STANDARD HANDBOOK OF BIOMEDICAL ENGINEERING AND DESIGN

STANDARD HANDBOOK OF BIOMEDICAL ENGINEERING AND DESIGN

Myer Kutz Editor in Chief

McGRAW-HILL
New York Chicago San Francisco Lisbon London Madrid
Mexico City Milan New Delhi San Juan Seoul
Singapore Sydney Toronto

The McGraw-Hill Companies

Library of Congress Cataloging-in-Publication Data

Standard handbook of biomedical engineering and design / Myer Kutz, editor in chief.
p. cm.
Includes index.
ISBN 0-07-135637-1
1. Biomedical engineering—Handbooks, manuals, etc. 2. Medical instruments and apparatus—Design and construction—Handbooks, manuals, etc. I. Kutz, Myer.

R856.15.S73 2003
610′.28—dc21 2002029356

1 2 3 4 5 6 7 8 9 0 DOC/DOC 0 9 8 7 6 5 4 3 2

ISBN 0-07-135637-1

The sponsoring editor for this book was Kenneth P. McCombs, the editing supervisor was David E. Fogarty, and the production supervisor was Sherri Souffrance. It was set in Times Roman by Progressive Information Technologies, Inc.

Printed and bound by R. R. Donnelley & Sons Company.

This book is printed on acid-free paper.

For Arlene, forever

CONTENTS

Part 3 Biomaterials

Part 4 Bioelectricity

Part 5 Design of Medical Devices and Diagnostic Instrumentation

CONTRIBUTORS

Ronald S. Adrezin *College of Engineering, University of Hartford, West Hartford, Conn.* (CHAPS. 7, 31)

Ronald E. Barr *Department of Mechanical Engineering, University of Texas, Austin, Texas* (CHAP. 6)

Christopher Batich *Materials Science and Engineering Department, University of Florida, Gainesville, Fla.* (CHAP. 11)

Ravi V. Bellamkonda *Department of Biomedical Engineering, Case Western Reserve University, Cleveland, Ohio* (CHAP. 16)

Anthony J. Brammer *National Research Council, Ottawa, Ontario, Canada and University of Connecticut Health Center, Farmington, Conn.* (CHAP. 10)

Thomas S. Buchanan *Center for Biomedical Research, University of Delaware, Newark, Del.* (CHAP. 5)

Albert M. Cook *Faculty of Rehabilitation Medicine, University of Alberta, Edmonton, Alberta, Canada* (CHAP. 30)

John Michael Currie *SmithGroup, Inc., Washington, D.C.* (CHAP. 38)

Alfred M. Dolan *Institute of Biomaterials and Biomedical Engineering, University of Toronto, Toronto, Canada* (CHAP. 36)

David Elad *Department of Biomedical Engineering, Tel Aviv University, Tel Aviv, Israel* (CHAP. 3)

Shmuel Einav *Department of Biomedical Engineering, Tel Aviv University, Tel Aviv, Israel* (CHAP. 3)

Laurie L. Fajardo *The Russell Morgan Department of Radiology and Radiological Science, Johns Hopkins Medical Institutions, Baltimore, Md.* (CHAP. 28)

Kenneth L. Gage *Department of Chemical Engineering, University of Pittsburgh, Pittsburgh, Penn.* (CHAP. 20)

Michele J. Grimm *Bioengineering Center, Wayne State University, Detroit, Mich.* (CHAP. 15)

James B. Grotberg *Department of Biomedical Engineering, University of Michigan, Ann Arbor, Mich.* (CHAP. 4)

Michael N. Helmus *Boston Scientific Corporation, Boston, Mass.* (CHAP. 14)

Arif Iftekhar *Department of Biomedical Engineering and Gleason Laboratory, University of Minnesota, Minneapolis, Minn.* (CHAP. 12)

Bhaskara R. Jasti *Department of Pharmaceutics and Medicinal Chemistry, University of the Pacific, Stockton, Calif.* (CHAP. 22)

Arthur T. Johnson *Department of Biological Resources Engineering, University of Maryland, College Park, Md.* (CHAP. 21)

Leo Joskowicz *Computer-Aided Surgery and Medical Image Processing Laboratory, The Hebrew University of Jerusalem, Israel* (CHAP. 29)

Tony M. Keaveny *Departments of Mechanical Engineering and Bioengineering, University of California, Berkeley, Calif. and Department of Orthopaedic Surgery, University of California, San Francisco, Calif.* (CHAP. 8)

David H. Kohn *Department of Biomedical Engineering, University of Michigan, Ann Arbor, Mich.* (CHAP. 13)

Patrick Leamy *Materials Science and Engineering Department, University of Florida, Gainesville, Fla.* (CHAP. 11)

Xiaoling Li *Department of Pharmaceutics and Medicinal Chemistry, University of the Pacific, Stockton, Calif.* (CHAP. 22)

Mark T. Madsen *Department of Radiology, University of Iowa, Iowa City, Iowa* (CHAP. 27)

Kurt. T. Manal *Center for Biomedical Research, University of Delaware, Newark, Del.* (CHAP. 5)

Nancy J. Meilander *Department of Biomedical Engineering, Case Western Reserve University, Cleveland, Ohio* (CHAP. 16)

Elise F. Morgan *Department of Mechanical Engineering, University of California, Berkeley, Calif.* (CHAP. 8)

Jit Muthuswamy *Department of Bioengineering, Arizona State University, Tempe, Ariz.* (CHAP. 18)

Patrick J. Nolan *DDL, Inc., Eden Prairie, Minn.* (CHAP. 23)

Michael D. Nowak *Department of Civil and Environmental Engineering, University of Hartford, West Hartford, Conn.* (CHAP. 9)

James P. O'Leary *Department of Mechanical Engineering, Tufts University, Medford, Mass.* (CHAP. 19)

Marcus Pandy *Department of Biomedical Engineering, University of Texas, Austin, Texas* (CHAP. 6)

Donald R. Peterson *University of Connecticut Health Center, Farmington, Conn.* (CHAPS. 7, 10)

Narender P. Reddy *Department of Biomedical Engineering, University of Akron, Akron, Ohio* (CHAP. 1)

David J. Reinkensmeyer *Department of Mechanical and Aerospace Engineering, University of California, Irvine, Calif.* (CHAP. 35)

Peter Rockett *Department of Engineering Science, Oxford University, Oxford, U.K.* (CHAP. 26)

Blair A. Rowley *Biomedical, Industrial and Human Factors Engineering, Wright State University, Dayton, Ohio* (CHAP. 34)

Daniel J. Schaefer *GE Medical Systems, Waukesha, Wisc.* (CHAP. 24)

David M. Shade *Johns Hopkins Pulmonary Laboratory, Johns Hopkins School of Medicine, Baltimore, Md.* (CHAP. 21)

Steve I. Shen *Department of Pharmaceutics and Medicinal Chemistry, University of the Pacific, Stockton, Calif.* (CHAP. 22)

M. Barbara Silver-Thorn *Department of Biomedical Engineering, Marquette University, Milwaukee, Wisc.* (CHAP. 33)

John Smith *Hospital for Sick Children, Toronto, Ontario, Canada* (CHAP. 37)

Roger W. Snyder *LVAD Technology, Inc., Detroit, Mich.* (CHAP. 14)

Ira Tackel *Department of Biomedical Instrumentation, Thomas Jefferson University Hospital, Philadelphia, Penn.* (CHAP. 39)

Russell Taylor *Computer Science Department, Johns Hopkins University, Baltimore, Md.* (CHAP. 29)

Kai E. Thomenius *Ultrasound Program, GE Corporate Research and Development, Schenectady, N.Y.* (CHAP. 25)

Bruce C. Towe *College of Engineering, Arizona State University, Tempe, Ariz.* (CHAP. 17)

William R. Wagner *Departments of Surgery, Bioengineering & Chemical Engineering, University of Pittsburgh, Pittsburgh, Penn.* (CHAP. 20)

Ge Wang *Department of Radiology, University of Iowa, Iowa City, Iowa* (CHAP. 26)

Richard F. ff. Weir *Prosthetics Research Laboratory and Rehabilitation Engineering Research Center, Northwestern University, Chicago, Ill.* (CHAP. 32)

Mark B. Williams *Department of Radiology, University of Virginia, Charlottesville, Va.* (CHAP. 28)

Oscar C. Yeh *Department of Mechanical Engineering, University of California, Berkeley, Calif.* (CHAP. 8)

Liang Zhu *Department Mechanical Engineering, University of Maryland, Baltimore County, Md.* (CHAP. 2)

PREFACE

How do important medical advances that change the quality of life come about? Sometimes, to be sure, they can result from the inspiration and effort of physicians or biologists working in remote, exotic locations or organic chemists working in the well-appointed laboratories of pharmaceutical companies with enormous research budgets. Occasionally, however, a medical breakthrough happens when someone with an engineering background gets a brilliant idea in less glamorous circumstances. One afternoon in the late 1950s, the story goes, when an electrical engineer named Wilson Greatbatch was building a small oscillator to record heart sounds, he accidentally installed the wrong resistor, and the device began to give off a steady electrical pulse. Greatbatch realized that a small device could regulate the human heart, and in two years he had developed the first implantable cardiac pacemaker, followed later by a corrosion-free lithium battery to power it. In the mid-1980s, Dominick M. Wiktor, a Cranford, New Jersey, engineer, invented the coronary stent after undergoing open heart surgery.

You often find that it is someone with an engineer's sensibility—someone who may or may not have engineering training, but does have an engineer's way of looking at, thinking about, and doing things—who not only facilitates medical breakthroughs, but also improves existing healthcare practice. This sensibility, which, I dare say, is associated in people's consciousness more with industrial machines than with the human body, manifests itself in a number of ways. It has a descriptive component, which comes into play, for example, when someone uses the language of mechanical engineering to describe blood flow, how the lungs function, or how the musculoskeletal system moves or reacts to shocks, or when someone uses the language of other traditional engineering disciplines to describe bioelectric phenomena or how an imaging machine works.

Medically directed engineer's sensibility also has a design component, which can come into play in a wide variety of medical situations, indeed whenever an individual, or a team, designs a new healthcare application, such as a new cardiovascular or respiratory device, a new imaging machine, a new artificial arm or lower limb, or a new environment for someone with a disability. The engineer's sensibility also comes into play when an individual or team makes an application that already exists work better—when, for example, the unit determines which materials would improve the performance of a prosthetic device, improves a diagnostic or therapeutic technique, reduces the cost of manufacturing a medical device or machine, improves methods for packaging and shipping medical supplies, guides tiny surgical tools into the body, improves the plans for a medical facility, or increases the effectiveness of an organization installing, calibrating, and maintaining equipment in a hospital. Even the improved design of time-released drug capsules can involve an engineer's sensibility.

The field that encompasses medically directed engineer's sensibility is, of course, called biomedical engineering. Compared to the traditional engineering disciplines, whose fundamentals and language it employs, this field is new and rather small. Although there are now over 80 academic programs in biomedical engineering in the United States, only 6500 undergraduates were enrolled in the year 2000. Graduate enrollment was just 2500. The U.S. Bureau of Labor Statistics reports total biomedical engineering employment in all industries in the year 2000 at 7221. The bureau estimates this number to rise by 31 percent to 9478 in 2010.

The effect this relatively young and small field has on the health and well being of people everywhere, but especially in the industrialized parts of the world that have the wherewithal to fund the field's development and take advantage of its advances, is, in my view, out of proportion to its age and size. Moreover, as the examples provided earlier indicate, the concerns of biomedical engineers are very wide-ranging. In one way or another, they deal with virtually every system and part in the human body. They are involved in all phases of healthcare—measurement and diagnosis, therapy and repair, and patient management and rehabilitation. While the work that biomedical engineers do involves the human body, their work is engineering work. Biomedical engineers, like other engineers in the more traditional disciplines, design, develop, make, and manage. Some work in traditional engineering settings—in laboratories, design departments, on the floors of manufacturing plants—while others deal directly with healthcare clients or are responsible for facilities in hospitals or clinics.

Of course, the field of biomedical engineering is not the sole province of practitioners and educators who call themselves biomedical engineers. The field includes people who call themselves mechanical engineers, materials engineers, electrical engineers, optical engineers, or medical physicists, among other names. The entire range of subjects that can be included in biomedical engineering is very broad. Some curricula offer two main tracks: biomechanics and bioinstrumentation. To some degree, then, there is always a need in any publication dealing with the full scope of biomedical engineering to bridge gaps, whether actually existing or merely perceived, such as the gap between the application of mechanical engineering knowledge, skills, and principles from conception to the design, development, analysis, and operation of biomechanical systems and the application of electrical engineering knowledge, skills, and principles to biosensors and bioinstrumentation.

The focus in the *Standard Handbook of Biomedical Engineering and Design* is on engineering design informed by description in engineering language and methodology. For example, the Handbook not only provides engineers with a detailed understanding of how physiological systems function and how body parts—muscle, tissue, bone—are constituted, it also discusses how engineering methodology can be used to deal with systems and parts that need to be assisted, repaired, or replaced.

I have sought to produce a practical manual for the biomedical engineer who is seeking to solve a problem, improve a technique, reduce cost, or increase the effectiveness of an organization. The Handbook is not a research monograph, although contributors have properly included lists of applicable references at the ends of their chapters. I want this Handbook to serve as a source of practical advice to the reader, whether he or she is an experienced professional, a newly minted graduate, or even a student at an advanced level. I intend the Handbook to be the first information resource a practicing engineer reaches for when faced with a new problem or opportunity—a place to turn to even before turning to other print sources or to sites on the Internet. (The Handbook is planned to be the core of an Internet-based update or current-awareness service, in which the Handbook chapters would be linked to news items, a bibliographic index of articles in the biomedical engineering research literature, professional societies, academic departments, hospital departments, commercial and government organizations, and a database of technical information useful to biomedical engineers.) So the Handbook is more than a voluminous reference or collection of background readings. In each chapter, the reader should feel that he or she is in the hands of an experienced consultant who is providing sensible advice that can lead to beneficial action and results.

I have divided the Handbook into eight parts. Part 1, which contains only a single chapter, is an introductory chapter on applying analytical techniques to biomedical systems. Part 2, which contains nine chapters, is a mechanical engineering domain. It begins with a chapter on the body's thermal behavior, then moves on to two chapters that discuss the mechanical functioning of the cardiovascular and respiratory systems. Six chapters of this part of the Handbook are devoted to analysis of bone and the musculoskeletal system, an area that I have been associated with from a publishing standpoint for a quarter-century, ever since I published David Winter's book on human movement.

Part 3 of the Handbook, the domain of materials engineering, contains six chapters. Three deal with classes of biomaterials—biopolymers, composite biomaterials, and bioceramics—and three deal with using biomaterials, in cardiovascular and orthopedic applications, and to promote tissue regeneration.

The two chapters in Part 4 of the Handbook are in the electrical engineering domain. They deal with measuring bioelectricity and analyzing biomedical signals, and they serve, in part, as an introduction to Part 5, which contains ten chapters that treat the design of therapeutic devices and diagnostic imaging instrumentation, as well as the design of drug delivery systems and the development of sterile packaging for medical devices, a deceptively robust and complex subject that can fill entire books on its own. Imaging also plays a role in the single-chapter Part 6 of the Handbook, which covers computer-integrated surgery.

The last two parts of the Handbook deal with interactions between biomedical engineering practitioners and both patients and medical institutions. Part 7, which covers rehabilitation engineering, includes chapters that treat not only the design and implementation of artificial limbs, but also ways in which engineers provide environments and assistive devices that improve a person's quality of life. Part 8, the last part of the Handbook, deals with clinical engineering, which can be considered the facilities-planning and management component of biomedical engineering.

Acknowledgments

The contributors to this Handbook work mainly in academia and hospitals. Several work in commercial organizations. Most work in the United States and Canada; a few work in Israel. What they all have in common is that what they do is useful and important: they make our lives better. That these busy people were able to find the time to write chapters for this Handbook is nothing short of miraculous. I am indebted to all of them. I am additionally indebted to multiple-chapter contributors Ron Adrezin of the University of Hartford and Don Peterson of the University of Connecticut School of Medicine for helping me organize the biomechanics chapters in the handbook, and for recruiting other contributors, Mike Nowak, a colleague at the University of Hartford and Anthony Brammer, now a colleague at the University of Connecticut Health Center. Also, contributor Alf Dolan of the University of Toronto was especially helpful in recommending contributors for the clinical engineering chapters.

Thanks to both of my editors at McGraw-Hill—Linda Ludwig, who signed the Handbook, and Ken McCombs, who saw the project to its completion. Thanks also to Dave Fogarty, who managed McGraw-Hill's editing process smoothly and expeditiously.

I want to give the final word to my wife Arlene, the family medical researcher and expert, in recognition of her patience and support throughout the life of this project, from development of the idea, to selection and recruiting of contributors, to receipt and editing of manuscripts: "It is our hope that this Handbook will not only inform and enlighten biomedical engineering students and practitioners in their present pursuits, but also provide a broad and sturdy staircase to facilitate their ascent to heights not yet scaled."

MYER KUTZ
Albany, New York

STANDARD HANDBOOK OF BIOMEDICAL ENGINEERING AND DESIGN

P • A • R • T • 1

BIOMEDICAL SYSTEMS ANALYSIS

CHAPTER 1
MODELING AND SIMULATION OF BIOMEDICAL SYSTEMS

Narender P. Reddy
University of Akron Akron, Ohio

1.1 MATHEMATICAL MODELING

Models are conceptual constructions that allow formulation and testing of hypotheses. A mathematical model attempts to duplicate the quantitative behavior of the system. Mathematical models are used in today's scientific and technological world because of the ease with which they can be used to analyze real systems. The most prominent value of a model is its ability to predict as yet unknown properties of the system. The major advantage of a mathematical or computer model is that the model parameters can be easily altered and the system performance can be simulated. Mathematical models allow the study of subsystems in isolation from the parent system. Model studies are often inexpensive and less time consuming than corresponding experimental studies. A model can also be used as a powerful educational tool, since it permits idealization of processes. Models of physiological systems often aid in the specification of design criteria for the design of procedures aimed at alleviating pathological conditions. Mathematical models are useful in the design of medical devices. Mathematical model simulations are first conducted in the evaluation of the medical devices before expensive animal testing and clinical trials. Models are often useful in the prescription of patient protocols for the use of medical devices. Pharmacokinetic models have been extensively used in the design of drugs and drug therapies.

There are two types of modeling approach: the black box approach and the building block approach. In the black box approach, a mathematical model is formulated on the basis of the input-output characteristic of the system without consideration of the internal functioning of the system. Neural network models and autoregressive models are some examples of the black box approach. In the building block approach, models are derived by applying the fundamental laws (governing physical laws) and constitutive relations to the subsystems. These laws together with physical constraints are used to integrate the models of subsystems into an overall mathematical model of the system. The building block approach is used when the processes of the system are understood. However, if the system processes are unknown or too complex, then the black box approach is used. With the building block approach, models can be derived at the microscopic or at the macroscopic levels. Microscopic models are spatially distributed and macroscopic models are spatially lumped and are rather global. The microscopic modeling often leads to partial differential equations, whereas the macroscopic or global modeling leads to a set of ordinary differential equations. For example, the microscopic approach can be used to derive the velocity profile for blood flow in an artery, but the global or macroscopic approach is needed to study the overall behavior of the circulatory system including the flow through arteries, capillaries, and the heart.

Models can also be classified into continuous time models and models lumped in the time domain. While the continuous time modeling leads to a set of differential equations, the models lumped in time are based on the analysis of discrete events in time and may lead to difference equations or sometimes into difference-differential equations. Random walk models and queuing theory models are some examples of discrete time models. Nerve firing in the central nervous system can be modeled by using such discrete time event theories. Models can be classified into deterministic and stochastic models. For example, in deterministic modeling, we could describe the rate of change of volume of an arterial compartment to be equal to the rate of flow in minus the rate of flow out of the compartment. However, in the stochastic approach, we regard the probability of increase in the volume of the compartment in an interval to be dependent on the probability of transition of a volume of fluid from the previous compartment and the probability of transition of a volume of fluid from the compartment to the next compartment. While the deterministic approach gives the mean or average values, the stochastic approach yields means, variances, and covariances. The stochastic approach may be useful in describing the cellular dynamics, cell proliferations, etc. However, in this chapter, we will consider only the deterministic modeling at the macroscopic level.

The real world is complex, nonlinear, nonhomogeneous, often discontinuous, anisotropic, multilayered, multidimensional, etc. The system of interest is isolated from the rest of the world by using a boundary. The system is then conceptually reduced to that of a mathematical model by using a set of simplifying assumptions. Therefore, the model results have significant limitations and are valid only in the regimes where the assumptions are valid.

1.2 COMPARTMENTAL MODELS

Compartmental models are lumped models. The concept of a compartmental model assumes that the system can be divided into a number of homogeneous well-mixed components called *compartments*. Various characteristics of the system are determined by the movement of material from one compartment to the other. Compartmental models have been used to describe blood flow distribution to various organs, population dynamics, cellular dynamics, distribution of chemical species (hormones and metabolites) in various organs, temperature distribution, etc.

As an example, let us consider a simple one-compartment model for the prescription of treatment protocols for dialysis by an artificial kidney device (Fig. 1.1). While the blood urea concentration (BUN) in the normal individual is usually 15 mg% (mg% = milligrams of the substance per 100 mL of blood), the BUN in uremic patients could reach 50 mg%. The purpose of the dialysis is to bring the BUN level closer to the normal. In the artificial kidney, blood flows on one side of the dialyzer membrane and dialysate fluid flows on the other side. Mass transfer across the dialyzer membrane occurs by diffusion due to concentration difference across the membrane. Dilysate fluid is a makeup solution consisting of saline, ions, and the essential nutrients that maintains zero concentration difference for these essential materials across the membrane. However, during the dialysis, some hormones also diffuse out of the dialyzer membrane along with the urea molecule. Too-rapid dialysis often leads to depression in the individual because of the rapid loss of hormones. On the other hand, too-slow dialysis may lead to unreasonable time required at the hospital.

Simple modeling can be used to calculate the treatment protocols. Let us consider a one-compartment model of the tissue where we assume that the blood and tissue are well mixed and that the concentration of urea is uniform throughout the body. Let C_o be the concentration of urea at the outlet of the body, i.e., at the inlet of the dialyzer in the arterial line that takes blood into the dialyzer, and let C_i be the concentration of urea at the inlet of the compartment, i.e., at the exit of the dialyzer

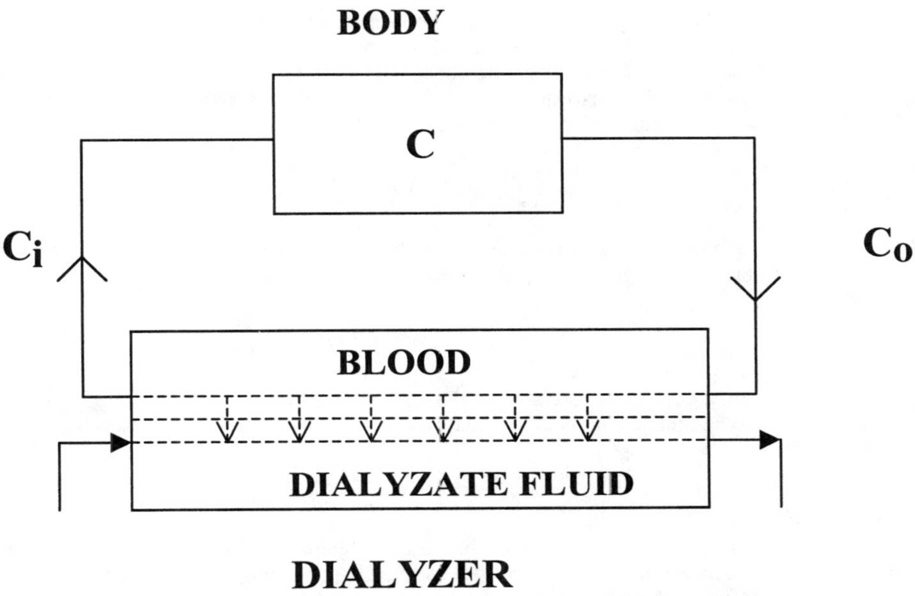

FIGURE 1.1 A one-compartment model of the human body to analyze the patient-dialyzer interactions.

in the venous line that brings the blood back to the body. Mass balance demands that the rate of change of mass in the body be equal to the net rate of mass coming into the body from the dialyzer, plus the metabolic production rate.

$$V\left(\frac{dC}{dt}\right) = Q(C_i - C_o) + m = -QC_o\left(1 - \frac{C_i}{C_o}\right) + m \tag{1.1}$$

where V = tissue volume plus the blood volume
Q = blood flow rate to the kidney
m = metabolic production rate of urea in the body

It should be noted that mass is equal to volume V times the concentration C.

Usually, the dialysate flow rate in the artificial kidney is much larger than that of the blood flow rate. Regardless of the type of the dialyzer (cocurrent, countercurrent, or mixed flow), the extraction ratio E can be expressed as (Cooney, 1980):

$$E = 1 - \exp\left(\frac{-kA}{Q}\right) \tag{1.2}$$

where A is the interfacial membrane surface area for mass transfer and k is the permeability of the membrane for that particular solute (urea in the present context). Since Q does not change during dialysis, and since k and A are design parameters, extraction ratio E remains a constant. Extraction can be further expressed in terms of the concentrations as follows:

$$E = 1 - \frac{C_i}{C_o} \tag{1.3}$$

It should be pointed out that C_o is the concentration at the outlet of the body and therefore at the inlet of the dialyzer, and C_i is the concentration in the blood coming into the body and therefore going out of the dialyzer. Also, it should be noted that the concentration C_o in the blood going out of the body is the same as the concentration in the body, since we assumed that the entire body (tissue and blood) constitutes a homogeneous, well-mixed compartment. Therefore, Eq. (1) can be rewritten as follows on substitution of Eq. (3)

$$V\left(\frac{dC}{dt}\right) = -QCE + m \tag{1.4}$$

When the dialyzer is turned on, metabolic production rate (m) can be neglected when compared to the other term in the equation, and upon integration will result in

$$C = C^0 \exp\left[-\left(\frac{QE}{V}\right)t\right] \tag{1.5}$$

where C^0 is the initial concentration of urea in the tissue.

When the patient is *not on* dialysis, then the blood flow to the dialyzer Q is zero, and therefore

$$V\left(\frac{dC}{dt}\right) = m \tag{1.6}$$

When the patient is *not on* dialysis, the concentration of urea will increase linearly if the metabolic production rate is constant or will increase exponentially if the metabolic production rate is a linear function of the concentration (first-order reaction). When the patient is *on* dialysis, the concentration would decrease exponentially. This way, the treatment protocol can be prescribed after simulating different on and off times (e.g., turn on the dialyzer for 4 hours every 3 days) to bring the BUN under control.

Now, let us examine the limitations of the one-compartment model. First, the entire blood and tissue are assumed to be in equilibrium. However, it is well known that intracellular urea concentration may be significantly different from the extracellular compartment. Moreover, urea may be preferentially produced in certain organs like the brain, heart, muscle, etc. An accurate treatment requires a multicompartment model.

Let us consider a two-compartment model (Fig. 1.2) consisting of an intracellular pool (compartment 1) and an extracellular pool (compartment 2). Urea is produced by the intracellular pool and is transported across the cell membrane into the interstitial fluids and then into the blood stream. Mass balance for these two compartments can be expressed as

$$V_1\left(\frac{dC_1}{dt}\right) = m - B(C_1 - C_2) \tag{1.7}$$

where $B\ (C_1 - C_2)$ = the interfacial transfer from compartment 1 to compartment 2 (from intracellular pool to extracellular pool)
C_1 and C_2 = concentrations of urea in compartments 1 and 2
B = a constant

The constant B is a product of permeability of the cellular membrane for urea and the interfacial surface area.

$$V_2\left(\frac{dC_2}{dt}\right) = B(C_1 - C_2) - QC_2E \tag{1.8}$$

Blood flow to the dialyzer (Q) is zero when the patient is not on the dialysis machine. Babb et al. (1967) simulated the two-compartmental model and the model results were in agreement with the experimental data (Fig. 1.3). However, the two compartmental model may not be sufficient if one wants to find the concentration of urea in the brain tissue. A multicompartment model involving

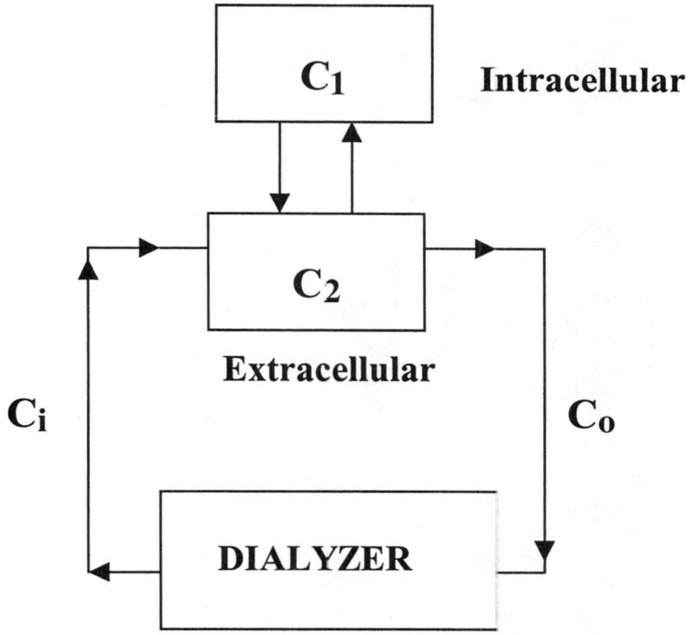

FIGURE 1.2 A two-compartmental model of the body used to simulate the patient-dialyzer interactions.

separate compartments for brain, heart, kidney, lean tissue, etc., may be needed to accurately determine the concentration of urea in critical organs. Similar compartmental models have been used in the development of metabolic modeling.

Now, let us consider an example of cellular dynamics in bone. Cellular dynamics plays an important role in natural and stress-induced bone remodeling. There are four kinds of functionally distinct cells: mesenchymal cells, osteoclasts, osteoblasts, and osteocytes. Mesenchymal cells have an outstanding capacity for proliferation and are capable of further differentiating into an osteoclast or an osteoblast. While the mesenchymal cell undergoes cell division, osteoblasts, osteoclasts, or the osteocytes do not undergo cell division. The osteoclast is responsible for bone resorption. The osteoblast is responsible for new bone formation. The osteoclast can transform into an osteoblast and vice versa. The osteocyte is a resident cell of the bone. When the osteoblast has surrounded itself with a matrix (after new bone formation), it becomes an osteocyte and it loses the characteristics of an actively secreting cell. Let us assume that these four types of cells can be divided into four different, well-mixed homogeneous compartments (Fig. 1.4). Let W be the number of mesenchymal cells, X be the number of osteoclasts, Y be the number of osteoblasts, and Z be the number of osteocytes at any given time. Let compartments 1 through 4, respectively, represent these four types of cells. The birth rate and death rate of cells in a compartment depend on the number of cells in that compartment. Let B be the birth rate of mesenchymal cells per individual cell per unit time. Let D_1 and D_2 represent the death rates of osteoclasts and osteocytes per cell per unit time. Equations representing the cell population dynamics can be expressed as follows by integrating the notions of bone cell physiology with the concepts of compartmental modeling.

The rate of change of the number of mesenchymal cells (W) can be expressed as

$$\frac{dW}{dt} = BW - C_1W - C_2W \tag{1.9}$$

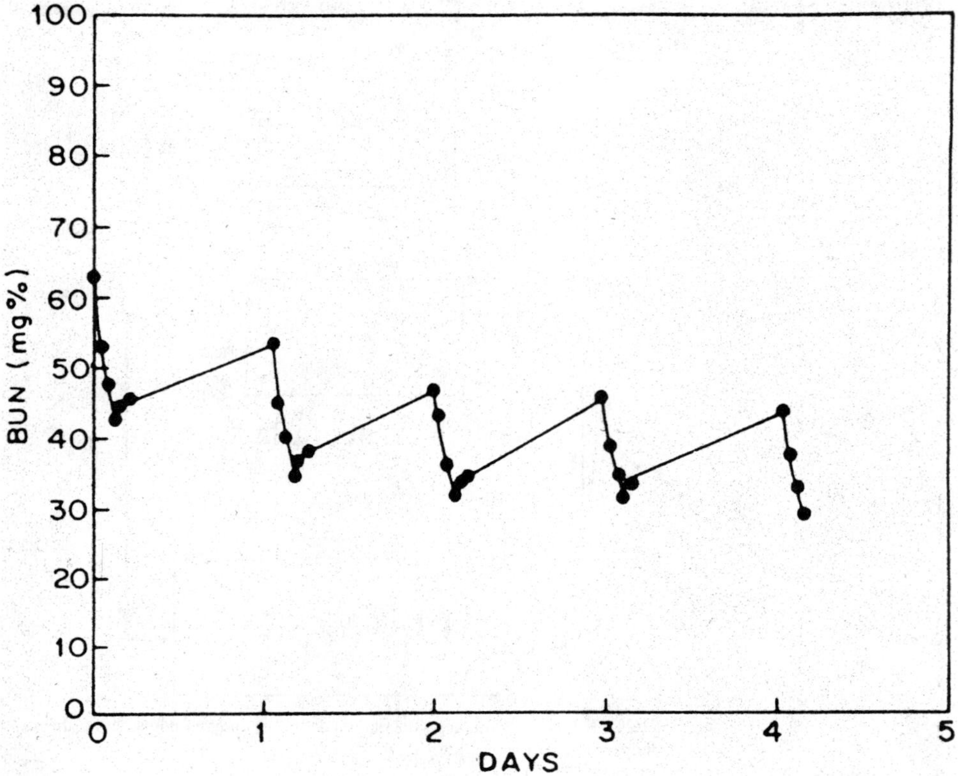

FIGURE 1.3 Model simulation results of the patient BUN (blood urea nitrogen levels) in mg% (mg% = mg of the substance per 100 mL of blood) plotted as a function of time. [*From Babb et al. (1967).*] Closed circles are experimental observations.

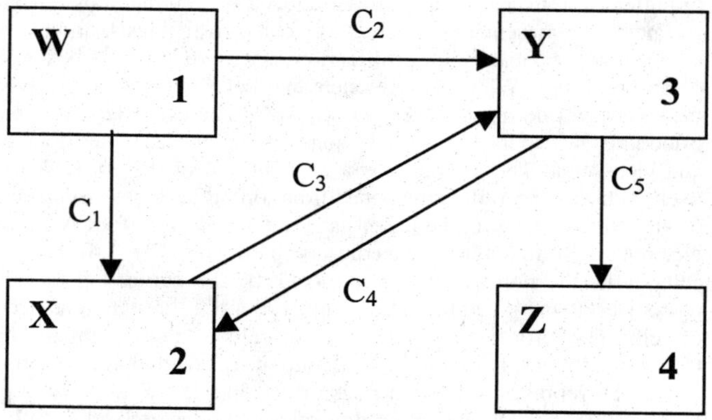

FIGURE 1.4 A four-compartment model of bone cells: Compartment 1 represents mesenchymal cells, compartment 2 represents osteoclasts, compartment 3 represents osteoblasts, and compartment 4 represents osteocytes. Mesenchymal cells reproduce and transform either into osteoclasts or into osteoblasts. The osteoclast can transform into an osteoblast and an osteoblast can transform into an osteoclast. The osteoblast can transform into an osteocyte.

The first term on the right-hand side represents the birth rate of mesenchymal cells, the second term represents the rate of number of mesenchymal cells transforming into an osteoclast, and the last term represents the rate of transformation of mesenchymal cells into osteoblasts.

The rate of change of the number of osteoclasts (X) as a function of time can be expressed as

$$\frac{dX}{dt} = C_1W - C_3X + C_4Y - D_1X \tag{1.10}$$

The first term on the right-hand side represents the rate of mesenchymal cells transforming into an osteoclast, the second term represents the rate of transformation of osteolasts into osteoblasts, the third term represents the rate of transformation of osteoblast into osteoclasts, and the last term represents the rate of death of osteoclasts.

The rate of change of the number of osteoblasts (Y) as a function of time can be expressed as

$$\frac{dY}{dt} = C_2W + C_3X - C_4Y - C_5Y \tag{1.11}$$

The first term on the right-hand side represents the rate of transformation of mesenchymal cells into osteoblasts, the second term represents the rate of transformation of osteoclasts into osteoblasts, the third term represents the rate of transformation of osteoblasts into osteoclasts, and the last term represents the rate of transformation of osteoblasts into osteocytes.

The rate of change of the number of osteocytes (Z) as a function of time can be expressed as

$$\frac{dZ}{dt} = C_5Y - C_6Z \tag{1.12}$$

The first term in the above equation represents the rate of osteoblast transformation into osteocytes, and the last term represents the rate of death of osteocytes. Reddy and Joshi (1987) simulated the stochastic compartmental model of bone cells in which the equation for the population means are the same as the above equations. In addition, the stochastic analysis provides information about the variations and covariences of cellular populations. Figure 1.5 shows the normalized number of osteoblasts plotted as a function of age of the individual when C_1 and C_2 are assumed to be sinusoidal functions in time. These simulation results of Reddy and Joshi (1987) are consistent with experimental observations of Frost (1963).

Compartmental models are used in the analysis of thermal interactions. Simon and Reddy (1992) formulated a mathematical model of the infant-incubator dynamics. Neonates who are born preterm often do not have the maturity for thermal regulation and do not have enough metabolic heat production. Moreover, these infants have a large surface area to volume ratio. Since these preterm babies can not regulate heat, they are often kept in an incubator until they reach thermal maturity. The incubator is usually a forced-convection heating system with hot air flowing over the infant. Incubators are usually designed to provide a choice of air control or skin control. In air control, the temperature probe is place in the incubator air space and the incubator air temperature is controlled. In the skin control operation, the temperature sensor is placed on the skin and the infant's skin temperature is controlled. Simon and Reddy (1992) used a four-compartment model (Fig. 1.6) to compare the adequacy of air control and skin control on the core temperature of the infant. They considered the infant, air, mattress, and the wall to be four separate, well-mixed compartments.

1.3 ELECTRICAL ANALOG MODELS

Electric analog models are a class of lumped models and are often used to simulate flow through the network of blood vessels. These models are useful in assessing the overall performance of a system or a subsystem. Integration of the fluid momentum equation (longitudinal direction, in cy-

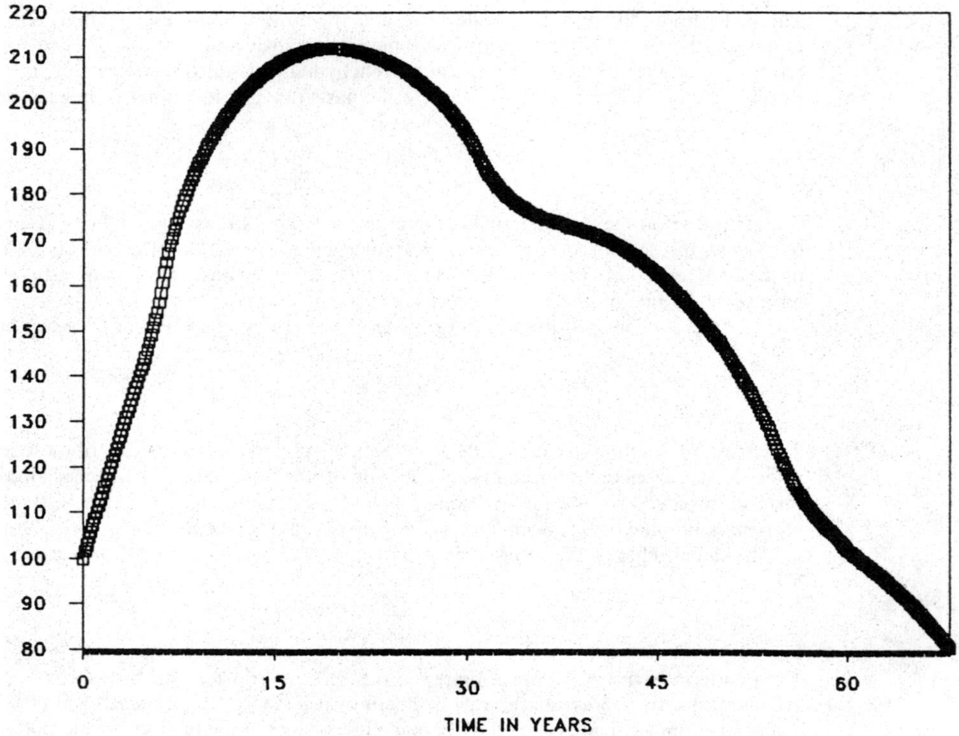

FIGURE 1.5 Simulation results of the normalized number of osteoblasts plotted as a function of age in years. [*From Reddy and Joshi (1987).*]

lindrical coordinates) across the cross section results in the following expression (Reddy, 1986; Reddy and Kesavan, 1989):

$$\rho \frac{dQ}{dt} = \frac{\pi a^2 \, \Delta P}{l} - 2a\tau_w \tag{1.13}$$

where ρ is the fluid density, Q is the flow rate, a is the wall radius, P is the pressure, l is the length, and τ_w is the fluid shear stress at the wall. If we assume that the wall shear stress can be expressed by using quasi-steady analysis, then the wall shear stress can be estimated by $\tau_w = 4\ \mu Q/a^3$. Substituting for the wall stress and rearranging results in

$$\left(\frac{\rho l}{\pi a^2}\right)\frac{dQ}{dt} = \Delta P - \left(\frac{8\mu l}{\pi a^4}\right)Q \tag{1.14}$$

which can be rewritten as

$$L\frac{dQ}{dt} = \Delta P - RQ \tag{1.15}$$

where $L = \rho l/(\pi a^2)$ and $R = 8\mu l/(\pi a^4)$.

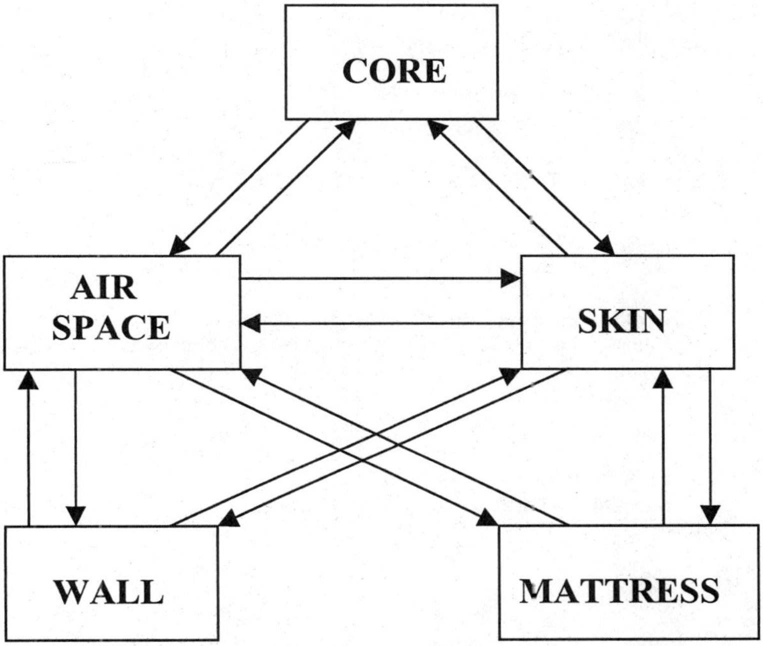

FIGURE 1.6 A lumped parameter model of the infant-incubator dynamics used by Simon and Reddy (1994) to simulate the effect of various control modes in a convectively heated infant incubator. Infants core, and skin are modeled as two separate compartments. The incubator air space, the incubator wall, and the mattress are treated as three compartments. Heat interactions occur between the core (infant's lungs) and the incubator air space through breathing. Skin-core heat interactions are predominantly due to blood flow to the skin. Heat transfer between the infant's skin and the incubator air are due to conduction and convection. Heat transfer from the skin to the mattress is via conduction, and heat transfer to the wall is via radiation from skin and convection from the air.

It can be easily observed that flow rate Q is analogous to electrical current i, and ΔP is analogous to the electrical potential drop (voltage) ΔE. In Eq. (1.15), L is the inductance (inertance) and R is the resistance to flow. Therefore, Eq. (1.15) can be rewritten as

$$L\frac{di}{dt} = \Delta E - Ri \tag{1.16}$$

The fluid continuity equation when integrated across the cross section can be expressed as

$$\frac{dV}{dt} = \Delta Q = Q_{\text{in}} - Q_{\text{out}} \tag{1.17}$$

where V is the volume. However, volume is a function of pressure (from the momentum balance for the vessel wall):

$$P = P_{\text{ext}} + (h/R_0)\sigma \tag{1.18}$$

where P_{ext} is the external pressure on the outside of the vessel wall, h is the wall thickness, and σ

is the hoop stress in the wall. The hoop stress is a function of wall radius R and modulus of elasticity Ω of the wall, and can be expressed as

$$\sigma = \Omega\left(\frac{R}{R_0} - 1\right) \tag{1.19}$$

Equations (1.17), (1.18), and (1.19) can be combined as

$$\frac{dP}{dt} = b\frac{dR}{dt} \tag{1.20}$$

where b is a constant equal to $\Omega h/(R_0^2)$.

Equation (1.20) can be expressed in terms of volume instead of the radius:

$$\frac{dV}{dt} = C\frac{dP}{dt} \tag{1.21}$$

C is often referred to as the *compliance* or *capacitance*.

Substituting Eq. (1.21) in Eq. (1.17) results in

$$C\frac{dP}{dt} = Q_{\text{in}} - Q_{\text{out}} \tag{1.22}$$

Equation 22 can be expressed in terms of an electrical equivalent as follows:

$$E = \left(\frac{1}{C}\right)\int i\,dt \tag{1.23}$$

Equations 16 and 23 [or (1.15) and (1.22)] can be used to simulate either a segment of a blood vessel or the entire blood vessel itself. In small blood vessels, the inductance L is very low compared to the resistance term R, and therefore the inductance term can be neglected in small arteries, arterioles, and capillaries. Since there is no oscillation of pressure in the capillaries, the inductance term can be neglected in vessels downstream of the capillary (i.e., venules, veins, and vena cava, etc.).

An electrical analog model of the circulation in the leg is illustrated in Fig. 1.7. Let us consider the flow from the femoral artery into the small leg arteries. There is no inductance in small leg arteries; there is only resistance. Since the small arteries are distensible, they have capacitance (compliance). The muscular pressure (P_{MP}) acts as the external pressure on the majority of small leg arteries. Consequently, P_{MP} is used as the reference pressure across the capacitor. The arterioles do not have inductance, but have a variable resistance, which is controlled by neurogenic and metabolic factors. In this model, the precapillary sphincters and the capillaries are lumped together. Since the capillaries are rather rigid, they do not have any capacitance (compliance), but the combined resistance of the sphincters and capillaries is variable subject to metabolic control. For instance, precapillary sphincters dilate in the presence of lactic acid and other end products of metabolism. Venules have resistance and a variable capacitance. This capacitance is subject to neurogenic control, since the diameter of the venule is under neurogenic control. From the venules, the flow goes into leg small veins which have a resistance and a variable capacitance subject to neurogenic control. In addition, the venules have valves that permit only unidirectional flow. These valves can be modeled as diodes. Again, the reference pressure for the capacitor is the muscle pressure P_{MP}. It is well known that the blood flow in the legs is aided by the muscle pump, which is essentially the external pressure oscillations on the blood vessel wall due to periodic skeletal muscle contractions during walking, etc. The muscle pump is absent in bedridden patients. Extremity pumps are used on such patients

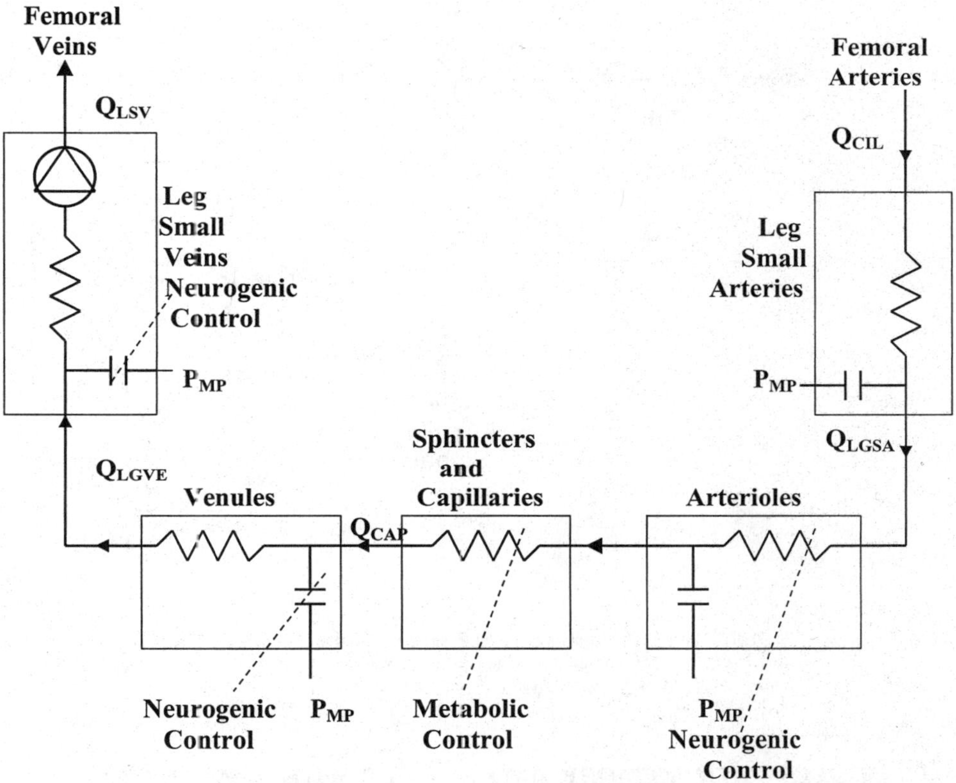

FIGURE 1.7 Electrical analog model of the circulation of the leg. P_{MP} is the muscle pump which exerts a periodic external pressure on the blood vessels. Q is the flow rate. Q_{LGSA} is the flow through the leg small arteries, Q_{CAP} is the flow rate through the capillary. Q_{LGVE} is the flow through the leg small veins. The elasticity is simulated with capacitance. The nonlinear capacitance of the leg small veins and the nonlinear resistance of arterioles and venules are under neurogenic control. The resistance of precapillary sphincters and capillaries is subject to metabolic control. The diaodes in the veins are used to simulate the valves which permit only the unidirectional flow.

to enhance blood flow to the legs. These extremity pumps provide a periodic, graded sequential external compression of the leg. The electrical analog model in Fig. 1.7 can be easily modified to simulate the effect of these extremity pumps.

Electrical analog models have been used in the study of cardiovascular, pulmonary, intestinal, and urinary system dynamics. Recently, Barnea and Gillon (2001) have used an electrical analog model to simulate flow through the urethra. Their model consisted of a simple *LRC* circuit with a variable capacitor. The time-varying capacitor simulated the time-dependent relaxation of the urethra. They used two types of resistance: a constant resistance to simulate Poiseouille-type viscous pressure drop and a flow dependant resistance to simulate Bernoulli-type pressure loss. With real-time pressure-flow data sets, Barnea and Gillon (2001) have used the model to estimate urethral resistance and changes in urethral compliance during voiding, and have suggested that the urethral elastance (inverse of compliance) estimated by the model provides a new diagnostic tool. Ventricular and atrial pumping can be modeled by similar techniques. The actual pump (pressure source) can be modeled as a variable capacitor. Figure 1.8 shows a model of the left heart with a multisegment representation of the ventricle (Rideout, 1991).

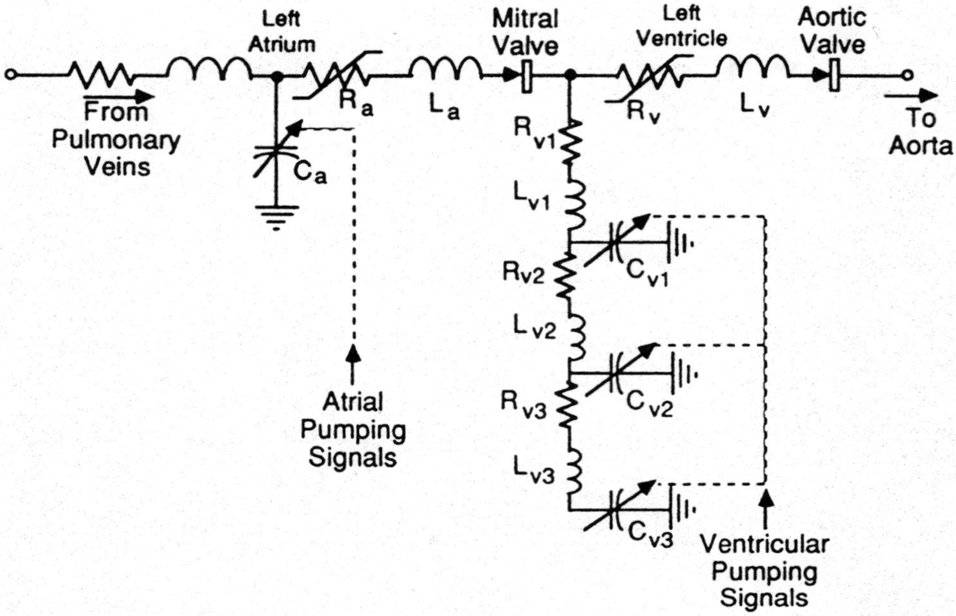

FIGURE 1.8 Electrical analog model to simulate atrial and ventricular pumping. Variable capacitances simulate the muscle contractions, and the filling and emptying through the ventricle can be simulated by a series of inductance and resistance elements. [*Adapted from Rideout (1991).*]

1.4 MODELS WITH MEMORY AND MODELS WITH TIME DELAY

Time delay and memory processes occur in several biomedical disciplines. An example of such an application occurs in modeling of the immune system (Reddy and Krouskop, 1978). In cellular immune response, lymphocytes are sensitized to a foreign material and have memory. The immune response is significantly enhanced if the similar material is reintroduced after a certain lag time. Another example is stress-induced bone remodeling. Modeling of the nervous system would involve time delays and memory. Similar hereditary functions are used to describe the material responses of viscoelastic materials. The effect of environmental pollutants can be modeled by using such hereditary functions. Stress-induced bone remodeling involves time lags between the actual application of stress and actual new bone formation, and also involves stress/strain histories. To illustrate the modeling of the effects of memory and time delay, let us consider a model of cellular immunity.

1.4.1 Modeling Cell-Mediated Immunity in Homograft Rejection

In cell-mediated immunity, lymphocytes in the tissue become sensitized to the target (graft) cells and travel to the regional lymph nodes where they initiate an immunological response by increasing the production of immunocompetent lymphocytes. The newly produced lymphocytes are then transported into the blood stream via the thoracic duct. Lymphocytes recirculate from the blood stream through the tissue and return to the blood stream via the lymphatic system. When foreign cells are introduced into the tissue, blood lymphocytes migrate into the tissue at an increased rate and bring about the destruction of the target cells. Lymphocytes have memory and they exhibit an increased secondary response; e.g., if, after the rejection of the first graft, a second graft is introduced into the host, the second graft is rejected much faster. A similar situation occurs in delayed hypersensitivity, which is another cell-mediated reaction. In this analysis, let us assume that blood and tissue are well-stirred compartments and that the newly produced lymphocytes are introduced into the blood compartment (Reddy and Krouskop, 1978).

For sensitization to occur, a lymphocyte has to come in contact with a target cell. The number of lymphocytes becoming sensitized at any given time ($L_s(t)$) is a function of the number of lymphocytes in the tissue ($L_T(t)$) and the number of target (foreign) cells ($g(t)$):

$$L_s(t) = C_1 L_T(t) g(t) \tag{1.24}$$

Certain lymphocytes, on encountering target cells, are transformed into memory cells. The memory cell formation depends on the number of lymphocytes in the tissue and the number of target cells. The number of memory cells formed at any time t may thus be expressed as

$$L_{m,s}(t) = C_1 L_T(t) g(t) \tag{1.25}$$

Sensitized lymphocytes stimulate the production of immunocompetent lymphocytes, and the effect of each sensitized cell lasts for a given period of time. For the purpose of the present analysis, it is assumed that the effect of each sensitized lymphocyte decays exponentially over a period of time. The production rate of blood lymphocytes at any time t due to the primary response $(dL_B/dt)_{\text{prim}}$ would then be equal to the sum of the residual effect of all the lymphocytes sensitized between time 0 and time $t - \Phi_1$, where Φ_1 is the time lag between sensitization and production of the lymphocytes.

The number of lymphocytes produced because of primary response between time t and time $(t - \Phi_1)$ would be

$$L_B(t) - L_B(t - \Delta t) = C_3\{\underbrace{L_s(t - \Phi_1)\,\Delta t}_{\substack{\text{Due to lymphocytes}\\ \text{sensitized at } t - \Phi_1}} + \underbrace{L_s(t - \Phi_1 - \Delta t)\,\Delta t}_{\substack{\text{Due to lymphocytes}\\ \text{sensitized at } t - \Phi_1 - \Delta t}} + \underbrace{L_s(t - \Phi_1 - 2\,\Delta t)e^{-K_1 \Delta t}\,\Delta t}_{\substack{\text{Due to lymphocytes}\\ \text{sensitized at } t - 2\Phi_1 - \Delta t}} +$$

$$+ \underbrace{L_s(t - \Phi_1 - r\,\Delta t)e^{-K_1 r \Delta t}\,\Delta t}_{\substack{\text{Due to lymphocytes}\\ \text{sensitized at } t - \Phi_1 - r\,\Delta t}} + \cdots\} \tag{1.26}$$

$$= C_3 \Sigma\, L_s(t - \Phi_1 - r\,\Delta t)e^{-K_1 r \Delta t}\,\Delta t \tag{1.27}$$

Dividing by Δt and taking the limits as $\Delta t \rightarrow 0$ changes the left-hand side to a derivative and the right-hand side to an integral in terms of the hereditary function:

$$\left(\frac{dL_B(t)}{dt}\right)_{\text{primary}} = C_3 \int_0^{t-\Phi_1} L_s(\tau)e^{-K_1(t-\Phi_1-\tau)}\,d\tau \tag{1.28}$$

Substituting for L_s in terms of L_T gives

$$\left(\frac{dL_B(t)}{dt}\right)_{\text{primary}} = k_2 \int_0^{t-\Phi_1} L_T(\tau)e^{-K_1(t-\Phi_1-\tau)}\,d\tau \tag{1.29}$$

For the secondary response to appear, a memory cell must encounter a target cell, and therefore the secondary response depends on the number of memory cells and the number of target cells. As in the above equation, Reddy and Krouskop (1978) expressed the secondary response in terms of a hereditary function

$$\left(\frac{dL_B(t)}{dt}\right)_{\text{secondary}} = k_3 \int_0^{t-\Phi_2} L_T(\tau)g(\tau)e^{-K_4(t-\Phi_2-\tau)}g(t-\Phi_3)\,d\tau \tag{1.30}$$

In developing the above equation, it is assumed that the effect of a memory cell also decays exponentially over a period of time. Thus the production rate of blood lymphocytes at time t due to secondary response $(dL_B/dt)_{\text{second}}$ is due to the sum of the residual effects of all the memory cells formed between 0 and time $t - \phi_2$, where ϕ_2 is the time lag between memory cell formation and the appearance of the secondary response.

The net rate in change of blood lymphocytes may then be described as

$$\frac{dL_B}{dt} = \underbrace{k_2 \int_0^{t-\Phi_1} L_T(\tau)e^{-K_1(t-\Phi_1-\tau)}\,d\tau}_{\text{Due to primary response}} + \underbrace{k_3 \int_0^{t-\Phi_2} L_T(\tau)g(\tau)e^{-K_4(t-\Phi_2-\tau)}g(t-\Phi_3)\,d\tau}_{\text{Due to secondary response}}$$
$$+ \underbrace{K_5L_T - K_6L_B}_{\text{Recirculation}} \quad \underbrace{-K_7L_B}_{\text{Death in the blood}} \quad \underbrace{-K_8L_Bg}_{\text{Migration into tissue due to target cell presence}} \tag{1.31}$$

The rates of change of tissue lymphocytes and the number of target cells can be described by a similar mass balance:

$$\frac{dL_T}{dt} = \underbrace{K_8L_Bg}_{\text{Increased migration}} \quad \underbrace{-K_5L_T + K_6L_B}_{\text{Recirculation}} \quad \underbrace{-K_9L_Tg}_{\text{Loss due to target cell destruction}} \tag{1.32}$$

$$\frac{dg}{dt} = \left(\frac{dg}{dt}\right)_{\text{input}} - k_{10}L_Tg \tag{1.33}$$

These equations were simulated by Reddy and Krouskop. Figure 1.9 shows the production rate of lymphocytes and the number of target cells when the target cells were introduced on day 0 and again on day 4. Figure 1.10 shows the production rate of lymphocytes and the number of target cells present in the tissue when the target cells were introduced continuously.

1.4.2 A Model to Predict the Number of Engineers in the United States

An easy to understand example of a deterministic model with time delay and memory is a model to predict the number of engineers in the United States at any given time. Let us restrict our analysis to a single discipline such as biomedical engineering. Let E be the number of engineers (biomedical) at any given time. The time rate of change of the number of engineers at any given time in the United States can be expressed as:

$$\frac{dE}{dt} = G + I - R - L - M \tag{1.34}$$

where G = number of graduates entering the profession (graduating from an engineering program) per unit time
I = number of engineers immigrating into the United States per unit time
R = number of engineers retiring per unit time
L = number of engineers leaving the profession per unit time (to become doctors, lawyers, managers, etc.)
M = number of engineers dying (before retirement) per unit time

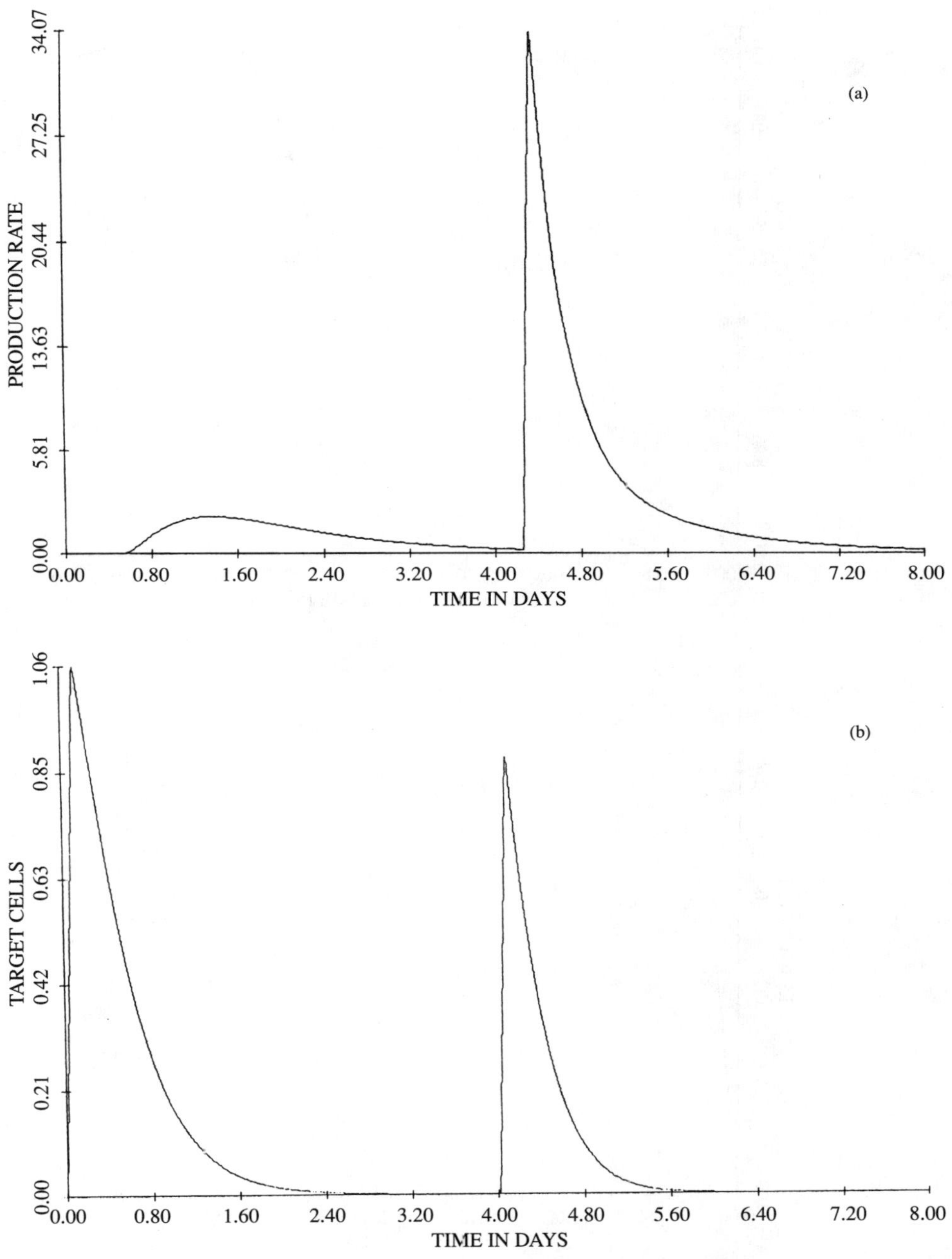

FIGURE 1.9 The simulation results of the production rate of lymphocytes (*a*), and the number of target cells or foreign cells (*b*) plotted as a function of time in days. In the simulation, the target cells were introduced on day 0 and again on day 4. [*From Reddy and Krouskop (1978).*] Note the increased secondary response.

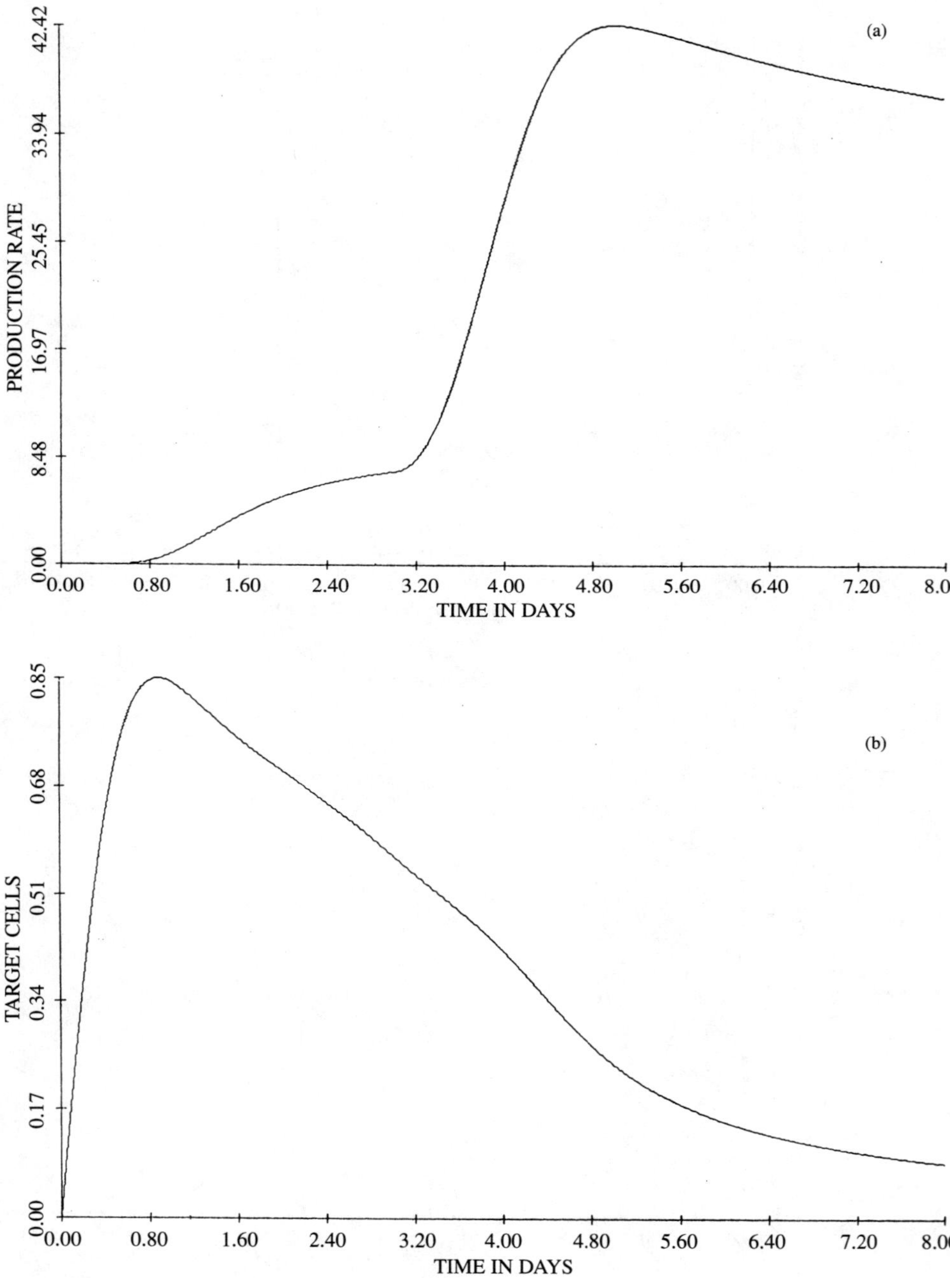

FIGURE 1.10 The simulation results of the production rate of lymphocytes (*a*), and the number of target cells or foreign cells (*b*) when the target cells were continuously introduced. [*From Reddy and Krouskop (1978).*]

In this equation, we have lumped the entire United States as a single region (a well-stirred compartment) with homogeneous distribution. In addition, we have not made any distinction with regard to age, sex, or professional level. We have considered the entire pool as a well-stirred homogeneous compartment. In reality, there is a continuous distribution of ages. Even with this global analysis with a lumped model, we could consider the age distribution with a series of compartments with each compartment representing engineers within a particular age group. Moreover, we have assumed that all engineering graduates enter the work force. A percentage of them go to graduate school and enter the workforce at a later time.

The number of graduates entering the profession is a function of the number of students entering the engineering school 4 years before:

$$G(t) = k_1 S(t - 4) \tag{1.35}$$

where $S(t)$ is the number of students entering the engineering school per unit time. The number of students entering the engineering school depends on the demand for the engineering profession over a period of years, i.e., on the demand history.

The number of engineers immigrating into the United States per unit time depends on two factors: demand history in the United States for engineers and the number of visas that can be issued per unit time. Assuming that the immigration visa policy is also dependent on the demand history, we can assume that I is dependent on demand history. Here we have assumed that immigrants from all foreign countries are lumped into a single compartment. In reality, each country should be placed in a separate compartment and intercompartmental diffusion should be studied.

The number of engineers retiring per unit time is proportional to the number of engineers in the profession at the time:

$$R(t) = k_2 E(t) \tag{1.36}$$

The number of engineers leaving the profession depends on various factors: the demand for the engineering profession at that time, demand for various other professions at that time, and several personal factors. For the purpose of this analysis, let us assume that the number of engineers leaving the profession in a time interval is proportional to the number of individuals in the profession at that time:

$$L(t) = k_3 E(t) \tag{1.37}$$

The number of engineers dying (before retirement) per unit time is proportional to the number of engineers at that time:

$$M(t) = k_4 E(t) \tag{1.38}$$

The demand for engineers at any given time is proportional to the number of jobs available at that time ($J(t)$) and is inversely proportional to the number of engineers available at that time:

$$D(t) = kJ(t)/E(t) \tag{1.39}$$

The number of jobs available depends on various factors, such as government spending for R&D projects, economic growth, sales of medical products, and number of hospitals. Let us assume in this case (biomedical engineering), the number of jobs is directly proportional to the sales of medical products (p), directly proportional to government spending for health care R&D (e), and directly proportional to the number of new medical product company start-ups (i):

$$J(t) = (k_6 e + k_7 c + k_8 I + k_9 + kp) \tag{1.40}$$

Although we assumed that the number of jobs at the present time is dependent on $e(t)$, $c(t)$, $h(t)$, $i(t)$, and $p(t)$, in reality the number of jobs at present may depend on previous values of these parameters, or may depend on the history of these parameters.

Let us now analyze the demand history. This history depends on the memory function. Let us assume that the effect of demand existing at a time decays exponentially (exponentially decaying memory). The net effect of demands from time 0 to t can be expressed as

$$H_1(t) = \int_{\tau=0}^{\tau=t} \{D(\tau) \exp[-k_{10} (t - \tau)]\} \, d\tau \tag{1.41}$$

The number of students entering engineering school per unit time is

$$S(t) = k_{11} H_1(t) \tag{1.42}$$

Immigration rate can similarly be expressed as:

$$I(t) = k_{12} H_2(t) \tag{1.43}$$

where

$$H_2(t) = \int_{\tau=0}^{\tau=t} \{D(\tau) \exp[-k_{13}(t - \tau)]\} \, d\tau \tag{1.44}$$

H_1 and H_2 are called *hereditary functions.* Instead of an exponential decay of memory, we could have a sinusoidal or some other functional form of memory decay, depending on the physical situation.

$$\frac{dE}{dt} = k_1 k_{10} H_1(t - 4) + k_{11} H_2(t) - (k_2 + k_3 + k_4) E(t) \tag{1.45}$$

In this analysis, making various assumptions, we have formulated a lumped-parameter deterministic model to predict the number of engineers (biomedical) present in the United States at any given time. If we want to know the geographical distribution, we can take two approaches. We can divide the entire United States into a number of compartments (e.g., northeast, east, west, etc.) and study the intercompartmental diffusion. Alternatively, we can make E a continuous variable in space and time, $I(x,y,t)$, and account for spatial diffusion.

1.5 ARTIFICIAL NEURAL NETWORK MODELS

Neural network models represent the black box type of model. These models are used where the precise functioning of the system is not understood but the sample input-output data are known. Neural networks represent a new generation of information processing systems that are artificially (virtually) constructed to make use of the organizational structure of the neuronal information processing in the brain. A neural network consists of several interconnecting neurons, also called *nodes.* These nodes are organized into an input layer, one or more hidden layers, and an output layer. The number of input layer nodes in the input layer depends on the number of input variables. The number of nodes in the output layer is determined by the number of output parameters. Each input parameter is represented by a node in the input layer, and each output parameter is represented by a node in the output layer. The number of nodes in the hidden layer could be variable. Each node in a feed-forward neural network is connected to every node in the next level of nodes. That is, each input node is connected to all the nodes in the hidden-layer neurons. Let us for simplicity consider only one hidden layer. Now, each node in the hidden layer is connected to all the nodes in the output layer. Figure 1.11 shows a network with four output nodes and five input nodes with four hidden-layer nodes. The connection strengths are determined by the weights.

Let us assume that $W_{i,j}$ represents the weight of the connection from jth node in the hidden layer to the ith node in the output layer, and let us assume that $w_{j,k}$ represents the weight of connection

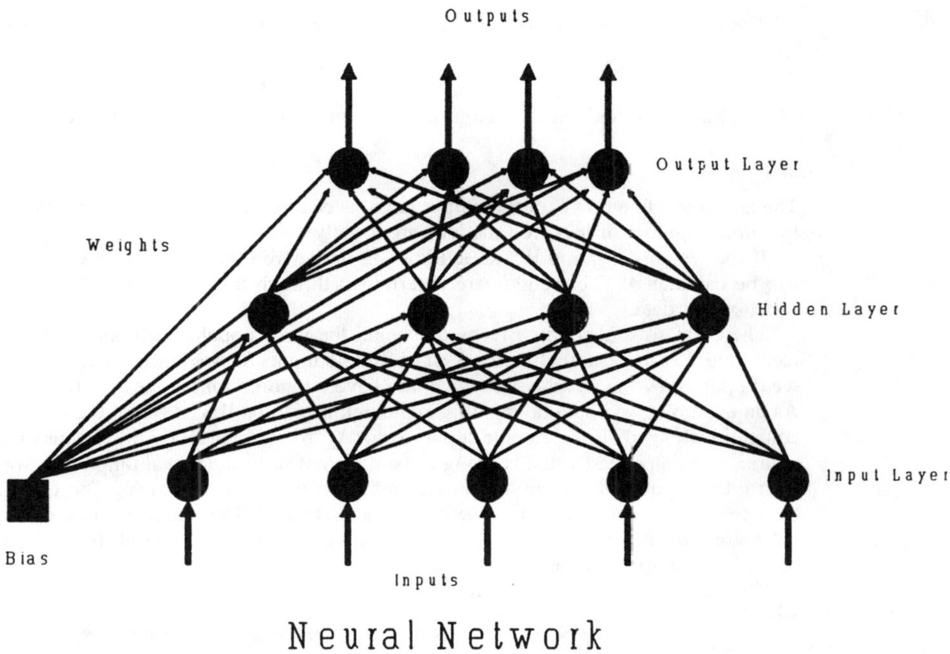

FIGURE 1.11 A neural network model consists of several input layer neurons (nodes), one or more neurons in the output layer, and one or more layers of hidden layer neurons each consisting of several neurons. Each neuron in the input layer corresponds to an input parameter, and each neuron in the output layer corresponds to an output parameter. Each neuron in a layer is connected to each of the neurons in the next level. In this example only one hidden layer is used. Each of the input neurons is connected with each neuron in the hidden layer, and each neuron in the hidden layer is connected to each neuron in the output layer. The connection strengths are represented by weights.

from kth input node to jth node in the hidden layer. Let X_k represent the value of kth input node. The sum of weighted inputs to the jth node in the hidden layer is

$$I_j = \Sigma w_{j,k} X_k \tag{1.46}$$

In other words,

$$I_1 = w_{1,1}X_1 + w_{1,2}X_2 + w_{1,3}X_3 + w_{1,4}X_4 \tag{1.47}$$

where X_1, X_2, X_3, and X_4 are the values of the four input parameters.

The output of a hidden-layer neuron is a function of its input:

$$H_j = f(I_j) \tag{1.48}$$

This function f is called the *activation function*. An example of this function is the sigmoid function

$$H_j = k\left[\frac{2}{1 + \exp(-aI_j + B)} - 1\right] \tag{1.49}$$

where k, a, and b are constants. B is called the *bias*, and can be zero. In general, any monotone, nondecreasing differentiable signal function can be used as the activation function.

The input G_i to the ith node in the output layer is the sum of its weighted inputs:

$$G_i = \Sigma W_{i,j} H_j \tag{1.50}$$

The output of the node in the output layer is some function of the input node.

$$Y_i = F(G_i) \tag{1.51}$$

The activation function F of the output neurons can be any monotone, nondecreasing differentiable function. Sigmoid or logistic functions are usually used.

If the weights $W_{j,k}$ and $W_{i,j}$ are all known, then given the inputs X_k, the output Y_i of the system can be calculated. The weights are determined through a training algorithm by using the sample input-output data.

There are several training techniques, and the most popular technique is the back propagation technique. Let us assume that, for a set of sample inputs X_k, we know the actual outputs d_i. Initially, we do not know the weights, but we could have a random initial guess of the weights $W_{j,k}$ and $W_{i,j}$. As an example, we could define all weights initially to be $W_{j,k} = W_{i,j} = 0.2$ or 0.5. Using the above equations along with the sample input vector $X_{k,}$ we can calculate the output of the system $Y_{i.}$ Of course, this calculated value is going to be different from the actual output (vector, if there is more than one output node) value d, corresponding to the input vector $X_{k.}$ The error is the difference between the calculated output value and the actual value. There are various algorithms to iteratively calculate the weights, each time changing the weights as a function of the error. The most popular of these is the gradient descent technique.

The sum of the error in the mth iteration is defined as

$$e_i^m = d_i - y_i^m = d_i - F(\Sigma W_{i,j}^m H_j) = d_i - F(\Sigma W_{i,j}^m f(\Sigma w_{j,k}^m X_k) \tag{1.52}$$

The instantaneous summed squared error at an iteration m corresponding to the sample data set n can be calculated as

$$E_n^m = \tfrac{1}{2}\, \Sigma (e_i^m)^2 \tag{1.53}$$

The total error E at each iteration, for all the sample data pairs (input-output), can be calculated as the sum of the errors E_n for the individual sample data.

Adjusting the weights for each iteration for connections between the hidden layer neurons and the output layer neurons $W_{i,j}$ can be calculated as

$$W_{i,j}^{m+1} = W_{i,j}^m - Z\left(\frac{\delta E^m}{\delta W_{i,j}}\right) \tag{1.54}$$

where Z is the learning rate.

The error gradient can be expressed as

$$\frac{\delta E^m}{\delta W_{i,j}} = \left(\frac{\delta E^m}{\delta F}\right)\left(\frac{dF}{dW_{i,j}}\right) \tag{1.55}$$

For a sigmoid activation function, it turns out that the differential is a simple function of the sigmoid:

$$dF = b(1 - F)F \tag{1.56}$$

where b is a constant. Thus,

$$\frac{dF}{dW_{i,j}} = b(1 - F(W_{i,j}))F(W_{i,j}) \tag{1.57}$$

For adjusting the weights for connections between the input and the hidden-layer neurons, the error

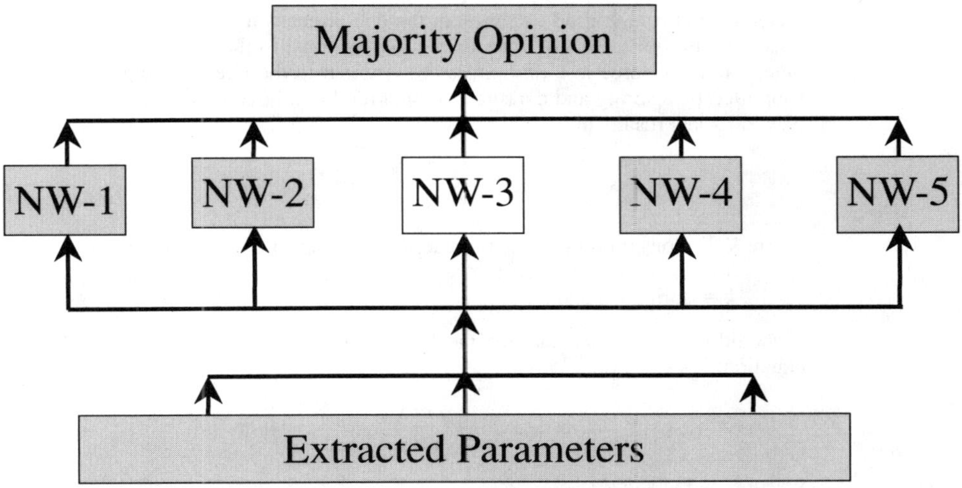

FIGURE 1.12 The committee of neural networks. Each of the input parameters is simultaneously fed to several networks working in parallel. Each network is different from the others in terms of initial training weights or the activation function (transfer function at the nodes). A majority opinion of the member networks provides the final decision of the committee. This committee of networks simulates the parliamentary process, and emulates a group of physicians making the decision. [*From Reddy and Buch (2000).*]

is back-propagated by calculating the partial derivative of the error E with respect to the weights $w_{j,k}$ similarly (Haykin, 1999).

The whole process of calculating the weights by using the sample data sets is called the *training process*. There is a Neural Network package in MATLAB that can be easily used in the training process. There are several algorithms in the package including back-propagation, modified back-propagation, etc., that the user can choose in the MATLAB software. Once the weights are calculated by MATLAB or any other software, it becomes a matter of obtaining the output vector for a given input vector by using matrix multiplications. The most important aspect of a neural network is that it should be tested with data not used in the training process.

Neural networks have been used for classification and control. For instance, Reddy et al. (1995) used neural networks to classify the degree of the disease in dysphagic patients by using noninvasive measurements (of throat acceleration, swallow suction, pressure, etc.) obtained from dysphagic patients during swallowing. These measurements were the inputs to the network and the outputs were normal, mild, moderate, and severe. Neural network performance depends on a black-box approach, and since the performance depends on the sample data, initial weights, etc., Reddy et al. (1995) trained several networks with various initial conditions and activation functions. On the basis of some initial testing with known data, they recruited the best five networks into a committee. A majority opinion of the committee was used as the final decision. Reddy and Buch (2000) and Das et al. (2001) obtained better results with a committee of neural networks (Fig. 1.12) when compared to a single network, and the majority opinion of the committee was in agreement with clinical or actual classification.

1.6 MECHANICAL MODELS

Mechanical models consisting of combinations of springs and dashpots are very popular in numerous disciplines. Spring-dashpot models have been used to model the mechanical behavior of viscoelastic materials and can be used to represent the one-dimensional behavior of tissue and other biological materials. In a linear spring, the force is proportional to the change in length or the strain. On the other hand, the force in a dashpot is proportional to the rate of change in strain. Consider a mass supported by a spring and a dashpot in parallel. Let a force F be acting on the mass. Application of Newton's law results in:

$$m\left(\frac{d^2X}{dt^2}\right) + b\left(\frac{dX}{dt}\right) + kX = F \tag{1.58}$$

where X = elongation or change in length with respect to the steady-state value
b = constant of the dashpot
k = spring constant

It should be pointed out that the mechanical equation (1.58) is similar to the following electrical equation:

$$L\left(\frac{di}{dt}\right) + Ri + \frac{1}{C}\int i\,dt = E \tag{1.59}$$

where L = inductance
R = resistance
i = current
E = voltage

This equation can be expressed in terms of the charge q instead of the current as

$$L\left(\frac{d^2q}{dt^2}\right) + R\left(\frac{dq}{dt}\right) + \frac{1}{C}q = E \tag{1.60}$$

Equations (1.58), (1.59), and (1.60) are similar. Therefore, mass is analogous to the inductor, the dashpot is analogous to the resistor, and the spring is analogous to the capacitor. The spring and the capacitor are storage units, whereas the dashpot and the resistor are dissipators of energy. The charge is analogous to the deformation or elongation, the current is similar to the velocity, and force is analogous to the voltage. Therefore, any electrical system can be modeled by mechanical analogs and any mechanical system can be modeled by electrical analogs.

Lumped mechanical models have been used to analyze impact dynamics. Generally, muscle is represented by a combination of a spring and a dashpot, whereas a ligament is modeled by a spring. Human body vibrations can be analyzed by similar models (Fritton et al., 1997).

1.7 MODEL VALIDATION

This chapter discussed the art of modeling, with a few examples. Regardless of the type of model developed, a mathematical model should be validated with experimental results. Validation becomes very important in the black-box type of models such as the neural network models. Moreover, the model results are valid only within certain regimes where the model assumptions are valid. Sometimes, any model can be fit to a particular set of data by adjusting the parameter values. The techniques of parameter estimation were not presented in this chapter; the presentation was limited to lumped-parameter analysis, or macroscopic modeling.

REFERENCES

Babb, A. L., et al. (1967). "Engineering Aspects of Artificial Kidney Systems," *Chemical Engineering in Medicine and Biology* (editor: D. Hershey), Plenum Press, New York, 289–332.

Barnea, O., and Gillon, G., (2001). "Model-based Estimation of Male Urethral Resistance and Elasticity Using Pressure-Flow Data," *Computers in Biology and Medicine*, **31**: 27–40.

Cooney, D. O. (1976). *Biomedical Engineering Principles: An Introduction to Fluid, Heat, and Mass Transport Processes*, M. Dekker, New York.

Chu, T. M., and Reddy, N. P. (1992). "A Lumped Parameter Mathematical Model of the Splanchnic Circulation," *ASME Journal of Biomechanical Engineering*, **114**, 222–226.

Das, A., Reddy, N. P., and Narayanan, J. (2001). "Hybrid Fuzzy Logic Committee Neural Networks for Recognition of Acceleration Signals Due to Swallowing," *Computer Methods and Programs in Biomedicine*, **64**, 87–99.

Fritton, J. C., Rubin, C. T., Qin, Y, and McLeod, K. J. (1997). "Whole Body Vibration in the Skeleton: Development of a Resonance-Based Testing Device," *Annals of Biomedical Engineering*, **25**: 831–839.

Frost, H. (1963). *Bone Remodeling Dynamics*, Little Brown, Boston.

Haykin, S. (1999). *Neural Networks: A Comprehensive Foundation*, 2d ed., Prentice Hall, Upper Saddle River, N. J.

Reddy, N. P., and Krouskop, T. A. (1978). "Modeling the Cell Mediated Immunity in Homograft Rejection," *International Journal of Biomedical Computing*, **9**, 327–340.

Reddy, N. P., and Kesavan, S. K. (1989). "Low Reynolds Number Liquid Propulsion in Contracting Tubular Segments Connected Through Valves," *Mathematical and Computer Modeling*, **12**, 839–844.

Reddy, N. P., and Buch, O. (2000). "Committee Neural Networks for Speaker Verification," *Intelligent Engineering Systems through Artificial Neural Networks*, Vol. 5, *Fuzzy Logic and Evolutionary Programming* (editors: C. Dagli, A. Akay, C. Philips, B. Fernadez, J. Ghosh), ASME Press, New York, 911–915.

Reddy, N. P., Prabhu, D., Palreddy, S., Gupta, V., Suryanarayanan, S., and Canilang, E. P. (1995). "Redundant Neural Networks for Medical Diagnosis: Diagnosis of Dysphagia," *Intelligent Engineering Systems through Artificial Neural Networks*, Vol. 5, *Fuzzy Logic and Evolutionary Programming* (editors: C. Dagli, A. Akay, C. Philips, B. Fernadez, J. Ghosh), ASME Press, New York, 699–704.

Reddy, N. P. (1986), "Lymph Circulation: Physiology, Pharmacology and Bioengineering." *CRC Critical Reviews in Biomedical Engineering*, **14**, 45–91, 1986.

Reddy, N. P., and Joshi, A. (1987). "A Stochastic Compartmental Model of Bone Cells," *International Journal of Biomedical Computing*, **23**, 163–174.

Rideout, V. C., (1991). *Mathematical and Computer modeling of Physiological Systems*, Prentice Hall, Englewood Cliffs, N. J.

Simon, B. N., Reddy, N. P., and Kantak, A. (1994). "A Theoretical Model of Infant Incubator Dynamics," *ASME Journal of Biomechanical Engineering*, **116**, 263–269.

P • A • R • T • 2

MECHANICS OF THE HUMAN BODY

CHAPTER 2
BIOHEAT TRANSFER

Liang Zhu
University of Maryland, Baltimore County, Baltimore, Maryland

2.1 INTRODUCTION

Over the past 100 years, the understanding of the thermal and mechanical properties of human tissues and the physics that governs biological processes has been greatly advanced by the utilization of fundamental engineering principles in the analysis of many biological heat and mass transfer applications. During the past decade, there has been an increasingly intense interest in bioheat transfer phenomena, with particular emphasis on therapeutic and diagnostic applications. Relying on advanced computational techniques, the development of complex mathematical models has greatly enhanced our ability to analyze various types of bioheat transfer process. The collaborations among physiologists, physicians, and engineers in the bioheat transfer field have resulted in improved treatment, preservation, and protection techniques for biological systems, including use of heat or cold treatments to destroy tumors and the protection of humans from extreme environmental conditions.

In this chapter we start with fundamental aspects of local blood tissue thermal interaction. Discussions on how the blood effect is modeled then follow. Different approaches to theoretically modeling the blood flow in the tissue are shown. In particular the assumptions and validity of several widely used continuum bioheat transfer equations are evaluated. Different techniques to measure temperature, thermophysical properties, and blood flow are then described. The final part of the chapter focuses on one of the medical applications of heat transfer, hyperthermia treatment for tumors.

2.2 FUNDAMENTAL ASPECTS OF BIOHEAT TRANSFER

One of the remarkable features of the human thermoregulatory system is that we can maintain a core temperature near 37°C over a wide range of environmental conditions and during thermal stress. The value of blood flow to the body varies over a wide range depending upon the need for its three primary functions:

1. **Mass transfer in support of body metabolisms.** Blood transports oxygen to the rest of the body and transports carbon dioxide and other waste from the cells.
2. **Regulation of systemic blood pressure.** The vascular system is a primary effector in the regulation of systemic blood pressure through its ability to alter the distribution of blood flow and regulate the cardiac output and thereby buffer systemic pressure fluctuations.
3. **Heat transfer for systemic thermoregulation.**

As for the third primary function, blood is known to have a dual influence on the thermal energy balance. First it can be a heat source or sink, depending on the local tissue temperature. During wintertime, blood is transported from the heart to warm the rest of the body. On the other hand, during hyperthermia treatment for certain diseases where the tissue temperature is elevated to as high as 45°C by external devices, the relatively cold blood forms cold tracks that can decrease the treatment efficacy. The second influence of the blood flow is that it can enhance heat dissipation from the inside of the body to the environment to maintain a normal body temperature. Theoretical study has shown that if the heat produced in the central areas of the body at rest condition could escape only by tissue conduction, the body temperature would not reach a steady state until it was about 80°C. A lethal temperature would be reached in only 3 hours. During exercise, our body temperature would typically rise 12°C in 1 hour if no heat were lost by blood flow. Maintaining a core temperature of 37°C during thermal stress or exercise is achieved by increasing the cardiac output by central and local thermoregulation, redistributing heat via the blood flow from the muscle tissue to the skin, and speeding the heat loss to the environment by evaporation of sweat.

Thermal interaction between blood and tissue can be studied either experimentally or theoretically. However, for the following reasons it is difficult to evaluate heat transfer in a biological system:

- The complexity of the vasculature. It is not practical to develop a comprehensive model that includes the effect of all thermally significant vessels in a tissue. Therefore, the most unusual and difficult basic problem of estimating heat transfer in living biologic systems is modeling the effect of blood circulation.
- Temperature response of the vasculature to external and internal effects is also a complex task. In a living system, the blood flow rate and the vessel size may change as a response to local temperature, local pH value, and the concentration of local O_2 and CO_2 levels.
- The small thermal length scale involved in the microvasculature. Thermally significant blood vessels are generally in a thermal scale of less than 300 μm. It has been difficult to build temperature-measuring devices with sufficient resolution to measure temperature fluctuation.

For the above reasons, even if the heat transfer function of the vascular system has been appreciated since the mid-nineteenth century, only in the past 2 decades, has there been a revolution in our understanding of how temperature is controlled at the local level, both in how local microvascular blood flow controls the local temperature field and how the local tissue temperature regulates local blood flow.

Until 1980, it was believed that, like gaseous transport, heat transfer took place in the capillaries because of their large exchange surface area. Several theoretical and experimental studies (Chato, 1980; Chen and Holmes, 1980; Weinbaum et al., 1984; Lemons et al., 1987) have been performed to illustrate how individual vessels participate in local heat transfer and thus, to understand where the actual heat transfer between blood and tissue occurs. In these analyses, the concept of thermal equilibration length was introduced. Thermal equilibration length of an individual blood vessel was defined as a distance over which the temperature difference between blood and tissue drops a certain percentage. For example if the axial variation of the tissue and blood temperature difference can be expressed as $\Delta T = \Delta T_0 \, e^{-x/L}$, where ΔT_0 is the temperature difference at the vessel entrance, then L and $4.6L$ are the thermal equilibration length over which ΔT decreases to 37 percent and 1 percent, respectively, of its value at the entrance. Blood vessels whose thermal equilibration length is comparable to their physical length are considered thermally significant.

Chato (1980) first theoretically investigated the heat transfer from individual blood vessels in three configurations: a single vessel, two vessels in counterflow, and a single vessel near the skin surface. It was shown that the Graetz number, proportional to the blood flow velocity and radius, is the controlling parameter determining the thermal equilibration between the blood and tissue. For blood vessels with very low Graetz number, blood quickly reaches the tissue temperature. It was also demonstrated that heat transfer between the countercurrent artery and vein was affected by the vessel center-to-center spacing and mass transport between them.

In an anatomic study performed on rabbit limbs, Weinbaum et al. (1984) identified three vascular layers (deep, intermediate, and cutaneous) in the outer 1-cm tissue layer. Subsequently, three fundamental vascular structures were derived from the anatomic observation: (1) an isolated vessel embedded in a tissue cylinder, as shown by the intermediate tissue layer; (2) a large artery and its countercurrent vein oriented obliquely to the skin surface, as shown in the deep tissue layer; and (3) a vessel or vessel pair running parallel to the skin surface in the cutaneous plexus. These three vascular structures served as the basic heat transfer units in the thermal equilibration analysis in Weinbaum et al. (1984).

As shown in Weinbaum et al. (1984), 99 percent thermal equilibration length of a single blood vessel embedded in a tissue cylinder was derived as

$$x_{cr} = 1.15a \Pr \mathrm{Re}\left[0.75 + K \ln\left(\frac{R}{a}\right)\right] \tag{2.1}$$

where a and R are the blood vessel and tissue cylinder radius, respectively, Pr and Re are the vessel Prandtl number and Reynolds number, respectively, and K is the ratio of blood conductivity to tissue conductivity. It is evident that x_{cr} is proportional to the blood vessel size and its blood flow velocity. Substituting the measured vascular geometry and the corresponding blood flow rate number for different blood vessel generations (sizes) from a 13-kg dog (Whitmore, 1968), one could calculate the thermal equilibration length as listed in Table 2.1.

TABLE 2.1 Thermal Equilibration Length in a Single Vessel Embedded in a Tissue Cylinder

Vessel radius a, μm	Vessel length L, cm	R/a	x_{cr}, cm
300	1.0	30	9.5
100	0.5	20	0.207
50	0.2	10	0.014
20	0.1	7	0.0006

Several conclusions were drawn from the comparison between x_{cr} and L. In contrast to previous assumptions that heat transfer occurs in the capillary bed, for blood vessels smaller than 50 μm in diameter, blood quickly reaches the tissue temperature; thus, all blood-tissue heat transfer must have already occurred before entering into these vessels. For blood vessels larger than 300 μm in diameter, there is little change in blood temperature in the axial direction because of their much longer thermal equilibration length compared with the vessel length. The medium-size vessels between 50 and 300 μm diameter are considered thermally significant because of their comparable thermal equilibration length and physical length. Those blood vessels are primary contributors to tissue heat transfer. Note that the conclusions are similar to that drawn by Chato (1980).

The most important aspect of the bioheat transfer analysis by Weinbaum and coinvestigators was the identification of the importance of countercurrent heat transfer between closely spaced, paired arteries and veins. The countercurrent heat exchange mechanism, if dominant, was suggested as an energy conservation means since it provides a direct heat transfer path between the vessels. It was observed that virtually all the thermally significant vessels (>50 μm in diameter) in the skeletal muscle were closely juxtaposed artery-vein pairs (Weinbaum et al., 1984). Thermal equilibration in the artery (50 to 300 μm diameter) in a countercurrent pair was estimated on the basis of a simple heat conduction analysis in the cross-sectional plane. It was noted that the thermal equilibration length in the countercurrent artery was at least 3 times shorter than that in a single vessel of the same size embedded in a tissue cylinder (Weinbaum et al., 1984). Significantly short thermal equilibration length in comparison with that of a single vessel suggests that the primary blood tissue heat exchange mechanism for vessels larger than 50 μm in the deep layer is the incomplete countercurrent heat exchange. Therefore, for modeling heat transfer in these tissue regions, reasonable assumptions related to the countercurrent heat exchange can be made to simplify the mathematical formulation.

Theoretical analysis of the thermal equilibration in a large vessel in the cutaneous layer (Chato, 1980; Weinbaum et al., 1984) demonstrated that its thermal equilibration length was much longer than its physical length during normal and hyperemic conditions despite the close distance from the skin surface. It was suggested that the large vessels in the cutaneous layer can be far from thermal equilibration and are therefore capable of delivering warm blood from the deep tissue to the skin layer. This superficial warm blood shunting is very important in increasing the normal temperature gradient at the skin surface and therefore plays an important role in losing heat during heavy exercise. On the contrary, during surface cooling there is rapid cutaneous vasoconstriction in the skin. The minimally perfused skin, along with the underlying subcutaneous fat, provides a layer of insulation, and the temperature gradient from the skin surface into the muscle becomes almost linear (Bazett, 1941) yielding the lowest possible heat transfer from the body.

2.3 BIOHEAT TRANSFER MODELING

The effects of blood flow on heat transfer in living tissue have been examined for more than a century, dating back to the experimental studies of Bernard in 1876. Since then, mathematical modeling of the complex thermal interaction between the vasculature and tissue has been a topic of interest for numerous physiologists, physicians, and engineers. A major problem for theoretical prediction of temperature distribution in tissue is the assessment of the effect of blood circulation, which is the dominant mode of heat removal and an important cause of tissue temperature inhomogeneity.

Because of the complexity of the vascular geometry, there are two theoretical approaches describing the effect of blood flow in a biological system. Each approach represents two length scales over which temperature variations may occur.

- *Continuum models*, in which the effect of blood flow in the region of interest is averaged over a control volume. Thus, in the considered tissue region, there is no blood vessel present; however, its effect is treated by either adding an additional term in the conduction equation for the tissue or changing some of the thermophysical parameters in the conduction equation. The continuum models are simple to use, since the detailed vascular geometry of the considered tissue region need not be known as long as one or two representative parameters related to blood flow are available. The shortcoming of the continuum model is that since the blood vessels disappear, no point-by-point variation in the blood temperature is available. Another shortcoming is associated with the assumptions introduced when the continuum model was derived. For different tissue regions and physiological conditions, these assumptions may not be valid.
- *Vascular models*, in which blood vessels are represented as tubes buried in tissue. Because of the complicated vascular geometry, one may only consider several blood vessels and neglect the others. Recent studies [Dorr and Hynynen, 1992; Crezee and Lagendijk, 1990; Roemer, 1990] have demonstrated that blood flow in large, thermally unequilibrated vessels is the main cause for temperature nonhomogeneity during hyperthermia treatment. Large blood vessels may significantly cool tissue volumes around them, making it very difficult to cover the whole tumor volume with therapeutic thermal exposure. In applications where point-to-point temperature nonuniform-

ities are important, vascular model has been proved to be necessary to predict accurately the tissue temperature field (Zhu et al., 1996a). In recent years, with the breakthrough of advanced computational techniques, vascular models [Raaymakers et al., 2000] for simulating vascular networks have grown rapidly and already demonstrated its great potential in accurate and point-to-point blood and tissue temperature mapping.

2.3.1 Continuum Models

In continuum models, blood vessels are not modeled individually. Instead, the traditional heat conduction equation for the tissue region is modified by either adding an additional term or altering some of the key parameters. The modification is relatively simple and is closely related to the local vasculature and blood perfusion. Even if the continuum models cannot describe the point-by-point temperature variation in the vicinity of larger blood vessels, they are easy to use and allow the manipulation of one or several free parameters. Thus, they have much wider applications than the vascular models. In the following sections, some of the widely used continuum models are introduced and their validity is evaluated on the basis of the fundamental heat transfer aspects.

Pennes Bioheat Transfer Model. It is known that one of the primary functions of blood flow in a biological system is its ability to heat or cool the tissue, depending on the relative local tissue temperature. The existence of a temperature difference between the blood and tissue is taken as evidence of its function to remove or release heat. On the basis of this speculation, Pennes [Pennes, 1948] proposed his famous bioheat transfer model, which is called Pennes' bioheat equation. Pennes suggested that the effect of blood flow in the tissue be modeled as as heat source or sink term added to the traditional heat conduction equation. The Pennes bioheat equation is given by

$$\rho C \frac{\partial T_t}{\partial t} = k_t \nabla^2 T_t + q_{\text{blood}} + q_m \tag{2.2}$$

where q_m is the metabolic heat generation in the tissue and the second term (q_{blood}) on the right side of the equation takes into account the contribution of blood flow to the local tissue temperature distribution. The strength of the perfusion source term can be derived as follows.

Figure 2.1 shows a schematic diagram of a small tissue volume perfused by a single artery and vein pair. The tissue region is perfused via a capillary network bifurcated from the transverse arterioles, and the blood is drained by the transverse venules. If one assumes that both the artery and

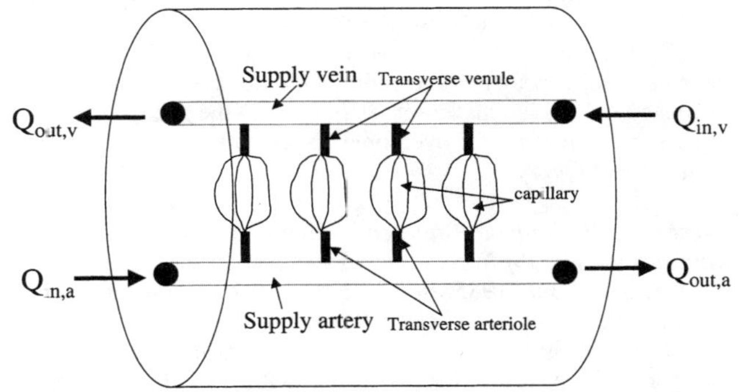

FIGURE 2.1 Schematic diagram of a tissue volume perfused by a single artery and vein pair.

vein keep a constant temperature when they pass through this tissue region, the total heat released is equal to the total amount of blood perfusing this tissue volume q multiplied by its density ρ_b, specific heat C_b, and the temperature difference between the artery and vein, and is given by

$$q\rho_b C_b(T_a - T_v) = (Q_{\text{in}} - Q_{\text{out}})\rho_b C_b(T_a - T_v) \tag{2.3}$$

The volumetric heat generation rate q_{blood}, defined as the heat generation rate per unit tissue volume, is then derived as

$$q_{\text{blood}} = [(Q_{\text{in}} - Q_{\text{out}})/V]\rho_b C_b(T_a - T_v) = \omega\rho_b C_b(T_a - T_v) \tag{2.4}$$

where ω is defined as the amount of blood perfused per unit volume tissue per second.

Note that both T_a and T_v in Eq. (2.4) are unknown. Applying the analogy with gaseous exchange in living tissue, Pennes believed that heat transfer occurred in the capillaries because of their large area for exchange. Thus, the local arterial temperature T_a could be assumed as a constant and equal to the body core temperature T_c. As for the local venous blood, it seems reasonable to assume that it equilibrates with the tissue in the capillary and enters the venules at the local tissue temperature. Then the Pennes bioheat equation becomes

$$\rho C \frac{\partial T_t}{\partial t} = k_t \nabla^2 T_t + \omega\rho_b C_b(T_c - T_t) + q_m \tag{2.5}$$

This is a partial differential equation for the tissue temperature. As long as an appropriate initial condition and boundary conditions are prescribed, the transient and steady-state temperature field in the tissue can be determined.

The limitations of the Pennes equation come from the basic assumptions introduced in this model. First it is assumed that the temperature of the arterial blood does not change when it travels from the heart to the capillary bed. As shown in Sec. 2.2, small temperature variations occur only in blood vessels with a diameter larger than 300 μm. Another assumption is that the venous blood temperature is approximated by the local tissue temperature. This is valid only for blood vessels with a diameter smaller than 50 μm. Thus, without considering the thermal equilibration in the artery and vein in different vessel generations, the Pennes source term obviously overestimates the effect of blood perfusion. To accurately model the effect of blood perfusion, the temperature variation along the artery and the heat recaptured by the countercurrent vein must be taken into consideration.

Despite the limitations of the Pennes bioheat equation, reasonable agreement between theory and experiment has been obtained for the measured temperature profiles in perfused tissue subject to various heating protocols. This equation is relatively easy to use, and it allows the manipulation of two blood-related parameters, the volumetric perfusion rate and the local arterial temperature, to modify the results. Pennes performed a series of experimental studies to validate his model. Over the years, the validity of the Pennes bioheat equation has been largely based on macroscopic thermal clearance measurements in which the adjustable free parameter in the theory, the blood perfusion rate [Xu and Anderson, 1999] was chosen to provide reasonable agreement with experiments for the temperature decay in the vicinity of the thermistor bead probe. Indeed, if the limitation of Pennes' bioheat equation is an inaccurate estimation of the strength of the perfusion source term, an adjustable blood perfusion rate will overcome its limitations and provide reasonable agreement between experiment and theory.

Weinbaum-Jiji Bioheat Equation. Since 1980, researchers (Chen and Holmes, 1980; Chato, 1980; Weinbaum et al., 1984) have begun to question the validity of the Pennes bioheat equation. Later, Weinbaum and Jiji (1985) developed a new equation for microvascular blood tissue heat transfer, based on the anatomic analysis (Weinbaum et al., 1984), that illustrated that the predominant mode of heat transfer in the tissue was the countercurrent heat exchange between a thermally significant artery and vein pair. The near-perfect countercurrent heat exchange mechanism implies that most of the heat leaving the artery is transferred to its countercurrent vein rather than released to the surrounding tissue. Once there is a tissue temperature gradient along the countercurrent vessel axes, the artery and vein will transfer a different amount of energy across a plane perpendicular to their axes even if there is no net mass flow. This gives rise to a net energy transfer that is equivalent to an enhancement in tissue conductivity in the axial direction of the vessels. In the Weinbaum-Jiji equation, the thermal effect of the blood perfusion is described by an enhancement in thermal conductivity k_{eff}, appearing in the traditional heat conduction equation,

$$\rho C \frac{\partial T_t}{\partial t} = k_{\text{eff}} \nabla^2 T_t + q_m \qquad k_{\text{eff}} = k_t[1 + f(\omega)] \tag{2.6}$$

It was shown that k_{eff} is a function of the local blood perfusion rate and local vascular geometry.

The main limitations of the Weinbaum-Jiji equation are associated with the importance of the countercurrent heat exchange. It was derived to describe heat transfer in peripheral tissue only, where its fundamental assumptions are most applicable. In tissue area containing a big blood vessel (>200 μm in diameter), the assumption that most of the heat leaving the artery is recaptured by its countercurrent vein could be violated, thus, it is not an accurate model to predict the temperature field. In addition, this theory was primarily developed for closely paired microvessels in muscle tissue, which may not always be the main vascular structure in other tissues, such as the renal cortex. Furthermore, unlike the Pennes equation, which requires only the value of blood perfusion, the estimation of the enhancement in thermal conductivity requires that detailed anatomical studies be performed to estimate the vessel number density, size, and artery-vein spacing for each vessel generation, as well as the blood perfusion rate (Zhu et al., 1995). These anatomic data are normally not available for most blood vessels in the thermally significant range.

A New Modified Bioheat Equation. The Pennes and Weinbaum-Jiji models represent two extreme situations of blood-vessel thermal interaction. In the original Pennes model, the arterial blood releases all of its heat to the surrounding tissue in the capillaries and there is no venous rewarming. Pennes did not realize that thermal equilibration was achieved in vessels at least an order of magnitude larger than the capillaries. In contrast, in the Weinbaum-Jiji model the partial countercurrent rewarming is assumed to be the main mechanism for blood-tissue heat transfer. The derivation of the Weinbaum-Jiji equation is based on the assumption that the heat transfer between the artery and the vein does not depart significantly from a perfect countercurrent heat exchanger. In other words, most of the heat lost by the artery is recaptured by its countercurrent vein rather than lost to the surrounding tissue. Subsequent theoretical and experimental studies have shown this is a valid assumption only for vessels less than 200 μm diameter (Charny et al., 1990; Zhu et al., 1996a).

Several theoretical studies have suggested that one way to overcome the shortcomings of both models was to introduce a "correction coefficient" in the Pennes perfusion term [Chato, 1980; Baish, 1994; Brinck and Werner, 1994; Weinbaum et al., 1997; Zhu et al., 2002]. In 1997, Weinbaum and coworkers (Weinbaum et al., 1997) modified the Pennes source term on the basis of the thermal analysis of a basic heat transfer unit of muscle tissue, a 1-mm-diameter tissue cylinder containing blood vessels smaller than 200 μm in diameter [Fig. 2.2]. The countercurrent heat exchange between the *s* artery and vein defined in the anatomical studies of Myrhage and Eriksson [1984] led to the estimation of the heat loss recaptured by the *s* vein. The strength of the source term was then rederived taking into account the rewarming of the countercurrent venous blood in the *s* tissue cylinder. The thermal equilibration analysis on the countercurrent *s* artery and vein in the tissue cylinder led to the following bioheat transfer equation:

$$\rho C \frac{\partial T_t}{\partial t} = k_t \nabla^2 T_t + \epsilon \omega \rho_b C_b (T_{a0} - T_t) + q_m \tag{2.7}$$

Note that the only modification to the Pennes model is a correction coefficient ϵ in the Pennes source term. This correction coefficient can be viewed as a weighting function to correct the overestimation of the original Pennes perfusion term. An easy-to-use closed-form analytic expression was derived

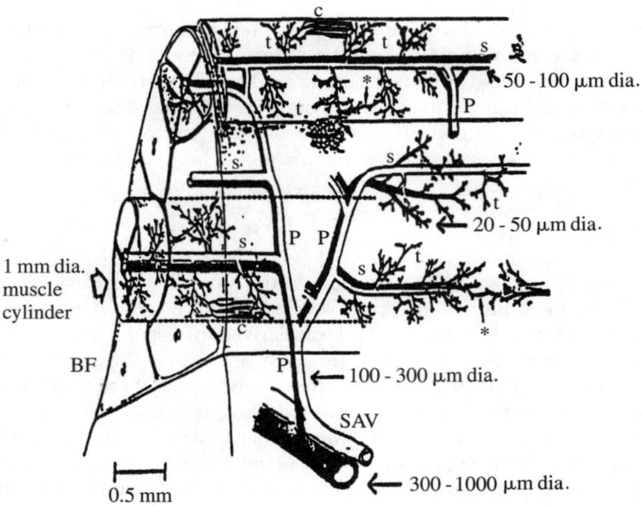

FIGURE 2.2 Macro- and microvascular arrangement in skeletal muscle. The blood supply for the muscle tissue cylinder comes from a branching countercurrent network of supply vessels. The primary (*P*) vessels, originating from the SAV vessels, run obliquely across the muscle tissue cylinders and then branch into the long secondary (*s*) vessels. [*From* Myrhage and Eriksson (1984), *with permission.*]

for this coefficient that depends on the vessel spacing and radius. From the anatomic studies of the vascular arrangements of various skeletal muscles, the correction coefficient was found to vary from 0.6 to 0.8 under normal physiological conditions, indicating that there is a 20 to 40 percent rewarming of the countercurrent vein. Note that it is close to neither unity (the Pennes model) nor zero (the Weinbaum-Jiji model). Thus, both the Pennes and the Weinbaum-Jiji equation are not valid for most muscle tissue. Furthermore, as shown in Zhu et al. (2002), the arterial temperature T_{a0} may not be approximated as the body core temperature unless the local blood perfusion is very high. In most physiological conditions, it is a function of the tissue depth, blood vessel bifurcation pattern, and the local blood perfusion rate.

2.3.2 Experimental and Theoretical Studies to Validate the Models

Pennes (1948) performed a series of experimental studies to validate his model. He inserted and pulled thermocouples through the arms of nine male subjects to measure the radial temperature profiles. He also measured the skin temperature distributions along the axis of the upper limb, as well as around the circumference of the forearm. Pennes then modeled the arm as a long cylinder and calculated the steady-state radial temperature profile. In this theoretical prediction, since the blood perfusion rate ω could not be directly measured, Pennes adjusted this parameter in his model to fit the solution to his experimental data for a fixed, representative ambient temperature and metabolic heating rate. The fitted value of blood perfusion rate ω was found to be between 1.2 and 1.8 mL blood/min/100g tissue, which is a typical range of values for resting human skeletal muscle. Recently, Wissler [1998] re-evaluated Pennes' original paper and analyzed his data. He found that the theoretical prediction agrees very well with Pennes' experimental results if the data were analyzed in a more rigorous manner.

Profound understanding of the heat transfer site and the countercurrent heat exchange between paired significant vessels was gained through the experiments [Lemons et al., 1987] performed on rabbit thigh to measure the transverse temperature profiles in the rabbit thigh using fine thermocouples. The experimental study was designed to achieve two objectives. The first is to examine whether there exists detectable temperature difference between tissue and blood for different-size vessels. Existing detectable blood-tissue temperature difference implies that blood has not reached thermal equilibration with the surrounding tissue. The second is to examine the temperature difference between the countercurrent artery and vein. If the countercurrent heat exchange is dominant in blood tissue heat transfer, the vein must recapture most of the heat leaving the artery; thus, the temperature difference between the countercurrent artery and vein should not vary significantly in the axial direction.

Experimental measurements (Lemons et al., 1987) revealed small temperature fluctuations of up to 0.5°C in the deep tissue. The irregularities in the tissue temperature profiles were closely associated with the existence of blood vessels in the vicinity of the thermocouple wire. It was shown that temperature fluctuation was observed in all the blood vessels larger than 500 μm, in 67 percent of the vessels between 300 and 500 μm, and in 9 percent of the vessels between 100 and 300 μm. No temperature fluctuation was observed in blood vessels less than 100 μm in diameter. This finding indicates that the assumption in the Pennes model that arterial blood reaches the capillary circulation without significant prior thermal equilibration is inaccurate for this vascular architecture, and thus most of the significant blood-tissue heat transfer occurs in the larger vessels upstream. It was also observed that the temperature field rarely exceeded 0.2°C in any countercurrent pair, even when the difference in temperature between the skin and the central part of the rabbit thigh exceeded 10°C. This implies the effectiveness of the countercurrent heat exchange process throughout the vascular tree.

In a series of experiments with an isolated perfused bovine kidney, Crezee and Lagendijk (1990) inserted a small plastic tube into the tissue of a bovine kidney and measured the disturbance temperature fields in a plane perpendicular to the tube while heated water was circulated through it, and the kidney cortex was perfused at different rates. They also used thermocouples to map the temperature distribution in the tissue of isolated perfused bovine tongues at various perfusion rates (Crezee et al., 1991). By examining the effect of increased perfusion on the amplitude and width of the thermal profile, they demonstrated that the temperature measurements agreed better with a perfusion-enhanced k_{eff} as opposed to the perfusion source term in the Pennes equation.

Charny (Charny et al., 1990) developed a detailed one-dimensional (1-D) three-equation model. Since this model was based on energy conservation and no other assumptions were introduced to simplify the analysis of the blood flow effect, it was viewed as a relatively more accurate model than both the Pennes and Weinbaum-Jiji equation. The validity of the assumptions inherent in the formulation of the Weinbaum-Jiji equation were tested numerically under different physiological conditions. In addition, the temperature profile predicted by the Pennes model was compared with that by the three-equation model and the difference between them was evaluated. The numerical simulation of the axial temperature distribution in the limb showed that the Weinbaum-Jiji equation provided very good agreement with the three-equation model for downstream vascular generations that are located in the outer layer in the limb muscle, while the Pennes model yielded better description of heat transfer in the upstream vessel generations. Considering that vessels bifurcate from approximately 1000 μm in the first generation to 150 μm in the last generation, one finds that the Pennes source term, which was originally intended to represent an isotropic heat source in the capillaries, is shown to describe instead the heat transfer from the largest countercurrent vessels, $>$500 μm in diameter. The authors concluded that this was largely attributed to the capillary bleedoff from the large vessels in these tissue regions. The capillary bleedoff appeared to result in a heat source type of behavior that matches the Pennes perfusion term. The Weinbaum-Jiji model, on the other hand significantly overestimated the countercurrent heat exchange in the tissue region containing larger blood vessels. The validity of the Weinbaum-Jiji equation requires that the ratio of the thermal equilibration length L_e of the blood vessel to its physical length L be less than 0.2. This criterion was found to be satisfied for blood vessels less than 300 μm in diameter under normothermic conditions.

2.4 TEMPERATURE, THERMAL PROPERTY, AND BLOOD FLOW MEASUREMENTS

2.4.1 Temperature

The control of human body temperature is a complex mechanism involving hormones, blood flow to the skin, respiration, and metabolic rate. The control mechanism can be altered by certain pathologic (fever) and external (hyperthermia treatment) events. Consequently, temperature is an important parameter in the diagnosis and treatment for many disease. Elevated local temperature can be an indication of excess or abnormal metabolic rates. Inflammation is the body's mechanism for removing foreign or diseased substances. Exercise also induces an increase in local temperature of skeletal muscles and joints. Some diagnostic procedures involve the measurement of temperature. Thermal images of the breasts have been used to detect the presence of malignant tumors. Temperature measurement is also critical in many therapeutic procedures involved in either hyperthermia or hypothermia.

Temperature-measuring devices can fall into two categories, invasive and noninvasive. Invasive temperature sensors offer the advantages of small size, fast response time, extreme sensitivity to temperature changes, and high stability. However, they have generally involved a limited number of measurement locations, uncertainties about the anatomic placement of thermometry devices, interaction with the energy applied, periodic rather than continuous temperature measurements, and, in some cases, surgical exposure of the target tissue for placement of the temperature probes.

Invasive temperature devices include thermocouples, thermistor beads, optical fiber sensors, etc. A thermocouple consists of two pieces of dissimilar metal that form two junctions. In the wire, an electric potential difference is formed if there exists a temperature difference between the two junctions. This potential difference can be measured with a voltmeter and translated to temperature with a fairly simple means of calibration. A thermocouple usually has a good long-term stability, responds very quickly to changes in temperature, and can be constructed in a manner that allows a good resolution. Another kind of invasive device, the thermistor bead, is made by depositing a small quantity of semiconductor paste onto closely spaced metal wires. The wire and beads are sintered at a high temperature when the material forms a tight bond. The wires are then coated with glass or epoxy for protection and stabilization. The resistors generally exhibit high thermal sensitivity. This characteristic sensitivity to temperature change can result in a thermistor resistance change of more than 50 Ω/°C. Unlike a thermocouple or thermistor bead, the fiber optic temperature probe does not interfere with an electromagnetic field. It has been used to measure tissue temperature rise induced by microwave and/or radio-frequency heating [Zhu et al., 1996b; 1998]. However, it is relatively big in size (~1.5 mm diameter) and has a lower temperature resolution (~0.2°C).

Noninvasive temperature-measuring techniques include MRI thermometry, infrared thermography, etc. Because of the theoretical sensitivity of some of its parameters to temperature, MRI has been considered to be a potential noninvasive method of mapping temperature changes during therapies using various forms of hyperthermia. MRI imaging has the advantage of producing three-dimensional anatomic images of any part of the body in any orientation. In clinical practice, MRI characteristic parameters such as the molecular diffusion coefficient of water, the proton spin-lattice (T_1) relaxation time (Parker et al., 1982), and the temperature-dependent proton resonance frequency (PRF) shift have been used to estimate the in vivo temperature distribution in tissues. MRI provides good spatial localization and sufficient temperature sensitivity. At the present time, it also appears to be the most promising modality to conduct basic assessments of heating systems and techniques. The disadvantages of MRI thermometry include limited temporal resolution (i.e., quasi-real time), high environmental sensitivity, high material expenditure, and high running costs.

Infrared thermography is based on Planck's distribution law describing the relationship between the emissive power and the temperature of a blackbody surface. The total radiative energy emitted by an object can be found by integrating the Planck equation for all wavelengths. This integration gives the Stefan-Boltzmann law $E(T) = \epsilon\sigma T^4$. The thermal spectrum as observed by an infrared-sensitive detector can be formed primarily by the emitted light. Hence, the formed thermal image is determined by the local surface temperature and the emissivity of the surface. If the emissivity of the object is known, and no intervening attenuating medium exists, the surface temperature can be quantified. Quantification of skin temperature is possible because the human skin is almost a perfect blackbody ($\epsilon = 0.98$) over the wavelengths of interest. Qualitative thermography has been successfully used in a wide range of medical applications (Jones, 1998) including cardiovascular surgery (Fiorini et al., 1982), breast cancer diagnoses (Gautherie and Gros, 1980; Lapayowker and Revesz, 1980), tumor hyperthermia (Cetas et al., 1980), laser angioplasty, and peripheral venous disease. Clinical studies on patients who had breast thermography demonstrated that an abnormal thermography was associated with an increased risk of breast cancer and a poorer prognosis for the breast cancer patients (Gautherie and Gros, 1980; Head et al., 1993). Infrared tympanic thermometry has also been developed and widely used in clinical practice and thermoregulatory research as a simple and rapid device to estimate the body core temperature (Matsukawa et al., 1996; Shibasaki et al., 1998).

2.4.2 Thermal Property (Thermal Conductivity and Thermal Diffusivity) Measurements

A knowledge of thermal properties of biological tissues is fundamental to understanding heat transfer processes in the biological system. This knowledge has increased importance in view of the concerns for radiological safety with microwave and ultrasound irradiation, and with the renewed interest in local and regional hyperthermia as a cancer therapy. The availability of a technique capable of accurately characterizing thermal properties of both diseased and normal tissue would greatly enhance diagnostic and therapeutic medicine.

The primary requirement in designing an apparatus to measure thermal conductivity and diffusivity is that the total energy supplied should be used to establish the observed temperature distribution within the specimen. For accurate measurements, a number of structural and environmental factors, such as undesired heat conduction to or from the specimen, convection currents caused by temperature-induced density variations, and thermal radiation, must be minimized. Biomaterials present additional difficulties. The existing literature on biological heat transfer bears convincing evidence of the complexity of heat transfer processes in living tissue. For example, thermal properties of living tissue differ from those of excised tissue. Clearly the presence of blood perfusion is a major factor in this difference. Relatively large differences in thermal conductivity exist between similar tissues and organs, and variation for the same organ, are frequently reported. Such variations suggest the importance of both defining the measurement conditions and establishing a reliable measurement technique.

The thermal property measurement techniques can be categorized as steady-state methods and transient methods. They can also be categorized as invasive and noninvasive techniques. In the following sections, we will focus on several widely used techniques for measuring tissue thermal conductivity or diffusivity. Their advantages and limitations will also be described.

Guarded Hot Plate. Thermal conductivity can be measured directly by using steady-state methods, such as the guarded hot plate. This method is invasive in that it requires the excision of the specimen for in vitro measurement. It typically involves imposing a constant heat flux through a specimen and measuring the temperature profile at specific points in the specimen after a steady-state temperature field has been established. Once a simple one-dimensional steady-state temperature field is established in the specimen, the thermal conductivity may be easily found by the expression

$$k = q'' \frac{L}{T_1 - T_2} \tag{2.8}$$

where q'' is the heat flux passing through the specimen, T_1 and T_2 are temperature values at any two measurement locations in the axial direction (or the direction of the heat flux), and L is the axial distance between these two temperature measurements.

Biological materials typically have moderate thermal conductivities and therefore require extensive insulation to ensure an unidirectional heat flow. The contact resistance between the specimen and the plate is also difficult to be minimized. In addition, this method cannot be used to obtain in vivo measurements. Once the tissue specimen is cut from the body, dehydration and temperature-dependent properties may need to be considered.

Flash Method. The transient flash method, first proposed by Parker et al. (1961), is the current standard for measuring the thermal diffusivity of solids. A schematic diagram of this method is shown in Fig. 2.3. The front face of a thin opaque solid, of uniform thickness, is exposed to a burst of intense radiant energy by either a high-energy flash tube or laser. The method assumes that the burst of energy is absorbed instantaneously by a thin layer at the surface of the specimen. Adiabatic boundary conditions are assumed on all other surfaces and on the front face during the measurement. The transient temperature at the rear surface is then measured by using thermocouples or an infrared detector.

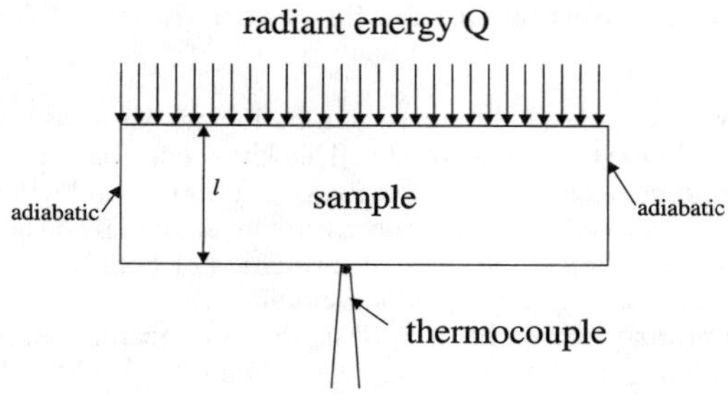

FIGURE 2.3 Schematic diagram of a flash apparatus for sample diffusivity measurements.

An analytic expression for the rear surface temperature transient is given by

$$T(l, t) - T(l, 0) = \frac{Q}{\rho C l}\left[1 + 2\int_{n=1}^{\infty}(-1)^n \exp\left(-\frac{n^2\pi^2}{l^2}\alpha t\right)\right] \tag{2.9}$$

where Q = absorbed radiant energy per unit area
ρ = mass density
C = specific heat
l = sample thickness
α = thermal diffusivity

The maximum temperature at the rear surface is determined by the volumetric heating as $T_{max} = T(l, 0) + Q/\rho Cl$. The thermal diffusivity in the direction of heat flow is usually calculated by the expression

$$\alpha = 1.38\frac{l^2}{\pi^2 t_{1/2}} \tag{2.10}$$

where $t_{1/2}$ is the time required for the rear surface to reach half of its maximum temperature.

The simplicity of the method described above is often offset by the difficulty in satisfying the required adiabatic boundary conditions. In order for this solution to be valid, the radiant energy incident on the front surface is required to be uniform, and the duration of the flash must be sufficiently short compared with the thermal characteristic time of the sample. In addition, it assumes that the sample is homogeneous, isotropic, and opaque, and that the thermal properties of the sample do not vary considerably with temperature.

Temperature Pulse Decay (TPD) Technique. Temperature pulse decay (TPD) technique is based on the approach described and developed by Arkin, Chen, and Holmes [Arkins et al., 1986; 1987]. This method needs no insulation, in contrast to some of the methods described above, since testing times are short, usually on the order of seconds. However, the determination of the thermal conductivity or the blood flow rate requires the solution of the transient bioheat transfer equation.

This technique employs a single thermistor serving as both a temperature sensor and a heater. Typically in this technique, either a thermistor is inserted through the lumen of a hypodermic needle, which is in turn inserted into the tissue, or the thermistor is embedded in a glass-fiber-reinforced epoxy shaft. Figure 2.4 shows the structure of a thermistor bead probe embedded in an epoxy shaft. Each probe can consist of one or two small thermistor beads situated at the end or near the middle of the epoxy shaft. The diameter of the finished probe is typically 0.3 mm, and the length can vary as desired. Because the end can be sharpened to a point, it is capable of piercing most tissues with very minimal trauma.

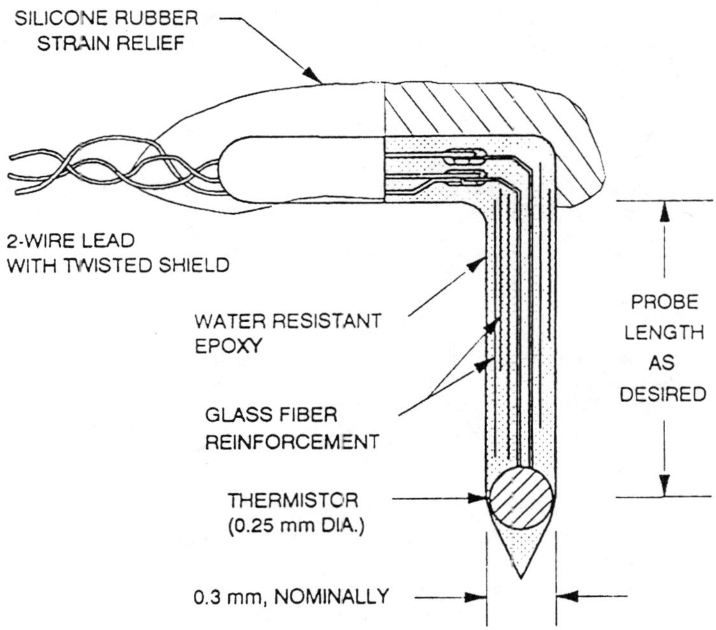

FIGURE 2.4 Sketch of a thermistor bead probe. [*From* Xu et al. (1998), *with permission.*]

During the experiment, a short heating pulse of approximately 3 seconds is delivered by the thermistor bead. The pulse heating results in a temperature rise in the area near the tip of the probe. After the pulse heating, the temperature of the probe will decrease. During the pulse heating and its subsequent temperature decay, the temperature at the tip of the probe is measured by the thermistor bead. To determine the thermal conductivity or blood flow rate, a theoretical prediction of the transient temperature profile is needed for the same tissue domain as in the experimental study. Typically, the theoretically predicted temperature profile is obtained by solving a bioheat transfer equation in which the blood flow rate and thermal conductivity have to be given as input to the model. The predicted temperature profile is then compared with the experimental measurements. The values of the blood flow rate and/or thermal conductivity will be adjusted to minimize the square difference between the predicted temperature profile and the experimental measurements using the linear-square residual fit. The values for the blood flow rate and thermal conductivity that give the best fit of the experimentally measured temperature profile are the calculated blood flow rate and thermal conductivity of the tissue sample.

Typically, the Pennes bioheat transfer equation is used to predict the temperature transient. It is assumed that the thermistor bead is small enough to be considered a point source inserted into the center of an infinitively large medium. The governing equation and initial condition for this thermal process are described as

$$\rho C \frac{\partial T_t}{\partial t} = k_t \frac{1}{r}\frac{\partial}{\partial r}\left(r \frac{\partial T_t}{\partial r}\right) + \omega \rho C (T_a - T_t) + q_p \qquad t = 0 \quad T_t = T_{ss}(r) \tag{2.11}$$

where q_p is the pulse heating deposited locally into the tissue through a very small thermistor bead probe; $q_p = P\,\delta(0)$ for $t \le t_p$; $q_p = 0$ for $t > t_p$. P is the deposited power, and $\delta(0)$ is the Dirac delta function. Before the measurement, the steady-state temperature distribution $T_{ss}(r)$ in the sample should satisfy the one-dimensional steady-state conduction equation without the pulse heating. The governing equation for $T_{ss}(r)$ is given by

$$0 = k_t \frac{1}{r}\frac{d}{dr}\left(r \frac{dT_{ss}}{dr}\right) + \omega \rho C (T_a - T_{ss}) \tag{2.12}$$

Subtracting Eq. (2.14) from Eq. (2.13), and introducing $\theta = T_t - T_{ss}$, one obtains

$$\rho C \frac{\partial \theta}{\partial t} = k_t \frac{1}{r}\frac{\partial}{\partial r}\left(r \frac{\partial \theta}{\partial r}\right) + \omega \rho C \theta + q_p \qquad t = 0 \quad \theta = 0 \tag{2.13}$$

For the limiting case of an infinitesimally small probe with an infinitesimally short heating pulse, the solution for Eq. (2.15) for the interval of temperature decay takes the form:

$$\theta = \lambda_2 \int_0^{t_p} (t - s)^{-1.5} e^{-\omega(t-s)} e^{-r^2/4\lambda_1(t-s)]}\, ds \tag{2.14}$$

where $\lambda_1 = P(\rho C)^{0.5}/(8\pi^{1.5})$ and $\lambda_2 = \alpha/(k_t^{1.5} t_p^{0.5})$. In this theoretical analysis, there are two unknowns, k_t and ω. A least squares residual fit allows one to find a set of values of k_t and ω that will lead to the best fit of the theoretical predictions to the experimentally measured temperature decay.

The temperature pulse decay technique has been used to measure both the in vivo and in vitro thermal conductivity and blood flow rate in various tissues (Xu et al., 1991; 1998). The specimen does not need to be cut from the body, and this method minimizes the trauma by sensing the temperature with a very small thermistor bead. For the in vitro experimental measurement, the measurement of thermal conductivity is simple and relatively accurate. The infinitively large tissue area surrounding the probe implies that the area affected by the pulse heating is very small in comparison with the tissue region. This technique also requires that the temperature distribution before the pulse heating should reach steady state in the surrounding area of the probe.

2.4.3 Blood Perfusion Measurement

Blood perfusion rate is defined as the amount of blood supplied to a certain tissue region per minute per 100 g tissue weight. In most of the situation, it is representing the nutrient need in that tissue area. High blood perfusion is also associated with heat dissipation during exercise or thermal stress. In humans, there are several tissue regions, such as kidney, heart, and choriocapillaris in the eye, possessing a high blood perfusion rate. The measured blood perfusion rate in the kidney is approximately 500 mL/min/100 g tissue (Holmes, 1997). In the heart, the blood perfusion rate is around 300 mL/min/100 g, which serves for the energy need of pumping the heart. The choriocapillaris in the eyes is a meshed structure within two thin sheets. Its blood perfusion rate is very high and can be as much as 8000 mL/min/100 g tissue. In addition to providing oxygen and other nutrients to the retina, the choriocapillaris also may play a role in stabilizing the temperature environment of the retina and retinal pigment epithelium (Aurer and Carpenter, 1980). In addition to its physiological role, blood perfusion measurement is important in theoretical modeling of the temperature distribution during various therapeutic and diagnostic applications.

Radiolabeled Microsphere Technique. Measurement of blood flow has become an integral part of the physiologic study of humans. While many methods have been utilized in measuring tissue blood flow, the one most often practiced today is dependent on injection of radioactively labeled microspheres. The reason for its overwhelming acceptance is due, in part, to the shortcomings of many of the alternative methods of blood flow determination.

In principle, small particles are uniformly mixed with blood and allowed to circulate freely until they impact in a vessel smaller in diameter than themselves. The tissue or organ is then removed and its radioactivity measured. In such a system, the number of particles impacted in a given tissue is assumed proportional to the volume of particle-containing blood perfusing that tissue. If the number of particles in the tissue sample is determined, and an adequate blood flow reference established, a tissue blood flow can be derived.

Calculating the blood flow rate is straightforward; it is based on the assumption that the number of microspheres in each organ should be directly proportional to blood flow to that organ, e.g.,

$$\frac{\text{Blood flow to organ A}}{\text{Microspheres in organ A}} = \frac{\text{blood flow to organ B}}{\text{microspheres in organ B}} = \frac{\text{cardiac output}}{\text{total microspheres injected}} \tag{2.15}$$

The cardiac output of the animal is obtained by another independent method.

Like any other experimental method, determination of blood flow by radioactive microspheres is subject to many sources of error, including individual variation among the sample population, the counting accuracy of the total microspheres in the tissue sample by the gamma counter, and the effect of arteriovenous shunting or the migration of the microspheres. Despite all the limitations of this method, the microsphere technique of blood flow determination has become the most powerful method available today and has been used to evaluate the accuracy of other techniques of blood flow measurement.

Doppler Ultrasound. Doppler ultrasound has been widely used to provide qualitative measurements of the average flow velocity in large to medium-size vessels if the vessel diameter is known. These include the extracranial circulation and peripheral limb vessels. It is also used in an assessment of mapped occlusive disease of the lower extremities. The frequency used for Doppler ultrasound is typically between 1 and 15 MHz. The basis of this method is the Doppler shift, which is the observed difference in frequency between sound waves that are transmitted from simple piezoelectric transducers and those that are received back when both transmitter and receiver are in relative motion. The average frequency shift of the Doppler spectrum is proportional to the average particulate velocity over the cross-sectional area of the sample. When used to measure blood flow, the transducers are stationary and motion is imparted by the flowing blood cells. In this event, red cell velocity V is described by the relationship

$$\frac{\delta F}{F} = \left(\frac{2V}{C}\right) \cos\theta \tag{2.16}$$

or

$$V = \frac{\delta F}{F}\left(\frac{C}{2\cos\theta}\right) \tag{2.17}$$

where δF = frequency change of the emitted wave
C = mean propagation velocity of ultrasound within tissues (about 1540 m/s)
θ = angle between the ultrasound beam and the flow velocity

The frequency shift is usually in the audible range and can be detected by an audible pitch variation or can be plotted graphically.

Attenuation of ultrasound increases nearly linearly with frequency in many types of tissue, causing high frequencies to be attenuated more strongly than low frequencies. The depth of penetration of the signal also depends on the density of the fluid; hence sampling of the velocity profile could be inaccurate in situations where this can vary. Determination of absolute flow/tissue mass with this technique has limited potential, since vessel diameter is not accurately measured and volume flow is not recorded. It is not possible, using currently available systems, to accurately measure the angle made by the ultrasonic beam and the velocity vector. Thus, Doppler flow measurements are semi-quantitative.

Laser Doppler Flowmetry. Laser Doppler flowmetry (LDF) offers the potential to measure flow in small regional volumes continuously and with repetitive accuracy. It is ideally suited to measure surface flow on skin or mucosa or following surgical exposure. LDF couples the Doppler principle in detecting the frequency shift of laser light imparted by moving red blood vessels in the blood stream. Incident light is carried to the tissue by fiber optic cables, where it is scattered by the moving red blood cells. By sampling all reflected light, the device can calculate flux of red blood cells within the sample volume. Depending on the light frequency, laser light penetrates tissue to a depth of less than approximately 3 mm.

The output from LDF is measured not in easily interpretable units of flow but rather in hertz. It would be ideal to define a single calibration factor that could be used in all tissues to convert laser output to flow in absolute units. Unfortunately, the calibration to determine an absolute flow is limited by the lack of a comparable standard and the lack of preset controlled conditions. This may be due to varying tissue optical properties affected by tissue density (Obeid et al., 1990). Further, LDF signals can be affected by movement of the probe relative to the tissue.

Despite its limitations, LDF continues to find widespread applications in areas of clinical research because of its small probe size, high spatial and temporal resolution, and entire lack of tissue contact if required. It has been suggested that LDF is well suited for comparisons of relative changes in blood flow during different experimental conditions (Smits et al., 1986). It is especially valuable to provide a direct measurement of cutaneous blood flow. In patients with Raynaud's phenomenon, it has been indicated that the abnormal cutaneous blood flow is related to the diminished fibrinolytic activity and increased blood viscosity (Engelhart and Kristensen, 1986). LDF has also been useful for assessing patients with fixed arterial obstructive disease of the lower extremity (Schabauer and Rooke, 1994). Other cutaneous uses of LDF include postoperative monitoring of digit reattachment, free tissue flap, and facial operations (Schabauer and Rooke, 1994). In the noncutaneous application of LDF, it has been reported to measure the retinal blood flow in patients with diabetes mellitus. LDF has also been used to monitor cerebral blood perfusion (Borgos, 1996). Recently, it was used to evaluate the brain autoregulation in patients with head injury (Lam et al., 1997).

Temperature Pulse Decay Technique. As described in Sec. 2.4 under "Temperature Pulse Decay (TPD) Technique," local blood perfusion rate can be derived from the comparison between the theoretically predicted and experimentally measured temperature decay of a thermistor bead probe. The details of the measurement mechanism have been described in that section. The temperature pulse decay technique has been used to measure the in vivo blood perfusion rates of different physical or physiological conditions in various tissues (Xu et al., 1991; 1998). The advantages of this technique are that it is fast and induces little trauma. Using the Pennes bioheat transfer equation, the intrinsic thermal conductivity and blood perfusion rate can be simultaneously measured. In some of the applications, a two-parameter least-square residual fit was first performed to obtain the intrinsic thermal conductivity of the tissue. This calculated value of thermal conductivity was then used to perform a one-parameter curve fit for the TPD measurements to obtain the local blood perfusion rate at the probe location. The error of blood perfusion measurement using the TPD technique is mainly inherited from the accuracy of the bioheat transfer equation. Theoretical study (Xu et al., 1993) has shown that this measurement is affected by the presence of large blood vessels in the vicinity of the thermistor bead probe. Further, poor curve fitting of the blood perfusion rate occurs if the steady state of the tissue temperature is not established before the heating (Xu et al., 1998).

2.5 HYPERTHERMIA TREATMENT FOR CANCERS AND TUMORS

2.5.1 Introduction

Within the past decade, there have been important advances in the use of hyperthermia in a wide variety of therapeutic procedures, especially for cancer treatment. Hyperthermia is used either as a singular therapy or as an adjuvant therapy with radiation and drugs in human malignancy. It has less complications and is preferable to more costly and risky surgical treatment (Dewhirst et al., 1997). The treatment objectives of current therapy are to raise tumor temperature higher than 43°C for periods of more than 30 to 60 minutes while keeping temperatures in the surrounding normal tissue below 43°C. It has been suggested that such elevated temperatures may produce a heat-induced cytotoxic response and/or increase the cytotoxic effects of radiation and drugs. Both the direct cell-killing effects of heat and the sensitization of other agents by heat are phenomena strongly dependent on the achieved temperature rise and the heating duration.

One of the problems encountered by physicians is that current hyperthermia technology can't deliver adequate power to result in effective tumor heating of all sites. The necessity of developing a reliable and accurate predictive ability for planning hyperthermia protocols is obvious. The treatment planning typically requires the determination of the energy absorption distribution in the tumor and normal tissue and the resulting temperature distribution. The heating patterns induced by various hyperthermia apparatus have to be studied to focus the energy on a given region of the body and provide a means for protecting the surrounding normal tissues. Over the past decade, optimization of the thermal dose is possible with known spatial and temporal temperature distribution during the hyperthermia treatment. However, large spatial and temporal variations in temperature are still observed because of the heterogeneity of tissue properties (both normal tissue and tumor), spatial variations in specific absorption rates, and the variations and dynamics of blood flow (Overgaard, 1987). It has been suggested that blood flow in large, thermally unequilibrated vessels is the main cause for temperature nonhomogeneity during hyperthermia treatment, since these large vessels can produce cold tracts in the heated volume. Thus, to heat the tissue and tumor volume effectively and safely, it is critical to experimentally or theoretically monitor the temporal and spatial temperature gradient during the hyperthermia treatment.

2.5.2 Temperature Monitoring during Thermal Treatment

One of the reasons hyperthermia has not yet become widely accepted as a mode of therapy is the lack of noninvasive and inexpensive temperature measurement technology for routine use. Invasive temperature devices have a number of restrictions when applied to temperature monitoring during hyperthermia. These restrictions include small representative tissue sample of the entire tissue and tumor volume, difficulty in inserting the devices into a deep-seated tumor, and discomfort to patients. Because of the problems associated with invasive temperature measurement techniques, there has been a strong demand for noninvasive temperature feedback techniques such as ultrasonic imaging and microwave radiometry imaging (MRI). In addition to their focusing and real-time capabilities, ultrasound-based techniques are capable of providing satisfactory temperature resolution as well as hot-spot localization in soft tissue. MRI was applied as a noninvasive thermometry method, but it has limited temperature resolution (~0.5°C) and spatial resolution (~1 cm) and therefore can provide only an estimate of the average temperature over a certain tissue volume. Further, MRI is a costly technique, and therefore, it does not comply with the clinical requirements of treatment monitoring for tissue temperature distribution.

2.5.3 Heating Patterns Induced by Hyperthermia Applicators

Ideal Treatment Volume and Temperature Distribution. Heating pattern or specific absorption rate (SAR) induced by external devices is defined as the heat energy deposited in the tissue or tumor per second per unit mass or volume of tissue. In optimal treatment planning, it is the temperature rather than the SAR distribution that is optimized in the treatment plan. The maximum temperature generally occurs in the tissue region with heat deposition. However, one should note that SAR and temperature distribution may not have the same profile, since temperature distribution can also be affected by the environment or imposed boundary conditions.

The thermal goal of a clinically practical hyperthermia treatment is to maximize the volume of tumor tissue that is raised to the therapeutic temperature. This maximization should be accomplished while keeping the volume of normal tissue at or below some clinically specific temperature level. There are difficulties to reaching the optimal temperature distribution with the presently available heating devices. Most clinical heating systems have had such fixed power deposition patterns that optimization was limited. In recent years, the cooperation between engineers and clinicians has resulted in a new generation of heating equipment. These heating devices have considerably more flexibility in their ability to deposit power in different patterns that help reach the treatment goal. Further, the ideal temperature distribution may be achieved by manipulating the geometrical consideration or regional blood flow. In most of the transurethral microwave catheters, circulated cold water is installed in the catheter to provide protection to the sensitive prostatic urethra. Manipulation of the flow rate and temperature of the water have been demonstrated to facilitate the achievement of high temperature penetrating deep in the transition zone (Liu et al., 2000). Preheating the large arterial blood to some extent before it enters the treatment region has also been shown to improve the temperature homogeneity in that area.

Currently Used Heating Devices. It is known that the size and location of the tumor have a significant impact on applicator design and type of heating. In most of the heating devices, heat is deposited in the tissue via electromagnetic wave absorption (microwave or radio frequency), electric conductive heating, ultrasound absorption, laser, and thermal seed, etc. In this section, different heating devices are introduced and their advantages and limitations are described.

High-energy dc shock (Scheinman et al., 1982) has been used as an implanted energy source for the treatment of drug-refractory supraventricular arrhythmias. During the dc catheter ablation, the peak voltage and current measured at the electrode-tissue interface are typically higher than 1000 V and 40 A, respectively. The high-voltage pulse results in a very high temperature at the electrode surface. Explosive gas formation and a shock wave can occur, which may cause serious complications, including ventricular fibrillation, cardiogenic shock, and cardiac peroration.

Alternating current in the radio-frequency range has been investigated as an alternative to dc shock for heating applicator (Huang et al., 1987; Nath et al., 1994; Nath and Haines, 1995; Wonnell, et al., 1992; Zhu and Xu, 1999). RF energy is readily controllable, and the equipment is relatively cheap. The standard RF generator used in catheter ablation produces an unmodulated sinusoidal wave alternating current at frequency of 200 to 1000 kHz. Two electrodes are needed to attach to the tissue and a current is induced between them. The passage of current through the tissue results in resistive or ohmic heating (I^2R losses). Resistive current density is inversely proportional to the square of the distance from the electrode. Thus, resistive heating decreases with the distance from the electrode to the fourth power. Maximal power occurs within a very narrow rim of tissue surrounding the electrodes. The heating typically leads to desiccation of tissue immediately surrounding the catheter electrodes, but diverges and decreases in between, which can cause broad variations of heating. Improved RF hyperthermia systems have been proposed to reduce the heterogeneity of the RF heating, including implanting a feedback power current system (Astrahan, 1982; Hartov et al., 1994) and using electrically insulating material around the electrodes (Cosset et al., 1986).

Microwave hyperthermia uses radiative heating produced by high-frequency power. High-frequency electromagnetic waves may be transmitted down an appropriately tuned coaxial cable and then radiated into the surrounding medium by a small antenna. The mechanism of electromagnetic heating from a microwave source is dielectric, rather than ohmic. The heating is due to a propagating electromagnetic wave that raises the energy of the dielectric molecules through which the field passes by both conduction and displacement currents. While maintaining alignment with the alternating electric field, neighboring energized dipole molecules collide with each other and the electromagnetic energy is transformed into kinetic energy (heat). The main limitation of microwave heating is that the energy is absorbed within a very narrow region around the microwave antenna. Typically, the generated heat decays fast and can be approximated as proportional to $1/r^2$. The highly absorptive nature of the water content of human tissue has limited the penetration of electromagnetic energy to 1 to 2 centimeters.

Laser photocoagulation is a form of minimally invasive thermotherapy in which laser energy is deposited into a target tissue volume through one or more implanted optical fibers. Laser is used in medicine for incision and explosive ablation of tumors and other tissues, and for blood vessel coagulation in various tissues. Laser light is nearly monochromatic. Most popular lasers utilized in the laboratory include Nd:YAG lasers operating at 1064 nm and diode lasers operating at 805 nm. Laser-beam power ranges from milliwatts to several watts. Usually the laser energy is focused on a small tissue area of a radius less than 300 μm, resulting in a very high radiance. Because there is minimal penetration of laser energy into the tissue, sufficient energy is delivered to heat tissues surrounding the point of laser contact to beyond 60°C or higher, leading to denaturation and coagulation of biomolecules. Because of the high temperature elevation in the target tissue, laser photocoagulation may produce vapor, smoke, browning, and char. A char is usually formed when temperature is elevated above 225°C (LeCarpentier et al., 1989; Thomsen, 1991; Torres et al., 1990; Whelan and Wyman, 1999).

Unlike the electromagnetic heating devices mentioned above, ultrasound heating is a mechanical hyperthermic technique. The acoustic energy, when absorbed by tissue, can lead to local temperature rise. Ultrasound offers many advantages as an energy source for hyperthermia because of its small wavelength and highly controllable power deposition patterns, including penetration depth control in human soft tissue. The depth of the resulting lesion could theoretically be increased or decreased by selecting a lower or higher ultrasound frequency, respectively. It has been shown that scanned focused ultrasound provides the ability to achieve more uniform temperature elevations inside tumors than the electromagnetic applicators. Moros and Fan (1998) have shown that the frequency of 1 MHz is not adequate for treating chest wall recurrences, since it is too penetrating. As for deep-seated tumor (3 to 6 cm deep), longer penetration depth is achieved by using relatively low frequency (1 MHz) and/or adjusting the acoustic output power of the transducer [Moros et al., 1996]. The practical problem associated with ultrasound heating is the risk of overheating the surrounding bone-tissue interface because of the high ultrasound absorption in bone.

Determination of SAR. Several methods are used to determine the SAR distribution induced by various heating applicators. The SAR distribution can be directly derived from the Maxwell equation of the electromagnetic field (Camart et al., 2000; Gentili et al., 1995; Ling et al., 1999; Stauffer et al., 1998; Strohbehn, 1984). The electrical field E and magnetic field B are first determined analytically or numerically from the Maxwell equation. The SAR (W/kg) is then calculated by the following equation (Sapozink et al, 1998):

$$\mathrm{SAR} = \frac{\sigma/2\rho}{E^2} \tag{2.18}$$

where ρ and σ represent the density and conductivity of the media, respectively. This method is feasible when the derivation of the electromagnetic field is not very difficult. It generally requires a rather large computational resource and a long calculation time, though it is flexible for modeling the applicators and the surrounding media.

Other methods in clinical and engineering applications are experimental determination of the SAR distribution based on the heat conduction equation. The experiment is generally performed on a tissue equivalent phantom gel. The applicability of the SAR distribution measured in the phantom gel to that in tissue depends on the electrical properties of the phantom gel. For energy absorption of ultrasound in tissue, the gel mimics tissue in terms of ultrasonic speed and attenuation/absorption properties. For heat pattern induced by microwave or radio frequency, the applicability requires that the phantom gel mimic the dielectric constant and electrical conductivity of the tissue. The electrical properties of various tissues at different wave frequencies have been studied by Stoy et al. (1982). It has been shown that in addition to the electromagnetic wave frequency, water content of the tissue is the most important factor determining the electrical properties. Thermal properties such as heat capacity and thermal conductivity of the gel are not required if no thermal study is conducted. The ingredients of the gel can be selected to achieve the same electrical characteristics of the tissue for a specific electromagnetic wavelength. As shown in Zhu and Xu (1999), the basic ingredients of the gel used for an RF heating applicator were water, formaldehyde solution, gelatin, and NaCl. Water was used to achieve a similar water content as the tissue. Formaldehyde and gelatin were the solidification agents. NaCl was added to obtain the desired electrical conductivity of tissue at that frequency. The resulted phantom gel was a semitransparent material that permits easy and precise positioning of the temperature sensors during the experimental study.

The simplest experimental approach to determining the SAR distribution is from the temperature transient at the instant of power on (Wong et al., 1993; Zhu et al., 1996b). In this approach, temperature sensors are placed at different spatial locations within the gel. Before the experiment, the gel is allowed to establish a uniform temperature distribution within itself. As soon as the initial heating power level is applied, the gel temperature is elevated and the temperatures at all sensor locations are measured and recorded by a computer. The transient temperature field in the gel can be described by the heat conduction equation:

$$\rho C \frac{\partial T}{\partial t} = k\nabla^2 T + \mathrm{SAR}(x,y,z) \qquad t = 0 \quad T = T_{\mathrm{env}} = \text{constant} \tag{2.19}$$

Within a very short period after the heating power is on, heat conduction can be negligible if the phantom gel is allowed to reach equilibration with the environment before the heating. Thus, the SAR can be determined by the slope of the initial temperature rise, i.e.,

$$\mathrm{SAR} = \rho C \left(\frac{\partial T}{\partial t} \right) \bigg|_{t=0} \tag{2.20}$$

Since the SAR at each spatial location is a constant during the heating, the temperature rise at each location is expected to increase linearly if heat conduction is negligible. Figure 2.5 gives the measured temperature rise at different radial locations from a cylindrical microwave antenna. Note that the temperatures at all three probe locations were very close to each other before the heating. The temperatures increased linearly once the power was on; however, after approximately 60 seconds, the plot became curved and heat conduction within the gel was no longer negligible. For convenience the loose SAR data are generally represented by an analytic expression with several unknown parameters. Then a least-square residual fit of the SAR measurement to the analytical expression is performed to determine the unknown parameters in the expression.

It is simple to determine the SAR distribution from the initial temperature transient. This method is fairly accurate as long as the temperature is uniform before the power is turned on. However, to obtain an accurate expression for the SAR distribution, enough temperature sensors should be placed in the region where the energy is absorbed. In the situation when the SAR decays rapidly in the radial direction because of the superficial penetration of the energy, and it is difficult to place many

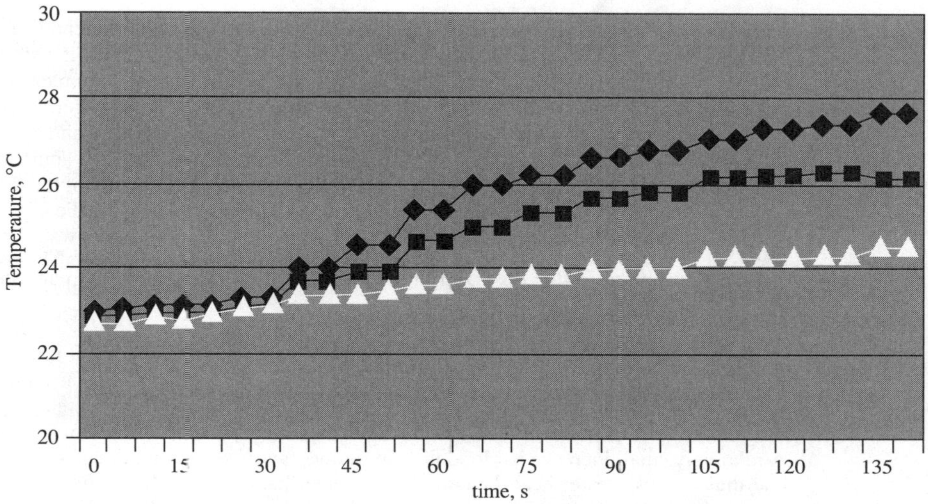

FIGURE 2.5 Measured transient temperatures at three probe locations in the phantom gel.

temperature sensors in the near field, the SAR distribution must be determined by only a few measurements in the near field, which increases the measurement error.

In the experimental study by Zhu et al. (1998), the heating pattern induced by a microwave antenna was quantified by solving the inverse problem of heat conduction in a tissue equivalent gel. In this approach, detailed temperature distribution in the gel is required and predicted by solving a two- or three-dimensional heat conduction equation in the gel. In the experimental study, all the temperature probes were not required to be placed in the near field of the catheter. Experiments were first performed in the gel to measure the temperature elevation induced by the applicator. An expression with several unknown parameters was proposed for the SAR distribution. Then, a theoretical heat transfer model was developed with appropriate boundary conditions and initial condition of the experiment to study the temperature distribution in the gel. The values of those unknown parameters in the proposed SAR expression were initially assumed and the temperature field in the gel was calculated by the model. The parameters were then adjusted to minimize the square error of the deviations theoretically predicted from the experimentally measured temperatures at all temperature sensor locations.

2.5.4 Dynamic Response of Blood Flow to Hyperthermia

As mentioned previously, blood flow plays a profound effect in the temperature field during hyperthermia treatment. Accurately measuring and monitoring blood flow in different tissue regions and at different heating levels is especially crucial to achieve the thermal goal. The distribution of blood flow is quite heterogeneous in the tissue. Blood flow rate may be higher in the skin than in the muscle. Blood flow in the tumor and normal tissue may also be quite different because of different vasculatures. Contrary to the general notion that blood flow is less in tumors than in normal tissues, blood flow in many tumors, particularly in small tumors, is actually greater than that in surrounding normal tissues at normothermic conditions. Even in the same tumor, blood flow generally decreases as the tumor grows larger, owing partially to progressive deterioration of vascular beds and to the rapid growth of tumor cell population relative to vascular bed.

The dynamic responses of the blood flow to hyperthermia in normal tissue and tumors are even more diversified than the blood flow heterogeneity. It is a well-known fact that heat induces a prompt increase in blood flow accompanied by dilation of vessels and an increase in permeability of the vascular wall in normal tissues. The degree of pathophysiological changes in the vascular system in normal tissue is, of course, dependent on the heating temperature, the heating duration, and the heating protocol. Experimental study by Song (1984) has shown how the vasculature changed in the skin and muscle of rodents at different time intervals after hyperthermia for varying heating temperatures at 42 to 45°C. It was shown that the blood flow in the skin increased by a factor of 4 and 6 upon heating at 43°C for 60 and 120 min, respectively. At 44°C the skin blood flow was about 12 times the control value within 30 minutes. At high heating temperature, there existed a critical time after which the blood flow decreased because of vasculature damage. This critical time was more quickly reached when the heating temperature was higher. The blood flow increase in the muscle was similar to that observed in the skin, a tenfold increase in the blood flow was noticed at the 45°C heating.

An indisputable fact emerging from various experimental data indicates that heat-induced change in the blood flow in some tumors is considerably different from that in normal tissue. As noted in Figure 2.6, there was a limited increase in blood flow in tumors during the initial heating period (Song, 1984). When the heating was prolonged, the tumor blood flow decreased progressively. The different responses of the normal tissue and tumors suggest the feasibility of selective tumor heating. A relatively small increase in blood flow in tumors favors retention of heat within the tumor volume, and thus causes greater heat damage. On the other hand, a large blood flow increase in the normal tissue by vascular dilation causes tissue cooling and high survival of cells.

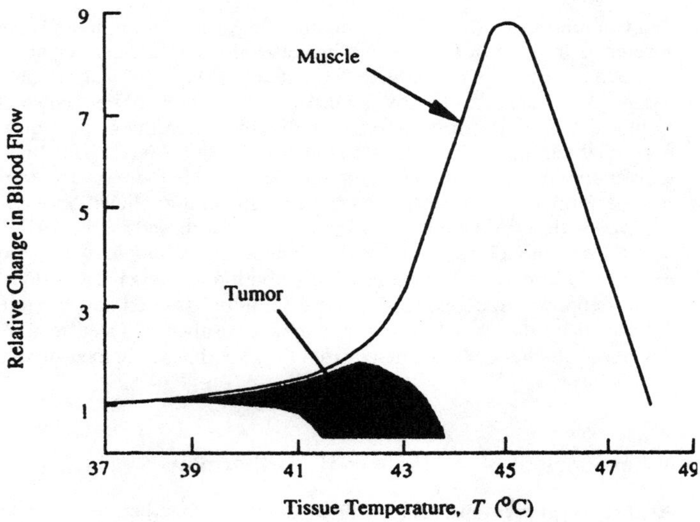

FIGURE 2.6 Temperature-dependent changes in the relative blood perfusion rates for muscle and animal tumors. [*From* Song (1984), *with permission.*]

Since blood flow is the major route of heat dissipation during hyperthermia, attempts have been made to modify the vascular responses in tumors and normal tissue to heat (Reinhold and Endrich, 1986; Song, 1991). Decreased tumor perfusion may induce changes in the tumor microenvironment, such as reduced pH and energy supply, and thus enhance the thermal cytotoxicity (Gerweck, 1977; Overgaard, 1976; Song et al., 1994). An injection of hydralazine to dogs was reported to decrease the blood flow by 50 percent in the tumor and increase the blood flow in the underlying muscle by a factor of 3 (Song, 1984). It was demonstrated that the use of vasoactive drugs led to intracellular acidification of the tumor environment (Song et al., 1994). It has been shown that 1 hour after an intravenous or intraperitoneal injection of KB-R8498, the blood flow in the SCK mammary carcinoma tumors of mice was reduced 30 to 60 percent (Griffin et al., 1998; Ohashi et al., 1998). The effect has also been made to induce a blood flow increase in the normal tissue. It was reported that preferentially dilating vessels in normal tissues using a vasodilator such as Na nitroprusside led to shunting of blood away from the tumor, thus reducing the cooling effect of the blood flow in a tumor during local hyperthermia (Jirtle, 1988; Prescott et al., 1992). Not surprisingly, radiation also altered the response of vasculatures to heat. It was reported that irradiation with 2000 R, given 1 hour before heating at 42°C for 30 min, enhanced the heat-induced vascular damage in the cervical carcinoma of hamsters. Another research showed that hyperthermia of 42°C for 1 hour, given several weeks after x-irradiation, enhanced the capacity of blood flow increase in skin and muscle (Song, 1984).

It has been increasingly evident that the response of vascular beds to heat in tumors differs considerably from that in normal tissues. The effective clinical use of hyperthermia depends on a careful application of these biological principles emerging from experimental work. More experimental measurements of temperature response are needed for different tumor types at different ages. It is also important to evaluate the applicability of the dynamic response measured in animal to human subjects. Another issue to be addressed is the hyperthermia-induced blood flow change in drug delivery, since the reduced tumor blood flow may decrease the drug delivered to the tumors.

2.5.5 Theoretical Modeling

In treatment planning, quantitative three-dimensional thermal modeling aids in the identification of power delivery for optimum treatment. Thermal modeling provides the clinician with powerful tools that improve the ability to deliver safe and effective therapy, and permits the identification of critical monitoring sites to assess tumor heating as well as to assure patient safety. Thermal modeling maximizes the information content of (necessarily sparse) invasive thermometry. The empirical temperature expression (if it is possible) can be used to develop an on-line reference for monitoring tissue temperatures and building a feedback control of the applied power to avoid overheatings in critical tissue areas during the hyperthermia therapy.

Tissue temperature distributions during hyperthermia treatments can be theoretically determined by solving the bioheat transfer equation (continuum model or vascular model), which considers the contributions of heat conduction, blood perfusion, and external heating. In addition to geometrical parameters and thermal properties, the following knowledge must be determined before the simulation. The SAR distribution induced by the external heating device should be determined first. The regional blood perfusion in the tissue and tumor and their dynamic responses to heating are also required. All this information, with appropriate boundary and initial conditions, allows one to calculate the temperature distribution of the tissue.

Analytical solution for the temperature field during the hyperthermia treatment is available for certain tissue geometries (Liu et al., 2000; Zhu and Xu, 1999). In most of the situations, the temperature field is solved by numerical methods because of the irregular tissue geometry and complicated dynamic response of blood flow to heating (Chatterjee and Adams, 1994; Charny et al., 1987; Clegg and Roamer, 1993).

Parametric studies can be performed to evaluate the influence of different parameters, such as heating level, tissue perfusion, and cooling fluid, on the temperature elevation. Extensive parametric studies can be performed quickly and inexpensively so that sensitive (and insensitive) parameters can be identified, systems can be evaluated, and critical experiments can be identified. This is especially important when the parameter is unknown. It is also possible to extract the ideal SAR distribution from the parametric study (Loulou and Scott, 2000) and design an improved heating applicator in the future. Knowledge of the expected thermal profiles could be used to guide experimental and clinical studies in using an optimum thermal dose to achieve a desired therapeutic effect. The optimal thermal dose needed in the treatment volume predicted by the theory helps physicians evaluate the effectiveness of the heating devices and their treatment protocols.

REFERENCES

Arkin, H., Holmes, K. R., Chen, M. M., and Bottje, W. G., 1986, "Thermal Pulse Decay Method for Simultaneous Measurement of Local Thermal Conductivity and Blood Perfusion: a Theoretical Analysis," *ASME Journal of Biomechanical Engineering*, **108**:208–214.

Arkin, H., Holmes, K. R., and Chen, M. M., 1987, "Computer Based System for Continuous On-line Measurements of Tissue Blood Perfusion," *Journal of Biomedical Engineering*, **9**:38–45.

Astrahan, M. A., and Norman, A., 1982, "A Localized Current Field Hyperthermia System for Use with 192 Iridium Interstitial Implants," *Med. Phys.*, **9**:419–424.

Aurer, C., and Carpenter, D. O., 1980, "Choroidal Blood Flow as a Heat Dissipating Mechanism in the Macula," *Am. J. Ophthalmology*, **89**:641–646.

Baish, J. W., 1994, "Formulation of a Statistical Model of Heat Transfer in Perfused Tissue," *ASME Journal of Biomechanical Engineering*, **116**:521–527.

Bazett, H. C., 1941, "Temperature Sense in Men," in *Temperature: Its Measurement and Control in Science and Industry*, Reinhold, New York, pp. 489–501.

Borgos, J., 1996, "Laser Doppler Monitoring of Cerebral Blood Flow," *Neurological Research*, **18**:251–255.

Brinck, H., and Werner, J., 1994, "Estimation of the Thermal Effect of Blood Flow in a Branching Countercurrent Network Using a Three-dimensional Vascular Model," *ASME Journal of Biomechanical Engineering*, **116**:324–330.

Camart, J. C., Despretz, D., Prevost, B., Sozanski, J. P., Chive, M., and Pribetich, J., 2000, "New 434 MHz Interstitial Hyperthermia System Monitored by Microwave Radiometry: Theoretical and Experimental Results," *Int. J. Hyperthermia*, **16**:95–111.

Cetas, T. C., Conner, W. G., and Manning, M. R., 1980, "Monitoring of Tissue Temperature During Hyperthermia Therapy," *Ann. New York Acad. Sci.*, **335**:281–297.

Charny, C. K., Hagmann, M. J., and Levin, R. L., 1987, "A Whole Body Thermal Model of Man during Hyperthermia," *IEEE Trans. Biomed. Engineering*, **34**(5):375–387.

Charny, C. K., Weinbaum, S., and Levin, R. L., 1990, "An Evaluation of the Weinbaum-Jiji Bioheat Equation for Normal and Hyperthermic Conditions," *ASME Journal of Biomechanical Engineering*, **112**:80–87.

Chato, J., 1980, "Heat Transfer to Blood Vessels," *ASME J. Biomech. Eng.*, **102**:110–118.

Chatterjee, I., and Adams, R. E., 1994, "Finite Element Thermal Modeling of the Human Body under Hyperthermia Treatment for Cancer," *Int. J. Computer Applications in Technology*, **7**:151–159.

Chen, M. M., and Holmes, K. R., 1980, "Microvascular Contributions to Tissue Heat Transfer," *Annals of the New York Academy of Science*, **335**:137–150.

Clegg, S. T., and Roemer, R. B., 1993, "Reconstruction of Experimental Hyperthermia Temperature Distributions: Application of State and Parameter Estimation," *ASME Journal of Biomech. Engineering*, **115**:380–388.

Cosset, J. M., Dutreix, J., Haie, C., Mabire, J. P., and Damia, E., 1986, "Technical Aspects of Interstitial Hyperthermia," in *Recent Results in Cancer Research*, Vol. 101, Berlin, Heidelberg: Springer Publishers, 56–60.

Crezee, J., and Lagendijk, J. J. W., 1990, "Experimental Verification of Bioheat Transfer Theories: Measurement of Temperature Profiles Around Large Artificial Vessels in Perfused Tissue," *Phys. Med. Biol.*, **35**(7):905–923.

Crezee, J., Mooibroek, J., Bos, C. K., and Lagendijk, J. J. W., 1991, "Interstitial Heating: Experiments in Artificially Perfused Bovine Tongues," *Phys. Med. Biol.*, **36**:823–833.

Dewhirst, M. W., Prosnitz, L., Thrall, D., Prescott, D., Clegg, S., Charles, C., MacFall, J., Rosner, G., Samulski, T., Gillette, E., and LaRue, S., 1997, "Hyperthermia Treatment of Malignant Diseases: Current Status and a View Toward the Future," *Seminars in Oncology*, **24**(6):616–625.

Dorr, L. N., and Hynynen, K., 1992, "The Effects of Tissue Heterogeneities and Large Blood Vessels on the Thermal Exposure Induced by Short High-power Ultrasound Pulses," *Int. J. Hyperthermia*, **8**:45–59.

Engelhart, M., and Kristensen, J. K., 1986, "Raynaud's Phenomenon: Blood Supply to Fingers During Indirect Cooling, Evaluated by Laser Doppler Flowmetry," *Clin. Physiol.*, **6**:481–488.

Fiorini, A., Fumero, R., and Marchesi, R., 1982, "Cardio-surgical Thermography," *Proc. SPIE*, **359**:249

Gautherie, M., and Gros, C. M., 1980, "Breast Thermography and Cancer Risk Prediction," *Cancer*, **45**(1):51–56.

Gentili, G. B., Leoncini, M., Trembly, B. S., Schweizer, S. E., 1995, "FDTD Electromagnetic and Thermal Analysis of Interstitial Hyperthermic Applicators: Finite-difference Time-domain," *IEEE Trans. Biomed. Engineering*, **42**(10):973–980.

Gerweck, L. E., 1977, "Modification of Cell Lethality at Elevated Temperatures: the pH Effect," *Radiation Research*, **70**:224–235.

Griffin, R. J., Ogawa, A., Shakil, A., and Song, C. W., 1998, "Local Hyperthermia Treatment of Solid Tumors: Interplay Between Thermal Dose and Physiological Parameters," ASME, HTD—Vol. 362, *Advances in Heat and Mass Transfer in Biological Systems*, pp. 67–69.

Hartov, A., Colacchio, T. A., Hoopes, P. J., and Strohbehn, J. W., 1994, "The Control Problem in Hyperthermia," ASME HTD—Vol. 288, *Advances in Heat and Mass Transfer in Biological Systems*, pp. 119–126.

Head, J. F., Wang, F., and Elliott R. L., 1993, "Breast Thermography Is a Noninvasive Prognostic Procedure That Predicts Tumor Growth Rate in Breast Cancer Patients," *Ann. New York Acad. Sci.*, **698**:153–158.

Holmes, K. R., 1997, "Biological Structures and Heat Transfer," in *Report from the Allerton Workshop on The Future of Biothermal Engineering*, pp. 14–37.

Huang, S. K., Bharati, S., Graham, A. R., Lev, M., Marcus, F. I., and Odell, R. C., 1987, "Closed Chest Catheter Desiccation of the Atrioventricular Junction Using Radiofrequency Energy—a New Method of Catheter Ablation," *J. Am. Coll. Cardiol.*, **9**:349–358.

Jirtle, R. L., 1988, "Chemical Modification of Tumor Blood Flow," *Int. J. Hyperthermia*, **4**:355–371.

Jones, B. F., 1998, "A Reappraisal of the Use of Infrared Thermal Image Analysis in Medicine," *IEEE Trans. Med. Imaging*, **17**(6):1019–1027.

Lam, J. M. K., Hsiang, J. N. K., and Poon, W. S., 1997, "Monitoring of Autoregulation Using Laser Doppler Flowmetry in Patients with Head Injury," *J. Neurosurg.*, **86**:438–455.

Lapayowker, M. S., and Revesz, G., 1980, "Thermography and Ultrasound in Detection and Diagnosis of Breast Cancer," *Cancer*, **46**(4 Suppl):933–938.

Le Carpentier, G. L., Motamédi, M., McMath, L. P., Rastegar, S., and Welch, A. J., 1989, "The Effect of Wavelength on Ablation Mechanism During cw Laser Irradiation: Argon Versus Nd: YAG (1.32 μm)," *Proceedings of the 11th Annual IEEE Engineering in Medicine and Biology Society Conference*, pp. 1209–1210.

Lemons, D. E., Chien, S., Crawshaw, L. I., Weinbaum, S., and Jiji, L. M., 1987, "The Significance of Vessel Size and Type in Vascular Heat Transfer," *Am. J. Physiol.*, **253**:R128–R135.

Ling, J. X., Hand, J. W., and Young, I. R., 1999, "Effect of the SAR Distribution of a Radio Frequency (RF) Coil on the Temperature Field Within a Human Leg: Numerical Studies," ASME HTD—Vol. 363, *Advances in Heat and Mass Transfer in Biotechnology*, pp. 1–7.

Liu, J., Zhu, L., and Xu, L. X., 2000, "Studies on the Three-Dimensional Temperature Transients in the Canine Prostate During Transurethral Microwave Thermal Therapy," *ASME Journal of Biomechanical Engineering*, **122**:372–379.

Loulou, T., and Scott, E. P., 2000 "2-D Thermal Dose Optimization in High Intensity Focused Ultrasound Treatments Using the Adjoint Method," ASME HTD—Vol. 368/BED—Vol. 47, *Advances in Heat and Mass Transfer in Biotechnology*—2000, pp. 67–71.

Moros, E. G., Fan, X., and Straube, W. L., 1996, "Penetration Depth Control with Dual Frequency Ultrasound," ASME HTD—Vol. 337/BED—Vol. 34, *Advances in Heat and Mass Transfer in Biotechnology*, pp. 59–65.

Moros, E. G., and Fan, X., 1998, "Model for Ultrasonic Heating of Chest Wall Recurrences," ASME HTD—Vol. 362/BED—Vol. 40, *Advances in Heat and Mass Transfer in Biotechnology*, pp. 27–33.

Matsukawa, T., Ozaki, M., Hanagata, K., Iwashita H., Miyaji, T., and Kumazawa, T., 1996, "A Comparison of Four Infrared Tympanic Thermometers with Tympanic Membrane Temperatures Measured by Thermocouples," *Can. J. Anaesth.*, **43**(12):1224–1228.

Myrhage, R., and Eriksson, E., 1984, "Arrangement of the Vascular Bed in Different Types of Skeletal Muscles," *Prog. Appl. Microcirc.*, **5**:1–14.

Nath, S., DiMarco, J. P., and Haines, D. E., 1994, "Basic Aspect of Radio Frequency Catheter Ablation," *J. Cardiovasc. Electrophysiol.*, **5**:863–876.

Nath, S., and Haines, D. E., 1995, "Biophysics and Pathology of Catheter Energy Delivery System," *Progress in Cardiovascular Diseases*, **37**(4):185–204.

Obeid, A. N., Barnett, N. J., Dougherty, G., and Ward, G., 1990, "A Critical Review of Laser Doppler Flowmetry," *J. Medical Engineering and Technology*, **14**(5):178–181.

Ohashi, M., Matsui, T., Sekida, T., Idogawa, H., Matsugi, W., Oyama, M., Naruse, K., Kotera, Y., Yoshino, K., Ikeda, Y., and Iuchi, K., 1998, "Anti-tumor Activity of a Novel Tumor Blood Flow Inhibitor, KB-R8498," *Proceedings of the American Association for Cancer Research*, Vol. 39, pp. 312.

Overgaard, J., 1976, "Influence of Extracellular pH on the Viability and Morphology of Tumor Cells Exposed to Hyperthermia," *J. National Cancer Institute*, **56**:1243–1250.

Overgaard, J., 1987, "Some Problems Related to the Clinical Use of Thermal Isoeffect Doses," *Int. J. Hyperthermia*, **3**:329–336.

Parker, D. L., Smith, V., Sheldon, P., Crooks, L. E., and Fussell, L., 1982, "Temperature Distribution Measurements in Two-dimensional NMR Imaging," *Med. Phys.*, **10**(3):321–325.

Parker, W. J., Jenkins, R. J., Butler, C. P., and Abbott, G. L., 1961, "Flash Method of Determining Thermal Diffusivity, Heat Capacity, and Thermal Conductivity," *J. Appli. Phys.*, **32**:1679–1684.

Pennes, H. H., 1948, "Analysis of Tissue and Arterial Blood Temperatures in the Resting Human Forearm," *J. Applied Physiology*, **1**:93–122.

Prescott, D. M., Samulski, T. V., Dewhirst, M. W., Page, R. L., Thrall, D. E., Dodge, R. K., and Oleson, J. R., 1992, "Use of Nitroprusside to Increase Tissue Temperature During Local Hyperthermia in Normal and Tumor-bearing Dogs," *Int. J. Radiation Oncology Biology Physics*, **23**:377–385.

Raaymakers, B. W., Crezee, J., and Lagendijk, J. J. W., 2000, "Modeling Individual Temperature Profiles from an Isolated Perfused Bovine Tongue," *Phys. Med. Biol.*, **45**:765–780.

Reinhold, H. S., and Endrich, B., 1986, "Tumor Microcirculation as a Target for Hyperthermia," *Int. J. Hyperthermia*, **2**:111–137.

Roemer, R. B., 1990, "Thermal Dosimetry," in *Thermal Dosimetry and Treatment Planning*, M. Gaytherie, ed., Berlin: Springer, pp. 119–214.

Sapozink, M. D., Cetas, T., Corry, P. M., Egger, M. J., and Fessenden, P., and The NCI Hyperthermia Equipment Evaluation Contractors' Group, 1988, "Introduction to Hyperthermia Device Evaluation," *Int. J. Hyperthermia*, **4**:1–15.

Schabauer, A. M. A., and Rooke, T. W., 1994, "Cutaneous Laser Doppler Flowmetry: Applications and Findings," *Mayo. Clin. Proc.*, **69**:564–574.

Scheinman, M. M., Morady, F., Hess, D. S., and Gonzalez, R., 1982, "Catheter-induced Ablation of the Atrioventricular Junction to Control Refractory Supraventricular Arrhythmias," *JAMA*, **248**:851–855.

Shibasaki, M., Kondo, N., Tominaga, H., Aoki, K., Hasegawa, E., Idota, Y., and Moriwaki, T., 1998, "Continuous Measurement of Tympanic Temperature with a New Infrared Method Using an Optical Fiber," *J. Applied Physiology*, **85**(3):921–926.

Smits, G. J., Roman, R. J., and Lombard, J. H., 1986, "Evaluation of Laser-doppler Flowmetry as a Measure of Tissue Blood Flow," *J. Appl. Physiol.*, **61**(2):666–672.

Song, C. W., 1984, "Effect of Local Hyperthermia on Blood Flow and Microenvironment: A Review," *Cancer Research*, **44**:4721s–4730s.

Song, C. W., 1991, "Role of Blood Flow in Hyperthermia," in *Hyperthermia and Oncology*, M. Urano and E. B. Douple, eds., VSP Publishers, Utrecht, The Netherlands, Vol. 3, pp. 275–315.

Song, C. W., Kim, G. E., Lyons, J. C., Makepeace, C. M., Griffin, R. J., Rao, G. H., and Cragon, E. J. Jr., 1994, "Thermosensitization by Increasing Intracellular Acidity with Amiloride and Its Analogs," *Int. J. Radiat. Oncol. Biol. Phys.*, **30**:1161–1169.

Stauffer, P. R., Rossetto, F., Leoncini, M., and Gentili, G. B., 1998, "Radiation Patterns of Dual Concentric Conductor Microstrip Antennas for Superficial Hyperthermia," *IEEE Trans. Biomed. Engineering*, **45**(5):605–613.

Stoy, R. D., Foster, K. R., and Schwan, H. P., 1982, "Dielectric Properties of Mammalian Tissues from 0.1 to 100 MHz: A Summary of Recent Data," *Phys. Med. Biol.*, **27**:501–513.

Strohbehn, J. W., 1984, "Calculation of Absorbed Power in Tissue for Various Hyperthermia Devices," *Cancer Research* (suppl.), **44**:4781s–4787s.

Thomsen, S., 1991, "Pathologic Analysis of Photothermal and Photomechanical Effects of Laser-Tissue Interactions," *Photochem. and Photobiol.*, **53**:825–835.

Torres, J. H., Motamedi, M., and Welch, A. J., 1990, "Disparate Absorption of Argon Laser Radiation by Fibrous Versus Fatty Plaque: Implications for Laser Angioplasty," *Lasers Surg. Med.*, **10**:149–157.

Weinbaum, S., Jiji, L. M., and Lemons, D. E., 1984, "Theory and Experiment for the Effect of Vascular Microstructure on Surface Tissue Heat Transfer—Part I: Anatomical Foundation and Model Conceptualization," *ASME J. Biomechanical Engineering*, **106**:321–330.

Weinbaum, S., and Jiji, L. M., 1985, "A New Simplified Bioheat Equation for the Effect of Blood Flow on Local Average Tissue Temperature," *ASME J. Biomech. Eng.*, **107**:131–139.

Weinbaum, S., Xu, L. X., Zhu, L., and Ekpene, A., 1997, "A New Fundamental Bioheat Equation for Muscle Tissue: Part I—Blood Perfusion Term," *ASME Journal of Biomechanical Engineering*, **119**:278–288.

Whelan, W. M., and Wyman, D. R., 1999, "A Model of Tissue Charring During Interstitial Laser Photocoagulation: Estimation of the Char Temperature," ASME HTD—Vol. 363/BED—Vol. 44, *Advances in Heat and Mass Transfer in Biotechnology*, pp. 103–107.

Whitmore, R. L., 1968, *Rheology of Circulation*, Pergamon Press, London.

Wissler, E. H., 1998, "Pennes' 1948 Paper Revisited," *Journal of Applied Physiology*, **85**:35–41.

Wong, T. Z., Jonsson, E., Hoopes, P. J., Trembly, B. S., Heaney, J. A., Douple, E. B., and Coughlin, C. T., 1993, "A Coaxial Microwave Applicator for Transurethral Hyperthermia of the Prostate," *The Prostate*, **22**: 125–138.

Wonnell, T. L., Stauffer, P. R., and Langberg, J. J., 1992, "Evaluation of Microwave and Radio Frequency Catheter Ablation in a Myocardium-equivalent Phantom Model," *IEEE Trans. Biomed. Engineering*, **39**:1086–1095.

Xu, L. X., Chen, M. M., Holmes, K. R., and Arkin, H., 1991, "The Theoretical Evaluation of the Pennes, the Chen-Holmes and the Weinbaum-Jiji Bioheat Transfer Models in the Pig Renal Cortex," ASME WAM, Atlanta, HTD—Vol. 189, pp. 15–22.

Xu, L. X., Chen, M. M., Holmes, K. R., and Arkin, H., 1993, "Theoretical Analysis of the Large Blood Vessel Influence on the Local Tissue Temperature Decay after Pulse Heating," *ASME Journal of Biomechanical Engineering*, **115**:175–179.

Xu, L. X., Zhu, L., and Holmes, K. R., 1998, "Thermoregulation in the Canine Prostate during Transurethral Microwave Hyperthermia, Part I: Temperature Response," *Int. J. Hyperthermia*, **14**(1):29–37.

Xu, L. X., and Anderson, G. T., 1999, "Techniques for Measuring Blood Flow in the Microvascular Circulation," in *Biofluid Methods and Techniques in Vascular, Cardiovascular, and Pulmonary Systems*, Vol. 4, Chap. 5, Gordon and Breach Science Publisher, Newark, NJ.

Zhu, L., Lemons, D. E., and Weinbaum, S., 1995, "A New Approach for Prediction of the Enhancement in the Effective Conductivity of Perfused Muscle Tissue Due to Hyperthermia," *Annals of Biomedical Engineering*, **23**:1–12.

Zhu, L., Weinbaum, S., and Lemons, D. E., 1996a, "Microvascular Thermal Equilibration in Rat Cremaster Muscle," *Annals of Biomedical Engineering*, **24**:109–123.

Zhu, L., Xu, L. X., Yuan, D. Y., and Rudie, E. N., 1996b, "Electromagnetic (EM) Quantification of the Microwave antenna for the Transurethral Prostatic Thermotherapy," ASME HTD—Vol. 337/BED—Vol. 34, *Advances in Heat and Mass Transfer in Biotechnology*, pp. 17–20.

Zhu, L., Xu, L. X., and Chencinski, N., 1998, "Quantification of the 3-D Electromagnetic Power Absorption Rate in Tissue During Transurethral Prostatic Microwave Thermotherapy Using Heat Transfer Model," *IEEE Trans. Biomed. Engineering*, 45(9):1163–1172.

Zhu, L., and Xu, L. X., 1999, "Evaluation of the Effectiveness of Transurethral Radio Frequency Hyperthermia in the Canine Prostate: Temperature Distribution Analysis," *ASME Journal of Biomechanical Engineering*, **121**(6):584–590.

Zhu, L., He, Q., Xu, L. X., and Weinbaum, S., 2002, "A New Fundamental Bioheat Equation for Muscle Tissue: Part II-Temperature of SAV Vessels," *ASME Journal of Biomechanical Engineering*, **124**:121–132.

CHAPTER 3
PHYSICAL AND FLOW PROPERTIES OF BLOOD

David Elad and Shmuel Einav
Tel Aviv University, Tel Aviv, Israel

3.1 PHYSIOLOGY OF THE CIRCULATORY SYSTEM

The circulatory transport system is responsible for oxygen and nutrient supply to all body tissues and removal of waste products. The discovery of the circulation of blood in the human body is related to William Harvey (1578–1657). The circulatory system consists of the *heart*—the pump that generates the pressure gradients needed to drive blood to all body tissues, the *blood vessels*—the delivery routes, and the *blood*—the transport medium for the delivered materials. The blood travels continuously through two separate loops; both originate and terminate at the heart. The *pulmonary circulation* carries blood between the heart and the lungs, whereas the *systemic circulation* carries blood between the heart and all other organs and body tissues (Fig. 3.1). In both systems blood is transported in the vascular bed because of a pressure gradient through the following subdivisions: arteries, arterioles, capillaries, venules, and veins. The *cardiac cycle* is composed of the *diastole*, during which the ventricles are filling with blood, and the *systole*, during which the ventricles are actively contracting and pumping blood out of the heart (Martini, 1995; Thibodeau and Patton, 1999).

The total blood volume is unevenly distributed. About 84 percent of the entire blood volume is in the systemic circulation, with 64 percent in the veins, 13 percent in the arteries, and 7 percent in the arterioles and capillaries. The heart contains 7 percent of blood volume and the pulmonary vessels 9 percent. At normal resting activities heart rate of an adult is about 75 beats/min with a stroke volume of typically 70 mL/beat. The *cardiac output*, the amount of blood pumped each minute, is thus 5.25 L/min. It declines with age. During intense exercise, heart rate may increase to 150 beats/min and stroke volume to 130 mL/beat, providing a cardiac output of about 20 L/min. Under normal conditions the distribution of blood flow to the various organs is brain, 14 percent; heart, 4 percent; kidneys, 22 percent; liver, 27 percent; inactive muscles, 15 percent bones, 5 percent; skin, 6 percent;

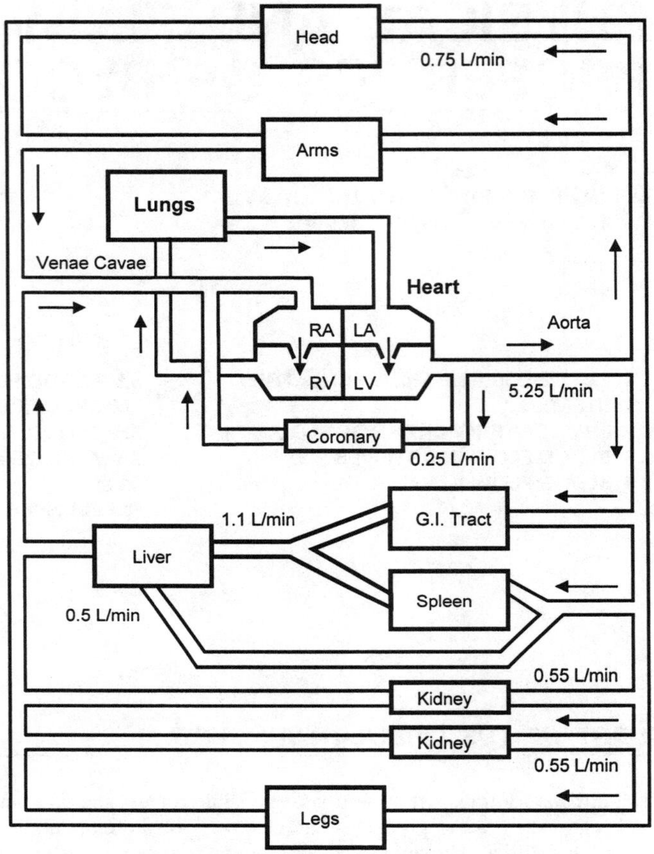

FIGURE 3.1 General organization of the circulatory system with averaged values of normal blood flow to major organs.

bronchi, 2 percent. The averaged blood velocity in the aorta (cross-sctional area of 2.5 cm^2) is 33 cm/s, while in the capillaries (cross-sectional area of 2500 cm^2) it is about 0.3 mm/s. The blood remains in the capillaries 1 to 3 seconds (Guyton and Hall, 1996; Saladin, 2001).

At normal conditions, the pulsatile pumping of the heart is inducing an arterial pressure that fluctuates between the systolic pressure of 120 mmHg and the diastolic pressure of 80 mmHg (Fig. 3.2). The pressure in the systematic capillaries varies between 35 mmHg near the arterioles to 10 mmHg near the venous end, with a functional average of about 17 mmHg. When blood terminates through the venae cavae into the right atrium of the heart, its pressure is about 0 mmHg. When the heart ejects blood into the aorta, a pressure pulse is transmitted through the arterial system. The traveling velocity of the pressure pulse increases as the vessel's compliance decreases; in the aorta it is 3 to 5 m/s, in the large arteries 7 to 10 m/s, and in small arteries 15 to 35 m/s. Figure 3.3 depicts an example of the variations in the velocity and pressure waves as the pulse wave travels toward peripheral arteries (Caro et al., 1978; Fung, 1984).

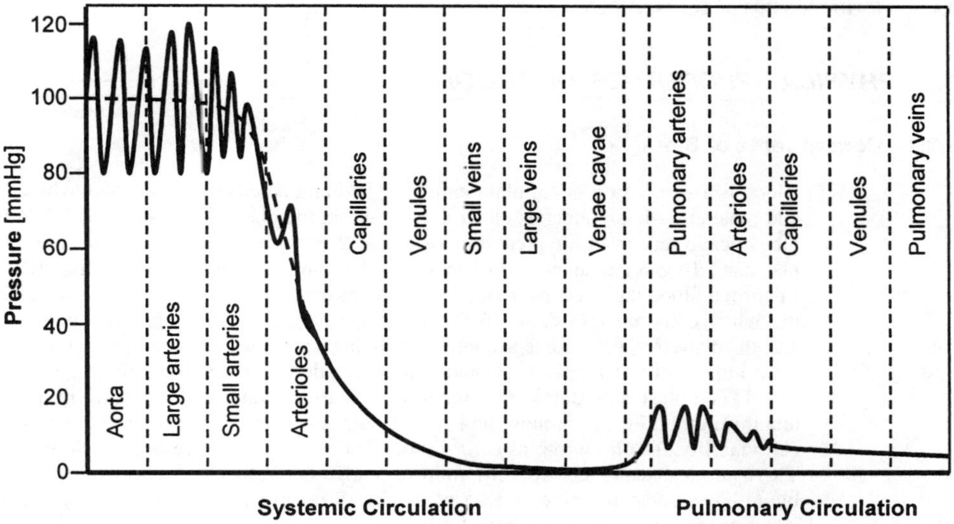

FIGURE 3.2 Variation of blood pressure in the circulatory system.

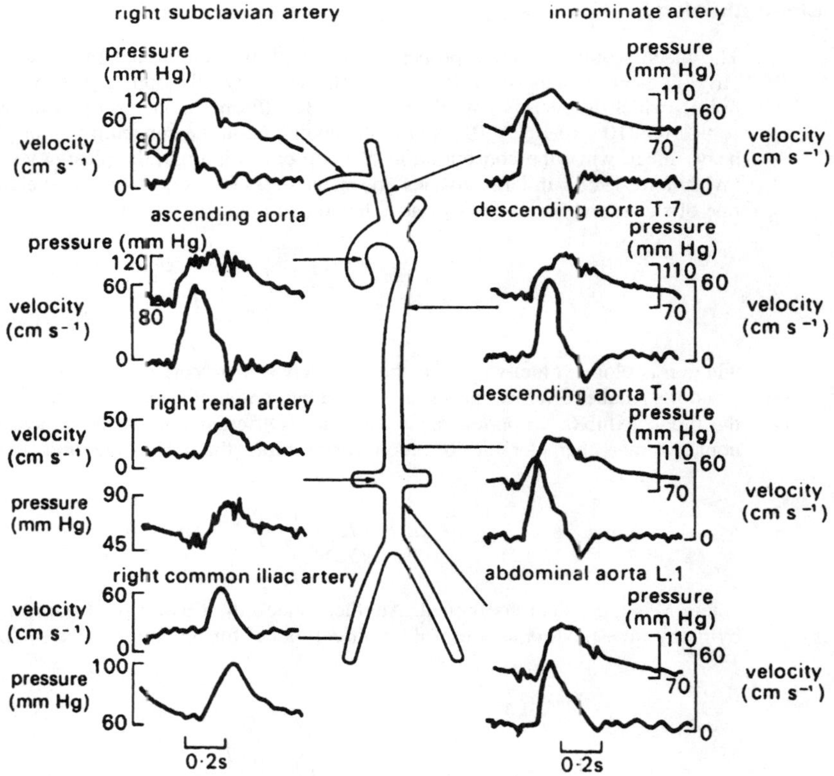

FIGURE 3.3 Pressure and flow waveforms in different arteries of the human arterial tree. [*From Mills et al. (1970) by permission.*]

3.2 PHYSICAL PROPERTIES OF BLOOD

3.2.1 Constituents of Blood

Blood is a suspension of cellular elements—red blood cells (erythrocytes), white cells (leukocytes), and platelets—in an aqueous electrolyte solution, the plasma. Red blood cells (RBC) are shaped as a biconcave saucer with typical dimensions of $2 \times 8\ \mu$m. Erythrocytes are slightly heavier than the plasma (1.10 g/cm^3 against 1.03 g/cm^3); thus they can be separated by centrifugation from the plasma. In normal blood they occupy about 45 percent of the total volume. Although larger than erythrocytes, the white cells are less than 1/600th as numerous as the red cells. The platelet concentration is 1/20th of the red cell concentration, and their dimensions are smaller (2.5 μm in diameter). The most important variable is the hematocrit, which defines the volumetric fraction of the RBCs in the blood. The plasma contains 90 percent of its mass in water and 7 percent in the principal proteins albumin, globulin, lipoprotein, and fibrinogen. Albumin and globulin are essential in maintaining cell viability. The lipoproteins carry lipids (fat) to the cells to provide much of the fuel of the body. The osmotic balance controls the fluid exchange between blood and tissues. The mass density of blood has a constant value of 1.05 g/cm^3 for all mammals and is only slightly greater than that of water at room temperature (about 1 g/cm^3).

3.2.2 Blood Rheology

The macroscopic rheologic properties of blood are determined by its constituents. At a normal physiological hematocrit of 45 percent, the viscosity of blood is $\mu = 4 \times 10^{-2}$ dyne · s/cm^2 (or poise), which is roughly 4 times that of water. Plasma alone (zero hematocrit) has a viscosity of $\mu = 1.1 \times 10^{-2}$ to 1.6×10^{-2} poise, depending upon the concentration of plasma proteins. After a heavy meal, when the concentration of lipoproteins is high, the plasma viscosity is quite elevated (Whitmore, 1968). In large arteries, the shear stress (τ) exerted on blood elements is linear with the rate of shear, and blood behaves as a newtonian fluid, for which,

$$\tau = \mu\left(-\frac{du}{dr}\right) \tag{3.1}$$

where u is blood velocity and r is the radial coordinate perpendicular to the vessel wall.

In the smaller arteries, the shear stress acting on blood elements is not linear with shear rate, and the blood exhibits a nonnewtonian behavior. Different relationships have been proposed for the nonnewtonian characteristics of blood, for example, the power-law fluid,

$$\tau = K\left(-\frac{du}{dr}\right)^n \qquad (n > 0) \tag{3.2}$$

where K is a constant coefficient. Another model, the Casson fluid (Casson, 1959), was proposed by many investigators as a useful empirical model for blood (Cokelet, 1980; Charm and Kurland, 1965),

$$\tau^{1/2} = K\left(-\frac{du}{dr}\right)^{1/2} + \tau_y^{1/2} \tag{3.3}$$

where τ_y is the fluid yield stress.

3.3 BLOOD FLOW IN ARTERIES

3.3.1 Introduction

The aorta and arteries have a low resistance to blood flow compared with the arterioles and capillaries. When the ventricle contracts, a volume of blood is rapidly ejected into the arterial vessels. Since the outflow to the arteriole is relatively slow because of their high resistance to flow, the arteries are inflated to accommodate the extra blood volume. During diastole, the elastic recoil of the arteries forces the blood out into the arterioles. Thus, the elastic properties of the arteries help to convert the pulsatile flow of blood from the heart into a more continuous flow through the rest of the circulation. Hemodynamics is a term used to describe the mechanisms that affect the dynamics of blood circulation.

An accurate model of blood flow in the arteries would include the following realistic features

1. The flow is pulsatile, with a time history containing major frequency components up to the eighth harmonic of the heart period.
2. The arteries are elastic and tapered tubes.
3. The geometry of the arteries is complex and includes tapered, curved, and branching tubes.
4. In small arteries, the viscosity depends upon vessel radius and shear rate.

Such a complex model has never been accomplished. But each of the features above has been "isolated," and qualitative if not quantitative models have been derived. As is so often the case in the engineering analysis of a complex system, the model derived is a function of the major phenomena one wishes to illustrate.

The general time-dependent governing equations of fluid flow in a straight cylindrical tube are given by the continuity and the Navier-Stokes equations in cylindrical coordinates,

$$\frac{\partial v}{\partial r} + \frac{v}{r} + \frac{\partial u}{\partial z} = 0 \tag{3.4}$$

$$\frac{\partial u}{\partial t} + v\frac{\partial u}{\partial r} + u\frac{\partial u}{\partial z} = F_z - \frac{1}{\rho}\frac{\partial P}{\partial z} + \frac{\mu}{\rho}\left(\frac{\partial^2 u}{\partial r^2} + \frac{1}{r}\frac{\partial u}{\partial r} + \frac{\partial^2 u}{\partial z^2}\right) \tag{3.5}$$

$$\frac{\partial v}{\partial t} + v\frac{\partial v}{\partial r} + u\frac{\partial v}{\partial z} = F_r - \frac{1}{\rho}\frac{\partial P}{\partial r} + \frac{\mu}{\rho}\left(\frac{\partial^2 v}{\partial r^2} + \frac{1}{r}\frac{\partial v}{\partial r} - \frac{v}{r^2} + \frac{\partial^2 v}{\partial z^2}\right) \tag{3.6}$$

Here, u and v are the axial and radial components of the fluid velocity, r and z are the radial and axial coordinates, and ρ and μ are the fluid density and viscosity, respectively. Eqs. (3.5) and (3.6) are the momentum balance equations in the z and r directions.

3.3.2 Steady Flow

The simplest model of steady laminar flow in a uniform circular cylinder is known as the Hagen-Poiseuille flow. For axisymmetric flow in a circular tube of internal radius R_0 and length l, the boundary conditions are

$$u(r = R_0) = 0 \qquad \text{and} \qquad \frac{\partial u}{\partial r}(r = 0) = 0 \tag{3.7}$$

For a uniform pressure gradient (ΔP) along a tube we get the parabolic Poiseuille solution

$$u(r) = -\frac{\Delta P}{4\mu l}(R_0^2 - r^2) \tag{3.8}$$

The maximal velocity $u_{max} = (R_0)^2 \Delta P/4\mu l$ is obtained at $r = 0$.

The Poiseuille equation indicates that the pressure gradient ΔP required to produce a volumetric flow $Q = uA$ increases in proportion to Q. Accordingly, the vascular resistance R will be defined as

$$R = \frac{\Delta P}{Q} \tag{3.9}$$

If the flow is measured in cm^3/s and P in dyn/cm^2, the units of R are $dyn \cdot s/cm^5$. If pressure is measured in mmHg and flow in cm^3/s, resistance is expressed in "peripheral resistance units," or PRU.

The arteries are composed of elastin and collagen fibers and smooth muscles in a complex circumferential organization with a variable helix. Accordingly, the arteries are compliant vessels, and their wall stiffness increases with deformation, as in all other connective tissues. Because of their ability to expand as transmural pressure increases, blood vessels may function to store blood volume under pressure. In this sense, they function as capacitance elements, similar to storage tanks. The linear relationship between the volume V and the pressure defines the capacitance of the storage element, or the vascular capacitance:

$$C = \frac{dV}{dP} \tag{3.10}$$

Note that the capacitance (or compliance) decreases with increasing pressure, and also decreases with age. Veins have a much larger capacitance than arteries and, in fact, are often referred to as capacitance or storage vessels.

Another simple and useful expression is the arterial compliance per unit length, C_u, that can be derived when the tube cross-sectional area A is related to the internal pressure $A = A(P, z)$. For a thin-wall elastic tube (with internal radius R_0 and wall thickness h), which is made of a hookean material (with Young modulus E), one can obtain the following useful relation,

$$C_u \equiv \frac{dC}{dz} \approx \frac{2\pi R_0^3}{hE} \tag{3.11}$$

3.3.3 Wave Propagation in Arteries

Arterial pulse propagation varies along the circulatory system as a result of the complex geometry and nonuniform structure of the arteries. In order to learn the basic facts of arterial pulse characteristics, we assumed an idealized case of an infinitely long circular elastic tube that contains a homogenous, incompressible, and nonviscous fluid (Fig. 3.4). In order to analyze the velocity of propagation of the arterial pulse, we assume a local perturbation, for example, in the tube cross-sectional area, that propagates along the tube at a constant velocity c.

The one-dimensional equations for conservation of mass and momentum for this idealized case are, respectively (Pedley, 1980; Fung, 1984),

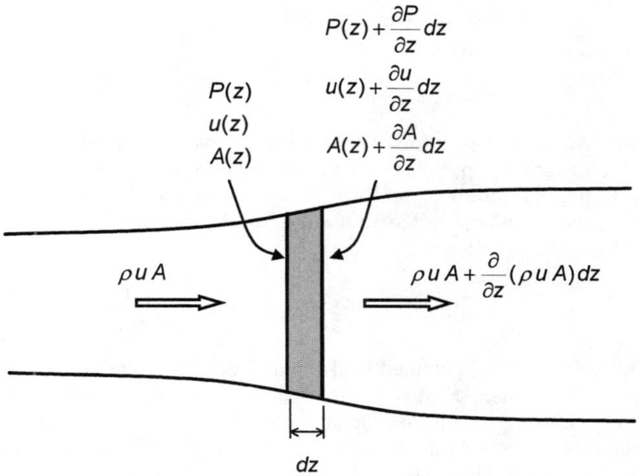

FIGURE 3.4 Cross section of artery showing the change of volume and volume flow.

$$\frac{\partial A}{\partial t} + \frac{\partial}{\partial z}(Au) = 0 \tag{3.12}$$

$$\frac{\partial u}{\partial t} + u\frac{\partial u}{\partial z} + \frac{1}{\rho}\frac{\partial P}{\partial z} = 0 \tag{3.13}$$

where $A(z)$ = tube cross-sectional area
$u(z)$ = uniform axial velocity of blood
$P(z)$ = pressure in the tube
ρ = fluid viscosity
z = axial coordinate
t = time

The elasticity of the tube wall can be prescribed by relationship between the local tube pressure and the cross-sectional area, $P(A)$.

We further assume that the perturbation is small, while the wave length is very large compared with the tube radius. Thus, the nonlinear inertia variables are negligible and the linearized conservation equation of mass and momentum become, respectively,

$$\frac{\partial A}{\partial t} = -A\frac{\partial u}{\partial z} \tag{3.14}$$

$$\frac{\partial u}{\partial t} = -\frac{1}{\rho}\frac{\partial P}{\partial z} \tag{3.15}$$

Next, we differentiate Eq. (3.14) with respect to t and Eq. (3.15) with respect to z, and upon adding the results we obtain the following wave equation:

$$\frac{\partial^2 P}{\partial z^2} = \frac{\rho}{A}\frac{\partial A}{\partial P}\frac{\partial^2 P}{\partial t^2} = c^{-2}\frac{\partial^2 P}{\partial t^2} \tag{3.16}$$

for which the wave speed c is given by

$$c^2 = \frac{A}{\rho}\frac{\partial P}{\partial A} \tag{3.17}$$

This suggests that blood pressure disturbances propagate in a wavelike manner from the heart toward the periphery of the circulation with a wave speed c. For a thin-wall elastic tube (with internal radius R_0 and wall thickness h), which is made of a hookean material (with Young modulus E) and subjected to a small increase of internal pressure, the wave speed c can be expressed as

$$c^2 = \frac{Eh}{2\rho R_0} \tag{3.18}$$

This equation was obtained by Thomas Young in 1808, and is known as the Moens-Kortweg wave speed. The Moens-Kortweg wave speed varies not only with axial distance but also with pressure. The dominant pressure-dependent term in Eq. (3.18) is E, the modulus of elasticity; it increases with increasing transmural pressure as the stiff collagen fibers bear more of the tension of the artery wall. The high-pressure portion of the pressure wave therefore travels at a higher velocity than the low-pressure portions of the wave, leading to a steepening of the pressure front as it travels from the heart toward the peripheral circulation (Fig. 3.5). Wave speed also varies with age because of the decrease in the elasticity of arteries.

The arteries are not infinitely long, and it is possible for the wave to reflect from the distal end and travel back up the artery to add to new waves emanating from the heart. The sum of all such propagated and reflected waves yields the pressure at each point along the arterial tree. Branching is clearly an important contributor to the measured pressures in the major arteries; there is a partial reflection each time the total cross section of the vessel changes abruptly.

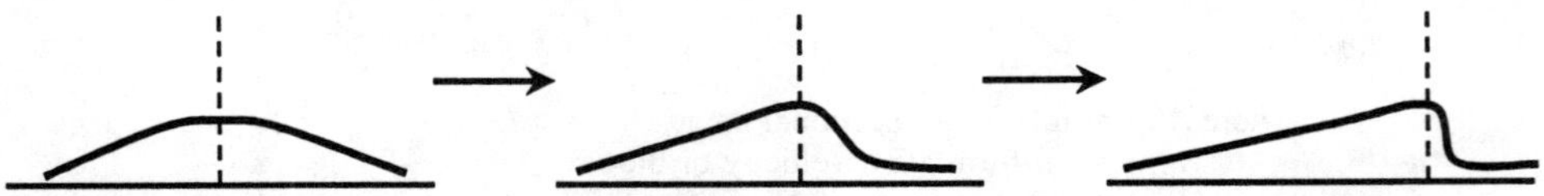

FIGURE 3.5 Steepening of a pressure pulse with distance along an artery.

3.3.4 Pulsatile Flow

Blood flow in the large arteries is driven by the heart, and accordingly it is a pulsating flow. The simplest model for pulsatile flow was developed by Womersley (1955a) for a fully developed oscillatory flow of an incompressible fluid in a rigid, straight circular cylinder. The problem is defined for a sinusoidal pressure gradient composed from sinuses and cosinuses,

$$\frac{\Delta P}{l} = Ke^{i\omega t} \tag{3.19}$$

where the oscillatory frequency is $\omega/2\pi$. Insertion of Eq. (3.19) into Eq. (3.5) yields

$$\frac{\partial^2 u}{\partial r^2} + \frac{1}{r}\frac{\partial u}{\partial r} - \frac{1}{\nu}\frac{\partial u}{\partial t} = -\frac{K}{\mu}e^{i\omega t} \tag{3.20}$$

The solution is obtained by separation of variables as follows:

$$u(r, t) = W(r) \cdot e^{i\omega t} \tag{3.21}$$

Insertion of Eq. (3.21) into (3.20) yields the Bessel equation,

$$\frac{d^2W}{dr^2} + \frac{1}{r}\frac{dW}{dr} + \frac{i^3\omega\rho}{\mu}W = -\frac{A}{\mu} \tag{3.22}$$

The solution for Eq. (3.22) is

$$W(r) = \frac{K}{\rho}\frac{1}{i\omega}\left\{1 - \frac{J_0\left(r\sqrt{\frac{\omega}{\nu}}\cdot i^{3/2}\right)}{J_0\left(R\sqrt{\frac{\omega}{\nu}}\cdot i^{3/2}\right)}\right\} \tag{3.23}$$

where J_0 is a Bessel function of order zero of the first kind, $\nu = \mu/\rho$ is the kinematic viscosity, and α is a dimensionless parameter known as the Womersley number and given by

$$\alpha = R_0\sqrt{\frac{\omega}{\nu}} \tag{3.24}$$

When α is large, the velocity profile becomes blunt (Fig. 3.6).

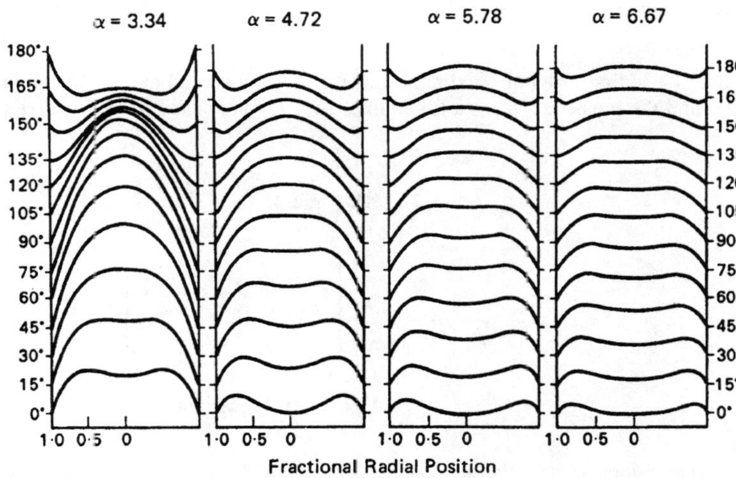

FIGURE 3.6 Theoretical velocity profiles of an oscillating flow resulting from a sinusoidal pressure gradient (cos ωt) in a pipe. α is the Womersley number. Profiles are plotted for intervals of $\Delta\omega t = 15°$. For $\omega t > 180°$, the velocity profiles are of the same form but opposite in sign. [*From Nichols and O'Rourke (1998) by permission.*]

Pulsatile flow in an elastic vessel is very complex, since the tube is able to undergo local deformations in both longitudinal and circumferential directions. The unsteady component of the pulsatile flow is assumed to be induced by propagation of small waves in a pressurized elastic tube. The mathematical approach is based on the classical model for the fluid-structure interaction problem, which describes the dynamic equilibrium between the fluid and the tube thin wall (Womersley, 1955b; Atabek and Lew, 1966). The dynamic equilibrium is expressed by the hydrodynamic equations (Navier-Stokes) for the incompressible fluid flow and the equations of motion for the wall of an elastic tube, which are coupled together by the boundary conditions at the fluid-wall interface. The motion of the liquid is described in a fixed laboratory coordinate system $(\hat{r}, \theta, \hat{z})$, and the dynamic equilibrium of a tube element in its deformed state is expressed in a lagrangian (material) coordinate system ($\hat{n}$, $\hat{t}$, θ), which is attached to the surface of the tube (Fig. 3.7).

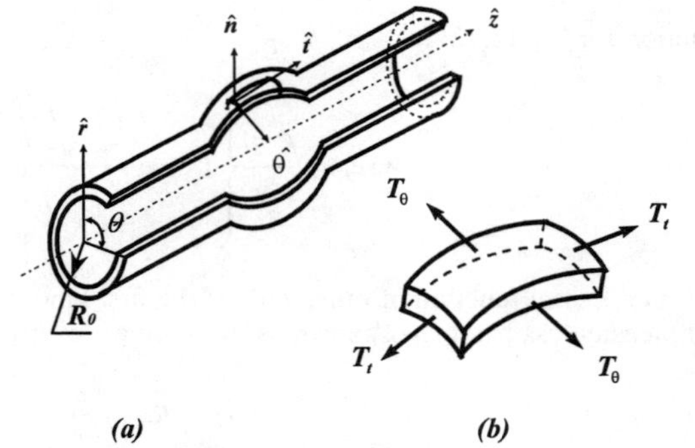

FIGURE 3.7 Mechanics of the arterial wall: (*a*) axisymmetric wall deformation; (*b*) element of the tube wall under biaxial loading. The *T*'s are longitudinal and circumferential internal stresses.

The first-order approximations for the axial (u_1) and radial (v_1) components of the fluid velocity, and the pressure (P_1) as a function of time (t) and space (r, z), are given by

$$u_1(r, z, t) = \frac{A_1}{c\rho_F}\left[1 + m\frac{J_0\left(\alpha_0 \frac{r}{R_0}\right)}{J_0(\alpha_0)}\right]\exp\left[i\omega\left(t - \frac{z}{c}\right)\right] \tag{3.25}$$

$$v_1(r, z, t) = \frac{A_1\beta}{c\rho_F}i\left[\frac{r}{R_0} + m\frac{J_1\left(\alpha_0 \frac{r}{R_0}\right)}{\alpha J_0(\alpha_0)}\right]\exp\left[i\omega\left(t - \frac{z}{c}\right)\right] \tag{3.26}$$

$$P_1(z, t) = A_1 \exp\left[i\omega\left(t - \frac{z}{c}\right)\right] \tag{3.27}$$

The dimensionless parameters m, x, k, τ_θ, and F_{10} are related to the material properties and defined as

$$m = \frac{2 + x[2\sigma - (1 - \tau_\theta)]}{x[(1 - \tau_\theta)F_{10} - 2\sigma]} \qquad x = \frac{Eh}{(1 - \sigma^2)R_0\rho_F c^2} \qquad k = \frac{\rho_T h}{\rho_F R_0}$$

$$c = \frac{2 \cdot c_0}{\{(k + 2) + [(k + 2)^2 - 8k(1 - \sigma^2)]^{1/2}\}^{1/2}} \qquad c_0 = \left(\frac{Eh}{2R_0\rho_F}\right)^{1/2} \tag{3.28}$$

$$\tau_\theta = T_{\theta_0}\frac{Eh}{1 - \sigma^2} \qquad T_{\theta_0} = P_0 R_0 \qquad F_{10} = \frac{2J_1(\alpha_0)}{\alpha_0 J_0(\alpha_0)}$$

$$\alpha = \frac{\omega R_0^2}{c} \qquad \alpha_0^2 = i^3\alpha \qquad \beta = \frac{\omega R_0}{c}$$

where c = wave speed
$\omega = 2\pi\text{HR}/60$ = angular frequency
HR = heart rate
A_1 = input pressure amplitude
J_0 and J_1 = Bessel functions of order 0 and 1 of the first kind
ρ_F and ρ_T = blood and wall densities
R_0 = undisturbed radius of the tube

Excellent recent summaries on pulsatile blood flow may be found in Nichols and O'Rourke (1998) and Zamir (2000).

3.3.5 Turbulence

Turbulence has been shown to exist in large arteries of a living system. It is especially pronounced when the flow rate increases in exercise conditions (Yamaguchi and Parker, 1983). Turbulence is characterized by the appearance of random fluctuations in the flow. The transition to turbulence is a very complex procedure, which schematically can be described by a hierarchy of motions: growth of two-dimensional infinitesimal disturbances to final amplitudes, three-dimensionality of the flow, and a maze of complex nonlinear interactions among the large-amplitude, three-dimensional modes resulting in a final, usually stochastically steady but instantaneously random, motion called turbulent flow (Akhavan et al., 1991; Einav and Sokolov, 1993).

In a turbulent flow field, all the dynamic properties (e.g., velocity, pressure, vorticity) are random functions of position and time. One thus looks at the statistical aspects of the flow characteristics (e.g., mean velocity, rms turbulent intensity). These quantities are meaningful if the flow is stochastically random (i.e., its statistics are independent of time) (Nerem and Rumberger, 1976). The time average of any random quantity is given by

$$\bar{f} \equiv \lim_{T\to\infty} \frac{1}{T}\int_0^\infty f(t)\, dt \tag{3.29}$$

One can thus decompose the instantaneous variables u and v as follows:

$$u(x, t) = \overline{U}(x) + u'(x, t) \tag{3.30}$$

$$v(y, t) = \overline{V}(y) + v'(y, t) \tag{3.31}$$

$$P(x, t) = \overline{P}(x) + p'(x, t) \tag{3.32}$$

We assume that u' is a velocity fluctuation in the x direction only and v' in the y direction only. The overbar denotes time average, so that by definition, the averages of u', v', and p' fluctuations are zero (stochastically random), and the partial derivatives in time of the mean quantities $\overline{U}, \overline{V}, \overline{P}$ are zeros (the Reynolds turbulence decomposition approach, according to which velocities and pressures can be decomposed to time-dependent and time-independent components).

By replacing u with $\overline{U} + u'$ etc. in the Navier-Stokes equation and taking time average, it can be shown that for the turbulent case the two-dimensional Navier-Stokes equation in cartesian coordinates becomes

$$\begin{gathered}\rho\overline{U}\frac{\partial\overline{U}}{\partial x} + \rho\overline{V}\frac{\partial\overline{U}}{\partial y} = -\frac{\partial\overline{P}}{\partial x} + \frac{\partial}{\partial y}\left[\mu\frac{\partial\overline{U}}{\partial y} - \rho\overline{u'v'}\right] \\ \frac{\partial\overline{U}}{\partial x} = 0 \to U = U(y) \qquad \overline{V} = 0\end{gathered} \tag{3.33}$$

and for the y direction one obtains

$$\frac{\partial}{\partial y}\left(\overline{P} + \rho\overline{v'^2} + \rho g z\right) = 0 \tag{3.34}$$

Integration yields

$$\overline{P} + \rho g z + \rho\overline{v'^2} = \text{constant} \tag{3.35}$$

As v' must vanish near the wall, the values of $P + \rho g z$ will be larger near the wall. That implies that, in a turbulent boundary layer, the pressure does not change hydrostatically (is not height or depth dependent), as is the case of laminar flow. Equation (3.35) implies that the pressure is not a function of y, and thus,

$$\frac{\partial \overline{P}}{\partial x} = \frac{\partial}{\partial y}\left(\mu \frac{d\overline{U}}{dy} - \rho\overline{u'v'}\right) \tag{3.36}$$

Since P is independent of y, integration in y yields

$$\mu \frac{d\overline{U}}{dy} - \rho\overline{u'v'} = y\frac{\partial \overline{P}}{\partial x} + C_1 \tag{3.37}$$

where $C_1 = \tau_0$ is the shear stress near the wall.

We see that in addition to the convective, pressure, and viscous terms, we have an additional term, which is the gradient of the nonlinear term $\rho u'v'$, which represents the average transverse transport of longitudinal momentum due to the turbulent fluctuations. It appears as a pseudo-stress along with the viscous stress $\mu \partial U/\partial y$, and is called the Reynolds stress. This term is usually large in most turbulent shear flows (Lieber and Giddens, 1988).

3.3.6 Flow in Curved Tubes

The arteries and veins are generally not straight uniform tubes but have some curved structure, especially the aorta, which has a complex three-dimensional curved geometry with multiplanar curvature. To understand the effect of curvature on blood flow, we will discuss the simple case of steady laminar flow in an in-plane curved tube (Fig. 3.8). When a steady fluid flow enters a curved pipe in the horizontal plane, all of its elements are subjected to a centripetal acceleration normal to their original directions and directed toward the bend center. This force is supplied by a pressure gradient in the plane of the bend, which is more or less uniform across the cross section. Hence, all the fluid elements experience approximately the same sideways acceleration, and the faster-moving elements with the greater inertia will thus change their direction less rapidly than the slower-moving ones. The net result is that the faster-moving elements that originally occupy the core fluid near the center of the tube are swept toward the outside of the bend along the diametrical plane, and their place is taken by an inward circumferential motion of the slower moving fluid located near the walls. Consequently, the overall flow field is composed of an outward-skewed axial component on which is superimposed a secondary flow circulation of two counterrotating vortices.

The analytical solution for a fully developed, steady viscous flow in a curved tube of circular cross section was developed by Dean in 1927, who expressed the ratio of centrifugal inertial forces to the viscous forces (analogous to the definition of Reynolds number Re) by the dimensionless Dean number,

$$\text{De} = \text{Re}\sqrt{\frac{r}{R_{\text{curve}}}} \tag{3.38}$$

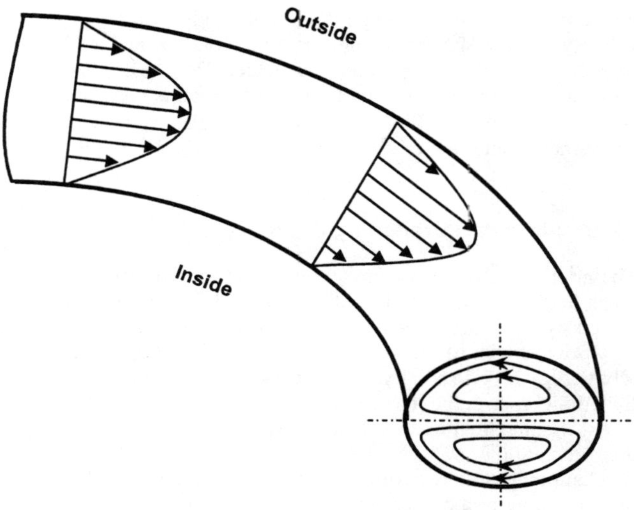

FIGURE 3.8 Schematic description of the skewed axial velocity profile and the secondary motions developed in a laminar flow in a curved tube.

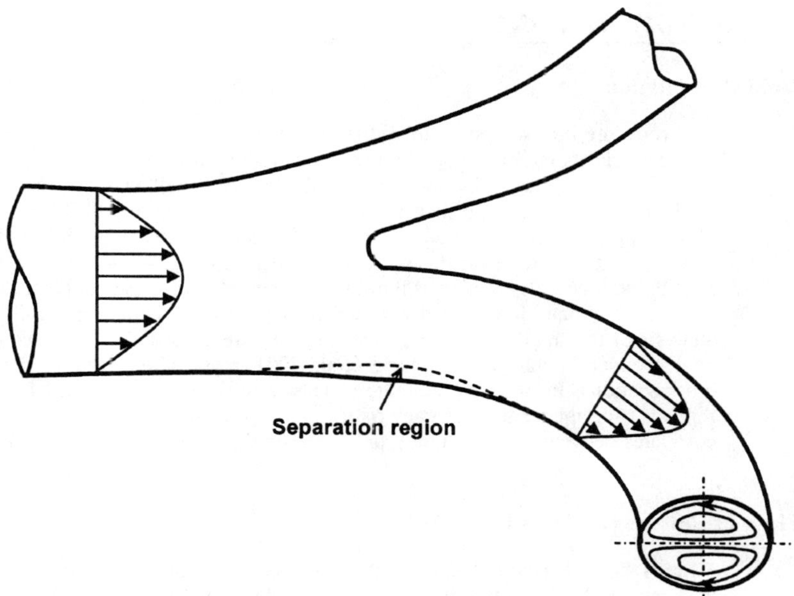

FIGURE 3.9 Qualitative illustration of laminar flow downstream of a bifurcation with a possible region of flow separation, secondary flow, and skewed axial profile.

where r is the tube radius and R_{curve} is the radius of curvature. As De increases, the maximal axial velocity is more skewed toward the outer wall. Dean's analytic solutions are limited to small ratios of radius to radius of curvature for which De < 96. However, numerical solutions extended the range up to 5000. Blood flow in the aortic arch is complex and topics such as entry flow from the aortic valve, pulsatile flow, and their influence on wall shear stress have been the subject of numerous experimental and numerical studies (Pedley, 1980; Berger et al., 1983; Chandran, 2001).

3.3.7 Flow in Bifurcating and Branching Systems

The arterial system is a complex asymmetric multigeneration system of branching and bifurcating tubes that distribute blood to all organs and tissues. A simplified arterial bifurcation may be represented by two curved tubes attached to a straight mother tube. Accordingly, the pattern of blood flow downstream of the flow divider (i.e., bifurcating region) is in general similar to flow in curved tubes (Fig. 3.9). Typically, a boundary layer is generated on the inside wall downstream from the flow divider, with the maximum axial velocity just outside the boundary layer. As in flow in curved tubes, the maximal axial velocity is skewed toward the outer curvature, which is the inner wall of the bifurcation.

Comprehensive experimental and computational studies were conducted to explore the pattern of blood flow in a branching vessel, energy losses, and the level of wall shear stress in the branch region (Ku and Giddens, 1987; Pinchak and Ostrach, 1976; Liepsch et al., 1989; Liepsch, 1993; Pedley, 1995; Perktold and Rappitsch, 1995). Of special interest are the carotid bifurcation and the lower extremity bypass graft-to-artery anastomosis whose blockage may induce stroke and walking inability, respectively. A recent review of computational studies of blood flow through bifurcating geometries that may aid the design of carotid endartectomy for stroke prevention and graft-to-artery configuration may be found in Kleinstreuer et al., 2001.

3.4 BLOOD FLOW IN VEINS

3.4.1 Vein Compliance

The veins are thin-walled tubular structures that may "collapse" (i.e., the cross-sectional area does not maintain its circular shape and becomes less than in the unstressed geometry) when subjected to negative transmural pressures P (internal minus external pressures). Experimental studies (Moreno et al., 1970) demonstrated that the structural performance of veins is similar to that of thin-walled elastic tubes (Fig. 3.10). Three regions may be identified in a vein subjected to a transmural pressure: When $P > 0$, the tube is inflated, its cross section increases and maintains a circular shape; when $P < 0$, the tube cross section collapses first to an ellipse shape; and at a certain negative transmural pressure, a contact is obtained between opposite walls, thereby generating two lumens. Structural analysis of the stability of thin elastic rings and their postbuckling shape (Flaherty et al., 1972), as well as experimental studies (Thiriet et al., 2001) revealed the different complex modes of collapsed cross sections. In order to facilitate at least a one-dimensional fluid flow analysis, it is useful to represent the mechanical characteristics of the vein wall by a "tube law" relationship that locally correlates between the transmural pressure and the vein cross-sectional area.

3.4.2 Flow in Collapsible Tubes

Venous flow is a complex interaction between the compliant structures (veins and surrounding tissues) and the flow of blood. Since venous blood pressure is low, transmural pressure can become negative, thereby resulting in blood flow through a partially collapsed tube. Early studies with a thin-walled elastic tube revealed the relevant experimental evidence (Conrad, 1969). The steady flow rate (Q) through a given length of a uniform collapsible tube depends on two pressure differences selected from among the pressures immediately upstream (P_1), immediately downstream (P_2), and external (P_e) to the collapsible segment (Fig. 3.11). Thus, the pressure-flow relationships in collapsible tubes are more complex than those of rigid tubes, where Q is related to a fixed pressure gradient, and may attain different shapes, depending on which of the pressures (e.g., P_1, P_2, P_e) are held fixed and which are varied. In addition, one should also consider the facts that real veins may be neither uniform nor straight, and that the external pressure is not necessarily uniform along the tube.

The one-dimensional theory for steady incompressible fluid flow in collapsible tubes (when $P - P_e < 0$) was outlined by Shapiro (1977) in a format analogous to that for gas dynamics. The governing equations for the fluid are that for conservation of mass,

$$\frac{\partial A}{\partial t} + \frac{\partial}{\partial z}(Au) = 0 \tag{3.39}$$

and that for conservation of momentum,

$$\frac{\partial u}{\partial t} + u\frac{\partial u}{\partial z} = -\frac{1}{\rho}\frac{\partial P}{\partial z} \tag{3.40}$$

where u = velocity
P = pressure in the flowing fluid
ρ = mass density of the fluid
A = tube cross-sectional area
t = time
z = longitudinal distance

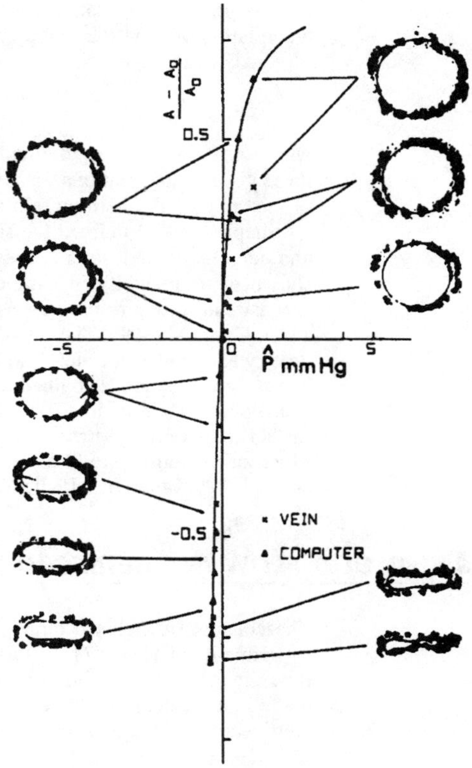

FIGURE 3.10 Relationship between transmural pressure, $P - P_e$, and normalized cross-sectional area, $(A - A_0)/A_0$, of a long segment of inferior vena cava of a dog. The solid line is a computer solution for a latex tube. [*From Moreno et al. (1970) by permission.*]

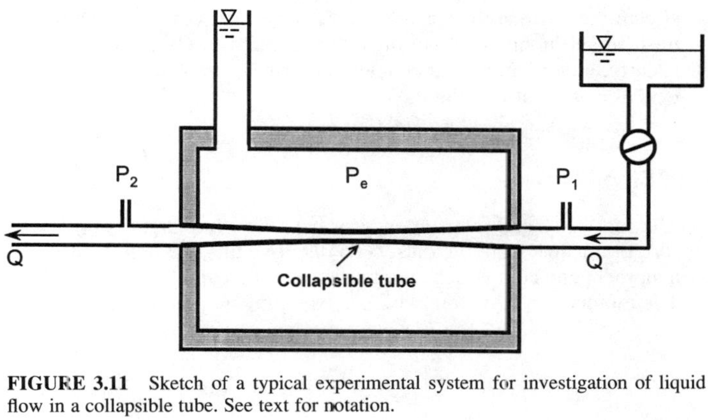

FIGURE 3.11 Sketch of a typical experimental system for investigation of liquid flow in a collapsible tube. See text for notation.

The governing equation for the tube deformation may be given by the tube law, which is also an equation of state that relates the transmural pressure to the local cross-sectional area,

$$P - P_e = K_p \cdot F\left(\frac{A}{A_0}\right) \tag{3.41}$$

where A_0 is the unstressed circular cross section and K_p is the wall stiffness coefficient. Solution of these governing equations for given boundary conditions provides the one-dimensional flow pattern of the coupled fluid-structure problem of fluid flow through a collapsible elastic tube.

Shapiro (1977) defined the speed index, $S = u/c$, similar to the Mach number in gas dynamics, and demonstrated different cases of subcritical ($S < 1$) and supercritical ($S > 1$) flows. It has been shown experimentally in simple experiments with compliant tubes that gradual reduction of the downstream pressure progressively increases the flow rate until a maximal value is reached (Holt, 1969; Conrad, 1969). The one-dimensional theory demonstrates that for a given tube (specific geometry and wall properties) and boundary conditions, the maximal steady flow that can be conveyed in a collapsible tube is attained for $S = 1$ (e.g., when $u = c$) at some position along the tube (Dawson & Elliott, 1977; Shapiro, 1977; Elad et al., 1989). In this case, the flow is said to be "choked" and further reduction in downstream pressure does not affect the flow upstream of the flow-limiting site. Much of its complexity, however, is still unresolved either experimentally or theoretically (Kamm et al., 1982; Kamm and Pedley, 1989; Elad et al., 1992).

3.5 BLOOD FLOW IN THE MICROCIRCULATION

The concept of a closed circuit for the circulation was established by Harvey (1578–1657). The experiments of Hagen (1839) and Poiseuille (1840) were performed in an attempt to elucidate the flow resistance of the human microcirculation. During the past century, major strides have been made in understanding the detailed fluid mechanics of the microcirculation and in depicting a concrete picture of the flow in capillaries and other small vessels.

3.5.1 The Microvascular Bed

We include in the term "microcirculation" those vessels with lumens (internal diameters) that are some modest multiple—say 1 to 10—of the major diameter of the unstressed RBC. This definition includes primarily the arterioles, the capillaries, and the postcapillary venules. The capillaries are of particular interest because they are generally from 6 to 10 μm in diameter, i.e., about the same size as the RBC. In the larger vessels, RBC may tumble and interact with one another and move from streamline to streamline as they course down the vessel. In contrast, in the microcirculation the RBC must travel in single file through true capillaries (Berman and Fuhro, 1969; Berman et al., 1982). Clearly, any attempt to adequately describe the behavior of capillary flow must recognize the particulate nature of the blood.

3.5.2 Capillary Blood Flow

The tortuosity and intermittency of capillary flow argue strongly that the case for an analytic description is lost from the outset. To disprove this, we must return to the Navier-Stokes equations for a moment and compare the various acceleration and force terms, which apply in the microcirculation. The momentum equation, which is Newton's second law for a fluid, can be written

$$\underset{(A)}{\rho \frac{\partial \mathbf{u}}{\partial t}} + \underset{(B)}{\rho(\mathbf{u} \cdot \nabla)\mathbf{u}} = \underset{(C)}{-\nabla P} + \underset{(D)}{F_{\text{shear}}} \tag{3.42}$$

Since we are analyzing capillaries in which the RBC are considered solid bodies traveling in a tube and surrounded by a waterlike fluid (plasma), a good representation of the viscous shear forces acting in the fluid phase is the newtonian flow,

$$F_{\text{shear}} = \mu \nabla^2 \mathbf{u} \tag{3.43}$$

We now examine the four terms in the momentum equation from the vantage point of an observer sitting on the erythrocytes. It is an observable fact that most frequently the fluid in the capillary moves at least 10 to 20 vessel diameters before flow ceases, so that a characteristic time for the unsteady term (A) is, say, 10 D/U. The distance over which the velocity varies by U is, typically, D. (In the gap between the RBC and the wall, this distance is, of course, smaller, but the sense of our argument is not changed.)

Dividing both sides of Eq. (3.42) by ρ, we have the following order-of-magnitude comparisons between the terms:

$$\begin{aligned} \frac{(A)}{(D)} &\approx \frac{U/(10D/U)}{(vU/D^2)} = \frac{UD}{10v} \\ \frac{(B)}{(D)} &\approx \frac{U^2/D}{(vU/D^2)} = \frac{UD}{v} \end{aligned} \tag{3.44}$$

The term UD/ν is the well-known Reynolds number. Typical values for human capillaries are U $\approx$ 500 μm/s, D $\approx$ 7 μm, $\nu \approx 1.5 \times 10^{-2}$ cm^2/s, so that the Reynolds number is about 2×10^{-3}. Clearly, the unsteady (A) and convective acceleration (B) terms are negligible compared to the viscous forces (Le-Cong and Zweifach, 1979; Einav and Berman, 1988).

This result is most welcome, because it allows us to neglect the acceleration of the fluid as it passes around and between the RBCs, and to establish a continuous balance between the local net pressure force acting on an element of fluid and the viscous stresses acting on the same fluid element. The equation to be solved is therefore

$$\nabla P = \mu \nabla^2 \mathbf{u} \tag{3.45}$$

subject to the condition that the fluid velocity is zero at the RBC surface, which is our fixed frame of reference, and U at the capillary wall. We must also place boundary conditions on both the pressure and velocity at the tube ends, and specify the actual shape and distribution of the RBC. This requires some drastic simplifications if we wish to obtain quantitative results, so we assume that all the RBC have a uniform shape (sphere, disk, ellipse, pancake, etc.) and are spaced at regular intervals. Then the flow, and hence the pressure, will also be subject to the requirement of periodicity, and we can idealize the ends of the capillary as being substantially removed from the region being analyzed. If we specify the relative velocity U between the capillary and the RBC, the total pressure drop across the capillary can be computed.

3.5.3 Motion of a Single Cell

For isolated and modestly spaced RBC, the fluid velocities in the vicinity of a red cell is schematically shown in Fig. 3.12. In the gap, the velocity varies from U to zero in a distance h, whereas in the "bolus" region between the RBC, the same variation is achieved over a distance of $D/4$. If $h < D/4$, as is often observed in vivo, then the viscous shear force is greatest in the gap region and tends to "pull" the RBC along in the direction of relative motion of the wall.

Counteracting this viscous force must be a net pressure, $P_u - P_d$, acting in a direction opposite to the sense of the shear force. This balance of forces is the origin of the parachutelike shape shown in Fig. 3.3 and frequently observed under a microscope.

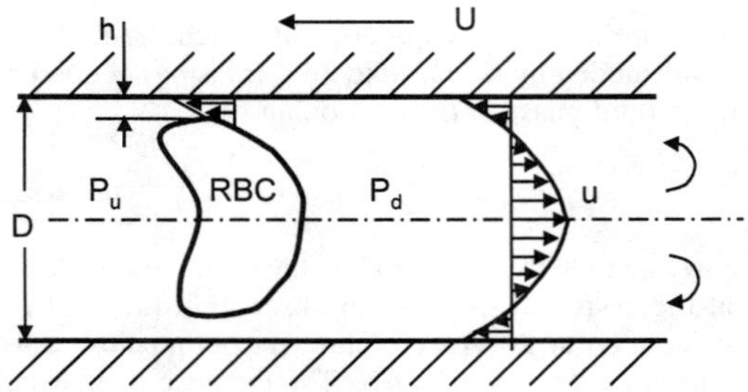

FIGURE 3.12 Diagram of the fluid pressure and velocity near the red blood cell within a capillary.

For $h \ll D/4$, we can approximate the net pressure,

$$(P_u - P_d)\frac{\pi D^2}{4} \approx \left(\mu \frac{U}{h}\right)(\pi D)\,(2b) \tag{3.46}$$

where 2b is the axial extent of the region of the gap. Suppose we use Eq. (3.46) to estimate the pressure drop across a typical capillary. Taking $h = 0.02D$, $b = 0.1D$, $U = 500\ \mu\text{m/s}$, $D = 7\ \mu\text{m}$, and $\mu = 1.4 \times 10^{-2}$ dyn · s/cm^2, then $P_u - P_d \approx 40$ dyn/cm^2.

3.6 BLOOD FLOW IN THE HEART

3.6.1 Flow in the Heart Ventricles

Under normal physiological conditions, systole and diastole occur in a definite coordination and constitute the *cardiac cycle*. Each cycle is considered to start with the atrial systole. The contraction begins a wave in that part of the right atrium where the orifices of the venae cavae are, and then involves both atria, which have a common musculature. With the cardiac rhythm of 75 contractions per minute, an atrial (auricular) systole lasts 0.1 second. As it ends, the ventricle systole begins, the atria then being in a state of diastole, which lasts 0.7 second. The contraction of the two ventricles occurs simultaneously, and their systole persists for about 0.3 second. After that, ventricular diastole begins and lasts about 0.5 second. One-tenth second before the end of the ventricular diastole, a new atrial systole occurs, and a new cycle of cardiac activity begins. The interconnection and sequence of the atrial and ventricular contractions depend upon where stimulation arises in the heart and how it spreads. Contraction of the ventricular myocardium ejects blood into the aorta and pulmonary arteries.

The heart valves are unidirectional valves, and in normal physiological conditions, blood flows in only one direction in the heart cavities: from the atria into the ventricles, and from the ventricles into the arterial system (Fig. 3.13). The ring-shaped muscle bundles of the atria, which surround the orifices, like a sphincter contract first during atrial systole, constricting these orifices so that blood flows from the atria only in the directions of the ventricles, and does not return into the veins. As the ventricles are relaxed during the atrial systole, and the pressure within them is lower than that in the contracting atria, blood enters them from the atria.

3.6.2 Flow through Heart Valves

The human heart contains four unidirectional valves that are anatomically grouped into two types: the atrioventricular valves and the semilunar valves. The tricuspid and mitral (bicuspid) valves

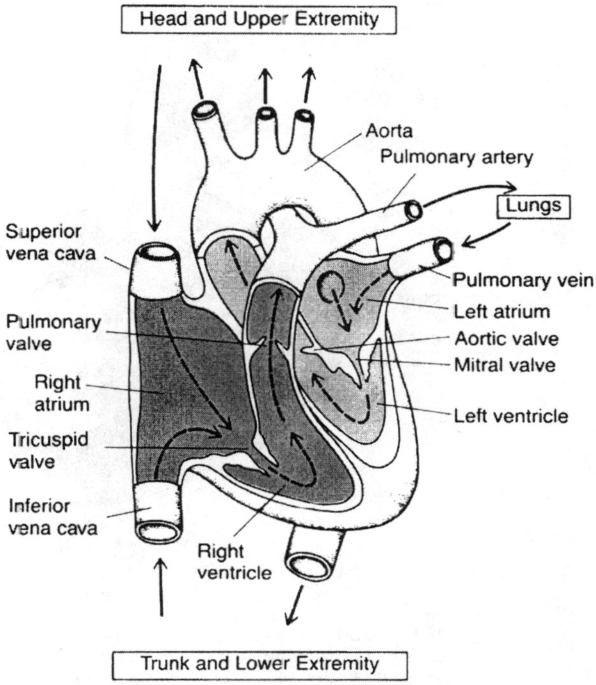

FIGURE 3.13 Structure of the heart and course of blood flow through the heart chambers. [*From Guyton and Hall (1996) by permission.*]

belong to the first type, whereas the pulmonic and aortic valves compose the second. Valves of the same type are not only structurally similar, but also functionally alike. For this reason, conclusions derived from studies on aortic and mitral valves are generally applicable to pulmonic and tricuspid valves, respectively.

One-way passages of blood from the ventricles into the main arteries is due to the heart valves. Though relatively simple in structure, heart valves are in their healthy state remarkably efficient, opening smoothly with little resistance to flow and closing swiftly in response to a small pressure difference with negligible regurgitation. Yet they are essentially passive devices that move in reaction to fluid-dynamical forces imposed upon them.

The motions of heart valves have drawn a considerable amount of attention from physicians and anatomists. One of the major reasons for this interest can be attributed to the frequent diagnosis of valvular incompetence in association with cardiopulmonary dysfunctionings. The first study exploring the nature of valve functioning, by Henderson and Johnson (1912), was a series of simple in vitro experiments, which provided a fairly accurate description of the dynamics of valve closure. Then came the Bellhouse and Bellhouse experiments (1969) and Bellhouse and Talbot (1969) analytical solution, which showed that the forces responsible for valve closure, are directly related to the stagnation pressure of the flow field behind the valve cusps.

Computational fluid dynamics models were developed over the years that include the effects of leaflet motion and its interaction with the flowing blood (Bellhouse et al., 1973; Mazumdar, 1992). Several finite-element structural models for heart valves were also developed in which issues such as material and geometric nonlinearities, leaflet structural dynamics, stent deformation, and leaflet coaptation for closed valve configurations were effectively dealt with (Bluestein and Einav, 1993; 1994). More recently, fluid-structure interaction models, based on the immersed boundary technique,

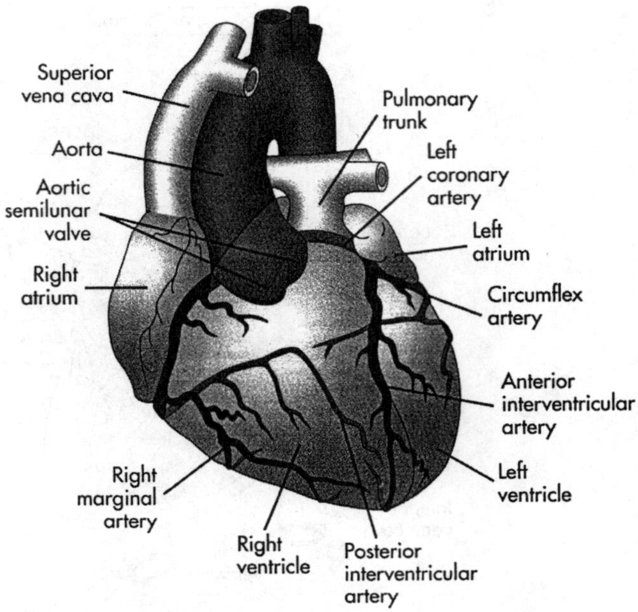

FIGURE 3.14 The arterial coronary circulation from anterior view. Arteries near the anterior surface are darker than those of the posterior surface seen through the heart. [*From Thibodeau and Patton (1999) by permission.*]

were used for describing valvular function (Sauob et al., 1999). In these models, strong coupled fluid-structure dynamic simulations were obtained. These models allowed for the inclusion of bending stresses, contact between adjacent leaflets when they coapted, and transient three-dimensional blood flow through the valve.

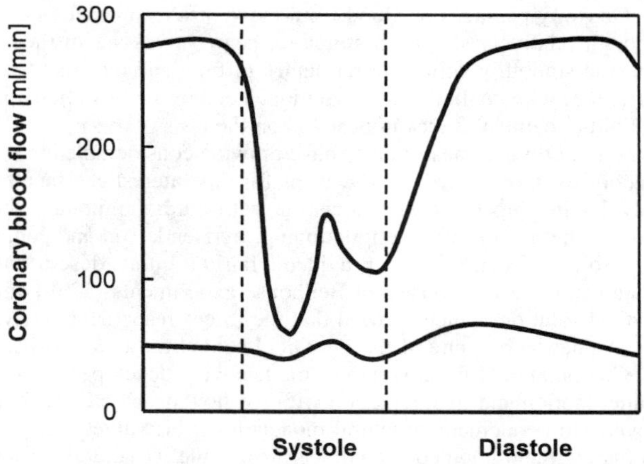

FIGURE 3.15 Blood flow through the coronary arteries: left (top) and right (bottom).

3.6.3 Coronary Blood Flow

The supply of blood to the heart is provided by the coronary vessels (Fig. 3.14), and occurs mainly during the diastole, opposite to other blood vessels. Coronary flow is low and even reversing during systole, while venous outflow is high. In diastole, coronary flow is high while venous flow is low. Contraction of the myocardium during the period of ventricular tension compresses the small arteries lying within it to such an extent that blood flow in the coronaries is sharply reduced (Fig. 3.15). Some blood from the cardiac veins enters the coronary sinus, which empties into the right atrium (Kajiya et al., 1990; Hoffman et al., 1985). The sinus receives blood mainly from the veins of the left ventricle, around 75 to 90 percent. A large amount of blood flows from the myocardium of the interatrial septum and from the right ventricle along numerous microvessels, and drains into the right ventricle (Krams et al., 1989). From 200 to 250 mL of blood flows through the coronaries of a human being per minute, which is about 4 to 6 percent of the minute volume of the heart. During physical exertion, coronary flow may rise to 3 or 4 L per minute (Bellhouse et al., 1968).

3.7 ANALOG MODELS OF BLOOD FLOW

3.7.1 Introduction

The circulatory system is a complex system of branching compliant tubes, which adjusts itself according to complex controllers. Mechanical and electrical analog models that can be solved analytically were devised to study either single arteries or the whole circulatory subsystem of specific organs (e.g., heart, kidney). Because of the complexity of the human system, there are a multitude of variables, which affect the functions, properties, and response of the circulatory system. Experimentally it is impossible to include all the known variables in a single system. However, analog models can deal with a group of variables at a time, and even make it possible to study the interaction between the variables. For extreme circulatory and physiological situations, it is simply too dangerous to study them experimentally, while the models are immune to studies.

3.7.2. Mechanical Models (Windkessel)

The *Windkessel* (air-cell in German) was suggested by Otto Frank (1899) to represent the cardiovascular system (Fig. 3.16*a*). In this linear mechanical model, the aorta and large blood vessels are

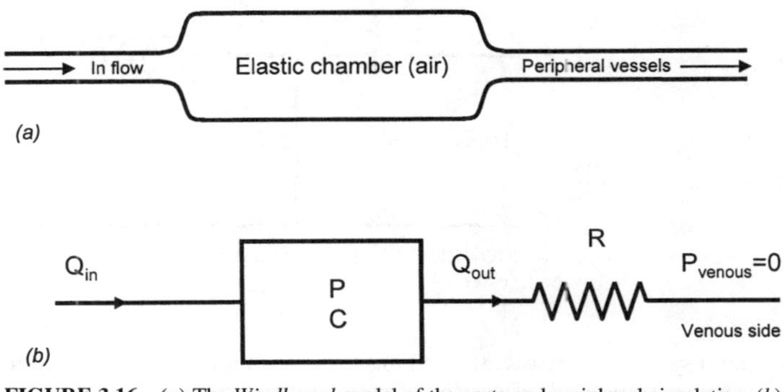

FIGURE 3.16 (*a*) The *Windkessel* model of the aorta and peripheral circulation. (*b*) Electrical model of the *Windkessel* model.

represented by a linear compliant air-cell and the peripheral vessels are replaced by a rigid tube with a linear resistance (Dinnar, 1981; Fung, 1984). Accordingly,

$$V = C \cdot P \qquad \text{and} \qquad Q_{\text{out}} = \frac{P}{R} \tag{3.47}$$

where V, P, and C = volume, pressure, and compliance of the air-cell and are linearly related
Q_{out} = outflow of the air-cell
R = linear resistance of the rigid tube
$P_{\text{venous}} = 0$

Conservation of mass requires that

$$Q_{\text{in}} = C\frac{dP}{dt} + \frac{P}{R} \tag{3.48}$$

where Q_{in} is the inflow to the air-cell and represents the cardiac stroke. This governing equation can be solved analytically for a set of given boundary conditions. The scheme of this analog is shown in Fig. 3.16*b*.

3.7.3. Electrical Models

A realistic model of blood flow in a compliant artery or vein requires a coupled solution of the governing equations for both the fluid and the elastic tube wall along with boundary conditions. These models are known as fluid-structure problems and can be solved by complex numerical techniques. In order to investigate linear approximations of complex blood flow phenomena, electrical analogs, which result in analogous linear differential equations, were developed (Dinnar, 1981). In these models the vessel resistance to blood flow is represented by a resistor R, the tube compliance by a capacitor C, and blood inertia by an inductance L. Assuming that blood flow and pressure are analogous to the electrical current and voltage, respectively, blood flow through a compliant vessel (e.g., artery or vein) may be represented by an electrical scheme (Fig. 3.17). Following Kirchhoff's law for currents, the governing differential equation becomes

$$\frac{d^2P}{dt^2} + RC\frac{dP}{dt} + P = L\frac{dQ}{dt} + RQ \tag{3.49}$$

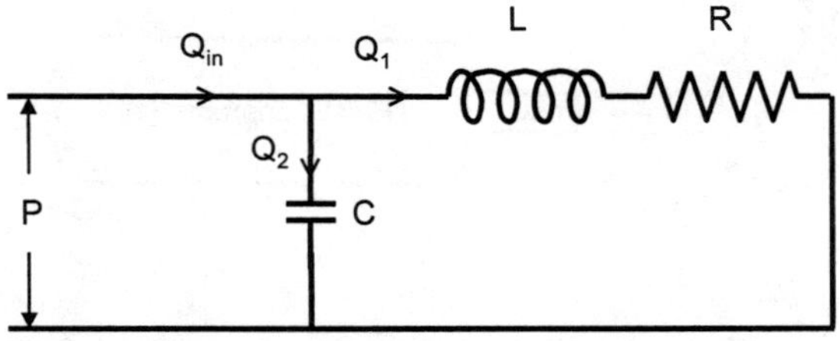

FIGURE 3.17 Electrical analog for blood flow in a compliant vessel.

For a system of vessels attached in series (or small units of a nonuniform vessel) a more complicated scheme, composed of a series of compartments whose characteristics are represented by R_i, C_i, and L_i (Fig. 3.18), is used. For this case, Kirchhoff's law applied to each loop will yield

$$P_i - P_{i-1} = R_i Q_i + L_i \frac{dQ_i}{dt} \tag{3.50}$$

$$P_i = \frac{V_i}{C_i} \tag{3.51}$$

$$Q_i - Q_{i-1} = \frac{dV_i}{dt} \tag{3.52}$$

The problem is reduced to a set of linear differential equations and can be solved for given boundary conditions. The *RCL* characteristics of a given problem may be derived from the known physical properties of blood and blood vessels (Dinnar, 1981; Van der Twell, 1957). This compartmental approach allowed for computer simulations of complex arterial circuits with clinical applications (McMahon et al., 1971; Clark et al., 1980; Barnea et al., 1990; Olansen et al., 2000; Westerhof and Stergiopulos, 2000; Ripplinger et al., 2001).

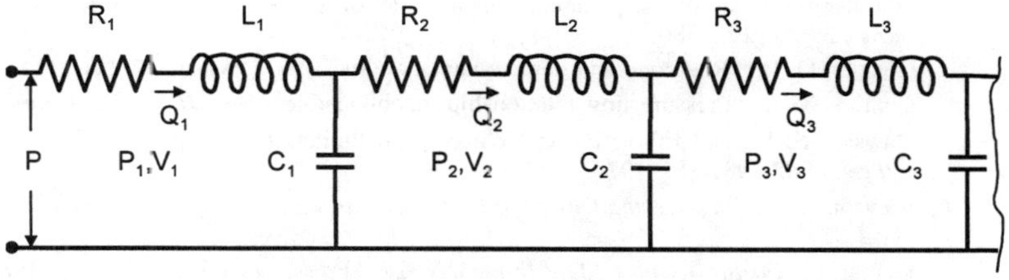

FIGURE 3.18 Lumped-parameter electrical model of a multicompartment arterial segment.

ACKNOWLEDGMENT

The authors are thankful to Dr. Uri Zaretsky for his technical assistance.

REFERENCES

Akhavan R. A, Kamm, R. D., and Shapiro, A. H., "An investigation of transition to turbulence in bounded oscillatory Stokes flows—Part 2: Numerical simulations," *J. Fluid Mech.*, **225**:423–444, 1991.

Atabek H. B., and Lew, H. S., "Wave propagation through a viscous incompressible fluid contained in a initially stressed elastic tube," *Biophys. J.*, **6**:481–503, 1966.

Barnea, O., Moore, T. W., Jaron, D., "Computer simulation of the mechanically-assisted failing canine circulation," *Ann. Biomed. Eng.*, **18**:263–283, 1990.

Bellhouse, B. J., Bellhouse, F. H., and Reid, K. G., "Fluid mechanics of the aortic root with application to coronary flow," *Nature*, **219**:1059–1061, 1968.

Bellhouse, B., and Bellhouse, F., "Fluid mechanics of model normal and stenosed aortic valves," *Circ. Res.*, **25**: 693–704, 1969.

Bellhouse, B., and Talbot, L., "Fluid mechanics of the aortic valve," *J. Fluid. Mech.*, **35**:721–735, 1969.

Bellhouse, B. J., Bellhouse, F., Abbott, J. A., and Talbot, L., "Mechanism of valvular incompetence in aortic sinus dilatation," *Cardiovasc. Res.*, **17**:490–494, 1973.

Berger, S. A., Goldsmith, W., and Lewis, E. R., *Introduction to Bioengineering*. Oxford University Press, 1996.

Berger, S. A., Talbot, L., and Yao, S., "Flow in curved tubes," *Ann. Rev. Fluid Mech.*, **15**:461–512, 1983.

Berman, H. J., and Fuhro, R. L., "Effect of rate of shear on the shape of the velocity profile and orientation of red blood cells in arterioles," *Bibl. Anat.*, **10**:32–37, 1969.

Berman, H. J., Aaron, A., and Behrman, S., "Measurements of red blood cell velocity in small blood vessels. A comparison of two methods: high-speed cinemicrography and stereopairs." *Microvas. Res.*, **23**:242, 1982.

Bluestein, D., and Einav, S., "Spectral estimation and analysis of LDA data in pulsatile flow through heart valves," *Experiments in Fluids*, **15**:341–353, 1993.

Bluestein, D., and Einav, S., "Transition to turbulence in pulsatile flow through heart valves: A modified stability approach," *J. Biomech. Eng.*, **116**:477–487, 1994.

Caro, C. G., Pedley T. J., Schroter R. C., and Seed, W. A., *The Mechanics of the Circulation*. Oxford University Press, Oxford, 1978.

Casson, M., in *Rheology of Dispersive Systems*, C. C. Mills, ed., Pergamon Press, Oxford, 1959.

Chandran, K. B., "Flow dynamics in the human aorta: Techniques and applications," in *Cardiovascular Techniques*, C. Leondes, ed., CRC Press, Boca Raton, Florida, Chap. 5, 2001.

Charm, S. E., and Kurland, G. S., "Viscometry of human blood for shear rates of 0–100,000 sec^{-1}," *Nature*, **206**:617–618, 1965.

Clark, J. W., Ling, R. Y., Srinivasan, R., Cole, J. S., and Pruett, R. C., "A two-stage identification scheme for the determination of the parameters of a model of left heart and systemic circulation," *IEEE Trans. Biomed. Eng.*, **27**:20–29, 1980.

Cokelet, G. R., "Rheology and hemodynamics," *Ann. Rev. Physiol.*, **42**:311–324, 1980.

Conrad, W. A., "Pressure-flow relationships in collapsible tubes," *IEEE Trans. Biomed. Eng.*, **16**:284–295, 1969.

Dawson, S. V., and Elliott E. A., "Wave-speed limitation on expiratory flow: A unifying concept," *J. Appl. Physiol.*, **43**:498–515, 1977.

Dawson, T. H., *Engineering Design of the Cardiovascular System of Mammals*, Prentice Hall, Englewood Cliffs, N. J., 1991.

Dinnar, U., *Cardiovascular Fluid Dynamics*, CRC Press, Boca Raton, Florida, 1981.

Einav, S., and Berman, H. J., "Fringe mode transmittance laser Doppler microscope anemometer. Its adaptation for measurement in the microcirculation," *J. Biomech. Eng.*, **10**:393–399, 1988.

Einav, S., and Sokolov, M., "An experimental study of pulsatile pipe flow in the transition range," *J. Biomech. Eng.*, **115**:404–411, 1993.

Elad, D., Sahar, M., Avidor, J. M., and Einav, S., "Steady flow through collapsible tubes: measurement of flow and geometry," *J. Biomech. Eng.*, **114**:84–91, 1992.

Elad, D., Kamm., R. D., and Shapiro, A. H., "Steady compressible flow in collapsible tubes: Application to forced expiration," *J. Fluid. Mech.*, **203**:401–418, 1989.

Flaherty, J. E., Keller, J. B., and Rubinow, S. I., "Post buckling behavior of elastic tubes and rings with opposite sides in contact," *SIAM J. Appl. Math.*, **23**:446–455, 1972.

Fung, Y. C., *Biodynamics: Circulation*, Springer-Verlag, New York, 1984.

Guyton, A. C., and Hall J. E., *Textbook of Medical Physiology*, 9th ed., W. B. Saunders, Philadelphia, 1996.

Henderson, Y., and Johnson, F. E., "Two models of closure of the heart valves," *Heart*, **4**:69, 1912.

Hoffman, H. E., Baer R. W., Hanley, F. L., Messina, L. M., and Grattan, M. T., "Regulation of transmural myocardial blood flow," *J. Biomech. Eng.*, **107**:2–9, 1985.

Holt, J. P., "Flow through collapsible tubes and through in situ veins," *IEEE Trans. Biomed. Eng.*, **16**:274–283, 1969.

Kajiya, F., Klassen, G. A., Spaan, J. A. E., and Hoffman J. I. E., *Coronary circulation. Basic mechanism and clinical relevance*, Springer, Tokyo, 1990.

Kamm, R. D., "Bioengineering studies of periodic external compression as prophylaxis against deep vein thrombosis—Part I: Numerical studies; Part II: Experimental studies on a stimulated leg," *J. Biomech. Eng.*, **104**: 87–95 and 96–104, 1982.

Kamm, R. D., and Pedley, T. J., "Flow in collapsible tubes: a brief review," *J. Biomech. Eng.*, **111**:177–179, 1989.

Kleinstreuer, C., Lei, M., and Archie, J. P., "Hemodynamics simulations and optimal computer-aided designs of branching blood vessels," in *Biofluid Methods in Vascular and Pulmonary Systems*, C. Leondes, ed., CRC Press, Boca Raton, Florida, Chap. 1, 2001.

Krams, R., Sipkema, P., Zegers, J., and Westerhof, N., "Contractility is the main determinant of coronary systolic flow impediment," *Am. J. Physiol.: Heart Circ. Physiol.*, **257**:H1936–H1944, 1989.

Ku, D. N., and Giddens, D. P., "Laser Doppler anemometer measurements of pulsatile flow in a model carotid bifurcation," *J. Biomech.*, **20**:407–421, 1987.

Le-Cong, P., and Zweifach, B. W., "In vivo and in vitro velocity measurements in microvasculature with a laser," *Microvasc. Res.*, **17**:131–141, 1979.

Lieber, B. B., and Giddens, D. P., "Apparent stresses in disturbed pulsatile flow," *J. Biomechanics*, **21**:287–298, 1988.

Liepsch, D., "Fundamental flow studies in models of human arteries," *Front. Med. Biol. Eng.*, **5**:51–55, 1993.

Liepsch, D., Poll, A., Strigberger, J., Sabbah, H. N., and Stein, P. D., "Flow visualization studies in a mold of the normal human aorta and renal arteries," *J. Biomech. Eng.*, **111**:222–227, 1989.

Martini, F. H., *Fundamentals of Anatomy and Physiology*. Prentice Hall, Englewood Cliffs, N. J., 1995.

Mazumdar, J. N., *Biofluid Mechanics*, World Scientific Pub., Singapore, 1992.

McDonald, D. A., The relation of pulsatile pressure to flow in arteries. *J. Physiol*, **127**:533–552, 1955.

McMahon, T. A., Clark, C., Murthy, V. S., and Shapiro, A. H., "Intra-aortic balloon experiments in a lumped-element hydraulic model of the circulation," *J. Biomech.*, **4**:335–350, 1971.

Mills, C. J., Gabe, I. T., Gault, J. H., Mason, D. T., Ross, J., Braunwald, E., and Shillingford, J. P., "Pressure-flow relationships and vascular impedance in man," *Cardiovasc. Res.*, **4**:405–417, 1970.

Moreno, A. H., Katz, I. A., Gold, L. D., Reddy, R. V., "Mechanics of distension of dog veins and other very thin-walled tubular structures," *Circ. Res.*, **27**:1069–1080, 1970.

Nerem, R. M., and Rumberger, J. A., "Turbulance in blood flow," *Recent Adv. Eng. Sci.*, **7**:263–272, 1976.

Nichols, W. W., O'Rourke, M. F., *McDonald's Blood Flow in Arteries: Theoretical, experimental, and clinical principles*, Arnold, London, 1998.

Olansen, J. B., Clark, J. W., Khoury, D., Ghorbel, F., and Bidani, A., "A closed-loop model of the canine cardiovascular system that includes ventricular interaction," *Comput. Biomed. Res.*, **33**:260–295, 2000.

Pedley, T. J., Schroter, R. C., and Sudlow, M. F., "Flow and pressure drop in systems of repeatedly branching tubes," *J. Fluid Mech.*, **46**:365–383, 1971.

Pedley, T. J., *The Fluid Mechanics of Large Blood Vessels*, Cambridge University Press: Cambridge, U. K., 1980.

Pedley, T. J., "High Reynolds number flow in tubes of complex geometry with application to wall shear stress in arteries," *Symp. Soc. Exp. Biol.*, **49**:219–241, 1995.

Perktold, K., and Rappitsch, G., "Computer simulation of local blood flow and vessel mechanics in a compliant carotid artery bifurcation model," *J. Biomech.*, **28**:845–856, 1995.

Pinchak, A. C., and Ostrach, S., "Blood flow in branching vessels," *J. Appl. Physiol.*, **41**:646–658, 1976.

Ripplinger, C. M., Ewert, D. L., and Koenig, S. C., "Toward a new method of analyzing cardiac performance," *Biomed. Sci. Instrum.*, **37**:313–318, 2001.

Saladin, K. S., *Anatomy and Physiology: The Unity of Form and Function*, 2d ed., McGraw-Hill, New York 2001.

Sauob, S. N., Rosenfeld, M., Elad, D., and Einav, S., "Numerical analysis of blood flow across aortic valve leaflets at several opening angles," *Int. J. Cardiovasc. Med. Sci.*, **2**:153–160, 1999.

Shapiro, A. H., "Steady flow in collapsible tubes," *J. Biomech. Eng.*, **99**:126–147, 1977.

Thibodeau, G. A., Patton, K. T., *Anatomy and Physiology*. 4th ed., Mosby, St Louis, 1999.

Thiriet, M., Naili, S., Langlet, A., and Ribreau, C., "Flow in thin-walled collapsible tubes," in *Biofluid Methods in Vascular and Pulmonary Systems*, C. Leondes, ed., CRC Press, Boca Raton, Florida, Chap. 10, 2001.

Van der Tweel, L. H., "Some physical aspects of blood pressure pulse wave, and blood pressure measurements," *Am. Heart J.*, **53**:4–22, 1957.

Westerhof, N., and Stergiopulos, N., "Models of the arterial tree," *Stud. Health Technol. Inform.*, **71**:65–77, 2000.

Whitmore, R. L., *Rheology of the Circulation*, Pergamon Press, Oxford, 1968.

Womersley, J. R., "Method for the calculation of velocity, rate of flow and viscous drag in arteries when the pressure gradient is known," *J. Physiol.*, **127**:553–563, 1955a.

Womersley, J. R., "Oscillatory motion of a viscous liquid in a thin-walled elastic tube: I. The linear approximation for long waves," *Philosophical Magazine*, Ser. 7, **46**:199–221, 1955b.

Yamaguchi, T., Parker K. H., "Spatial characteristics of turbulence in the aorta," *Ann. N. Y. Acad. Sci.*, **404**: 370–373, 1983.

Zamir, M., *The Physics of Pulsatile Flow*, Springer, New York, 2000.

CHAPTER 4
RESPIRATORY MECHANICS AND GAS EXCHANGE

James B. Grotberg
University of Michigan, Ann Arbor, Michigan

4.1 ANATOMY

4.1.1 General

As shown in Fig. 4.1, the pulmonary system consists of two lungs; each lung is conical in shape, with an inferior border (or base) that is concave as it overlies the diaphragm and abdominal structures and a superior border (or apex) that is convex and extends above the first rib. The anterior, lateral, and posterior lung surfaces are adjacent to the rib cage, while the medial surface is adjacent to the mediastinum, which contains the heart, great vessels, and esophagus. All of the lung surface is covered by the visceral pleural membrane, while the inside of the chest wall, mediastinum, and diaphragm are covered by the parietal pleural membrane. The two pleural membranes are separated by a thin liquid film (~20 to 40 μm thick) called the pleural fluid which occupies the pleural space, Fig. 4.1. This fluid lubricates the sliding motion between the lung and chest wall, and its pressure distribution determines lung inflation and deflation. Because the air-filled lung is surrounded by the pleural liquid, it experiences a buoyancy force that contributes to the overall force including lung weight and fluid pressures.[17]

4.1.2 Airway Divisions

The right main bronchus branches from the end of the trachea to the right lung, while the left main bronchus branches to the left. This division of a "parent" airway into two "daughter" airways is called a *bifurcation*. Each level of branching is called a *generation* and is given an integer value n, according to the Weibel symmetric model[35] shown in Table 4.1. The trachea is designated as $n = 0$, the main bronchi, $n = 1$, etc. For a symmetrically bifurcating geometry, then, there would be 2^n airways at each generation level. Generations $0 \leqslant n \leqslant 16$ constitute the conducting zone

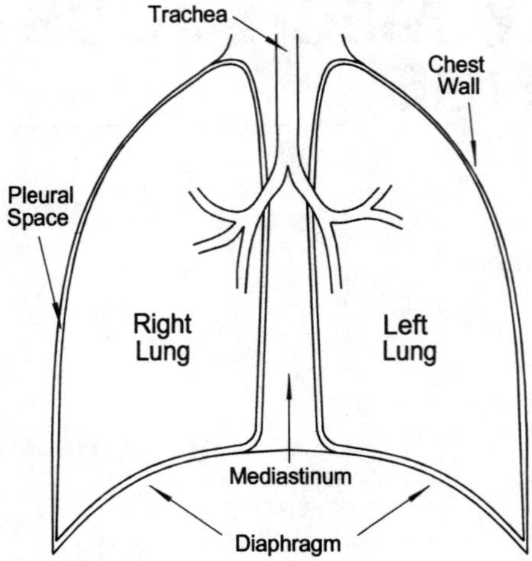

FIGURE 4.1 Sketch of the pulmonary system.

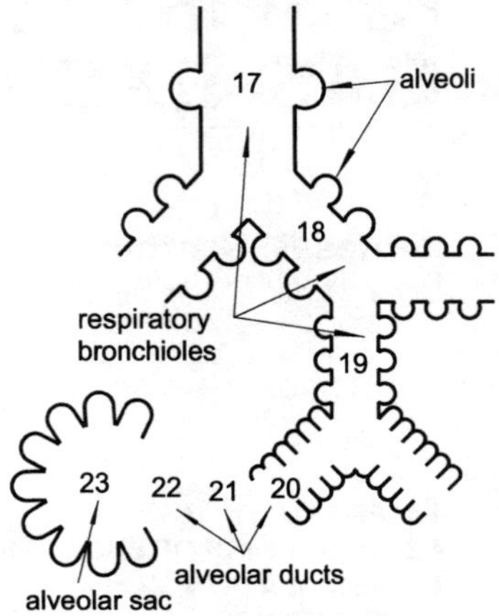

FIGURE 4.2 The respiratory zone of the airway network.

(trachea to terminal bronchiole) whose airways have no alveoli and, hence, do not participate in gas exchange with the pulmonary blood circulation. Each terminal bronchiole enters an acinus or respiratory zone unit for $17 \leqslant n \leqslant 23$; see Fig. 4.2. The respiratory bronchioles occupy generations $17 \leqslant n \leqslant 19$ and are tubes with partially alveolated surfaces. Generations $20 \leqslant n \leqslant 22$ are the alveolar ducts, tubes with totally alveolated surfaces. The alveoli are polyhedral in shape, but, viewed as nearly spherical caps, they each have a diameter of $\sim$200 μm at ¾ inflation.

4.1.3 Pulmonary Circulation

The pulmonary circulation consists of mixed venous blood entering the pulmonary artery from the right ventricle of the heart. After passing through the pulmonary capillary bed, the blood returns to the left atrium via the pulmonary vein. Flow through portions of the capillary bed that are in ventilated alveolar regions of the lung will discharge CO_2 and absorb O_2. Flows that go to unventilated alveoli will not exchange gas and are called the *shunt flow*. Typical values of pulmonary arterial blood pressures are 20 mmHg systolic/12 mmHg diastolic, about one-sixth the systemic vascular pressures. Because the pulmonary system is a low-pressure system, the vessel walls are much thinner than their systemic counterparts. The pulmonary arterial and venous systems have accompanying divisions, forming a triad with each airway division. The pulmonary capillaries, with a vessel diameter of $\sim$6 μm, form a multivessel mesh surrounding each alveolus and are contained in the alveolar walls. The blood-gas barrier is composed of (starting from the gas side): the alveolar liquid lining, alveolar epithelial cell, basement membrane, and capillary endothelial cell. These combined layers sum to approximately 0.5 to 1.0 μm, in thickness. This is the distance gas must diffuse to enter or leave the blood under normal conditions.

The conducting airways are fed by a second blood supply, the bronchial circulation, which comes from the systemic circulation via the bronchial arteries. Blood returns via the bronchial veins, which empty partly to the systemic veins and partly to the pulmonary veins. The latter case is a circuit that bypasses the lungs entirely and makes the systemic cardiac output slightly larger than the pulmonary cardiac output or lung perfusion $\dot{Q}$.

4.1.4 Lymphatics and Nerves

To maintain homeostasis, the continuous transudation of fluid and solutes from the pulmonary capillary bed into the surrounding interstitium and alveolar space is balanced by lymphatic drainage out of the lung. The lymphatic flow is directed toward the hilum from the pleural surfaces. From lymph nodes in the hilum, the lymph travels to the paratracheal nodes and then eventually into the venous system via the thoracic duct. The lung has nerve fibers from both the vagal nerves (parasympathetic) and the sympathetic nerves. The efferent fibers go to the bronchial musculature and the afferents come from the bronchi and alveoli.

4.2 MECHANICS OF BREATHING

4.2.1 Chest Wall

The rib cage and its muscles form the chest wall, which protects the vital thoracic organs while keeping the lungs inflated. During inspiration, the ribs swing on an axis defined by their articulation with the vertebrae,[20] dashed lines in Fig. 4.3. The result is that upper ribs, such as rib 1, swing forward and up, like a pump handle, increasing the anterior-posterior diameter of the upper chest wall. However, lower ribs swing primarily laterally, like a bucket handle pinned at the spine and sternum. Their motion increases the lateral diameter of the thorax.

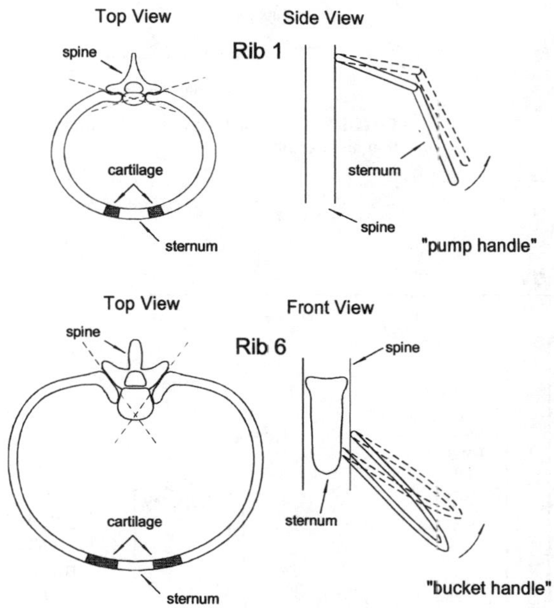

FIGURE 4.3 Rib motion during inspiration.

4.2.2 Muscles of Inspiration and Expiration

The major muscles involved in inspiration and expiration include the diaphragm and the intercostal muscles. During inspiration, the diaphragm contracts, pulling the inferior lung surface (via the pleural fluid) downward, while the external intercostal muscles contract, lifting the ribs; see Fig. 4.4. Expiration is primarily a passive event; the elastic structures simply return to their original, less-stretched, state as the diaphragm and external intercostals relax. At a normal breathing rate of 15 breaths per minute (bpm), inspiration may occupy one-third of the 4-second breathing cycle, while passive expiration occupies the rest, an inspiratory to expiratory time ratio of 1:2. Forceful expiration is accomplished by contraction of the internal intercostal muscles and the abdominal wall muscles that squeeze the abdominal contents hard enough to push them upward. The normal inspiratory to expiratory time ratio can increase with forceful expiration, as may occur during exercise, or decrease with prolonged expiration, as one finds in obstructive airways disease like asthma or emphysema.

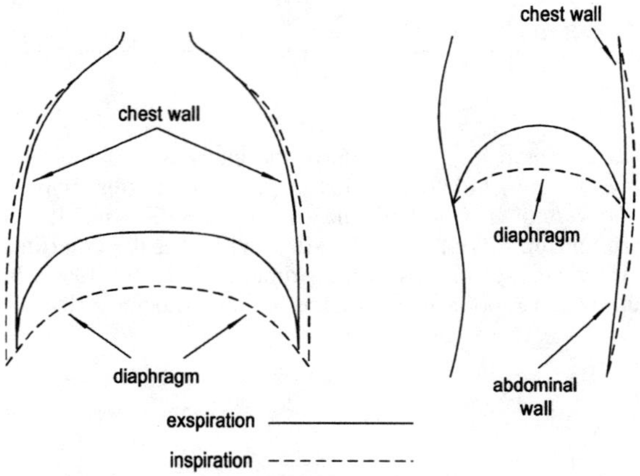

FIGURE 4.4 Diaphragm and abdominal muscles during inspiration and expiration.

4.3 VENTILATION

4.3.1 Lung Volumes

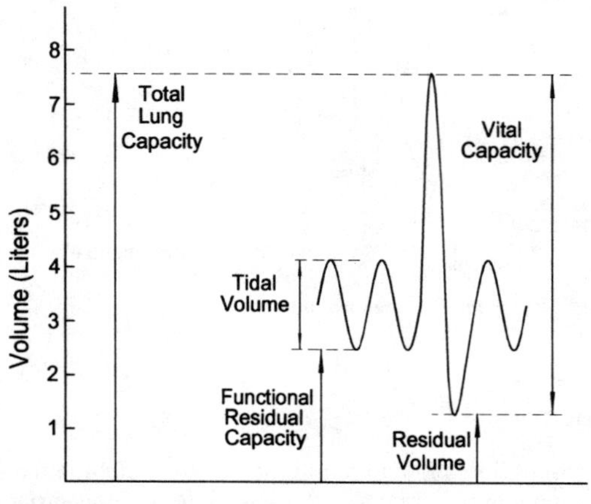

FIGURE 4.5 Lung volumes and definitions.

There is common terminology for different lung volume measurements, as shown in Fig. 4.5. The maximum volume is total lung capacity (TLC), which can be measured by dilution of a known amount of inspired helium gas whose insolubility in tissue and blood prevents it from leaving the air spaces. The minimum is residual volume (RV). Normal ventilation occurs within an intermediate range and has a local minimum called functional residual capacity (FRC). The volume swing from FRC to the end of inspiration is the tidal volume V_T. The vital capacity (VC) is defined by VC = TLC − RV. Within the lung there are also important volume concepts. The anatomic dead space V_D is the summed volume of all conducting airways, measured by the Fowler method, while the physiologic dead space is that portion of the lung that does not transfer CO_2 from the capillaries, measured by the Bohr method. While the methods used to determine these two values are different, normally they yield essentially the same result, approximately 150 cm^3 in an adult male. In some disease states, however, the Bohr method may be affected by abnormalities of the ventilation-perfusion relationship. The alveolar volume V_A, where gas exchange occurs, is the lung volume minus the dead space.

4.3.2 Air Flow and Resistance

A typical V_T of 500 cm^3 at a rate of 15 bpm yields a total ventilation $\dot{V}$ of (0.5 liters $\times$ 15 breaths/min) = 7.5 liters/min. Assuming that air flows in the conducting airways like a plug, for each 500 cm^3 breath inspired the tail 150 cm^3 fills the dead space while the front 350 cm^3 expands the alveoli, where it mixes with the alveolar gas previously retained at FRC. This makes the alveolar ventilation $\dot{V}_A = (0.5 - 0.15)$ liters $\times$ 15 bpm = 5.25 liters/min. The plug flow assumption is not correct, of course, but for resting ventilation it is a useful simplification. It fails, for example, in high-frequency ventilation,[3,24] where the mechanical ventilator can operate at 15 Hz with $V_T \leqslant 5$ cm^3; i.e., the tidal volumes can be smaller than the dead space. Adequate gas exchange occurs under these circumstances, because of the Taylor dispersion mechanism,[33] which is a coupling of curvilinear, axial velocity profiles with radial diffusion applied to a reversing flow.[5] The properties of airway branching, axial curvature, and flexibility can modify the mechanism considerably.[6,8,14]

The ventilatory flow rate $\dot{\mathbf{V}}$ results from the pressure boundary conditions imposed at the trachea and alveoli. In lung physiology, the flow details are often neglected and simply lumped into a resistance to air flow defined as the ratio of the overall pressure drop $\mathbf{\Delta P}$ to the flow rate; i.e., $\Delta P/\dot{V}$ = Res. A typical value is Res = 0.2 cmH_2O-s/L for resting breathing conditions. The appeal of defining airway resistance this way is its analogy to electrical circuit theory and Ohm's law, and other elements such as capacitance (see compliance, below) and inertance may be added.[7,27,32] Then an airway at generation n has its own resistance, Res_n, that is in series with its $n - 1$ and $n + 1$ connections but in parallel with the others at n.

The detailed fluid dynamics can be viewed as contributing to either the pressure drop due to kinetic energy changes, ΔP_K, or that due to frictional, viscous, effects, ΔP_F, so that

$$\Delta P = \Delta P_K + \Delta P_F \tag{4.1}$$

For a frictionless system, $\Delta P_F = 0$, Eq. (4.1) becomes the Bernoulli equation, from which we learn that, in general, $\Delta P_K = \frac{1}{2}\,\rho(u_{\mathrm{out}}^2 - u_{\mathrm{in}}^2)$, where $u_{\mathrm{out}}(u_{\mathrm{in}})$ is the outlet (inlet) velocity. For a given flow rate $\dot{V}$, the average air velocity at generation n is $u_n = \dot{V}/A_n$, where A_n is the total cross-sectional area at generation n. Because $A_0 \ll A_{23}$, air velocities diminish significantly from the tracheal value to the distal airways. This implies that $\Delta P_K < 0$ for inspiration while $\Delta P_K > 0$ for expiration.

The frictional pressure drop ΔP_F is always positive, but its value depends on the flow regime in each airway generation. For fully developed, laminar tube flow (Poiseuille flow) $\Delta P_F = (128\,\mu L/\pi d^4)\dot{V}$, where μ is the fluid (gas) viscosity, d is the tube diameter, and L is the axial distance. This formula is helpful in understanding some of the general trends for R, like its strong dependence on airway diameter, which can decrease in asthma. Airways are aerodynamically short tubes under many respiratory situations, i.e., not long enough for Poiseuille flow to develop. Then the viscous pressure drop is governed by energy dissipation in a thin boundary layer region near the airway wall where the velocity profile is curvilinear. For this entrance flow, $\Delta P_F \sim \dot{V}^{3/2}$, so resistance now is flow dependent. The local Reynolds number is given by $\mathrm{Re}_n = u_n d_n/\nu$, where $\nu \approx 0.15$ cm^2/s is the kinematic viscosity of air. For $\mathrm{Re}_n \geqslant 2300$, (see Table 4.1), the flow is turbulent and ΔP_F in those airways has an even stronger dependence on ventilation, $\Delta P_F \sim \dot{V}^{7/4}$ smooth wall, $\sim \dot{V}^2$ rough wall. Expressing these three cases for a single tube (airway) in terms of the friction coefficient, $C_F = \Delta P_F / \frac{1}{2}\rho u^2$. The comparison is shown in Eq. (4.2):

$$C_F = 64\frac{L}{d}\mathrm{Re}^{-1} \qquad C_F = 6\left(\frac{L}{d}\right)^{1/2}\mathrm{Re}^{-1/2} \qquad C_F = 0.32\frac{L}{d}\mathrm{Re}^{-1/4} \qquad C_F = k\frac{L}{d} \tag{4.2}$$

Poiseuille flow — entrance flow — turbulent flow smooth wall — turbulent flow rough wall where k depends on the wall roughness; a typical value is $k = 0.04$. Experiments of inspiratory flow with a multigeneration cast of the central airways[31] show that Poiseuille effects dominate for $350 \leq \text{Re}_0 \leq 500$, entrance effects for $500 \leq \text{Re}_0 \leq 4000$, and rough-walled turbulent effects for $4000 \leq \text{Re}_0 \leq 30{,}000$, approximately. C_F tends to be higher during expiration than inspiration.

As shown from measurements in airway models[19] or theories using modifications of the friction coefficients,[28] the vast majority of the pressure drop across the entire lung occurs within the large airways, say $0 \leq n \leq 8$. Thus clinical evaluation of total airway resistance can miss diseases of the small airways whose diameters are less than 2 mm, the so-called silent zone of the network. Pressure drop measurements and models are also important for the design of ventilators and other respiratory assist or therapeutic devices that interface with lung mechanics.

4.3.3 Gas Transport

Transport of O_2, CO_2, anesthetics, toxins, or any other soluble gas occurs by convection, diffusion, and their interplay. The ability of one gas species to diffuse through another is quantified by the molecular diffusivity D for the gas pair. For O_2 diffusing through air, the value of D is $D_{O_2-\text{air}} = 0.22$ cm²/s (at atmospheric pressure and 37°C), whereas for CO_2 the value is less, $D_{CO_2-\text{air}} = 0.17$ cm²/s. A characteristic time for diffusion to provide effective transport of a concentration front over a distance L is $T_d = L^2/D$, where L can be the airway length L_n. By contrast, a characteristic time for convective transport over the same distance is $T_c = L/U$, where $U = u_n$ is the average flow speed of the bulk gas mixture in an airway. The ratio of the two time scales tells us which mechanism, diffusion, or convection, may dominate (i.e., occur in the shortest time) in a given airway situation. This ratio is the Peclet number, $\text{Pe} = T_d/T_c = UL/D$, where $\text{Pe} \gg 1$ indicates convection-dominated transport while $\text{Pe} \ll 1$ indicates diffusion-dominated transport. Table 4.1 shows how Pe_n for O_2 transport varies through the airway tree when $\dot{V} = 500$ cm³/s. Convection dominates in the conducting airways, where negligible O_2 is lost, so the O_2 concentration in these airways on inspiration is essentially the ambient inspired level, normally $P_{I_{O_2}} = 150$ mmHg. At approximately generation 15, where Pe is close to unity and both convection and diffusion are important, the axial concentration gradient in O_2 develops over the next ~0.5 cm in path length to reach the alveolar level ($P_{A_{O_2}} = 100$ mmHg). CO_2 has the reverse gradient during inspiration from its alveolar level ($P_{A_{CO_2}} =$

TABLE 4.1 Airway Geometry

n generation	Name	2^n number	d_n, diameter, cm	L_n, length, cm	A_n, total cross-sectional area, cm²	V_n, volume of generation, cm³	X_n, distance from carina, cm	Re_n*	Pe_n* (O_2)
0	Trachea		1.80	12.00	2.54	30.50	0	2,362	1,611
1	Main bronchus	1	1.22	4.76	2.33	11.25	4.76	1,745	1,190
2	Lobar bronchus	4	0.83	1.90	2.13	3.97	6.66	1,299	886
3	Segmental bronchus	8	0.56	1.76	2.00	1.52	7.42	933	636
4	Subsegmental bronchus	16	0.45	1.27	2.48	3.46	8.69	605	412
5	Small bronchus	32	0.35	1.07	3.11	3.30	9.76	375	256
10	Small bronchus	1,024	0.13	0.46	13.40	6.21	13.06	32	22
16	Terminal bronchiole	65,536	0.06	0.17	180.00	29.70	14.46	1.11	0.79
17	Respiratory bronchiole	131,072	0.054	0.141	300.00	41.80	14.55	0.60	0.43
20	Alveolar duct	1,048,576	0.045	0.083	1,600.00	139.50	14.77	0.09	0.06
23	Alveolar sac	8,388,608	0.041	0.050	11,800.00	591.00	14.98	0.01	0.01

*Re_n and Pe_n for O_2 transport are evaluated for $\dot{V} = 500$ cm³/s.
Source: Data from Weibel (1963) (Ref. 35).

40 mmHg) to the ambient level of near zero. During expiration the alveolar levels of both gases are forced out through the airway tree.

Once in the alveoli, gas transport through the alveolar wall and endothelium into the blood (or vice versa) is diffusive transport, which is quantified by the concept of the lung's diffusing capacity. As shown in Fig. 4.6, with wall thickness h, total alveolar membrane surface area A_m, and a gas concentration or partial pressure difference of P_1 (alveolar) − P_2 (blood serum), the mass flow rate across is $\dot{V}_{gas} = A_m D_{gas\text{-}tiss}(P_1 - P_2)/h = D_L(P_1 - P_2)$. $D_{gas\text{-}tiss}$ is the diffusivity of the gas (O_2, CO_2) in tissue and is proportional to solubility/(molecular weight)$^{1/2}$, making $D_{CO_2\text{-tiss}} \sim 20\, D_{O_2\text{-tiss}}$ since CO_2 has far greater solubility in tissue. The details of the area, thickness, and diffusivity are lumped into an overall coefficient called the *diffusing capacity* D_L. The diffusion equation can be simplified for transport of carbon monoxide, CO, which is so rapidly taken up by hemoglobin within the red blood cells that the serum P_2 value is essentially zero. Then we find, after rearranging, that the diffusing capacity is given by

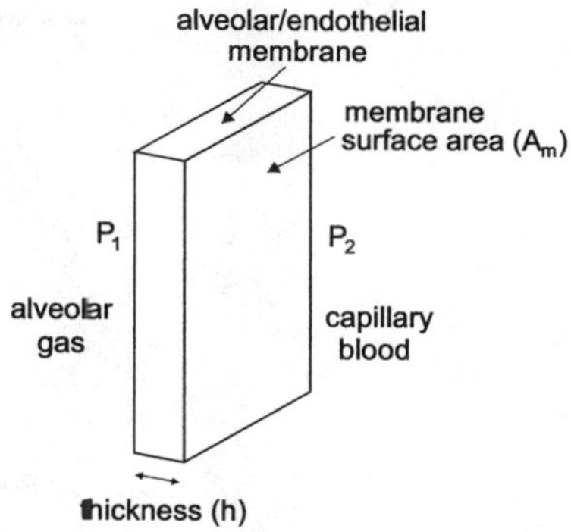

FIGURE 4.6 Diffusion across the alveolar/endothelial membrane.

$$D_L = \frac{\dot{V}_{CO}}{P_{A_{CO}}} \tag{4.3}$$

and there are both single-breath and steady-state methods employed in pulmonary function testing to evaluate D_L by breathing small concentrations of CO. Diseases that thicken the membrane wall, like pulmonary fibrosis, or reduce the available surface area, like emphysema, can cause abnormally low D_L values.

4.4 ELASTICITY

4.4.1 Pressure-Volume Relationship of Isolated Lung

The overall mechanical properties of the isolated lung are demonstrated by its pressure-volume (P-V) curve. Lung inflation involves both inflating open alveoli and recruiting those which are initially closed because of liquid plugs or local collapse. Deflation from TLC starts with a relatively homogenous population of inflated alveoli. This irreversible cycling process is partly responsible for the inflation and deflation limbs of the curve being different. Inflation requires higher pressures than deflation for any given volume; see the air-cycled (solid line) curve of Fig. 4.7 adapted from Ref. 26. Irreversible cycling of a system exhibits its hysteresis, and the hatched area within the loop created by the two limbs is called the hysteresis area. The magnitude of the area is the work per cycle required to cycle the lung. The overall slope of the curve, obtained by connecting the endpoints with a straight line, is the total compliance $C_{tot} = \Delta V/\Delta P$, which measures the total volume change obtained for the total pressure increase. Stiffer lungs have lower compliance, for example. Often one wants a local measure of compliance, which is the local slope of the $V(P)$ curve, defined as the dynamic

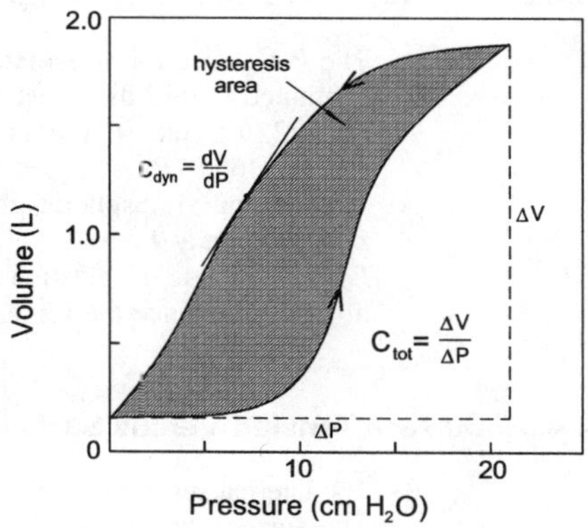

FIGURE 4.7 Air cycling of isolated dog lungs.

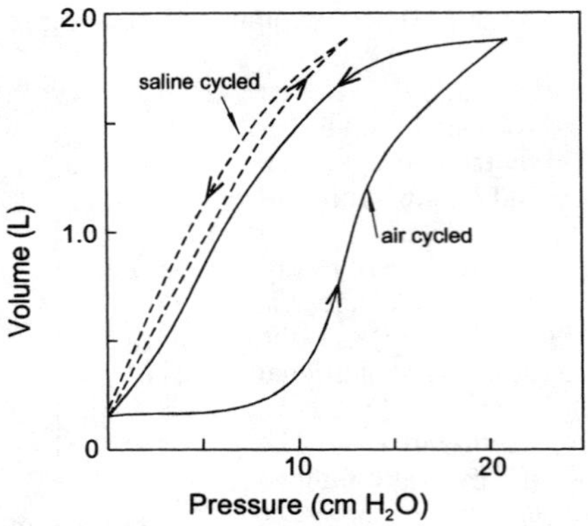

FIGURE 4.8 Saline cycling compared to air cycling.

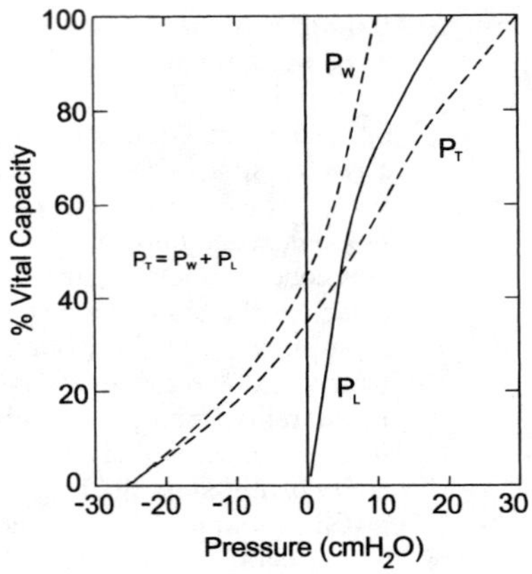

FIGURE 4.9 Pressure-volume curves for lung (P_L), chest wall (P_w), and total respiratory system (P_T).

compliance $C_{\text{dyn}} = dV/dP$. Other useful parameters are the specific compliance $V^{-1}\, dV/dP$, the elastance $1/C$, and the specific elastance $V\, dP/dV$.

Shown in Fig. 4.8 is the curve for a lung inflated and deflated with saline rather than air. Saline cycling was first noted by Von Neergaard[34] to reduce inflation pressures, as can be seen in the resulting curve that is shifted to the left, because of increased compliance, and has lost almost all of its hysteresis area. Because it removes the air-liquid interface that alveoli normally possess, saline cycling reveals that the surface tension forces exerted by this interface have very significant effects on the lung's mechanical response. Saline cycling allows us to understand the elasticity and hysteresis properties of the lung tissue by removing the surface tension forces and their contributions to hysteresis and compliance.

4.4.2 Pressure-Volume Relationship of the Respiratory System

The *P-V* curve for an isolated lung is only part of the total picture. The intact pulmonary system is contained within the chest cavity, which has its own *P-V* characteristics. In Fig. 4.9 adapted from Ref. 22, the static *P-V* curves for an isolated lung (P_L), a passive chest with no lung contents (P_W), and the total ($P_T = P_L + P_W$) are shown. When $P_T = 0$, the system is in equilibrium with the surrounding atmospheric pressure and the graph indicates that the lung is stretched from its near-zero volume at $P_L = 0$ while the chest cavity is compressed from its higher equilibrium volume at $P_W = 0$. Hence the elastic recoil of the chest wall is pulling outward while the lung tissue is pulling inward to balance the forces, which are in series.

4.4.3 Surface Tension Versus Surface Area

The internal surface of the lung is coated with a thin liquid film, which is in contact with the resident air. The surface tension arising at this air-liquid interface has significant effects on the overall lung mechanical response, as shown in Fig. 4.8, when it is removed. Alveolar Type II cells produce important surface-active substances called *surfactants*, which reduce the interfacial tension. A major component of lung surfactant is dipalmitoylphosphatidylcholine (DPPC). Lung surfactants reduce alveolar surface tension from the surfactant-free value for air-water, 70 dyn/cm, to 1 to 5 dyn/cm in the alveoli depending on the concentration and state of lung inflation. Figure 4.10 shows the relationship between surface tension and surface area for cycling of an air-liquid interface containing lung surfactants. Note that there is a hysteresis area and average slope, or compliance, that characterize the loop and have significant effects on their counterparts in Fig. 4.7.

Insufficient surfactant levels can occur in premature neonates whose alveolar cells are not mature enough to produce sufficient quantities, a condition leading to respiratory distress syndrome, also called hyaline membrane disease.[2] Instilling liquid mixtures, containing either natural or man-made surfactants, directly into airways via the trachea has developed from early work in animal models[9] into an important clinical tool called surfactant replacement therapy. The movement of these liquid boluses through the network relies on several mechanisms, including air-blown liquid plug flow dynamics, gravity, and surface tension and its gradients.[4,10,18]

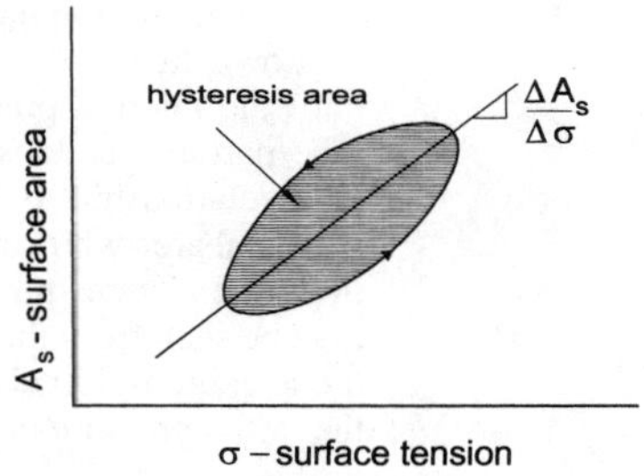

FIGURE 4.10 Surface area A_s versus surface tension σ for a cycled interface containing lung surfactant. There is a hysteresis area and mean slope or compliance to the curve; compare to Fig. 4.7.

4.5 VENTILATION, PERFUSION, AND LIMITS

The lung differs from many other organs in its combination of gas-phase and liquid-phase constituency. Because the densities of these two phases are so different, gravity plays an important role in determining regional lung behavior, both for gas ventilation and for blood perfusion. In an upright adult, the lower lung is compressed by the weight of the upper lung and this puts the lower lung's alveoli on a more compliant portion of their regional *P-V* curve; see Fig. 4.7. Thus inhaled gas tends to be directed preferentially to the lower lung regions, and the regional alveolar ventilation $\dot{V}_A$ decreases in a graded fashion moving upward in the gravity field.

Blood flow $\dot{Q}$ is also preferentially directed toward the lower lung, but for different reasons. The blood pressure in pulmonary arteries, P_a, and veins, P_v, sees a hydrostatic pressure gradient in the upright lung. These vessels are imbedded within the lung's structure, so they are surrounded by alveolar gas pressure, P_A, which is essentially uniform in the gravity field since its density is negligible. Thus there is a region called Zone III in the upper lung where $P_a < P_A$. This difference in pressures squeezes the capillaries essentially shut and there is relatively little pulmonary blood flow there. In the lower lung called Zone I, the hydrostatic effect is large enough to keep $P_a > P_v > P_A$ and blood flow is proportional to the arterial-venous pressure difference, $P_a - P_v$. In between these two zones is Zone II where $P_a > P_A > P_v$. Here there will be some length of the vessel that is neither fully closed nor fully opened. It is partially collapsed into more of a flattened, oval shape. The physics of the flow, called the *vascular waterfall*[29] or *choked flow*[30] or *flow limitation*[23], dictates that Q is no longer dependent on the downstream pressure, but is primarily determined by the pressure difference $P_a - P_A$.

Figure 4.11*a* shows this interesting type of flexible

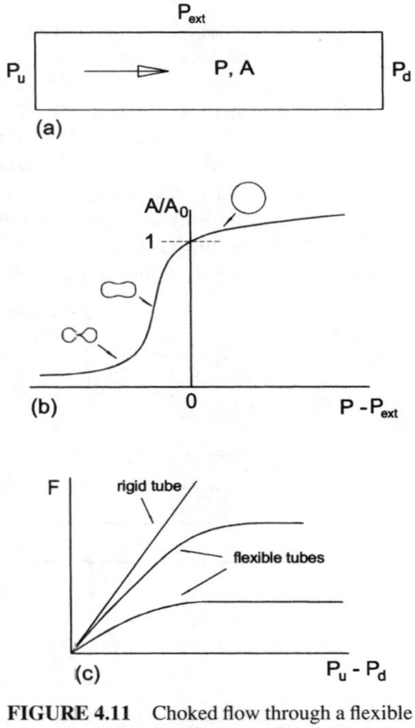

FIGURE 4.11 Choked flow through a flexible tube. (*a*) upstream pressure P_u, downstream pressure P_d, external pressure P_{ext}, flow F, pressure P, cross-sectional area A. (*b*) A/A_0 versus transmural pressure with shapes indicated. (*c*) Flow versus pressure drop, assuming P_d is decreased and P_u is fixed.

tube configuration and flow limitation phenomena where $P_a = P_u$, $P_A = P_{ext}$, $P_v = P_d$ and $\dot{Q} = F$. As downstream pressure P_d is decreased and upstream pressure P_u is kept fixed, the flow increases until the internal pressure of the tube drops somewhat below the external pressure P_{ext}. Then the tube partially collapses, decreasing in cross-sectional area according to its pressure-area or "tube law" relationship $A(P)$, shown in Fig. 4.11*b*. As P_d is further decreased, the tube reduces its cross-sectional area while the average velocity of the flow increases. However, their product, the volumetric flow rate F, does not increase as shown in Fig. 4.11*c*. A simplified understanding of this behavior may be seen from the conservation of momentum equation for the flow, a Bernoulli equation, where the average fluid velocity U is defined as the ratio of flow to cross-sectional area, $U = F/A$, all of the terms pressure dependent.

$$P_{\text{res}} = P + \frac{1}{2}\rho\left(\frac{F(P)}{A(P)}\right)^2 \tag{4.4}$$

where P_{res} = pressure in a far upstream reservoir where fluid velocity is negligibly small
ρ = fluid density

P_{res} is the alveolar air pressure, for example, when applied to limitation of air flow discussed below. Taking the derivative of Eq. (4.4) with respect to P and setting the criterion for flow limitation as $dF/dP = 0$ gives

$$U_c = \left(\frac{A}{\rho}\frac{dP}{dA}\right)^{1/2} = \left(\frac{E}{\rho}\right)^{1/2} \tag{4.5}$$

where E is the specific elastance of the tube. The quantity $(E/\rho)^{1/2}$ is the "wave speed" of small pressure disturbances in a fluid-filled flexible tube, and flow limitation occurs when the local fluid speed equals the local wave speed. At that point, pressure information can no longer propagate upstream, since waves carrying the new pressure information are all swept downstream.

The overall effect of nonuniform ventilation and perfusion is that both decrease as one progresses vertically upward in the upright lung. But perfusion decreases more rapidly so that the dimensionless ratio of ventilation to perfusion, $\dot{V}_A/\dot{Q}$, decreases upward, and can vary from approximately 0.5 at the lung's bottom to 3 or more at the lung's top.[36] Extremes of this ratio are ventilated regions with no blood flow, called *dead space*, where $\dot{V}_A/\dot{Q} \rightarrow \infty$, and perfused regions with no ventilation, called *shunt*, where $\dot{V}_A/\dot{Q} \rightarrow 0$.

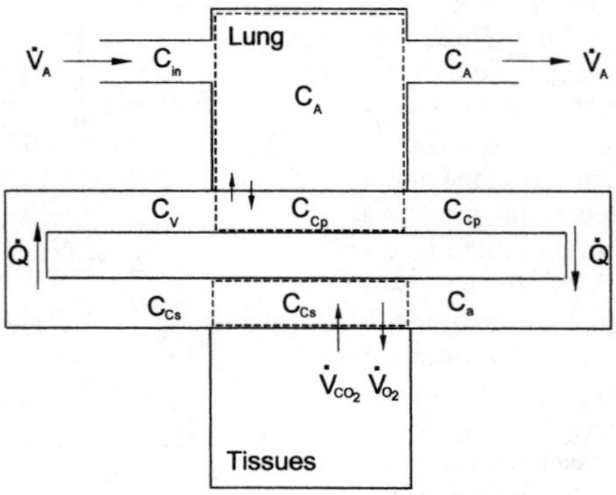

FIGURE 4.12 Schematic for ventilation and perfusion in the lung and tissues. Dashed line indicates control volume for mass balance. Assume instantaneous and homogenous mixing. Subscripts indicate gas concentrations for systemic (*s*) or pulmonary (*p*) end capillary C_c, venous C_v, alveolar C_A, and arterial C_a.

The steady-state gas concentrations within an alveolus reflect the balance of inflow to outflow, as shown in the control volumes (dashed lines) of Fig. 4.12. For CO_2 in the alveolar space, net inflow by perfusion must equal the net outflow by ventilation, $\dot{Q}(C_{vCO_2} - C_{cCO_2}) = \dot{V}_A(C_{ACO_2} - C_{inCO_2})$ where C indicates concentration. In the tissue compartment, the CO_2 production rate from cellular metabolism, $\dot{V}_{CO_2}$, is balanced by net inflow versus outflow in the tissues; i.e., $\dot{V}_{CO_2} = \dot{Q}(C_{vCO_2} - C_{cCO_2})$. Noting that $C_{inCO_2} \approx 0$ and combining the two equations, while converting concentrations to partial pressures, leads to the alveolar ventilation equation

$$\dot{V}_A = 8.63\frac{\dot{V}_{CO_2}}{P_{ACO_2}} \tag{4.6}$$

where the constant 8.63 comes from the units conversion. The inverse relationship between alveolar ventilation and alveolar CO_2 is well known clinically. Hyperventilation drops alveolar, and hence arterial, CO_2 levels whereas hypoventilation raises them. Using a similar approach for oxygen consumption $\dot{V}_{O_2}$ and using the results of the CO_2 balance yields the ventilation-perfusion equation

$$\frac{\dot{V}_A}{\dot{Q}} = \frac{8.63R(C_{a_{O_2}} - C_{V_{O_2}})}{P_{A_{CO_2}}} \tag{4.7}$$

Equation 4.7 uses the definition of the respiratory exchange ratio, $R = \dot{V}_{CO_2}/\dot{V}_{O_2}$, which usually has a value of $R \approx 0.8$ for $\dot{V}_A/\dot{Q} = 1$. It also replaces the end capillary concentration with the systemic arterial value, $C_{a_{O_2}} = C_{c_{O_2}}$, assuming equilibration. From Eq. 4.7, the extreme limits of $\dot{V}_A/\dot{Q}$, mentioned earlier, may be recognized. Intermediate solutions are more complicated, however, since there are nonlinear relationships between gas partial pressure and gas content or concentration in the blood. Equation 4.7 also demonstrates that higher $\dot{V}_A/\dot{Q}$, as occurs in the upper lung, is consistent with a higher end capillary and alveolar oxygen level. It is often thought that tuberculosis favors the upper lung for this reason.

The $\dot{V}/\dot{Q}$ variation leads to pulmonary venous blood having a mix of contributions from different lung regions. Consequently, there is a difference between the lung-average alveolar $P_{A_{O_2}}$ and the average or systemic arterial $P_{a_{O_2}}$, sometimes called the *A-a* gradient of O_2. An average $P_{A_{O_2}}$ can be derived from the alveolar gas equation,

$$P_{A_{O_2}} = P_{I_{O_2}} - \frac{P_{A_{CO_2}}}{R} + f \tag{4.8}$$

which derives from the mass balance for O_2 in Fig. 4.12. Here $P_{I_{O_2}}$ is the inspired value and f is a small correction normally ignored. Clinically, an arterial sample yields $P_{a_{CO_2}}$, which can be substituted for $P_{A_{CO_2}}$ in Eq. 4.8. The *A-a* gradient becomes abnormally large in several lung diseases that cause increased mismatching of ventilation and perfusion.

4.6 AIRWAY FLOW, DYNAMICS, AND STABILITY

4.6.1 Forced Expiration and Flow Limitation

A common test of lung function consists of measuring flow rate by a spirometer apparatus. When the flow signal is integrated with time, the lung volume is found. Important information is contained in the volume versus time curves. The amount of volume forcefully exhaled with maximum effort in 1 second, FEV1, divided by the maximal volume exhaled or forced vital capacity, FVC, is a dimensionless ratio used to separate restrictive and obstructive lung disease from normal lungs. FEV1/FVC is normally 80 percent or higher, but in obstructed lungs (asthma, emphysema) the patient cannot exhale very much volume in 1 second, so FEV1/FVC drops to diagnostically low levels, say 40 percent. The restricted lung (fibrosis) has smaller than normal FVC, though the FEV1/FVC ratio may fall in the normal range because of the geometric scaling as a smaller lung.

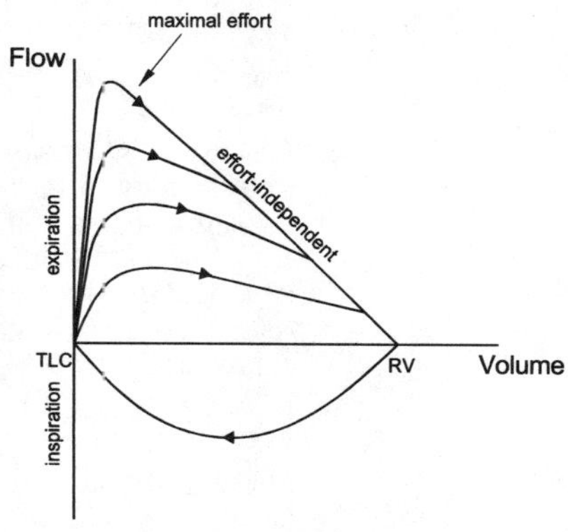

FIGURE 4.13 Flow-volume curves for increasing effort level during expiration, including maximal effort. Note effort-independent portion of curves.

Flow and volume are often plotted against one another as in Fig. 4.13. The flow-volume curves shown are for increasing levels of effort during expiration. The maximal effort curve sets a portion common to all of the curves, the effort-independent region. Expiratory flow limitation is another example of the choked flow phenomenon discussed earlier in the context of blood flow. Similar flow versus driving pressure curves shown in Fig. 4.11 can be extracted from Fig. 4.13 by choosing flows at the same lung volume and measuring the pressure drop at that instant, forming the isovolume pressure-flow curve. Since airway properties and E vary along the network, the most susceptible place for choked flow seems to be the proximal airways. For example, we expect gas speeds prior to choke to be largest at generation $n = 3$ where the total cross-sectional area of the network is minimal; see Table 4.1. So criticality is likely near that generation. An interesting feature during choked flow in airways is the production of flutter oscillations, which are heard as wheezing breath sounds,[13,15,16] so prevalent in asthma and emphysema patients whose maximal flow rates are significantly reduced, in part because E and U_c are reduced.

4.6.2 Airway Closure and Reopening

Most of the dynamics during expiration, so far, have been concerned with the larger airways. Toward the very end of expiration, smaller airways can close off as a result of plug formation from the liquid lining, a capillary instability,[11,21] or from the capillary forces pulling shut the collapsible airway,[25] or from some combination of these mechanisms. Surfactants in the airways help to keep them open by both static and dynamic means. The lung volume at which airway closure occurs is called the closing volume, and in young healthy adults it is ~10 percent of VC as measured from a nitrogen washout test. It increases with aging and with some small airway diseases. Reopening of closed airways was mentioned earlier as affecting the shape of the *P-V* curve in early inspiration as the airways are recruited. When the liquid plugs break and the airway snaps open, a crackle sound, or cascade of sounds from multiple airways, can be generated and heard with a stethoscope.[1,12] Diseases that lead to increased intra-airway liquid, such as congestive heart failure, are followed clinically by the extent of lung crackles, as are certain fibrotic conditions that affect airway walls and alveolar tissues.

REFERENCES

1. Alencar, A. M., Z. Hantos, F. Petak, J. Tolnai, T. Asztalos, S. Zapperi, J. S. Andrade, S. V. Buldyrev, H. E. Stanley, and B. Suki, "Scaling behavior in crackle sound during lung inflation," *Phys. Rev. E*, **60**: 4659–4663, 1999.
2. Avery, M. E., and J. Mead, "Surface properties in relation to atelectasis and hyaline membrane disease," *Am. J. Dis. Child.*, **97**:517–523, 1959.
3. Bohn, D. J., K. Miyasaka, E. B. Marchak, W. K. Thompson, A. B. Froese, and A. C. Bryan, "Ventilation by high-frequency oscillation," *J. Appl. Physiol.*, **48**:710–16, 1980.
4. Cassidy, K. J., J. L. Bull, M. R. Glucksberg, C. A. Dawson, S. T. Haworth, R. B. Hirschl, N. Gavriely, and J. B. Grotberg, "A rat lung model of instilled liquid transport in the pulmonary airways," *J. Appl. Physiol.*, **90**:1955–1967, 2001.
5. Chatwin, P. C., "On the longitudinal dispersion of passive contaminant in oscillatory flows in tubes," *J. Fluid Mech.*, **71**:513–527, 1975.
6. Dragon, C. A., and J. B. Grotberg, "Oscillatory flow and dispersion in a flexible tube," *J. Fluid Mech.*, **231**: 135–155, 1991.
7. Dubois, A. B., A. W. Brody, D. H. Lewis, and B. F. Burgess, "Oscillation mechanics of lungs and chest in man." *J. Appl. Physiol.* **8**:587–594, 1956.
8. Eckmann, D. M., and J. B. Grotberg, "Oscillatory flow and mass transport in a curved tube," *J. Fluid Mech.*, **188**:509–527, 1988.
9. Enhorning, G., and B. Robertson, "Expansion patterns in premature rabbit lung after tracheal deposition of surfactant," *Acta Path. Micro. Scand. Sect. A—Pathology*, **A79**:682, 1971.

10. Espinosa, F. F., and R. D. Kamm, "Bolus dispersal through the lungs in surfactant replacement therapy," *J. Appl. Physiol.*, **86**:391–410, 1999.

11. Everett, D. H., and J. M. Haynes, "Model studies of capillary condensation: 1. Cylindrical pore model with zero contact angle," *J. Colloid, Inter. Sci.*, **38**:125–137, 1972.

12. Forgacs, P., "Functional basis of pulmonary sounds," *Chest*, **73**:399–405, 1978.

13. Gavriely, N., K. B. Kelly, J. B. Grotberg, and S. H. Loring, "Forced expiratory wheezes are a manifestation of airway flow limitation," *J. Appl. Physiol.*, **62**:2398–2403, 1987.

14. Godleski, D. A., and J. B. Grotberg, "Convection-diffusion interaction for oscillatory flow in a tapered tube," *Journal of Biomechanical Engineering—Transactions of ASME*, **110**:283–291, 1988.

15. Grotberg, J. B., and S. H. Davis, "Fluid-dynamic flapping of a collapsible channel: sound generation and flow limitation," *J. Biomech. Eng.*, **13**:219–230, 1980.

16. Grotberg, J. B., and N. Gavriely, "Flutter in collapsible tubes: a theoretical model of wheezes," *J. Appl. Physiol.*, **66**:2262–2273, 1989.

17. Haber, R., J. B. Grotberg, M. R. Glucksberg, G. Miserocchi, D. Venturoli, M. D. Fabbro, and C. M. Waters, "Steady-state pleural fluid flow and pressure and the effects of lung buoyancy," *J. Biomech. Eng.*, **123**:485–492, 2001.

18. Halpern, D., O. E. Jensen, and J. B. Grotberg, "A theoretical study of surfactant and liquid delivery into the lung," *J. Appl. Physiol.*, **85**:333–352, 1998.

19. Isabey, D., and H. K. Chang, "Steady and unsteady pressure-flow relationships in central airways," *J. Appl. Physiol.*, **51**:1338–1348, 1981.

20. Jordanog, J., "Vector analysis of rib movement," *Respir. Physiol.*, **10**:109, 1970.

21. Kamm, R. D., and R. C. Schroter, "Is airway closure caused by a thin liquid instability?" *Respir. Physiol.*, **75**:141–156, 1989.

22. Knowles, J. H., S. K. Hong, and H. Rahn, "Possible errors using esophageal balloon in determination of pressure-volume characteristics of the lung and thoracic cage," *J. Appl. Physiol.*, **14**:525–530, 1959.

23. Lambert, R. K., and T. A. Wilson, "Flow limitation in a collapsible tube," *Journal of Applied Physiology*, **33**:150–153, 1972.

24. Lunkenheimer, P. P., W. Rafflenbeul, H. Kellar, I. Frank, H. H. Dickhut, and C. Fuhrmann, "Application of tracheal pressure oscillations as a modification of 'Diffusional Respiration'," *Br. J. Anaesth.*, **44**:627, 1972.

25. Macklem, P. T., D. F. Proctor, and J. C. Hogg. The stability of peripheral airways. *Respir. Physiol.* **8**:191–203, 1970.

26. Mead, J., J. L. Whittenberger, and E. P. Radford, Jr., "Surface tension as a factor in pulmonary volume-pressure hysteresis," *J. Appl. Physiol.*, **10**:191–196, 1957.

27. Otis, A. B., C. B. McKerrow, R. A. Bartlett, J. Mead, M. B. Mcllroy, N. J. Selverstone, and E. P. Radford, "Mechanical factors in distribution of pulmonary ventilation," *J. Appl. Physiol.*, **8**:427–443, 1956.

28. Pedley, T. J., R. C. Schroter, and M. F. Sudlow, "The prediction of pressure drop and variation of resistance within the human bronchial airways," *Respir. Physiol.*, **9**:387–405, 1970.

29. Permutt, S., and R. L. Riley, "Hemodynamics of collapsible vessels with tone—vascular waterfall," *J. Appl. Physiol.*, **18**:924, 1963.

30. Shapiro, A. H., "Steady flow in collapsible tubes," *J. Biomech. Eng.*, **99**:126–147, 1977.

31. Slutsky, A. S., G. G. Berdine, and J. M. Drazen, "Steady flow in a model of human central airways," *J. Appl. Physiol.*, **49**:417–423, 1980.

32. Suki, B., F. Petak, A. Adamicza, Z. Hantos, and K. R. Lutchen, "Partitioning of airway and lung-tissue properties—comparison of in-situ and open-chest conditions," *J. Appl. Physiol.*, **79**:861–869, 1995.

33. Taylor, G. I., "Dispersion of solute matter in solvent flowing through a tube," *Proc. Roy. Soc. Lond.*, **A291**: 186–203, 1953.

34. Von Neergaard, K., "Neue Auffassungen über einen Grundbegriff der Atemmechanik. Die Retraktionskraft der Lunge, abhängig von der Oberflächenspannung in den Alveolen," *Z. Gesampte Exp. Med.*, **66**:373–394, 1929.

35. Weibel, E. R., *Morphometry of the human lung*, New York: Academic Press, p. 151, 1963.

36. West, J. B., "Regional differences in gas exchange in the lung of erect man," *J. Appl. Physiol.*, **17**:893–898, 1962.

BIBLIOGRAPHY

Fredberg, J. J., N. Wang, D. Stamenovic, and D. E. Ingber, "Micromechanics of the lung: from the parenchyma to the cytoskeleton," *Complexity in Structure and Function of the Lung (Lung Biol. Health Dis. Ser.)*, 99–122, 1998.

Grotberg, J. B., "Respiratory fluid mechanics and transport processes," *Annu. Rev. Biomed. Engr.*, **3**:421–457, 2001.

Kamm, R. D., "Airway wall mechanics," *Annu. Rev. Biomed. Eng.*, **1**:47–72, 1999.

The Lung: Scientific Foundations, R. G. Crystal, and J. B. West, eds., New York: Raven Press, 1991.

West, J. B., *Respiratory Physiology: The Essentials*, 6th ed., Baltimore: Lippincott Williams & Wilkins, 2000.

CHAPTER 5
BIOMECHANICS OF HUMAN MOVEMENT

Kurt T. Manal and Thomas S. Buchanan
University of Delaware, Newark

5.1 WHY STUDY HUMAN MOVEMENT?

The biomechanics of human motion is a fascinating field. Who among us has never marveled at the graceful motions of a dancer or the rapid finger movements of a musician? From the time of Aristotle onward there have been countless books written on the topic of movement in animals and humans. Despite the great minds that have considered the topic, it is just recently that much advancement has been made experimentally. Historically, the study of human movement has been costly and very time consuming. This is because, in order to study them, movements are almost always discretized and then analyzed step by step, with the size of the steps determined by the speed of the movement (and the questions being asked). Whether it be frames of film from a movie camera or digitized samples from an electrogoniometer, most movements are recorded as series of static images that are then reassembled to provide kinematic and kinetic information.

There has been a tremendous growth in the study of human movement in the past 2 decades because of the low cost of digital data acquisition systems that make possible the storage and analysis of massive amounts of data that are required to accurately characterize complex motion. This growing interest in the study of human movement is coming from five predominate groups:

1. Basic scientists are interested in the control of human movement. How the nervous system controls the large number of degrees of freedom necessary to produce smooth, complex movements (or even simple ones!) is poorly understood. The study of the coordination of movement can be compared to the inverse problem faced by the roboticist. The roboticist develops computer programs to produce coordinated movements in a robot. On the other hand, the motor control researcher measures coordinated movements in order to understand what the "neural program" is.

2. Human movements are studied to understand and treat pathologies. For example, gait analysis is often used to help guide the physician contemplating surgery for children with cerebral palsy. The best choice for a tendon transfer or muscle lengthening surgery can be predicted by using combinations of movement analysis and biomechanical modeling (e.g., Delp et al., 1996). Gait analysis can also be used to monitor the progression of the disease and the efficacy of the treatment.
3. The study of human athletic performance has been revolutionized by motion analysis equipment and software that make it possible to readily analyze complex three-dimensional movements. From cricket bowling to figure skating to swimming to pole vaulting, the kinematics and kinetics have been examined with an aim to improve human performance.
4. There is substantial interest in human movement from those studying ergonomics and human factors related to military applications. Both the development of human-machine interfaces for high-tech weapons and the minimization of industrial injuries require knowledge of human kinematics and kinetics.
5. The kinematics of human movement has been studied by animators interested in making computer-generated characters move in realistic ways. By recording actors while they perform choreographed dances and movements, it is possible to get complex kinematic data into a computer, which can then be used to animate a computer-generated image.

5.2 FORWARD VERSUS INVERSE DYNAMICS

There are two fundamentally different approaches to studying the biomechanics of human movement: forward dynamics and inverse dynamics. Either can be used to determine joint kinetics (e.g., estimate joint moments during movements).

5.2.1 Forward Dynamics

In a forward dynamics approach to the study of human movement, the input to the system is the neural command (Fig. 5.1). This specifies the level of activation to the muscles. The neural command can be estimated by optimization models (Zajac, 1989; Pandy and Zajac, 1991) or from electromyograms (EMGs). The neural command is the sum of the neuronal signals from the α-motorneurons (that originate in the spinal cord) to the fibers of each muscle. This can be represented by a single value (at any given time) for each muscle, which we will call muscle activation α_i, and which we will represent mathematically as a value between zero and one. Hence, if it is to be estimated from EMGs, additional steps are needed to transform EMGs to muscle activation.

Musculotendon dynamics govern the transformation of muscle activation α_i, to muscle force F_i. Once the muscle begins to develop force, the tendon (in series with the muscle) begins to carry load as well. Depending upon the kinetics of the joint, the relative length changes in the tendon and the muscle may be very different. For example, this is certainly the case for a "static contraction." (This commonly used name is an oxymoron, as something cannot *contract*, i.e., shorten, and be *static* at the same time. Hence, the tendon must lengthen as the muscle shortens if the joint is not to move!).

The force in each musculotendonous unit contributes toward the total moment about the joint. The musculoskeletal geometry determines the moment arms of the muscles. (Since muscle force is dependent upon muscle length, i.e., the classic muscle "length-tension curve," there is feedback between joint angle and musculotendon dynamics.) It is important to note that the moment arms of muscles are not constant values, but change as a function of joint angles. Also, one needs to keep in mind the multiple degrees of freedom of each joint, as a muscle may have multiple actions at a joint, depending on its geometry. Finally, it is important to note that the joint moment T_i is determined

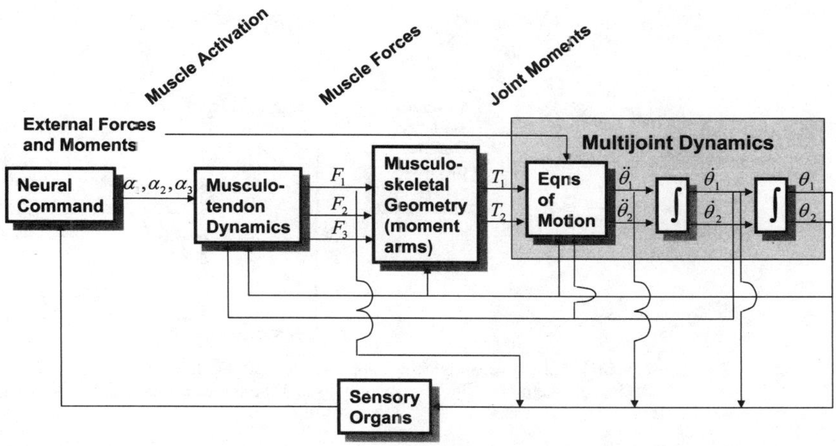

FIGURE 5.1 Forward dynamics approach to studying human movement. This simplified figure depicts the neural command and forces for three muscles and the moments and joint angles for a two-joint system. See text for details.

from the sum of the contributions for each muscle. If not all muscles are included in the process, the joint moment will be underestimated. The output of this transformation is a moment for each joint (or, more precisely, each degree of freedom).

From the joint moments, multijoint dynamics can be used to compute the accelerations, velocities, and angles for each joint of interest. On the feedback side, the neural command is influenced by muscle length (via muscle spindles) and tendon force (via Golgi tendon organs). Many other sensory organs play a role in this as well, but these two are generally the most influential.

There are several limitations of the forward dynamics approach. First, it requires estimates of muscle activation. EMG methods have been used to this end, but the high variability in EMG signals has made this difficult, especially during dynamic conditions. Second, the transformation from muscle activation to muscle force is difficult, as it is not completely understood. Most models of this (e.g., Zajac, 1989) are based on phenomenological models derived from A. V. Hill's classic work (Hill, 1938) or the more complex biophysical model of Huxley's (Huxley, 1957; Huxley and Simmons, 1971), such as Zahalack's models (Zahalack, 1986; 2000). One way around the problem of determining force from EMGs is to employ optimization methods to predict muscle forces directly (bypassing these first two limitations). However, the choice of a proper cost function is a matter of great debate. Scientists doing research in human motor control find it surprising that biomechanical engineers replace their entire line of study (and indeed, the entire central nervous system) with a simple, unverified equation. Nevertheless, some cost functions provide reasonable fits of the data in addressing specific questions. Another limitation is that of determining musculoskeletal moment arms. These are difficult to measure in cadavers and even harder to determine with any accuracy in a living person. Finally, joint moments can easily be underestimated. Using forward dynamics, small errors in joint torques can lead to large errors in joint position.

5.2.2 Inverse Dynamics

Inverse dynamics approaches the problem from the opposite end. Here we begin by measuring position and the external forces acting on the body (Fig. 5.2). In gait analysis, for example, the position of tracking targets attached to the segments can be recorded by using a camera-based system and the external forces can be recorded by using a force platform.

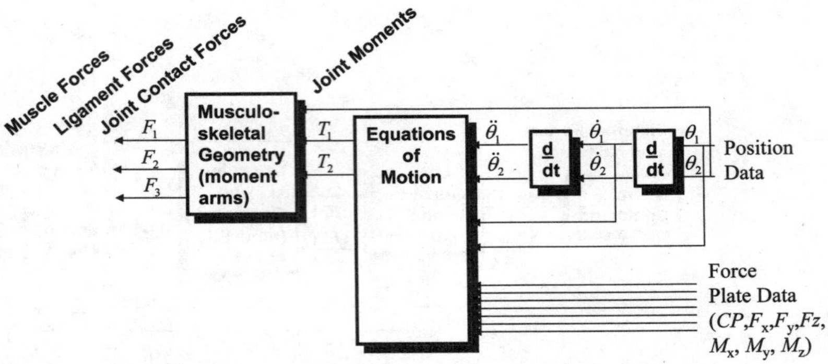

FIGURE 5.2 Inverse dynamics approach to studying human movement. This simplified figure depicts the angular position for two joints and the forces for three muscles. See text for details.

The relative position of tracking targets on adjacent segments is used to calculate joint angles. These data are differentiated to obtain velocities and accelerations.

The accelerations and the information about other forces exerted on the body (e.g., the recordings from a force plate) can be input to the equations of motion to compute the corresponding joint reaction forces and moments.

If the musculoskeletal geometry is included, muscle forces can then be estimated from the joint moments, and from these it may be possible to estimate ligament and joint compressive forces.

As with forward dynamics, inverse dynamics has important limitations. First, in order to estimate joint moments correctly, one must know the inertia of each body segment (this is embedded in the equations of motion). These parameters are difficult to measure and must be estimated. Typically, they are estimated by using established values from cadavers and scaled by simplistic scaling rules, the accuracies of which are rarely verified. Second, the resultant joint reaction forces and moments are net values. This is important to keep in mind if an inverse dynamics approach is used to predict muscle forces. For example, if a person activates his hamstrings generating a 30-N · m flexion moment and at the same time activates the quadriceps generating a 25-N · m extension moment, the inverse dynamics method (if it is perfectly accurate) will yield a net knee flexion moment of 5 N · m. Since the actual contribution of the knee flexor muscles was 6 times greater, this approach is grossly inaccurate and inappropriate for estimating the role of the knee flexors during this task. This is strongly stated because cocontraction of muscles is very common, yet this approach is widely used to estimate muscular contributions. Another limitation of the inverse dynamics approach occurs when one tries to estimate muscle forces. Since there are multiple muscles spanning each joint, the transformation from joint moment to muscle forces yields an infinite number of solutions. Choosing the proper solution requires some sort of optimization analysis, requiring the use of a cost function whose validity is sure to be challenged. Finally, if one wishes to examine muscle activations, there is no current model available that will do this inverse transformation. However, this is rarely the goal of an inverse dynamics analysis.

5.2.3 Comparing Forward and Inverse Dynamics Methods

Given the limitations of each method, which should be used: forward or inverse dynamics? That depends on the question being asked. If one's primary interest is in joint kinematics, it makes more sense to start with a measurement of position as in the inverse dynamics approach. If one is primarily interested in muscle forces, one could argue that forward dynamics has more advantages. For estimating joint moments during movements, inverse dynamics is probably the best bet, depending upon the specific application.

For the remainder of this chapter, we will concentrate on the inverse dynamics approach for the study of human movement. Inverse dynamics is more commonly used than forward dynamics for studying human movement. A forward dynamics approach will be addressed in a subsequent chapter (Chap.6, "Biomechanics of the Musculoskeletal System," by Marcus G. Pandy and Ronald E. Barr).

5.3 TOOLS FOR MEASURING HUMAN MOVEMENT

In this section we will discuss three of the more common methods used to collect human movement data: electrogoniometers, electromagnetic tracking devices, and optoelectronic measuring systems. Of these distinctly different measuring tools, optoelectronic systems are the most common registration method, and therefore most of this section will focus on video-based motion analysis.

5.3.1 Electrogoniometers

Electrogoniometers are devices that convert joint angle to a voltage. The voltage can be sampled continuously, making electrogoniometers ideal for measuring dynamic movement. There are basically two designs, both of which fall under the category of resistive transducers. These devices, namely potentiometers and strain gauges, output a voltage related to the angular position of the joint. The voltage is converted to an angle by using a manufacturer supplied scale factor specific to each transducer. The joint angle can be displayed in real time and/or stored on a computer equipped with an analog-to-digital data acquisition card.

Potentiometers. A potentiometer is nothing more than a variable resistor that is sensitive to changes in angular position. Two arms, one fixed to the outer casing of the potentiometer and the other to the rotating shaft, can be used to mount the device to the segments on either side of a joint. The potentiometer is placed over the joint axis of rotation with the arms secured to the segments by medical tape or elasticized wraps. Changes in joint angle will cause the *wiper* (i.e., sliding contact) of the potentiometer to slide across the resistor, resulting in an output voltage linearly related to the joint angle. It is important that the potentiometer be placed over the axis of rotation; otherwise, movement of the joint will be restricted. The electrogoniometer is ideally positioned when the rotating shaft of the potentiometer and the joint axis of rotation are aligned. More elaborate mounting methods have been designed to house mutually perpendicular potentiometers in multiple degree of freedom exoskeletal linkages (e.g., Chao, 1980; Shiavi et al., 1987). These devices are no longer commonly used, but are mentioned for historical purposes since they have played an important role in many previous studies.

Strain Gauges. Strain gauges can also be used to detect changes in joint angular position. An example of a 1 degree of freedom electrogoniometer is illustrated in Fig. 5.3. Two and three degrees of freedom electrogoniometers of this type are also available. Strain-sensitive wires are mounted within a connecting element and electrically connected to form a Wheatstone bridge. Each strain-sensitive wire is, in effect, a resistor and is sensitive to strains in particular directions. Hence, when the electrogoniometer is forced to rotate about the axis of rotation as drawn in the figure, the bridge circuitry becomes unbalanced. This unbalance is noted as a change in the output voltage of the bridge and is proportional to the amount of rotation. The design is clever because pure rotation about axes perpendicular to the axis of rotation depicted in Fig. 5.3 will *not* unbalance the bridge. Another interesting and practical characteristic of this device is that it does not have to be positioned over the joint axis of rotation as is the case for rotatory potentiometers. Note, however, that the base of the mounting blocks must lie in the plane of rotation without necessarily being centered over the axis of rotation.

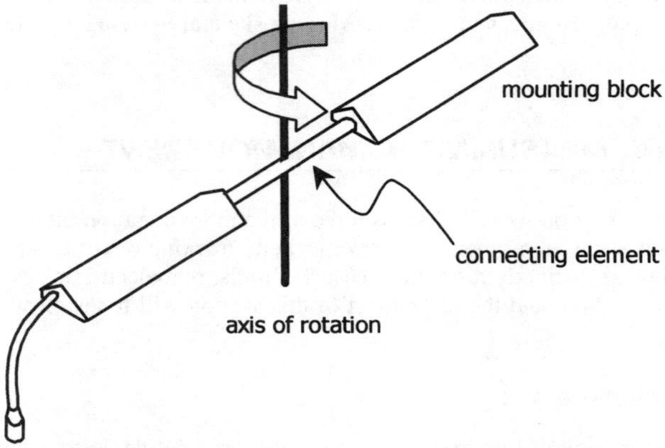

FIGURE 5.3 Single degree of freedom strain gauge for measuring joint angular position. Strain-sensitive wires are fixed to the connecting element between the mounting blocks. The mounting blocks are secured to both segments on either side of a joint. Changes in joint angle are output as a voltage proportional to the amount of rotation about the axis of rotation.

Electrogoniometers of this type can be configured to display joint angles in real time and/or interfaced with a computer for data storage. Additionally, data can be saved to a storage unit (i.e., data logger) strapped to the subject. The data logger is ideal for recording dynamic movements in the field. The stored data can be uploaded to a computer at a later time and converted to joint angles for analysis.

There are advantages and disadvantages associated with the use of electrogoniometers. In their favor are ease of use and cost. On the other hand, they are less accurate than other systems used to record movement. In addition, both designs (i.e., potentiometer and strain gauge) require placement over the joint, which may interfere with the natural kinematics because of cumbersome cabling and/ or method of attachment. Another drawback of these devices is that while they provide a *relative* measure of joint angular position, the data do not lend themselves to an inverse dynamics analysis in which joint reaction forces and moments are of interest, the computation of which requires knowledge of the absolute positions of the body segments.

5.3.2 Electromagetic Tracking Systems

Electromagnetic tracking technology originated in the defense industry and has since become widely used in the entertainment industry (e.g., motion pictures, animation, and gaming). The use of electromagnetic tracking has become increasingly popular in the academic environment as evinced by the growing number of research publications using this technology.

Electromagnetic tracking is based on Faraday's law of magnetic induction. That is, electrons in a conductor experience a spontaneous magnetic force when moved through a magnetic field. The magnitude of the induced force (i.e., electromotive force or EMF) is proportional to the strength of the magnetic field through which the conductor is moved. The magnitude of the EMF (i.e., voltage) is also related to the speed the conductor is moving. If the conductor is in the shape of a loop, the same principles apply with an induced EMF proportional to the strength of the magnetic field perpendicular to the cross-sectional area of the loop. The induced EMF is related to the magnetic flux Φ_B:

$$\text{EMF} = -\frac{d\Phi_B}{dt} \tag{5.1}$$

Conceptually, the strength and direction of a magnetic field can be thought of as the density and direction of magnetic field lines. The magnetic flux will vary as the conductor moves closer to/farther from the source of the magnetic field and also as it rotates relative to the magnetic field lines. The more field lines passing through the loop of the conductor, the greater the induced EMF. This principle forms the basis for electromagnetic tracking. That is, the general idea is to move a conducting sensor through a magnetic field and record the induced voltage.

The basic components of an electromagnetic tracking system are an active transmitter and passive sensors. The transmitter is stationary and contains three orthogonal coils (i.e., antennas) that are activated in sequence, with only one antenna generating a magnetic field at a time. Interestingly, if the subject (and therefore the sensors attached to the subject) stops moving within the magnetic field, we might think the induced voltage in each sensor would remain constant. However, revisiting Eq. (5.1) shows that this is not the case, because the magnetic flux must change with respect to time or the EMF goes to zero. There are two things that can be controlled to ensure a changing flux: (1) make sure the subject never stops moving or (2) change the strength and direction of the magnetic field. Electromagnetic tracking systems use the latter strategy to ensure the magnetic flux changes. The transmitter not only emits a magnetic field, but also serves as a fixed reference about which position and orientation of each sensor is reported.

Each receiving sensor contains three orthogonal coils used to detect the magnetic field emitted by the transmitter according to the principles of magnetic induction. The receiving coils are contained within a 1-cubic-inch plastic housing that protects and provides a convenient method for attaching the sensor to the subject. The sensors are generally secured by double-sided tape and wrapped with an elasticized band. Proprietary signal processing takes place in real time, compensating for the strength of the earth's magnetic field. Individual coil signals can be used to determine the orientation of the sensor relative to the antenna generating the magnetic field. Each coil within the sensor detects three signals from the transmitter (i.e., one for each antenna of the transmitter) for a total of nine signals. These nine signals suffice to locate the position and orientation of the sensor relative to the transmitter. For example, the receiving coil most parallel to the currently active transmitting antenna will experience the largest EMF, while the more orthogonal the coil, the smaller the induced voltage. Because each coil within a sensor is the same distance from the transmitter, it is possible to determine the distance and orientation of the sensor relative to the currently active antenna by comparing the strength of the induced EMF in each coil to the strength of the emitted magnetic field.

There are two types of electromagnetic tracking systems that are used for the study of human movement. The biggest difference between these systems is that one (e.g., from Polhemus Incorporated) uses an ac magnetic field, while the other type (e.g., from Ascension Technology Corporation) uses a pulsed dc magnetic field. The precision and accuracy of electromagnetic tracking systems is affected by metallic objects, low-frequency electronic noise, and the distance of the sensor from the transmitting antennas. The radius within which precise and accurate data are sampled depends on the particular system and strength of the transmitter. However, when used in an ideal environment, the precision and accuracy of both systems is more than adequate for studying human movement. With proper setup, static accuracy of less than 2 mm rms and 0.5° rms for both systems is possible. Real-time data can be sampled at a rate of up to 144 Hz, depending on the system, the number of sensors tracked, and the type of data communication interface with the computer. These systems do not suffer from line-of-sight problems typical of optoelectronic systems and are therefore ideal for capturing complex movements.

5.3.3 Optical Methods: Camera-Based Systems

The most common method of recording human movement involves "filming" the motion of interest. Historically, images were stored on conventional media such as 16-mm film or on videotape. Today's standard is based on digital technology, bypassing physical media per se, sending the data directly to the computer. Data are commonly sampled at a rate between 50 and 240 frames per second, depending on the movement of interest. For example, natural cadence walking is often sampled at a rate of 60 Hz, while running and arm movements tend to be sampled at 100 Hz or faster. There are two types of high-speed video-based systems that are used for studying human movement. The fundamental difference between designs is related to their use of active or passive tracking of targets.

Active-tracking target systems use infrared light-emitting diodes (LEDs) to indicate the position of the target in space. The diodes are pulsed so that only one target is illuminated (i.e., active) at a time. Thus, if a target is not detected by a camera and then suddenly reappears in the camera's field of view, it will automatically be identified by its order in the pulse sequence. Active target systems are subject to line-of-sight problems common to all optical-based tracking systems. That is, the target must be "seen" by a camera to be detected. Active targets emit a restricted angle of light that may not be detected by the camera if the target rotates relative to the segment to which it is attached.

A limitation of active target systems is that on-board electronics must be strapped to the subject with leads to each of the diodes. These wires, combined with the subject being tethered to the cameras, can interfere with certain movements. Tetherless systems (i.e., telemetered systems) are available; however, wires to each diode are still necessary.

In contrast, passive tracking targets merely reflect projected light and do not actively communicate their position in space. It is therefore important that the tracking targets reflect more light than surrounding objects. To promote this, tracking targets are covered with a highly reflective material, most often in the form of retroreflective tape; however, reflective ink or paint can also be used. In addition, a ring of stroboscopic LEDs mounted around the camera lens housing is used to illuminate the tracking targets (see right panel of Fig. 5.4).

Passive tracking targets typically range between 10 and 40 mm in diameter, with the size of the target usually related to the field of view in which the movement takes place and the accuracy of the experimental setup. There is a trade-off in target size, since overly large targets may obscure the detection of other targets, while too small a target may not reflect sufficient light to be detected by the cameras. A reasonable rule of thumb is that the diameter of the tracking targets should be between 1 and 2 percent of the largest dimension of the calibrated workspace. The workspace may be thought of as the region in which the movement will take place, and is generally defined during a system calibration process (discussed in "Camera Calibration," below).

The Role of the Video Camera. The determination of the three-dimensional coordinates of tracking targets from multiple two-dimensional camera views is often taken for granted or treated as a *black box*. In the sections that follow, we will discuss the basic principles of reconstructing three-dimensional target coordinates from multiple camera images. It is advantageous to describe how the process works for a single tracking target prior to discussing how three-dimensional kinematics of segmental motion are calculated.

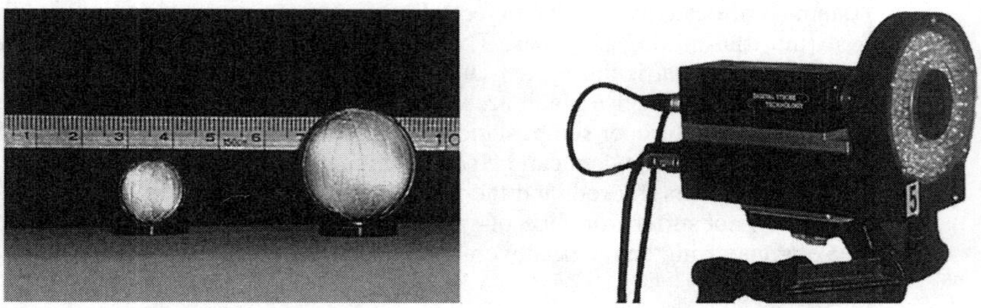

(a) (b)

FIGURE 5.4 (*a*) 10- and 25-mm retroreflective-tape-covered tracking targets. Note how the targets are mounted on lightweight plastic pedestals. The pedestals make it easier to attach the targets to the segment. (*b*) High-speed digital video camera used to "film" the position of the tracking targets. Note the stroboscopic ring of LEDs around the lens of the camera.

For the purposes of this discussion, we assume the image plane of our high-speed video camera is a charge-coupled device (CCD) sensor. The image plane may be thought of as the *exposure medium* onto which real-world *object-space* is projected. The term *object-space* will be used to describe the X, Y, Z inertial reference system in which the tracking targets move. Individual elements of the sensor are arranged in a series of rows and columns, with each element responsible for converting the intensity of light to a voltage such that the greater the intensity of light striking the element, the greater the voltage. This is particularly relevant because the tracking targets should reflect more light than all other objects detected by the camera. The matrix arrangement of light-sensitive elements is illustrated schematically in Fig. 5.5. Note the internal u, v coordinate system of the image plane.

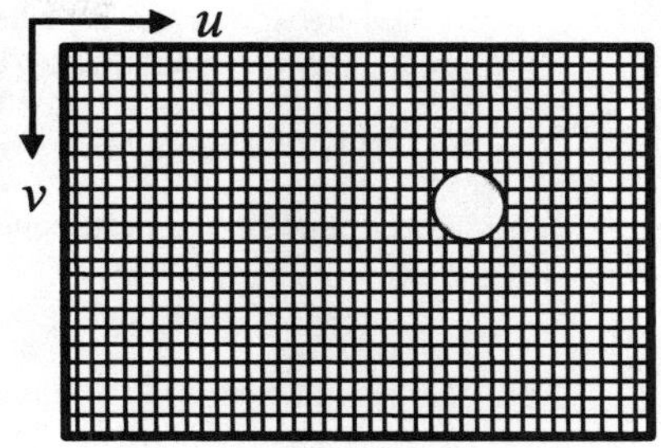

FIGURE 5.5 Schematic representation of the light-sensitive elements of the imaging sensor. The circle is the projection of a target onto the image plane. Note the direction of the u, v reference axes.

Consider the case where a single tracking target is detected by a camera and no other light-reflecting objects are visible. The circle in Fig. 5.5 is used to depict the projection of the target onto the imaging sensor. Clearly, other elements of the sensor would be excited to varying degrees depending on the ambient lighting, but have been turned off for the sake of this example. Figure 5.5 is also helpful in depicting the effect of using binary thresholding to suppress background light sources or reflective objects other than the tracking targets. The idea is to suppress all voltages below a user-specified threshold, which has the effect of moving the light-reflective tracking targets to the foreground. This is ideal since we are ultimately concerned with locating the center of a prospective target in u, v coordinates and do not want other sources of light affecting the location of the computed center. One method of determining the center of a target is to scan the matrix of light-sensitive elements for transitions in voltage (i.e., edge detection) and fit a circle of best fit to the resulting "edges." While this approach is conceptually straight forward, binary thresholding as described here would eliminate useful information that could otherwise be used to determine the center of the projected target at greater, subpixel accuracy. For example, rather than simply treating each element of the sensor as being on or off, we could use a weighted average of sensor element voltages and a geometric constraint that the active elements form a circle. The center of the target in the image plane is assumed to lie at the center the circle.

If one were to draw a line from the center of the target in the image plane to the X, Y, Z coordinates of the target in object-space, it would be clear that the mapping between these spaces is not unique, since all targets lying on this line would map to the same u, v coordinates. From this, it is evident that the location of a target in object-space cannot be determined with only one camera. This raises a subtle but important distinction regarding the role of the camera in video-based motion analysis. The camera does not actually record the location of a target in object-space, but rather the role of the camera is to define a ray in the direction of the target. When multiple cameras view the same target, the location of the target in object-space is assumed to lie at the intersection of the directed rays from each camera. The cameras must first be calibrated before the intersection of these rays can be calculated.

Camera Calibration. Each camera must be calibrated before it can contribute to locating a target in object-space. Camera calibration defines a *mapping* from three-dimensional object-space into the two-dimensional u, v coordinates of the camera. This mapping is expressed in Eq. (5.2) in terms of homogeneous coordinates:

$$\begin{pmatrix} \lambda u \\ \lambda v \\ \lambda \end{pmatrix} = \mathbf{A} \begin{pmatrix} X \\ Y \\ Z \\ 1 \end{pmatrix} \tag{5.2}$$

where λ = scale factor relating the spaces
u, v = image plane coordinates of a target
$\mathbf{A}$ = 3×4 transformation matrix
X, Y, Z = coordinates of a target in object-space

Expanding the right-hand side of equation 2 results in the following set of equations.

$$\lambda u = \alpha_{11}X + \alpha_{12}Y + \alpha_{13}Z + \alpha_{14} \tag{5.3}$$

$$\lambda v = \alpha_{21}X + \alpha_{22}Y + \alpha_{23}Z + \alpha_{24} \tag{5.4}$$

$$\lambda = \alpha_{31}X + \alpha_{32}Y + \alpha_{33}Z + \alpha_{34} \tag{5.5}$$

Substituting λ into Eq. (5.3) and (5.4), and introducing the following equation:

$$\beta_{ij} = \alpha_{ij} / \alpha_{34} \tag{5.6}$$

leads to two convenient expressions relating the coordinates of the center of the target in the image plane and the location of the target in the object-space.

$$u = \beta_{11}X + \beta_{12}Y + \beta_{13}Z + \beta_{14} - u\beta_{31}X - u\beta_{32}Y - u\beta_{33}Z \tag{5.7}$$

$$v = \beta_{21}X + \beta_{22}Y + \beta_{23}Z + \beta_{24} - v\beta_{31}X - v\beta_{32}Y - v\beta_{33}Z \tag{5.8}$$

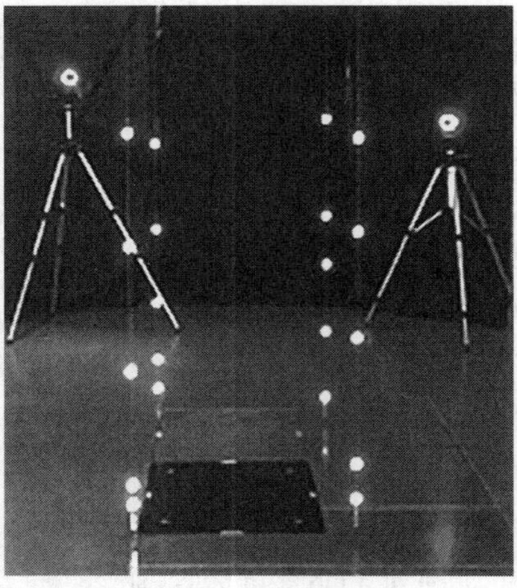

FIGURE 5.6 The X, Y, Z coordinates of the control points (i.e., reflective targets) are known relative to the origin of the object-space. Once the cameras have been calibrated, the hanging strings with the control points are removed from the field of view of the cameras. The black rectangle flush with the floor is a force platform (see Sec. 5.4.5).

Note that the u, v coordinates of the target are known. Therefore, if the X, Y, Z coordinates of the target also are known, we are left with 11 unknowns (i.e., transformation parameters) in two equations. The unknown betas (i.e., β_{ij}) can be determined if the X, Y, Z coordinates of at least six *control points* are detected by the camera. That is, each control point provides two equations that can be used to solve for the 11 unknown betas. The term *control point* is used to make clear that the X, Y, Z coordinates for these targets are known, having been accurately measured relative to the origin of the object-space. The control points are used solely for the purpose of calibrating the cameras and are removed from the field of view once the cameras have been calibrated. The distribution of the $n \geqslant 6$ control points must not be colinear, and the control points should encompass the volume within which the movement will take place. This volume is often referred to as the *workspace*. One method of defining the workspace is to hang four strings with a number of control points attached to each string, as shown in Fig. 5.6.

The direct linear transformation (DLT) proposed by Abdel-Aziz and Karara (1971) is perhaps the most well known method of calibrating the cameras among those conducting video-based motion analysis. The unknown betas for each camera are related to internal and external camera parameters. Examples of internal parameters include the principal distance from the center of the camera lens to the image plane and the u, v coordinates of the principal point. (The principal point lies at the intersection of the principal axis and the image plane.) Although the number of internal parameters can vary depending on the accuracy of the geometric representation of the camera, the number of external parameters remains fixed at six. The six external parameters (i.e., three position and three orientation) describe the relationship between the internal camera coordinate system and the object-space.

Prior to development of the DLT method, the six external parameters were measured manually. This was a painstaking process and subject to errors. The simple act of bumping a camera or repositioning the cameras for a new experimental setup involved remeasuring the external parameters. The DLT greatly facilitated video-based motion analysis, providing a convenient method of solving for the external camera parameters and determining the mapping from object-space to the u, v coordinates of the image plane.

This discussion on camera calibration is not meant to be comprehensive. However, it does provide the basic background for understanding how and why cameras are calibrated. Additional terms can be added to the basic 11-parameter DLT model to correct for symmetrical and asymmetrical lens distortions. These errors can be treated, in part, during camera calibration, and may also be accounted for by using lens correction maps provided by the manufacturer.

In recent years, the so-called wand or dynamic calibration method has become widely used in place of hanging strings with control points. In a dynamic calibration, two retroreflective targets attached to a wand are moved throughout the entirety of the volume in which the movement will take place. The targets on the wand are not control points *per se* because their locations in object-space are not known *a priori*. However, since the coordinates of the two wand targets can be measured with respect to the u, v coordinates of each camera and because the distance between targets should remain constant, the cameras can be calibrated in an iterative manner until the length of the wand as detected by the cameras matches the true length of the wand (i.e., distance between targets). Although the length of the wand can be reconstructed very accurately by this method, the direction of the object-space reference axes does not have to be known for determining the length. A static frame with a predefined origin and control points arranged to define the object-space reference axes is placed in the field of view of the "calibrated" cameras to establish the direction of the X, Y, Z object-space axes.

Calculating Object-Space Coordinates. Once the cameras have been calibrated and a set of betas for each camera is known, the opposite approach can be used to locate the position of a target in object-space. The term *reconstruction* is often used to describe the process of calculating three-dimensional coordinates from multiple ($n \geqslant 2$) camera views. Consider the example illustrated in Fig. 5.7, where two cameras have a unique perspective of the same tracking target. The u, v coordinates of the target in each camera view are known, as are the betas for both cameras as a result of the calibration. The unknowns in this case are the X, Y, Z coordinates of the target in the object-space. Rearranging Eq. (5.7) and (5.8) and adding two more equations for the second camera leads to the following:

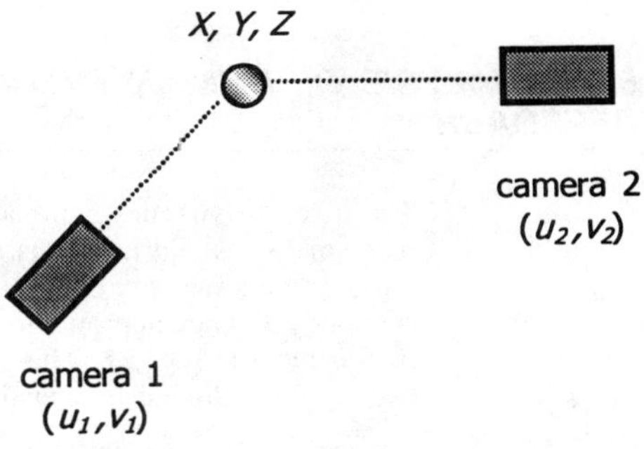

FIGURE 5.7 Cameras 1 and 2 each have a unique perspective of the tracking target in object-space (i.e., silver circle). The X, Y, Z coordinates of the target can be calculated by using the u, v coordinates and the betas determined during calibration.

$$u_1 = (\beta'_{11} - u_1\beta'_{31})X + (\beta'_{12} - u_1\beta'_{32})Y + (\beta'_{13} - u_1\beta'_{33})Z + \beta'_{14} \tag{5.9}$$

$$v_1 = (\beta'_{21} - v_1\beta'_{31})X + (\beta'_{22} - v_1\beta'_{32})Y + (\beta'_{23} - v_1\beta'_{33})Z + \beta'_{24} \tag{5.10}$$

$$u_2 = (\beta''_{11} - u_2\beta''_{31})X + (\beta''_{12} - u_2\beta''_{32})Y + (\beta''_{13} - u_2\beta_{33}'')Z + \beta''_{14} \tag{5.11}$$

$$v_2 = (\beta''_{21} - v_2\beta''_{31})X + (\beta''_{22} - v_2\beta''_{32})Y + (\beta''_{23} - v_2\beta''_{33})Z + \beta''_{24} \tag{5.12}$$

where the subscript on u and v indicates camera 1 or 2, with β'_{ij} and β''_{ij} used to identify the betas for cameras 1 and 2, respectively. We can express Eq. (5.9) to (5.12) compactly if we let C_{ij} be the terms in parentheses, where i indicates row and j column [see Eq. (5.13) for an example], and let L_i

be the combination of the left-hand side and the lone beta on the right-hand side of Eqs. (5.9) to (5.12) [see Eq. (5.14) for an example]. Equation (5.15) reflects this compact notation.

$$C_{11} = (\beta'_{11} - u_1\beta'_{31}) \tag{5.13}$$

$$L_3 = (u_2 - \beta''_{14}) \tag{5.14}$$

$$\mathbf{L} = \mathbf{C}\begin{pmatrix} X \\ Y \\ Z \end{pmatrix} \tag{5.15}$$

Since **C** is not a square matrix (it is 4 × 3), the unknown X, Y, Z coordinates can be solved by using the Moore-Penrose generalized inverse, as follows:

$$\begin{pmatrix} X \\ Y \\ Z \end{pmatrix} = (\mathbf{C}^T\mathbf{C})^{-1}\mathbf{C}^T\mathbf{L} \tag{5.16}$$

which, in essence, yields a least-squares solution for the X, Y, Z coordinates of the tracking target. The solution is easily expanded to account for $n > 2$ cameras.

While only two cameras are necessary to reconstruct the three-dimensional coordinates of a tracking target in object-space, more than two cameras are recommended to help ensure that a minimum of two cameras see the target every point in time. The cameras should be positioned so that each has a unique perspective of the workspace. Ideally the cameras should be placed at an angle of 90° with respect to one another. In practice this may not be possible, and every effort should be taken to maintain a minimum separation angle of at least 60°.

5.4 ANALYSIS OF HUMAN MOTION: AN INVERSE DYNAMICS APPROACH

The inverse dynamics approach is the most commonly used method to solve for unknown joint reaction forces and moments. The analysis begins with the most distal segment, moving upward through the kinematic chain, requiring that all external forces acting on the system are known. A free-body diagram appropriate for a two-dimensional inverse dynamics analysis of the foot and shank is illustrated in Fig. 5.8. This can be expressed mathematically in a generalized form suitable for a two- or three-dimensional analysis of n segments.

$$\Sigma\mathbf{M}_i = I_i\, d\boldsymbol{\omega}_i/dt \tag{5.17}$$

$$\Sigma\mathbf{F}_i = m_i\, d\mathbf{v}_i/dt \tag{5.18}$$

where $\Sigma\mathbf{M}_i$ = sum of the moments acting on segment i
I_i = inertia tensor for segment i about its center of mass (COM)
$\boldsymbol{\omega}_i$ = angular velocity of the segment

Forces acting on segment i, mass, and linear velocity of the segment correspond to $\mathbf{F}_i$, m_i and $\mathbf{v}_i$, respectively.

The three types of measurement *tools* described in Sec. 5.3 all provide kinematic data of some form. For example, goniometers provide an estimate of joint angular position, while electromagnetic tracking systems output the relative position and orientation of the sensors attached to the segments. In the case of video-based motion analysis, output data are in the form of target coordinates expressed in the object-space. It is important to note that output from all of these devices contains some degree of error. That is, the sampled signal is actually a combination of "true" signal and "noise." This is

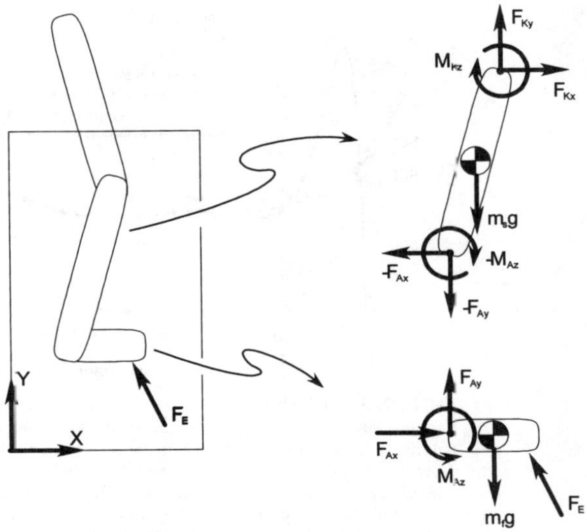

FIGURE 5.8 Two-dimensional FBD of the foot and shank segments. Note how the forces and moment acting at the distal end of the shank are equal in magnitude but opposite in direction to the forces and moment acting at the ankle. *F* and *M* are used to indicate force and moment respectively, with *A* and *K* used to distinguish between the ankle and knee joints. The foot and shank segments are represented by *f* and *s*, while F_E is used to denote the external force acting on the foot.

an important consideration because the equations of motion contain linear and angular acceleration terms, values that are obtained by numerically differentiating the position data. Differentiating the raw data will have the undesirable effect of magnifying the noise, which can severely compromise the integrity of the results. An excellent discussion of this topic can be found in Winter (1990). The point we wish to make is that the raw data should be treated to reduce or eliminate the amount of contaminating noise before the data are used in subsequent calculations. This process of treating the raw data is commonly referred to as *data smoothing*.

5.4.1 Data Smoothing

As previously stated, all motion capture data contains some degree of noise. For example, consider the plot in Fig. 5.9. The dotted line is the trajectory of a tracking target attached to the shank of a subject walking at a natural cadence. The data are limited to the stance phase of gait and have not been treated (i.e., these data are "raw"). The thick line passing through the raw data is a smoothed form of the original signal. The purpose of this section is to introduce several methods that are commonly used in biomechanical studies to smooth motion capture data.

For the purpose of this discussion, consider a single tracking target whose position has been determined by video-based motion analysis. We begin by assuming the data were collected at an adequate sampling rate to prevent aliasing of the signal. The sampling theorem states that the data should be sampled at a rate of at least 2 times greater than the highest-frequency component in the signal being sampled. This minimum rate is commonly described as the *Nyquist limit*. It is not unreasonable to sample the data at 5 times the Nyquist limit to ensure the integrity of the data in both the frequency and time domains. Historically, sampling at such a high rate was prohibitive

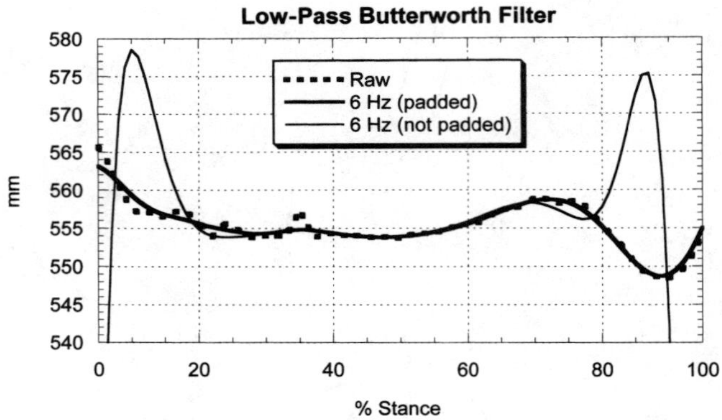

FIGURE 5.9 The bold black squares are raw data for the *X* coordinate of a tracking target attached to the shank of a subject during the stance phase of natural cadence walking, where *Y* is the forward direction and *Z* is the vertical direction. The thick solid line is the result of filtering the raw data in the forward and reverse direction using a fourth-order Butterworth low-pass digital filter set at a cutoff frequency of 6 Hz. The front and back ends of the raw data were padded prior to filtering. Note how the thick line fits through the original raw data. The thin line is the result of filtering the raw data using exactly the same Butterworth filter, with the only difference being that the raw data were not padded prior to filtering. Clearly, the smoothed data represented by the thin line are not suitable for analysis.

because of constraints on disk space. Storage space, however, is no longer an issue and sampling data well above the Nyquist limit is recommended.

Two general approaches of smoothing data include curve fitting and digital filtering. Both approaches can yield similar results; however, the underlying theory behind each approach is different.

Curve Fitting. Curve fitting, as the name implies, involves fitting a function or a series of functions through the raw data with a goodness of fit generally based on a least-squares difference. For example, polynomial regression and piecewise polynomial approximation are methods of curve fitting, the latter of which is more commonly used in studying human movement. Cubic and quintic splines are the most common of the piecewise approximation methods. These splines require that a smoothing parameter be specified to determine how closely the smoothed data fit through the original data points. The goal is to select a smoothing parameter that does not over- or undersmooth the raw data. In practice, it may be difficult to determine an ideal smoothing parameter. Algorithms have been created in which an ideal smoothing parameter can be determined by a statistical procedure known as *generalized cross validation* (GCV). The GCVSPL package (Woltring, 1986) is one such program that uses GCV to identify an ideal smoothing parameter for the spline. A description of the GCVSPL package and source code is available for download from the International Society of Biomechanics (1999).

Digital Filtering. Digital filtering is another method that is used to smooth biomechanical data. The concept is based on the fact that any signal, if sampled at an adequate rate, can be recreated from a series of sine and cosine waveforms of varying frequency. This principle can be used to reduce the amount of noise, if the frequency content of the noise is known. For the sake of this example we assume the acquired data are contaminated with high-frequency noise. Although a variety of digital filters exist, we focus our attention on the Butterworth filter because it is perhaps the most widely used filter in biomechanics research.

A low-pass Butterworth filter is designed to attenuate frequencies above a specified cutoff frequency, while allowing frequencies below the cutoff to pass through the filter unattenuated. Butterworth filters are not infinitely sharp. The *order* of the filter characterizes the sharpness, or how much the signal is attenuated in the vicinity of the cutoff frequency. The higher the order, the sharper the filter response. Computer implementation of a Butterworth filter is straightforward, which may be in part why it is so widely used. Although we described the role of the Butterworth filter as extracting particular frequencies, the raw signal is actually filtered in the time domain as seen below.

$$Y(t) = a_0X(t) + a_1X(t-1) + a_2X(t-2) + b_1Y(t-1) + b_2Y(t-2) \tag{5.19}$$

where $Y(t)$ and $X(t)$ are the filtered and raw data at time (t). The $(t - 1)$ and $(t - 2)$ notation is used to indicate data at 1 and 2 samples prior to the current time. Equation (5.19) is for a second-order recursive filter; higher-order filters require additional recursive terms. The a and b coefficients for a Butterworth low-pass filter can be determined from the following equations:

$$\omega_c = \tan\left(\frac{\pi f_c}{f_s}\right) \tag{5.20}$$

$$K_1 = \sqrt{2}\omega_c \tag{5.21}$$

$$K_2 = \omega_c^2 \tag{5.22}$$

$$K_3 = \frac{2a_0}{K_2} \tag{5.23}$$

$$a_0 = a_2 = \frac{K_2}{1 + K_1 + K_2} \tag{5.24}$$

$$a_1 = 2a_0 \tag{5.25}$$

$$b_1 = -2a_0 + K_3 \tag{5.26}$$

$$b_2 = 1 - 2a_0 - K_3 \tag{5.27}$$

where f_c and f_s are the cutoff frequency and the sampling frequency expressed in hertz, respectively.

Several practical considerations should be noted when using a Butterworth digital filter. First, we see from Eq. (5.19) that the filtered data at time t are related in a recursive manner to raw and filtered data at times $t - 1$ and $t - 2$. This can cause problems at the beginning of the data set unless the front end of the data is padded with extra data points. The dotted line in Fig. 5.9 illustrates the consequence of not padding the data set prior to filtering. Clearly the smoothed data at the beginning (and also at the end!) of the data set are erroneous. Two methods of padding the front end of the data involve reflecting the first n data points (10 or more generally work well for a second-order filter) about data point 1, or simply by collecting more data than is actually needed. It should be noted that this type of digital filter introduces a phase lag in the smoothed signal. The easiest method of correcting for this phase lag is to refilter the already filtered data in the reverse direction. This will shift the data in an equal and opposite direction, realigning the raw and filtered data temporally. Note that filtering the already-filtered data in the reverse direction will increase the sharpness of the filter response. If the data are filtered in the reverse direction, it is advisable to pad the back end of the data set for the reasons cited earlier.

The thick line in Fig. 5.9 is the result of filtering the raw data in the forward and reverse directions using a fourth-order, low-pass Butterworth filter set at a cutoff frequency of 6 Hz (note that the front and back ends of the raw data were padded). This raises an interesting question, that is, how do we identify an appropriate cutoff frequency for the filter? There are a number of methods that can be used to help select an appropriate cutoff frequency. A fast Fourier transform (FFT) can be used to examine the content of the signal in the frequency domain, or one of several residual analysis methods can be used (Jackson, 1979; Winter, 1990).

5.4.2 Tracking Motion of the Segment and Underlying Bone

We continue with our example of how motion data collected with a video-based tracking system is used in an inverse dynamics analysis. Calculating joint kinetics from the observed kinematics and the external forces acting on the body requires knowledge of how the bones are moving. In this section we describe how tracking targets attached to the segments can be used to track motion of the underlying bones. We assume the target coordinates have been smoothed by an appropriate method.

The first step in calculating joint and segmental kinematics is to define orthogonal anatomical coordinate systems (ACSs) embedded in each segment. Because it is the kinematics of the underlying bones that is most often of interest, we must define a set of reference axes that are anatomically meaningful for the purposes of describing the motion. An anatomical coordinate system (i.e., ACS) is constructed for each segment in the kinematic chain. Retroreflective targets positioned over anatomical sites (hence the term *anatomical targets*) are used to define the ACS for each segment. Consider the case in Fig. 5.10*a* where anatomical targets are positioned over the malleoli and femoral condyles. These targets are used to define an ACS for the shank (ACS_{shank}). The frontal plane of the shank is defined by fitting a plane through the four anatomical targets (Fig. 5.10*b*). The next step is to define the ankle and knee joint centers, which are assumed to lie midway between the malleoli and femoral condyle targets, respectively. The longitudinal axis of the shank lies in the previously defined plane, originating at the distal joint center (i.e., the ankle joint) and pointing in the direction of the knee joint center. The origin of the ACS_{shank} is set at the COM of the segment. The COM lies along the longitudinal axis, at a location generally determined by using anthropometric lookup tables

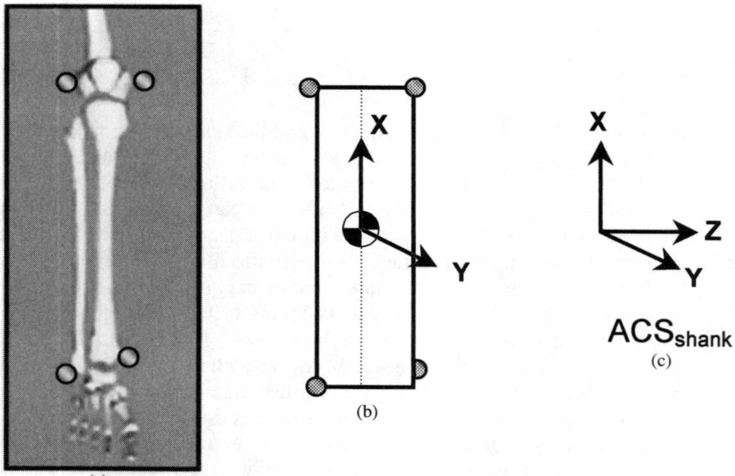

FIGURE 5.10 Retroreflective targets are placed over the medial and lateral malleoli and femoral condyles (*a*). These anatomical targets are used to define the frontal plane of the shank (*b*). The *X* axis projects from the ankle joint center toward the knee joint center. The *Y* axis lies perpendicular to the frontal plane, with the *Z* axis given by the cross product of *X* and *Y*. The orthogonal axes in the right panel represent the ACS_{shank}, which is located at the COM of the segment.

(see Sec. 5.4.4). The unit vector X_s, originating at the COM, will be used to define the direction of the longitudinal axis of the ACS_{shank}. An anteroposterior (AP) axis lies perpendicular to the frontal plane of the shank, with a mediolateral (ML) axis formed by the cross product of the longitudinal and AP axes. Unit vectors Y_s and Z_s originating at the COM of the shank are used to define the direction of the AP and ML axes, respectively. The X_s, Y_s and Z_s unit vectors are orthonormal by way of construction and form the basis of the ACS_{shank} (Fig. 5.10*c*).

A similar approach can be used to construct ACSs for other segments in the kinematic chain, given suitable placement of anatomical targets (see Cappozzo et al., 1995). Figure 5.11 illustrates ACSs for the shank and thigh without specifying the exact details of how the ACS_{thigh} was constructed.

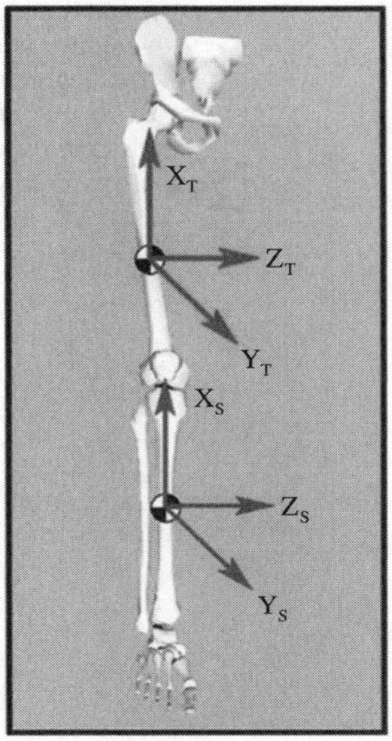

FIGURE 5.11 ACSs for the shank (subscript *S*) and thigh (subscript *T*) segments. The ACSs originate at the COM of each segment. Note that changes in the knee angle will cause the relative orientation between the ACS_{shank} and ACS_{thigh} to change.

Although anatomical targets are used to construct the ACSs, and it is motion of the ACSs that is of interest, it is not practical to track motion of the anatomical targets because they are prone to being knocked off, and in many cases pose line-of-sight problems. The medial malleolus and femoral condyle targets are especially prone to these problems. From a data collection perspective, it is easier to track targets attached to a segment that have been positioned for optimal viewing by the cameras than it is to track targets over anatomical sites. If we define a relationship between the tracking targets and the ACS, we can estimate how the bones are moving by tracking motion of targets on the segment. The easiest way to do this is to construct a set of orthogonal axes using three tracking targets and a series of vector cross products [Eqs. (5.28) to (5.30)]. The resulting orthogonal axes are illustrated in Fig. 5.12. These axes will be referred to as a *local coordinate system.*

$$\mathbf{i} = A - C \tag{5.28}$$

$$\mathbf{j} = (B - C) \times (A - C) \tag{5.29}$$

$$\mathbf{k} = \mathbf{i} \times \mathbf{j} \tag{5.30}$$

where the *X*, *Y*, *Z* coordinates of tracking targets *A*, *B*, and *C* are known in the object-space.

The relative position and orientation of the local coordinate system and the ACS can be represented as a translation vector relating their origins and a rotation matrix of direction cosines. Moreover, the relative position and orientation between the local coordinate system and the ACS should not change if we assume the segment is rigid. This relationship can be used to estimate the position and orientation of the ACS at any point in time by tracking motion of targets attached to the segment. This idea is easily expanded to multiple segments and forms the basis for comparing relative motion between adjacent bones (i.e., ACSs).

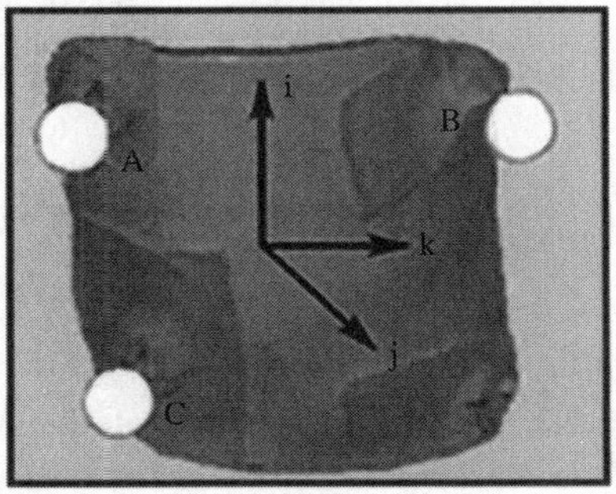

FIGURE 5.12 Three tracking targets *A*, *B*, and *C* are fastened to a contour molded shell. The shell is attached to the segment by elasticized wrap or some other convenient method. Targets *A*, *B*, and *C* are used to construct an orthogonal local coordinate system as in Eqs. (5.28) to (5.30). The origin of the local coordinate system has been drawn in the middle of the figure for convenience.

Constructing a local coordinate system for the purposes of estimating motion of the ACS is a straightforward and convenient method. However, the position and orientation of the local coordinate system is generally sensitive to errors in the coordinates of the tracking targets, and therefore, the estimated position and orientation of the ACS will also be affected. For this reason, it is generally advantageous to use more then three targets per segment and a least-squares method to track motion of the segment and underlying bone. The singular value decomposition (SVD) method has been used to this end with good success (Soderkvist and Wedin, 1993; Cheze et al., 1995). The SVD method *maps* all of the tracking targets ($n \geqslant 3$) from position a to position b using a least-squares approximation. This is illustrated schematically in Fig. 5.13 and represented algebraically in Eq. (5.31).

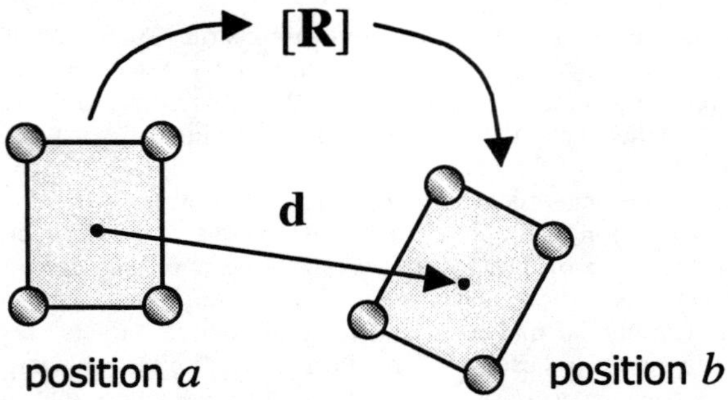

FIGURE 5.13 Least-squares mapping of the tracking targets from position a to position b. **R** is a 3×3 rotation matrix and **d** is a displacement vector.

$$\min \sum_{i=1}^{n} \| Ra_i + d - b_i \|^2 \tag{5.31}$$

where n represents the number of targets attached to the segment, with a_i and b_i used to indicate the object-space coordinates of the individual tracking targets. R is a 3×3 rotation matrix, while d is a displacement vector that, when combined with R, maps all targets in a least-squares sense from their position in a to their position in b. Because the coordinates of the tracking targets are also known relative to the ACS, the same least-squares approach can be used to determine how the ACS moved between position a and position b. Note that although this example maps the targets on the same segment, this idea can also be used to determine relative kinematics between adjacent segments (cf. Soderkvist and Wedin, 1993).

5.4.3 Joint Kinematics: Relative Motion between Adjacent Anatomical Coordinate Systems

It is clear from Fig. 5.11 that changing the knee angle will affect the relative orientation between the ACS_{shank} and the ACS_{thigh}. The orientation at any point in time can be represented by a 3×3 matrix of direction cosines. The nine elements of the direction cosine matrix are related to an ordered sequence of rotations about a particular set of axes. This can be visualized by starting out with the ACS_{shank} and ACS_{thigh} initially aligned, moving the ACS_{shank} into its final orientation relative to the ACS_{thigh} by rotating about the Z, Y', X'' axes of a moving reference frame. The ACS_{shank} is the moving reference in our example. The prime superscripts indicate that the orientation of the primed axes is related to a previous rotation. The first rotation in the Z, Y', X'' sequence takes place about the ML axis of the ACS_{thigh} (or equivalently about the Z axis of the ACS_{shank} because both ACSs are aligned at the onset!). The Y' axis about which the second rotation occurs is perpendicular to both the ML axis of the thigh and the longitudinal axis of the shank. This mutually perpendicular axis is often called the *line of nodes* (or *floating axis* in joint coordinate system terminology). The line of nodes is formed by the vector cross product of the ML axis of the thigh and the longitudinal axis of the shank. The final rotation takes place about the longitudinal axis of the shank (i.e., X''). Note the double superscript indicating the orientation of the longitudinal axis has been influenced by two previous rotations about the Z and Y' axes. These ordered rotations are known as Euler Z, Y', X'' angles. The Euler angles described here are commonly reported in biomechanical studies (e.g., Grood and Suntay, 1983) because the rotations take place about clinically meaningful axes corresponding to joint flexion-extension (Z), abduction-adduction (Y'), and internal-external rotation (X''). The Z, Y', X'' sequence of rotations is expressed in matrix notation in Eq. (5.32), with the elements of the individual matrices shown in Eq. (5.33).

$$\mathbf{R} = [R_z][R_{y'}][R_{x''}] \tag{5.32}$$

$$R_{x''} = \begin{bmatrix} 1 & 0 & 0 \\ 0 & \cos\psi & -\sin\psi \\ 0 & \sin\psi & \cos\psi \end{bmatrix} \quad R_{y'} = \begin{bmatrix} \cos\theta & 0 & \sin\theta \\ 0 & 1 & 0 \\ -\sin\theta & 0 & \cos\theta \end{bmatrix}$$

$$R_z = \begin{bmatrix} \cos\phi & -\sin\phi & 0 \\ \sin\phi & \cos\phi & 0 \\ 0 & 0 & 1 \end{bmatrix} \tag{5.33}$$

where ϕ, θ, and φ are the Euler angles about the Z, Y', X'' axes, respectively. Expanding Eq. (5.32) by using the matrices from Eq. (5.33) leads to the rotation matrix $\mathbf{R}$ in Eq. (5.34).

$$\mathbf{R} = \begin{bmatrix} \cos(\phi)\cos(\theta) & \cos(\phi)\sin(\theta)\sin(\psi) - \sin(\phi)\cos(\psi) & \cos(\phi)\sin(\theta)\cos(\psi) + \sin(\phi)\sin(\psi) \\ \sin(\phi)\cos(\theta) & \sin(\phi)\sin(\theta)\sin(\psi) + \cos(\phi)\cos(\psi) & \sin(\phi)\sin(\theta)\cos(\psi) - \cos(\phi)\sin(\psi) \\ -\sin(\theta) & \cos(\theta)\sin(\psi) & \cos(\theta)\cos(\psi) \end{bmatrix} \tag{5.34}$$

It is easy to show that a different sequence of rotations can be used to move the $\text{ACS}_{\text{shank}}$ from its initially aligned position to its final orientation relative to the $\text{ACS}_{\text{thigh}}$. Because matrix multiplication is not commutative in general, the terms of $\mathbf{R}$ in Eq. (5.34) will differ depending on the sequence of rotations selected. Equations (5.35) to (5.37) can be used to determine the Euler angles for this particular sequence of rotations:

$$\phi = \arctan\left(\frac{r_{21}}{r_{11}}\right) \tag{5.35}$$

$$\theta = \arctan\left(\frac{-r_{31}}{\sqrt{r_{11}^2 + r_{21}^2}}\right) \tag{5.36}$$

$$\psi = \arctan\left(\frac{r_{32}}{r_{33}}\right) \tag{5.37}$$

where r_{ij} is the element in the ith row and jth column of matrix $\mathbf{R}$.

The methods outlined above can also be used to calculate segmental kinematics. For example, instead of calculating the relative orientation between the shank and thigh at time 1, we can use Euler angles to determine the relative orientation between the $\text{ACS}_{\text{shank}}$ at times 1 and 2.

5.4.4 Body Segment Parameters

Re-examining Eqs. (5.17) and (5.18), we see that estimates for mass (m) and the inertia tensor (I) for each segment are required to determine the right-hand side of the equations. Several other terms, including the location of the center of mass and the distances from the distal and proximal joint centers to the COM, are embedded in the left-hand side of Eq. (5.18). The term *body segment parameters* (BSPs) is used to describe this collection of anthropometric information.

There are essentially two approaches that are used for estimating BSP values. The more exact approach is to measure the BSP values experimentally. In practice, this is rarely done because the process is tedious, subject to error, and certainly not practical to perform for every subject. Because the BSP values are difficult to measure accurately, they are generally estimated by using anthropometric lookup tables and/or regression equations (e.g., Dempster, 1957; Winter, 1990; Zatsiorsky and Seluyanov, 1985).

For the case of a two-dimensional analysis, when motion is assumed planar, the moment of inertia in Eq. (5.18) takes on a single value. In the case of a three-dimensional analysis, I becomes a 3×3 inertia tensor. The main diagonal of the inertia tensor is constant and the off-diagonal elements vanish when the principle axis of inertia is aligned with the axes of the ACS. The diagonal matrix in Eq. (5.38) reflects this alignment and is the form used in Eq. (5.18) for a three-dimensional analysis in which the moments are expressed in the ACS of the segment.

$$\mathbf{I} = \begin{bmatrix} I_{xx} & 0 & 0 \\ 0 & I_{yy} & 0 \\ 0 & 0 & I_{zz} \end{bmatrix} \tag{5.38}$$

If we assume that each segment is a homogeneous solid of known geometry, we can use standard formulas for calculating mass moment of inertia about the X, Y, and Z axes.

5.4.5 Force Transducers

The role of a force transducer is to record external forces acting on the body. Force plates used in gait and postural studies to measure ground reaction forces are perhaps the most familiar type of force transducer used in biomechanics. A force platform is sensitive to the load a subject applies to the plate, with the plate exerting an equal and opposite load on the subject (hence the term *ground reaction forces*). Although we will limit our discussion to force plates, we wish to point out that other types of force transducers are used in the study of human movement. For example, multiaxial load cells are used to investigate the motor control of arm movements (e.g., Buchanan et al., 1993; 1998).

Commercially available force platforms use one of two different measuring principles to determine the applied load. The first type of force plate uses strain gauge technology to indirectly measure the force applied to the plate (e.g., AMTI and Bertec), while the second type uses piezoelectric quartz (e.g., Kistler). Piezoelectric materials produce an electrical charge directly proportional to the magnitude of the applied load. In this section we focus on how the output of a force platform is used in an inverse dynamics analysis, without considering how the forces and moments detected by the plate are calculated.

Force platforms are used to resolve the load a subject applies to the ground. These forces and moments are measured about X, Y, and Z axes specific to the force platform. In general, the orientation of the force platform axes will differ from the orientation of the reference axes of the object-space. This is illustrated schematically in Fig. 5.14. Thus, it is necessary that the ground reaction forces be transformed into the appropriate reference system before they are used in subsequent calculations. For example, the ground reaction forces acting on the foot should be transformed into the foot coordinate system, if ankle joint forces and moments are expressed in an anatomically meaningful reference system (i.e., about axes of the ACS_{foot}).

Another variable that must be considered is the location of the external force acting on the system. For the case of a subject stepping on a force platform, the location of the applied load is assumed to act at the center of pressure (COP). The term is aptly named, since the subject really applies a

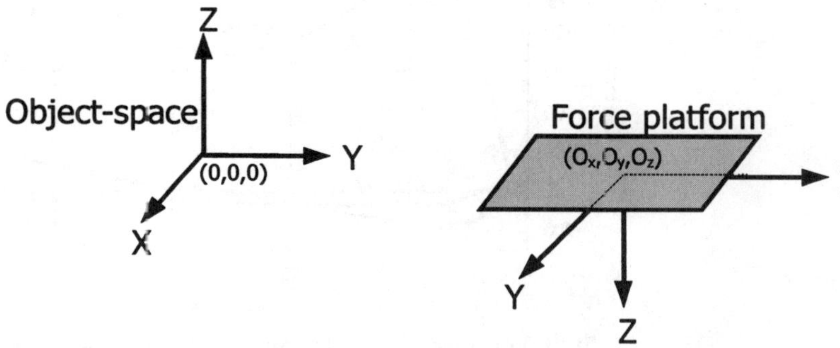

FIGURE 5.14 Relationship between the reference axes of the force platform and the reference axes of the object-space. In general, the reference axes will not be aligned. The force platform is located at a distance O_x, O_y, O_z from the origin of the object-space.

distributed pressure to the top surface of the force plate that is treated as an equivalent point force. As with the forces and moments, the location of the COP in the force platform system should be transformed into the appropriate reference system. Other devices such as pressure insoles and mats can measure pressure distributions, but are not suitable for three-dimensional motion analysis because they do not provide a complete 6 degree of freedom history of the applied load. If the data from a force platform are used in an inverse dynamics analysis, the data must be synchronized with the kinematic observations. Generally, analog data sampled from a force platform are collected at an integer multiple of the video collection rate.

5.4.6 Example: Results from an Inverse Dynamics Analysis

We close this section by presenting several examples of joint kinematic and kinetic data calculated using the methods outlined above. The ground reaction forces for a subject walking at a natural cadence are illustrated in Fig. 5.15. The vertical component of the ground reaction force (GRF) is by far the largest, with the peak AP component of the GRF next largest in magnitude. Notice how the AP force component has both a negative phase and a positive phase corresponding to *braking* and *propulsive* phases during stance. The first hump of the vertical component of the GRF corresponds to a deceleration of the whole body COM during weight acceptance (note how this corresponds with the AP braking force). The second hump in the vertical component of the GRF and the positive phase of the AP force component accelerate the body COM upward and forward as the subject prepares for push-off at the end of stance.

The curve in Fig. 5.16 is the sagittal plane ankle moment that was calculated using the data from the force platform and Eqs. (5.17) and (5.18). At the ankle, we see a small dorsiflexion moment shortly after contact. This moment prevents the foot from slapping down during initial contact with the ground (i.e., the dorsiflexion moment "pulls" the toes toward the shank). As the subject moves into the latter half of stance, a sizable plantarflexion moment is generated as a main contributor to the body's forward progression. This increase in plantarflexion moment is due to the gastrocnemius and soleus muscles contracting, essentially "pushing" the foot into the ground.

Also toward the end of the stance phase, the knee joint flexion angle increases in preparation for push-off. (Think what would happen if the knee did not flex as the leg begins to swing.) The weight acceptance period shortly after initial contact is mediated, in part, by the knee joint, which undergoes a brief period of flexion (Fig. 5.17). During this initial period of stance, the knee acts as a spring, resisting the force of impact. Hence, in Fig. 5.18 we see that a substantial knee extension moment is generated by the quadriceps muscle group to control knee flexion during this time.

Reporting the results of an inverse dynamics analysis in graphical form as we have done here demonstrates the interdependence of the kinematic and kinetic variables. The figures are helpful

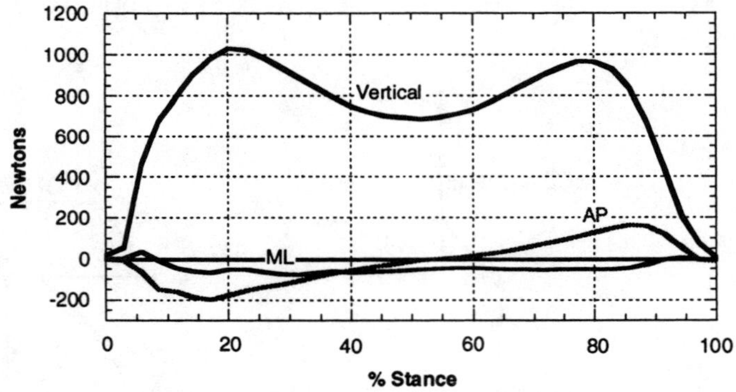

FIGURE 5.15 Ground reaction forces during the stance phase of natural cadence walking. The stance phase begins at foot strike and ends when the foot leaves the ground. ML = mediolateral, AP = anteroposterior.

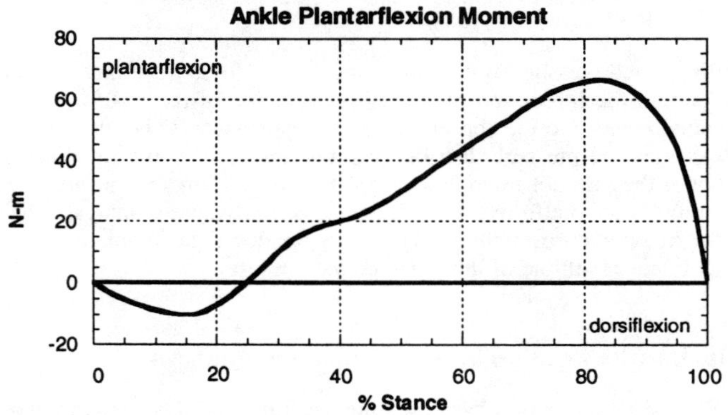

FIGURE 5.16 Sagittal plane ankle moment during the stance phase of natural cadence walking. Notice the small dorsiflexion moment during the first 20 percent of stance. This prevents the foot from slapping the ground shortly after contact. The large plantarflexion moment during the latter half of stance helps propel the body upward and forward.

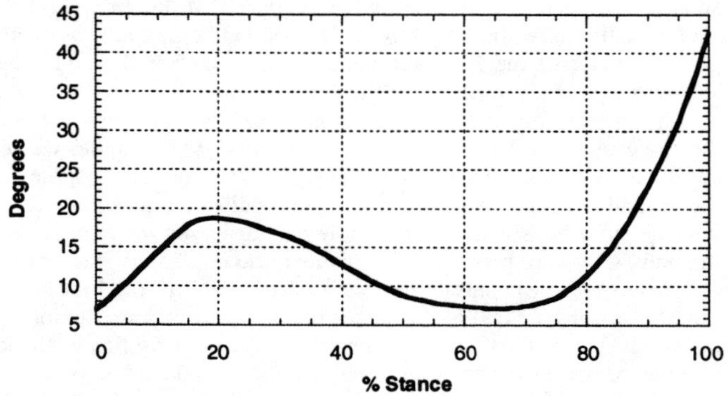

FIGURE 5.17 Knee flexion angle during the stance phase of natural cadence walking. The initial period of flexion (0 to 20 percent stance) helps absorb the shock of impact when the foot hits the ground. A second period of flexion begins at approximately 70 percent of stance, increasing rapidly in preparation for the swing phase.

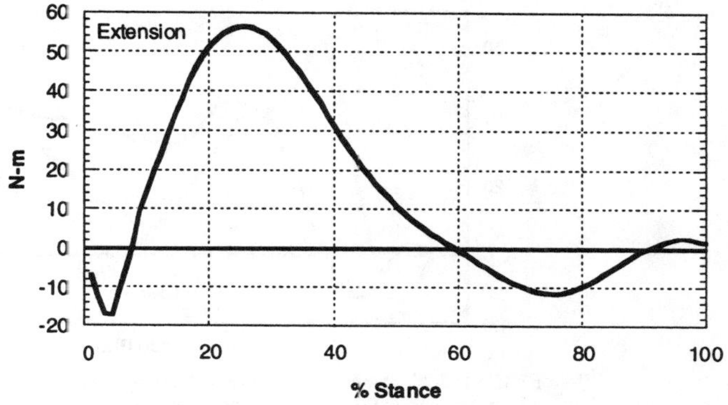

FIGURE 5.18 Knee extension moment during the stance phase of natural cadence walking. The net extension moment reveals that the quadriceps muscles are the dominant group during stance. The initial small flexion moment is caused by the vertical ground reaction force when the foot hits the ground.

when communicating with an athlete or trainer in breaking down a movement pattern to determine how performance might be improved. Inverse dynamics is also a valuable tool that is used to plan surgical treatment and assess the outcome. For example, consider the case in Fig. 5.19. The right panel depicts a varus aligned knee (i.e., *genu varum*). Genu varum is more commonly known as bow-leggedness. Because the net moment is the result of all muscles acting about the joint, it is a reasonable assumption that the medial compartment forces will be greater for the genu varum knee than forces for the normally aligned knee. These increased forces coupled with reduced joint contact

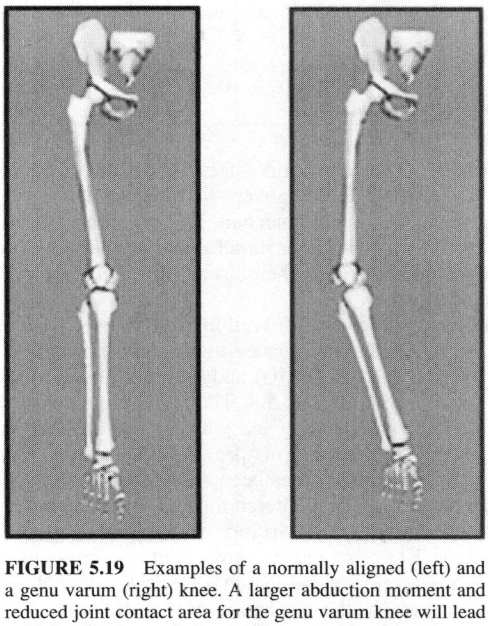

FIGURE 5.19 Examples of a normally aligned (left) and a genu varum (right) knee. A larger abduction moment and reduced joint contact area for the genu varum knee will lead to higher than normal stresses during stance.

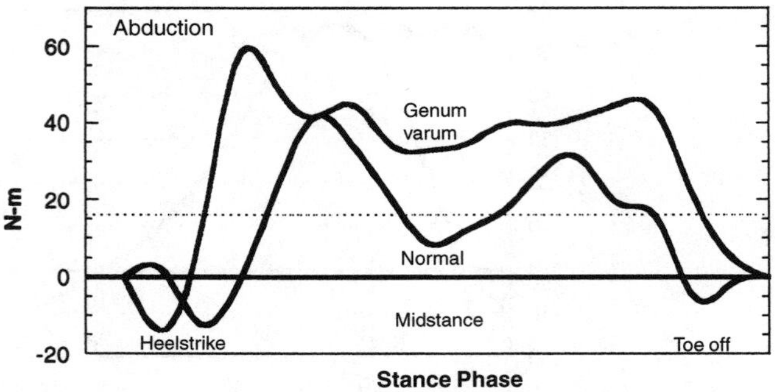

FIGURE 5.20 Stance phase knee abduction moment for a normally aligned and a genu varum knee. Note the difference during midstance. The dashed line is the postsurgical mean data for patients undergoing high tibial osteotomy to correct the genu varum. The mean postsurgical midstance abduction moment (mean = 16 N · m) has returned to a near normal level. Data taken from Weidenhielm et al. (1995).

area (the area being substantially less, as it is mostly over the medial side) will lead to greater stresses in the joint, which may predispose an individual to knee osteoarthritis. The abduction moments for an unimpaired and a genu varum knee are shown in Fig. 5.20. Note how the midstance abduction moment is significantly greater for the presurgical genu varum knee. The dashed line is the average postsurgical midstance moment for patients who underwent high tibial osteotomy surgery, which is a surgical procedure in which a wedge of bone is removed to better align the joint. Note that the midstance knee abduction moment for these subjects has returned to near normal following surgery.

These examples demonstrate how, by using human movement analysis, we can calculate the kinematics and kinetics during complex motions and how these, in turn, can be used to provide information about clinical efficacy and athletic performance.

5.5 CONCLUDING REMARKS

In this chapter we have examined forward and inverse dynamics approaches to the study of human motion. We have outlined the steps involved in using the inverse approach to studying movement with a particular focus on human gait. This is perhaps the most commonly used method for examining joint kinetics. The forward or direct dynamics approach requires that one start with knowledge of the neural command signal, the muscle forces, or, perhaps, the joint torques. These are then used to compute kinematics.

Before concluding, a brief word might be said for hybrid approaches that combine both forward and inverse dynamics approaches to meet in the middle. These methods record both the neural command signal (i.e., the EMG) and the joint position information using standard motion analysis methods as described in Sec. 5.4. The EMG is processed so as to determine muscle forces, which are then summed together to yield joint moments. These same joint moments can also be computed from the inverse dynamics. This provides a means by which to calibrate the EMG to muscle force relationships. This method has been shown to work well with gait studies (Bessier, 2000) and has great potential for studying altered muscle function associated with pathological gait, which cannot be readily examined by optimization techniques.

The biomechanics of human movement is a growing field, spanning many disciplines. As new techniques are developed and shared across these disciplines, the field will continue to grow, allowing us to peer deeper into the mechanics of movement.

REFERENCES

Abdel-Aziz, Y. I., and Karara, H. M. (1971). "Direct linear transformation from comparator coordinates into object-space coordinates," *Close-range photogrammetry*. American Society of Photogrammetry, Falls Church, Va.

Bessier, T. F. (2000). "Examination of neuromuscular and biomechanical mechanisms of non-contact knee ligament injuries," doctoral dissertation, University of Western Australia.

Buchanan, T. S., Moniz, M. J., Dewald, J. P., and Zev Rymer, W. (1993). "Estimation of muscle forces about the wrist joint during isometric tasks using an EMG coefficient method," *Journal of Biomechanics*, **26**:547–560.

Buchanan, T. S., Delp, S. L., and Solbeck, J. A. (1998) Muscular resistance to varus and valgus loads at the elbow. *Journal of Biomechanical Engineering*, **120**(5), 634–639.

Cappozzo, A., Catani, F., Della Croce, U., and Leardini, A. (1995). "Position and orientation in space of bones during movement: anatomical frame definition and determination." *Clinical Biomechanics*, **10**(4), 171–178.

Chao, E. Y. S. (1980). "Justification of triaxial goniometer for the measurement of joint rotation," *Journal of Biomechanics*, **13**:989–1006.

Cheze, L., Fregly, B. J., and Dimnet, J. (1995). "A solidification procedure to facilitate kinematic analyses based on video system data," *Journal of Biomechanics*, **28**(7), 879–884.

Delp, S. L., Arnold, A. S., Speers, R. A., Moore, C. A. (1996). "Hamstrings and psoas lengths during normal and crouch gait: implications for muscle-tendon surgery," *Journal of Orthopaedic Research*, **14**:144–151.

Dempster, W. T. (1955) "Space Requirements of the Seated Operator: Geometrical, Kinematic, and Mechanical Aspects of the Body with Special Reference to the Limbs," Technical Report (55–159) (AD 87892), Wright Air Development Center, Air Research and Development Command, Wright-Patterson Air Force Base, Ohio.

Grood, E. S., and Suntay, W. J. (1983). "A joint coordinate system for the clinical description of three-dimensional motions: application to the knee," *Journal of Biomechanical Engineering*, **105**:136–144.

Hill, A. V. (1938). "The heat of shortening and the dynamic constants of muscle," *Proceedings of the Royal Society of London Series B*, **126**:136–195.

Huxley, A. F. (1957). "Muscle structure and theories of contraction," *Progress in Biophysical Chemistry*, **7**:255–318.

Huxley, A. F., and Simmons, R. M. (1971). "Proposed mechanism of force generation in striated muscle," *Nature*, **233**:533–538.

International Society of Biomechanics (1999). *ISB software sources*, http://isb.ri.ccf.org/software.

Jackson, K. M. (1979). "Fitting of mathematical functions to biomechanical data," *IEEE Transactions on Biomedical Engineering*, BME, **26**(2):122–124.

Pandy, M. G., and Zajac, F. E. (1991). "Optimal muscular coordination strategies for jumping," *J. Biomech.*, **24**:1–10.

Shiavi, R., Limbird, T., Frazer, M., Stivers, K., Strauss, A., and Abramovitz J. (1987). "Helical motion analysis of the knee—I. Methodology for studying kinematics during locomotion," *Journal of Biomechanics*, **20**(5): 459–469.

Soderkvist, I., and Wedin, P. (1993). "Determining the movements of the skeleton using well-configured markers," *Journal of Biomechanics*, **26**(12), 1473–1477.

Weidenhielm, L., Svensson, O. K., and Brostrom, L-A. (1995). "Change of Adduction Moment About the Hip, Knee, and Ankle Joints After High Tibial Osteotomy in Osteoarthrosis of the Knee," *Clinical Biomechanics*, **7**:177–180.

Winter, D. A. (1990), *Biomechanics and Motor Control of Human Movement*, 2d ed., John Wiley & Sons, New York.

Woltring, H.J. (1986). "A FORTRAN package for generalized, cross-validatory spline smoothing and differentiation," *Advances in Engineering Software*, **8**(2):104–113.

Zahalak, G. I. (1986). "A comparison of the mechanical behavior of the cat soleus muscle with a distribution-moment model," *Journal of Biomechanical Engineering*, **108**:131–140.

Zahalak, G. I. (2000). "The two-state cross-bridge model of muscle is an asymptotic limit of multi-state models," *Journal of Theoretical Biology*, **204**:67–82.

Zajac, F. E. (1989). "Muscle and tendon: properties, models, scaling and application to the biomechanics of motor control," in Bourne, J. R. (ed.), *Critical Reviews in Biomedical Engineering*, **17,** CRC Press, pp. 359–411.

Zatsiorsky, V., and Seluyanov, V. (1985). "Estimation of the mass and inertia characteristics of the human body by means of the best predictive regression equation," in Winter, D., Norman, R., Wells, R., Hayes, K., and Patla, A. (eds.), *Biomechanics IX-B* (pp. 233–239), Human Kinetics, Champaign, Ill.

CHAPTER 6
BIOMECHANICS OF THE MUSCULOSKELETAL SYSTEM

Marcus G. Pandy and Ronald E. Barr
University of Texas at Austin

6.1 INTRODUCTION

As the nervous system plans and regulates movement, it does so by taking into account the mechanical properties of the muscles, the mass and inertial properties of the body segments, and the external forces arising from contact with the environment. These interactions can be represented schematically as in Figure 6.1, which suggests that the various elements of the neuromusculoskeletal system can be compartmentalized and modeled independently.

Muscles provide the forces needed to make movement possible; they transmit their forces to tendons, whose forces in turn cause rotation of the bones about the joints. Muscles, however, are not simple force generators: the force developed by a muscle depends not only on the level of neural excitation provided by the central nervous system (CNS), but also on the length and speed at which the muscle is contracting. Thus, muscles are the interface between the neuromuscular and musculoskeletal systems, and knowledge of their force-producing properties is crucial for understanding how these two systems interact to produce coordinated movement.

In this chapter, we review the structure and properties of the neuromusculoskeletal system, and show how the various components of this system can be idealized and described in mathematical terms. Section 6.2 begins with an overview of the mechanical properties of muscle, tendon, ligament, and cartilage. In Secs. 6.3 and 6.4, we focus on the structure of the body-segmental (skeletal) system, emphasizing how musculoskeletal geometry (i.e., muscle moment arms) converts linear actuation (musculotendon forces) into rotary (joint) motion. How motor output from the CNS is converted to muscle activation and ultimately muscle force is described in Sec. 6.5. Section 6.6 presents two methods commonly used to determine musculoskeletal loading during human movement. Repre-

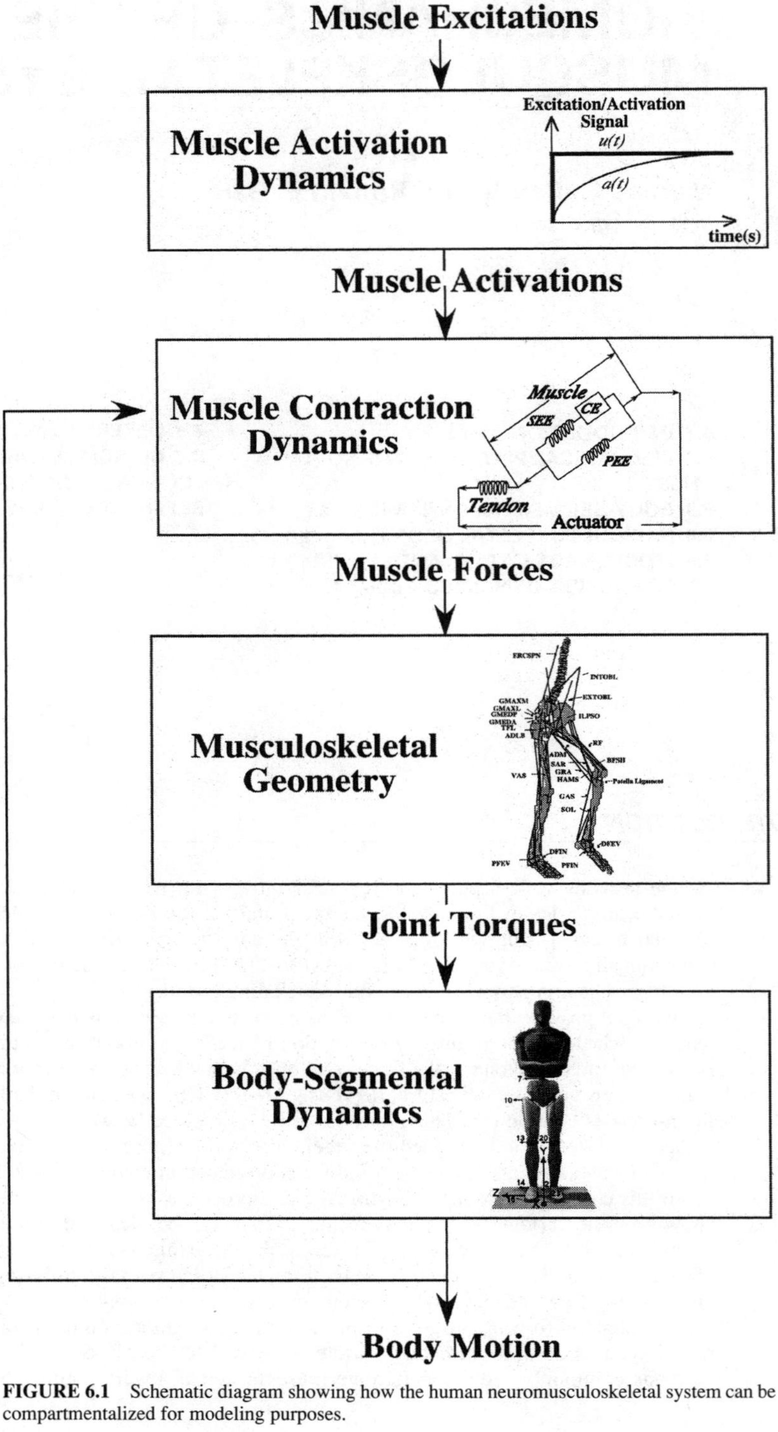

FIGURE 6.1 Schematic diagram showing how the human neuromusculoskeletal system can be compartmentalized for modeling purposes.

sentative results of muscle, ligament, and joint-contact loading incurred during exercise and daily activity are given in Sec. 6.7.

6.2 MECHANICAL PROPERTIES OF SOFT TISSUE

We focus our description of the mechanical properties of soft tissue on muscle, tendon, ligament, and cartilage. The structure and properties of bone are treated elsewhere in this volume.

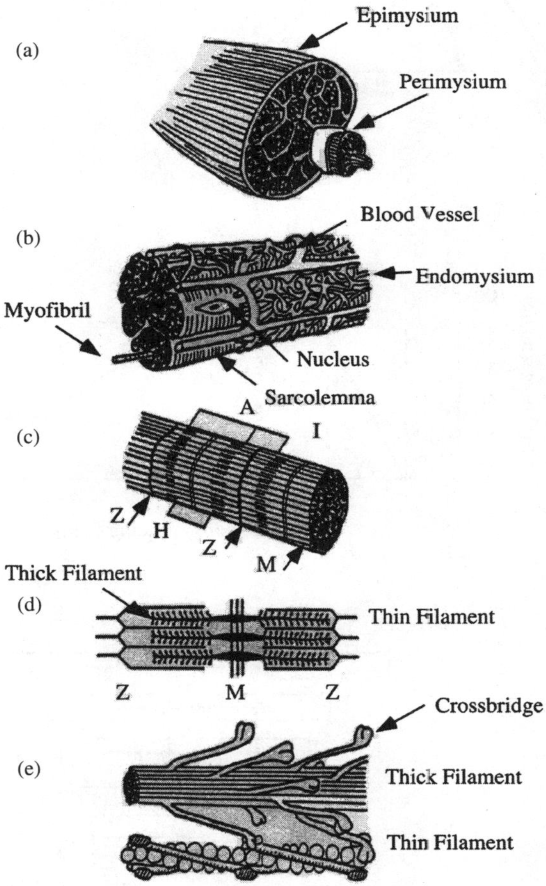

FIGURE 6.2 Structural organization of skeletal muscle from macro- to microlevel. Whole muscle (*a*), bundles of myofibrils (*b*), single myofibril (*c*), sarcomere (*d*), and thick (myosin) filament and thin (actin) filament (*e*). All symbols are defined in the text. [*Modified from Enoka* (*1994*).]

6.2.1 Muscle

Gross Structure. Muscles are molecular machines that convert chemical energy into force. Individual muscle fibers are connected together by three levels of collagenous tissue: endomysium, which surrounds individual muscle fibers; perimysium, which collects bundles of fibers into fascicles; and epimysium, which encloses the entire muscle belly (Fig. 6.2*a*). This connective tissue matrix connects muscle fibers to tendon and ultimately to bone.

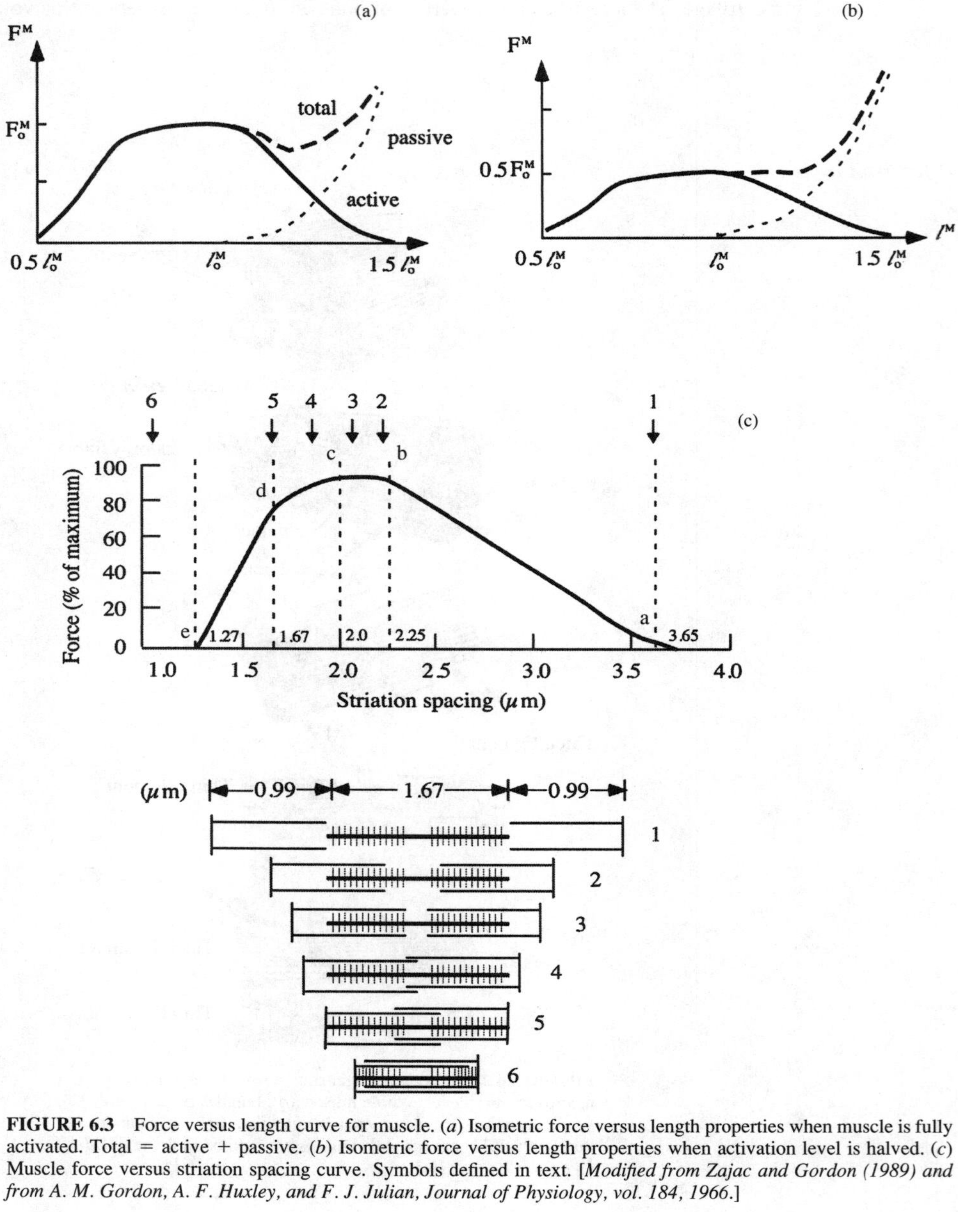

FIGURE 6.3 Force versus length curve for muscle. (*a*) Isometric force versus length properties when muscle is fully activated. Total = active + passive. (*b*) Isometric force versus length properties when activation level is halved. (*c*) Muscle force versus striation spacing curve. Symbols defined in text. [*Modified from Zajac and Gordon (1989) and from A. M. Gordon, A. F. Huxley, and F. J. Julian, Journal of Physiology, vol. 184, 1966.*]

Whole muscles are composed of groups of muscle fibers, which vary from 1 to 400 mm in length and from 10 to 60 μm in diameter. Muscle fibers, in turn, are composed of groups of myofibrils (Fig. 6.2*b*), and each myofibril is a series of sarcomeres added end to end (Fig. 6.2*c*). The sarcomere is both the structural and functional unit of skeletal muscle. During contraction, the sarcomeres are shortened to about 70 percent of their uncontracted, resting length. Electron microscopy and biochemical analysis have shown that each sarcomere contains two types of filaments: thick filaments, composed of myosin, and thin filaments, containing actin (Fig. 6.2*d*). Near the center of the sarcomere, thin filaments overlap with thick filaments to form the AI zone (Fig. 6.2*e*).

In the discussion below, the force-length and force-velocity properties of muscle are assumed to be scaled-up versions of the properties of muscle fibers, which in turn are assumed to be scaled-up versions of properties of sarcomeres.

Force-Length Property. The steady-state property of muscle is defined by its isometric force-length curve, which is obtained when activation and fiber length are both held constant. When a muscle is held isometric and is fully activated, it develops a steady force. The difference in force developed when the muscle is activated and when the muscle is passive is called the active muscle force (Fig. 6.3*a*). The region where active muscle force is generated is (nominally) $0.5l_o^M < l^M < 1.5l_o^M$, where l_o^M is the length at which active muscle force peaks; that is, $F^M = F_o^M$ when $l^M = l_o^M$; l_o^M is called muscle fiber resting length or *optimal muscle fiber length* and F_o^M is the maximum isometric force developed by the muscle (Zajac and Gordon, 1989). In Fig. 6.3*a* passive muscle tissue bears no force at length l_o^M. The force-length property of muscle tissue that is less than fully activated can be considered to be a scaled-down version of the one that is fully activated (Fig. 6.3*b*). Muscle tissue can be less than fully activated when some or all of its fibers are less than fully activated.

The shape of the active force-length curve (Fig. 6.3) is explained by the experimental observation that active muscle force varies with the amount of overlap between the thick and thin filaments within a sarcomere (see also "Mechanism of Muscle Contraction" in Sec. 6.5.2). The muscle force versus striation spacing curve given in Fig. 6.3*c* shows that there is minimal overlap of the thick and thin filaments at a sarcomere length of 3.5 μm, whereas at a length of about 2.0 μm there is

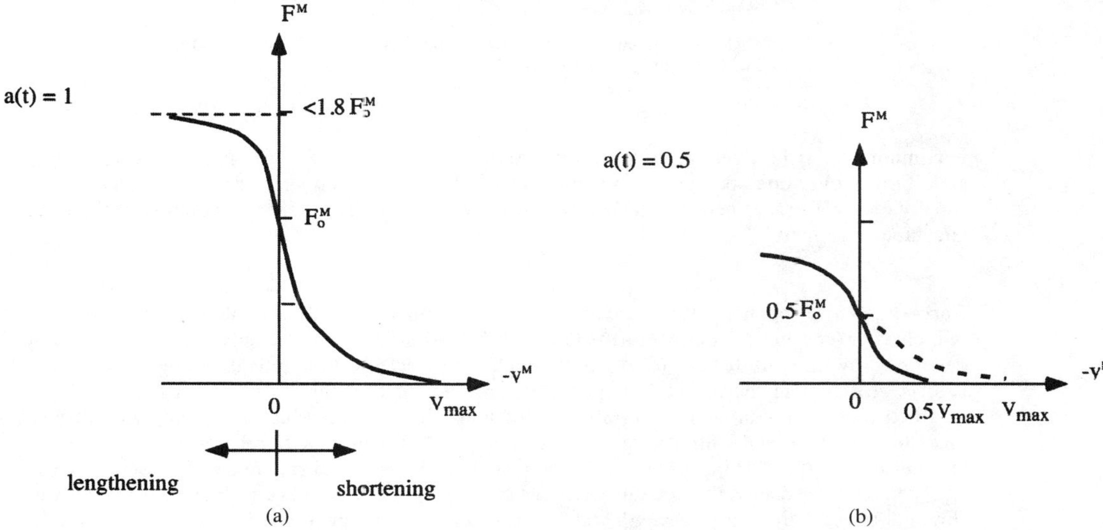

FIGURE 6.4 Force versus velocity curve for muscle when (*a*) muscle tissue is fully activated and (*b*) activation is halved. Symbols defined in text. [*Modified from Zajac and Gordon (1989).*]

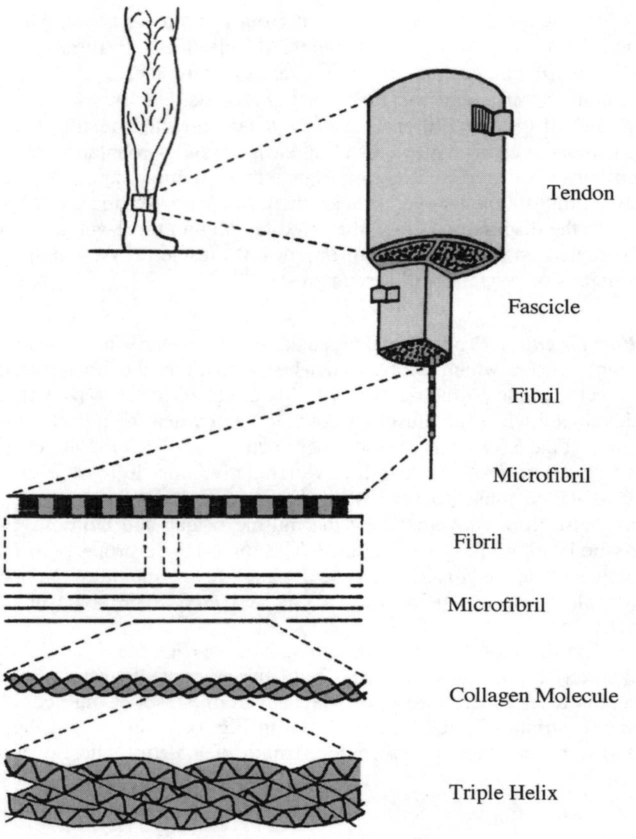

FIGURE 6.5 Structural organization of tendon or ligament from macro- to microlevel. Modified from Enoka (1994).

maximum overlap between the cross-bridges. As sarcomere length decreases to 1.5 μm, the filaments slide farther over one another, and the amount of filament overlap again decreases. Thus, muscle force varies with sarcomere length because of the change in the number of potential cross-bridge attachments formed.

Force-Velocity Property. When a constant load is applied to a fully activated muscle, the muscle will shorten (concentric contraction) if the applied load is less than the maximum isometric force developed by the muscle for the length at which the muscle is initially contracting. If the applied load is greater than the muscle's maximum isometric force at that length, then the muscle will lengthen (eccentric contraction). From a set of length trajectories obtained by applying different loads to shortening and lengthening muscle, an empirical force-velocity relation can be derived for any muscle length l^M (Fig. 6.4*a*). At the optimal fiber length l_o^M, a maximum shortening velocity v_{max}, can be defined such that the muscle tissue cannot resist any load even when it is fully activated (Fig. 6.4*a*). As with the force-length curve, it is commonly assumed that the force-velocity relation scales proportionally with activation, although some studies have found that the maximum shortening velocity v_{max} is also a function of muscle length (Fig. 6.4*b*).

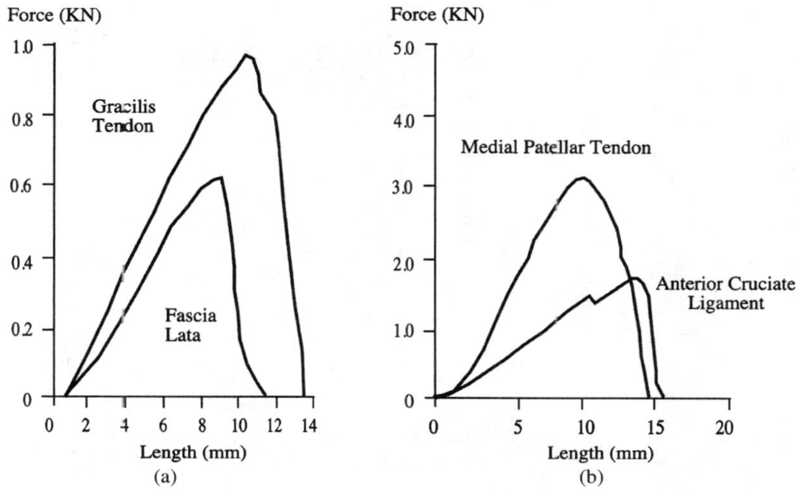

FIGURE 6.6 Force versus length curves for human medial patellar tendon, gracilis tendon, fascia lata tendon, and the anterior cruciate ligament.

6.2.2 Tendon and Ligament

Gross Structure. Tendon connects muscle to bone, whereas ligament connects bone to bone. The main difference between the structure of tendon and ligament is the organization of the collagen fibril. In tendon, the fibrils are arranged longitudinally in parallel to maximize the resistance to tensile (pulling) forces exerted by muscle. In ligament, the fibrils are generally aligned in parallel with some oblique or spiral arrangements to accommodate forces applied in different directions.

Tendons and ligaments are dense connective tissues that contain collagen, elastin, proteoglycans, water, and fibroblasts. Approximately 70 to 80 percent of the dry weight of tendon and ligament consists of Type I collagen, which is a fibrous protein. Whole tendon is composed of bundles of fascicles, each made up in turn of bundles of fibrils (Fig. 6.5). The collagen fibril is the basic load-bearing unit of tendon and ligament. The fibril consists of bundles of microfibrils held together by biochemical bonds (called *cross-links*) between the collagen molecules. Because these cross-links bind the microfibrils together, the number and state of the cross-links are thought to have a significant effect on the strength of the connective tissue.

Stress-Strain Property. When a tensile force is applied to tendon or ligament at its resting length, the tissue stretches. Figure 6.6 shows the force-length curves for three different muscle tendons and one knee ligament found in humans. These data show that the medial patellar tendon is much stronger than either the gracilis or fascia lata tendons. Interestingly, the anterior cruciate ligament is also stronger than the gracilis and fascia lata tendons, and is slightly more compliant as well.

Force-length curves can be normalized to subtract out the effects of geometry; thus, force can be normalized by dividing by the cross-sectional area of a tissue, while length can be normalized by dividing by the initial length of the tendon or ligament. The resulting stress-strain curve displays three characteristic regions: the toe region, the linear region, and the failure region (Fig. 6.7). The toe region corresponds to the initial part of the stress-strain curve and describes the mechanical behavior of the collagen fibers as they are being stretched and straightened from the initial, resting zigzag pattern. The linear region describes the elastic behavior of the tissue, and the slope of the curve in this region represents the elastic modulus of the tendon or ligament. The failure region describes plastic changes undergone by the tissue, where a small number of fibrils first rupture, followed by ultimate failure of the whole tissue.

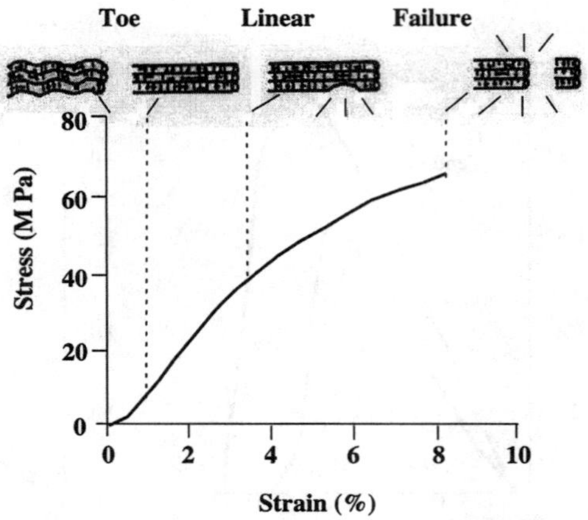

FIGURE 6.7 Idealized stress versus strain curve for tendon or ligament. [*Modified from Butler et al. (1978).*]

6.2.3 Articular Cartilage

Articular cartilage is a white, dense connective tissue, which appears as a layer anywhere from 1 mm to 5 mm thick on the bony articulating ends of a diarthrodial joint. Cartilage is multiphasic, nonlinearly permeable, and viscoelastic. It consists of two phases: a solid organic matrix, composed predominantly of collagen fibrils and proteoglycan macromolecules, and a movable interstitial fluid phase, composed primarily of water. The ratio by wet weight of these two phases is approximately 4:1, with water accounting for roughly 80 percent of the total mass.

In mathematical models of healthy human joints, cartilage is often represented as a single-phase, elastic material with homogeneous and isotropic properties. This approximation is valid, provided that only the short-term response of the tissue is of interest; when cartilage is loaded for 1 to 5 seconds, its response is more or less elastic (Hayes and Bodine, 1978; Hori and Mockros, 1976; Mak, 1986). In the long term, however, say more than 1 minute, the response of the tissue is dominated by the nonlinear, viscoelastic properties of creep and stress relaxation (Hayes and Mockros, 1971; Mow et al., 1984).

Kempson (1980) performed uniaxial compression tests on an isolated cylinder of cartilage. Values of the elastic (Young's) modulus for articular cartilage of the human knee ranged from $E = 8.4$ to $E = 15.4$ MPa when measured 0.2 s after application of the force. Hori and Mockros (1976) estimated the values of elastic modulus and Poisson's ratio at 1 s of loading by conducting torsion and confined compression tests on cartilage of the human tibia. The elastic modulus E of normal cartilage was reported to lie in the range 5.6 to 10.2 MPa, and that of degenerate cartilage in the range 1.4 to 9.3 MPa. Poisson's ratio ν for normal and degenerate cartilage was also found to lie in the range 0.42 to 0.49.

6.3. BODY-SEGMENTAL DYNAMICS

6.3.1 Generalized Coordinates and Degrees of Freedom

The minimum number of coordinates needed to specify the position and orientation of a body in space is the set of generalized coordinates for the body. The number of degrees of freedom (dof) of the body is equal to the number of generalized coordinates minus the number of kinematic constraints acting on the body at the instant under consideration (Kane and Levinson, 1985). For example, the

FIGURE 6.8 Number of degrees of freedom (dof) of a joint. See text for explanation. [*Modified from Zajac and Gordon (1989).*]

number of dof of a joint is at most 6: 3 associated with the orientation of one segment relative to the other (θ_1, θ_2, θ_3 in Fig. 6.8*a*), and 3 associated with the position of a point on one segment relative to the other segment (x, y, z in Fig. 6.8*a*).

In coordination studies, models of joints are kept as simple as possible. In studies of walking, for example, the hip is often assumed to be a ball-and-socket joint, with the femur rotating relative to the pelvis only (Anderson and Pandy, 2001b). Thus, the position of the femur relative to the pelvis can be described by three angles, as illustrated in Fig. 6.8*b*. In jumping, pedaling, and rising from a chair, the hip may even be assumed to be a simple hinge joint, in which case the position of the femur relative to the pelvis is described by just one angle, as shown in Fig. 6.8*c*.

Contact with the environment serves to constrain the motion of the body segments during a motor task. Consider the task of vertical jumping as illustrated in Fig. 6.9. Assuming that movement of the body segments is constrained to the sagittal plane, Figure 6.9*a* indicates that the motor task has 6 dof when the body is in the air (x_p, y_p, which specify the position of the metatarsals plus θ_1, θ_2, θ_3, θ_4, which specify the orientation of the foot, shank, thigh, and trunk, respectively). When the metatarsals touch the ground, only 4 generalized coordinates are needed to specify the position and orientation of the body segments relative to any point on the ground, and accordingly the motor task has only 4 dof (Figure 6.9*b*). Similarly, when the feet are flat on the ground, the number of generalized coordinates and the number of dof are each correspondingly reduced by 1 (Fig. 6.9*c*).

6.3.2 Equations of Motion

Once a set of generalized coordinates has been specified and a kinematic model of the motor task has been defined, the governing equations of motion for the motor task can be written. The number of dynamical equations of motion is equal to the number of dof of the motor task; thus, if the number of dof changes during a motor task (see Fig. 6.9), so too will the structure of the equations of motion.

Different methods are available to derive the dynamical equations of motion for a motor task. In the Newton-Euler method (Pandy and Berme, 1988), free-body diagrams are constructed to show the external forces and torques acting on each body segment. The relationships between forces and

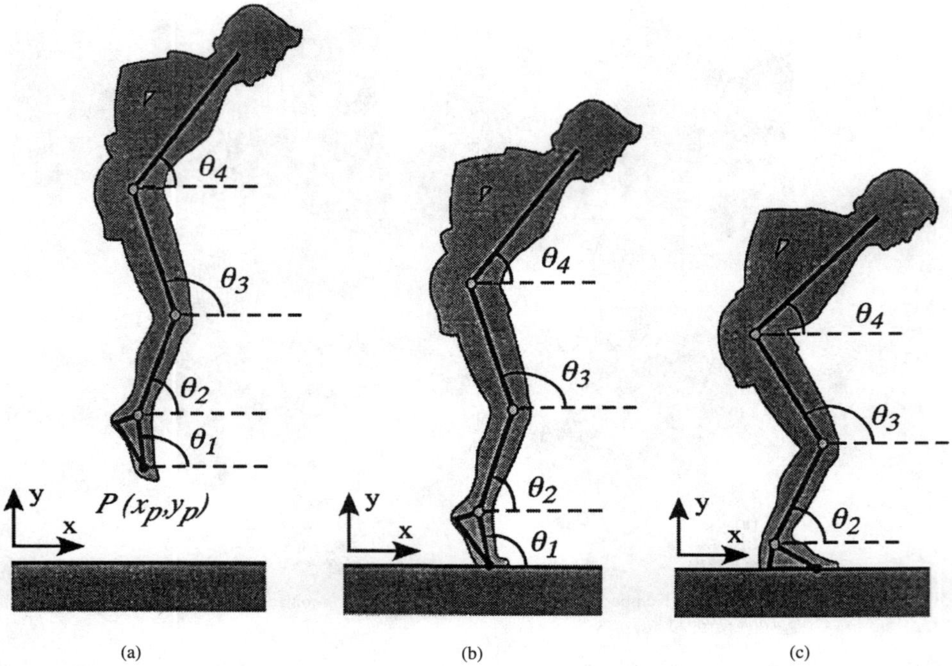

FIGURE 6.9 Number of dof of a motor task such as landing from a vertical jump. [*Modified from Zajac and Gordon (1989).*]

linear accelerations of the centers of mass of the segments are written according to Newton's second law ($\Sigma \mathbf{F} = m\mathbf{a}$), and the relationships between torques and angular accelerations of the segments are written using Euler's equation ($\Sigma \mathbf{T} = I\boldsymbol{\alpha}$).

The Newton-Euler method is well suited to a recursive formulation of the kinematic and dynamic equations of motion (Pandy and Berme, 1988); however, its main disadvantage is that all of the intersegmental forces must be eliminated before the governing equations of motion can be formed. In an alternative formulation of the dynamical equations of motion, Kane's method (Kane and Levinson, 1985), which is also referred to as Lagrange's form of D'Alembert's principle, makes explicit use of the fact that constraint forces do not contribute directly to the governing equations of motion. It has been shown that Kane's formulation of the dynamical equations of motion is computationally more efficient than its counterpart, the Newton-Euler method (Kane and Levinson, 1983).

The governing equations of motion for any multijoint system can be expressed as

$$M(\mathbf{q})\ddot{\mathbf{q}} + C(\mathbf{q})\dot{\mathbf{q}}^2 + \mathbf{G}(\mathbf{q}) + R(\mathbf{q})\mathbf{F}^{MT} + \mathbf{E}(\mathbf{q},\dot{\mathbf{q}}) = \mathbf{0}, \tag{6.1}$$

where $\mathbf{q}, \dot{\mathbf{q}}, \ddot{\mathbf{q}}$ = vectors of the generalized coordinates, velocities, accelerations, respectively
$M(\mathbf{q})$ = system mass matrix
$M(\mathbf{q})\ddot{\mathbf{q}}$ = vector of inertial forces and torques
$C(\mathbf{q})\dot{\mathbf{q}}^2$ = vector of centrifugal and Coriolis forces and torques
$\mathbf{G}(\mathbf{q})$ = vector of gravitational forces and torques
$R(\mathbf{q})$ = matrix of muscle moment arms (see Sec. 6.4.3)
$\mathbf{F}^{MT}$ = vector of musculotendon forces
$R(\mathbf{q})\mathbf{F}^{MT}$ = vector of musculotendon torques
$\mathbf{E}(\mathbf{q}, \dot{\mathbf{q}})$ = vector of external forces and torques applied to the body by the environment

If the number of dof of the motor task is greater than, say, 4, a computer is needed to obtain Eq. (6.1) explicitly. A number of commercial software packages are available for this purpose including AUTOLEV by On-Line Dynamics Inc., SD/FAST by Symbolic Dynamics Inc., ADAMS by Mechanical Dynamics Inc., and DADS by CADSI.

6.4 MUSCULOSKELETAL GEOMETRY

6.4.1 Modeling the Paths of Musculotendinous Actuators

Two different methods are used to model the paths of musculotendinous actuators in the body: the straight-line method and the centroid-line method. In the straight-line method, the path of a musculotendinous actuator (muscle and tendon combined) is represented by a straight line joining the centroids of the tendon attachment sites (Jensen and Davy, 1975). Although this method is easy to implement, it may not produce meaningful results when a muscle wraps around a bone or another muscle (see Fig. 6.10). In the centroid-line method, the path of the musculotendinous actuator is represented as a line passing through the locus of cross-sectional centroids of the actuator (Jensen and Davy, 1975). Although the actuator's line of action is represented more accurately in this way, the centroid-line method can be difficult to apply because (1) it may not be possible to obtain the locations of the actuator's cross-sectional centroids for even a single position of the body, and (2) even if an actuator's centroid path is known for one position of the body, it is practically impossible to determine how this path changes as body position changes.

One way of addressing this problem is to introduce effective attachment sites or via points at specific locations along the centroid path of the actuator. In this approach, the actuator's line of action is defined by either straight-line segments or a combination of straight-line and curved-line segments between each set of via points (Brand et al., 1982; Delp et al., 1990). The via points remain fixed relative to the bones even as the joints move, and muscle wrapping is taken into account by making the via points active or inactive, depending on the configuration of the joint. This method works quite well when a muscle spans a 1-dof hinge joint, but it can lead to discontinuities in the calculated values of moment arms when joints have more than 1 rotational dof (Fig. 6.10).

6.4.2 Obstacle-Set Method

An alternative approach, called the obstacle-set method, idealizes each musculotendinous actuator as a frictionless elastic band that can slide freely over the bones and other actuators as the configuration of the joint changes (Garner and Pandy, 2000). The musculotendinous path is defined by a series of straight-line and curved-line segments joined together by via points, which may or may not be fixed relative to the bones.

To illustrate this method, consider the example shown in Fig. 6.11, where the path of the actuator is constrained by a single obstacle, which is fixed to bone A. (An *obstacle* is defined as any regular-shaped rigid body that is used to model the shape of a constraining anatomical structure such as a bone or another muscle.) The actuator's path is defined by the positions of the five fixed via points and the positions of the two obstacle via points. (A *fixed via point* remains fixed in a bone reference frame and is always active; an *obstacle via point* is not fixed in any reference frame, but is constrained to move on the surface of an underlying obstacle.) Via points P_1 and P_2 are fixed to bone A, while S_1, S_2, and S_3 are fixed to bone B. For a given configuration of the joint, the paths of all segments of the muscle are known, except those between the bounding-fixed via points, P_2 and S_3. The path of the actuator is known once the positions of the obstacle via points Q and T have been found.

Figure 6.12 shows a computational algorithm that can be used to model the centroid path of a musculotendinous actuator for a given configuration of the joints in the body. There are four steps in the computational algorithm. Given the relative positions of the bones, the locations of all fixed via points are known and can be expressed in the obstacle reference frame (Fig. 6.12, Step 1). The locations of the remaining via points in the actuator's path, the obstacle via points, can be calculated

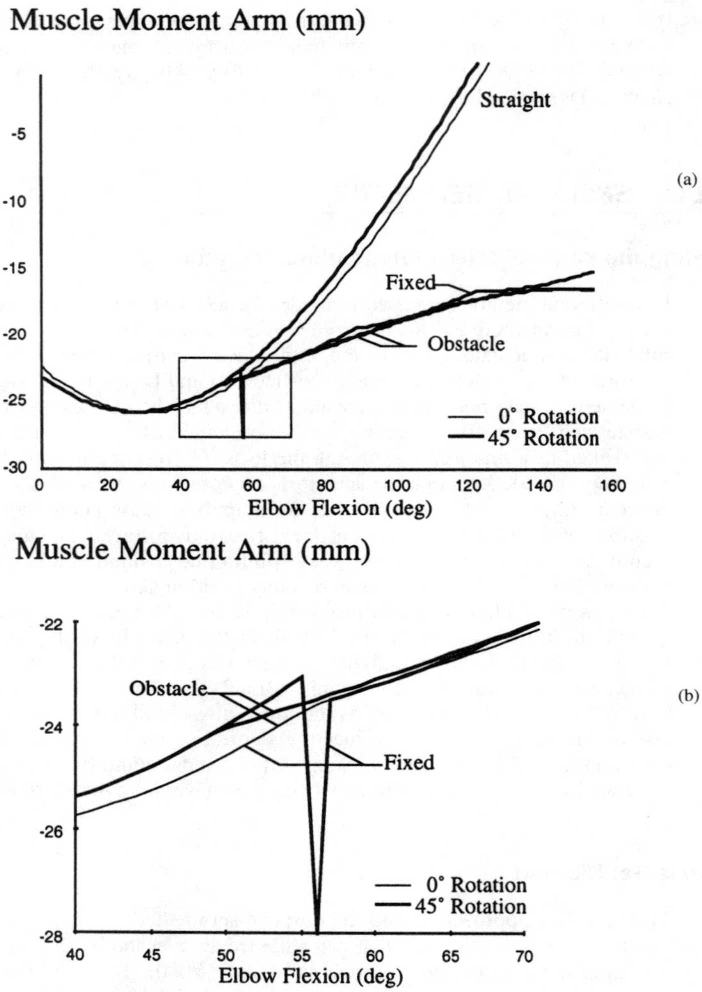

FIGURE 6.10 (*a*) Comparison of moment arms for the long head of triceps obtained using the straight-line model (Straight), fixed-via-point model (Fixed), and the obstacle-set model (Obstacle). For each model, the moment arms were calculated over the full range of elbow flexion, with the humerus positioned alongside the torso in neutral rotation (solid lines) and in 45° internal rotation (dotted lines). (*b*) Expanded scale of the graph above, where the moment arms obtained using the fixed-via-point and obstacle-set models are shown near the elbow flexion angle where the muscle first begins to wrap around the obstacle (cylinder) placed at the elbow [see part (*a*)]. The fixed-via-point model produces a discontinuity in moment arm when the shoulder is rotated 45° internally (Fixed, dotted line). [*Modified from Garner and Pandy (2000).*]

using one or more of the four different obstacle sets given in Appendices C to F in Garner and Pandy (2000) (Fig. 6.12, Step 2). Once the locations of the obstacle via points are known, a check is made to determine whether any of these points should be inactive (Fig. 6.12, Step 3). If an obstacle via point should be inactive, it is removed from the actuator's path and the locations of the remaining obstacle via points are then recomputed (Fig. 6.12, repeat Steps 2 and 3). Finally, the lengths of

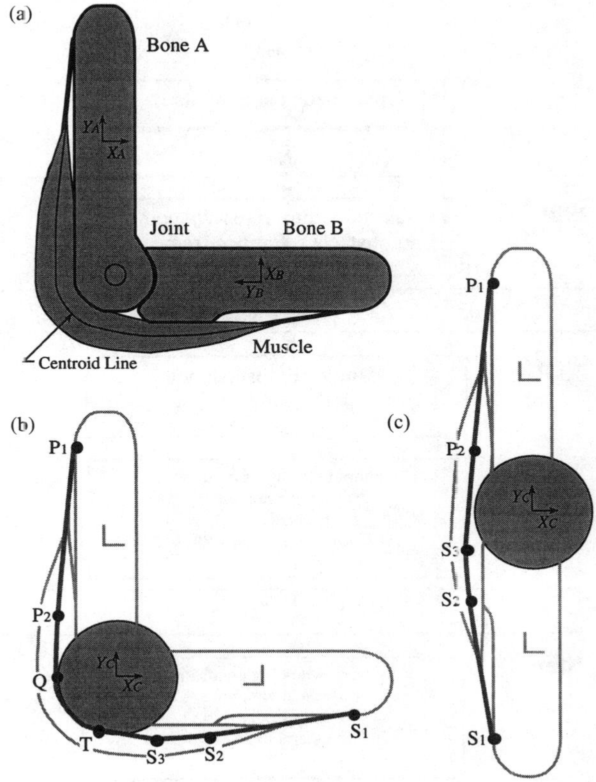

FIGURE 6.11 Schematic of a hypothetical muscle (*a*) represented by the obstacle-set model [(*b*) and (*c*)]. The muscle spans a single joint that connects bones A and B. Reference frames attached to the bones are used to describe the position and orientation of the bones relative to each other. At certain joint configurations, the muscle contacts the bones and causes the centroid path of the muscle to deviate from a straight line (*a*). The muscle path is modeled by a total of seven via points; five of these are fixed via points (P_1, P_2, S_1, S_2, S_3); the remaining two are obstacle via points (Q, T), which move relative to the bones and the obstacle. Points P1 and S1 are the origin and insertion of the muscle. Points P_2 and S_3 are bounding-fixed via points. The obstacle is shown as a shaded circle, which represents the cross section of a sphere or cylinder. An obstacle reference frame positions and orients the obstacle relative to the bones. The obstacle is made larger than the cross sections of the bones in order to account for the thickness of the muscle belly. The obstacle set is defined by the obstacle, the four via points (P_2, Q, T, S_3), and the segments of the muscle path between the via points P_2 and S_3. At some joint configurations, the muscle path can lose contact with the obstacle, and the obstacle via points then become inactive (*c*). See Garner and Pandy (2000) for details of the obstacle-set method. [*Modified from Garner and Pandy (2000).*]

the segments between all of the active via points along the actuator's path are computed (Fig. 6.12, Step 4).

The computational algorithm shown in Fig. 6.12 was used to model the paths of the triceps brachii muscle in the arm (Garner and Pandy, 2000). The muscle was separated into three segments:

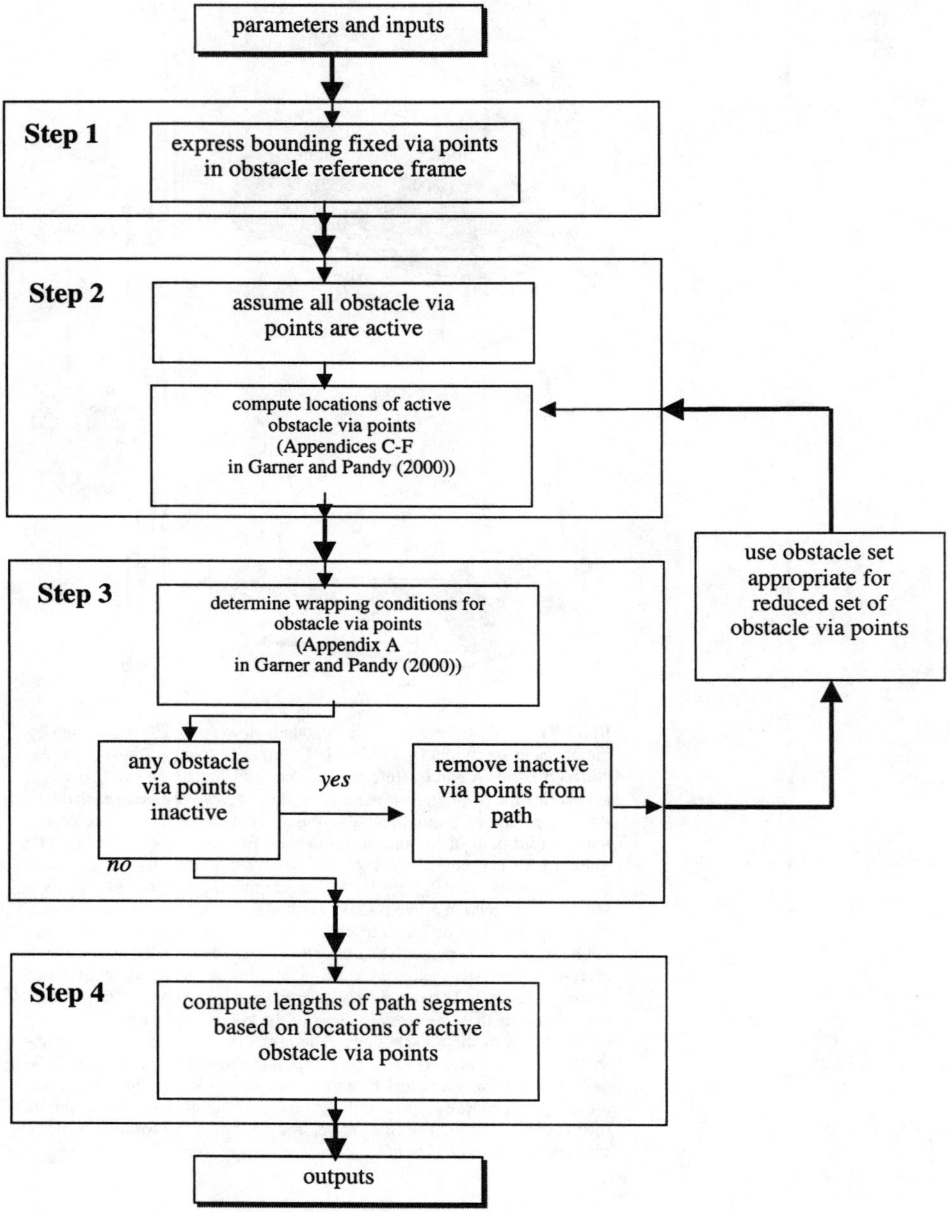

FIGURE 6.12 Flowchart of the obstacle-set algorithm. See text for details.

the medial, lateral, and long heads (Fig. 6.13). The medial head [Fig. 6.13 (1)] and lateral head [Fig. 6.13 (3)] were each modeled by a single-cylinder obstacle set (not shown in Fig. 6.13). The long head [Fig. 6.13 (2)] was modeled by a double-cylinder obstacle set as illustrated. The locations of the attachment sites of each muscle segment and the locations and orientations of the obstacles were chosen to reproduce the centroid paths of each head as accurately as possible. The geometry of the bones and the centroid paths of the muscle segments were obtained from three-dimensional reconstructions of high-resolution medical images obtained from the National Library of Medicine's Visible Human Male dataset (Garner and Pandy, 2001). Because the path of a muscle is not improperly constrained by contact with neighboring muscles and bones, the obstacle-set method produces not only accurate estimates of muscle moment arms, but also smooth moment arm–joint angle curves, as illustrated in Fig. 6.10.

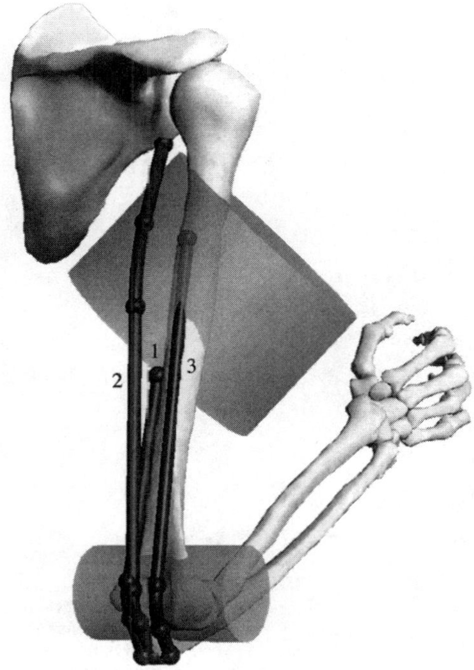

FIGURE 6.13 Posterolateral view of the obstacle-set model used to represent the paths of the triceps brachii in a model of the arm. The medial head (1) and lateral head (3) were each modeled by a single-cylinder obstacle set, which is not shown in the diagram. The long head (2) was modeled by a double-cylinder obstacle set as illustrated here. The locations of the attachment sites of the muscle and the locations and orientations of the obstacles were chosen to reproduce the centroid paths of each portion of the modeled muscle. The geometry of the bones and the centroid paths of the muscles were based on three-dimensional reconstructions of high-resolution medical images obtained from the National Library of Medicine's Visible Human Male dataset. [*Modified from Garner and Pandy (2000).*]

6.4.3 Muscle Moment Arms

Muscles develop forces and cause rotation of the bones about a joint. The tendency of a musculotendinous actuator to rotate a bone about a joint is described by the actuator's moment arm. Two methods are commonly used to measure the moment arm of an actuator—the geometric method and the tendon excursion method.

In the geometric method, a finite center of rotation is found by x-rays, computed tomography, or magnetic resonance imaging, and the moment arm is found by measuring the perpendicular distance from the joint center to the line of action of the muscle (Jensen and Davy, 1975). In the tendon excursion method, the change in length of the musculotendinous actu ator is measured as a function of the joint angle, and the moment arm is obtained by evaluating the slope of the actuator-length versus joint-angle curve over the full range of joint movement (An et al., 1983).

Consider two body segments, A and B, which articulate at a joint (Fig. 6.14). By considering the instantaneous power delivered by a musculotendinous actuator to body B, it can be shown that the moment arm of the actuator can be written as (Pandy, 1999):

$$\mathbf{r}^M = ({}^A\hat{\boldsymbol{\omega}}^B \cdot \mathbf{r}^{OQ} \times \hat{\mathbf{F}}^M)\,{}^A\hat{\boldsymbol{\omega}}^B \tag{6.2}$$

where $\mathbf{F}^M$ = musculotendinous force directed from the effective insertion of the actuator on body B (point Q) to the effective origin of the actuator on body A (point P)

${}^A\hat{\boldsymbol{\omega}}^B$ = unit vector that specifies the direction of the angular velocity of body B in body A [i.e., parallel to the instantaneous axis of rotation (ISA) of body B relative to body A]

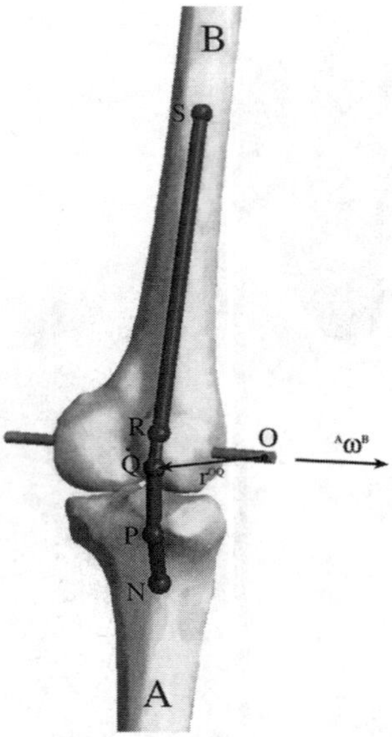

FIGURE 6.14 Two bodies, A and B, shown articulating at a joint. Body A is fixed in an inertial reference frame, and body B moves relative to it. The path of a generic muscle is represented by the origin S on B, the insertion N on A, and three intermediate via points P, Q, and R. Q and R are via points arising from contact of the muscle path with body B. Via point P arises from contact of the muscle path with body A. The ISA of B relative to A is defined by the angular velocity vector of B in A (${}^A\boldsymbol{\omega}^B$). [*Modified from Pandy (1999).*]

$\mathbf{r}^{OQ}$ = position vector directed from any point O on the ISA to any point Q on the line of action of the actuator between its origin and insertion sites

$\hat{\mathbf{F}}^M$ = unit vector in the direction of the actuator force F^M (see Fig. 6.14)

In Eq. (6.2), $\mathbf{r}^M$ is the moment-arm vector, which is equal to the *moment* of the musculotendinous force per unit of musculotendinous force. The direction of the moment-arm vector is along the instantaneous axis of rotation of the body B relative to body A whose direction is described by the unit vector ${}^A\hat{\boldsymbol{\omega}}^B$.

In the tendon excursion method, Euler's theorem on rotation is used to define an equivalent relation for the moment arm of a musculotendinous force. Thus, if every change in the relative orientation of body A and body B can be produced by means of a simple rotation of B in A about the ISA, then the moment arm of a musculotendinous force can also be written as (Pandy, 1999):

$$\mathbf{r}^M = -\frac{dl^M}{d\theta}\,{}^A\hat{\boldsymbol{\omega}}^B \tag{6.3}$$

where l^M = length of the musculotendinous actuator between the effective origin and insertion sites (P to Q in Fig. 6.14)

θ = angle associated with the simple rotation of body B about the ISA

$\dfrac{dl^M}{d\theta}$ = total derivative of musculotendinous length with respect to the angle of rotation θ of body B relative to body A

Equations (6.2) and (6.3) have the same geometric interpretation: the moment arm of a muscle force r^M, given either by ${}^A\hat{\boldsymbol{\omega}}^B \cdot \mathbf{r}^{OQ} \times \hat{\mathbf{F}}^M$ from Eq. (6.2) or by $dl^M/d\theta$ from Eq. (6.3), is equal to the perpendicular (shortest) distance between the ISA of B relative to A and the line of action of the muscle force, multiplied by the sine of the angle between these two lines (Pandy, 1999).

6.5 MUSCLE ACTIVATION AND CONTRACTION DYNAMICS

6.5.1 Muscle Activation Dynamics

Neural Excitation of Muscle. Voluntary contraction of human muscle initiates in the frontal motor cortex of the brain, where impulses from large pyramidal cells travel downward through corticospinal tracts that lead out to peripheral muscles. These impulses from the motor cortex are called *action potentials*, and each impulse is associated with a single motor neuron. The principle structure of a motor neuron is shown in Fig. 6.15. The action potential initiates in the cell body, or soma, and travels down a long efferent trunk, called the axon, at a rate of about 80 to 120 m/s. The action potential waveform is the result of a voltage depolarization-repolarization phenomenon across the neuron cell membrane. The membrane ionic potential at rest is disturbed by a surrounding stimulus, and Na^+ ions are allowed to momentarily rush inside. An active transport mechanism, called the $Na^+{-}K^+$ pump, quickly returns the transmembrane potential to rest. This sequence of events, which lasts about 1 ms, stimulates a succession of nerve impulses or waves that eventually reach muscle

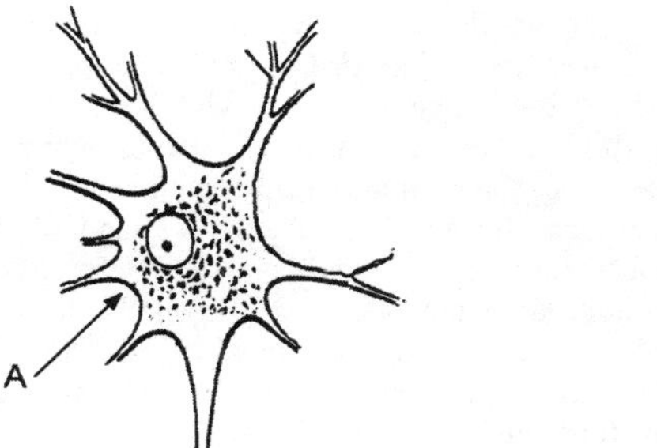

FIGURE 6.15 Motor neuron. (A) cell body, (B) axon, (C) neuromuscular junction, and (D) muscle fiber. The action potential travels downward, from the cell body towards the neuromuscular junction.

tissue. When an impulse reaches the muscle, it conducts over muscle tissue as a motor unit action potential (MUAP). Motor unit action potentials can have a variety of general shapes (Fang et al., 1997), as depicted in Fig. 6.16.

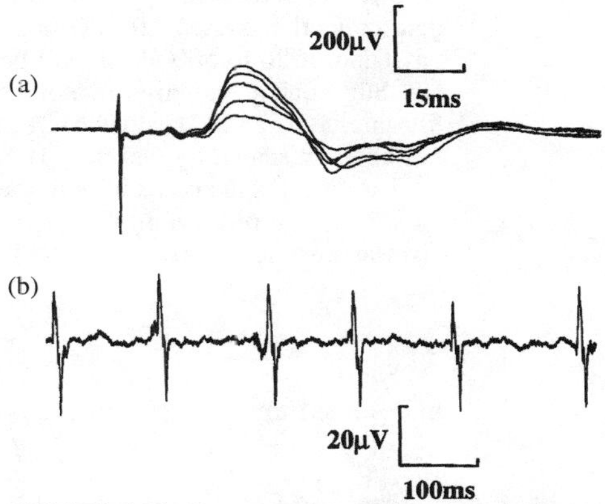

FIGURE 6.16 Representative motor unit action potentials (MUAPs). (*a*) Change in shapes due to increase in stimulation, (*b*) Sequence of MUAPs during voluntary contraction. [*Source: Fang et al., 1997.*]

The connecting site between the motor neuron and the muscle is called the *neuromuscular junction*. One motor neuron can have branches to many muscle fibers, and these together are called a motor unit. When the nerve impulse reaches the end of the nerve fiber, a neurotransmitter called *acetylcholine* is released into the motor end plate of the muscle. This in turn causes the release of Ca^{++} ions deep into the muscle fiber. The presence of Ca^{++} ions causes cross-bridges to be formed between actin and myosin filaments in the sarcomeres. The actin filaments slide inward along the myosin filaments, causing the muscle fiber to contract.

In addition to cortical motor neurons, control of the musculoskeletal system also relies on afferent receptors or sensory neurons that carry information from the periphery to the brain or spinal cord. The simplest nerve pathways lead directly from sensory neurons to motor neurons, and are known as *reflex arcs*. One such example is the withdrawal reflex. When skin receptors sense something is hot or sharp (a pin prick, for example), a sensory impulse is sent to the spinal cord, where an interneuron integrates the information and relays it to a motor neuron. The motor neuron in turn transmits a signal to the appropriate flexor muscle, which contracts and thus completes the reflex loop.

Muscle Electromyography. While motor unit action potentials form the cellular origin for muscle activation, they are normally not observable in routine clinical applications. The electromyogram (EMG) is the primary tool to study muscle activation in clinical and research settings using both surface and indwelling electrodes (Basmajian and DeLuca, 1985). The electromyogram is the summation of all motor unit action potentials at a given location during muscle contraction. Hence it represents a "gross" measure of the strength of a muscle contraction, since the number of muscle fibers contracting is directly related to the number of motor units firing.

The EMG appears as a random series of bursts that represent periods of muscle contraction and relaxation. Figure 6.17 shows a series of EMG bursts. The EMG signal is acquired by both invasive and noninvasive techniques. In invasive detection, a needle or fine wire is inserted directly into the muscle to a depth of several centimeters. In noninvasive techniques, also known as surface electromyography, a recording electrode is pasted onto the skin approximately halfway between the muscle's origin and insertion sites. In either case, because of the low voltage of the EMG signal (100 μV to several millivolts), some amplification of the signal is needed before it enters the digital data collection system. In addition, to reduce unwanted noise in the signal, an analog filter with passband of 20 to 500 Hz should be used. This necessitates a typical digital sampling rate of 1000 Hz. Subsequent reduction of noise in the EMG signal can be handled computationally by digital filtering (Barr and Chan, 1986). A set of 16 recommendations for EMG acquisition and processing has been presented by DeLuca (1997).

The EMG is analyzed in both the amplitude and frequency domains. For amplitude analysis, the most common processing technique is to compute a running root-mean-square (rms) of the signal. Over a short observational period T, the rms of the signal $x(t)$ can be expressed as

$$\mathrm{EMG}_{\mathrm{rms}} = \sqrt{\frac{1}{T}\int_0^T [x(t)]^2\, dt} \tag{6.4}$$

In discrete form, over N contiguous samples, the rms can be expressed as:

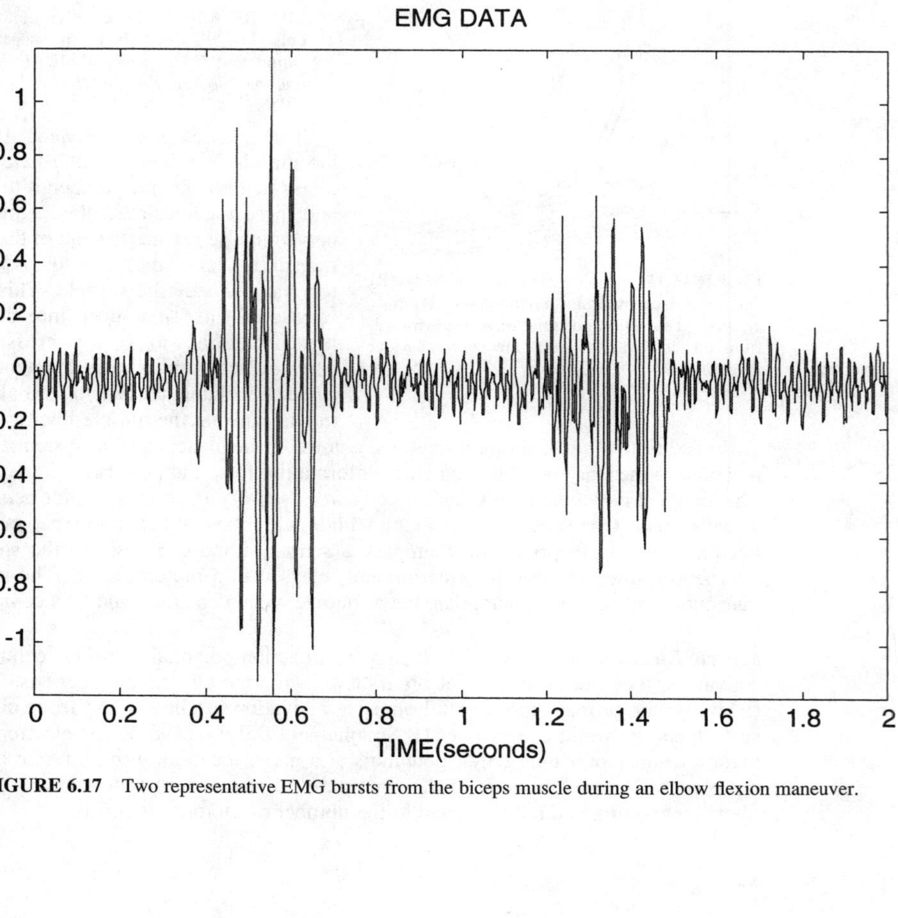

FIGURE 6.17 Two representative EMG bursts from the biceps muscle during an elbow flexion maneuver.

$$\text{EMG}_{\text{rms}} = \sqrt{\frac{1}{N}\sum_{i=1}^{N}[x_i(t)]^2} \tag{6.5}$$

An alternative, but similar, approach to EMG amplitude analysis is to rectify and then integrate the signal over the short period T (called *integrated EMG*, or IEMG). This approach has the advantage in that a simple electronic circuit can be used and no numerical computation is necessary. In either case, the objective is to obtain an estimation of the underlying muscle contraction force, which is roughly proportional to the EMG_{rms} or IEMG. In recent efforts to use the EMG signal as a myoelectric control signal (Evans et al., 1994), a three-stage digital processing approach has arisen; it is illustrated in Fig. 6.18. The process consists of (1) band-pass filtering (20 to 200 Hz) of the signal, (2) full-wave rectification, and (3) low-pass filtering (10-Hz cutoff) to produce a smooth signal.

In the frequency domain, the most common tool is to apply a fast Fourier transform (FFT) to the EMG signal. A typical result is shown in Fig. 6.19. One can see that most of the power in the signal is confined to the 20- to 200-Hz range, and that the center frequency or mean power point is typically below 100 Hz. One use of EMG frequency analysis is in the study of muscle fatigue. It has been shown that as the muscle fatigues, the median power frequency of the EMG shifts downward (Bigland-Ritchie et al., 1983).

Modeling Activation Dynamics. As noted is Sec. 6.5.1 ("Neural Excitation of Muscle"), muscle cannot activate or relax instantaneously. The delay between excitation and activation (or the development of muscle force) is due mainly to the time taken for calcium pumped out of the sarcoplasmic reticulum to travel down the T-tubule system and bind to troponin (Ebashi and Endo, 1968). This delay is often modeled as a first-order process (Zajac and Gordon 1989; Pandy et al., 1992):

$$\dot{a}^m = \left(\frac{1}{\tau_{\text{rise}}}\right)(u^2 - ua^m) + \left(\frac{1}{\tau_{\text{fall}}}\right)(u - a^m) \qquad u = u(t) \qquad a^m = a^m(t) \tag{6.6}$$

where u represents the net neural drive to muscle and a^m the activation level. Other forms of this relation are also possible; for example, an equation that is linear in the control u has been used to simulate jumping (Pandy et al., 1990) as well as pedaling (Raasch et al., 1997). Implicit in the formulation of Eq. (6.6) is the assumption that muscle activation depends only on a single variable u. Other models assume that a depends on two inputs, u_1 and u_2 say, which represent the separate effects of recruitment and stimulation frequency (Hatze, 1978). In simulations of multijoint movement, whether or not both recruitment and stimulation frequency are incorporated in a model of excitation-contraction (activation) dynamics is probably not as important as the values assumed for the time constants τ_{rise} and τ_{fall}. Values of these constants range from 12 to 20 ms for rise time, τ_{rise}, and from 24 to 200 ms for relaxation time, τ_{fall} (Pandy, 2001). Changes in the values of τ_{rise} and τ_{fall} within the ranges indicated can have a significant effect on predictions of movement coordination (F. C. Anderson and M. G. Pandy, unpublished results).

6.5.2 Muscle Contraction Dynamics

Mechanism of Muscle Contraction. Our understanding of how a muscle develops force is based on the sliding-filament theory of contraction. This theory, formulated by A. F. Huxley in 1957, proposes that a muscle shortens or lengthens because the thick and thin filaments slide past each other without the filaments themselves changing length (Huxley, 1957). The central tenet of the theory is that adenosine triphosphate (ATP) dependent interactions between thick filaments (myosin proteins) and thin filaments (actin proteins) generate a force that causes the thin filaments to slide past the thick filaments. The force is generated by the myosin heads of thick filaments, which form cross-bridges to actin thin filaments in the AI zone, where the two filaments overlap (see Fig. 6.2*e*). Subsequent structural changes in these cross-bridges cause the myosin heads to "walk" along an

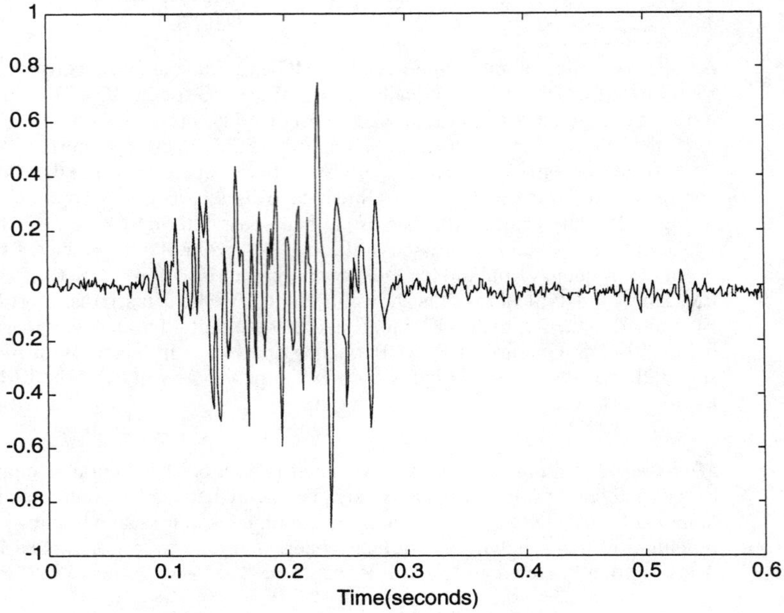

(a)

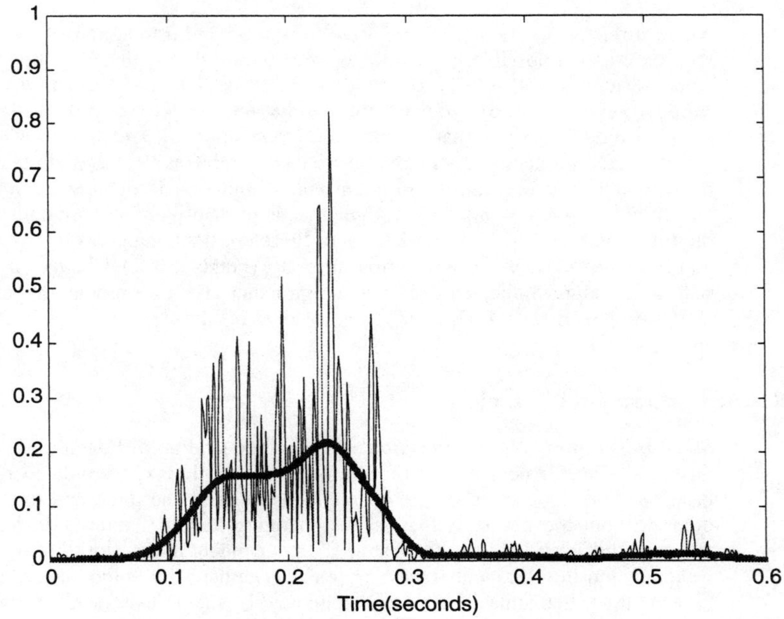

(b)

FIGURE 6.18 A method for processing EMG data for a myoelectric control signal. (*a*) the raw EMG data and (*b*) the signal after being rectified and low pass filtered (10 Hz cutoff).

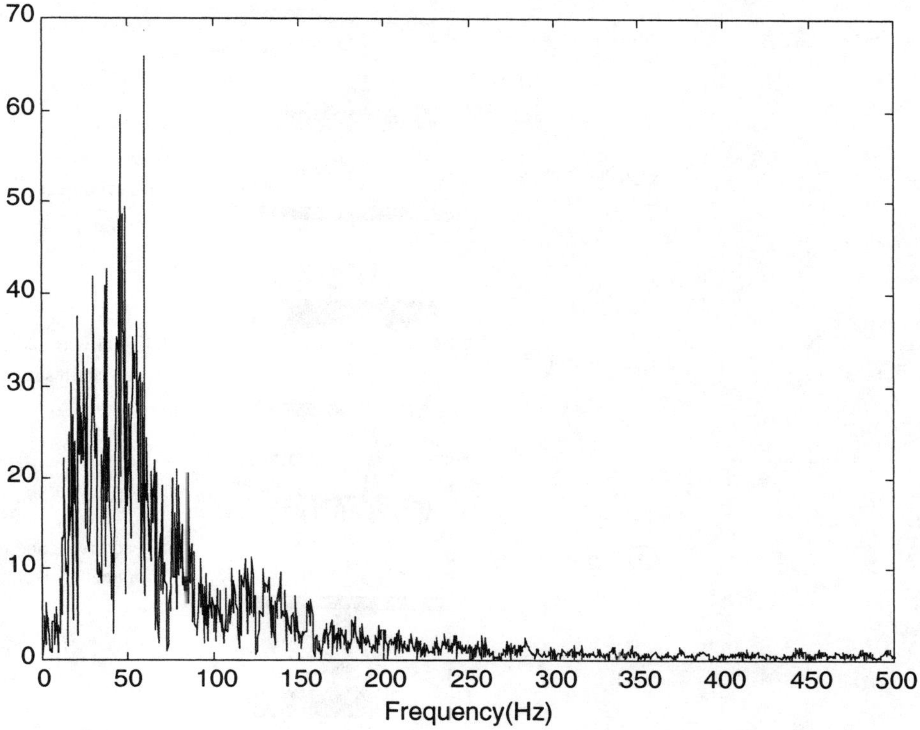

FIGURE 6.19 The fast Fourier transform (FFT) of an EMG sample. Most of the signal power is in the 20- to 200-Hz range.

actin filament. The sliding-filament theory correctly predicts that the force of contraction is proportional to the amount of overlap between the thick and thin filaments (see Fig. 6.3).

To understand the mechanism of muscle contraction, consider Figure 6.20, which shows the various steps involved in the interaction between one myosin head and a thin filament, also often referred to as the *cross-bridge cycle*. In the absence of ATP, a myosin head binds tightly to an actin filament in a "rigor" state. When ATP binds (Step 1), it opens the cleft in the myosin head, which weakens the interaction with the actin filament. The myosin head then reacts with ATP (Step 2), causing a structural change in the head that moves it to a new position, closer to the end of the actin filament, or Z disk, where it then rebinds to the filament. Phosphate is then released from the ATP-binding pocket on the myosin head (Step 3), and the head then undergoes a second structural change, called the power stroke, which restores myosin to its rigor state. Because the myosin head is bound to the actin filament, this second structural change exerts a force that causes the myosin head to move the actin filament (Fig. 6.20).

Modeling Contraction Dynamics. A. F. Huxley developed a mechanistic model to explain the structural changes at the sarcomere level that were seen under the electron microscope in the late 1940s and early 1950s. Because of its complexity, however, this (cross-bridge) model is rarely, if ever, used in studies of coordination. Instead, an empirical model, proposed by A. V. Hill, is used in virtually all models of movement to account for the force-length and force-velocity properties of muscle (Hill, 1938) (Fig. 6.21).

In a Hill-type model, muscle's force-producing properties are described by four parameters

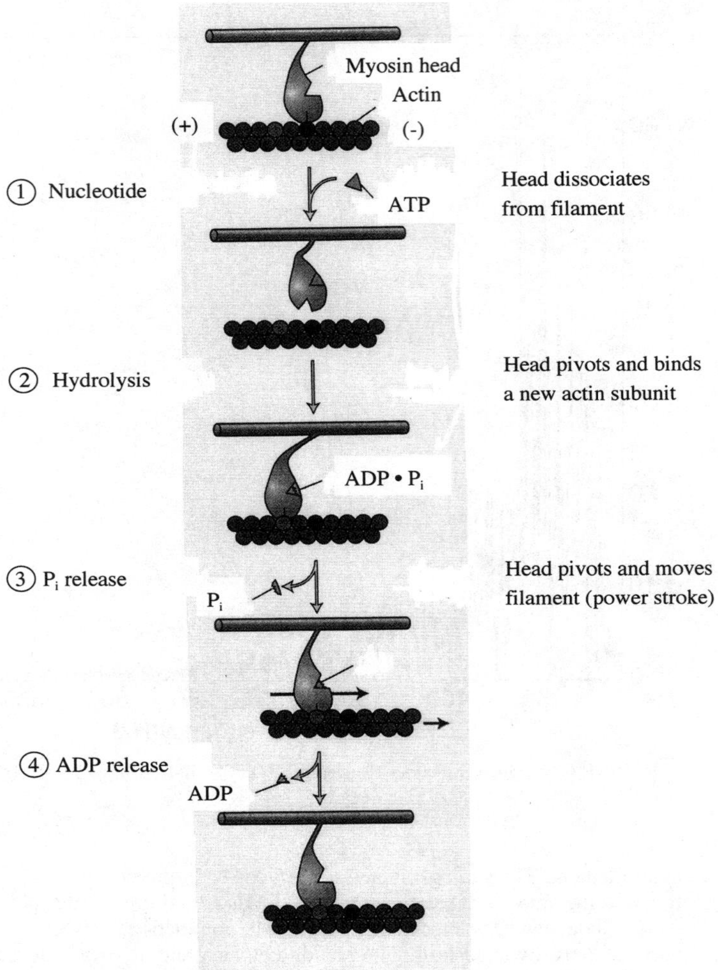

FIGURE 6.20 Schematic diagram illustrating the mechanism of force development in muscle. [*Modified from Lodish et al. (2000).*]

(Zajac, 1989): muscle's peak isometric force (F_o^m) and its corresponding fiber length (l_o^m) and pennation angle (α), and the intrinsic shortening velocity of muscle (v_{max}). F_o^m is usually obtained by multiplying muscle's physiological cross-sectional area by a generic value of specific tension. Values of optimal muscle fiber length, l_o^m, and α, the angle at which muscle fibers insert on tendon when the fibers are at their optimal length, are almost always based on data obtained from cadaver dissections (Freiderich and Brand, 1990). v_{max} is often assumed to be muscle independent; for example, simulations of jumping (Pandy et al., 1990), pedaling (Raasch et al., 1997), and walking (Anderson and Pandy, 2001b) assume a value of $v_{max} = 10\ s^{-1}$ for all muscles, which models the summed effect of slow, intermediate, and fast fibers (Zajac, 1989). Very few studies have examined the sensitivity of model simulations to changes in v_{max}, even though a change in the value of this parameter has been found to affect performance nearly as much as a change in the value of F_o^m (Pandy, 1990).

Tendon is usually represented as elastic (Pandy et al., 1990; Anderson and Pandy, 1993). Even though force varies nonlinearly with a change in length as tendon is stretched from its resting length

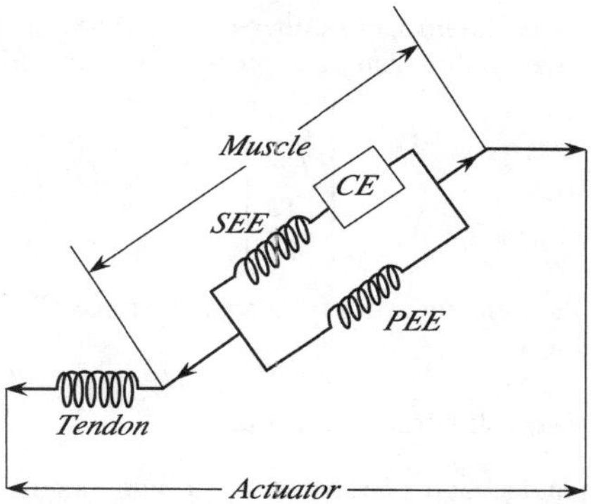

FIGURE 6.21 Schematic diagram of a model commonly used to simulate musculotendon actuation. Each musculotendon actuator is represented as a three-element muscle in series with an elastic tendon. The mechanical behavior of muscle is described by a Hill-type contractile element (CE) that models muscle's force-length-velocity property, a series-elastic element (SEE) that models muscle's active stiffness, and a parallel-elastic element (PEE) that models muscle's passive stiffness. The instantaneous length of the actuator is determined by the length of the muscle, the length of the tendon, and the pennation angle of the muscle. In this model the width of the muscle is assumed to remain constant as muscle length changes. [*Modified from Zajac and Gordon (1989) and Pandy et al. (1990).*]

l_s^T (see Fig. 6.6), a linear force-length curve is sometimes used (Anderson and Pandy, 1993). This simplification will overestimate the amount of strain energy stored in tendon, but the effect on actuator performance is not likely to be significant, because tendon force is small in the region where the force-length curve is nonlinear. Values of the four muscle parameters plus tendon rest length for a number of musculotendinous actuators in the human arm and leg can be found in Garner and Pandy (2001) and Anderson and Pandy (1999), respectively.

For the actuator shown in Fig. 6.21, musculotendon dynamics is described by a single, nonlinear, differential equation that relates musculotendon force (F^{MT}), musculotendon length (l^{MT}), musculotendon shortening velocity (v^{MT}), and muscle activation (a^m) to the time rate of change in musculotendon force:

$$\dot{F}^{MT} = f(F^{MT}, l^{MT}, v^{MT}, a^m) \qquad 0 \leq a^m \leq 1 \tag{6.7}$$

Given values of F^{MT}, l^{MT}, v^{MT}, and a^m at one instant in time, Eq. (6.7) can be integrated numerically to find musculotendon force at the next instant.

6.6 DETERMINING MUSCLE FORCE

6.6.1 Muscle Forces and Joint Torques

The torque of a musculotendinous force is equal to the magnitude of the actuator force F^{MT}, multiplied by the moment arm of the actuator r^{MT}. Thus, the torque exerted by actuator i about joint j is

$$T_j^{MT} = r_i^{MT} F_i^{MT} \tag{6.8}$$

where r_i^{MT} can be found from either Eq. (6.2) or Eq. (6.3). For a system with m actuators and n joints, the above relation can be expressed in matrix form [see also Eq. (6.1)]:

$$\begin{bmatrix} T_1^{MT} \\ T_2^{MT} \\ \cdot \\ T_n^{MT} \end{bmatrix} = \begin{bmatrix} r_{11} & r_{12} & \cdot & r_{1m} \\ r_{21} & r_{22} & \cdot & r_{2m} \\ \cdot & \cdot & \cdot & \cdot \\ r_{n1} & r_{n2} & \cdot & r_{nm} \end{bmatrix} \begin{bmatrix} F_1^{MT} \\ F_2^{MT} \\ \cdot \\ F_m^{MT} \end{bmatrix} \tag{6.9}$$

where r_{11} is the moment arm of the actuator force F_1^{MT} that exerts a torque T_1^{MT} about joint 1, etc.

6.6.2 Indeterminate Problem in Biomechanics

Both agonist and antagonist muscles contribute (unequally) to the net torque developed about a joint. In fact, for any given joint in the body, there are many more muscles crossing the joint than there are dof prescribing joint movement. The knee, for example, has at most 6 dof, yet there are at least 14 muscles that actuate this joint. One consequence of this arrangement is that the force developed by each muscle cannot be determined uniquely. Specifically, there are more unknown musculotendinous actuator forces than net actuator torques exerted about the knee; that is, $m > n$ in Eq. (6.9), which means that the matrix of muscle moment arms is not square and therefore not invertible. This is the so-called indeterminate problem in biomechanics, and virtually all attempts to solve it are based on the application of optimization theory (see also the Chap. 5 by Manal and Buchanan).

6.6.3 Inverse-Dynamics Method

In the inverse-dynamics method, noninvasive measurements of body motions (position, velocity, and acceleration of each segment) and external forces are used as inputs in Eq. (6.1) to calculate the net actuator torques exerted about each joint (see Fig. 6.22). This is a determinate problem because the number of net actuator torques is equal to the number of equations of motion of the system. Specifically, from Eq. (6.1), we can write

$$\mathbf{T}^{MT}(\mathbf{q}) = -[M(\mathbf{q})\ddot{\mathbf{q}} + C(\mathbf{q})\dot{\mathbf{q}}^2 + \mathbf{G}(\mathbf{q}) + \mathbf{E}(\mathbf{q},\dot{\mathbf{q}})] \tag{6.10}$$

where Eq. (6.9) has been used to replace $\mathbf{R}(\mathbf{q})\mathbf{F}^{MT}$ with $\mathbf{T}^{MT}(\mathbf{q})$ in Eq. (6.1). The right-hand side of Eq. (6.10) can be evaluated by noninvasive measurements of body motions ($\mathbf{q}$, $\dot{\mathbf{q}}$, $\ddot{\mathbf{q}}$) and external forces $\mathbf{E}(\mathbf{q}, \dot{\mathbf{q}})$. This means that all quantities on the left-hand side of Eq. (6.9) are known. The matrix of actuator moment arms on the right-hand side of Eq. (6.9) can also be evaluated if the origin and insertion sites of each musculotendinous actuator and the relative positions of the body segments are known at each instant during the movement (Sec. 6.4).

However, Eq. (6.9) cannot be solved for the m actuator forces because $m > n$ (i.e., the matrix of actuator moment arms is nonsquare). Static optimization theory is usually used to solve this indeterminate problem (Seireg and Arvikar, 1975; Hardt, 1978; Crowninshield and Brand, 1981). Here, a cost function is hypothesized, and an optimal set of actuator forces is found, subject to the equality constraints defined by Eq. (6.9) plus additional inequality constraints that bound the values of the actuator forces. If, for example, actuator stress is to be minimized, then the static optimization problem can be stated as follows (Seireg and Arvikar, 1975; Crowninshield and Brand, 1981): Find the set of actuator forces that minimizes the sum of the squares of actuator stresses:

$$J = \sum_{i=1}^{m} (F_i^{MT}/F_{oi}^{MT})^2 \tag{6.11}$$

subject to the equality constraints

$$\mathbf{T}^{MT} = R(\mathbf{q})\mathbf{F}^{MT} \tag{6.12}$$

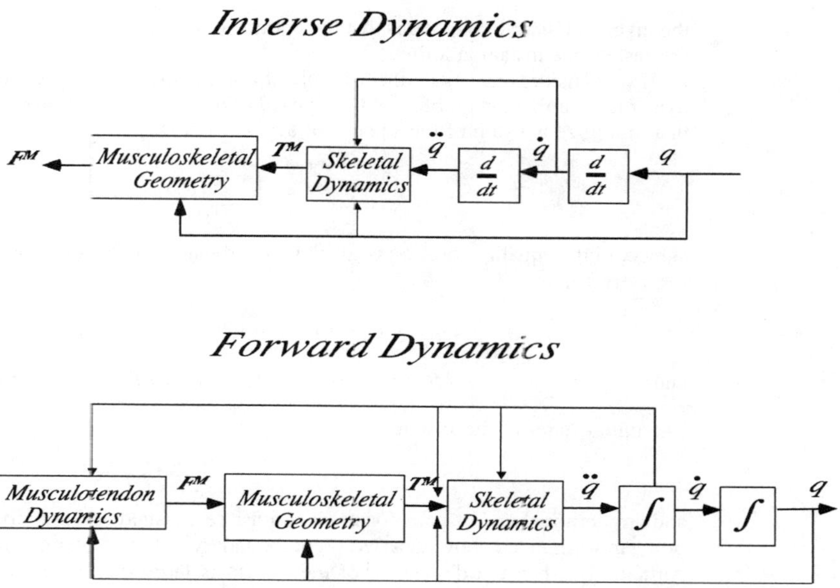

FIGURE 6.22 Comparison of the forward- and inverse-dynamics methods for determining muscle forces during movement. Top: Body motions are the inputs and muscle forces are the outputs in inverse dynamics. Thus, measurements of body motions are used to calculate the net muscle torques exerted about the joints, from which muscle forces are determined using static optimization. Bottom: Muscle excitations are the inputs and body motions are the outputs in forward dynamics. Muscle force (F^M) is an intermediate product (i.e., output of the model for musculotendon dynamics).

and the inequality constraints

$$0 \leqslant \mathbf{F}^{MT} \leqslant \mathbf{F}_0^{MT} \tag{6.13}$$

F_{oi}^{MT} is the peak isometric force developed by the ith musculotendinous actuator, a quantity that is directly proportional to the physiological cross-sectional area of the ith muscle. Equation (6.12) expresses the n relationships between the net actuator torques $\mathbf{T}^{MT}$, the matrix of actuator moment arms $\mathbf{R}(\mathbf{q})$, and the unknown actuator forces $\mathbf{F}^{MT}$. Equation (6.13) is a set of m equations that constrains the value of each actuator force to remain greater than zero and less than the peak isometric force of the actuator defined by the cross-sectional area of the muscle. Standard nonlinear programming algorithms can be used to solve this problem [e.g., sequential quadratic programming (Powell, 1978)].

6.6.4 Forward-Dynamics Method

Equations (6.1) and (6.7) can be combined to form a model of the musculoskeletal system in which the inputs are the muscle activation histories (**a**) and the outputs are the body motions ($\mathbf{q}, \dot{\mathbf{q}}, \ddot{\mathbf{q}}$) (Fig. 6.22). Measurements of muscle EMG and body motions can be used to calculate the time histories of the musculotendinous forces during movement (Hof et al., 1987; Buchanan et al., 1993). Alternatively, the goal of the motor task can be modeled and used, together with dynamic optimization theory, to calculate the pattern of muscle activations needed for optimal performance of the task (Hatze, 1976; Pandy et al., 1990; Raasch et al., 1997; Pandy, 2001). Thus, one reason why the forward-dynamics method is potentially more powerful for evaluating musculotendinous forces than the inverse-dynamics method is that the optimization is performed over a complete cycle of the task, not just at one instant at a time.

If we consider once again the example of minimizing muscle stress (cf. Sec. 6.6.3), an analogous dynamic optimization problem may be posed as follows. Find the time histories of all actuator forces that minimize the sum of the squares of actuator stresses:

$$J = \int_0^{t_f} \sum_{i=1}^{m} (F_i^{MT}/F_{oi}^{MT})^2 \tag{6.14}$$

subject to the equality constraints given by the dynamical equations of motion [Eqs. (6.7) and (6.1), respectively]:

$$\dot{F}^{MT} = f(F^{MT}, l^{MT}, v^{MT}, a^m) \qquad 0 \leq a^m \leq 1$$

and

$$M(q)\ddot{q} + C(q)\dot{q}^2 + G(q) + R(q)F^{MT} + E(q,\dot{q}) = 0$$

The initial states of the system,

$$x(0) = x_o \qquad x = \{q, \dot{q}, F^{MT}\} \tag{6.15}$$

and any terminal and/or path constraints, must be satisfied additionally. The dynamic optimization problem formulated above is a two-point, boundary-value problem, which is often difficult to solve, particularly when the dimension of the system is large (i.e., when the system has many dof and many muscles).

A better approach involves parameterizing the input muscle activations (or controls) and converting the dynamic optimization problem into a parameter optimization problem (Pandy et al., 1992). The procedure is as follows. First, an initial guess is assumed for the control variables **a**. The system dynamical equations [Eq. (6.7) and (6.1)] are then integrated forward in time to evaluate the cost function in Eq. (6.14). Derivatives of the cost function and constraints are then calculated and

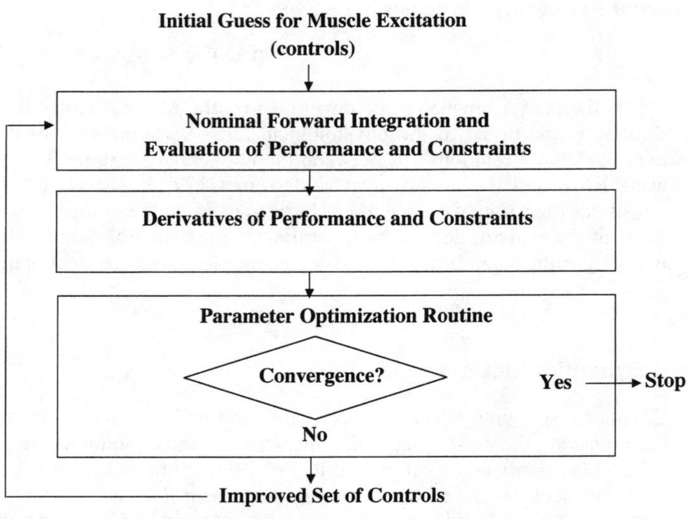

FIGURE 6.23 Computational algorithm used to solve dynamic optimization problems in human movement studies. The algorithm computes the muscle excitations (controls) needed to produce optimal performance (e.g., maximum jump height). The optimal controls are found by parameter optimization.

used to find a new set of controls that will improve the values of the cost function and the constraints in the next iteration (see Fig. 6.23). The computational algorithm shown in Fig. 6.23 has been used to find the optimal controls for vertical jumping (Anderson and Pandy, 1993), rising from a chair (Pandy et al., 1995), pedaling (Fregly and Zajac, 1996), and walking (Anderson and Pandy, 2001b).

If accurate measurements of body motions and external forces are available, then inverse dynamics should be used to determine musculotendinous forces during movement, because this method is much less expensive computationally. If, instead, the goal is to study how changes in body structure affect function and performance of a motor task, then the forward-dynamics method is preferred, for measurements of body motions and external forces are a priori not available in this instance.

6.7 MUSCLE, LIGAMENT, AND JOINT-CONTACT FORCES

Because muscle, ligament, and joint-contact forces cannot be measured noninvasively in vivo, estimates of these quantities have been obtained by combining mathematical models with either the inverse-dynamics or the forward-dynamics approach (Sec. 6.6). Below we review the levels of musculoskeletal loading incurred in the lower-limb during rehabilitation exercises, such as isokinetic knee extension, as well as during daily activity such as gait.

6.7.1 Knee Extension Exercise

The quadriceps is the strongest muscle in the body. This can be demonstrated by performing an isometric knee-extension exercise. Here, the subject is seated comfortably in a Cybex or Biodex dynamometer with the torso and thigh strapped firmly to the seat. The hip is flexed to 60°, and the leg is strapped to the arm of the machine, which can be either fixed or allowed to rotate at a constant angular velocity (see Fig. 6.24). Locking the machine arm in place allows the muscles crossing the

FIGURE 6.24 Photograph and schematic diagram showing the arrangement commonly used when people perform a knee-extension exercise on a Biodex or Cybex dynamometer. Notice that the strap fixed on the machine arm is attached distally (near the ankle) on the subject's leg.

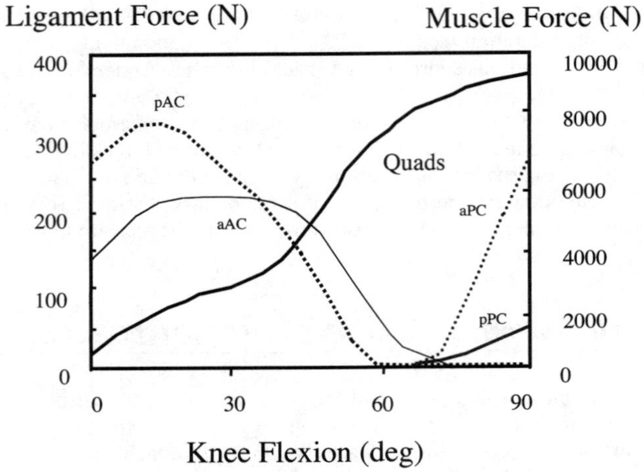

FIGURE 6.25 Muscle and knee-ligament forces incurred during a maximum isometric knee-extension exercise. The results were obtained from a two-dimensional mathematical model of the knee joint, assuming the quadriceps muscles are fully activated and there is no cocontraction in the flexor muscles of the knee (Shelburne and Pandy, 1997). The thick solid line represents the resultant force acting in the quadriceps tendon. The thin lines are the forces transmitted to the cruciate ligaments (aAC, black solid line; pAC, black dashed line; aPC, gray dashed line; pPC, gray solid line). The forces in the collateral ligaments are nearly zero. [*Modified from Shelburne and Pandy (1997).*]

knee to contract isometrically, because the knee angle is then held fixed. Under these conditions, the quadriceps muscles can exert up to 9500 N when fully activated. As shown in Fig. 6.25, peak isometric force is developed with the knee bent to 90°, and decreases as the knee is moved toward extension (Fig. 6.25, quads). Quadriceps force decreases as knee-flexion angle decreases because the muscle moves down the ascending limb of its force-length curve as the knee extends (see Fig. 6.3).

Quadriceps force increases monotonically from full extension and 90° of flexion, but the forces borne by the cruciate ligaments of the knee do not (Fig. 6.25, ACL). Calculations obtained from a mathematical model of the knee (Shelburne and Pandy, 1997; Pandy and Shelburne, 1997; Pandy and Sasaki, 1997) indicate that the ACL is loaded from full extension to 80° of flexion during knee-extension exercise. The model calculations also show that the resultant force in the ACL reaches 500 N at 20° of flexion, which is lower than the maximum strength of the human ACL (2000 N) (Noyes and Grood, 1976).

The calculations show further that load sharing within a ligament is not uniform. For example, the force borne by the anteromedial bundle of the ACL (aAC) increases from full extension to 20° of flexion, where peak force occurs, and aAC force then decreases as the knee flexion increases (Figure 6.25, aAC). The changing distribution of force within the ACL suggests that single-stranded reconstructions may not adequately meet the functional requirements of the natural ligament.

For isokinetic exercise, in which the knee is made to move at a constant angular velocity, quadriceps force decreases as knee-extension speed increases. As the knee extends more quickly, quadriceps force decreases because the muscle shortens more quickly, and, from the force-velocity property, an increase in shortening velocity leads to less muscle force (see Fig. 6.4). As a result, ACL force also decreases as knee-extension speed increases (Figure 6.26), because of the drop in shear force applied to the leg by the quadriceps (via the patellar tendon) (Serpas et al., in press).

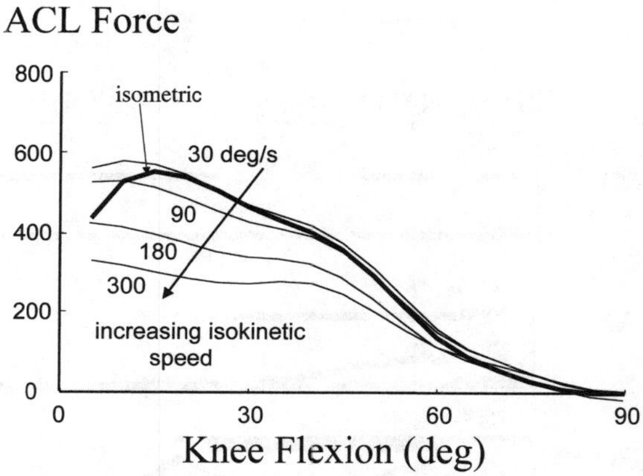

FIGURE 6.26 Resultant force in the ACL for isometric (thick line) and isokinetic (30, 90, 180, and 300 deg/sec) knee-extension exercises. The results were obtained from a two-dimensional model of the knee joint, assuming the quadriceps are fully activated and there is no cocontraction in the flexor muscles of the knee (Serpas et al., in press). The model results show that exercises in the studied speed range can reduce the force in the ACL by as much as one-half. [*Modified from Serpas et al. (in press).*]

The forces exerted between the femur and patella and between femur and tibia depend mainly on the geometry of the muscles that cross the knee. For maximum isometric extension peak forces transmitted to the patellofemoral and tibiofemoral joints are around 11,000 N and 6500 N, respectively (i.e., 15.7 and 9.3 times body weight, respectively) (Fig. 6.27). As the knee moves faster during isokinetic extension exercise, joint-contact forces decrease in dire proportion to the drop in quadriceps force (T. Yanagawa and M. G. Pandy, unpublished results).

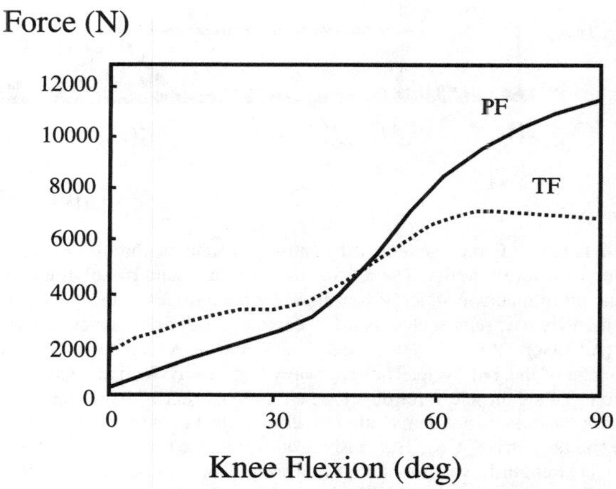

FIGURE 6.27 Resultant forces acting between the femur and patella (PF) and between the femur and tibia (TF) during maximum isometric knee-extension exercise. See Fig. 6.25 for details.

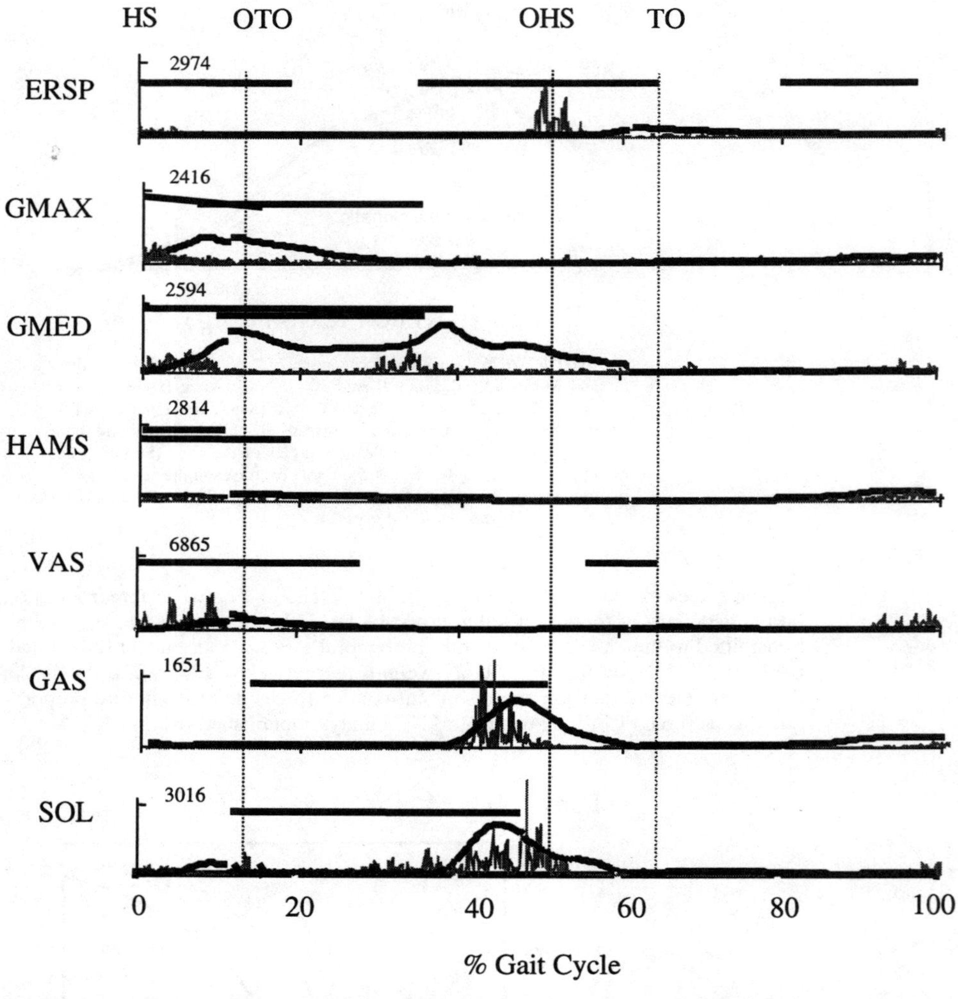

FIGURE 6.28 Forces generated by muscles spanning the hip, knee, and ankle (heavy black lines) during normal walking over level ground. The muscle forces were found by solving a dynamic optimization problem that minimized the amount of metabolic energy consumed by the muscles in the body per meter walked. In the model used to simulate gait, the body was represented as a 10-segment, 23-dof articulated linkage, and was actuated by 54 musculotendinous units (Anderson, 1999; Anderson and Pandy, 2001a; Anderson and Pandy, 2001b). The time scale is represented as a percentage of the gait cycle. The gray wavy lines are EMG data recorded from one subject, and are given to show that the timing of the muscle-force development is consistent with the measured EMG activity of the muscles. The following gait cycle landmarks are demarcated by thin vertical lines: opposite toe-off (OTO), opposite heel-strike (OHS), toe-off (TO), and heel-strike (HS). The muscle-force plots are scaled to each muscle's maximum isometric strength (e.g., peak isometric strength in vasti is 6865 N). Muscle abbreviations are as follows: erector spinae (ERSP), gluteus maximus (GMAX), gluteus medius (GMED), hamstrings (HAMS), vasti (VAS), gastrocnemius (GAS), soleus (SOL). [*Modified from Anderson and Pandy (2001a).*]

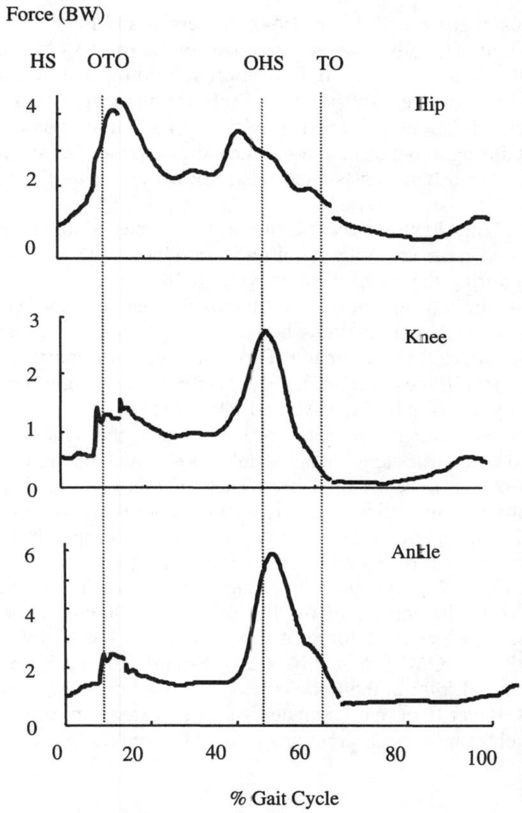

FIGURE 6.29 Joint contact forces acting at the hip, knee, and ankle during gait. The results were obtained by solving a dynamic optimization problem for normal walking (see Fig. 6.28 for details). [*Modified from Anderson and Pandy (2001a).*]

The results of Fig. 6.25 to 6.27 have significant implications for the design of exercise regimens aimed at protecting injured or newly reconstructed knee ligaments. For maximum isolated contractions of the quadriceps, Fig. 6.25 shows that the ACL is loaded at all flex angles less than 80°. Quadriceps-strengthening exercises should therefore be limited to flexion angles greater than 80° if the ACL is to be protected from load. In this region, however, very high contact forces can be applied to the patella (Fig. 6.27), so limiting this exercise to large flexion angles may result in patellofemoral pain. This scenario, known as the "paradox of exercise," is the reason why surgeons and physical therapists now prescribe so-called closed-chain exercises, such as squatting, for strengthening the quadriceps muscles subsequent to ACL reconstruction.

6.7.2 Gait

Muscle and joint loading are much lower during gait than during knee-extension exercise. Various studies have used inverse-dynamics or static optimization (Hardt, 1978; Crowninshield and Brand, 1981; Glitsch and Baumann, 1997) and forward-dynamics or dynamic optimization (Davy and Audu,

1987; Yamaguchi and Zajac, 1990; Anderson and Pandy, 2001b) to estimate muscle forces during normal gait. There is general agreement in the results obtained from these modeling studies.

Analysis of a dynamic optimization solution has shown that the hip, knee, and ankle extensors are the prime movers of the lower limb during normal walking. Specifically, gluteus maximus and vasti provide support against gravity during initial stance; gluteus medius provides most of the support during midstance; and soleus and gastrocnemius support and simultaneously propel the body forward during terminal stance (also known as *push-off*) (Anderson, 1999; Anderson and Pandy, 2001a).

The plantarflexors generate the largest forces of all the muscles in the leg during normal gait. Soleus and gastrocnemius, combined, produce peak forces of roughly 3000 N (or 4 times body weight) during the push-off phase (Fig. 6.28, SOL and GAS prior to OHS). By comparison, the hip extensors, gluteus maximus, and gluteus medius combined produce peak forces that are slightly less (around 2500 N or 3.5 times body weight during initial stance) (Fig. 6.28, GMAXL, GMAXM, GMEDP, and GMEDA near OTO). Finally, the quadriceps, vasti, and rectus femoris combined develop peak forces that are barely 2 times body weight near the transition from double support to single support (Fig. 6.28, VAS and RF at OTO).

Muscles dominate the forces transmitted to the bones for most activities of daily living. Thus, the loading histories applied at the ankle, knee, and hip are in close correspondence with the predicted actions of the muscles that cross each of these joints (Fig. 6.29). For example, the peak force transmitted by the ankle is considerably higher than the peak forces transmitted by the hip or knee, which is consistent with the finding that the ankle plantarflexors develop the largest forces during normal gait (compare joint-contact force at ankle in Fig. 6.29 with forces produced by SOL and GAS in Fig. 6.28). Similarly, peak hip-contact force is around 4 times body weight near OTO, which results from the actions of the hip extensors, gluteus maximus, and gluteus medius, at this time (compare joint-contact force at hip in Fig. 6.29 with forces developed by GMAXL, GMAXM, GMEDP, and GMEDA in Fig. 6.28). Finally, the peak articular-contact force transmitted by the tibio-femoral joint at the knee is less than 3 times body weight for gait. This level of joint loading is much lower than that estimated for knee-extension exercise, where forces approaching 10 times body weight have been predicted to act between the femur and tibia (Fig. 6.27).

REFERENCES

An, K. N., Ueba, Y., Chao, E. Y., et al. (1983). "Tendon excursion and moment arm of index finger muscles," *Journal of Biomechanics*, **16**:419–425.

Anderson, F. C. (1999). "A dynamic optimization solution for a complete cycle of normal gait," Ph.D. dissertation, University of Texas at Austin.

Anderson, F. C., and Pandy, M. G. (2001b). "Dynamic optimization of human walking," *Journal of Biomechanical Engineering*, **123**:381–390.

Anderson, F. C., and Pandy, M. G. (2001a). "Static and dynamic optimization solutions for gait are practically equivalent," *Journal of Biomechanics*, **34**:153–161.

Anderson, F. C., and Pandy, M. G. (1999). "A dynamic optimization solution for vertical jumping in three dimensions." *Computer Methods in Biomechanics and Biomedical Engineering*, **2**:201–231.

Anderson, F. C., and Pandy, M. G. (1993). "Storage and utilization of elastic strain energy during jumping." *Journal of Biomechanics*, **26**:1413–1427.

Barr, R., and Chan, E. (1986). "Design and implementation of digital filters for biomedical signal processing," *Journal of Electrophysiological Techniques*, **13**:73–93.

Basmajian, J., and DeLuca, C. (1985). *Muscles Alive: Their Function Revealed by Electromyography*, Williams and Wilkens, Baltimore.

Bigland-Ritchie, R., Johansson, R., Lippold, O., et al. (1983). "Changes in motor neuron firing rates during sustained maximal voluntary contractions," *Journal of Physiology*, **340**:335–346.

Brand, R. A., Crowninshield, R. D., Wittstock, C. E., et al. (1982). "A model of lower extremity muscular anatomy," *Journal of Biomechanical Engineering*, **104**:304–310.

Buchanan, T. S., Moniz, M. J., Dewald, J. P., and Rymer, W. Z. (1993). "Estimation of muscle forces about the wrist joint during isometric tasks using an EMG coefficient method," *Journal of Biomechanics*, **26**:547–560.

Butler, D. L., Grood, E. S., Noyes, F. R., and Zernicke, R. F. (1978). "Biomechanics of ligaments and tendons," *Exercise and Sport Sciences Reviews*, **6**:125–181.

Crowninshield, R. D., and Brand, R. A. (1981). "A physiologically-based criterion of muscle force prediction in locomotion," *Journal of Biomechanics*, **14**:793–801.

Davy, D. T., and Audu, M. L. (1987). "A dynamic optimization technique for predicting muscle forces in the swing phase of gait," *Journal of Biomechanics*, **20**:187–201.

Delp, S. L., Loan, J. P., Hoy, M. G., Zajac, F. E., Topp, E. L., and Rosen, J. M. (1990). "An interactive graphics-based model of the lower extremity to study orthopaedic surgical procedures," *IEEE Transactions on Biomedical Engineering*, **37**:757–767.

DeLuca, C. (1997). "The use of surface electromyography in biomechanics," *Journal of Applied Biomechanics*, **13**:135–163.

Ebashi, S., and Endo, M. (1968). "Calcium ion and muscle contraction," *Progress in Biophysics and Molecular Biology*, **18**:125–183.

Enoka, R. M. (1994). *Neuromechanical Basis of Kinesiology*, 2d ed., Human Kinetics, New York.

Evans, H., Pan, Z., Parker, P., and Scott, R. (1994). "Signal processing for proportional myoelectric control," *IEEE Transactions on Biomedical Engineering*, **41**:207–211.

Fang, J., Shahani, B., and Graupe, D. (1997). "Motor unit number estimation by spatial-temporal summation of single motor unit potentials," *Muscle and Nerve*, **20**:461–468.

Fregly, B. J., and Zajac, F. E. (1996). "A state-space analysis of mechanical energy generation, absorption, and transfer during pedaling," *Journal of Biomechanics*, **29**:81–90.

Friederich, J., and Brand, R. A. (1990). "Muscle fiber architecture in the human lower limb," *Journal of Biomechanics*, **23**:91–95.

Garner, B. A., and Pandy, M. G. (2001). "Musculoskeletal model of the human arm based on the Visible Human Male dataset," *Computer Methods in Biomechanics and Biomedical Engineering*, **4**:93–126.

Garner, B. A., and Pandy, M. G. (2000). "The obstacle-set method for representing muscle paths in musculoskeletal models," *Computer Methods in Biomechanics and Biomedical Engineering*, **3**:1–30.

Glitsch, U., and Baumann, W. (1997). "The three-dimensional determination of internal loads in the lower extremity," *Journal of Biomechanics*, **30**:1123–1131.

Hardt, D. E. (1978). "Determining muscle forces in the leg during human walking: An application and evaluation of optimization methods," *Journal of Biomechancal Engineering*, **100**:72–78.

Hatze, H. (1978). "A general myocybernetic control model of skeletal muscle," *Biological Cybernetics*, **28**:143–157.

Hatze, H. (1976). "The complete optimization of human motion," *Mathematical Biosciences*, **28**:99–135.

Hayes, W. C., and Bodine, A. J. (1978). "Flow-independent viscoelastic properties of articular cartilage matrix," *Journal of Biomechanics*, **11**:407–419.

Hayes W. C., and Mockros, L. F. (1971). "Viscoelastic properties of human articular cartilage." *Journal of Applied Physiology*, **31**:562–568.

Hill, A. V. (1938). "The heat of shortening and the dynamic constants of muscle," *Proceedings of the Royal Society (London)*, Series B, **126**:136–195.

Hof, A. L., Pronk, C. A. N., J. A. van Best, (1987). "Comparison between EMG to force processing and kinetic analysis for the calf muscle moment in walking and stepping," *Journal of Biomechanics*, **20**:167–178.

Hori, R. Y., Mockros, L. F. (1976). "Indention tests of human articular cartilage," *Journal of Applied Physiology*, **31**:562–568.

Huxley, A. F. (1957). "Muscle structure and theories of contraction," *Progress in Biophysics and Biophysical Chemistry*, **7**:255–318

Jensen, R. H., and Davy, D. T. (1975). "An investigation of muscle lines of action about the hip: A centroid line approach vs the straight line approach," *Journal of Biomechanics*, **8**:103–110.

Kane, T. R., and Levinson, D. A. (1983). "The use of Kane's dynamical equations in robotics," *International Journal of Robotics Research*, **2**:3–21.

Kane, T. R., and Levinson, D. A. (1985). *Dynamics: Theory and Applications*, McGraw-Hill, New York.

Kempson, G. E. (1980) *The Joints and Synovial Fluid*, Vol. **2**, Chap. 5, pp. 177–238, Academic Press, New York.

Lodish, H., Berk, A., Zipursky, S. L., et al. (2000). *Molecular Cell Biology*, 4th ed., W. H. Freeman and Company, New York.

McMahon, T. A. (1984). *Muscles, Reflexes, and Locomotion*, Princeton University Press, New Jersey.

Mak, A. F. (1986). "The apparent viscoelastic behavior of articular cartilage: The contributions from the intrinsic matrix viscoelasticity and interstitial fluid flow," *Journal of Biomechanical Engineering*, **108**:123–130.

Mow, V. C., Holmes, M. H., and Lai, W. M. (1984). "Fluid transport and mechanical properties of articular cartilage: a review," *Journal of Biomechanics*, **17**:377–394.

Noyes, F. R., and Grood, E. S. (1976). "The strength of the anterior cruciate ligament in humans and rhesus monkeys," *Journal of Bone and Joint Surgery*, **58-A**:1074–1082.

Pandy, M. G. (2001). "Computer modeling and simulation of human movement," *Annual Review of Biomedical Engineering*, **3**:245–273.

Pandy, M. G. (1999). "Moment arm of a muscle force," *Exercise and Sport Sciences Reviews*, **27**:79–118.

Pandy, M. G., and Sasaki, K. (1998). "A three-dimensional musculoskeletal model of the human knee joint. Part II: Analysis of ligament function." *Computer Methods in Biomechanics and Biomedical Engineering*, **1**:265–283.

Pandy, M. G., Sasaki, K., and Kim, S. (1997). "A three-dimensional musculoskeletal model of the human knee joint. Part I: Theoretical construction." *Computer Methods in Biomechanics and Biomedical Engineering*, **1**: 87–108.

Pandy, M. G., and Shelburne, K. B. (1997). "Dependence of cruciate-ligament loading on muscle forces and external load," *Journal of Biomechanics*, **30**:1015–1024.

Pandy, M. G., Garner, B. A., and Anderson, F. C. (1995). "Optimal control of non-ballistic muscular movements: A constraint-based performance criterion for rising from a chair." *Journal of Biomechanical Engineering*, **117**: 15–26.

Pandy, M. G., Anderson, F. C., and Hull, D. G. (1992). "A parameter optimization approach for the optimal control of large-scale musculoskeletal systems," *Journal of Biomechanical Engineering*, **114**:450–460.

Pandy, M. G., and Zajac, F. E. (1991). "Optimal muscular coordination strategies for jumping," *Journal of Biomechanics*, **24**:1–10.

Pandy, M. G., Zajac, F. E., Sim, E., and Levine, W. S. (1990). "An optimal control model for maximum-height human jumping." *Journal of Biomechanics*, **23**:1185–1198.

Pandy, M. G. (1990). "An analytical framework for quantifying muscular action during human movement," in Winters J. M., and Woo, S. L-Y. (eds.), *Multiple Muscle Systems—Biomechanics and Movement Organization*, Springer-Verlag, New York, pp. 653–662.

Pandy, M. G., and Berme, N. (1988). "A numerical method for simulating the dynamics of human walking," *Journal of Biomechanics*, **21**:1043–1051.

Powell, M. J. D. (1978). "A fast algorithm for nonlinearly constrained optimization calculations," in Matson, G. A. (ed.): *Numerical Analysis: Lecture Notes in Mathematics*, Vol. 630, pp. 144–157, Springer-Verlag, New York.

Raasch, C. C., Zajac, F. E., Ma, B., and Levine, W. S. (1997). Muscle coordination of maximum-speed pedaling. *Journal of Biomechanics* **6**:595–602.

Seireg, A., and Arvikar, R. J. (1973). "A mathematical model for evaluation of forces in the lower extremities of the musculoskeletal system," *Journal of Biomechanics*, **6**:313–326.

Serpas, F., Yanagawa, T., and Pandy, M. G. (2002). "Forward-dynamics simulation of anterior cruciate ligament forces developed during isokinetic dynamometry." *Computer Methods in Biomechanics and Biomedical Engineering*, **5**:33–43.

Shelburne, K. B., and Pandy, M. G. (1998). "Determinants of cruciate-ligament loading during rehabilitation exercise," *Clinical Biomechanics*, **13**:403–413.

Shelburne, K. B., and Pandy, M. G. (1997). "A musculoskeletal model of the knee for evaluating ligament forces during isometric contractions," *Journal of Biomechanics*, **30**:163–176.

Yamaguchi, G. T., and Zajac, F. E. (1990). "Restoring unassisted natural gait to paraplegics via functional neuromuscular stimulation: A computer simulation study." *IEEE Transactions on Biomedical Engineering*, **37**: 886–902.

Zajac, F. E. (1993). "Muscle coordination: a perspective," *Journal of Biomechanics*, **26** (suppl. 1):109–124.

Zajac F. E., and Gordon, M. E. (1989). "Determining muscle's force and action in multi-articular movement," *Exercise and Sport Sciences Reviews*, **17**:187–230.

CHAPTER 7

BIODYNAMICS: A LAGRANGIAN APPROACH

Donald R. Peterson
Biodynamics Laboratory at the Ergonomic Technology Center, University of Connecticut Health Center, Farmington, Connecticut

Ronald S. Adrezin
University of Hartford, College of Engineering, West Hartford, Connecticut

7.1 MOTIVATION

Athletic performance, work environment interaction, and medical sciences involving rehabilitation, orthotics, prosthetics, and surgery rely heavily on the analysis of human performance. Analyzing human motion allows for a better understanding of the anatomical and physiological processes involved in performing a specific task, and is an essential tool for accurately modeling intrinsic and extrinsic biodynamic behaviors (Peterson, 1999). Analytical dynamic models (equations of motion) of human movement will assist the researcher in identifying key forces, movements, and movement patterns to measure, providing a base or fundamental model from which an experimental approach can be determined and the efficacy of initially obtained data can be evaluated. At times, the fundamental model may be all that is required if laboratory or field-based measurements prove to be costly and impractical. Finally, it permits the engineer to alter various assumptions and/or constraints of the problem and compare the respective solutions, ultimately gaining an overall appreciation for the nature of the dynamics.

As an example, consider the motion of an arm-forearm system illustrated in Fig. 7.1. The corresponding equation of motion for the elbow joint (point C), or a two-link, multibody system is given in Eq. (7.1).

$$I\ddot{\theta} + mgl\cos\theta + M_{\text{Applied}} = M_{\text{Elbow}} \tag{7.1}$$

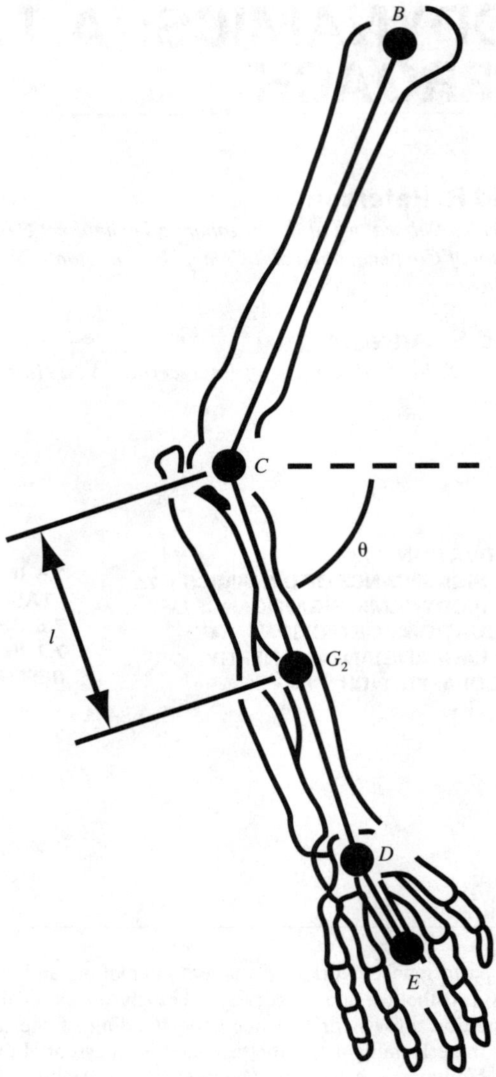

FIGURE 7.1 The bones of the arm, forearm, and hand with line segments superimposed. (Note that the line segments and points coincide with Figs. 7.6, 7.7, and 7.8. Point G_2 represents the center of gravity for segment CD, or the forearm.)

where I = mass moment of inertia of the forearm
m = mass of the forearm
M_{Elbow} = moment at the elbow created by the muscles
M_{Applied} = moment due to externally applied loads
l = distance from the elbow to the center of mass (G_2) of the forearm
θ = angle between the forearm and the horizontal plane

By solving for the moment created by the muscles at the elbow, an estimation of the effort required during a forearm movement activity can be determined.

In order to yield estimates for each term within the equations of motion for a biodynamic system, an experimental approach involving the simultaneous collection of information from several modalities would be required. More specifically, sensors such as accelerometers (to measure acceleration or vibration), inclinometers (to measure displacement or position), load cells or force sensitive resistors (to measure applied load or grip forces), and electrogoniometers (to measure anatomic angles) should be considered. The selection of modalities, and their respective sensors, will depend upon the nature of the terms given in the equations of motion. In addition, estimates of anthropometric quantities involving mass, segment length, location of the center of mass, and the mass moment of inertia of each body segment are also required. It should be understood that these modalities might be costly to purchase and operate, and can generate large volumes of data that must be analyzed and modeled properly in order to yield practical estimates for each desired term. The results for each term can then be substituted into the equation of motion to model the dynamic behavior of the system. This approach to modeling the biodynamics of the human musculoskeletal system has proved to be extremely valuable for investigating human motion characteristics in settings of normal biomechanical function and in settings of disease (Peterson, 1999).

This chapter presents a straightforward approach to developing dynamic analytical models of multirigid body systems that are analogous to actual anatomic systems. These models can yield overall body motion and joint forces from estimated joint angles and applied loads, and even begin to structure dynamic correlations such as those between body segment orientations and body segment kinematics and/or kinetics. The applications of these equations to clinical or experimental scenarios will vary tremendously. It is left to the reader to utilize this approach, with the strong suggestion of reviewing the current literature to identify relevant correlations significant to their applications. Listing and discussing correlations typically used would be extensive and beyond the scope of this chapter.

7.2 THE SIGNIFICANCE OF DYNAMICS

The theory and applications of engineering mechanics is not strictly limited to nonliving systems. The principles of statics and dynamics, the two fundamental components within the study of engineering mechanics, can be applied to any biological system. They have proved to be equally effective in yielding a relatively accurate model of the mechanical state of both intrinsic and extrinsic biological structures (Peterson, 1999). In fact, nearly all of the dynamic phenomena observed within living and nonliving systems can be modeled by using the principles of rigid body kinematics and dynamics. Most machines and mechanisms involve multibody systems where coupled dynamics exist between two or more rigid bodies. Mechanized manipulating devices, such as a robotic armature, are mechanically analogous to the human musculoskeletal system, which is an obvious multibody system. Consequently, the equations of motion governing the movement of the armature will closely resemble the equations derived for the movement of an extremity (i.e., arm or leg).

Careful steps must be taken in structuring the theoretical approach, particularly in identifying the initial assumptions required to solve the equations of motion. By varying any of the initial assumptions (and the justification for making them), the accuracy of the model is directly affected. For example, modeling a human shoulder joint as a mechanical ball and socket neglects the shoulder's ability to translate and thus prohibits any terms describing lateral motion exhibited by the joint. Therefore, any such assumption or constraint should be clearly defined on the basis of the desired approach or theoretical starting point for describing the dynamics of the system. Elasticity is another example of such a constraint, and is present to some degree within nearly all of the dynamics of a biological system. Ideally, elasticity should not be avoided and will have direct implications on determining the resulting dynamic state of the system.

7.3 THE BIODYNAMIC SIGNIFICANCE OF THE EQUATIONS OF MOTION

The equations of motion are dynamic expressions relating kinematics with forces and moments. In a musculoskeletal biodynamic system, the forces and moments will consist of joint reactions; internal forces, such as muscle, tendon, or ligament forces; and/or externally applied loads. Consequently, the equations of motion can provide a critical understanding of the forces experienced by a joint and effectively model normal joint function and joint injury mechanics. They can yield estimates for forces that cannot be determined by direct measurement. For example, muscle forces, which are typically derived from other quantities such as external loads, center of mass locations, and empirical data including anatomical positioning and/or electromyography, can be estimated.

In terms of experimental design, the equations of motion can provide an initial, theoretical understanding of an actual biodynamic system and can aid in the selection of the dynamic properties of the actual system to be measured. More specifically, the theoretical model is an initial basis that an experimental model can build upon to determine and define a final predictive model. This may involve comparative and iterative processes used between the theoretical and actual models, with careful consideration given to every assumption and defined constraint.

Biodynamic models of the human musculoskeletal system have direct implications on device/tool design and use and the modeling of normal and/or abnormal (or undesired) movements or movement patterns (the techniques with which a device or tool is used). Applications of the models can provide a better understanding for soft and hard tissue injuries, such as repetitive strain injuries (RSI), and can be used to identify and predict the extent of a musculoskeletal injury (Peterson, 1999).

7.4 THE LAGRANGIAN (AN ENERGY METHOD) APPROACH

The equations of motion for a dynamic system can be determined by any of the following four methods:

1. Newton-Euler method
2. Application of Lagrange's equation
3. D'Alembert's method of virtual work
4. Principle of virtual power using Jourdain's principle or Kane's equation

Within this chapter, only the first two methods are discussed. For a detailed discussion of other methods, consult the references.

The Newton-Euler (Newtonian) approach involves the derivation of the equations of motion for a dynamic system using the Newton-Euler equations, which depend upon vector quantities and accelerations. This dependence, along with complex geometries, may promote derivations for the equations of motion that are timely and mathematically complex. Furthermore, the presence of several degrees of freedom within the dynamic system will only add to the complexity of the derivations and final solutions.

The energy method approach uses Lagrange's equation (and/or Hamilton's principle, if appropriate) and differs from the newtonian approach by the dependence upon scalar quantities and velocities. This approach is particularly useful if the dynamic system has several degrees of freedom and the forces experienced by the system are derived from potential functions. In summary, the energy method approach often simplifies the derivation of the equations of motion for complex multibody systems involving several degrees of freedom as seen in human biodynamics.

Within this section, several applications of the Lagrangian approach are presented and discussed. In particular, Lagrange's equation is used to derive the equations of motion for several dynamic systems that are mechanically analogous to the musculoskeletal system. A brief introduction of Lagrange's equation is provided, however, the derivation and details are left to other sources identified in the references. Also note that an attempt was made to be consistent with the symbols used from figure to figure to allow the reader to correlate the demonstrated examples.

7.4.1 Brief Introduction to Lagrange's Equation

The application of Lagrange's equation to a model of a dynamic system can be conceptualized in six steps (Baruh, 1999):

1. Draw all free-body diagrams.
2. Determine the number of degrees of freedom in the system and select appropriate independent generalized coordinates.
3. Derive the velocity for the center of mass of each body and any applicable virtual displacements.
4. Identify both conservative and nonconservative forces.
5. Calculate the kinetic and potential energy and the virtual work.
6. Substitute these quantities into Lagrange's equation and solve for the equations of motion.

A system determined to have n degrees of freedom would correspondingly have n generalized coordinates denoted as q_k, where k may have values from 1 to n. A generalized, nonconservative force corresponding to a specific generalized coordinate is represented by Q_k, where k again may range from 1 to n. The derivative of q_k with respect to time is represented as $\dot{q}_k$. Equation (7.2) shows the general form of Lagrange's equation.

$$\frac{d}{dt}\left(\frac{\partial L}{\partial \dot{q}_k}\right) - \frac{\partial L}{\partial q_k} = Q_k \qquad k = 1,2, \ldots ,n \tag{7.2}$$

The Lagrangian L is defined as the difference between the kinetic energy T and the potential energy V:

$$L = T - V \tag{7.3}$$

After determining the Lagrangian, differentiating as specified in (7.2) will result in a set of n scalar equations of motion due to the n degrees of freedom.

Since the Lagrangian approach yields scalar equations, it is seen as an advantage over a Newtonian approach. Only the velocity vector $\mathbf{v}$, not the acceleration, of each body is required and any coordinate system orientation desired may be chosen. This is a result of the kinetic energy expressed in terms of a scalar quantity as demonstrated in Eq. (7.4).

$$\begin{aligned} T &= T_{\text{Translation}} + T_{\text{Rotation}} \\ T_{\text{Translation}} &= \tfrac{1}{2}m(\mathbf{v} \cdot \mathbf{v}) = \tfrac{1}{2}mv^2 \\ T_{\text{Rotation}} &= \tfrac{1}{2}\{\omega\}^T\{I_G\}\{\omega\} \end{aligned} \tag{7.4}$$

For certain problems where constraint forces are considered, a Newtonian approach, or the application of both techniques, may be necessary. The following sections will present some applications of Lagrange's equation as applicable to human anatomical biodynamics. Other mechanically based examples are easily found within the cited literature (Baruh, 1999; Moon, 1998; Wells, 1967).

7.4.2 Biodynamic Models of Soft Tissue

Single Viscoelastic Body with Two Degrees of Freedom. Consider a single viscoelastic body pendulum that consists of a mass suspended from a pivot point by a spring of natural length l and a dashpot, as seen in Fig. 7.2. The system is constrained to move within the r-θ plane, and is acted upon by a gravitational field of magnitude g, which acts in the negative vertical direction; r and θ are determined to be the only two generalized coordinates, since the motion of the mass is limited to movement in the radial and transverse (r or θ respectively) directions only. A model of this type can be used as a mechanical analogy to structure an approach to investigate the dynamics of soft tissues such as a muscle.

The velocity of the mass is given by

$$\mathbf{v} = \dot{r}\mathbf{e}_r + r\dot{\theta}\mathbf{e}_\theta$$

in the r-θ frame of reference or, in terms of the $\mathbf{b}_1$, $\mathbf{b}_2$, $\mathbf{b}_3$ frame of reference,

$$\mathbf{v} = \dot{r}\mathbf{b}_1 + r\dot{\theta}\mathbf{b}_2 \tag{7.5}$$

The origin of this reference frame is considered to be located at the pivot point, as seen in Fig. 7.2.

The kinetic energy of the system can be represented in vector form as

$$T = \tfrac{1}{2}m(\mathbf{v} \cdot \mathbf{v}) \tag{7.6}$$

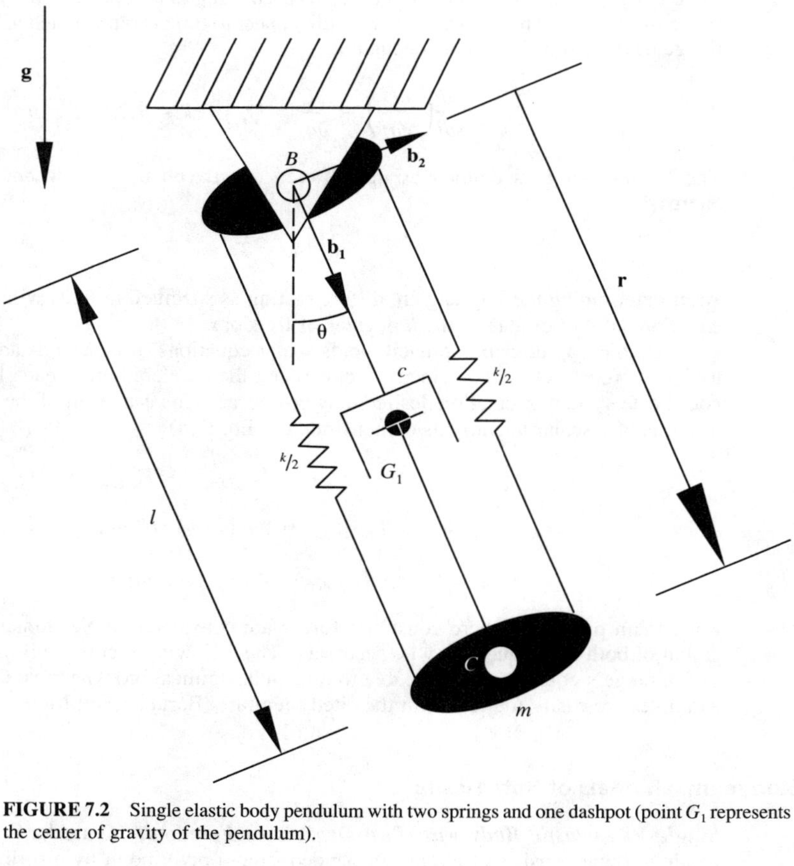

FIGURE 7.2 Single elastic body pendulum with two springs and one dashpot (point G_1 represents the center of gravity of the pendulum).

where m is mass and $\mathbf{v}$ is the velocity vector. Substituting Eq. (7.5) into Eq. (7.6) yields

$$T = \frac{m}{2}[\dot{r}^2 + (r\dot{\theta})^2] \tag{7.7}$$

The potential energy of the system is determined to be

$$V = -mgr\cos\theta + \frac{k}{2}(r - l)^2 \tag{7.8}$$

Rayleigh's dissipation function is applied in order to properly handle the viscous component of the given dynamic system. In this case, it is assumed that the viscous nature of the dashpot will be linearly dependent upon the velocity of the mass. More specifically, the viscous damping force is considered to be proportional to the velocity of the mass and is given by the following relation,

$$\mathbf{F}_d = -c\dot{\mathbf{r}} \tag{7.9}$$

Equation (7.9) is expressed in terms of the generalized coordinate r, where c is defined as the coefficient of proportionality for viscous damping. Rayleigh's dissipation function is defined as

$$\mathcal{F} = -\tfrac{1}{2}c\dot{r}^2 \tag{7.10}$$

Further information concerning the definition and use of Rayleigh's dissipation function can be found in the references.

Lagrange's equation can be modified to accommodate the consideration of the viscous damping forces within the dynamic system. Subsequently, Eq. (7.2) can be rewritten as

$$\frac{d}{dt}\left(\frac{\partial L}{\partial \dot{q}_k}\right) - \frac{\partial L}{\partial q_k} + \frac{\partial \mathcal{F}}{\partial \dot{q}_k} = Q_k \tag{7.11}$$

where the Lagrangian, $L = T - V$, is determined by using Eqs. (7.7) and (7.8),

$$L = \frac{m}{2}[\dot{r}^2 + (r\dot{\theta})^2] + mgr\cos\theta - \frac{k}{2}(r - l)^2 \tag{7.12}$$

By applying Lagrange's equation (7.11), and using the first generalized coordinate r, where

$$q_1 = r \tag{7.13}$$

each term of Eq. (7.11) can easily be identified by differentiation of Eq. (7.12). The resulting relations are

$$\frac{\partial L}{\partial \dot{r}} = m\dot{r} \tag{7.14}$$

so that

$$\frac{d}{dt}\left(\frac{\partial L}{\partial \dot{r}}\right) = m\ddot{r} \tag{7.15}$$

and

$$\frac{\partial L}{\partial r} = mr\dot{\theta}^2 + mg\cos\theta - k(r - l) \tag{7.16}$$

Similarly, the term for the dissipation function can be shown to be

$$\frac{\partial \mathcal{F}}{\partial \dot{r}} = -c\dot{r} \tag{7.17}$$

and the generalized force, $Q_r = 0$, since there are no externally applied forces acting on the system in the radial direction. By insertion of Eqs. (7.15), (7.16), and (7.17) into Eq. (7.11), the equation of motion in the r direction is

$$m\ddot{r} - mr\dot{\theta}^2 - mg\cos\theta + k(r - l) - c\dot{r} = 0 \tag{7.18}$$

Lagrange's equation must now be solved by using the second generalized coordinate θ, where

$$q_2 = \theta \tag{7.19}$$

By using the same approach as seen for the first generalized coordinate, the resulting differential relations are

$$\frac{\partial L}{\partial \dot{\theta}} = mr^2\dot{\theta} \tag{7.20}$$

so that

$$\frac{d}{dt}\left(\frac{\partial L}{\partial \dot{\theta}}\right) = 2mr\dot{r}\dot{\theta} + mr^2\ddot{\theta} \tag{7.21}$$

and

$$\frac{\partial L}{\partial \theta} = -mgr\sin\theta \tag{7.22}$$

The dissipation function in terms of the second generalized coordinate is shown to be

$$\frac{\partial \mathcal{F}}{\partial \dot{\theta}} = 0 \tag{7.23}$$

and the generalized force, $Q_\theta = 0$, since there are no externally applied torques acting on the system in the θ direction. By insertion of Eqs. (7.21), (7.22), and (7.23) into Eq. (7.11), the equation of motion in the θ direction is

$$2mr\dot{r}\dot{\theta} + mr^2\ddot{\theta} + mgr\sin\theta = 0 \tag{7.24}$$

Through simplification and rearranging of terms, Eqs. (7.18) and (7.24) can be rewritten as

$$mg\cos\theta - k(r - l) + c\dot{r} = m(\ddot{r} - r\dot{\theta}^2) \tag{7.25}$$

and

$$-mg\sin\theta = m(2\dot{r}\dot{\theta} + r\ddot{\theta}) \tag{7.26}$$

respectively, yielding the equations of motion for the spring-dashpot pendulum system. Note that Eqs. (7.25) and (7.26) are also the equations of motion obtained by using Newton's laws.

Single Elastic Body with Two Degrees of Freedom. Next, consider a single elastic body pendulum consisting of a mass suspended from a pivot point. Assume that the mass is suspended solely by a spring of natural length l, as seen in Fig. 7.3, and that the system experiences no viscous damping. As before, the system is constrained to move within the vertical, or r-θ plane, and is acted upon by a gravitational field g, which acts in the negative vertical direction; r and θ are again determined to be the only two generalized coordinates, since the motion of the mass is constrained to movement

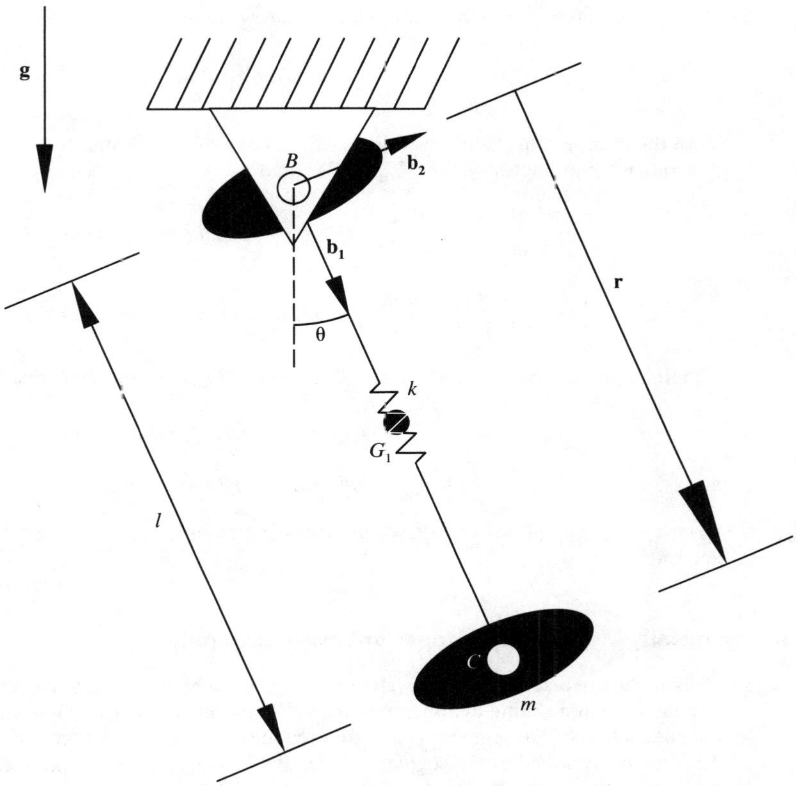

FIGURE 7.3 Single elastic body pendulum with one spring (point G_1 represents the center of gravity of the pendulum).

within the r-θ plane only. The spring pendulum system of Fig. 7.3 is another mechanically analogous system that can also be used to initially model the dynamic behavior of soft tissues.

The steps needed to determine the equations of motion for this system are essentially the steps for the previous system in Fig. 7.2. In this example, the effects of viscous damping are neglected, thereby omitting Rayleigh's dissipation function within the solution. This is a simple omission where removal of the dissipation function from Lagrange's equation will have an impact on the final solutions.

The velocity of the mass remains the same as before and is given by

$$\mathbf{v} = \dot{r}\mathbf{b}_1 + r\dot{\theta}\mathbf{b}_2 \tag{7.27}$$

The kinetic energy of the system,

$$T = \frac{m}{2}[\dot{r}^2 + (r\dot{\theta})^2] \tag{7.28}$$

and the potential energy of the system,

$$V = -mgr\cos\theta + \frac{k}{2}(r - l)^2 \tag{7.29}$$

are also the same, and therefore, the lagrangian will not change,

$$L = T - V = \frac{m}{2}[\dot{r}^2 + (r\dot{\theta})^2] + mgr\cos\theta - \frac{k}{2}(r - l)^2 \tag{7.30}$$

With the omission of the dissipation function, Lagrange's equation is solved for each of the chosen generalized coordinates: $q_1 = r$, Eq. (7.31), and $q_2 = \theta$, Eq. (7.32).

$$\frac{d}{dt}\left(\frac{\partial L}{\partial \dot{r}}\right) - \left(\frac{\partial L}{\partial r}\right) = Q_r \Rightarrow m\ddot{r} - mr\dot{\theta}^2 - mg\cos\theta + k(r - l) = 0 \tag{7.31}$$

$$\frac{d}{dt}\left(\frac{\partial L}{\partial \dot{\theta}}\right) - \left(\frac{\partial L}{\partial \theta}\right) = Q_\theta \Rightarrow 2mr\dot{r}\dot{\theta} + mr^2\ddot{\theta} + mgr\sin\theta = 0 \tag{7.32}$$

Simplificating and rearranging Eqs. (7.31) and (7.32) yields the two equations of motion:

$$mg\cos\theta - k(r - l) = m(\ddot{r} - r\dot{\theta}^2) \tag{7.33}$$

and

$$-mg\sin\theta = m(2\dot{r}\dot{\theta} + r\ddot{\theta}). \tag{7.34}$$

Once again, Eqs. (7.33) and (7.34) are the same equations of motion obtained by the Newtonian approach.

7.4.3 Biodynamically Modeling the Upper or Lower Extremity

Consider the two-segment system shown in Fig. 7.4, where a single segment of length l_3 is connected to a large nontranslating cylinder at point B. For initial simplicity, it is assumed that for this system, the connection of the segment is that of a revolute (or hinge) joint having only one axis of rotation. The free-moving end of the segment is identified as point C, while point G_1 identifies the center of gravity for this segment. The large cylinder of height l_1 and radius l_2 (e.g., torso) is fixed in space and is free to rotate about the vertical axis at some angular speed Ω. A moving coordinate system $\mathbf{b}_1$, $\mathbf{b}_2$, $\mathbf{b}_3$ is fixed at point B and is allowed to rotate about the $\mathbf{b}_3$ axis so that the unit vector $\mathbf{b}_1$ will always lie on segment BC. This particular system considers only one segment, which can represent the upper arm or thigh, and is presented as an initial step for dynamically modeling the human upper or lower extremity. To complete the extremity model, additional segments are subsequently added to this initial segment. The derivations of the models involving additional segments are presented later within this section.

The position vectors designating the locations of point B, G, and C are given as

$$\mathbf{r}_B = \mathbf{l}_1 + \mathbf{l}_2 \tag{7.35}$$

$$\mathbf{r}_{G_1} = \frac{1}{2}\mathbf{l}_3 + \mathbf{r}_B \tag{7.36}$$

and

$$\mathbf{r}_C = \mathbf{r}_B + \mathbf{l}_3 \tag{7.37}$$

respectively.

The angular velocity vector of segment BC is determined to be

$$\boldsymbol{\omega}_b = -\Omega\cos\theta_1\mathbf{b}_1 + \Omega\sin\theta_1\mathbf{b}_2 + \dot{\theta}_1\mathbf{b}_3 \tag{7.38}$$

where θ_1 is the angle between the segment and the vertical and $\dot{\theta}_1$ is the time rate of change of that angle.

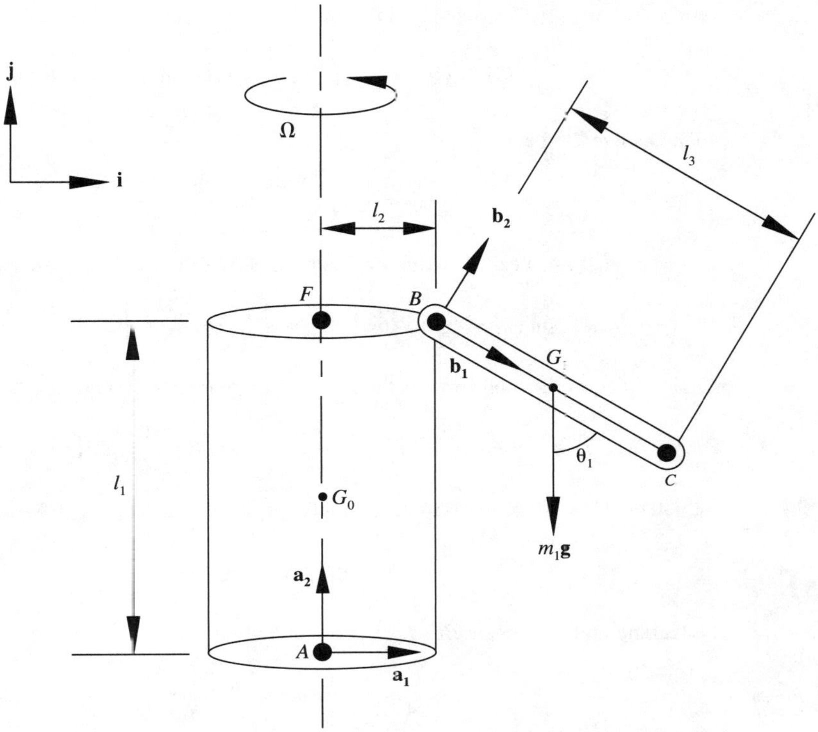

FIGURE 7.4 Two-segment model: one rigid segment connected to a nontranslating cylinder.

The components for the mass moment of inertia about point G_1 in the $\mathbf{b}_1$, $\mathbf{b}_2$, $\mathbf{b}_3$ frame of reference are

$$I_{b_1} = 0 \tag{7.39}$$

and

$$I_{b_2} = I_{b_3} = \tfrac{1}{12} m_1 l_3^2 \tag{7.40}$$

The kinetic energy of segment BC is defined by the equation

$$T_1 = \tfrac{1}{2}(I_{b_1}\omega_{b_1}^2 + I_{b_2}\omega_{b_2}^2 + I_{b_3}\omega_{b_3}^2) + \tfrac{1}{2} m_1 \mathbf{v}_{G_1} \cdot \mathbf{v}_{G_1} \tag{7.41}$$

where $\mathbf{v}_{G_1}$ is the velocity vector of segment taken at the center of mass. This vector is determined by using the relative velocity relation

$$\mathbf{v}_{G_1} = \mathbf{v}_B + \boldsymbol{\omega}_b \times \mathbf{r}_{G_1/B} \tag{7.42}$$

where $\mathbf{v}_B$ is the velocity vector for point B and is

$$\mathbf{v}_B = -l_2 \Omega \mathbf{b}_3 \tag{7.43}$$

and $\mathbf{r}_{G_1/B}$ is the relative position vector for point G_1 as defined from point B and is

$$\mathbf{r}_{G_1/B} = \frac{l_3}{2}\mathbf{b}_1 \tag{7.44}$$

Substitution of Eqs. (7.43) and (7.44) into Eq. (7.42) yields

$$\mathbf{v}_{G_1} = -l_2\Omega\mathbf{b}_3 + (-\Omega\cos\theta_1\mathbf{b}_1 + \Omega\sin\theta_1\mathbf{b}_2 + \dot{\theta}_1\mathbf{b}_3) \times \frac{l_3}{2}\mathbf{b}_1 \tag{7.45}$$

which is solved to be

$$\mathbf{v}_{G_1} = \frac{l_3}{2}\dot{\theta}_1\mathbf{b}_2 - \Omega\left(\frac{l_3\sin\theta_1}{2} + l_2\right)\mathbf{b}_3 \tag{7.46}$$

Therefore, Eq. (7.41) can be written by using Eqs. (7.38), (7.39), (7.40), and (7.46):

$$T_1 = \frac{1}{2}\left(\frac{1}{12}m_1l_3^2\Omega^2\sin^2\theta_1 + \frac{1}{12}m_1l_3^2\dot{\theta}_1^2\right) + \frac{1}{2}m_1\left\{\left(\frac{l_3}{2}\dot{\theta}_1\right)^2 + \left[-\Omega\left(\frac{l_3\sin\theta_1}{2} + l_2\right)\right]^2\right\} \tag{7.47}$$

and simplified to the final form of the kinetic energy for the system:

$$T_1 = \frac{1}{24}m_1l_3^2(\Omega^2\sin^2\theta_1 + \dot{\theta}_1^2) + \frac{1}{8}m_1l_3^2\dot{\theta}_1^2 + \frac{1}{2}m_1\Omega^2\left(\frac{l_3\sin\theta_1}{2} + l_2\right)^2 \tag{7.48}$$

The potential energy of the system is simply

$$V_1 = -m_1g\frac{l_3}{2}\cos\theta_1 \tag{7.49}$$

The lagrangian for segment BC, L_1, is subsequently determined to be

$$\begin{aligned} L_1 = &\frac{1}{24}m_1l_3^2(\Omega^2\sin^2\theta_1 + \dot{\theta}_1^2) + \frac{1}{8}m_1l_3^2\dot{\theta}_1^2 \\ &+\frac{1}{2}m_1\Omega^2\left(\frac{l_3\sin\theta_1}{2} + l_2\right)^2 + m_1g\frac{l_3}{2}\cos\theta_1 \end{aligned} \tag{7.50}$$

Applying Lagrange's equation (7.11) and using the only generalized coordinate for this system, θ_1,

$$q_1 = \theta_1 \tag{7.51}$$

each term of Eq. (7.11) can easily be identified by differentiation of Eq. (7.50). The derivative of L_1 with respect to θ_1 is solved accordingly:

$$\begin{aligned} \frac{\partial L_1}{\partial\theta_1} = &\frac{1}{12}m_1l_3^2\Omega^2\sin\theta_1\cos\theta_1 + \frac{1}{4}m_1l_3^2\Omega^2\sin\theta_1\cos\theta_1 \\ &+\frac{1}{2}m_1l_2l_3\Omega^2\cos\theta_1 - \frac{1}{2}m_1gl_3\sin\theta_1 \end{aligned} \tag{7.52}$$

which reduces to

$$\frac{\partial L_1}{\partial\theta_1} = \frac{1}{3}m_1l_3^2\Omega^2\sin\theta_1\cos\theta_1 + \frac{1}{2}m_1l_2l_3\Omega^2\cos\theta_1 - \frac{1}{2}m_1gl_3\sin\theta_1 \tag{7.53}$$

The derivative of L_1 with respect to $\dot{\theta}_1$ is

$$\frac{\partial L_1}{\partial\dot{\theta}_1} = \frac{1}{12}m_1l_3^2\dot{\theta}_1 + \frac{1}{4}m_1l_3^2\dot{\theta}_1 = \frac{1}{3}m_1l_3^2\dot{\theta}_1 \tag{7.54}$$

So that

$$\frac{d}{dt}\left(\frac{\partial L_1}{\partial \dot{\theta}_1}\right) = \frac{1}{3}m_1 l_3^2 \ddot{\theta}_1 \tag{7.55}$$

The appropriate terms can be substituted into Lagrange's equation (7.11) to give

$$\frac{1}{3}m_1 l_3^2 \ddot{\theta}_1 - \frac{1}{3}m_1 l_3^2 \Omega^2 \sin\theta_1 \cos\theta_1 - \frac{1}{2}m_1 l_2 l_3 \Omega^2 \cos\theta_1 + \frac{1}{2}m_1 g l_3 \sin\theta_1 = 0 \tag{7.56}$$

since there are no externally applied torques acting on the system in the θ_1 direction. The resulting equation of motion for the one-segment system is solved as

$$\ddot{\theta}_1 - \Omega^2 \sin\theta_1 \cos\theta_1 - \frac{3}{2}\frac{l_2}{l_3}\Omega^2 \cos\theta_1 + \frac{3}{2}\frac{g}{l_3}\sin\theta_1 = 0 \tag{7.57}$$

Next, consider an additional segment added to the two-segment system, in the previous example, as seen in Fig. 7.5. Assume that the additional segment added adjoins to the first segment at point C by way of a revolute joint. The new segment is of length l_4, with point D defining the free-moving end of the two-segment system and point G_2 identifies the center of gravity for the second segment. An additional moving body-fixed, coordinate system $\mathbf{c}_1$, $\mathbf{c}_2$, $\mathbf{c}_3$ is defined at point C and is allowed to rotate about the $\mathbf{c}_3$ axis so that the unit vector $\mathbf{c}_1$ will always lie on segment CD.

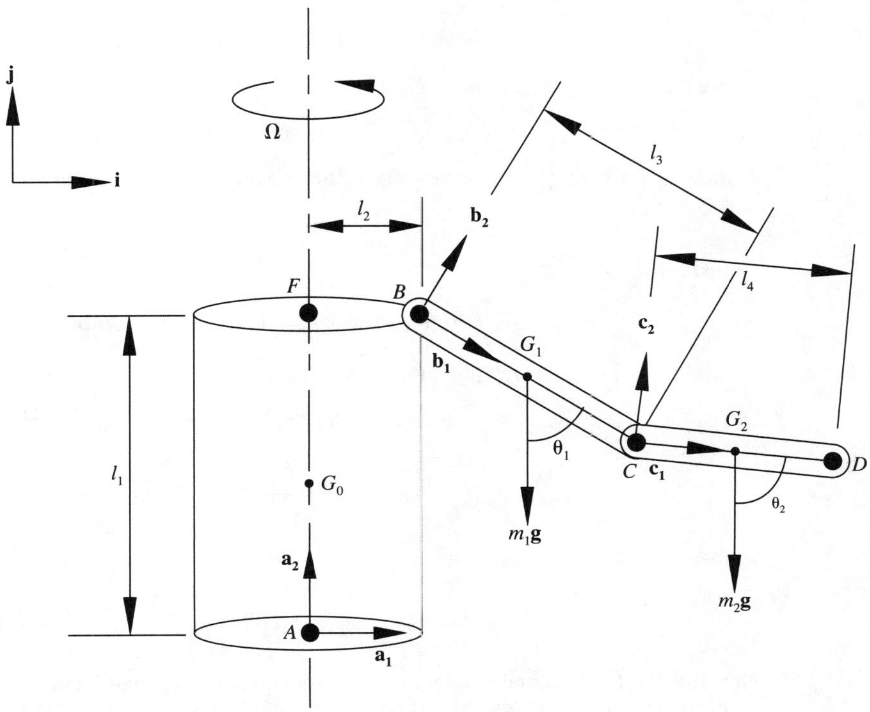

FIGURE 7.5 Two rigid segments connected to a nontranslating cylinder.

New position vectors designating the locations of point G_2 and D are given as

$$\mathbf{r}_{G_2} = \frac{1}{2}\mathbf{l}_4 + \mathbf{r}_C \tag{7.58}$$

and

$$\mathbf{r}_D = \mathbf{r}_C + \mathbf{l}_4 \tag{7.59}$$

respectively for the new segment.

When solving a three-segment system as the one seen in Fig. 7.5, the kinetic and potential energy must be determined independently for each segment. Since the kinetic and potential energy was solved previously for segment BC and is given in Eqs. (7.48) and (7.49), consideration needs to be given only to segment CD.

In a similar manner as before, the angular velocity vector of segment CD is determined to be

$$\boldsymbol{\omega}_c = -\Omega \cos \theta_2 \mathbf{c}_1 + \Omega \sin \theta_2 \mathbf{c}_2 + \dot{\theta}_2 \mathbf{c}_3 \tag{7.60}$$

where θ_2 is the angle between the segment and the vertical and $\dot{\theta}_2$ is the time rate of change of the angle. Also, the components for the mass moment of inertia about point G_2 in the $\mathbf{c}_1$, $\mathbf{c}_2$, $\mathbf{c}_3$ frame of reference are

$$I_{c_1} = 0 \tag{7.61}$$

and

$$I_{c_2} = I_{c_3} = \tfrac{1}{12} m_2 l_4^2 \tag{7.62}$$

and the kinetic energy of segment CD is defined by the equation

$$T_2 = \tfrac{1}{2}(I_{c_1}\omega_{c_1}^2 + I_{c_2}\omega_{c_2}^2 + I_{c_3}\omega_{c_3}^2) + \tfrac{1}{2} m_2 \mathbf{v}_{G_2} \cdot \mathbf{v}_{G_2} \tag{7.63}$$

where the velocity vector at point G_2 is

$$\mathbf{v}_{G_2} = \mathbf{v}_C + \boldsymbol{\omega}_c \times \mathbf{r}_{G_2/C} \tag{7.64}$$

In order to solve Eq. (7.64), the velocity vector at point C and the cross product must be determined.

$$\mathbf{v}_C = \mathbf{v}_B + \boldsymbol{\omega}_b \times \mathbf{r}_{C/B}$$

so that

$$\mathbf{v}_C = -l_2 \Omega \mathbf{b}_3 + (-\Omega \cos \theta_1 \mathbf{b}_1 + \Omega \sin \theta_1 \mathbf{b}_2 + \dot{\theta}_1 \mathbf{b}_3) \times (l_3 \mathbf{b}_1)$$

$$\mathbf{v}_C = l_3 \dot{\theta}_1 \mathbf{b}_2 - \Omega(l_2 + l_3 \sin \theta_1)\mathbf{b}_3 \tag{7.65}$$

and

$$\boldsymbol{\omega}_C \times \mathbf{r}_{G_2/C} = (-\Omega \cos \theta_2 \mathbf{c}_1 + \Omega \sin \theta_2 \mathbf{c}_2 + \dot{\theta}_2 \mathbf{c}_3) \times \left(\frac{l_4}{2}\mathbf{c}_1\right) \tag{7.66}$$

$$\boldsymbol{\omega}_C \times \mathbf{r}_{G_2/C} = \frac{l_4}{2}\dot{\theta}_2 \mathbf{c}_2 - \frac{l_4}{2}\Omega \sin \theta_2 \mathbf{c}_3$$

so that

$$\mathbf{v}_{G_2} = l_3 \dot{\theta}_1 \mathbf{b}_2 - \Omega(l_3 \sin \theta_1 + l_2)\mathbf{b}_3 + \frac{l_4}{2}\dot{\theta}_2 \mathbf{c}_2 - \frac{l_4}{2}\Omega \sin \theta_2 \mathbf{c}_3 \tag{7.67}$$

Note that Eq. (7.67) contains velocity terms from both moving coordinate systems $\mathbf{b}_1$, $\mathbf{b}_2$, $\mathbf{b}_3$ and $\mathbf{c}_1$, $\mathbf{c}_2$, $\mathbf{c}_3$ and should be represented entirely in terms of $\mathbf{c}_1$, $\mathbf{c}_2$, $\mathbf{c}_3$. Therefore, a coordinate transformation matrix is defined that will manage this.

$$\begin{vmatrix} \mathbf{c}_1 \\ \mathbf{c}_2 \\ \mathbf{c}_3 \end{vmatrix} = \begin{vmatrix} \cos(\theta_2 - \theta_1) & \sin(\theta_2 - \theta_1) & 0 \\ -\sin(\theta_2 - \theta_1) & \cos(\theta_2 - \theta_1) & 0 \\ 0 & 0 & 1 \end{vmatrix} \begin{vmatrix} \mathbf{b}_1 \\ \mathbf{b}_2 \\ \mathbf{b}_3 \end{vmatrix} \tag{7.68}$$

From Eq. (7.68), the velocity at point G_2 becomes

$$\begin{aligned} \mathbf{v}_{G_2} &= l_3\dot{\theta}_1 \sin(\theta_2 - \theta_1)\mathbf{c}_1 + \left[l_3\dot{\theta}_1 \cos(\theta_2 - \theta_1) + \frac{l_4}{2}\dot{\theta}_2 \right]\mathbf{c}_2 \\ &\quad - \left[\Omega(l_3 \sin\theta_1 + l_2) + l_2\Omega + \frac{l_4}{2}\Omega \sin\theta_2 \right]\mathbf{c}_3 \end{aligned} \tag{7.69}$$

Substituting Eqs. (7.61), (7.62), and (7.69) into Eq. (7.63), the equation for the kinetic energy of segment CD, and solving gives

$$\begin{aligned} T_2 &= \tfrac{1}{2}m_2 l_3^2\dot{\theta}_1^2 + \tfrac{1}{2}m_2 l_3 l_4 \dot{\theta}_1\dot{\theta}_2 \cos(\theta_2 - \theta_1) + \tfrac{1}{6}m_2\, l_4^2\dot{\theta}_2^2 \\ &\quad + \tfrac{1}{2}m_2 l_3^2\Omega^2 \sin^2\theta_1 + \tfrac{1}{2}m_2 l_3 l_4 \Omega^2 \sin\theta_1 \sin\theta_2 + m_2 l_2 l_3 \Omega^2 \sin\theta_1 \\ &\quad + \tfrac{1}{2}m_2 l_2^2\Omega^2 + \tfrac{1}{2}m_2 l_2 l_4 \Omega^2 \sin\theta_2 + \tfrac{1}{6}m_2 l_4^2\Omega^2 \sin^2\theta_2 \end{aligned} \tag{7.70}$$

and the potential energy of segment CD is determined to be

$$V_2 = -m_2 g l_3 \cos\theta_1 - m_2 g \frac{l_4}{2} \cos\theta_2 \tag{7.71}$$

The total kinetic and potential energy of the system can be determined by summation of

$$T = T_1 + T_2 \tag{7.72}$$

and

$$V = V_1 + V_2 \tag{7.73}$$

respectively, where the lagrangian for the system is given by

$$L = (T_1 + T_2) - (V_1 + V_2) = T - V \tag{7.74}$$

It is left to the reader to apply Lagrange's equation (7.11), using the two generalized coordinates of the two-segment system, $q_1 = \theta_1$ and $q_2 = \theta_2$, to work through the mathematics involved in solving Eq. (7.11) to yield the equations of motion. For the first generalized coordinate, $q_1 = \theta_1$, the equation of motion is solved to be

$$\begin{aligned} &(\tfrac{1}{3}m_1 + m_2)l_3^2\ddot{\theta}_1 + \tfrac{1}{2}m_2 l_3 l_4 \cos(\theta_2 - \theta_1)\ddot{\theta}_2 - (\tfrac{1}{3}m_1 + m_2)l_3^2\Omega^2 \sin\theta_1 \cos\theta_1 \\ &\qquad - (\tfrac{1}{2}m_1 + m_2)l_2 l_3 \Omega^2 \cos\theta_1 + \tfrac{1}{2}m_2 l_3 l_4 \dot{\theta}_1\dot{\theta}_2 \sin\theta_1 \cos\theta_2 \\ &\qquad - \tfrac{1}{2}m_2 l_3 l_4(\dot{\theta}_1\dot{\theta}_2 + \Omega^2)\cos\theta_1 \sin\theta_2 + (\tfrac{1}{2}m_1 + m_2)g l_3 \sin\theta_1 = 0 \end{aligned} \tag{7.75}$$

For the second generalized coordinate, $q_2 = \theta_2$, the equation of motion is solved to be

$$\begin{aligned} &\tfrac{1}{3}m_2 l_4^2\ddot{\theta}_2 + \tfrac{1}{2}m_2 l_3 l_4 \cos(\theta_2 - \theta_1)\ddot{\theta}_1 + \tfrac{1}{2}m_2 l_3 l_4 \dot{\theta}_1\dot{\theta}_2 \cos\theta_1 \sin\theta_2 \\ &\qquad - \tfrac{1}{2}m_2 l_3 l_4(\dot{\theta}_1\dot{\theta}_2 + \Omega^2)\sin\theta_1 \cos\theta_2 \\ &\qquad - \tfrac{1}{3}m_2 l_4^2\Omega^2 \sin\theta_2 \cos\theta_2 - \tfrac{1}{2}m_2 l_2 l_4 \Omega^2 \cos\theta_2 + \tfrac{1}{2}m_2 g l_4 \sin\theta_2 = 0 \end{aligned} \tag{7.76}$$

To complete the hand-arm or foot-leg system, consider yet another additional segment added to the three-segment system, as seen in Fig. 7.6. As before, it is assumed that the additional segment

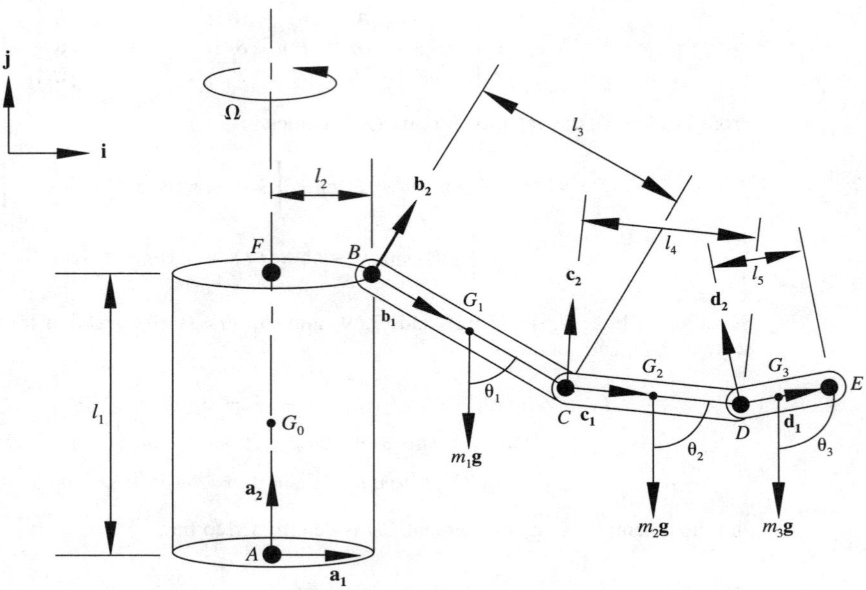

FIGURE 7.6 Three rigid segments connected to a nontranslating cylinder.

added adjoins to the second segment at point D by way of a revolute joint. The new segment is of length l_5, with point E defining the free-moving end of the four-segment system and point G_3 identifying the center of gravity for the third segment. An additional moving coordinate system $\mathbf{d}_1$, $\mathbf{d}_2$, $\mathbf{d}_3$ is defined at point D and is allowed to rotate about the $\mathbf{d}_3$ axis so that the unit vector $\mathbf{d}_1$ will always lie on segment DE.

The solution to the four-segment system is obtained in a similar manner to that of the three-segment system with careful consideration of the coordinate transformations. It is not provided because of its bulky content and is left for the reader to solve.

7.5 INTRODUCTION TO THE KINEMATICS TABLE METHOD

The kinematics table, shown in Table 7.1, introduces a method (referred to within this chapter as the table method) for efficiently managing the mathematics involved in analyzing multibody and multiple coordinate system problems, and can be used in either the Lagrangian or the Newton-Euler approach. A schematic diagram, which defines an inertial or body-fixed coordinate system, must accompany every kinematics table. The purpose of the schematic is to identify the point on the body at which the absolute velocity and acceleration is to be determined. The corresponding schematic for Table 7.1 is shown in Fig. 7.7.

The kinematics table is divided into three sections.

1. Top section (rows a, b, and c): Acts as a worksheet to assist in the writing of the expressions for the next two sections.

TABLE 7.1 Blank Kinematics Table

	i	j	k
$\boldsymbol{\omega}_B$ (*a*) Angular velocity of coordinate system with its origin at *B*			
$\boldsymbol{\alpha}_B = \dot{\boldsymbol{\omega}}_B$ (*b*) Angular acceleration of coordinate system with its origin at *B*			
$\mathbf{r}_{P/B}$ (*c*) Relative displacement of point *P* with respect to *B*			
$\mathbf{v}_B$ (*d*) Base velocity of origin *B*			
$\mathbf{v}_{P/B} = \dot{\mathbf{r}}_{P/B}$ (*e*) Relative velocity of point *P* with respect to *B*			
$\boldsymbol{\omega}_B \times \mathbf{r}_{P/B}$ (*f*) Transport velocity			
$\mathbf{a}_B$ (*g*) Base acceleration			
$\mathbf{a}_{P/B} = \ddot{\mathbf{r}}_{P/B}$ (*h*) Relative acceleration			
$\boldsymbol{\alpha}_B \times \mathbf{r}_{P/B}$ (*i*) Acceleration due to angular acceleration of rotation frame			
$\boldsymbol{\omega}_B \times (\boldsymbol{\omega}_B \times \mathbf{r}_{P/B})$ (*j*) Centripetal acceleration			
$2\boldsymbol{\omega}_B \times \dot{\mathbf{r}}_{P/B}$ (*k*) Coriolis acceleration			

2. Middle section (rows *d*, *e*, and *f*): Summed to yield the absolute velocity (velocity with respect to an inertial frame of reference) of point *P*.
3. Bottom section (rows *g*, *h*, *i*, *j*, and *k*): Summed to yield the absolute acceleration (acceleration with respect to an inertial frame of reference) of point *P*.

All terms in the first column are vectors and are resolved into their vector components in columns 2, 3, and 4. Note that the unit vectors of the selected coordinate system are written at the top of columns 2, 3, and 4.

For a multibody system, each body would require a kinematics table and a corresponding schematic. The following examples illustrate the steps required for solving problems by the table method. Note that one example includes the expressions for acceleration to demonstrate the use of the table method with the Newton-Euler approach, while all other examples consider only the velocity.

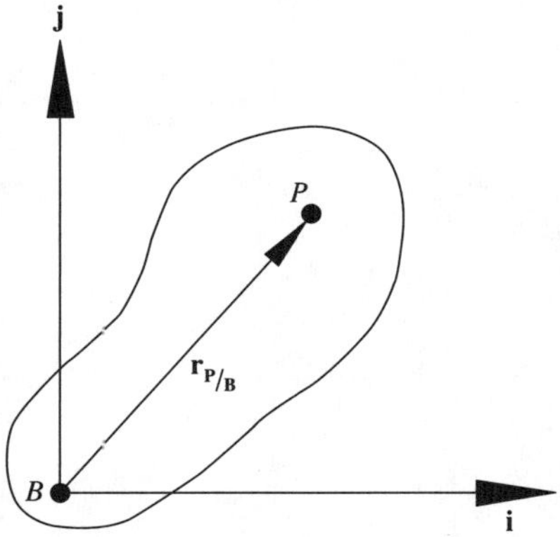

FIGURE 7.7 Schematic to accompany the kinematics table (Table 7.1).

7.5.1 Single Rigid Body with a Single Degree of Freedom

The upper arm, simplified as a single rigid body, is shown in Fig. 7.8. The velocity and acceleration for the center of mass of the arm are derived and presented in two coordinate systems. Table 7.2 presents the kinematics in an inertial coordinate system, while Table 7.3 utilizes a body-fixed, moving coordinate system. For this system, not unlike the two-segment system of Fig. 7.4, a moving coordinate system $\mathbf{b}_1$, $\mathbf{b}_2$, $\mathbf{b}_3$ is fixed at point B and is allowed to rotate about the $\mathbf{b}_3$ axis so that the unit vector $\mathbf{b}_1$ will always lie on segment BC.

From Tables 7.2 and 7.3, the absolute velocity of the center of gravity, or point G_1, is

$$\mathbf{v}_{G_1} = r_{G_1}\dot{\theta}_1 \cos \theta_1\mathbf{i} + r_{G_1}\dot{\theta}_1 \sin \theta_1\mathbf{j}$$
$$\mathbf{v}_{G_1} = r_{G_1}\dot{\theta}_1\mathbf{b}_3 \tag{7.77}$$

in the $\mathbf{i}$, $\mathbf{j}$, $\mathbf{k}$ and $\mathbf{b}_1$, $\mathbf{b}_2$, $\mathbf{b}_3$ frame of references respectively.

The absolute acceleration is

$$\mathbf{a}_{G_1} = (r_{G_1}\ddot{\theta}_1 \cos \theta_1 - r_{G_1}\dot{\theta}_1^2 \sin \theta_1)\mathbf{i} + (r_{G_1}\ddot{\theta}_1 \sin \theta_1 + r_{G_1}\dot{\theta}_1^2 \cos \theta_1)\mathbf{j}$$
$$\mathbf{a}_{G_1} = -r_{G_1}\dot{\theta}_1^2\mathbf{b}_1 + r_{G_1}\ddot{\theta}_1\mathbf{b}_2 \tag{7.78}$$

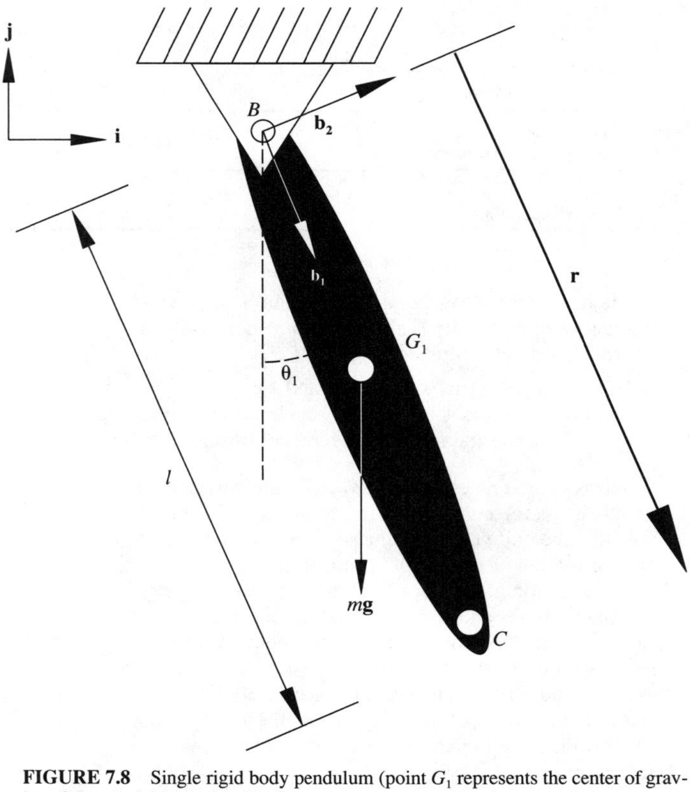

FIGURE 7.8 Single rigid body pendulum (point G_1 represents the center of gravity of the pendulum and is fixed).

TABLE 7.2 Kinematics Table for the Single Rigid Body in an Inertial Coordinate System

	$\mathbf{i}$	$\mathbf{j}$	$\mathbf{k}$
$\boldsymbol{\omega}_B$	0	0	0
$\boldsymbol{\alpha}_B = \dot{\boldsymbol{\omega}}_B$	0	0	0
$\mathbf{r}_{G_1/B} = \mathbf{r}_{G_1}$	$r_{G_1} \sin \theta_1$	$-r_{G_1} \cos \theta_1$	0
$\mathbf{v}_B$	0	0	0
$\mathbf{v}_{G_1} = \dot{\mathbf{r}}_{G_1}$	$r_{G_1}\dot{\theta}_1 \cos \theta_1$	$r_{G_1}\dot{\theta}_1 \sin \theta_1$	0
$\boldsymbol{\omega}_B \times \mathbf{r}_{G_1}$	0	0	0
$\mathbf{a}_B$	0	0	0
$\mathbf{a}_{G_1} = \ddot{\mathbf{r}}_{G_1}$	$r_{G_1}\ddot{\theta}_1 \cos \theta_1$ $- r_{G_1}\dot{\theta}_1^2 \sin \theta_1$	$r_{G_1}\ddot{\theta}_1 \sin \theta_1$ $+ r_{G_1}\dot{\theta}_1^2 \cos \theta_1$	0
$\boldsymbol{\alpha}_B \times \mathbf{r}_{G_1}$	0	0	0
$\boldsymbol{\omega}_B \times (\boldsymbol{\omega}_B \times \mathbf{r}_{G_1})$	0	0	0
$2\boldsymbol{\omega}_B \times \dot{\mathbf{r}}_{G_1}$	0	0	0

TABLE 7.3 Kinematics Table for the Single Rigid Body in a Body-Fixed Coordinate System

	$\mathbf{b}_1 = \mathbf{e}_r$	$\mathbf{b}_2 = \mathbf{e}_\theta$	$\mathbf{b}_3$
$\boldsymbol{\omega}_B$	0	0	$\dot{\theta}_1$
$\boldsymbol{\alpha}_B = \dot{\boldsymbol{\omega}}_B$	0	0	$\ddot{\theta}_1$
$\mathbf{r}_{G_1/B} = \mathbf{r}_{G_1}$	r_{G_1}	0	0
$\mathbf{v}_B$	0	0	0
$\mathbf{v}_{G_1} = \dot{\mathbf{r}}_{G_1}$	0	0	0
$\boldsymbol{\omega}_B \times \mathbf{r}_{G_1}$	0	$r_{G_1}\dot{\theta}_1$	0
$\mathbf{a}_B$	0	0	0
$\mathbf{a}_{G_1} = \ddot{\mathbf{r}}_{G_1}$	0	0	0
$\boldsymbol{\alpha}_B \times \mathbf{r}_{G_1}$	0	$r_{G_1}\ddot{\theta}_1$	0
$\boldsymbol{\omega}_B \times (\boldsymbol{\omega}_B \times \mathbf{r}_{G_1})$	$-r_{G_1}\dot{\theta}_1^2$	0	0
$2\boldsymbol{\omega}_B \times \dot{\mathbf{r}}_{G_1}$	0	0	0

in the **i**, **j**, **k** and $\mathbf{b}_1$, $\mathbf{b}_2$, $\mathbf{b}_3$ frame of references, respectively. In applying either form of Eq. (7.77), the kinetic energy of the arm is found to be

$$T = \tfrac{1}{2}m\mathbf{v}_{G_1} \cdot \mathbf{v}_{G_1} + \tfrac{1}{2}I_{G_1}\dot{\theta}_1^2 = \tfrac{1}{2}mr_{G_1}^2\dot{\theta}_1^2 + \tfrac{1}{2}I_{G_1}\dot{\theta}_1^2 \tag{7.79}$$

The gravitational potential energy of the arm is

$$V = -mgr_{G_1}\cos\theta_1 \tag{7.80}$$

Substituting Eqs. (7.79) and (7.80) into the lagrangian, Eq. (7.3), results in

$$L = T - V = \tfrac{1}{2}mr_{G_1}^2\dot{\theta}_1^2 + \tfrac{1}{2}I_{G_1}\dot{\theta}_1^2 + mgr_{G_1}\cos\theta_1 \tag{7.81}$$

Substituting Eq. (7.81) into Eq. (7.2) with $k = \theta_1$ results in the equation of motion for the upper arm (assuming that the torso is fixed), as shown in Eq. (7.82). Q_θ would include all nonconservative or externally applied torques.

$$mr_{G_1}^2\ddot{\theta}_1 + I_{G_1}\ddot{\theta}_1 + mgr_{G_1}\sin\theta_1 = Q_\theta \tag{7.82}$$

where $I_B = mr_{G_1}^2 + I_{G_1}$, which can be found by using the parallel axis theorem.

In a similar manner to the derivation of the three-segment system of Fig. 7.5, a coordinate transformation matrix is defined in order to convert between coordinate systems. A transformation between the b frame and the inertial frame for a single rotation, θ_1, about the $\mathbf{b}_3$ axis in matrix form is shown in Eq. (7.83).

$$\begin{vmatrix} \mathbf{b}_1 \\ \mathbf{b}_2 \\ \mathbf{b}_3 \end{vmatrix} = \begin{vmatrix} \sin\theta_1 & -\cos\theta_1 & 0 \\ \cos\theta_1 & \sin\theta_1 & 0 \\ 0 & 0 & 1 \end{vmatrix} \begin{vmatrix} \mathbf{i} \\ \mathbf{j} \\ \mathbf{k} \end{vmatrix} \tag{7.83}$$

This results in the following equations:

$$\begin{aligned} \mathbf{b}_1 &= \sin\theta_1\mathbf{i} - \cos\theta_1\mathbf{j} \\ \mathbf{b}_2 &= \cos\theta_1\mathbf{i} + \sin\theta_1\mathbf{j} \\ \mathbf{b}_3 &= \mathbf{k} \\ \mathbf{i} &= \sin\theta_1\mathbf{b}_1 + \cos\theta_1\mathbf{b}_2 \\ \mathbf{j} &= -\cos\theta_1\mathbf{b}_1 + \sin\theta_1\mathbf{b}_2 \end{aligned} \tag{7.84}$$

In transferring a velocity or acceleration from one table to the next, a conversion between frames, as shown in Eq. (7.68), or a conversion from the b frame to the inertial frame as in Eqs. (7.83) and (7.84), and then from the inertial into the c frame, may be used.

7.5.2 Single Elastic Body with Two Degrees of Freedom

Table 7.4 is the kinematics table for the single elastic body pendulum shown in Fig. 7.3. The absolute velocity of the center of mass G_1 is required to complete the Lagrangian approach, and the Newtonian approach utilizes the absolute acceleration of point G_1, Eq. (7.86), in $\mathbf{F} = m\mathbf{a}_{G_1}$ for problems of constant mass.

The absolute velocity and acceleration of G_1 are expressed in terms of the body-fixed coordinate system, $\mathbf{b}_1$, $\mathbf{b}_2$, $\mathbf{b}_3$, which is fixed to the pendulum and rotates about the $\mathbf{b}_3$ axis as before. Although it is equivalent to expressing within the inertial frame of reference, **i**, **j**, **k**, the body-fixed coordinate system, $\mathbf{b}_1$, $\mathbf{b}_2$, $\mathbf{b}_3$, uses fewer terms. The velocity and acceleration for G_1 are respectively as follows:

TABLE 7.4 Kinematics Table for the Single Elastic Body Pendulum

	$\mathbf{b}_1 = \mathbf{e}_r$	$\mathbf{b}_2 = \mathbf{e}_\theta$	$\mathbf{b}_3$
$\boldsymbol{\omega}_B$	0	0	$\dot{\theta}_1$
$\boldsymbol{\alpha}_B = \dot{\boldsymbol{\omega}}_B$	0	0	$\ddot{\theta}_1$
$\mathbf{r}_{G_1/B} = \mathbf{r}_{G_1}$	r_{G_1}	0	0
$\mathbf{v}_B$	0	0	0
$\mathbf{v}_{G_1} = \dot{\mathbf{r}}_{G_1}$	$\dot{r}_{G_1}$	0	0
$\boldsymbol{\omega}_B \times \mathbf{r}_{G_1}$	0	$r_{G_1}\dot{\theta}_1$	0
$\mathbf{a}_B$	0	0	0
$\mathbf{a}_{G_1} = \ddot{\mathbf{r}}_{G_1}$	$\ddot{r}_{G_1}$	0	0
$\boldsymbol{\alpha}_B \times \mathbf{r}_{G_1}$	0	$r_{G_1}\ddot{\theta}_1$	0
$\boldsymbol{\omega}_B \times (\boldsymbol{\omega}_B \times \mathbf{r}_{G_1})$	$-r_{G_1}\dot{\theta}_1^2$	0	0
$2\boldsymbol{\omega}_B \times \dot{\mathbf{r}}_{G_1}$	0	$2\dot{r}_{G_1}\dot{\theta}_1$	0

$$\mathbf{v}_{G_1} = \dot{r}_{G_1}\mathbf{b}_1 + r_{G_1}\dot{\theta}_1\mathbf{b}_2 \tag{7.85}$$

and

$$\mathbf{a}_{G_1} = (\ddot{r}_{G_1} - r_{G_1}\dot{\theta}_1^2)\mathbf{b}_1 + (r_{G_1}\ddot{\theta}_1 + 2\dot{r}_{G_1}\dot{\theta}_1)\mathbf{b}_2 \tag{7.86}$$

7.5.3 Biodynamically Modeling the Upper or Lower Extremity by the Table Method

It is clear that the multilink systems of Figs. 7.4, 7.5, and 7.6 are applicable to many biodynamic scenarios. They can represent a torso with an upper or lower extremity, as well as several other combinations of multibody problems, as can be seen within the cited references.

If the multibody system represented in Fig. 7.6 is considered to represent a human torso, upper arm, forearm, and hand, then Table 7.5 results in the velocity at the shoulder (point B) expressed in the body-fixed coordinate system of the torso segment, $\mathbf{a}_1$, $\mathbf{a}_2$, $\mathbf{a}_3$. Similarly, the $\mathbf{b}_1$, $\mathbf{b}_2$, $\mathbf{b}_3$ coordinate system is body-fixed to the upper arm segment, the $\mathbf{c}_1$, $\mathbf{c}_2$, $\mathbf{c}_3$ system to the forearm segment, and

TABLE 7.5 The Absolute Velocity of Point B as Expressed Relative to $\mathbf{a}_1$, $\mathbf{a}_2$, $\mathbf{a}_3$

	$\bar{\mathbf{a}}_1$	$\bar{\mathbf{a}}_2$	$\bar{\mathbf{a}}_3$
$\boldsymbol{\omega}_A$	$\omega_{A_1} = -\dot{\psi}_1^a \cos\psi_2^a \sin\psi_3^a + \dot{\psi}_2^a \cos\psi_3^a$	$\omega_{A_2} = \dot{\psi}_1^a \sin\psi_2^a + \dot{\psi}_3^a$	$\omega_{A_3} = \dot{\psi}_1^a \cos\psi_2^a \cos\psi_3^a + \dot{\psi}_2^a \sin\psi_3^a$
$\mathbf{r}_{B/A}$	0	$r_{F/A}$	$r_{B/F}$
$\mathbf{v}_A$	0	0	0
$\mathbf{v}_{B/A} = \dot{\mathbf{r}}_{B/A}$	0	0	0
$\boldsymbol{\omega}_A \times \mathbf{r}_{B/A}$	$r_{B/F}\omega_{A_2} - r_{F/A}\omega_{A_3}$	$-r_{B/F}\omega_{A_1}$	$r_{F/A}\omega_{A_2}$

the $\mathbf{d}_1$, $\mathbf{d}_2$, $\mathbf{d}_3$ system to the hand segment. Tables 7.6, 7.7, and 7.8 are the results for the velocities at the elbow (point C), wrist (point D), and metacarpophalangeal joint of the third digit (point E), respectively. The results from these tables will yield the velocities at the end points of each segment considered. The end point velocities are required in order to determine the velocities at the centers of gravity for each segment. In Table 7.6, the velocity at the shoulder (point B), $\mathbf{v}_B$, is found from Table 7.5 by following these steps:

1. Sum the terms to yield the velocity in the a frame.
2. Either apply a coordinate transformation between the a and b frames directly, or convert from the a frame to the inertial frame and then from the inertial frame to the b frame.
3. Once the velocity is expressed in the b frame, it may be inserted into Table 7.6.

This process is then repeated for Tables 7.7 and 7.8 to complete the tables with the velocities of points C and D, respectively. The derivations of the velocity equations within these tables are lengthy and left as an exercise for the reader.

To determine the velocity at the center of gravity for each segment, additional tables can be subsequently constructed that correspond respectively to Tables 7.5, 7.6, 7.7, and 7.8. Each new table must take into account the position vector that defines the location of G_j for each segment, where j ranges from 0 to $n - 1$ and n is the number of segments considered within the system.

The kinetic energy of the entire system is determined by expanding the expression for the kinetic energy in Eq. (7.4) and is given as

$$\begin{aligned} T = {} & \tfrac{1}{2} m_A \mathbf{v}_{G_0} \cdot \mathbf{v}_{G_0} + \tfrac{1}{2} m_B \mathbf{v}_{G_1} \cdot \mathbf{v}_{G_1} \\ & + \tfrac{1}{2} m_C \mathbf{v}_{G_2} \cdot \mathbf{v}_{G_2} + \tfrac{1}{2} m_D \mathbf{v}_{G_3} \cdot \mathbf{v}_{G_3} \\ & + \tfrac{1}{2}\{\omega_A\}^T\{I_{G_0}\}\{\omega_A\} + \tfrac{1}{2}\{\omega_B\}^T\{I_{G_1}\}\{\omega_B\} \\ & + \tfrac{1}{2}\{\omega_C\}^T\{I_{G_2}\}\{\omega_C\} + \tfrac{1}{2}\{\omega_D\}^T\{I_{G_3}\}\{\omega_D\} \end{aligned} \tag{7.87}$$

where $\{I_{G_j}\}$ is a 3×3 matrix containing the mass moment of inertia about each axis and, again, j is from 0 to $n - 1$, where n is the number of segments considered within the system. (For this system, 0 designates the torso segment, 1 the upper arm segment, 2 the forearm segment, and 3 the hand segment.) Each matrix within the $\tfrac{1}{2}\{\omega_i\}^T\{I_{G_n}\}\{\omega_i\}$ terms must be expressed in the same coordinate system. In general, it is practical to select a coordinate system with an axis along the length of a body segment (e.g., upper arm, or forearm). This is demonstrated by the use of the b, c, and d frames in the figures. A transformation may be performed on the inertia matrix if another coordinate system is desired.

Euler angles are used in order to define the movement of each segment. More specifically, the angular velocities $\boldsymbol{\omega}_i$ of points B, C, D, and E are determined by using a 3-1-2 coordinate transfor-

TABLE 7.6 The Absolute Velocity of Point C as Expressed Relative to $\mathbf{b}_1$, $\mathbf{b}_2$, $\mathbf{b}_3$

	$\mathbf{b}_1$	$\mathbf{b}_2$	$\mathbf{b}_3$
$\boldsymbol{\omega}_B$	$\omega_{B_1} = -\dot{\psi}_1^b \cos\psi_2^b \sin\psi_3^b + \dot{\psi}_2^b \cos\psi_3^b$	$\omega_{B_2} = \dot{\psi}_1^b \sin\psi_2^b + \dot{\psi}_3^b$	$\omega_{B_3} = \dot{\psi}_1^b \cos\psi_2^b \cos\psi_3^b + \dot{\psi}_2^b \sin\psi_3^b$
$\mathbf{r}_{C/B} = \mathbf{r}_C$	0	r_C	0
$\mathbf{v}_B$			
$\mathbf{v}_{C/B} = \dot{\mathbf{r}}_{C/B}$	0	0	0
$\boldsymbol{\omega}_B \times \mathbf{r}_{C/B}$	$-r_C\omega_{B_3}$	0	$r_C\omega_{B_1}$

TABLE 7.7 The Absolute Velocity of Point D as Expressed Relative to $\mathbf{c}_1$, $\mathbf{c}_2$, $\mathbf{c}_3$

	$\mathbf{c}_1$	$\mathbf{c}_2$	$\mathbf{c}_3$
$\boldsymbol{\omega}_C$	$\omega_{C_1} = -\dot{\psi}_1^c \cos\psi_2^c \sin\psi_3^c + \dot{\psi}_2^c \cos\psi_3^c$	$\omega_{C_2} = \dot{\psi}_1^c \sin\psi_2^c + \dot{\psi}_3^c$	$\omega_{C_3} = \dot{\psi}_1^c \cos\psi_2^c \cos\psi_3^c + \dot{\psi}_2^c \sin\psi_3^c$
$\mathbf{r}_{D/C} = \mathbf{r}_D$	0	r_D	0
$\mathbf{v}_C$			
$\mathbf{v}_{D/C} = \dot{\mathbf{r}}_{D/C}$	0	0	0
$\boldsymbol{\omega}_C \times \mathbf{r}_{D/C}$	$-r_D\omega_{C_3}$	0	$r_D\omega_{C_1}$

mation, which is otherwise referred to as using Cardan angles (Allard et al., 1997). This transformation is commonly used within biodynamics to represent human segment movement and is chosen for its ability to define the movement of three-dimensional joints with limited ranges of motion in terms of anatomical movement (flexion/extension, adduction/adduction, and internal/external rotations). For the multisegment systems of Fig. 7.4, 7.5, and 7.6, ω_i is determined within Tables 7.5, 7.6, 7.7, and 7.8, assuming that each segment link is that of a ball-and-socket joint, or a globular or spherical pair. Unlike the revolute joint, the ball-and-socket joint has three axes of rotation and allows ω_i to have components in any direction. This assumption also prompts the introduction of a new set of symbols, which are somewhat different from the ones used previously, to describe the motion of each segment.

The $\mathbf{e}_1$, $\mathbf{e}_2$, $\mathbf{e}_3$ coordinate system is defined to generalize the discussion of the angular velocity derivations and represents the inertial frame of reference. The 3-1-2 transformation follows an initial rotation about the third axis, $\mathbf{e}_3$, by an angle of ψ_1 to yield the $\mathbf{e}_1'$, $\mathbf{e}_2'$, $\mathbf{e}_3'$ coordinate system. Then a second rotation is performed about the $\mathbf{e}_1'$ axis by an angle of ψ_2, yielding the $\mathbf{e}_1''$, $\mathbf{e}_2''$, $\mathbf{e}_3''$ system. Finally, a third rotation is performed about the $\mathbf{e}_2''$ axis by ψ_3 to yield the final $\mathbf{e}_1'''$, $\mathbf{e}_2'''$, $\mathbf{e}_3'''$ body frame of reference. This defines the transformation from the $\mathbf{e}_1$, $\mathbf{e}_2$, $\mathbf{e}_3$ system to the $\mathbf{e}_1'''$, $\mathbf{e}_2'''$, $\mathbf{e}_3'''$ system. To supplement the kinematics tables, an expression for the angular velocity vector is defined from this transformation as

$$\boldsymbol{\omega}_i = \dot{\psi}_1\mathbf{e}_3 + \dot{\psi}_2\mathbf{e}_1' + \dot{\psi}_3\mathbf{e}_2'' \tag{7.88}$$

where $\mathbf{e}_2'' = \mathbf{e}_2'''$ by nature of the rotations, and $\mathbf{e}_3$ and $\mathbf{e}_1'$ must be defined in terms of the $\mathbf{e}_1'''$, $\mathbf{e}_2'''$, $\mathbf{e}_3'''$ coordinate system. This is typically accomplished by using a figure that demonstrates the rotations and respective orientations of the coordinate systems. Euler angles and coordinate transformations are discussed in greater detail in the cited references and are also presented in a slightly different manner in Chap. 5, "Biomechanics of Human Movement."

TABLE 7.8 The Absolute Velocity of Point E as Expressed Relative to $\mathbf{d}_1$, $\mathbf{d}_2$, $\mathbf{d}_3$

	$\mathbf{d}_1$	$\mathbf{d}_2$	$\mathbf{d}_3$
$\boldsymbol{\omega}_D$	$\omega_{D_1} = -\dot{\psi}_1^d \cos\psi_2^d \sin\psi_3^d + \dot{\psi}_2^d \cos\psi_3^d$	$\omega_{D_2} = \dot{\psi}_1^d \sin\psi_2^d + \dot{\psi}_3^d$	$\omega_{D_3} = \dot{\psi}_1^d \cos\psi_2^d \cos\psi_3^d + \dot{\psi}_2^d \sin\psi_3^d$
$\mathbf{r}_{E/D} = \mathbf{r}_E$	0	r_E	0
$\mathbf{v}_D$			
$\mathbf{v}_{E/D} = \dot{\mathbf{r}}_{E/D}$	0	0	0
$\boldsymbol{\omega}_D \times \mathbf{r}_{E/D}$	$-r_E\omega_{D_3}$	0	$r_E\omega_{D_1}$

The gravitational potential energy of the system, expressed in vector form, is given as

$$V = (m_A \mathbf{r}_{G_0} + m_B \mathbf{r}_{G_1} + m_C \mathbf{r}_{G_2} + m_D \mathbf{r}_{G_3}) \cdot (-g)\mathbf{j} \qquad (7.89)$$

The unit vector **j**, according to Figs. 7.4, 7.5, and 7.6, is in the inertial coordinate system and is always directed upward. Taking a dot product between any quantity and this unit vector results in the vertical component of the quantity. As a result of the dot products in both Eqs. (7.87) and (7.89), the resulting kinetic and potential energies are scalar quantities. As before, these quantities can be incorporated into Lagrange's equation to determine the equations of motion for the system.

7.6 BRIEF DISCUSSION

7.6.1 Forces and Constraints

Forces play an integral role in the dynamic behavior of all human mechanics. In terms of human movement, forces can be defined as intrinsic or extrinsic. For example, a couple about a particular joint will involve the intrinsic muscle and frictional forces as well as any extrinsic loads sustained by the system. If the influence of intrinsic muscle activity within the system is to be considered, the location of the insertion points for each muscle must be determined to properly solve the equations of motion.

Conservative forces due to gravity and elasticity are typically accounted for within the terms defining the potential energy of the system, while inertial forces are derived from the kinetic energy. Forces due to joint friction, tissue damping, and certain external forces are expressed as nonconservative generalized forces.

In biodynamic systems, motions that occur between anatomical segments of a joint mechanism are not completely arbitrary (free to move in any manner). They are constrained by the nature of the joint mechanism. As a result, the structure of the joint, the relative motions the joint permits, and the distances between successive joints must be understood in order to properly determine the kinematics of the system.

The Lagrangian approach presented within this section has been limited to unconstrained systems with appropriately selected generalized coordinates that match the degrees of freedom of the system. For a system where constraints are to be considered, Lagrange multipliers are used with the extended Hamilton's principle (Baruh, 1999). Each constraint can be defined by a constraint equation and a corresponding constraint force. For any dynamic system, the constraint equations describe the geometries and/or the kinematics associated with the constraints of the system. For a biodynamic joint system, the contact force between the segments linked by the joint would be considered a constraint force. Constraint forces may also involve restrictions on joint motion due to orientations and interactions of the soft tissues (e.g., ligamentous, tendonous, and muscular structures) that surround the joint.

7.6.2 Hamilton's Principle

In cases where equations of motion are desired for deformable bodies, methods such as the extended Hamilton's principle may be employed. The energy is written for the system and, in addition to the terms used in Lagrange's equation, strain energy would be included. Application of Hamilton's principle will yield a set of equations of motion in the form of partial differential equations as well as the corresponding boundary conditions. Derivations and examples can be found in other sources (Baruh, 1999; Benaroya, 1998). Hamilton's principle employs the calculus of variations, and there are many texts that will be of benefit (Lanczos, 1970).

7.7 IN CLOSING

This chapter is presented as an introduction to the use of the Lagrangian approach to biodynamically model human mechanics. Several important aspects of dynamics are briefly introduced and discussed, and may require a review of the literature for more detailed explanations and additional examples. Assumptions were made within each example to simplify the solution and provide a clear presentation of the material.

Further applications may consider dynamic systems that involve adding two or more viscoelastic or elastic bodies to the single-body pendulum examples. As a result, solutions defining the dynamic behavior of a multisegment pendulum problem would be determined. Combinations of viscoelastic and elastic segments may also be linked together, but may add to the complexity of the solutions because of the elasticity variations between segments. Other applications may include various combinations of spring and dashpot systems, such as a Maxwell model or a Kelvin body, to further study the effects of viscoelasticity on a dynamic system.

The multisegment extremity model demonstrated the ability to subsequently add segments to a base model and determine the equations of motion with each addition. These models were derived with the assumption that the links between segments were revolute joints. Further modifications of this example may involve combinations of revolute and ball-and-socket joints to more accurately model an actual biodynamic system. The final example (Tables 7.5, 7.6, 7.7, and 7.8) begins to solve a system that assumes all links to be of a ball-and-socket type. If one those links is assumed to be a revolute joint (e.g., point C, the elbow), then the appropriate angles ψ and angular velocities $\dot{\psi}$ for the adjoining segments would be negligible on the basis of the constraints of a revolute joint.

REFERENCES

Allard, P., Cappozzo, A., Lundberg, A., and Vaughan, C. L., *Three-dimensional Analysis of Human Locomotion*, John Wiley and Sons, New York, 1997.

Baruh, H., *Analytical Dynamics*, McGraw-Hill, New York, 1999.

Benaroya, H., *Mechanical Vibration: Analysis, Uncertainties, and Control*, Prentice Hall, Englewood Cliffs, N. J. 1998.

Harrison, H. R., and Nettleton, T., *Advanced Engineering Dynamics*, John Wiley and Sons, New York, 1997.

Lanczos, C., *The Variational Principles of Mechanics*, Dover, New York, 1970.

Meirovitch, L: *Methods of Analytical Dynamics*, McGraw-Hill, New York, NY, 1970

Moon, F. C., *Applied Dynamics with Applications to Multibody and Mechatronic Systems*, J. Wiley and Sons, New York, 1998.

Peterson, D. R., "A Method for Quantifying the Biodynamics of Abnormal Distal Upper Extremity Function: Application to Computer Keyboard Typing," Ph.D. Dissertation, University of Connecticut, 1999.

Wells, D. A., *Theory and Problems of Lagrangian Dynamics*, McGraw-Hill, New York, 1967.

CHAPTER 8
BONE MECHANICS

Tony M. Keaveny
University of California, San Francisco, California and University of California, Berkeley, California

Elise F. Morgan
University of California, Berkeley

Oscar C. Yeh
University of California, Berkeley

8.1 INTRODUCTION

Bone is a complex tissue that is continually being torn down and replaced by biological remodeling. As the main constituent in whole bones (which as organs contain other tissues such as bone marrow, nerves, and blood vessels), the two types of bone tissue—cortical and trabecular bone—have the functional task of withstanding substantial stress during the course of locomotion and strenuous activities such as lifting heavy weights or fast running. Since bones are loaded both cyclically and statically, fatigue and creep responses are important aspects of their mechanical behavior. Indeed, there is evidence that a primary stimulus for bone remodeling is the repair of damage that accumulates from habitual cyclic loading.[1,2] With aging, however, the balance between bone loss and gain is disrupted, and bone deteriorates, leading to a variety of devastating clinical problems. In modern populations, fractures from osteoporosis are becoming increasingly common, the spine, hip, and wrist being the primary sites. Implantation of orthopedic prostheses for conditions such as disc degeneration and osteoarthritis require strong bone for optimal fixation, a difficult requirement for sites such as the aged spine or hip, where bone strength can be greatly compromised. The goal of this chapter is to summarize the highlights of what is known about the mechanical behavior of bone as a material. With a focus on the behavior of human bone, we review the mechanics of cortical bone, trabecular bone, and trabecular tissue material. Rather than attempt to provide an encyclopedic review of the literature, our intent is to provide a focused summary that will be most useful as input for biomechanical analyses of whole bone and bone-implant systems.

8.2 COMPOSITION OF BONE

At the nanometer scale, bone tissue is composed of inorganic and organic phases and water. On a weight basis, bone is approximately 60 percent inorganic, 30 percent organic, and 10 percent water,[3] whereas on a volume basis, these proportions are about 40 percent, 35 percent, and 25 percent, respectively. The inorganic phase of bone is a ceramic crystalline-type mineral that is an impure form of naturally occurring calcium phosphate, most often referred to as *hydroxyapatite*: $Ca_{10}(PO_4)_6(OH)_2$.[4] Bone hydroxyapatite is not pure hydroxyapatite because the tiny apatite crystals (2- to 5-nm-thick × 15-nm-wide × 20- to 50-nm-long plates) contain impurities such as potassium, magnesium, strontium, and sodium (in place of the calcium ions), carbonate (in place of the phosphate ions), and chloride or fluoride (in place of the hydroxyl ions). The organic phase of bone consists primarily of type I collagen (90 percent by weight), some other minor collagen types (III and VI), and a variety of noncollagenous proteins such as osteocalcin, osteonectin, osteopontin, and bone sialoprotein.[5] The collagen molecules are arranged in parallel with each other head to tail with a gap or "hole zone" of approximately 40 nm between each molecule.[6] Mineralization begins in the hole zones and extends into other intermolecular spaces, resulting in a mineralized fibril. The three-dimensional arrangement of collagen molecules within a fibril is not well understood. However, collagen fibrils in bone range from 20 to 40 nm in diameter, suggesting that there are 200 to 800 collagen molecules in the cross section of a fibril.

8.3 BONE AS A HIERARCHICAL COMPOSITE MATERIAL

At the micron scale and above, bone tissue is a hierarchical composite (Fig. 8.1). At the lowest level ($\approx$0.1-μm scale), it is a composite of mineralized collagen fibrils. At the next level (1- to 10-μm scale), these fibrils can be arranged in two forms, either as stacked thin sheets called *lamellae* (about 7 μm thick) that contain unidirectional fibrils in alternating angles between layers or as a block of randomly oriented *woven* fibrils. Lamellar bone is most common in adult humans, whereas woven bone is found in situations of rapid growth, such as in children and large animals, as well as during the initial stages of fracture healing.

Lamellar bone is found in different types of histological structures at the millimeter scale. Primary lamellar bone is new tissue that consists of large concentric rings of lamellae that circle the outer 2 to 3 mm of diaphyses similar to growth rings on a tree. The most common type of cortical bone in adult humans is osteonal or Haversian bone, where about 10 to 15 lamellae are arranged in concentric cylinders about a central Haversian canal, a vascular channel about 50 μm in diameter that contains blood vessel capillaries, nerves, and a variety of bone cells (Fig. 8.2*a*). The substructures of concentric lamellae including the Haversian canal is termed an *osteon*, which has a diameter of about 200 μm and lengths of 1 to 3 mm. Volkmann's canals are about the same diameter as Haversian canals but run transverse to the diaphyseal axis, providing a radial path for blood flow within the bone. Osteons represent the main discretizing unit of human adult cortical bone and are continually being torn down and replaced by the bone remodeling process. Over time, the osteon can be completely removed, leaving behind a resorption cavity that is then filled in by a new osteon. Typically, there are about 13 Haversian canals per square millimeter of adult human cortical bone. Since mineralization of a new osteon is a slow process that can take months, at any point in time there is a large distribution of degree of mineralization of osteons in any whole-bone cross section. A *cement*

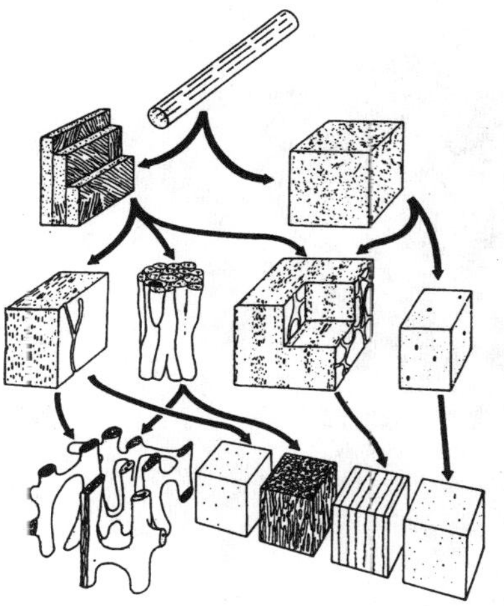

FIGURE 8.1 The four levels of bone microstructure, from the level of mineralized collagen fibrils to cortical and trabecular bone. It is generally assumed that at the former level, all bone is equal, although there may be subtle differences in the nature of the lamellar architecture and degree of mineralization between cortical and trabecular bone. (*Adapted from Ref. 145.*)

line, which is a thin layer of calcified mucopolysaccharides with very little collagen and low mineral content,[7] remains around the perimeter of each newly formed osteon. The cement line is thought to represent a weak interface between the osteon and the surrounding interstitial bone.[8] These weak interfaces are thought to improve the fatigue properties of cortical bone by providing avenues for dissipation of energy during crack propagation.[7]

The bone matrix that comprises lamellar and woven bone contains another level of porosity on the order of 5 to 10 μm that is associated with the bone cells (see Fig. 8.2*a*, *b*, *c*). Osteocytes, the most common type of bone cell, are surrounded by a thin layer of extracellular fluid within small ellipsoidal holes (5 μm minor diameter, 7 to 8 μm major diameter) called *lacunae*, of which there are about 25,000 per μm^3 in bone tissue. The lacunae are generally arranged along the interfaces between lamellae. However, the lacunae also have a lower-scale porosity associated with them. Each osteocyte has dendritic processes that extend from the cell through tiny channels ($\approx$0.5 μm diameter, 3 to 7 μm long) called *canaliculi* to meet at cellular gap junctions with the processes of surrounding cells.* There are about 50 to 100 canaliculi per single lacuna and about 1 million per mm^3 of bone.

At the highest hierarchical level (1 to 5 mm), there are two types of bone: cortical bone, which comes as tightly packed lamellar, Haversian, or woven bone; and trabecular bone, which comes as a highly porous cellular solid. In the latter, the lamellae are arranged in less well-organized "packets" to form a network of rods and plates about 100 to 300 μm thick interspersed with large marrow spaces. Many common biological materials, such as wood and cork, are cellular solids.[9]

The distinction between cortical and trabecular bone is most easily made based on porosity. Cortical bone can be defined as bone tissue that has a porosity P of less than about 30 percent or, equivalently, a volume fraction V_f of greater than about 0.70 ($V_f = 1 - P$). Volume fraction is the ratio of the volume of actual bone tissue to the bulk volume of the specimen. In the field of bone mechanics, porosity measures usually ignore the presence of lacunae and canaliculi. Porosity of adult

* Gap junctions are arrays of small pores in the cell membrane that make connections between the interiors of neighboring cells, allowing direct passage of small molecules such as ions from one cell to another.

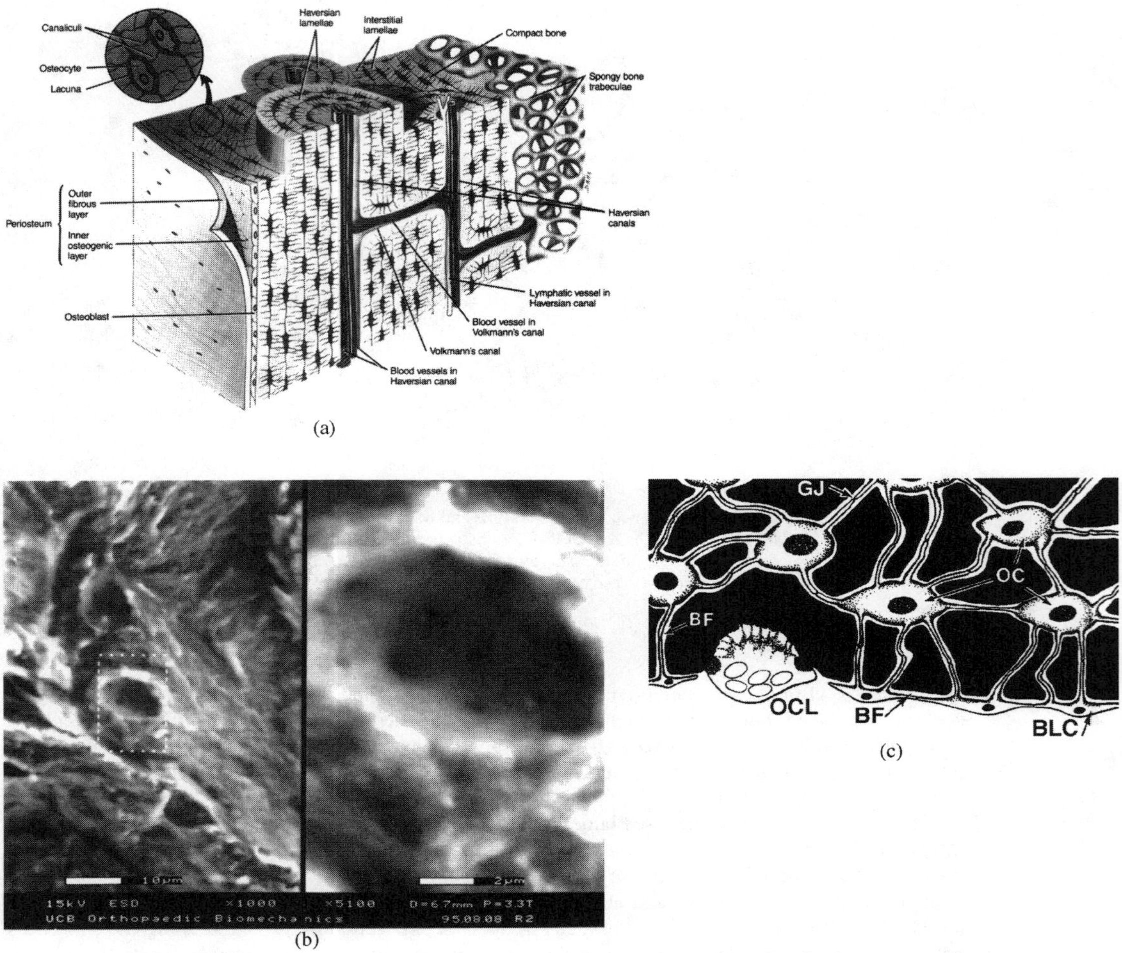

FIGURE 8.2 (*a*) Diagram of a sector of the shaft of a long bone showing the different types of cortical bone, trabecular bone, and the various channels. (*From Figure 5–1d of Ref. 146.*) (*b*) Environmental scanning electron micrograph of a fracture surface of a piece of cortical bone showing a fractured lacuna at low (*left*) and high (*right*) magnifications. Note the elliptical shape of the lacuna and the multiple smaller canaliculi. (*c*) A schematic depicting the interconnection of osteocytes (OC) via the cell processes that extend along the canaliculi and meet at gap junctions (GJ). Bone-lining cells (dormant osteoblasts, BLC) lie on each exposed bone surface, where osteoclasts (OCL) can be found removing bone as part of the ongoing remodeling process. A thin layer of bone fluid (BF) surrounds the cells and their processes.

human femoral cortical bone, for example, can vary from as low as 5 percent at age 20 up to almost 30 percent above age 80.[10] Porosity of trabecular bone can vary from 70 percent in the femoral neck[11] up to about 95 percent in the elderly spine.[12]

Two other common measures of bone density in biomechanical studies are termed *tissue* and *apparent densities*. Tissue density ρ_{tiss} is defined as the ratio of mass to volume of the actual bone tissue. It is similar for cortical and trabecular bone, varies little in adult humans, and is about 2.0 g/cm^3. Apparent density ρ_{app} is defined as the ratio of the mass of bone tissue to the bulk volume of the specimen, including the volume associated with the vascular channels and higher-level porosity. Volume fraction, tissue density, and apparent densities are related as follows:

$$\rho_{app} = \rho_{tiss} V_f$$

Typically, mean values of apparent density of hydrated cortical bone are about 1.85 g/cm^3, and this does not vary much across anatomic sites or species. By contrast, the average apparent density of trabecular bone depends very much on anatomic site. It is as low as 0.10 g/cm^3 for the spine,[13] about 0.30 g/cm^3 for the human tibia,[14] and up to about 0.60 g/cm^3 for the load-bearing portions of the proximal femur.[11] After skeletal maturity (around ages 25 to 30), trabecular bone density decreases steadily with aging, at a rate of about 6 percent per decade.[15]

Spatially, the relatively high porosity of trabecular bone is in the form of a network of interconnected pores filled with bone marrow. The trabecular tissue forms an irregular lattice of small rods and plates that are called *trabeculae* (Fig. 8.3). Typical thicknesses of individual trabeculae are in the range 100 to 300 μm, and typical intertrabecular spacing is on the order of 500 to 1500 μm.[16] The spatial arrangement of the trabeculae is referred to as the *trabecular architecture*. Architectural type varies across anatomic site and with age. Bone from the human vertebral body tends to be more rodlike, whereas bone from the bovine proximal tibia consists almost entirely of

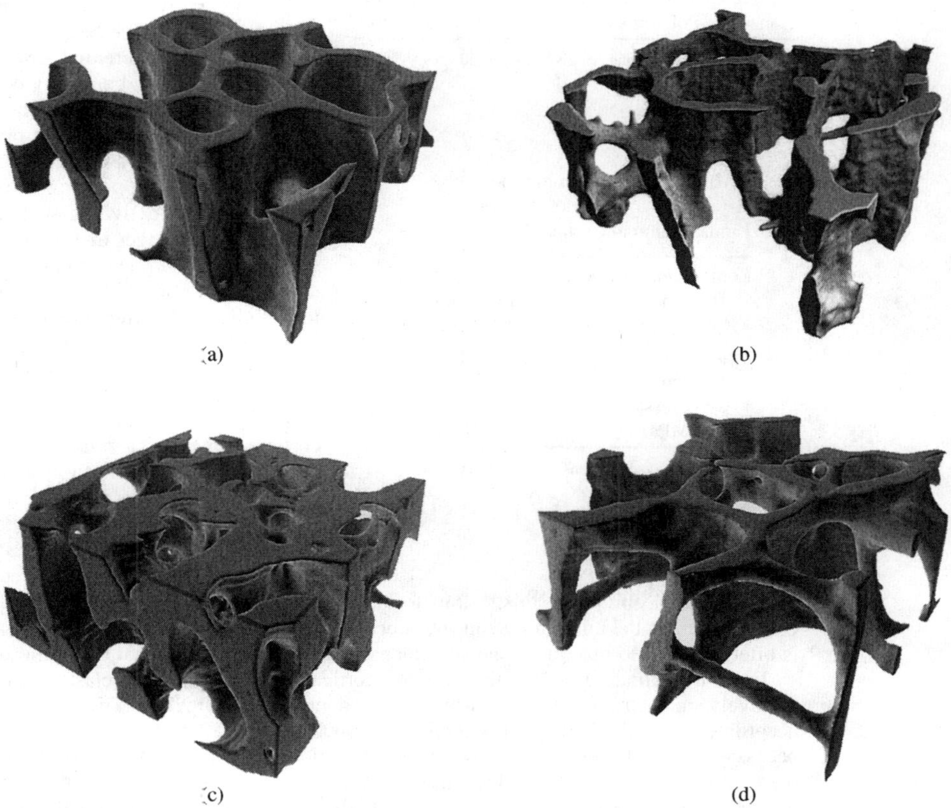

(a) (b) (c) (d)

FIGURE 8.3 Three-dimensional reconstructions of trabecular bone from the (*a*) bovine proximal tibia, (*b*) human proximal tibia, (*c*) human femoral neck, (*d*) human vertebra. Each volume is 3 × 3 × 1 mm^3. (*From Ref. 142.*)

plates. As age increases and volume fraction decreases, the architecture becomes increasingly rodlike, and these rods become progressively thin and can be perforated. Quantification of trabecular architecture with the intent of understanding its role in the mechanical behavior of trabecular bone has been the subject of intense research. In addition to calculating trabecular thickness, spacing, and surface-to-volume ratio, stereological and three-dimensional methods may be used to determine the mean orientation (main grain axis) of the trabeculae, connectivity, and the degree of anisotropy.[17] While earlier studies used two-dimensional sections of trabecular bone to perform these architectural analyses,[18,19] more recent investigations use three-dimensional reconstructions generated by micro-computed tomography and other high-resolution imaging techniques.[16,20–22]

8.4 MECHANICAL PROPERTIES OF CORTICAL BONE

TABLE 8.1 Anisotropic Elastic Properties of Human Femoral Cortical Bone

Longitudinal modulus (MPa)	17,900 (3900)*
Transverse modulus (MPa)	10,100 (2400)
Shear modulus (MPa)	3,300 (400)
Longitudinal Poisson's ratio	0.40 (0.16)
Transverse Poisson's ratio	0.62 (0.26)

*Standard deviations are given in parentheses.
Source: Data from Ref. 150.

TABLE 8.2 Anisotropic and Asymmetrical Ultimate Stresses of Human Femoral Cortical Bone

Longitudinal (MPa)	
Tension	135 (15.6)*
Compression	205 (17.3)
Transverse (MPa)	
Tension	53 (10.7)
Compression	131 (20.7)
Shear (MPa)	65 (4.0)

*Standard deviations are given in parentheses.
Source: Data from Ref. 150.

Reflecting the anisotropy of its microstructure, the elastic and strength properties of human cortical bone are anisotropic. Cortical bone is both stronger and stiffer when loaded longitudinally along the diaphyseal axis compared with the radial or circumferential "transverse" directions (Table 8.1). Comparatively smaller differences in modulus and strength have been reported between the radial and circumferential directions, indicating that human cortical bone may be treated as transversely isotropic. This is probably a reflection of its evolutionary adaptation to produce a material that most efficiently resists the largely uniaxial stresses that develop along the diaphyseal axis during habitual activities such as gait. Cortical bone is also stronger in compression than in tension (Table 8.2). The percentage strength-to-modulus ratio for cortical bone is about 1.12 and 0.78 for longitudinal compression and tension, respectively. Compared with high-performance engineering metal alloys such as aluminum 6061-T6 and titanium 6Al-4V with corresponding ratios of about 0.45 and 0.73, respectively, it is seen that cortical bone has a relatively large strength-to-modulus ratio. In this sense, it can be considered a relatively high-performance material, particularly for compression. It should be noted that these properties only pertain to its behavior when loaded along the principal material direction. If the specimen is loaded oblique to this, a transformation is required to obtain the material constants. This consequence of the anisotropy can introduce technical challenges in biomechanical testing since it is often difficult to machine bone specimens in their principal material orientations.

From a qualitative perspective, human cortical bone is a linearly elastic material that fails at relatively small strains after exhibiting a marked yield point (Fig. 8.4). This yield point is determined according to standard engineering definitions such as the 0.2 percent offset technique and does not necessarily reflect plasticity. However, when cortical bone is loaded to close to its yield point and then unloaded, permanent residual strains develop (Fig. 8.5). Unlike the ultimate stresses, which are higher in compression, ultimate strains are higher in tension for longitudinal loading. These longitudinal tensile ultimate strains can be up to 5 percent in young adults but decrease to about 1 percent in the elderly.[10] Cortical bone is relatively weak in shear but is weakest when loaded transversely in tension (see Table 8.2). An example of such loading is the circumferential or "hoop" stress that

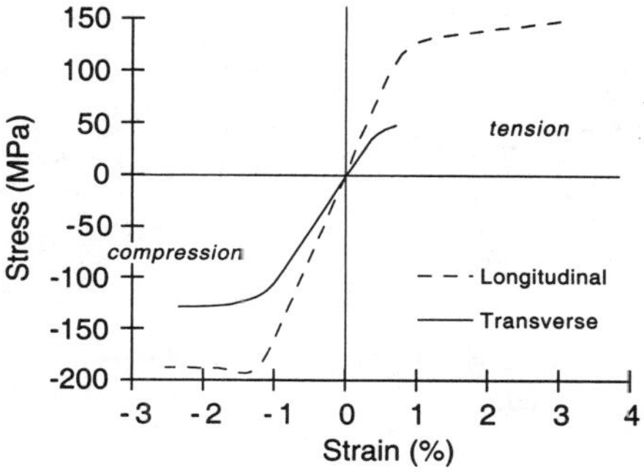

FIGURE 8.4 Typical stress-strain behavior for human cortical bone. The bone is stiffer in the longitudinal direction, indicative of its elastic anisotropy. It is also stronger in compression than in tension, indicative of its strength asymmetry (modulus is the same in tension and compression). (*From Ref. 9.*)

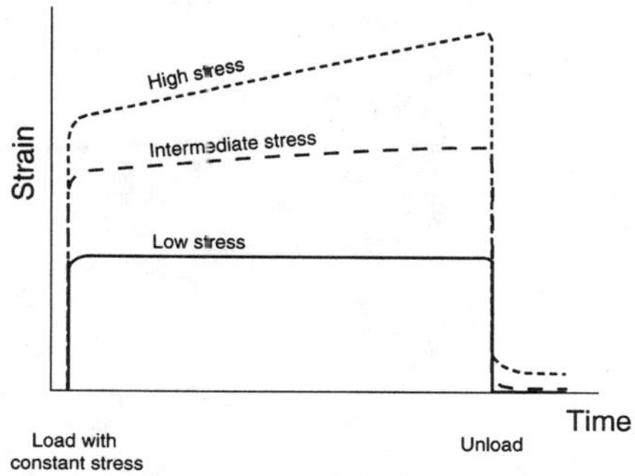

FIGURE 8.5 Creep response of cortical bone for three different stress levels. When a low stress is applied to the bone, the strain remains constant over time, and there is no permanent deformation after unloading. For stresses just below yield, strains increase with time at a constant rate, and a small permanent deformation exists after unloading. As the magnitude of the stress is increased, the rate of creep increases, and a larger permanent deformation exists after unloading. (*From Ref. 108.*)

can develop when large intramedullary tapered implants such as uncemented hip stems are impacted too far into the diaphysis.

While it is often appropriate to assume average properties for cortical bone, as shown in Tables 8.1 and 8.2, it may be necessary in some cases to account for the heterogeneity that can arise from variations in microstructural parameters such as porosity and percentage mineralization. Both modulus and ultimate stress can halve when porosity is increased from 5 to 30 percent[10,23] (Fig. 8.6*a*). Small increases in percentage mineralization cause large increases in both modulus and strength (see Fig. 8.6*b*), and while this parameter does not vary much in adult humans,[10] it can vary substantially across species.[24]

Aging also affects the mechanical properties of cortical bone. Tensile ultimate stress decreases at a rate of approximately 2 percent per decade[25] (Fig. 8.7*a*). Perhaps most important, tensile ultimate strain decreases by about 10 percent of its "young" value per decade, from a high of almost 5 percent

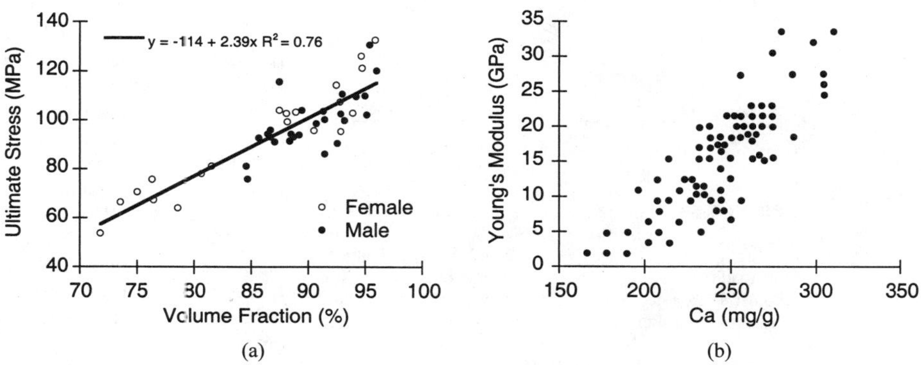

FIGURE 8.6 (*a*) Dependence of the ultimate tensile stress of human cortical bone on volume fraction (expressed as a percentage). Ages of the specimens were in the range 20 to 100 years. (*From Ref. 10.*) (*b*) Modulus versus calcium content (in mg/gm of dehydrated bone tissue) for cortical bone taken from 18 different species. (*From Ref. 24.*)

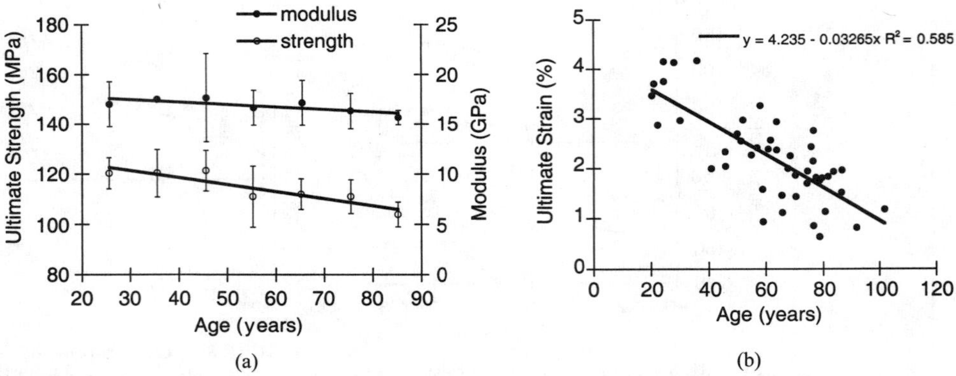

FIGURE 8.7 Reductions of human cortical bone mechanical properties with age. (*a*) Modulus is not reduced much, if at all, whereas strength is reduced more, at a rate of about 2 percent per decade. (*From Ref. 25.*) (*b*) Ultimate strain decreases markedly with age, at a rate of about 10 percent of its young value per decade. (*From Ref. 10.*)

strain at ages 20 to 30 years to a low of less than 1 percent strain above age 80 years[10] (see Fig. 8.7*b*). Thus the energy to fracture, given by the total area under the stress-strain curve before fracture, is much less for old bone than for younger bone. As discussed below, fracture mechanics studies also show a decrease in the fracture toughness with aging. For these reasons, old cortical bone is more brittle than young bone. It is not currently clear if this age-related brittleness arises from hypermineralization or collagen changes, although it appears that the latter is more plausible, since mineralization does not change much in adult humans with aging.[10] Many of these age-related changes in mechanical properties are to be expected, since porosity increases with age. However, there are concurrent changes in other aspects of the tissue microstructure and composition such that porosity is not simply a surrogate measure of age. For example, although strength and ductility clearly decrease with age in adults, there is controversy over whether elastic modulus changes with age.[10,25,26]

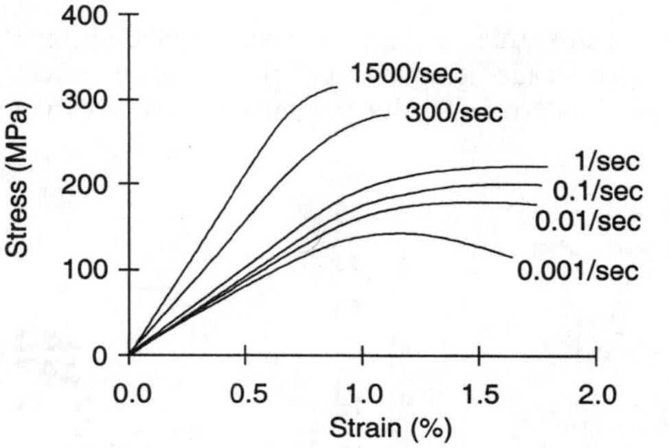

FIGURE 8.8 Strain-rate sensitivity of cortical bone for longitudinal tensile loading. Typically, modulus and strength increase only by factors 2 and 3, respectively, as the loading rate is increased by 6 orders of magnitude. The higher strain rates shown here may occur in vehicular accidents or gunshot wounds. (*Data from Ref. 27.*)

Although cortical bone is viscoelastic, the effect of loading rate on modulus and strength is only moderate. Over a six order of magnitude increase in strain rate, modulus only changes by a factor of 2 and strength by a factor of 3[27] (Fig. 8.8). Thus, for the majority of physiological activities that tend to occur in a relatively narrow range of strain rates (0.01 to 1.0 percent strain per second), the monotonic response of cortical bone reasonably can be assumed to have minor rate effects. Similarly, dynamic sinusoidal experiments indicate that the loss tangent attains a broad minimum (0.01 to 0.02) over the range of physiological frequencies.[28,29] These values, which are lower than those for polymers by a factor of 10, indicate that significant mechanical damping does not occur within this frequency range. Clearly, in extraordinary situations such as high-speed trauma, strength properties can increase by a factor of 2 to 3, and this should be recognized. Additionally, it has been found that loading rate has a significant effect on the accumulation of damage within the tissue. Slower loading rates produce higher numbers of acoustic emission events, but these events are of lower amplitude than those emitted at faster rates.[30]

Multiaxial failure properties of cortical bone are not well understood, although it is clear that simple isotropic and symmetrical criteria such as the von Mises are not capable of describing the multiaxial strength properties of this tissue. The Tsai-Wu criterion, commonly used for fiber-reinforced composite materials, has been applied to cortical bone using both transversely isotropic[31] and orthotropic[32] treatments. The transversely isotropic case works quite well for axial-shear-loading configurations,[31] but neither this case nor the orthotropic one has been validated across the full range of multiaxial stresses. Regardless, this criterion accounts for the difference in tensile and compressive strengths, as well as the low shear strength with respect to the tensile strength, and in this sense is the most suitable criterion currently available.

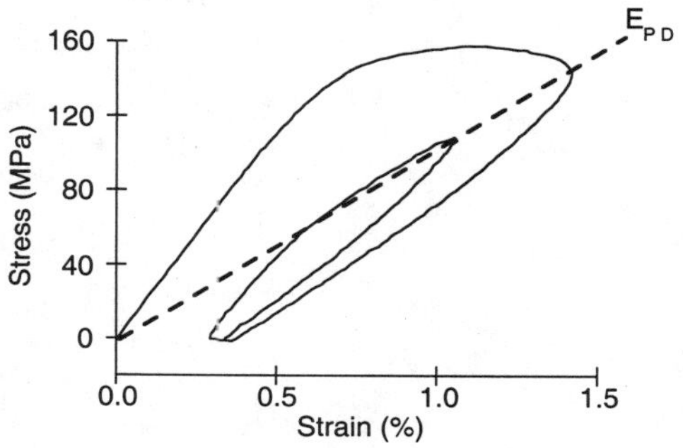

FIGURE 8.9 Comparison of loading and reloading tensile stress-strain curves for human cortical bone. On reloading, the modulus is similar to that for initial loading, but it is quickly reduced to a value that is close to the *perfect damage modulus*, the secant modulus at the unloading point. Substantial residual strains are evident even after a 1- to 2-minute hold between loading cycles. (*Data from Ref. 148.*)

Cortical bone exhibits mechanical property degradations characteristic of a damaging material. When cortical bone is loaded beyond its yield point, unloaded, and reloaded, its modulus is reduced[33,34] (Fig. 8.9). This evidence of mechanical damage does not occur for metals, for which the reloading modulus after yield is the same as the initial modulus. Studies using acoustic emissions to monitor structural changes in the tissue during monotonic loading to failure support the idea that the postyield behavior of cortical bone is damage-related.[35,36] Fatigue loading can also induce modulus reductions, and these reductions are accompanied by increases in energy dissipation per cycle.[37,38] Similar to engineering composites, the secant modulus exhibits a gradual reduction in stiffness until the final 5 percent of fatigue life, at which point the stiffness declines sharply until complete fracture.[38] However, there may be a load threshold below which this fatigue damage does not occur.[39] Cortical bone has a greater resistance to fatigue failure in compression than in tension, and the effect of mean strain on fatigue life is negligible.[37,40] For strain amplitude controlled tests, the following life prediction has been reported for human femoral cortical bone[37]:

$$N_f = (2.94 \times 10^{-9})\Delta\epsilon^{-5.342} \qquad n = 68$$

where N_f is the number of cycles to failure, and $\Delta\epsilon$ is the applied strain amplitude. The standard error of the estimate for this regression on the log-transformed data is 0.4085.[37] Interestingly, creep appears to be an intrinsic component of the fatigue behavior. With increasing numbers of cycles, increasing creep strains can be observed.[38] When fatigue and creep behaviors are expressed as functions of stress/modulus versus time to failure, fatigue life is independent of frequency (0.2- to 2.0-Hz range), and substantial similarities appear between the fatigue and creep behaviors[40,41] (Fig. 8.10).

Microscopy techniques have established the presence of histological damage in cortical bone in vivo. Collectively termed *microdamage*, the patterns of damage include longitudinal and transverse microcracks, diffuse damage, and cross-hatched shear band patterns. It appears that histological damage increases with age[42,43] and is more pronounced in women.[43,44] These correlations have fueled a large body of research attempting to determine a relationship between mechanical property degradations and microdamage. True cause-and-effect relationships have not been established and have been controversial. The ability to detect microdamage at a high enough resolution, as well as to quantify it unambiguously, has been proposed as a confounding factor.

Damage may have direct biological consequences since the underlying cells will undergo structural damage as the surrounding bone matrix permanently deforms and sustains microdamage. This cellular damage may induce a biological response, perhaps prompting the bone cells to repair the subtle matrix damage. This is an important point when interpreting fatigue or creep properties be-

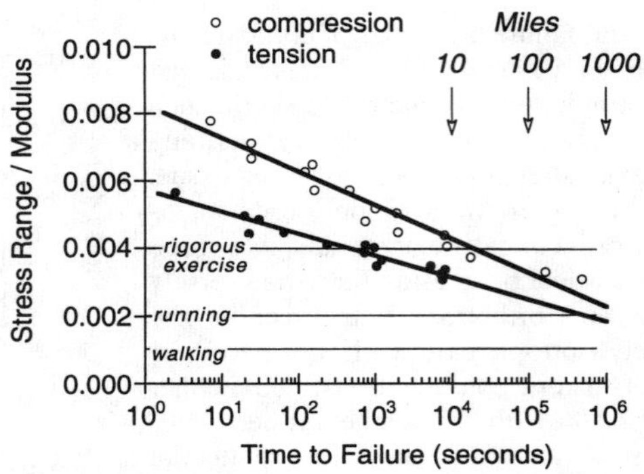

FIGURE 8.10 Fatigue and creep behaviors of human cortical bone versus time to failure. For fatigue loading, the ordinate on this graph can be converted to number of cycles by multiplying the time to failure by the frequency, which is typically one cycle per second for normal walking. Note that both creep and fatigue resistance are lower in tension, consistent with monotonic behavior. *(Data from Refs. 37 and 41.)*

cause it should be realized that no biological healing can occur during in vitro experiments. Thus the preceding fatigue characteristics are best considered as lower bounds on the in vivo fatigue life (see Fig. 8.10). It is unlikely that high-cycle (low-stress) fatigue failure occurs in vivo since the resulting fatigue damage would be healed biologically before large enough cracks could develop that would cause overt fracture of the bone. However, it should also be noted that the increase in porosity associated with the initial stages of the bone remodeling process may actually weaken the bone tissue even as the process attempts to strengthen it.

Fracture mechanics has been applied to cortical bone to determine its resistance to crack initiation and propagation. Various experimental techniques involving single-edge-notch (SEN), center-notched-cylindrical (CNC), and compact-tension (CT) specimens have been used to measure critical stress intensity factor K_c and critical energy release rate G_c. Size requirements of standard fracture toughness tests cannot be strictly met due to limitations on the size of available human tissue. Therefore, experimentally determined values of fracture toughness depend on specimen geometry and do not reflect a true material property. Additionally, plane-strain conditions and the associated relationships between K_c and G_c cannot be assumed. Theoretical developments[45] and tests on larger bones (see Ref. 46 for review), such as bovine tibiae, have been used to determine correction factors that are used to account for specimen geometry. Comparisons of reported values should be made with care because some studies do not attempt to correct for specimen geometry but rather report values in a comparative sense only consistent with the specific study.

Average values for K_c and G_c range from 2 to 6 $MNm^{-3/2}$ and 50 to over 1000 N/m, respectively, for specimens oriented such that the crack propagated along the longitudinal axis of the long bone

TABLE 8.3 Fracture Toughness Values per Anatomic Site for Human Cortical Bone

	G_{Ic} (N/m)		G_{IIc} (N/m)	
Anatomic site	Mean	SD	Mean	SD
Femoral neck	1128	344	5642	1272
Fermoral shaft	706	288	1817	1090
Tibial shaft	816	327	5570	1749

Source: Data from Ref. 45.

(Table 8.3). These values are similar, for example, to those of polystyrene. This orientation has been found to be the weaker in terms of crack resistance relative to the transverse orientation (ratio of 1: 1.75 for bovine bone[46]). Fracture toughness decreases with age in the diaphyses of long bones[26,47] at a rate of 4.1 percent per decade in the femoral shaft.[26] Fracture toughness in mode II loading can be greater by as much as fivefold than that in mode I.[45,48] The most significant factors that are correlated with fracture toughness are age and a variety of porosity-related measures (including apparent density, water content, and osteon density).[47,48]

A number of studies have developed micromechanical models for the elastic and strength properties of cortical bone. This work has been motivated by observations that changes in the mechanical properties with age and disease are accompanied by alterations in the tissue microstructure.[49] While it is generally agreed on that bone behaves mechanically as a composite material, the complex hierarchy of composite structure has led to disagreements over which scale or scales are most relevant for a particular aspect of the mechanical behavior. For instance, ultrastructural models have focused on the role of the mineral phase as reinforcement for the collagen matrix,[50,51] whereas macrostructural models have treated the Haversian systems as fibers embedded in a matrix of cement lines and interstitial lamellae.[52–54] On an intermediate scale, the individual mineralized collagen fibers have been modeled as reinforcement for the noncollagenous mineralized matrix.[55,56] Still another class of models has used a hierarchical approach to synthesize two or more of these different length scales.[57–59] These modeling efforts rely extensively on accurate information regarding the constituent and interface properties. Obtaining this information has proven challenging not only because of the extremely small scale involved but also because, unlike the case with engineering composites, isolating the different phases of bone tissue often involves processes that may alter the properties being measured.[60] Recent studies using nanoindentation,[61,62] for example, have begun to address the issue of scale by providing the ability to measure the elastic modulus and microhardness of various microstructures within cortical tissue. Thus efforts are ongoing to develop micromechanical constitutive models for cortical bone.

It is also of keen interest to determine the role of mechanical stimuli on bone cells and the interaction between cell biology and bone mechanical properties. Biological processes that are active throughout an individual's lifetime can alter the bone tissue on many scales. It has often been suggested that osteocytes act as sensors of mechanical loading and initiators of the bone-adaptation processes.[63–66] Whether these processes add or remove bone, or whether they are activated at all, is thought to depend on the level of mechanical loading[67] and on the presence or absence of tissue damage.[1,2] Many studies have suggested that strain or strain rate is the appropriate measure of mechanical stimulus,[68–70] although others have used stress or strain energy.[71,72] In addition, other characteristics of the mechanical loading, such as mode, direction, frequency, duration, and distribution, have also been identified as important. Using this collective empirical evidence, several theories of mechanical adaptation have been proposed (see Ref. 73 for a comprehensive treatment of this topic). When applied to finite-element models of whole bones, some of these theories have been able to predict the density pattern of the trabecular bone in the proximal femur[74,75] and the loss of bone in the regions surrounding a hip implant.[76] This remains an area of active research since experimentally validated mechanobiological constitutive relations of the remodeling process have yet to be developed.

8.5 MECHANICAL PROPERTIES OF TRABECULAR BONE

Although trabecular bone—also referred to as *cancellous* or *spongy bone*—is nonlinearly elastic even at small strains,[77] it is most often modeled as linearly elastic until yielding. It yields in compression at strains of approximately 1 percent, after which it can sustain large deformations (up to 50 percent strain) while still maintaining its load-carrying capacity. Thus trabecular bone can absorb substantial energy on mechanical failure. A heterogeneous porous cellular solid, trabecular bone has anisotropic mechanical properties that depend on the porosity of the specimen as well as the architectural arrangement of the individual trabeculae. Its apparent (whole-specimen) level properties also depend on the tissue-level material properties of the individual trabeculae. An overwhelming portion

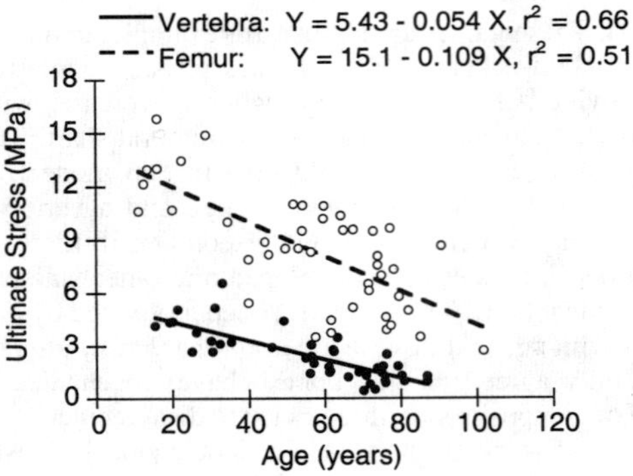

FIGURE 8.11 Dependence of ultimate stress on age for trabecular bone from the human vertebra and femur. For both anatomic sites, strength decreases approximately 10 percent per decade. (*Data from Refs. 15 and 149.*)

of trabecular bone mechanics research has been devoted to improving our understanding of the relative contributions and interplay of porosity, architecture, and tissue properties in the apparent level properties.

The elastic and strength properties of trabecular bone display substantial heterogeneity with respect to donor age and health, anatomic site, loading direction (with respect to the principal orientation of the trabeculae), and loading mode. Both modulus and strength decrease with age, falling approximately 10 percent per decade[15,78] (Fig. 8.11). Pathologies such as osteoporosis, osteoarthritis, and bone cancer are also known to affect mechanical properties.[79,80] Young's modulus can vary 100-fold within a single epiphysis[81] and 3-fold depending on loading direction.[82–85] Typically, the modulus of human trabecular bone is in the range 10 to 3000 MPa depending on the preceding factors; strength, which is linearly and strongly correlated with modulus,[11,81,82] is generally 2 orders of magnitude lower than modulus and is usually in the range 0.1 to 30 MPa.

In compression, the anisotropy of trabecular bone strength increases with age[78] and decreasing density (Fig. 8.12). The strength also depends on loading mode, being highest in compression and lowest in shear.[86,87] Ratios of compressive to tensile strength and compressive to shear strength are not constant but rather depend on modulus[87] and density (see Fig. 8.12). Both modulus and strength depend heavily on apparent density, yet these relationships vary for different types of trabecular bone because of the anatomic site-, age-, and disease-related variations in trabecular architecture. Linear and power-law relationships* can be used to describe the dependence of modulus and compressive strength on apparent density (Tables 8.4 and 8.5), with typical coefficients of determination (r^2 values) in the range 0.5 to 0.9.

Interestingly, the failure (yield and ultimate) strains of human trabecular bone have only a weak dependence, if any, on apparent density and modulus.[11,13,78,88–91] A recent study designed to test for intersite differences found that yield strains were approximately uniform within anatomic sites, with standard deviations on the order of one-tenth the mean value, but mean values could vary across sites[11] (Fig. 8.13). Thus, for analysis purposes, yield strains can be considered constant within sites but heterogeneous across sites. Regardless of anatomic site, however, yield stains are higher in compression than in tension.[11] Ultimate strains are typically in the range of 1.0 to 2.5 percent. Evidence from experiment on bovine bone indicates that yield strains are also isotropic[92,93] despite substantial anisotropy of modulus and strength.

* Differences in the predictive power between the various linear and power laws are usually negligible within a single anatomic site because the range of apparent density exhibited by trabecular bone is less than 1 order of magnitude.

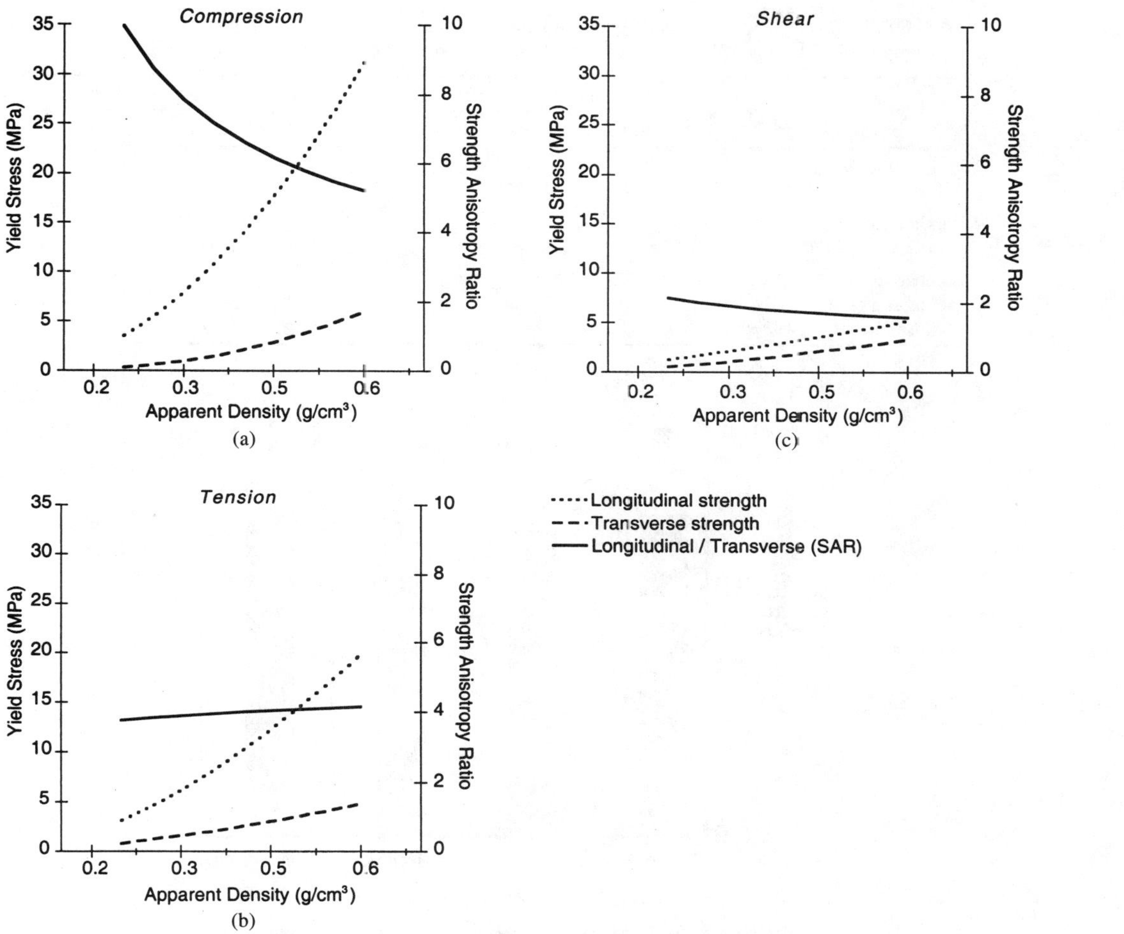

FIGURE 8.12 Dependence of yield stress in (*a*) compression, (*b*) tension, and (*c*) torsion on apparent density for bovine tibial trabecular bone specimens oriented both longitudinally (along) and transverse to the principal trabecular orientation. Overall, strength is greatest in compression and least in shear. In compression, the strength-anisotropy ratio [SAR = (longitudinal strength)/(transverse strength)] increases with decreasing density. (*Data from Ref. 102.*)

TABLE 8.4 Power-Law Regressions Between Modulus E (in MPa) and Apparent Density ρ (in g/cm^3) for Human Trabecular Bone Specimens from a Range of Anatomic Sites

	Cadavers			$\sigma = a\rho^b$		
Study	Number	Age, years	No. of specimens	a	b	r^2
Vertebra (T_{10}–L_5)	25	20–90	61	4,730	1.56	0.73
Proximal tibia	16	40–85	31	15,490	1.93	0.84
Femoral greater trochanter	21	49–101	23	15,010	2.18	0.82
Femoral neck	23	57–101	27	6,850	1.49	0.85

Source: Data from Ref. 11.

TABLE 8.5 Power-Law Regressions Between Ultimate Stress σ (in MPa) and Apparent Density ρ (in g/cm^3) for Compressive Loading of Human Trabecular Bone Specimens from a Range of Anatomic Sites

Study	Cadavers		No. of specimens	$\sigma = a\rho^b$		
	Number	Age, years		a	b	r^2
Proximal tibia						
Linde et al., 1989[151]	9	59–82	121	34.2	1.56	0.79
Proximal femur						
Lotz et al., 1990[131]	4	25–82	49	25.0	1.80	0.93
Lumbar spine						
Hansson et al., 1987[88]	3	71–84	231	50.3	2.24	0.76
Mosekilde et al., 1987[78]	42	15–87	40	24.9	1.80	0.83

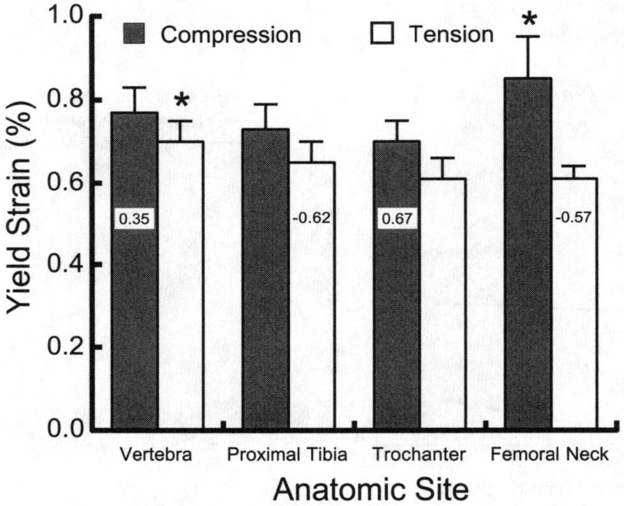

FIGURE 8.13 Mean yield strain per anatomic site in both compression and tension. Error bars indicate 1 standard deviation. Yield strain is only weakly dependent on apparent density for four of the groups, as indicated by the Pearson correlation coefficient r in the bar. Compressive yield strains are higher than tensile yield strains for each anatomic site. Intrasite scatter is on the order of one-tenth the mean values. (*From Ref. 11.*)

The advent of high-resolution finite-element modeling[94] has led to enormous progress in determining elastic stiffness matrices, multiaxial failure behavior, and as will be seen later, trabecular tissue properties. Finite-element models of individual specimens, developed using microcomputed tomography[95,96] and other types of microscopic imaging,[17,97] have been used to compute the full set of elastic constants for specimens from multiple anatomic sites. Results indicate that trabecular bone can be considered orthotropic[98,99] or, in some cases, transversely orthotropic.[100] Poisson's ratios, which are difficult to measure experimentally for trabecular bone, range from 0.03 to 0.60.[22,99]

Given the complexity of in vivo loading conditions, there is a need to develop a multiaxial failure criterion for trabecular bone. Results have been reported for bovine bone only.[101–103] These studies indicate that the von Mises criterion does not work well and that expression of the criterion in terms of strain (or nondimensional measures of stress divided by modulus) greatly simplifies the mathematical form of the criterion since it eliminates the dependence of the criterion on specimen density.

Criteria such as the Tsai-Wu criterion have only a limited ability to describe multiaxial failure of trabecular bone for arbitrary stress states. Coupling between normal strengths in different directions (longitudinal versus transverse, for example) appears to be minimal.[104] At present, it is recommended for multiaxial failure analysis that the tensile-compressive-shear-strength asymmetry be recognized, as well as the strength anisotropy. If properties are not available for a specific site under analysis, failure strains should be used from sites that have a similar density range and architecture.

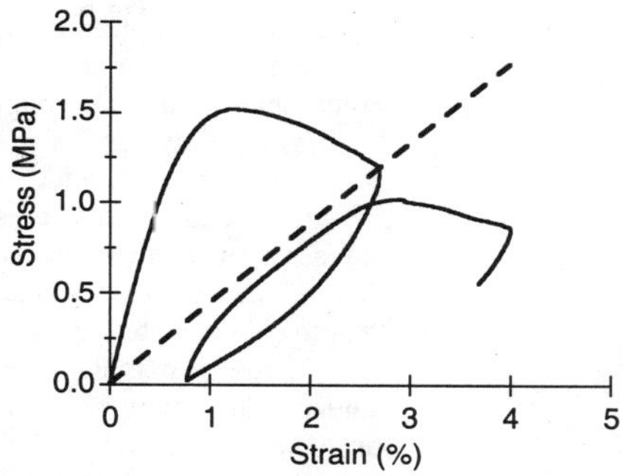

FIGURE 8.14 Compressive load-unload-reload behavior of human vertebral trabecular bone. Similar to cortical bone tested in tension, an initial overload causes residual strains and a reloading curve whose modulus quickly reduces from a value similar to the intact modulus to a value similar to the perfect damage modulus. (*From Ref. 105.*)

When trabecular bone is loaded in compression beyond its elastic range, unloaded, and reloaded, it displays loss of stiffness and development of permanent strains[105] (Fig. 8.14). In particular, it reloads with an initial modulus close to its intact Young's modulus but then quickly loses stiffness. The residual modulus is statistically similar to the perfect-damage modulus (a secant modulus from the origin to the point of unloading). In general, the reloading stress-strain curve tends to reload back toward the extrapolated envelope of the original curve. Phenomenologically, trabecular bone therefore exhibits elements of both plasticity and damage. The magnitudes of stiffness loss $\%\Delta E$ and residual strain $\epsilon_{\text{RESDIUAL}}$ for human vertebral trabecular bone are highly correlated with the applied strain ϵ_{TOTAL} (all expressed in percent) in the initial overload:

$$\epsilon_{\text{RESIDUAL}} = -0.046 + 0.104\epsilon_{\text{TOTAL}} + 0.073\epsilon_{\text{TOTAL}}^2 \qquad r^2 = 0.96$$

$$\%\Delta E = 178 - \frac{496}{\epsilon_{\text{TOTAL}} + 2.58} \qquad r^2 = 0.94$$

Also, at any given strain, modulus on reloading is reduced more than strength:

$$\%\Delta S = 17.1 + 19.1\epsilon_{\text{TOTAL}} - 149\rho_{\text{APP}} \qquad r^2 = 0.62$$

where $\%\Delta S$ is the percentage change in strength, and ρ_{APP} is the apparent density. These relations are for applied strains on the order of 1 to 5 percent only; residual behavior for much greater applied strains has not yet been reported. The percent modulus reductions and residual strains do not depend on volume fraction because similar trends have been reported for bovine bone, which is much more dense and platelike.[106,107] Furthermore, the trabecular behavior is qualitatively similar to that for cortical bone loaded in tension.[34,108] This suggests that the dominant physical mechanisms for damage behavior act at the nanometer scale of the collagen and hydroxyapatite.

Regarding time-dependent behavior, trabecular bone is only slightly viscoelastic when tested in vitro, with both compressive strength and modulus being related to strain rate raised to a power of 0.06.[109,110] The stiffening effect of marrow is negligible except at very high strain rates (10 strains/s), although there is evidence that the constraining effects of a cortical shell may allow hydraulic stiffening of whole bones in vivo under dynamic loads.[111] Minor stress relaxation has been shown to occur[112] and depends on the applied strain level,[113] indicating that human trabecular bone is nonlinearly viscoelastic. Little else is known about its time-dependent properties, including creep and fatigue for human bone. Fatigue and creep studies on bovine bone have revealed the following power law relations between compressive normalized stress (stress divided by modulus, expressed in percent) and time to failure (frequency for fatigue loading was 2 Hz):

Fatigue: $$t_f = 1.71 \times 10^{-24}(\Delta\sigma/E_o)^{-11.56} \qquad r^2 = 0.77$$

Creep: $$t_f = 9.66 \times 10^{-33}(\sigma/E_o)^{-16.18} \qquad r^2 = 0.83$$

Failure in these experiments was defined by a 10 percent reduction in secant modulus compared with the initial Young's modulus.

It should be noted that the in vitro mechanical test methods most often used to date on trabecular bone are known to introduce substantial errors in the data as a result of the porous anisotropic trabecular structure. Damage preferentially occurs at the ends of machined specimens when they are compressed between loading platens due to disruption of the trabecular network at the specimen boundaries.[114–116] In addition, friction at the platen-specimen interface creates a triaxial stress state that may result in an overestimation of modulus.[114,115,117] If strains are computed from the relative displacement of the platens, substantial systematic and random errors in modulus on the order of 20 ± 12 percent can occur.[118] Strength and failure strain data are also affected.[119,120] The trabecular anisotropy indicates that experimental measurements of the orthotropic elastic constants should be done in the principal material coordinate system. Since off-axis moduli are functions of multiple material elastic constants, substantial errors can be introduced from misalignment[121] if not corrected for. Much of the data in the literature have been obtained in various types of off-axis configurations, since specimens are often machined along anatomic rather than material axes. The difficulty in interpreting these off-axis measurements is heightened when intersite and interstudy comparisons are attempted. For all these reasons, and since the in vitro test boundary conditions rarely replicate those in vivo, interpretation and application of available data must be done with care.

An important development for structural analysis of whole bones is the use of quantitative computed tomography (QCT) to generate anatomically detailed models of whole bone[122–127] or bone-implant[128,129] systems. At the basis of such technology is the ability to use QCT to noninvasively predict the apparent density and mechanical properties of trabecular bone. Some studies have reported excellent predictions ($r^2 \geqslant 0.75$) of modulus and strength from QCT density information for various anatomic sites.[82,89,130,131] Since the mechanical properties depend on volume fraction and architecture, it is important to use such relations only for the sites for which they were developed; otherwise, substantial errors can occur. Also, since QCT data do not describe any anisotropic properties of the bone, trabecular bone is usually assumed to be isotropic in whole-bone and bone-implant analyses. Typically, in such structural analyses, cortical bone properties are not assigned from QCT since it does not have the resolution to discern the subtle variations in porosity and mineralization that cause variations in cortical properties. In these cases, analysts typically assign average properties, sometimes transversely isotropic, to the cortical bone.

8.6 MECHANICAL PROPERTIES OF TRABECULAR TISSUE MATERIAL

While most biomechanical applications at the organ level require knowledge of material properties at the scales described earlier, there is also substantial interest in the material properties of trabecular tissue because this information may provide additional insight into diseases such as osteoporosis and drug treatments designed to counter such pathologies. Disease- or drug-related changes could be most obvious at the tissue rather than apparent or whole-bone level, yet only recently have researchers begun to explore this hypothesis. This is so primarily because the irregular shape and small size of individual trabecula present great difficulties in measuring the tissue material properties.

Most of the investigations of trabecular tissue properties have addressed elastic behavior. Some of the earlier experimental studies concluded that the trabecular tissue has a modulus on the order of 1 to 10 GPa.[132–135] Later studies using combinations of computer modeling of the whole specimen and experimental measures of the apparent modulus have resulted in a wide range of estimated values for the effective modulus of the tissue (Table 8.6) such that the issue became controversial. However, studies using ultrasound have concluded that values for elastic modulus are about 20 percent lower than for cortical tissue.[136,137] This has been supported by results from more recent nanoindentation studies.[61,62,138] The combined computer-experiment studies that successfully eliminated end artifacts in the experimental protocols also found modulus values more typical of cortical bone than the much lower values from the earlier studies.[139] Thus an overall consensus is emerging that the elastic modulus of trabecular tissue is similar to, and perhaps slightly lower than, that of cortical bone.[140] Regarding failure behavior, it appears from the results of combined computational-

TABLE 8.6 Trabecular Tissue Moduli Using a Variety of Experimental and Computational Techniques

Study	Testing method	Anatomic site	No. of specimens	Tissue modulus (GPa) Mean	SD
Ashman and Rho, 1988[136]	Ultrasound	Human femur	53	13.0	1.5
Rho et al., 1993[137]	Ultrasound	Human tibia	20	14.8	1.4
	Microtensile			10.4	3.5
Rho et al., 1997[61]	Nanoindentation	Human vertebra	2	13.4	2.0
Zysset et al., 1998[152]	Nanoindentation	Human femoral neck	8	11.4	5.6
Hou et al., 1998[153]	FEM	Human vertebra	28	5.7	1.6
Ladd et al., 1998[154]	FEM	Human vertebra	5	6.6	1.0
Turner et al., 1999[138]	Nanoindentation	Human distal femur	1	18.1	1.7
	Ultrasound			17.5	1.1
Niebur et al., 2000[139]	FEM	Bovine tibia	7	18.7	3.4

experimental studies that the yield strains for trabecular tissue are similar to those for cortical bone, being higher in compression than in tension.[139] Experimental studies on machined microbeams have shown that the fatigue strength of trabecular tissue is lower than that of cortical tissue.[141]

8.7 CONCLUDING REMARKS

The field of bone mechanics has evolved to a very sophisticated level where mechanical properties of cortical and trabecular bone are available for many anatomic sites. Studies have also reported on the effects of bone density, aging, and disease on these properties, enabling researchers to perform highly detailed specimen-specific analyses on whole bone and bone-implant systems. We have reviewed here much of that literature. Our focus was on data for human bone, although we reported bovine data when no other reliable data were available. One important theme in bone mechanics is to account for the substantial heterogeneity in bone properties that can occur for both cortical and trabecular bone, particularly for the latter. The heterogeneity results from aging, disease, and natural interindividual biological variation and thus occurs longitudinally and cross-sectionally in populations. The heterogeneity also exists spatially within bones. Successful structural analysis depends on appreciation of this heterogeneity so that appropriate material properties are used for the analysis at hand. Improved understanding of the micromechanics and damage behaviors of bone is also leading to unique insight into mechanisms of disease and their treatment as well as biological remodeling and tissue engineering. While a number of excellent texts are available for more detailed study of these topics and many of those presented here,[73,142–144] it is hoped that this review will provide a concise basis for practical engineering analysis of bone.

ACKNOWLEDGMENTS

Support is gratefully acknowledged from NIH (AR41481, AR43784), NSF (BES-9625030), and The Miller Institute for Basic Research in Science, Berkeley, Calif.

REFERENCES

1. Burr, D. B., Forwood, M. R., Fyhrie, D. P., Martin, R. B., Schaffler, M. B., and Turner, C. H. (1997), Bone microdamage and skeletal fragility in osteoporotic and stress fractures, *J. Bone Miner. Res.* **12**(1):6–15.

2. Burr, D. B., Martin, R. B., Schaffler, M. B., and Radin, E. L. (1985), Bone remodeling in response to in vivo fatigue microdamage, *J. Biomech.* **18**(3):189–200.
3. Gong, J. K., Arnold, J. S., and Cohn, S. H. (1964), Composition of trabecular and cortical bone, *Anat. Rec.* **149**:325–332.
4. Lowenstam, H. A., and Weiner, S. (1989), *On Biomineralization*, Oxford University Press, New York.
5. Herring, G. (1972), The organic matrix of bone, in G. Bourne (ed.), *The Biochemistry and Physiology of Bone*, 2d ed., Vol. 1, pp.127–189, Academic Press, New York.
6. Hodge, A. J., Petruska, J. A. (1963), Recent studies with the electron microscope on ordered aggregates of the tropocollagen molecule, in G. N. Ramachandran (ed.) *Aspects of protein structure*, Academic Press, London, pp 289–300.
7. Burr, D. B., Schaffler, M. B., and Frederickson, R. G. (1988), Composition of the cement line and its possible mechanical role as a local interface in human compact bone, *J. Biomech.* **21**(11):939–945.
8. Lakes, R., and Saha, S. (1979), Cement line motion in bone, *Science* **204**(4392):501–503.
9. Gibson, L. J., and Ashby, M. F. (1997), *Cellular Solids: Structures & Properties*, 2d ed. Pergamon Press, Oxford, U.K.
10. McCalden, R. W., McGeough, J. A., Barker, M. B., and Court-Brown, C. M. (1993), Age-related changes in the tensile properties of cortical bone: The relative importance of changes in porosity, mineralization, and microstructure, *J. Bone Joint Surg.* **75A**(8):1193–1205.
11. Morgan, E. F., and Keaveny, T. M., (2001), Dependence of yield strain of human trabecular bone on anatomic site, *J. Biomech.* **34**(5):569–577.
12. Snyder, B. D., Piazza, S., Edwards, W. T., and Hayes, W. C. (1993), Role of trabecular morphology in the etiology of age-related vertebral fractures, *Calcif. Tissue Int.* **53S**(1):S14–S22.
13. Kopperdahl, D. L., and Keaveny, T. M. (1998), Yield strain behavior of trabecular bone, *J. Biomech.* **31**(7): 601–608.
14. Linde, F., Hvid, I., and Pongsoipetch, B. (1989), Energy absorptive properties of human trabecular bone specimens during axial compression, *J. Orthop. Res.* **7**(3):432–439.
15. McCalden, R. W., McGeough, J. A., and Court-Brown, C. M. (1997), Age-related changes in the compressive strength of cancellous bone: The relative importance of changes in density and trabecular architecture, *J. Bone Joint Surg.* **79A**(3):421–427.
16. Hildebrand, T., Laib, A., Müller, R., Dequeker, J., and Rüegsegger, P. (1999), Direct three-dimensional morphometric analysis of human cancellous bone: microstructural data from spine, femur, iliac crest, and calcaneus, *J. Bone Miner. Res.* **14**(7):1167–1174.
17. Odgaard, A. (1997), Three-dimensional methods for quantification of cancellous bone architecture, *Bone* **20**(4):315–328.
18. Mosekilde, L. (1989), Sex differences in age-related loss of vertebral trabecular bone mass and structure: Biomechanical consequences, *Bone* **10**(6):425–432.
19. Whitehouse, W. J. (1974), The quantitative morphology of anisotropic trabecular bone, *J. Microsc.*, **2**:153–168.
20. Goldstein, S. A., Goulet, R., and McCubbrey, D. (1993), Measurement and significance of three-dimensional architecture to the mechanical integrity of trabecular bone, *Calcif. Tissue Int.* **53S**(1):S127–S133.
21. Majumdar, S., Kothari, M., Augat, P., Newitt, D. C., Link, T. M., Lin, J. C., Lang, T., Lu, Y., and Genant, H. K. (1998), High-resolution magnetic resonance imaging: three-dimensional trabecular bone architecture and biomechanical properties, *Bone* **22**(5):445–454.
22. Ulrich, D., Van Rietbergen, B., Laib, A., and Rueegsegger, P. (1999), The ability of three-dimensional structural indices to reflect mechanical aspects of trabecular bone, *Bone* **25**(1):55–60.
23. Schaffler, M. B., and Burr, D. B., (1988), Stiffness of compact bone: Effects of porosity and density, *J. Biomech.* **21**(1):13–16.
24. Currey, J. D. (1988), The effect of porosity and mineral content on the Young's modulus of elasticity of compact bone, *J. Biomech.* **21**(2):131–139.
25. Burstein, A. H., Reilly, D. T., and Martens, M. (1976), Aging of bone tissue: Mechanical properties, *J. Bone Joint Surg.* **58A**(1):82–86.
26. Zioupos, P., and Currey, J. D. (1998), Changes in the stiffness, strength, and toughness of human cortical bone with age, *Bone* **22**(1):57–66.

27. McElhaney, J. H., and Byars, E. F. (1965), Dynamic response of biological materials, in *Proc. Amer. Soc. Mech. Eng.*, ASME 65-WA/HUF-9:8, Chicago.

28. Lakes, R. S. (1982), Dynamical study of couple stress effects in human compact bone, *J. Biomech. Eng.* **104**(1):6–11.

29. Lakes, R. S., Katz, J. L., and Sternstein, S. S. (1979), Viscoelastic properties of wet cortical bone: I. Torsional and biaxial studies, *J. Biomech.* **12**(9):657–678.

30. Fischer, R. A., Arms, S. W., Pope, M. H., and Seligson, D. (1986), Analysis of the effect of using two different strain rates on the acoustic emission in bone, *J. Biomech.* **19**(2):119–127.

31. Cezayirlioglu, H., Bahniuk, E., Davy, D. T., and Heiple, K. G. (1985), Anisotropic yield behavior of bone under combined axial force and torque, *J. Biomech.* **18**(1):61–69.

32. Cowin, S. C. (1989), *Bone Mechanics*, CRC Press, Boca Raton, Fla.

33. Courtney, A. C., Hayes, W. C., and Gibson, L. J. (1996), Age-related differences in post-yield damage in human cortical bone: Experiment and model, *J. Biomech.* **29**(11):1463–1471.

34. Fondrk, M. T., Bahniuk, E. H., and Davy, D. T. (1999), A damage model for nonlinear tensile behavior of cortical bone, *J. Biomech. Eng.* **121**:533–541.

35. Wright, T. M., Vosburgh, F., and Burstein, A. H. (1981), Permanent deformation of compact bone monitored by acoustic emission, *J. Biomech.* **14**(6):405–409.

36. Zioupos, P., Currey, J. D., and Sedman, A. J. (1994), An examination of the micromechanics of failure of bone and antler by acoustic emission tests and laser scanning confocal microscopy, *Med. Eng. Phys.* **16**(3): 203–212.

37. Carter, D. R., Caler, W. E., Spengler, D. M., and Frankel, V. H. (1981), Fatigue behavior of adult cortical bone: The influence of mean strain and strain range, *Acta Orthop. Scand.* **52**(5):481–490.

38. Pattin, C. A., Caler, W. E., and Carter, D. R. (1996), Cyclic mechanical property degradation during fatigue loading of cortical bone, *J. Biomech.* **29**(1):69–79.

39. Schaffler, M. B., Radin, E. L., and Burr, D. B. (1990), Long-term fatigue behavior of compact bone at low strain magnitude and rate, *Bone* **11**(5):321–326.

40. Carter, D. R., Caler, W. E., Spengler, D. M., and Frankel, V. H. (1981), Uniaxial fatigue of human cortical bone: The influence of tissue physical characteristics, *J. Biomech.* **14**(7):461–470.

41. Caler, W. E., and Carter, D. R. (1989), Bone creep-fatigue damage accumulation, *J. Biomech.* **22**(6–7): 625–635.

42. Frost, H. M. (1960), Presence of microscopic cracks in vivo in bone, *Bull. Henry Ford Hosp.* **8**:27–35.

43. Schaffler, M. B., Choi, K., and Milgrom, C. (1995), Aging and matrix microdamage accumulation in human compact bone, *Bone* **17**(6):521–525.

44. Norman, T. L., and Wang, Z. (1997), Microdamage of human cortical bone: Incidence and morphology in long bones, *Bone* **20**(4):375–379.

45. Brown, C. U., Yeni, Y. N., and Norman, T. L. (2000), Fracture toughness is dependent on bone location: A study of the femoral neck, femoral shaft, and the tibial shaft, *J. Biomed. Mater. Res.* **49**(3):380–389.

46. Melvin, J. W. (1993), Fracture mechanics of bone, *J. Biomech. Eng.* **115**(4B):549–554.

47. Yeni, Y. N., Brown, C. U., and Norman, T. L. (1998), Influence of bone composition and apparent density on fracture toughness of the human femur and tibia, *Bone* **22**(1):79–84.

48. Yeni, Y. N., Brown, C. U., Wang, Z., and Norman, T. L. (1997), The influence of bone morphology on fracture toughness of the human femur and tibia, *Bone* **21**(5):453–459.

49. Landis, W. J. (1995), The strength of a calcified tissue depends in part on the molecular structure and organization of its constituent mineral crystals in their organic matrix, *Bone* **16**(5):533–544.

50. Katz, J. L. (1971), Hard tissue as a composite material: I. Bounds on the elastic behavior, *J. Biomech.* **4**(5): 455–473.

51. Mammone, J. F., and Hudson, S. M. (1993), Micromechanics of bone strength and fracture, *J. Biomech.* **26**(4–5):439–446.

52. Gottesman, T., and Hashin, Z. (1980), Analysis of viscoelastic behaviour of bones on the basis of microstructure, *J. Biomech.* **13**(2):89–96.

53. Katz, J. L. (1981), Composite material models for cortical bone, in S. C. Cowin (ed.), *Mechanical Properties of Bone: Proceedings of the Joint ASME-ASCE Applied Mechanics, Fluids Engineering, and Bioengineer-*

ing Conference, Boulder, Colorado, June 22–24, 1981, Vol. 45, pp. 171–184, American Society of Mechanical Engineers, New York.

54. Krajcinovic, D., Trafimow, J., and Sumarac, D. (1987), Simple constitutive model for cortical bone, *J. Biomech.* **20**(8):779–784.
55. Braidotti, P., Branca, F. P., and Stagni, L. (1997), Scanning electron microscopy of human cortical bone failure surfaces, *J. Biomech.* **30**(2):155–162.
56. Pidaparti, R. M., Chandran, A., Takano, Y., and Turner, C. H. (1996), Bone mineral lies mainly outside collagen fibrils: Predictions of a composite model for osteonal bone, *J. Biomech.* **29**(7):909–916.
57. Akiva, U., Wagner, H. D., and Weiner, S. (1998), Modelling the three-dimensional elastic constants of parallel-fibred and lamellar bone, *J. Mater. Sci.* **33**(6):1497–1509.
58. Sasaki, N., Ikawa, T., and Fukuda, A. (1991), Orientation of mineral in bovine bone and the anisotropic mechanical properties of plexiform bone, *J. Biomech.* **24**(1):57–61.
59. Wagner, H. D., and Weiner, S. (1992), On the relationship between the microstructure of bone and its mechanical stiffness, *J. Biomech.* **25**(11):1311–1320.
60. McCutchen, C. W. (1975), Do mineral crystals stiffen bone by straitjacketing its collagen? *J. Theor. Biol.* **51**(1):51–58.
61. Rho, J. Y., Tsui, T. Y., and Pharr, G. M. (1997), Elastic properties of human cortical and trabecular lamellar bone measured by nanoindentation, *Biomaterials* **18**(20):1325–1330.
62. Zysset, P. K., Guo, X. E., Hoffler, C. E., Moore, K. E., and Goldstein, S. A. (1999), Elastic modulus and hardness of cortical and trabecular bone lamellae measured by nanoindentation in the human femur, *J. Biomech.* **32**(10):1005–1012.
63. Cowin, S. C., Moss-Salentijn, L., and Moss, M. L. (1991), Candidates for the mechanosensory system in bone, *J. Biomech. Eng.* **113**(2):191–197.
64. Lanyon, L. E. (1993), Osteocytes, strain detection, bone modeling and remodeling, *Calcif. Tissue Int.* **53**(Suppl. 1):S102–106; see also Discussion, pp. S106–107.
65. Marotti, G., Ferretti, M., Muglia, M. A., Palumbo, C., and Palazzini, S. (1992), A quantitative evaluation of osteoblast-osteocyte relationships on growing endosteal surface of rabbit tibiae, *Bone* **13**(5):363–368.
66. Mullender, M. G., and Huiskes, R. (1995), Proposal for the regulatory mechanism of Wolff's law, *J. Orthop. Res.* **13**(4):503–512.
67. Frost, H. M. (1987), Bone "mass" and the "mechanostat": A proposal, *Anat. Rec.* **219**(1):1–9.
68. Biewener, A. A., Thomason, J., Goodship, A., and Lanyon, L. E. (1983), Bone stress in the horse forelimb during locomotion at different gaits: A comparison of two experimental methods, *J. Biomech.* **16**(8):565–576.
69. Lanyon, L. E. (1984), Functional strain as a determinant for bone remodeling, *Calcif. Tissue Int.* **36**(Suppl. 1):S56–61.
70. Turner, C. H., Owan, I., and Takano, Y. (1995), Mechanotransduction in bone: Role of strain rate, *Am. J. Physiol.* **269**(3 Pt. 1):E438–442.
71. Fyhrie, D. P., and Carter, D. R. (1986), A unifying principle relating stress to trabecular bone morphology, *J. Orthop. Res.* **4**(3):304–317.
72. Weinans, H., Huiskes, R., and Grootenboer, H. J. (1992) The behavior of adaptive bone-remodeling simulation models, *J. Biomech.* **25**(12):1425–1441.
73. Martin, R. B., and Burr, D. B. (1989), *Structure, Function, and Adaptation of Compact Bone*, Raven Press, New York.
74. Carter, D. R., Orr, T. E., and Fyhrie, D. P. (1989), Relationships between loading history and femoral cancellous bone architecture, *J. Biomech.* **22**(3):231–244.
75. Mullender, M. G., Huiskes, R., and Weinans, H. (1994), A physiological approach to the simulation of bone remodeling as a self-organizational control process, *J. Biomech.* **27**(11):1389–1394.
76. Huiskes, R., Weinans, H., and van Rietbergen, R. (1992), The relationship between stress shielding and bone resorption around total hip stems and the effects of flexible materials, *Clin. Orthop.* **274**:124–134.
77. Morgan, E. F., Yeh, O. C., Chang, W. C., and Keaveny, T. M. (2001), Non-linear behavior of trabecular bone at small strains, *J. Biomech. Eng.* **123**(1):1–9.

78. Mosekilde, L., Mosekilde, L., and Danielsen, C. C. (1987), Biomechanical competence of vertebral trabecular bone in relation to ash density and age in normal individuals, *Bone* **8**(2):79–85.

79. Hipp, J. A., Rosenberg, A. E., and Hayes, W. C. (1992), Mechanical properties of trabecular bone within and adjacent to osseous metastases, *J. Bone Miner. Res.* **7**(10):1165–1171.

80. Pugh, J. W., Radin, E. L., and Rose, R. M. (1974), Quantitative studies of human subchondral cancellous bone: Its relationship to the state of its overlying cartilage, *J. Bone Joint Surg.* **56A**(2):313–321.

81. Goldstein, S. A., Wilson, D. L., Sonstegard, D. A., and Matthews, L. S. (1983), The mechanical properties of human tibial trabecular bone as a function of metaphyseal location, *J. Biomech.* **16**(12):965–969.

82. Ciarelli, M. J., Goldstein, S. A., Kuhn, J. L., Cody, D. D., and Brown, M. B. (1991), Evaluation of orthogonal mechanical properties and density of human trabecular bone from the major metaphyseal regions with materials testing and computed tomography, *J. Orthop. Res.* **9**(5):674–682.

83. Goulet, R. W., Goldstein, S. A., Ciarelli, M. J., Kuhn, J. L., Brown, M. B., and Feldkamp, L. A. (1994), The relationship between the structural and orthogonal compressive properties of trabecular bone, *J. Biomech.* **27**(4):375–389.

84. Linde, F., Pongsoipetch, B., Frich, L. H., and Hvid, I. (1990), Three-axial strain controlled testing applied to bone specimens from the proximal tibial epiphysis, *J. Biomech.* **23**(11):1167–1172.

85. Townsend, P. R., Raux, P., Rose, R. M., Miegel, R. E., and Radin, E. L. (1975), The distribution and anisotropy of the stiffness of cancellous bone in the human patella, *J. Biomech.* **8**(6):363–367.

86. Ford, C. M., and Keaveny, T. M. (1996), The dependence of shear failure properties of bovine tibial trabecular bone on apparent density and trabecular orientation, *J. Biomech.* **29**:1309–1317.

87. Keaveny, T. M., Wachtel, E. F., Ford, C. M., and Hayes, W. C. (1994), Differences between the tensile and compressive strengths of bovine tibial trabecular bone depend on modulus, *J. Biomech.* **27**:1137–1146.

88. Hansson, T. H., Keller, T. S., and Panjabi, M. M. (1987), A study of the compressive properties of lumbar vertebral trabeculae: Effects of tissue characteristics, *Spine* **12**(1):56–62.

89. Hvid, I., Bentzen, S. M., Linde, F., Mosekilde, L., and Pongsoipetch, B. (1989), X-ray quantitative computed tomography: The relations to physical properties of proximal tibial trabecular bone specimens, *J. Biomech.* **22**(8–9):837–844.

90. Hvid, I., Jensen, N. C., Bunger, C., Solund, K., and Djurhuus, J. C. (1985), Bone mineral assay: Its relation to the mechanical strength of cancellous bone, *Eng. Med.* **14**:79–83.

91. Rohl, L., Larsen, E., Linde, F., Odgaard, A., and Jorgensen, J. (1991), Tensile and compressive properties of cancellous bone, *J. Biomech.* **24**(12):1143–1149.

92. Chang, W. C. W., Christensen, T. M., Pinilla, T. P., and Keaveny, T. M. (1999), Isotropy of uniaxial yield strains for bovine trabecular bone, *J. Orthop. Res.* **17**:582–585.

93. Turner, C. H. (1989), Yield behavior of bovine cancellous bone, *J. Biomech. Eng.* **111**(3):256–260.

94. Van Rietbergen, B., Weinans, H., Huiskes, R., and Odgaard, A. (1995), A new method to determine trabecular bone elastic properties and loading using micromechanical finite element models, *J. Biomech.* **28**(1):69–81.

95. Kinney, J. H., Lane, N. E., and Haupt, D. L. (1995), In vivo three-dimensional microscopy of trabecular bone, *J. Bone Miner. Res.* **10**(2):264–270.

96. Kuhn, J. L., Goldstein, S. A., Feldkamp, L. A., Goulet, R. W., and Jesion, G. (1990), Evaluation of a microcomputed tomography system to study trabecular bone structure, *J. Orthop. Res.* **8**(6):833–842.

97. Beck, J. D., Canfield, B. L., Haddock, S. M., Chen, T. J. H., Kothari, M., and Keaveny, T. M. (1997), Three-dimensional imaging of trabecular bone using the computer numerically controlled milling technique, *Bone* **21**:281–287.

98. Yang, G., Kabel, J., Van Rietbergen, B., Odgaard, A., Huiskes, R., and Cowin, S. (1999), The anisotropic Hooke's law for cancellous bone and wood, *J. Elasticity* **53**:125–146.

99. Zysset, P. K., Goulet, R. W., and Hollister, S. J. (1998), A global relationship between trabecular bone morphology and homogenized elastic properties, *J. Biomech. Eng.* **120**(5):640–646.

100. Odgaard, A., Kabel, J., an Rietbergen, B., Dalstra, M., and Huiskes, R. (1997), Fabric and elastic principal directions of cancellous bone are closely related, *J. Biomech.* **30**(5):487–495.

101. Fenech, C. M., and Keaveny, T. M. (1999), A cellular solid criterion for predicting the axial-shear failure properties of trabecular bone, *J. Biomech. Eng.* **121**:414–422.

102. Keaveny, T. M., Wachtel, E. F., Zadesky, S. P., and Arramon, Y. P. (1999), Application of the Tsai-Wu quadratic multiaxial failure criterion to bovine trabecular bone, *J. Biomech. Eng.* **121**:99–107.

103. Stone, J. L., Beaupre, G. S., and Hayes, W. C. (1983), Multiaxial strength characteristics of trabecular bone, *J. Biomech.* **16**(9):743–752.

104. Niebur, G. L., and Keaveny, T. M. (2001), Trabecular bone exhibits multiple failure surfaces in biaxial loading, *Trans. Orthop. Res. Soc.* **26**:517.

105. Keaveny, T. M., Wachtel, E. F., and Kopperdahl, D. L. (1999), Mechanical behavior of human trabecular bone after overloading, *J. Orthop. Res.* **17**:346–353.

106. Keaveny, T. M., Wachtel, E. F., Guo, X. E., and Hayes, W. C. (1994), Mechanical behavior of damaged trabecular bone, *J. Biomech.* **27**(11):1309–1318.

107. Zysset, P. K., and Curnier, A. (1996), A 3D damage model for trabecular bone based on fabric tensors, *J. Biomech.* **29**(12):1549–1558.

108. Fondrk, M., Bahniuk, E., Davy, D. T., and Michaels, C. (1988), Some viscoplastic characteristics of bovine and human cortical bone, *J. Biomech.* **21**(8):623–630.

109. Carter, D. R., and Hayes, W. C. (1977), The compressive behavior of bone as a two-phase porous structure, *J. Bone Joint Surg.* **59A**:954–962.

110. Linde, F., Norgaard, P., Hvid, I., Odgaard, A., and Soballe, K. (1991), Mechanical properties of trabecular bone: Dependency on strain rate, *J. Biomech.* **24**(9):803–809.

111. Ochoa, J. A., Sanders, A. P., Kiesler, T. W., Heck, D. A., Toombs, J. P., Brandt, K. D., and Hillberry, B. M. (1997), In vivo observations of hydraulic stiffening in the canine femoral head, *J. Biomech. Eng.* **119**(1):103–108.

112. Zilch, H., Rohlmann, A., Bergmann, G., and Kolbel, R. (1980), Material properties of femoral cancellous bone in axial loading: II. Time dependent properties, *Arch. Orthop. Trauma Surg.* **97**(4):257–262.

113. Deligianni, D. D., Maris, A., and Missirlis, Y. F. (1994), Stress relaxation behaviour of trabecular bone specimens, *J. Biomech.* **27**(12):1469–1476.

114. Linde, F., Hvid, I., and Madsen, F. (1992), The effect of specimen geometry on the mechanical behaviour of trabecular bone specimens, *J. Biomech.* **25**:359–368.

115. Odgaard, A., and Linde, F. (1991), The underestimation of Young's modulus in compressive testing of cancellous bone specimens, *J. Biomech.* **24**(8):691–698.

116. Zhu, M., Keller, T. S., and Spengler, D. M. (1994), Effects of specimen load-bearing and free surface layers on the compressive mechanical properties of cellular materials, *J. Biomech.* **27**(1):57–66.

117. Keaveny, T. M., Borchers, R. E., Gibson, L. J., and Hayes, W. C. (1993), Theoretical analysis of the experimental artifact in trabecular bone compressive modulus, *J. Biomech.* **26**(4–5):599–607.

118. Keaveny, T. M., Pinilla, T. P., Crawford, R. P., Kopperdahl, D. L., and Lou, A (1997), Systematic and random errors in compression testing of trabecular bone, *J. Orthop. Res.* **15**:101–110.

119. Keaveny, T. M., Borchers, R. E., Gibson, L. J., and Hayes, W. C. (1993), Trabecular bone modulus and strength can depend on specimen geometry, *J. Biomech.* **26**:991–1000.

120. Keaveny, T. M., Guo, X. E., Wachtel, E. F., McMahon, T. A., and Hayes, W. C. (1994), Trabecular bone exhibits fully linear elastic behavior and yields at low strains, *J. Biomech.* **27**(9):1127–1136.

121. Turner, C. H., and Cowin, S. C. (1988), Errors introduced by off-axis measurements of the elastic properties of bone, *J. Biomech.* **110**:213–214.

122. Cody, D. D., Gross, G. J., Hou, F. J., Spencer, H. J., Goldstein, S. A., and Fyhrie, D. P. (1999), Femoral strength is better predicted by finite element models than QCT and DXA, *J. Biomech.* **32**(10):1013–1020.

123. Keyak, J. H., Meagher, J. M., Skinner, H. B., and Mote, C. D., Jr. (1990), Automated three-dimensional finite element modelling of bone: a new method, *J. Biomed. Eng.* **12**(5):389–397.

124. Keyak, J. H., Rossi, S. A., Jones, K. A., and Skinner, H. B. (1998), Prediction of femoral fracture load using automated finite element modeling, *J. Biomech.* **31**:125–133.

125. Keyak, J. H., and Skinner, H. B. (1992), Three-dimensional finite element modelling of bone: effects of element size, *J. Biomed. Eng.* **14**(6):483–489.

126. Lotz, J. C., Cheal, E. J., and Hayes, W. C. (1991), Fracture prediction for the proximal femur using finite element models: I. Linear analysis, *J. Biomech. Eng.* **113**(4):353–360.

127. Lotz, J. C., Cheal, E. J., and Hayes, W. C. (1991), Fracture prediction for the proximal femur using finite element models: II. Nonlinear analysis, *J. Biomech. Eng.* **113**(4):361–365.

128. Cheal, E. J., Spector, M., and Hayes, W. C. (1992), Role of loads and prosthesis material properties on the mechanics of the proximal femur after total hip arthroplasty, *J. Orthop. Res.* **10**(3):405–422.

129. Keaveny, T. M., and Bartel, D. L. (1995), Mechanical consequences of bone ingrowth in a hip prosthesis inserted without cement, *J. Bone Joint Surg.* **77A**:911–923.

130. Kopperdahl, D. L., and Keaveny, T. M. (2002), Quantitative computed tomography estimates of the mechanical properties of human vertebral trabecular bone, *J. Orthop. Res.* (in press).

131. Lotz, J. C., Gerhart, T. N., and Hayes, W. C. (1990), Mechanical properties of trabecular bone from the proximal femur: A quantitative CT study, *J. Comput. Assist. Tomogr.* **14**(1):107–114.

132. Choi, K., Kuhn, J. L., Ciarelli, M. J., and Goldstein, S. A. (1990), The elastic moduli of human subchondral, trabecular, and cortical bone tissue and the size-dependency of cortical bone modulus, *J. Biomech.* **23**(11): 1103–1113.

133. Kuhn, J. L., Goldstein, S. A., Choi, K., London, M., Feldkamp, L. A., and Matthews, L. S. (1989), Comparison of the trabecular and cortical tissue moduli from human iliac crests, *J. Orthop. Res.* **7**(6):876–884.

134. Mente, P. L., and Lewis, J. L. (1989), Experimental method for the measurement of the elastic modulus of trabecular bone tissue, *J. Orthop. Res.* **7**(3):456–461.

135. Ryan, S. D., and Williams, J. L. (1989), Tensile testing of rodlike trabeculae excised from bovine femoral bone, *J. Biomech.* **22**(4):351–355.

136. Ashman, R. B., and Rho, J. Y. (1988), Elastic modulus of trabecular bone material, *J. Biomech.* **21**(3): 177–181.

137. Rho, J. Y., Ashman, R. B., and Turner, C. H. (1993), Young's modulus of trabecular and cortical bone material: Ultrasonic and microtensile measurements, *J. Biomech.* **26**:111–119.

138. Turner, C. H., Rho, J., Takano, Y., Tsui, T. Y., and Pharr, G. M. (1999), The elastic properties of trabecular and cortical bone tissues are similar: Results from two microscopic measurement techniques, *J. Biomech.* **32**(4):437–441.

139. Niebur, G. L., Feldstein, M. J., Yuen, J. C., Chen, T. J., and Keaveny, T. M. (2000), High-resolution finite element models with tissue strength asymmetry accurately predict failure of trabecular bone, *J. Biomech.* **33**:1575–1583.

140. Guo, X. E., and Goldstein, S. A. (1997), Is trabecular bone tissue different from cortical bone tissue? *Forma* **12**:185–196.

141. Choi, K., and Goldstein, S. A. (1992), A comparison of the fatigue behavior of human trabecular and cortical bone tissue, *J. Biomech.* **25**(12):1371–1381.

142. Cowin, S. (2001), *Bone Mechanics Handbook*, 2d ed., CRC Press, Boca Raton, Fla.

143. Currey, J. (1984), *The Mechanical Adaptations of Bones*, Princeton University Press, Princeton, N.J.

144. Martin, R., Burr, D., and Sharkey, N. (1998), *Skeletal Tissue Mechanics*, Springer-Verlag, New York.

145. Wainwright, S., Gosline, J., and Biggs, W. (1976), *Mechanical Design in Organisms*, Halsted Press, New York.

146. Tortora, G. (1983), *Principles of Human Anatomy* 3d ed., Harper & Row, New York.

147. Ross, M., and Romrell, L. (1989), *Histology*, 2d ed., Williams & Wilkins, Baltimore.

148. Fondrk, M. T. (1989), An experimental and analytical investigation into the nonlinear constitutive equations of cortical bone, Ph.D. thesis, Case Western Reserve University, Cleveland.

149. Mosekilde, L., and Mosekilde, L. (1986), Normal vertebral body size and compressive strength: Relations to age and to vertebral and iliac trabecular bone compressive strength, *Bone* **7**:207–212.

150. Reilly, D. T., and Burstein, A. H. (1975), The elastic and ultimate properties of compact bone tissue, *J. Biomech.* **8**:393–405.

151. Linde, F., and Hvid, I. (1989), The effect of constraint on the mechanical behaviour of trabecular bone specimens, *J. Biomech.* **22**(5):485–490.

152. Zysset, P. K., Guo, X. E., Hoffler, C. E., Moore, K. E., and Goldstein, S. A. (1998), Mechanical properties of human trabecular bone lamellae quantified by nanoindentation, *Technol. Health Care* **6**(5–6):429–432.

153. Hou, F. J., Lang, S. M., Hoshaw, S. J., Reimann, D. A., and Fyhrie, D. P. (1998), Human vertebral body apparent and hard tissue stiffness, *J. Biomech.* **31**(11):1009–1015.

154. Ladd, A. J., Kinney, J. H., Haupt, D. L., and Goldstein, S. A. (1998), Finite-element modeling of trabecular bone: Comparison with mechanical testing and determination of tissue modulus, *J. Orthop. Res.* **16**(5): 622–628.

CHAPTER 9
FINITE-ELEMENT ANALYSIS

Michael D. Nowak
University of Hartford, West Hartford, Connecticut

9.1 INTRODUCTION

In the realm of biomedical engineering, computer modeling in general and finite-element modeling in particular are powerful means of understanding the body and the adaptations that may be made to it. Using the appropriate inputs, a better understanding of the interrelation of the components of the body can be achieved. In addition, the effects of surgical procedures and material replacement can be evaluated without large numbers of physical trials. The "what if" and iterative aspects of computer modeling can save a great deal of time and money, especially as compared with multiple bench testing or series of live trials.

In attempting to understand the human body, much can be learned from material and fluids testing and cadaveric examination. These processes do not, in general, determine the relative forces and interactions between structures. They are also not able to determine the stresses within hard or soft tissue nor the patterns of flow due to the interaction of red blood cells within the vascular system.

Every aspect of the human body and devices to aid or replace function fall within the realm of computer modeling. These models range from the more obvious arenas based on orthopedic and vascular surgery to trauma from accident (Huang et al., 1999) to the workings of the middle ear (Ferris and Prendergast, 2000).

There are an increasing number of journals that include articles using finite-element analysis (FEA) in evaluation of the body and implants. These range from the journals dedicated to the engineering evaluation of the body (such as *The Journal of Biomechanics* and *The Journal of Biomechanical Engineering*) to those associated with surgical and other specialties (such as *The Journal of Orthopaedic Research, The Journal of Prosthetic Dentistry*, and *The Journal of Vascular Surgery*).

As with any use of FEA results, the practitioner must have some understanding of the actual structure to determine if the model is valid. One prime example of error when overlooking the end use in the FEA realm is that of the femoral component of hip implants. If one only examines loading patterns, the best implant would be placed on the exterior of the femoral bone, since the bone transfers load along its outer material. Doing so in reality would lead to failure because the nutrient supply to the bone would be compromised, and the bone would resorb (bone mass would be lost).

This chapter will give an overview of the requirements and uses of FEA and similar codes for biomedical engineering analysis. The examples that will be used refer to FEA, but the techniques will be the same for other systems, such as finite difference and computational fluid dynamics. The literature cited in this chapter gives a flavor of the breadth of information available and the studies being undertaken. This listing is far from exhaustive because of the large number of ongoing efforts. Numerous search engines are available to find abstracts relating to the subjects touched on in this chapter.

A large number of software packages and a wide range of computational power are used in FEA of the human body, ranging from basic personal computer (PC) programs and simplified constructs to high-powered nonlinear codes and models that require extensive central processing unit (CPU) time on supercomputers. Most of the examples in this chapter will be those run on desktop PCs.

The remainder of this chapter will focus on three basic concerns of a useful finite-element model: the geometry, the material properties, and the boundary conditions.

9.2 GEOMETRIC CONCERNS

9.2.1 Two Dimensions versus Three Dimensions

The very first question is two-dimensional versus three-dimensional. While the human body is three-dimensional (3D), many situations lend themselves to a successful two-dimensional (2D) analysis.

First approximations for implants (such as the hip) may include 2D analysis. If the leg is in mid-stance (not stair climbing), the loading pattern is 2D. The head and proximal end of the femur are placed in compression and bending, but there is minimal out-of-plane loading. Long bone fracture fixation may require the use of a plate and screws, such as noted in Fig. 9.1. Since the plate is axial and not expected to be subjected to off-axis motion, a 2D model reasonably models the system. Loading for this model is single-leg stance, so the weight is almost directly above the section shown (at approximately 10 degrees to vertical). There is no torque applied to the femur at this point in walking. If one wishes to examine the full walking cycle or a position in early or late stance where the leg is bent forward or back, a 3D analysis would be better suited (Chu et al., 2000; Kurtz et al., 1998).

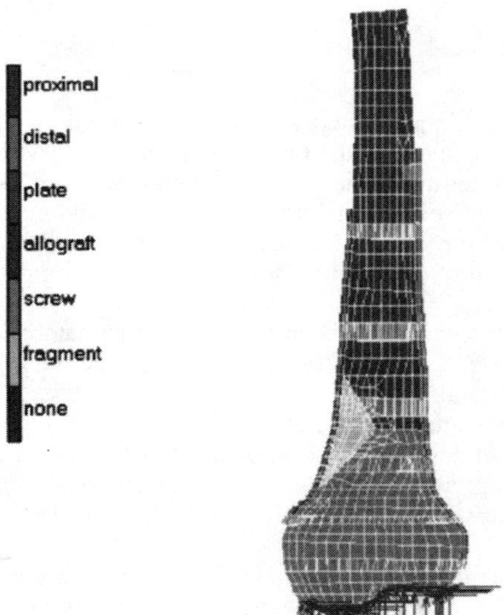

FIGURE 9.1 Two-dimensional distal femur with plate, screws, and bone allograft. A fracture with butterfly bone separation is shown (butterfly bone segment is the light colored triangle on the left). The plate is shown at the right, with screws through overdrilled holes on the near side and full attachment to the cortical bone on the left side. The additional bone graft is on the far left of the bone.

The analysis of dental implants follows a similar pattern. For simple biting, the loading on a tooth is basically 2D in nature. An implant using an isotropic material such as metal may also be evaluated in 2D (Maurer et al., 1999), whereas a composite implant in general will require a 3D analysis to include the out-of-plane material properties (Augerean et al., 1998; Merz et al., 2000).

Many examinations of single ligament or tendon behavior may also be considered 2D. For example, the carpal arch or the wrist (noted in cases of carpal tunnel syndrome) may be modeled in 2D (Nowak and Cherry, 1995). Multiple attachment sites or changes in orientation would necessitate a shift to 3D.

In the cardiovascular realm, due to symmetry, the evaluation of single aortic valve leaflets may be considered 2D for a first approximation (de Hart et al., 2000). The full pattern of valve motion or flow in the proximal aorta would require a 3D analysis (de Hart et al., 1998; Grande et al., 2000).

Fluid boundary layer separation at vascular bifurcations or curves (as found in the carotid and coronary arteries) may be considered 2D if evaluating centerline flow (Steinman and Ethier, 1994). Along this plane, the secondary flows brought on by the vessel cross-sectional curvature will not affect the flow patterns. These models may be used to evaluate boundary layer separation in the carotid artery, the coronaries, and graft anastomoses.

9.2.2 Model Detail

In addition to the number of dimensions is the detail of the model. As with 2D versus 3D models, detail varies with area of interest. If you are examining a plate along a long bone, the bone surface may be considered smooth. If you are assuming complete integration of the screw threads into the bone, this interface may not include the screw teeth (but rather a "welded" interface). Braces and orthotics (if precast) may also be considered smooth and of uniform thickness. Thermoformed materials (such as ankle-foot orthotics) may be considered of a uniform thickness to start, although the heavily curved sections are generally thinner. Large joints, such as the knee, hip, and shoulder may also be considered smooth, although they may have a complex 2D shape.

Bones such as the spine require varying levels of detail depending on the analysis of interest. A general model examining fixation of the spine via a rod or clamps would not require fine detail of vertebral geometries. A fracture model of the spinous processes would require a greater level of detail.

Joints such as the ankle and wrist consist of many small bones, and their surfaces must be determined accurately. This may be done via cadaveric examination or noninvasive means. The cadaveric system generally consists of the following (or a modification): The ligaments and tendons are stained to display the attachments (insertions) into the bones; the construct is then embedded in Plexiglas, and sequential pictures or scans are taken as the construct is either sliced or milled. Slices on the order of a fraction of a millimeter are needed to fully describe the surfaces of wrist carpal bones. Noninvasive measures, such as modern high-sensitivity computed tomographic (CT) scans, may also be able to record geometries at the required level of detail.

Internal bone and soft tissue structures may generally be considered uniform. A separation should be made with bone to distinguish the hard cortical layer from the spongier cancellous material, but it is generally not required to model the cell structure of bone. If one is interested in specifically modeling the head of the femur in great detail, the trabecular architecture may be required. These arches provide a function similar to that of buttresses and flying buttresses in churches. While work continues in detailed FEA studies of these structural alignments down to the micron level, this is not required if one only seeks to determine the general behavior of the entire structure.

Similar approximations may be made in the biofluids realm. Since the size and orientation of vessels vary from person to person, a simple geometry is a good first step. The evaluation of bifurcations can be taken to the next level of complexity by varying the angle of the outlet sides. Cadaveric studies are helpful in determining general population data. As mentioned earlier, CT studies lend exact data for a single subject or a small group of subjects [such as the abdominal aortic aneurysm work of Raghaven et al. (2000)].

9.3 MATERIAL PROPERTIES

Hard tissue should be modeled as orthotropic, even when using 2D analysis. The differences in properties are on the order of those seen in wood, with the parallel-to-the-grain direction (vertical along the tree trunk) being the stiffest. Basic mechanical properties can be found in Fung's text on biomechanics (1996). As can be seen, isotropic modeling will produce significant deviation from expected clinical results. Most FEA codes allow orthotropic material properties in 2D or 3D. This being stated, it must also be noted that since people vary widely in shape and size, material properties may have standard deviations of 30 percent or more. Once a determination of what material properties are going to be used is made, it may be beneficial to vary these parameters to ensure that changes of properties within the standards will not significantly alter the findings.

Distinction must be made between cortical and cancellous bone structures. The harder cortical outer layer carries the bulk of the loading in bones. Although one should not neglect the cancellous bone, it must be recalled that during femoral implant procedures for hip implants, the cancellous canal is reamed out to the cortical shell prior to fitting the metal implant.

Joint articular cartilage is a particular area of concern when producing an FEA model. Work continues on fully describing the mechanical behavior of these low-friction cushioning structures that allow us to move our joints freely. As a first approximation, it is generally assumed that these surfaces are hard and frictionless. If the purpose of the model is to investigate general bone behavior away from the joint, this will be a reasonable approximation. To isolate the cartilage properties, one should perform a data search of the current literature. Of course, implant modeling will not require this information because the cartilage surface is generally removed.

When evaluating intact joints, after the articular cartilage has been modeled, one is forced to examine the synovial fluid. This material is a major component to the nearly frictionless nature of joints. Long molecular chains and lubricants interact to reduce friction to levels far below that of the best human-made materials. At a normal stress of 500 kPa, a sample of a normal joint with synovial fluid has a friction coefficient of 0.0026, whereas a Teflon-coated surface presents coefficients in the range of 0.05 to 0.1 (Fung, 1996). In addition, the articular cartilage reduces friction by exuding more synovial fluid as load increases, effectively moving the articular surfaces further apart. When the load is reduced, the fluid is reabsorbed by the articular materials.

Implants require their own careful study as pertains to the implant-bone interface. If an uncemented implant is to be investigated, the question is one of how much bone integration is expected. If one examines the typical implant, e.g., the femoral component of a hip implant or the tibial component of a knee implant, one will see that a portion of the surface is designed for bone ingrowth through the use of either a beaded or wire-mesh style of surface. These surfaces do not cover the entire implant, so it should not be assumed that the entire implant will be fixed to the bone after healing. Cemented implants are expected to have a better seal with the bone, but two additional points must be made. First, the material properties of bone cement must be included. These are isotropic and are similar to the acrylic grout that is the primary component of bone cement. This brings up the second point, which is that bone cement is not an actual cement that bonds with a surface but rather a tight-fitting grout. As a first approximation, this bonding may be considered a rigid linkage, but it must be recalled that failure may occur by bone separation from the implant (separation from the bone is less likely as the bone cement infiltrates the pores in the bone).

Ligaments and tendons may be modeled as nonlinear springs or bundles of tissue. In the simplest case, one may successfully model a ligament or tendon as one or more springs. The nonlinear behavior is unfortunately evident as low force levels for ligaments and tendons. The low force segment of their behavior, where much of the motion occurs, is nonlinear (forming a toe or J in the stress-strain curve) due to straightening of collagen fibers of different lengths and orientations (Nowak, 1993; Nowak and Logan, 1991). The upper elastic range is quasi-linear. As multifiber materials, ligaments and tendons do not fail all at once but rather through a series of single-fiber microfailures. If one is interested in evaluating behavior in the permanent deformation region, this ropelike behavior must be accounted for.

Dental implants area also nonisotropic in nature, as are many of the implant materials, especially composite-reinforced materials (Nowak et al., 1999). Fiber-reinforced composites generally require a 3D model for analysis, and the properties must be applied using a local coordinate system. This is so because the composites are generally wrapped around a post, and the lengthwise properties no longer run simply in one direction. Choosing a local coordinate system around curves will allow the modeler to keep the high-modulus direction along the cords of the reinforcements.

Although skin is an anisotropic material having different properties in each direction, Bischoff et al. (2000) have presented an interesting model as a second step beyond isotropic material selection.

They used an anisotropic stress loading on the material and achieved a more anisotropic material response. While an anisotropic, nonlinear elastic-plastic model would be best to model skin, the preceding may be used as an intermediate step in FEA.

Vascular flow models (i.e., arterial, venous, and cardiac) and airways models have to be concerned with vessel wall elasticity. As a first step, a rigid wall may be useful when investigating simply the flow patterns. As one progresses toward a more realistic model, elastic properties (e.g., radial, circumferential, and axial) may be required. Texts such as that by Fung (1996) present some of these properties. Another issue is that of tissue tethering. Most mechanical property evaluations have used material removed from its surrounding tissue. Vascular tissue is intimately connected with its surrounding tissue, which may reduce the elastic component slightly. Commercial silicone tubing is a reasonable material on which to model vessel walls if one is evaluating the material as isotropic. If one varies vessel compliance (elasticity) from rigid to that of thin-walled silicone tubing (and beyond), one can achieve a reasonable cross section of possible flow patterns. This is of particular importance when evaluating pulsatile flows. The downside to adding vessel compliance is the dramatic increase in computational time. Steinman and Ethier (1994) evaluated an end-to-side anastomosis pulsatile flow model in two dimensions with either a rigid or elastic wall. The computational time on a workstation was 3 to 4 hours for the rigid model versus 78 hours for the same model when elastic walls were included.

The next topic when discussing material behavior in vascular flows is the nonlinear flow material itself (blood). As a first approximation, a Newtonian fluid (water) may be used to model blood. This is generally the first material used when evaluating flows by bench testing. Flows in the larger arteries and veins, such as the carotid artery, may be successfully evaluated in this manner. Whole-blood viscosity is 0.035 cP, with a density just slightly above that of water (Fung, 1996). This model is reasonable for larger vessels because the solid components of blood do not interact enough to greatly vary the results. The flow in the boundary layer near the vessel walls tends to have fewer red cells in the large vessels because the particles tend to shift toward the higher-velocity core flow. Smaller-diameter vessels, such as the coronaries, will demonstrate greater deviation from Newtonian flow as the red blood cells become a dominant factor in the flow. These flows are best modeled as a combination of Bingham material and power-law flow. The flow is often modeled as a Casson's flow, including a pluglike flow in the vessel core and a combination of the preceding two flows nearer the wall. A finite shear stress is required for flow to exist, and then the viscosity changes as a power-law fit to the shear rate. At higher shear rates, this tends to become a straight line and would be well modeled as Newtonian. Capillary flows should not be modeled as Newtonian or power-law flows because they are much more complex. The normal red blood cell is larger than the diameter of a capillary and must be "forced" through the vessel by deforming the cell in either a balloonlike form or a so-called slipper-zipper form that forces the cell along one side of the vessel.

The decision to "accurately" model blood flow becomes difficult when evaluating the red blood cell component. Although the red blood cell begins as the biconcave form seen in most anatomy and physiology texts, it deforms under the shears seen in flow. The cell becomes elongated (looking much like a football), and the cell membrane slides in a tank-treading motion (Sutera et al., 1989). The interior of the cell is also a viscoelastic fluid, making the modeling of this system difficult. While the preceding is correct for normal red blood cells, disease states, such as sickle cell anemia, modify the mechanical structure of the cell membrane, which in turn alters the flow characteristics of the blood.

9.4 BOUNDARY CONDITIONS

The final important area of FEA is that of boundary conditions. What are the loadings and constraints of the systems? For hard and soft tissues, these include loading sites and multiple attachments. For biofluids, these include pressures, velocities, and pulsatile conditions.

The primary concern when evaluating hard tissues is to determine how many ligament and tendon attachments are required. For example, the wrist is made up of a vast network of ligaments but only a few major ones (An et al., 1991). In a two-dimensional model of the wrist or even in a 3D model, a reasonably accurate model may be produced by only adding a few ligament attachments per carpal bone. Ligament attachment sites may also be approximated as a first step, although point attachments such as might be made by a single spring are cause for concern.

As noted in Sec. 9.2, a 3D model of a joint will be more accurate than a 2D model. If one is evaluating a long bone model in isolation, it must be decided whether the joint area is of importance and whether the model is to be constructed for static or dynamic evaluation. In a static case, if one is concerned with structures away from the articular cartilage, a simple loading or constraint system may be used successfully to represent the joint. A fully dynamic model or one that examines the joint itself will require a more complex 3D model with fixed constraints and loadings further from the joint surfaces.

Soft tissue analysis requires an evaluation of insertion geometries and paths for complete evaluation. As tendons travel from muscle to bone they often travel through sheaths and are partially constrained. While these issues are generally not of importance to general modeling of bone interactions and muscle forces, they may be needed if the desired result is a clinically accurate model of a given tendon.

Blood flow requires a knowledge of both the velocity and pressure conditions and the time-based changes in these parameters. While venous flow is basically steady state, it is readily apparent that arterial flow is pulsatile. Even in arterial flows, a mean flow is considered reasonable as a first modeling mechanism. Higher and lower steady flow velocities and pressures begin to produce a fuller picture of behavior. In the more complex models, a fully pulsatile flow is desirable, although generally at a large computational expense (hence the use of supercomputers for many of these models). As noted in Sec. 9.3, vessel wall behavior is an important boundary condition for blood flow.

Airways flow has a boundary condition of diffusion only at the terminal level within the alveolar sacs. While air is moved in and out in the upper respiratory system, diffusion is the means by which material is shifted near the lung-blood interfaces. A flow model of the lower respiratory system should be such that air does not physically move in and out of the alveolar sacs.

9.5 CASE STUDIES

The first study is an evaluation of an ankle-foot orthosis (AFO) or brace such as is used for patients with "drop foot" (Abu-Hasaballah et al., 1997). Although the orthosis is made of lightweight thermoplastic, many patients still find it heavy and uncomfortable, especially in the summer when the orthosis may become sweat covered.

A 3D model with appropriate geometries and material properties was developed from a physical brace (Fig. 9.2). After verifying the model by comparison to actual AFO behavior, a series of design modifications was made, by element removal, to reduce weight while retaining structural integrity. Figure 9.3 presents one of the final versions, with segments along the calf removed. The weight reduction was approximately 30 percent, and the large openings along the calf would reduce sweat buildup in this region.

It should be noted that a single AFO cost \$200 or more, so FEA is an inexpensive means by which many design changes may be evaluated before any actual brace has to be built. Computational times were reasonably low, from a few minutes to a few hours on a PC.

As a second example of the process that may be followed when modeling the human body with FEA, let us consider blood flow through the carotid artery. This is the main blood supply to the brain (through the internal carotid artery) and is the site of stenoses. Plaque formation (artherosce-losis) narrows the common carotid artery at the origin of the external carotid artery (which flows toward the face) until blood flow is reduced to the brain.

To first evaluate this flow, many simplifications will be made to reduce the opportunities for error in modeling. Each model result should be evaluated based on available bench and clinical data prior

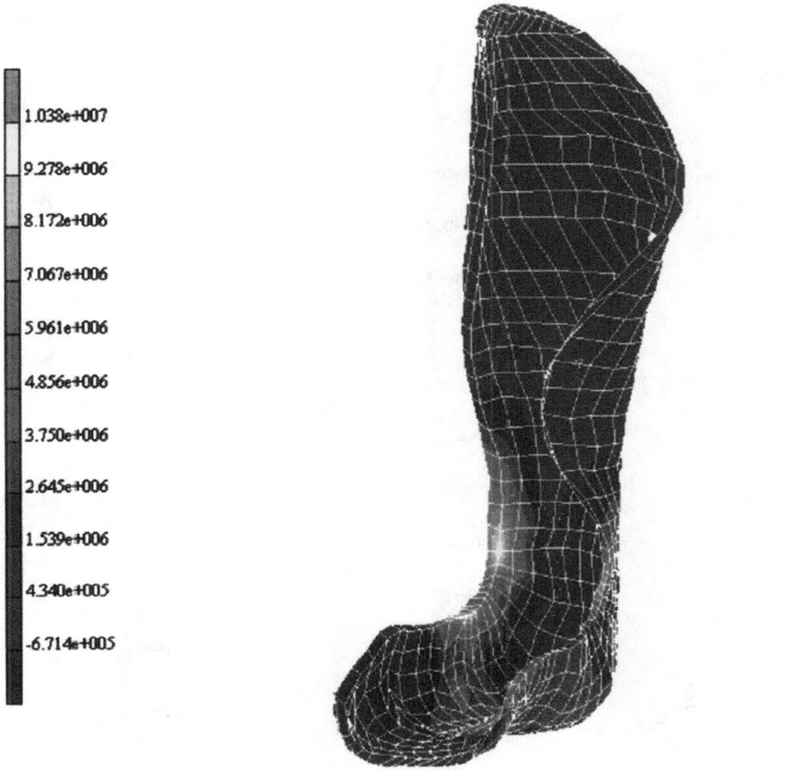

FIGURE 9.2 Finite-element model of an isentropic ankle-foot orthosis demonstrating high-stress regions near inner ankle. The material is thermoplastic.

to adding the next level of complexity. These checks may also be used to determine which simplifications do not affect the relevant results.

As a first approximation of geometry, a Y bifurcation using straight rigid tubing may be used. The blood may be modeled as a Newtonian fluid without particles. As boundary conditions, the inlet flow may either be considered Poiseuille or uniform across the tube (if a sufficient entry length is included to produce a fully developed flow profile). Steady flow will be used for this model.

The next sequence in modeling will be to improve the geometry. The bulb shape of the carotid sinus will be added, as well as the geometry of the external carotid inlet. For this second model, a Y shape may still be used, or a more anatomically correct angle may be used. Liquid and inlet conditions will be maintained from the first model, and the vessel walls will remain rigid. Differences in flow behavior may be apparent between these first two models.

Once the second model is functional, we will turn our attention to the material in the flow. A combination of Bingham and power-law fluid will be used to better model the non-Newtonian characteristics of whole blood. RBCs will not be added for this model because the fluid is now a reasonable approximation of large-vessel flow. We will still use steady flow in this model.

Once the third version solves smoothly, pulsatile flow will be added. As a first step, a sinusoidal flow pattern will be added to the constant-flow conditions. Subsequent modifications will add user-defined patterns resembling the actual pulse forms found in the human carotid.

The fourth generation of the model will seek to include the nonrigid vessel wall properties. As a first step, the walls will include varying amounts of elasticity, such as might be found in silicone

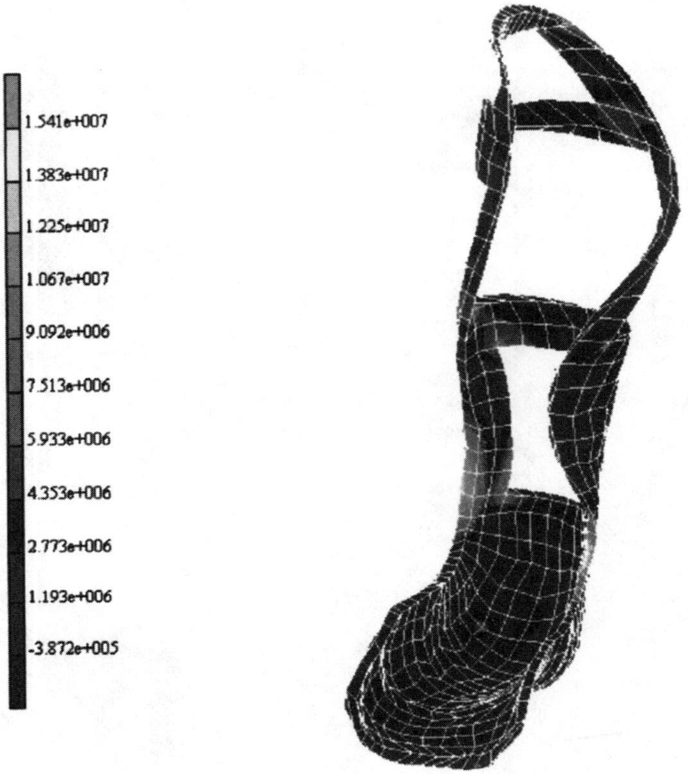

FIGURE 9.3 Ankle-foot orthosis optimized manually to reduce material in regions of low stress. Total weight reduction is approximately 30 percent.

tubing. The viscoelastic behavior of actual arterial wall may be investigated during the final model or two in the process.

The fifth version of the model will now seek to address the issues of the red blood cells. Using software allowing for the addition of user-defined particles, rigid disks may first be used. After success with this version, deformable cells may be used to better approximate red blood cells.

At each point in the process (each generation of the model), the user should compare the findings with those of available bench and clinical studies, along with the results from previous model versions. Little change from previous models may suggest either that there is no need for a more complex model or that all the modifications have not been taken account of.

It should also be noted that varying the simpler models may encompass the behavior of more complex models. For example, varying the static flow parameters may account for many of the behavior modifications seen in the pulsatile model. For similar reasons, varying the elastic properties of the vessel wall may encompass behavioral changes brought about by viscoelastic modeling at a great savings of computational time and power.

9.5.1 Case Studies Conclusions

As can be seen from these examples, multiple generations of modeling are often used to evaluate all the aspects of biological systems. Many simplifications are made initially to determine a first-generation solution. The number of subsequent model generations will vary depending on the sophistication of results desired. At this point in time it is perhaps unlikely that all nonlinear parameters of the human body can be included in an FEA model, mainly due to the fact that all properties are not yet known. Comparison with bench and clinical findings will demonstrate, however, that similar behavioral results are possible.

The researcher should not be overly concerned with the minutiae of the model parameters. Considering the variances between people, an overview covering most of the noted effects should be the goal. The purpose of the models is to examine system behavior when changes are made, such as the effects of geometric and material modifications. Although the actual behavior may not be exact, the variances due to changes in the model may well mimic those of the device or human system.

9.6 CONCLUSIONS

The main points to be taken from this chapter are that many simplifications must be made initially when evaluating human-related structures with FEA, but this may not be a major problem. Each of the three main areas (geometry, material properties, and boundary conditions) is of equal importance and should be considered separately. Depending on the detail of the construct modeled, many geometric components may be simplified. Care should be taken when using isentropic models for solids, but even these may be used for general evaluations of tissue regions. Soft tissues are difficult to fully model without the use of nonlinear properties. Newtonian flows are reasonable for large-vessel flows (such as the aorta or carotid artery) as long as the aspects of flows have been shown not to be affected by the particular components. Boundary conditions may begin with steady-state values, but dynamic components will add the complexities of the true system.

In closing, a computer model of portions of the human body or an added component may be as simple or complex as the user desires. One should use properties to be found in this and other sources and always confirm findings with clinical or bench values. The future is unlimited in this field as we learn more about the body and seek to best aid or replace its many functions.

REFERENCES

Abu-Hasaballah, K. S., Nowak, M. D., and Cooper, P. S. (1997), Enhanced solid ankle-foot orthosis design: Real-time contact pressures evaluation and finite element analysis, *1997 Advances in Bioengineering,* pp. 285–286.

An, K.-N., Berger, R. A., and Cooney, W. P. (eds.) (1991), *Biomechanics of the Wrist Joint,* Springer-Verlag, New York.

Augereau, D., Pierrisnard, L., Renault, P., and Barquins, M. (1998), Prosthetic restoration after coronoradicular resection: Mechanical behavior of the distal root remaining and surrounding tissue, *J. Prosthet. Dent.* **80**:467–473.

Bischoff, J. E., Arruda, E. M., and Grosh, K. (2000), Finite element modeling of human skin using an isotropic, nonlinear elastic constitutive model, *J. Biomech.* **33**:645–652.

Chu, Y. H., Elias, J. J., Duda, G. N., Frassica, F. J., and Chao, E. Y. S. (2000), Stress and micromotion in the taper lock joint of a modular segmental bone replacement prosthesis, *J. Biomech.* **33**:1175–1179.

Ferris, P., and Prendergast, P. J. (2000), Middle-ear dynamics before and after ossicular replacement, *J. Biomech.* **33**:581–590.

Fung, Y. C. (1996), *Biomechanics: Mechanical Properties of Living Tissues,* 2d ed., Springer-Verlag, New York.

Grande, K. J., Cochran, R. P., Reinhall, P. G., and Kunzelman, K. S. (2000), Mechanisms of aortic valve incompetence: Finite element modeling of aortic root dilation, *Ann. Thorac. Surg.* **69**:1851–1857.

Hart, J. de, Cacciola, G., Schreurs, P. J. G., and Peters, G. W. M. (1998), A three-dimensional analysis of a fiber-reinforced aortic valve prosthesis, *J. Biomech.* **31**:629–638.

Hart, J. de, Peters, G. W. M., Schreurs, P. J. G., and Baaijens, F. P. T. (2000), A two-dimensional fluid-structure interaction model of the aortic valve, *J. Biomech.* **32**:1079–1088.

Huang, H. M., Lee, M. C., Chiu, W. T., Chen, C. T., and Lee, S. Y. (1999). Three-dimensional finite element analysis of subdural hematoma, *J. Trauma* **47**:538–544.

Klinnert, J., Nowak, M., and Lewis, C. (1994), Addition of a medial allograft to stabilize supracondylar femur fractures, *1994 Advances in Bioengineering,* pp. 193–194.

Kurtz, S. M., Ochoa, J. A., White, C. V., Srivastav, S., and Cournoyer, J. (1998), Backside nonconformity and locking restraints affect liner/shell load transfer mechanisms and relative motion in modular acetabular components for total hip replacement, *J. Biomech.* **31**:431–437.

Maurer, P., Holwig, S., and Schubert, J. (1999), Finite-element analysis of different screw diameters in the sagittal split osteotomy of the mandible, *J. Craniomaxillofac. Surg.* **27**:365–372.

Merz, B. R., Hunenbaart, S., and Belser, U. C. (2000), Mechanics of the implant-abutment connection: An 8-degree taper compared to a butt joint connection, *Int. J. Oral Maxillofac. Implants* **15**:519–526.

Nowak, M. D. (1993), Linear versus nonlinear material modeling of the scapholunate ligament of the human wrist, in H. D. Held, C. A. Brebbia, R. D. Ciskowski, H. Power (eds), *Computational Biomedicine*, pp. 215–222, Computational Mechanics Publications, Boston.

Nowak, M. D., and Cherry, A. C. (1995), Nonlinear finite element modeling of the distal carpal arch, *1995 Advances in Bioengineering,* pp. 321–322.

Nowak, M. D., Haser, K., and Golberg, A. J. (1999), Finite element analysis of fiber composite dental bridges: The effect of length/depth ratio and load application method, *1999 Advances in Bioengineering* pp. 249–250.

Nowak, M. D., and Logan, S. E. (1991), Distinguishing biomechanical properties of intrinsic and extrinsic human wrist ligaments, *J. Biomech. Eng.* **113**:85–93.

Raghavan, M. L., Vorp, D. A., Federle, M. P., Makaroun, M. S., and Webster, M. W. (2000), Wall stress distribution on three-dimensionally reconstructed models of human abdominal aortic aneurysm, *J. Vasc. Surg.* **31**:760–769.

Steinman, D. A., and Ethier, C. R. (1994), The effect of wall distensibility on flow in a two-dimensional end-to-side anastomosis, *J. Biomech. Eng.* **116**:294–301.

Sutera, S. P., Pierre, P. R., and Zahalak, G. I. (1989), Deduction of intrinsic mechanical properties of the erythrocyte membrane from observations of tank-treading in the rheoscope, *Biorheology* **26**:177–197.

CHAPTER 10
VIBRATION, MECHANICAL SHOCK, AND IMPACT

Anthony J. Brammer
Biodynamics Laboratory at the Ergonomic Technology Center, University of Connecticut Health Center, Farmington, Connecticut and Institute for Microstructural Sciences, National Research Council, Ottawa, Ontario, Canada

Donald R. Peterson
Biodynamics Laboratory at the Ergonomic Technology Center, University of Connecticut Health Center, Farmington, Connecticut

10.1 INTRODUCTION

Time-varying forces and accelerations occur in daily life, and are commonly experienced, for example, in an elevator and in aircraft, railway trains, and automobiles. All of these situations involve motion of the whole body transmitted through a seat, or from the floor in the case of a standing person, where the human response is commonly related to the relative motion of body parts, organs, and tissues. The vibration, shocks, and impacts become of consequence when activities are impaired (e.g., writing and drinking on a train, or motion sickness), or health is threatened (e.g., a motor vehicle crash). Equally important are exposures involving a localized part of the body, such as the hand and arm (e.g., when operating a hand tool), or the head (e.g., impacts causing skull fracture or concussion).

In this chapter, methods for characterizing human response to vibration, shock, and impact are considered in order to prescribe appropriate countermeasures. The methods involve data from experiments on humans, animals, and cadavers, and predictions using biodynamic models and manikins. Criteria for estimating the occurrence of health effects and injury are summarized, together with methods for mitigating the effects of potentially harmful exposures. There is an extensive literature on the effects of vibration, shocks, and impacts on humans (Griffin, 1990; von Gierke et al., 2002; von Gierke, 1997; Nahum et al., 1993).

10.1.1 Definitions and Characterization of Vibration, Mechanical Shock, and Impact

Vibration and Mechanical Shock. *Vibration* is a time-varying disturbance of a mechanical, or biological, system from an equilibrium condition for which the long-term average of the motion will tend to zero, and on which may be superimposed either translations or rotations, or both. The mechanical forces may be distributed, or concentrated over a small area of the body, and may be applied at an angle to the surface (e.g., tangential or normal). Vibration may contain random or deterministic components, or both; they may also vary with time (i.e., be nonstationary). Deterministic vibration may contain harmonically related components, or pure tones (with sinusoidal time dependence), and may form "shocks." A mechanical *shock* is a nonperiodic disturbance characterized by suddenness and severity with, for the human body, the maximum forces being reached within a few tenths of a second, and a total duration of up to about a second. An *impact* occurs when the body, or body part, collides with an object. When considering injury potential, the shape of the object in contact with, or impacting, the body is important, as is the posture. In addition, for hand tools, both the compressive (grip) and thrust (feed) forces employed to perform the manual task need to be considered.

Although vibration, shock, and impact may be expressed by the *displacement* of a reference point from its equilibrium position (after subtracting translational and rotational motion), they are more commonly described by the *velocity* or *acceleration*, which are the first and second time derivatives of the displacement.

Vibration Magnitude. The magnitude of vibration is characterized by second, and higher even-order mean values, as the net motion expressed by a simple, long-term time average will be zero. For an acceleration that varies with time t, as $a(t)$, the higher-order mean values are calculated from:

$$a_{RM} = \left[\frac{1}{T}\int_0^T [a(t)]^m \, dt\right]^{1/r} \tag{10.1}$$

where the integration is performed for a time T, and m and r are constants describing the moment and root of the function. By far the most common metric used to express the magnitude of whole-body or hand-transmitted vibration is the *root mean square* (RMS) acceleration a_{RMS}, which is obtained from Eq. (10.1) with $m = r = 2$; i.e.,

$$a_{RMS} = \left[\frac{1}{T}\int_0^T [a(t)]^2 \, dt\right]^{1/2} \tag{10.2}$$

Other metrics used to express the magnitude of vibration and shock include the *root mean quad* (RMQ) acceleration a_{RMQ}, with $m = r = 4$ (and higher even orders, such as the *root mean sex* (RMX) acceleration a_{RMX}, with $m = r = 6$).

The RMS value of a continuous random vibration with a gaussian distribution corresponds to the magnitude of the 68th percentile of the amplitudes in the waveform. The higher-order means correspond more closely to the peak values of the waveform, with the RMQ corresponding to the 81st percentile and the RMX to the 88th percentile of this amplitude distribution. The relationships between these metrics depend on the amplitude distribution of the waveform, wherein they find application to characterize the magnitude of shocks entering, and objects impacting, the body. This can be inferred from the following example, where the RMS value corresponds to 0.707 of the amplitude of a sinusoidal waveform, while the RMQ value corresponds to 0.7825 of the amplitude.

EXAMPLE 10.1 *Calculate the RMS and RMQ accelerations of a pure-tone (single-frequency) vibration of amplitude A and angular frequency ω.*

Answer*: The time history (i.e., waveform) of a pure-tone vibration of amplitude A can be expressed as $a(t) = A \sin \omega t$, so that, from Eq. 10.(2):*

$$a_{RMS} = \left[\frac{1}{T}\int_0^T [A \sin(\omega t)]^2 \, dt\right]^{1/2} \quad \text{or} \quad a_{RMS} = \left[\frac{A^2}{2T}\int_0^T [1 - \cos(2\omega t)] \, dt\right]^{1/2}$$

Let us integrate for one period of the waveform, so that $T = 2\pi/\omega$ (any complete number of periods will give the same result). Then:

$$a_{\text{RMS}} = \frac{A}{\sqrt{2}} = 0.707A$$

From Eq. (10.1):

$$a_{\text{RMQ}} = \left[\frac{1}{T}\int_0^T [A \sin(\omega t)]^4\, dt\right]^{1/4} \quad \text{or} \quad a_{\text{RMQ}} = \left[\frac{A^4}{4T}\int_0^T [1 - \cos(2\omega t)]^2\, dt\right]^{1/4}$$

$$a_{\text{RMQ}} = \left[\frac{A^4}{4T}\int_0^T \left[\frac{3}{2} - 2\cos(2\omega t) - \frac{1}{2}\cos(4\omega t)\right] dt\right]^{1/4}$$

Again, integrating for one period of the waveform:

$$a_{\text{RMQ}} = \left[\frac{3A^4}{8}\right]^{1/4} = 0.7825A$$

Equinoxious Frequency Contours. Human response to vibration, shock, and impact depends on the frequency content of the stimulus, as well as the magnitude. This may be conveniently introduced electronically, by filtering the time history of the stimulus signal, and has led to the specification of vibration magnitudes at different frequencies with an equal probability of causing a given human response or injury, so defining an *equinoxious frequency contour*. The concept, while almost universally employed, is strictly only applicable to linear systems. The biomechanic and biodynamic responses of the human body to external forces and accelerations commonly depend nonlinearly on the magnitude of the stimulus, and so any equinoxious frequency contour can be expected to apply only to a limited range of vibration, shock, or impact magnitudes.

Equinoxious frequency contours may be estimated from epidemiological studies of health effects, or from the response of human subjects, animals, cadavers, or biodynamic models to the stimuli of interest. Human subjects cannot be subjected to injurious accelerations and forces for ethical reasons, and so little direct information is available from this source. Some information has been obtained from studies of accidents, though in most cases the input acceleration-time histories are poorly known.

Frequency Weighting. The inverse frequency contour (i.e., reciprocal) to an equinoxious contour should be applied to a stimulus containing many frequencies to produce an overall magnitude that appropriately combines the contributions from each frequency. The frequency weightings most commonly employed for whole-body and hand-transmitted vibration are shown in Fig. 10.1 (ISO 2631-1, 1997; ISO 5349-1, 2001). The range of frequencies is from 1 to 80 Hz for whole-body vibration, and from 8 to 1250 Hz for vibration entering the hand. A frequency weighting for shocks may also be derived from a biodynamic model (see "Dynamic Response Index (DRI)" in Sec. 10.3.1).

Vibration Exposure. Health disturbances and injuries are related to the magnitude of the stimulus, its frequency content, and its duration. A generalized expression for exposure may be written

$$E(a_w, T)_{m,r} = \left[\int_0^T [F(a_w(t))]^m\, dt\right]^{1/r} \tag{10.3}$$

where $E(a_w, T)_{m,r}$ is the exposure occurring during a time T to a stimulus function that has been frequency weighted to equate the hazard at different frequencies, $F(a_w(t))$. In general, $F(a_w(t))$ may be expected to be a nonlinear function of the frequency-weighted acceleration-time history $a_w(t)$.

Within this family of exposure functions, usually only those with *even* integer values of m are of interest. A commonly used function is the so-called *energy-equivalent* vibration exposure for which $F(a_w(t)) \equiv a_w(t)$ and $m = r = 2$:

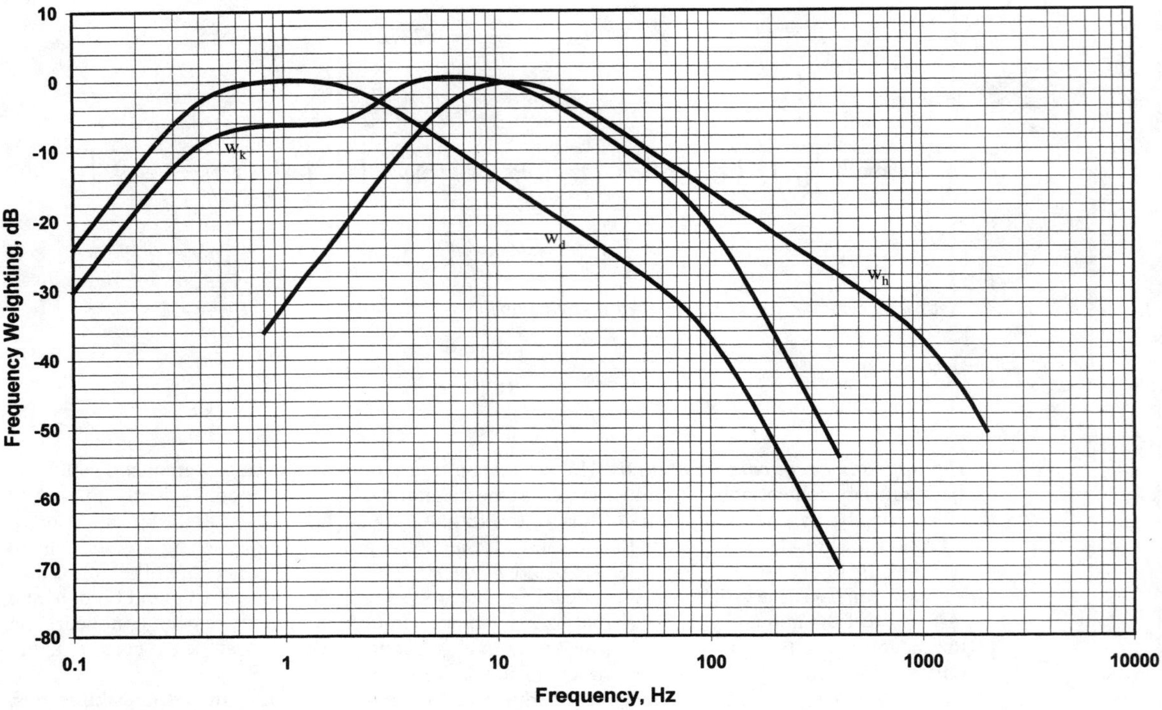

FIGURE 10.1 Frequency weightings for whole-body (W_k and W_d) and hand-transmitted (W_h) vibration. W_k and W_d are for the z direction and x and y directions, respectively, and are applicable to seated and standing persons (see Fig. 10.2). W_h is for all directions of vibration entering the hand. The filters are applied to acceleration-time histories $a(t)$. (*ISO 2631-1, 1997; ISO 5349-1, 2001.*)

$$E(a_w, T)_{2,2} = \left[\int_0^T [F(a_w(t))]^2 \, dt \right]^{1/2} \tag{10.4}$$

For an exposure continuing throughout a working day, $T \equiv T_{(8)} = 28{,}800$ s, and Eq. (10.4) can be written [using Eq. (10.2)]:

$$E(a_w, T)_{2,2} = T_{(8)}^{1/2} \left[\frac{1}{T_{(8)}} \int_0^{T_{(8)}} [F(a_w(t))]^2 \, dt \right]^{1/2} = T_{(8)}^{1/2} a_{\mathrm{RMS}(8)} \tag{10.5}$$

where $a_{\mathrm{RMS}(8)}$ is the 8-hour, energy-equivalent, frequency-weighted RMS acceleration.

A second function, used for exposure to whole-body vibration, is the *vibration dose value*, VDV, for which $F(a_w(t)) \equiv a_w(t)$ and $m = r = 4$. The function is thus:

$$\mathrm{VDV} = E(a_w, T)_{4,4} = \left[\int_0^T [a_w(t)]^4 \, dt \right]^{1/4} \tag{10.6}$$

which is more influenced by the large amplitudes in a fluctuating vibration than the energy-equivalent exposure. A related function, the *severity index* for which $F(a_w(t)) \equiv a_w(t)$, $m = 2.5$, and $r = 1$, is sometimes used for the assessment of head impact, though it cannot be applied to continuous acceleration-time histories owing to the value of m.

10.1.2 Human Response to Vibration, Mechanical Shock, and Impact

Mechanical damage can occur at large vibration magnitudes, which are usually associated with exposure to shocks, and to objects impacting the body (e.g., bone fracture, brain injury, organ hemorrhage, and tearing or crushing of soft tissues). At moderate magnitudes there can be physiological effects leading to chronic injury, such as to the spine, and disorders affecting the hands. At all magnitudes above the threshold for perception there can be behavioral responses ranging from discomfort to interference with tasks involving visual or manual activities.

Injury from Vibration. *Whole-Body Vibration.* Small animals (e.g., mice and dogs) have been killed by intense vibration lasting only a few minutes (see Griffin, 1990). The internal injuries observed on postmortem examination (commonly heart and lung damage, and gastro-intestinal bleeding) are consistent with the organs beating against each other and the rib cage, and suggest a *resonance* motion of the heart, and lungs, on their suspensions. In man, these organ suspension resonances are at frequencies between 3 and 8 Hz.

Chronic exposure to whole-body vibration may result in an increased risk of low back pain, sciatic pain, and prolapsed or herniated lumbar disks compared to control groups not exposed to vibration. These injuries occur predominantly in crane operators, tractor drivers, and drivers in the transportation industry (Bovenzi and Hulshof, 1998). However, it is difficult to differentiate between the roles of whole-body vibration and ergonomic risk factors, such as posture, in the development of these disorders.

Hand-Transmitted Vibration. Chronic injuries may be produced when the hand is exposed to vibration. Symptoms of numbness or paresthesia in the fingers or hands are common. Reduced grip strength and muscle weakness may also be experienced, and episodic finger blanching, often called colloquially "white fingers," "white hand," or "dead hand," may occur in occupational groups (e.g., operators of pneumatic drills, grinders, chipping hammers, riveting guns, and chain saws). The blood vessel, nerve, and muscle disorders associated with regular use of hand-held power tools are termed the *hand-arm vibration syndrome* (HAVS) (Pelmear et al., 1998). An exposure-response relationship has been derived for the onset of finger blanching (Brammer, 1986). Attention has also recently been drawn to the influence of vibration on task performance and on the manual control of objects (Martin et al., 2001).

Repeated flexing of the wrist can injure the tendons, tendon sheaths, muscles, ligaments, joints and nerves of the hand and forearm. These *repetitive strain injuries* commonly occur in occupations involving repeated hand-wrist deviations (e.g., keyboard and computer operators), and frequently involve nerve compression at the wrist (e.g., *carpal tunnel syndrome*) (Cherniack, 1999).

Injury from Shock and Impact. Physiological responses to shocks and objects impacting the body include those discussed for whole-body vibration. For small contact areas, the injuries are often related to the elastic and tensile limits of tissue (Haut, 1993; von Gierke et al., 2002). The responses are critically dependent on the magnitude, direction, and time history of the acceleration and forces entering the body, the posture, and on the nature of any body supports or restraints (e.g., seat belt or helmet).

Vertical Shocks. Exposure to single shocks applied to a seated person directed from the seat pan toward the head ("headward") has been studied in connection with the development of aircraft ejection seats, from which the conditions for spinal injury and vertebral fractures have been documented (Anon., 1950; Eiband, 1959). Exposure to intense *repeated* vertical shocks is experienced in some off-the-road vehicles and high-performance military aircraft, where spinal injury has also been reported. A headward shock with acceleration in excess of $g = 9.81$ m/s^2 (the acceleration of gravity) is likely to be accompanied by a downward ("tailward") impact, when the mass of the torso returns to being supported by the seat.

Horizontal Shocks. Exposure to rapid decelerations in the horizontal direction has been extensively studied in connection with motor vehicle and aircraft crashes ("spineward" deceleration). Accident statistics indicate that serious injuries to the occupants of motor vehicles involved in frontal collisions are most commonly to the head, neck, and torso, including the abdomen (AGARD-AR-330, 1997).

Injuries to the head usually involve diffuse or focal brain lesions either with or, commonly, without skull fracture. The former consists of brain swelling, concussion, and *diffuse axonal injury*, that is, mechanical disruption of nerve fibers; the latter consists of localized internal bleeding and contusions (coup and contrecoup).

The most common neck injury is caused by rearward flexion and forward extension ("whiplash"), which may result in dislocation or fracture of the cervical vertebrae, and compression of the spinal cord.

10.2 PHYSICAL MEASUREMENTS

The complexity of a living organism, and its ability to modify its mechanical properties (e.g., in response to mechanical or physiological demands or muscle tension), necessitates the careful design of experiments. There is a large variability in response between individuals. Also, the direct attachment of vibration and shock sensors to soft tissues produces a mechanical load that influences tissue motion. With appropriate measurement methods and instrumentation (ISO 8041, 1990), mechanical responses to vibration can be determined for tissues, body segments, and the whole body.

10.2.1 Methods and Apparatus

Tissue Vibration. Noncontact methods are preferred for measuring the surface motion of tissues. Laser vibrometers are commercially available with sufficient bandwidth and resolution for most studies. A direct mass load on the skin, together with the skin's elasticity, forms a mechanical low-pass filter (see "Simple Lumped Models" in Sec. 10.3.1). If a device is to be mounted directly on the skin, it must be of low weight (e.g., <3 g) and possess a comparatively large attachment area (e.g., $>5\ cm^2$), in order for vibration to be recorded without attenuation of the motion at 80 Hz. An upper frequency limit of 200 Hz is theoretically achievable (-3dB) with a transducer and skin mount weighing 3 g and an attachment area of 1.8 cm^2.

The measurement of hand-transmitted vibration requires a bandwidth extending to at least 1250 Hz, which is unobtainable by localized, skin-mounted transducers. Attempts to strap the transducer to a bony prominence (e.g., by a "watch strap" at the wrist) have demonstrated that the upper frequency limit for this method of measurement is about 200 Hz (Boileau et al., 1992). A distributed sensor, such as a pressure-sensitive lightweight film, would permit the motion of a large skin area to be monitored with acceptable mass loading, and could respond to vibration at higher frequencies.

Interface between Body and Vibrating Surface. Devices have been developed to measure the motion of the interface between the skin and a source of vibration in contact with the body, such as a vehicle seat pan or tool handle. The former consists of a flexible disk of rubberlike material thickened at the center, where an accelerometer is mounted. The dimensions have been standardized (ISO 7096, 1982). Attempts have been made to design transducer mounts for the palm of the hand to measure the vibration entering the hand from tool handles (see, for example, ISO 10819, 1997), and also the static compressive force (i.e., the combined grip and thrust forces exerted by the hand on the tool handle). The frequency response of one such device extends to more than 1 kHz (Gillmeister et al., 2001).

Structures in Contact with the Body. The vibration of structures in contact with the body, such as seats and tool handles, is conveniently measured by accelerometers rigidly attached to a structural element. Accelerometers are designed to respond to the vibration in a given direction and may be grouped to record simultaneously accelerations in three orthogonal directions. They are commercially available in a wide range of sizes and sensitivities, and may be attached by screws or adhesives to the structure of interest.

Orientation of Sensors. Tissue-, interface-, and structure-mounted sensors may be aligned as closely as possible with the axes of a *biodynamic coordinate system* (see, for example, Fig. 10.2), even though these may have inaccessible origins that are anatomical sites within the body. In practice, sensors are commonly oriented to record the component accelerations defined by the *basicentric coordinate systems* shown in Fig. 10.2 (ISO 2631–1, 1997; ISO 5349–1, 2001), which have origins at the interface between the body and the vibrating surface. The location of accelerometers to record the handle vibration of specific power tools is described in an international standard (ISO 5349-2, 2001).

Errors in Shock and Impact Measurement. Care must be taken when accelerometers employing piezoelectric elements are used to measure large-magnitude shocks and impacts, as they are subject to internal crystalline changes that result in dc shifts in the output voltage. Results containing such shifts should be considered erroneous. This limitation of piezoelectric transducers may be overcome by mounting the sensor on a mechanical low-pass filter (see "Simple Lumped Models" in Sec. 10.3.1), which is, in turn, attached to the structure of interest. Such filters possess a resilient element that serves to reduce the transmission of vibration at high frequencies. The filter cutoff frequency is selected to be above the maximum vibration frequency of interest but below the internal mechanical resonance frequency of the accelerometer.

Data Recording. The signal produced by vibration, shock, and impact sensors is first conditioned to remove bias voltages or other signals required for the device to function, and then amplified and buffered for output to a data recording system. The output may be stored on high-quality magnetic tape (e.g., a DAT recorder), or by a digital data acquisition system. The latter should possess low-pass, antialiasing filters (with cutoff frequency typically one-half the sampling frequency), and an analog-to-digital (A/D) converter with sufficient dynamic range (commonly 12 or 16 bits). The data acquisition system should be capable of recording time histories at sampling frequencies of at least 2500 Hz for hand-transmitted vibration, or 160 Hz for whole-body vibration, *per component and measurement site* (e.g., palm and wrist, or seat surface and seat back).

10.2.2 Small-Amplitude Response of the Human Body

Tissue Properties. The properties of human tissues when the body is considered a linear, passive mechanical system are summarized in Table 10.1 (von Gierke et al., 2002; Goldstein et al., 1993). The values shown for soft tissues are typical of muscle tissue, while those for bone depend on the structure of the specific bone. Cortical bone is the dominant constituent of the long bones (e.g., femur, tibia), while trabecular bone, which is more elastic and energy absorbent, is the dominant constituent of the vertebrae. The shear viscosity and bulk elasticity of soft tissue are from a model for the response in vivo of a human thigh to the vibration of a small-diameter piston (von Gierke et al., 1952)

The nonlinear mechanical properties of biological tissues have been studied extensively in vitro, including deviations from Hooke's law (Fung, 1993; Haut, 1993).

Mechanical Impedance of Muscle Tissue. The (input) *mechanical impedance* is the complex ratio between the dynamic force applied to the body and the velocity at the interface where vibration enters the body. The real and imaginary parts of the mechanical impedance of human muscle in vivo are shown as a function of frequency in Fig. 10.3 (von Gierke et al., 1952). In this diagram the measured resistance (open circles) and reactance (diamonds) are compared with the predictions of a model, from which some tissue properties may be derived (see Table 10.1). It should be noted that

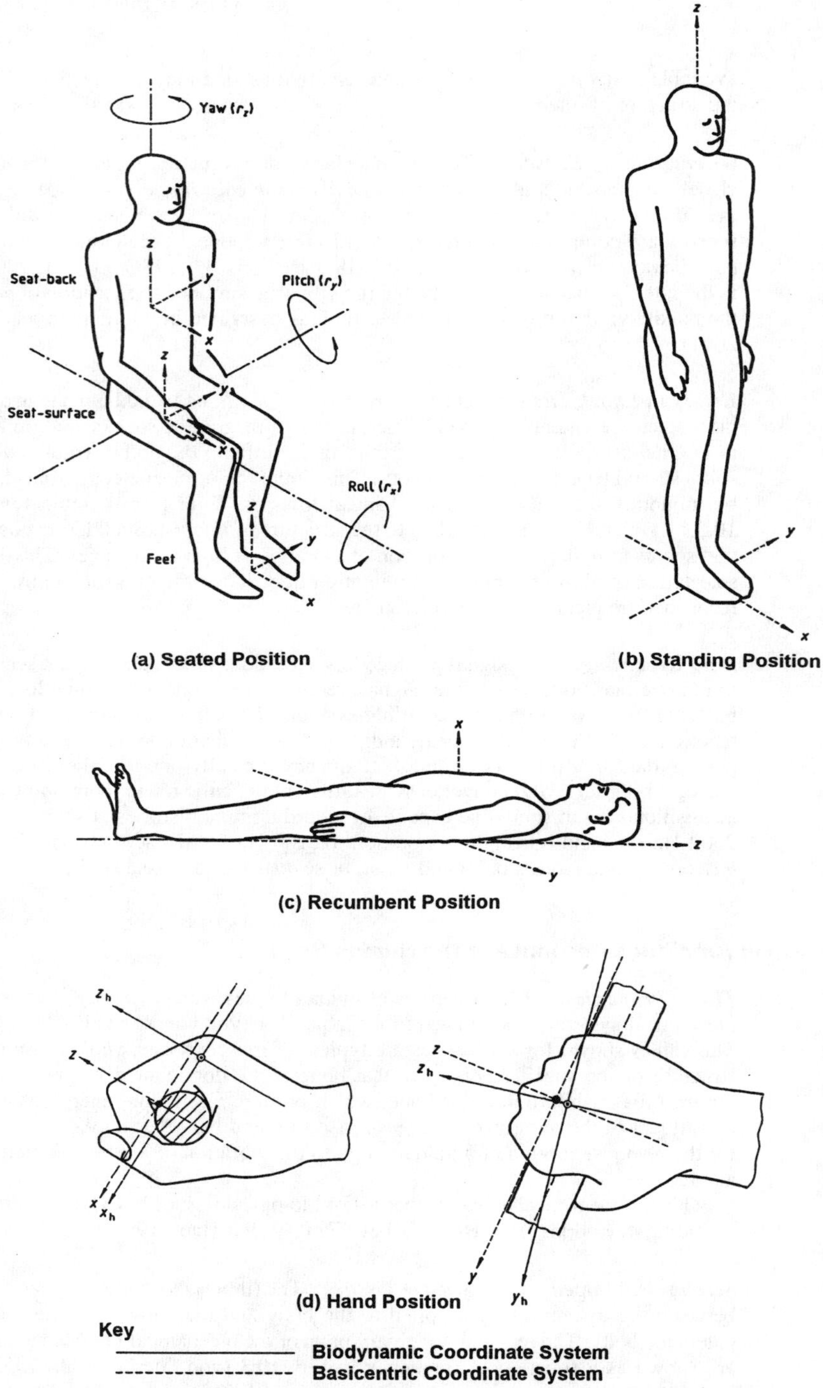

FIGURE 10.2 Basicentric coordinate axes for translational (x, y, and z) and rotational (r_x, r_y, and r_z) whole-body vibration, and basicentric (filled circles and dashed lines) and biodynamic (open circles and continuous lines) axes for hand-transmitted vibration. The biodynamic coordinates axes for the hand are x_h, y_h, and z_h. *(ISO 2631-1, 1997; ISO 5349-1, 2001.)*

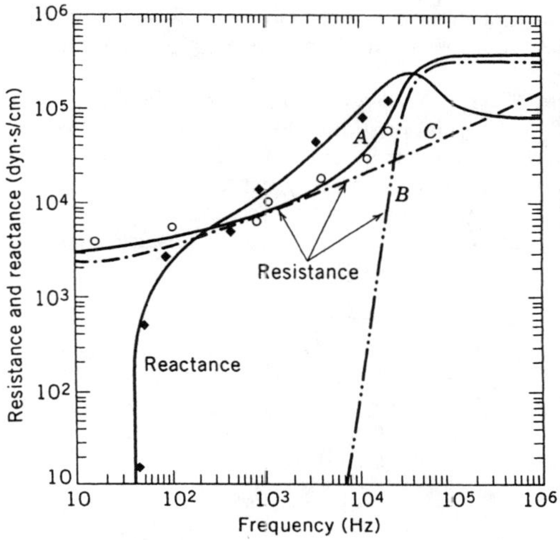

FIGURE 10.3 Mechanical resistance and reactance of soft thigh tissue (2 cm in diameter) in vivo from 10 Hz to 1 MHz. The measured values (open circles—resistance; diamonds—reactance) are compared with the calculated resistance and reactance of a 2-cm-diameter sphere vibrating in a viscous, elastic compressible medium with properties similar to soft human tissue (continuous lines, curves A). The resistance is also shown for the sphere vibrating in a frictionless compressible fluid (acoustic compression wave, curve B) and an incompressible viscous fluid (curve C). *(von Gierke et al., 1952.)*

TABLE 10.1 Typical Physical Properties of Human Tissues at Frequencies Less than 100 kHz

Property	Soft tissues	Bone (wet)	Bone (dry)
Density, kg/m³	$1–1.2 \times 10^3$	$1.9–2.3 \times 10^3$	1.9×10^3
Young's modulus, Pa	7.5×10^3	$1.6–2.3 \times 10^{10}$	1.8×10^{10}
Shear modulus,* Pa	2.5×10^3†	$2.9–3.4 \times 10^9$	7.1×10^9
Bulk modulus, Pa	2.6×10^9†		1.3×10^{10}
Shear viscosity, Pa · s	15†		
Sound velocity, m/s	$1.5–1.6 \times 10^3$	3.4×10^3	
Acoustic impedance, Pa · s/m	1.7×10^6	6×10^6	6×10^6
Tensile strength, Pa			
Cortical bone		$1.3–1.6 \times 10^8$	1.8×10^8
Compressive strength, Pa			
Cortical bone		$1.5–2.1 \times 10^8$	
Trabecular bone (vertebrae)		$0.4–7.7 \times 10^6$	
Shear strength, Pa			
Cortical bone		$7.0–8.1 \times 10^7$	

*Lamé constant.
†From soft tissue model (von Gierke et al., 1952).
Source: After von Gierke et al., 2002; and Goldstein et al., 1993.

the mechanical stiffness and resistance of soft tissues approximately triple in magnitude when the static compression of the surface increases by a factor of three. The relationship, however, is not linear.

Apparent Mass of Seated Persons. The *apparent mass* is often used to describe the response of the body at the point of stimulation rather than the mechanical impedance, and is the complex ratio between the dynamic force applied to the body and the acceleration at the interface where vibration enters the body. It is commonly expressed as a function of frequency, and is equal to the static weight of a subject in the limiting case of zero frequency when the legs are supported to move in unison with the torso. The influence of posture, muscle tension, and stimulus magnitude on the

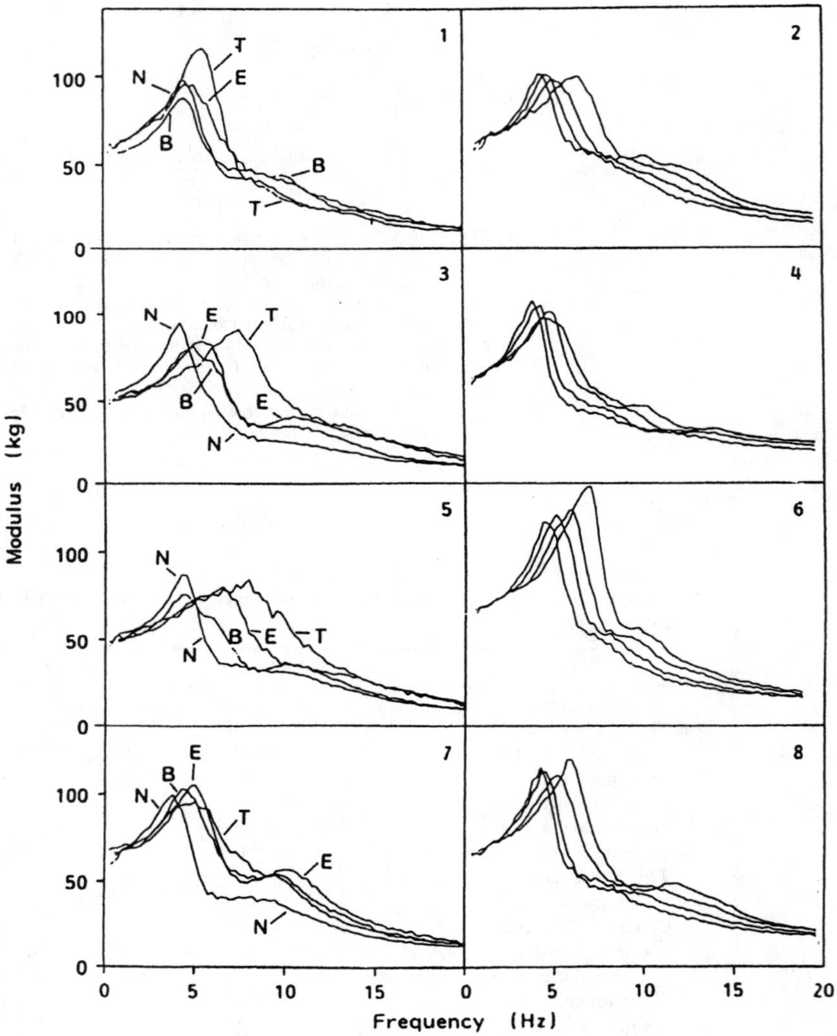

FIGURE 10.4 Effect of posture (N—"normal"; E—erect; B—with backrest), muscle tension (T—tensed muscles), and stimulus magnitude (0.25, 0.5, 1.0, and 2.0 m/s^2) on the apparent mass of a seated person for four subjects (see text for explanation). (*Fairley et al., 1989.*)

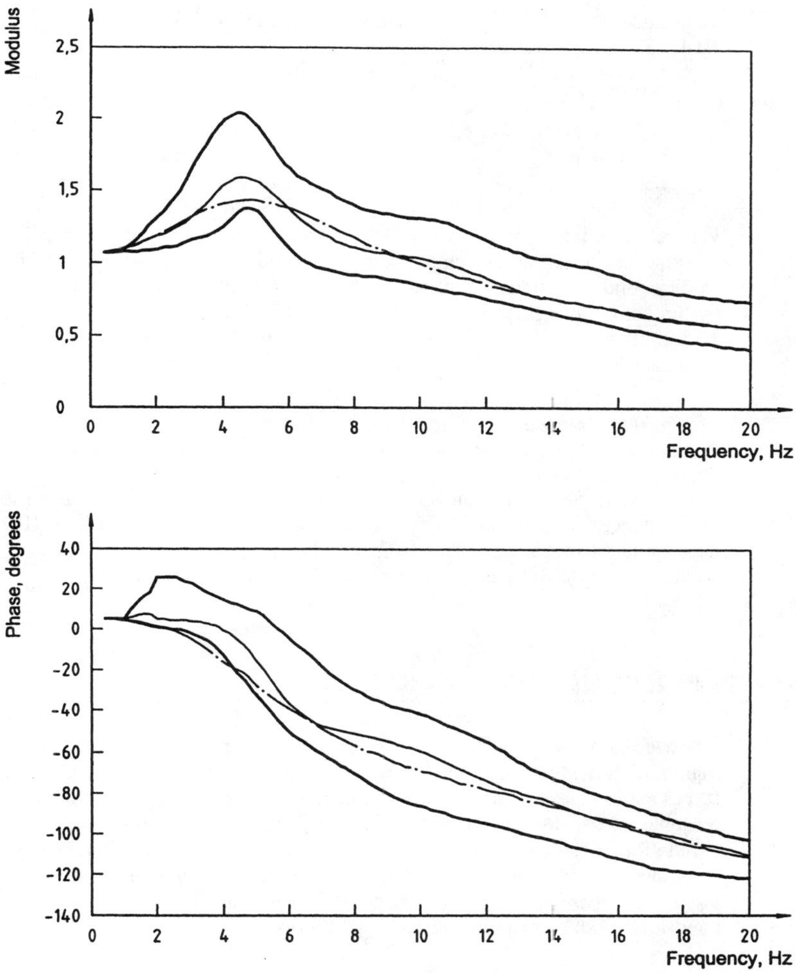

FIGURE 10.5 Idealized values for the modulus and phase of the seat-to-head transmissibility of seated persons subjected to vertical vibration. The envelopes of the maximum and minimum mean values of studies included in the analysis are shown by thick continuous lines, and the mean of all data sets is shown by the thin line. The response of a biodynamic model (see text and Fig. 10.8) is plotted as a dash-dot line. (*ISO 5982, 2001.*)

apparent mass of seated persons, in the vertical direction, is shown for four subjects in Fig. 10.4 (Fairley et al., 1989). The column of graphs to the left of the diagram shows the modulus of the apparent mass measured with a comfortable, "normal," upright posture and muscle tension (labeled N), with this posture but an erect torso and the shoulders held back (E), with all muscles in the upper body tensed (T), and, finally, with the subject leaning backward to rest against a rigid backrest (B). The largest variation in apparent mass between these conditions was associated with tensing the back muscles, which clearly increased the frequency of the characteristic peak in the response (at around 5 Hz). In some subjects the frequency of this peak could be changed by a factor of 2 by muscle tension. A variation in the apparent mass could also be induced by changing the stimulus magnitude, as is shown for four RMS accelerations (0.25, 0.5, 1.0, and 2.0 ms^{-2}) to the right of Fig. 10.4. Within this range of stimulus magnitudes, the frequency of the characteristic peak in the apparent mass was found to decrease with increasing stimulus magnitude, for each subject.

Seat-to-Head Transmissibility. The *transmissibility* expresses the response of one part of a mechanical system (e.g., the head or hand) to steady-state forced vibration of another part of the system (e.g., the buttocks), and is commonly expressed as a function of frequency. A synthesis of measured values for the seat-to-head transmissibility of seated persons has been performed for vibration in the vertical direction, to define the *idealized* transmissibility. The idealized transmissibility attempts to account for the sometimes large and unexplained variations in the results from different experimental studies conducted under nominally equivalent conditions. The results of one such analysis are shown by the continuous lines in Fig. 10.5 (ISO 5982, 2001). It can be seen by comparing Figs. 10.4 and 10.5 that the characteristic peak of the apparent mass remains in the modulus of the idealized transmissibility.

Mechanical Impedance of the Hand-Arm System. The idealized mechanical input impedance of the hand-arm system when the hand is gripping a cylindrical or oval handle has been derived for the three directions of the basicentric coordinate system shown in Fig. 10.2 (ISO 10068, 1998). The transmissibility of vibration through the hand-arm system has not been measured with noncontact or lightweight transducers satisfying the conditions described in Sec. 10.2.1. However, it has been demonstrated that the transmissibility from the palm to the wrist when a hand grips a vibrating handle is unity at frequencies up to 150 Hz (Boileau et al., 1992).

10.3 MODELS AND HUMAN SURROGATES

Knowledge of tolerable limits for human exposure to vibration, shock, and impact is essential for maintaining health and performance in the many environments in which man is subjected to dynamic forces and accelerations. As already noted, humans cannot be subjected to injurious stimuli for ethical reasons, and so little direct information is available from this source. In these circumstances, the simulation of human response to potentially life-threatening dynamic forces and accelerations is desirable, and is commonly undertaken using *biodynamic models*, and *anthropometric* or *anthropomorphic manikins*. They are also used in the development of vehicle seats and, in the case of hand-arm models, of powered hand-held tools.

10.3.1 Biodynamic Models

Simple Lumped Models. At frequencies up to several hundred hertz, the biodynamic response of the human body can be represented theoretically by point masses, springs, and dampers, which constitute the elements of *lumped* biodynamic models. The simplest one-dimensional model consists of a mass supported by a spring and damper, as sketched in Fig. 10.6, where the system is excited at its base. The equation of motion of a mass m when a spring with stiffness k and damper with resistance proportional to velocity, c, are base driven with a displacement $x_0(t)$ is:

$$ma_1(t) + c(\dot{x}_1(t) - \dot{x}_0(t)) + k(x_1(t) - x_0(t)) = 0 \qquad (10.7)$$

where the displacement of the mass is $x_1(t)$, its acceleration is $a_1(t)$, and differentiation with respect to time is shown by dots.

For this simple mechanical system, the apparent mass may be expressed as a function of frequency by (Griffin, 1990):

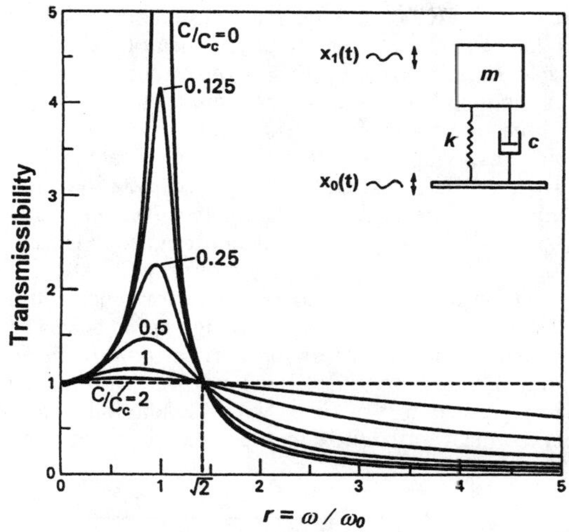

FIGURE 10.6 Single-degree-of-freedom, lumped-parameter biodynamic model. The mass m is supported by a spring with stiffness k and viscous damper with resistance c. The transmissibility of motion to the mass is shown as a function of the frequency ratio r ($=\omega/\omega_0$) when the base is subjected to a displacement $x_0(t)$. *(After Griffin, 1990.)*

$$M(\omega) = \frac{m(k + i\omega c)}{k - \omega^2 m + i\omega c} \tag{10.8}$$

where ω is the angular frequency ($= 2\pi f$), $i = (-1)^{1/2}$, and the transmissibility from the base to the mass is

$$H(\omega) = \frac{k + i\omega c}{k - \omega^2 m + i\omega c} \tag{10.9}$$

The modulus of the transmissibility may then be written

$$|H(\omega)| = \left[\frac{1 + (2\xi r)^2}{(1 - r^2)^2 + (2\xi r)^2}\right]^{1/2} \tag{10.10}$$

where r is the ratio of the angular excitation frequency to the angular *resonance frequency* of the system, ω/ω_0, and

$$\omega_0 = 2\pi f_0 = \left(\frac{k}{m}\right)^{1/2} \tag{10.11}$$

In Eq. (10.10), the damping is expressed in terms of the damping ratio $\xi = c/c_c$, where c_c is the critical viscous damping coefficient [$= 2(mk)^{1/2}$]. The transmissibility of the system is plotted as a function of the ratio of the angular excitation frequency to the natural (resonance) frequency in Fig. 10.6. It can be seen from the diagram that, at excitation frequencies less than the resonance frequency (i.e., $r << 1$), the motion of the mass is the same as that of the base. At frequencies greater than the resonance frequency, however, the motion of the mass becomes progressively less than that of the base. At angular excitation frequencies close to the resonance frequency ω_0, the motion of the mass exceeds that of the base. This response is that of a *low-pass mechanical filter*.

Dynamic Response Index (DRI). The response of the spine to shocks in the vertical (headward) direction has long been modeled in connection with the development of ejection seats for escape from high-performance aircraft. The simple model shown in Fig. 10.6 has been used to simulate the maximum stress within the vertebral column by calculating the maximum dynamic deflection of the spring, $|x_1(t) - x_0(t)|_{max}$, for a given input acceleration–time history to the model. The potential for spinal injury is estimated by forming the *dynamic response index* (DRI), which is defined as $(\omega_0)^2|x_1(t) - x_0(t)|_{max}/g$, where the natural frequency is 52.9 rad/s (i.e., $f_0 = 8.42$ Hz), the damping ratio is 0.224, and g is the acceleration of gravity (9.81 ms^{-2}). The success of the model has led to its adoption for specifying ejection seat performance and its extension to a metric for exposure to repeated shocks and for ride comfort in high-speed boats (Allen, 1978; Brinkley, 1990; Payne 1976).

Whole-Body Apparent Mass for Vertical (z-Direction) Vibration. The apparent mass of the seated human body may be described by a variant of the simple biodynamic model of Fig. 10.6. A satisfactory prediction is obtained with the addition of a second mass, m_0, on the seat platform to represent the mass of the torso and legs that does not move relative to the seat (i.e., to the base of

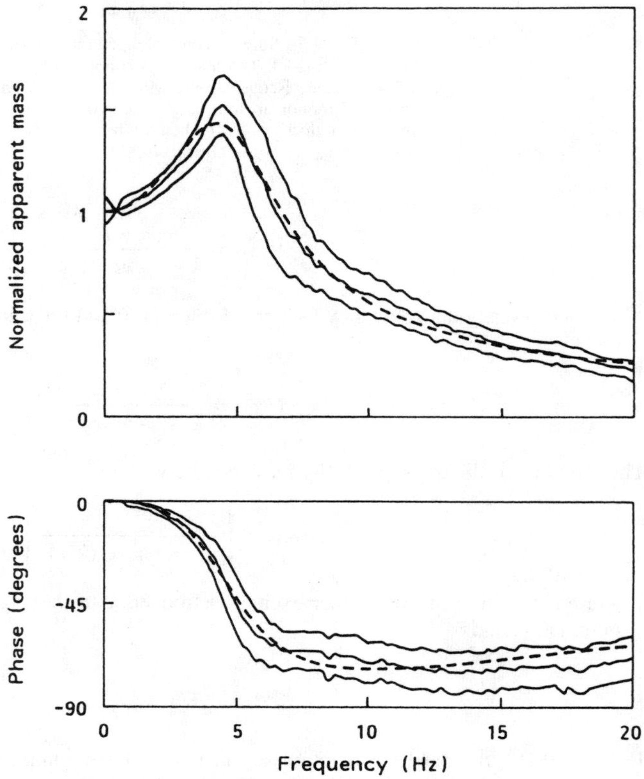

FIGURE 10.7 Comparison between predicted and observed apparent mass. The mean, normalized apparent masses of 60 subjects, ± 1 standard deviation, are shown by the continuous lines, and the predictions of a single-degree-of-freedom, lumped parameter biodynamic model by the dashed line. (*Fairley et al., 1989.*)

the model). The apparent mass measured on 60 subjects (24 men, 24 women, and 12 children) can be well represented by this model when the legs are supported to move in phase with the seat, and the data are normalized for the different weights of individuals using the value of apparent mass recorded at 0.5 Hz (Fairley and Griffin, 1989). A comparison of the observed values for the magnitude, and phase, of the normalized apparent mass (continuous lines) and the predicted values (dashed lines) is shown in Fig. 10.7. To obtain this agreement, the natural angular frequency of the model is 31.4 rad/s (i.e., f_0 = 5 Hz), and the damping ratio is 0.475. These values differ considerably from those of the DRI model (see above), reflecting the stimulus magnitude dependent, nonlinear response of the human body to vibration.

Whole-Body Impedance, Apparent Mass, and Transmissibility for Vertical (z-direction) Vibration. A model has been developed to predict idealized values for the input impedance, apparent mass, *and* transmissibility of the seated human body when subjected to vertical vibration. The model, shown in Fig. 10.8, comprises elements forming three of the models sketched in Fig. 10.6 (ISO 5982, 2001). It contains elements similar to those used to predict apparent mass on the left-hand side of the diagram with, in addition, two basic models in series to the right of the diagram (i.e., one on top of the other), in order to represent "head" motion (i.e., motion of mass m_2). The model predictions for the transmissibility from the seat to the head are shown by the dash-dot lines in Fig. 10.5, and should be compared with the target mean values (the thin continuous lines). While the model meets the considerable challenge of predicting three biodynamic responses with one set of model values, the parameters of the model do not possess precise anatomical correlates, a common failing of simple lumped models, and the predictions are applicable only to a limited range of stimulus magnitudes.

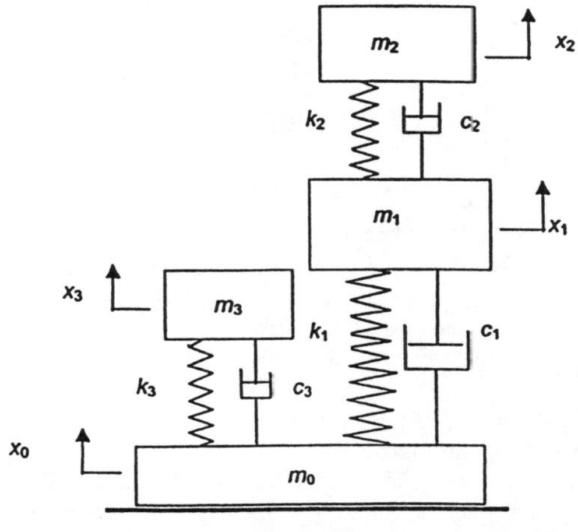

FIGURE 10.8 Three-degree-of-freedom, lumped-parameter biodynamic model of the seated human body for estimating mechanical impedance, apparent mass, and transmissibility, for vertical vibration. The model is driven at the base (x_0), and the transmissibility is calculated to the "head," mass m_2. *(ISO 5982, 2001.)*

Impedance Model of the Hand-Arm System. A 4-degree-of-freedom model consisting of four of the models sketched in Fig. 10.6, arranged in series (i.e., on top of each other), has been used to predict idealized values for the input impedance to the hand (ISO 10068, 1998). As is the case for the lumped whole-body models, the parameters of the hand-arm model do not possess direct anatomical correlates.

Rigid, Multielement, and Finite-Element (FE) Models. More complex models have been developed to incorporate more realistic descriptions of individual body parts and to predict the motion of one body part relative to another. One of the first such models consisted of a seven-segment numerical model for describing the motion of a vehicle occupant in a collision. The segments consisted of the head and neck, upper torso, lower torso, thighs, legs, upper arms, and forearms and hands. Of the many multielement models that have since been developed, the *Articulated Total Body* (ATB) model and the *MAthematical DYnamical MOdel* (MADYMO) are the most widely used. Each employs a combination of rigid bodies, joints, springs, and dampers to represent the human, or in some cases manikin, and sets up and solves numerically the equations of motion for the system. The applied forces and torques on these components are established by using different routines for contact with external surfaces and for joint resistance, the effect of gravity, and restraining the body (e.g., by a seat belt).

ATB and MADYMO Models. The ATB and MADYMO computer models commonly employ 15 ellipsoidal segments with masses and moments of inertia appropriate for the body parts they represent, as determined from anthropomorphic data for adults and children. The connections between these segments are flexible, and possess elastic and resistive properties based on measurements on human joints. The environment to be simulated may include contact surfaces (e.g., seats and vehicle dashboard), seat belts, and air bags. The models may include advanced harness systems and wind forces to model seat ejection from aircraft and can also describe aircraft crash. An example of the use of these models to predict human response in crashlike situations is shown in Fig. 10.9 (von Gierke, 1997). In this diagram, the motion of an unrestrained child standing on the front seat of an automobile is shown in response to panic breaking. The output of the model has been calculated at the onset of breaking and after 500 and 600 ms. It can be seen from the diagram that the model predicts the child will slide down and forward along the seat until its head impacts the dashboard of the vehicle.

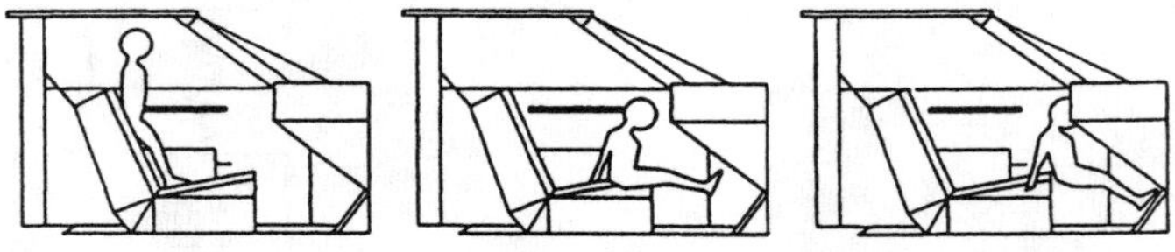

FIGURE 10.9 Articulated total body (ATB) model prediction for the response of an unrestrained standing child to panic braking in an automobile at 500 and 600 ms after commencing braking. (*von Gierke, 1997.*)

Finite-Element (FE) Models. The detailed response of selected parts of the body and the environment being simulated (e.g., head and neck, spine, and vehicle seating) have been modeled by using finite elements (FEs). In the MADYMO model, the FEs can interact with the multibody model elements. Examples of human body subsystems that have been modeled with FEs include the spine, to predict the injury potential of vertebral compression and torsional loads, and the head and neck, to predict rotation of the head and neck loads during rapid horizontal deceleration (see Fig. 10.12) (RTO-MP-22, 1999).

10.3.2 Anthropometric Manikins

Mechanically constructed manikins, or dummies, are used extensively in motor vehicle crash testing and for evaluating aircraft escape systems and seating. Several have been developed for these purposes (AGARD AR-330, 1997), some of which are commercially available.

Hybid III Manikin. The Hybrid III manikin, shown in Fig. 10.10, was originally developed by General Motors for motor vehicle crash testing, and has since become the de facto standard for

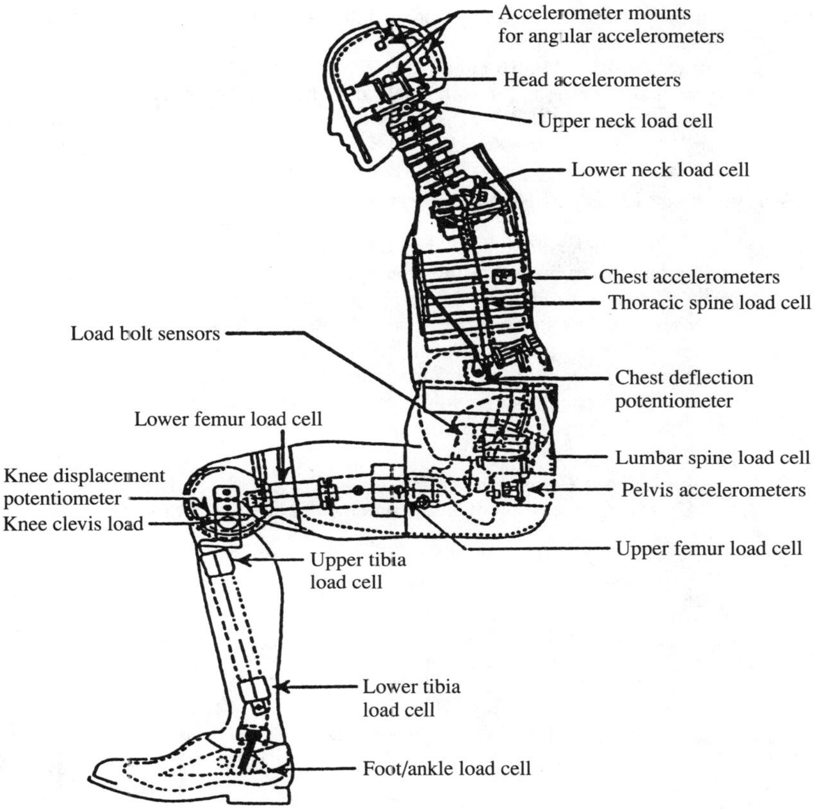

FIGURE 10.10 Hybrid III anthropometric dummy designed for use in motor-vehicle frontal crash tests, showing elements of construction and sensors. (*AGARD-AR-330, 1997.*)

simulating the response of motor vehicle occupants to frontal collisions and for tests of occupant safety restraint systems. The manikin approximates the size, shape, and mass of the 50th percentile North American adult male, and consists of metal parts to provide structural strength and define the overall geometry. This "skeleton" is covered with foam and an external vinyl skin to produce the desired shape. The manikin possesses a rubber lumbar spine, curved to mimic a sitting posture. The head, neck, chest, and leg responses are designed to simulate the following human responses during rapid deceleration: head acceleration resulting from forehead and side-of-the-head impacts; fore-and-aft, and lateral, bending of the neck; deflection of the chest to distributed forces on the sternum; and impacts to the knee (Mertz, 1993). The instrumentation required to record these responses together with local axial compressional loads is shown in Fig. 10.10. Hybrid III dummies are now available for small (fifth percentile) adult females, and large (95th percentile) adult males, as well as for infants and children. A related *side impact dummy* (SID) has been developed for the U.S. National Highway Traffic Safety Administration (NHTSA), for crash tests involving impacts on the sides of motor vehicles.

ADAM. ADAM (Advanced Dynamic Anthropomorphic Manikin) is a fully instrumented manikin primarily used in the development of aircraft ejection systems. Its overall design is conceptually similar to that of the Hybrid III dummy (see Fig. 10.10), in that ADAM replicates human body segments, surface contours, and weight. In addition to a metal skeleton, the manikin posseses a sandwich skin construction of sheet vinyl separated by foamed vinyl to mimic the response of human soft tissue. ADAM also attempts to replicate human joint motion and the response of the spine to vertical accelerations for both small-amplitude vibration and large impacts. The spine consists of a mechanical spring-damper system, which is mounted within the torso.

10.3.3 Biodynamic Fidelity of Human Surrogates

Animals, cadavers, manikins, and computer models have been used to predict human responses to potentially injurious or life-threatening stimuli. To evaluate the biofidelity of the surrogate or model, it is necessary to identify the response characteristics that are most relevant. For biodynamic responses, the time histories of the acceleration, velocity, displacement, and forces provide the most meaningful comparisons, though point-by-point comparisons can be misleading if the system response to the stimulus of interest is extremely nonlinear. In these circumstances, evaluating peak values in the time history, impulses calculated from the acceleration or contact forces, or energy absorption may be more appropriate.

Manikins and Computer Models. The current state of the art is illustrated in Figs. 10.11 and 10.12, where several biodynamic parameters of the response of the human head and neck to rapid

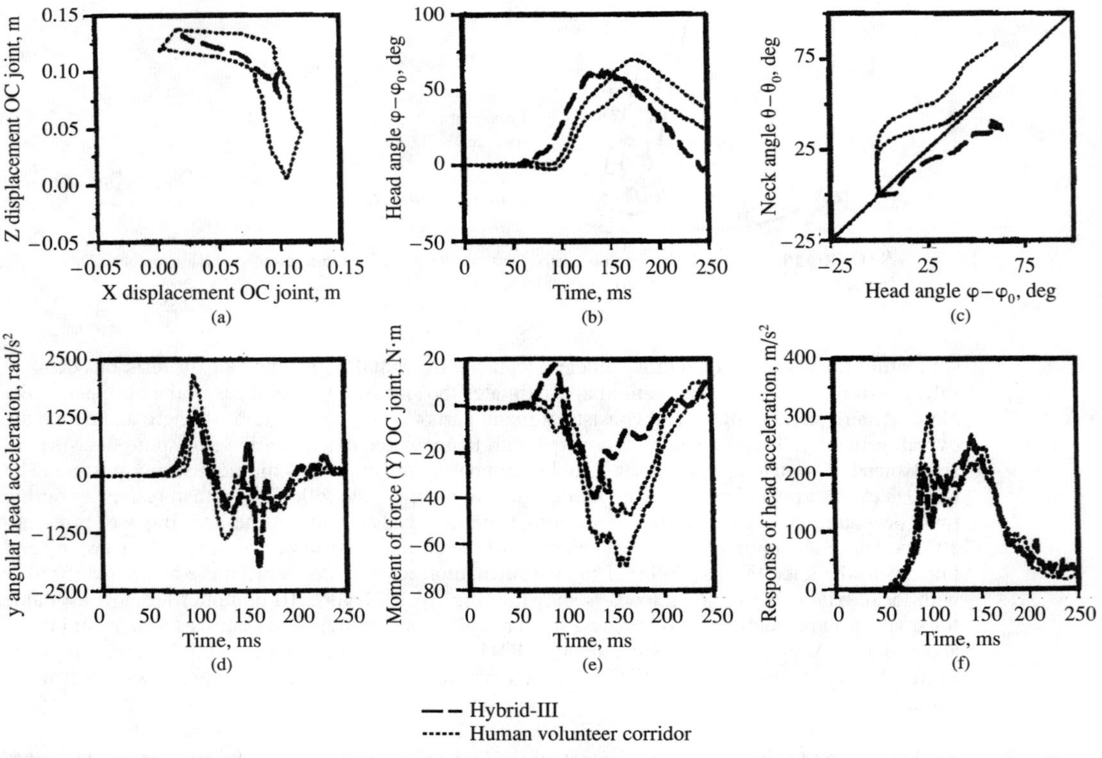

FIGURE 10.11 Response of human volunteers and the Hybrid III head and neck to 15 g spineward deceleration. The range of responses from human subjects is shown by the dotted lines, and the response of the manikin by the dashed lines (see text for explanation of the motions plotted). (*RTO-MP-20, 1999.*)

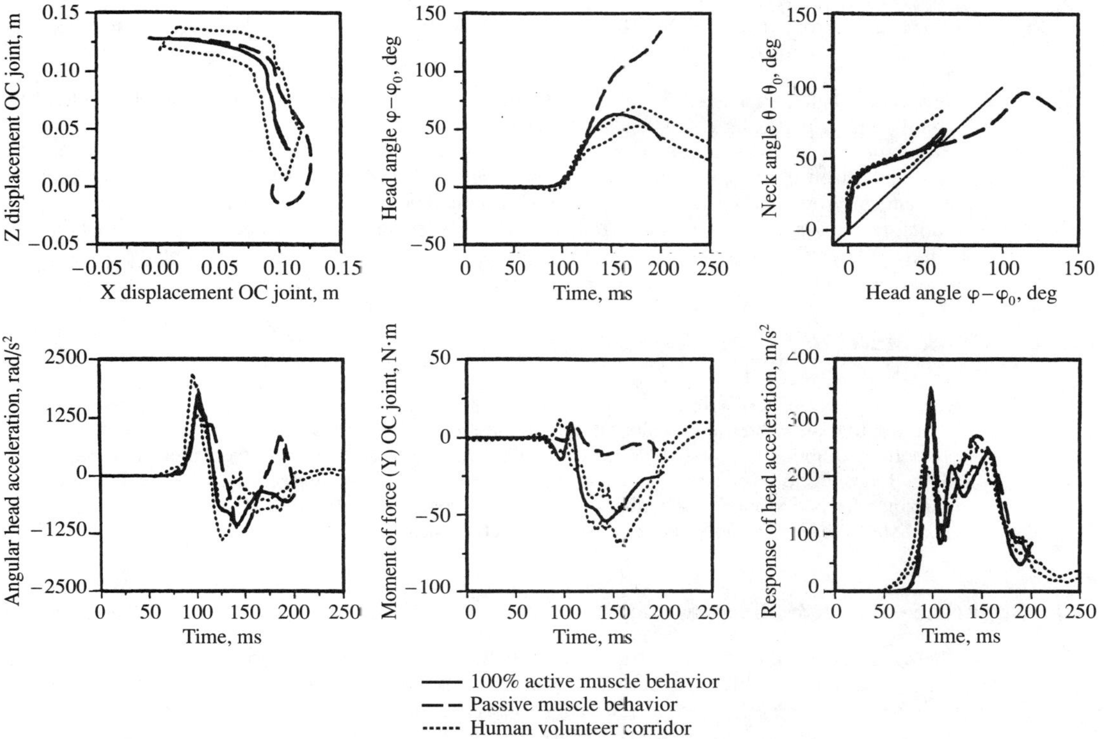

FIGURE 10.12 Response of human volunteers and the 3D MADYMO model of the head and neck to spineward decelerations. The range of responses from human subjects is shown by the dotted lines, and the response of the model with passive and active muscles is shown by the dashed and continuous lines, respectively (see text for explanation of the motions plotted). (*RTO-MP-20, 1999.*)

horizontal decelerations are compared. The comparisons are between the responses of human volunteers, the limits of which are shown by dotted lines, with those of the Hybrid III manikin at the same deceleration (Fig. 10.11), and of a three-dimensional head and neck for the MADYMO computer model (Fig. 10.12) (RTO-MP-20, 1999).

While the Hybrid III manikin can reproduce some human responses, the head and neck system does not introduce appropriate head rotation lag (see neck angle versus head angle in Fig. 10.11*c*) or torque at the occipital condyles joint (see moment of force OC joint versus time (Fig. 10.11*e*)]. Note, however, that the linear acceleration of the center of gravity of the head is well reproduced by the manikin, except for the peak acceleration at 100 ms [see response of head acceleration versus time (Fig. 10.11*f*)]. In contrast, the three-dimensional head and neck model for MADYMO can be seen to reproduce most human responses when active muscle behavior is included (the continuous lines in Fig. 10.12). In this computer model, the neck muscles are represented by simple cords between anatomical attachment points on the head and the base of the neck.

Animals. Employing the results of experiments with animals to predict biodynamic responses in humans introduces uncertainties associated with interspecies differences. Of particular concern is the difference in size of body parts and organs, which influences resonance frequencies. For this reason, most animal research on the limit of exposure to rapid horizontal deceleration and to vertical acceleration, which commonly involves shock and impact, has employed mammals of roughly similar size and mass to man (i.e., pigs and chimpanzees). Research on pathophysiological mechanisms is less subject to concerns with different body size, and more with the biological equivalence of the systems being studied.

Cadavers. Human cadavers have also been used for experiments involving potentially injurious stimuli, and in particular to relate skull fracture to frontal head impact. Such studies resulted in the formulation of the Wayne State concussion tolerance curve, which has been widely used to define survivable head impacts in motor vehicle collisions (SAE J885, 1986). Cadavers lack appropriate mechanical properties for tissues and muscle tension. The latter is important for obtaining realistic human responses, as can be seen from Fig. 10.12 by comparing the results with and without active muscle behavior.

10.4 COUNTERMEASURES

Reducing the effects of vibration, shock, and impact on humans is effected in several ways: (1) by isolation, to reduce the transmission of dynamic forces and accelerations to the body; (2) by personal protective equipment, to distribute the dynamic forces over as large a surface area as possible; and (3) by redesign, to reduce the source's vibration intensity. These subjects are considered after summarizing the occurrence of health effects from exposure to vibration, shock, and impact, which establishes the performance required of ameliorative measures.

10.4.1 Occurrence of Health Effects and Injury

There is extensive literature on the occurrence of health effects and injury from exposure to vibration, shock, and impact, which has been reviewed recently (von Gierke et al., 2002) and serves as the basis for the present discussion. Estimates of exposures necessary for common human responses and health effects are summarized in Table 10.2 for healthy adults; the interested reader is directed to the references given in the table for more complete information, or to the recent review article cited. Included in Table 10.2 are the metric used for assessing the exposure, the frequency weighting of the stimulus, and a *representative* value for the response or health effect under consideration. As already noted, there are large variations between individuals in response, and susceptibility, to vibration, shock, and impact.

Vibration Perception. The perception of vibration depends on the body site and on the stimulus frequency. The thresholds in Table 10.2 are typical of those for healthy adults, and are expressed as instantaneous RMS accelerations [i.e., with $T \approx 1$ s in Eq. (10.2)]. The value for whole-body vibration is given in terms of a frequency-weighted acceleration, and so is applicable to vibration at frequencies from 1 to 80 Hz. The values for hand-transmitted vibration are for sinusoidal stimuli applied to the fingertips of males (M) and females (F) at the specified frequencies.

Thresholds for Health Effects. Thresholds for the onset of health effects have been estimated for regular, near-daily exposure to hand-transmitted and to whole-body vibration. The metrics employed, however, differ. For hand-transmitted vibration, the assessment is in terms of the magnitude of the 8-hour, energy-equivalent, frequency-weighted, RMS, vector acceleration sum, $a_{\mathrm{WAS}(8)}$. This metric employs values of the RMS component accelerations averaged over 8 hours [i.e., $T = 28{,}800$ s in Eq. 10.2] which have been frequency-weighted using W_h in Fig. 10.1. The components are determined for each of the directions of the basicentric, or biodynamic, coordinate system shown in Fig. 10.2, $a_{X,\mathrm{RMS}(8)}$, $a_{Y,\mathrm{RMS}(8)}$, and $a_{Z,\mathrm{RMS}(8)}$, respectively. Thus:

$$a_{\mathrm{WAS}(8)} = [a^2_{X,\mathrm{RMS}(8)} + a^2_{Y,\mathrm{RMS}(8)} + a^2_{Z,\mathrm{RMS}(8)}]^{1/2} \tag{10.12}$$

A reduction in the metric will occur for a given acceleration magnitude if the duration of the exposure is reduced, as is illustrated by the following example (see also "Vibration Exposure") in Sec. 10.1.1).

TABLE 10.2 Estimates of Health and Injury Criteria for Healthy Adults.

Human response	Metric	Frequency or weighting	Value	Source
Perception (population mean, threshold):				
Whole Body (vertical)	RMS accel.	W_k	0.015 m/s^2	ISO 2631-1, 1997
Fingertip (selected frequencies)	RMS accel.	4 Hz	0.0075 m/s^2	ISO/FDIS 13091-2, 2002
	RMS accel.	31 Hz	M—0.10 m/s^2	ISO/FDIS 13091-2, 2002
	RMS accel.	31 Hz	F—0.12 m/s^2	ISO/FDIS 13091-2, 2002
	RMS accel.	125 Hz	M—0.25 m/s^2	ISO/FDIS 13091-2, 2002
	RMS accel.	125 Hz	F—0.32 m/s^2	ISO/FDIS 13091-2, 2002
Health effects (threshold):				
Hand (HAVS)	$a_{\mathrm{WAS}(8)}$	W_h	1 m/s^2	ISO 5349-1, 2001
Whole body				
Z direction	VDV	W_k	8.5 m/s$^{1.75}$	ISO 2631-1, 1997
X and Y directions	VDV	W_d	6.1 m/s$^{1.75}$	ISO 2631-1, 1997
Risk of injury (5–10% probability):				
Hand (HAVS)	$a_{\mathrm{WAS}(8)}$	W_h	3 m/s^2	ISO 5349-1, 2001
Whole-body vibration, Z (direction)	VDV	W_k	17 m/s$^{1.75}$	ISO 2631-1, 1997
Shocks to the body (headward):*				
1 shock/day ($n_q = 1$)	$E(\mathrm{DRI}_q, n_q)$	DRI model	$\mathrm{DRI}_{\max} = 14$	After Allen, 1978, and Payne, 1976
10 shocks/day ($n_q = 10$)	$E(\mathrm{DRI}_q, n_q)$	DRI model	$\mathrm{DRI}_{\max} = 8$	Allen, 1978
100 shocks/day ($n_q = 100$)	$E(\mathrm{DRI}_q, n_q)$	DRI model	$\mathrm{DRI}_{\max} = 6$	Allen, 1978
Survivable single shock or impact:				
To body (headward)*	DRI	DRI model	20	After Allen, 1978
To head (manikin)	HIC	None	1000	NHTSA FMVSS 208†

M-male
F-female
*When body, is restrained.
†U.S. National Highway Traffic Safety Administration Federal Motor Vehicle Safety Standard 208.

EXAMPLE 10.2 *A worker uses a percussive rock drill for 3 hours daily. Measurement of the handle vibration indicates a frequency-weighted RMS acceleration of 18 m/s^2 along the drill axis, and a frequency-weighted RMS acceleration of 5 m/s^2 perpendicular to this axis. Estimate the 8-hour daily exposure.*

Answer: *From the (limited) available information, we obtain the following approximate values:*

$$a_{Z,\mathrm{RMS}(3)} = 18 \qquad a_{X,\mathrm{RMS}(3)} = a_{Y,\mathrm{RMS}(3)} = 5$$

So

$$a_{\mathrm{WAS}(3)} = (5^2 + 5^2 + 18^2)^{1/2} = 19.3$$

Now an exposure for 3 hours ($T_{(3)}$) can be expressed as an equivalent 8-hour exposure using Eq. (10.5):

$$a_{\mathrm{WAS}(8)} = a_{\mathrm{WAS}(3)}\left(\frac{T_{(3)}}{T_{(8)}}\right)^{1/2}$$

so that

$$a_{\mathrm{WAS}(8)} = 19.3\left(\frac{10{,}800}{28{,}800}\right)^{1/2}$$

$$a_{\mathrm{WAS}(8)} = 11.8 \text{ m/s}^2$$

The assessment of whole-body vibration employs the VDV averaged over 8 hours [i.e., T = 28,800 s in Eq. (10.6)], with frequency weighting W_k for vertical vibration and frequency weighting W_d for horizontal vibration. It is believed that the higher-power metrics, as recommended here, better represent the hazard presented by motion containing transient events, particularly when these become small-magnitude shocks or impacts. The most appropriate metric, however, remains a subject for research.

Risk of Injury. Exposures to vibration magnitudes between the threshold for perception and that for health effects commonly occur in daily life. Near-daily exposures to values of $a_{\mathrm{WAS(8)}}$ and VDV in excess of those estimated to result in 5 to 10% injury in Table 10.2 occur in numerous occupations (involving some 8 million persons in the United States) and lead to the symptoms described in Sec. 10.1.2.

There have been several attempts to relate shocks and impacts to risk of injury, as they can be life threatening (Brinkley et al., 1990; Eiband, 1959; SAE J885, 1986). For headward accelerations, the method proposed by Allen is included here (Allen, 1978). This is based on the DRI biodynamic model (see Sec. 10.3.1), which is extended to include multiple shocks or impacts. The assessment of the risk of spinal injury depends on the number of shocks or impacts entering the buttocks of a seated person, as listed in Table 10.2. For exposures consisting of multiple headward shocks or impacts of differing magnitudes (e.g., as in off-the-road vehicles), if there are n_q shocks or impacts of magnitude DRI_q, where $q = 1, 2, 3, \ldots, Q$, then the exposure is believed to be acceptable if

$$E(\mathrm{DRI}_q, n_q) = \sum_{q=1}^{Q}\left[\frac{\mathrm{DRI}_q}{(\mathrm{DRI}_{\mathrm{max}})_q}\right] \leq 1 \tag{10.13}$$

In this expression, the denominator is the maximum allowable DRI corresponding to the *observed* number of shocks or impacts n_q with magnitude DRI_q and is derived from Table 10.2.

Survivable Shocks and Impacts. Estimates for *survivable* exposures to single impacts and shocks are given for both headward acceleration and spineward deceleration in Table 10.2. For headward acceleration, experience with nonfatal ejections from military aircraft suggests that a DRI of 18 is associated with a 5 to 10 percent probability of spinal injury among healthy, *young* males who are restrained in their seats. For spineward deceleration, the estimate is based on the *head injury criterion* (HIC), which was developed from the severity index (see Sec. 10.1.1). The metric is defined as:

$$\mathrm{HIC} = \left\{(t_2 - t_1)\left[\frac{1}{(t_2 - t_1)}\int_{t_1}^{t_2} a(t)\, dt\right]^{2.5}\right\}_{\mathrm{max}} \tag{10.14}$$

where t_1 and t_2 are the times between which the HIC attains its maximum value and $a(t)$ is measured at the location of the center of gravity of the head. The HIC is applied to instrumented crash test dummies; its ability to rank order, by magnitude, the severity of injuries in humans has been questioned. The time interval $(t_2 - t_1)$ is commonly chosen to be 36 ms, a value prescribed by the NHTSA, though a shorter interval (e.g., 15 ms) has been suggested as more appropriate (SAE J885, 1986). The value listed in Table 10.2 is the maximum allowable by the NHTSA in frontal collisions. The assessment is related to the Wayne State concussion curve (see Sec. 10.3.3) and is applicable to head injury in vehicle and aircraft crash. Survivable injuries to the neck and chest, and fracture limits for the pelvis, patella, and femur in frontal and side impacts have also been proposed for manikins (Mertz, 1993).

10.4.2 Protection against Whole-Body Vibration

Vibration Isolation. Excessive whole-body vibration is most commonly encountered in transportation systems, where it predominantly affects seated persons. In consequence, an effective remedial measure is to reduce the vertical component of vibration transmitted through seats (and, where applicable, vehicle suspension systems), by means of low-pass mechanical filters. The transmissibility of a vibration-isolated seat or vehicle suspension (neglecting tire flexure) can be modeled in the vertical direction by the mechanical system shown in Fig. 10.6. For a vibration-isolated seat, the spring and damper represent mechanical elements supporting the seat pan and person, which are represented by a single mass *m*. For a vehicle suspension, the mechanical elements support the passenger compartment. With this model, it is evident that the vibration transmitted to the body will be reduced at frequencies above the resonance frequency of the mechanical system defined by the spring, damper, and total supported mass [i.e., for $r > \sqrt{2}$ in Fig. 10.6 and Eq. (10.10)]. The frequency weightings of Fig. 10.1 suggest that effective vibration isolation for humans will require a resonance frequency of 1 Hz or less. The low resonance frequency can be achieved by using a soft coiled spring, or an air spring, and a viscous damper. Vehicle suspensions with these properties are commonly employed. So-called *suspension seats* are commercially available, but are limited to applications in which the vertical displacement of the seat pan that results from the spring deflection is acceptable. A situation can be created in which the ability of a driver to control a vehicle is impaired by the position, or motion, of the person sitting on the vibration-isolated seat relative to the (nonisolated) controls.

Active Vibration Reduction. An *active vibration control system* consists of a hydraulic or electrodynamic actuator, vibration sensor, and electronic controller designed to maintain the seat pan stationary irrespective of the motion of the seat support. Such a control system must be capable of reproducing the vehicle motion at the seat support, which will commonly possess large displacement at low frequencies, and supply a phase-inverted version to the seat pan to counteract the vehicle motion in real time. This imposes a challenging performance requirement for the control system and vibration actuator. Also, the control system must possess safety interlocks to ensure it does not erroneously generate harmful vibration at the seat pan. While active control systems have been employed commercially to adjust the static stiffness or damping of vehicle suspensions, to improve the ride comfort on different road surfaces, there are currently no active seat suspensions.

10.4.3 Protection against Hand-Transmitted Vibration

Vibration-Isolated Tool Handles. Vibration isolation systems have been applied to a range of powered hand tools, often with dramatic consequences. For example, the introduction of vibration-isolated handles to gasoline-powered chain saws has significantly reduced the incidence of HAVS among professional saw operators. Unfortunately, such systems are not provided for the handles of all consumer-grade chain saws. The principle is the same as that described for whole-body vibration isolation, but in this case the angular resonance frequency can be ~350 rad/s (i.e., $f_0 \approx 55$ Hz) and still effectively reduce chain-saw vibration. The higher resonance frequency results in a static deflection of the saw tip relative to the handles that, with skill, does not impede the utility of the tool.

Tool Redesign. Some hand and power tools have been redesigned to reduce the vibration at the handles. Many are now commercially available (Linqvist, 1986). The most effective designs counteract the dynamic imbalance forces at the source—for example, a two-cylinder chain saw with 180° opposed cylinders and synchronous firing. A second example is a pneumatic chisel in which the compressed air drives both a cylindrical piston into the chisel (and workpiece) and an opposing counterbalancing piston; both are returned to their original positions by springs. A third is a rotary grinder in which the rotational imbalance introduced by the grinding wheel and motor is removed by a *dynamic balancer*. The dynamic balancer consists of a cylindrical enclosure, attached to the motor spindle, containing small ball bearings that self-adjust with axial rotation of the cylinder to positions on the walls that result in the least radial vibration—the desired condition.

Gloves. There have been attempts to apply the principle of vibration isolation to gloves, and so called antivibration gloves are commercially available. However, none has yet demonstrated a capability to reduce vibration at the frequencies most commonly responsible for HAVS, namely

200 Hz and below (an equinoxious frequency contour for HAVS is the inverse of frequency weighting W_h in Fig. 10.1). Performance requirements for antivibration gloves are defined by an international standard (ISO 10819, 1997). No glove has satisfied the modest transmissibility requirements, namely <1 at vibration frequencies from 31.5 to 200 Hz, and <0.6 at frequencies from 200 to 1000 Hz. An extremely soft spring is needed for the vibration isolation system because of the small dynamic mass of the hand if the resonance frequency is to remain low [see Eq. (10.11)]. An air spring formed from several air bladders sewn into the palm, fingers, and thumb of the glove appears most likely to fulfill these requirements (Reynolds, 2001).

10.4.4 Protection against Mechanical Shocks and Impacts

Protection against potentially injurious shocks and impacts is obtained by distributing the dynamic forces over as large a surface area of the body as possible and transferring the residual forces preferably to the pelvic region of the skeleton (though not through the vertebral column). Modifying the impact-time history to involve smaller peak forces lasting for a longer time is usually beneficial. Progressive crumpling of the passenger cabin floor and any forward structural members while absorbing horizontal crash forces, as well as extension of a seat's rear legs, are all used for this purpose.

Seat Belts and Harnesses. For seated persons, lap belts, or combined lap and shoulder harnesses, are used to distribute shock loads, and are routinely installed in automobiles and passenger aircraft. In addition, the belts hold the body against the seat, which serves to strengthen the restrained areas. Combined lap and shoulder harnesses are preferable to lap belts alone for forward-facing passengers, as the latter permit the upper body to flail in the event of a spineward deceleration, such as occurs in motor vehicle and many aircraft crashes. Harnesses with broader belt materials can be expected to produce lower pressures on the body and consequently less soft tissue injury. For headward accelerations the static deformation of the seat cushion is important, with the goal being to spread the load uniformly and comfortably over as large an area of the buttocks and thighs as possible.

A significant factor in human shock tolerance appears to be the acceleration-time history of the body immediately before the transient event. A *dynamic preload* imposed immediately before and/or during the shock, and in the same direction as the impending shock forces (e.g., vehicle braking before crash), has been found experimentally to reduce body accelerations (Hearon et al., 1982).

Air Bags. Although originally conceived as an alternative to seat belts and harnesses, air bags are now recognized to provide most benefit when used with passive restraints, which define the position of the body. The device used in automobiles consists, in principle, of one or more crash sensors (accelerometers) that detect rapid decelerations, and a controller that processes the data and initiates a pyrotechnic reaction to generate gas. The gas inflates a porous fabric bag between the decelerating vehicle structure and the occupant within about 25 to 50 ms, to distribute the shock and impact forces over a large surface of the body.

An example of the use of the MADYMO model to simulate air bag inflation and occupant response to the frontal collision of an automobile is shown in Fig. 10.13. In this diagram, the response of a person wearing a shoulder and lap seat belt has been calculated at 25-ms time intervals following the initiation of air bag inflation. The forward rotation of the head is clearly visible and is arrested before it impacts the chest. Also, the flailing of the arms can be seen.

Air bags can cause injury if they impact someone positioned close to the point of inflation, which may occur in the event of unrestrained, or improperly restrained, vehicle occupants (e.g., small children; see Fig. 10.9).

Helmets. Impact-reducing helmets surround the head with a rigid shell to distribute the dynamic forces over as large an area of the skull as possible. The shell is supported by energy absorbing material formed to the shape of the head, to reduce transmission of the impact to the skull. The shell of a helmet must be as stiff as possible consistent with weight considerations, and must not deflect sufficiently on impact for it to contact the head. The supporting webbing and energy absorbing foam

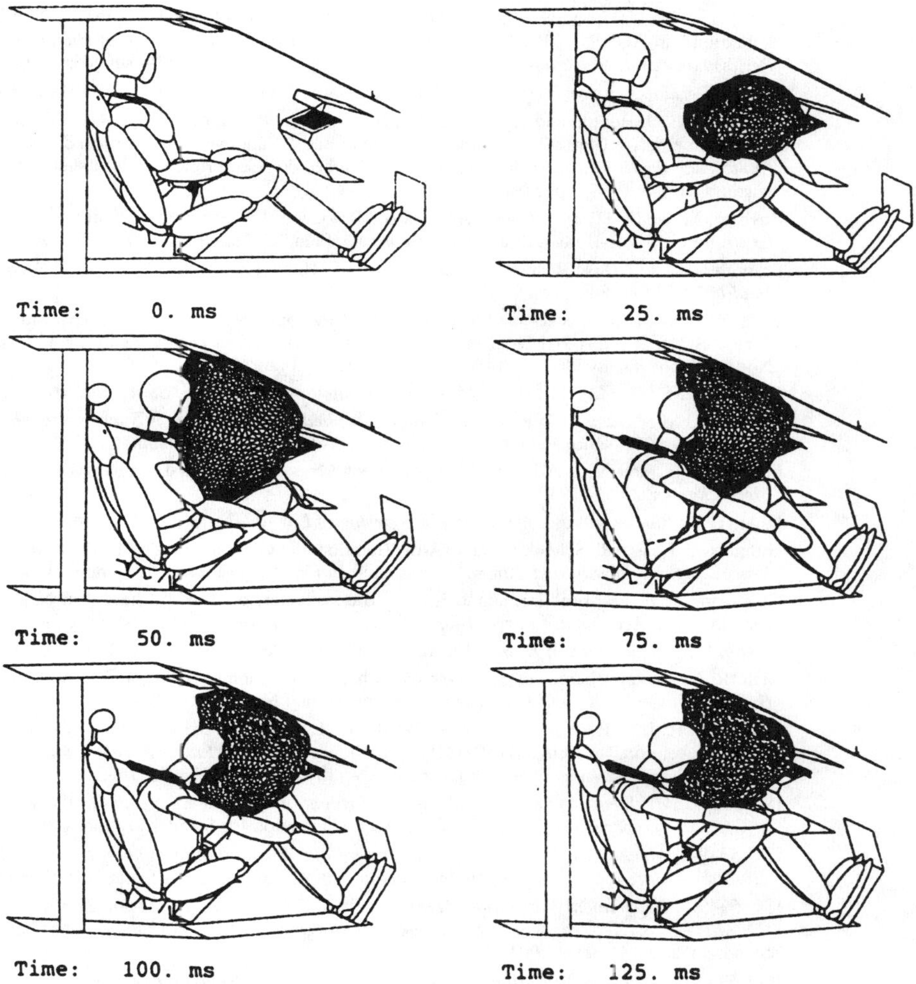

FIGURE 10.13 MADYMO simulation of the response of a person wearing a shoulder and lap seat belt to the inflation of an air bag in a frontal motor vehicle collision. The model predicts the body position every 25 ms after the collision. Note the time for the air bag to inflate (between 25 and 50 ms), the rotation of the head, the flailing of the arms, and the bending of the floor. (*AGARD-AR-330, 1996.*)

plastic must maintain the separation between the shell and the skull, and not permit shell rotation on the head, to avoid the edges of the helmet impacting the neck or face.

Most practical helmet designs are a compromise between impact protection and other considerations (e.g., bulk and weight, visibility, comfort, ability to communicate). Their efficacy has been demonstrated repeatedly by experiment and accident statistics.

REFERENCES

AGARD-AR-330, *Anthropomorphic Dummies for Crash and Escape System Testing*, North Atlantic Treaty Organization, Neuilly sur Seine, France, 1997.

Allen, G.: "The Use of a Spinal Analogue to Compare Human Tolerance to Repeated Shocks with Tolerance to Vibration," in AGARD-CP-253, *Models and Analogues for the Evaluation of Human Biodynamic Response, Performance and Protection*, North Atlantic Treaty Organization, Neuilly sur Seine, France, 1978.

Anon., "*German Aviation Medicine, World War II*," Vol. 2, Government Printing Office, Washington, DC, 1950.

Boileau, P.-E., J. Boutin, and P. Drouin: "Experimental Validation of a Wrist Measurement Method using Laser Interferometry," in Christ E., H. Dupuis, A. Okada, and J. Sandover (eds.), *Proceedings of the 6th International Conference on Hand-Arm Vibration*, Hauptverbandes der Gewerblichen Berufsgenossenschaften, Sankt Augustin, Germany, 1992, p. 519.

Bovenzi, M., and C. T. J. Hulshof: "An Updated Review of Epidemiologic Studies of the Relationship Between Exposure to Whole-Body Vibration and Low back Pain," *J. Sound Vib.*, **215**:595 (1998).

Brammer, A. J.: "Dose-Response Relationships for Hand-Transmitted Vibration," *Scand. J. Work Environ. Health*, **12**:284 (1986).

Brinkley, J. W., L. J. Specker, and S. E. Mosher: "Development of Acceleration Exposure Limits for Advanced Escape Systems," in AGARD-CP-472: *Implications of Advanced Technologies for Air and Spacecraft Escape*, North Atlantic Treaty Organization, Neuilly sur Seine, France, 1990.

Cherniack, M. G. (ed.): "Office Ergonomics," *Occupational Medicine: State of the Art Reviews*, **14**:1 (1999).

Eiband, A. M.: "*Human Tolerance to Rapidly Applied Accelerations: A Summary of the Literature*," NASA Memo 5-19-59E, National Aeronautics and Space Administration, Washington, D.C., 1959.

Fairley, T. E., and M. J., Griffin: "The Apparent Mass of the Seated Human Body: Vertical Vibration," *J. Biomechanics*, **22**:81 (1989).

Fung, Y. C.: *Biomechanics—Mechanical Properties of Living Tissues*, 2d ed., Springer-Verlag, New York, 1993.

Gillmeister, F., and T. Schenk.: "Hand-Arm Vibration and Coupling Force Measurement with a New Adapter," *Proceedings of the 9th International Conference on Hand-Arm Vibration*, Nancy, France, 2001.

Goldstein, S., E. Frankenburg, and J. Kuhn: "Biomechanics of Bone," Chap. 9 in Nahum, A. M., and J. W. Melvin (eds.), *Accidental Injury—Biomechanics and Prevention*, Springer-Verlag, New York, 1993.

Griffin, M. J.: *Handbook of Human Vibration*, Academic Press, London, 1990.

Haut, R. C.: "Biomechanics of Soft Tissues," Chap. 10 in Nahum, A. M., and J. W. Melvin (eds.): *Accidental Injury—Biomechanics and Prevention*, Springer-Verlag, New York, 1993.

Hearon, B. F., J. A. Raddin, Jr., and J. W. Brinkley, "Evidence for the Utilization of Dynamic Preload in Impact Injury Prevention," in AGARD-CP-322; *Impact Injury Caused by Linear Acceleration: Mechanisms, Prevention and Cost*, North Atlantic Treaty Organization, Neuilly sur Seine, France, 1982.

ISO 2631-1, *Mechanical Vibration and Shock—Evaluation of Human Exposure to Whole Body Vibration—Part 1: General Requirements*, 2d ed., International Organization for Standardization, Geneva, 1997.

ISO 5349-1, *Mechanical Vibration—Measurement and Evaluation of Human Exposure to Hand-Transmitted Vibration—Part 1: General Guidelines*, International Organization for Standardization, Geneva, 2001.

ISO 5349-2, *Mechanical Vibration—Measurement and Evaluation of Human Exposure to Hand-Transmitted Vibration—Part 2: Practical Guidance for Measurement at the Workplace*, International Organization for Standardization, Geneva, 2001.

ISO 5982, *Mechanical Vibration and Shock—Range of Idealized Values to Characterize Seated Body Biodynamic Response under Vertical Vibration*, International Organization for Standardization, Geneva, 2001.

ISO 7096, *Earth Moving Machinery—Operator Seat—Measurement of Transmitted Vibration*, International Organization for Standardization, Geneva, 1982.

ISO 8041, *Human Response to Vibration—Measuring Instrumentation*, International Organization for Standardization, Geneva, 1990.

ISO 10068, *Mechanical Vibration and Shock—Free Mechanical Impedance of the Human Hand-Arm System at the Driving-Point*, International Organization for Standardization, Geneva, 1998.

ISO 10819, *Mechanical Vibration and Shock—Hand-Arm Vibration—Method for the Measurement and Evaluation of the Vibration Transmissibility of Gloves at the Palm of the Hand*, International Organization for Standardization, Geneva, 1997.

ISO/FDIS 13091-2, *Mechanical Vibration—Vibrotactile Perception Thresholds for the Assessment of Nerve Dysfunction—Part 2: Analysis and Interpretation of Measurements at the Fingertips*, International Organization for Standardization, Geneva, 2002.

Linqvist, B. (ed.): *Ergonomic Tools in Our Time*, Atlas-Copco, Stockholm, Sweden, 1986.

Martin, B. J., and D. Adamo: "Effects of Vibration on Human Systems: A Neurophysiological Perspective," *Can. Acoust. Assoc. J.*, **29**:3, 10 (2001).

Mertz, H. J.: "Anthropometric Test Devices", Chap. 3 in Nahum, A. M., and J. W. Melvin (eds.): *Accidental Injury—Biomechanics and Prevention*, Springer-Verlag, New York, 1993.

Nahum, A. M., and J. W. Melvin (eds.): *Accidental Injury—Biomechanics and Prevention*, Springer-Verlag, New York, 1993.

Payne, P. R., "*On Quantizing Ride Comfort and Allowable Accelerations*," Paper 76-873, AIAA/SNAME Advanced Marine Vehicles Conf., Arlington, Va., American Institute of Aeronautics and Astronautics, New York, 1976.

Pelmear, P. L., and D. E. Wasserman (eds.): *Hand-Arm Vibration*, 2d ed., OEM Press, Beverly Farms, Mass., 1998.

Peterson, D. R., and Cherniack, M. G.: "Vibration, Grip Force, Muscle Activity, and Upper Extremity Movement of a Manual Hammering Task," *Proceedings of the 9th International Conference on Hand-Arm Vibration*, Nancy, France, 2001.

Peterson, D. R., and Cherniack, M. G.: "Repetitive Impact from Manual Hammering: Physiological Effects on the Hand-Arm System," *Can. Acoust. Assoc. J.* **29**:3; 12 (2001).

Reynolds, D. D.: "Design of Antivibration Gloves," *Can. Acoust. Assoc. J.* **29**:3; 16 (2001).

RTO-MP-20: "Models for Aircrew Safety Assessment: Uses, Limitations and Requirements," North Atlantic Treaty Organization, Neuilly sur Seine, France, 1999.

SAE J885: "Human Tolerance to Impact Conditions as Related to Motor Vehicle Design," Society of Automotive Engineers, Warrendale, Pa. 1986.

von Gierke, H. E.: "Effects of Vibration and Shock on People," Chap. 145 in Crocker, M. (ed.), *Encyclopedia of Acoustics*, Academic Press, New York, 1997,

von Gierke, H. E., and Brammer, A. J.: "Effects of Shock and Vibration on Humans," Chap. 42 in Harris, C. M., and A. G. Piersol (eds.), *Harris' Shock and Vibration Handbook*, 5th ed. McGraw-Hill, New York, 2002.

von Gierke, H. E., H. L. Oestreicher, E. K. Franke, H. O. Parrach, and W. W. von Wittern: "Physics of Vibrations in Living Tissues," *J. Appl. Physiol.*, **4**:886 (1952).

P • A • R • T • 3

BIOMATERIALS

CHAPTER 11
BIOPOLYMERS

Christopher Batich and Patrick Leamy
University of Florida, Gainesville, Florida

11.1 INTRODUCTION

Polymers are large molecules synthesized from smaller molecules called *monomers*. Most polymers are organic compounds with carbon as the base element. Plastics are polymers that are rigid solids at room temperature and generally contain additional additives. Some common plastics used in biomedical applications are polymethyl methacrylate for intraocular lenses, braided polyethylene terephthalate for vascular grafts, and ultrahigh-molecular-weight polyethylene for the articulating surfaces of orthopedic implants. Polymers and biopolymers in particular encompass a much broader spectrum than plastics alone. Biopolymers include synthetic polymers and natural polymers such as proteins, polysaccharides, and polynucleotides. This chapter will only cover the most commonly used examples in each class but will provide references to more specific sources.

Many useful polymers are water-soluble and are used as solutions. Hyaluronic acid is a naturally occurring high-molecular-weight polymer found in connective tissues and is used to protect the iris and cornea during ophthalmic surgery. Polyvinyl pyrrolidinone is a synthetic polymer used as a binder or additive in 25 percent of all pharmaceuticals.[1] Hydrogels are another class of polymers that has many biomedical applications. Hydrogels are polymers that swell in water but retain their overall shape. They are therefore soft and moist and mimic many natural tissues. The most well-known hydrogel series is poly(hydroxyethyl methacrylate) (PHEMA) and PHEMA copolymers, which are used in soft contact lenses.

Gelling polymers are hydrogels that can be formed in situ using chemical or physical bonding of polymers in solution. Alginates, for instance, are acidic polysaccharides that can be cross-linked using divalent cations such as calcium. Other examples of gelling polymers are the poloxamers that can gel with an increase in temperature. Alginates are widely used in cell immobilization, and poloxamers are used as coatings to prevent postsurgical adhesions.

Elastomers are low-modulus polymers that can reversibly deform up to many times (some over 500 percent) their original size. Silicones and polyurethanes are common elastomeric biopolymers. Polyurethane is used as a coating for pacemaker leads and for angioplasty balloons. Silicones are used for a variety of catheters, soft contact lenses, and foldable intraocular lenses.

The next section begins with an overview of polymer science topics, including synthesis, structure, and mechanical properties. The remainder of this chapter (Sec. 11.3) will discuss individual polymers, including their applications and properties. The polymers are presented in the following order: water-soluble polymers, gelling polymers, hydrogels, elastomers, and finally, rigid polymers. These five categories are roughly ordered from low to high modulus (i.e., high to low compliance). Water-soluble polymers in solution do not have an elastic modulus since they are fluids, so these are presented first. In fact, most polymers do not have a true elastic modulus since they are viscoelastic and exhibit solid and viscous mechanical behavior depending on the polymer structure, strain rate, and temperature.

Natural tissues are continuously repaired and remodeled to adjust to changes in the physiologic environment. No current synthetic biomaterial or biopolymer can mimic these properties effectively. Consequently, the ideal biomaterial or biopolymer performs the desired function, eventually disappears, and is replaced by natural tissue. Therefore, degradable polymers are of great interest to the biomedical engineering community. Polylactides and their copolymers are currently used as bone screws and sutures since they have good mechanical properties and degrade by hydrolysis so that they can, under optimum conditions, be replaced by natural tissue.

In addition to classification as water-soluble polymers, gelling polymers, hydrogels, elastomers, and rigid polymers, polymers can also be classified as bioinert, bioerodable, and biodegradable. Bioinert polymers are nontoxic in vivo and do not degrade significantly even over many years. Polymers can degrade by simple chemical means or under the action of enzymes. For the purposes of this chapter, bioerodable polymers such as polylactide are those that degrade by simple chemical means, and biodegradable polymers are those that degrade with the help of enzymes. Most natural polymers (proteins, polysaccharides, and polynucleotides) are biodegradable, whereas most synthetic degradable polymers are bioerodable. The most common degradation reactions for bioerodable polymers are hydrolysis and oxidation.

11.2 POLYMER SCIENCE

11.2.1 Polymer Synthesis and Structure

Polymers are frequently classified by their synthesis mechanism as either step or chain polymers. Step polymers are formed by stepwise reactions between functional groups. Linear polymers are formed when each monomer has two functional groups (functionality = 2). The second type of polymerization is chain polymerization, where monomers are added one at a time to the growing polymer chain.

Most polymerization techniques yield polymers with a distribution of polymer molecular weights. Polymer molecular weight is of great interest since it affects mechanical, solution, and melt properties of the polymer. Figure 11.1 shows a schematic diagram for a polymer molecular weight distribution. Number average molecular weight M_n averages the molecular weight over the number of molecules, whereas weight average molecular weight M_w averages over the weight of each polymer chain. Equations (11.1) and (11.2) defined M_w and M_n.

$$M_n = \frac{\Sigma N_i M_i}{\Sigma N_i} \tag{11.1}$$

$$M_w = \frac{\Sigma N_i M_i^2}{\Sigma N_i M_i} \tag{11.2}$$

where N_i is the number of polymer chains with molecular weight M_i.

The polymerization mechanism is a useful classification because it indicates the likely low-molecular-weight contaminants present. Chain-growth polymers frequently contain unreacted monomers, whereas step-growth polymers have low-molecular-weight oligomers (short chains) present.

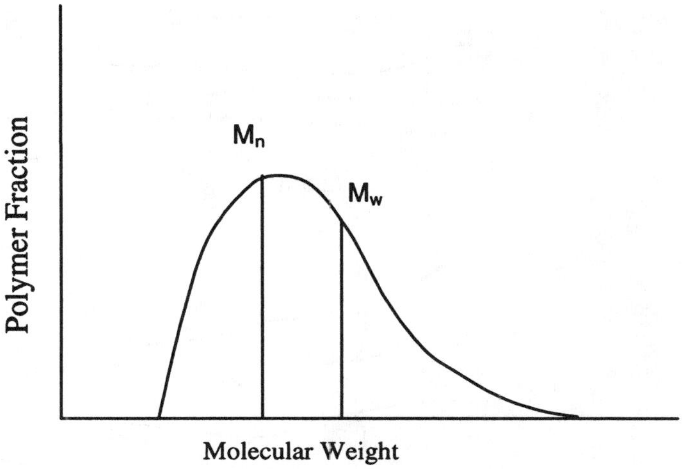

FIGURE 11.1 Typical polymer molecular weight distribution.

These low-molecular-weight species are more mobile or soluble than polymers and hence more likely to have physiologic effects. For instance, the monomer of PMMA causes a lowering of blood pressure and has been associated with life-threatening consequences when present (e.g., in some bone cements). Furthermore, the same polymer can be prepared by both mechanisms, leading to different impurities. For instance, polylactide is usually prepared by a chain-growth mechanism involving ring opening of a cyclic dimer (lactide) rather than the condensation of lactic acid.

As with all materials, a polymer's properties can be predicted and explained by understanding the polymer's structure on the atomic, microscopic, and macroscopic scales. Polymers can be roughly classified into two different classes, thermoplastic and thermoset. Thermoplastic polymers are made of individual polymer chains that are held together by relatively weak van der Waals and dipole-dipole forces. Thermoplastic polymers can be processed into useful products by melt processing, namely, injection molding and extrusion. They can also be dissolved in solvents and cast to form films and other devices. Although they often degrade or denature before melting, most proteins and polysaccharides can be considered thermoplastics since they are made of individual chains and can be dissolved in solvents. Finally, thermoplastics can be linear or branched.

Thermosetting polymers contain cross-links between polymer chains. Cross-links are covalent bonds between chains and can be formed using monomers with functionalities of greater than 2 during synthesis. Some polyurethanes and many silicones are formed using monomers with functionalities greater than 2. Cross-links also can be created after the polymer is formed. An example of this is vulcanization, which was discovered by Charles Goodyear in 1839 to toughen natural rubber. Vulcanization uses sulfur as a cross-linking agent.

Thermosets are, in essence, one giant molecule since all the polymer chains are connected through the cross-links. Thermosets cannot be melted after they are formed and cannot be dissolved in solvents. Depending on the cross-link density, thermosets can swell in certain solvents. When a cross-linked polymer solidifies or gels, it usually has some linear or unconnected polymer present, which sometimes can be extracted after implanation. Figure 11.2 shows a schematic diagram for linear, branched, and cross-linked polymers.

Polymers in the solid state have varying degrees of crystallinity. No polymer is truly 100 percent crystalline, but some are purely amorphous. Figure 11.3 is a simple model depicting a crystalline polymer. Polymer chains folding over themselves form crystalline regions. Amorphous regions of disordered polymer connect the crystals. Polymer chains are packed tighter in crystalline regions, leading to higher intermolecular forces. This means that mechanical properties such as modulus and

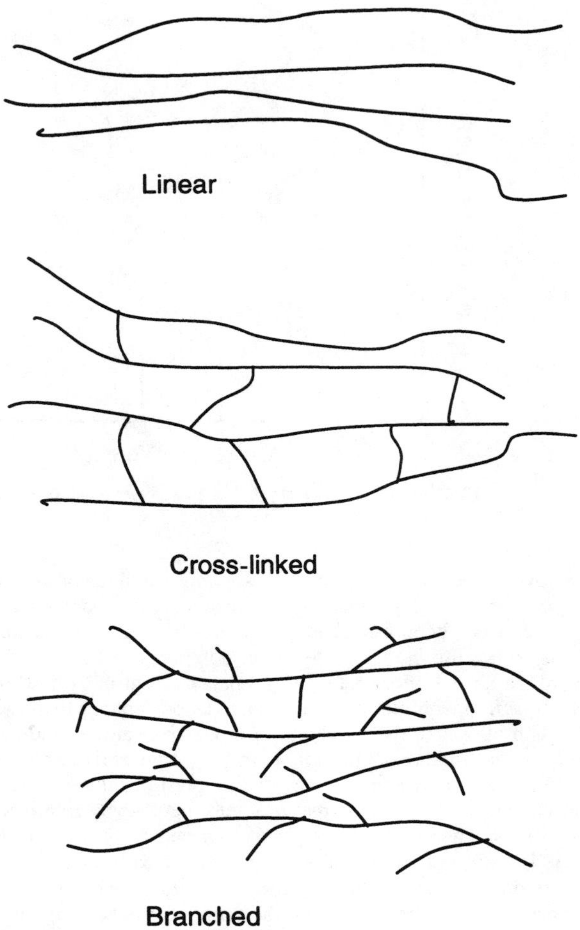

FIGURE 11.2 Schematic diagram showing different polymer structures.

strength increase with crystallinity. Ductility decreases with crystallinity since polymer chains have less room to slide past each other.

The primary requirement for crystallinity is an ordered repeating chain structure. This is why stereoregular polymers are often crystalline and their irregular counterparts are amorphous. Stereoregular polymers have an ordered stereostructure, either isotactic or syndiotactic. Isotactic polymers have the same configuration at each stereo center, whereas configuration alternates for syndiotactic polymers (Fig. 11.4). Atactic polymers have no pattern to their stereostructure. Polypropylene (PP) is a classic example of a polymer whose crystallinity and properties change drastically depending on stereostructure. Syndiotactic and isotactic PP have a high degree of crystallinity, whereas atactic polypropylene is completely amorphous. Isotactic PP has excellent strength and flexibility due to this regular structure and makes excellent sutures. Atactic PP is a weak, gumlike material. Recent advances in polymer synthesis have made available new polymers with well-controlled tacticity based on olefins. It is likely that they will find use as biomaterials in the future.

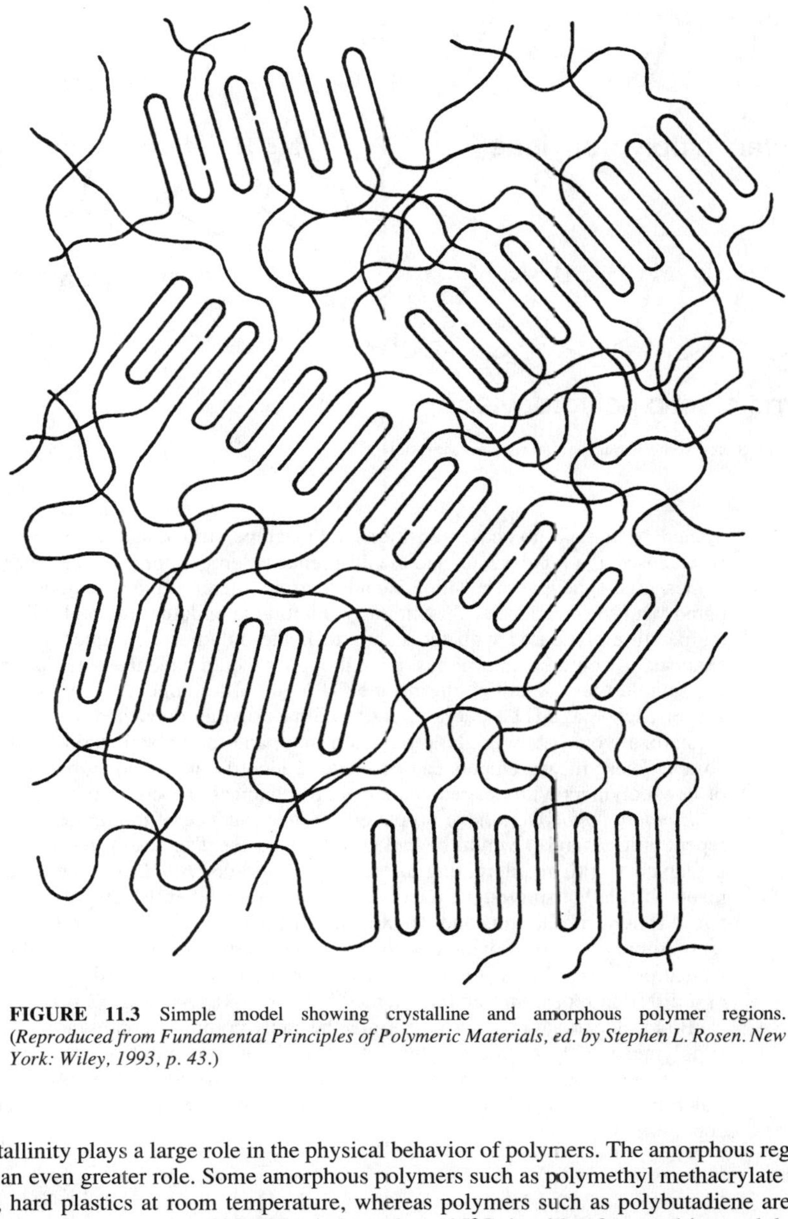

FIGURE 11.3 Simple model showing crystalline and amorphous polymer regions. (*Reproduced from Fundamental Principles of Polymeric Materials, ed. by Stephen L. Rosen. New York: Wiley, 1993, p. 43.*)

Crystallinity plays a large role in the physical behavior of polymers. The amorphous regions play perhaps an even greater role. Some amorphous polymers such as polymethyl methacrylate (PMMA) are stiff, hard plastics at room temperature, whereas polymers such as polybutadiene are soft and flexible at room temperature. If PMMA is heated to 105°C, it will soften, and its modulus will be reduced by orders of magnitude. If polybutadiene is cooled at to −73°C, it will become stiff and hard. The temperature at which this hard-to-soft transformation takes place is called the *glass transition temperature* T_g.

Differential thermal analysis (DTA) or a similar technique called differential scanning calorimetry (DSC) can be used to determine the temperature at which phase transitions such as glass transition temperature and melting temperature T_m occur. DTA involves heating a polymer sample along with

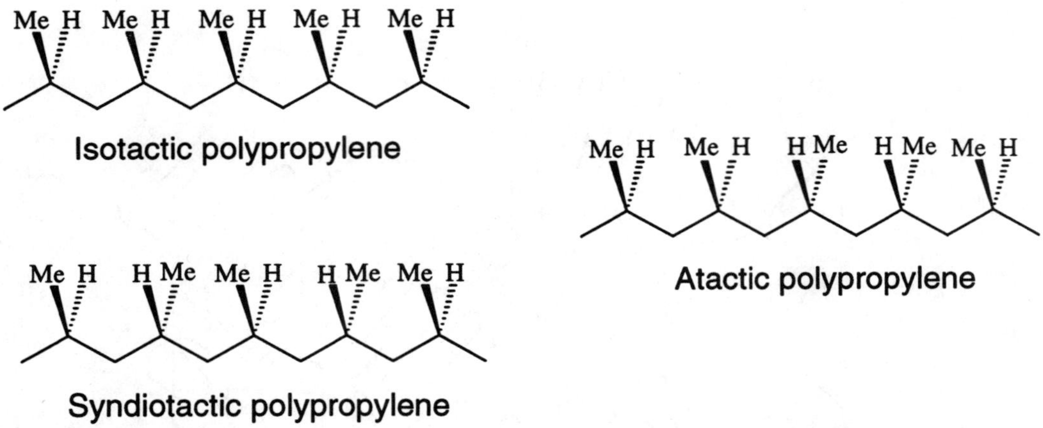

FIGURE 11.4 Stereoisomerism in polypropylene (Me$=$$CH_3$).

a standard that has no phase transitions in the temperature range of interest. The ambient temperature is increased at a regular rate, and the difference in temperature between the polymer and the standard is measured. The glass transition is endothermic; therefore, the polymer sample will be cooler compared with the standard at T_g. Similarly, melting is endothermic and will be detected as a negative temperature compared with the standard. If the polymer were quenched from melt prior to DTA analysis, it would be amorphous, even though it would have the potential to crystallize. In this case, the sample will crystallize during the DTA run at a temperature between T_g and T_m. Figure 11.5 shows a schematic DTA curve for a crystalline polymer quenched for melt. A polymer that does not crystallize would show a glass transition only, and the crystallization and melting peaks would be absent. These measurements can be made to identify an unknown plastic or to aid in the synthesis of new polymers with desired changes in mechanical properties at certain temperatures.

Random copolymers have no pattern to the sequence of monomers. A random copolymer using repeat units A and B would be called poly(A-*co*-B). The term *alternating copolymer* is fairly self-explanatory with an alternating pattern of repeat units. *Block copolymers* consist of long-chain segments (blocks) of single-repeat units attached to each other. Block polymers most commonly employ two different repeat units and contain two or three blocks. Block copolymers are named poly(A-*b*-B) or simply AB for polymers with two blocks (diblock polymer). A triblock copolymer would be named poly(A-*b*-B-*b*-A) or simply ABA. *Graft copolymers* consist of a backbone with side chains of a different repeat unit and are named poly(A-*g*-B). (See Fig. 11.6.)

Block and random copolymers are the most common copolymers. An example of a random copolymer is poly(lactide-*co*-glycolide), also known as poly(lactic-*co*-glycolic acid) depending on the synthesis route. Note that the structure for poly(lactide-*co*-glycolide) does not specify the type (random, alternating, block, or graft) and must be accompanied by the structure name to specify copolymer type.

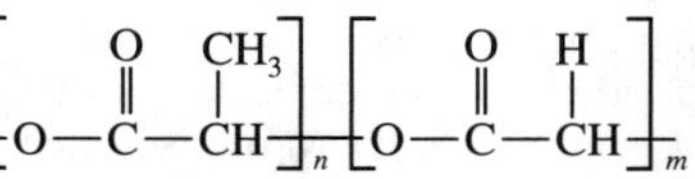

Poly(lactide-*co*-glycolide)

Block copolymers often phase segregate into an A-rich phase and a B-rich phase. If one repeat unit (or phase) is a soft phase and the other is a hard glassy or crystalline phase, the result can be a *thermoplastic elastomer*. The crystalline or hard glassy phase acts as a physical cross-link. The

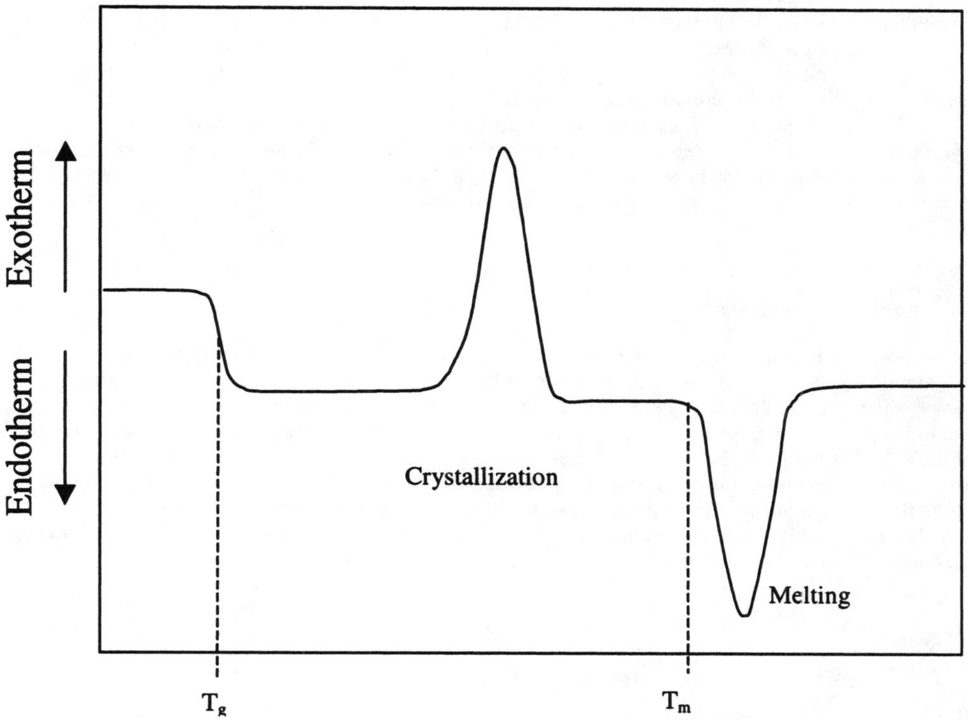

FIGURE 11.5 Schematic representation of a DTA curve for crystalline polymer quenched from melt prior to analysis.

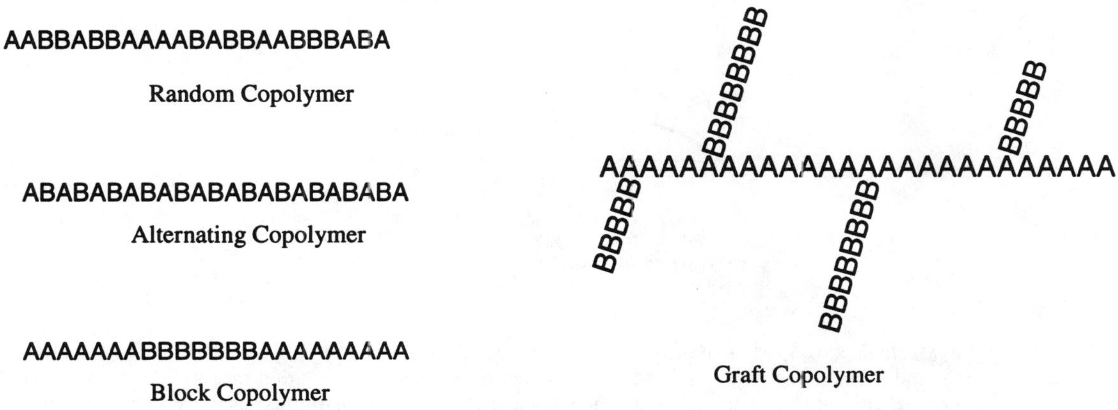

FIGURE 11.6 Schematic diagram showing classes of copolymer.

advantage of thermoplastic elastomers, unlike chemically cross-linked elastomers, is that they can be melt or solution processed. Many polyurethanes are thermoplastic elastomers. They consist of soft segments, either a polyester or a polyether, bonded to hard segments. The hard segments are ordinarily synthesized by polymerizing diisocyanates with glycols.

Similarly, thermoplastic hydrogels can be synthesized by using a hydrophilic A block and a hydrophobic B block. Poly(ethylene oxide-*b*-lactides) (PEO-*b*-PLA) are biodegradable hydrogel polymers that are being developed for drug-delivery applications.[2–6] PEO is a water-soluble polymer that promotes swelling in water, and PLA is a hard, degradable polymer that acts as a physical cross-linker.

11.2.2 Polymer Mechanical Properties

Solid polymer mechanical properties can be classified into three categories: brittle, ductile, and elastomeric (Fig. 11.7). Brittle polymers are polymers with a T_g that is much higher than room temperature, such as PMMA. These polymers have a high modulus and high ultimate tensile strength but low ductility and toughness. Ductile polymers are semicrystalline polymers such as polyethylene and PTFE that have a T_g below room temperature for the amorphous polymer content. The crystals lend strength, but the rubbery amorphous regions offer toughness. These polymers have lower strength and modulus but greater toughness than brittle polymers. Elastomers have low moduli since they have a T_g well below room temperature, but they can return to their original shape following high extensions since cross-links prevent significant polymer chain translations.

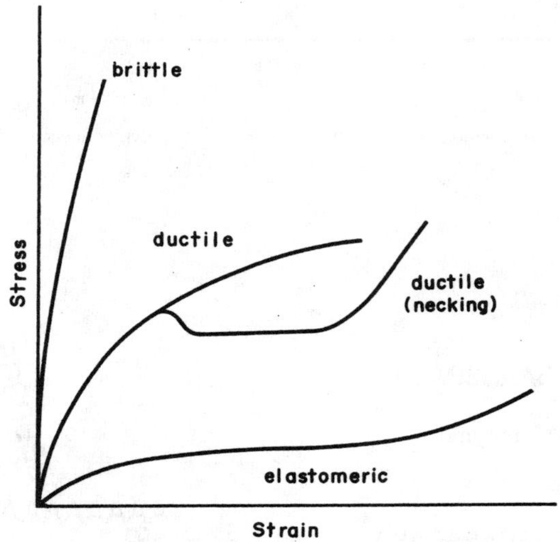

FIGURE 11.7 Mechanical behavior of polymers. (*Reproduced from Encyclopedia of Materials Science and Engineering, ed. by M. B. Bever. Cambridge, MA: MIT Press, 1986, p. 2917.*)

Mechanical properties of polymers, unlike those of other engineering materials, are highly strain rate and temperature dependent. Modulus increases with increasing strain rate and decreasing temperature (Fig. 11.8). The strain-rate dependence for mechanical properties shows that polymers exhibit viscous behavior in addition to solid or elastic behavior.

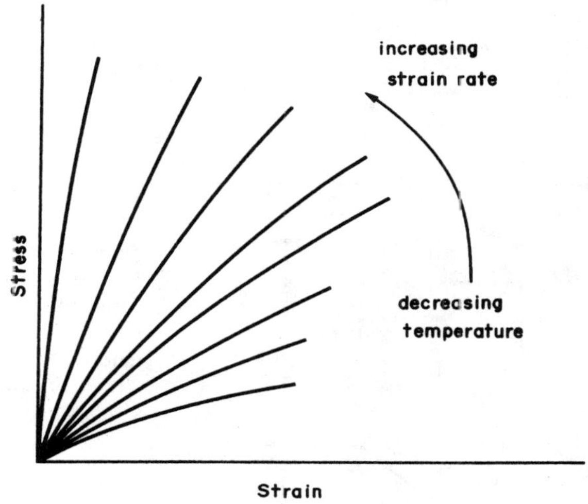

FIGURE 11.8 Schematic diagram showing strain rate and temperature dependence of polymer mechanical properties. *(Reproduced from Encyclopedia of Materials Science and Engineering, ed. by M. B. Bever. Cambridge, MA: MIT Press, 1986, p. 2917.)*

For an elastic solid, stress σ is a linear function of the applied strain ϵ, and there is no strain-rate dependence. Elastic modulus E is the slope of the stress versus strain curve. An elastic material can be modeled as a spring, whereas viscous materials can be modeled as a dashpot. For a fluid (viscous material), stress is proportional to strain rate ($d\epsilon/dt$) and unrelated to strain. Viscosity η is the slope of the stress versus strain rate curve. Figure 11.9 shows the stress/strain relationship for elastic solids and the stress/strain-rate relationships for viscous liquids.

Polymers can exhibit both viscous and solid mechanical behavior; this phenomenon is called *viscoelasticity*. For a given polymer, the degree of viscous behavior depends on temperature. Below T_g, polymers will behave more or less as elastic solids with very little viscous behavior. Above T_g,

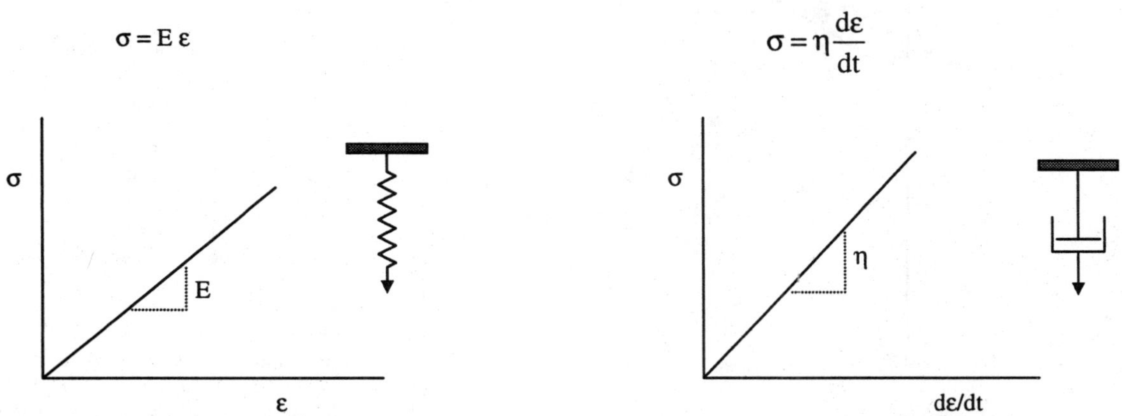

FIGURE 11.9 Stress/strain relationship for elastic solids and the stress/strain rate relationships for viscous liquids.

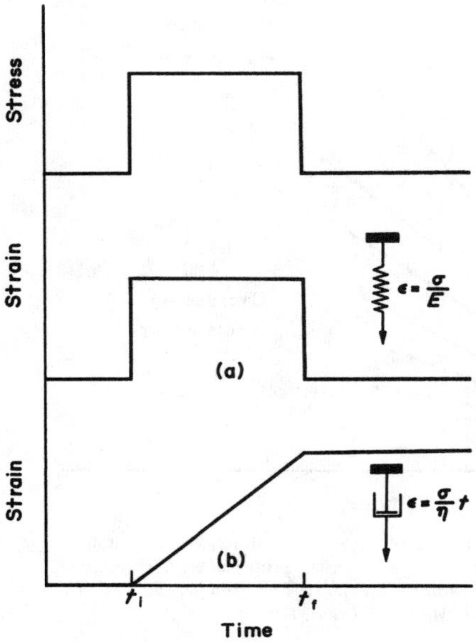

FIGURE 11.10 Response of elastic model (a) and viscous model (b) to a constant stress applied from t_i to t_f. (*Reproduced from Encyclopedia of Materials Science and Engineering, ed. by M. B. Bever. Cambridge, MA: MIT Press, 1986, p. 2919.*)

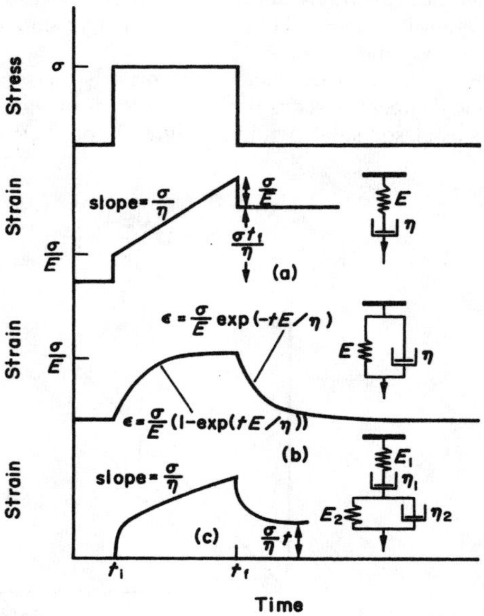

FIGURE 11.11 Creep response for (a) Maxwell model, (b) Voight-Kelvin model, and (c) four-parameter model for constant stress applied at t_i and removed at t_f.

polymers exhibit viscoelastic behavior until they reach their melting temperature, where they behave as liquids.

$$\sigma = E\epsilon \tag{11.3}$$

When designing with polymers, it is important to keep in mind that many polymers deform over time when they are under a continuous load. This deformation with time of loading is called *creep*. Ideal elastic solids do not creep since strain (deformation) is proportional to stress, and there is no time dependence. Viscous materials (liquids) deform at a constant rate with a constant applied stress. Equation (11.6) describes the strain in a viscous material under constant load or stress σ.

$$\sigma = \eta \frac{d\epsilon}{dt} \tag{11.4}$$

$$\int_0^\epsilon d\epsilon = \frac{\sigma}{\eta}\int_0^t dt \tag{11.5}$$

$$\epsilon = \frac{\sigma}{\eta} t \tag{11.6}$$

Figure 11.10 shows the strain with time of constant stress for a viscous and elastic material. The stress is applied at t and removed at t_f. The elastic model shows an instantaneous deformation when stress is applied at t_i, a constant deformation with time, and then a return to its original length when the load is removed. Therefore, the elastic solid does not creep. The viscous (dashpot) model deforms continuously (creeps) from t_i to t_f and remains permanently deformed after removal of the load.

Adding spring and dashpot models in series and parallel creates viscoelastic models. Several models have been proposed. Figure 11.11 shows the creep behavior for four viscoelastic models. *Stress relaxation* is a similar phenomenon and is defined as a reduction in stress during a constant deformation. One example of stress relaxation is the use of plastic washers between a nut and bolt. After the screw is secured, the washer deformation is constant, but the stress in the washer diminishes with time (stress relaxation), and the screw is therefore more likely to loosen with time.

Therefore, creep and stress relaxation should be accounted for when designing with polymers. The strain rate for a given application must also be known since modulus, ductility, and strength are strain rate dependent.

11.3 SPECIFIC POLYMERS

11.3.1 Water-Soluble Polymers

Water-soluble polymers are used for a variety of applications. They can be adsorbed or covalently bound to surfaces to make them more hydrophilic, less thrombogenic, and more lubricious. They can be used as protective coatings to prevent damage during surgery. Hyaluronic acid solutions are used in ophthalmic surgery to prevent damage to the cornea and iris. They can be cross-linked to form hydrogels for soft-tissue replacement and for drug-delivery applications. There are numerous water-soluble biopolymers. The polymers discussed below are some of the more common and useful examples.

Poly(N-Vinyl-Pyrrolidinone). Degradation: bioinert.

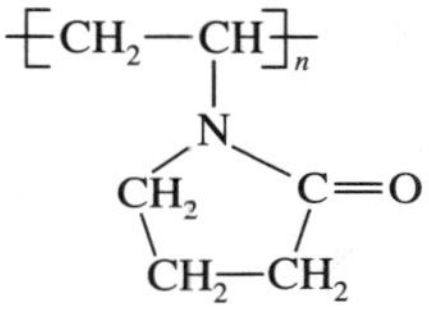

Poly(*N*-vinyl-pyrrolidinone) (or PVP) is a widely used water-soluble polymer. Similar to dextran, it has been used as a plasma volume expander to replace lost blood in mass-casualty situations. PVP can also be used as a detoxifying agent many toxic compounds form nontoxic complexes with PVP, which the kidneys eventually excrete. PVP is also used extensively as a binder in the pharmaceutical industry.

Polyethylene Glycol. Degradation: bioinert.

$$-\!\left[CH_2-CH_2-O \right]_n\!-$$

Polyethylene glycol (PEG), also known as polyethylene oxide (PEO), is used primarily to make hydrophobic surfaces more hydrophilic. These hydrophilic coatings are known to drastically reduce bacterial adhesion to substrates, making the surfaces antimicrobial.[7–9] PEO also can be coated or grafted onto the surfaces of microparticles to aid in colloidal stability.[10–12] Microparticles for drug-delivery applications are quickly recognized and cleared from the circulation by the reticuloendothelial system (RES). PEO coatings help particles elude the RES, thereby increasing their residence time in the circulation.[13–16]

Hyaluronic Acid. Degradation: biodegradable.

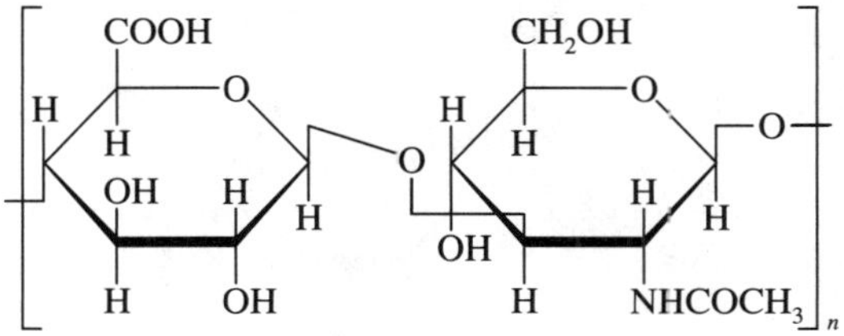

β-D-Glucoronic acid   *N*-Acetyl-β-D-glucosamine

Hyaluronic acid (HA) is a very lubricious high-molecular-weight water-soluble polymer found in connective tissue and the sinovial fluid that cushions the joints. HA is also found in the vitreous and aqueous humors of the eye. Solutions are injected in the eye during intraocular lens surgery to protect the cornea and the iris from damage during surgery. Table 11.1 shows data on HA concentration,

TABLE 11.1 Data for Commercial HA Solutions Used in Ophthalmic Surgery

Product	Concentration, %	Molecular weight, g/mol × 10^6	Viscosity, mPa · s
Microvisc plus	1.4	7.9	n.i.*
Morcher Oil Plus	1.4	7.9	1×10^6
Morcher Oil	1	6.1	1×10^6
Microvisc	1	5	n.i.
Healon GV	1.4	5	2×10^6
Viscorneal Plus	1.4	5	500,000
Allervisc Plus	1.4	5	500,000
Viscorneal	1	5	200,000
Allervisc	1	5	200,000
Healon5	2.3	4	7×10^6
HSO Plus	1.4	4	4.8×10^6
HSO	1	4	1×10^6
Healon	1	4	200,000
Dispasan Plus	1.5	>3	2.5×10^6
Visko Plus	1.4	3	500,000
BioLon	1	3	115,000
Dispasan	1	>2	35,000
Visko	1	2	300,000
Hya-Ophtal	2	2	n.i.
Amvisc Plus	1.6	1.5	60,000
IALUM	1.2	1.2	10,000
IALUM-F	1.8	1.2	22,000
Provisc	1	>1.1	50,000
Rayvisc	3	0.8	50,000
AMO Vitrax	3	0.5	40,000
Viscoat	3	>0.5	*ca.* 40,000

*n.i. = not investigated.
Source: Reproduced from H. B. Dick and O. Schwenn, *Viscoelastics in Ophthalmic Surgery*. Berlin: Springer-Verlag, 2000, p. 34.

molecular weight, and viscosity for some commercially available HA solutions. HA is currently being investigated to prevent postoperative adhesions. Since HA has many functional groups (OH, carboxylate, acetamido), it can be cross-linked by a variety of reagents. Therefore, HA may have applications as a hydrogel drug-delivery matrix.[17]

Dextran. Degradation: biodegradable.

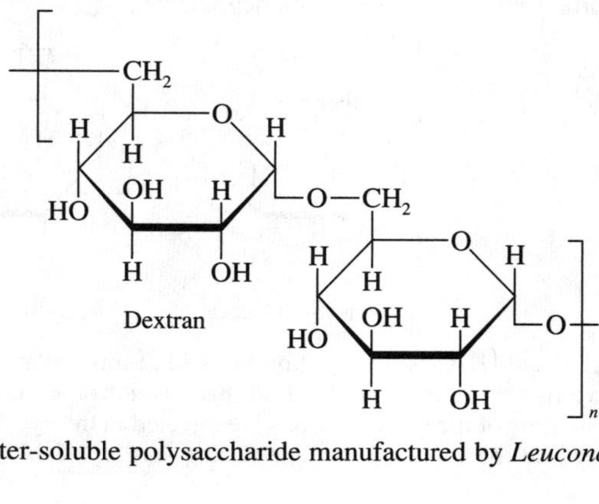

Dextran

Dextran is a simple water-soluble polysaccharide manufactured by *Leuconostoc mesenteroides* and

L. dextranicum (Lactobacteriaceae). Its structure is shown as a linear polymer, but some branching occurs at the three remaining OH groups. The native form of dextran has a high molecular weight near 5×10^8 g/mol. Dextran is depolymerized to yield a variety of molecular weights depending on the application. Similar to polyvinyl pyrrolidinone, dextran solutions can be used as a blood plasma extender for mass-casualty situations. Dextran of between 50,000 and 100,000 g/mol is used for this application. Like many of the water-soluble polymers, cross-linked dextran can be used as a drug-delivery matrix in whole or microsphere form. Dextran-coated magnetite (Fe_3O_4) nanoparticles are finding use as a magnetic resonance imaging (MRI) contrast agent. The dextran adsorbs onto the particle surfaces and provides a steric barrier to prevent agglomeration of the nanoparticles.

Starch. Degradation: biodegradable.

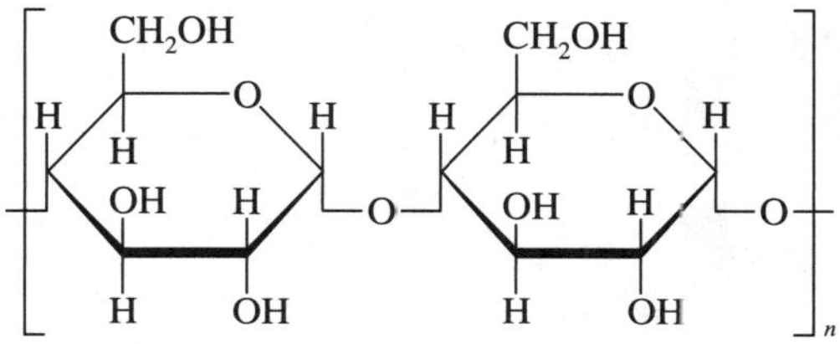

Amylose: Poly(1,4′-α-D-glucopyranose)

Starch is the primary source of carbohydrate in the human diet. Starch is composed of two monosaccharides: amylose and amylopectin. Amylose is a linear polymer that varies in molecular weight between 100,000 and 500,000 g/mol. Amylopectin is similar to amylose, having the same backbone structure but with 4 percent branching. Starch is insoluble in water but can be made soluble by treating with dilute HCl. Soluble starch has similar properties to dextran and therefore has similar applications.

11.3.2 Gelling Polymers

Gelling polymers are polymers in solution that transform into relatively rigid network structures with a change in temperature or by addition of ionic cross-linking agents. This class of polymers is useful because hydrogels can be formed at mild conditions. These polymers can therefore be used for cell immobilization and for injectable materials that gel in vivo. They are also used as coatings for drug tablets to control release in vivo.

Poloxamers. Degradation: bioinert.

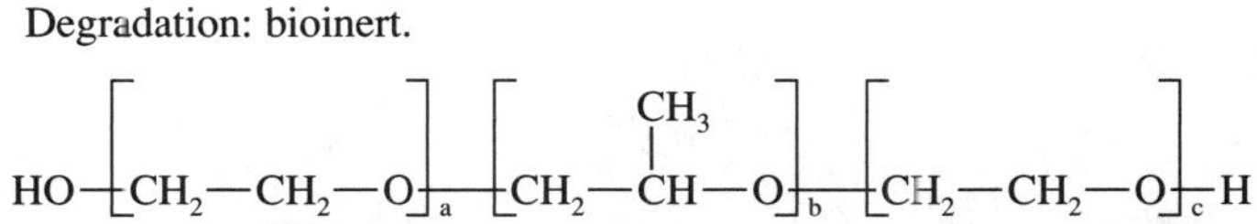

Poloxamers consist of two polyethylene oxide (PEO) blocks attached on both sides of a polypropylene oxide (PPO) block. The polymers are water-soluble, but increasing the temperature or concentration can lead to gel formation. The gelling properties are a function of the polypropylene content and the block lengths. Figure 11.12 shows the viscosity as a function of temperature for poloxamer 407. For a given concentration of poloxamer, the viscosity increases by several orders of magnitude at a transition temperature. The transition temperature decreases as polymer concentration increases.

The unique gelling properties of poloxamers make them useful as a coating to prevent postsurgical adhesions. They can be applied as a liquid since they gel at body temperature to provide a strong barrier for the prevention of adhesions. Similarly, poloxamers are being investigated for use as an injectable drug depot. Drug can be mixed with an aqueous poloxamer solution that thermally gels

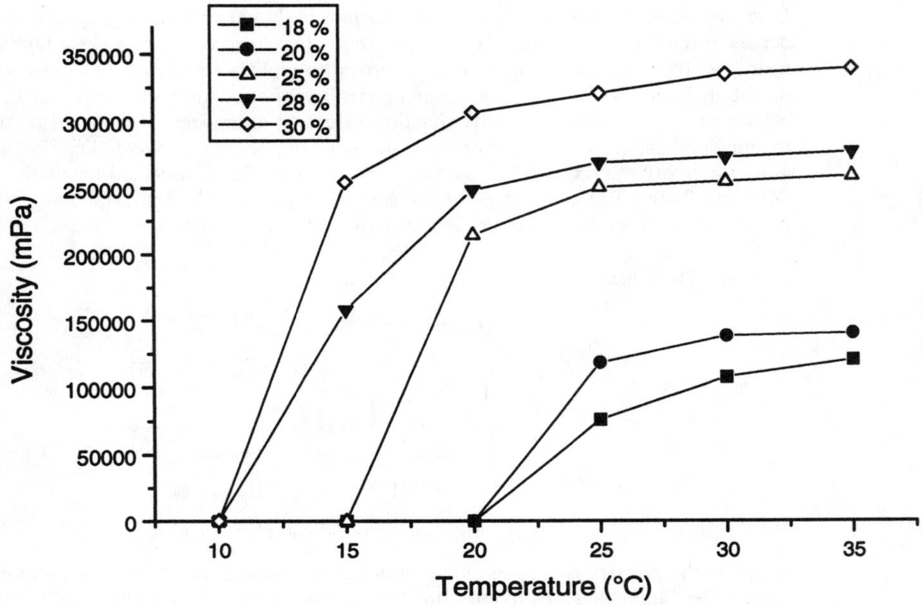

FIGURE 11.12 Viscosity of poloxamer solutions as a function of temperature and polymer concentration. (*Reproduced from L. E. Reeve, "Poloxamers: Their chemistry and applications," in Handbook of Biodegradable Polymers, A. J. Domb, J. K. Kost, and D. M. Wiseman (eds.) London: Harwood Academic Publishers, 1997, p. 235.*)

in the body and provides a matrix for sustained release. Another research area for poloxamers is for coating hydrophobic polymer microspheres. The PPO block adsorbs to the hydrophobic microsphere, whereas the PEO blocks extend into the solution and provide steric repulsion to prevent coagulation. The PEO blocks also prolong circulation after intravenous injection since the hydrophilic PEO retards removal by the reticuloendothelial system.

Alginate. Degradation: slow or nondegradable.

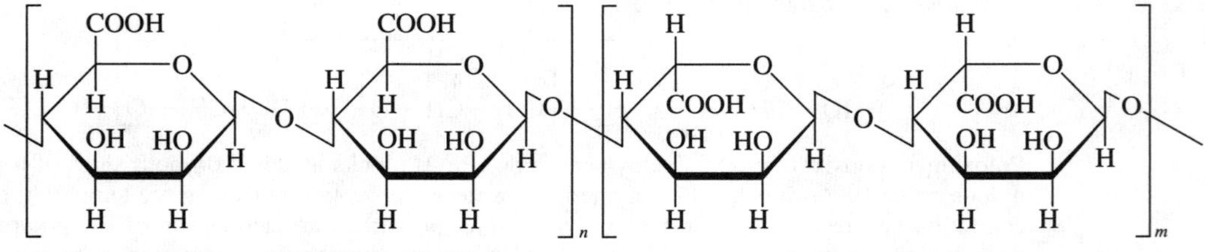

D-Mannuronic acid units L-Guluronic acid units

As this structure shows, alginate is a copolymer of guluronic and mannuronic acids. Alginate is a natural polysaccharide that is readily cross-linked using divalent or trivalent cations. Cross-linking occurs between acid groups of adjacent mannuronic acid units. Ca^{2+} is commonly used as a cross-linking agent. The sodium salt of alginate (sodium alginate) is used rather than the plain alginate since the acidic alginate can be harmful to cells and tissues.

Since cross-linking is chemically mild and easily accomplished, calcium cross-linked alginate is commonly used for cell immobilization. Cells are immobilized to prevent immune response in vivo and to prevent them from traveling from the desired location in vivo. Immobilization is most often accomplished by adding cells to a sodium alginate solution, followed by dripping the solution into a calcium chloride solution to cross-link the alginate and entrap cells.

Gelatin. Degradation: biodegradable. Gelatin is a protein prepared by hydrolyzing type I collagen using aqueous acids or bases. Collagen is discussed further in the section on hydrogels. Hydrolysis involves disruption of the collagen tertiary triple helix structure and reduction of molecular weight to yield gelatin that is soluble in warm water. Following hydrolysis, gelatin is purified and dried to yield a powder. Contrary to the poloxamers, gelatin solutions ($>$0.5 wt %) gel with a reduction in temperature. Gelatin gels melt between 23 and 30°C, and gelatin solutions set at around 2 to 5°C lower than the melting point. Gelatin is used as a tablet coating or capsule materials to control the release rate of drugs. Gelatin sponges are similar to collagen sponges and are used as hemostatic agents.

Fibrin. Degradation: biodegradable. Fibrin is the monomer formed from fibrinogen in the blood when a clot is formed. It is a protein that first polymerizes and then cross-links during clot formation and has been isolated and used as a biological adhesive and matrix for tissue engineering. The gel formation involves mixing fibrinogen with the gelling enzyme (thrombin) and a second calcium-containing solution. Speed of gellation is controlled by concentrations. Biodegradation occurs fairly rapidly due to natural enzymatic activity (fibrinolysis) resulting from plasmin in tissue. Fibrin is used as a soft-tissue adhesive and is used in tissue scaffolds.

11.3.3 Hydrogels

Hydrogels are materials that swell when placed in aqueous environments but maintain their overall shape. Hydrogels can be formed by cross-linking nearly any water-soluble polymer. Many natural materials such as collagen and chitosan (derived from chitin) absorb significant amounts of water and can be considered to be hydrogels. Hydrogels are compliant since the polymer chains have high mobilities due to the presence of water. Hydrogel mechanical properties are dependent on water content. Modulus and yield strength decrease with water content, whereas elongation tends to increase. Hydrogels are lubricious due to their hydrophilic nature. Hydrogels resist protein absorption and microbial attack due to their hydrophilicity and dynamic structure.

Poly(hydroxyethyl methacrylate). Degradation: bioinert.

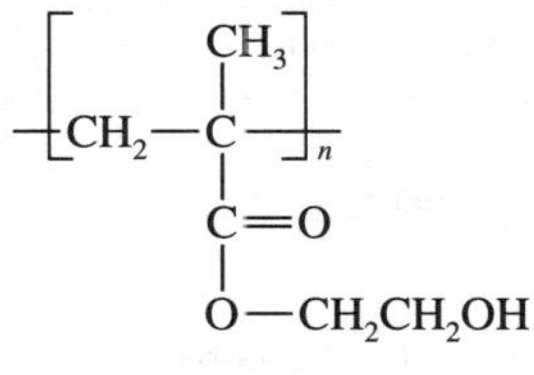

Poly(hydroxyethyl methacrylate) (PHEMA) is a hydrogel generally cross-linked with ethylene glycol dimethacrylate (which is normally present as a contaminant in the monomer). PHEMA's hydrogel properties such as resistance to protein adsorption and lubricity make it an ideal material for contact lenses. Hydrated PHEMA gels have good oxygen permeability, which is necessary for the health of the cornea. PHEMA is copolymerized with polyacrylic acid (PAA) or poly(N-vinyl pyrrolidinone) to increase its water-absorbing capability.

Chitosan. Degradation: biodegradable.

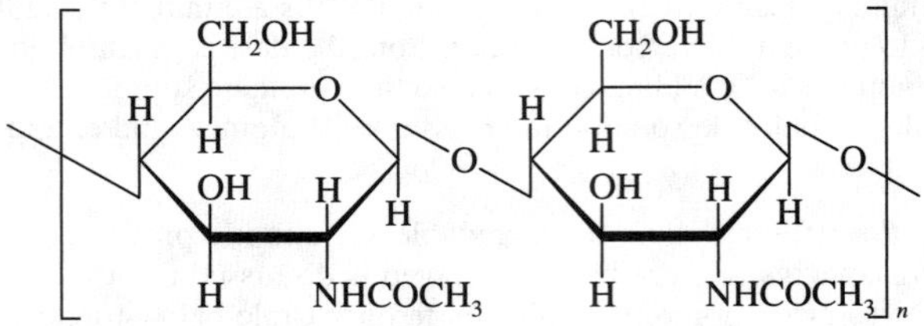

Chitin: Poly(1,4′-β-*N*-acetyl-2-amino-2-deoxy-D-glucopyranose)

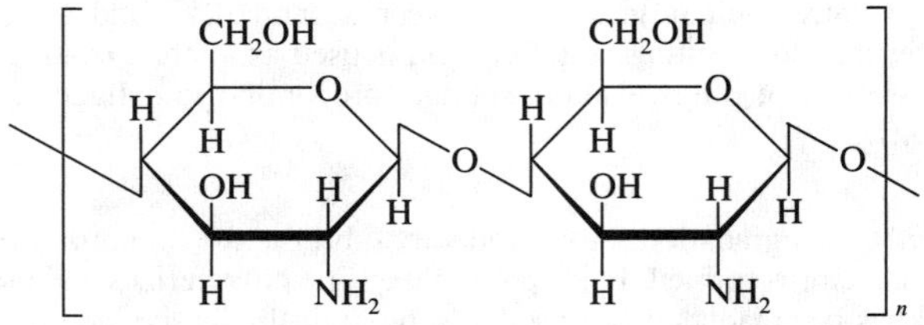

Chitosan: Poly(1,4′-β-2-amino-2-deoxy-D-glucopyranose)

Chitin is a polysaccharide that is the major component of the shells of insects and shellfish. Chitosan is deacetylated chitin. Deacetylation is accomplished using basic solutions at elevated temperatures. Chitin is not 100 percent acetylated, and chitosan is not 100 percent deacetylated. The degree of acetylation has a large influence on properties, in particular solubility. Chitin is difficult to use as a biomaterial since it is difficult to process. It cannot be melt processed and is insoluble in most aqueous and organic solutions. It is soluble only in strong acid solutions. Chitosan, on the other hand, is soluble in dilute organic acids; acetic acid is most commonly used. Chitosan has a positive charge due to the primary amines in its structure. The positive charge is significant because most tissues are negatively charged. Chitosan has been used for artificial skin and sutures and as a drug-delivery matrix.[18]

Chitosan absorbs a significant amount of water when placed in aqueous solution. Equilibrium water content of 48 percent was determined by immersing chitosan films in deionized water. Tensile testing on these wet films resulted in an ultimate tensile stress of approximately 1600 psi with 70 percent elongation at break.[19]

Collagen. Degradation: biodegradable. Collagen is the major structural protein in animals and exists in sheet and fibrillar forms. Collagen fibrils consist of a triple helix of three protein chains. Type I collagen is a fibrillar form of collagen that makes up 25 percent of the protein mass of the human body. Due to its prevalence and ability to be separated from tissues, type 1 collagen is most often used in medical devices. Collagen fibrils are strong and biodegradable, and collagen is hemostatic, making it useful in a variety of applications. Table 11.2 shows many of the applications for collagen. Collagen is usually obtained from bovine corium, the lower layer of bovine hide. Bovine collagen is nonimmunogenic for most people, but immune response may be triggered in those with allergies to beef. [20]

Both water-soluble and water-insoluble collagen can be extracted from animal tissues. Water-soluble collagen can be extracted from collagen using salt solutions, organic acids, or a combination of organic acids and proteases. Proteases break down cross-links and nonhelical ends, yielding more soluble collagen than acid alone or the salt solutions. Water-soluble collagen finds little use in the preparation of materials and devices since it quickly resorbs in the moist environment of the body. Water-insoluble collagen, however, is routinely used in the manufacture of medical devices. Water-insoluble collagen is ground and purified to yield a powder that can be later processed into materials and devices. Collagen cannot be melt processed and is therefore processed by evaporating water

TABLE 11.2 Medical Applications of Collagen

Specialty	Application
Cardiology	Heart valves
Dermatology	Soft-tissue augmentation
Dentistry	Oral wounds Biocoating for dental implants Support for hydroxyapatite Periodontal attachment
General surgery	Hemostasis Hernia repair IV cuffs Wound repair Suture
Neurosurgery	Nerve repair Nerve conduits
Oncology	Embolization
Orthopedic	Bone repair Cartilage reconstruction Tendon and ligament repair
Ophthalmology	Corneal graft Tape or retinal attachment Eye shield
Plastic surgery	Skin replacement
Urology	Dialysis membrane Sphincter repair
Vascular	Vessel replacement Angioplasty
Other	Biocoatings Drug delivery Cell culture Organ replacement Skin test

Source: Reproduced from F. H. Silver and A. K. Garg, "Collagen characterization, processing, and medical applications," in *Handbook of Biodegradable Polymers*, A. J. Domb, J. Krost, and D. M. Wiseman (eds.). London: Harwood Academic Publishers, 1997, Chap. 17, p. 336.

from collagen suspensions. Insoluble collagen disperses well at pH between 2 and 4. Evaporating 1% suspensions forms collagen films. Freezing suspensions followed by lyophilizing (freeze drying) forms sponges. Ice crystals form during freezing and result in porosity after water is removed during lyophilizing. Freezing temperature controls ice crystal size, and 14-μm pores result from freezing at $-80°C$ and 100-μm pores at $-30°C$. Fibers and tubes are formed by extruding collagen suspensions into aqueous solutions buffered at pH 7.5.[20]

Collagen absorbs water readily in the moist environment of the body and degrades rapidly; therefore, devices are often cross-linked or chemically modified to make them less hydrophilic and to reduce degradation. Viswwanadham and Kramer[21] showed that water content of untreated collagen hollow fibers (15 to 20 μm thick, 400 μm outer diameter) is a function of humidity. The absorbed water plasticizes collagen, lowering both the modulus and yield strength. Table 11.3 summarizes these results. Cross-linking the fibers using ultraviolet (UV) radiation increased the modulus of the fibers.[21]

TABLE 11.3 Water Absorption and Its Effect on Modulus *E* and Yield Strength of Collagen Hollow Fibers[21]

Humidity, %	Water absorption, g/100 g collagen
Wet	240
90	50
80	25
60	17
30	10
Humidity, %	**Yield stress, psi**
Wet	3000
90	5200
66	13,000
36	19,000
8	24,000
Humidity, %	***E*, ksi***
Wet	44
90	450
75	750
8	970

*1 ksi = 1000 psi.

Albumin. Degradation: biodegradable. Albumin is a globular, or soluble, protein making up 50 percent of the protein content of plasma in humans. It has a molecular weight of 66,200 and contains 17 disulfide bridges.[22] Numerous carboxylate and amino (lysyl) groups are available for cross-linking reactions, providing for a very broad range of mechanical behavior. Heating is also an effective cross-linking method, as seen in ovalbumin (egg white cooking). This affords another gelling mechanism and is finding increasing use in laser welding of tissue, where bond strengths of 0.1 MPa have been achieved.[23]

As with collagen, the most common cross-linking agent used is glutaraldehyde, and toxic byproducts are of concern. Careful cleaning and neutralization with glycine wash have provided biocompatible albumin and collagen structures in a wide variety of strengths up to tough, very slowly degradable solids. It should be noted that albumin and collagen solidification generally is different from that of fibrin, which gels by a normal biological mechanism. The glutaraldehyde methods yield a variety of nonbiologic solids with highly variable mechanical properties. This has led to an extensive literature and a very wide range of properties for collagen and albumin structures, which are used for tissue substitutes and drug-delivery vehicles.

Oxidized Cellulose. Degradation: bioerosion. Oxidized cellulose is one of the fastest-degrading polymers at physiologic pH. It is classified as bioerodable since it degrades without the help of enzymes. It is relatively stable at neutral pH, but above pH 7, it degrades. Oxidized cellulose disappears completely in 21 days when placed in phosphate-buffered saline (PBS). Similarly, it dissolves 80 percent after 2 weeks in vivo. Cellulose is oxidized using nitrogen tetroxide (N_2O_4). Commercially available oxidized cellulose contains between 0.6 and 0.93 carboxylic acid groups per glucose unit, which corresponds to between 16 and 24 wt /% carboxylic acid.[24]

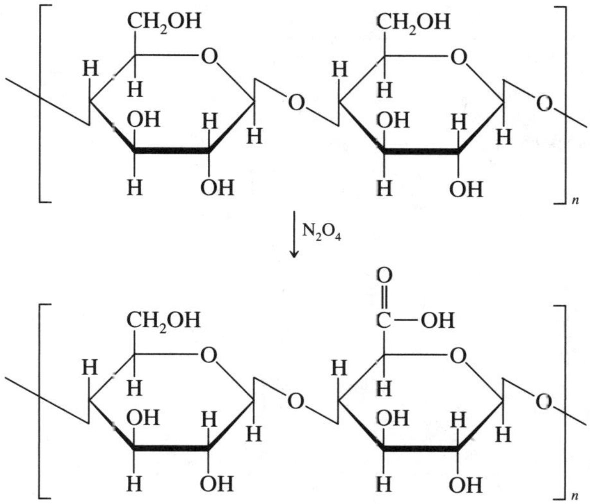

Cellulose

Oxidized cellulose

Oxidized cellulose is used as a degradable hemostatic agent. The acid groups promote clotting when placed in wounds. Furthermore, oxidized cellulose swells with fluid to mechanically close damaged vessels. Oxidized cellulose sheets are placed between damaged tissues following surgery to prevent postsurgical adhesions. The sheets separate tissue during healing and dissolve in a few weeks after healing occurs.[24]

11.3.4 Elastomers

Silicones and polyurethanes are the two classes of elastomers used for in vivo applications. Both are versatile polymers with a wide range of mechanical properties. Polyurethanes tend to be stiffer and stronger than silicones, whereas silicones are more inert and have the advantage of being oxygen-permeable. Polyurethanes are more versatile from a processing standpoint since many polyurethanes are thermoplastics, whereas silicones rely on covalent cross-linking and are therefore thermosets.

Polyurethane elastomers. Degradation: bioinert or slow bioerosion.

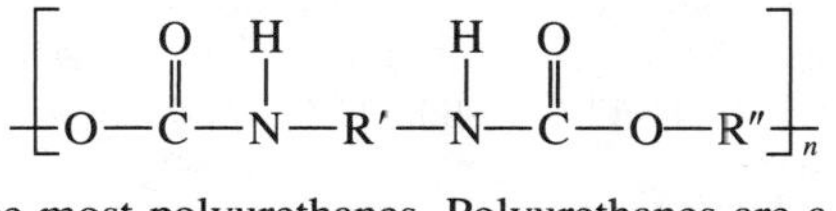

This repeat unit can describe most polyurethanes. Polyurethanes are a versatile class of block copolymers consisting of a hard block (R′) and a soft block (R″). The hard block is a glassy polymer (T_g above room temperature) often synthesized by polymerizing diisocyanates with glycols. R″ is a low-T_g (T_g << room temperature) polyester or polyether. Polyurethanes with polyester soft blocks are degradable, whereas those with polyether blocks degrade very slowly. Polyurethanes are usually elastomers since hard and soft blocks are present. Rubbers of different hardness or durometer can be prepared by varying the ratio of R′ to R″. Covalently cross-linked polymers can be prepared by using monomers with functionalities greater than 2. However, the most useful polyurethanes for medical applications are the thermoplastic elastomers since these can be melt processed or solution cast. Polyurethanes have good fatigue strength and blood compatibility and are used for pacemaker lead insulation, vascular grafts, and ventricular assist device/artificial heart membranes.[25] Table 11.4 shows properties for thermoplastic polyurethane elastomers available from Cardio Tech International, Inc. These values are for the Chronoflex C series of polymers.

TABLE 11.4 Properties of Chronoflex Thermoplastic Polyurethanes Available from CardioTech

Property	ASTM procedure	Typical values		
Hardness (Shore durometer)	ASTM D-2240	80A	55D	75D
Ultimate tensile strength (psi)	ASTM D-638	5500–6500	6000–7500	7000–8000
Elongation at break (%)	ASTM D-638	400–490	365–440	255–320
100% secant modulus (psi)	ASTM D-638	770–1250	1850–2200	5300–5700
300% secant modulus (psi)	ASTM D-638	700–1400	1700–2000	2700–3200
Flexural strength (psi)	ASTM-D790	350	550	10,000
Flexural modulus (psi)	ASTM D-790	5500	9300	300,000
Melt index (g/10 min), 210°C, 2.17 kg	ASM D-1238	8	5	3
Vicat softening point (F/C)	ASTM D-1525	160/70	180/80	—
Water absorption (%)	ASTM D-5170	1.2	1	0.8
Specific gravity	ASTM D-792	1.2	1.2	1.2
Coefficient of friction	ASTM D-1894	1.5	0.8	0.64
Abrasion resistance (% loss at 1000 cycles)	ASTM D-1044	0.008	0.035	0.053
Melt processing temp. (°F/°C)		375–430/190–220		
Recommended sterilization		Gamma, e-beam, ethylene oxide		
Class VI biocompatability test	U.S.P. XXII	Pass	Pass	Pass

Silicone Elastomers (Polysiloxanes). Degradation: bioinert.

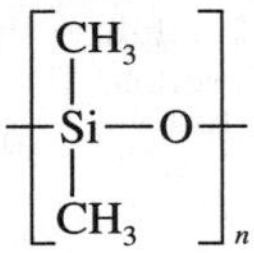

Poly(dimethyl siloxane)

Silicone elastomers are cross-linked derivatives of poly(dimethyl siloxane) (PDMS). Polysiloxane liquids with functional endgroups such as OH or sidegroups such as $CH{=}CH_2$ can be molded at room temperature and cross-linked to form elastomers using various cross-linking agents. Silicone elastomer kits consisting of the polysiloxane precursor liquids and cross-linking agents are commercially available from corporations such as GE Bayer Silicones (Table 11.5). Silicones cross-linked at room temperature are called *room-temperature vulcanized* (RTV) *elastomers*, and those requiring elevated temperatures are called *heat-cured silicone elastomers*.

Silicones are more flexible and of lower strength than polyurethanes. However, they are more chemically stable and are used for artificial finger joints, blood vessels, heart valves, breast implants, outer ears, and chin and nose implants. Silicones have high oxygen permeability and are used for membrane oxygenators and soft contact lenses.[26]

11.3.5 Rigid Polymers

Most bioinert rigid polymers are commodity plastics developed for nonmedical applications. Due to their chemical stability and nontoxic nature, many commodity plastics have been used for implantable materials. This subsection on rigid polymers is separated into bioinert and bioerodable materials. Table 11.6 contains mechanical property data for bioinert polymers and is roughly ordered by elastic modulus. Polymers such as the nylons and poly(ethylene terephthalate) slowly degrade by hydrolysis of the polymer backbone. However, they are considered bioinert since a significant decrease in properties takes years.

Most rigid degradable polymers degrade without the aid of enzymes and are therefore bioerodable. Table 11.7 shows mechanical property data for bioerodable polymers.

TABLE 11.5 Mechanical Properties of Cured Silicone Available from GE Bayer Silicones

Product	Shore hardness	Tensile strength, psi	Elongation at break, %	Tear strength, lb/in
Silopren HV 3/322	30A	800	600	86
Silopren HV 3/822	80A	1300	400	142
Silopren HV 4/311	35A	1000	600	57
Silopren HV 4/811	80A	1450	400	114
Silopren LSR 4020	22A	940	1000	86
Silopren LSR 4070	70A	1300	400	114

TABLE 11.6 Literature Values for Physical Properties of Bioinert Plastics

Material	Specific gravity	Tensile strength, psi	Elastic modulus, ksi	T_g, °C	T_m, °C
Cellulose acetate	1.22–1.34[a]	1900–9000[a]	580[b]		230[b]
Nylon 6,6	1.13–1.15[a]	11,000†–13,700*[a]	230*–550[a]	50[c]	255–265[a]
			230*–500[a]		
Poly(methyl methacrylate)	1.17–1.20[a]	7000–10,500[a]	325–470[a]	85–105[a]	Amorphous
Nylon 6	1.12–1.14[a]	6000†–24,000*[a]	380*–464[a]	40–52[c]	210–220[a]
			100†–247[a]		
Poly(ethylene terephthalate)	1.30–1.38[a]	8200–8700[a]	280–435[a]	62–77[c]	220–267[a]
Poly(vinyl chloride)	1.39–1.43[b]	7200[d]	363[d]	75–85[b]	120–210[b]
Polypropylene	0.90–0.91[a]	4500–6000[a]	165–225[a]	−20[a]	160–175[a]
High-density polyethylene	0.95–0.97[a]	3200–4500[a]	155–158[a]	−125[c]	130–137[a]
Poly(tetrafluoroethylene)	2.14–2.20[a]	3000–5000[a]	58–80[a]	−80, 126[c]	327[a]
Perfluorinated ethylene-propylene	2.12–2.17[a]	2700–3100[a]	50[a]	−90, 70[c]	275[a]

*Tested at 0.2 percent moisture content.
†Tested after conditioning at 50 percent relative humidity.
Sources:
[a]Data from *Modern Plastics Encyclopedia*, New York: McGraw-Hill, 1999.
[b]Data from *Encyclopedia of Polymer Science and Engineering*, 2d ed. New York: Wiley, 1985.
[c]Data from *Polymer Handbook*, 3d ed. New York: Wiley, 1989.
[d]Data from *Encyclopedia of Materials Engineering*. Cambridge, MA: MIT Press, 1986.

TABLE 11.7 Physical Properties of Degradable Polyesters Available from Birmingham Polymers, Inc.

Material	Specific gravity	Tensile strength, psi	Elastic modulus, ksi	T_g, °C	T_m, °C
Polyglycolide	1.53	10,000	1000	35–40	225–230
Poly (L-lactide)	1.24	8000–12,000	600	60–65	173–178
50/50 Poly (DL-lactide-*co*-glycolide)	1.34	6000–8000	400	45–50	Amorphous
Poly (DL-lactide)	1.25	4000–6000	400	55–60	Amorphous
Polycaprolactone	1.11	3000–5000	50	−65–−60	58–63

Cellulose. Degradation: bioinert.

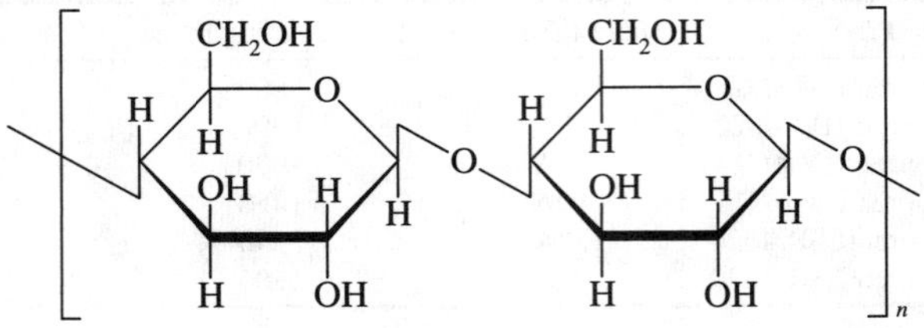

Cellulose: Poly(1,4'-β-D-glucopyranose)

Cellulose is a partially crystalline polysaccharide and is the chief constituent of plant fiber. Cotton is the purest natural form of cellulose, containing 90 percent cellulose. Cellulose decomposes before melting and therefore cannot be melt processed. It is insoluble in organics and water and can only be dissolved in strong basic solutions. Regenerated cellulose, also known as *rayon,* is cellulose that has been precipitated form a basic solution. Cellulose is used in bandages and sutures. Cuprophan is cellulose precipitated from copper hydroxide solutions to form hemodialysis membranes.

Cellulose acetate. Degradation: bioinert.

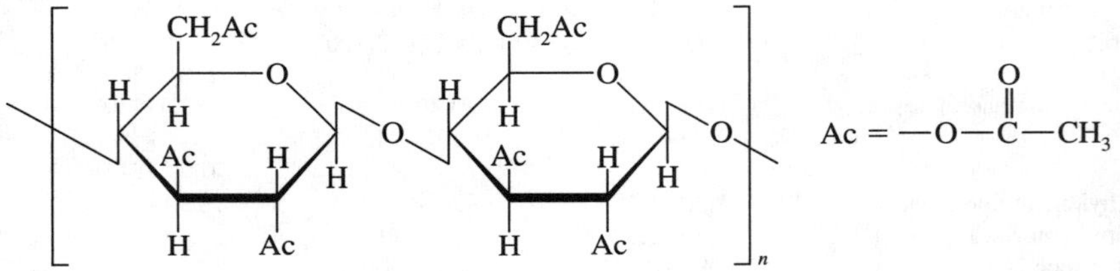

Cellulose acetate

Cellulose acetate is a modified cellulose that can be melt processed. Cellulose acetate membranes are used for hemodialysis.

Nylon 6,6. Degradation: slow bioerosion.

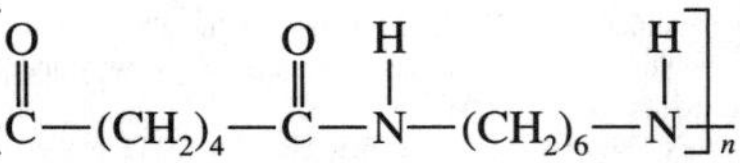

Poly (hexamethylene adipimide) is also known as *Nylon 6,6* since its repeat unit has two six-carbon sequences. Nylon is tough, abrasion resistant, and has a low coefficient of friction, making it a popular suture material.[27] Nylon 6,6 is hydrophilic and absorbs water when placed in tissues or in humid environments (9 to 11 percent water when fully saturated[28]). Absorbed water acts as a plasticiser, increasing the ductility and reducing the modulus of Nylon 6,6. Nylon bioerodes at a very slow rate. Nylon 6,6 implanted in dogs lost 25 percent of its tensile strength after 89 days and 83 percent after 725 days.[29]

Nylon 6: Poly(caprolactam). Degradation: slow bioerosion.

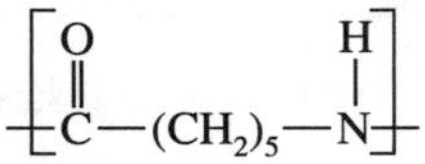

Nylon 6 has similar properties to Nylon 6,6, the primary difference being that Nylon 6 has a lower melting temperature and its properties are more moisture-sensitive.

Poly(ethylene terephthalate). Degradation: very slow bioerosion.

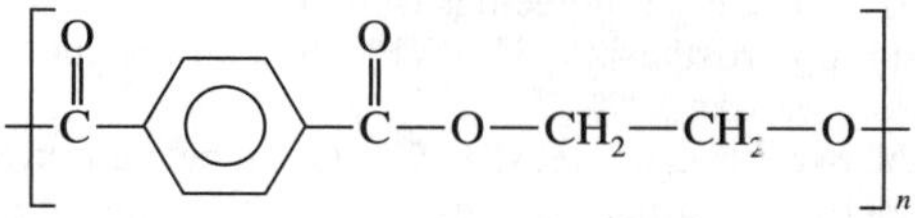

Poly(ethylene terephthalate) (PET), also known simply as *polyester* or *Dacron*, is a rigid semicrystalline polymer. It is widely used as a material for woven large-diameter vascular grafts. PET is usually considered to be stable, but it undergoes very slow bioerosion in vivo.

Poly(methyl methacrylate) (PMMA). Degradation: bioinert.

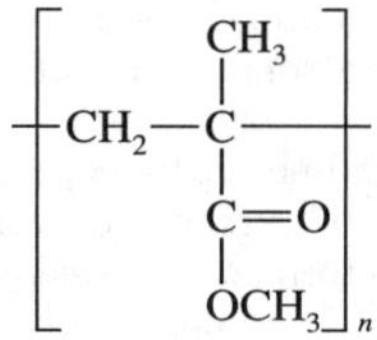

PMMA is an amorphous polymer with a high T_g around 100°C. PMMA is a stiff, hard, transparent material with a refractive index of 1.5 and is therefore used for intraocular lenses and hard contact lenses. PMMA is very bioinert but less so than PTFE due to possible hydrolysis of ester sidegroups.

PMMA is a thermoplastic that can be formed by injection molding or extrusion. Casting monomer or monomer-polymer syrup and polymerizing can also form PMMA. PMMA plates, commonly known as *Plexiglass* or *Lucite*, are formed this way.

Polyvinyl chloride. Degradation: nondegradable.

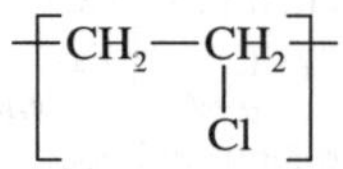

Polyvinyl chloride (PVC) a rigid glassy polymer that is not used in vivo because it causes a large inflammatory response probably due to metal stabilizers and residual catalysts. However, PVC softened with plasticizers such as dioctyl phthalate is used for medical tubing and blood bags. PVC is a thermoplastic and can be melt processed.

Polypropylene (PP). Degradation: bioinert.

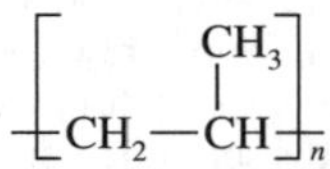

Commercial polypropylene is isotactic since atactic polypropylene has poor mechanical properties and isotactic is difficult to synthesize. Polypropylene has similar structure and properties as HDPE, except that it has superior flex-fatigue properties and a higher melting point. Polypropylene is commonly used for nondegradable sutures. Like PE, polypropylene can be melt processed.[26]

Polyethylene (PE). Degradation: bioinert.

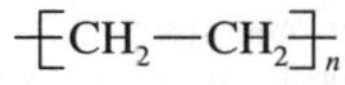

Polyethylene is a flexible polymer with a T_g of around −125°C. It is available in three different forms: low density (LDPE), linear low density (LLDPE), and high density (HDPE). LDPE and LLDPE are typically not used in vivo since they cannot be autoclaved. HDPE can be autoclaved and is used in tubing for drains and as a catheter material. Ultrahigh-molecular-weight polyethylene (UHMWPE) has a high molecular weight ($\sim 2 \times 10^6$ g/mol) and is used for the articulating surfaces of knee and hip prosthesis. UHMWPE, like all PE, has a low coefficient of friction but is very hard and abrasion-resistant.[30]

With the exception of UHMWPE, PE can be melt processed. UHMWPE has a high melting point and, like PTFE, is formed by pressing and sintering of powders.

Polytetrafluoroethylene (PTFE). Degradation: very bioinert.

$$\left[CF_2 - CF_2 \right]_n$$

Polytetrafluoroethylene (PTFE, Teflon, Fluorel) is best known for its excellent chemical stability and low coefficient of friction. Expanded PTFE (ePTFE) contains micropores created by stretching PTFE film and is used in small-diameter vascular grafts and for artificial heart valve sewing rings (e.g., Gore-Tex).

PTFE is highly crystalline (92 to 98 percent) and degrades near its melting temperature of 327°C; therefore, it cannot be melt processed even though it is a thermoplastic. Due to its inability to be melt processed, PTFE is formed by pressing PTFE powder, followed by heating to sinter the powder, or it is heated and pressed simultaneously (pressure sintered). [31]

Perfluorinated Ethylene-Propylene Polymer (FEP). Degradation: bioinert.

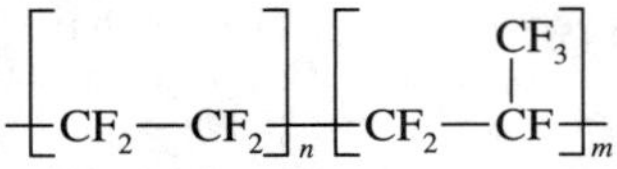

Perfluorinated ethylene-propylene polymer (FEP) is a copolymer of tetrafluoroethylene (TFE) and hexafluoropropylene (HFP). FEP has similar properties to PTFE, but its lower melting temperature of 275°C allows it to be melt processed.[31]

Polylactide, Polyglycolide, and Copolymers. (Also known as polylactic acid and polyglycolic acid.) Degradation: bioerosion; polylactide: R$=$$CH_3$; polyglycolide: R$=$H.

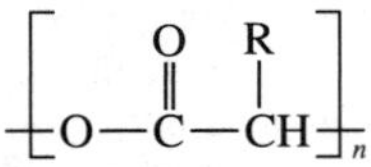

Polylactide and polyglycolide are the most widely used synthetic degradable biopolymers. They are popular since they have good mechanical properties and degrade to nontoxic metabolites (glycolic or lactic acid). Polylactide, polyglycolide, and copolymers of the two find clinical use in degradable sutures and orthopedic pins and screws. Recent research has focused on their use as a drug-delivery matrix since sustained release of drugs can be achieved as the materials degrade. Drug-delivery matrices include monoliths and microspheres. Microspheres are routinely prepared by dissolution of polymer and drug in chloroform (or dichloromethane), suspension in aqueous polyvinyl alcohol (to form an oil-in-water emulsion), and evaporation to form drug-entrapped microspheres or nanospheres. Primarily stirring speed and polymer-drug concentration in the oil phase (chloroform or dichloromethane solution) control sphere size.

Polylactide differs from polyglycolide in that R is a methyl group (CH_3) for polylactide and a hydrogen for polyglycolide. Polylactide and polyglycolide are usually synthesized from lactide and glycolide cyclic monomers using initiators such as stannous 2-ethyl hexanoate (stannous octoate). As with polypropopylene, the stereochemistry of the repeat unit has a large effect on the structure and properties of polylactide. Poly(DL-lactide) is atactic, meaning that it has no regular stereostructure and as a result is purely amorphos. Poly(D-lactide) and poly(L-lactide) are isotactic and consequntly are approximatley 35 percent crystalline. Poly(D-lactide) is seldom used commercially since D-lactic acid (degradation product of D-lactide) does not occur naturally in the human body, whereas L-lactic acid is a common metabolite. Poly(L-lactide) has a higher modulus and tensile strength than the amorphous poly (DL-lactide). Similarly, the crystalline poly(L-lactide) degrades completely in vivo in 20 months to 5 years, whereas poly(DL-lactide) degrades much faster, in 6 to 17 weeks.[32]

Copolymers of glycolide and lactide [poly(lactide-*co*-glycolide)] are amorphous and have similar mechanical properties and degradation rates as poly(DL-lactide). Pure polyglycolide is very strong and stiff yet has similar degradation as the poly(DL-lactides) and the lactide-glycolide copolymers. Polyglycolide is highly crystalline, with crystallinities between 35 and 70 percent.[32] Figures 11.13 and 11.14 show degradation rates for polyglycolide and poly(L-lactide).

Polylactide, polyglycolide, and poly(lactide-*co*-glycolide) are often called *polylactic acid, polglycolic acid*, and *poly(lactic-co-glycolic acid)* since their structures can be deduced by the direct condensation of lactic and glycolic acid. Although it is rare, synthesis of polylactic and glycolic acids can be achieved by direct condensation, but this results in a low-molecular-weight polymer (on the order of 2000 g/mol) with poor mechanical properties but increased degradation rates.

Polycaprolactone. Degradation: bioerosion.

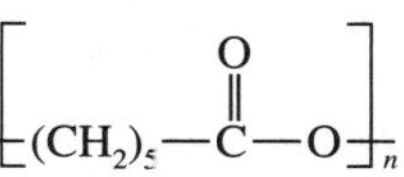

Polycaprolactone (PCL) is a biodegradable semicrystalline polyester that is synthesized from caprolactone using stannous octoate in a similar manner to polylactide or polyglycolide. PCL has a very low modulus of around 50 ksi since it has a low T_g of −60°C. PCL degrades very slowly, and therefore, it is usually not used as a homopolymer. Caprolactone, however, is copolymerized with glycolide to make a flexible suture material (trade name Monocryl).[33]

Poly(Alkylcyanoacrylates). Degradation: bioerosion.

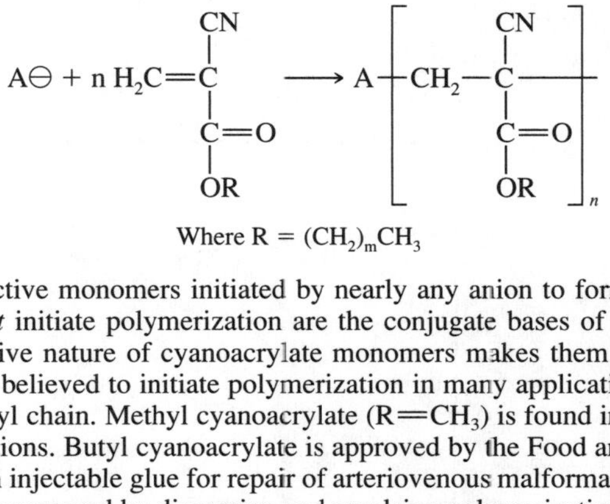

Where R = $(CH_2)_mCH_3$

Cyanoacrylates are reactive monomers initiated by nearly any anion to form a rigid polymer. The only anions that *cannot* initiate polymerization are the conjugate bases of strong acids (e.g., Cl^-, NO_3^-, SO_4^{2-}). The reactive nature of cyanoacrylate monomers makes them useful adhesives. OH^- from adsorbed water is believed to initiate polymerization in many applications. R in the preceding figure represents an alkyl chain. Methyl cyanoacrylate ($R{=}CH_3$) is found in commercial adhesives for nonmedical applications. Butyl cyanoacrylate is approved by the Food and Drug Administration (FDA) and is used as an injectable glue for repair of arteriovenous malformations. Microspheres and nanospheres can also be prepared by dispersion and emulsion polymerization and loaded with drugs for drug-delivery applications.

Degradation is slow at neutral or acidic conditions, but above pH 7, polycyanoacrylates degrade faster. Formaldehyde is one of the degradation products (especially for methyl cyanoacrylate); therefore, there is some question as to the safety of polycyanoacrylates.[34–36] Degradation rates increase with increasing alkyl chain length (R) since hydrophobicity increases with alkyl chain length. Degradation occurs at the polymer surface; therefore, surface degradation rates are highly surface area dependent. For example, poly (ethyl cyanoacrylate) microspheres[37] degrade completely in PBS (pH 7.4) in 4 to 20 hours. Depending on polymerization conditions, smaller-sized poly (methyl cyanoacrylate) nanospheres degrade completely in 20 minutes in PBS at pH 7.4 and 1 hour in fetal calf serum.[38] The longer alkyl chain poly (isobutyl cyanoacrylate) and poly (isohexyl cyanoacrylate) nanospheres take over 24 hours to degrade completely.[38] Larger poly (ethyl cyanoacrylate) particles,

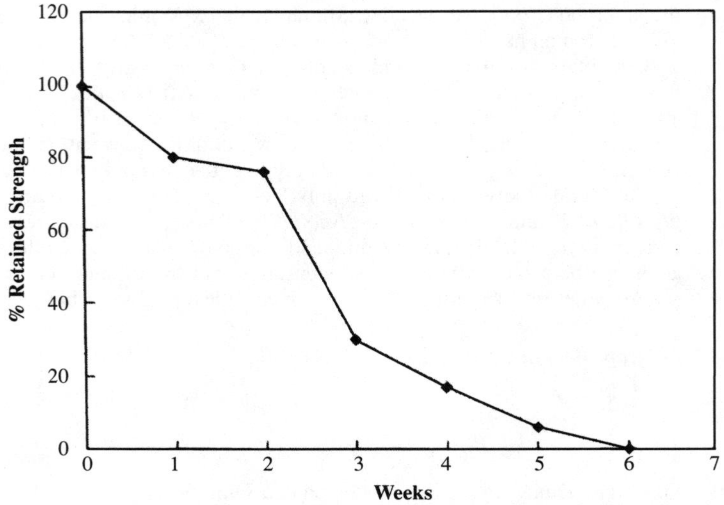

FIGURE 11.13 In vitro degradation of polyglycolide. Retained tensile strength versus time. *(Reproduced from D. E. Perrin and J. P. English, "Polyglycolide and polylactide," in Handbook of Biodegradable Polymers, A. J. Domb, J. K. Kost, and D. M. Wiseman (eds.). London: Harwood Academic Publishers, 1997, p. 12.)*

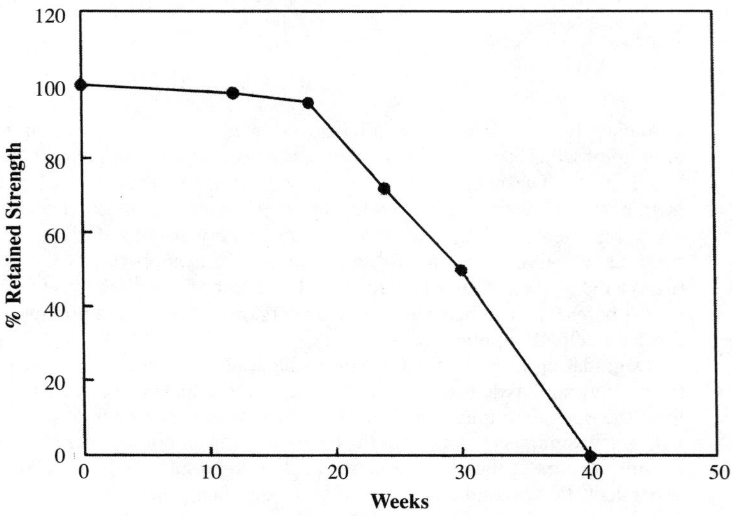

FIGURE 11.14 In vitro degradation of poly (L-lactide). Retained tensile strength versus time. *(Reproduced from D. E. Perrin and J. P. English, "Polyglycolide and polylactide," in Handbook of Biodegradable Polymers, A. J. Domb, J. K. Kost, and D. M. Wiseman (eds.). London: Harwood Academic Publishers, 1997, p. 12.)*

near 100 μm in size, take weeks to degrade completely at pH 7.4 due to their small surface area compared with microspheres and nanospheres.[35]

REFERENCES

1. *Encyclopedia of Polymer Science and Engineering*, vol. 17. New York: Wiley, 1985, pp. 198–257.
2. Chen, X. H., S. P. McCarthy, and R. A. Gross, "Synthesis and characterization of [L]-lactide-ethylene oxide multiblock copolymers," *Macromolecules*, **30**(15):4295–4301 (1997).
3. Cohn, D., and H. Younes, "Biodegradable Peo/Pla block copolymers," *Journal of Biomedical Materials Research*, **22**(11):993–1009 (1988).
4. Hu, D. S. G., and H. J. Liu, "Structural analysis and degradation behavior in polyethylene-glycol poly (L-lactide) copolymers," *Journal of Applied Polymer Science*, **51**(3):473–482 (1994).
5. Li, Y. X., and T. Kissel, "Synthesis and properties of biodegradable aba triblock copolymers consisting of poly(L-lactic acid) or poly(L-lactic-*co*-glycolic acid) *a*-blocks attached to central poly(oxyethylene) *b*-blocks," *Journal of Controlled Release*, **27**(3):247–257 (1993).
6. Yamaoka, T., et al., "Synthesis and properties of multiblock copolymers consisting of poly(L-lactic acid) and poly(oxypropylene-*co*-oxyethylene) prepared by direct polycondensation," *Journal of Polymer Science [A]*, **37**(10):1513–1521 (1999).
7. Razatos, A., et al., "Force measurements between bacteria and poly(ethylene glycol)-coated surfaces," *Langmuir*, **16**(24):9155–9158 (2000).
8. Vigo, T. L., and K. K. Leonas, "Antimicrobial activity of fabrics containing cross-linked polyethylene glycols," *Textile Chemist and Colorist & American Dyestuff Reporter*, **1**(1):42–46 (1999).
9. Park, K. D., et al., "Bacterial adhesion on PEG modified polyurethane surfaces," *Biomaterials*, **19**(7–9): 851–859 (1998).
10. Guo, Y. Q., and S. W. Hui, "Poly(ethylene glycol)-conjugated surfactants promote or inhibit aggregation of phospholipids," *Biochimica et Biophysica Acta*, **1323**(2):185–194 (1997).
11. Slepushkin, V. A., et al., "Sterically stabilized pH-sensitive liposomes: Intracellular delivery of aqueous contents and prolonged circulation in vivo," *Journal of Biological Chemistry*, **272**(4):2382–2388 (1997).
12. Woodle, M. C., M. S. Newman, and J. A. Cohen, "Sterically stabilized liposomes: Physical and biological properties," *Journal of Drug Targeting*, **2**(5):397–403 (1994).
13. Dunn, S. E., et al., "Polystyrene-poly(ethylene glycol) (PS-PEG 2000) particles as model systems for site-specific drug-delivery: 2. The effect of PEG surface-density on the in vitro cell-interaction and in vivo biodistribution," *Pharmaceutical Research*, **11**(7):1016–1022 (1994).
14. Verrecchia, T., et al., "Non-stealth (poly(lactic acid albumin)) and stealth (poly(lactic acid-polyethylene glycol)) nanoparticles as injectable drug carriers," *Journal of Controlled Release*, **36**(1–2):49–61 (1995).
15. Maruyama, K., et al., "Immunoliposomes bearing polyethyleneglycol-coupled Fab′ fragment show prolonged circulation time and high extravasation into targeted solid tumors in vivo," *FEBS Letters*, **413**(1): 177–180 (1997).
16. Vittaz, M., et al., "Effect of PEO surface density on long-circulating PLA-PEO nanoparticles which are very low complement activators," *Biomaterials*, **17**(16):1575–1581 (1996).
17. Kost, J., and R. Goldbart, "Natural and modified polysaccharides," in *Handbook of Biodegradable Polymers*, A. J. Domb, J. Kost, and M. W. Wiseman (eds.) London: Harwood Academic Publishers, 1997, pp. 285–286.
18. Kost, J., and R. Goldbart, "Natural and modified polysaccharides," in *Handbook of Biodegradable Polymers*, A. J. Domb, J. Kost, and M. W. Wiseman (eds.). London: Harwood Academic Publishers, 1997, pp. 282–284.
19. Qurashi, M. T., H. S. Blair, and S. J. Allen, "Studies on modified chitosan membranes; 1. Preparation and characterization," *Journal of Applied Polymer Science*, **46**(2):255–261 (1992).
20. Silver, F. H., and A. K. Garg, "Collagen: Characterization, processing, and medical applications," in *Handbook of Biodegradable Polymers*, A. J. Domb, J. Kost, and M. W. Wiseman (eds.) London: Harwood Academic Publishers, 1997, pp. 319–346.

21. Viswanadham, R. K., and E. J. Kramer, "Elastic properties of reconstituted collagen hollow fiber membranes," *Journal of Materials Science*, **11**(7):1254–1262 (1976).
22. Mathews, C., and K. Holde, *Biochemistry*. San Francisco: Benjamin Cummings, 1990, p. 578.
23. Chivers, R., "In vitro tissue welding using albumin solder: Bond strengths and bonding temperatures," *International Journal of Adhesion and Adhesives*, **20**:179–187 (2000).
24. Stillwell, R. L., et al., "Oxidized cellulose: Chemistry, processing and medical applications," in *Handbook of Biodegradable Polymers*, A. J. Domb, J. Kost, and M. W. Wiseman (eds.). London: Harwood Academic Publishers, 1997, pp. 291–306.
25. Ratner, B. D., et al. (eds.), *Biomaterials Science: An Introduction to Materials in Medicine*. New York: Academic Press, 1996, p. 60.
26. Ratner, B. D., et al. (eds.), *Biomaterials Science: An Introduction to Materials in Medicine*. New York: Academic Press, 1996, p. 59.
27. *Encyclopedia of Polymer Science and Engineering*, Vol. 11. New York: Wiley, 1985, pp. 315–381.
28. Ratner, B. D., et al. (eds.), *Biomaterials Science: An Introduction to Materials in Medicine*. New York: Academic Press, 1996, p. 247.
29. Zaikov, G. E. " Quantitative aspects of polymer degradation in the living body," *Journal of Macromolecular Science–Reviews in Macromolecular Chemistry and Physics*, **C25**(4):551–597 (1985).
30. Ratner, B. D. et al. (eds.), *Biomaterials Science: An Introduction to Materials in Medicine*. New York: Academic Press, 1996, p. 58.
31. *Encyclopedia of Polymer Science and Engineering*, Vol. 17. New York: Wiley, 1985, pp. 577–647.
32. Perrin, D. E., and J. P. English, "Polyglycolide and polylactide," in *Handbook of Biodegradable Polymers*, A. J. Domb, J. Kost, and M. W. Wiseman (eds.). London: Harwood Academic Publishers, 1997, pp. 2–27.
33. Perrin, D. E., and J. P. English, "Polycaprolactone," in *Handbook of Biodegradable Polymers*, A. J. Domb, J. Kost, and M. W. Wiseman (eds.). London: Harwood Academic Publishers, 1997, pp. 63–77.
34. Wade, C., and F. Leonard, "Degradation of poly(methyl 2-cyanoacrylates)," *Journal of Biomedical Materials Research*, **6**:215–220 (1992).
35. Vezin, W. R., and T. Florence, "In vitro heterogenious degradation of poly(*n*-alkyl α-cyanoacrylates)," *Journal of Biomedical Materials Research*, **14**:93–106 (1980).
36. Leonard, F., et al., "Synthesis and degradation of poly(alkyl α-cyanoacrylates)," *Journal of Applied Polymer Science*, **10**:259–272 (1966).
37. Tuncel, A., et al., "Monosize poly(ethylcyanoacrylate) microspheres: Preparation and degradation properties," *Journal of Biomedical Materials Research*, **29**:721–728 (1995).
38. Muller, R. H., et al., "In vitro model for the degradation of alkylcyanoacrylate nanoparticles," *Biomaterials*, **11**:590–595 (1990).

CHAPTER 12
BIOMEDICAL COMPOSITES

Arif Iftekhar
University of Minnesota, Minneapolis

12.1 INTRODUCTION

A common debate about the definition of composite materials among composite engineers and materials scientists continues to this day. More recently, biomedical engineers have used the term *composite* prolifically for newly developed biomaterials, and it might be argued that not every usage of the term *composite* for a biomaterial would satisfy the traditional composite engineer, who is used to thinking in terms of fibers, matrices, and laminates. That said, defining composites a certain way in this chapter is not meant to preclude its use outside this definition.

The difficulty lies, on the one hand, in the depth of the material to which the definition refers. Practically everything is a composite material in some sense, except for pure elements. For example, a common piece of metal is a composite (polycrystal) of many grains (or single crystals). Thus alloys, ceramics, steels, etc., would be considered composites if the definition refers to the microstructure. However, if it is the macrostructure that concerns us, then we get the traditional treatment of composites as a *materials system* of different macroconstituents.[1]

On the other hand, there is also a question in this definition regarding how these macroconstituents are brought together and for what purpose. For instance, thin coatings on a material do not make it a typical composite, and the same could be said about adding resin-extending fillers to plastics, although the constituents exist at the macrostructure. Furthermore, a structure that is assembled with components made of different materials does not qualify it to be a composite. Thus a pacemaker lead that has a metallic core and a polymeric sheath would not be considered a composite in the strict sense, whereas a catheter tube polymer reinforced with embedded braided metal wires would. In addition, foams and porous coatings on materials will not be considered composites in this discussion.

The following is an operational definition for the purpose of this chapter:

A *composite material* consists of two or more physically and/or chemically distinct, suitably arranged or distributed materials with an interface separating them. It has characteristics that are not depicted by any of the components in isolation, these specific characteristics being the purpose of combining the materials.

Composite materials have a bulk phase, which is continuous, called the *matrix*, and one or more dispersed, noncontinuous phases, called the *reinforcement*, which usually has superior mechanical or thermal properties to the matrix. The region between the two can be simply a surface, called an *interface*, or a third phase, called an *interphase*. An example of the latter is a layer of coupling agent coated on glass fibers that facilitates adhesion of the glass to the matrix polymer.

The essence of the concept of composites is this: the bulk phase accepts the load over a large surface area and transfers it to the reinforcement phase, which, being different, changes the mechanical properties of the composite suitably, whether it is strength, stiffness, toughness, or fatigue resistance. For instance, in structural polymer composites, the reinforcement is much stiffer than the matrix, making the composite several times stiffer than the bulk polymer and resulting in a reduction in bulk strain on deformation, as seen in Fig. 12.1. The significance here lies in the fact that there are numerous matrix materials and as many reinforcement types that can be combined in countless ways to produce just the desired properties.

The concept of composite materials is ancient: to combine different materials to produce a new material with performance and efficiency unattainable by the individual constituents. An example is adding straw to mud for building stronger mud walls. Some more recent examples, but before engineered materials became prominent, are steel rods in concrete, cement and asphalt mixed with sand, fiberglass in resin, etc. In nature, examples abound: a palm leaf, cellulose fibers in a lignin matrix (wood), collagen fibers in an apatite matrix (bone), etc.

Most research in engineered composite materials has been done since the mid-1960s. Today, given the most efficient design of, say, an aerospace structure, a boat, or a motor, we can make a composite material that meets or exceeds the performance requirements. The benefits are mostly in weight and cost, measured in terms of ratios such as stiffness/weight, strength/weight, etc. Advances in biomedical composites have been focused on the design of dental and orthopedic implants, which are mainly structural applications. However, tremendous stiffness and strength improvement is not always the concern in the design of biomedical composites and even less so for large weight savings. Other concerns, such as biocompatibility, precise property matching, mimicking natural structures, etc., can become more important, yet these too are areas where composites offer much promise in device design. Engineers and materials scientists who are used to working with traditional materials such as metal alloys, ceramics, and plastics are increasingly challenged to design with composites that have different physical characteristics, mechanical behaviors, and processing methods.

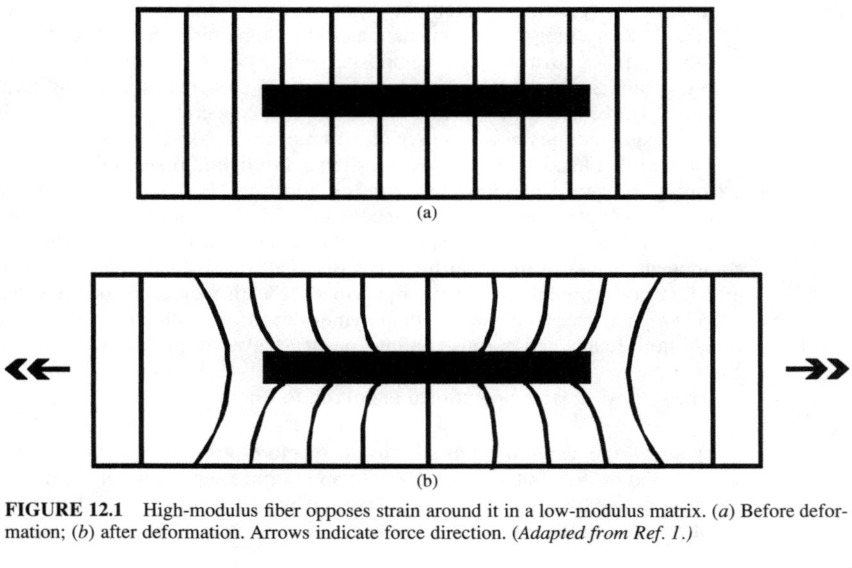

FIGURE 12.1 High-modulus fiber opposes strain around it in a low-modulus matrix. (*a*) Before deformation; (*b*) after deformation. Arrows indicate force direction. (*Adapted from Ref. 1.*)

In the design of composite problem solutions, particularly load-bearing implants, it is important to base the design on a firm theoretical understanding of composites to avoid creating new problems while trying to solve an existing one. Three factors need to be considered rationally: material selection for the bulk and reinforcement phases, internal and external structure of the device, and a suitable processing method.[2]

12.2 CLASSIFICATION

The factors that most contribute to the engineering performance of the composite include[1]

1. Materials that make up the individual components
2. Quantity, form, and arrangement of the components
3. Interaction between the components

Of these, the reinforcement system in a composite material strongly determines the properties achievable in a composite. It is thus convenient and common to classify composites according to the characteristics of the reinforcement. These can include the shape, size, orientation, composition, distribution, and manner of incorporation of the reinforcement. For the purposes of a discussion of biomedical composites, this results in two broad groups, namely, fiber-reinforced and particle-reinforced composites. Figure 12.2 shows further divisions within these groups.

Composite materials can also be broadly classified based simply on the matrix material used. This is often done more for processing than for performance purposes. Thus there are polymer-matrix composites (PMCs), ceramic-matrix composites (CMCs), or metal-matrix composites (MMCs). The last type is an advanced composite uncommon in biomedical applications and is mostly used for high-temperature applications.

12.3 CONSTITUENTS

12.3.1 Matrices

The matrix in a composite is the continuous bulk phase that envelopes the reinforcement phase either completely or partially. It serves several important functions.[1] It holds the fibers or particles in place, and in oriented composites, it maintains the preferred direction of fibers. The matrix transfers the applied load to the reinforcement and redistributes the stress. When used with brittle fibers, the matrix helps increase fracture toughness because it is typically of a lower stiffness material and can tolerate greater elongation and shear forces than the reinforcement. The matrix also determines the environmental durability of the composite by resisting chemical, hygroscopic, and thermal stresses and protecting the reinforcement from these stresses. The matrix also greatly influences the processing characteristics of a composite.

Common matrices in biomedical composites are listed in Table 12.1. In nonmedical applications, thermosets make up the bulk of the matrix materials, particularly in structural and aerospace applications, where high stiffness and temperature resistance are very important requirements. In most medical applications, however, thermoplastics are the matrix materials of choice due to their nonreactive nature, processing flexibility, and generally greater toughness. Also, thermosets are nondegradable, whereas some thermoplastics can be designed to be biodegradable. Instead of using off-the-shelf materials, new matrices are constantly being developed for medical applications that have designed reactivity, flexibility, and strength. Resorbable matrices are useful when a composite is not permanently needed once implanted, but it is challenging to design a stiff reinforcing material that

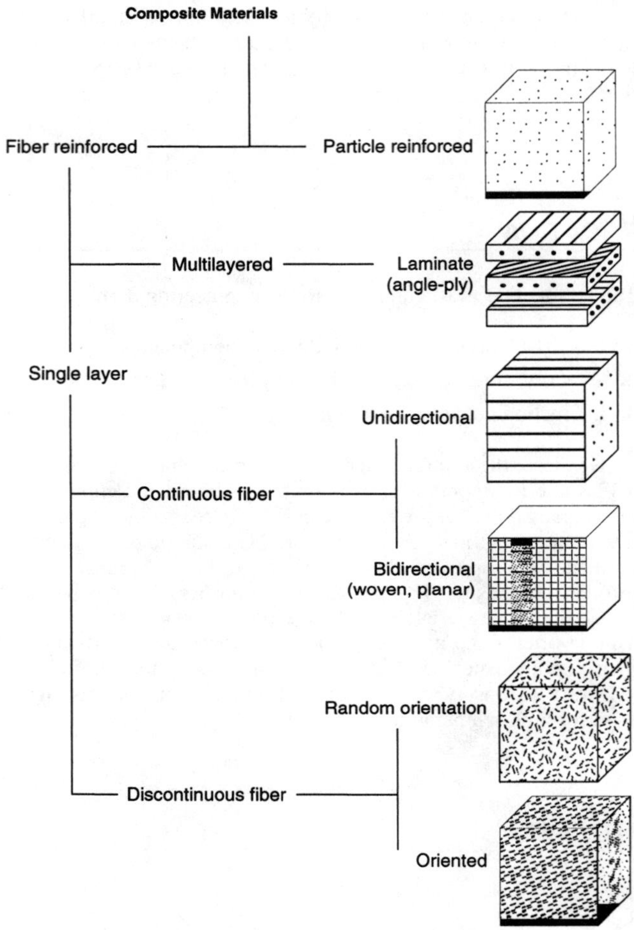

FIGURE 12.2 Classification of composite materials.

has a comparable degradation rate to such matrices. Ceramic matrices are used for their compressive properties and bioactive possibilities but suffer from poor fracture toughness.

12.3.2 Fibers

A great majority of materials is stronger and stiffer in the fibrous form than in any other form. This explains the emphasis on using fibers in composite materials design, particularly in structural applications, where they are the principal load-carrying component. Fibers have a very high aspect ratio of length to diameter compared with particles and whiskers, and the smaller the diameter, the greater is the strength of the fiber due to a reduction in surface flaws. Many properties of a composite are determined by the length, orientation, and volume fraction of fibers of a given type.

Fibers are often manufactured as continuous filaments, with diameters in the range of 5 to 50 μm, and then they are arranged to produce tows, yarns, strands, rovings, mats, etc. These are

TABLE 12.1 Constituents of Biomedical Composites

Matrix	Fibers	Particles
Thermosets	*Polymers*	*Inorganic*
Epoxy	Aromatic polyamides (aramids)	Glass
Polyacrylates	UHMWPE	Alumina
Polymethacrylates	Polyesters	*Organic*
Polyesters	Polyolefins	Polyacrylate
Silicones	PTFE	Polymethacrylate
Thermoplastics	*Resorbable polymers*	
Polyolefins (PP, PE)	Polylactide, and its copolymers with polyglyocolide	
UHMWPE	Collagen	
Polycarbonate	Silk	
Polysulfones	*Inorganic*	
Poly(ether ketones)	Carbon	
Polyesters	Glass	
Inorganic	Hydroxyapatite	
Hydroxyapatite	Tricalcium phosphate	
Glass ceramics		
Calcium carbonate ceramics		
Calcium phosphate ceramics		
Carbon		
Steel		
Titanium		
Resorbable polymers		
Polylactide, polyglycolide and their copolymers		
Polydioxanone		
Poly(hydroxy butyrate)		
Alginate		
Chitosan		
Collagen		

used to fabricate continuous-fiber composites, often for large structural applications. Filaments can be chopped to form short fibers ranging in length from 3 to 50 mm that are used to make discontinuous or short-fiber composites, more commonly for low-cost applications or small intricate parts. Whiskers are fibers made of single crystals with very small diameters around 10 μm, but their aspect ratio is high (>100). They have very high strengths but also high manufacturing cost.[3] Compared with continuous-fiber composites, short-fiber composites are less efficient in the use of fibers and in achieving a desired orientation, but they are also less limited in design and processing possibilities and can come very close to achieving their theoretical strength.[1]

Both continuous and short fibers can be oriented in one, two, or three dimensions, resulting in unidirectional, planar, and random reinforcement systems. The volume fraction of fibers oriented in a given direction strongly affects the physical properties of a composite in that direction. Unidirectional and planar reinforced composites exhibit anisotropy; i.e., their properties vary depending on the axis of measurement. The third composite type is isotropic, having equal properties in all directions. As mentioned earlier, it is difficult to orient short fibers, particularly in mold-filling processes, and the resulting composites tend to be isotropic. Laminate composites are a type of fiber-reinforced composite consisting of anisotropic layers or plies bonded together that can differ in relative fiber orientation and volume fraction. This allows high-fiber-volume fractions and three-dimensional orientation not achievable in isotropic short-fiber composites.

Table 12.1 lists some fibers common in biomedical composites. There are many naturally occurring fibers, such as cotton, flax, collagen, jute, wood, hemp, hair, wool, silk, etc., but these have extremely varying properties and present many processing challenges. Among these, collagen fibers have been successfully utilized in tissue engineering of skin and ligament. Borosilicate glass fiber is ubiquitous in the composites industry but not common in biomedical composites, where, instead, adsorbable bioglass fibers made from calcium phosphate have found some applications. Carbon fiber is as strong as glass fiber but is several times stiffer owing to its fine structure of axially aligned graphite crystallites and is also lighter than glass. It is used extensively to make high-strength lightweight composites in prosthetic structural components, where the fatigue resistance of carbon-fiber composites is also an advantage.[4] Carbon fibers tend to be brittle and are anisotropic, particularly in their thermal properties. They also add electrical conductivity to a composite, which can have corrosive effects next to metallic implants. Among polymers, highly oriented aramid fibers such as Kevlar are used in orthopedic applications because of their high resistance to impact fracture. However, Kevlar has very poor compressive properties, making it unsuitable for bending applications, and it is difficult to process due to its strong cut-through resistance. Teflon and polyester (Dacron) fibers are used to make vascular prostheses that are flexible. Polylactide and polyglycolide and their copolymers are used to make fiber composites in which adsorbability is more important than mechanical properties.

12.3.3 Particles

Particles can be added to a matrix to improve mechanical properties such as toughness and hardness. Other properties, such as dimensional stability, electrical insulation, and thermal conductivity, can also be controlled effectively by particles, especially when added to polymer matrices. Particulate reinforcement is randomly distributed in a matrix, resulting in isotropic composites. Particles can either strengthen or weaken a matrix depending on its shape, stiffness, and bonding strength with the matrix. Spherical particles are less effective than platelet- or flakelike particles in adding stiffness. Hard particles in a low-modulus polymer increase stiffness, whereas compliant particles such as silicone rubber, when added to a stiff polymer matrix, result in a softer composite. Fillers are nonreinforcing particles such as carbon black and glass microspheres that are added more for economic and not performance purposes.

Particulate reinforcement in biomedical composites is used widely for ceramic matrices in dental and bone-analogue applications. The most common such particle form is hydroxyapatite, a natural component of bone where it exists in a composite structure with collagen. Hydroxyapatite particles have very poor mechanical properties and may serve more as a bioactive than reinforcement component.

12.3.4 Interface

The transfer and distribution of stresses from the matrix to the fibers or particles occur through the interface separating them. The area at the interface and the strength of the interfacial bond greatly affect the final composite properties and long-term property retention.[1] A low interfacial area denotes poor wetting of the fiber with the matrix material. Wetting can be enhanced by processing methods in which there is greater pressure (metal matrices) or lower-viscosity flow (polymer matrices). When mechanical coupling is not sufficient, coupling agents are often used to coat fibers to improve chemical compatibility with the matrix.

Interfacial shear strength determines the fiber-matrix debonding process and thus the sequence and relative magnitude of the different failure mechanisms in a composite. Strong interfaces common in polymer matrix composites make ductile matrices very stiff but also lower the fracture toughness. Weak interfaces in ceramic matrix composites make brittle matrices tough by promoting matrix crack but also lower strength and stiffness.[5]

12.4 PROCESSING

12.4.1 Polymer-Matrix Composites

Continuous-fiber composites can be made in a variety of ways, such as manual layup of laminates, filament winding, pultrusion, and resin transfer molding. Manual layup involves stacking preimpregnated (or prepreg) tapes and sheets of parallel-fiber filaments held together by thermoplastic resin or partially cured thermoset resin, followed by autoclaving. This is not very expensive and not suitable for small medical implants. Like layup, filament winding allows for high-fiber-volume fraction and control of properties but is limited to tubular shapes, and the fibers cannot be oriented along the axis of the component. Pultrusion is ideal for very stiff composite rods and beams and can be used for making orthodontic archwires. Resin transfer molding allows very complex shapes and short cycle times but requires expert design of preforms and molds. Compression molding is another method suitable for both thermosets and thermoplastics.

Short-fiber thermoplastic composites are typically injection molded, which allows rapid, high-volume, and economical production but requires expensive tooling. The fiber length and volume fraction are limited by this method, and fiber orientation and distribution are difficult to control.

12.4.2 Ceramic-Matrix Composites

Ceramic-matrix composites are manufactured commonly by either pressing methods or infiltration methods. In the former, the reinforcement is mixed with a powder of the matrix, which is densified by hot pressing or hot isotactic pressing (HIP). Near-zero porosity can be achieved, but the simultaneous high pressure and temperature can degrade fibers or result in a strong interfacial bond that may reduce fracture toughness. In infiltration methods, a sintered fibrous or particulate preform is filled by the matrix by such methods as chemical vapor deposition, glass melt infiltration, and preceramic polymer or sol infiltration. Alumina-glass dental composites are made by this method. This method has the advantage of complex shape capability, low pressure, and flexible fiber architecture. Disadvantages are high matrix porosity (>10 percent) and long fabrication cycles. Particulate reinforcement can also be used through tape-casting and slip-casting the initial preform.

12.5 PHYSICAL PROPERTIES

Composite materials can be designed to have a wide range of physical and biochemical properties, making it important to develop predictive models to aid in designing with the many complex variables that face the composites engineer. Although these models tend be complicated due to the microstructure of composites, it is important to resist the temptation to treat a composite as a black box with gross properties because this macroscopic approach does not have the predictive power crucial for failure analysis. There have been three general approaches to predicting the basic mechanical properties of a composite[3,5]:

1. Mechanics of materials models
2. Theory of elasticity models
3. Semiempirical models

12.5.1 Mechanics of Materials Models

The mechanics of materials model uses simple analytical equations to arrive at effective properties of a composite, using simplifying assumptions about the stress and strain distribution in a representative volume element of the composite. This approach results in the common rule of mixtures equations for composites, where properties are relative to the volume fraction of the fibers and matrix.

Physical properties such as density are easily calculated by the following equations:

$$V_f + V_m + V_v = 1 \tag{12.1}$$

$$\rho_c = \rho_f V_f + \rho_m V_m \tag{12.2}$$

where V_f, V_m, and V_v are the volume fractions of the fiber, matrix, and voids, respectively, and similarly, ρ_c, ρ_f, and ρ_m are densities of the composite, fiber, and matrix.

The rule of mixtures is useful in roughly estimating upper and lower bounds of mechanical properties of an oriented fibrous composite, where the matrix is istropic and the fiber orthotropic, with coordinate 1 the principal fiber direction and coordinate 2 transverse to it. For the upper bound, the Voight model is used (Fig. 12.3), where it is assumed that the strain is the same in the fiber and matrix. For the lower bound, the Reuss model is used, where the stress is assumed to be the same. This gives the following equations for composite moduli:

$$E_{1c} = E_{1f} V_f + E_m V_m \tag{12.3}$$

$$\frac{1}{E_{2c}} = \frac{V_f}{E_{2f}} + \frac{V_m}{E_m} \tag{12.4}$$

$$\frac{1}{G_{12c}} = \frac{V_f}{G_{12f}} + \frac{V_m}{G_m} \tag{12.5}$$

where E and G are the Young's modulus and shear modulus, respectively. The equations for transverse modulus and shear modulus are known as the inverse law of mixtures. Some fibers, such as carbon, have different properties along their longitudinal and transverse axes that the preceding equations can take into account.

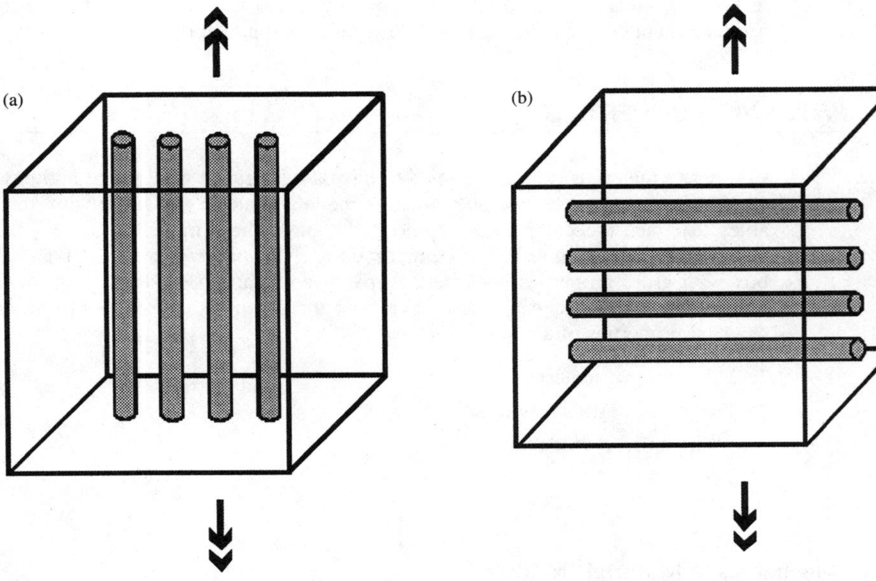

FIGURE 12.3 Composite stress models: (*a*) Voight or isostrain; (*b*) Reuss or isostress. Arrows indicate tension force direction.

The rule of mixtures equations have several drawbacks. The isostrain assumption in the Voight model implies strain compatibility between the phases, which is very unlikely because of different Poisson's contractions of the phases. The isostress assumption in the Reuss model is also unrealistic since the fibers cannot be treated as a sheet. Despite this, these equations are often adequate to predict experimental results in unidirectional composites. A basic limitation of the rule of mixtures occurs when the matrix material yields, and the stress becomes constant in the matrix while continuing to increase in the fiber.

The ultimate tensile strength of a fibrous composite σ_c^* depends on whether failure is fiber-dominated or matrix-dominated. The latter is common when V_f is small. One result of such treatment is

$$\sigma_c^* = \frac{\sigma_m^*}{E_m} E_{1f} V_f + \sigma_m^* V_m \tag{12.6}$$

where σ_m^* is the fracture strength of the matrix.

Other results from this simple analytical approach for orthotropic composites are

$$\alpha_{1c} = \frac{E_{1f}\alpha_{1f}V_f + E_m\alpha_m V_m}{E_{1f}V_f + E_m V_m} \tag{12.7}$$

$$C_c = \frac{1}{\rho_c}(\rho_f C_f V_f + \rho_m C_m V_m) \tag{12.8}$$

$$K_{1c} = K_f V_f + K_m V_m \tag{12.9}$$

where α is the coefficient of thermal expansion, C is the specific heat, and K is the thermal conductivity. The coefficient of hygroscopic expansion β can be found by substituting α with β above. These results are for the longitudinal directions only.

12.5.2 Theory of Elasticity Models

In this approach, no assumptions are made about the stress and strain distributions per unit volume. The specific fiber-packing geometry is taken into account, as is the difference in Poisson's ratio between the fiber and matrix phases. The equations of elasticity are to be satisfied at every point in the composite, and numerical solutions generally are required for the complex geometries of the representative volume elements. Such a treatment provides for tighter upper and lower bounds on the elastic properties than estimated by the rule of mixtures, as is described in the references used in this section.

One illustrative result for the longitudinal Young's modulus is

$$E_{1c} = \frac{[2(\nu_f - \nu_m)^2 E_f V_f E_m V_m]}{\{E_m V_m(1 - \nu_f - 2\nu_f^2) + E_f[V_f(1 - \nu_m - 2\nu_m^2) + (1 - \nu_m)]\}} + E_m + (E_f - E_m)V_f \tag{12.10}$$

where ν_f and ν_m are the Poisson's ratios of the fiber and matrix, respectively. Very small differences in the Poisson's ratios of the phases cause such equations to be simplified to the rule of mixtures.

12.5.3 Semiempirical Models

Curve-fitting parameters are used in semiempirical and generalized equations to predict experimental results. The most common model was developed by Halpin and Tsai, and it has been modified for aligned discontinuous fiber composites to produce such results as the following for the longitudinal modulus:

$$\frac{E_{1c}}{E_m} = \frac{1 + \xi\eta V_f}{1 - \eta V_f} \tag{12.11}$$

where
$$\eta = \frac{E_{1f}/E_m - 1}{E_{1f}/E_m + \eta} \tag{12.12}$$

and the Halpin-Tsai curve-fitting parameter is assumed to be $\xi = 2L/d$, where L is the length and d the diameter of the fiber. The expression can be simply substituted to also obtain E_{2c} and G_{12} using experimental values for ξ.

For two-dimensional randomly oriented fibers in a composite, approximating theory of elasticity equations with experimental results yielded this equation for the planar isotropic composite stiffness and shear modulus in terms of the longitudinal and transverse moduli of an identical but aligned composite system with fibers of the same aspect ratio:

$$E_c = \tfrac{3}{8}E_1 + \tfrac{5}{8}E_2 \tag{12.13}$$

$$G_c = \tfrac{1}{8}E_1 + \tfrac{1}{4}E_2 \tag{12.14}$$

For a three-dimensional random orientation of fibers, a slightly different equation is proposed for the isotropic tensile modulus:

$$E_c = \tfrac{1}{5}E_1 + \tfrac{4}{5}E_2 \tag{12.15}$$

The stiffness of particulate composites can be predicted depending on the shape of the particles.[6] For a dilute concentration of rigid spherical particles, the composite stiffness is approximated by

$$E_c = \frac{5(E_p - E_m)V_p}{3 + 2E_p/E_m} + E_m \tag{12.16}$$

where E_p and V_p are the stiffness and volume fraction of particles, respectively.

It is important to note that any model for composite behavior requires experimental validation and may prove to be quite inaccurate for not taking into account many irregularities typical in composite design and processing. In addition, these results are usually valid for static and short-term loading.

12.6 FRACTURE AND FATIGUE FAILURE

Failure of fiber-reinforced composites is generally preceded by an accumulation of different types of internal damage that slowly or catastrophically renders a composite structure unsafe. The damage can be process-induced, such as nonuniform curing control of a dental resin, or it can be service-induced, such as undesired water absorption. Fiber breaking, fiber bridging, fiber pullout, matrix cracking, and interface debonding are the failure mechanisms common in all composites, but the sequence and interaction of these mechanisms depend on the type of loading and the properties of the constituents, as well as on the interfacial shear strength (Fig. 12.4). Various fracture mechanics theories are available for failure analysis of composites, among them the maximum stress theory and the maximum strain theory described in more detail in Ref. 5.

Energy absorption and crack deflection during fracture lead to increased toughness of the composite. The two most important energy-absorbing failure mechanisms in a fiber-reinforced composite are debonding at the fiber-matrix interface and fiber pullout. If the interface bonds relatively easily, the crack propagation is interrupted by the debonding process, and instead of moving through the fiber, the crack is deflected along the fiber surface, allowing the fiber to carry higher loads. Fiber

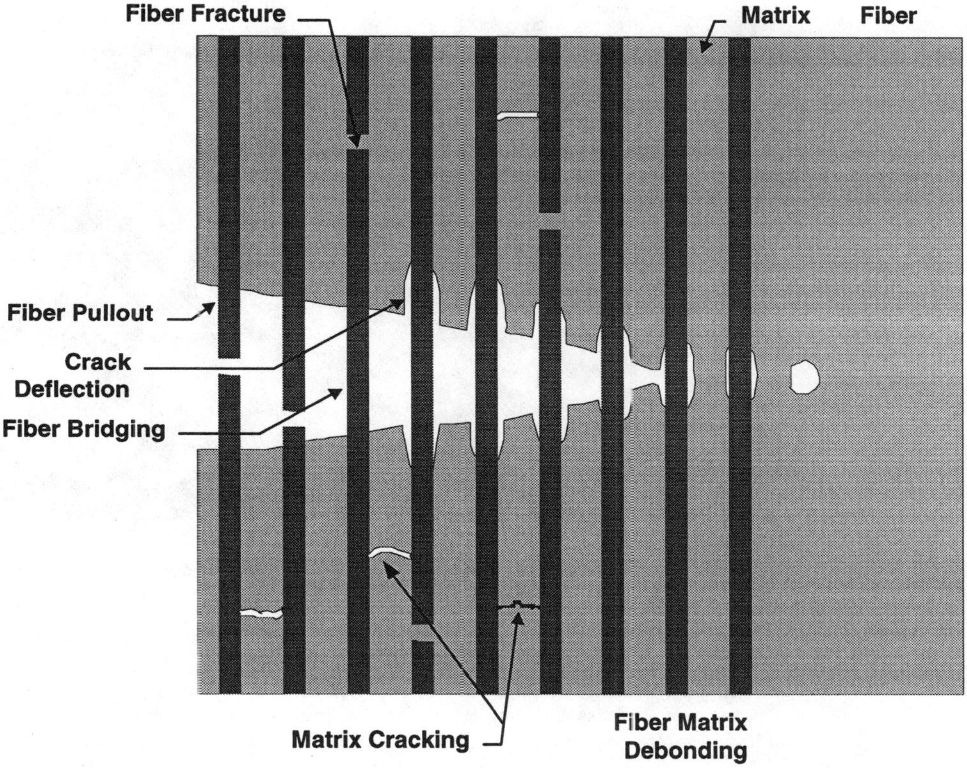

FIGURE 12.4 Failure mechanisms in a unidirectional fiber composite. (*Adapted from Ref. 3.*)

pullout occurs because fibers do not all break at the crack plane but at random locations away from this plane. If the pullout occurs against high frictional forces or shear stresses at the interface, there maybe a significant increase in fracture toughness.[3] The same holds true for particulate reinforcements, although here crack deflection is more common than bridging and pullout due to the smaller aspect ratio. In an alumina-glass composite, as shown in Fig. 12.5, indentation cracks in the glass matrix are deflected around the angular alumina granules in (*a*) but not as much in (*b*), where the cracks propagate more through the granules, indicating a stronger interface but more likely lower toughness. In laminate composites, delamination loads between the layers from in-plane shear loads are a common failure mechanism, initiating at the free edge of the plate or a hole, e.g., where screws would go in a carbon-reinforced bone plate.

Long-term durability of composites is a challenging and still developing area of study. Unlike many metallic biomaterials, the static strength of most load-bearing composite materials does not correlate well with their long-term performance, especially under cyclic loading. The reasons for this include fatigue damage, matrix creep, and stress relaxation, as well as environmental effects at the implant site.[2] Fatigue damage and implant life are difficult to predict, unlike for metals, which have a distinct endurance stress limit below which the material can be loaded an infinite number of times without failure. Many composites do not exhibit an endurance limit. In addition, the loads on implants are highly variable both in direction and magnitude and between patients, and the most damaging loads may occur randomly as a result of accidents. Water absorption in polymer matrices can cause swelling and have a plasticizing effect, which is problematic in dental composites and

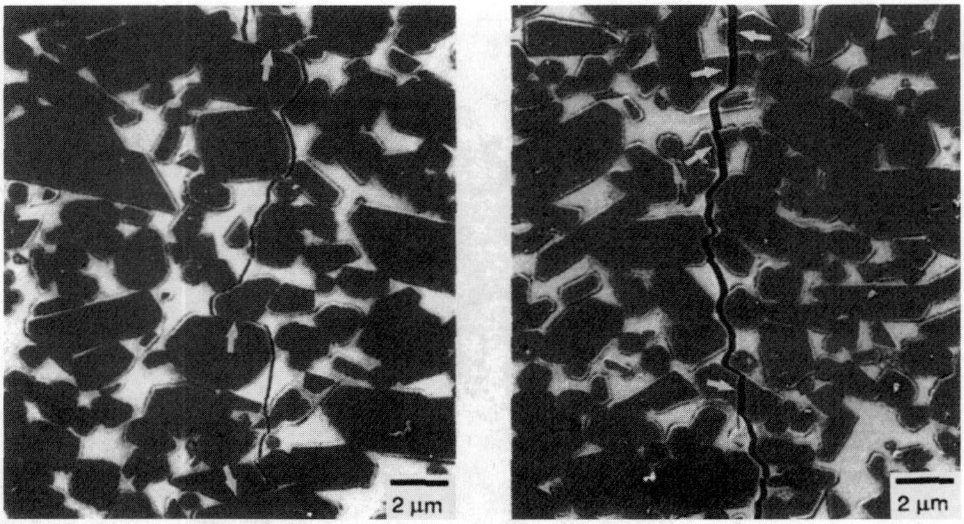

(a) (b)

FIGURE 12.5 Surface traces of indentation cracks in alumina-glass dental composite by backscattered scanning electron microscopy (SEM). Grain bridging sites are near the arrows in (*a*) and transgranular fracture sites in (*b*). (*From Ref. 21.*)

bone cements, although this is a very slow phenomenon. Visocelastic effects such as the strain-rate dependence of stiffness and long-term creep are proportional to the column fraction of matrix such that highly reinforced continuous-fiber composites are less prone than short-fiber composites.

12.7 BIOLOGIC RESPONSE

In designing biomedical composites and predicting their performance, several issues must be considered regarding the biological response. As the number of constituent materials in a composite increases, so can the variations in the host response. Additional tests are necessary to establish that while the individual materials may be by themselves biocompatible, their specific composition, arrangement, and interaction are also biocompatible. This has implications for both the flexibility of design and obtaining regulatory approval. The potential of composite design to obtain the desired set of properties can be restricted by being conservative in the choice and number of materials used. Even if all the materials used may be approved by the Food and Drug Administration (FDA), their particular combination in a composite may require additional approval.

Materials can elicit a different host response in the bulk form than in the fibrous or particulate form. For instance, UHMWPE, as in an acetabular cup of a hip prosthesis, is generally biocompatible, whereas its fibrous form, as in a finely woven fabric, has been shown to produce a different, more adverse reaction. Furthermore, when the discontinuous phase is particles, whiskers, platelets, or microspheres with dimensions on a cellular scale, the inflammatory response can include their ingestion by immune cells and transport to other parts of the body. This can be accompanied by the release of enzymes that can adversely affect the performance of the composite, such as by altering the degradation kinetics of a biodegradable composite.[7] The composite can be designed in such a way that the fibers or particles are not exposed to the host, but this is challenging because it involves elimination of all voids at the fiber-matrix or particle-matrix interface during processing. In addition, friction in a moving part, such as in orthopedic or dental composites, can cause abrasion of the matrix and produce new voids at the interface, exposing the reinforcing material to the host.

The interaction of materials at the interface is integral to composite performance, and this can be affected by the tissue response in various ways, such as filling and swelling of interfacial voids with fluid and fibrous tissue, altering the interfacial adhesion strength between matrix and reinforcement, and delamination of laminate composites. Such effects can lead to failure of a composite, particularly in structural applications.

Thermosetting polymers, although uncommon in biomedical implants, may contain unreacted monomer and cross-linking agents, particularly in laminated composites made from prepreg layers. In both thermosetting and thermoplastic polymer composites, the sizing applied to glass and carbon fibers is another compound that may be present, and some residual solvents may also leach from the matrix if they are not completely removed during processing. These trace amounts may not be an issue if the application is external to the body, as in prosthetic limbs.

12.8 BIOMEDICAL APPLICATIONS

The use of composite materials in biomedical implants and devices is illustrated by the following examples of structural applications.

12.8.1 Orthopedic

Composite materials have found wide use in orthopedic applications, as summarized by Evans,[2] particularly in bone fixation plates, hip joint replacement, bone cement, and bone grafts. In total hip replacement, common materials for the femoral-stem component such as 316L stainless steel, Co-Cr alloys, and Ti-6A1-4V titanium alloy have very high stiffness compared with the bone they replace. Cortical bone has a stiffness of 15 GPa and tensile strength of 90 MPa.[8] Corresponding values for titanium are 110 GPa and 800 MPa, which are clearly very high. This produces adverse bone remodeling and stress shielding, which over the long term leads to reduction in bone mass and implant loosening, specially in the proximal region. Fiber composites can be tailored to match the specific mechanical properties of the adjacent bone. Carbon-fiber composites in PEEK or polysulfone matrices can be fabricated with stiffness in the range 1 to 170 GPa and tensile strength from 70 to 900 MPa.[9] Examples are press-fit femoral stems made from laminated unidirectional carbon fibers in PEEK, polysulfone, liquid crystalline polymer (LCP),[10] and polyetherimide (PEI).[11] These composites are difficult to fabricate and have not had very encouraging durability, but they continue to be developed for the inherent advantages of tailorability, flexibility, noncorrosiveness, and radiolucency.[12] Problems with biocompatibility due to particulate carbon debris from these composites have been addressed by polishing and coating with hydroxyapatite (Fig. 12.6) or carbon-titanium alloy.[13]

For fracture fixation, a fully resorbable bone plate is desirable to avoid the need for a second operation to remove the implant after healing. This in the form of a tailored low-stiffness composite also avoids the problem of stress shielding described earlier. The rate of degradation must be controlled to maintain the mechanical properties such that strength loss in the implant mirrors strength increase in the healing. In addition, the degradation by-products must be nontoxic. A summary of the design of adsorbable fixation devices is provided by Pietrzak.[14] Examples of composite bone plates include laminated continuous carbon fiber in a polylactide (PLA) matrix, which is partially adsorbable, and calcium-phosphate glass fibers also in PLA, which is fully resorbable.[15] Continuous poly (L-lactide) fibers in a PLA matrix also produced a fully resorbable composite.[16] These composites, however, did not have adequate mechanical properties and degraded quite rapidly. Nonresorbable carbon-epoxy bone plates with sufficient strength and fatigue properties are available from such manufacturers as Orthodesign, Ltd.

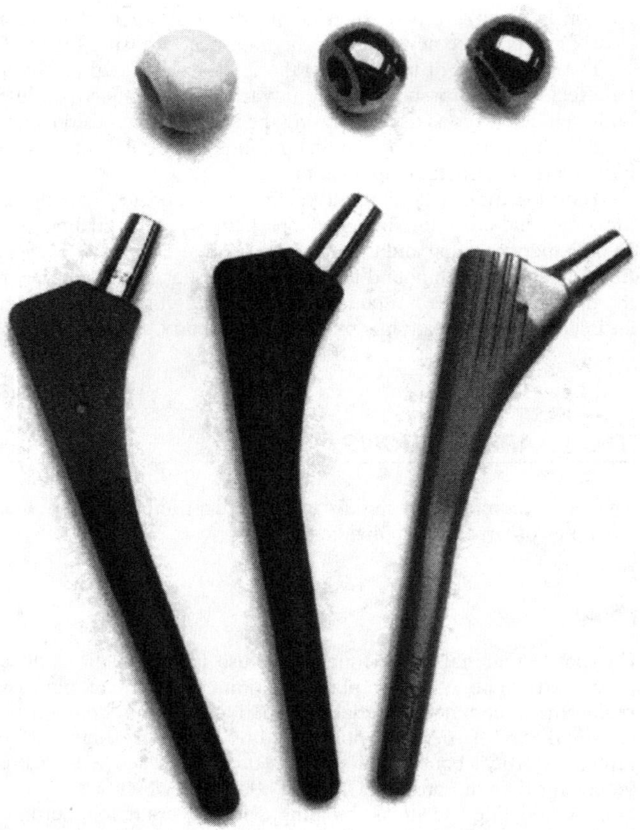

FIGURE 12.6 Carbon-fiber-reinforced thermoplastic hip prosthesis, with and without hydroxyapatite coating, shown alongside conventional titanium equivalent. (*Courtesy of Orthodesign, Ltd., Christchurch, U.K.*)

Bone cements used to fill the void and improve adhesion between implants and the host bone tissue have been reinforced with various fibers to prevent loosening and enhance shear strength. The typical bone cement is PMMA powder mixed with a methacrylate-type monomer that is polymerized during fixation. Low volume fractions of graphite, carbon, and Kevlar fibers have been added to PMMA matrices to increase fatigue life and reduce creep deformation.[17]

12.8.2 Dental

Composites have been by far the most successful in dental applications by meeting several stringent design requirements difficult to achieve with homogeneous materials such as ceramics and metal alloys. Whether it is preparation of crowns, repair of cavities, or entire tooth replacement, the product needs to be aesthetically matched in color and translucence with other teeth and retain its gloss. It must match the hardness of the opposing tooth and be resistant to wear or fatigue fracture. It must be dimensionally stable and withstand the largely varying thermal stresses in the mouth. It also has to have short processing time and near-net shape. Particulate composites used to repair cavities have a polymer resin matrix filled with stiff inorganic or organic inclusions. The resin monomer is typically a methacrylate or a urethane dimethacrylate ester derivative such as bis-GMA cured on site by cross-linkers, ultraviolet light (UV), or light-emitting diodes. The stiff filler particles that increase strength and impart wear resistance can be glass ceramics, calcium silicate, calcium fluoride, crystalline quartz, and silicon nitride whiskers. These are usually silane treated or fused with silica particles to improve retention in the matrix, especially in stress-bearing restorations and to reduce water absorption. The filler can be from 50 percent to 80 percent by volume of the composite and have varying sizes from 20 nm to 50 μm. In a microfilled dental resin, fused silica particles 20 to 40 nm in size can be incorporated to modest volume fractions up to 40 percent to produce a composite that is translucent and can be polished to a high gloss but not mechanically strong enough for posterior teeth and difficult to handle because of low viscosity. Hybrid dental resins have particle sizes of different orders of magnitude from around 0.1 to around 10 μm, allowing for higher filler volume, up to 80 percent and higher viscosity for easier handling, as well as lower water absorption compared with microfilled resins. Commercial dental composite resins have polymerization shrinkages varying from 1.6 to 2.5 percent,[18] followed by water absorption of up to 1.5 percent,[19] causing dimensional changes. They also have poor adhesion to dentin,[20] making it important to use bonding agents to prevent fitting and leakage problems.

All-ceramic dental composites are used for stress-bearing restorations of dental crowns and bridges. Fracture toughness is a very important concern and is addressed by designing crack deflection and bridging mechanisms into the composite. A common type is the alumina-glass composite known as In-Ceram, in which a skeleton of alumina particles is slip cast and sintered, followed by melt infiltration of glass into the porous core. The composition in one example is 75 percent by volume α-alumina particles of average 3 μm size and 25 percent glass.[21] Thermal expansion mismatch between the alumina and glass was shown not to affect fracture toughness significantly. Recent development of aqueous tape-casting of the alumina core has made it easier to conform the composite to a tooth model.[22] The composite can be many times harder than the enamel of the opposing teeth and can thus wear them out. They are coated with either alkali aluminosilicate dental porcelains or calcium phosphate composites to reduce the surface hardness.[23]

Another application of dental composites is orthodontic archwires.[24] One example is a unidirectional pultruded S2-glass-reinforced dimethacrylate thermoset resin.[25] Depending on the yarn of glass fiber used, the fiber volume fraction varied from 32 to 74 percent. The strength and modulus were comparable with those of titanium wires. Orthodontic brackets were also made from composites with a polyethylene matrix reinforced with ceramic hydroxyapatite particles,[26] resulting in isotropic properties and good adhesion to enamel.

12.8.3 External Prosthetics and Orthotics

Traditional prosthetic and orthotic materials such as wood, aluminum, and leather have been largely replaced by high-performance composites and thermoplastics.[27] The requirements of low weight, durability, size reduction, safety, and energy conservation have made fiber-reinforced plastics very attractive in this area. This is one application where traditional composites design and manufacturing, particularly using thermosets, are common because the products are structural and external to the body. For transtibial (TT) and transfemoral (TF) prostheses, composites have been used for the socket-frame component that interfaces with the residual limb and transmits the load and for the shank component that supports the load over the ground. TT and TF prostheses have a target weight limit of 1 and 2 kg, respectively, which makes carbon-fiber-reinforced (CFR) composites ideal. Socket frames in the ISNY system have been made from carbon fiber tape and acrylic resin produced by laminate layup. The matrix used is a blend of rigid and flexible methyl methacrylate resins for tailoring the stiffness of the socket. CFR epoxy tubing has been used to replace stainless steel in artificial arms. Satin-weave carbon-fiber cloth in epoxy prepreg has been used to make the shank of the Endolite TF prosthesis. Hybrid composites for the shank with carbon and nylon fibers in polyester resins had better impact resistance than just carbon- or nylon-only composites. Using thermoplastics, CFR Nylon 6,6 made easily by injection molding has excellent vibration-damping characteristics important for shock absorption. However, the part will need to be heavier because a simple analysis reveals that the wall thickness of a tubular molding in 30 percent discontinuous CFR Nylon 6,6 needs to be seven times that in 60 percent continuous CFR epoxy at constant tube radius. This can be overcome by using more advanced thermoplastics such as PEEK and longer fibers in injection molding. For the prosthetic foot unit, various composites have been used with the objective of storing energy and returning it during motion. The Flex-Foot unit has a long CFR epoxy composite beam that stores flexural energy along the entire length of the prosthesis rather than just the foot unit and is tailored to the individual patient characteristics and activity level. In the Carbon Copy 2 foot, the keel is made of a posterior Kevlar-reinforced nylon block and anterior CFR plastic leaf springs, providing two-stage resistance to flexion. Finally, a nylon-reinforced silicone elastomer has been used to make a durable cover for the flexible foam that encases the prosthesis. In the area of orthotics designed to support injured tissue, composites find use in new bandage-form splinting materials replacing the old cotton fabric and plaster of paris method. These are commonly laminated fiberglass or polyester knitted fabrics in partially cured polyurethane matrices. Besides strength, they have the advantages of better x-ray transmission and lower water adsorption. However, these casts also produce more dust during sawing and are harder to remove.

12.8.4 Soft-Tissue Engineering

Cross-linked hydrogel networks are suitable as a scaffold for skin regeneration due to their high water content and barrier properties, but they have poor mechanical properties. The tensile strength and break point of the hydrogel poly(2-hydroxyethyl methacrylate) (pHEMA) were dramatically enhanced by reinforcing it with Spandex and gauze fibers, resulting in gels that withstood greater forces before tearing. In cartilage repair, where high compressive and shear properties are desirable, low-density linear polyethylene was melt coated onto a woven three-dimensional fabric of UHMWPE fibers to produce a composite that had compressive behavior approximating that of natural cartilage.[28] Fibrous poly(glycolic acid) (PGA) felts and poly(lactide-*co*-glycolide) (PLGA) fibers have been used as surfaces and scaffolds for cartilage cell growth, where the cells produce the bulk matrix when implanted at the site of the cartilage defect. The fiber diameter, interfiber distance, and biodegradation rate were important parameters that affected the quality of the neocartilage formed from such constructs. In the area of vascular grafts, woven-fiber tubes made from polyester (Dacron) have high permeability during implantation, resulting in severe blood leakage through the graft walls. To avoid the lengthy preclotting time needed to overcome this leakage problem, various impermeable materials are coated on such grafts to form a composite, where the matrix functions as the sealant rather than the load bearer. For example, alginate and gelatin have been used to thoroughly wet Dacron fibers and seal the vascular prosthesis, and over time they are biodegraded. For hemodialysis vascular access, where the graft has to be resilient to frequent needle puncture, a technique employed in the DIASTAT graft involves PTFE fibers sandwiched between layers of porous expanded PTFE. The fibers are pushed aside where the needle enters, and after needle withdrawal, the fibers create a baffle effect to reduce leaking blood velocity and improve resealing.[29]

REFERENCES

1. M. M. Schwartz, *Composite Materials Handbook*, 2d ed. New York: McGraw-Hill, 1992.
2. S. L. Evans and P. J. Gregson, "Composite technology in load-bearing orthopaedic implants," *Biomaterials* **19**:1329–1342 (1998).
3. P. K. Mallick, *Composites Engineering Handbook*. New York: Marcel Dekker, 1997.

4. D. L. Wise, *Human Biomaterials Applications*. Totowa, N. J.: Humana Press, 1996.
5. E. E. Gdoutos, K. Pilakoutas, and C. A. Rodopoulos, *Failure Analysis of Industrial Composite Materials*. New York: McGraw-Hill, 2000.
6. R. Christensen, *Mechanics of Composite Materials*. New York: Wiley, 1979.
7. B. D. Ratner, *Biomaterials Science: An Introduction to Materials in Medicine*. San Diego: Academic Press, 1996.
8. J. Katz, "Orthopedic applications," in *Biomaterials Science*, B. D. Ratner (ed.). San Diego: Academic Press, 1966, pp. 335–346.
9. H. Yildiz, S. K. Ha, and F. K. Chang, "Composite hip prosthesis design: 1. Analysis," *Journal of Biomedical Materials Research* **39**:92–101 (1998).
10. J. Kettunen, E. A. Makelaa, H. Miettinen, et al., "Mechanical properties and strength retention of carbon fiber-reinforced liquid crystalline polymer (LCP/CF) composite: An experimental study on rabbits," *Biomaterials* **19**:1219–1228 (1998).
11. R. De Santis, L. Ambrosio, and L. Nicolais, "Polymer-based composite hip prostheses," *Journal of Inorganic Biochemistry* **79**:97–102 (2000).
12. S. Srinivasan, J. R. de Andrade, S. B. Biggers, Jr., and R. A. Latour, Jr., "Structural response and relative strength of a laminated composite hip prosthesis: effects of functional activity," *Biomaterials* **21**:1929–1940 (2000).
13. L. Bacakova, V. Start, O. Kofronova, and V. Lisa, "Polishing and coating carbon fiber-reinforced carbon composites with a carbon-titanium layer enhances adhesion and growth of osteoblastlike MG63 cells and vascular smooth muscle cells in vitro," *Journal of Biomedical Materials Research* **54**:567–578 (2001).
14. W. S. Pietrzak, "Principles of development and use of absorbable internal fixation," *Tissue Engineering* **6**: 425–433 (2000).
15. H. Alexander, "Composites," in *Biomaterials Science*, B. D. Ratner (ed.). San Diego: Academic Press, 1996, pp. 94–105.
16. M. Dauner, H. Planck, L. Caramaro, et al., "Resorbable continuous-fiber reinforced polymers for osteosynthesis," *Journal of Materials Science: Materials in Medicine* **9**:173–179 (1998).
17. A. Kelly, R. W. Cahn, and M. B. Bever, *Concise Encyclopedia of Composite Materials*, Revised Edition. New York: Pergamon Press, 1994.
18. W. D. Cook, M. Forrest, and A. A. Goodwin, "A simple method for the measurement of polymerization shrinkage in dental composites," *Dental Materials* **15**:447–449 (1999).
19. M. W. Beatty, M. L. Swartz, B. K. Moore, et al., "Effect of microfiller fraction and silane treatment on resin composite properties," *Journal of Biomedical Materials Research* **40**:12–23 (1998).
20. J. W. Nicholson, "Adhesive dental materials: A review," *International Journal of Adhesion and Adhesives* **18**:229–236 (1998).
21. W. D. Wolf, K. J. Vaidya, and L. Falter Francis, "Mechanical properties and failure analysis of alumina-glass dental composites," *Journal of the American Ceramic Society* **79**:1769–1776 (1996).
22. D. J. Kim, M. H. Lee, and C. E. Kim, "Mechanical properties of tape-cast alumina-glass dental composites," *Journal of the American Ceramic Society* **82**:3167–3172 (1999).
23. L. F. Francis, K. J. Vaidya, H. Y. Huang, and W. D. Wolf, "Design and processing of ceramic-based analogs to the dental crown," *Materials Science and Engineering [C]* **3**:63–74 (1995).
24. R. P. Kusy, "A review of contemporary archwires: Their properties and characteristics," *Angle Orthodontist* **67**:197–207 (1997).
25. D. W. Fallis and R. P. Kusy, "Variation in flexural properties of photo-pultruded composite archwires: Analyses of round and rectangular profiles," *Journal of Materials Science: Materials in Medicine* **11**:683–693 (2000).
26. M. Wang, C. Berry, M. Braden, and W. Bonfield, "Young's and shear moduli of ceramic particle filled polyethylene," *Journal of Materials Science: Materials in Medicine* **9**:621–624 (1998).
27. R. Hanak and E. S. Hoffman, "Specification and fabrication details for the ISNY above-knee socket system," *Orthotics and Prosthetics* **40**:38–42 (1986).
28. B. Seal and A. Panitch, "Polymeric biomaterials for tissue and organ regeneration," *Materials Science and Engineering [R]* **262**:1–84 (2001).
29. J. M. Lohr, K. V. James, A. T. Hearn, and S. A. Ogden, "Lessons learned from the DIASTAT vascular access graft," *American Journal of Surgery* **172**:205–209 (1996).

CHAPTER 13
BIOCERAMICS

David H. Kohn
University of Michigan, Ann Arbor

13.1 INTRODUCTION

The clinical goal when using ceramic biomaterials, as is the case with any biomaterial, is to replace lost tissue or organ structure and/or function. The rationale for using ceramics in medicine and dentistry was initially based on the relative biological inertness of ceramic materials compared with metals. However, in the past two decades, this emphasis has shifted more toward the use of bioactive ceramics, materials that not only elicit normal tissue formation but may also form an intimate bond with bone tissue. Most recently, bioceramics have been utilized in conjuction with more biological therapies. In other words, the ceramic, usually resorbable (i.e., a greater degree of bioactivity than surface-reactive materials), facilitates the delivery and function of a biological agent (i.e., cells, proteins, and/or genes) with an end goal of eventually regenerating a full volume of functional tissue.

Ceramic biomaterials are processed to yield one of four general types of surfaces and associated mechanisms of tissue attachment (Kohn and Ducheyne, 1992): (1) fully dense, relatively inert crystalline ceramics that attach to tissue by either a press fit, tissue ongrowth onto a roughened surface, or a grouting agent, (2) porous, relatively inert ceramics into which tissue ingrowth occurs, creating a mechanical attachment, (3) fully dense, surface-active ceramics that attach to tissue via a chemical bond, and (4) resorbable ceramics that integrate with tissue and eventually are replaced by host tissue. Ceramics may therefore be classified by their macroscopic surface characteristics (smooth, fully dense, roughened, or porous) or their chemical stability (inert, surface reactive, or bulk reactive/resorbable). The integration of biological (i.e., inductive) agents with ceramics further expands the clinical potential of these materials.

Relatively inert ceramics elicit minimal tissue response and lead to a thin layer of fibrous tissue immediately adjacent to the surface. Surface-active ceramics are partially soluble, resulting in ion-exchange and the potential to lead to a direct chemical bond with bone. Bulk bioactive ceramics are fully resorbable, have much greater solubility than surface-active ceramics, and may ultimately be replaced by an equivalent volume of regenerated tissue. The relative level of bioactivity mediates the thickness of the interfacial zone between the biomaterial surface and host tissue (Fig. 13.1). There are, however, no standardized measures of reactivity, but the most common are pH changes, ion solubility, tissue reaction, and any number of assays that assess some parameter of cell function.

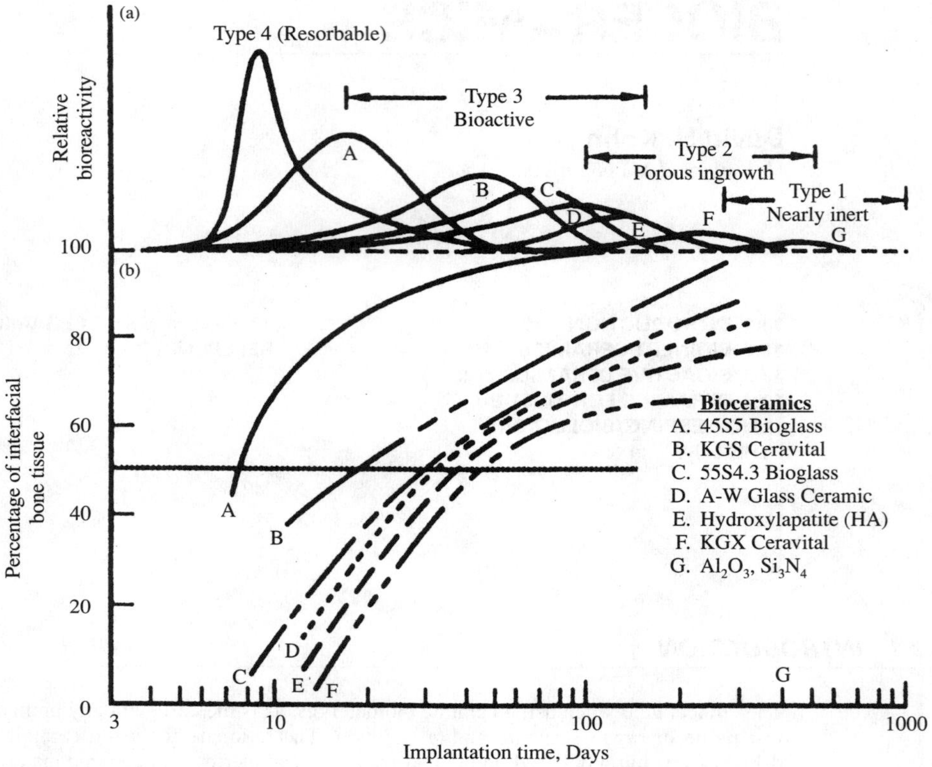

FIGURE 13.1 Bioactivity spectra for selected bioceramics: (*a*) relative magnitudes and rates of bioactivity; (*b*) time dependence of formation of bone at bioceramic surface and ceramic-bone bonding. (*From Hench, 1996, with permission.*)

Five main ceramic materials are used in musculoskeletal reconstruction/regeneration: carbon (Christel et al., 1987; Haubold et al., 1981; Huttner and Huttinger, 1984), alumina (Al_2O_3) (Kohn and Ducheyne, 1992; Boutin et al., 1988; Heimke et al., 1978; Hulbert et al., 1970; Webster et al., 2000; Zreiqat et al., 1999), zirconia (ZrO_2) (Kohn and Ducheyne, 1992; Cales and Stefani, 1995; Christel et al., 1989; Filiaggi et al., 1996a, 1996b), bioactive glasses and glass ceramics (Kohn and Ducheyne, 1992; Ducheyne, 1985, El-Ghannam et al., 1997; Gross and Strunz, 1980; Hench et al., 1972; Nakamura et al., 1985; Effah Kaufmann et al., 2000), and calcium-phosphates (Kohn and Ducheyne, 1992; Murphy et al., 2000a; Ducheyne, 1987; Ducheyne et al., 1980; Koeneman et al., 1990; Van Raemdonck et al., 1984). Alumina, zirconia, and carbon are considered bioinert, whereas glasses and calcium phosphates are bioactive ceramics. Bioactive ceramics are primarily used as permanent or temporary scaffolds (with or without in vitro expanded cells), coatings on more structurally sound substrates, and vehicles for drug delivery.

In this chapter the three main classifications of bioceramics (i.e., bioinert, bioactive, and bioactive/biological constructs) are discussed, with a focus on musculoskeletal applications. A materials science approach is taken to address design issues of importance to a biomedical engineer. In other words, the processing-structure-composition-property synergy is discussed for each material, and then critical properties important to the design and clinical success of each class of bioceramic are presented. Within the discussion of the processing-composition-structure synergism, issues of material selection, service conditions, fabrication routes, and important characterization methodologies are discussed.

13.2 BIOINERT CERAMICS

Ceramics are fully oxidized materials and are therefore chemically very stable. Thus ceramics are less likely to elicit an adverse biological response than metals, which only oxidize at their surface. Three types of inert ceramics are of interest in musculoskeletal applications: carbon, alumina, and zirconia.

13.2.1 Carbon

The benign biological reaction elicited by carbon-based materials, along with the similarity in stiffness and strength between carbon and bone, made carbon a candidate material for musculoskeletal reconstruction (Bokros et al., 1972). Carbon has a hexagonal crystal structure that is formed by strong covalent bonds. Graphite has a planar hexagonal array structure with a crystal size of approximately 1000 Å (Bokros, 1978). The carbon-carbon bond energy within the planes is large (114 kcal/mol), whereas the bond between the planes is weak (4 kcal/mol) (Hench and Ethridge, 1982). Therefore, carbon derives its strength from the strong in-plane bonds, whereas the weak bonding between the planes results in a low modulus, near that of bone (Bokros, 1978).

Isotropic carbon, on the other hand, has no preferred crystal orientation and hence possesses isotropic material properties. There are three types of isotropic carbon: pyrolytic, vitreous, and vapor-deposited carbon. Pyrolytic carbons are formed by the deposition of carbon from a fluidized bed onto a substrate. The fluidized bed is formed from pyrolysis of hydrocarbon gas at between 1000 and 2500°C (Hench and Ethridge, 1982). Low-temperature isotropic (LTI) carbons are formed at temperatures below 1500°C. LTI pyrolytic carbon possesses good frictional and wear properties, and incorporation of silicon can further increase hardness and wear resistance (Bokros, 1978).

Vitreous carbon is a fine-grained polycrystalline material formed by slow heating of a polymer. On heating, the more volatile components diffuse from the structure, and only carbon remains (Hench and Ethridge, 1982). Since the process is diffusion-mediated and potentially volatile, heating must be slow, and the dimensions of the structure are therefore limited to approximately 7 mm (Bokros, 1978). Salient properties of all three forms of carbon are summarized in Table 13.1.

Attempts have been made at depositing LTI coatings onto metallic substrates (Hench and Ethridge, 1982). The limiting factor in these systems was the brittleness of the carbon coating and the propensity for coating fracture and coating-substrate debonding. Carbon may also be vapor deposited onto a substrate by the evaporation of carbon atoms from a high-temperature source and subsequent condensation onto a low-temperature substrate (Hench and Ethridge, 1982). Vapor-deposited coatings are typically about 1 μm thick. As a result, the bulk properties of the substrate are retained. More recently, diamondlike carbon (DLC) coatings have been studied, primarily as a means of improving wear and corrosion resistance of articulating components of joint replacements (Allen et al., 2001). Carbon-based thin films are produced from solid carbon or liquid/gaseous hydrocarbon sources using ion-beam or plasma-deposition techniques. Resulting coatings are metastable amorphous with properties intermediate to those of graphite and diamond (Allen et al., 2001).

13.2.2 Alumina

High-density, high-purity polycrystalline alumina is used for femoral stems, femoral heads, acetabular components, and dental implants (Kohn and Ducheyne, 1992; Boutin et al., 1988; Heimke et al., 1978). More recently, ion-modified and nanostructured forms of Al_2O_3 have been synthesized in attempts to make these traditionally inert bioceramics more bioactive (Webster et al., 2000; Zreiqat et al., 1999). The attributes of alumina, aside from its chemical stability and relative biologic inertness, are its hardness and excellent friction and wear resistance. As a result, a main motivation for using alumina in reconstructive surgery has been for tribologic improvements, and many total hip replacements are now designed as modular devices; i.e., an alumina femoral head is press-fit onto

TABLE 13.1 Physical and Mechanical Properties of Bioceramics

Material	Porosity, %	Density, mg/m^3	Modulus, GPa	Compressive strength MPa	Tensile strength MPa	Flexural strength MPa	K_{Ic}, $MPa{\cdot}m^{1/2}$
Graphite (isotropic)	7	1.8	25	—	—	140	—
	12	1.8	20–24	65–95	24–30	45–55	—
	16–20	1.6–1.85	6–13.4	18–58	8–19	14–27	—
	30	1.55	7.1	—	—	—	—
	—	0.1–0.5	—	2.5–30	—	—	—
Pyrolytic graphite, LTI	2.7	2.19	28–41	—	—	—	—
	—	1.3–2	17–28	900	200	340–520	—
	—	1.7–2.2	17–28	—	—	270–550	—
Glassy (vitreous) carbon	—	1.4–1.6	—	—	—	70–205	—
	—	1.45–1.5	24–28	700	70–200	150–200	—
	—	1.38–1.4	23–29	—	—	190–255	—
	≤50	<1.1	7–32	50–330	13–52	—	—
Bioactive ceramics and glass ceramics	—	—	—	—	56–83	—	—
	—	2.8	—	500	—	100–150	—
	31–76	0.65–1.86	2.2–21.8	—	—	4–35	—
Hydroxyapatite	0.1–3	3.05–3.15	7–13	350–450	38–48	100–120	—
	10	2.7	—	—	—	—	—
	30	—	—	120–170	—	—	—
	40	—	—	60–120	—	15–35	—
	2.8–19.4	2.55–3.07	44–48	310–510	—	60–115	—
	2.5–26.5	—	55–110	≤800	—	50–115	—
Tetracalcium-phosphate	Dense	3.1	—	120–200	—	—	—
Tricalcium-phosphate	Dense	3.14	—	120	—	—	—
Other calcium phosphates	Dense	2.8–3.1	—	70–170	—	—	—
Al_2O_3	0	3.93–3.95	380–400	4000–5000	350	400–500	5–6
	25	2.8–3.0	150	500	—	70	—
	35	—	—	200	—	55	—
	50–75	—	—	80	—	6–11.4	—
ZrO_2, stabilized (~ 3% Y_2O_3)	0	4.9–5.6	150–200	1750	—	150–900	4–12
	1.5	5.75	210–240	—	—	280–450	—
	5	—	150–200	—	—	50–500	—
	28	3.9–4.1	—	<400	—	50–65	—

Source: Modified from Kohn and Ducheyne (1992) with permission.

the neck of a metallic femoral stem. The alumina head then articulates in either a polyethylene or alumina acetabular cup.

High-purity (e.g., 99.5 percent) alumina powder, prepared by calcining alumina trihydrate, is typically isostatically compacted and shaped. Subsequent sintering at temperatures in the range 1600 to 1800°C transforms a preform into a dense polycrystalline solid having a grain size of less than 5 μm (Boutin et al., 1988). Addition of trace amounts of MgO aids in sintering and limits grain growth. If processing is kept below 2050°C, α-Al_2O_3, which is the most stable phase, forms. Alternatively, single crystals (sapphire) may be grown by feeding powder onto a seed and allowing buildup.

The physical and mechanical properties (i.e., ultimate strength, fatigue strength, fracture toughness, and wear resistance) of α-alumina are a function of purity, grain size, grain size distribution, porosity, and inclusions (Kohn and Ducheyne, 1992; Boutin et al., 1988; Dorre and Dawihl, 1980) (see Table 13.1). Both grain size d and porosity P ($0 \leq P \leq 1$) affect strength σ via well-characterized relations [Eqs. (13.1) and (13.2)], where σ_0 is the strength of the dense ceramic and A, n and B are material constants, experimentally determined, with n typically approximately 0.5.

$$\sigma = Ad^{-n} \tag{13.1}$$

$$\sigma_p = \sigma_0 e^{-BP} \tag{13.2}$$

For example, decreasing grain size from 4 to 7 μm increases strength by approximately 20 percent (Dorre and Dawihl, 1980). Advanced ceramics processing now makes it possible to fabricate alumina with grain sizes on the order of 1 μm and a small grain size distribution, material characteristics that result in increased strength. The elastic modulus of fully dense alumina is two- to fourfold greater than that of metals commonly used in bone and joint reconstruction.

Sintering and annealing cycles and the subsequent cooling can lead to residual stresses in alumina. Microcrack nucleation occurs at locally high residual stresses and at pores, inclusions, segregated grain boundaries, and grain-boundary triple points (Boutin et al., 1988). Furthermore, if there is a wide distribution of grain sizes, the material anisotropy leads to a larger range of thermal expansions and higher residual stresses (Boutin et al., 1988). The long-term mechanical properties of ceramics, however, may be predicted through the use of fracture mechanics and statistical methods.

Environment also affects the mechanical properties of alumina. For example, fatigue strength is reduced in an aqueous environment due to subcritical crack growth, which is enhanced by water adsorption through hydrogen bonding of water molecules with oxygen in the Al_2O_3 lattice (Ritter et al., 1979; Dawihl et al., 1979). The threshold stress for subcritical crack growth is also a function of processing and purity. For example, thermal cycling can lead to microcracking, whereas CaO impurities compromise mechanical integrity (Boutin et al., 1988).

The amount of wear in alumina-alumina couples can be as much as 10 times less than in a metal-polyethylene system depending on experimental conditions and surface roughness of the ceramic (Davidson, 1993; Kumar et al., 1991). The coefficient of friction of both alumina-alumina and alumina-polyethylene is less than that of metal-polyethylene because of alumina's low surface roughness and wettability (Boutin et al., 1988; Semlitsch et al., 1977).

The major limitation of alumina is that it possesses relatively low tensile and bending strengths and fracture toughness and, as a consequence, is sensitive to stress concentrations and overloading. Clinically retrieved alumina total hip replacements exhibit damage thought to be caused by fatigue, impact, or overload (Walter and Lang, 1986). Elevated contact and shear stresses, which lead to subsurface damage accumulation at microstructural defects and grain boundaries, have been implicated in the failure process (Walter and Lang, 1986). In general, a large number of ceramic failures can be attributed to materials processing or design deficiencies and can be minimized through better materials choice and quality control.

Based on clinical experience and retrieval analyses, implant geometries should have a sphericity of less than 1 μm and a radius tolerance between components of 7 to 10 μm (Boutin et al., 1988). These specifications are based on the rationale that too small a gap between the components does not provide sufficient room for necessary lubrication or an escape route for alumina particles, whereas too large a gap increases contact pressures. To design around the low mechanical properties, a 32-mm femoral head and an acetabular cup with a minimum 44-mm outside diameter are recommended (Boutin et al., 1998).

13.2.3 Zirconia

Yttrium oxide partially stabilized zirconia (YPSZ) has been advocated as an alternative to alumina (Christel et al., 1989; Cales and Stefani, 1995). This class of ceramic has a higher toughness than alumina since it can be transformation toughened and is used in bulk form or as a coating (Filiaggi et al., 1996a, 1996b). There are currently approximately 150,000 zirconia components in clinical use (Cales and Stefani, 1995).

At room temperature, pure zirconia has a monoclinic crystal symmetry and a density of 5.5 g/cm³. On heating, it transforms to a tetragonal phase (density ~ 6.1 g/cm³) at approximately 1000 to 1100°C and then to a cubic phase at approximately 2000°C (Fig. 13.2). A partially reversible volumetric shrinkage (density increase) of 3 to 10 percent occurs during the monoclinic-to-tetragonal transformation (Christel et al., 1989). The volumetric changes resulting from the phase transformations can lead to residual stresses and cracking. Furthermore, because of the large volume reduction, pure zirconia cannot be sintered. However, sintering and phase transformations can be controlled via the addition of stabilizing oxides. Yttrium oxide (Y_2O_3) serves as a stabilizer for the tetragonal phase such that on cooling the tetragonal crystals are maintained in a metastable state and do not transform back to a monoclinic structure. The normal tetragonal-to-monoclinic transformation

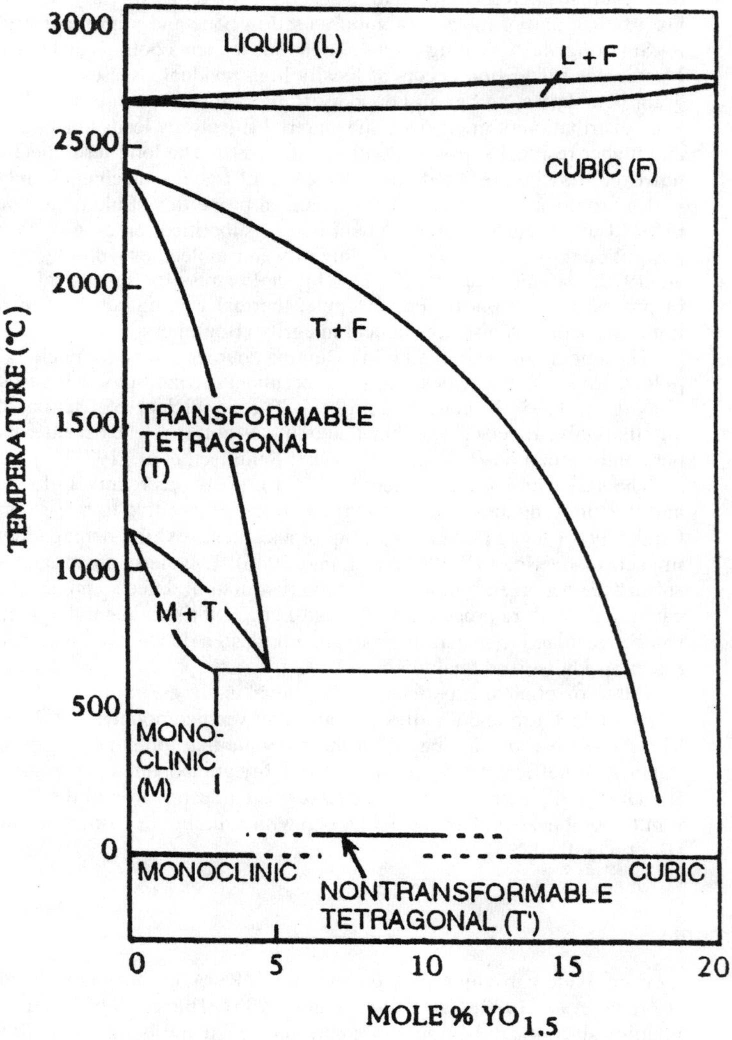

FIGURE 13.2 Schematic phase diagram of the ZrO_2-Y_2O_3 system. *(From Miller et al., 1981, with permission.)*

and volume change is additionally prevented by neighboring grains inducing compressive stresses on one another (Christel et al., 1989).

The modulus of partially stabilized zirconia is approximately half that of alumina, whereas the bending strength and fracture toughness are two to three and two times greater, respectively (see Table 13.1). The relatively high strength and toughness are a result of transformation toughening, a mechanism that manifests itself as follows (Fig. 13.3): crack nucleation and propagation lead to locally elevated stresses and energy in the tetragonal crystals surrounding the crack tip. The elevated energy induces the metastable tetragonal grains to transform into monoclinic grains in this part of the microstructure. Since the monoclinic grains are larger than the tetragonal grains, there is a local volume increase, compressive stresses are induced, more energy is needed to advance the crack, and crack blunting occurs.

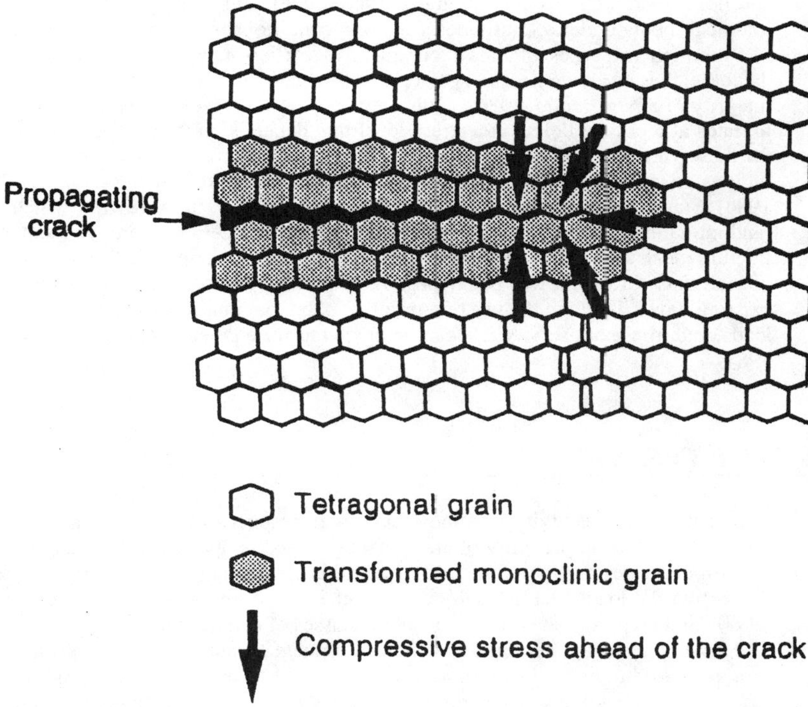

FIGURE 13.3 Schematic of microstructure in yttrium partially stabilized zirconia (YPSZ) bioceramic undergoing transformation toughening at a crack tip. (*From Cales and Stefani, 1995, with permission.*)

The wear rate of YPSZ on UHMWPE can be five times less than the wear rate of alumina on UHMWPE, depending on experimental conditions (Kumar et al., 1991; Davidson, 1993; Derbyshire et al., 1994). Wear resistance is a function of grain size, surface roughness, and residual compressive stresses induced by the phase transformation. The increased mechanical properties may allow for smaller-diameter femoral heads to be used in comparison with alumina.

Partially stabilized zirconia is typically shaped by cold isostatic pressing and then densified by sintering. Sintering may be performed with or without a subsequent hot isostatic pressing (HIP-ing) cycle. The material is usually presintered until approximately 95 percent dense and then HIP-ed to remove residual porosity (Christel et al., 1989). Sintering can be performed without inducing grain growth, and final grain sizes can be less than 1 μm.

13.2.4 Critical Properties of Bioinert Ceramics

In addition to wear resistance and minimal biological response, other properties of bioinert ceramics important for their long-term efficacy are stiffness, strength, and toughness. Stiffness represents one gauge of the mechanical interaction between an implant and its surrounding tissue; it is one determinant of the stress magnitude and distribution in the biomaterial as well as the tissue, and it also dictates the efficacy of stress transfer and potential for stress shielding (Kohn and Ducheyne, 1992; Ko et al., 1995). Although the effect of biomaterial stiffness has never been truly assessed in a functionally loaded prosthesis under controlled conditions in which stiffness was the only variable, it is generally acknowledged that stiffness is an important design parameter.

Load-bearing biomaterials must also be conservatively designed to ensure that they maintain their structural integrity, i.e., designed to be fail-safe at stresses above peak in-service stresses for a lifetime greater than the expected service life of the prosthesis. Thus the static (tensile, compressive, and flexural strength), dynamic (high-cycle fatigue), and toughness properties of ceramics in physiological media under a multitude of loading conditions and rates must be well characterized.

Although the types of data recommended earlier are an important aspect of any bioceramic database, these data are necessary but not sufficient input for designing bioceramics. The mechanical integrity of a bioceramic also depends on its processing, size, and shape. Failure of ceramics usually initiates at a critical defect at a stress level that depends on the geometry of the defect. To account for these variables and minimize the probability of failure, fracture mechanics and statistical distributions are recommended to predict failure probability at different load levels (Soltesz and Richter, 1984). Statistical methods are useful to account for scatter in properties, which is largely due to the random nature of the defects in these brittle materials. Ceramics are well suited for applying these fracture mechanics concepts because of the validity of linear elastic fracture mechanics for this class of materials. Crack growth behavior may be predicted based on stresses and the size and shape of existing flaws. Through the use of a proof test, the maximum allowable flaw sizes, minimum failure loads, and minimum service life can be predicted for a given set of conditions (Soltesz and Richter, 1984).

13.3 BIOACTIVE CERAMICS

The concept of bioactivity was introduced with respect to bioactive glasses via the following hypothesis: the biocompatibility of an implant material is optimal if the material elicits the formation of normal tissues at its surface and in addition if it establishes a contiguous interface capable of supporting the loads that normally occur at the site of implantation (Hench et al., 1972; Ducheyne, 1987). Under appropriate conditions, three classes of ceramics may fulfill these requirements: bioactive glasses and glass ceramics, calcium phosphate ceramics, and composites of these glasses and ceramics. Additionally, incorporation of inductive factors into each of these classes may enhance bioactivity. Collectively, these different classes of materials/biological constituents are used clinically and experimentally in a wide variety of applications, including bulk implants (surface-active), coatings on metallic or ceramic implants (surface-active), permanent bone augmentation devices/scaffold materials (surface-active), temporary tissue engineering devices (surface- or bulk-active), fillers such as in cements (surface- or bulk-active), and drug-delivery vehicles (bulk-active).

It is important to reemphasize that the nature of the biomaterial-tissue interface and the reactions (e.g., ion exchange) at the ceramic surface and in the tissues dictate the resulting mechanical, chemical, physical, and biological processes that occur. In general, four factors determine the long-term effect of bioactive ceramic implants: (1) the site of implantation, (2) tissue trauma, (3) the bulk and surface properties of the material, and (4) the relative motion at the implant-tissue interface (Ducheyne et al., 1987). For resorbable materials, additional design requirements include the following: the strength/stability of the material-tissue interface needs to be maintained during the period of degradation and replacement by host tissue; material resorption and tissue repair/regeneration rates should be matched; and the resorbable material should consist only of metabolically acceptable species.

13.3.1 Bioactive Glasses and Glass Ceramics

Bioactive glasses were first developed by Hench (1972), who synthesized several glasses containing mixtures of silica, phosphate, calcia, and soda. These materials are used as bulk implants, coatings on metallic or ceramic implants, and scaffolds for guiding biological therapies (Kohn and Ducheyne, 1992; El-Ghannam et al., 1997; Gross and Strunz, 1980; Hench et al., 1972; Effah Kaufmann et al., 2001; Nakamura et al., 1985) (Table 13.2).

Chemical reactions are limited to the surface (~300 to 500 μm) of the glass, and bulk properties are not affected by surface reactivity. The degree of activity and physiological response depends on the chemical composition of the glass and may vary by over an order of magnitude (Kohn and Ducheyne, 1992). For example, the substitution of CaF for CaO decreases the solubility of the glass, whereas the addition of B_2O_3 increases the solubility of the glass (Hench and Ethridge, 1982).

Ceravital, a variation of Bioglass, is a glass ceramic. The material is first quench-melted to form a glass and then heat-treated to form nuclei for crystal growth and subsequent transformation from a glass to a ceramic. Compared with Bioglass, Ceravital has a different alkali oxide concentration—small amounts of alkaline oxides are added to control dissolution rates (see Table 13.2)—but the physiological response of both these glasses is similar (Gross and Strunz, 1980). It is hypothesized that the general biologic response to both glasses is the nucleation of hydroxyapatite crystals at the implant surface within an ordered collagen matrix, followed by the formation of mineralized bone (Kohn and Ducheyne, 1992).

A glass ceramic containing crystalline oxyapatite and fluorapatite $[Ca_{10}(PO_4)_6(O,F_2)]$ and β-Wollastonite (SiO_2-CaO) in a MgO-CaO-SiO_2 glassy matrix (denoted glass-ceramic A-W) represents a third bioactive glass (Kitsugi et al., 1986; Kokubu et al., 1990a, 1990b; Nakamura et al., 1985). This material is formed by heating a glass powder compact, having a composition listed in Table 13.2, at 1050°C. A-W glass ceramic bonds to living bone through a thin calcium- and phosphorus-rich layer that is formed at the surface of the glass ceramic (Kitsugi et al., 1986, 1987; Nakamura et al., 1985). If the physiological environment is correctly simulated in terms of ion concentration, pH, and temperature, this layer consists of small carbonated hydroxyapatite crystallites with a defective structure, and the composition and structural characteristics are similar to those of bone (Kokubu et al., 1990a).

The physical/chemical basis for glass and glass-ceramic interfacing with the biologic milieu is as follows: ceramics are susceptible to surface changes in an aqueous medium. Lower-valence ions tend to segregate to surfaces and grain boundaries, leading to concentration gradients and ion exchange. These reactions depend on the local pH and reactive cellular constituents (Hench and Ethridge, 1982). It is important to note that these reactions can be either biologically beneficial or adverse and therefore must be well controlled and characterized as a function of microstructure and surface state (Kohn and Ducheyne, 1992).

TABLE 13.2 Composition (Weight Percent) of Bioactive Glasses and Glass Ceramics

Material	45S5 Bioglass	45S5-F Bioglass	45S5-B5 Bioglass	52S4.6 Bioglass	Ceravital	Stabilized Ceravital	A-W glass ceramic
SiO_2	45.0	45.0	45.0	52.0	40–50	40–50	34.2
P_2O_5	6.0	6.0	6.0	6.0	10–15	7.5–12.0	16.3
CaO	24.5	12.3	24.5	21.0	30–35	25–30	44.9
Na_2O	24.5	24.5	24.5	21.0	5–10	3.5–7.5	—
B_2O_3	—	—	5.0	—	—	—	—
CaF_2	—	12.3	—	—	—	—	0.5
K_2O	—	—	—	—	0.5–3.0	0.5–2.0	—
MgO	—	—	—	—	2.5–5.0	1.0–2.5	4.6
Al_2O_3	—	—	—	—	—	5.0–15.0	—
TiO_2	—	—	—	—	—	1.0–5.0	—
Ta_2O_5	—	—	—	—	—	5.0–15.0	—

Source: From Kohn and Ducheyne (1992) with permission.

When placed in a physiological medium, bioactive glasses leach Na^+ ions and subsequently also K^+, Ca^{2+}, P^{5+}, Si^{4+}, and SiOH species. These ionic species are replaced with H_3O^+ ions from the medium through an ion-exchange reaction that produces a silica-rich gel surface layer (Hench and Ethridge, 1982). In an in vitro setting at least, the depletion of H^+/H_3O^+ ions in solution causes a pH increase, which further enhances glass dissolution. With increasing time of exposure to the medium, the high-surface-area silica-rich surface gel chelates calcium and phosphate ions, and a Ca-P-rich amorphous apatite layer forms on top of the silica-rich layer. This Ca-P-rich layer may form after as little as 1 hour in physiological solution (Hench and Ethridge, 1982). The amorphous Ca-P layer eventually crystallizes, and CO_3^{2-} substitutes for OH^- in the apatite lattice, leading to the formation of a carbonated apatite layer. Depending on animal species, anatomic site, and time of implantation, steady-state thickness of the Ca-P-rich and Si-rich zones can range from 30 to 70 μm and 60 to 230 μm respectively (Hench and Ethridge, 1982).

In parallel with these physical/chemical-mediated reactions, in an in vivo setting, proteins adsorb/desorb from the silica gel and carbonated layers. The bioactive surface and subsequent preferential protein adsorption that can occur can enhance attachment, differentiation, and proliferation of osteoblasts and secretion of the cells' own extracellular matrix. Crystallization of carbonated apatite within an ordered collagen matrix leads to an interfacial bond.

The overall rate of change of the glass surface R may be quantified as the sum of the reaction rates of each stage of the reaction (Hench, 1996):

$$R = -k_1 t^{0.5} - k_2 t^{1.0} + k_3 t^{1.0} + k_4 t^y + k_5 t^z \qquad (13.3)$$

where k is the rate constant for each stage and represents, respectively, the rate of exchange between alkali cations in glass and H^+/H_3O^+ in solution (k_1), interfacial SiO_2 network dissolution (k_2), repolymerization of SiO_2 (k_3), carbonate precipitation and growth (k_4), and other precipitation reactions (k_5). Using these rates, the following design criterion may be established: The kinetics of each stage, especially stage 4, should match the rate of biomineralization in vivo. For $R \gg$ in vivo rates, resorption will occur, whereas if $R \ll$ in vivo rates, the glass will be nonbioactive (Hench, 1996).

The degree of activity and physiological response (e.g., rate of formation of the Ca-P surface and ultimate glass-tissue bond) therefore depends on the composition of the glass as well as on time and is mediated by the material, the solution, and the cell. The reactivity and rate of bond formation depend on composition and can be expressed by the ratio of the network former to the network modifier: $SiO_2/(CaO + Na_2O + K_2O)$ (Hench and Clark, 1982). The higher this ratio is, the less soluble is the glass and the slower is the rate of bone formation. Inspection of a SiO_2-Na_2O-CaO ternary diagram (Fig. 13.4) enables a quantitative association between composition and biological response (Hench, 1996). In general, the ternary diagram may be divided into three zones of biological interest: zone A, bioactive bone bonding [glasses are characterized by CaO/P_2O_5 ratios > 5 and $SiO_2/(CaO + Na_2O)$ ratios < 2]; zone B, nearly inert [bone bonding does not occur (only fibrous tissue formation occurs at the surface) because the SiO_2 content is too high and reactivity is too low; these high-SiO_2 glasses develop only a surface hydration layer or too dense a silica-rich layer to enable further dissolution and ion exchange]; and zone C, resorbable glasses (no bone bonding occurs because reactivity is too high and SiO_2 undergoes rapid selective alkali ion exchange with protons or H_3O^+, leading to a thick but porous unprotected SiO_2-rich film that dissociates at a high rate).

The level of bioactivity has been related to the physiological process of bone formation via an index of bioactivity I_B that is related to the amount of time it takes for 50 percent of the interface to be bonded (Hench, 1996):

$$I_B = 100/t_{0.5BB} \qquad (13.4)$$

The compositional dependence of the biological response may therefore be understood by viewing iso-I_B contours superposed onto the ternary diagram (see Fig. 13.4). The cohesion strength of the glass-tissue interface will be a function of the surface area, thickness, and stiffness of the inerfacial zone and is optimum for $I_B \sim 4$ (Hench, 1996).

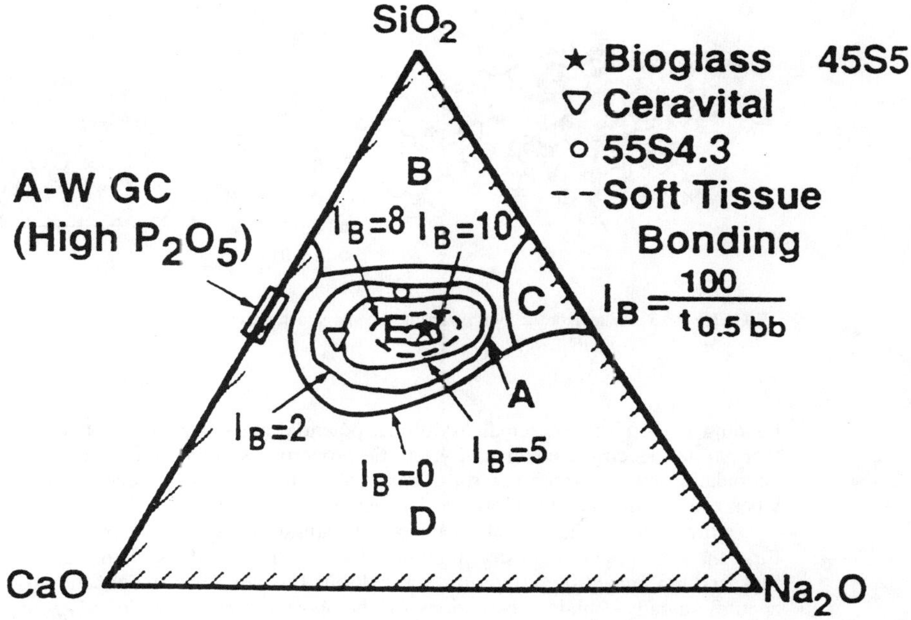

FIGURE 13.4 Ternary diagram (SiO_2-Na_2O-CaO, at fixed 6 percent P_2O_5) showing the compositional dependence (in weight percent) of bone bonding and fibrous tissue bonding to the surfaces of bioactive glasses and glass ceramics: zone A, bioactive bone bonding ceramics; zone B, nearly inert ceramics (bone bonding does not occur at the ceramic surface; only fibrous tissue formation occurs); zone C, resorbable ceramics (no bone bonding occurs because reactivity is too high). I_B = index of bioactivity for bioceramics in zone A. *(From Hench, 1996, with permission.)*

13.3.2 Calcium-Phosphate Ceramics

Calcium phosphate ceramics are ceramics with varying calcium-to-phosphate ratios. Among them, the apatite ceramics, defined by the chemical formula $M_{10}(XO_4)_6Z_2$, have been studied most. The apatites form a range of solid solutions as a result of ionic substitution at the M^{2+}, XO_4^{3-}, or Z^- sites. In general, apatites are nonstoichiometric and contain less than 10 mol of M^{2+} ions, less than 2 mol of Z^- ions, and exactly 6 mol of XO_4^{3-} ions (Van Raemdonck et al., 1984). The M^{2+} species is typically a bivalent metallic cation, such as Ca^{2+}, Sr^{2+}, Ba^{2+}, Pb^{2+}, or Cd^{2+}. The XO_4^{3-} species is typically one of the following trivalent anions: AsO_4^{3-}, VO_4^{3-}, CrO_4^{3-}, or MnO_4^{3-}. The monovalent Z^- ions are usually F^-, OH^-, Br^-, or C_2^- (Van Raemdonck et al., 1984).

More complex ionic structures may also exist. For example, replacing the two monovalent Z^- ions with a bivalent ion, such as CO_3^{2-}, results in the preservation of charge neutrality, but one anionic position becomes vacant. Similarly, the M^{2+} positions may also have vacancies. In this case, charge neutrality is maintained by vacancies at the Z^- positions or by substitution of some trivalent PO_4^{3-} ions with bivalent ions (Van Raemdonck et al., 1984).

The most common apatite used in medicine and dentistry is hydroxyapatite (HA), a material with a hexagonal crystal lattice; an ideal chemical formula, i.e., $Ca_{10}(PO_4)_6(OH)_2$; ideal weight percents of 39.9 percent Ca, 18.5 percent P, and 3.38 percent OH; and an ideal calcium-phosphate ratio of 1.67 (Fig. 13.5). The crystal structure and crystallization behavior of HA is strongly dependent on the substitutional nature of the ionic species.

The impetus for using synthetic HA as a biomaterial stems from the perceived advantage of using a material similar to the mineral phase in bone and teeth for replacing these materials. Better tissue

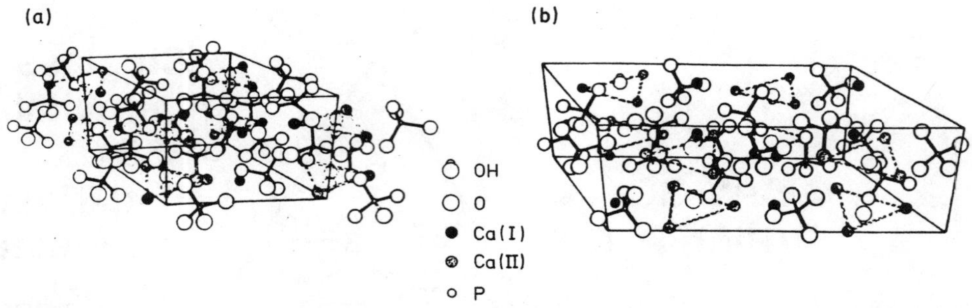

FIGURE 13.5 Schematic of hydroxyapatite crystal structure: (*a*) hexagonal; (*b*) monoclinic. (*From Kohn and Ducheyne, 1992, with permission.*)

bonding is therefore expected. Additional potential advantages of bioactive ceramics include low thermal and electrical conductivity, elastic properties similar to those of bone, control of in vivo degradation rates through control of material properties, and the ceramic functioning as a barrier when coated onto a metal substrate (Koeneman et al., 1990).

Two aspects of the crystal chemistry of natural and synthetic apatites need to be recognized. First, the HA in bone is nonstoichiometric, has a Ca/P ratio of less than 1.67, and contains carbonate ions, sodium, magnesium, fluorine, and chlorine (Posner, 1985a). Second, most synthetic hydroxyapatites actually contain substitutions for the phosphate and/or hydroxyl groups and vary from the ideal stoichiometry and Ca/P ratios. Oxyhydroxyapatite [$Ca_{10}(PO_4)_6O$], α-tricalcium phosphate (α-TCP), β-tricalcium phosphate (β-TCP) or β-Whitlockite [$Ca_3(PO_4)_2$], tetracalcium phosphate ($Ca_4P_2O_9$), and octocalcium phosphate [$Ca_8(HPO_4)_2(PO_4)_4{\cdot}5H_2O$] have all been detected via x-ray diffraction (XRD), Fourier transform infrared spectroscopy (FTIR), and chemical analyses (Kohn and Ducheyne, 1992; Ducheyne et al., 1986, 1990; Koch et al., 1990). These compounds are not apatites per se since the crystal structure differs from that of actual apatite.

Numerous processing techniques have been developed for producing synthetic apatites, including hydrolysis, hydrothermal synthesis and exchange, sol-gel techniques, wet chemistry, and conversion of natural bone and coral (Koeneman et al, 1990). Because of the variability in processing strategies, it is critical to appreciate that differences in the structure, chemistry, and composition of apatites arise from differences in material processing techniques, time, temperature, and atmosphere. Understanding the processing-composition-structure-processing synergy for calcium phosphates is therefore critical to understanding the in vivo function of these materials.

As stoichiometric HA is heated from room temperature, it may become dehydrated. Between 25 and 200°C, adsorbed water is reversibly lost. Between 200 and 400°C, lattice-bound water (from H_2O or $2HPO_4^{2-}$ substituting OH^- or PO_4^{3-} ions) is irreversibly lost, causing a contraction of the crystal lattice. Lattice water is only present when synthetic apatite is prepared from an aqueous system. At temperatures greater than 850°C, a reversible weight loss occurs, indicating another reversible dehydration reaction. Should pyrolysis occur, an oxyhydroxyapatite forms. Above 1050°C, HA may decompose into β-TCP and tetracalcium phosphate, by the following reaction: $Ca_{10}(PO_4)_6(OH)_2 \rightarrow 2\beta Ca_3(PO_4)_2 + Ca_4P_2O_9 + H_2O$ (Van Raemdonck et al., 1984). At temperatures above 1350°C, β-TCP transforms into α-TCP, which is retained on cooling. Analogous reactions occur with nonstoichiometric hydroxyapatite, but the reaction products differ as a function of the Ca/P ratio (Van Raemdonck et al., 1984).

The dissolution behavior of HA in an aqueous medium and subsequent biological reactions also depend on the chemical composition of the crystal. Ion exchange occurring at the apatite surface depends on (1) the rate of formation and dissolution of the various phases, (2) the powder-weight-to-liquid-volume ratio, (3) pH, (4) specific surface area, (5) crystal defects, impurities, and vacancies, and (6) substitutional ions.

The general mechanism of biological bonding to calcium phosphates is as follows (de Bruijn et al., 1995). Differentiated osteoblasts secrete a mineralized matrix at the ceramic surface, resulting in a narrow, amorphous electron-dense band approximately 3 to 5 μm thick. Collagen bundles form between this zone and the cells. Bone mineral crystals nucleate within this amorphous zone, first in the form of an octocalcium phosphate precursor phase, and, ultimately they undergo a conversion to HA. As the healing site matures, the bonding zone shrinks to approximately 0.05 to 0.2 μm, and bone attaches through a thin epitaxial bonding layer to the bulk implant as the growing bone crystals align with the apatite crystals of material.

Calcium-phosphate-based bioceramics have also been used as coatings on dense implants and porous surface layers to accelerate and enhance fixation of a substrate biomaterial to tissue (Kohn and Ducheyne, 1992; Cook et al., 1992; Ducheyne et al., 1980; Oonishi et al., 1994). Results of these studies vary with respect to bond strength, solubility, and overall in vivo function, suggesting a window of material variability in parallel with a window of biologic variability (Kohn and Ducheyne, 1992).

Numerous processing techniques are used to deposit and bond Ca-P powders to substrates, inclusing plasma and thermal spraying (de Groot et al., 1987; Koch et al., 1990), ion-beam and other sputter-deposition techniques (Ong et al., 1991; Wolke et al., 1994), electrophoretic deposition (Ducheyne et al., 1986, 1990), sintering (de Groot, 1983; Ducheyne et al., 1986, 1990), sol-gel techniques (Chai et al., 1998), pulsed laser deposition (Garcia et al., 1998), and chemical vapor deposition (Gao et al., 1999).

During plasma spraying, powder is sprayed onto a metal core. Plasma temperatures as high as 10,000°C can be reached. The ceramic particles form a loosely bonded coating on impact. Due to the rapid cooling of the particles, the temperature of the metal remains low, and the structure and properties of the bulk metal remain unchanged. However, porosity and phase changes are induced in the ceramic. For example, mixtures of HA, TCP, and tetracalcium phosphate evolve as a result of plasma spraying nominally pure HA (Koch et al., 1990; Radin and Ducheyne, 1992).

During sputter deposition, a beam (ion) sputters off atoms from a target to form a thin coating on a substrate. During electrophoretic deposition, the calcium-phosphate ceramic (CPC) is precipitated out of suspension onto a substrate. An advantage of this technique, as well as low-temperature sol-gel techniques, is that it is more conducive to coating the internal surfaces of a porous substrate. Both sintering and electrophoretic deposition of hydroxyapatite onto titanium result in an interfacial layer rich in Ti and P (Ducheyne et al., 1986). The interfacial zone has a different composition from that of the bulk Ca-P coating.

The different structures and compositions of Ca-P coatings resulting from different processing modulate the subsequent biological reactions. For example, increased Ca/P ratios, fluorine and carbonate contents, and degree of crystallinity all lead to greater stability of the biological precipitate (Posner, 1985b; Van Raemdonck et al., 1984). In general, calcium phosphates with Ca/P ratios in the range 1.5 to 1.67 (tricalcium phosphate and HA, respectively) yield the most beneficial tissue response. Given the range of chemical compositions available in bioactive ceramics and the resulting fact that pure HA is used rarely, the broader term *calcium-phosphate ceramics* has been proposed in lieu of the more specific term *hydroxyapatite* (Ducheyne, 1987). Each individual CPC is defined by its own unique set of chemical and physical properties.

13.3.3 Bioactive Ceramic Composites

Bioactive ceramics typically exhibit low strength and toughness. The reason for the mechanical deficiency is that the design requirement of bioactivity supersedes any mechanical property requirement, and as a result, mechanical properties are restricted. Bioceramic composites therefore have been synthesized as a means of creating a material with superior properties to that of the individual constituents. Three separate design objectives are used as criteria for developing bioceramic composites: (1) use the beneficial biological response to bioceramics and reinforce the ceramic with a second phase to strengthen the ceramic, (2) use bioceramic materials as the second phase to achieve desirable strengths and stiffnesses, and (3) synthesize scaffold materials for tissue (re)generation (Ducheyne, 1987).

Bioactive glass composites have been synthesized via thermal treatments that create a second phase (Gross and Strunz, 1980, 1985; Kitsugi et al., 1986). By altering the firing temperature and composition of the bioactive glasses, stable multiphase bioactive glass composites have been produced. Adding oxyapatite, fluorapatite, β-Wollastonite, and/or β-Whitlockite results in bending strengths two to five times greater than that of unreinforced bioactive glasses (Kitsugi et al., 1986). Calcium phosphates have been strengthened via incorporation of glasses, alumina, and zirconia (Ioku et al., 1990; Knowles and Bonfield, 1993; Li et al., 1995).

13.3.4 Critical Properties of Bioactive Ceramics

Important issues in bioactive ceramics research and development still include characterization of the processing-composition-structure-property synergy, characterization of in vivo function, and establishing the relationship between in vitro and in vivo function (Kohn and Ducheyne, 1992). Parameters important to understanding and improving the ceramic-tissue bond and reactions at the ceramic surface and in the tissues have been outlined (Ducheyne, 1987): (1) characterization of surface activity, including surface analysis, biochemistry, and ion transport, (2) physical chemistry, pertaining to strength and degradation, stability of the tissue-ceramic interface, and tissue resorption, and (3) biomechanics, as related to strength, stiffness, design, wear, and tissue remodeling. It is essential to note that these properties are time-dependent and therefore must be characterized as a function of loading and environmental history. The importance of this characterization is underscored by the variability in experimental results, discussed earlier, and the potential limitations of bioactive materials, such as fatigue, fracture, and coating-substrate failure and unwanted material degradation.

Specific physicochemical properties that are important to characterize and relate to in vitro and in vivo biologic response include powder particle size and shape; pore size, shape, and distribution; specific surface area; phases present; crystal structure and size; grain size; density; coating thickness; hardness; and surface roughness (Kohn and Ducheyne, 1992).

Starting powders may be identified for their particle size, shape, and distribution via sifting techniques or more quantitative stereological imaging techniques. Pore size, shape, and distribution, important properties with respect to strength and bioreactivity, also may be quantified via stereological methods and/or scanning electron microscopy (SEM). Specific surface area, important in understanding the dissolution and precipitation reactions at the ceramic-fluid interface, may be characterized by the Brunnauer, Emmet, and Teller (BET) technique.

Phase identification may be accomplished via XRD. FTIR is recommended as a complementary technique because it allows identification of phase amounts and structures not readily detectable with XRD (Ducheyne, 1990). Grain sizes may be determined through either optical microscopy, SEM, or transmission electron microscopy (TEM), depending on the order of the grain size. Additionally, TEM is useful to characterize second phases, crystal structure, and lattice imperfections. Auger electron spectroscopy (AES) and x-ray photoelectron spectroscopy (XPS) may also be utilized to determine surface and interfacial compositions. Chemical stability and surface activity may be analyzed via XPS and measurements of ionic fluxes and zeta potentials. It is assumed that two different pathways of activity exist: solution and cell-mediated (Jarcho, 1981).

An additional parameter that should be considered in evaluating chemical stability and surface activity of bioceramics is the simulated in vivo environment. Factors such as the type and concentration of electrolytes in solution and the presence of proteins or cells may influence in vitro immersion results. For example, in a study on glass-ceramic A-W that can be generalized to other bioceramics, a solution with constituents, concentrations, and pH equivalent to human plasma most accurately reproduces in vivo surface structural changes, whereas more standard buffers do not reproduce these changes (Kokubo et al., 1990b). Such control of the ionic environment also forms the basis for biomimetic approaches discussed in the last section of this chapter.

The physiologic response at a biomaterial-tissue interface depends on both the implant and the tissue. Therefore, analyses of both these constituents must be well posed: the implant surface must be analyzed, and the species released into the environment and tissues also must be determined. Analysis can be accomplished with solution chemical methods, such as atomic absorption spectroscopy; physical methods, such as thin-film XRD, electron microprobe analysis (EMP), energy dispersive x-ray analysis (EDXA), and FTIR; and surface-sensitive methods, such as AES, XPS, and secondary ions mass spectroscopy (SIMS) (Fig. 13.6). The type of loading is also important in analyzing bone-bonding behavior. The integrity of the implant-tissue interface is strongly dependent on the specific nature of the loading pattern since loading may alter the chemical and mechanical behavior of the interface (Kohn and Ducheyne, 1992).

The major factor limiting expanded use of bioactive ceramics is their low tensile strength and fracture toughness. Their use in bulk form is therefore limited to functions in which only compressive loads are applied. To use ceramics as self-standing implants that are able to withstand tensile stresses is a primary engineering design objective. Four general approaches have been used to achieve this objective: (1) use of the bioactive ceramic as a coating on a metal or ceramic substrate (Ducheyne et al., 1980), (2) strengthening of the ceramic, such as via crystallization of glass (Gross et al., 1981); (3) use of fracture mechanics as a design approach (Ritter et al., 1979); and (4) reinforcing of the ceramic with a second phase (Ioku et al., 1990; Kitsugi et al., 1986; Knowles and Bonfield, 1993; Li et al., 1995).

No matter which of these four strategies is used, the resulting bioactive ceramic must be stable, both chemically and mechanically, until it fulfills its intended function(s). It should be noted, however, that the specific properties needed depend on the specific application (Kohn and Ducheyne, 1992). For example, if a smooth total hip prosthesis is to be fixed by coating the metal stem with a CPC coating, then the ceramic-metal bond must remain intact throughout the service life of the

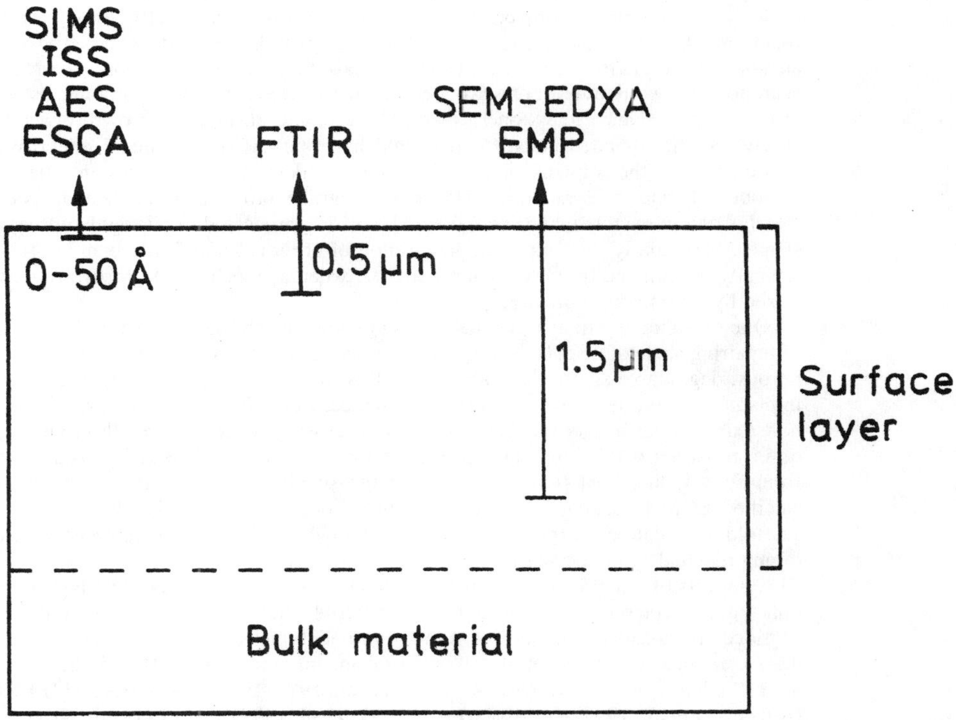

FIGURE 13.6 Schematic of sampling depths for different surface analysis techniques used to characterize bioceramics. (*From Kohn and Ducheyne, 1992, with permission.*)

prosthesis. However, if the CPC coating is used on a porous-coated prosthesis with the intent of accelerating ingrowth into the pores of the metal, then the ceramic-metal bond need only be stable until tissue ingrowth is achieved.

In either scenario, mechanical testing of the ceramic-metal bond is a critical area (Filiaggi et al., 1991; Mann et al., 1994). A number of tests are available, including pullout, lap-shear, three- and four-point-bending, double-cantilever-beam, double-torsion, indentation, and scratch tests (Koeneman et al., 1990). However, there are inherent difficulties in performing each of these tests, particularly for thin, brittle coatings. A more fundamental fracture mechanics approach has been taken through the use of an interfacial fracture toughness test with a resulting reduction in data variability (Filiaggi and Pilliar, 1991).

Ex vivo pushout tests indicate that the ceramic-metal bond fails before the ceramic-tissue bond and is the weak link in the system (Kohn and Ducheyne, 1992). There is therefore reason for concern about the weak ceramic-metal bond and integrity of that interface over a lengthy service life of functional loading. Material function under dynamic loading represents a critical parameter and a much-needed research area (Koeneman et al., 1990; Lemons, 1988).

13.4 CERAMICS FOR TISSUE ENGINEERING/BIOLOGICAL THERAPY

An ideal tissue substitute would possess the biological advantages of an autograft and supply advantages of an allograft (Laurencin et al., 1996) but alleviate the complications to which each of these types of grafts is subject. An ideal bone substitute would also satisfy the following design requirements (Yaszemski et al., 1996): (1) biocompatibility, (2) osteoconductivity (it should provide an appropriate environment for attachment, proliferation, and function of osteoblasts, leading to secretion of osteoid or new bone extracellular matrix), (3) ability to incorporate osteoinductive factors to direct and enhance new bone growth, (4) it should allow for ingrowth of vascularity to ensure survival of transplanted cells, (5) it should have sufficient mechanical integrity to support loads encountered at the implant site, (6) it should be degradable, with a controlled, predictable, and reproducible rate of degradation, (7) it should degrade into nontoxic molecular species that are easily metabolized or excreted, and (8) it should be readily processed into irregular three-dimensional (3D) shapes. Particularly difficult is the integration of criteria 4 and 5 into one design since transport is typically maximized by maximizing porosity, whereas mechanical properties are frequently maximized by minimizing porosity.

One new strategy to achieve these design goals, which may ultimately lead to an improved way of repairing skeletal defects, is to create a composite graft in which autogenous cells are seeded onto porous, degradable scaffolds. Following cell seeding, cell-scaffold constructs may be immediately implanted or cultured further and then implanted. In the latter case, the cells proliferate and secrete new extracellular matrix (ECM) and factors necessary for tissue growth in vitro, and the scaffold-tissue construct is then implanted as a graft. Once implanted, the scaffold is also populated by cells from surrounding host tissue. Ideally, for the specific case of bone regeneration, secretion of a calcified ECM by osteoblasts and subsequent bone growth and development are concurrent with scaffold degradation. In the long term, a functional ECM and tissue are regenerated and are devoid of any residual synthetic scaffold.

Synthesis of bonelike tissue has been pursued by culturing cells capable of expressing the osteoblast phenotype onto synthetic or natural materials that mimic aspects of natural ECMs. A number of bioceramics satisfy the eight design requirements listed earlier, including bioactive glasses and glass ceramics (Ducheyne et al., 1994; El-Ghannam et al., 1997); HA, TCP, and coral (Ohgushi et al., 1990; Kuznetsov et al., 1997; Krebsbach et al., 1997, 1998, 1999); HA and HA/TCP plus collagen (Kuznetsov et al., 1997; Krebsbach et al., 1997, 1998, 1999); and PLGA-apatite composites (Murphy et al., 2000a; Attawia et al., 1995; Thomson et al., 1998). An important consideration is that varying the scaffold, even subtly, can lead to a significant variation in biologic effect in vitro (e.g., osteoblast attachment, osteoblast proliferation, collagen and noncollagenous protein synthesis, and RNA transcription) (Puleo et al., 1991; El-Ghannam et al., 1997; Zreiqat et al., 1999). The nature of the scaffold can also significantly affect the outcome in vivo procedures (e.g., rate of new bone formation, intensity or duration of any transient or sustained inflammatory response, and progenitor cell differentiation to osteoblasts) (Ohgushi et al., 1990; Kuznetsov et al., 1997; Krebsbach et al., 1997, 1998, 1999; James et al., 1999).

13.5 BIOMIMETIC CERAMICS

Through millions of years of evolution, the skeleton has evolved into a near-optimally designed system that performs the functions of load bearing, organ protection, and chemical balance efficiently and with a minimum expenditure of energy. Traditional engineering approaches might have accomplished these design goals by using materials with much greater mass. However, nature has designed the skeleton to be an efficient, relatively lightweight system. The reasons why the skeletal system can perform its functions using a minimum amount of mass are that nature has used several important design approaches. First is the ability to adapt to environmental cues; i.e., physiologic systems are smart. Second, tissues are hierarchical composites consisting of elegant interdigitations of organic and inorganic constituents that are synthesized via solution chemistry under benign conditions. Third, nature has optimized the orientation of the constituents and developed functionally graded materials; i.e., the organic and inorganic phases are heterogeneously distributed to accommodate variation in anatomic needs.

A biomimetic surface could promote preferential absorption of biologic molecules that regulate cell function, serving to promote events leading to biomineralization. The rationale for this strategy is based on the mechanisms of action of bioactive materials discussed in Sec. 13.3. Bioactive glasses and glass ceramics bond to bone through a layer of bonelike apatite, which forms on the surfaces of these materials in vivo and is characterized by a carbonate-containing apatite with small crystallites and defective structure (Ducheyne, 1987; Nakamura et al., 1985). This type of apatite is not observed at the interface between nonbioactive materials and bone. It is therefore hypothesized that the essential requirement for a biomaterial to bond to living bone is the formation of a biologically active bonelike apatite layer (Kohn and Ducheyne, 1992; Ducheyne, 1987; Nakamura et al., 1985). The bonelike apatite layer is hypothesized to form by a mechanism in which Ca and Si ions dissolve from the surfaces of glass or glass ceramics and react with PO_4 ions from the surrounding body fluid. The dissolved Ca ions increase the degree of supersaturation of the surrounding body fluid with respect to apatite, whereas the hydrated silica formed on the glass surfaces provides favorable sites for apatite nucleation (Ohtsuki et al., 1992; de Bruijn et al., 1995). Once apatite nuclei form, they spontaneously grow, consuming Ca and PO_4 ions from the surrounding fluid.

On the basis of these findings, it has been hypothesized that a bonelike apatite layer could be formed in vitro at normal pressure and temperature (Murphy et al., 2000a; Abe et al., 1990; Li et al., 1992; Bunker et al., 1994; Campbell et al., 1996; Tanahashi et al., 1995; Yamamoto et al., 1997; Wu et al., 1997; Wen et al., 1997). The basis for in vitro mineralization processes lies in the observation that in nature organisms use various macromolecules to control nucleation and growth of mineral phases (Weiner, 1986; Bunker et al., 1994). These macromolecules usually contain functional groups that are negatively charged at the crystallization pH (Weiner, 1986). It is hypothesized that these groups chelate ionic species present in the surrounding media, stimulating crystal nucleation (Bunker et al., 1994). The presence (or absence) of a bioapatite may determine the specific location of cellular activity on a biomaterial surface. The key requirement is to chemically modify a substrate to produce a surface that induces heterogeneous nucleation from an aqueous solution (Bunker et al., 1994). Such modifications may be achieved via a number of processes, including grafting, irradiation, NaOH treatment, or simple hydrolysis (Murphy et al., 2000a, 2000b, 2000c; Tanahashi et al., 1995; Yamamoto et al., 1997; Wu et al., 1997; Hanawa et al., 1998). Such biomimetic strategies have primarily been utilized with metals to accelerate osseointegration (Kohn, 1998; Abe et al., 1990; Campbell et al., 1996; Wen et al., 1997; Hanawa et al., 1998) and, more

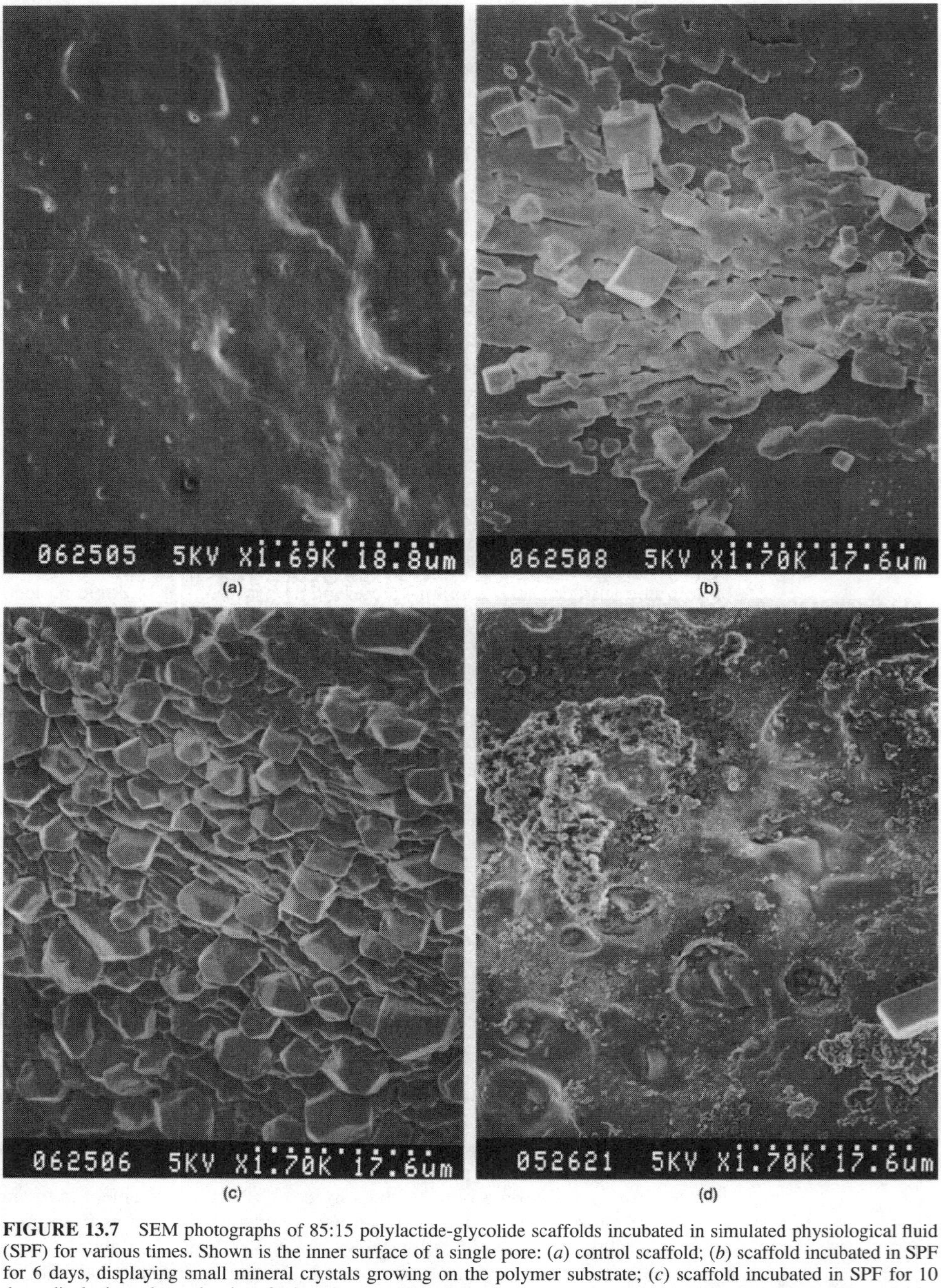

FIGURE 13.7 SEM photographs of 85:15 polylactide-glycolide scaffolds incubated in simulated physiological fluid (SPF) for various times. Shown is the inner surface of a single pore: (*a*) control scaffold; (*b*) scaffold incubated in SPF for 6 days, displaying small mineral crystals growing on the polymer substrate; (*c*) scaffold incubated in SPF for 10 days, displaying a large density of mineral crystals growing on the polymer substrate; (*d*) scaffold incubated in SPF for 16 days, displaying continuous mineral film grown over the polymer substrate. All photographs at 1700×. (*From Murphy et al., 2000a, with permission.*)

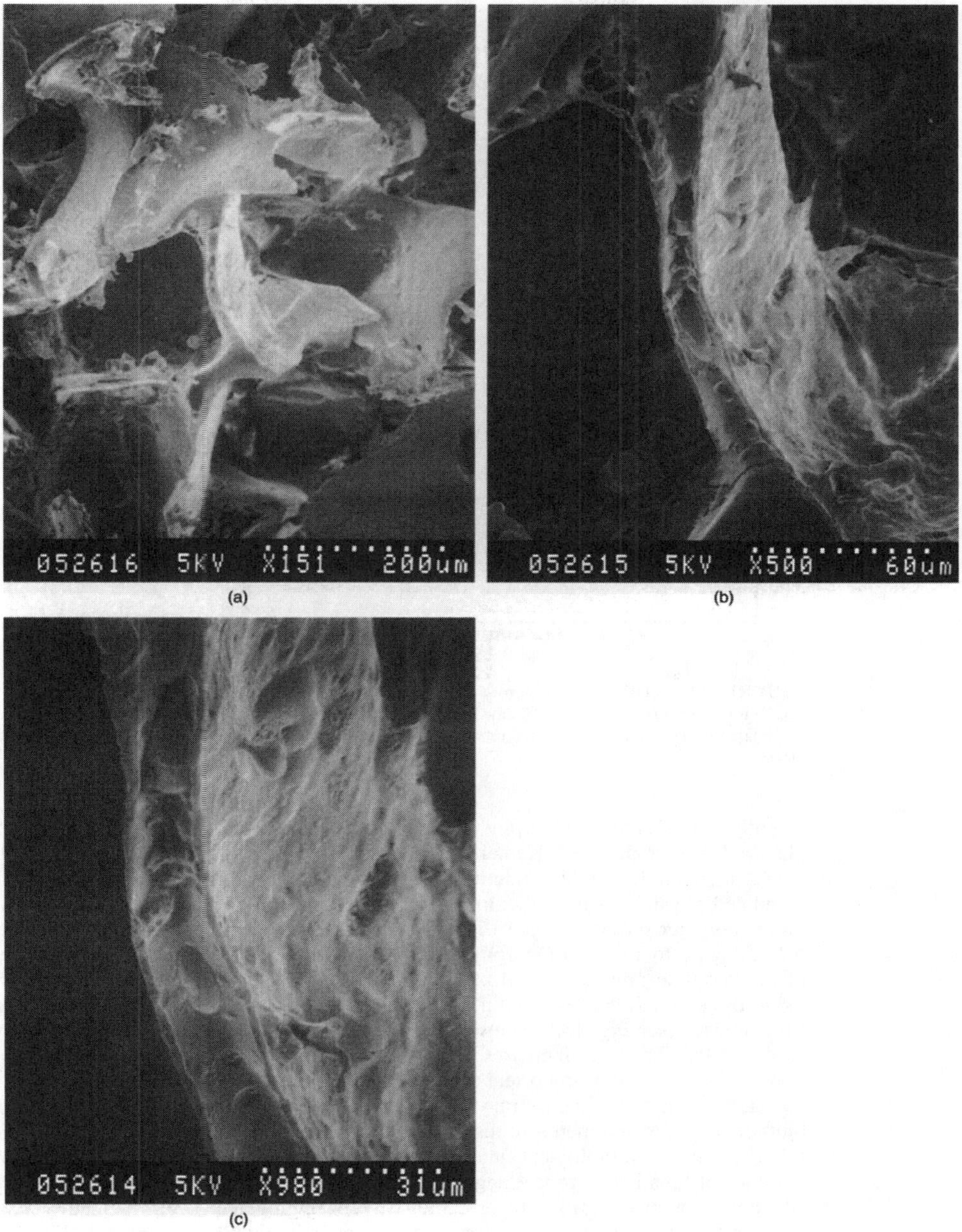

FIGURE 13.8 SEM photographs of the cross section of a porous 85:15 polylactide-glycolide scaffold incubated in simulated physiological fluid (SPF) for 16 days: (*a*) large pore structure of scaffold (150×); (*b*) cross section of mineralized pore wall (500×); (*c*) cross section of mineralized pore wall (1000×). Mineralization does not significantly reduce the porosity of the scaffold. (*From Murphy et al., 2000a, with permission.*)

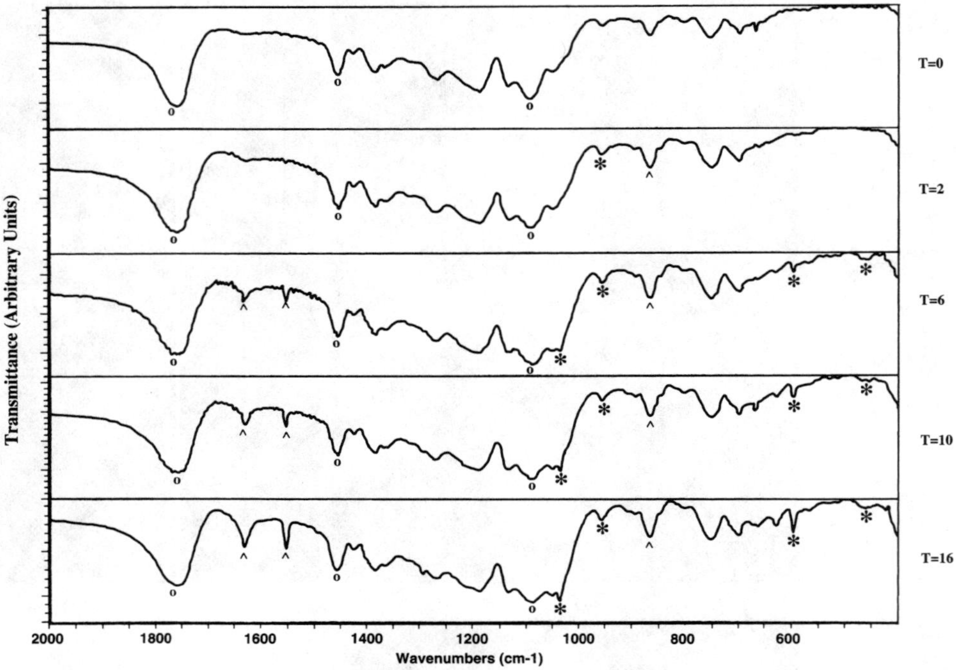

FIGURE 13.9 FTIR spectra of porous 85:15 polylactide-glycolide scaffold incubated in simulated physiological fluid (SPF) for various times ($T = 0-16$ days). The spectra show the development of phosphate (*) and carbonate (Λ) peaks as incubation time increases. Peaks representing polymer (o) are also indicated. *(From Murphy et al., 2000a, with permission.)*

recently, with polymers (Murphy et al., 2000a, 2000b, 2000c; Tanahashi et al., 1995; Yamamoto et al., 1997; Wu et al., 1997; Kamei et al., 1997; Du et al., 1999; Taguchi et al., 1999).

As an example of a biomimetic strategy, porous polylactide-co-glycolide scaffolds incubated in simulated physiological fluid exhibit nucleation and growth of a continuous bonelike apatite layer on the polymer surfaces (Fig. 13.7) and within the pores (Fig. 13.8) after relatively short incubation times (e.g., 7 to 14 days) (Murphy et al., 2000a, 2000b, 2000c). FTIR analyses confirm the nature of the bonelike mineral (Fig. 13.9). There was also a five-fold increase in compressive modulus (Murphy, et al., 2000a) without a significant reduction in scaffold porosity due to mineral growth. SEM images (see Fig. 13.8) demonstrate that the mineral sheath is only 1 to 10 μm thick (Murphy et al., 2000a). The large increases in mechanical properties with the addition of only a thin bonelike mineral are extremely important in light of the competing design requirements of transport and mechanics, which frequently may only be balanced by choosing an intermediate porosity. Such a biomimetic approach therefore results in a powerful model system for analyzing the effects of material chemistry on biological function without the usual concurrent architectural changes. The importance of biomimetic processing conditions (e.g., room temperature, atmospheric pressure) is that attachment or incorporation of osteoinductive or angiogenic factors is readily achievable without concern for denaturing, thus enabling a dual conductive/inductive approach (Fig. 13.10) (Murphy et al., 2000b, 2000c). In general, such biomimetic processes are guided by the pH and ionic concentration of the microenvironment, and conditions conducive to heterogeneous nucleation will support epitaxial growth of mineral (Fig. 13.11).

It should be noted that these biomimetic approaches do not completely mimic nature; rather, just selected biological aspects are mimicked. However, if selected biomimetic aspects are rationally

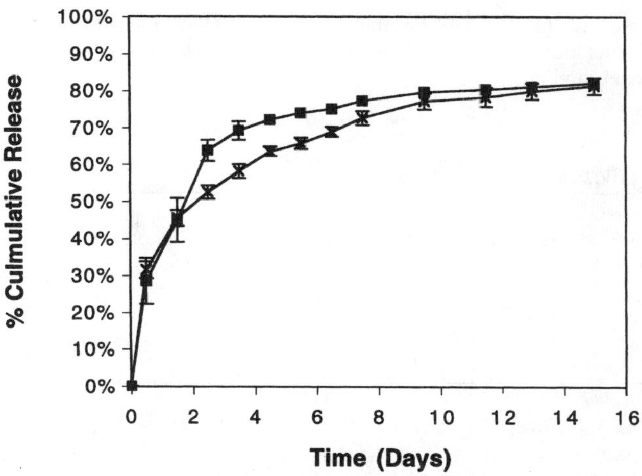

FIGURE 13.10 Cumulative release of vascular endothelial growth factor (VEGF) from mineralized (X) and nonmineralized (■) porous polylactide-glycolide scaffolds. Values represent the mean and standard deviation of five samples. *(From Murphy et al., 2000c, with permission.)*

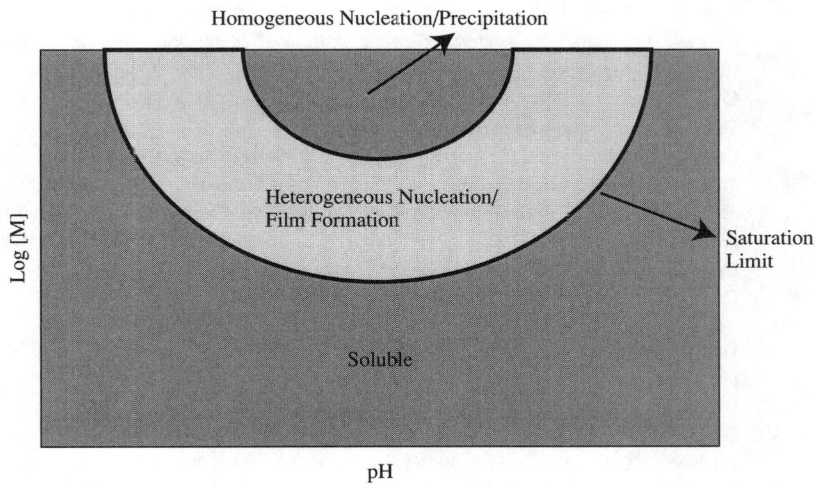

$$\Delta G = -RT \ln S + \sigma_{cl}A_{cl} + (\sigma_{cl} - \sigma_{sl})A_{cs}$$

FIGURE 13.11 Schematic depicting a design space for biomimetic processing of materials. Variations in ionic concentration and pH modulate mineral nucleation. Heterogeneous nucleation and film formation on a substrate are desirable. The free energy for crystal nucleation ΔG is a function of the degree of solution supersaturation S, temperature T, crystal interfacial energy σ, and crystal surface area A Subscripts c, s, and l denote interfaces involving the crystal, solid substrate, and liquid, respectively.

designed into biomaterial, then the biological system will be able to respond in a more controlled, predictable, and efficient manner, providing an exciting new arena for biomaterials research and development.

ACKNOWLEDGMENT

Parts of the author's research discussed in this chapter were supported by NIH/NIDCR Grant DE-13380.

REFERENCES

Abe, Y., Kokubo, T., Yamamuro, T., *J. Mater. Sci.: Mater. Med.* **1**:233–238, 1990.

Allen, M., Myer, B., Rushton, N., *J. Biomed. Mater. Res. (Appl. Biomat.)* **58**:319–328, 2001.

Attawia, M. A., Devin, J. E., Laurencin, C. T., *J. Biomed. Mater. Res.* **29**:843–848, 1995.

Bokros, J. C., *Trans. Biomed. Mater. Res. Symp.* **2**:32–36, 1978.

Bokros, J. C., LaGrange, L. D., Schoen, G. J., in *Chemistry and Physics of Carbon*, Walker, P. L., ed. New York: Dekker, 1972, vol. 9, pp. 103–171.

Boutin, P., Christel, P., Dorlot, J. M., et al., *J. Biomed. Mater. Res.* **22**:1203–1232, 1988.

Bunker, B. C., Rieke, P. C., Tarasevich, B. J., et al., *Science* **264**:48–55, 1994.

Cales, B., Stefani, Y., in *Biomedical Engineering Handbook*, Bronzino, J. D., ed. Boca Raton, Fla.: CRC Press, 1995, pp 415–452.

Campbell, A. A., Fryxell, G. E., Linehan, J. C., Graff, G. L., *J. Biomed. Mater. Res.* **32**:111–118, 1996.

Chai, C. S., Gross, K. A., Ben-Nissan, B., *Biomaterials* **19**:2291–2296, 1998.

Christel, P., Meunier, A., Leclercq, S., et al., *J. Biomed. Mater. Res.: Appl. Biomat.* **21**(A2):191–218, 1987.

Christel, P., Meunier A., Heller, M., et al., *J. Biomed. Mater. Res.* **23**:45–61, 1989.

Cook, S. D., Thomas, K. A., Dalton, J. E., et al., *J. Biomed. Mater. Res.*, **26**:989–1001, 1992.

Davidson, J. A., *Clin. Orthop.* **294**:361–178, 1993.

Dawihl, W., Mittelmeier, H., Dorre, E., et al., *Med. Orthop. Tech.* **99**:114–118, 1979.

de Bruijn, J. D., van Blitterswijk, C. A., Davies, J. E., *J. Biomed. Mater. Res.* **29**:89–99, 1995.

de Groot, K., ed., *Bioceramics of Calcium Phosphate*. Boca Raton, Fla.: CRC Press, 1983.

de Groot, K., Geesink, R. G. T., Klein, C. P. A. T., Serekian, P., *J. Biomed. Mater. Res.* **21**:1375–1381, 1987.

Derbyshire, B., Fisher, J., Dowson, D., et al., *Med. Eng. Phys.* **16**:229–236, 1994.

Dorre, E., Dawihl, W., in *Mechanical Properties of Biomaterials*, Hastings, G. W., Williams, D. F., eds. New York, Wiley, 1980, pp. 113–127.

Du, C., Cui, F. Z., Zhu, X. D., de Groot, K., *J. Biomed. Mater. Res.* **44**:407–415, 1999.

Ducheyne, P., *J. Biomed. Mater. Res.* **19**:273–291, 1985.

Ducheyne, P., *J. Biomed. Mater. Res.: Appl. Biomat.* **21**(A2):219–236, 1987.

Ducheyne, P., Hench, L. L., Kagan, A. II, et al., *J. Biomed. Mater. Res.* **14**:225–237, 1980.

Ducheyne, P., Van Raemdonck, W., Heughebaert, J. C., Heughebaert, M., *Biomaterials* **7**:97–103, 1996.

Ducheyne, P., Radin, S., Heughebaert, M., Heughebaert, J. C., *Biomaterials* **11**:244–254, 1990.

Ducheyne, P., El-Ghannam A., Shapiro I., *J. Cell Biochem.* **56**:162–167, 1994.

Effah Kaufmann, E. A. B., Ducheyne, P., Shapiro, I. M., *J. Biomed. Mater. Res.* **52**:783–796, 2001.

El-Ghannam, A., Ducheyne, P., Shapiro, I. M., *J. Biomed. Mater. Res.* **36**:167–180, 1997.

Filiaggi, M. J., Coombs, N. A., Pilliar, R. M., *J. Biomed. Mater. Res.* **25**:1211–1229, 1991.

Filiaggi, M. J., Pilliar, R. M., Yakubovich, R., Shapiro, G., *J. Biomed. Mater. Res.: Appl. Biomat.* **33**:225–238, 1996a.

Filiaggi, M. J., Pilliar, R. M., Abdulla, D., *J. Biomed. Mater. Res.: Appl. Biomat.* **33**:239–256, 1996b.

Gao, Y., in *Biomedical Materials: Drug Delivery, Implants and Tissue Engineering*, Neenan, T., Marcolongo, M., Valentini, R. F., eds. Warrendale, Pa.: Materials Research Society, 1999, pp. 361–366.

Garcia, F., Arias, J. L., Mayor, B., et al., *J. Biomed. Mater. Res.: Appl. Biomat.* **43**:69–76, 1998.

Gross, U., Strunz, V., *J. Biomed. Mater. Res.* **14**:607–618, 1980.

Gross, U., Strunz, V., *J. Biomed. Mater. Res.* **19**:251–271, 1985.

Gross, U., Brandes, J., Strunz, V., et al., *J. Biomed. Mater. Res.* **15**:291–305, 1981.

Hanawa, T., Kon, M., Ukai, H., et al., *J. Biomed. Mater. Res.* **41**:227–236, 1998.

Haubold, A. D., Shim, H. S., Bokros, J. C., in *Biocompatibility of Clinical Implant Materials*, Williams, D. F., ed. Boca Raton, Fla.: CRC Press, 1981, vol. II, pp. 3–42.

Heimke, G., Jentschura, G., Werner, E., *J. Biomed. Mater. Res.* **12**:57–65, 1978.

Hench L. L., in *Biomaterials Science: An Introduction to Materials in Medicine*, Ratner, B. D., Hoffman, A. S., Schoen, F. J., Lemons, J. E., eds. San Diego, Academic Press, 1996, pp. 73–84.

Hench, L. L., Clark, A. E., in *Biocompatibility of Orthopaedic Implants*, Williams, D. F., ed. Boca Raton, Fla.: CRC Press, 1982, vol. II, pp. 129–170.

Hench, L. L., Ethridge, E. C., *Biomaterials: An Interfacial Approach.* New York: Academic Press, 1982.

Hench, L. L., Splinter, R. J., Allen, W. C., Greenlee, T. K. Jr., *J. Biomed. Mater. Res. Symp.* **2**:117–141, 1972.

Hulbert, S. F., Young, F. A., Mathews, R. S., et al., *J. Biomed. Mater. Res.* **4**:433–456, 1970.

Huttner, W., Huttinger, K. J., in *The Cementless Fixation of Hip Endoprostheses*, Morscher, E., ed. Berlin: Springer-Verlag, 1984, pp. 81–94.

Ioku, K., Yoshimura, M., Somiya, S., *Biomaterials* **11**:57–61, 1990.

James, K., Levene, H., Parson, J. R., Kohn, J., *Biomaterials* **20**:2203–2212, 1999.

Jarcho, M., *Clin. Orthop.* **157**:259–278, 1981.

Kamei, S., Tomita, N., Tamai, S., et al., *J. Biomed. Mater. Res.* **37**:384–393, 1997.

Kitsugi, T., Yamamuro, T., Nakamura, T., et al., *J. Biomed. Mater. Res.* **20**:1295–1307, 1986.

Kitsugi, T., Nakamura, T., Yamamuro, T., et al, *J. Biomed. Mater. Res.* **21**:1255–1271, 1987.

Knowles, J. C., Bonfield, W., *J. Biomed. Mater. Res.* **27**:1591–1598, 1993.

Ko, C. C., Kohn, D. H., Hollister, S. J., *J. Mater. Sci. Mater. Med.* **7**:109–117, 1995.

Koch, B., Wolke, J. G. C., de Groot, K., *J. Biomed. Mater. Res.* **24**:655–667, 1990.

Koeneman, J., Lemons, J., Ducheyne, P., et al., *J. Appl. Biomat.* **1**:79–90, 1990.

Kohn, D. H., *Curr. Opin. Solid State Mater. Sci.* **3**:309–316, 1998.

Kohn, D. H., Ducheyne, P., in *Medical and Dental Materials*, D. F. Williams, ed. Berlin: VCH Verlagsgesellschaft, 1992, pp. 29–109.

Kokubo, T., Ito, S., Huang, Z. T., et al., *J. Biomed. Mater. Res.* **24**:331–343, 1990a.

Kokubo, T., Kushitani, H., Sakka, S., et al., *J. Biomed. Mater. Res.* **24**:721–734, 1990b.

Krebsbach, P. H., Kuznetsov, S. A., Satomura, K., et al., *Transplantation* **63**:1059–1069, 1997.

Krebsbach, P. H., Mankani, M. H., Satomura, K., et al., *Transplantation* **66**:1272–1278, 1998.

Krebsbach, P. H., Kuznetsov, S. A., Bianco, P., Gehron-Robey, P., *Crit. Rev. Oral Biol. Med.* **10**:165–181, 1999.

Kumar, P., Oka, M., Ikeuchi, K., et al., *J. Biomed. Mater. Res.* **25**:813–828, 1991.

Kuznetsov, S. A., Krebsbach, P. H., Satomura, K., et al., *J. Bone Min. Res.* **12**:1335–1347, 1997.

Laurencin, C. T., El-Amin, S. F., Ibim, S. E., et al., *J. Biomed. Mater. Res.* **30**:133–138, 1996.

Lemons, J. E., *Clin. Orthop.* **235**:220–223, 1988.

Li, P., Ohtsuki, C., Kokubo, T., et al., *J. Am. Ceram. Soc.* **75**:2094–2097, 1992.

Li, J., Fartash, B., Hermansson, L., *Biomaterials* **16**:417–422, 1995.

Mann, K. A., Edidin, A. A., Kinoshita, R. K., Manley, M. T., *J. Appl. Biomat.* **5**:285–291, 1994.

Miller, R. A., Smialek, R. G., and Garlick, in *Advances in Ceramics*, vol. 3, *Science and Technology of Zirconia*, Westerville, Ohio: American Ceramic Society, 1981, p. 241.

Murphy, W. L., Kohn, D. H., Mooney, D. J., *J. Biomed. Mater. Res.* **50**:50–58, 2000a.

Murphy, W. L., Gilhool, K. A., Kohn, D. H., Mooney, D. J., in *Mineralization in Natural and Synthetic Biomaterials*, P. Li, P. Calvert, R. J. Levy, T. Kokubo, C. R. Schneid, eds. Warrendale, Pa.: Materials Research Society, 2000b, pp 347–352.

Murphy, W. L., Peters, M. C., Kohn, D. H., Mooney, D. J., *Biomaterials* **21**:2521–2527, 2000c.
Nakamura, T., Yamamuro, T., Higashi, S., et al., *J. Biomed. Mater. Res.* **19**:685–698, 1985.
Ogino, M., Ohuchi, F., Hench, L. L., *J. Biomed. Mater. Res.* **14**:55–64, 1980.
Ohgushi, H., Okumura, M., Tamai, S., et al. *J. Biomed. Mater. Res.* **24**:1563–1570, 1990.
Ohtsuki, C., Kokubo, T., Yamamuro, T., *J. Non-Cryst. Solids* **143**:84–92, 1992.
Ong, J. L., Harris, L. A., Lucas, L. C., et al., *J. Am. Ceram. Soc.* **74**:2301–2304, 1991.
Oonishi, H., Noda, T., Ito, S., et al., *J. Appl. Biomat.* **5**:23–37, 1994.
Posner, A. S., *Clin. Orthop.* **200**:87–99, 1985a.
Posner, A. S., *J. Biomed. Mater. Res.* **19**:241–250, 1985b
Puleo, D. A., Holleran, L. A., Doremus, R. H., Bizios, R., *J. Biomed. Mater. Res.* **25**:711–723, 1991.
Radin, S., Ducheyne, P., *Mater. Med.* **3**:33–42, 1992.
Ritter, J. E., Jr., Greenspan, D. C., Palmer, R. A., Hench, L. L., *J. Biomed. Mater. Res.* **13**:251–263, 1979.
Semlitsch, M., Lehmann, M., Weter, H., et al., *J. Biomed. Mater. Res.* **11**:537, 1977.
Soltesz, U., Richter, H., in *Metal and Ceramic Biomaterials*, vol. II: *Strength and Surface*, Ducheyne, P., Hastings, G. W., eds. Boca Raton, Fla.: CRC Press, 1984, pp 23–61.
Taguchi, T., Shiraogawa, M., Kishida, A., Arashi, M., *J. Biomater. Sci. Polymer. Edn.* **10**:19–31, 1999.
Tanahashi, M., Yao, T., Kokubo, T., et al., *J. Biomed. Mater. Res.* **29**:349–357, 1995.
Thomson, R. C., Yaszemski, M. J., Powers, J. M., Mikos, A. G., *Biomaterials* **19**:1935–1943, 1998.
Van Raemdonck W, Ducheyne P, De Meester P, in *Metal and Ceramic Biomaterials*, vol. II: *Strength and Surface*, Ducheyne, P., Hastings, G. W., eds. Boca Raton, Fla.: CRC Press, 1984, pp 143–166.
Walter, A., Lang, W., in *Biomedical Materials—Mater. Res. Soc. Symp. Proc.* vol. 55, Williams, J. M., Nichols, M. F., Zingg, W., eds. Pittsburgh, Pa.: Materials Research Society, 1986, pp. 181–190.
Webster, T. J., Ergun, C., Doremus, R. H., et al., *J. Biomed. Mater. Res.* **51**:475–483, 2000.
Weiner, S., *CRC Crit. Rev. Biochem.* **20**:365–408, 1986.
Wen, H. B., de Wijn, J. R., Cui, F. Z., de Groot, K., *J. Biomed. Mater. Res.* **35**:93–99, 1997.
Wolke, J. G. C., van Dijk, K., Schaeken, H. G., et al., *J. Biomed. Mater. Res.* **28**:1477–1484, 1994
Wu, W., Zhuang, H., Nancollas, G. H., *J. Biomed. Mater. Res.* **35**:93–99, 1997.
Yamamoto, M., Kato, K., Ikada, Y., *J. Biomed. Mater. Res.* **37**:29–36, 1997.
Yaszemski, M. J., Payne, R. G., Hayes, W. C., et al., *Biomaterials*, **17**:175–185, 1996.
Zreiqat, H., Evans, P., Howlett, C. R., *J. Biomed. Mater. Res.* **44**:389–396, 1999.

CHAPTER 14
CARDIOVASCULAR BIOMATERIALS

Roger W. Snyder
L. VAD Technology, Inc., Detroit, Michigan

Michael N. Helmus
Boston Scientific Corporation, Boston, Massachusetts

14.1 INTRODUCTION

Numerous definitions for *biomaterials* have been proposed. One of the more inclusive is "any substance (other than a drug) or combination of substances synthetic or natural in origin, which can be used for any period of time, as a whole or part of a system which treats, augments, or replaces tissue, organ, or function of the body," proposed by a Biomaterials Consensus Committee meeting at the National Institutes of Health (NIH).[1] This definition must be extended because biomaterials are currently being utilized as scaffolds for tissue-engineered devices (hybrids of synthetic or biologic scaffolds and living cells and tissue for vessels, heart valves, and myocardium). Completely resorbable scaffolds can result in new organs without a trace of the original biomaterial. Efforts are also underway to use coatings that release bioactive agents to prevent thrombosis, infection, and hyperplastic reactions (excessive tissue formation).

The cardiovascular system consists of the heart and all the blood vessels. Cardiovascular biomaterials may contact blood (both arterial and venous), vascular endothelial cells, fibroblasts, and myocardium, as well as a number of other cells and acellular matrix material that make up all biological tissue. This chapter will consider a wide range of biomaterials that interact with the heart, blood, and blood vessels.

Biomaterials used in the cardiovascular system are susceptible to a number of failure modes. Like all materials, mechanical failure is possible, particularly in implants. Although typical loads are low (as compared with orthopedic implants, for example), implant times are expected to exceed 10 years. At a typical heart rate of 90 beats per minute, 10 years of use would require more than 470 million cycles.

Thrombosis is a unique failure mode for cardiovascular biomaterials. The resulting clots may occlude the device or may occlude small blood vessels, resulting in heart attacks, strokes, paralysis, failures of other organs, etc. On the other hand, devices can also damage blood cells. Hemolysis can diminish the oxygen-carrying capacity of the blood. Hemolysis can occur as a reaction to the material or its degradation products or as a result of shear due to the relative motion between the material surface and the blood.

Cardiovascular biomaterials are also in contact with other tissues. Another common failure mode of these devices is excessive growth of the tissues surrounding the device. This can be caused by reaction to the material (the natural encapsulation reaction to any foreign body), stresses on surrounding tissues caused by the device, or reaction to material degradation products. Vascular grafts (in particular, smaller-diameter grafts) are subject to anastomotic hyperplasia, which reduces the diameter of the graft at the anastomosis. A similar hyperplastic response occurs around endovascular stents used to keep vessels open after angioplasty or as a de novo treatment. Heart valves can fail if tissue grows into the space occupied by the moving disk. Finally, tissue surrounding a device can die. As in hemolysis, this can be as a result of reaction with the material or its degradation products or as a result of continuous micromotion between the device and the tissue.

Biomaterials that have been used in the cardiovascular system include processed biological substances, metals, and polymers (see Table 14.1 for typical materials and applications). Materials of biologic origin include structures such as pericardia, arteries and veins, and heart valves. Devices can also include biological substances, e.g., as coatings, such as collagen and heparin.

Metals such as titanium, stainless steel, nitinol, cobalt-chrome alloys, etc., are used in many devices. Generally, these are metals with passive surfaces or surfaces that can be passivated. Silver has been used as a coating designed to resist infection. Glassy carbons have also been used as coatings to render surfaces thromboresistant. Pyrolytic carbon structures or coatings on graphite have been utilized in the fabrication of bileaflet heart valves. These are the most popular mechanical valves in use today.

Polymeric materials that have been used in the cardiovascular system include polytetrafluorethylene, polyethylene terephthalate, polyurethane, polyvinyl chloride, etc. Textiles based on polytetrafluorethylene and polyethylene terephthalate are used extensively as fabrics for repair of vasculature and larger-vessel replacement (greater than 6 mm in diameter). Stent-grafts are hybrid stent grafts placed by catheter to treat aortic aneurysms nonsurgically and are fabricated of the same metallic alloys used in stents and textiles similar to those used in vascular grafts. Table 14.1 lists many of the biomaterials currently used in the cardiovascular system.

Biomaterials are used throughout the cardiovascular system in both temporary and permanent devices. Cardiovascular devices can be divided into three categories: temporary external devices, temporary internal devices, and permanent internal devices. These categories are useful in determining the type of testing required.

Temporary external devices range from simple tubing (for bypass or hemodialysis) to more complicated devices such as oxygenators, arterial filters, and hemodialysis equipment. For the purposes of this chapter, we will consider devices that contact blood only as external devices. Temporary internal devices include a wide range of catheters used for diagnostics and treatment. These also include guidewires and introducers for use with catheters and cannulae for use in bypass circuits.

Vascular grafts and patches, as well as heart valves, are among the oldest of cardiovascular implants. More recently, permanent internal devices include pacemakers, defibrillators, stents, left ventricular assist devices, and artificial hearts.

14.2 MATERIALS

14.2.1 Metals

Metals are utilized for applications requiring high strength and/or endurance, such as structural components of heart valves, endovascular stents, and stent-graft combinations. Commonly used alloys include austenitic stainless steels (SS), cobalt-chrome (Co-Cr) alloys including molybdenum-based alloys, tantalum (Ta), and titanium (Ti) and its alloys. Elgiloy, a cobalt-nickel-chrome-iron

TABLE 14.1 Biomaterials Used in the Cardiovascular System

Typical Materials	Applications
Hydrogels	
Hydrocolloids, hydroxyethyl-methacrylate, poly(acrylamide), poly(ethylene oxide), poly(vinylalcohol), poly(vinyl-pyrrolidone)	Slippery coatings for catheters, vascular sealants, antiadhesives, thromboresistant coatings, endovascular paving, drug-delivery coatings
Elastomers	
Latex rubber, poly(amide) elast, poly(ester) elast, poly(olefin) elast, poly(urethanes), biostable poly(urethanes), poly(vinylchloride) silicones, styrene butadiene copolymers	Central venous catheters, intraaortic balloon pump balloons (polyurethanes), artificial heart bladders (polyurethanes), carrier for drug-delivery coatings, insulators for pacemaker leads, vascular grafts (e.g., biostable polyurethanes), heart-valve components (silicones), extracorporeal tubing
Plastics	
Acrylics, cyanoacrylates, fluorocarbons, ethylene-tetrafluoroethylene, ethylene-chloro-tri-fluoroethylene, fluorinated ethylene propylene, poly(tetrafluoro-ethylene), poly(vinylidene fluoride), poly(amides) poly(carbonates), poly(esters), poly(methyl pentene), poly(ethylene), poly(propylene), poly(urethane), poly(vinylchloride)	Housings for extracorporeal devices [acrylics, poly(carbonates), poly(methylpentane)], catheters, angioplasty balloons, sutures vascular grafts (polyester textiles, expanded PTFE), medical tubing oxygenator and hemdialysis membranes
Engineering plastics and thermosets	
Epoxies, poly(acetals), poly(etherketones), poly(imides), poly(methylmethacrylate), poly(olefin), high-crystallinity poly(sulfones)	Structural components for bioprosthetic heart valves [poly(acetals)], artificial heart housings, catheter components, two-part systems for adhesives (e.g., epoxies)
Bioresorbables	
Poly(amino acids), poly(anhydrides), poly(caprolactones), poly(lactic/glycolic) acid copolymers, poly(hydroxybutyrates), poly(orthoesters), tyrosine-derived polycarbonates	Sutures, scaffolds for tissue engineering, nanoparticles for treatment of blood vessels to prevent restenosis, drug-delivery coatings
Biologically derived materials	
Bovine vessels, bovine pericardium, human umbilical vein, human heart valves, porcine heart valve	Vascular grafts, pericardial substitute, heart valves
Bioderived macromolecules	
Albumin, cellulose acetates, cuprammonium cellulose, chitosans, collagen, fibrin, elastin, gelatin, hyaluronic acid, phospholipids, silk	Vascular graft coatings, hemodialysis membranes, experimental coatings, lubricious coatings (e.g., hyaluronic acid), controlled-release coatings, scaffolds for tissue engineering, tissue sealants, antiadhesives, nanoparticles for intravascular drug delivery, thromboresistant coatings, sutures
Passive coatings	
Albumin, alkyl chains, fluorocarbons, hydrogels, silica-free silicones, silicone oils	Thromboresistance, lubricious coatings for catheters, cannulas, needles
Bioactive coatings	
Anticoagulants, e.g., heparin and hirudin, antimicrobials, cell adhesion peptides, cell adhesion proteins, negative surface charge plasma polymerized coating, thrombolytics	Thromboresistance, infection resistance, enhanced cell adhesion, enhanced vascular healing
Tissue Adhesives	
Cyanoacrylates, fibrin, molluscan glue, PEG-based systems	Microsurgery for anastomosing vessels, vascular graft coating, enhancement of cell adhesion
Metals and metallic alloys	
Cobalt chrome alloys, gold alloys, mercury amalgams, nickel-chrome alloys, nitinol alloys (shape memory and superelastic), stainless steels, tantalum, titanium, and titanium alloys	Guidewires, mechanical heart-valve housings and struts, biologic heart-valve stents, vascular stents, vena cava umbrellas, artificial heart housings, pacemaker leads, leads for implantable electrical stimulators, surgical staples, supereleastic properties of some nickel-titanium formulations, shape memory properties of some Ni titanium formulation, radiopaque markers

TABLE 14.1 Biomaterials Used in the Cardiovascular System *(Continued)*

Typical Materials	Applications
Ceramics, inorganics, and glasses	
Bioactive glasses, bioactive glass/ceramics, high-density alumina, hydroxylapatite, single-crystal alumina, zirconia	Hermetic seals for pacemakers, enhanced cell adhesion, limited vascular applications, experimental heart-valve components.
Carbons	
Pyrolytic (low-temperature isotropic) carbon, ultralow-temperature isotropic carbon, pyrolized polymers for carbon-carbon composites, pyrolized fibers for fiber composites	Heart valves, coatings, fibers for carbon-fiber-reinforced plastics, or carbon-carbon composites
Composites	
Carbon-fiber-based: Epoxy, poly(ether keytones), poly(imide), poly(sulfone), radiopacifiers ($BaSO_4$, $BaCl_2$, TiO_2) blended into poly(olefins), poly(urethanes), silicones	Heart-valve housing and struts and stents, housings for artificial heart, composites to control torque and steering of catheters, Radiopaque fillers in polymers to identify location on x-ray

alloy, has been used in fine-wire devices such as self-expanding endovascular stents. The shape memory or superelastic properties of nickel-titanium alloys are used in stents.

Noble metals such as platinum-iridium are also utilized in pacemaker electrodes. In addition to the noble metals, stainless steel and tantalum can also be used in sensing (nonpacing) electrodes. Stainless steel has also been used as wire braids and reinforcements in catheters, particularly in high-pressure catheters, such as those used for radiopaque dye injection.

Fatigue life is critical in many of the applications for which metallic alloys are used. Therefore, processing and joining methods must be such that crack initiation and propagation are mitigated.

14.2.2 Carbons and Ceramics

Carbons and glassy carbons have been widely used as heart valve components, particularly as pyrolytic carbon in the leaflets and housings of mechanical valves.[10] They demonstrate good biocompatibility and thromboresistance, as well as high lubricity and resistance to wear, in this application. Graphite is used as the substrate for many of the pyrolytic carbon coatings. Strength and durability are imparted by the pyrolytic coatings. The use of a graphite substrate reduces residual stresses that become significant in thick pyrolytic coatings. The substrate has the potential to act as a barrier to crack propagation within the pyrolytic coating. Low-temperature isotropic coatings (LTI) can be used to coat more heat-sensitive polymeric substrates. Sapphires have also been utilized as bearings in high-rpm implantable rotary blood pumps.

Ceramics have had limited application in cardiovascular devices except for hermetic seals on pacemakers and for insulation in radioablation catheters. Potentially, bioactive ceramics and glasses could have uses for enhanced cell and tissue adhesion. Experimental heart valves have been fabricated from ceramics, such as single-crystal sapphire leaflets for heart valves. Ceramic coatings of heart-valve components to improve their wear properties, particularly by chemical vapor deposition methods, e.g., diamondlike coatings, are another potential application.

14.2.3 Polymers

In the late 1800s, autologous venous grafts and homologous grafts were used to close arterial defects.[4] However, the supply of these materials was limited. Long-term results were not promising, with many of the grafts developing aneurysms. In the early 1900s, solid-wall tubes of glass, methyl methacrylate, and various metals were tried. These were largely unsuccessful due to thrombosis and anastomotic aneurysms.

During World War II and the Korean War, great progress was made in vascular surgery. Based on observations of sutures placed in the aorta, a textile was shown to have the ability to retain a fibrin layer, which then organized into a fibrous tissue layer. A number of materials were tested. Selection of these materials was based on two criteria: (1) minimal tissue reactivity and (2) availability in a textile form.

Materials such as polyvinyl alcohol, polyamide, polyacrylonitrile, polyethylene terephthalate, and polytetrafluorethylene were all tried. As long as these textile tubes were implanted in the aorta, there was little clinical difference among the materials. Most of the differences in results were due to the different textile structures used. However, biostability and availability of commercial yarns did depend on the polymer chosen.

Polyvinyl alcohol was soon abandoned due to excessive ruptures. Polyamide and polyacrylonitrile were discovered to be biodegradable, although it took 12 to 24 months to occur. Thus polyethylene terephthalate (polyester) and polytetrafluorethylene (PTFE) became the polymers of choice. Both these materials have demonstrated their longevity as an implant.[5]

Clinical results are excellent when the devices are implanted in high-flow, large-diameter arteries such as the aorta. However, patency rates decrease significantly when these devices are implanted below the aortic bifurcation. Thus other materials and structures have been investigated for low-flow, small-diameter arteries.

In the mid-1970s, PTFE in a different form was introduced. Expanded PTFE is formed by compressing PTFE with a carrier medium and extruding the mixture. This is termed a *paste extrusion* because PTFE is not a thermomelt polymer. The resulting extrudate is then heated to near the glass transition temperature and stretched. Finally, the stretched material is sintered at a higher temperature. The resulting structure is microscopically porous with transverse plates of PTFE joined by thin PTFE fibers. This form of PTFE was indicated for use in smaller arteries with lower flow rates. However, the patency results obtained with this vascular graft were not significantly higher than those obtained with a polyester textile.

The PTFE textile graft is no longer commercially available. In general, the handling characteristics of that device were not as good as those of the polyester textile because commercially available PTFE fibers were larger in diameter than polyester fibers.

However, polyester and PTFE textiles, as well as expanded PTFE, are available as flat sheets. The textile materials are available as knits, weaves, and felts. These materials are used for patches and suture buttresses.

Silicone is a rubberlike polymer. It is normally cross-linked in a mold or during extrusion. The most common silicone used is room-temperature-vulcanizing (RTV) silicone. In general, tissue does not adhere to silicone. The first commercially viable heart valve used a silicone ball in a cage. However, when first used, these silicone balls absorbed lipids and swelled, causing premature failure of the valves. These problems were corrected, and a small number of these valves are still implanted today. It is not uncommon that explants are recovered 30 years after implantation showing no degradation, minor wear, and only discoloration of the silicone ball. In current cardiovascular devices, silicone may be used as percutaneous drive lines.

Synthetic bioresorbable materials have had a wide application as suture materials, although they have not generally been used in vascular anastomoses. They are being investigated for scaffolds for tissue-engineered heart valves and blood vessels. They are also being investigated as drug-release coatings on vascular prostheses and stents (to prevent thrombosis, infection, and excessive tissue formation) and as nanoparticles to deliver drugs to prevent restenosis.

14.2.4 Biological Materials

Materials of biological origin are used as cardiovascular devices and as coatings. Most of the devices commercially available rely on collagen as the structural material. Collagen is a macromolecule that exists as a triple helical structure of several amino acids. Procollagen is expressed by cells. The ends of the procollagen molecule are enzymatically trimmed, allowing the trimmed helical strands to self-assemble into the collagen molecule. At least 19 different types of collagen have been identified,[5,13] with types I and III predominating in cardiovascular structures.

The collagen molecule can be cross-linked by a number of techniques to improve its structural integrity and biostability. An early example of this is the tanning of skin to make leather. The use of formaldehyde to preserve biological samples is another example.

In 1969, porcine aortic valves were cross-linked with gluteraldehyde and used to replace human aortic valves. Gluteraldyhde cross-linking was shown to yield a more biostable structure than cross-linking with formaldehyde.[5] These valves have been very successful in older patients and do not require the anticoagulation regimen needed for mechanical heart valves. Significant effort has been made to reduce the calcification of bioprosthetic heart valves, both porcine aortic valve prostheses and bovine pericardial valve prostheses. Calcification and degradation mechanisms limit the use of these devices in young patients and children. Reduction of calcification entails modification of the surface, e.g., binding amino oleic acid or treatments with surfactants to remove lipids and other agents that can be a nidus for calcification.[7] Their use in patients under 60 years of age is increasing and will continue to increase with the development of new treatments to reduce calcification.

Cross-linked bovine arteries and human umbilical veins have been used as vascular replacements. Cross-linked pericardium has been used as a patch material, primarily between the myocardium and the pericardial sac to prevent adhesions.

Biological materials can also be used as coatings. Textile vascular grafts are porous and must be sealed (preclotted) prior to use. Research[6] suggests that a four-step procedure, including a final coat with heparinized blood, can improve patency results. However, surgeons typically take nonheparinized blood and coat the graft in one or two applications.

Precoated grafts are commercially available. Cross-linked collagen and gelatin (a soluble form of collagen), as well as cross-linked albumin, can be used to seal porous materials. The rate of degradation of the coating will depend on the material chosen, as well as the degree of cross-linking.

Significant effort is now focusing on tissue-engineered vessels and heart valves. The history of this effort is found in the seeding or culturing of endothelium on synthetic vascular prostheses. The clinical outcomes did not justify continued development. However, new technology allows vascular tissue to be formed on scaffolds of either synthetic or resorbable materials. The awareness that endothelium alone was not suitable has led to the evolution of techniques to recreate the vascular tissue utilizing multiple cells types.[8,9] This approach utilizes the cell types expected in final structure, e.g., endothelium, smooth muscle cells, and fibroblasts, or pluripotential cells such as stem cells.

14.3 TESTING

The testing program for any medical device can be divided into five phases: (1) biocompatibility, (2) short-term bench (or in vitro) tests, (3) long-term bench tests, (4) animal (or in vivo) studies, and (5) human clinical studies.[11] For each of these five phases, the type of device and length of time it will be used must be considered in developing test protocols.

14.3.1 Biocompatibility

Biocompatibility testing must measure the effects of the material on blood and tissue, as well as the effects of the organism on the material. International standards exist for demonstrating biocompatibility. These standards prescribe a series of tests, the selection of which depends on the length of time that the device will be in contact with the body. In these standards, any use less than 30 days is considered short term. Whether or not the device is external and will only contact blood or will be internal and in contact with tissue and blood also dictates which tests are necessary.

Biocompatibility will be affected by surface contamination. Surface contamination can occur as a result of processing. Process aids, cleaning agents, finger oils, etc., can all have an impact on compatibility. Residues can also result from sterilization. Residual sterilants or sterilant by-products (such as ethylene oxide), modification of the material surface from radiation, and toxic debris from microbes can have an impact on biocompatibility. Materials such as solvents, plasticizers, unreacted monomers, and low-molecular-weight polymers can diffuse from polymers. Certain cleaning or thermal processes can accelerate diffusion.

Therefore, all samples for biocompatibility testing should be from completed devices that have seen the complete process, including sterilization. If there is a chance that contaminates could continue to diffuse from the material, testing samples after storage should be considered.

Since the cost of doing some of these tests (including the cost of the samples) can be significant, screening tests can be performed on any new material (or process). These tests are subsets of the standard tests. Some materials manufacturers provide biocompatibility data of this type.

Once the materials used in the device have been shown to be biocompatible, consideration must be given to the function of the device. For example, devices subjected to flowing blood must be tested to document the lack of damage to blood components. Devices that rely on tissue ingrowth must be tested for this feature. These types of tests could be part of animal testing, which will be discussed later.

The Blue Book Memo[2] issued by the U.S. Food and Drug Administration (FDA), tabulates the tests required to demonstrate biocompatibility. These tables are based on an International Standards Organization (ISO) Document ISO-10993.[3] For implant devices contacting blood for more than 30 days, the following tests are required: cytotoxicity, sensitization, irritation or intracutaneous reactivity, acute system toxicity, subchronic toxicity, genotoxicity, implantation, and hemocompatibility. For devices in contact with blood for less than 24 hours, subchronic toxicity and genotoxicity are not required. The international standard also has a category for devices that are in use for between 24 hours and 30 days. This standard does not require the subchronic toxicity testing. The FDA may require this type of testing, however. The tests required for implanted cardiovascular devices that do not contact blood require the same type of testing program except for the hemocompatibility requirement and for the implantation tests for devices in use for less than 24 hours.

The tests for external devices contacting blood are also the same as for implanted devices, although implantation tests are noted as "may be applicable." For long-term devices, either external or implants, chronic toxicity and carcinogenicity testing may also be required.

Many manufacturers can provide biocompatibility data either in their literature or as an FDA Master File. Often material manufacturers will advertise that a material meets class IV biocompatibility requirements. Class VI requirements are an old set of tests published in the U.S. Pharmacopeia that were developed for testing food packaging. They are similar to the cytotoxicity, acute toxicity, and subchronic toxicity tests. However, the data provided by a materials manufacturer are on samples that have not seen the processing and storage of the device. The data simply are an indication that the material can pass the initial set of biocompatibility tests if processed appropriately.

There are a wide variety of tests in the literature addressing these various requirements. Protocols for many of these tests have been issued as ISO standards. The American Society for Testing and Materials (ASTM) has also developed protocols for demonstrating biocompatibility. Since these standard protocols are recognized by many regulatory agencies, their use will often aid in the device approval process.

Collaborating with a laboratory that specializes in these types of tests and is familiar with the regulatory requirements will generally produce the best data to demonstrate biocompatibility.

14.3.2 Short-Term Bench Testing

Short-term bench (in vitro) testing includes material identification, surface characterization, mechanical properties, etc. Material identification tests characterize the bulk properties of the material. Tests chosen depend on the type of material. Chemical formula, molecular weight, percentage crystallinity, melting or softening point, and degree of cross-linking may all be important to characterize a polymer. Composition, grain size, and contamination levels may define metallic materials. Composition, molecular weight, cross-linking, shrinkage temperature, and purity may define materials of a biological origin.

Surface properties will affect the reaction between the material and tissue. Material composition at the surface may differ from the bulk composition. Coatings, either deliberately applied or as contaminates, will modify the biological response. Extraction studies, perhaps performed as part of the development of the cleaning process, could identify any inadvertent contamination. The coating should have an identification test. The geometry of the surface, such as surface roughness, will also modify this response. The dimensions of the surface roughness can be measured microscopically. For smaller features, scanning electron microscopy or atomic force microscopy can be utilized to characterize the surface roughness.

Mechanical properties of the material will determine if a device will suffer an early failure. Tensile strength and elastic modulus can be measured by a simple tensile test. If the device could be subjected to impact loading, an impact type test can be performed. Tear tests are important for materials in sheets such as fabrics and films. This is particularly true for tear-sensitive materials such as silicone.

The ASTM has published protocols for mechanical tests and for operating test equipment.

14.3.3 Long-Term Bench Testing

For long-term devices, endurance (or fatigue) testing is required. In general, simple tensile or bending tests can be performed on the basic material. Frequently, however, the device itself is tested in some simulated loading condition. Such a test includes the effects of processing, sterilization, and shelf life on the material. It also allows the designer to calculate reliability. There are test and reliability standards for some cardiovascular devices. Vascular grafts, heart valves, stents, and left-ventricular assist devices, among others, have reliability requirements and recommendations for the types of tests that can be employed. However, since there is a wide variation in these types of devices, tests that fit the device must be developed.

Materials can be tested in tension, compression, or bending. Using the material in a sheet form, biaxial loading can be applied. For larger stress or strain ranges (and thus a lower number of cycles), the same equipment used to test for material strength can be used. However, for smaller loads and higher numbers of cycles, specialized equipment is required to complete the tests in a reasonable time. Loading devices using rotating cam shafts will apply fixed strain ranges. Fixed stress ranges can be applied using pressure-actuated devices. For very small stress or strain loads approaching the endurance limit, electronic mechanisms similar to those used to drive audio speakers can be used to drive a material at very high speeds.

If one plots a variable such as stress range or strain range versus number of cycles, the resulting curve will approach a limit known as the *endurance limit*. Below this limit, the number of cycles that a material can withstand is theoretically infinite. Above this limit, the number of cycles that a material can withstand under a variable load can be calculated from Miner's rule:

$$\frac{n_1}{N_1} + \frac{n_2}{N_2} + \frac{n_3}{N_3} + \quad . \; . \; . \quad + \frac{n_n}{N_n} = 1$$

where n_i is the number of cycles for a given load and N_i is the total number of cycles to failure under that load. Thus, if the stresses on a device can be calculated, the fatigue life of a device can be estimated.

However, regulatory agencies prefer that the life of a device, or its reliability, be measured rather than calculated. Therefore, it is common practice to perform reliability testing on the device as it will be used in a patient. Although it is usually not possible to use blood or other biological fluids in a long-term test setup due to the difficulty in preserving the fluid, if the environment will affect the material, then a reasonable substitute must be found. In general, saline has been accepted as a substitute test media. If necessary, the saline can be buffered to an appropriate pH.

Since cardiovascular devices, particularly implants, are expected to function for multiple years, tests to demonstrate reliability must be accelerated. At the same time, the test setup should apply loads that are as close to the actual usage as possible. Although some forms of degradation can be accelerated by an increase in temperature, it is common practice to test materials and devices at normal body temperature. Thus reliability tests are normally accelerated by increasing the cyclic rate.

The upper limit of this type of acceleration is determined by several factors. First, the normal range of load and motion must be duplicated. For larger device parts, such as heart-valve disks, inertia will limit the cyclic rate. Second, any increase in temperature due to accelerating bending must not change the material. Internal temperatures of materials with glass transition temperatures above body temperature must remain below these transition points. Finally, the rate of loading must not affect, either negatively or positively, the amount of creep (or viscoelastic deformation) experienced by the material in actual use.

Reliability can be calculated by several methods. One method used in other industries (such as the automotive industry) is to test 12 samples to the proposed life of the device. If all 12 samples survive, then the sample size is adequate to demonstrate a reasonable risk of failure using a binomial model. To determine failure modes, the stress or strain on the device can be increased by 10 percent for 10 percent of the proposed life of the device. If there are no failures at 110 percent of the proposed life, the stress or strain range is increased another 10 percent. This stair-step method continues until a failure mode is demonstrated.

A more common method for medical devices is to run the life test until failure occurs. Then an exponential model can be used to calculate the percentage survivability. Using a chi-square distribution, limits of confidence on this calculation can be established. These calculations assume that a failure is equally likely to occur at any time. If this assumption is unreasonable (e.g., if there are a number of early failures), it may be necessary to use a Weibull model to calculate the mean time to failure. This statistical model requires the determination of two parameters and is much more difficult to apply to a test that some devices survived. In the heart-valve industry, lifetime prediction based on S-N (stress versus number of cycles) or damage-tolerant approaches is required. These methods require fatigue testing and ability to predict crack growth.[10,11,14]

Another long-term bench test required for these devices is shelf life. Some cardiovascular biomaterials degrade on the shelf. Thus typical devices, having seen the standard process, are packaged and aged before testing. Generally, aging can be accelerated by increasing the temperature of the storage conditions. As a general rule, it is accepted that the rate of degradation doubles for every 8°C increase in temperature. Some test labs also include variations in humidity in the protocol and may include a short period of low-temperature storage.

Products that have been packaged and aged can then be tested to determine if they still meet the performance criteria. At the same time, these products can be tested for sterility, thus demonstrating that the packaging material and packaging process also yield the appropriate shelf life.

Polymeric and biologic-based devices may also need to be evaluated on the basis of the biostability of the materials. This could include hydrolytic and enzymatic stability, requiring a combination of testing that examines hydrolytic stability under simulated physiologic stresses as well as evaluation in animals. Stress tends to accelerate many of these degradative mechanisms, and materials that look stable under static conditions may not perform well when stressed. Soft grades of polyether polyurethane are an example of a material than can undergo oxidative degradation when stressed due to the presence of oxidative enzymes present in biologic systems.

14.3.4 Animal Studies

There are few standard protocols for animal studies. Each study is typically designed to take into account the function and dimensions of the device. There are two approaches to developing a protocol. First, one could use a model of the condition being treated to demonstrate the function of the device. For example, vascular grafts can be implanted as replacements for arterial segments. The second approach is to design a test that will demonstrate the functioning of the device but not treat an abnormal condition. For example, a left ventricular assist device can be implanted in normal animals. The protocol would then consist of operating the device and monitoring the effect of the device on the blood. In addition, the effect of the biological environment on the device could be documented.

The first step in developing an appropriate protocol is to determine the purpose of the test. Is the purpose to demonstrate the functionality of the device, to demonstrate that the device is effective in treating a medical condition, or to test long-term biocompatibility and biostability of the device? The purpose of the test and the proposed use of the device (i.e., biological environment and length of use) will determine the model and length of time required for the test.

If the device can be miniaturized, or if nonfunctioning devices are appropriate for the purpose of the test, then smaller animals can be used. Of course, the life span of the animal must be adequate for the implant time required. If, however, the purpose of the test requires a full-sized, functioning device, a larger animal model will have to be selected.

14.4 MATERIAL PROCESSING AND DEVICE DESIGN

Processing methods can have a major impact on the success or failure of a cardiovascular biomaterial. As described previously, surface features (either deliberately introduced or as the result of machining or tool imperfections), residues (from cleaning, handling, or sterilization), or process aids (either as surface residues or as bulk material diffusing from the biomaterial) can change the biological results.

There are three methods currently used to decrease thrombosis: (1) use of a smooth surface to prevent thrombi from adhering, (2) use of a rough surface (usually a fiberlike surface) to encourage the formation of a neointima, and (3) use of a coating to prevent coagulation or platelet adherence. All these methods have been used in cardiovascular medical devices with varying degrees of success.

As a general rule, the slower the flow of blood, the more likely thrombi will form. Thus the design of a device should avoid areas of stasis or low flow. A suitable surface must be either smooth, avoiding any small features that might cause microeddies in the flow, or of sufficient roughness so as to allow the resulting coagulation products to securely anchor to the surface.

The thickness of the resulting layer is limited by the need to provide nutrients to the underlying tissue. Without the formation of blood vessels, this is generally about 0.7 mm. Should the material itself cause the formation of a thicker layer or should parts of the underlying structure move or constrict the tissue, the tissue will die. If this occurs continuously, the tissue will calcify or the underlying biomaterial will remain unhealed.

Cleanliness of a material will also affect the biologic outcome. All processing aids should be completely removed. This includes any surfactant used in the preliminary cleaning steps. Surfactants can cause cell lysis or pyrogenic reactions. Solvents can diffuse into plastics and diffuse out slowly after implantation, causing a local toxic reaction. Some plastics may retain low-molecular-weight polymer or even monomers from their formation. These can also be toxic to cells. These may also leach out slowly after implantation. Plasticizers (used to keep some polymers pliable) can also damage blood components. The oxidation by-products of some metals can also be locally toxic.

Thus it is important to establish a processing method prior to final evaluation of a cardiovascular biomaterial for use in a medical device.

14.5 MATERIAL REPLACEMENT

A number of major suppliers of materials no longer sell their products for long-term (more than 30 days) implantation in humans. In the past, the market for such materials was small (compared with the commercial market) and the liability high. Even if the material manufacturer could demonstrate that it was not liable for a particular use of its material, the legal cost was still quite high compared with possible sales. Even after Congress passed the Biomaterials Access Assurance Act of 1998 that shielded manufacturers of these materials from liability, many manufacturers remained reluctant to allow their materials to be used for long-term implants. In addition, materials may be totally withdrawn from the market if other commercial uses decline significantly.

Many implant device manufacturers were forced to find replacement materials for their devices. Sometimes it was possible to locate an alternate supplier. At the same time, a number of material suppliers to the medical device industry have been started. In particular, new suppliers of silicones and polyurethanes have been established. Following the withdrawal of silicone from the breast implant market, the FDA modified its procedures, not requiring that replacement materials be tested as new materials but allowing manufacturers to demonstrate equivalency to previously used materials.

Equivalency can be established through a series of data. Replacement materials can be compared with the original material in terms of chemical (structure, molecular weight, thermal properties, etc.) and mechanical properties (strength, tensile modulus, endurance, etc.). Data from actual devices fabricated from the new material should be compared with those from the original device. Biocompatibility data may be available from the material manufacturer, although the agency will request that additional data be supplied if the process used to fabricate the test samples differs from the process used to fabricate the device.

REFERENCES

1. Williams, D. F., ed., "Definitions in biomaterials," *Progress in Biomedical Engineering* **4**:67 (1987).
2. Blue Book Memorandum G#95-1, U.S. Food and Drug Administration, 1995. See http://www.fda.gov/cdrh.
3. *Biological Evaluation of Medical Devices,* Part 1: *Evaluation and Testing,* International Standards Organization (ISO) Document Number 10993. Geneva: Iso, 1997.
4. Weslowski, S. A., *Evaluation of Tissue and Prosthetic Vascular Grafts.* Springfield, IL: Charles C. Thomas, 1963.
5. Guidoin, R. C., Snyder, R. W., Awad, J. A., and King, M. W., "Biostability of vascular prostheses," in *Cardiovascular Biomaterials*, Hastings, G. W. (ed.). New York: Springer-Verlag, 1991.
6. Greisler, H. P., *New Biologic and Synthetic Vascular Prostheses.* Austin, Tex.: R. G. Landes Co., 1991.
7. Schoen, F. J., and Levy, R. J., "Founder's Award, 25th Annual Meeting of the Society for Biomaterials, Perspectives. Providence, RI, April 28–May 2, 1999: Tissue heart valves: Current challenges and future research perspectives," *Journal of Biomedical Materials Research* **15**(47, 4):439–465 (1999).
8. Helmus, M. N., "Introduction/general perspective," Frontiers of Industrial Research, International Society for Applied Cardiovascular Biology, (Abstract), *Cardiovascular Pathology* **7**(5):281 (1998).
9. Helmus, M. N., "From bioprosthetic tissue engineered constructs for heart valve replacement," in *First International Symposium, Tissue Engineering for Heart Valve Bioprostheses, Satellite Symposium of the World Symposium on Heart Valve Disease,* Friday, June 11, 1999, Westminster, London (Abstract), pp. 35–36.
10. Ritchie, R. O., "Fatigue and fracture of pyrolytic carbon: A damage-tolerant approach to structural integrity and life prediction in 'ceramic' heart valve prostheses," *Journal of Heart Valve Disease* **5**:(Suppl. 1): S9–S31 (1996).
11. Helmus, M. N., ed., *Biomaterials in the Design and Reliability of Medical Devices.* Georgetown, Tex.: Landes Bioscience, 2001.
12. von Recum, A. F., ed., *Handbook of Biomaterials Evaluation: Scientific, Technical, and Clinical Testing of Implant Materials,* 2d ed. Philadelphia: Taylor & Francis, 1999.
13. Prockop, D. J., and Kivirikko, K. I., "Collagens: Molecular biology, diseases, and potentials for therapy," *Annual Revue of Biochemistry* **64**:403–434 (1995).
14. Kafesjian, R., and Schmidt, P., "Life analysis and testing: Short course, evaluation and testing of cardiovascular devices," Society for Biomaterials, Course Notebook, Cerritos, Calif., 1995.

CHAPTER 15
ORTHOPEDIC BIOMATERIALS

Michele J. Grimm
Wayne State University, Detroit, Michigan

15.1 INTRODUCTION

Before the circulation of blood was discovered by Harvey in 1628 (Lee, 2000) and before Vesalius systematically documented the anatomy of the human body in 1543 (Venzmer, 1968), the structural function of the skeletal system was understood. Bone protected organs such as the brain and provided the frame on which the soft tissues of the body were formed. Based on this basic understanding, the first medical interventions to replace bone—removed due to damage or underlying injury—were seen at least as far back as the time of the Aztecs, who are known to have used gold and silver to replace pieces of the skull following craniotomies (Sanan and Haines, 1997). The fact that bone was a living material that could heal itself was documented over 5000 years ago when the ancient Egyptians recorded techniques for setting fractures on papyrus (Peltier, 1990), and this knowledge has led to interventions designed to manipulate the fracture-healing properties of the tissue.

Our greater understanding of the overall physiology of bone did not develop until much more recent history. The complex and important role of the cellular component of bone, although only a small fraction of the overall material volume, is still being investigated. While the general properties of bone, ligament, tendon, and cartilage have been well characterized over the past century, knowledge of how these properties can be best mimicked or taken advantage of to promote tissue healing remains in its infancy.

Orthopedic tissues are affected by both the stresses that they experience on a daily basis and as a result of trauma and disease processes. Many of these injuries or pathologies require medical intervention that may be assisted through the use of engineered materials. The science behind the selection of these materials has moved from the realm of trial and error to one based on scientific theory and understanding. This chapter will give a brief overview of the natural orthopedic biomaterials—bone, cartilage, tendon, and ligament—before proceeding to a discussion of the historical development and current technology in engineered biomaterials for orthopedic applications.

15.2 NATURAL MATERIALS

15.2.1 Bone

Bone has a diverse set of physiological roles ranging from the obvious structural support and protection to maintenance of calcium homeostasis and hematopoesis, the production of red blood cells by the bone marrow. As such, both the material and cellular characteristics of bone must be understood in order to fully appreciate the complexity of the tissue. However, to initiate this understanding, it is easier to examine the material and cellular components of bone separately at first.

Bone's Material Components. From a structural standpoint, bone is essentially a composite of organic and inorganic components—namely, collagen and hydroxyapatite. Collagen is a protein with a high tensile strength and viscoelastic properties, whereas hydroxyapatite is a calcium phosphate compound with properties similar to that of a ceramic. Hydroxyapatite crystals, needlelike structures with a size on the order of an angstrom, are embedded in the sides of long collagen fibers. The collagen fibers are then arranged in sheets as parallel structures, which in turn are layered in concentric circles with the collagen fiber orientation varying between layers. The dimension about which these concentric layers of composite, or lamellae, are formed differs with the type of bone involved.

Cortical bone, or compact bone, is the dense form of the tissue that is generally called to mind when an image of *bone* is produced. It is found on the outer surface of all bones and comprises the majority of the shaft (or diaphysis) of long bones such as the femur. Two basic forms of cortical bone exist in humans: osteonal and lamellar. Lamellar bone is formed when the concentric layers of collagen-mineral composite are wrapped around the inner (endosteal) or outer (periosteal) surfaces of a whole bone structure. Osteonal bone involves a more complex microstructure, with the composite layers wrapped in concentric circles about a vascular, or Haversian, canal (Fig. 15.1). A group of these lamellae with its central Haversian canal form an osteon, the diameter of which can range from 150 to 250 μm for secondary (or remodeled) osteons, whereas primary osteons tend to be smaller. The axis of the osteon is generally oriented along the direction of primary loading in a bone.

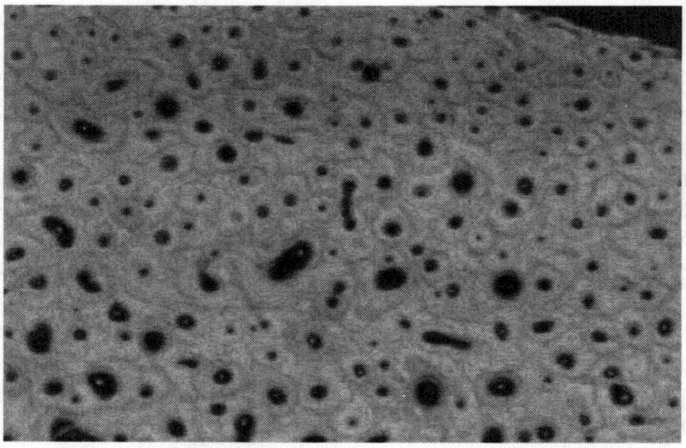

FIGURE 15.1 Scanning acoustic microscopy image of cortical bone from a human femur. Note the circular arrangement of the lamellae around the central haversian canal.

Trabecular bone is formed through a different arrangement of lamellae. An individual trabeculum is a tube of wrapped lamellae on the order of 150 to 300 μm in diameter. Trabeculae can also form in the shape of plates, which have a slightly larger dimension but are again composed of parallel layers of the collagen-mineral composite. The trabecular plates, beams, and struts are arranged into a three-dimensional structure that mimics the internal skeleton of a modern skyscraper (Fig. 15.2). The beams and plates are generally arranged in the direction of primary loading, whereas the struts provide supporting structures in an off-axis direction in order to minimize buckling. Healthy trabecular bone is "designed" to have an improved strength-to-weight ratio compared with cortical bone—it can carry a substantial amount of load without contributing added weight to the body. It is found at the ends of long bones, in the metaphyseal and epiphyseal regions, as well as in the inner portions of bones such as the vertebrae of the spine, the carpal bones of the wrist, and the flat bones of the ribs and skull.

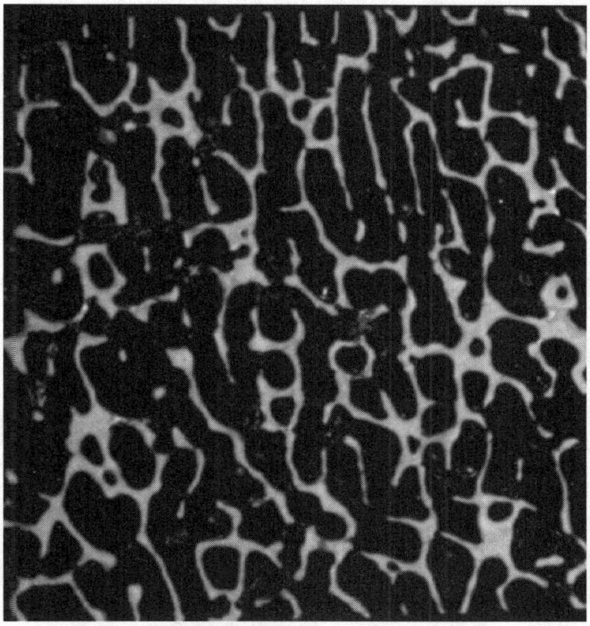

FIGURE 15.2 Scanning acoustic microscopy image of vertebral trabecular bone from a young individual. The vertical beams and the horizontal struts form a three-dimensional network to maximize the mechanical properties while minimizing weight.

The mechanical and material properties of bone have been extensively characterized, and representative properties are listed in Table 15.1. The structural and material properties of cortical bone are approximately equal due to its low porosity. However, since the porosity and arrangement of trabecular bone play an important role in the structural properties of this phase, the material modulus may be up to 3 orders of magnitude higher than the structural modulus. It must be noted, however, that unlike those of traditional engineering materials, the properties of bone are not constant. The strength, modulus, and density can vary between individuals, between anatomic locations, and as a result of age or disease processes. Variations in the properties of bone may be a function of changes in either the structure of the tissue (e.g., how many trabeculae are present and how they are arranged) or the material of the tissue (e.g., the properties of the collagen-mineral composite itself). In healthy tissue, the material of bone changes very little, with the mineral density fairly constant at a level of 1.8 to 1.9 g/cc (Kaplan et al., 1994) and the mineral-to-collagen ratio set to about 1:1 by volume. Disease processes such as osteomalacia or osteogenesis imperfecta can affect the collagen or mineral components of bone and, as such, have a profound effect on the underlying properties of the tissue.

TABLE 15.1 Representative Properties of Cortical and Trabecular Bone

Cortical bone	Compressive strength, MPa	131–224 longitudinal
		106–133 transverse
	Tensile strength, MPa	80–172 longitudinal
		51–56 transverse
	Shear strength, MPa	53–70
	Elastic modulus, GPa	11–20 longitudinal
Trabecular bone	Tissue compressive strength, MPa	0.5–50
	Tissue elastic modulus, MPa	5–150
	Material elastic modulus, GPa	1–11

Sources: An (2000), Cowin (1989), and Hayes and Bonxsein (1997).

The structural properties of bone, even at the microscopic level, can also vary due to anatomic location (which can be seen as a design variation), age, or disease. A prime example of this is the loss of trabecular bone seen in all individuals after age 35 and exacerbated by osteoporosis. It has been shown that in the vertebrae, for example, osteoporosis results in the selective resorption of the horizontal, supporting trabeculae. The trabecular bone, which makes up all but a small fraction of the volume of the vertebral centrum, is weakened as each of the load-bearing beams is then defined by a larger characteristic length. Based on Euler's theories, these trabeculae will be more susceptible to buckling—and hence failure—at lower loads. Figure 15.3 shows a buckled trabeculae in an image of vertebral trabecular bone from a 75-year-old.

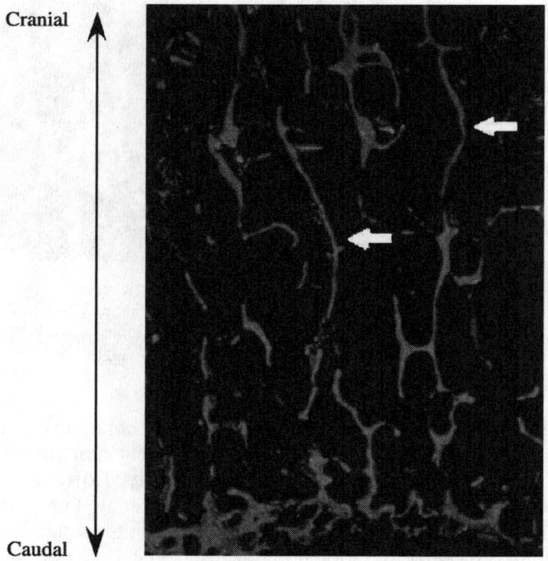

FIGURE 15.3 Scanning acoustic microscopy image of vertebral trabecular bone from a 75-year-old man. The arrows indicate the location of trabeculae that have begun to buckle under the superoinferiorly (craniocaudally) directed physiologic load due to a loss of supporting struts.

Finally, the properties of a whole bone will be affected by the amounts of trabecular and cortical bone present and their geometric arrangement. As will be discussed later, bone is a living tissue that can adapt to its loading environment. The loss of cross-sectional area in the diaphysis of a long bone, the reduction in trabecular volume fraction, and the change in shape of a bone all will affect a bone's overall properties and likelihood of fracture.

Bone's Living Components. Like liver, kidney, and muscle, bone is a living tissue that responds to its environment. Two basic processes take place in bone as it responds to physiological demands. Bone modeling occurs primarily in children and young adults and results in bone growth—both in length and in cross-sectional area. The growth of bones through the addition of material to the endosteum or periosteum, which is the result of the modeling process, can also continue throughout life. Bone remodeling involves the removal and—in general—replacement of bone. This process allows for the continual recycling of bone, and in healthy tissue it prevents the accumulation of microcracks that could lead to fatigue failure of the structure. The same general processes are seen in fracture healing.

The cellular component of bone consists of three cell types: osteoblasts, osteoclasts, and osteocytes. Osteoblasts are the cells in bone that will lay down new collagen matrix, which is then mineralized to form the lamellae of bone. Osteoclasts remove bone during the normal remodeling process, which is then replaced through osteoblastic activity. Osteoclasts also act to remove bone due to changes in the loading environment. This response in bone, which has tremendous implications in implant design and use, will be discussed further below in the section on Wolff's law. Osteocytes are the general cells of bone, acting as communication systems from one location in bone to another. Connected through cellular processes in the canaliculi of osteonal bone, osteocytes are thought to act as transducers that sense the mechanical and chemical environment around bone and then relay this information to the osteoclasts and osteoblasts in order to elicit the necessary cellular response.

Wolff's Law. Developed in 1892 by Professor Wolff (Wolff, 1892), this theory of bone behavior remains the governing principle behind our understanding of bone physiology. After observing that the structural orientation of trabeculae in the head and neck of the femur resembled the principal stress trajectories of a Cullman crane (a mechanical structure with a similar shape and loading pattern), Wolff hypothesized that bone develops in response to the loading environment that it experiences. Through the last 100 years, this hypothesis has been reinforced through empirical and experimental data. Thus bones that are not loaded sufficiently will lose tissue mass, whereas bones that are loaded at a greater level than previously will add bone in order to reduce the stress experienced. This response does require a time-averaged response—a single day spent in bed or lifting weights will not change the structure of bone. However, extended periods in a hypogravity environment, such as the space shuttle, will result in bone loss and therefore a reduction in whole-bone strength. In loading-related bone remodeling, the changes in bone mass are due to increases or decreases in the structural arrangement of bone, not a change in the amount of mineral per unit volume of collagen at the material level.

15.2.2 Cartilage

From an orthopedic material viewpoint, the type of cartilage of interest is articular cartilage—located at the bearing surfaces of the joints. Cartilage provides a covering surface on the ends of bones that meet to form an articulation, such as the femur and tibia at the knee. It acts to provide a smooth, low-friction bearing surface, as well as to absorb some of the energy transferred through the joints during normal activities.

Cartilage is a soft tissue composed of a proteoglycan matrix reinforced with collagen. The orientation of the collagen varies through the thickness of the structure, with fibers oriented perpendicular to the articular surface at the deepest level (furthest from the point of joint contact) and parallel to the surface in the uppermost region (Mankin et al., 1994). Between 65 and 80 percent of the total tissue weight is due to the water contained within the tissue matrix (Mankin et al., 1994). Cartilage is predominantly loaded in compression and is viscoelastic in nature. Under initial loading, the water within the proteoglycan matrix is extruded, and the stiffness of the material is a function of the tissue permeability. In fact, the fluid pressure within the matrix supports approximately 20 times more load than the underlying material during physiological loading (Mankin et al., 1994). Under extended, noncyclic loading, the collagen and proteoglycan matrix will determine the material behavior after the water has been forced from the tissue. Table 15.2 shows representative values for cartilage properties.

TABLE 15.2 Representative Properties for Human Articular Cartilage Taken from the Lateral Condyle of the Femur

Property	Value
Poisson's ratio	0.10
Compressive modulus, MPa	0.70
Permeability coefficient, $m^4/N{\cdot}s$	1.18×10^{-15}

Source: Mankin et al., (1994).

The low-friction environment provided by healthy cartilage as an articulating surface is also due to the fluid that is forced from the structure under compressive loading. As the tissue is loaded in compression, the water released from the proteoglycan matrix provides fluid-film lubrication between the two surfaces. During the unloading portion of a motion cycle, the cartilage resorbs a portion of the water, returning it to the matrix.

The principal cell in cartilage is the chondrocyte. Responsible for matrix production during growth and maintenance of the matrix in mature tissue, chondrocytes occupy only about 10 percent of the overall tissue volume (Mankin et al., 1994). Due to the avascular nature of articular cartilage, the provision of metabolites to the cells is assumed to occur via diffusion from the synovial fluid or, to a lesser extent, the underlying bone (Mankin et al., 1994). However, the lack of blood supply severely diminishes the ability of cartilage to heal once it has been damaged.

15.2.3 Ligaments and Tendons

Although different structures with different physiological functions, ligaments and tendons often are examined together due to their similar tensile loading patterns. Ligaments connect bones to each other across a joint, whereas tendons attach muscles to bone and provide the anchor necessary for muscles to cause movement. Each is composed of a combination of collagen and elastin fibers arranged primarily in parallel along the axis of loading. However, in the unloaded state, the fibers are slightly crimpled. Therefore, initial tensile loading of the structure acts only to straighten out the component fibers—resulting in a region of low stiffness. Once the fibers have been completely straightened, the individual fiber stiffness dictates the overall structural stiffness. The resulting load-deformation curve (Fig. 15.4) exhibits a characteristic low-stiffness toe region followed by a region of increasing stiffness. If loading continues, failure of individual fibers within the structure will result in decreasing overall stiffness followed by rupture.

Table 15.3 shows typical values for the tensile properties of ligament and tendon. Tendon tends to be slightly stiffer than ligament due to the higher concentration of collagen. Both tissues are highly viscoelastic and will fail at lower extensions when loaded at high rates. This behavior explains why a slow stretch will not injure a tendon or ligament, whereas a rapid motion may result in rupture. The properties of both types of tissue vary based on anatomic location, indicating that the structures develop to match normal physiological demands.

Tendons tend to be avascular if they are surrounded by a tendon sheath to direct passage around a sharp prominence of bone, such as is seen in the flexor tendons of the hand. However, the remaining tendons tend to have a reasonable blood supply through surrounding connective tissue (Woo et al., 1994). Ligaments have a very limited blood supply through the insertion sites. In all cases, tendons

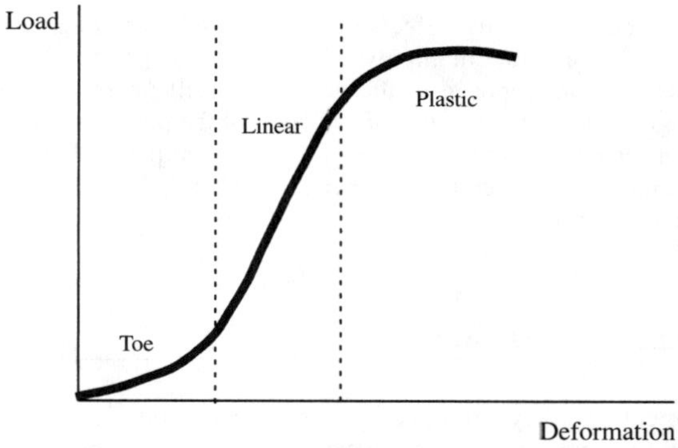

FIGURE 15.4 Schematic diagram of a typical stress-strain curve for collagenous soft tissues such as ligament and tendon. The initial toe region results from the straightening and aligning of the collagen fibers. The middle region, with increased stiffness, indicates participation from the majority of the fibers in the tissue and is relatively linear in behavior. Finally, as individual fibers begin to fail, the modulus again drops, and deformation proceeds under lower forces until rupture occurs.

TABLE 15.3 Representative Properties of Human Tendon and Ligament under Tensile Loading

Tissue	Property	Value
Tendon	Elastic modulus, GPa	1.2–1.8
	Ultimate strength, MPa	50–105
	Ultimate strain, %	9–35
	Energy absorbed during elastic deformation	4–10% per cycle
Ligament	Stiffness, N/mm	150
	Ultimate load, N	368
	Energy absorbed to failure, N-mm	1330

Note: Ligament properties were measured for rabbit anterior cruciate ligament with its bony attachments (Woo et al., 1994, 1997).

and ligaments have a small population of cells (fibroblasts) within the collagen and elastin fibers. The vascular supply that does exist is necessary for the maintenance of tissue properties. Periods of immobilization, such as occur when a limb is casted, result in a decrease in both stiffness and strength in ligaments. The ligament substance can recover in a period of time approximately equal to that of immobilization. However, the strength of the insertion has been seen to reach only 80 to 90 percent of its original strength after 12 months of recovery following 9 weeks of non-weight bearing (Woo et al., 1994).

15.2.4 Autografts and Allografts

In many cases, natural tissues can be used to replace damaged or diseased tissue structures. Natural tissue that is obtained from an individual and will be implanted into the same person is termed an *autograft*. If the donor is a different individual, the material is referred to as an *allograft*. Bone grafts, used to fill bony defects or replace whole sections of bone, can range from morselized bone fragments to an intact hemipelvis. The larger the graft, the more likely the need to obtain it through a tissue bank as opposed to the patient himself or herself. Soft tissue grafts are more likely to be autologous in nature. The use of a portion of the patellar tendon to replace a ruptured anterior cruciate ligament is one example. Tissue grafts face unique problems in terms of viability, tissue matching, and damage to the donor site (for autografts and allografts from living donors) that are not seen with artificial materials.

15.3 ENGINEERED MATERIALS

Treatment of many orthopedic injuries or pathologies includes the introduction of an engineered material to replace a portion of tissue or to augment the structure to assist in healing. These interventions may be permanent or temporary in nature. For the selection of any material for biomedical applications, both the function of the implant and the material's biocompatibility must be considered. The general concerns of corrosion, leaching, absorption, and mutagenicity must be addressed for orthopedic biomaterials, as they are for other applications. The following subsections provide a brief history of the selection of various material types for orthopedic implant use. The material considerations are discussed, including biocompatibility issues that are specific to orthopedic tissue replacement. The clinical success of an implant depends not only on the material choice but also on the overall implant design; however, clinical studies into implant efficacy have not generally been included.

15.3.1 Hard Tissue

The most common biomaterials applications for the replacement or augmentation of bone are used to treat injuries, particularly fractures. A much smaller proportion of implants are used in the treatment of bony diseases, such as replacing bone resected due to osteosarcoma. Total-joint replacement, such as the hip, knee, or shoulder, can be used to treat both bony fractures and joint disease.

Stress Shielding. Beyond the traditional biocompatibility issues, hard tissue biomaterials must also be designed to minimize a phenomenon known as *stress shielding*. Due to the response of bone remodeling to the loading environment, as described by Wolff's law, it is important to maintain the stress levels in bone as close to the preimplant state as possible. When an implant is in parallel with bone, such as in a bone plate or a hip stem, the engineered material takes a portion of the load—which then reduces the load, and as a result, the stress, in the remaining bone. When the implant and bone are sufficiently well bonded, it can be assumed that the materials deform to the same extent and therefore experience the same strain. In this isostrain condition, the stress in one of the components of a two-phase composite can be calculated from the equation:

$$\sigma_1 = \frac{E_1 P}{E_1 A_1 + E_2 A_2} \tag{15.1}$$

where P is the total load on the structure and E and A are the Young's modulus and cross-sectional area of each of the components. Thus the fraction of the load carried by each material, and the resulting stress, is related to its Young's modulus and cross-sectional area as compared with those of the other components of the composite structure. The stiffer materials in the composite will carry a greater proportion of the load per unit cross-sectional area.

If bone in its natural state is compared with bone with a parallel implant, the effect of this intervention on the stress in the bone, and therefore its remodeling response, can be estimated from Eq. (15.1). The applied load can be assumed to be the same before and after implantation, which yields the following equations for the stress in the bone in the two configurations:

Preimplantation ($E_{implant} = 0$; $A_{implant} = 0$):

$$\sigma_{bone} = \frac{E_{bone}P}{E_{bone}A_{bone}} = \frac{P}{A_{bone}} \tag{15.2a}$$

Postimplantation:

$$\sigma_{bone} = \frac{E_{bone}P}{E_{bone}A_{bone} + E_{implant}A_{implant}} \tag{15.2b}$$

Thus the amount of the stress reduction in bone when an implant is included depends on the modulus and cross-sectional area of the implant. Implants with a higher modulus and a larger cross-sectional area will shield the bone from a greater proportion of its normal physiological stress, resulting in bone loss according to Wolff's law.

An ideal implant would match the modulus of bone and occupy no greater cross-sectional area than the tissue replaced while meeting all the other design requirements of the implant. Since such a constraint generally cannot be met by current materials or designs, it is necessary to construct an implant that will minimize—if not entirely eliminate—stress shielding.

Metals for Bone Applications. Due to the structural role of bone, metals—with their high strength and modulus (Table 15.4)—are an obvious choice for replacement or augmentation of the tissue. The first metals implanted into the body for bony replacement were used in prehistoric times in a nonstructural role to replace cranial defects (Sanan and Haines, 1997). Gold, though of lower modulus than most metals, proved to be a suitable selection for this application because of its lack of reactivity within the body. Structural augmentation of bone using metals to assist fracture healing began in the nineteenth century, when common materials such as silver wires, iron nails, and galvanized steel plates were used to hold fragments of bone together (Peltier, 1990). In the case of metals susceptible to oxidation, such as steel and iron, corrosion led to premature mechanical failure and severe tissue reactions. In 1912, Sherman developed a steel alloy that contained vanadium and chromium (Sherman, 1912), providing it with higher strength and ductility than the previously used tool steels or crucible steels. Studies on cytotoxicity that began in the 1920s (Zierold, 1924) and 1930s (Jones and Lieberman, 1936) led to a reduction in the types of metals used in implants, focusing attention on gold, lead, aluminum, and specific formulations of steels (Peltier, 1990).

TABLE 15.4 Summary of Properties of Metals Currently Used in Hard Tissue Implants in Comparison with Bone

Material	Elastic modulus, GPa	Compressive strength, MPa
Stainless steel (316L)	200	505–860
Cast Co-Cr-Mo	200	655
Wrought CoNiCrMo	200	600–1790
Titanium alloy (Ti6A14V)	110	860
Porous tantalum	NA	Compressive: 63 Tensile: 60
Cortical bone	11–20	Compressive: 106–224 Tensile: 51–172

Note: The properties for porous tantalum are taken from a sample composition and will vary substantially based on porosity and structure.

Sources: ASTM-F136 (1998), ASTM-F562 (2000), Cowin (1989).

The first metallic implant for a hip replacement, introduced in 1940, was constructed of Vitallium, a form of cobalt-chromium alloy (Rang, 2000). Along with stainless steel, this became a standard metal for large-joint replacement and internal fracture fixation. Both materials showed good biocompatibility and excellent structural properties. The choice between the two often depended on the individual opinions of the designing physicians as they balanced biocompatibility and mechanical performance. Multiple medical grades of stainless steel were developed and standardized, including the most common formulation in use today—316L (containing iron, chromium, nickel, molybdenum, and manganese in decreasing concentrations, with additional trace elements). In addition to the cast alloy that is Vitallium (Co-30Cr-6Mb), a wrought alloy was also introduced (Co-20Cr-10Ni-15Tu) that possesses improved tensile strength and ductility (Brettle and Jordan, 1971).

One of the keys to the chemical biocompatibility of stainless steel and cobalt chromium was the formation of a passivation layer in vivo, thus minimizing the amount of corrosion that occurs to the implant. However, as indicated in Table 15.4, while the strength of these two metals reduced the chance for failure within the implant, their elastic moduli are an order of magnitude higher than that seen in healthy cortical bone. This resulted in the occurrence of stress shielding and concomitant bone loss in many patients with large implants.

In the 1940s, the aerospace industry introduced titanium and its alloys into the market. The high strength-to-weight ratio and comparatively low modulus attracted the attention of surgeons and implant designers. Titanium also proved to be chemically biocompatible, forming its passivation layer in air before implantation—thus further reducing the chemical reactions occurring at the implant interface. Despite the higher cost of the bulk material and the difficulty encountered in machining, due to their tendency to seize when in contact with other metals, titanium alloys have proven to be an effective choice for large-joint replacement and some techniques of fracture fixation, including compression plates. The most common titanium alloy used in orthopedic surgery is T318 (Ti-6Al-4V). The strength of the alloy is greater than that of pure titanium, and it maintains titanium's good biocompatibility (Brettle and Jordan, 1971). Titanium has been shown to promote good bone apposition to its surface when it is implanted, and porous surfaces have proven to be receptive to bone ingrowth. Neither of these features is as apparent in ferrous or cobalt-based alloys.

The latest metal to hit the orthopedic market is tantalum. The benefit of tantalum is the ability to form it into porous foams with a structure on the order of trabecular bone, providing a scaffold that is optimum for bone ingrowth. The mechanical properties of this novel metal depend on its porosity and structure but are sufficient to provide mechanical support during the period of bony integration (Zardiackas et al., 2001). Bone ingrowth into the porous structure after 4 weeks of implantation into cortical bone provided stronger fixation than observed in many other porous structures and progressed to fill over 60 percent of the pores by 16 weeks of implantation (Bobyn et al., 1999). In addition to its strong mechanical attributes, both in terms of initial stability and bony fixation, tantalum has been shown to be virtually inert, provoking a minimal tissue response (Black, 1994). This combination of properties has lead to the development of tantalum structures for the backing of acetabular cups and spinal fusion cages. It shows great promise for future implant development and is sure to be the subject of substantial research and development during the coming decade.

In addition to bulk implants, metals have been used to form the ingrowth surface for total-joint replacements. The design goal of these implants, which use a porous surface on all or part of the bone-contacting portion of the implant, is to better transfer the load from the implant through to the bone. Various companies have developed porous surface systems based on sintered particles, sintered wires, or rough, plasma-sprayed surfaces. The common goal in these systems is to produce a pore size into which bone will grow and become firmly fixed. Due to the substantially increased surface area of the metal in these implants, corrosion becomes a point of increased concern. In addition, it is necessary to maintain a strong bond between the porous surface and the underlying bulk metal in order to allow full load transfer to occur.

Ceramics for Bone Applications. Since bone is a composite consisting essentially of ceramic and polymeric components, and due to the essential inertness of many ceramics, this class of materials was looked to to find truly biocompatible materials for structural applications. However, the brittle nature and low tensile strength of ceramics have led to some concerns regarding the fracture behavior of these materials, whereas the high modulus again raises the specter of stress shielding for implants with large geometries (Table 15.5).

TABLE 15.5 Summary of Mechanical Properties of Some Ceramics Used in Orthopedic Applications

Material	Elastic modulus, GPa	Compressive strength, MPa	Tensile strength, MPa
Alumina	380	4500	270
Dense calcium phosphate	40–117	294	—
Bioglass	—	—	100–200

Note: Calcium phosphate properties vary depending on formulation (e.g., tricalcium phosphate versus hydroxyapatite).

Sources: Boutin et al. (1988) and Park and Lakes (1992).

Ceramics, particularly alumina, were first introduced as structural orthopedic biomaterials in the late 1960s (Boutin, 1972). However, limitations in processing technology and lack of quality control led to materials with higher than desired levels of impurities and imperfections, including high porosity levels. These defects caused a further reduction in the strength of ceramics in tensile or shear loading, resulting in premature failure in a number of clinical cases (Holmer and Nielsen, 1993; Peiro et al., 1991).

Processing techniques for ceramics improved by 1977, resulting in smaller and less variable grain sizes. As processing technologies improved, the true chemical biocompatibility of these materials caused them to be reexamined for use in orthopedic applications. Alumina and zirconia have become the most popular ceramics for use in total-joint replacement. Zirconia was introduced in an attempt to further reduce the risks of component fracture and wear-particle production (Jazwari et al., 1998). In general, the low tensile strength of both materials has precluded their use in structures subjected to substantial bending, such as the femoral stem of a total-hip replacement. However, highly polished ceramics have shown good success as articulating components in total-joint arthroplasty—with articulation against either a polymer or another ceramic both possible. Implants constructed predominantly of ceramics, particularly for total-knee replacement, are currently being investigated. These designs are particularly useful in patients with demonstrated metal sensitivities, which often precludes the use of a standard implant design.

At the opposite end of the spectrum to the ceramics investigated for their inert nature are a group of ceramic materials that are designed to induce a reaction from the surrounding tissue. These bioactive materials take advantage of the tissue's cellular physiology and structural component materials to induce bone remodeling, growth, and integration into the implant. An ideal bioactive ceramic would actually spur bone growth adjacent to the implant, promote integration of the bone with the implant structure, and gradually biodegrade as healthy bone tissue replaces the artificial structure. Two general categories of bioactive ceramics have been developed: calcium-based ceramics, such as calcium phosphate, calcium sulfate, and hydroxyapatite; and bioglasses, mineral-rich structures that can be tailored to optimize the tissue response. Bioactive materials such as these can have either *osteoinductive* or *osteoconductive* properties. The former refers to the ability of a material to trigger bone cell differentiation and remodeling in locations where bone cell proliferation and healing would not normally occur (such as a large defect), whereas the latter defines a material that promotes bony ingrowth and vascularization, allowing for integration and remodeling to take place.

Calcium-based composites rely on their similarity to the mineral component of natural bone—hydroxyapatite (HA). The theory behind their use is that the body will see these materials as tissues that need to be remodeled, allowing them to be integrated with and then replaced by bone. Tricalcium phosphate [TCP, $Ca_3(PO_4)_2$], calcium sulfate [plaster of paris, $CaSO_4$], and hydroxyapatite [$Ca_{10}(PO_4)_6(OH)_2$] are all currently being used to fill bony deflects and stimulate or direct bone formation. Calcium sulfate has been used for over a century due to its ready availability and biocompatibility (Tay et al., 1999). The crystal size (nanometers) of biological HA is much smaller than can be produced in synthetic versions of the material (Cooke, 1992); however, it has still been shown to be more osteoconductive in nature than TCP (Klein et al., 1983).

TCP, calcium sulfate, and HA can be inserted into a defect in the cortical or trabecular bone in the form of pellets or particles. The high surface-to-volume ratios of these implants, used in areas where immediate structural support is maintained through remaining bone or fracture fixation, allows for more rapid integration and remodeling of the material. The calcium sulfate formulation has been shown to resorb in only 6 to 8 weeks (Ladd and Pliam, 1999).

Less frequently, blocks of calcium-based ceramics are used to replace large segments of bone that have been resected due to injury or disease. These implants have not proven to become fully replaced by living bone but serve as a continued structural support that is integrated with the surrounding bone surface. The blocks can be made porous, to mimic the structure of trabecular bone, and this has been shown to increase bone ingrowth into the material. One brand of porous hydroxyapatite has been manufactured from the tricalcium phosphate laid down by marine coral and has been shown to possess substantial osteoconductive properties, filling over 50 percent of the porosity volume with bone within 3 months (White and Shors, 1986). Hydroxyapatite has also been combined with polymethyl methacrylate bone cement (see below) with the goal of inducing bone growth into the cement through the remodeling of these small particles of calcium-based ceramic. Ceramic bone grafts can be augmented with biological molecules designed to increase their osteoinductive nature, including transforming growth factor β (TGF-β) and bone morphogenic protein (BMP) (Ladd and Pliam, 1999). These materials then begin to bridge the gap to tissue engineering.

One of the newest techniques for applying a calcium-based material to assist with fracture fixation is through the injection of a viscous "cement" that then fully hardens in vivo. Norian SRS (Skeletal Replacement System), an injectable calcium phosphate material, was introduced in 1995 (Constantz et al., 1995). It has been shown to reduce the immobilization time required during fracture fixation (Kopylov et al., 1999) because it carries a portion of the load during bone healing. After 12 hours in vivo, Norian has cured to between 85 and 95 percent of its ultimate properties, with a final compressive strength of 55 MPa (Constantz et al., 1995). Successful clinical applications have included the reduction and stablization of unstable or intraarticular radial fractures (Kopylov et al., 1999; Yetkinler et al., 1999), complex calcaneal fractures (Schildhauer et al., 2000), and vertebral compression fractures (Bai et al., 1999) and augmentation of hip screw fixation in unstable fractures of the the intertrochanteric region of the femur (Elder et al., 2000).

In addition to their use in bulk form, calcium-based ceramics can be coated onto metallic implants to improve fixation. The ceramic can be applied through plasma spraying, creating a rough or porous surface approximately 50 μm thick (Cooke, 1992). Hydroxyapatite-coated titanium has shown firm fixation to bone in implant conditions both in mechanically stable and mechanically unstable conditions (Soballe et al., 1999), with the fixation occurring at a faster rate than in implants where the porous coating is manufactured from titanium itself (Thomas, 1994). HA coatings degrade with time and are replaced with natural bone, allowing close apposition with the underlying implant material. Clinical studies have shown that inclusion of the additional material layer does not promote increased wear or osteolysis in a properly designed implant (Capello et al., 1998). While cemented implants and metallic porous coatings remain the predominant designs of choice for total-joint replacements, ceramic-coated designs have garnered increasing interest.

Bioglass was introduced to the scientific world in the late 1960s by Dr. Hench. These glass-ceramics, which contained varied proportions of SiO, Na_2O, CaO, P_2O_3, CaF_2, and B_2O_3, were designed to interact with the normal physiology of bone to allow strong bone bonding (Ducheyne, 1985). Initial work by Greenspan and Hench (1976) indicated that an alumina implant coated with Bioglass showed substantially improved attachment to bone and new bone formation when implanted in rats compared with alumina-only controls. The bonding mechanism was found to depend on the composition of the glass, and this has sparked the development of other variations of glass-ceramics. These include Ceravital (which contains K_2O and MgO in place of CaF_2 and B_2O_3) (Ducheyne, 1985) and a form containing apatite and wollastonite (Nishio et al., 2001).

Glass-ceramics have low tensile strength and fracture toughness, limiting their use in bulk form to applications subject to purely compressive loading. Attempts have been made to use these materials as part of composite structures to increase their application. The most common method is to coat a ceramic or metallic implant with the glass to create an osteoinductive surface. The coating may be applied in a pure layer of glass or as an enamel coating with embedded glass particles (Ducheyne, 1985). For the enamel systems, it is important to ensure that the components of the enamel do not interfere with the bone-formation process (Ducheyne, 1985). The glass coating is still a brittle material and must be handled with care—any substantial impact may lead to failure of the entire coating system.

Glass composites have also been investigated using stainless steel fibers (50 to 200 μm thick) to reinforce the glass-ceramic (Ducheyne and Hench, 1982). The goal of these composites follows that of other fiber-reinforced materials—to increase their resistance to fracture by blunting crack growth and introducing a residual compressive stress within the material (Ducheyne, 1985). This procedure was found to make the material significantly more ductile and stronger, thus reducing its tendency to fail catastrophically. In addition, the elastic modulus was reduced from that of the pure glass (Ducheyne, 1985), bringing it closer to the ideal properties for bony replacement.

To date, glass-ceramics have been used clinically for only limited applications. These include material for filling bony defects along the lines of a bone graft (Pavek et al., 1994), reconstruction of the ossicular bones (Hughes, 1987), spine reconstruction (Yamamuro and Shimizu, 1994), and dental reconstruction (Kudo et al., 1990; Yukna et al., 2001).

Polymers for Bone Applications. Until recently, the only polymer used to replace or augment bone itself (as opposed to the articulating surfaces, which are actually cartilage replacement) was polymethyl methacrylate (PMMA), or bone cement. This material was introduced to the world of orthopedics in 1951 (Rang, 2000), becoming widely used in the 1960s, and provided good clinical success at maintaining the fixation of a total-joint implant within a medullary canal. Bone cement does not act as an adhesive but rather a space filler. It fills the void left between the stem of an implant and the endosteum of the bone, interdigitating with both the implant and the natural surfaces. This minimizes the need for an exact fit between the implant and the bone, required with press-fit implants, and provides an immediate source of fixation, as opposed to porous-coated implants that require several weeks for bone to grow into the implant surface. Bone cement has been used for over 50 years with little change in its composition and is still the preferred fixation method for some implants—particularly those to be used in patients with poor bone quality. Three negative factors affect the use of bone cement. First, it polymerizes in vivo through an exothermic reaction that elevates the temperature of the surrounding tissues. The effect of this high temperature on cells has not been fully established. Second, it can deteriorate through fatigue and biological processes, resulting in the production of wear debris. These particles of cement can cause osteolysis (bone loss) of the femoral bone or enter the articulating region, promoting third-body wear of the acetabular and/or femoral head components. This latter process would then further exacerbate any debris-related bone loss. Finally, the cement provides an additional material and an additional interface (bone-cement-implant versus bone-implant) at which macroscopic failure can occur. This can result in a reduced life span for the implant.

In the 1990s, researchers and clinicians began to look at polymers for fracture fixation. This work built on the idea of epoxy–carbon fiber composite plates introduced in the preceding decade (Ali et al., 1990). While they do not possess the same mechanical strength seen in the metals traditionally used for bone plates and screws (Table 15.6), they do have some properties that may outweigh this lack of strength. First, the bone plates are less stiff—resulting in reduced stress shielding and less bone loss compared with current plates. Second, the polymers can be designed to degrade with time, allowing the healing bone to eventually take over the entire load-bearing role while avoiding a second surgery for plate removal. While the fixed constructs (bone plus plate) are generally stronger during the initial healing when a metal plate is used, the loss of bone due to the higher stress shielding of stainless steel causes a substantial reduction in bone strength at extended time points (Hanafusa et al., 1995). Degradable plates and screws are typically constructed of poly(lactide-co-glycolide) (PLGA), poly(L-lactide) (PLA), or polyglycolic acid (PGA). The polymer matrix can be augmented with hydroxyapatite to improve the mechanical strength or bonding with bone (Furukawa et al., 2000; Hasirci et al., 2000). Clinical results with these new constructs appear promising for certain

TABLE 15.6 Characteristic Properties of Polymers Used in Orthopedic Implant Applications

Material property	Young's modulus, GPa	Tensile strength, MPa	Compressive strength, MPa	Elongation, %
UHMWPE	0.4–1.2	44	15.2–24.8	400–500
PMMA bone cement	1.35	45.5	89.6	4.6
PLA	1.2–3.0	28–48	NA	1.8–3.7

Sources: Dumbleton and Black (1975) and Engelberg and Kohn (1991).

applications, such as malleolar fractures (Bostman, 1987), and they are sure to be used with increasing frequency in the near future.

Tissue-Engineered Bone Replacements. The phrase *tissue engineering* has been applied to bone for a wide range of developments. Interventions can be as straightforward as delivery of osteoinductive factors such as TGF-β and BMP to the surrounding tissue through a porous scaffold. The more complicated designs include cultured bone cells within a three-dimensional matrix. Due to bone's hard tissue nature, both hard (ceramic) and soft (polymer) scaffolds are being investigated for this application (Burg et al., 2000). In general, all the calcium-based ceramics and the degradable polymers—including natural collagen—have been the subject of research interest for this application. Some polymers may need to be reinforced to provide adequate mechanical stability (Burg et al., 2000). These scaffolds have been seeded with chondrocytes, periosteal osteoblasts, and marrow progenitor cells to determine the best cell type to promote osteogenesis when implanted into a defect site (Burg et al., 2000).

15.3.2 Soft Tissue

As with bone, replacement or augmentation of orthopedic soft tissues can be used to treat injury-or disease-based degradation to the original tissue. Osteoarthritis—characterized by degradation of the articular cartilage that progresses to the bony surfaces themselves—is one of the most common pathologies experienced by the aging population, with up to 20 percent of the aging population showing signs of degenerative joint disease (DJD) (Felson et al., 2000). Ligament damage is generally the result of injury, often (though not exclusively) caused by athletic activity. Many ligaments are designed with redundant systems—the failure of a single ligament need not result in complete instability in a joint. One of the first questions that must be asked following ligament damage is whether a repair is needed or if (given the activity level of the individual) conservative treatment and bracing will provide the needed support to the joint. Tendons are damaged much less frequently than other orthopedic soft tissues and are not the site of common implant-based repair. Thus they will not be addressed in this section.

Polymers for Cartilage Replacement. Given the relatively low stiffness of cartilage and the need for low coefficients of friction, polymers have been the principal material of choice for replacement of articulating joint surfaces or at least one of the surfaces of an articulating joint. Replacements of large joints, such as the hip, knee, and shoulder, are generally designed with a metal or ceramic component articulating against a polymer surface. For smaller joints, such as those of the fingers, polymeric pieces have been used as spacers and hinges.

Silicone, polyethylene (PE), and polyolefin have all been used as a flexible hinge to replace a joint of the hand damaged through injury or arthritis. The most widely accepted implant for this application was designed by Swanson in the 1960s and continues to be used today (Linscheid, 2000). Constructed of Silastic, a form of silicone rubber, it achieves fixation through the planned formation of a fibrous capsule around the implant. Such capsular formation is a standard response to implanted structures in the body, but in many cases it has been determined to be contraindicated for optimal implant performance. In the case of Swanson's joint replacement, the implant is designed to move within the medullary canal of the connected bones, promoting an enhanced fibrous capsule formation. This fixation avoids the problems seen in the hand with screw fixation or porous coatings designed to promote bony ingrowth (Linscheid, 2000).

Beyond the traditional biocompatibility concerns, which include the effects of leaching and absorption, the greatest obstacle to the use of polymers in the role of articulating surfaces has been wear. The cyclic motion of an opposing implant component or bone against the polymer may produce substantial amounts of wear debris that can then precipitate bone loss and implant failure.

When Charnley introduced his low-friction arthroplasty in the late 1950s, he originally selected Teflon (PTFE, polytetrafluoroethylene) for the acetabular component. However, within a few years, he realized that while Teflon possessed a very low coefficient of friction, its wear resistance was poor (Charnley, 1970). While the observations made on these implants provided substantial information regarding wear processes for plastics in vivo, it was obvious that another material was required. A "filled" Teflon (given the name Fluorosint) was investigated, in which glass fibers or synthetic mica was added to improve the wear resistance of the artificial joint. While laboratory tests using a water lubricant showed that the newly formulated material had a 20-fold reduction in wear, clinical studies showed that the filled Teflon suffered wear at the same rate as the pure version. The clinical picture was worsened, however, because it was discovered that the particles used in the new formulation acted as an abrasive against the stainless steel femoral head (Charnley, 1970). This difference emphasizes the need to conduct laboratory tests in conditions that mimic the physiological environment as closely as possible before progressing to animal and human trials. Charnley hypothesized that the difference in results was due to the action of the extracellular fluids on the Teflon, preventing the formation of a protective surface layer (Charnley, 1970).

After the failure of Teflon, high-density polyethylene (HDP) was investigated as a bearing material. It was shown to be substantially more resistant to wear than PTFE, although the particles produced by the wear that was still expected to occur were a concern of Charnley's back in 1970 (Charnley, 1970). The creep behavior of HDP under compressive loading was also a concern because this would alter the shape of the articulating surfaces. New or modified materials were thus investigated. In order to counter the problem of creep, Delrin 150 was introduced and used clinically in Europe. This is a high-viscosity extruded polymer that is biocompatible, significantly harder than HDP, and resistant to creep—a property that is extremely important for sites such as the tibia (Fister et al., 1985). Polyester was also examined in the early 1970s for use in trunion designs of implants. However, wear proved to be the downfall of these materials as well (Clarke, 1992; Havelin et al., 1986; Sudmann et al., 1983). Similarly, composites of carbon fiber–reinforced PE were also developed for use as a joint surface with the goal of reducing wear. It proved to be as biocompatible as PE alone (Tetik et al., 1974). However, while the laboratory studies showed improved wear resistance, clinical results proved to be substantially worse (Busanelli et al., 1996; Clarke, 1992).

Today, the density of polyethylene has been increased further from that first used by Charnley, and joint bearings are now typically constructed from ultra-high-molecular-weight polyethylene (UHMWPE). The material has proven to provide good articulation, with the main concern being long-term wear. The problem with wear is not only the mechanical impingement that can occur as a result of a change in the articulating surface geometry but, more importantly, the effect of wear debris on the surrounding tissue. Bone, as a living material, is affected by inflammatory processes. The body reacts to the presence of foreign debris by triggering the immune system; in an attempt to rid the body of this unwanted material, phagocytotic processes are set in motion that eventually produce chemicals that adversely affect the surrounding bone. This process of osteolysis and the resulting loss of bone are principal causes of implant failure in the absence of infection. Substantial efforts are still underway to develop an implant system that minimizes the production of wear debris and protects the surrounding tissue.

Metals and Ceramics for Cartilage Replacement. Due to the problems encountered with wear debris from the polymeric components of large-joint implants, a number of designs have appeared that use highly polished hard materials on both articulating surfaces. The initial designs for hard-bearing surfaces may have been abandoned in part due to the high frictional torques and early failures that were caused by problems in both implant design and material processing (Amstutz and Grigoris, 1996; Boutin et al., 1988). Second-generation metal-metal and ceramic-ceramic bearings generally have similar coefficients of friction to joints with UHMWPE components (Table 15.7). They have proved to be clinically feasible, and studies indicate good long-term survival rates (Boutin et al., 1988; Dorr et al., 2000; Wagner and Wagner, 2000). In small-joint replacement, components manufactured from pyrolitic carbon—a material proven to have exceptional biocompatibility—have also shown good preliminary results in clinical trials (Cook et al., 1999). Both ceramic-ceramic and metal-metal designs have been shown to produce substantially reduced volumes of wear (Boutin et al., 1988; Schmalzried et al., 1996; Wagner and Wagner 2000); however, in both cases, the particles are substantially smaller than those produced from a metal or ceramic articulating against polyethylene (Boutin et al., 1988; Shahgaldi et al., 1995; Soh et al., 1996). In fact, the number of particles produced per step is about the same for cobalt-chromium articulating with either UHMWPE or itself (Wagner and Wagner, 2000). Thus further research must be conducted into the local and systemic effects of these smaller, easily transported particles within the body and on long-term outcomes. Despite questions that still deserve to be addressed, hard-bearing implants for total-joint replacement have gained increasing amounts of interest, especially for application in younger patients for whom the lifetime accumulation of wear debris is of greater concern.

TABLE 15.7 Coefficients of Friction for Sample Material Combinations Used in Total-Hip Replacement

Material combination	Coefficient of friction
Cartilage-cartilage	0.002
CoCr-UHMWPE	0.094
Zirconia-UHMWPE	0.09–0.11
Alumina-UHMWPE	0.08–0.12
CoCr-CoCr	0.12
Alumina-alumina	0.05–0.1

Note: UHMWPE ultra-high-molecular-weight polyethylene; CoCr, cobalt chromium alloy (Park and Lakes, 1992; Streicher et al., 1992).

Tissue-Engineered Cartilage Replacements. The ideal replacement material would be one that would mimic all the functions of the original tissue, including those attributed to the cellular components. Artificial biomaterials cannot meet this goal. However, the new technologies of tissue engineering have opened the door to the development of living replacement tissues that can be "manufactured" in the laboratory. Thus these are not allografts or autografts, with their inherent problems, but materials that can either be banked for use when necessary or grown to meet the needs of a particular individual. Since the majority of past interventions for replacement of cartilage (e.g., not part of a total-joint replacement) have not proved to be successful, tissue-engineered cartilage holds great promise.

The premise behind an engineered tissue is to manufacture a scaffold from a biocompatible and possibly biodegradable material and then to seed this material with appropriate cells. The scaffold supports the cells, allowing them to grow, proliferate, and become integrated with the surrounding healthy tissue. In the case of cartilage, chondrocytes must be harvested and allowed to reproduce in the laboratory to provide the required number of cells. These can be taken from healthy cartilage (articular cartilage or the epiphysis) or isolated as more primitive cells that can be directed to differentiate into the desired form (mesenchymal stem cells or bone marrow stromal cells) (Suh and Fu, 2000). The choice of scaffold is equally challenging, with the goal being to match the property of the normal cartilage matrix. In the case of cartilage, research is being conducted into the construction and application of scaffolds based on collagen, polyglycolic acid (PGA) and poly(L-lactic) acid

(PLLA) (both alone and as copolymers), hyaluronic acid, and polysaccharide-based hydrogels (Suh and Fu, 2000; Suh and Matthew, 2000; Temenoff and Mikos, 2000b). A three-dimensional scaffold is required to prevent the chondrocytes from dedifferentiating and losing some of their needed properties (Temenoff and Mikos, 2000a). Injectable materials that can deliver chondrocytes to the area of interest without an invasive surgery are also being investigated. Fibrinogen and thrombin can be combined in vivo to provide the necessary stability to the cells (Temenoff and Mikos, 2000a). This research is in its infancy, but it promises great advances during the next decades.

Polymers and Ceramics for Ligament Replacement and Augmentation. The most frequently damaged ligament is the anterior cruciate (ACL), located in the knee. Therefore, much of the work that has been done on ligament repair, replacement, and augmentation has examined this anatomic location. However, the knowledge gained through decades of work on the ACL can be transferred to other sites in the body as long as new designs undergo appropriate application-specific testing.

Four schools of thought exist when it comes to repair of damaged ligaments:

1. If sufficient joint stability exists, do nothing and allow collateral structures to maintain the mechanical function of the joint.
2. Use autologous structures to replace the damaged ligament, such as a section of the patellar tendon for the ACL.
3. Provide a bridge that the damaged structure or implanted replacement (allograft or autograft) can use as it heals. This augmentation device also carries a significant portion of the tensile load until the ligament has healed sufficiently.
4. Replace the ligament completely with an artificial material or allograft material.

Much of the debate in this field comes from the healing behavior of ligaments. Because they posess a minimal vascular supply, ligaments heal and remodel slowly. During this healing process, they are not able to carry the normal amount of tensile load. However, ligaments—like bone—also require regular cyclic loading beyond some threshold value to regain and maintain their mechanical properties. Most initial repairs of the ACL involve autograft tissue taken from the patellar tendon, the iliotibial band, or other similar tissues (Schepsis and Greenleaf, 1990). However, donor-site morbidity and the occasional failure of these grafts have driven the need for the development of other implant options. For artificial augmentation or replacement implants, polymeric fabrics have become the material of choice.

The goals for a ligament prosthesis or augmentation device must be to provide the necessary mechanical stability to the joint without premature degradation or failure. Table 15.8 provides a summary of mechanical properties for a number of synthetic grafts in comparison with normal ACL tissue.

TABLE 15.8 Representative Properties for Normal ACL and Devices Designed to Replace the Ligament or Augment Healing of an Allograft or Autograft

Material	Yield force, N	Stiffness, kN/m
Normal ACL	1750	182
GoreTex prosthesis	5000	320
Polypropylene LAD	1500–1730	330

Note: LAD, ligament augmentation device (Schepsis and Greenleaf, 1990).

At the beginning of the twentieth century, silk was applied as the first artificial material for ACL replacement (Alwyn-Smith, 1918); however, these implants failed within a few months of implantation. Use of synthetic materials for this application was virtually abandoned until the 1970s, when UHMWPE rods were introduced (Schepsis and Greenleaf, 1990). This design, along with the Proplast rod of propylene copolymer, had a short life span before fracture or elongation of the prosthetic occurred (Ahlfeld et al., 1987; Schepsis and Greenleaf, 1990). Carbon fiber was investigated as a potential material for prosthetic ligaments (Jenkins, 1978, 1985); however, its brittle nature and tendency to fail became more problematic than the benefits of the biodegradable nature of the fibers. The fibers were then coated with PLA to improve handling in the operating room as well as prevent failure in vivo (Alexander et al., 1981), but this has not gained wide use.

PTFE ligaments have been seen in clinical studies to provide higher levels of patient satisfaction than the Proplast structures (Ahlfeld et al., 1987); however, the failure rate is still higher than desirable (Schepsis and Greenleaf, 1990). The most recent material additions to the field of prosthetic ligaments have been Dacron (nylon) and a polyethylene braid; results using these implants are mixed (Schepsis and Greenleaf, 1990). Despite their promise in terms of mechanical stability and long-term outcomes, artificial ligaments have proven to be controversial. There have been substantial numbers of cases reported in which the artificial material produced a synovitis—inflammation of the synovial fluid in the joint—or failed completely (Christel, 1994). While they have gained acceptance for revision surgery for chronically unstable knees—such as may result from failure of a graft—prosthetic ligaments have not yet met the performance of autografts for primary repairs.

The advent of ligament augmentation devices (LADs) was the result of the observation that autografts or allografts experienced a period of decreased mechanical strength and stiffness soon after implantation (Kumar and Maffulli, 1999). This degradation results from the natural remodeling process that takes place to fully integrate the biologic structure into its new surroundings. One implant designed to minimize the chance of failure for the healing graft is constructed of diamond-braided polypropylene (Kumar and Maffulli, 1999). Other designs have included PLA-coated carbon fiber (Strum and Larson, 1985), knitted Dacron (Pinar and Gillquist, 1989), and polydioxanone (Puddu et al., 1993). Despite expectations based on laboratory studies, clinical results have not shown an improvement in outcomes when LADs have been used to supplement the biologic reconstruction of the ACL (Kumar and Maffulli, 1999). There is concern that an LAD will stress shield a healing ligament graft (Schepsis and Greenleaf, 1990), therefore reducing its mechanical properties and increasing the likelihood of graft failure.

The state of the art in ligament replacement remains the application of allografts. The use of artificial materials in this application is in its relative adolescence compared with fracture fixation and total-joint replacement. While artificial structures for total-ligament replacement or graft augmentation have not been fully optimized to date, they have proven to be effective in secondary repair situations—where a primary graft has failed—or cases of chronic instability. Future developments in materials, particularly composites, may produce a structure that can meet the mechanical and fixation requirements for ligament replacement with improved clinical outcomes.

15.4 CONCLUSION

Orthopedic injuries and pathologies are among the most common medical conditions. While fractures are no longer an anticipated part of a child's medical history, over 6.2 million fractures occur each year in the U.S. population (Tay et al., 1999). Osteoarthritis affects 1 in 5 individuals over the age of 70 (Felson et al., 2000). These and other conditions—including rheumatoid arthritis, osteosarcoma, and ligament tears—often require an intervention that includes the replacement of some portion of tissue. Autografts and allografts have proven problematic based primarily on material availability. Artificial materials have shown tremendous success in a number of applications, particularly total-joint replacement, but are not without their downsides, including long-term biocompatibility and survivability issues. As is the case in many areas of tissue and organ replacement, the future of tissue engineering holds great promise. If orthopedic surgeons are able to replace diseased or damaged tissue with a material that will immediately take over the original structural and mechanical function while it becomes completely integrated with the natural tissue with time, the current gap between natural and engineered materials may be bridged. Until the time when tissue engineering of complex structural tissues is perfected, the need for the continued development, refinement, and investigation of artificial materials for orthopedic applications continues. The material selected plays a large role in the overall success of an implant design—as past failures have so dramatically shown. However, a poor result with a material cannot be divorced from the implant design itself. Boutin provided good evidence of this in his historical discussion of alumina acetabular components—in which loosening rates were reduced from 26 to 0.5 percent based on the position of the stabilizing pegs alone (Boutin, 1988). Thus materials should not be deleted from the database of potential choices until it is determined if a failure was due to material selection or mechanical design. It is also important to have a thorough understanding of the underlying physiological processes in orthopedic tissues in both normal and pathological conditions so that implants of any type can be designed to be optimally integrated into the body from a mechanical and biological standpoint. The combined efforts of materials scientists, tissue physiologists, biomechanists, clinicians, and others will continue to offer developments in the field of orthopedic tissue replacement and enhancement—a discipline that already dates back thousands of years.

REFERENCES

Ahlfeld, S. K., Larson R. L., and Collins, H. R. (1987), Anterior cruciate reconstruction in the chronically unstable knee using an expanded polytetrafluoroethylene (PTFE) prosthetic ligament, *Am. J. Sports Med.* **15**(4):326–330.

Alexander, H., Parsons, J. R., and Strauchler, I. D. (1981), Canine patellar tendon replacement with a polyactic acid polymer-filamentous carbon tissue scaffold, *Orthop. Rev.* **10**:41–51.

Ali, M. S., French, T. A., Hastings, G. W., Rae, T., Rushton, N., Ross, E. R., and Wynn-Jones, C. H. (1990), Carbon fibre composite bone plates: Development, evaluation and early clinical experience, *J. Bone Joint Surg.* **72B**(4):586–591.

Alwyn-Smith, S. (1918), The diagnosis and treatment of injuries to the crucial ligaments, *Br. J. Surg.* **6**:176.

Amstutz, H. C., and Grigoris, P. (1996), Metal on metal bearings in hip arthroplasty, *Clin. Orthop.*, **329S**: 11–34.

An, Y. H. (2000), Mechanical properties of bone, in Y. H. An, and R. A. Draughn (eds.), *Mechanical Testing of Bone and the Bone-Implant Interface*, pp. 41–63, CRC Press, Boca Raton, FL.

ASTM-F136 (1998), *Standard Specification for Wrought Titanium-6 Aluminum-4 Vanadium ELI (Extra Low Interstitial) Alloy (UNS R56401) for Surgical Implant Applications*, American Society for Testing and Materials, W. Conshohocken, Pa.

ASTM-F562 (2000), *Standard Specification for Wrought Cobalt-35 Nickel-20 Chromium-10 Molybdenum Alloy for Surgical Implant Applications*, American Society for Testing and Materials, W. Conshohocken, Pa.

Bai, B., Jazrawi, L. M., Kummer, F. J., and Spivak, J. M. (1999), The use of an injectable, biodegradable calcium phosphate bone substitute for the prophylactic augmentation of osteoporotic vertebrae and the management of vertebral compression fractures, *Spine* **24**(15):1521–1526.

Black, J. (1994), Biological performance of tantalum, *Clin. Mater.* **16**(3):167–173.

Bobyn, J. D., Stackpool, G. J., Hacking, S. A., Tanzer, M., and Krygier, J. J., (1999), Characteristics of bone ingrowth and interface mechanics of a new porous tantalum biomaterial, *J. Bone Joint Surg.*, **81B**(5):907–914.

Bostman, O., Vainionpaa, S., Hirvensalo, E., Makela, A., Vihtonen, K., Tormala, P., and Rokkanen, P. (1987), Biodegradable internal fixation for malleolar fractures: A prospective randomised trial, *J. Bone Joint Surg.* **69B**(4):615–619.

Boutin, P. (1972), Arthroplastie totale de la hanche par prostheses en aluminine fritte, *Rev. Chir. Orthop.* **58**: 230–246.

Boutin, P., Christel, P., Dorlot, J.-M., Meunier, A., deRoquancourt, A., Blanquaert, D., Herman, S., Sedel, L., and Witvoet, J. (1988), The use of dense alumina-alumina ceramic combination in total hip replacement, *J. Biomed. Mater. Res.* **22**:1203–1232.

Brettle, J., Hughes, A. N., and Jordan, B. A. (1971), Metallurgical aspects of surgical implant materials, *Injury* **2**, (3):225–234.

Burg, K. J., Porter, S., and Kellam, J. F. (2000), Biomaterial developments for bone tissue engineering, *Biomaterials* **21**(23):2347–2359.

Busanelli, L., Squarzoni, S., Brizio, L., Tigani, D., and Sudanese, A. (1996), Wear in carbon fiber–reinforced polyethylene (poly-two) knee prostheses, *Chir. Organi. Mov.* **81**(3):263–267.

Capello, W. N., D'Antonio, J. A., Manley, M. T., and Feinberg, J. R. (1998), Hydroxyapatite in total hip arthroplasty: Clinical results and critical issues, *Clin. Orthop.* **16**(355):200–211.

Charnley, J. (1970), Total hip replacement by low-friction arthroplasty, *Clin. Orthop.* **72**:7–21.

Christel, P. (1994), Prosthetic replacement of the anterior cruciate ligament: A challenge, *Clin. Mater.* **15**(1): 3–13.

Clarke, I. C. (1992), Role of ceramic implants: Design and clinical success with total hip prosthetic ceramic-to-ceramic bearings, *Clin. Orthop.* **16**(282):19–30.

Constantz, B. R., Ison, I. C., Fulmer, M. T., Poser, R. D., Smith, S. T., VanWagoner, M., Ross, J., Goldstein, S. A., Jupiter, J. B., and Rosenthal, D. I. (1995), Skeletal repair by in situ formation of the mineral phase of bone, *Science* **267**(5205):1796–1799.

Cook, S. D., Beckenbaugh, R. D., Redondo, J., Popich, L. S., Klawitter, J. J., and Linscheid, R. L. (1999), Long-term follow-up of pyrolytic carbon metacarpophalangeal implants, *J. Bone Joint Surg.* **81A**(5):635–648.

Cooke, F. W. (1992), Ceramics in orthopedic surgery, *Clin. Orthop.* **16**(276):135–146.

Cowin, S. C. (1989), Mechanical properties of cortical bone, in S. C. Cowin (ed.), *Structure and Function of Cortical Bone,* pp. 97–127, CRC Press, Boca Raton, Fla.

Dorr, L. D., Wan, Z., Longjohn, D. B., Dubois, B., and Murken, R. (2000), Total hip arthroplasty with use of the Metasul metal-on-metal articulation: Four to seven-year results, *J. Bone Joint Surg.* **82A**(6):789–798.

Ducheyne, P. (1985), Bioglass coatings and Bioglass composites as implant materials, *J. Biomed. Mater. Res.* **19**:273–291.

Ducheyne, P., and Hench, L. L. (1982), The processing and static mechanical properties of metal fiber–reinforced Bioglass, *J. Mater. Sci.* **17**:595–606.

Dumbleton, J., and Black, J. (1975), *An Introduction to Orthopaedic Materials*, Charles C. Thomas, Publishers, Springfield, Ill.

Elder, S., Frankenburg, E., Goulet, J., Yetkinler, D., Poser, R., and Goldstein, S. (2000), Biomechanical evaluation of calcium phosphate cement-augmented fixation of unstable intertrochanteric fractures, *J. Orthop. Trauma* **14**(6):386–393.

Engelberg, I., and Kohn, J. (1991), Physicomechanical properties of degradable polymers used in medical applications: A comparative study, *Biomaterials* **12**:292–304.

Felson, D. T., Lawrence, R. C., Dieppe, P. A., Hirsch, R., Helmick, C. G., Jordan, J. M., Kington, R. S., Lane, N. E., Nevitt, M. C., Zhang, Y., Sowers, M., McAlindon, T., Spector, T. D., Poole, A. R., Yanovski, S. Z., Ateshian, G., Sharma, L., Buckwalter, J. A., Brandt, K. D., and Fries, J. F. (2000), Osteoarthritis: New insights; part 1: The disease and its risk factors, *Ann. Intern. Med.* **133**(8):635–646.

Fister, J. S., Memoli, V. A., Galante, J. O., Rostoker, W., and Urban, R. M. (1985), Biocompatibility of Delrin 150: A creep-resistant polymer for total joint prostheses, *J. Biomed. Mater. Res.* **19**:519–533.

Furukawa, T., Matsusue, Y., Yasunaga, T., Nakagawa, Y., Shikinami, Y., Okuno, M., and Nakamura, T. (2000), Bone bonding ability of a new biodegradable composite for internal fixation of bone fractures, *Clin. Orthop.* **16**(379):247–258.

Greenspan, D. C., and Hench, L. L. (1976), Chemical and mechanical behavior of Bioglass-coated alumina, *J. Biomed. Mater. Res. Symp.* **7**:503–509.

Hanafusa, S., Matsusue, Y., Yasunaga, T., Yamamuro, T., Oka, M., Shikinami, Y., and Ikada, Y. (1995), Biodegradable plate fixation of rabbit femoral shaft osteotomies: A comparative study, *Clin. Orthop.* **16**(315): 262–271.

Hasirci, V., Lewandrowski, K. U., Bondre, S. P., Gresser, J. D., Trantolo, D. J., and Wise, D. L. (2000), High-strength bioresorbable bone plates: Preparation, mechanical properties and in vitro analysis, *Biomed. Mater. Eng.* **10**(1):19–29.

Havelin, L. I., Gjerdet, N. R., Lunde, O. D., Rait, M., and Sudmann, E. (1986), Wear of the Christiansen hip prosthesis, *Acta Orthop. Scand.* **57**(5):419–422.

Hayes, W. C., and Bouxsein, M. L. (1997), Biomechanics of cortical and trabecular bone: Implications for assessment of fracture risk, in V. C. Mow and W. C. Hayes (eds.) *Basic Orthopaedic Biomechanics,* 2d ed., Lippincott-Raven Publishers, Philadelphia.

Holmer, P., and Nielsen, P. T. (1993), Fracture of ceramic femoral heads in total hip arthroplasty, *J. Arthroplasty* **8**(6):567–571.

Hughes, G. B. (1987), Ossicular reconstruction: A comparison of reported results, *Am. J. Otol.* **8**(5):371–374.

Jazrawi, L. M., Bogner, E., Della Valle, C. J., Chen, F. S., Pak, K. I., Stuchin, S. A., Frankel, V. H., and Di Cesare, P. E. (1999), Wear rates of ceramic-on-ceramic bearing surfaces in total hip implants: A 12-year follow-up study, *J. Arthroplasty* **14**(7):781–787.

Jazrawi, L. M., Kummer, F. J., and DiCesare, P. E. (1998), Alternative bearing surfaces for total joint arthroplasty, *J. Am. Acad. Orthop. Surg.* **6**(4):198–203.

Jenkins, D. H. R. (1978), The repair of cruciate ligaments with flexible carbon fibers, *J. Bone Joint Surg.* **60B**(520):520.

Jenkins, D. H. R. (1985), Ligament induction by filamentous carbon fiber, *Clin. Orthop.* **196**:88–89.

Jones, L., and Lieberman, B. A. (1936), Interaction of bone and various metals, *AMA Arch. Surg.* **32**:990–1006.

Kaplan, F. S., Hayes, W. C., Keaveny, T. M., Boskey, A., Einhorn, T. A., and Iannotti, J. P. (1994), Form and function of bone, in S. R. Simon (ed.), *Orthopaedic Basic Science,* American Academy of Orthopaedic Surgeons, Chicago.

Klein, C. P. A. T., Driessen, A. A., deGroot, K., and van den Hoof, A. (1983), Biodegradation behavior of various calcium phosphate materials in bone tissue, *J. Biomed. Mater. Res.* **17**:769.

Kopylov, P., Runnqvist, K., Jonsson, K., and Aspenberg, P. (1999), Norian SRS versus external fixation in redisplaced distal radial fractures: A randomized study in 40 patients, *Acta Orthop. Scand.* **70**(1):1–5.

Kudo, K., Miyasawa, M., Fujioka, Y., Kamegai, T., Nakano, H., Seino, Y., Ishikawa, F., Shioyama, T., and Ishibashi, K. (1990), Clinical application of dental implant with root of coated Bioglass: Short-term results, *Oral Surg. Oral Med. Oral Pathol.* **70**(1):18–23.

Kumar, K., and Maffulli, N. (1999), The ligament augmentation device: An historical perspective, *Arthroscopy* **15**(4):422–432.

Ladd, A. L., and Pliam, N. B. (1999), Use of bone-graft substitutes in distal radius fractures, *J. Am. Acad. Orthop. Surg.* **7**(5):279–290.

Lee, H. S. J. (2000), *Dates in Cardiology: A Chronological Record of Progress in Cardiology over the Last Millennium,* Parthenon, New York.

Linscheid, R. L. (2000), Implant arthroplasty of the hand: Retrospective and prospective conditions, *J. Hand Surg.* **25A**(5):796–816.

Mankin, H. J., Mow, V. C., Buckwalter, J. A., Iannotti, J. P., and Ratcliffe, A. (1994), Form and function of articular cartilage, in S. R. Simon (ed.) *Orthopaedic Basic Science,* American Academy of Orthopaedic Surgeons, Chicago.

Nishio, K., Neo, M., Akiyama, H., Okada, Y., Kokubo, T., and Nakamura, T. (2001), Effects of apatite and Wollastonite containing glass-ceramic powder and two types of alumina powder in composites on osteoblastic differentiation of bone marrow cells, *J. Biomed. Mater. Res.* **55**(2):164–176.

Park, J. B., and Lakes, R. S. (1992), *Biomaterials: An Introduction,* Plenum Press, New York.

Pavek, V., Novak, Z., Strnad, Z., Kudrnova, D., and Navratilova, B. (1994), Clinical application of bioactive glass-ceramic BAS-O for filling cyst cavities in stomatology, *Biomaterials* **15**(5):353–358.

Peiro, A., Pardo, J., Navarrete, R., Rodriguez-Alonso, L., and Martos, F. (1991), Fracture of the ceramic head in total hip arthroplasty: Report of two cases, *J. Arthroplasty* **6**(4):371–374.

Peltier, L. F. (1990), *Fractures: A History and Iconography of Their Treatment,* Norman Publishing, San Francisco.

Pinar, H., and Gillquist, J. (1989), Dacron augmentation of a free patellar tendon graft: A biomechanical study, *Arthroscopy* **5**:328–330.

Puddu, G., Gipolla, M., Cerullo, G., Franco, V., and Gianni, E. (1993), Anterior cruciate ligament reconstruction and augmentation with PDS graft, *Clin. Sports Med.* **12**:13–24.

Rang, M. (2000), *The Story of Orthopaedics,* Saunders, Philadelphia.

Sanan, A., and Haines, S. J. (1997), Repairing holes in the head: A history of cranioplasty, *Neurosurgery* **40**(3):588–603.

Schepsis, A. A., and Greenleaf, J. (1990), Prosthetic materials for anterior cruciate ligament reconstruction, *Orthop. Rev.* **19**(11):984–991.

Schildhauer, T. A., Bauer, T. W., Josten, C., and Muhr, G. (2000), Open reduction and augmentation of internal fixation with an injectable skeletal cement for the treatment of complex calcaneal fractures, *J. Orthop. Trauma* **14**(5):309–317.

Schmalzried, T. P., Peters, P. C., Maurer, B. T., Bragdon, C. R., and Harris, W. H. (1996), Long-duration metal-on-metal total hip arthroplasties with low wear of the articulating surfaces, *J. Arthroplasty* **11**(3):322–331.

Shahgaldi, B. F., Heatley, F. W., Dewar, A., and Corrin, B. (1995), In vivo corrosion of cobalt-chromium and titanium wear particles, *J. Bone Joint Surg.* **77B**(6):962–966.

Sherman, W. D. (1912), Vanadium steel plates and screws, *Surg. Gynecol. Obstet.* **14**:629.

Soballe, K., Overgaard, S., Hansen, E. S., Brokstedt-Rasmussen, H., Lind, M., and Bunger, C. (1999), A review of ceramic coatings for implant fixation, *J. Long Term Eff. Med. Implants* **9**(1–2):131–151.

Soh, E. W., Blun, G. W., Wait, M. E., et al. (1996), Size and shape of metal particles from metal-on-metal total hip replacements, *Trans. Orthop. Res. Soc.* **21**:462.

Streicher, R. M., Senlitsch, M., and Schön, R. (1992), Articulation of ceramic surfaces against polyethylene, in A. Ravaglioli and A. Krajewsk (eds.) *Bioceramics and the Human Body,* pp. 118–123, Elsevier Applied Science, New York.

Strum, G. M., and Larson, R. L. (1985), Clinical experience and early results of carbon fiber augmentation of anterior cruciate reconstruction of the knee, *Clin. Orthop.* **196**:124–138.

Sudmann, E., Havelin, L. I., Lunde, O. D., and Rait, M. (1983), The Charnley versus the Christiansen total hip arthroplasty: A comparative clinical study, *Acta Orthop. Scand.* **54**(4):545–552.

Suh, J. K., and Fu, F. H. (2000), Application of tissue engineering to cartilage repair, in *Gene Therapy and Tissue Engineering in Orthopaedic and Sports Medicine,* Birkäuser, Boston.

Suh, J. K., and Matthew, H. W. (2000), Application of chitosan-based polysaccharide biomaterials in cartilage tissue engineering: A review, *Biomaterials* **21**(24):2589–2598.

Tay, B. K., Patel, V. V., and Bradford, D. S. (1999), Calcium sulfate– and calcium phosphate–based bone substitutes: Mimicry of the mineral phase of bone, *Orthop. Clin. North Am.* **30**(4):615–623.

Temenoff, J. S., and Mikos, A. G. (2000a), Injectable biodegradable materials for orthopedic tissue engineering, *Biomaterials* **21**(23):2405–2412.

Temenoff, J. S., and Mikos, A. G. (2000b), Review: Tissue engineering for regeneration of articular cartilage, *Biomaterials* **21**(5):431–440.

Tetik, R. D., Galante, J. O., and Rostoker, W. (1974), A wear-resistant material for total joint replacement: Tissue biocompatibility of an ultra-high-molecular-weight (UHMW) polyethylene-graphite composite, *J. Biomed. Mater. Res.* **8**(5):231–250.

Thomas, K. A. (1994), Hydroxyapatite coatings, *Orthopedics* **17**(3):267–278.

Venzmer, G. (1968), *Five Thousand Years of Medicine,* Taplinger, New York.

Wagner, M., and Wagner, H. (2000), Medium-term results of a modern metal-on-metal system in total hip replacement, *Clin. Orthops.* **379**:123–133.

White, E., and Shors, E. C. (1986), Biomaterial aspects of Interpore-200 porous hydroxyapatite, *Dent. Clin. North Am.* **30**(1):49–67.

Wolff, J. (1892), *Das Gesetz der Transformation der Knochen,* A. Hirchwald, Berlin.

Woo, S. L. Y., An, K. N., Arnoczky, S. P., Wayne, J. S., Fithian, D. C., and Myers, B. S. (1994), Anatomy, biology, and biomechanics of tendon, ligament and meniscus, in S. R. Simon (ed), *Orthopaedic Basic Science,* American Academy of Orthopaedic Surgeons, Chicago.

Woo, S. L. Y., Livesay, G. A., Runco, T. J., and Young, E. P. (1997), Structure and function of tendons and ligaments, in V. C. Mow and W. C. Hayes (eds), *Basic Orthopaedic Biomechanics,* Lippincott-Raven, Philadelphia.

Yamamuro, T., and Shimizu, K. (1994), Clinical application of AW glass ceramic prosthesis in spinal surgery, *Nippon Seikeigeka Gakkai Zasshi* **68**(7):505–515.

Yetkinler, D. N., Ladd, A. L., Poser, R. D., Constantz, B. R., and Carter, D. (1999), Biomechanical evaluation of fixation of intraarticular fractures of the distal part of the radius in cadavera: Kirschner wires compared with calcium-phosphate bone cement, *J. Bone Joint Surg.* **81A**(3):391–399.

Yukna, R. A., Evans, G. H., Aichelmann-Reidy, M. B., and Mayer, E. T. (2001), Clinical comparison of bioactive glass bone replacement graft material and expanded polytetrafluoroethylene barrier membrane in treating human mandibular molar class II furcations, *J. Periodontol.* **72**(2):125–133.

Zardiackas, L. D., Parsell, D. E., Dillon, L. D., Mitchell, D. W., Nunnery, L. A., and Poggie, R. (2001), Structure, metallurgy, and mechanical properties of a porous tantalum foam, *J. Biomed. Mater. Res.* **58**(2):180–187.

Zierold, A. A. (1924), Reaction of bone to various metals, *AMA Arch. Surg.* **9**:365–412.

CHAPTER 16
BIOMATERIALS TO PROMOTE TISSUE REGENERATION

Nancy J. Meilander and Ravi V. Bellamkonda
Case Western Reserve University, Cleveland, Ohio

16.1 BACKGROUND

Historically, biomaterial implants have been designed to replace a specific function, usually mechanical, and were considered ideal if they elicited little or no response in vivo. For instance, synthetic vascular grafts have typically been made of inert materials such as expanded polytetrafluoroethylene (Teflon) with the necessary mechanical strength to support the forces created by pulsatile blood flow. Likewise, materials for orthopedic screws and plates for bone fixation were usually metals, chosen for their mechanical strength. However, synthetic small-diameter vascular grafts fail by thrombosis, and orthopedic implants can weaken existing bone due to stress shielding or fail due to implant loosening. In addition, most surface-modification approaches to improve integration of the implant with the host tissue have not been very successful over the last 20 years. These failures demonstrate that synthetic materials alone cannot fully replace all the functions (structural, mechanical, biochemical, metabolic, etc.) that the original tissue provided.

Due to new understanding from a molecular biology perspective, it is clear that most tissue has two categories of components: (1) structural and (2) biochemical. The field of tissue engineering considers both these components in the design of implants to promote regeneration and facilitate better integration of implants. With a growth rate of 22.5 percent per annum (based on the annual spending in 1994 and 1997), tissue engineering is a rapidly expanding field. In 1997, over $450 million was allocated for tissue engineering to fund the research efforts of nearly 2500 scientists and support staff (Lysaght et al., 1998). Tissue-engineered skin products (Apligraf from Organogenesis, Transcyte from Advanced Tissue Sciences) and bioengineered cartilage (Carticel from Genzyme Tissue Repair) are examples of commercially available treatments, and many other tissue regeneration products are currently in clinical trials.

Using the tissue-engineering strategy, materials designed for tissue regeneration attempt to control the physiological response to the implanted biomaterial by mimicking the two-component structure of natural tissue. A variety of techniques have been implemented in this approach to the design of regenerative biomaterials, as listed in Table 16.1. Tissue engineers typically use synthetic or

TABLE 16.1 Parameters to Optimize Biomaterials for Tissue Regeneration

Structural components	Biochemical components
Physical properties	Immobilized signals
Shape	Matrix proteins
Size	Adhesive peptides
Porosity	Growth factors
Three-dimensionality	Diffusible signals
Surface topography	One-component system
Mechanical properties	Two-component system
Reinforced composites	Gene delivery
Mechanical loading	Living component
Electrical stimuli	Cell seeding
Chemical properties	Differentiated cells
Hydrophilicity	Scaffold prevascularization
Charge	Cells for neuroregeneration
Biodegradation	Combination therapy
	Stem cells
	Commercial, cell-based products

natural biomaterials to achieve the structural design parameters. The design space for optimizing the structural parameters includes the physical, mechanical, and chemical characteristics of the material. The biochemical parameters are incorporated via immobilized signals, diffusible chemoattractive agents, or living cells. The critical factors that promote tissue growth in vivo can be determined by examining environments that stimulate tissue generation and regeneration in the body, such as those found during development, tissue remodeling, and wound healing. By providing similar cues, regenerative materials can function in a supportive role, aiding the biological processes when necessary, rather than permanently replacing a lost function. The materials discussed in this chapter are listed in Table 16.2.

16.2 STRUCTURAL COMPONENT

In order to properly guide tissue regeneration, a biomaterial ought to satisfy the structural requirements of the native tissue. For each type of tissue and application, the implant must have the correct physical architecture with the appropriate mechanical and chemical properties. Table 16.3 provides examples of approaches that utilize these properties for tissue regeneration.

16.2.1 Physical Properties

In tissue regeneration, both natural and synthetic materials contribute to the spectrum of physical properties displayed in biomaterials today. Collagen, a naturally occurring bioresorbable protein, is commonly used as a scaffold for tissue regeneration. The most frequently used collagen is type I collagen, but other types may also be appropriate. However, natural materials such as collagen can only provide a limited range of properties, and they may be immunogenic (Madsen and Mooney, 2000). Synthetic materials, including the biodegradable aliphatic polyesters, namely, poly(L-lactic acid) (PLLA), poly(D,L-lactic acid) (PDLLA), poly(glycolic acid) (PGA), and poly(lactide-co-glycolide) copolymers (PLGA), can be formed to provide a range of physical properties and may therefore offer greater flexibility. These materials are often produced in a less expensive, more

TABLE 16.2 Example Materials Used for Tissue Regeneration

Material	
Agarose	Poly(D,L-lactic acid)
Alginate	Polyester
Calcium phosphate	Poly(ethylene glycol)
Cellulose	Poly(glycolic acid)
Chitosan	Poly(L-lactic acid)
Collagen	Poly(L-lactide-co-6-caprolactone)
Collagen-glycosaminoglycan, chitosan-linked	Poly(lactide-glycolide)
Dextran	Polylysine
Dicalcium phosphate dihydrate	Poly-*N*-(2-hydroxypropyl)-methacrylamide
Expanded polytetrafluoroethylene	Polyorthoesters
Fibrin	Polypropylene
Fluorinated ethylene propylene	Poly(propylene fumarate)
Gelatin	Polypyrrole
Gelfoam matrices	Polystyrene
Hyaluronic acid	Poly(styrenesulfonate)
Hydroxyethyl methacrylate-methyl methacrylate	Polyurethane
Hydroxyapatite	Polyvinyl alcohol
Matrigel	Quartz
Nanophase alumina	Silica
Poly-4-hydroxybutyrate	Titanium
Poly(acrylonitrile-co-vinyl chloride)	Tricalcium phosphate
Polyanhydrides	Vinylidene fluoride–trifluoroethylene copolymer
Polycarbonate	

reproducible manner relative to natural materials. However, processing often involves the use of high temperatures or harsh organic solvents that may preclude the incorporation of sensitive proteins or nucleic acids. Also, residual solvent in the final product may detrimentally affect the in vivo response to the material. Consequently, while evaluating the physical properties of an implant, including shape, size, porosity, three-dimensionality, and surface topography, other innate characteristics of the material should also be considered.

Shape. The overall shape of a biomaterial implant is often dictated by the application. PLGA polymers (Giordano et al., 1997; Lu et al., 1998; Thomson et al., 1996) and collagen (Tiller et al., 2001) have been formed into thin films to facilitate the growth of cell monolayers, including epithelium. For perivascular treatment of intimal hyperplasia in blood vessels, alginate hydrogels containing PLGA drug-delivery systems were fabricated into perivascular wraps (Edelman et al., 2000). In the nervous system, polymeric degradable rods made of PDLLA foams mixed with poly(ethylene oxide)-block-poly(D,L-lactide) (PELA) have been used for spinal cord regeneration in rats (Maquet et al., 2001), and porous conduits of PLGA, PLLA (Widmer et al., 1998), or poly(L-lactide-co-6-caprolactone) (Giardino et al., 1999) have been fabricated to facilitate regeneration of peripheral nerves (Evans et al., 1999a). To form tubular shapes for applications such as intestinal tissue engineering, PLGA copolymers have been made into porous films, formed into tubular structures, and implanted in vivo to develop tubular fibrovascular tissue (Mooney et al., 1994). Tissues with complex three-dimensional (3D) shapes require a more sophisticated technique to fabricate appropriately shaped implants. One study produced a dome-shaped bone formation in a rabbit calvarial defect using PLLA–tricalcium phosphate (TCP) matrices molded into a dome-shaped structure (Lee et al., 2001). For joint repair, the specific 3D shapes of the metacarpal-phalangeal pieces have been replicated with PLLA and PGLA using a lamination technique to precisely control the profile (Mikos et al., 1993). These studies illustrate the variety of methods available to create almost any desired shape.

TABLE 16.3 Examples of Structural Techniques for Designing Regenerative Biomaterials

Design parameter	Biomaterial	Application	References
		Physical properties	
Shape	PLGA thin film	Retinal pigment epithelium sheets	Giordano et al., 1997
	Fibrin microbeads	Wound healing	Gorodetsky et al., 1999
Size	PGA fibers	SMC transplantation	Eiselt et al., 1998
	Agar substrate	Spheroid growth in 3D	Folkman et al., 1973
Porosity	Polycarbonate membranes	Corneal epithelial tissue migration	Steele et al., 2000
	PDLLA-PELA with longitudinal pores	Nerve regeneration	Maquet et al., 2001
Three-dimensionality	Alginate beads	3D hepatocyte culture	Selden et al., 1999
	Collagen microcarriers	3D chondrocyte culture	Frondoza et al., 1996
Surface topography	Textured polyurethane	BAEC spreading	Goodman et al., 1996
	Grooved acrylic substrates	Directed neurite extension	Clark et al., 1990
		Mechanical properties	
Reinforced composites	PLGA with hydroxyapatite	Bone regeneration	Devin et al., 1996
	PPF with β-TCP	Bone regeneration	Yaszemski et al., 1996
Mechanical loading	PGA/poly-4-hydroxybutyrate	Vascular grafts	Hoerstrup et al., 2001
	Agarose	Chondrocyte differentiation	Elder et al., 2000
Electrical stimuli	P(VDF-TrFE)	Peripheral nerve regeneration	Fine et al., 1991
	Polypyrrole	Enhanced neurite extension	Schmidt et al., 1997
		Chemical properties	
Hydrophilicity	PLGA with varying hydrophilicity	Fibroblast adhesion	Khang et al., 1999
	Silica films with PEG	Inhibit cell adhesion	Alcantar et al., 2000
Charge	PLLA and PLGA with polylysine	Increase chondrocyte attachment	Sittinger et al., 1996
	Agarose hydrogel with chitosan	Induce neurite extension	Dillon et al., 2000
Biodegradation	Polyanhydrides	Adjustable degradation rates	Burkoth et al., 2000
	PLGA in fluid flow	Adjustable degradation rates	Agrawal et al., 2000

In addition to large, continuous shapes, smaller, discontinuous materials have also been used for tissue regeneration, either for delivery of growth factors or as a scaffold to culture and transport cells to diseased or injured areas. To act as a drug-delivery system, microparticles made of PLGA and poly(ethylene glycol) (PEG) were loaded with transforming growth factor-β1 (TGF-β1). The loaded microparticles increased the proliferation of rat marrow stromal cells (Peter et al., 2000). When designed as cell scaffolds, microcarriers made of collagen have been used to culture human chondrocytes (Frondoza et al., 1996), and fibrin microbeads have been used to transport fibroblasts for skin wound-healing applications (Gorodetsky et al., 1999). Glass and collagen-coated dextran microspheres have been used to transplant cells for neuroregeneration (Borlongan et al., 1998; Saporta et al., 1997). Also for neuroregeneration, PC12 cells have been microencapsulated in 75:25 hydroxyethyl methacrylate-methyl methacrylate copolymers (Vallbacka et al., 2001), as well as in alginate-polylysine-alginate (Winn et al., 1991). Small microparticles are advantageous because, unlike large implants, suspensions of microparticles are usually injectable.

Size. The overall size of an implant is important, especially with respect to the diffusion of oxygen and nutrients within the implant. It is essential for regenerating tissue to have access to the appropriate nutrients as well as a method of waste removal, and the oxygen supply is often the limiting factor. Cells typically require an oxygen partial pressure (P_{O_2}) of 1 to 3 mmHg for basic cellular metabolism (Hunter et al., 1999). Due to the capillary P_{O_2} and the oxygen diffusion coefficient in tissue, cells are usually found no more than 50 μm from a capillary in vivo (Guyton and Hall, 1996b). When biomaterial scaffolds have a thickness significantly greater than 100 μm, oxygen deficits may occur, depending on the diffusive properties of the scaffold. When PLLA sponges infiltrated with polyvinyl alcohol (PVA) were formed into 1-mm-thick samples for the seeding and delivery of hepatocytes, the cells on the interior of the sponges necrosed in vivo (Mooney et al., 1995), probably because the oxygen and nutrient demands of the cells exceeded the supply. Another study seeded smooth muscle cells (SMCs) onto PGA fibers for transplantation into rats. After 18 days, viable SMCs were found at the edges of the implant, but cells more than 100 μm into the implant did not survive (Eiselt et al., 1998). Consequently, if a cellular implant is sufficiently large, it will require enhanced oxygen transport, perhaps via an internal blood supply, to support the cells deep within the implant.

Blood vessels naturally invade foreign materials as part of the wound-healing response, so macroporous scaffolds will eventually vascularize. However, this process is lengthy, taking over 3 weeks even for relatively small (5 mm) PLLA scaffolds (Voigt et al., 1999). To avoid the development of a necrotic core, implant size should be based on the diffusive properties of the material as well as the metabolic needs of the cells. Multicell spheroids cultured in soft agar enlarged until they reached a critical size, after which they ceased to expand. Although the critical size ranged from 2.4 to 4.0 mm, the first cells to necrose were greater than 150 to 200 μm from the surface, probably due to the limitation of oxygen diffusion (Folkman and Hochberg, 1973). In this study, the allowable distance between the viable cells within the spheroids and the nutrient/oxygen source was larger than the typical in vivo distance between cells and blood vessels (50 μm), most likely due to the diffusive properties of the agar and the density and metabolism of the cells. Nevertheless, diffusion alone was not sufficient to sustain cells deep within the spheroids. To overcome limitations such as this, a biochemical stimulus to enhance nutrient influx can be included in the material, as discussed later.

Porosity. A biomaterial for tissue regeneration must be porous enough to allow cells to invade the material, with ample surface area for an appropriate quantity of cells to adhere and proliferate. Studies have shown that creating pores in polycarbonate and polyester membranes enhanced corneal epithelial tissue migration (Steele et al., 2000), and the introduction of pores into PLGA foams increased rat hepatocyte adhesion (Ranucci and Moghe, 1999). The surface area for cell attachment is correlated with the porosity of the polymer, so highly porous PLGA foams have the advantage of increased available surface area to support cellular attachment and growth (Zheng et al., 1998). The porosity of a biomaterial also affects the diffusion of oxygen and nutrients within the scaffold, as well as the rate of neovascularization (Eiselt et al., 1998). Larger pores in the biomaterial may allow for improved mass transport and neovascularization (Zhang and Ma, 2000). In addition, pore shape can be adjusted to enhance tissue regeneration. Biodegradable PDLLA-PELA foams have been made with longitudinally oriented pores to facilitate spinal cord regeneration (Maquet et al., 2001).

Often, a specific pore size enhances cellular activity, and the optimal pore diameter depends on the cell type. In one set of studies, the nominal pore size of porous polycarbonate membranes affected epithelial tissue migration (Steele et al., 2000) and the formation of a continuous basement membrane (Evans et al., 1999b). Likewise, when the average pore radius of agarose hydrogels (Fig. 16.1) was below a critical threshold, neurite extension from rat embryonic striatal cells and embryonic chick dorsal root ganglia (DRGs) was inhibited (Bellamkonda et al., 1995; Dillon et al., 1998). Also, cellulose membranes with 0.2-μm pores were unable to support cell ingrowth, but when the pore size was increased to 8 μm, cells were able to invade the membranes (Padera and Colton, 1996). Interestingly, rat hepatocyte attachment to 3D PLGA foams of varying porosity showed no preference for any given porosity. However, when protein secretion was evaluated, cells cultured on foams with supracellular-sized (67 μm) pores had a significantly increased secretion of albumin as compared with cells on foams with intermediate-sized (17 μm) or subcellular-sized (3 μm) pores. This effect may have been due to increased cell-cell contacts in the scaffolds with supracellular-sized pores (Ranucci and Moghe, 1999). These results indicate that while pore size and shape should be optimized for each application, it may not be sufficient to use only one fundamental parameter, such as cell attachment, to determine the best pore morphology.

Since the optimal pore size depends on the application, it is important to have control over the porosity of materials. Novel methods of fabricating a porous matrix include using a 3D paraffin mold to create a spherical pore network (Ma and Choi, 2001); using sugar particles, disks, and fibers to produce a variety of controlled porous PLLA structures (Zhang and Ma, 2000); and using high-

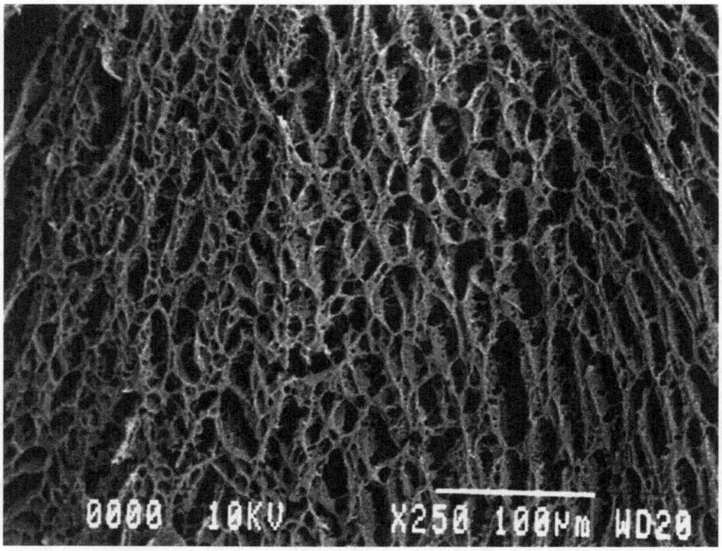

FIGURE 16.1 Scanning electron micrograph showing the porosity of a critical-point-dried 1.5% (wt/vol) agarose hydrogel (bar = 100 μm).

pressure gas foaming in combination with particulate leaching to create porous scaffolds from PLGA copolymers without the use of organic solvents (Murphy and Mooney, 1999). The porosity of PLGA-PEG blends, fabricated by solvent casting and particulate leaching methods, was significantly dependent on the initial salt fraction as well as the ratio of PLGA to PEG (Wake et al., 1996). Similarly, the porosity of collagen-chitosan matrices was controlled by changing the component ratios, with increased amounts of chitosan resulting in smaller pores. As the chitosan concentration increased, proliferation of a human hemopoietic cell line decreased, perhaps due to the smaller pore size. Notably, cell viability remained unchanged (Tan et al., 2001). These collagen-chitosan matrices may be suitable for applications such as cell encapsulation, where it is desirable for cells to remain viable but not highly proliferative to avoid overcrowding the scaffold. Even with a thorough, controlled design of initial porosity as described in these studies, processes including fibrous encapsulation, scaffold degradation, cellular ingrowth, and extracellular matrix (ECM) production must be considered because they may alter the scaffold porosity after implantation.

Three-Dimensionality. Since tissues in vivo are generally 3D constructs, cells cultured in 3D substrates are more likely to reflect in vivo scenarios. Culturing cells in vitro on two-dimensional (2D) substrates rather than in three dimensions has been shown to affect cell phenotype. Hepatocytes cultured in alginate beads (400 μm diameter) secreted more albumin, fibrinogen, α_1-antitrypsin, α_1-acid glycoprotein, and prothrombin than a similar cell density in a 2D monolayer culture. Moreover, the levels produced by the cells on the alginate beads approached those of normal hepatocytes in vivo, perhaps because the 3D cell architecture mimicked that found in vivo (Selden et al., 1999). Primary human chondrocytes cultured on 2D substrates transitioned to a fibroblastoid cell phenotype. When cultured on 3D collagen microcarriers, they reverted back to the original chondrocytic phenotype (Frondoza et al., 1996). Similarly, when neurons were cultured on 2D agarose cultures, they did not adhere and grow. However, when embedded in 3D agarose hydrogels with the identical chemistry, the neurons were viable and extended long processes (Bellamkonda et al., unpublished observations). Photomicrographs of growth cones extending from primary neurons illustrate the difference between the growth-cone morphology on 2D tissue culture substrates and the 3D morphology in agarose hydrogel (Fig. 16.2). These examples show that cell phenotype and morphology depend on the dimensionality of the culture conditions.

Despite this fact, many in vitro studies related to cell migration and proliferation have been performed on 2D tissue culture substrates due to the relative simplicity of these experiments. Some groups have performed transmigration assays (Gosiewska et al., 2001) and proliferation assays (Chu et al., 1995; Tan et al., 2001) with 3D cultures. However, 3D analysis of cultures can be difficult because cells cannot be manipulated or imaged as easily as those on 2D substrates. Live imaging of cells cultured in 3D scaffolds via light microscopy is challenging due to the thickness and non-transparent nature of the 3D material.

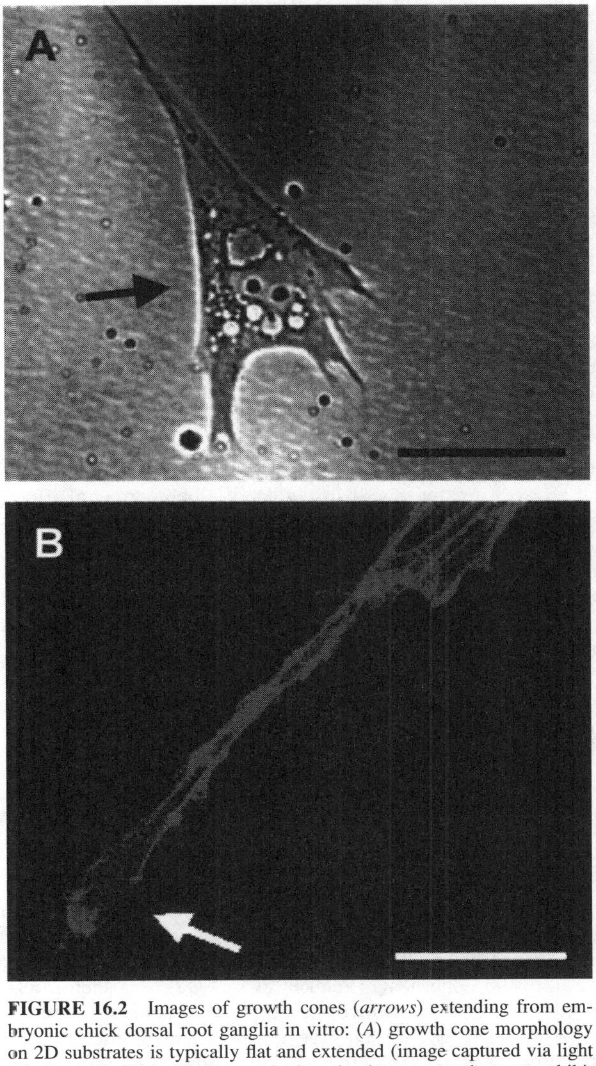

FIGURE 16.2 Images of growth cones (*arrows*) extending from embryonic chick dorsal root ganglia in vitro: (*A*) growth cone morphology on 2D substrates is typically flat and extended (image captured via light microscopy); (*B*) in 3D agarose hydrogel cultures, growth cones exhibit a 3D ellipsoidal shape (image is one optical slice through a 3D growth cone captured via confocal microscopy) (bar = 25 μm).

In addition, techniques such as immunostaining require adaptation for 3D cultures in vitro. Often, when in vitro experiments are performed using 3D scaffolds, the cultures are either fixed and sectioned prior to staining (Gosiewska et al., 2001; Schreiber et al., 1999) or else stained in situ and then sectioned for analysis (Attawia et al., 1995; Chu et al., 1995; Holy et al., 2000). Recently, a method for staining PLGA scaffolds in situ and then performing a histological analysis has been published (Holy and Yakubovich, 2000). Immunostaining neurites and growth cones is not as straightforward. Three dimensional cultures of primary neurons in agarose hydrogels can be imaged via light microscopy because the hydrogels are transparent and the cell density is low (Dillon et al., 2000). Nevertheless, staining and sectioning the cultures are not recommended because the fine neuronal processes may be destroyed. To maintain the integrity of the neuronal processes, a modified staining method is under development in our laboratory. This method involves partially digesting the hydrogel to increase the porosity and allow the unimpeded diffusion of antibodies and fluorophores in and out of the hydrogel. While culturing in 3D substrates can be useful in maintaining normal phenotypes, significant challenges with cultures, qualitative visualization, and assays remain.

Surface Topography. Modifying the surface topography and surface roughness of a material is another approach to enhancing interactions between the implant and the tissue. When the topography of the subendothelial extracellular matrix was replicated with biomedical polyurethane, bovine aortic endothelial cells (BAECs) spread more rapidly and appeared more like cells in a native environment as compared with cells on nontextured polyurethane (Goodman et al., 1996). To determine the effect of surface roughness on bone augmentation, smooth and grit-blasted (textured) titanium cylinders were implanted in rabbits. Although both types of surfaces resulted in similar amounts of trabecular bone formation, the grit-blasted titanium had a larger area of bone directly contacting the implant (Lundgren et al., 1999). As these studies illustrate, cells sometimes prefer a textured surface to a smooth surface, perhaps due to the increased surface area available for cell attachment and tissue integration.

Grooves have also been used to adjust the surface topography of materials. By creating microgrooves on tissue culture polystyrene, a significantly higher number of rat dermal fibroblasts attached onto and aligned actin filaments with 1-μm-wide grooves relative to smooth surfaces and surfaces with grooves greater than 2 μm wide (Walboomers et al., 2000). Similarly, chick embryo cerebral neurons and their processes have been shown to follow micron- but not nanometer-scale grooves faithfully in vitro (Clark et al., 1990, 1991). *Xenopus* spinal cord neurons cultured on grooved quartz extended neurites parallel to the grooves, regardless of groove dimensions, but rat hippocampal neurons extended neurites parallel to deep, wide grooves and perpendicular to shallow, narrow grooves (Rajnicek et al., 1997). Also, when polymeric guidance channels were used for rat sciatic nerve regeneration, the topography of the luminal surface affected the regeneration (Guenard et al., 1991). Only channels with a smooth inner surface produced a discrete nerve cable with microfascicles consisting of myelinated axons (Aebischer et al., 1990). Due to the critical influence of surface topography on tissue organization and response, topography should be considered when designing regenerative biomaterials.

16.2.2 Mechanical Properties

Biomaterials to regenerate load-bearing tissues, including bone, cartilage, and blood vessels, have obvious requirements for mechanical properties such as modulus of elasticity, tensile and shear strength, and compliance. For instance, biodegradable bone cement designed to aid in the repair of broken bones or to fill bone voids must have mechanical properties similar to those of the native bone until the bone heals (Peter et al., 1998a). For such implants, composite reinforced materials may be necessary to achieve the desired mechanical properties. Moreover, in the regeneration of soft tissue such as cartilage, the development of the appropriate mechanical properties may be critically dependent on the mechanical conditioning imposed in vitro.

Mechanical properties are also important for other applications where mechanical strength is not the primary function. One such application is electrodes for stimulation or recording in the brain. Silicon microelectrodes inevitably seem to elicit astroglial scar due to mechanical mismatch with the host neural tissue and lack of integration between the implant and the brain tissue (Maynard et al., 2000), and this fibrous scarring can lead to electrode failure (Williams et al., 1999). It has also been directly demonstrated that the rate of neurite extension is affected by the stiffness of the material in which the neuron is growing (Balgude et al., 2001). Soft tissues like the nervous system should not be overlooked when considering mechanical properties.

Reinforced Composites. Bone regeneration and fixation require strong materials, so applicable biomaterials must have the strength to sustain the mechanical loads experienced in bone. However, many materials with desirable physical and chemical properties, such as biodegradable polymers, are often unable to provide this strength by themselves. By combining two different materials, reinforced composites with the joint properties of both materials can be fabricated.

PLLA, PGA, and PLGA are common scaffolds for a number of tissue-regeneration applications, including bone regeneration. The Bioscrew, a PLLA screw, has been found to perform comparably with metal screws as a fixation device for patellar tendon autografts. Moreover, the Bioscrew is completely degradable and has the ability to compress slightly and conform to its surroundings when inserted into bone (Barber et al., 2000). However, most forms of PLLA, PDLLA, PGA, and PLGA are not high-strength materials, and adjusting physical parameters to enhance tissue ingrowth, such as by increasing the pore size, leads to a further decrease in the mechanical strength (Ma and Choi, 2001). Reinforcing techniques via the addition of mineral and ceramic components have been developed to strengthen these and other materials for load-bearing applications.

Hydroxyapatite (HA), a mineral component found in bone, is commonly used to reinforce polymers for bone regeneration because it can enhance mechanical properties as well as provide osteoinductive properties. Absorbable high-strength composites of PLLA reinforced with HA are being considered for fixation devices in bone. These composite rods have an initial bending strength exceeding that of human cortical bone and have promoted greater bone contact and better bone integration than PLLA without HA in a rabbit bone defect model (Furukawa et al., 2000). PLGA has also been strengthened with HA to form PLGA-HA composites with an elastic modulus and yield strength similar to those of cancellous bone (Devin et al., 1996). Short HA fibers have been used to reinforce PLGA and produce stronger composite foams for bone regeneration (Thomson et al., 1998). Based on these and other studies, it can be concluded that HA incorporated into biodegradable polymers can increase the strength of the composite to a level useful for orthopedic implants.

The bioactive ceramic β-tricalcium phosphate (β-TCP) has also been used to develop composite materials for bone regeneration. Poly(propylene fumarate) (PPF) has been strengthened by incorporating β-TCP to result in a biodegradable composite material with sufficient mechanical strength to temporarily replace trabecular bone (Yaszemski et al., 1996). It is notable that the mechanical properties of the composite actually increased during degradation, maintaining a compressive strength of at least 5 MPa and a compressive modulus greater than 50 MPa for 3 (Peter et al., 1998b) to 12 (Yaszemski et al., 1996) weeks depending on the composition. Materials that maintain their mechanical strength in this manner are useful for tissue regeneration because they allow a gradual, smoother replacement of the temporary biomaterial with the host matrix and cells. If the mechanical properties of a biomaterial decrease very rapidly, it may be beneficial to overengineer the initial implant to compensate for the loss.

Reinforcements are used for nonpolymeric materials as well. HA coatings on titanium bone fixtures increased contact between the native bone and the implant and provided greater shear and antitorque strength (Meffert, 1999). Microporous calcium phosphate ceramics (75 percent HA and 25 percent β-TCP) generally have weak mechanical properties under compression. When the pores were filled with calcium phosphate cement consisting of β-TCP and dicalcium phosphate dihydrate, the mechanical strength of the ceramic improved without compromising the bone healing and regeneration process (Frayssinet et al., 2000). By reinforcing available materials with components that provide strength as well as osteoinductive properties, bone regeneration can be enhanced.

Mechanical Conditioning. Tissue regeneration has been shown to benefit from external mechanical stimuli. One study demonstrated that constant mechanical tension elongated the axon bundles of synapsed primary embryonic rat cortical neurons in vitro (Smith et al., 2001). Moreover, certain tissues require mechanical loads for the generation of proper cell phenotypes. When in vitro culture conditions emulate in vivo loading, native cell phenotypes can be maintained.

Blood vessels in vivo are mechanically loaded by the pulsatile flow of the bloodstream. To simulate these conditions, tissue-engineered blood vessels are often grown under pulsatile flow conditions. Tissue-engineered arteries grown in vitro under pulsatile flow appeared more similar to native arteries than vessels not cultured under flow (Niklason et al., 1999), and cell-seeded vascular grafts cultured under pulsatile flow demonstrated superior mechanical strength relative to static control grafts (Hoerstrup et al., 2001). Thus in vitro cultures of blood vessels benefit from mechanical conditioning similar to the in vivo mechanical loading.

Cartilage is another tissue that experiences mechanical loads in vivo. Explants of healthy cartilage respond to compressive loads and in particular to the release of those loads. The release of a static compressive load stimulated the chondrocytes, and continuous dynamic loading was able to stimulate or inhibit the biosynthesis of proteins depending on the amplitude and frequency of the compression (Sah et al., 1989; Steinmeyer and Knue, 1997). This behavior of native cartilage extends to chondrocytes cultured in vitro. Chondrocytes cultured on agarose disks were evaluated under static and dynamic compression. Dynamic compression resulted in an increase in proteoglycan and protein synthesis with time, whereas little change in synthesis was seen with static compression (Buschmann et al., 1995; Lee and Bader, 1997). Likewise, cyclic loading was able to stimulate chick mesenchymal stem cells cultured on agarose hydrogels to differentiate into chondrocytes, making them useful for cartilage regeneration applications (Elder et al., 2000). Moreover, the chondrogenesis depended on the frequency and duration of the cyclic compressive load, indicating the similarity between these constructs and native cartilage (Elder et al., 2001). These and other studies have established the importance of mechanical stimulation for chondrocytes and offer a possible approach to enhance cartilage generation in vitro and regeneration in vivo.

Electrical Stimuli. Electrically active materials have also been used to encourage tissue growth. The use of piezoelectric materials made of vinylidene fluoride–trifluoroethylene copolymer [P(VDF-TrFE)] enhanced peripheral nerve regeneration in vivo (Fine et al., 1991), and when PC12 cells were cultured on oxidized polypyrrole, the application of an electrical stimulus resulted in enhanced neurite extension (Schmidt et al., 1997), as shown in Fig. 16.3. Implant vascularization was enhanced when bilayer films of polypyrrole–hyaluronic acid and polypyrrole-poly(styrenesulfonate) were implanted subcutaneously (Collier et al., 2000). These types of electrical stimuli can be used in conjunction with a biomaterial to promote tissue regrowth.

16.2.3 Chemical Properties

The chemistry of a material affects the interactions that occur between the implant and its surrounding environment, including protein adsorption, cellular response, and bioresorption. In particular, the hydrophilicity, charge, and degradability of the material can impact tissue regeneration.

Hydrophilicity. One aspect of material chemistry that affects cell behavior is the hydrophilicity of the biomaterial. Most mammalian cells are anchorage-dependent and only viable when attached to a substrate in a receptor-mediated fashion. Since they have no receptors for most synthetic biomaterials, cells will not attach to bare materials in a receptor-mediated manner. However, protein adsorption onto biomaterials results in a permissible surface for cell attachment. It has been shown that moderately hydrophilic surfaces, in the presence of serum proteins, often support greater cell attachment (Chang et al., 1999; Khang et al., 1999; van Wachem et al., 1985), spreading (Webb et al., 1998), and normal phenotypes (McClary et al., 2000) relative to hydrophobic or highly hydrophilic surfaces. Focal contacts and stress fibers in cells on this moderately hydrophilic surfaces are well defined, indicating active binding and outside-in signaling due to the proteins adsorbed onto the surface (McClary et al., 2000). The increased cell attachment on moderately hydrophilic surfaces could be due to preferential binding of cells to ECM proteins that adsorbed to the surfaces (Khang et al., 1999).

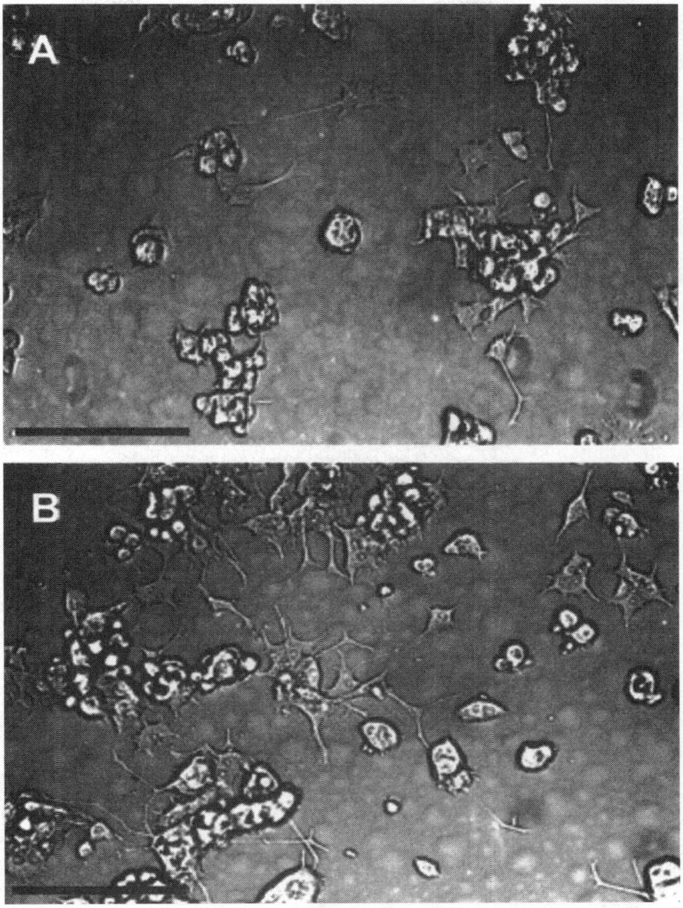

FIGURE 16.3 PC12 cell differentiation on polypyrrole (PP) without (*A*) and with (*B*) application of an electric potential. PC12 cells were grown on PP for 24 h in the presence of nerve growth factor and then exposed to electrical stimulation (100 mV) across the polymer film (*B*). Images were acquired 24 h after stimulation. Cells grown for 48 h but not subjected to electrical stimulation are shown for comparison (*A*) (bar = 100 μm). (*Courtesy of C. Schmidt, Ph.D.*)

Spatial control of cells is achieved by including zones that permit cell attachment with adjoining zones that inhibit attachment. Very hydrophilic surfaces are relatively resistant to protein adsorption and can be used as an inhibitory surface to generate patterns of attached cells on a biomaterial. High-molecular-weight PEG (above 18,500 g/mol) is known to be resistant to protein adsorption and cell adhesion and can be modified for the covalent coupling of other molecules (Desai and Hubbell, 1991; Gombotz et al., 1991). When PEG chains were grafted onto silica films, significantly less protein adsorption was observed as compared with unmodified surfaces (Alcantar et al., 2000). PEG chains can also reduce cell adhesion. Collagen, known for its ability to promote cell attachment, can be modified into a nonpermissive substrate by attaching PEG to the collagen (PEG-collagen) (Tiller et al., 2001). These and other methods can be used to create regions of differing hydrophilicity to control cell micropatterns on surfaces.

Charge. In addition to preferring moderately hydrophilic surfaces, cell attachment is improved on positively charged surfaces. Polymers that carry a positive charge at physiological pH augment cell attachment and growth in the presence of proteins (McKeehan and Ham, 1976). In the absence of proteins, cells still prefer positively charged hydrophilic surfaces to those with a neutral or negative charge (Webb et al., 1998). When polylysine (a positively charged protein) was coated onto PLLA and PGLA nonwoven structures, chondrocyte attachment increased (Sittinger et al., 1996). When fluorinated ethylene propylene films were surface modified with patterns of amine groups, preadsorption with albumin resulted in neuroblastoma cell attachment along the amine patterns (Ranieri et al., 1993). Also, neurite extension, an important element of nerve tissue regeneration, has been shown to depend on the polarity of the substrate in which the cells are cultured. When embryonic chick DRGs were cultured in 3D agarose hydrogels, negatively charged dermatan sulfate coupled to the hydrogel inhibited neurite extension, whereas positively charged chitosan coupled the hydrogel enhanced neurite extension (Dillon et al., 2000). To determine the effect of charge on the ability of collagen to support cells, collagen was modified to have either a net positive or negative charge. However, the charge difference had minimal effect, with 90 percent of mouse fibroblasts and endothelial cells attaching after 60 minutes, regardless of the substrate's net charge. When PEG-collagen, which did not support cell attachment, was given a net positive charge through amination, fibroblast and endothelial cell attachment returned to the levels of attachment on unmodified collagen (Tiller et al., 2001). As demonstrated by these studies, positive charges added to biomaterial scaffolds can induce tissue growth.

Biodegradation. Biomaterials for some applications such as total-hip replacements and synthetic vascular grafts are designed to be long-lived and even permanent, if possible. However, many tissue-regeneration implants are best designed using materials that can completely degrade in vivo to be replaced with autologous tissue. The lack of a synthetic remnant is advantageous because it reduces the likelihood of infection and chronic inflammation that are often seen with permanent implants. Biodegradable materials are also ideal for tissues that are still growing and developing, such as bones and heart valves in pediatric patients.

The chemistry of certain materials allows them to degrade in vivo, typically due to cleavable bonds. The most commonly used degradable polyester polymers include PLLA, PDLLA, PGA, and PLGA. These polymers undergo bulk degradation into lactic and glycolic acid by hydrolysis of the ester bonds. The degradation by-products are nontoxic and easily metabolized, and the rate of degradation can be controlled by the ratio of PLLA to PGA. Polyanhydrides are another type of biodegradable polymer used for drug delivery and tissue engineering. Due to the hydrophobicity of the polymer chains, they degrade by predictable surface erosion rather than bulk degradation, as observed with the polyesters. A variety of other synthetic and natural polymers are also biodegradable.

The time course of degradation should be planned and controlled. A degradable scaffold must support cell growth and provide the appropriate mechanical properties until the tissue is capable of fulfilling these functions. In the vasculature, a human artery typically experiences an average pressure of at least 100 mmHg (Guyton and Hall, 1996a). A material implanted to promote vessel regeneration must therefore supply the necessary strength and compliance until the native smooth muscle cells and fibroblasts are able to support the load. After the tissue is self-sufficient, it is optimal for the implanted material to completely degrade. The degradation timing can be controlled by adjusting polymer properties, such as the type and location of the degradable bonds. New methods of modifying polyanhydride networks have allowed for controlled variation of porosity, rate of degradation, and interactions with cells (Burkoth et al., 2000).

One must also consider the fact that the rate of tissue regeneration and polymer degradation will vary among individuals. The environment will affect the degradation rate as well. It has been shown that events such as fluid flow around a degradable PLGA scaffolds will decrease the degradation rate (Agrawal et al., 2000). A novel approach to bioresorbable scaffolds involves enzyme-dependent degradation, currently under development by Hubbell and coworkers (Schense et al., 2000; Ye et al., 2000). Rather than relying on the chemistry of the environment to degrade the material, this approach depends on the cells to control degradation. Until the cells enzymatically trigger degradation, the scaffold retains its mechanical and structural properties. This promising method has the potential to allow for individualized degradation that is completely dependent on the tissue-regeneration process.

Polymer degradation usually produces small degradation products. Some materials may degrade into small toxic by-products, even though the original material as a whole is nontoxic. When PLGA degrades into lactic acid and glycolic acid, the local pH may drop if the area is not well perfused (Wake et al., 1998). Also, if the by-products are immunogenic, the immune system may attack the area of desired tissue regeneration, and successful regeneration will not occur. To ensure biocompatibility, the toxicity of the degradation products ought to be evaluated.

16.3 BIOCHEMICAL COMPONENT

Once a material has been optimized based on its structural properties, bioactive agents can be incorporated to mimic normal regenerative environments found in vivo and further enhance tissue regeneration. These bioactive agents could consist of peptides, proteins, chemicals, oligonucleotides, genes, or living cells and can be included by two methods: (1) immobilization to the material and (2) release as diffusible agents. A combination of these two approaches could result in a diffusible signal that is chemoattractive and an immobilized signal that encourages cell attachment and proliferation. Table 16.4 provides a sampling of biochemical components incorporated into materials to enhance tissue growth.

16.3.1 Presentation of Immobilized Signals on the Scaffold

A scaffold material can provide cues that will direct and control the cell-matrix interactions to promote and regulate tissue regeneration. Biochemical signals such as matrix proteins, adhesive peptides, and growth factors can be incorporated into a material by nonspecific adsorption or through covalent immobilization. Since many mammalian cells require adhesion for cell viability, these molecules can provide the essential cues for cell attachment.

Matrix Proteins for Cell Adhesion. Cell adhesion in a receptor-mediated fashion is key for cellular processes such as intracellular signaling, synthesis of ECM proteins, and mitosis. Specific cell adhesion to scaffolds often occurs via ECM molecules that are presented on the scaffold surface. One adhesion study confirmed that smooth muscle cells adhere via integrins to extracellular matrix proteins (Nikolovski and Mooney, 2000). By immobilizing matrix proteins to the material, cell adhesion can be facilitated.

The ECM protein collagen includes the Arg-Gly-Asp (RGD) amino acid sequence that can facilitate specific interactions with cells via integrin receptors (Culp et al., 1997). Collagen has been covalently immobilized on PLGA surfaces to increase cell attachment (Zheng et al., 1998) and on polycarbonate and polyester membranes to enhance corneal epithelial tissue migration (Steele et al., 2000). A porous collagen membrane placed around a titanium implant resulted in increased bone repair in a model of alveolar ridge deficiency as compared with uncoated titanium implants (Zhang et al., 1999b). Collagen has also been used in combination with other ECM components to improve a polypropylene mesh for hernia repair. A collagen-glycosaminoglycan (GAG) matrix was placed around the mesh and shown to facilitate tissue growth and vascularization (Butler et al., 2001). The numerous applications for abundant ECM proteins such as collagen illustrate the usefulness of matrix proteins in tissue regeneration.

Laminin, another ECM protein, has also been used to modify materials, particularly for neural applications. When laminin was covalently immobilized on 3D agarose hydrogels, neurite extension from embryonic chick DRGs was stimulated. Further, immobilization was necessary, since laminin simply mixed into the agarose gel was unable to stimulate neurite extension (Yu et al., 1999). Uniform laminin substrates have also been used for the attachment of Schwann cells, but no preferential cell orientation was shown. When micropatterns of laminin and albumin were fabricated on

TABLE 16.4 Examples of Biochemical Components Incorporated into Biomaterials for Tissue Regeneration

Design parameter	Biomaterial	Biochemical agent	Application	References
		Presentation of immobilized signals		
Matrix proteins	Polypropylene mesh	Collagen-GAG	Hernia repair	Butler et al., 2001
Adhesive peptides	Hyaluronic matrix	RGD peptide	Skin regeneration	Hansbrough et al., 1995
Growth factors	PEG hydrogel	TGF-β1	SMC matrix production	Mann et al., 1999
		Presentation of chemoattractive, diffusible signals		
One-component system	Collagen sponges	bFGF	Cartilage regeneration	Fujisato et al., 1996
	PLLA/PGA	Tetracycline	Bone regeneration	Park et al., 2000
Two-component system	Gelatin and collagen	bFGF	Angiogenesis	Kawai et al., 2000
	Agarose and lipid microtubules	NGF	Nerve regeneration	Yu et al., 1999
Gene delivery	PLGA matrices	PDGF-B plasmids	Wound healing	Shea et al., 1999
	Collagen carrier	BMP-7 adenovirus	Bone regeneration	Franceschi et al., 2000
		Presentation of a living component		
Cell seeding	PLGA foams	Stromal osteoblasts	Osteogenic activity	Ishaug et al., 1997
	PGA fiber matrix	Smooth muscle cells	Cell density	Kim et al., 1998
Differentiated cell-scaffold systems	Expanded PTFE	Endothelial cells	Vascular lining regeneration	Ahlswede et al., 1994
	Collagen sponge	Fibroblasts	Skin regeneration	Kuroyanagi et al., 2001
Cell scaffolds for neuroregeneration	Glass beads	Adrenal chromaffin cells	Neuroregeneration	Cherksey et al., 1996
	PAN/PVC	Baby kidney cells	Neuroregeneration	Emerich et al., 1994
Combination therapy	Matrigel 3D scaffolds	Myoblasts and bFGF	Myogenesis and angiogenesis	Barbero et al., 2001
	PLGA-HA scaffold	Stromal cells with BMP-2 genes	Bone regeneration	Laurencin et al., 2001
Stem cell-scaffold systems	Hyaluronan-based 3D scaffolds	Mesenchymal stem cells	Osteochondral repair	Radice et al., 2000
	Collagen gel	Mesenchymal stem cells	Tendon regeneration	Young et al., 1998
Cell-based products	Collagen	Fibroblasts and keratinocytes	Venous ulcer treatment	Eaglstein et al., 1999

glass coverslips, Schwann cells attached only on the laminin surfaces and were oriented with the laminin patterns (Thompson and Buettner, 2001). These studies demonstrate how a more specialized protein such as laminin can also be used to enhance and direct tissue outgrowth.

Protein adsorption studies have demonstrated the importance of surface protein conformation. For effective interactions between cells and ECM proteins, the active domains of the proteins should be available for cell recognition. However, when a protein is adsorbed onto a surface to promote cell adhesion, the adhesion sequences will not necessarily be available. Changes in fibronectin conformation due to adsorption to glass and silane surfaces decreased endothelial cell adhesion strength and spreading (Iuliano et al., 1993). On the contrary, adsorption of vitronectin onto nanophase alumina resulted in a conformational change of the protein that increased the adhesion of rat osteoblasts, perhaps due to an increased exposure of the adhesive epitopes (Webster et al., 2001). Thus conformational changes should be considered when designing immobilization strategies.

Cellular attachment and proliferation are not the only parameters of regeneration to evaluate when optimizing tissue regeneration. It has been shown that increased cell adhesion due to adhesion-promoting surface modifications can result in a decrease of ECM production (Mann et al., 1999). When a biodegradable biomaterial is used, the cells must be self-sufficient when the material is gone, and this includes making their own ECM. A biomaterial with strong adhesive properties may be detrimental to overall tissue function. This example illustrates the importance of balancing cellular activities, such as cell adhesion and matrix production.

Adhesive Peptides. Immobilizing adhesive peptides to the material can also encourage cell attachment to biomaterials. Protein adsorption can result in a loss of quaternary and tertiary structure, but peptides should not denature or change conformation when immobilized to a surface. The most commonly used amino acid sequence for cell adhesion is the RGD sequence found in a variety of ECM proteins, including collagen, fibronectin, and vitronectin. Other peptide sequences used include Tyr-Ile-Gly-Ser-Arg (YIGSR) and Ile-Lys-Val-Ala-Val (IKVAV) from laminin, as well as Arg-Glu-Asp-Val (REDV) from fibronectin. All these sequences are short, so they are often contained within a slightly larger peptide to facilitate immobilization on scaffolds as well as interactions with cells.

RGD peptides work via interactions with cell surface integrin receptors, as shown by the receptor-mediated attachment of human umbilical vein endothelial cells (HUVECs) to silica substrates modified with RGD peptides (Porte-Durrieu et al., 1999). When immobilized on a surface, the RGD peptides typically localized in clusters rather than distributing uniformly (Kouvroukoglou et al., 2000). Clustering is advantageous since it has been shown to decrease the average peptide density required to support migration (Maheshwari et al., 2000). Moreover, the peptide surface density should be at or below the saturation level of the cellular receptors because peptide densities beyond that level have shown no further increase in cell attachment, spreading, or matrix production (Rezania and Healy, 2000). Consequently, there is an optimum surface density for RGD peptides that depends on the application.

Peptides are often used to enhance specific cellular activity with the materials. RGD peptides were covalently linked to quartz surfaces to examine their effect on rat calvaria osteoblast-like cell (RCO) adhesion. Although RCOs adhered and spread on both modified and unmodified quartz in the presence of serum proteins, the RGD-modified surfaces stimulated an increase in mineralization from the RCOs (Rezania and Healy, 1999). The RGD peptide also has been coupled to alginate hydrogels. Since these hydrogels are highly hydrophilic, they have minimal interaction with cells. C2C12 mouse skeletal myoblasts were cultured on 2D membranes of alginate with and without RGD-containing peptides. Only with the RGD peptides present did the myoblasts attach and spread (Rowley et al., 1999). Other studies have immobilized RGD peptides to a hyaluronic matrix to accelerate wound healing and regeneration in mice (Cooper et al., 1996) and skin healing and regeneration for pediatric burn patients (Hansbrough et al., 1995). Immobilizing RGD to polystyrene via Pluronic F108 allowed for increased fibroblast adhesion and spreading (Neff et al., 1998). In addition, bovine pulmonary artery endothelial cells were shown to have an increased migration on glass surfaces modified with either RGD or YIGSR containing peptides (Kouvroukoglou et al., 2000). The versatility of RGD peptides demonstrated by these studies suggests that RGD peptides can play an influential role in enhancing tissue regeneration.

Neurite extension can be controlled using immobilized peptides. RGD peptides linked to PLLA-PEG-biotin surfaces via a streptavidin linker were shown to spatially control the adhesion and the direction of neurite extension from PC12 cells (Patel et al., 1998). Neurite extension from PC12 cells and NG108-15 neuroblastoma cells was also spatially controlled using laminin-derived peptides YIGSR and IKVAV bound to fluorinated ethylene propylene films (Ranieri et al., 1994). These studies demonstrate the potential impact of peptides on the regeneration of nerve tissues.

Another method of immobilizing peptides to a surface is to use self-assembled monolayers (SAMs) to modify a surface. Using the microcontact printing technique to create SAMs that both promoted (via RGD peptide) and inhibited (via PEG) cell adhesion, patterns of cell adhesion were established (Zhang et al., 1999a). A different self-assembly method used a molecule with a hydrophobic tail and a peptide head group that contains a collagenlike structural motif. These peptide-amphiphiles were able to promote the adhesion and spreading of human melanoma cells on tissue culture plates (Fields et al., 1998). Whether the immobilization technique involves SAMs or covalent chemistry, adhesive peptides are useful for modifying biomaterials to induce cell adhesion.

Growth Factor Presentation. In addition to ECM-related molecules, growth factors have been coupled to materials to provide a biochemical stimulus. Vascular endothelial growth factor (VEGF) immobilized to fibrin matrices had a greater mitogenic effect on human endothelial cells in vitro than soluble VEGF (Zisch et al., 2001). Also, basic fibroblast growth factor (bFGF) immobilized onto gas-plasma-modified polystyrene accelerated the formation of a confluent HUVEC layer in vitro, whereas soluble bFGF in culture medium was ineffective (Bos et al., 1999). Studies such as these indicate that immobilized growth factors can sometimes provide a more potent stimulation than soluble growth factors. In another study, TGF-β1 was covalently immobilized to PEG hydrogels and found to increase the matrix production of vascular smooth muscle cells (Mann et al., 2001). Peptides based on the active domains of growth factors have also been utilized as biochemical cues. Alginate hydrogels have been covalently modified with a 20-amino-acid sequence derived from bone morphogenic protein-2 (BMP-2) to induce bone formation in vivo. After 8 weeks, the gel was vascularized, and osteoblasts had formed new trabecular bone within the alginate pores (Suzuki et al., 2000). Therefore, by immobilizing growth factors to biomaterials, a range of cell responses can be achieved.

16.3.2 Presentation of Chemoattractive, Diffusible Signals on the Scaffold

In addition to immobilized signals, the body also creates chemoattractive gradients using diffusible factors to guide cell migration. To imitate this approach, a bioactive agent can be incorporated into the biomaterial to diffuse away and provide a biochemical cue. For agent release, the biomaterial could consist of one component that fulfills both the role of the scaffold to support the cells and the role of the delivery system to release the biochemicals. Alternatively, a two-component system using one material for the scaffold and a different material for the delivery system could be employed.

Scaffold-Drug Delivery: One-Component System. Using the biomaterial scaffold as a drug-delivery device is a simple means of incorporating a diffusible agent into the material. A bioactive agent incorporated into the biomaterial will slowly diffuse away after implantation to create a concentration gradient. Diffusible agents have been included in matrices by adding them either during or after the fabrication process.

Biodegradable polyesters are among the materials usually loaded by adding the biochemicals during fabrication. Acidic fibroblast growth factor (aFGF) was incorporated into PDLLA-PELA copolymers for nerve regeneration (Maquet et al., 2001), and tetracycline was added to a PLLA film/PGA mesh to stimulate bone regeneration and marrow formation in a rat craniotomy defect (Park et al., 2000). PLLA and PLLA-TCP membranes designed to release platelet-derived growth factor-BB (PDGF-BB) enhanced bone formation and allowed for complete bony reunion in a critical-sized defect in rats (Lee et al., 2001; Park et al., 1998). Also, titanium coated with PDLLA that contained insulinlike growth factor and TGF-β1 significantly accelerated fracture healing in rodents as compared with implants without growth factors (Schmidmaier et al., 2001). Fabricating scaffolds in the presence of bioactive agents is an effective method to include diffusible factors in biomaterials.

Another method for loading a biomaterial is to add the bioactive agent after the matrix has already been formed, such as by soaking a matrix in the agent solution. This method is frequently utilized for loading collagen scaffolds. An absorbable, macroporous collagen sponge soaked in a recombinant human bone morphogenic protein (BMP) solution increased the osteoinductive activity of the material (Uludag et al., 1999). To promote cartilage formation, collagen sponges were soaked in bFGF. They induced tissue regeneration at 1 week and mature chondrocytes at 4 weeks postimplantation, whereas sponges without bFGF did not show significant cartilage regeneration (Fujisato et al., 1996). These studies suggest that growth factor release can be a potent stimulator of tissue regeneration.

The release of angiogenic factors to encourage neovascularization allows for increased tissue survival. Methods to support this facet of tissue regeneration have included the slow release of angiogenic factors such as VEGF (Eiselt et al., 1998). PLGA foam scaffolds fabricated to release VEGF have been shown to increase the proliferation of human dermal microvascular endothelial cells as compared with scaffolds without VEGF (Murphy et al., 2000). Thus various aspects of regeneration can be enhanced by appropriately selecting the material and the bioactive agent to be released.

Scaffold-Drug Delivery: Two-Component System. An assortment of drug-delivery systems is available and can be incorporated into a scaffold for tissue regeneration. This two-component method of delivering agents is particularly useful for biomaterials that cannot be loaded with agents by the previously described methods or are not capable of providing adequate slow release. One designated delivery system is the PLGA microsphere system (Fig. 16.4). As loaded microspheres degrade, they release the particles trapped within and provide the diffusible agent. PLGA microspheres loaded with heparin, known for its antiproliferative and anticoagulative activity, were embedded in alginate films. These films were then wrapped around vein grafts and denuded carotid arteries in rats and shown to reduce intimal hyperplasia as compared with control films without heparin (Edelman et al., 2000). PLGA microspheres have also been loaded with VEGF to induce the proliferation of human dermal microvascular endothelial cells (Eiselt et al., 1998). To promote angiogenesis, gelatin microspheres were loaded with bFGF and incorporated into an artificial dermis consisting of an outer silicone layer and an inner collagen sponge layer. Implants with the bFGF-loaded microspheres were shown to accelerate fibroblast proliferation and capillary formation in vivo (Kawai et al., 2000). Another delivery system used to provide chemical gradients is the lipid microtubule delivery system (Meilander et al., 2001). Lipid microtubules are hollow tubules that can be loaded with a bioactive agent and embedded into the biomaterial. Due to the concentration gradient between the tubules and the external environment, the agent is released. Lipid microtubules loaded with nerve growth factor (NGF) and incorporated into agarose hydrogels have been shown to directionally stimulate neurite extension from 3D cultures of chick embryonic DRGs (Yu et al., 1999), as depicted in Fig. 16.5. These studies demonstrate that a specialized delivery system is an effective means to provide a diffusible biochemical signal within a scaffold.

Gene-Delivery Scaffolds. Gene therapy can also provide a biochemical component to promote tissue regeneration. Instead of releasing a protein, the gene for that protein can be released to transduce cells for local protein synthesis. This is especially useful for sensitive proteins with short in vivo half-lives. One application that has benefited from gene delivery is wound healing. Gas-foamed PLGA matrices loaded with plasmid DNA encoding PDGF-B increased the formation of granulation tissue and vascularization following subcutaneous implantation in rodents (Shea et al., 1999). Likewise, collagen matrices containing adenoviral or plasmid DNA for PDGF-A or PDGF-B increased granulation tissue, reepithelialization, and neovascularization to result in accelerated wound healing (Chandler et al., 2000).

Gene delivery has also been used to enhance bone regeneration in vivo. An adenovirus encoding bone morphogenic protein-7 (BMP-7) was mixed with a collagen carrier and implanted in mice to

FIGURE 16.4 Scanning electron microscope image of PLGA microspheres of various sizes. After microsphere fabrication, sieves can be used to select a specific size range (bar = 10 μm). *(Courtesy of J. Gao., Ph. D.)*

produce ectopic formation of cortical and trabecular bone with a marrow cavity after 4 weeks (Franceschi et al., 2000). Similarly, plasmid DNA encoding a fragment of parathyroid hormone (PTH) was incorporated in a collagen gel to reproducibly promote bone regeneration in a canine model of bone injury. Since the DNA was shown to be incapable of diffusing out of the initial form of the gel, the DNA remained localized at the injury site until cells arrived and degraded the collagen

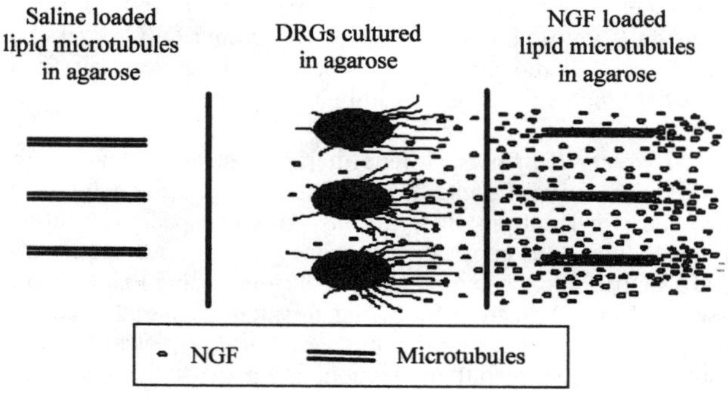

FIGURE 16.5 Schematic of directional neurite extension from chick DRGs in a 3D agarose culture. NGF concentration gradients were established by the microtubule delivery system, and neurite extension was directed preferentially toward the NGF-loaded microtubules (diagram not to scale).

matrix (Bonadio et al., 1999). To test the effects of using two plasmids simultaneously, collagen sponges were soaked in a solution of plasmid DNA encoding either the osteoinductive protein bone morphogenic protein-4 (BMP-4), a fragment of PTH, or both BMP-4 and the PTH fragment. The loaded sponges were implanted in critical defects in the femoral diaphysis of adult rats. Defects with either the BMP-4 or PTH fragment plasmid healed completely and demonstrated the same mechanical strength as the unoperated femur. Moreover, combination gene therapy, with a collagen sponge containing plasmid DNA encoding both BMP-4 and the PTH fragment, was found to further enhance bone regeneration (Fang et al., 1996). As these examples illustrate, gene therapy is a useful method to stimulate tissue regeneration, especially with the use of multiple genes. Advances in the field of gene therapy will extend the use of this promising technique to a variety of tissue-regeneration applications.

16.3.3 Presentation of a Living Component

A more recent approach to enhance the regenerative effort incorporates cells into tissue-engineered biomaterials. Rather than relying entirely on the migration of host cells into the implant, cells can be seeded in the biomaterial. Differentiated cells, as well as stem cells, are being used for regenerative therapies. In addition, cells can be used in combination with other biochemical therapies to further enhance regeneration.

Cell Seeding. The seeding density of cells on the implanted material can be adjusted for each application. It was shown that the final cell concentration of smooth muscle cells on PGA fiber matrices was proportional to the seeding density (Kim et al., 1998). Likewise, the osteogenic activity of stromal osteoblasts cultured on 3D PLGA foams was found to depend on the seeding density of the cells (Ishaug et al., 1997). An appropriate seeding density can be chosen based on the rate of implanted cell proliferation and host tissue proliferation. An implant with a low density of cells allows for the proliferation and migration of both implanted and host cells. If the cellular density on the implant is similar to that of the tissue in vivo, minimal proliferation of the host tissue is necessary. This could be useful in tissues, such as central nervous system tissue, that only have minimal regenerative capacity. In addition, the seeding method affects the cell distribution. Dynamic seeding methods result in a larger number of adherent cells as well as a more uniform distribution of the cells (Kim et al., 1998). Accordingly, the effective use of cells to stimulate tissue regeneration will involve choosing an appropriate cell seeding density and method.

Differentiated Cell-Scaffold Systems. A number of differentiated cell types have been used to aid in the regeneration of tissue in vivo. Endothelial cells are often included for vascular tissue regeneration or for neovascularization of scaffolds. Endothelial cells were perivascularly transplanted in Gelfoam matrices around balloon-denuded rat carotid arteries to control the tissue healing response and reduce intimal hyperplasia (Nathan et al., 1995). Endothelial cells have also been used in the sodding of expanded polytetrafluoroethylene (ePTFE) vascular grafts to enhance the regeneration of an endothelial cell lining in a rat aortic graft model (Ahlswede and Williams, 1994). Endothelial cells and vascular myofibroblasts were seeded onto polyglycolic-acid–poly-4-hydroxybutyrate copolymers to create small-diameter vascular grafts (Hoerstrup et al., 2001). To improve scaffold vascularization, endothelial cells and dermal fibroblasts were added to a skin equivalent consisting of a chitosan-linked collagen-GAG sponge containing keratinocytes. The endothelial cells and fibroblasts promoted the formation of a network of capillarylike structures in vitro and may serve as a framework to facilitate vascularization upon implantation (Black et al., 1998). Other studies have incorporated fibroblasts on spongy collagen to aid in the healing of skin defects (Kuroyanagi et al., 2001), chondrocytes in fibrin glues to result in actively proliferating and ECM-producing cells (Sims et al., 1998), keratinocytes on collagen-coated dextran microcarriers to reconstitute epithelium (Voigt et al., 1999), and hepatocytes on polyanhydrides, polyorthoesters, or PLGA fibers for possible liver regeneration (Vacanti et al., 1988). These studies suggest that numerous cells types have the potential to provide a living component in biomaterials.

Scaffold Prevascularization. Vascularization is important in implants seeded with cells because these cells will immediately require a source of nourishment, most likely via the bloodstream. One approach to ensure an adequate vascular supply when cells are implanted is to implant a cell-free scaffold to allow for fibrovascular tissue ingrowth prior to seeding the cells. Using a PVA foam, the timing of in vivo fibrovascular tissue formation was determined. It was concluded that cells should be seeded into the material as soon as the fibrovascular tissue has reached the center of the implant (Wake et al., 1995). A similar method implanted an encapsulation chamber prior to adding the cells with the intent of increasing vasculature around the chamber, since normal wound healing results in vessel regression during the second week. Cellulose membranes with 8-μm pores were able to maintain the vascularization of the fibrous capsule that surrounded the membrane beyond 2 weeks, allowing for a sustained vascular supply (Padera and Colton, 1996). Such methods may provide improved local vascularization and mass transport at the time of cell implantation. However, these strategies may have limited application because they require two implantation procedures, one for the scaffold and one for the cells.

Cell-Scaffold Constructs for Neuroregeneration. Neurodegenerative diseases are prime candidates for cell therapy approaches to tissue regeneration. One disease that has received considerable attention for cell therapy approaches is Parkinson's disease. In one study, rat adrenal chromaffin cells (ACC) were seeded on collagen-coated dextran or glass beads and injected into the brains of hemiparkinsonian rats. These animals showed significant behavioral recovery over the 12-month study. Animals receiving only the ACC without the microcarriers showed improvement at 1 month but reverted to the original diseased state by 2 months, demonstrating that the scaffold was necessary for prolonged treatment and recovery (Borlongan et al., 1998; Cherksey et al., 1996). Bovine chromaffin cells have also been used to treat hemiparkinsonian rats. Using alginate-polylysine-alginate as a microencapsulator, the cells were implanted in hemiparkinsonian rats and shown to decrease apomorphine-induced rotation (Xue et al., 2001). Likewise, human and rat fetal ventral mesencephalon cells have also been seeded onto collagen-coated dextran microcarriers as a possible treatment for Parkinson's disease. The cells had an increased survival when attached to the microcarriers versus cell suspensions (Saporta et al., 1997). To supply dopamine, PC12 cells capable of secreting dopamine were encapsulated in agarose–poly(styrene sulfonic acid) and grafted into the corpus striatum of guinea pigs. The cells were stained with a tyrosine hydroxylase antibody, suggesting that they continued to secrete dopamine in the brain. In addition, there was no immunological rejection or tumor formation when the biomaterial was used (Date et al., 1996). Cell therapy also has potential to treat Alzheimer's disease. In one such effort, baby hamster kidney cells were genetically engineered to overexpress NGF. They were then mixed with vitrogen and infused into poly(acrylonitrile-co-vinyl chloride) copolymers (PAN/PVC) to reduce the degeneration of basal forebrain cholinergic neurons (Emerich et al., 1994). These findings regarding Parkinson's disease and Alzheimer's disease suggest that biomaterials, either as a scaffold or as a means of encapsulation, play an important role when using cells as a treatment for neurodegenerative diseases.

Combination Therapy. In vivo, cells interpret and react to numerous signals simultaneously. Taking advantage of this ability, multiple biochemical cues can be provided for tissue regeneration. One study evaluated human skin fibroblasts grown on fibrin microbeads. The microbeads with cells were shown to enhance wound healing as compared with fibroblasts or microbeads by themselves. However, when the diffusible growth factor PDGF-BB was included with the fibroblast-fibrin microbeads, the healing response was further improved (Gorodetsky et al., 1999). Growth factors that encourage neovascularization can also be included with the cells. In one such approach, murine myoblasts were cultured in 3D scaffolds of Matrigel containing either bFGF or hepatocyte growth factor to promote angiogenesis. When the scaffolds were implanted in a model of ectopic muscle regeneration, the presence of the growth factors increased angiogenesis, resulting in improved cell viability and enhanced myogenesis (Barbero et al., 2001).

Gene therapy and cell therapy have been combined to result in murine stromal cells transduced to produce BMP-2. The cells were seeded on a PLGA-HA matrix and found to induce heterotopic bone regeneration (Laurencin et al., 2001). Likewise, rat dermal cells were infected with a retrovirus containing the PDGF-B gene and seeded onto PGA scaffolds to modulate wound healing (Breitbart et al., 1999). In another study, fibroblasts engineered to express neurotrophic factors were seeded onto poly-*N*-(2-hydroxypropyl)-methacrylamide hydrogels with RGD-containing peptides. This therapy, incorporating cells, DNA, and adhesive peptides into one material, was shown to enhance optic nerve regeneration. In addition, when the cells produced two neurotrophic factors, axonal growth into the hydrogels was greater than with either growth factor alone (Loh et al., 2001). These examples illustrate the positive regenerative effects when a biomaterial is designed to utilize several parameters in concert.

Stem Cell–Scaffold Systems. Stem cells, also known as *progenitor cells,* are pluripotent cells with the ability to differentiate into a variety of cell types. The most widely studied stem cell is the mesenchymal stem cell (MSC) derived from bone marrow. In the embryo, these cells give rise to skeletal tissues, including bone, cartilage, tendon, ligament, marrow stroma, adipocytes, dermis, muscle, and connective tissue (Caplan, 1991).

Mesenchymal stem cells are relevant for tissue regeneration because of their ability to differentiate, a trait they share with many cells that facilitate tissue repair in vivo. Harnessing this capability by combining MSCs, biomaterials, and the appropriate cues for differentiation should allow for regeneration of the skeletal tissues just mentioned. A composite matrix of gelatin and esterified hyaluronic acid matrix seeded with MSCs enabled osteochondrogenic cell differentation when implanted subcutaneously in mice (Angele et al., 1999). Similarly, MSCs were harvested from rabbit bone marrow and seeded on a hyaluronan-based 3D scaffold to fill an osteochondral defect in rabbits. While the plain scaffold was able to enhance osteochondral repair, the morphology of the repair improved when progenitor cells were included (Radice et al., 2000). For cartilage regeneration, MSCs were seeded onto collagen sponges and shown to augment meniscus cartilage regeneration in rabbits (Walsh et al., 1999). Another study demonstrated the effectiveness of MSCs for bone regeneration. MSCs were seeded onto coral scaffolds and implanted in a bone defect in sheep. The MSC scaffolds were more likely to develop into new cortical bone with a medullary canal for clinical union than the coral scaffold alone (Petite et al., 2000). MSCs have also been suspended in collagen gels and implanted into tendon defects to improve tendon regeneration and biomechanical properties (Awad et al., 1999; Young et al., 1998). These studies with MSCs are just beginning to realize the potential of stem cells as tools for tissue regeneration.

As a combination therapy, stem cells were used in conjunction with gene therapy to further enhance regeneration. Human bone marrow–derived mesenchymal stem cells were infected with an adenovirus containing the BMP-2 gene, cultured on a collagen matrix, and transplanted in murine radial defects. Unlike the sham-infected MSCs, BMP-2–infected MSCs were able to differentiate into chondrocytes in vivo, regenerate bone, and bridge the radial gap (Turgeman et al., 2001). In a similar model, genetically engineered pluripotent mesenchymal cells expressing BMP-2 were compared with a nonprogenitor cell line also genetically engineering to express BMP-2. The cells were seeded onto collagen sponges and transplanted into a bone defect in mice. Even though the nonprogenitor cells had a much higher BMP-2 expression in vitro, they showed no organized bone formation in vivo. The progenitor cells, despite a lower gene expression in vitro, induced orderly formation of bone and cartilage (Gazit et al., 1999). Again, materials designed with more than one factor to enhance regeneration proved superior to single-factor designs.

Commercial Cell-Based Tissue-Engineered Products. Tissue-engineered products containing cells are beginning to appear on the commercial market. Apligraf, a bilayered product developed and manufactured by Organogenesis (Canton, MA) and marketed by Novartis Pharmaceuticals Corporation (East Hanover, NJ), is composed of neonatal-derived dermal fibroblasts, keratinocytes, and bovine collagen. It has been approved for the treatment of venous ulcers in the United States (Eaglstein et al., 1999). Carticel, from Genzyme Tissue Repair (Cambridge, MA), consists of healthy autologous chondrocytes that are harvested, proliferated in vitro, and implanted into sites of injured cartilage to improve the tissue. Dermagraft (Advanced Tissue Sciences), a dermal substitute consisting of dermal fibroblasts on a resorbable substrate, has recently received Food and Drug Administration (FDA) approval in the United States. In addition, several other tissue-engineered products are available in Europe and other countries but are not yet approved for marketing in the United States. These products include Epicel (Genzyme Tissue Repair, Cambridge, MA) and VivoDerm (ER Squibb and Sons, Inc., Princeton, NJ; ConvaTec, Ltd., authorized user).

16.4 CONCLUSIONS

Tissue regeneration allows for the development of self-renewing, responsive tissue that can remodel itself and its matrix. A number of methods to encourage tissue regeneration are currently available, most of which place the host tissue in the primary role and the biomaterial in a supportive role, mimicking permissive in vivo environments. Since a barrage of structural and biochemical cues orchestrates tissue regeneration in vivo, the design of biomaterials for tissue regeneration must progress to include a combination of design parameters. As knowledge of cell and molecular biology continues to advance, more parameter combinations will be discovered, and the effectiveness of biomaterials designed for tissue regeneration will undoubtedly improve and extend to a variety of organs.

ACKNOWLEDGMENTS

We wish to thank Professor Jinming Gao (Case Western Reserve University, Cleveland OH) and Professor Christine Schmidt (University of Texas at Austin, Austin TX) for their generous contributions to the figures in the chapter.

REFERENCES

Aebischer, P., Guenard, V., and Valentini, R. F. (1990), The morphology of regenerating peripheral nerves is modulated by the surface microgeometry of polymeric guidance channels, *Brain Res.* **531**(1–2):211–218.

Agrawal, C. M., McKinney, J. S., Lanctot, D., and Athansiou, K. A. (2000), Effects of fluid flow on the in vitro degradation kinetics of biodegradable scaffolds for tissue engineering, *Biomaterials* **21**(23):2443–2452.

Ahlswede, K. M., and Williams, S. K. (1994), Microvascular endothelial cell sodding of 1-mm expanded polytetrafluoroethylene vascular grafts, *Arterioscler Thromb.* **14**(1):25–31.

Alcantar, N. A., Aydil, E. S., and Israelachvili J. N. (2000), Polyethylene glycol-coated biocompatible surfaces, *J. Biomed. Mater. Res.* **51**(3):343–351.

Angele, P., Kujat, R., Nerlich, M., Yoo, J., Goldberg, V., and Johnstone, B. (1999), Engineering of osteochondral tissue with bone marrow mesenchymal progenitor cells in a derivatized hyaluronan-gelatin composite sponge, *Tissue Eng.* **5**(6):545–554.

Attawia, M. A., Herbert, K. M., and Laurencin, C. T. (1995), Osteoblast-like cell adherance and migration through 3-dimensional porous polymer matrices, *Biochem. Biophys. Res. Commun.* **213**(2):639–644.

Awad, H. A., Butler, D. L., Boivin, G. P., Smith, F. N., Malaviya, P., Huibregtse, B., and Caplan A. I. (1999), Autologous mesenchymal stem cell-mediated repair of tendon, *Tissue Eng.* **5**(3):267–277.

Balgude, A. P., Yu, X., Szymanski, A., and Bellamkonda, R. V. (2001), Agarose gel stiffness determines rate of DRG neurite extension in 3D cultures, *Biomaterials* **22**(10):1077–1084.

Barber, F. A., Elord, B. F., McGuire, D. A., and Paulos, L. E. (2000), Bioscrew fixation of patellar tendon autografts. *Biomaterials* **21**(24):2623–2629.

Barbero, A., Benelli, R., Minghelli, S., Tosetti, F., Dorcaratto, A., Ponzetto, C., Wernig, A., Cullen, M. J., Albini, A., and Noonan, D. M. (2001), Growth factor supplemented matrigel improves ectopic skeletal muscle formation: A cell therapy approach, *J. Cell Physiol.* **186**(2):183–192.

Bellamkonda, R., Ranieri, J. P., Bouche, N., and Aebischer, P. (1995), Hydrogel-based three-dimensional matrix for neural cells, *J. Biomed. Mater. Res.* **29**(5):663–671.

Black, A. F., Berthod, F., L'Heureux, N., Germain, L., and Auger, F. A. (1998), In vitro reconstruction of a human capillary-like network in a tissue-engineered skin equivalent. *FASEB J.* **12**(13):1331–1340.

Bonadio, J., Smiley, E., Patil, P., and Goldstein, S. (1999), Localized, direct plasmid gene delivery in vivo: Prolonged therapy results in reproducible tissue regeneration, *Nature Med.* **5**(7):753–759.

Borlongan, C. V., Saporta, S., and Sanberg, P. R. (1998), Intrastriatal transplantation of rat adrenal chromaffin cells seeded on microcarrier beads promote long-term functional recovery in hemiparkinsonian rats, *Exp. Neurol.* **151**(2):203–214.

Bos, G. W., Scharenborg, N. M., Poot, A. A., Engbers, G. H., Beugeling, T., van Aken, W. G., and Feijen, J. (1999), Proliferation of endothelial cells on surface-immobilized albumin-heparin conjugate loaded with basic fibroblast growth factor, *J. Biomed. Mater. Res.* **44**(3):330–340.

Breitbart, A. S., Mason, J. M., Urmacher, C., Barcia, M., Grant, R. T., Pergolizzi, R. G., and Grande D. A. (1999), Gene-enhanced tissue engineering: Applications for wound healing using cultured dermal fibroblasts transduced retrovirally with the PDGF-B gene, *Ann. Plast. Surg.* **43**(6):632–639.

Burkoth, A. K., Burdick, J., and Anseth, K. S. (2000), Surface and bulk modifications to photocrosslinked polyanhydrides to control degradation behavior, *J. Biomed. Mater. Res.* **51**(3):352–359.

Buschmann, M. D., Gluzband, Y. A., Grodzinsky, A. J., and Hunziker, E. B. (1995), Mechanical compression modulates matrix biosynthesis in chondrocyte/agarose culture, *J. Cell Sci.* **108**(4):1497–1508.

Butler, C. E., Navarro, F. A., and Orgill, D. P. (2001), Reduction of abdominal adhesions using composite collagen-GAG implants for ventral hernia repair, *J. Biomed. Mater. Res.* **58**(1):75–80.

Caplan, A. I. (1991), Mesenchymal stem cells, *J. Orthop. Res.* **9**(5):641–650.

Chandler, L. A., Gu, D. L., Ma, C., Gonzalez, A. M., Doukas, J., Nguyen, T., Pierce, G. F., and Phillips, M. L. (2000), Matrix-enabled gene transfer for cutaneous wound repair, *Wound Repair Regen.* **8**(6):473–479.

Chang, S. J., Kuo, S. M., Lan, J. W., and Wang, Y. J. (1999), Amination of polycarbonate surface and its application for cell attachment, *Artif. Cells Blood Substit. Immobil. Biotechnol.* **27**(3):229–244.

Cherksey, B. D., Sapirstein, V. S., and Geraci, A. L. (1996), Adrenal chromaffin cells on microcarriers exhibit enhanced long-term functional effects when implanted into the mammalian brain, *Neuroscience* **75**(2):657–664.

Chu, C. R., Monosov, A. Z., and Amiel, D. (1995), In situ assessment of cell viability within biodegradable polylactic acid polymer matrices, *Biomaterials* **16**(18):1381–1384.

Clark, P., Connolly, P., Curtis, A. S., Dow, J. A., and Wilkinson, C. D. (1990), Topographical control of cell behavior: II. Multiple grooved substrata, *Development* **108**(4):635–644.

Clark, P., Connolly, P., Curtis, A. S., Dow, J. A., and Wilkinson, C. D. (1991), Cell guidance by ultrafine topography in vitro, *J. Cell Sci.* **99**(1):73–77.

Collier, J. H., Camp, J. P., Hudson, T. W., and Schmidt, C. E. (2000), Synthesis and characterization of polypyrrole-hyaluronic acid composite biomaterials for tissue engineering applications, *J. Biomed. Mater. Res.* **50**(4):574–584.

Cooper, M. L., Hansbrough, J. F., and Polarek, J. W. (1996). The effect of an arginine-glycine-aspartic acid peptide and hyaluronate synthetic matrix on epithelialization of meshed skin graft interstices, *J. Burn. Care Rehabil.* **17**(2):108–116.

Culp, L. A., O'Conner, K. L., and Lechner, R. (1997), Extracellular matrix adhesion: Biological, molecular, and pathogenic mechanisms, in Bittar E. E., Bittar N. (eds.), *Principles of Medical Biology*, pp. 573–607, JAI Press, Greenwich, Conn.

Date, I., Miyoshi, Y., Ono, T., Imaoka, T., Furuta, T., Asari, S., Ohmoto, T., and Iwata H. (1996), Preliminary report of polymer-encapsulated dopamine-secreting cell grafting into the brain, *Cell Transplant* **5**(5 suppl. 1): S17–19.

Desai, N. P., and Hubbell, J. A. (1991), Biological responses to polyethylene oxide modified polyethylene terephthalate surfaces, *J. Biomed. Mater. Res.* **25**(7):829–843.

Devin, J. E., Attawia, M. A., and Laurencin, C. T. (1996), Three-dimensional degradable porous polymer-ceramic matrices for use in bone repair, *J. Biomater. Sci. Polym. Ed.* **7**(8):661–669.

Dillon, G. P., Yu, X., and Bellamkonda, R. V. (2000), The polarity and magnitude of ambient charge influences three-dimensional neurite extension from DRGs, *J. Biomed. Mater. Res.* **51**(3):510–519.

Dillon, G. P., Yu, X., Sridharan, A., Ranieri, J. P., and Bellamkonda, R. V. (1998), The influence of physical structure and charge on neurite extension in a 3D hydrogel scaffold, *J. Biomater. Sci. Polym. Ed.* **9**(10):1049–1069.

Eaglstein, W. H., Alvarez, O. M., Auletta, M., Leffel, D., Rogers, G. S., Zitelli, J. A., Norris, J. E., Thomas, I., Irondo, M., Fewkes, J., Hardin-Young, J., Duff, R. G., and Sabolinski, M. L. (1999), Acute excisional wounds treated with a tissue-engineered skin (Apligraf), *Dermatol. Surg.* **25**(3):195–201.

Edelman, E. R., Nathan, A., Katada, M., Gates, J., and Karnovsky, M. J. (2000), Perivascular graft heparin delivery using biodegradable polymer wraps, *Biomaterials* **21**(22):2279–2286.

Eiselt, P., Kim, B. S., Chacko, B., Isenberg, B., Peters, M. C., Greene, K. G., Roland, W. D., Loebsack, A. B., Burg, K. J., Culberson, C., Halberstadt, C. R., Holder, W. D., and Mooney, D. J. (1998), Development of technologies aiding large-tissue engineering, *Biotechnol. Prog.* **14**(1):134–140.

Elder, S. H., Goldstein, S. A., Kimura, J. H., Soslowsky, L. J., and Spengler, D. M. (2001), Chondrocyte differentiation is modulated by frequency and duration of cyclic compressive loading, *Ann. Biomed. Eng.* **29**(6): 476–482.

Elder, S. H., Kimura, J. H., Soslowsky, L. J., Lavagnino, M., and Goldstein, S. A. (2000), Effect of compressive loading on chondrocyte differentiation in agarose cultures of chick limb-bud cells, *J. Orthop. Res.* **18**(1): 78–86.

Emerich, D. F., Winn, S. R., Harper, J., Hammang, J. P., Baetge, E. E., and Kordower, J. H. (1994), Implants of polymer-encapsulated human NGF-secreting cells in the nonhuman primate: Rescue and sprouting of degenerating cholinergic basal forebrain neurons, *J. Comp. Neurol.* **349**(1):148–164.

Evans, G. R., Brandt, K., Widmer, M. S., Lu, L., Meszlenyi, R. K., Gupta, P. K., Mikos, A. G., Hodges, J., Williams, J., Gurlek, A., Nabawi, A., Lohman, R., and Patrick, C. W., Jr. (1999a), In vivo evaluation of poly(L-lactic acid) porous conduits for peripheral nerve regeneration, *Biomaterials* **20**(12):1109–1115.

Evans, M. D., Dalton, B. A., and Steele, J. G. (1999b), Persistent adhesion of epithelial tissue is sensitive to polymer topography, *J. Biomed. Mater. Res.* **46**(4):485–493.

Fang, J., Zhu, Y. Y., Smiley, E., Bonadio, J., Rouleau, J. P., Goldstein, S. A., McCauley, L. K., Davidson, B. L., and Roessler, B. J. (1996), Stimulation of new bone formation by direct transfer of osteogenic plasmid genes, *Proc. Natl. Acad. Sci. USA* **93**(12):5753–5758.

Fields, G. B., Lauer, J. L., Dori, Y., Forns, P., Yu, Y. C., and Tirrell, M. (1998), Protein-like molecular architecture: Biomaterial applications for inducing cellular receptor binding and signal transduction, *Biopolymers* **47**(2):143–151.

Fine, E. G., Valentini, R. F., Bellamkonda, R., and Aebischer, P. (1991), Improved nerve regeneration through piezoelectric vinylidenefluoride-trifluoroethylene copolymer guidance channels, *Biomaterials* **12**(8):775–780.

Folkman, J., and Hochberg, M. (1973), Self-regulation of growth in three dimensions, *J. Exp. Med.* **138**(4):745–753.

Franceschi, R. T., Wang, D., Krebsbach, P. H., and Rutherford, R. B. (2000), Gene therapy for bone formation: In vitro and in vivo osteogenic activity of an adenovirus expressing BMP7, *J. Cell Biochem.* **78**(3):476–486.

Frayssinet, P., Mathon, D., Lerch, A., Autefage, A., Collard, P., and Rouquet, N. (2000), Osseointegration of composite calcium phosphate bioceramics, *J. Biomed. Mater. Res.* **50**(2):125–130.

Frondoza, C., Sohrabi, A., and Hungerford, D. (1996), Human chondrocytes proliferate and produce matrix components in microcarrier suspension culture, *Biomaterials* **17**(9):879–888.

Fujisato, T., Sajiki, T., Liu, Q., and Ikada, Y. (1996), Effect of basic fibroblast growth factor on cartilage regeneration in chondrocyte-seeded collagen sponge scaffold, *Biomaterials* **17**(2):155–162.

Furukawa, T., Matsusue, Y., Yasunaga, T., Nakagawa, Y., Okada, Y., Shikinami, Y., Okuno, M., and Nakamura, T. (2000), Histomorphometric study on high-strength hydroxyapatite/poly(L-lactide) composite rods for internal fixation of bone fractures, *J. Biomed. Mater. Res.* **50**(3):410–419.

Gazit, D., Turgeman, G., Kelley, P., Wang, E., Jalenak, M., Zilberman, Y., and Moutsatsos, I. (1999), Engineered pluripotent mesenchymal cells integrate and differentiate in regenerating bone: A novel cell-mediated gene therapy. *J. Gene Med.* **1**(2):121–133.

Giardino, R., Fini, M., Nicoli Aldini, N., Giavaresi, G., and Rocca, M. (1999), Polylactide bioabsorbable polymers for guided tissue regeneration, *J. Trauma* **47**(2):303–308.

Giordano, G. G., Thomson, R. C., Ishaug, S. L., Mikos, A. G., Cumber, S., Garcia, C. A., and Lahiri-Munir, D. (1997), Retinal pigment epithelium cells cultured on synthetic biodegradable polymers, *J. Biomed. Mater. Res.* **34**(1):87–93.

Gombotz, W. R., Wang, G. H., Horbett, T. A., and Hoffman, A. S. (1991), Protein adsorption to poly(ethylene oxide) surfaces, *J. Biomed. Mater. Res.* **25**(12):1547–1562.

Goodman, S. L., Sims, P. A., and Albrecht, R. M. (1996), Three-dimensional extracellular matrix textured biomaterials, *Biomaterials* **17**(21):2087–2095.

Gorodetsky, R., Clark, R. A., An, J., Gailit, J., Levdansky, L., Vexler, A., Berman, E., and Marx, G. (1999), Fibrin microbeads (FMB) as biodegradable carriers for culturing cells and for accelerating wound healing, *J. Invest. Dermatol.* **112**(6):866–872.

Gosiewska, A., Rezania, A., Dhanaraj, S., Vyakarnam, M., Zhou, J., Burtis, D., Brown, L., Kong, W., Zimmerman, M., and Geesin, J. C. (2001), Development of a three-dimensional transmigration assay for testing cell-polymer interactions for tissue engineering applications, *Tissue Eng.* **7**(3):267–277.

Guenard, V., Valentini, R. F., and Aebischer, P. (1991), Influence of surface texture of polymeric sheets on peripheral nerve regeneration in a two-compartment guidance system, *Biomaterials* **12**(2):259–263.

Guyton, A. C., and Hall, J. E. (1996a), *Textbook of Medical Physiology*, 9th ed., pp. 175–176, Saunders, Philadelphia.

Guyton, A. C., and Hall, J. E. (1996b), *Textbook of Medical Physiology*, 9th ed., pp. 513–523, Saunders, Philadelphia.

Hansbrough, J. F., Herndon, D. N., Heimbach, D. M., Solem, L. D., Gamelli, R. L., and Tompkins, R. G. (1995), Accelerated healing and reduced need for grafting in pediatric patients with burns treated with arginine-glycine-aspartic acid peptide matrix: RGD Study Group, *J. Burn Care Rehabil.* **16**(4):377–387.

Hoerstrup, S. P., Zund, G., Sodian, R., Schnell, A. M., Grunenfelder, J., and Turina, M. I. (2001), Tissue engineering of small caliber vascular grafts, *Eur. J. Cardiothorac. Surg.* **20**(1):164–169.

Holy, C. E., Shoichet, M. S., and Davies, J. E. (2000), Engineering three-dimensional bone tissue in vitro using biodegradable scaffolds: Investigating initial cell-seeding density and culture period, *J. Biomed. Mater. Res.* **51**(3):376–382.

Holy, C. E., and Yakubovich, R. (2000), Processing cell-seeded polyester scaffolds for histology, *J. Biomed. Mater. Res.* **50**(2):276–279.

Hunter, S. K., Kao, J. M., Wang, Y., Benda, J. A., and Rodgers, V. G. (1999), Promotion of neovascularization around hollow fiber bioartificial organs using biologically active substances, *ASAIO J.* **45**(1):37–40.

Ishaug, S. L., Crane, G. M., Miller, M. J., Yasko, A. W., Yaszemski, M. J., and Mikos, A. G. (1997), Bone formation by three-dimensional stromal osteoblast culture in biodegradable polymer scaffolds, *J. Biomed. Mater. Res.* **36**(1):17–28.

Iuliano, D. J., Saavedra, S. S., and Truskey, G. A. (1993), Effect of the conformation and orientation of adsorbed fibronectin on endothelial cell spreading and the strength of adhesion, *J. Biomed. Mater. Res.* **27**(8):1103–1113.

Kawai, K., Suzuki, S., Tabata, Y., Ikada, Y., and Nishimura, Y. (2000), Accelerated tissue regeneration through incorporation of basic fibroblast growth factor-impregnated gelatin microspheres into artificial dermis, *Biomaterials* **21**(5):489–499.

Khang, G., Lee, S. J., Lee, J. H., Kim, Y. S., and Lee, H. B. (1999), Interaction of fibroblast cells on poly(lactide-co-glycolide) surface with wettability chemogradient, *Biomed. Mater. Eng.* **9**(3):179–187.

Kim, B. S., Putnam, A. J., Kulik, T. J., and Mooney, D. J. (1998), Optimizing seeding and culture methods to engineer smooth muscle tissue on biodegradable polymer matrices, *Biotechnol. Bioeng.* **57**(1):46–54.

Kouvroukoglou, S., Dee, K. C., Bizios, R., McIntire, L. V., and Zygourakis, K. (2000), Endothelial cell migration on surfaces modified with immobilized adhesive peptides, *Biomaterials* **21**(17):1725–1733.

Kuroyanagi, Y., Yamada, N., Yamashita, R., and Uchinuma, E. (2001), Tissue-engineered product: Allogeneic cultured dermal substitute composed of spongy collagen with fibroblasts, *Artif. Organs* **25**(3):180–186.

Laurencin, C. T., Attawia, M. A., Lu, L. Q., Borden, M. D., Lu, H. H., Gorum, W. J., and Lieberman, J. R. (2001), Poly(lactide-co-glycolide)/hydroxyapatite delivery of BMP-2-producing cells: A regional gene therapy approach to bone regeneration, *Biomaterials* **22**(11):1271–1277.

Lee, D. A., and Bader, D. L. (1997), Compressive strains at physiological frequencies influence the metabolism of chondrocytes seeded in agarose, *J. Orthop. Res.* **15**(2):181–188.

Lee, S. J., Park, Y. J., Park, S. N., Lee, Y. M., Seol, Y. J., Ku, Y., and Chung, C. P. (2001), Molded porous poly(L-lactide) membranes for guided bone regeneration with enhanced effects by controlled growth factor release, *J. Biomed. Mater. Res.* **55**(3):295–303.

Loh, N. K., Woerly, S., Bunt, S. M., Wilton, S. D., and Harvey, A. R. (2001), The regrowth of axons within tissue defects in the CNS Is promoted by implanted hydrogel matrices that contain BDNF and CNTF producing fibroblasts, *Exp. Neurol.* **170**(1):72–84.

Lu, L., Garcia, C. A., and Mikos, A. G. (1998), Retinal pigment epithelium cell culture on thin biodegradable poly(DL-lactic-co-glycolic acid) films, *J. Biomater. Sci. Polym. Ed.* **9**(11):1187–1205.

Lundgren, A. K., Lundgren, D., Wennerberg, A., Hammerle, C. H., and Nyman, S. (1999), Influence of surface roughness of barrier walls on guided bone augmentation: Experimental study in rabbits, *Clin. Implant Dent. Relat. Res.* **1**(1):41–8.

Lysaght, M. J., Nguy, N. A., and Sullivan, K. (1998), An economic survey of the emerging tissue engineering industry, *Tissue Eng.* **4**(3):231–238.

Ma, P. X., and Choi, J. W. (2001), Biodegradable polymer scaffolds with well-defined interconnected spherical pore network, *Tissue Eng.* **7**(1):23–33.

Madsen, S-K., and Mooney, D. J. (2000), Delivering DNA with polymer matrices: Applications in tissue engineering and gene therapy, *Pharmaceut. Sci. Technol. Today* **3**(11):381–384.

Maheshwari, G., Brown, G., Lauffenburger, D. A., Wells, A., and Griffith, L. G. (2000), Cell adhesion and motility depend on nanoscale RGD clustering, *J. Cell. Sci.* **113**(10):1677–1686.

Mann, B. K., Schmedlen, R. H., and West, J. L. (2001), Tethered-TGF-beta increases extracellular matrix production of vascular smooth muscle cells, *Biomaterials* **22**(5):439–444.

Mann, B. K., Tsai, A. T., Scott-Burden, T., and West J. L. (1999), Modification of surfaces with cell adhesion peptides alters extracellular matrix deposition, *Biomaterials* **20**(23–24):2281–2286.

Maquet, V., Martin, D., Scholtes, F., Franzen, R., Schoenen, J., Moonen, G., and Jerome, R. (2001), Poly(D,L-lactide) foams modified by poly(ethylene oxide)-block-poly(D,L- lactide) copolymers and a-FGF: In vitro and in vivo evaluation for spinal cord regeneration, *Biomaterials* **22**(10):1137–1146.

Maynard, E. M., Fernandez, E., and Normann, R. A. (2000), A technique to prevent dural adhesions to chronically implanted microelectrode arrays, *J. Neurosci. Methods* **97**(2):93–101.

McClary, K. B., Ugarova, T., and Grainger, D. W. (2000), Modulating fibroblast adhesion, spreading, and proliferation using self-assembled monolayer films of alkylthiolates on gold, *J. Biomed. Mater. Res.* **50**(3):428–439.

McKeehan, W. L., and Ham, R. G. (1976), Stimulation of clonal growth of normal fibroblasts with substrata coated with basic polymers, *J. Cell. Biol.* **71**(3):727–734.

Meffert, R. M. (1999), Ceramic-coated implant systems, *Adv. Dent. Res.* **13**:170–172.

Meilander, N. J., Yu, X., Ziats, N. P., and Bellamkonda, R. V. (2001), Lipid-based microtubular drug delivery vehicles, *J. Control Release* **71**(1):141–152.

Mikos, A. G., Sarakinos, G., Leite, S. M., Vacanti, J. P., and Langer, R. (1993), Laminated three-dimensional biodegradable foams for use in tissue engineering, *Biomaterials* **14**(5):323–330.

Mooney, D. J., Organ, G., Vacanti, J. P., and Langer, R. (1994), Design and fabrication of biodegradable polymer devices to engineer tubular tissues, *Cell Transplant* **3**(2):203–210.

Mooney, D. J., Park, S., Kaufmann, P. M., Sano, K., McNamara, K., Vacanti, J. P., and Langer, R. (1995), Biodegradable sponges for hepatocyte transplantation, *J. Biomed. Mater. Res.* **29**(8):959–965.

Murphy, W. L., and Mooney, D. J. (1999), Controlled delivery of inductive proteins, plasmid DNA and cells from tissue engineering matrices, *J. Periodontal. Res.* **34**(7):413–419.

Murphy, W. L., Peters, M. C., Kohn, D. H., and Mooney, D. J. (2000), Sustained release of vascular endothelial growth factor from mineralized poly(lactide-co-glycolide) scaffolds for tissue engineering, *Biomaterials* **21**(24):2521–2527.

Nathan, A., Nugent, M. A., and Edelman, E. R. (1995), Tissue-engineered perivascular endothelial cell implants regulate vascular injury, *Proc. Natl. Acad. Sci. USA* **92**(18):8130–8134.

Neff, J. A., Caldwell, K. D., and Tresco, P. A. (1998), A novel method for surface modification to promote cell attachment to hydrophobic substrates, *J. Biomed. Mater. Res.* **40**(4):511–519.

Niklason, L. E., Gao, J., Abbott, W. M., Hirschi, K. K., Houser, S., Marini, R., and Langer, R. (1999), Functional arteries grown in vitro, *Science* **284**(5413):489–493.

Nikolovski, J., and Mooney, D. J. (2000), Smooth muscle cell adhesion to tissue engineering scaffolds, *Biomaterials* **21**(20):2025–2032.

Padera, R. F., and Colton, C. K. (1996), Time course of membrane microarchitecture-driven neovascularization, *Biomaterials* **17**(3):277–284.

Park, Y. J., Ku, Y., Chung, C. P., and Lee, S. J. (1998), Controlled release of platelet-derived growth factor from porous poly(L-lactide) membranes for guided tissue regeneration, *J. Control Release* **51**(2–3):201–211.

Park, Y. J., Lee, Y. M., Park, S. N., Lee, J. Y., Ku, Y., Chung, C. P., and Lee, S. J. (2000), Enhanced guided bone regeneration by controlled tetracycline release from poly(L-lactide) barrier membranes, *J. Biomed. Mater. Res.* **51**(3):391–397.

Patel, N., Padera, R., Sanders, G. H., Cannizzaro, S. M., Davies, M. C., Langer, R., Roberts, C. J., Tendler, S. J., Williams, P. M., and Shakesheff, K. M. (1998), Spatially controlled cell engineering on biodegradable polymer surfaces, *FASEB J.* **12**(14):1447–1454.

Peter, S. J., Lu, L., Kim, D. J., Stamatas, G. N., Miller, M. J., Yaszemski, M. J., and Mikos, A. G. (2000), Effects of transforming growth factor beta1 released from biodegradable polymer microparticles on marrow stromal osteoblasts cultured on poly(propylene fumarate) substrates, *J. Biomed. Mater. Res.* **50**(3):452–462.

Peter, S. J., Miller, M. J., Yasko, A. W., Yaszemski, M. J., and Mikos, A. G. (1998a), Polymer concepts in tissue engineering, *J. Biomed. Mater. Res.* **43**(4):422–427.

Peter, S. J., Miller, S. T., Zhu, G., Yasko, A. W., and Mikos, A. G. (1998b), In vivo degradation of a poly(propylene fumarate)/beta-tricalcium phosphate injectable composite scaffold, *J. Biomed. Mater. Res.* **41**(1):1–7.

Petite, H., Viateau, V., Bensaid, W., Meunier, A. de, Pollak, C., Bourguignon, M., Oudina, K., Sedel, L., and Guillemin, G. (2000), Tissue-engineered bone regeneration, *Nature Biotechnol.* **18**(9):959–963.

Porte-Durrieu, M. C., Labrugere, C., Villars, F., Lefebvre, F., Dutoya, S., Guette, A., Bordenave, L., and Baquey, C. (1999), Development of RGD peptides grafted onto silica surfaces: XPS characterization and human endothelial cell interactions, *J. Biomed. Mater. Res.* **46**(3):368–375.

Radice, M., Brun, P., Cortivo, R., Scapinelli, R., Battaliard, C., and Abatangelo, G. (2000), Hyaluronan-based biopolymers as delivery vehicles for bone-marrow-derived mesenchymal progenitors, *J. Biomed. Mater. Res.* **50**(2):101–109.

Rajnicek, A., Britland, S., and McCaig, C. (1997), Contact guidance of CNS neurites on grooved quartz: Influence of groove dimensions, neuronal age and cell type, *J. Cell Sci.* **110**(23):2905–2913.

Ranieri, J. P., Bellamkonda, R., Bekos, E. J., Gardella, J. A., Jr., Mathieu, H. J., Ruiz, L., and Aebischer, P. (1994), Spatial control of neuronal cell attachment and differentiation on covalently patterned laminin oligopeptide substrates, *Int. J. Dev. Neurosci.* **12**(8):725–735.

Ranieri, J. P., Bellamkonda, R., Jacob, J., Vargo, T. G., Gardella, J. A., and Aebischer, P. (1993), Selective neuronal cell attachment to a covalently patterned monoamine on fluorinated ethylene propylene films, *J. Biomed. Mater. Res.* **27**(7):917–925.

Ranucci, C. S., and Moghe, P. V. (1999), Polymer substrate topography actively regulates the multicellular organization and liver-specific functions of cultured hepatocytes, *Tissue Eng.* **5**(5):407–420.

Rezania, A., and Healy, K. E. (1999), Biomimetic peptide surfaces that regulate adhesion, spreading, cytoskeletal organization, and mineralization of the matrix deposited by osteoblast-like cells, *Biotechnol. Prog.* **15**(1): 19–32.

Rezania, A., and Healy, K. E. (2000), The effect of peptide surface density on mineralization of a matrix deposited by osteogenic cells, *J. Biomed. Mater Res.* **52**(4):595–600.

Rowley, J. A., Madlambayan, G., and Mooney, D. J. (1999), Alginate hydrogels as synthetic extracellular matrix materials, *Biomaterials* **20**(1):45–53.

Sah, R. L., Kim, Y. J., Doong, J. Y., Grodzinsky, A. J., Plaas, A. H., and Sandy, J. D. (1989), Biosynthetic response of cartilage explants to dynamic compression, *J. Orthop Res.* **7**(5):619–636.

Saporta, S., Borlongan, C., Moore, J., Mejia-Millan, E., Jones, S. L., Bonness, P., Randall, T. S., Allen, R. C., Freeman, T. B., and Sanberg, P. R. (1997), Microcarrier enhanced survival of human and rat fetal ventral mesencephalon cells implanted in the rat striatum, *Cell Transplant* **6**(6):579–584.

Schense, J. C., Bloch, J., Aebischer, P., and Hubbell, J. A. (2000), Enzymatic incorporation of bioactive peptides into fibrin matrices enhances neurite extension, *Nature Biotechnol.* **18**(4):415–419.

Schmidmaier, G., Wildemann, B., Bail, H., Lucke, M., Fuchs, T., Stemberger, A., Flyvbjerg, A., Haas, N. P., and Raschke, M. (2001), Local application of growth factors (insulin-like growth factor-1 and transforming growth factor-beta1) from a biodegradable poly(D,L=lactide) coating of osteosynthetic implants accelerates fracture healing in rats, *Bone* **28**(4):341–350.

Schmidt, C. E., Shastri, V. R., Vacanti, J. P., and Langer R. (1997), Stimulation of neurite outgrowth using an electrically conducting polymer, *Proc. Natl. Acad. Sci. USA* **94**(17):8948–8953.

Schreiber, R. E., Dunkelman, N. S., Naughton, G., and Ratcliffe, A. (1999), A method for tissue engineering of cartilage by cell seeding on bioresorbable scaffolds, *Ann N.Y., Acad. Sci.* **875**:398–404.

Selden, C., Shariat, A., McCloskey, P., Ryder, T., Roberts, E., and Hodgson, H. (1999), Three-dimensional in vitro cell culture leads to a marked upregulation of cell function in human hepatocyte cell lines: An important tool for the development of a bioartificial liver machine, *Ann. N.Y. Acad. Sci.* **875**:353–363.

Shea, L. D., Smiley, E., Bonadio, J., and Mooney, D. J. (1999), DNA delivery from polymer matrices for tissue engineering, *Nature Biotechnol.* **17**(6):551–554.

Sims, C. D., Butler, P. E., Cao, Y. L., Casanova, R., Randolph, M. A., Black, A., Vacanti, C. A., and Yaremchuk, M. J. (1998), Tissue-engineered neocartilage using plasma-derived polymer substrates and chondrocytes, *Plast. Reconstr. Surg.* **101**(6):1580–1585.

Sittinger, M., Reitzel, D., Dauner, M., Hierlemann, H., Hammer, C., Kastenbauer, E., Planck, H., Burmester, G. R., and Bujia, J. (1996), Resorbable polyesters in cartilage engineering: Affinity and biocompatibility of polymer fiber structures to chondrocytes, *J. Biomed. Mater. Res.* **33**(2):57–63.

Smith, D. H., Wolf, J. A., and Meaney, D. F. (2001), A new strategy to produce sustained growth of central nervous system axons: Continuous mechanical tension, *Tissue Eng.* **7**(2):131–139.

Steele, J. G., Johnson, G., McLean, K. M., Beumer, G. J., and Griesser, H. J. (2000), Effect of porosity and surface hydrophilicity on migration of epithelial tissue over synthetic polymer, *J. Biomed. Mater. Res.* **50**(4): 475–482.

Steinmeyer, J., and Knue, S. (1997), The proteoglycan metabolism of mature bovine articular cartilage explants superimposed to continuously applied cyclic mechanical loading, *Biochem. Biophys. Res. Commun.* **240**(1): 216–221.

Suzuki, Y., Tanihara, M., Suzuki, K., Saitou, A., Sufan, W., and Nishimura, Y. (2000), Alginate hydrogel linked with synthetic oligopeptide derived from BMP-2 allows ectopic osteoinduction in vivo, *J. Biomed. Mater. Res.* **50**(3):405–409.

Tan, W., Krishnaraj, R., and Desai, T. A. (2001), Evaluation of nanostructured composite collagen-chitosan matrices for tissue engineering, *Tissue Eng.* **7**(2):203–210.

Thompson, D. M., and Buettner, H. M. (2001), Schwann cell response to micropatterned laminin surfaces, *Tissue Eng.* **7**(3):247–265.

Thomson, R. C., Giordano, G. G., Collier, J. H., Ishaug, S. L., Mikos, A. G., Lahiri-Munir, D., and Garcia, C. A. (1996), Manufacture and characterization of poly(alpha-hydroxy ester) thin films as temporary substrates for retinal pigment epithelium cells, *Biomaterials* **17**(3):321–327.

Thomson, R. C., Yaszemski, M. J., Powers, J. M., and Mikos, A. G. (1998), Hydroxyapatite fiber reinforced poly(alpha-hydroxy ester) foams for bone regeneration, *Biomaterials* **19**(21):1935–1943.

Tiller, J. C., Bonner, G., Pan, L. C., and Klibanov, A. M. (2001), Improving biomaterial properties of collagen films by chemical modification, *Biotechnol. Bioeng.* **73**(3):246–252.

Turgeman G., Pittman, D. D., Muller, R., Kurkalli, B. G., Zhou, S., Pelled, G., Peyser, A., Zilberman, Y., Moutsatsos, I. K., and Gazit, D. (2001), Engineered human mesenchymal stem cells: A novel platform for skeletal cell mediated gene therapy, *J. Gene. Med.* **3**(3):240–251.

Uludag, H., Friess, W., Williams, D., Porter, T., Timony, G., D'Augusta, D., Blake, C., Palmer, R., Biron, B., and Wozney, J. (1999), rhBMP-collagen sponges as osteoinductive devices: Effects of in vitro sponge characteristics and protein pI on in vivo rhBMP pharmacokinetics, *Ann. N.Y. Acad. Sci.* **875**:369–378.

Vacanti, J. P., Morse, M. A., Saltzman, W. M., Domb, A. J., Perez-Atayde, A., and Langer, R. (1988), Selective cell transplantation using bioabsorbable artificial polymers as matrices, *J. Pediatr Surg.* **23**(1 pt. 2):3–9.

Vallbacka, J. J., Nobrega, J. N., and Sefton, M. V. (2001) Tissue engineering as a platform for controlled release of therapeutic agents: Implantation of microencapsulated dopamine producing cells in the brains of rats, *J. Control Release* **72**(1–3):93–100.

van Wachem, P. B., Beugeling, T., Feijen, J., Bantjes, A., Detmers, J. P., and van Aken, W. G. (1985), Interaction of cultured human endothelial cells with polymeric surfaces of different wettabilities, *Biomaterials* **6**(6):403–408.

Voigt, M., Schauer, M., Schaefer, D. J., Andree, C., Horch, R., and Stark, G. B. (1999), Cultured epidermal keratinocytes on a microspherical transport system are feasible to reconstitute the epidermis in full-thickness wounds, *Tissue Eng.* **5**(6):563–572.

Wake, M. C., Gerecht, P. D., Lu, L., and Mikos, A. G., (1998), Effects of biodegradable polymer particles on rat marrow-derived stromal osteoblasts in vitro, *Biomaterials* **19**(14):1255–1268.

Wake, M. C., Gupta, P. K., and Mikos, A. G. (1996), Fabrication of pliable biodegradable polymer foams to engineer soft tissues, *Cell Tranplant* **5**(4):465–473.

Wake, M. C., Mikos, A. G., Sarakinos, G., Vacanti, J. P., and Langer, R. (1995), Dynamics of fibrovascular tissue ingrowth in hydrogal foams, *Cell Transplant* **4**(3):275–279.

Walboomers, X. F., Ginsel, L. A., and Jansen, J. A. (2000), Early spreading events of fibroblasts on microgrooved substrates, *J. Biomed. Mater. Res.* **51**(3):529–534.

Walsh, C. J., Goodman, D., Caplan, A. I., and Goldberg, V. M. (1999), Meniscus regeneration in a rabbit partial meniscectomy model, *Tissue Eng.* **5**(4):327–337.

Webb, K., Hlady, V., and Tresco, P. A. (1998), Relative importance of surface wettability and charged functional groups on NIH 3T3 fibroblast attachment, spreading, and cytoskeletal organization, *J. Biomed. Mater. Res.* **41**(3):422–430.

Webster, T. J., Schadler, L. S., Siegel, R. W., and Bizios, R. (2001), Mechanisms of enhanced osteoblast adhesion on nanophase alumina involve vitronectin, *Tissue Eng.* **7**(3):291–301.

Widmer, M. S., Gupta, P. K., Lu, L., Meszlenyi, R. K., Evans, G. R., Brandt, K., Savel, T., Gurlek, A., Patrick, C. W., Jr., and Mikos, A. G. (1998), Manufacture of porous biodegradable polymer conduits by an extrusion process for guided tissue regeneration, *Biomaterials* **19**(21):1945–1955.

Williams, J. C., Rennaker, R. L., and Kipke, D. R. (1999), Long-term neural recording characteristics of wire microelectrode arrays implanted in cerebral cortex, *Brain Res. Brain Res. Protoc.* **4**(3):303–313.

Winn, S. R., Tresco, P. A., Zielinski, B., Greene, L. A., Jaeger, C. B., and Aebischer, P. (1991) Behavioral recovery following intrastriatal implantation of microencapsulated PC 12 cells, *Exp. Neurol.* **113**(3):322–329.

Xue, Y., Gao, J., Xi, Z., Wang, Z., Li, X., Cui, X., Luo, Y., Li., C., Wang, L., Zhou, D., Sun, R., and Sun, A. M. (2001), Microencapsulated bovine chromaffin cell xenografts into hemiparkinsonian rats: A drug-induced rotational behavior and histological changes analysis, *Artif. Organs* **25**(2):131–135.

Yaszemski, M. J., Payne, R. G., Hayes, W. C., Langer, R., and Mikos, A. G. (1996), In vitro degradation of a poly(propylene fumarate)–based composite material, *Biomaterials* **17**(22):2127–2130.

Ye, Q., Zund, G., Benedikt, P., Jockenhoevel, S., Hoerstrup, S. P., Sakyama, S., Hubbell, J. A., and Turina, M. (2000), Fibrin gel as a three dimensional matrix in cardiovascular tissue engineering, *Eur. J. Cardiothorac. Surg.* **17**(5):587–591.

Young, R. G., Butler, D. L., Weber, W., Caplan, A. I., Gordon, S. L., and Fink, D. J. (1998), Use of mesenchymal stem cells in a collagen matrix for Achilles tendon repair, *J. Orthop. Res.* **16**(4):406–413.

Yu, X., Dillon, G. P., and Bellamkonda, R. B. (1999), A laminin and nerve growth factor-laden three-dimensional scaffold for enhanced neurite extension, *Tissue Eng.* **5**(4):291–304.

Zhang, Q., Yao, K., Liu, L., Sun, Y., Xu, L., Shen, X., and Lai, Q. (1999b), Evaluation of porous collagen membrane in guided tissue regeneration, *Artif. Cells Blood Substit. Immobil. Biotechnol.* **27**(3):245–253.

Zhang, R., and Ma, P. X. (2000), Synthetic nanofibrillar extracellular matrices with predesigned macroporous architectures, *J. Biomed. Mater. Res.* **52**(2):430–438.

Zhang, S., Yan, L., Altman, M., Lassle, M., Nugent, H., Frankel, F., Lauffenburger, D. A., Whitesides, G. M., and Rich, A. (1999a), Biological surface engineering: a simple system for cell pattern formation, *Biomaterials* **20**(13):1213–1220.

Zheng, J., Northrup, S. R., and Hornsby, P. J. (1998), Modification of materials formed from poly(L-lactic acid) to enable covalent binding of biopolymers: Application to high-density three- dimensional cell culture in foams with attached collagen, *In Vitro Cell Dev. Biol. Anim.* **34**(9):679–684.

Zisch, A. H., Schenk, U., Schense, J. C., Sakiyama-Elbert, S. E., and Hubbell, J. A. (2001), Covalently conjugated VEGF-fibrin matrices for endothelialization, *J. Control Release* **72**(1–3):101–113.

P · A · R · T · 4

BIOELECTRICITY

CHAPTER 17
BIOELECTRICITY AND ITS MEASUREMENT

Bruce C. Towe
Arizona State University, Tempe, Arizona

17.1 INTRODUCTION

Bioelectricity is fundamental to all of life's processes. Indeed, placing electrodes on the human body, or on any living thing, and connecting them to a sensitive voltmeter will show an assortment of both steady and time-varying electric potentials depending on where the electrodes are placed. These biopotentials result from complex biochemical processes, and their study is known as *electrophysiology*. We can derive much information about the function, health, and well-being of living things by the study of these potentials. To do this effectively, we need to understand how bioelectricity is generated, propagated, and optimally measured.

17.2 THE NATURE OF BIOELECTRICITY

In an electrical sense, living things can be modeled as a bag of water having various dissolved salts. These salts ionize in solution and create an electrolyte where the positive and negative ionic charges are available to carry electricity. Positive electric charge is carried primarily by sodium and potassium ions, and negative charges often are carried by chloride and hydroxyl. On a larger scale, the total numbers of positive and negative charges in biological fluids are equal, as illustrated in Fig. 17.1. This, in turn, maintains net electroneutrality of living things with respect to their environment.

Biochemical processes at the cell molecular level are intimately associated with the transfer of electric charge. From the standpoint of bioelectricity, living things are analogous to an electrolyte container filled with millions of small chemical batteries. Most of these batteries are located at cell membranes and help define things such as cell osmotic pressure and selectivity to substances that

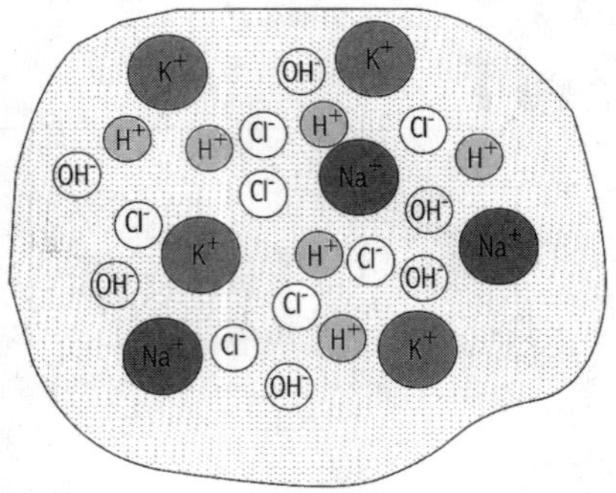

FIGURE 17.1 All living things have a net electroneutrality, meaning that they contain equal numbers of positive and negatively charged ions within their biological fluids and structure.

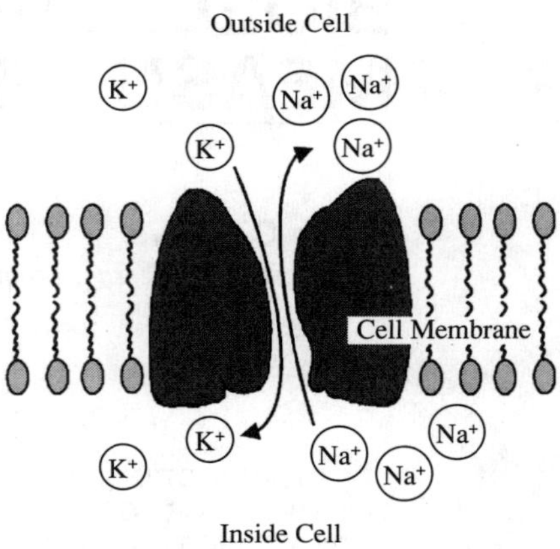

FIGURE 17.2 Concept of a sodium-potassium ionic pump within the phospholipid structure of the cell membrane. The molecular pump structures are not exactly known.

cross the membrane. These fields can vary transiently in time and thus carry information and so underlie the function of the brain and nervous system.

The most fundamental bioelectric processes of life occur at the level of membranes. In living things, there are many processes that create segregation of charge and so produce electric fields within cells and tissues. Bioelectric events start when cells expend metabolic energy to actively transport sodium outside the cell and potassium inside the cell. The movement of sodium, potassium, chloride, and, to a lesser extent, calcium and magnesium ions occurs through the functionality of molecular pumps and selective channels within the cell membrane.

Membrane ion pumps consist of assemblies of large macromolecules that span the thickness of the phospholipid cell membrane, as illustrated in Fig. 17.2. The pumps for sodium and potassium ion are coupled and appear to be a single structure that acts somewhat like a turnstile. Their physical construction transports sodium and potassium in opposite directions in a 3:2 atomic ratio, respectively.

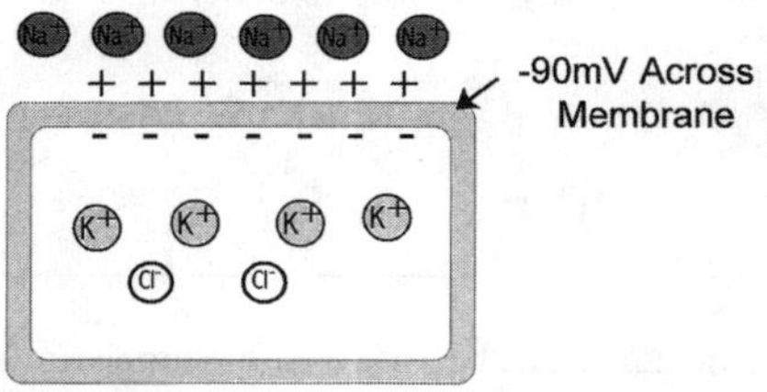

FIGURE 17.3 Electric field across the membrane due to action of the sodium-potassium pump.

As shown in Fig. 17.3, the unbalance of actively transported charge causes a low intracellular sodium ion concentration and relatively higher potassium concentration than the extracellular environment. Chloride is not actively transported. The unbalance in ionic charge creates a negative potential in the cell interior relative to the exterior and gives rise to the resting transmembrane potential (TMP).

This electric field, appearing across an approximately 70-Å-thick cell membrane, is very large by standards of the macro physical world. For example, a cellular TMP of -70 mV creates an electric field of 10^7 V/m across the membrane. In air, electric breakdown would occur and would produce lightning-bolt-sized discharges over meter-order distances. Electric breakdown is resisted in the microcellular environment by the high dielectric qualities of the cell membrane. Even so, the intensity of these fields produces large forces on ions and other charged molecules and is a major factor in ionic transfer across the cell membrane. When a cell is weakened or dies by lack of oxygen, its transmembrane field also declines or vanishes. It both regulates and results from the life and function of the cell.

In addition to ion pumps, the membrane has ion channels. These are partially constructed of helical-shaped proteins that longitudinally align to form tubes that cross the cell membrane. These proteins are oriented such that parts of their charged structure are aligned along the inside of the channels. Due to this alignment and the specific diameter of the formed channel lumen, there results selectivity for specific ionic species. For example, negatively charged protein carboxylic acid groups lining the channel wall will admit positively charged ions of a certain radius, such as sodium, and exclude negatively charged ions, such as chloride.

Some membrane ionic channels respond to changes in the surrounding membrane electric field by modulating their ionic selectivity and conductance. These are known as *voltage-gated ion channels*. Their behavior is important in excitable cells, such as nerve and muscle, where the membrane permeability and its electric potential transiently change in response to a stimulus. This change accompanies physiological events such as the contraction of muscle cells and the transmission of information along nerve cells.

17.2.1 Bioelectric Currents

Electric fields can arise from a static distribution of charge, whereas electric currents are charges in motion. Since current flow in a volume of fluid is not confined to a linear path as in a wire, currents can flow in many directions. When describing bioelectric currents, we often use a derivation of Ohm's law:

$$J = \sigma E$$

where J = current density, amps/m^2
σ = medium conductivity, siemens/m
E = electric field, volts/m

Thus, within the conductivity of living tissue, electric current flows are driven by electric field differences.

Electric currents in a wire always move from a source to a sink. Ionic currents in an electrolyte flow in a similar way. Charged ions, such as sodium, move from some origin to a destination, thereby producing an electric current flow. The magnitude of these currents is directly proportional to the number of ions flowing. This idea is shown in Fig. 17.4.

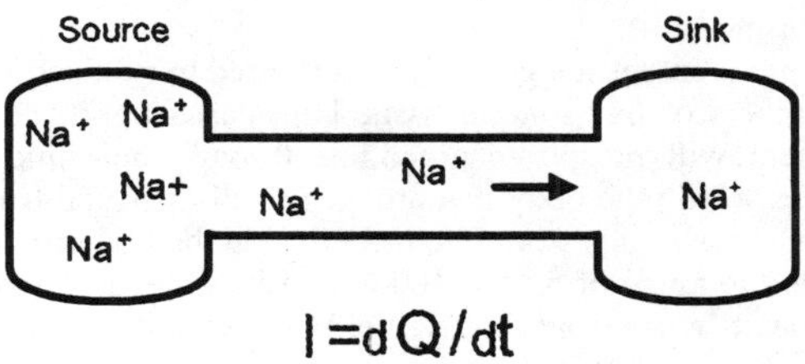

FIGURE 17.4 Diffusion of charged ions from a region of high concentration (a source) to low (a sink). This constitutes an electric current that can be expressed in terms of amperes.

Current flow I in amperes is defined as the time rate of charge Q movement:

$$I = dQ/dt$$

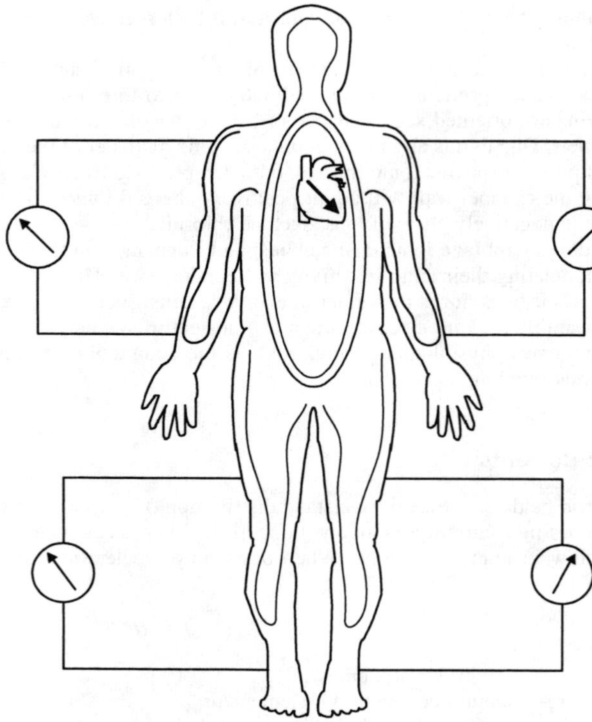

FIGURE 17.5 The volume conductivity of the body means that bioelectric current generators such as the heart produce electric fields detectable elsewhere on the body.

Thus, given the charge of 1.6×10^{-23} C, approximately 10^{23} sodium ions, for example, moving in one direction each second will produce an electric current of 1 A. Such large numbers of ions do not ordinarily flow in one direction over a second in biologic organisms. Currents in the milliampere region, however, are associated with physiologic processes such as muscular contraction and the beating of the heart.

Biological tissues are generally considered to be electric volume conductors. This means that the large numbers of free ions in tissue bulk can support the conduction of currents. Even bones, to some extent, will transport charged ions through their fluid phase. On a macroscopic scale, there are few places within the body that are electrically isolated from the whole. This is reflected in the fact that electric resistance from one place inside the body to almost any other will show a finite value, usually in the range of 500 to 1000 Ω.[1] Likewise, a current generator within the body, such as the beating heart, will create electric field gradients that can be detected from most parts of the body, as illustrated in Fig. 17.5.

17.2.2 Nernst Potentials

The origin and equilibrium of electric fields at the cellular level can be understood in terms of Nernst potentials. These potentials arise where any kind of membrane, not just living ones, selectively passes certain ions and blocks others.

If, for example, a membrane selective for potassium separates a compartment of high-concentration potassium chloride solution from another of lower concentration, there is a diffusion-driven current J_{diff} across the membrane. The value of this is given by a version of Fick's law:

$$J_{diff} = -\frac{RT}{f}\left[\frac{dC}{dx}\right]$$

where dC/dx = ion concentration gradient
R = universal gas constant
T = absolute temperature, kelvins
f = frictional constant related to the membrane permeability

In this situation, it might be expected that potassium ions would continue to diffuse until their concentration is the same on both sides of the membrane. However, since potassium is charged and the membrane prevents charged counter-ions such as chloride from also moving across, there arises an electric field across the membrane. This electric field creates forces that tend to inhibit further potassium diffusion. The electric field tends to cause a flux of potassium J_{elect} back across the membrane. It can be shown that the flux due to this electric field dV/dx is given by

$$J_{elect} = -\left[\frac{dV}{dx}\right] nCF\frac{1}{f}$$

where n = charge on each ion
C = concentration of ions
F = Faraday's constant

At some point of equilibrium, the forward flux of ions moving under forces of diffusion equals the reverse flux of ions driven by the electric field. The *Nernst potential* of a membrane is thus defined as the electric field where the diffusive flux of ions exactly counterbalances the electric field flux such that

$$J_{diff} = -J_{elect}$$

The negative sign on the electric field term is needed for balance, since the positive direction of ionic diffusion is from a high to a low concentration. Thus an equilibrium condition is established where the forces of diffusion are exactly balanced by the forces of the electric field. This situation is illustrated in Fig. 17.6.

Substituting and solving the equation by integration for the potential difference $V_{in} - V_{out}$ across a membrane gives

$$V_{in} - V_{out} = -\frac{RT}{nF}\ln\left[\frac{C_{in}}{C_{out}}\right]$$

where C_{in} and C_{out} are the respective ionic concentrations inside and outside the cell membrane. At a body temperature of 37°C and for univalent ions, the value of $RT/nF \times 2.303$ (to convert the natural log to log base 10) gives a result of 61 mV per decade of concentration ratio.

This Nernst relationship can be applied to living cell membranes. Cells are most permeable to potassium, and if we substitute known concentration values of potassium inside and outside the cell, we find that the cell trans-membrane potential is near this value, but not exactly.

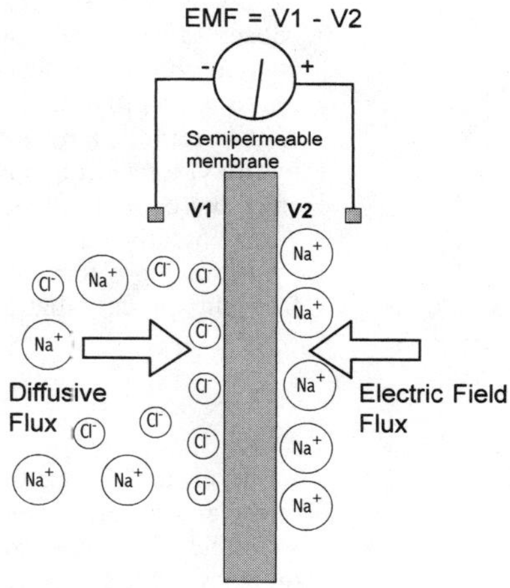

FIGURE 17.6 Nernst potentials arise across semipermeable membranes due to opposing forces of electric field and diffusion.

17.2.3 The Goldman Equation

Actually, in living cells there are many ionic species, with the important ones being potassium, sodium, and chloride. All these individually have their own Nernst potential that contributes to the total membrane potential. The fraction to which each adds to the total TMP depends on the membrane's permeability for the specific ion. Each ionic potential might be imagined as a battery in series with a specific electric conductance (related to the membrane's ionic permeability) and all connected across the membrane. Each ion species contributes according to the membrane's intrinsic permeability for that ion like batteries in parallel.

The Goldman equation accounts for multiple ionic contributors to the cell membrane potential. It assumes that the membrane permeability P is equivalent to the internal resistance of an ionic battery. Thus, with multiple ionic species that all contribute, the cell transmembrane potential is given by

$$\text{TMP} = V_{\text{in}} - V_{\text{out}} = (-RT/F)\ln\frac{\{P_{\text{Na}^+}[\text{Na}^+]_{\text{in}} + P_{\text{K}^+}[\text{K}^+]_{\text{in}} + P_{\text{Cl}^-}[\text{Cl}^-]_{\text{out}}\}}{\{P_{\text{Na}^+}[\text{Na}^+]_{\text{out}} + P_{\text{K}^+}[\text{K}^+]_{\text{out}} + P_{\text{Cl}^-}[\text{Cl}^-]_{\text{in}}\}}$$

where the quantities in brackets are the concentrations of the ionic species. Chloride ion is negatively charged, and hence its concentration ratio in the equation is inverted compared with those of sodium and potassium.

This equation produces a result that is very close to the measured cell TMP. The activity of membrane pumps is also known to contribute a few millivolts. Together, all these sources can account for the cell TMP.

17.2.4 Diffusion Potentials

Tissue sometimes is found to be electrically polarized even where there is no membrane that would give rise to Nernst or Goldman potentials. The origin of this polarization can often be attributed to what are called *diffusion potentials*. These arise from differences in ionic mobilities in the tissue volume conductivity and because metabolic processes produce a range of different kinds of free ions.

Living processes of metabolism and biochemistry constantly liberate ions that are higher in concentration locally than at more distant locations. Small ions, e.g., hydrogen carrying a positive charge, will diffuse faster away from a site of their production than will larger, negatively charged ions such as hydroxyl. This ionic separation process, however, is self-limiting. As hydrogen ions diffuse rapidly outward, this creates an electric field in tissue relative to the negatively charged hydroxyl ions left behind. The increasing electric field slows the outward diffusion of hydrogen, but since there are no physical barriers, both ions continue to move.

Eventually, equilibrium occurs, where the rapid outward diffusion of charged fast ions is balanced by a growing electric field from slower, oppositely charged ions. The potential magnitude resulting from this process that exists across some region of tissue is given by

$$V = RT/nF\left(\frac{\mu_a - \mu_c}{\mu_a + \mu_c}\right)\ln\left(\frac{C_1}{C_2}\right)$$

where C_1 and C_2 are the differential concentrations, respectively, and μ_a and μ_c are the mobilities of the anions and cations involved.

Table 17.1 shows some representative values of ionic mobilities. Hydrogen has a much larger mobility than any other ion. Therefore, local tissue pH changes are a prime source of diffusion potentials. Some ions such as sodium tend to collect water molecules around them, resulting in a larger hydrated radius, and even though they are small, they diffuse more slowly than might be expected from their size.

TABLE 17.1 Mobilities of Various Ions[2]

Ion	Mobility of ions at infinite dilution in water, (cm/s)/(V/cm)
Hydrogen	$\mu_c = 36 \times 10^{-4}$
Hydroxyl	$\mu_a = 20.5 \times 10^{-4}$
Sodium	$\mu_c = 5.2 \times 10^{-4}$
Potassium	$\mu_c = 7.6 \times 10^{-4}$
Chloride	$\mu_a = 7.9 \times 10^{-4}$

Unlike Nernst potentials, where ions are in equilibrium and create a static electric field, diffusion potentials result from ions in motion, making them electric currents in tissue. If diffusion is isotropic, there would be an outward-directed electric field and current flow from their source. This situation is shown in Fig. 17.7.

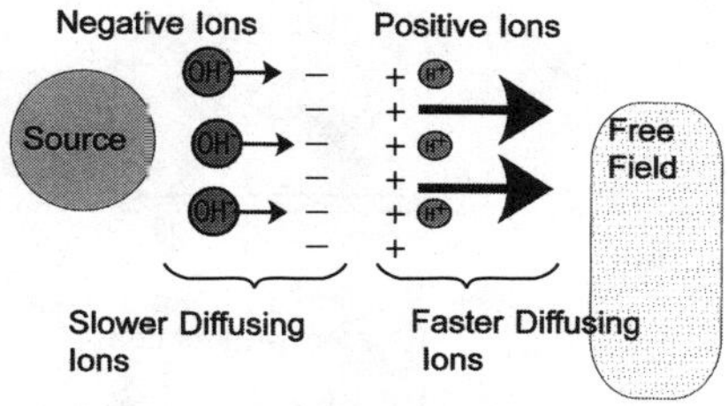

FIGURE 17.7 Fast-diffusing positively charged hydrogen ions move away from slower negatively charged hydroxyl ions, creating an electric field known as the *diffusion potential.*

Sustained diffusion currents depend on the active generation of free ions. Otherwise, the currents will decay as ionic concentration gradients move toward equilibrium. In biological organisms, diffusion currents often flow over time scales that are minutes, hours, or continuous if the source of production of the ions is continuous. Diffusion currents, in principle, can be differentiated from steady electric fields in tissue since currents can be detected using biomagnetometers that respond to ionic flow rather than to an electric field.

Diffusion currents are often associated with biological processes such as growth, development, wound healing, and tissue injury of almost any sort. For example, depriving heart tissue of oxygen, which occurs during a heart attack, produces an immediate diffusion current, known as a *current of injury*. This current is believed to result mostly from potassium ions released from inside the heart cells. Without aerobic metabolism, cells can no longer support the membrane pumps that maintain the cellular TMP. Potassium ions then flow out of the cells and diffuse outward from the injured tissues. Electrodes used to map bioelectricity during coronary vessel blockage will show a steady current flowing radially outward from the affected site. Likewise, it is known that skin wounds, bruises, other tissue traumas, and even cancer[3] can produce injury currents.

Nernst potentials, Goldman potentials, membrane ion pumps, and diffusion potentials are thus primary drivers for bioelectric currents that flow in organisms. Since natural bioelectric generators have a maximum of about −100 mV at the cellular level, most bioelectric events measured by electrodes do not exceed this millivoltage. There are a few exceptions. Certain species of electric fish, such as the electric eel, have tissues that effectively place the TMP of cells in electric series. Some remarkably high bioelectric voltages can be achieved by this arrangement, in the range of hundreds of volts. On synchronized cellular depolarization, large eels are known to discharge currents in the ampere range into seawater.

17.3 ACTION EVENTS OF NERVE

Excitable cells are those that can respond to variously electric, chemical, optical, and mechanical stimuli by sudden changes in their cell TMP. Ordinarily, a loss of a cell's membrane potential is lethal. However, fast transient changes in cell TMP, called *action events*, are part of the natural function of excitable cells.

Action events involve a multiphase biochemical-bioelectric process. A localized stimulus to an excitable cell can launch a series of cascading molecular events affecting the membrane's ionic permeability. The accompanying changes in TMP feed back on the membrane by way of voltage-gated channels and magnify the effect of the stimulus. If the stimulus amplitude reaches a threshold value, this causes further and more dramatic changes in the membrane's ionic permeability.

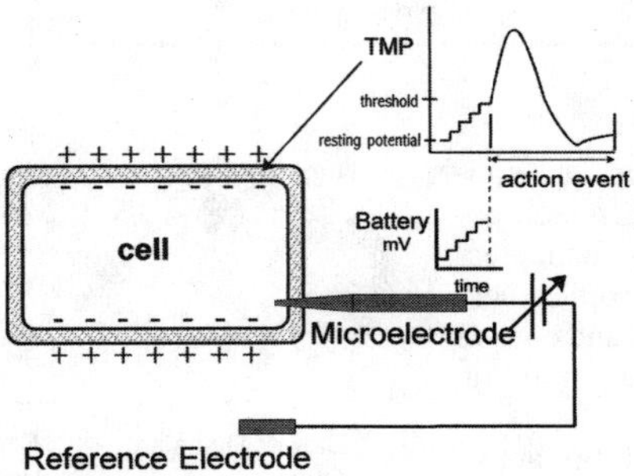

FIGURE 17.8 Illustration of a cell punctured by a microelectrode connected to a stepwise current source that drives the cell membrane potential above threshold.

The threshold behavior of excitable cells can be observable by placing a tiny (usually a 1-μm tip diameter) microelectrode inside a cell. Sufficient charge injection from an external voltage source can drive the transmembrane potential toward zero. Now if the transmembrane potential is moved stepwise more positive an action event will occur above some threshold. At this point, the membrane potential will suddenly jump of its own accord from a negative TMP to a positive value. Figure 17.8 shows a setup where a battery is used to stepwise drive the cell TMP more positive.

During an action event, membrane sodium conductance abruptly increases allowing the higher concentration of sodium in the extracellular medium to rush into the cell. The net negative electric field inside the cell is thereby reduced toward zero through the positive charge on sodium. This is known as *depolarization*. Shortly after the sodium inrush, there is an increase in membrane potassium conductance. This allows the higher concentration of potassium ions inside the cell to move outward. Because of a flow of positive charge to the outside of the cell, the net negative charge inside the cell is restored. This is known as *repolarization*. These changes in sodium and potassium membrane conductance are illustrated in Fig. 17.9.

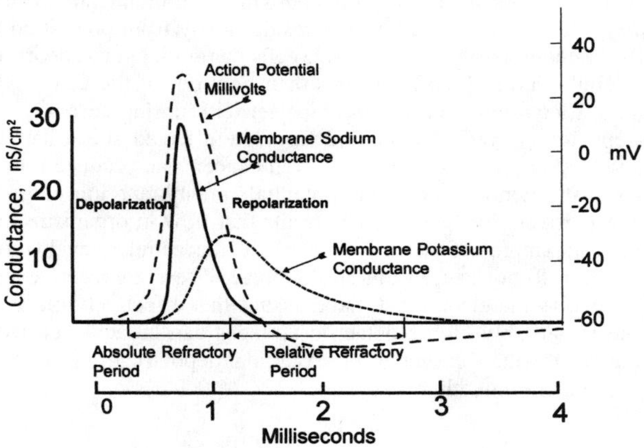

FIGURE 17.9 Illustration of the cycle of ionic conductances associated with action events.

The depolarization and repolarization phases of action events can occur quickly over intervals of tens of microseconds, although the actual durations depend very much on the cell type. During the time when the cell is depolarized, it cannot be restimulated to another action event. This interval is known as the cell's *absolute refractory period*. The cell's *relative refractory period* is the interval from partial to complete repolarization. During this time, the cell can be restimulated, but a higher stimulus is required to produce an action potential event, and it is lower in magnitude.

17.3.1 Membrane Bioelectrical Models

Cell membranes separate charges, and this causes a potential across their thickness. This aspect of the membrane can be modeled as a charged capacitor that follows the relationship

$$V = Q/C$$

where V = potential
Q = charge separation
C = capacitance

In an action potential event, the total amount of charge Q transferred across the membrane by movements of sodium and potassium is relatively little, just sufficient to charge the membrane capacitance C_m. The amplitude of potential change and time course of an action event depend on the type of excitable cell being studied.

Membranes also have a finite and distributed electric resistance both across their thickness and along their length. These resistances tend to short circuit the membrane capacitor and cause its stored charge to decay to zero if the membrane ion pumps cease. There are several different models of the cell membrane that describe its static and transient bioelectric behavior. Models of the cell membrane are useful because they help explain the propagation of action events, such as that along a nerve.

The electric nature of the membrane is usually modeled as distributed capacitors and resistors, as shown in Fig. 17.10. The value of R_m is the membrane equivalent electric resistance. R_e and R_i are the resistances of the external and internal membrane environments. The stored charge across the membrane is electrically represented by a capacitance C_m, and the action of the ionic pumps is represented by E_m.

With local membrane depolarization caused by some stimulus, the longitudinal membrane conduction causes the local rise in TMP to spread along the adjacent regions of the membrane. Adjacent voltage-gated membrane channels are driven above threshold, propagating the depolarization event. This effect on adjacent membrane is illustrated in Fig. 17.11 for a nerve axon. By this mechanism, bioelectric action events spread outward from the point of stimulus, move along the cell membrane, and cease when the entire membrane is depolarized.

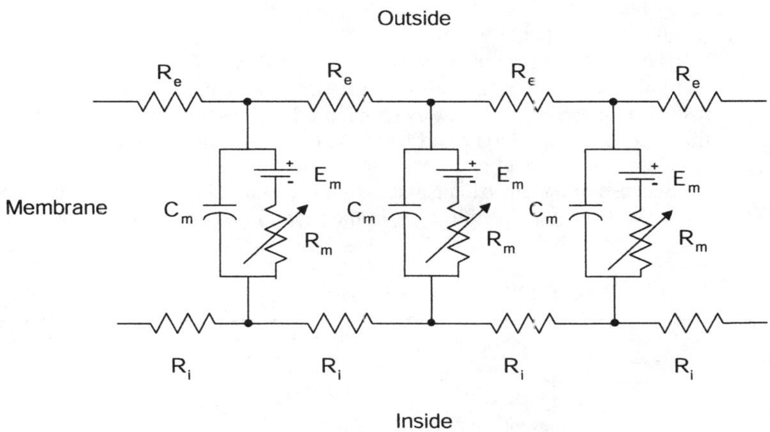

FIGURE 17.10 Model of the electrical equivalent circuit of a nerve membrane.

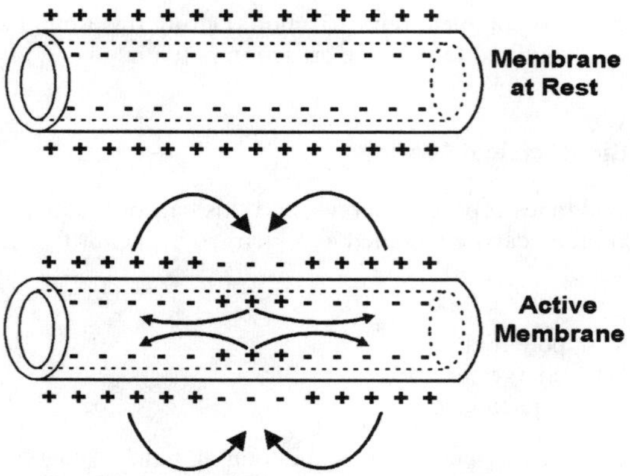

FIGURE 17.11 Depolarization of a nerve cell membrane creating a circulating flow of ionic currents.

17.3.2 Propagation of Action Events

Nerves in higher biological organisms are bundles of long, excitable cells that can extend to meter-order lengths. A sufficient stimulus to a point on a nerve cell membrane will rapidly conduct along its length. Bioelectric action currents generated in a single nerve fiber by the depolarization wavefront are relatively small, in the tens of nanoampere region. They can range into the microampere region, though, with synchronous depolarizations of many fibers within large nerve bundles. As these currents travel along the nerve length, they can produce detectable electric fields at nearby recording electrodes. The time course of these recorded biopotentials depends largely on the velocity of the action current wave as it passes by the point of measurement.

The velocity of action event propagation depends on the nerve diameter d as well as on passive electric properties of the membrane. Assuming that these electric properties are constant along the membrane, it can be shown that the depolarization wave velocity of a nerve is proportional to

$$\text{Action potential velocity (m/s)} \approx \sqrt{d}$$

Thus larger-diameter nerves propagate the depolarization wave faster. In a given bundle of individual nerve fibers, there are clusters of different fiber diameters, and hence propagation speeds within a bundle are distributed.

The smallest nerve fibers are classified as C-type. These have diameters on the order of a micron and have conduction velocities on the order of 1 m/s. A-type fibers are coated with a layer of an insulating material known as *myelin*, a product of Schwann cells that wrap around them. Myelinated fibers are generally larger and faster with a conduction velocity median around 50 m/s. Table 17.2 compares the physical and electric properties of A-type and C-type fibers.

Because of the electric insulating quality of myelin, bioelectric depolarization is prevented along the nerve except at specific gaps in the myelin known as the *nodes of Ranvier*. At these points the

TABLE 17.2 Nerve Properties[4]

Property	A-type myelinated nerve	C-type unmyelinated nerve
Diameter	1–22 μm	0.3–1.3 μm
Velocity	5–120 m/s	0.7–2.3 m/s
Action event duration	0.4–0.5 ms	2 ms
Refractory time	0.4–10 ms	2 ms

extracellular sodium and potassium ionic currents flow during action events. The nodes confine the bioelectric current sources to relatively small areas on the nerve membrane, and the current flows appear to hop from node to node down the length of the fiber. An example of current flow in a myelinated fiber is shown in Fig. 17.12.

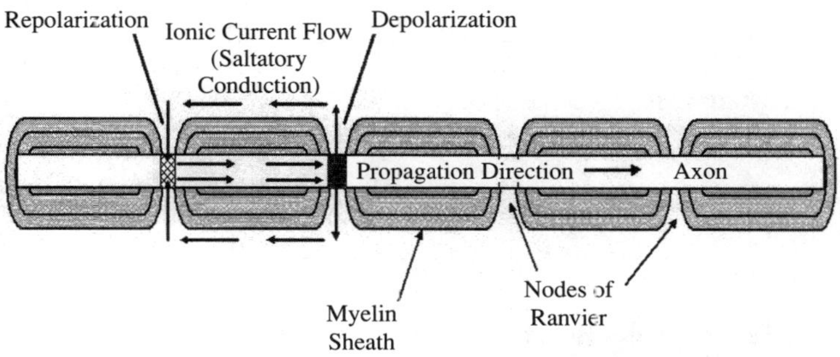

FIGURE 17.12 Myelin sheath around a nerve that limits action currents to the nodes of Ranvier.

However, there are still electrode-detectable bioelectric fields from myelinated nerves during action events, since volume currents flow, as shown in Fig. 17.13. These bioelectric currents are dipolar in nature, meaning that ionic currents flow spatially from the source at the leading edge of depolarization to a sink at the repolarization edge. Separation distances between these nodes in mammals are on the order of a millimeter.

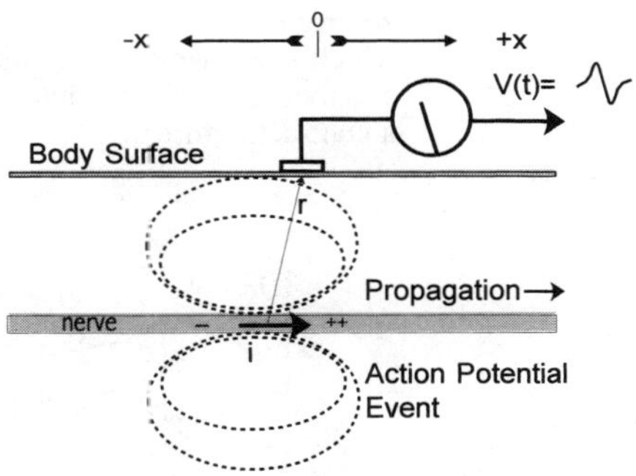

FIGURE 17.13 Electric dipole equivalent of a traveling depolarization event in nerve.

17.3.3 Action Events of Muscle

Skeletal muscles produce some of the larger bioelectric signals within the body. Muscle is composed of many small and elongated myofibers of about 50 μm in diameter. Myofibers are made of myofibril cells, which contain filaments of contractile proteins—actin and myosin. Actin and myosin are oriented longitudinal to the direction of their contraction. These proteins are believed to engage in a walk-along movement by a sliding filament process that coincides with internal fluxes of calcium ions triggered by an action event. Myofibers are in turn organized into innervated functional groups known as *motor units*. Motor units are bundled together and interwoven with others and are then called *fascicles*, ultimately forming the whole muscle.

Single motor units are actuated by a nerve fiber that terminates at the synaptic cleft of the myoneural junction. The neurotransmitter acetylcholine opens up membrane ion channels, allowing the influx of sodium and causing an action event followed by a short, twitchlike mechanical contraction.

Both an increase in recruitment of individual motor units and an increase in their firing rates control muscle force. Single-motor-unit action events produce bioelectric currents of 3 to 15 ms in duration. Under sustained nervous stimulation, these can occur with repetition rates of 5 to 30 per second. Even though the myofibril bioelectric events are pulsatile, muscle contraction can be smooth because at high firing frequencies there is a blending of the forces together. Increases in skeletal muscle contractile force are accompanied by an increase in the recruitment of myofibrils. The forces

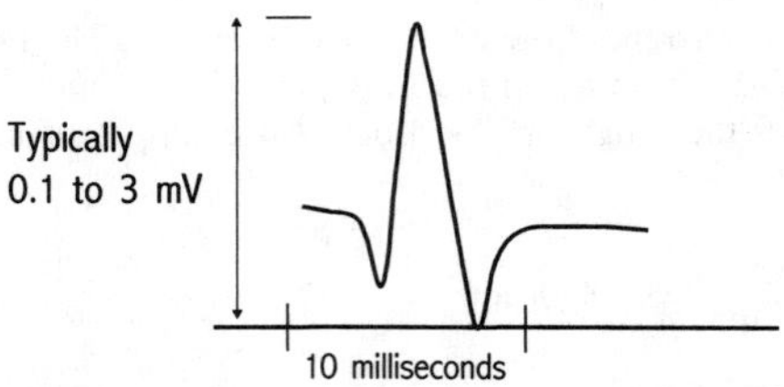

FIGURE 17.14 Illustration of the general form of an action potential event of a single motor unit as recorded by a needle electrode.

of the individual motor units sum together, producing a relatively larger force. When monitored by a local invasive needle electrode, motor unit action events produce a triphasic extracellular waveform in the range of a few tens of microvolts to 1 mV. A typical motor unit bioelectric waveform is shown in Fig. 17.14.

Complex, random-looking electromyographic (EMG) waveforms are detected by electrodes placed near skeletal muscle. These waveforms result from the time and spatial superposition of thousands of motor unit events. During a strong contraction of a large skeletal muscle, monitored skin bioelectric waveforms can reach millivolt amplitudes.

17.3.4 Action Events of the Heart

Heart cells are unique compared with skeletal muscle cells because they are in electric contact with each other through electric tight junctions between cell membranes and so propagate bioelectric depolarization events from one cell to the next. By contrast, motor units of skeletal muscle are controlled by nerves, and each unit has a separate junction. Heart cells are also physically intertwined, which is known as being *intercalated*. The overall result is that the heart acts as both an electrical and mechanical syncytium. That is, the heart cells act together to conduct bioelectricity and produce a contractile force.

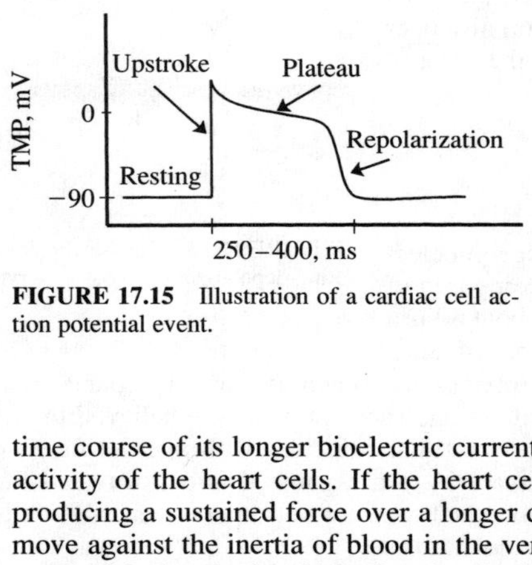

FIGURE 17.15 Illustration of a cardiac cell action potential event.

The action potential event of the heart is different from that of skeletal muscle. It has a prolonged depolarization phase that varies from tens of milliseconds for cells of the atria to about 250 to 400 ms for cells of the ventricle. The general waveform shape for an action event of the ventricle is shown in Fig. 17.15.

The prolongation of the heart action event results from the activity of slow calcium channels in the cell membranes. These act in concert with the fast sodium channels to maintain a longer depolarization phase of the heart. This prolongation is important in the physiological function of the heart since it defines the timing and duration of the mechanical events required for a contraction.

The mechanical contraction of the heart follows the time course of its longer bioelectric current flows. Bioelectric currents initiate and define the motor activity of the heart cells. If the heart cells simply twitched during an action event rather than producing a sustained force over a longer duration, there would not be time for the myocardium to move against the inertia of blood in the ventricles.

The bioelectric depolarization that leads to contraction starts in the pacemaker cells in the sinoatrial node. There it passes as a narrow band of depolarization from the atria toward the ventricles. Although the process of depolarization is complex and not spatially uniform, it can be thought of as a moving band of current flow from top to bottom of the heart. For example, assuming a depolarization velocity of 25 m/s and a sodium channel-opening rise time of 30 μs, a traveling band of current approximately 0.75 mm wide would move through the heart during the contraction phase (systole). This simplified notion is illustrated in Fig. 17.16. In actuality, specialized nodal tracts of the atria, the bundle of His, and the Purkinje fibers in the ventricles direct the propagation wave and affect its dipolar length.

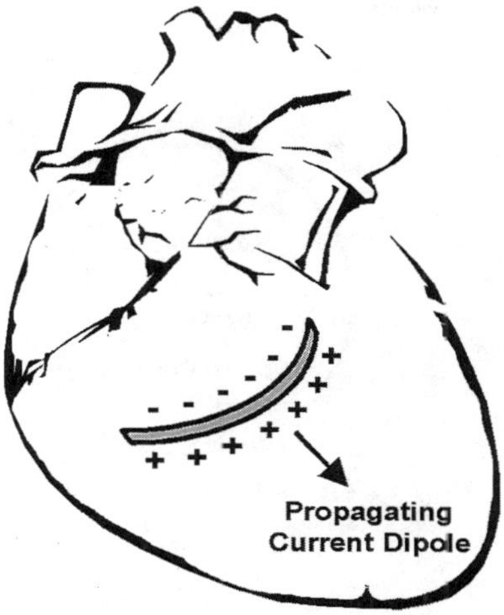

FIGURE 17.16 Illustration showing the simplified concept of a traveling band of depolarization creating a net current dipole in the heart.

Subsequently, heart cells repolarize during the resting (diastole) phase of the heart cycle. The repolarization wave event is again a cellular current flow creating a bioelectric potential. In humans, many millions of cells participate in this depolarization. Each cell is a tiny bioelectric current generator. The sum of the generators produces biopotentials on the order of 1 to 3 mV on the chest wall surface. A recording of these cardiac potentials is known as an *electrocardiogram* (ECG, or sometimes EKG, from the German, *electrokardiogram*).

17.4 VOLUME CONDUCTOR PROPAGATION

Living things create and conduct current within their tissues and so produce remotely detectable electric fields. Excitable tissue function is both initiated and followed by bioelectric events. Thus there exists the possibility of monitoring physiological function through the use of remote and noninvasive potential-sensing electrodes. Measurement of the bioelectric magnitude and time course may allow us to gain an understanding of biological function and performance. Indeed, changes in monitored biocurrent flows have been found to be useful indicators of disease. Unfortunately, it is not straightforward to use noninvasive surface electrodes to infer the location and time courses of biocurrent flows inside tissue.

17.4.1 Dipolar Current Sources in Volume Conductors

When discussing bioelectric currents, electrophysiologists usually use the term *dipole moment p*. This is defined as

$$p = id$$

where i is the ionic current in milliamperes and d is the separation distance between the current's

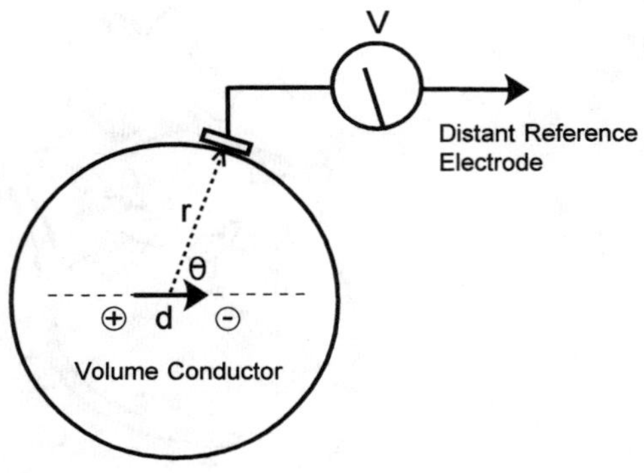

FIGURE 17.17 Model of a dipolar current within a large spherical volume conductor.

source and sink. The reason for this usage is that the magnitude of potential remotely detected from an ionic current flow depends not only on the current amplitude but also on its source-sink separation.

A model situation often used is a dipolar current source immersed in a large spherical volume conductor. The surface potential is monitored with respect to a distant electrode, as seen in Fig. 17.17. Due to spherical symmetry, the potential V at a point on the sphere surface is given by a fairly simple volume-conductor relationship:

$$V = \frac{id \cos \theta}{4\pi\sigma r^2}$$

where θ is the angle formed by the current vector to the remote monitoring point. This equation defines the fraction of the electric potential seen by remote monitoring electrodes. Biological tissues have widely varying conductivities σ at low biopotential frequencies (10 Hz) ranging from 0.6 S/m for blood,[5] which is a relatively good conductor, to 0.04 S/m for fat, which is a relatively poor one.

Bioelectric dipole moments range from milliampere-millimeter currents associated with the heart and skeletal muscle to nanoampere-millimeter for nerve currents. The volume-conductor relation shows that the detection of these current moments depends on a number of variables besides their strength. For example, surface potentials depend both on the dipole distance and on the orientation of the dipole moment with respect to the electrode.

17.4.2 The Problem of Biocurrent Localization

The task of the electrophysiologist or the physician is often to interpret recorded bioelectric waveform shapes in terms of physiological events. Bioelectric field propagation in biological objects, however, is complex partly because of nonuniformities in tissue conductivity. Therefore, skin surface potentials often do not directly follow the pattern distribution of the internal currents that produce them.

If we know the distribution of biocurrents in tissue and know or can estimate tissue conductivities, it is possible to calculate the biopotential distribution on the body surface. In electrophysiology, this is known as the *forward problem*. Conversely, inferring the magnitude and distribution of biocurrent generators from surface biopotentials is called the *inverse problem*. Indeed, solving this problem would be very useful for medical diagnostics and to allow us to gain a greater understanding of physiology. Unfortunately, the inverse problem is known to be mathematically ill-posed, bearing no direct solution. This derives from the fact that there are too many variables and too few measurements possible to define a three-dimensional current from potentials on a two-dimensional surface.

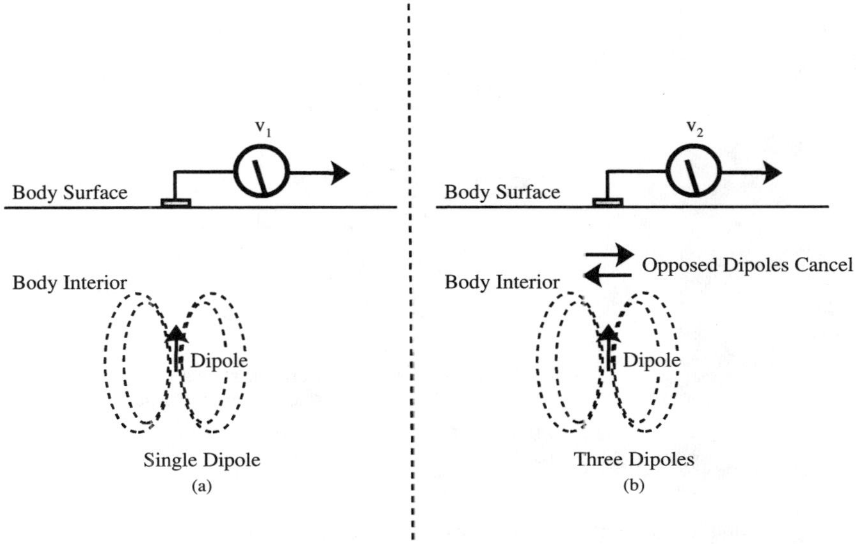

FIGURE 17.18 Illustration of the bioelectrical inverse problem.

For example, multiple biocurrent vector directions and variable dipole distances can create a situation where, if they are close together and have opposite vector directions, they can cancel, rendering no trace of their existence. Figure 17.18 illustrates the problem of vector cancellation. When far enough away from the skin, one current source in (*a*) creates the same surface potential as the three sources seen in (*b*). This is just one of many different geometric possibilities of dipole orientation and body surface distance that create identical surface potentials.

There are, however, constraints that can be put on the inverse problem to make it somewhat more tractable. Estimates of tissue conductivity, knowledge of how currents flow in biological tissues, and knowledge of the normal biocurrent time course can simplify the problem. Complex computer models have been developed to derive medical diagnostic and other electrophysiological data from maps of surface potentials. These are still the subject of research.

Another example of ambiguity in what a waveform shows is that the biocurrent dipole amplitude may appear to be changing when really it is moving. A model situation is where a biopotential electrode is placed on the surface of a volume conductor and a dipole moment p moves past it. The distance from the center of the dipole to the electrode is r, and θ is the angle that the dipole makes with respect to the vertical. Plotting the potential V at the electrode measurement point as the current dipole moves past, assuming that the dipole is relatively far away from the measuring point compared with its length, we see that motion of the dipole produces a complex biopotential, as plotted in Fig. 17.19. It has a biphasic component where, in fact, there is no change in amplitude or reversal of the biocurrent source.

Another factor that makes it difficult to use skin biopotential amplitudes as quantitative indicators of physiologic events is that there are significant variations between people relative to body fat, muscle distribution, and size and position of tissue biocurrent sources.

In practice, simple observation of the biopotential waveform is often very useful. For example, a common biopotential recording is the fetal ECG. Electrodes are placed on the mother's abdomen to bring the bioelectric monitoring point close to the source; however, the fetal current generators are still relatively weak. Furthermore, the mother's ECG is also captured in the recording. Nonetheless, the two waveforms are easily discriminated by a trained eye. The fetal ECG is typically faster,

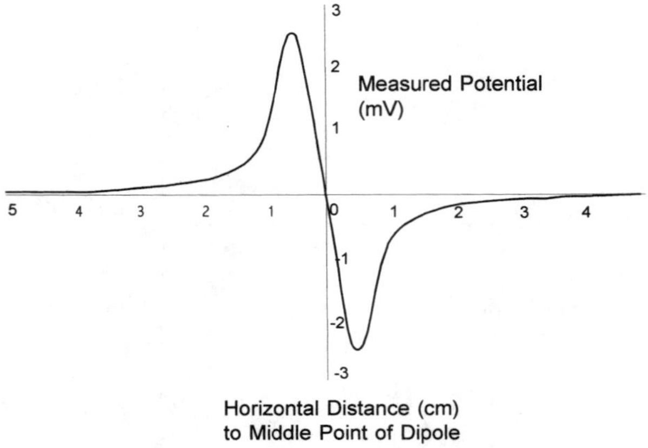

FIGURE 17.19 Plot of a biocurrent dipole moving past the measuring point.

smaller in amplitude, and may exhibit a simpler waveform than the maternal ECG, making the detected waveform rather obvious to the physician.

17.5 DETECTION OF BIOELECTRIC EVENTS

17.5.1 Bioelectrodes and the Electrode-Skin Interface

Biopotentials from the human body are generally low-level signals, maximally in the millivolt range; however, in many cases, microvolt-level signals are of significant interest. Although these potentials are not difficult to amplify, the detection and measurement of bioelectric potentials from tissue or skin are not straightforward processes. Many people have found through hard experience that the placement of electrodes on the skin often creates an assortment of biopotential measurement problems. These include baseline drifts, motion artifacts, and interference originating from power wiring and radiofrequency sources. Some of these problems originate from a lack of understanding of electrode characteristics. Careful electrode selection and placement, with a solid understanding of electrode use, are important in achieving good-quality biopotential recordings.

Electrode metals used to monitor low-level biopotentials perform the function of transducing ionic electrochemical reactions into a current carried by electrons in wires. Electrochemical reactions must occur at the interface of the electrode surface. Otherwise, no charge transfer will occur to the electrode wire, and a recording apparatus would not measure a biopotential.

17.5.2 Electrode-Electrolyte Interface

When metals are placed in electrolytes, such as in the saline environment of tissue or skin, atoms of the metal slightly dissolve into solution. This slight dissolution of metal is accompanied by a loss of electrons to the atoms that remain with the parent metal. It leaves the parent metal with a net negative charge and the dissolved metal ions with a positive charge. The created electric field tends to draw the ions back to the vicinity of the metal surface; however, energy is required for their recombination, so the ions are just held close to the metal surface. An interfacial double layer of charged ions is formed, as seen in the model in Fig. 17.20.

This model is attributable to the great nineteenth-century scientist Herman von Helmholtz, although there are several other models. The dissolved metal ions line up in a double layer of atomically thin sheets immediately next to the parent metal. This layer has electric qualities of a battery, a capacitor, and an electric resistor. These qualities profoundly affect bioelectrode performance and its ability to accurately report biopotentials.

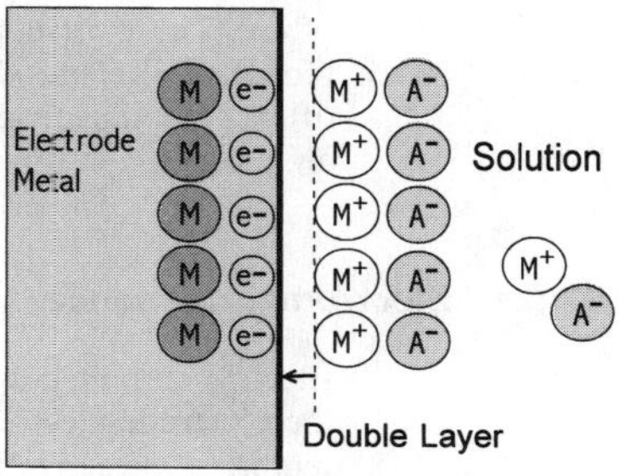

FIGURE 17.20 The Helmholtz layer is an electric double layer of charged metal ions (M^+) and anions (A^-) facing each other at the electrode-electrolyte interface.

17.5.3 Electrode Half-Cell Potential

Two metal electrodes placed into an electrolytic solution constitute what is called an *electrochemical cell.* A potential arises at the solution interface of each electrode due to the separation of charged ions across its double layer. This potential is known as the *electrode half-cell potential.* It cannot be measured directly since it requires some reference potential for comparison.

Two dissimilar metal electrodes placed in an electrolyte will show a potential difference that is a function of the metals used, the electrolyte composition, and the temperature. Half-cell potential differences can be up to several volts and are the basis of the electric batteries used to power appliances. Theoretically, the cell potential difference between two electrodes E_{cell} can be determined by knowing the half-cell potential of each electrode E_{hc1} and E_{hc2} with respect to some common reference electrode. Thus the cell potential difference is given by

$$E_{cell} = E_{hc1} - E_{hc2}$$

The value of this electric field depends on the electrochemical potentials of the electrode metals and is related to their position in the electrochemical series. Table 17.3 shows some values for some common electrode metals at room temperature. Electrochemists have by convention adopted the hydrogen electrode as a standard of reference potential and assigned it a value of 0 V. All other metals have a nonzero potential with respect to it. Metal half-cell potentials depend on their electrochemical oxidation state, and they are usually arranged in a table showing their activity relative to others, such as seen in Table 17.3.

TABLE 17.3 Half-Cell Potentials of Some Common Metals at 25°C[6]

Metal and reaction	Electrode potential E_{hc}
$Al \rightarrow Al^{3+} + 3e^-$	−1.67
$Fe \rightarrow Fe^{2+} + 2e^-$	−0.441
$H_2 \rightarrow 2H^+ + 2e^-$	0 by definition
$Cu \rightarrow Cu^{2+} + 2e^-$	+0.340
$Ag^+ + Cl^- \rightarrow AgCl$	+0.223
$Ag \rightarrow Ag^+ + e^-$	+0.799

In theory, if two electrodes are of the same metal, their individual half-cell potentials are the same, and so the cell potential difference should equal zero. This is desirable for biopotential monitoring. Otherwise, the electrode pair would add an offset potential to the biopotential being monitored.

Unfortunately, unbalanced half-cell potentials are often the case with biopotential electrodes. Primarily, this is due to different electrode metal surface states that occur through oxidation occurring with air exposure, tarnishing, metallurgical preparation, previous electrolyte exposure, or past history of use. These offset potentials typically range from several tens of millivolts with commercial silver biopotential electrodes to hundreds of millivolts with electrodes such as stainless steel. Electrode offset potentials tend to be unstable and drift in amplitude with time and accumulated use. Electrodes are a major contributor to slow baseline drifts in biopotential measurements.

17.5.4 Electrode Impedance

Electric current flow though electrodes occurs through the agency of electrochemical reactions at the electrode-electrolyte interface layer. These reactions do not always occur readily and usually require some activation energy, and so the electrode interface can be a source of electrical resistance. Depending on the electrode composition and its area, electrode-electrolyte resistance is on the order of a few hundred ohms with typical ECG skin electrodes and thousands to millions of ohms with small wire electrodes and microelectrodes. Electrode interface resistance is usually not large compared with other resistance in the electrode-biological circuit.

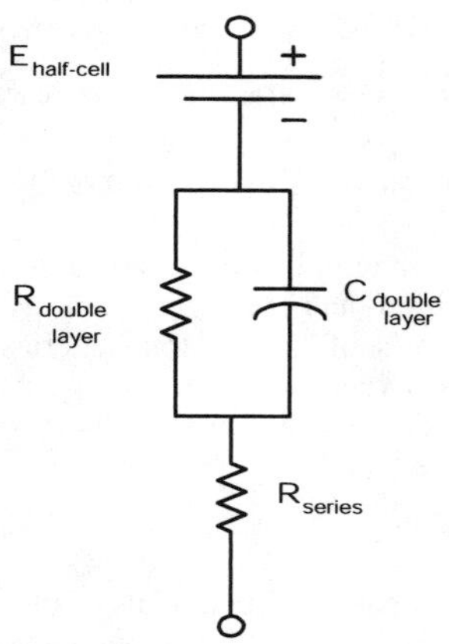

FIGURE 17.21 Electric equivalent circuit of a biopotential electrode.

The separation and storage of charge across the electrode double layer are equivalent to an electric capacitance. This capacitance is relatively high per unit area, on the order of 10 to 100 $\mu F/cm^2$ with many metals in physiologic electrolytes. Together the electrode resistance and capacitance form a complex impedance to electric current flows. A simple diagram of an equivalent electric circuit is given in Fig. 17.21. The electrode-electrolyte interface is somewhat more complicated than linear models can easily show. For example, the equivalent capacitance and resistance of an electrode R_e and C_e are not fixed but fall in value roughly as a square-root function of frequency.

The result is that bioelectrodes more easily pass alternating or time-varying currents at frequencies above a few hundred hertz than they do with steady currents or low frequencies. Fortunately, this has little impact on most bioelectric recording applications, such as ECG or electroencephalography (EEG). Even higher-frequency electromyographic (EMG) recordings are only nominally impacted. This is so because high-input-impedance bioelectric amplifiers are relatively insensitive to moderate changes in electrode impedance that occur as a function of frequency.

The frequency-dependent behavior of electrode impedance is of more concern in intracellular microelectrode recording applications, where electrode resistances are very high. In conjunction with the electrode capacitance and that of the recording circuit, a low-pass filter is formed at the microelectrode interface that tends to distort high-frequency portions of action events.

Frequency-dependent electrode impedance is also important in bioelectric stimulation applications, where relatively larger currents and complex waveforms are often used to stimulate excitable tissues.

17.5.5 Electrode Electrochemistry

In most metal electrodes, a process of chemical oxidation and reduction transfers current. At the anode, oxidation occurs, and electrons are given up to the parent metal. Reduction occurs at the cathode, where metal ions in solution accept electrons from the metal.

Actually, this is a process of mass transfer since, with oxidation, metal ions leave the electrode and move into solution and, with reduction, metal ions form a metal precipitate on the electrode. Ordinarily, current flow in biopotential recording electrodes is very small. Therefore, only minuscule quantities of the electrode metal are transferred to or from electrode surfaces. Biocurrent flows are also generally oscillating, so the transference of ions is cyclic between the electrodes with no net mass transfer. This process of electrode mass transfer through dissolution and precipitation of metal ions is depicted in Fig. 17.22.

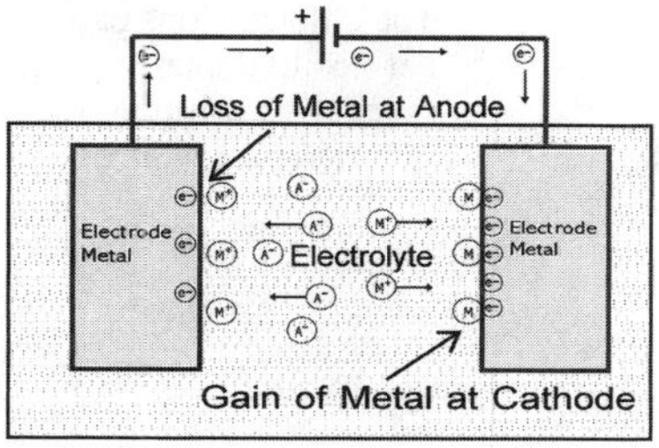

FIGURE 17.22 Transfer of charge due to transfer of ionized atoms of the electrode mass.

There are exceptions to electric conduction by electrode mass transfer. Noble metals such as platinum, iridium, and gold are electrochemically inert. When placed into solution, they usually do not significantly lose their substance to ionization. They are still useful as bioelectrodes, however, because they catalyze electrochemical reactions in solution by donating or accepting electrons without actually contributing their mass to the electrochemical process. They participate by catalyzing reactions at their surface but are not consumed. They are often electrically modeled as capacitors, but this is not a perfect model since they will pass steady currents. Although unconsumed, noble metal electrodes are generally not as electrically stable as other kinds of electrodes used in monitoring applications.

17.5.6 Electrode Polarization

Metal electrodes present a complex electrochemical interface to solution that often does not allow electric charge to traverse with equal ease in both current directions. For example, in some metal-electrolyte systems, such as iron-saline solutions, there is a greater tendency for iron to oxidize than to reduce.

Common experience reveals that iron easily oxidizes (rusts) but resists reversal back to the iron base metal. In electrochemistry, this process is known as *electrode polarization.* A result of polarization is higher electrode resistance to current flow in one direction versus the other. With polarization, the electrode half-cell potential value also tends to vary from table values and depends on the direction and magnitude of electrode current flow. It is desirable for electrodes to be electrochemically reversible since this prevents the process of polarization. Electrode polarization is a problem in biopotential measurements because it is associated with electric instability. It gives rise to offset potentials between electrodes, electrode noise, and high resistance.

Electrodes made from electrically conductive metals such as pure silver, copper, tin, or aluminum are not electrochemically reversible and so do not provide the best or lowest-resistance pathway to electrolytic solutions or to tissue. These metals are electrochemically reactive (corrode) in (physiologic) electrolytes, even with no current flow.

The best bioelectric interfaces are combinations of metals and their metallic salts, usually chlorides. The metal salt is used as a coating on the base metal and acts as an intermediary in the electrode-electrolyte processes. Silver in combination with a chloride coating is the most widely used biopotential recording electrode. Silver chemically reacts in chloride-bearing fluids such as saline, skin sweat, and body fluids containing potassium and sodium chloride. After a few hours of immersion, a silver electrode will become coated with a thin layer of silver chlorides. This occurs by the spontaneous reaction:

$$\mathrm{Ag} + \mathrm{NaCl} \rightarrow \mathrm{AgCl} + \mathrm{Na}^+$$

The outer layer of silver chloride is highly reversible with chloride ions in physiologic electrolytes, and the silver is highly reversible with its salts. Thus current can reversibly flow across the electrode by a two-stage chemical reaction:

$$\mathrm{Ag} \leftrightarrow \mathrm{Ag}^+ + e^-$$

$$\mathrm{Ag}^+ + \mathrm{Cl}^- \leftrightarrow \mathrm{AgCl}$$

The chloride ions carry net negative charges in chloride-bearing electrolytes, and silver ions carry a net positive charge. This two-stage process of electric charge transfer greatly lowers electrode resistance to solution, reduces electrode polarization to near zero, improves electrode offset potential stability, and reduces electrochemical noise.

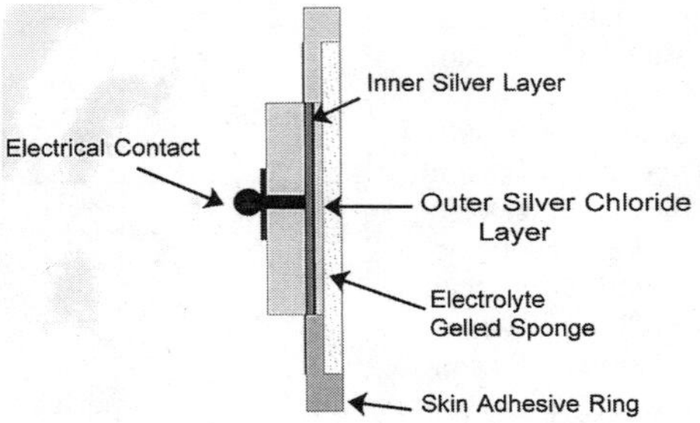

FIGURE 17.23 Silver–silver chloride electrode construction.

To conserve silver metal, commercial silver–silver chloride electrodes, sold for clinical and other high-volume biopotential monitoring, consist of thin layers of silver deposited over an underlying steel electrode. A thin layer of silver chloride is then formed electrochemically on the silver surface by the manufacturer. It may also be formed during recording by a chemical reaction with an applied gel containing high concentrations of potassium chloride. Usually, a thin sponge pad holds a gel electrolyte in contact with the silver, as shown in Fig. 17.23. It provides a suitable wick for electrolyte contact with the skin. Thus coupling of the silver chloride with the skin occurs through the electrolyte gel rather than by direct contact.

Thin-layer silver electrodes are usually designed for single-use application, not chronic or continuous monitoring applications, since the silver surface and chloride layer are easily scratched. When the underlying silver base metal is exposed, the silver electrode becomes polarizable, degrading its performance. An example of a single-use silver–silver chloride electrode can be seen in Fig. 17.24*a*. The pure-silver electrodes seen in Fig. 17.24*b* are sometimes used in EEG recording. These electrodes develop a chloride layer with the application of a salty gel during their use and can be cleaned afterwards for reuse.

The most rugged and reusable electrodes are made of a matrix of fused silver and silver chloride powders. These composite electrodes form a hard and durable pellet. They are relatively expensive but offer superior performance. A photograph of this type is shown in Fig. 17.24*c*. If the surface of this composite electrode is scratched or damaged, it can be resurfaced with abrasion to provide a renewed composition of silver–silver chloride. Furthermore, these reusable electrodes exhibit low polarization, low electrical noise, and low electrical impedance.

17.5.7 Stimulating Bioelectrodes

Stimulating electrodes pass significant currents to biological fluids for evoking action potential events and for other purposes. They must satisfy a different set of performance requirements than those for biopotential monitoring. For stimulation, silver–silver chloride electrodes are not usually the best choice, particularly when stimulating current pulses are not symmetrical in polarity through the electrode. High-current monophasic pulses, for example, can cause significant amounts of silver to be lost from the anode, and silver or other metal ions in solution will precipitate on the cathode. This leads to an asymmetry in chemical composition of the electrodes, differential impedance, and the usually undesirable introduction of silver metal into biological media, which may have toxic consequences.

To avoid injection of metal ions into biological tissues, electrodes using the noble metals are the best choice. Although they are expensive, their resistance to corrosion and good biocompatibility often outweigh the expense. Like all metal-electrolyte systems, their electric impedance and ability to transfer charge to solution depends on the current frequency. At higher frequencies, the high capacitive nature of noble electrodes in solution permits good charge transfer.

Stainless steel and certain other noncorrodable alloys are often used for cost reasons when the electrode surface area is relatively large. Also, graphite powder dispersed in a polymer binder has been used as a low-cost, large-area electrode for such applications as skeletal muscle stimulation.

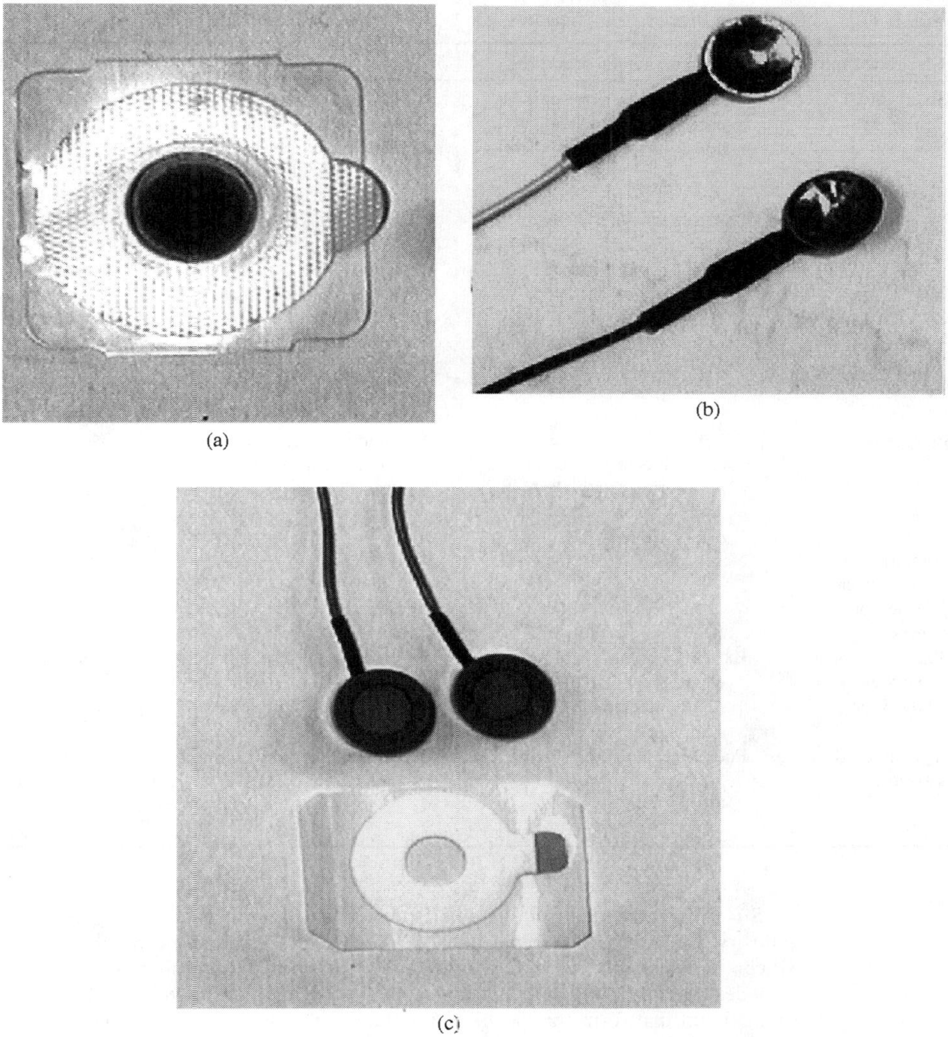

(a) (b) (c)

FIGURE 17.24 (*a*) A conventional thin-film disposable ECG silver chloride electrode from 3M Corp. (*b*) A pure-silver EEG electrode from Teca Corp. (*c*) A pressed-pellet silver–silver chloride electrode from IVM Corp.

Although stainless steel and graphite electrodes generally have high impedance per unit area, this becomes less of a concern when electrodes have large (several square centimeter) surface areas because the overall impedance is low. Tungsten is not a noble metal, but it is minimally reactive in physiologic fluids when current densities are low and the stimulating current is biphasic. Tungsten is often used for penetrating microelectrodes since it has a high stiffness and good mechanical qualities.

The reversible electrode tin–tin chloride is available commercially at about half the cost of silver-based electrodes. Tin electrodes are approved in the United States for EEG monitoring applications; however, they seem not as durable as silver electrodes and are not as widely used.

TABLE 17.4 Comparison Bioelectrode Properties

Electrode type	For biopotential recording	For bioelectrical stimulation	For cellular microelectrodes
Silver–silver chloride	Excellent for general-purpose use; low electrochemical noise, low drift, low offset potentials	Moderately useful with biphasic, low-level stimulation; loss of silver chloride on anode with monophasic stimulation	Works well for placement inside glass barrel liquid-coupled microelectrodes; silver itself lacks sufficient stiffness for cell penetration
Platinum, gold, tungsten	Can be somewhat noisy at low frequencies; often exhibit drifting electrode potentials	Excellent for long-term stimulation, implantable electrodes, either monophasic or biphasic current pulses	Mechanical and physical qualities are useful and often outweigh problems of higher noise; tungsten has high stiffness for cellular penetration
Iridium–iridium oxide	Noisy	Good; a particularly high charge-transfer capability per unit area; expensive	Excellent for microstimulation
Stainless steel	Suitable mostly for noncritical applications and for monitoring relatively high amplitude bioelectric signals such as ECG	Suitable for short-term, noncritical applications; low cost	Moderately useful for stimulation; less expensive
Graphite, pyrolytic carbon loaded polymers	Noisy	Good	Poor mechanical qualities
Glass micropipette, electrolyte-filled	Exhibits low electrical noise when silver–silver chloride contact made to filling solution	Often used, but accompanied by small amounts of ionophoresis of filling electrolyte	Good for low-noise intracellular recordings
Capacitive dielectric	Poor, electrically noisy	Moderately useful in some applications	Little used due to the difficulty of making thin dielectrics that maintain a high resistivity in saline

Silicon is convenient for microelectrode applications since it is widely used in the semiconductor industry and can be fabricated into complex mechanical and electrical structures. Unfortunately, silicon is chemically reactive in body fluids and corrodes, forming an insulating glass (SiO_2) that impedes current flow. For this reason, silicon is usually coated with thin layers of other metals such as platinum that form the electrical interface with tissue. Table 17.4 compares different types of bioelectrodes for different applications.

17.5.8 Electrode-Skin Interface

Perhaps the greatest problems encountered in biopotential recording are noise, motion artifacts, and a general instability of the biopotential monitored signal. These problems are most severe when electrodes are placed on dry skin. They are less of an issue with invasive, implanted, or intracellular electrodes.

Placement of electrodes directly on the skin will show a large electrical resistance, often in the megohm (10^6) region. This is due mostly to the surface layer of dead and dehydrated skin cells on the epidermis, known as the *corneum.* Below the epidermis, the next sublayer is the living part of the skin, known as the *dermis.* It exhibits a complex electrical behavior having a resistance, capacitance, and potential generators similar in some respects to that of electrodes.

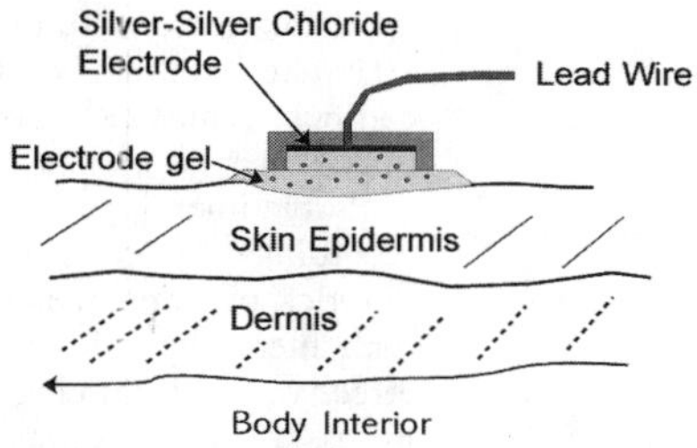

FIGURE 17.25 Electrode-skin coupling using a gel electrolyte.

More stable bioelectric recordings from the skin are achieved with application of a liquid or gel electrolyte between the skin and electrode, as shown in Fig. 17.25. This bridges the electrode surface to the skin. The gel is an aqueous chloride-bearing solution that will hydrate the skin, reduce the impedance of the corneum, and produce a more uniform medium for charge transfer. Skin-electrode impedance can drop below 5 kΩ across the hydrated skin and abraded corneum. High impedances in the range of 20 to 100 kΩ are not uncommon if the skin is not adequately prepared, i.e., abraded and clean.

Gel electrolytes provide a convenient method of coupling between the silver–silver chloride electrodes to the skin surface. The gel also helps protect the thin silver chloride layer on the electrode from abrasion with the skin surface as well as to hydrate the skin. After several minutes, gel electrolytes saturate the outer resistive layers of dead skin and form a low-impedance pathway to the body interior.

Commercial electrode gels usually use relatively high concentrations of potassium or sodium chlorides at a neutral pH. Since these concentration levels can irritate the skin, there are different types of gels are on the market offering trade-offs of low resistance versus gentleness to the skin.

The living nature of the dermis also creates an electrochemical potential called a *skin battery*.[7] This battery is associated with the function of the sweat glands. It is very pressure-sensitive. Small movements or forces on the electrodes cause transient millivolt-order changes in the skin-battery magnitude.[8] These changes appear as motion artifacts and wandering baselines on biopotential traces.

The skin potential due to electrolytes placed on the skin E_{skin} and the intrinsic battery potential generated by the sweat glands E_{sweat} are usually modeled as two batteries in parallel with internal resistances, as shown in Fig. 17.26.

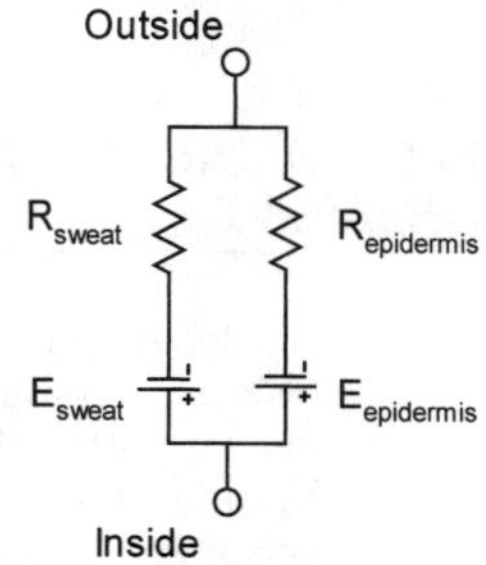

FIGURE 17.26 Illustration of the electric nature of the skin.

The electric sources in the skin vary with time, temperature, hydration, and pressure. For example, a buildup of sweat under the electrodes will cause slow changes in the monitored dc potential. Pressure on the electrodes causes deformation of the skin and consequent resistance change that shunts the skin battery and changes its magnitude. These effects are particularly frustrating to the recording of low-frequency biopotentials since they produce a wandering baseline.

The skin battery is partially modulated by sympathetic nervous system activity. This interesting effect is known as the *skin potential response* (SPR). It arises due to a varying electric shunt resistance across the skin that occurs with both capillary vasodilatation and changes in sweat duct activity that are under autonomic nervous control.[9,10] Fast-changing emotional states such as excitement or alarm can produce millivolt-order changes in the skin potential that are then superimposed on the monitored biopotential. This effect is similar to the better-known *galvanic skin response* (GSR), which directly measures the changes in skin resistance as a function of emotional stress. GSR recording is sometimes used in forensic applications, such as lie detectors.

17.5.9 Skin Preparation

The key to fast electrode equilibration with the skin is removing or abrogating the skin corneum. This is usually performed by skin abrasion. Fine-grit abrasive materials are sometimes incorporated into commercial electric coupling creams so that outer layers of dead skin will be removed on application. Conventional 320- to 400-grit sandpaper applied with light pressure using about 6 to 10 strokes will barely redden the skin but dramatically reduce the skin resistance.[11–13] Alcohol wipes can help remove skin oils in preparation for gel application but are not used alone to reduce skin impedance.

Sometimes it is important to reduce skin resistance to the lowest possible value, as in EEG recordings, to detect microvolt-level signals. In this case, the skin barrier is punctured by using a pinprick or sometimes drilling with small abrasive tools.[14] The abraded region or puncture fills up with interstitial fluids and immediately forms a low-resistance pathway to the body interior. Although effective for improving bioelectrode performance, it is often not popular (for obvious reasons) with most subjects or patients.

Commercial skin electrodes are usually packaged in metal foil wrappers that are impervious to moisture. These wrappers help prevent evaporation of the gel and should not be removed until the electrodes are to be used. The electrodes should always be examined for adequate gel prior to application.

Although electrode gels are needed for good skin contact, hydrating electrolytes produce another source of electric potential, known as a *skin diffusion potential*, in addition to the skin battery. This diffusion potential arises because skin membranes are semipermeable to ions. The magnitude of this skin potential varies as a function of the salt composition in the electrolyte.

Thus there are multiple electric potentials in addition to the biopotential detected by skin electrodes. These electrode and skin potentials drift with skin temperature, pressure, equilibration time, and emotional excitement. Whether these potentials are significant in recording depends on the frequency of the biopotential signal. The diagnostic ECG, for example, requires the recording of relatively low-frequency (0.05 Hz) signals and is vulnerable to these effects. The higher-frequency EMG recordings can largely filter these out.

17.6 ELECTRICAL INTERFERENCE PROBLEMS IN BIOPOTENTIAL MEASUREMENT

Bioelectric potentials are relatively low-level electric signals and must be substantially amplified before a recording can be easily displayed. Modern integrated-circuit amplifiers are well suited to this task since they have high gains and contribute little noise to the biopotential measurement. Unfortunately, skin-monitored biopotentials typically show a complexity of interfering electric signals that may overwhelm the desired signals. This is so because the size and bulk of animals and humans contribute to a type of antenna effect for environmental electric fields.

17.6.1 Power-Line Interference

In modern society we are immersed in a complex electromagnetic environment originating from power wiring in buildings, radio transmitters of various sorts, radiofrequency-emitting appliances such as computers, and natural sources such as atmospheric electricity. Within the home, environmental electric fields will typically induce in the human body a few tens of millivolts with respect to a ground reference. These fields can be much larger, as high as a few volts, if someone is using a cell phone or is located within a foot of power-line wiring or near a radio station. These induced voltages are thousands of times larger than the bioelectric signals from the heart as recorded on the chest surface.

Electric coupling of the body to environmental sources is usually due to proximity capacitance and, to a lesser extent, inductive (magnetic) fields. Capacitive coupling between two objects results from the electric field between them. The space or air gap separating the objects acts as a dielectric. The coupling can be simply modeled as a parallel-plate capacitor, which can be determined by

$$C = \frac{\epsilon A}{d}$$

where $\epsilon = \epsilon_0 \epsilon_r$, where the individual ϵ dielectric constants are for free space and air, respectively
A = area of mutual conductor plate interception
d = separation distance

For example, let A be equal to the surface area of the human body intercepted by a power line and d be a distance less than 1 m from it. In this situation, only fractions of a picofarad (10^{-12} F) coupling results. However, due to the closer proximity to the floor, a person usually will have greater coupling to ground than to the power line, perhaps by a factor of 100 or more. As a result, a voltage divider is created, and the body acquires a potential, known as a *floating potential* V_{float}. Its magnitude, as measured by a voltmeter with a ground reference, is determined by the ratio of impedances of the body to the power line Z_{line}, where $Z_{\text{line}} = 2\pi f C_{\text{line}}$, and ground Z_{ground}, where $Z_{\text{ground}} = 2\pi f C_{\text{ground}}$. This arrangement is shown in Fig. 17.27 and ignores any ground resistance since it is usually negligible. The body itself is a volume conductor with such a low relative resistance that it is not significant to this analysis.

In this case, a subject's floating potential is given by the impedance divider relationship:

$$V_{\text{float}} = V_{\text{line}} \frac{Z_{\text{ground}}}{Z_{\text{line}} + Z_{\text{ground}}}$$

The floating potential can be a relatively large induced potential. Take, for example, a person sitting in a metal chair. If the person's proximity capacitance to the power line is assumed to be 1 pF, the ground impedance Z_{ground} is 10^7 Ω, and the line voltage is 120 V_{rms} (338 $V_{\text{pk-pk}}$) at 60 Hz (in the USA), then the calculated floating potential is

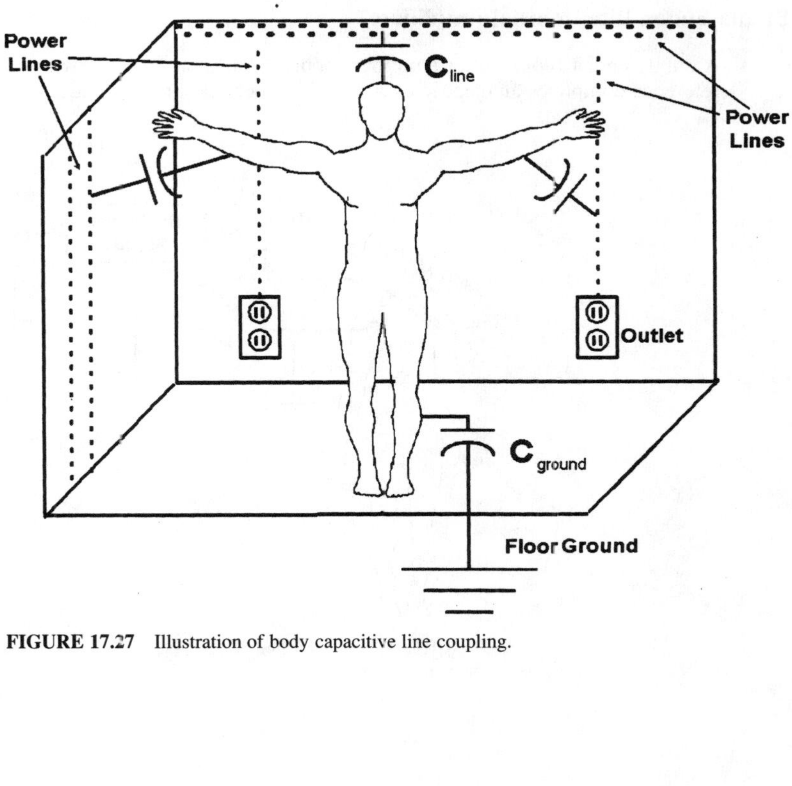

FIGURE 17.27 Illustration of body capacitive line coupling.

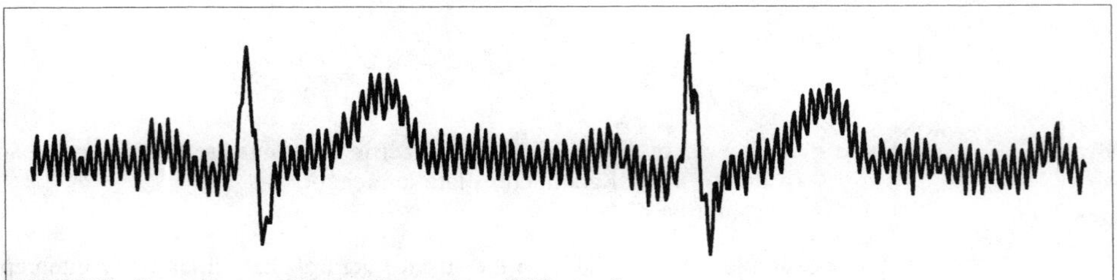

FIGURE 17.28 Power-line coupling interference in an ECG waveform.

$$Z_{\text{line}} = \frac{1}{2\pi fC} = \frac{1}{2\pi(60)(10^{-12})} = 2.65 \times 10^9 \Omega$$

$$V_{\text{float}} = 338(10^7/2.65 \times 10^9 + 10^7) = 1.27\ \text{V}_{\text{pk-pk}}$$

This V_{float} is present uniformly over the body and will sum with a skin biopotential V_{bio}. The output of the amplifier V_{out} is the sum of all potentials at its input. Thus the floating potential is summed with the biopotential:

$$V_{\text{out}} = V_{\text{float}} + V_{\text{bio}}$$

In the preceding example, the line-induced interference is more than 10,000 times greater than could be tolerated in a recording. It is an artifact with a sinusoidal waveform characteristic. Figure 17.28 provides an example of this type of power-line interference in an ECG recording.

17.6.2 Single-Ended Biopotential Amplifiers

A single-ended biopotential amplifier monitors an input voltage with respect to its reference. Its reference completes an electric circuit to the object of study. Because they are simple, single-ended

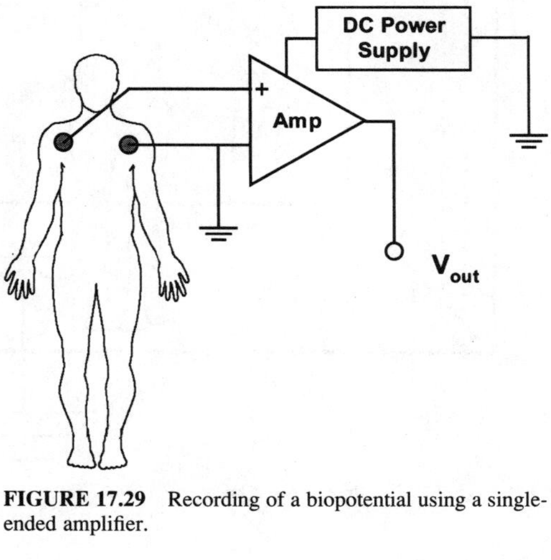

FIGURE 17.29 Recording of a biopotential using a single-ended amplifier.

amplifiers, they are sometimes used in biopotential monitoring. They need only two electrodes, a single monitoring electrode and a reference electrode. This kind of amplifier should not to be confused with a differential amplifier, where there are two inputs that subtract from one another. Figure 17.29 shows a schematic of a single-ended amplifier where its reference is connected to both the subject and ground.

Environmental line-frequency interference coupled to the subject is a major challenge for this amplifier configuration. One approach to reduce interference is to ground the body with a reference electrode to an earth ground. In principle, grounding a biological object will reduce V_{float} to zero if Z_{ground} is zero. With this idealized configuration, a biopotential V_{bio} would be amplified without concern for environmental noise. In practice, bioelectrodes have significant electrical impedances. This means that the divider ratio defined by Z_{line} and Z_{ground} produces a V_{float} value that is not reduced to zero by the grounding electrode. Therefore, to achieve quality single-ended amplifier recordings, it is essential to minimize coupling capacitance and ensure low-impedance reference electrodes.

Low line-coupling capacitance reduces noise and can often be achieved with small biological specimens by (1) removing the work area from power-wire proximity and/or (2) shielding the specimen with a metal enclosure or copper mesh connected to ground. Capacitive coupling does not occur through grounded conductive enclosures unless line-frequency power wires to equipment are allowed to enter the enclosure.

It is more difficult to shield the human body than biological specimens because of the body's bulk and greater surface area. Skin-electrode resistance is also greater than that of invasive electrodes used in biological specimens, and this causes higher floating potentials on the human body. Except under certain conditions discussed later, these circumstances can create frustration in using single-ended amplifiers for human recording.

For example, if a subject's electrode-skin impedance $Z_{\text{electrode}}$ is 20 kΩ, this value is low enough that it dominates over his or her ground capacitance such that essentially $Z_{\text{ground}} = Z_{\text{electrode}}$. Assuming the same line-coupling capacitance as before (1 pF) and using the voltage-divider relation given earlier, the calculated floating potential V_{float} is now much lower:

$$V_{\text{float}} = 338 \frac{20\ \text{k}\Omega}{20\ \text{k}\Omega + 2.65 \times 10^9 \Omega}$$

$$= 2.5\ \text{mV at 60 Hz}$$

Under these electric capacitive coupling circumstances (which are rather severe), this example shows that a single-ended amplifier referenced to a grounded body experiences power-main interference 250 percent larger than a 1-mV ECG. The ECG will appear to be undesirably riding on a large oscillatory artifact.

17.6.3 Biotelemetry

Isolating the amplifier and recording system from earth ground can significantly reduce problems with line-noise interference. Remember, a subject's floating potential appears with respect to an earth-ground reference since it originates from capacitive coupling to power lines. Power lines from the electric service are referenced to earth ground in the electric distribution system. Thus, isolating the biopotential amplifier and its reference from earth ground would eliminate the appearance of floating potentials across the amplifier inputs. An isolated system, however, still must have an amplifier reference on the subject's body in order to complete the electric circuit.

Biotelemetry is one way to provide this kind of isolation. In conjunction with battery-powered amplifiers, biotelemetry can be a very effective way to reduce electromagnetic interference in biopotential recordings since it can be isolated from ground reference. This is depicted in Fig. 17.30. Here, the isolated system reference is shown by the triangle symbol rather than the segmented lines of earth ground.

An isolated amplifier is referenced only to the body under measurement and so in principle does

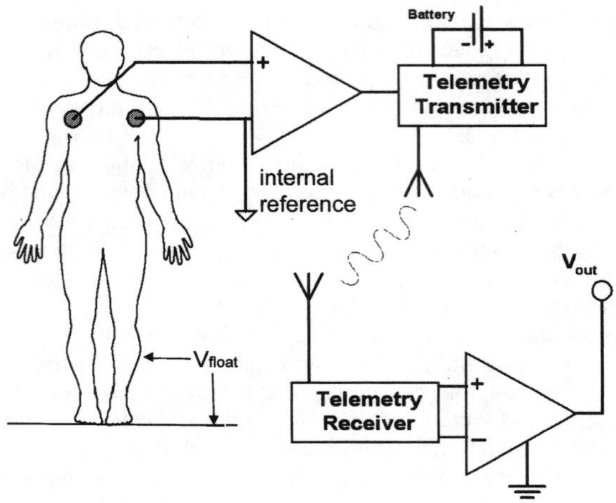

FIGURE 17.30 Biotelemetry recording system.

not detect any of the V_{float}. It is not that V_{float} is zero; it is just that it does not appear at the amplifier input and so is not in the equations. Thus, for an amplifier gain of $A = 1$,

$$V_{\text{out}} = V_{\text{bio}}$$

Biotelemetry also has the significant advantage of allowing the use of single-ended amplifiers that only require two electrodes for measurement. Three electrodes are required by differential amplifier recording techniques. Fewer electrodes simplify measurement and require less labor in electrode application and maintenance.

In practice, biopotential recording by battery-powered biotelemetry works quite well in interference rejection. The remaining sources of signal interference have to do with magnetically coupled potentials induced into the lead wires, and these problems are specific to certain environments, such as those in close proximity to high-current-carrying wires or high-power appliances.

Miniature telemetry systems are becoming more widely used for biopotential recording applications. However, the added cost and complexity are often a disadvantage. Their greatest utility has so far been in patient ambulatory applications in the clinic, such as ECG monitoring. However, small wireless systems for multichannel monitoring the EMG in biomechanics applications are often useful, as are microsized systems for neural recordings in brain research on animals. In this latter case, the wireless nature of the link can essentially eliminate power-line interferences that are a problem in recording the microvolt-level signals from EEG.

Battery-powered amplifiers are isolated and convenient for portable applications but are not the same as biotelemetry. Unfortunately, connection of a battery-powered amplifier to a recording system without any other isolation in the connecting cables unavoidably interconnects the amplifier reference to the recording instrument reference, the latter of which itself is usually connected to a power-line ground. This abrogates the advantage of battery isolation on the power supply. Solutions can consist of using optical isolators on the signal output leads or using commercial isolation amplifiers.

17.6.4 Differential Biopotential Amplifiers

The use of differential amplifiers is common in biopotential measurements because of a greater ability to reject environmental interference compared with ground-referenced single-ended amplifiers. Differential amplifiers subtract the electric potential present at one place on the body from that of another. Both potentials are measured with respect to a third body location that serves as a common point of reference.

Differential amplifiers are useful because biopotentials generated within the body vary over the body surface, but line-coupled noise does not. For example, subtraction of the heart electric potential at two points on the chest surface will produce a resulting potential since the local biopotential amplitudes and wave shapes at each electrode are different. Environmental electric fields from the power line are more remote and couple such that they are present uniformly over the body. This is partly due to the distributed nature of capacitive coupling. It is also because the low 50- to 60-Hz line frequencies have electric field wavelengths so long (hundreds of meters) that a person's body can be considered to be, in some sense, an antenna in the uniform near field of an electric field source.

The induced body potential V_{float} is present at both inputs of a difference amplifier, and as used here, it is also known as the common-mode potential V_{cm}. This is so because it is common (equal) in amplitude and phase at each of the two amplifier inputs. Thus, for our differential amplifier,

$$V_{\text{cm}} = V_{\text{float}}$$

The connection of a differential amplifier always requires three electrodes since a common reference point for the amplifier is needed. This point can be anywhere on the body, but by convention in ECG measurements, for example, the right leg is used as this reference. Figure 17.31 illustrates this situation.

Differential amplifiers of gain A perform the following operation:

$$V_{\text{out}} = A(V_1 - V_2)$$

where V_1 and V_2 are the signal levels on each of the noninverting and inverting inputs of the amplifier, respectively, with respect to a reference. In practice, differential amplifiers very closely approach this ideal. Modern differential amplifiers can have common-mode rejection ratios of 120 dB or better, meaning that they perform the subtraction to better than one part per million.

The grounding electrode, as in the preceding single-ended case, also reduces the common-mode potential on the body surface. Even if the ground were not effective due to a large electrode resis-

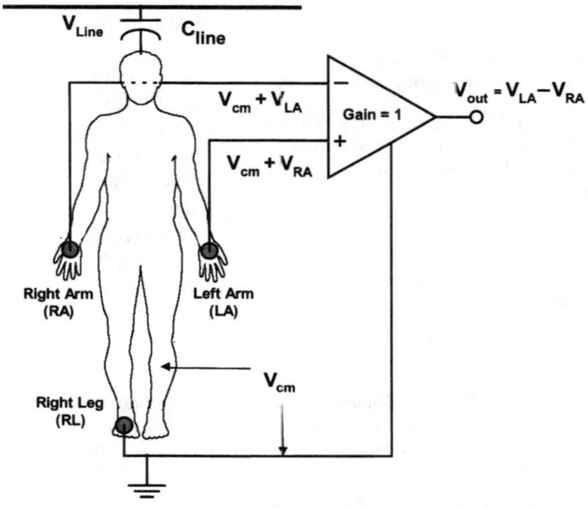

FIGURE 17.31 Diagram of the placement of differential electrodes on a subject's body. The capacitively coupled potential is common mode to the amplifier and sums with the biopotential.

tance, within a fairly large range of common-mode levels, the differential amplifier would be capable of near-perfect cancellation of the common-mode signal.

In ECG monitoring, for example, the common-mode interference on the body sums with the arm-lead potentials V_{RA} and V_{LA}. If we calculate V_{out} under these circumstances, we get

$$V_{out} = A[(V_{cm} + V_{LA}) - (V_{cm} + V_{RA})] = A(V_{LA} - V_{RA})$$
$$= AV_{ECG}$$

The quantity $V_{LA} - V_{RA}$ is the definition of ECG lead I.

In practice, this interference cancellation process works fairly well; however, the assumption that the power-line-induced signal is common mode (uniform) over the body does not hold perfectly in all situations. Slight differences in its phase or amplitude over the subject's body when in close proximity to some electric field source or unbalanced electrode impedances can cause this cancellation process to be less than perfect. Some line noise may still pass through into the recording. Usually, improvements in skin electrode preparation can remedy this problem.

17.6.5 Bandpass Filtering

An understanding of the frequency content of bioelectric signals can be helpful in maximizing the biopotential recording quality. Cell membrane sodium channel activation is probably the fastest of bioelectric events in living things. In heart cells, for example, these occur over time scales on the order of 30 μs. In practice, most bioelectric recordings of fast body surface potentials such as the EMG and cell neural spikes show no significant difference in wave shape with amplifier bandwidths at about 5 kHz. Wider bandwidths than this tend to admit disproportionately more noise than signal.

Both high-pass and low-pass filters are generally needed in biopotential recordings. Setting the filter cutoffs implies knowledge of the biopotential frequency spectrum of interest. In general, it is desirable to narrow amplifier bandwidths as much as possible to reduce power-line noise, radio-station interference, digital computer noise, and electric fields in the tens of kilohertz range such as those associated with the cathode-ray-tube monitors.

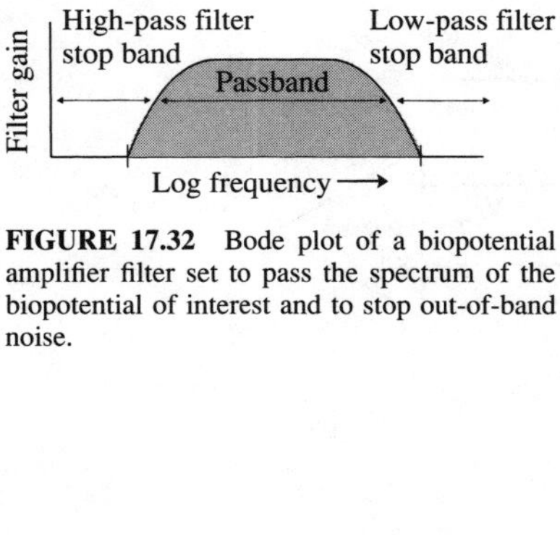

FIGURE 17.32 Bode plot of a biopotential amplifier filter set to pass the spectrum of the biopotential of interest and to stop out-of-band noise.

Figure 17.32 shows a Bode plot of the desirable situation. The roll-off of the filters brackets the frequency spectrum of interest in the biopotential. This produces the greatest rejection of out-of-band noise. Simple single-pole filters are usually employed that give a relatively gentle cutoff with increasing frequency. Low-pass filtering the ECG at 100 Hz, for example, can greatly reduce EMG artifacts whose frequencies extend up into the kilohertz region. High-pass filtering at 0.5 or 1 Hz helps reject slow baseline wander from electrode potentials associated with offset potential and skin hydration drifts. Table 17.5 gives a rule of thumb for the spectral content of some common biopotentials.

TABLE 17.5 Characteristics of Some Common Biopotentials

Biopotential	Typical amplitude	Frequency spectrum
Electrocardiogram (heart)	1 mV	0.05–100 Hz (diagnostic)
Electroencephalogram (brain)	1–100 μV	0.1–50 Hz
Electromyogram (muscle)	10 μV–1 mV	10 Hz–5 kHz

It is also tempting to use low-pass filters set at 30 Hz and below to reduce the level of 50- or 60-Hz line-frequency interference, but the monitored biopotential waveform shape may change. The ECG is probably the best example where waveform shape is fundamental to its utility as a medical diagnostic. Subtle changes in the shape of the QRS or ST segment, for example, can be indicative of myocardial pathology or infarction. Any waveform filtering or processing operation that affects the waveform shape can be misleading to a physician trained to interpret subtle ECG waveform changes as clues to pathology.

Because the bioelectric wave shape is so important in diagnosis, ECG machines are required in the United States to have filtering of 0.05 Hz high pass and 100 Hz low pass at the −3-dB breakpoints. Some commercial machines have 150-Hz bandpass capability.

If exact waveform fidelity does not need to be maintained, as is often the case in the EMG and even occasionally in the EEG, bandwidths can be narrowed with concomitant improvements in recording noise rejection. Even with the ECG, not all monitoring circumstances require high-fidelity waveform reproduction. In some clinical ECG machines, a front-panel switch can change the waveform filtering to a nondiagnostic monitor mode where the bandpass is 1 to 30 Hz. These filter breakpoints substantially reduce problems with electrode motion artifact at the low-frequency end and reduce EMG skeletal muscle signals from the electrodes at the high-frequency end. The result is more ECG waveform distortion but with the advantage of less false heart-rate and loss-of-signal alarms for clinical staff and a more stable and easier-to-read waveform display. The effect of 3-Hz highpass filtering on ECG baseline wander can be seen in Fig. 17.33. It effectively eliminates the large drifts of the waveform trace.

Sharp cutoff filters should be avoided in biopotential measurements where the bioelectric waveform shape is of interest. Filtering can greatly distort waveforms where waveform frequencies are near the filter breakpoints. Phase and amplitude distortions are more severe with higher-order sharp-cutoff filters. Filters such as the Elliptic and the Tchebyscheff exhibit drastic phase distortion that can seriously distort bioelectric waveforms. Worse still for biopotential measurements, these filters have a tendency to "ring" or overshoot in an oscillatory way with transient events. The result can be addition of features in the bioelectric waveform that are not really present in the raw signal, time delays of parts of the waveforms, and inversion of phase of the waveform peaks. Figure 17.34 shows that the sharp cutoff of a fifth-order elliptical filter applied to an ECG waveform produces a dramatically distorted waveform shape.

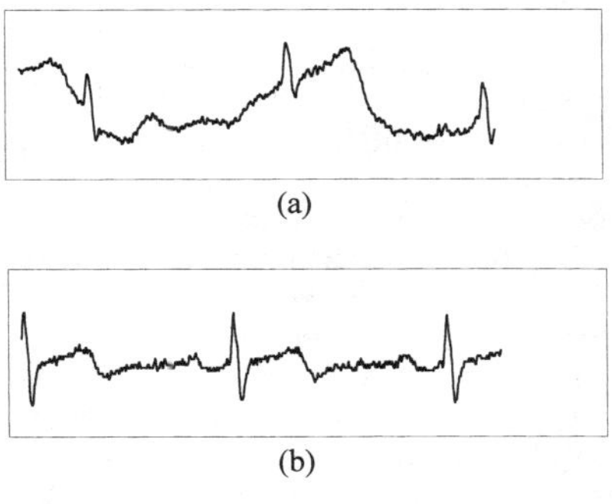

FIGURE 17.33 (*a*) Electrocardiogram showing muscle noise and baseline wander. (*b*) The effect of introducing an analog 3-Hz (−3-dB) high-pass filter into the signal path. There is a marked reduction in baseline wander.

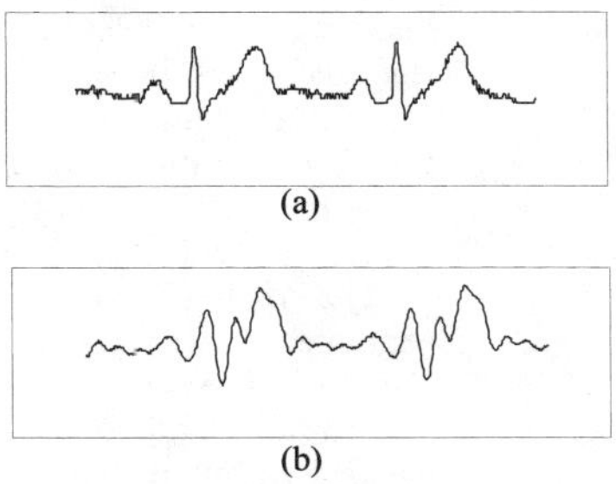

FIGURE 17.34 (*a*) Electrocardiogram with slight muscle tremor. (*b*) Electrocardiogram with a strong fifth-order analog Elliptic filter introduced into the signal path. Considerable amplitude and phase distortion of the waveform in (*b*) is evident, obscuring most of the ECG features and introducing spurious waveform artifacts.

17.6.6 Electrode Lead Wires

Some care must be taken with lead wires to biopotential recording electrodes. Commercial bioelectrodes are often sold with 100-cm lengths of an insulated but unshielded signal-carrying wire. Unshielded wires increase capacitive coupling, in proportion to their length, to environmental sources of electrical interference. In electrically quiet environments or with low electrode impedances to the skin, these may work satisfactorily without coupling excessive noise. Keeping the electrode wires close to grounded objects or near the body can be helpful in reducing line-frequency noise pickup.

In more demanding applications, the electrode lead wires should be shielded, preferably by purchasing the type that has a fine coaxial braided copper wire around the central signal-carrying lead. The shield should be connected to the amplifier reference but not connected to the electrode signal wire itself since this would abrogate its shielding effect.

Interference by ac magnetic induction into the electrode lead wires is a possible but often not a serious problem in most recording environments. The exceptions are where nearby machinery draws large current (many amperes) or where electrode lead wires pass close to current-carrying wiring. In these situations, the amplitude of the induced electric field is defined by Ampere's law and is roughly proportional to the open area defined by the perimeter of the loop formed by two electrode wires. Twisting the electrode lead wires together can reduce this area to near zero, and this approach can be an effective method of reducing this source of electromagnetic noise.

17.6.7 Biopotential Amplifier Characteristics

Modern integrated-circuit amplifiers, and even most inexpensive op-amps, are capable of giving good-quality recordings of millivolt-level biopotentials such as from the heart and muscles. Higher-quality integrated-circuit instrumentation amplifier chips are more suited to measuring microvolt-level signals such as associated with the electroneurography (ENG), EEG, and intracortical events.

Some of the desirable characteristics of bioelectric amplifiers are

- Gains of 10^3 to 10^4
- Amplifier broad-spectrum noise less than 20 nV/Hz
- Input impedance greater than $10^8\ \Omega$
- Common-mode rejection ratio greater than 100 dB

Most of these characteristics are easily found in generic amplifier chips. The aspects of circuit design that separate biopotential amplifiers from other applications mostly concern the ease of selecting the right bandpass filtering.

Instrumentation amplifiers are differential voltage amplifiers that combine several operational-type stages within a single package. Their use can simplify the circuitry needed for biopotential amplification because they combine low-noise, high-impedance buffer input stages and high-quality differential amplification stages.

The Analog Devices, Inc. (Norwood, Mass.) AD624 instrumentation amplifier is used here as an example, but there are other chips by other manufacturers that would serve as well. This chip has high-impedance ($>10^9\ \Omega$) input buffers that work well with even high-impedance bioelectrodes. The amplifier gain is easily selected from several standard gains by interconnection of pins on the package or, alternatively, by adjusting the value of a single external resistor.

Figure 17.35 shows its configuration, and it can be used to monitor many different biopotentials. Its gain is conveniently determined by adjusting R_G to select an amplifier gain A according to the relation $A = (40k/R_G) + 1$. Gains of 1 to 1000 can be easily selected; however, it is desirable to keep the gain of this part of the circuit less than about 50 since higher gains will tend to saturate the amplifier if electrode dc offset potentials become large.

The instrumentation amplifier is connected to a bandpass filter consisting of simple single-pole high-pass and low-pass filter. The component values on the filter stage are variable and selected for

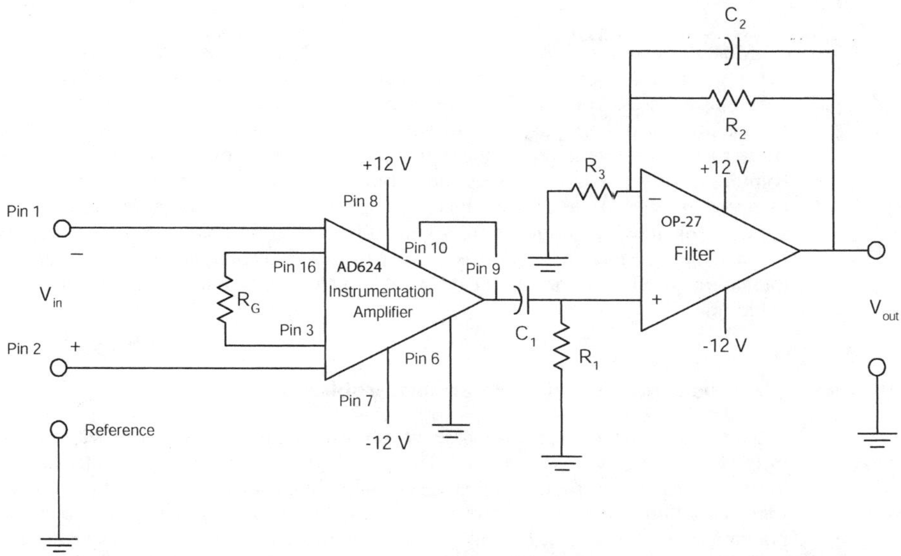

FIGURE 17.35 A simple biopotential amplifier design based on an integrated-circuit instrumentation amplifier.

frequency cutoffs suitable for a specific biopotential monitoring application. Generic but good-quality operational amplifiers such as the industry-standard OP-27 are used in the filter stage.

The −3-dB low-pass and high-pass cutoff frequencies of the filter, formed by R_2 and C_2 and R_1 and C_1, respectively, are given by $f_c = 1/(2\pi RC)$. Typical values for EMG recording, for example, using cutoffs approximately at 10 Hz high pass and 3 kHz low pass would be $R_1 = 10\ \text{k}\Omega$, $C_1 = 1.6\ \mu\text{F}$, $R_2 = 50\ \text{k}\Omega$, and $C_2 = 1$ nF. $R_3 = 10\ \text{k}\Omega$ for an in-band gain of 5. Overall, this circuit with an R_G of 800 Ω provides an in-band gain of 250, meaning that a 1-mV EMG signal will give a 0.25-V output. Good-quality electrodes are required for optimal performance.

Improvements in this simple design might include protection on the amplifier inputs by using series input lead resistors of 50 kΩ or more and capacitive input couplings. Zener diodes across the input leads offer further protection from defibrillator and static electricity by shorting transients to ground. Selectable gains and added filtering options would give the amplifier more flexibility in application.

17.6.8 Other Sources of Measurement Noise

In addition to power-line coupled noise, computers and cathode-ray-tube monitors can be a troublesome source of interference in many clinical and laboratory situations. Digital logic electronics often involve fast switching of large currents, particularly in the use of switching-type solid-state power supplies. These produce and can radiate harmonics that spread over a wide frequency spectrum. Proximity of the amplifier or the monitored subject to these devices can sometimes result in pulselike biopotential interference.

Cathode-ray-tube monitors internally generate large electric fields (tens of thousands of volts) at frequencies of 15.75 to 70 kHz needed to drive the horizontal beam deflection on the cathode-ray tube. These fields are strong and have a high frequency, so they would have the possibility of being a major problem in biopotential recording. However, since the 1970s, the radiated electric field emissions from these devices have been regulated by government agencies in most countries, and this considerably reduces the likelihood of their interference.

Modern designs of electronic equipment usually minimize radiofrequency interference (RFI) by the use of metal cases, internal metal shielding, and/or conductive nickel coatings on the insides of their plastic cases. This largely confines the electric fields to within the packaging. Even so, sensitive biopotential amplifiers placed on or very near operating microprocessor-based instrumentation can often suffer interference from these sources. Since capacitive coupling is usually the problem, the solution is often the simple separation of biopotential amplifiers (and the subject) from the proximity of operating digital electronics. Placing the offending equipment within a grounded metal box or surrounding it with grounded foils generally can stop capacitively induced fields. Shielding the bioamplifiers is also possible. It is usually of little consolation, however, that removing the instrumentation to an outdoor setting or powering the total system with batteries can be an effective solution.

17.6.9 Analog Versus Digital Ground Reference Problems

Other interference problems can arise when sensitive high-gain amplifiers are connected to computers through data-acquisition interfaces. These ports often have two differently labeled grounds, digital ground and analog ground (sometimes-called *signal ground*). Small currents arising from digital logic-switching transients present on the computer digital ground reference can pass through a biopotential amplifier circuit by way of its ground reference connection. Current flows between these two different grounds, digital and analog, can create millivolt levels of digital noise that in the biopotential amplifier create a high-frequency noise problem in the recording.

This problem is particularly severe with high-gain bioamplifiers, such as those used for microvolt-level EEG measurements. This problem is alleviated in most commercial instrumentation designs by internally isolating the analog signal ground from digital ground.

17.6.10 Ground Loops

Ground loops can create seemingly intractable problems with line-frequency (50 or 60 Hz) interference in low-level biopotential recordings. They often occur when biopotential amplifiers are connected to signal processing or recording systems such as filters, oscilloscopes, or computer-based data-acquisition systems. The root cause is often that the reference wire of signal cables interconnect the ground references of all the instruments. However, each instrument is also referenced to a power-line ground through its third-wire grounding pin on the power-line plug, as required by electrical safety codes.

Interference arises from the fact that all power grounds and signal-ground references are not equal. Line-powered instruments and appliances are only supposed to draw current through the hot and neutral wires of the line. However, the third-pin instrument power ground conducts small capacitive and resistive leakage currents from its power supply to wall-outlet ground. These ground leakage currents originate mostly in the instrument power-line transformer winding capacitance to ground and in resistive leakage from less than perfect wire insulation. Appliances having heavy motors, such as refrigerators, air conditioners, etc., that draw high currents tend to have greater leakage currents.

Currents flowing in the power-line ground cause voltage drops in the resistance of the building wiring ground. As a result, a voltmeter will often show several tens of millivolts between one power-outlet ground and another even in the same room. Since recording amplifier grounds are usually referenced to the instrument power-line ground, millivolt-level potential differences can create circulating currents between power-ground and instrument-ground references. By the same process, there can also arise current flows between various instruments through their individual ground references.

The result is that small line-frequency currents flow in loops between different instruments interconnected by multiple signal and ground references. Line-frequency voltage drops in the interconnecting reference path wiring can appear as a signal across the amplifier inputs.

The solutions are often varied and cut and dry for each application. Some typical solutions include

- Plugging all instruments into the same power bus feeding off a single power socket
- Using battery-powered equipment where possible or feasible
- Earth grounding the amplifier input stages
- Independent ground wires from the subject to the bioamplifier
- Identifying and replacing certain offending instruments that contribute disproportionately to induced noise
- Use of isolation modules on the amplifier
- Grounding the work tables, patient beds, or surrounding area

Within modern work environments, it is not infrequent for interference in biopotential recordings to appear from the following sources when they are near (perhaps a meter) the recording site:

- Computers, digital electronics, digital oscilloscopes
- Cathode-ray-tube displays
- Brush-type electric motors
- Fluorescent light fixtures
- Operating televisions, VCRs

Less frequent but also possible sources include

- Local radio stations, cell phone repeaters, amateur radio transmitters, military radars
- Arc welding equipment, medical diathermy, and HVAC systems
- Digital cell phones
- Institutional carrier-current control signals sent through power wiring to synchronize clocks and remotely control machinery

Not usually a problem are portable radios, CD players, stereos, and regular telephones.

17.6.11 Electrical Safety

Whenever a person is connected to an electrical device by a grounded conductive pathway that is low resistance, such as through biopotential electrodes, there is a concern about electrical safety. Electric voltages that ordinarily would be harmless to casual skin contact can become dangerous or even lethal if someone happens to be well grounded. Wetted skin can provide low resistance, and from this stems the old adage about the inadvisability of standing in pools of water when around electric appliances.

Electric shock hazard is particularly a concern in clinical biopotential monitoring where the patient may be in a vulnerable state of health. Bioelectrodes are desirably low-resistance connections to the body for amplifier performance reasons. As discussed earlier, they similarly can offer a low-resistance pathway to ground for electric fault currents.

These fault currents could arise from many sources, including within the amplifier itself through some internal failure of its insulation or from a patient coming into contact with defective electric devices in the immediate environment. These might include such things as a table lamp, a reclining bed controller, electric hand appliances (e.g., hair dryers), and computers, radios, television sets, etc. All have potential failure modes that can place the hot lead of the power main to the frame or case of the device. Current flows of about 6 mA or greater at line frequencies cause muscle paralysis, and the person may be in serious straits indeed and unable to let go. In some failure modes, the device can still appear to function quite normally.

Ground-Fault Circuit Interrupters. Commercial ground-fault circuit interrupters (GFCIs) detect unbalanced current flows between the hot and neutral sides of a power line. Small fault current flows through a person's body to ground will trip a GFCI circuit breaker, but large currents through a normally functioning appliance will not. Building codes for power outlets near water in many countries now requires these devices. GFCI devices can significantly increase safety when used with bioinstrumentation. These devices are often incorporated into the wall socket itself and can be identified since they usually show a red test-reset button near the electrical socket.

A GFCI can shut off the power supply to a device or bioinstrumentation amplifier within 15 to 100 ms of the start of a fault current to ground. This interruption process is so fast that a person may not experience the sensation of a fault shock and even may then wonder why the instrument is not working.

It should be noted that GFCI devices are not foolproof, nor are they absolute protection against electric shock from an instrument. They only are effective with fault currents that follow a path to ground. It is still possible to receive a shock from an appliance fault where a person's body is across the hot-to-neutral circuit, requiring a simultaneous contact with two conductors. This is a more unusual situation and implies exposed wiring from the appliance.

Isolation Amplifiers. Isolation amplifiers increase safety by preventing fault currents from flowing through the body by way of a ground-referenced biopotential electrode. Isolation amplifiers effectively create an electric system similar to biotelemetry. They eliminate any significant resistive coupling between the isolated front-end stages of the amplifier and the output stages to the recorder. They also reduce the amplifier capacitive proximity coupling to ground to a few dozen picofarads or less, depending on manufacture. This provides very high isolation impedance at the power frequency. In principle, isolation amplifiers give many of the advantages of biotelemetry in terms of safety.

Isolated amplifiers are universally used in commercial biopotential monitoring instruments and are mandated in many countries by government regulatory agencies for ECG machines. There are a number of commercial prepackaged modules that conveniently perform this function, so the researcher or design engineer can concentrate on other areas of system design. Isolation amplifiers can be small, compact modules of dimensions 2 by 7.5 cm, as seen in Fig. 17.36.

From an electronic design standpoint, isolation requires considerable sophistication in the bioamplifier. Isolation amplifiers use special internal transformers that have a low interwinding capacitance. They employ a carrier-frequency modulation system for signal coupling between isolated stages and use a high-frequency power-induction system that is nearly equivalent to battery power in isolation. The isolation module effectively creates its own internal reference known as a *guard* (triangle symbol in Fig. 17.37) that connects to the biopotential reference electrode on the subject. The functional components in an isolated amplifier, are seen in Fig. 17.37.

In addition to an improvement in electrical safety, isolated amplifiers in principle should eliminate line-frequency noise in biopotential recordings in the same way as does biotelemetry. The biopo-

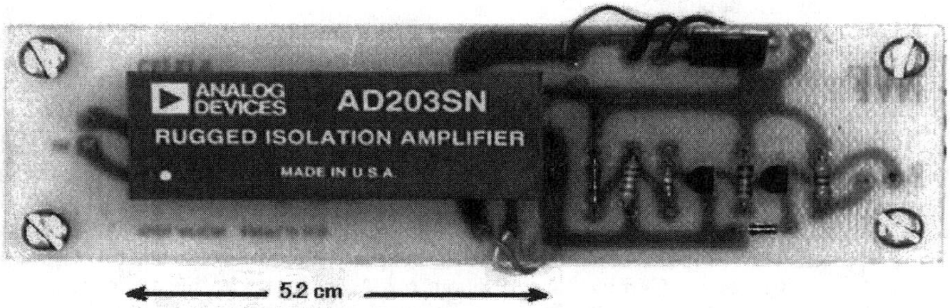

FIGURE 17.36 Photograph of a commercial isolation amplifier module.

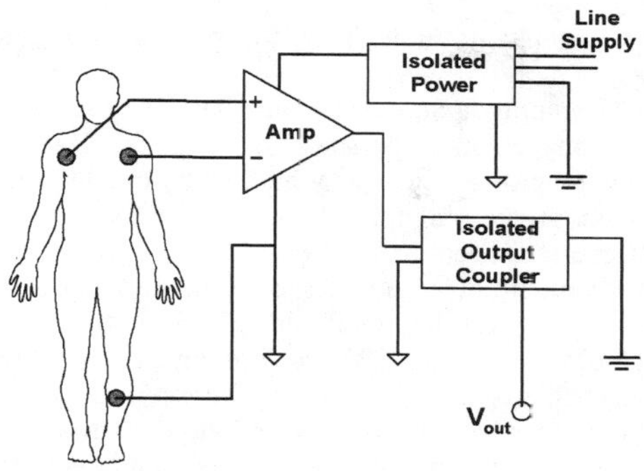

FIGURE 17.37 An isolation amplifier configuration.

tential amplifier reference electrode provides the needed current return path, but because it is isolated, however, it does not reduce the patient's floating potential with respect to ground. This means that V_{float} does not appear between the amplifier input and guard but only with respect to ground, and the bioamplifier has no connection to ground. V_{float} then does not appear in any current nodal analysis, and the isolated bioamplifier effectively ignores this potential.

In practice, the performance advantages are not quite as great as what might be expected. The few picofarads of capacitive coupling that remain to ground as a result of imperfect isolation still can allow the amplifier to be ground referenced to a slight degree. For this reason, isolated bioamplifiers are not totally immune to capacitively coupled power-line interference. There is often a small floating potential present at the amplifier input; however, its amplitude is greatly reduced over the nonisolated case.

17.7 BIOPOTENTIAL INTERPRETATION

17.7.1 Electrocardiography

A recording of the heart's bioelectrical activity is called an electrocardiogram (ECG). It is one of the most frequently measured biopotentials. The study of the heart's bioelectricity is a way of understanding its physiological performance and function. Dysfunction of the heart as a pump often shows up as a characteristic signature in the ECG because its mechanical events are preceded and initiated by electrical events.

Measurement of the ECG is performed using a differential amplifier. Two limb electrodes at a time are selected for the input to the differential amplifier stage. ECG amplifiers typically have gains of about 1000 and so increase the nominally 1-mV biopotential to about 1 V for driving a strip-chart recorder, cathode-ray-tube display, or computer data-acquisition card. Chart speeds in ECG recording have been standardized at 25 and 50 mm/s.

The heart's depolarization is an electric vector called the *cardiac equivalent vector*. The placement of two electrodes creates a direction between them called the *electrode lead*. The electrode-detected bioelectric signal V results from a dot-product relationship of the cardiac equivalent vector **m** with the electrode lead **a** such that

$$V = \mathbf{m} \cdot \mathbf{a} = \text{ma} \cos \theta$$

One of the contributors to the ECG waveform morphology is the changing direction of the cardiac depolarization front. Placing electrodes on the chest such that they are in line with the general direction of the cardiac biocurrent flow, from the sinoatrial node to the heart apex, provides the largest ECG signal amplitude. This is also the direction of a line connecting the right shoulder and the left leg.

The ECG amplitude and waveshape detected from the chest surface change with electrode position and from person to person. This is so because the biocurrents in the heart are relatively localized and so produce relatively high electric field gradients on the chest surface. Small errors in electrode position are magnified by differences in the size, shape, and position of the heart in the chest from one individual to another. Moving a biopotential electrode as little as a centimeter or so in the high-gradient-potential region over the sternal area of the chest, for example, can substantially change and even invert the monitored ECG waveform.

A Dutch physician named Willem Einthoven in 1902 proposed a measurement convention where the ECG electrodes were placed on the limbs. At this position, the monitored electric activity waveshape is more reproducible from one individual to the next because they are in a relative far field to the heart where electric gradients are lower. Einthoven used electrodes placed on the right and left arms and left leg and made the assumption that the heart was at the electric center of a triangle defined by the electrodes.

Einthoven's electrode placement has been adopted and standardized into the ECG Lead system known as the *Einthoven triangle* and is shown in Fig. 17.38. Electric potential differences are measured between the three limb electrodes along the line between the electrode placements, and their potentials are called lead I, II, and III such that

$$\text{Lead I} = V_{\text{LA}} - V_{\text{RA}}$$

$$\text{Lead II} = V_{\text{LL}} - V_{\text{RA}}$$

$$\text{Lead III} = V_{\text{LL}} - V_{\text{LA}}$$

These three leads create an electrical perspective of the heart from three different vectorial directions.

The letter designations of the major features of the ECG wave are derived from the Dutch. The first wave (the preliminary) resulting from atrial contraction was designated as P and is followed in sequence by the triphasic ventricular depolarization called the QRS, and the electric repolarization wave follows as the T wave. Lead II is often used in monitoring heart rhythm because it gives the largest biopotential during ventricular contraction. The normal ECG is seen in Fig. 17.39.

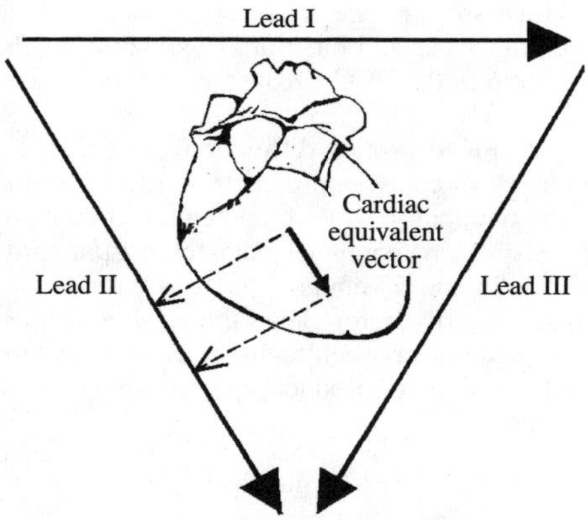

FIGURE 17.38 The ECG lead vectors are based on the Einthoven triangle placement of electrodes.

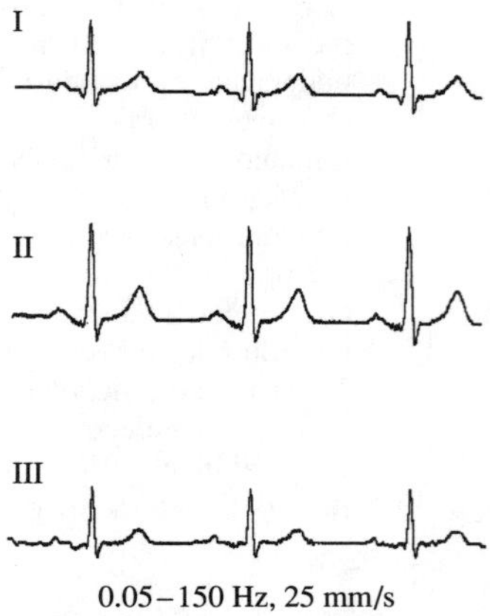

FIGURE 17.39 The normal ECG for leads I, II, and III.

The right-leg electrode is an important electrode even though it does not explicitly appear in the ECG Lead system. It provides the common reference point for the differential amplifier, and in older ECG machines, it is a grounding point for the body to reduce electric noise.

Other Lead systems are also possible and have been standardized into what is known as the *12-lead system* that constitutes the clinical ECG. In addition to leads I, II, and III, there are three others known as the augmented leads aVR, aVL, and aVF. These leads use an electric addition process of the limb-electrode signals to create a virtual signal reference point in the center of the chest known as the *Wilson central terminal* (WCT). From the WCT reference point, the augmented lead vectors point to the right arm, left arm, and left leg, respectively.

There are also six precordial leads, V_1 to V_6, that track along the rib cage over the lower heart region that are used to give more localized ECG spatial information. This total of 12 leads are used clinically to allow the cardiologist to view the electric activity of the heart from a variety of electrical perspectives and to enable a diagnosis based on a more comprehensive data set.

In the plane of the surface electrodes, the signals from the various electrode leads provide sufficient information to define the time-varying biocurrent amplitude and direction in the heart. When this is done, the recording is called a *vectorcardiogram*.

ECG Waveforms in Pathology. ECG waveform interpretation has over the years evolved into a powerful tool for assessing pathologies of the heart. Although its diagnostic interpretation requires considerable training, gross waveform changes that occur in common pathologies can be easily seen even by the inexperienced. Myocardial infarction (heart attack), for example, often is accompanied by an ST-segment elevation. The usually isoelectric (flat) part of the ECG between the S wave and the onset of the T wave is elevated from zero, as can be seen in the sequential waveform comparison in Fig. 17.40.

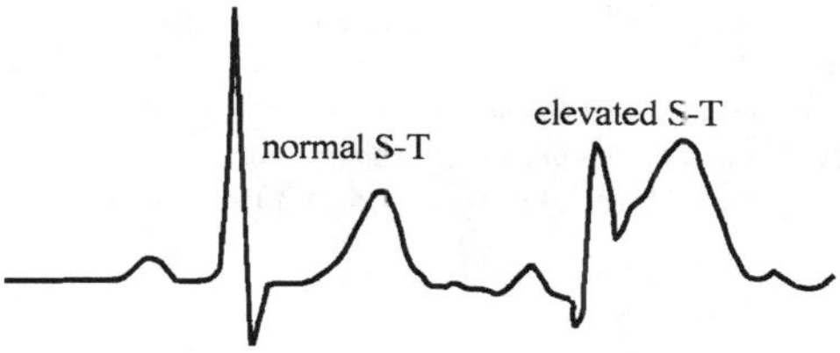

FIGURE 17.40 A dramatic ST-segment elevation subsequent to a myocardial infarction.

This ST-segment displacement is known to be a result of a bioelectric injury current flowing in the heart. This injury current arises from oxygen deprivation of the heart muscle and is usually due to blockage of one or more of the coronary vessels. The elevated ST segment can be 20 percent or more of the ECG amplitude and is sometimes taken as an indicator, along with other factors, of the relative degree of cardiac injury.

After infarction, this injury current slowly subsides over a period of hours or days as the affected cells either live or die. Some cells near the border of the affected region continue to survive but are on the threshold of ischemia (poor blood flow resulting in oxygen deprivation). When stressed, these cells are particularly vulnerable to increased demands placed on the heart, and they are sources of injury currents during exercise. Treadmill exercise, for example, will often stress these cells, and the presence or absence of ST-segment elevation can give insight into amount of ischemic tissue and so indicate the degree of healing in the patient's recovery from infarction. ST-segment monitoring can also be useful for gaining insight into a progressive coronary vessel blockage that creates the pain in the chest known as *angina*.

17.7.2 Electromyography

Skin-surface electrodes placed over contracting skeletal muscles will record a complex time-varying biopotential waveform known as an *electromyogram* (EMG). This electric signal arises from the summed action potential events of individual muscle motor units. Because a large number of cells are undergoing action events during muscle contraction, bioelectric current flows are relatively large and can produce skin potentials as high as 10 mV, although more usually it is in the range of a few millivolts.

The amplitude of the EMG as measured on the skin surface is related, although not linearly, to muscle contraction force. In general, an increasing frequency and complexity of the myogram biopotential follows increasing muscular contraction. This results from both the increased motor unit firing rate and the increased recruitment of muscle fibrils.

The EMG waveform looks random and is fairly well represented by random noise having a Gaussian distribution function. The peak energy is roughly in the 30- to 150-Hz range with some low-energy components as high as 500 Hz. As the muscle fatigues, the EMG frequency spectrum shifts toward the lower frequencies, and the bioelectric amplitude decreases.

Many attempts have been made to use the skin-surface-detected EMG as an indicator of the strength of voluntary muscle force generation, but with often less than hoped-for success. The biopotential frequency and amplitude tend to change little over a range of low contractile force, rise quickly with small changes in moderate force, and vary relatively little with progressively larger forces. In applications where the EMG is used for control of prosthetics, there have been problems due to a relatively little range of proportional control.

Even so, the EMG is widely monitored and used for many applications in research and medicine partly because it is so accessible and easily recorded. For example, the EMG is often used in biomechanics research where it is desired to monitor which muscle groups are active and/or their timing and sequence of contraction. Frequency-spectrum analysis of the EMG can be used to detect the degree of muscle fatigue and to gain insight into muscle performance. EMG monitoring of jaw and neck muscle tension is employed in commercial biofeedback monitors designed to encourage relaxation.

Biopotential amplifiers used for EMG monitoring are fairly simple since standard instrumentation amplifiers or even inexpensive operational amplifiers can be used in a differential mode. An electrode placement and differential amplifier setup used to measure the EMG is shown in Fig. 17.41.

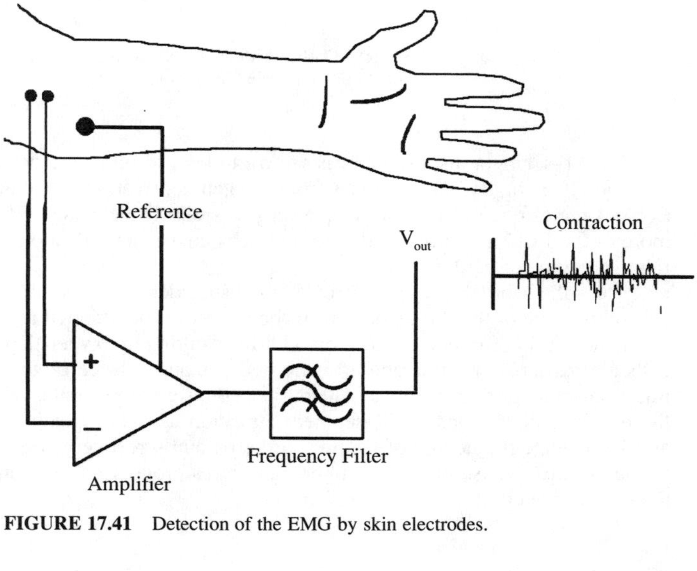

FIGURE 17.41 Detection of the EMG by skin electrodes.

Frequency filtering can be an important aspect of EMG biopotential measurement. For example, electrode movement artifacts can appear with strong muscle contractions, and these add to the EMG waveform, giving the illusion of a larger signal than is really present. With some relatively poor electrode placements and skin preparations, electrode motion artifacts can be the largest signal measured. High-pass filtering helps reject electrode motion artifacts that can accompany forceful skeletal muscle contraction. Analog filters set for 20 to 500 Hz can often be adequate, but some special research applications set the filter as low as 1 to 2 Hz to monitor low-frequency muscle tremors in some cases associated with disease states.

The raw EMG waveform is usually not directly utilized in applications but rather is first subjected to signal processing. One method consists of EMG signal full wave rectification and then low-pass filtering to give the envelope of the muscle-effort waveform. Threshold detectors, slope proportional controllers, and integral control can be implemented using the resulting envelope waveform. Multiple channels derived from different voluntary muscle groups can be used for more complex applications.

For example, Fig. 17.42 shows the results of an experiment using the bicep muscle EMG as a control signal. Two gelled silver–silver chloride electrodes were applied to the upper arm of a volunteer near the bicep about 2 cm apart, and a third reference electrode was placed near the wrist. The EMG biopotentials resulting from two strong voluntary contractions of the bicep were amplified with a gain about 500 in a bandpass of 10 Hz to 1 kHz. The amplified signal was then applied to the data-acquisition interface of a computer. The analog EMG signal was sampled at 10 kHz to avoid aliasing and to capture a high-fidelity EMG waveform. Figure 17.42*a* shows this raw digitized EMG signal. It has a complex waveform, whereas the average amplitude of its envelope rises and falls twice as a result of two discrete muscle efforts over a period of 5 seconds.

To use this signal as a control, we need to integrate the EMG signal over time since its frequency components are far too high. First, it is necessary to take the absolute value of the waveform to provide full wave rectification; otherwise, the alternating waveform integrates to zero. This operation is conveniently done in the digital domain, as shown in Fig. 17.42*b*. Rectification can also be accomplished by using an analog electronic system before the waveform is digitally sampled. Next, calculating a running average of the amplitudes digitally smoothes the rectified waveform. This result is shown in Fig. 17.42*c*. The degree of smoothing determines the response time of the system, as well as its sensitivity to false triggering. The user can display the smoothed EMG waveform to set a threshold. When the computer detects a threshold crossing, it then performs subsequent control steps according to a program.

Low-pass filtering can also be performed prior to computer sampling by conventional analog filtering. This allows a designer to be less concerned about the speed of computer processing needed for real-time signals.

EMG Waveforms in Pathology. Various forms of needle electrodes introduced into the muscle are used clinically to detect biopotentials from the firing of single motor units. Although invasive, the biopotentials monitored from needle electrodes contain medically diagnostic information in a waveshape that is otherwise obscured by the placement of electrodes on the surface of the skin. Needle EMG is used to help detect loss of neurons innervating a muscle, which may be seen in cases of nerve root compression, disk herniation, and peripheral nerve injury, all leading to axonal degeneration.

In muscular disease, needle EMG waveforms vary in strength, waveshape, and frequency content. For example, in muscular dystrophy, the EMG action events are low in amplitude, short in duration (1 to 2 ms), and high in frequency (up to 40 per second). When the bioelectric signals are introduced into a loudspeaker so that they can be heard, there is a high-pitched characteristic sound. These diseased muscles fatigue easily, and this is reflected in their action events that are greatly reduced with sustained contraction.

When the innervation of muscles is severed by accident or disease, spontaneous action events known as *fibrillation waves* appear several weeks after the trauma. These waves result from weak random contraction and relaxation of individual motor units and produce no muscle tension. Fibrillation potentials persist as long as denervation persists and as long as the muscles are present. The

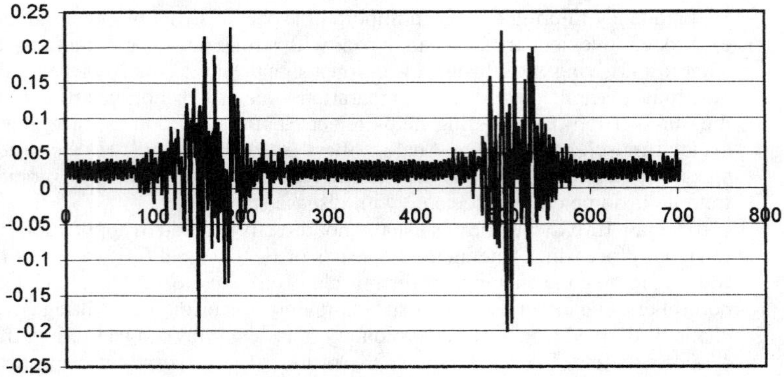

(a)

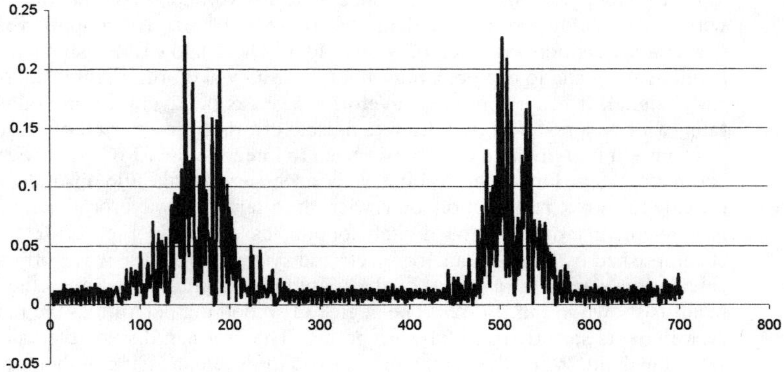

(b)

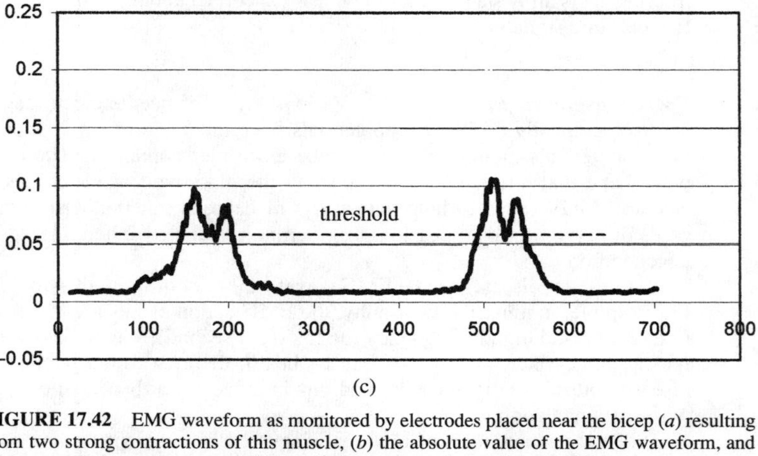

(c)

FIGURE 17.42 EMG waveform as monitored by electrodes placed near the bicep (*a*) resulting from two strong contractions of this muscle, (*b*) the absolute value of the EMG waveform, and (*c*) a smoothed moving average that could be used to trigger a threshold circuit to actuate an EMG-controlled prosthetic device.

waves slowly fade away, as the muscle atrophies. If reinnervation occurs as the injury heals, low-amplitude EMG events can be seen long before the muscle produces signs of visible contraction.

17.7.3 Electroencephalography

Biopotential recordings from the head reflect the bioelectric function of the brain. This recording is known as an *electroencephalogram* (EEG). The first systematic recording of the human EEG is attributed to the Austrian psychiatrist Dr. Hans Berger, who published his results in 1929. By using a primitive galvanometer and surface electrodes placed on his son's scalp, he showed the EEG as a rhythmic pattern of electrical oscillation.

The EEG is produced by electrical dipoles in the outer brain cortex. The waveform is too low in frequency to be the summed result of fast action potential events. Instead, the electric signal is believed to be attributable to the aggregate of excitatory and inhibitory postsynaptic potentials (EPSPs/IPSPs).

The EEG monitored on the top of the head originates primarily from the pyramidal cell layer of the neocortex. These cells have a dipolar nature and are organized with their dendrites in vertical columns. This sets up what is called a *dendrosomatic dipole*, which oscillates in amplitude with the arrival of excitatory or inhibitory postsynaptic potentials. Changes in membrane polarization, as well as EPSPs/IPSPs, impress voltages that are conducted throughout the surrounding medium. A highly conductive layer of cerebrospinal fluid (CSF) between the dura and the skull tends to spatially diffuse biopotentials from the brain cortex. Furthermore, the skull bone is a relatively poor electric conductor. The properties of both the CSF and the skull together reduce millivolt-level bioelectric events in the brain to only microvolts of biopotential on the brain surface.

EEG waveforms are typically 1 to 50 μV in amplitude with frequencies of 2 to 50 Hz. During brain disease states, such as in epileptic seizures, the EEG amplitudes can be much greater, nearing 1000 μV. In some research applications, their frequency spectrum of interest can extend above 100 Hz. Alternatively, invasive electrodes can be placed directly on or in the brain. This electric measurement is known as an *electrocortiogram*, and it has frequency components that extend into the kilohertz range.

The EEG is known to be a composite of many different bioelectric generators within the brain. However, it has not proved to be a powerful technique in studying brain function. The brain is a complex organ that has enormous numbers of bioelectrically active neurons and glial cells. Millions of these cells fire at any given time, and their potentials do not organize well or sum in ways that are characteristic of an activity of a specific part of the brain. Even so, spectral analysis of the EEG shows certain peaks, and study of the EEG suggests that characteristics of the waveforms can be associated with certain mental states. Table 17.6 shows the major brain rhythms categorized according to their major frequency component.

TABLE 17.6 The Major Brain Wave Frequencies

Brain rhythm	Frequency range, Hz
Alpha	8–13
Beta	14–30
Theta	4–7
Delta	<3.5

From wakefulness to deep sleep, there is a progression of EEG activity, slowing from beta wave activity (about 18 Hz) to theta-delta wave activity (3.5 to 8 Hz). Figure 17.43 shows the appearances of the major brain wave frequencies. The delta waves occur in deep sleep or coma. Theta waves are

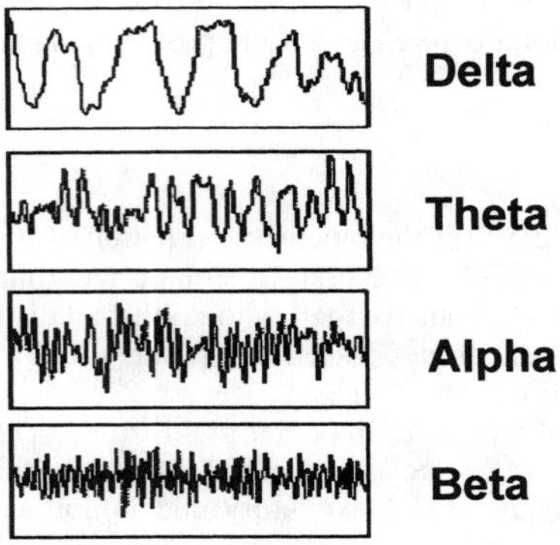

FIGURE 17.43 Waveform appearances of four brain rhythms in order of increasing frequency.

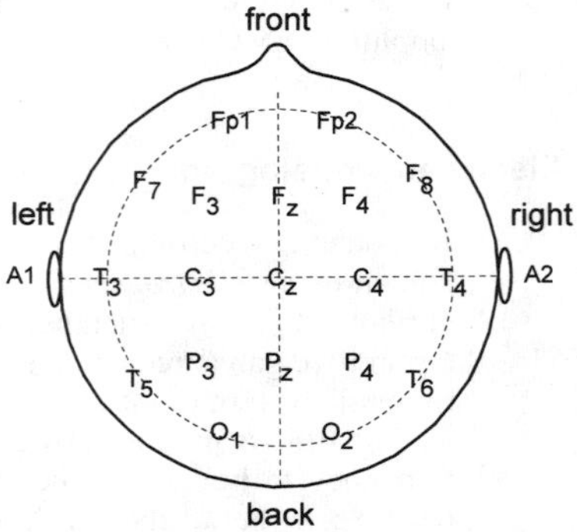

FIGURE 17.44 The 10-20 lead system is the standard for placement of EEG electrodes.

around 7 Hz and are believed to reflect activity in the hippocampus and limbic system. They appear to mediate and/or motivate adaptive, complex behaviors, such as learning and memory.

Alpha waves are around 10 Hz and strongest over the occipital and frontal lobes of the brain. They occur whenever a person is alert and not actively processing information. Alpha waves have been linked to extraversion and creativity, since subjects manifest alpha waves when coming to a solution for creative problems. During most of waking consciousness, the active brain produces beta waves at around 12 to 18 Hz. There also is a high-frequency EEG wave band known as gamma that is above 35 Hz. These waves are relatively small in amplitude and carry little power. The origin of these waves is unclear, and they are not listed in many reference books.

The EEG is widely used in the study of sleep patterns, the various stages of sleep, and the effects of various pharmaceuticals on sleep. The EEG is also used in the study of epilepsy, both in its diagnosis and in research applications that seek to predict epileptic seizure onset.

The clinical EEG is generally monitored using high-quality silver–silver chloride electrodes attached to the scalp in a specific arrangement. This arrangement is known as the *10-20 lead system* and is shown in Fig. 17.44. This lead system gains its name from the fact that electrodes are spaced apart by either 10 or 20 percent of the total distances from landmarks on the skull. Along a line front to back, this is between the bridge of the nose, the nasion, and the bump at the rear base of the skull, the inion. Side to side, the ears form the landmarks.

The electrodes are labeled by letters according to their positions on the scalp, such as frontal, parietal, temporal, or occipital regions. Intermediate positions are labeled by two letters, such as Fp1, meaning the first of a frontal-parietal electrode placement. EEG electrodes are often, but not exclusively, used in differential pairs. This allows a greater sensitivity to biopotentials located between a specific pair of electrodes in an attempt to gain a greater spatial specificity to the generators within the brain. The various electrode placements also allow a greater sensitivity to specific rhythm generators. The alpha rhythm, for example, is known to be the strongest in the occipital electrodes, whereas the beta activity is typically greatest in the frontal electrodes.

Mapping of the brain electric activity using more densely spaced sets of electrodes has been accomplished in combination with computer data-acquisition systems. Maps of the instantaneous distribution of bioelectricity over the head are a research tool used to help study the functionality of the brain.

Evoked EEG Potentials. When a stimulus is applied to the brain through the sensory system, such as a tone burst in the ear, a flash of light to the eye, or a mechanical or electric stimulus to the skin or nerve, there is a subtle response in the EEG. This is known as an *evoked response*, and it reflects the detection and processing of these bioelectric sensory events within the brain. The amplitude of evoked potentials is quite variable but is typically in the range of 3 to 20 μV.

Ordinarily, the brain response to these stimuli is so subtle that they are undetectable in the noise of other brain electric activity. However, if the stimuli is repetitive, then a powerful computer-based technique known as *signal averaging* can be used to extract the corresponding bioelectric events from the rest of the larger but uncorrelated EEG brain activity.

The most well known of these evoked potentials are the auditory evoked response (AER), the visual evoked response (VER), and the somatosensory evoked response (SSER). The somatosensory system evoked signal can result from multiple types of stimuli to the body—light touch, pain, pressure, temperature, and proprioception (also called joint and muscle position sense).

The procedures for evoked waveform monitoring and signal extraction are all basically the same. In each case, electrodes are placed over or near the brain structures most involved in processing the sensory signal. The pulsed stimulus is applied at rates that do not produce fatigue or accommodation of the brain or nervous system. Computer data acquisition and processing create an averaged waveform that has noise suppressed and provides a display of the evoked potential waveform. A physician then evaluates the recording. Experience has shown that certain waveform shapes are associated with normal function, and so deviations, such as different propagation delays or amplitudes, may indicate pathology and be of clinical interest.

Evoked EEG Waveforms in Pathology. The VER is used clinically as a test for optic neuritis or other demyelinating events along the optic nerve or further back along the optic pathways. The test involves stimulating the visual pathways by having a patient watch a black and white checkered pattern on a TV screen in a darkened room. The black and white squares alternate on a regular cycle that generates electrical potentials along the optic nerve and the brain. Electrodes are placed over the visual centers on the occipital scalp. The pattern reversal stimulates action events along the optic nerve, the optic chiasm, and the optic tract through complex processing pathways that ultimately feed to the occipital cortex.

The whole trip of the bioelectric wave propagation (the latency) normally takes about 100 ms. The VER averaged waveforms take on morphology having characteristic peaks such as the P100. This waveform is given this designation because it is a positive-polarity electric event that is delayed by 100 ms after the visual stimulus. White-matter brain lesions anywhere along this pathway will slow or even stop the signal.

Figure 17.45 shows two overlying VER waveforms. Each waveform is the average of 128 individual EEG responses from a visual stimulus. The slight variation of the two waveforms gives an indication of the noise in the measurement that is not fully averaged out, as well as the physiologic variability from trial to trial.

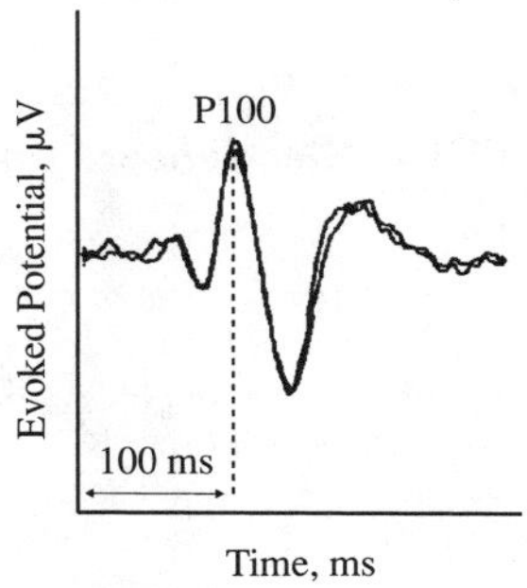

FIGURE 17.45 A visual evoked response waveform showing the P100 peak. It is delayed 100 ms from the onset of a visual flash stimulus in healthy people.

Signal Processing of Evoked Biopotential Waveforms. Due to the relatively low amplitude of evoked waveforms, bandpass filtering is not sufficient to show their responses against the background noise. Noise originates from uncorrelated EEG activity, low-level EMG activity of the head and neck, and sometimes even cardiac bioelectricity.

The signal-to-noise ratio (SNR) is an indicator of the quality of the biopotential recording in terms of being able to identify the biopotential of interest apart from the noise:

$$\text{SNR} = \frac{\text{signal power}}{\text{noise power}}$$

An SNR of 1 means that the signal and noise are of equal power, and hence the signal waveform is indistinguishable from the noise. SNRs of at least 3 to 5 are needed to reliably see the presence of biopotential waveforms, and much higher SNRs, perhaps 10 to 20, are needed for diagnosis.

Signal averaging is a very powerful method of improving the SNR and extracting low-level repetitive signals from uncorrelated noise. It can be employed as long as one can repetitively evoke the biopotential. Signal theory provides that when the interfering noise is random and thus uniformly distributed in the frequency spectrum, signal averaging increases the ratio of the signal amplitude to noise amplitude by

$$\text{SNR ratio improvement} \propto \sqrt{n}$$

where n is the number of waveforms averaged.

EMG noise in evoked bioelectric recordings often does not strictly meet the criteria of randomness and flat frequency spectrum since it is somewhat peaked at certain frequencies. Line-frequency interference has a single, large-frequency peak but is otherwise uncorrelated with the evoked waveform. Despite these variations from the optimum, signal averaging can still be quite useful in improving the SNR when bioelectric responses can be repetitively evoked.

A computer-based data-acquisition system has an analog-to-digital converter that is triggered to sample the bioelectric waveform in synchrony with, and just following, the sensory stimulus. The evoked biopotential along with the noise is digitally sampled and stored in memory as a sequence of numbers representing the signal's instantaneous amplitude over time. Successively sampled waveforms are then numerically added in a pointwise fashion with the previous corresponding time intervals for each past event. The resulting ensemble of added waveforms are then digitally scaled in amplitude by the number of events, and the averaged waveform is displayed. Many commercial signal-processing software packages can perform this function.

The improvement in the ability to see the evoked waveform is limited only by the degree to which the noise deviates from the ideal, the practical limitations on the time one can wait to accumulate evoked events, and physiological variability. Generally, averaging as few as 64 to perhaps greater than 1024 evoked events can provide the needed waveform fidelity.

17.7.4 Electroneurography

The electroneurogram (ENG) is a bioelectric recording of a propagating nerve action potential event. This biopotential is usually evoked by an electric stimulus applied to a portion of the nerve proximal to the spinal cord. The ENG is of interest in assessing nerve function, particularly after injury to a nerve. As the nerve heals and regrows, the ENG amplitude becomes detectable and increases over time. Recording this waveform can give insight into the progress of nerve healing.

The propagated bioelectric response is detected with skin bioelectrodes some distance away from the stimulus. Nerve action events are transient and not synchronous with other fibers within a nerve bundle, so these action currents do not sum together in the same way as they do, for example, in the heart. Detection of the skin surface potential created by an action event from a single nerve fibril within a larger bundle is very difficult, perhaps impossible.

The exception is when a large external electric stimulus is applied to the nerve bundle causing a synchronous depolarization of many fibers. This causes a relatively larger action potential that is more detectable; however, the skin potential signals are still only in the range of 10 to 30 μV. Even with total suppression of line-frequency noise, this signal can be easily obscured by small EMG activity from unrelaxed local muscles. The solution is to evoke the ENG by repetitive stimulation of the nerve. This allows computer-based signal averaging, which increases the ENG signal-to-noise ratio.

An example method of recording the ENG involves stimulation of the ulnar nerve on the upper forearm under the bicep. This stimulation produces a twitching of the muscles in the fingers. The ENG is detected as the efferent action potential passes two biopotential electrodes straddling the nerve at the wrist. Figure 17.46 shows the placement of electrodes and an illustration of the propagation delay in the ENG waveform.

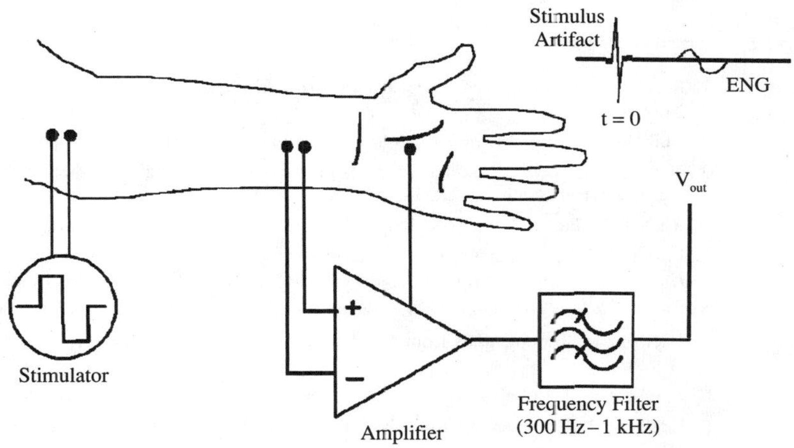

FIGURE 17.46 Illustration of the instrumentation setup for monitoring the ENG.

The ENG in this case arises from the depolarization of many myelinated nerve fibrils within the bundle. ENG amplitude is usually on the order of a few tens of microvolts and creates a transient, sinusoidal-looking waveform as it passes the electrodes. The finite propagation velocity causes a several-millisecond delay between the stimulus and the electrode recording. The nerve action potential conduction velocity V can be determined from

$$\text{Nerve conduction velocity } V = \frac{d}{t}$$

where d is the separation distance of the electrodes and t is the propagation delay time. The propagation velocity is slowed in nerve trauma and stretch injuries. Monitoring nerve conduction through the affected region of the nerve can help a physician assess the progress of nerve repair.

17.7.5 Other Biopotentials

Bioelectrodes placed on or near many major organs having bioelectrically active tissue will often register a bioelectric potential associated with their function. For example, electrodes placed over the stomach will register a slowly varying biopotential associated with digestion. This recording is known as an *electrogastrogram* (EGG).

Electrodes placed on the skin around the eye will detect a biopotential due to a dipolar current flow from the cornea to the retina. The dot product of this biocurrent dipole with an electrode pair will change in amplitude according to the gaze angle. This is known as an *electroocculogram* (EOG), and it can be used to determine the eye's angular displacement.

Finally, there are also bioelectric currents flowing in the retina. They are detectable using corneal electrodes, and the monitored biopotential is known as an *electroretinogram* (ERG). It changes in amplitude and waveshape with pulsed illumination. A strobe light, for example, will produce a transient ERG biopotential of nearly 1 mV.

ACKNOWLEDGMENT

I would like to acknowledge the help and contributions of Mr. William Phillips in the creation of the illustrations and in manuscript editing.

REFERENCES

1. Reilly, J. Patrick, *Applied Bioelectricity: From Electrical Stimulations to Electropathology*, Springer, New York, 1998.
2. Bull H. B., *An Introduction to Physical Chemistry*, F. A. Davis, Philadelphia, 1971.
3. Nordenstrom, B. E., Biokinetic impacts on structure and imaging of the lung: The concept of biologically closed electric circuits, *A.J.R.* **145**:447–467 (1985).
4. Ruch T. C., Patton J. W., Woodbury J. W., and Towe A. L., *Neurophysiology*, Saunders, Philadelphia, 1968.
5. Foster K. R., and Schawn R. P., Dielectric properties of tissues, in C. Polk and E. Postow (eds.), *CRC Handbook of Effects of Electromagnetic Fields*, pp. 25–102, CRC Press, Boca Raton, Fla., 1986.
6. Lide, D. R. (ed.), *Handbook of Chemistry and Physics*, CRC Press, Boca Raton, Fla., 1972.
7. Edelberg R., Biopotentials from the skin surface: The hydration effect, *Ann. N.Y. Acad Sci* **148**:252–262 (1968).
8. Edelberg R., Local electrical response of the skin to deformation, *J. Appl. Physiol.* **34**(3):334–340 (1973).
9. Edelberg R., Biopotentials from the skin surface: The hydration effect, *Ann. N.Y. Acad. Sci.* **148**:252–262 (1968).
10. Wilcott R. C., Arousal sweating and electrodermal phenomena, *Psychol. Bull.* **67**(1):58–72 (1967).
11. Tam H. W., and Webster J. G., Minimizing electrode motion artifact by skin abrasion. *IEEE Trans. Biomed. Eng.* **24**(2):134–139 (1977).
12. Burbank D. P., and Webster J. G., Reducing skin potential motion artifact by skin abrasion, *Med. Biol. Eng. Comput.* **16**:31–38 (1978).
13. Thakor N. V., and Webster J. G., The origin of skin potential and its variations, presented at the 31st Annual Conference on Engineering in Medicine and Biology (ACEMB), Georgia, Alliance for Engineering and Biology, Bethesda, 1978.
14. Shackel B., Skin drilling: A method of diminishing galvanic skin potentials, *Am. Jo. Physiol.* **72**:114–121 (1959).

CHAPTER 18
BIOMEDICAL SIGNAL ANALYSIS

Jit Muthuswamy
Department of Bioengineering, Arizona State University, Tempe, Arizona

18.1 INTRODUCTION

Any signal transduced from a biological or medical source could be called a *biosignal*. The signal source could be at the molecular level, cell level, or a systemic or organ level. A wide variety of such signals are commonly encountered in the clinic, research laboratory, and sometimes even at home. Examples include the electrocardiogram (ECG), or electrical activity from the heart; speech signals; the electroencephalogram (EEG), or electrical activity from the brain; evoked potentials (EPs, i.e., auditory, visual, somatosensory, etc.), or electrical responses of the brain to specific peripheral stimulation; the electroneurogram, or field potentials from local regions in the brain; action potential signals from individual neurons or heart cells; the electromyogram (EMG), or electrical activity from the muscle; the electroretinogram from the eye; and so on.

Clinically, biomedical signals are primarily acquired for monitoring (detecting or estimating) specific pathological/physiological states for purposes of diagnosis and evaluating therapy. In some cases of basic research, they are also used for decoding and eventual modeling of specific biological systems. Furthermore, current technology allows the acquisition of multiple channels of these signals. This brings up additional signal-processing challenges to quantify physiologically meaningful interactions among these channels.

Goals of signal processing in all these cases usually are noise removal, accurate quantification of signal model and its components through analysis (system identification for modeling and control purposes), feature extraction for deciding function or dysfunction, and prediction of future pathological or functional events as in prosthetic devices for heart and brain. Typical biological applications may involve the use of signal-processing algorithms for more than one of these reasons. The monitored biological signal in most cases is considered an additive combination of signal and noise. Noise can be from instrumentation (sensors, amplifiers, filters, etc.), from electromagnetic interference (EMI), or in general, any signal that is asynchronous and uncorrelated with the underlying physiology of interest. Therefore different situations warrant different assumptions for noise characteristics, which will eventually lead to an appropriate choice of signal-processing method.

The focus of this chapter is to help the biomedical engineer or the researcher choose the appropriate representation or analysis of the signal from the available models and then guide the engineer toward an optimal strategy for quantification. This chapter is not meant to be an exhaustive review of biosignals and techniques for analyzing. Only some of the fundamental signal-processing techniques that find wide application with biosignals are discussed in this chapter. It is not structured to suit the reader who has a scholarly interest in biomedical signal-processing techniques. For a more detailed overview of biomedical signal-processing techniques, the reader is referred to Refs. 1 and 2. This chapter will not deal with measurement issues of the signal. The reader is assumed to have acquired the signal reliably and is poised to make decisions based on the signals. This chapter will help the reader navigate his or her way from the point of signal acquisition to the point where it is useful for decision making.

A general classification of biomedical signals is attempted in Sec. 18.2. This will enable the reader (user) to place his or her signal of interest in the appropriate class. Subsequently, the sections are outlined according to different techniques for signal analysis. As far as possible, the first paragraph of each section generally outlines the class(es) of signals for which the corresponding technique is best suited. Toward the end of each section, appropriate MATLAB functions useful for analysis are indicated. Each section is then illustrated by an application.

18.2 CLASSIFICATIONS OF SIGNALS AND NOISE

The biomedical signal sources can be broadly classified into *continuous processes* and *discrete-time* or *point processes*. Each of these types of signals could be deterministic (or predictable), stochastic (or random), fractal, or chaotic. The continuous processes are typically encountered in one of the following situations.

18.2.1 Deterministic Signals in Noise

Examples of this type are ECG or single-fiber EMG signals in noise. The measured signal $x(t)$ can be represented as follows:

$$x(t) = s(t) + n(t)$$

where $s(t)$ is the actual deterministic signal and $n(t)$ is the additive noise. A segment of blood pressure shown in Fig. 18.1*a* is an example of a deterministic periodic signal. A Gaussian white noise assumption is valid in many biological cases. The goal in many situations is feature extraction under noisy (could be EMI, ambient, or instrumentation noise) conditions and subsequently correlating with the underlying physiological or pathological state.

18.2.2 Deterministic Signals (Synchronized to Another Stimulus Signal or Perturbation) in Noise

Examples of this type include all the different evoked responses (auditory, somatosensory, visual, etc.) and event-related potentials recorded in response to controlled stimuli administered to the body (or any biological system in general). These signals usually reveal functional characteristics of specific pathways in the body. For instance, evoked responses to peripheral somatosensory stimulation reveal the performance of the somatosensory pathway leading to the sensory cortex. A segment of cortical somatosensory evoked potential is shown in Fig. 18.1*b* that was obtained after averaging 100 stimulus-response pairs. Evoked responses or event-related potentials are usually superimposed

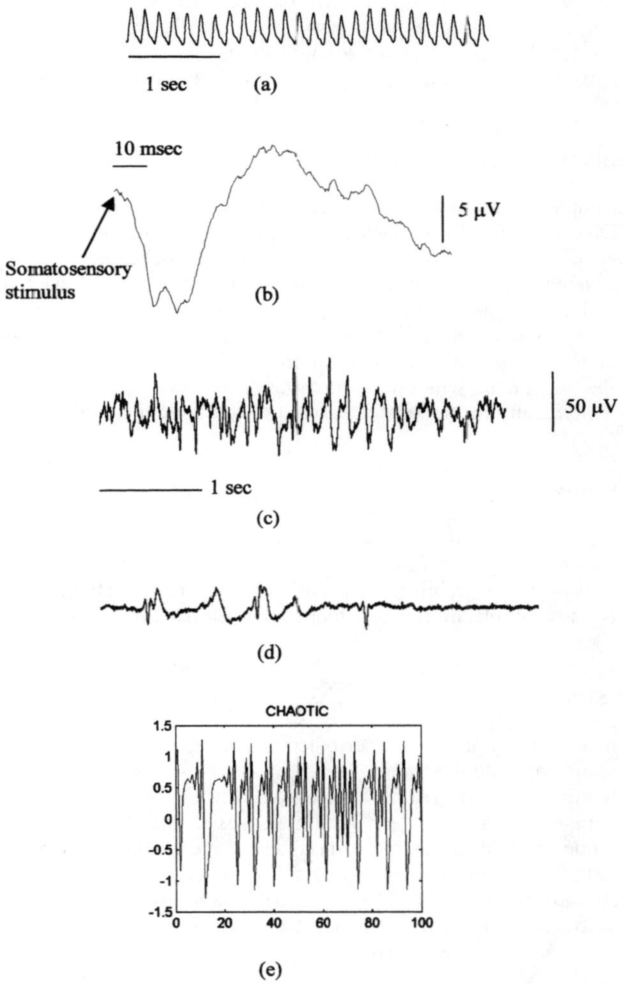

FIGURE 18.1 Types of biosignals: (*a*) *deterministic signal in noise*, illustrated by a segment of blood pressure signal recorded using a fluid-filled catheter in the femoral artery; (*b*) *deterministic signal (in noise) synchronized to an external cue or perturbation*, illustrated by an epoch of somatosensory evoked potential recorded from the somatosensory cortex in response to an electrical stimulus to the forelimb; (*c*) *stationary stochastic signal*, illustrated by a segment of EEG from the cortex that shows no specific morphology or shape, but the statistics of the signals are more or less stable or stationary in the absence of any physiological perturbations; (*d*) *nonstationary stochastic signal*, illustrated by a segment of EEG recorded from an adult rat recovering from a brain injury showing bursts of activity, but the statistics of this signal change with time; (*e*) *a chaotic signal* that was artificially generated resembles a stochastic signal but is actually generated by a deterministic dynamical system.

over spontaneous background electrical activity that is unrelated and hence asynchronous to the administered stimulation or perturbation. Therefore, signal-processing efforts have been directed toward extraction of evoked responses from the spontaneous background activity (could also be considered "noise" in this case), noise (could be interfering signals such as ECG, EMG, or ambient noise) removal, and analysis of the evoked responses to quantify different components.

18.2.3 Stochastic or Random Signals

Examples of this type include EEGs, EMGs, field potentials from the brain, and R-R intervals from ECGs. Random signals lack the morphology of the signals found in the preceding two categories. Depending on the underlying physiology, the stochastic biosignals could be *stationary* (statistics of the signal do not change with time) or *nonstationary* (fluctuations in the signal statistics due to physiological perturbations such as drug infusion or pathology or recovery). A segment of EEG signal (random signal) that is stationary within the window of observation is shown in Fig. 18.1*c*, and an EEG signal that is nonstationary with alternate patterns of bursts and suppressions in amplitudes within the window of observation is shown in Fig. 18.1*d*. Signal-analysis techniques have been typically for noise removal and for accurate quantification and feature extraction.

18.2.4 Fractal Signals

Fractal signals and patterns in general are self-replicating, which means that they look similar at different levels of magnification. They are therefore scale-invariant. There is evidence to suggest that heart rate variability is fractal in nature. The branching of the airway into bronchioles seems to have a self-replicating nature that is characteristic of a fractal.

18.2.5 Chaotic Signals

Chaotic signals are neither periodic nor stochastic, which makes them very difficult to predict beyond a short time into the future. The difficulty in prediction is due to their extreme sensitivity to initial conditions, characteristic of these nonlinear systems. While fractal theory details the spatial characteristics of the nonlinear systems, chaos theory describes the temporal evolution of the system parameters or the dynamical variables. The essential problem in nonlinear biosignal analysis is to determine whether a given biosignal (a time series) is a deterministic signal from a dynamical system. Subsequently, signal analysis is usually done to determine the dimensionality of the signal and quantification of the dynamical states of the system. An example of a chaotic signal is shown in Fig. 18.1*e* that resembles a random signal but is actually generated by a deterministic dynamical system.

18.2.6 Multichannel Signals

Multichannel signals could include signals of any of the preceding five types but acquired using multichannel recording technology. Analysis and interpretation usually involve a matrix formulation of the single-channel analysis technique. A segment of four-channel multiunit activity (extracellular action potential signals) from thalamic and cortical structures is shown in Fig. 18.2. The goals of signal analysis are usually to identify correlation and hence synchrony among different channels, to achieve feature extraction under noisy conditions, and to identify underlying physiology.

The same six broad types of signals are found among point processes as well. The most common example of a point process is the action potential traces (recorded from either the extracellular or the intracellular space). Derived measures such as the neuronal firing rate histograms or cross-correllograms that are continuous in time are often used for analysis of these point processes.

The following sections are organized as follows. Section 18.3 deals with analysis of deterministic and stationary stochastic signals. Analysis of nonstationary signals is discussed in Sec.18.4. Sub-

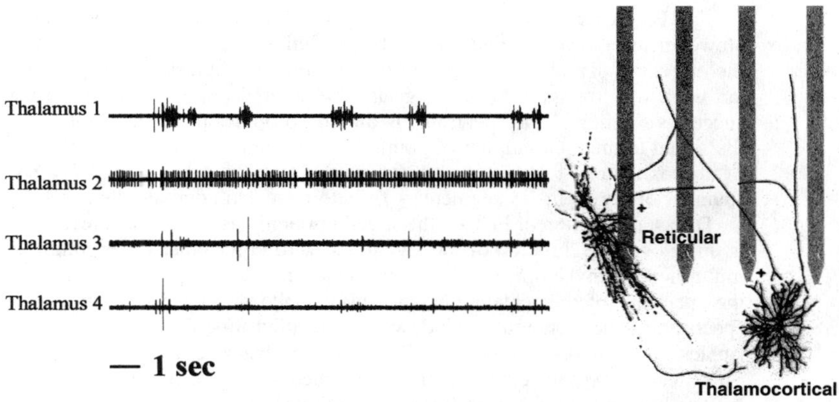

FIGURE 18.2 An example of a *multichannel biosignal* acquired using a multichannel microelectrode implanted in the thalamus and cortex of an adult rat. Each channel is a trace of multiunit activity (action potentials from a few neurons in the vicinity of each microelectrode as illustrated in the adjacent cartoon).

sequently, Sec. 18.5 deals with alternative orthogonal basis of representing biosignals that are optimal in certain physiological situations. Techniques for dealing with pairs of signals (both stochastic or a combination of stochastic and deterministic) are discussed in Sec. 18.6. Finally, Sec. 18.7 deals with analysis of fractal and chaotic signals. The statistics of the estimates are discussed wherever appropriate. By discussing these techniques separately, I am not suggesting that each of these techniques should be used in isolation. My hope is that the reader will get a deeper appreciation for each technique and will be driven toward creative solutions based on the data that could very well include a combination of several techniques.

18.3 SPECTRAL ANALYSIS OF DETERMINISTIC AND STATIONARY RANDOM SIGNALS

One of the most common analysis techniques that is used for biological signals is aimed at breaking down the signal into its different spectral (or frequency) components.

18.3.1 Fast Fourier Transforms

Signal Types—Deterministic Biosignals (with or without Noise). Fast Fourier transform (FFT) is commonly used in analyzing the spectral content of any deterministic biosignal (with or without noise). I will discuss the issue of estimating the spectrum of the signal under noisy conditions in the following subsections. Discrete Fourier transform (DFT) allows the decomposition of discrete time signals into sinusoidal components whose frequencies are multiples of a fundamental frequency. The amplitudes and phases of the sinusoidal components can be estimated using the DFT and is represented mathematically as

$$X(k) = \frac{1}{N} \sum_{n=0}^{N-1} x(n) e^{-j(2\pi kn/NT)} \tag{18.1}$$

for a given biosignal $x(n)$ whose sampling period is T with N number of total samples (NT is therefore the total duration of the signal segment). The spectrum $X(k)$ is estimated at multiples of f_s/N, where f_s is the sampling frequency.

Fast Fourier transform (FFT) is an elegant numerical approach for quick computation of the DFT. However, users need to understand the resolution limitations (in relation to the signal length) and the effects of signal windowing on the accuracy of the estimated spectrum. In general, FFT does not work well for short-duration signals. The spectral resolution (or the spacing between ordinates of successive points in the spectrum) is directly proportional to the ratio of sampling frequency f_s of the signal to the total number of points N in the signal segment. Therefore, if we desire a resolution of approximately 1 Hz in the spectrum, then we need to use at least 1-s duration of the signal (number of points in 1-s segment $= f_s$) before we can compute the spectral estimates using FFT.

Direct application of FFT on the signal implicitly assumes that a rectangular window whose value is unity over the duration of the signal and zero at all other times multiplies the signal. However, multiplication in the time domain results in a convolution operation in the frequency domain between the spectrum of the rectangular window and the original spectrum of the signal $x(n)$. Since the spectrum of the rectangular window is a so-called Sinc function consisting of decaying sinusoidal ripples, the convolved spectrum $X(f)$ can be a distorted version of the original spectrum of $x(n)$. Specifically, spectral contents from one frequency component (usually the dominating spectral peak) tend to leak into neighboring frequency components due to the convolution operation. Therefore, it is often advisable to window the signals (particularly when one expects a dominating spectral peak adjacent to one with lower amplitude in the signal spectrum). Several standard windows, such as Hamming, Hanning, Kaiser, Blackman-Tukey, etc., are available in any modern signal-processing toolbox, each with its own advantages and disadvantages. For a thorough review on different windows and their effect on the estimation of spectral contents, the reader is referred to Ref. 3.

18.3.2 Periodogram Approach

Signal Types—Deterministic Signals in Noise or Pure Stationary Random Signals. For deterministic signals in noise or pure random signals, Welch's periodogram approach can be used for estimating the power spectrum of the signal. The given length of the signal N is segmented into several (K) overlapping or nonoverlapping epochs each of length M. The power spectrum of each epoch is evaluated using FFT and averaged over K epochs to obtain the averaged periodogram. An implicit assumption in this method is that the statistical properties of the noise or the signal do not change over the length of the given sequence of data (assumption of *stationarity*.) Periodogram leads to statistically unbiased estimates (the mean value of the *estimate* of the power spectrum equals the true value of the power spectrum). Further, the variance (or uncertainty) in the individual estimates is inversely proportional to K. Therefore, it is desirable to increase the number of segments K in order to decrease the uncertainty in the estimates. However, increasing K will also decrease the resolution in the power spectra for reasons discussed in the preceding subsection. The power spectra using Welch's periodogram approach can be estimated using *spectrum* function in MATLAB. Since the power spectral values are estimates, one has to specify the confidence in these estimates for a complete description. When the epochs are nonoverlapping, the ratio of the power spectral estimate at each frequency to its actual value can be approximated to be a χ^2_{2K} random variable with $2K$ degrees of freedom.[4] Therefore, the interval within which the actual value is likely to lie can be estimated easily given the desired level of confidence. The function *spectrum* in MATLAB has options to estimate the confidence interval as well.

18.3.3 Parametric Methods

Signal Types—Short-Duration Signals and Stationary Random Signals In Noise. When a high-resolution spectrum is desired and only a short-duration signal is available, parametric approaches outlined in this subsection provide a better alternative (as long as the parametric model is accurate) to FFT-based approaches for power spectral estimation in the case of *random or stochastic signals in noise*. For deterministic or periodic signals (with or without noise), Fourier transform–based approaches will still be preferable.[5] Examples of parametric modeling include the autoregressive (AR) model, the autoregressive moving-average (ARMA) model, and the moving-average (MA) model. So far there is no automatic method to choose from AR, ARMA, or MA. The given signal is treated as the output of a linear time-invariant system (more advanced adaptive methods can be used to track time-varying model parameters) driven by a white Gaussian noise. The AR model can also be treated as an attempt to predict the current signal sample based on p past values of the signal weighted by constant coefficients. We estimate the best model by trying to minimize the mean squared error between the signal sample predicted by the model and the actual measured signal sample.

In an AR model, the signal $x(n)$ is represented in terms of its prior samples as follows:

$$x(n) = e(n) - a_1x(n-1) - a_2x(n-2) - \cdots - a_px(n-p) \tag{18.2}$$

where $e(n)$ = assumed to be zero mean white Gaussian noise with a variance of σ^2
p = order of the AR model
$x(n-i)$ = signal sample i time periods prior to the current sample at n
a_i = coefficients or parameters of the AR model

This representation can also be seen as a system model in which the given biosignal is assumed to be the output of a linear time-invariant system that is driven by a white noise input $e(n)$. The coefficients or parameters of the AR model a_i become the coefficients of the denominator polynomial in the transfer function of the system and therefore determine the locations of the poles of the system model. As long as the biosignal is stationary, the estimated model coefficients can be used to reconstruct any length of the signal sequence. Theoretically, therefore, power spectral estimates of any desired resolution can be obtained. The three main steps in this method are

1. Estimation of approximate model order (p)
2. Estimation of model coefficients (a_i)
3. Estimation of the power spectrum using the model coefficients or parameters

It is critical to estimate the right model order because this determines the number of poles in the model transfer function (between the white noise input and the signal output). If the model order is too small, then the power spectral estimate tends to be biased more toward the dominant peaks in the power spectrum. If the model order is larger than required, it often gives rise to spurious peaks in the power spectral estimate of the signal.

Several asymptotic model order selection methods use a form of generalized information criteria (GIC) that could be represented as $\text{GIC}(\alpha, p) = N\ln(\rho_p) + \alpha p$, where p is the model order, α is a constant, N is the number of data points, and ρ_p is the variance in the residual or error for model p.[6] The error or residual variance ρ_p can be determined using a (forward) prediction error $e_p(n)$ defined as

$$e_p(n) = \sum_{k=0}^{p} a_p(k)x(n-k),$$

where $a_p(k)$ is a parameter in the AR model of order p. The error variance is then simply a summation of squared forward prediction errors given by

$$\rho_p = \sum_{n=0}^{N-1} |e_p(n)|^2$$

When the value of α is 2 in the expression for GIC, it takes the form of the Akaike information criterion (AIC).[7,8] The optimum model order p is one that minimizes the generalized or Akaike information criterion.[8] More recent methods to estimate model order for signal sequences with a finite number of sample points include predictive least squares (PLS) and finite sample information criteria (FSIC).[6]

Having determined the optimal model order for the given segment of signal, the model can be estimated using one of the following MATLAB functions: *arburg, arcov* (uses a covariance approach), *armcov* (uses a modified covariance approach), or *aryule* (uses the Yule-Walker equations for estimation). The *arburg* function uses Burg's method[9] that minimizes both forward and backward prediction errors to arrive at the estimates of the AR model (similar to the modified covariance approach in *armcov*). The AIC model order estimation works well in combination with the Burg's method. Each of these functions uses different numerical procedures to arrive at the minimum mean square estimate of the AR model. The *aryule* function uses the autocorrelation estimates of the signal to determine the AR coefficients, which could potentially degrade the resolution of the spectrum. A brief discussion of the different AR parameter estimators and their application in ultrasonic tissue characterization is found in Ref. 10. Once the AR model has been estimated, the power spectra can be estimated by using the following expression:

$$P(f) = \frac{\sigma_p^2 T}{\left| 1 + \sum_{k=0}^{p} a_p(k) \exp(-j2\pi fkT) \right|^2} \tag{18.3}$$

where $P(f)$ = power spectral estimate at the frequency f
σ_p^2 = variance of the white noise input to the model
T = sampling period of the signal

The power spectrum of the signal can be estimated using the *pburg, pcov, pmcov*, or *pyulear* functions in MATLAB.

Statistics of the Spectral Estimates. The exact results for the statistics of the AR spectral estimator are not known. For large samples of stationary processes, the spectral estimates have approximately a Gaussian probability density function and are asymptotically unbiased and consistent estimates of the power spectral density. The variance of the estimate is given by

$$\mathrm{Var}[P(f)] = \begin{cases} \dfrac{4p}{N} P^2(f) & f = 0 \text{ and } \dfrac{f_s}{2} \\ \dfrac{2p}{N} P^2(f) & \text{otherwise} \end{cases} \tag{18.4}$$

where f_s = sampling frequency of the signal
N = number of samples
p = model order[4]

In general, parametric methods are preferred for stochastic signals, provided the estimations of model order and the model parameters are done carefully.

APPLICATION *An 8000-point EEG signal segment sampled at 278 Hz is shown in the Fig. 18.3a. The power spectrum of the signal as evaluated using Welch's periodogram method is shown in Fig. 18.3b. The MATLAB function* spectrum *was used with a Hanning window of length 512, allowing for 256-point overlap between successive segments. The resolution of the spectrum is the ratio of the sampling frequency to the number of points, which turns out to be 0.54 Hz. The corresponding 95 percent confidence interval shown (upper curve) along with the power spectral estimate (lower curve) indicates the slightly higher uncertainty in the estimates around 5 Hz. An AR model order of 40 was chosen using the AIC shown in Fig. 18.3c. The power spectral estimate from the AR model using Burg's method in Fig. 18.3d shows a much higher resolution.*

18.4 SPECTRAL ANALYSIS OF NONSTATIONARY SIGNALS

A signal is nonstationary when the statistics of the signal (mean, variance, and higher-order statistics) change with time. The traditional spectral estimation methods just outlined will only give an

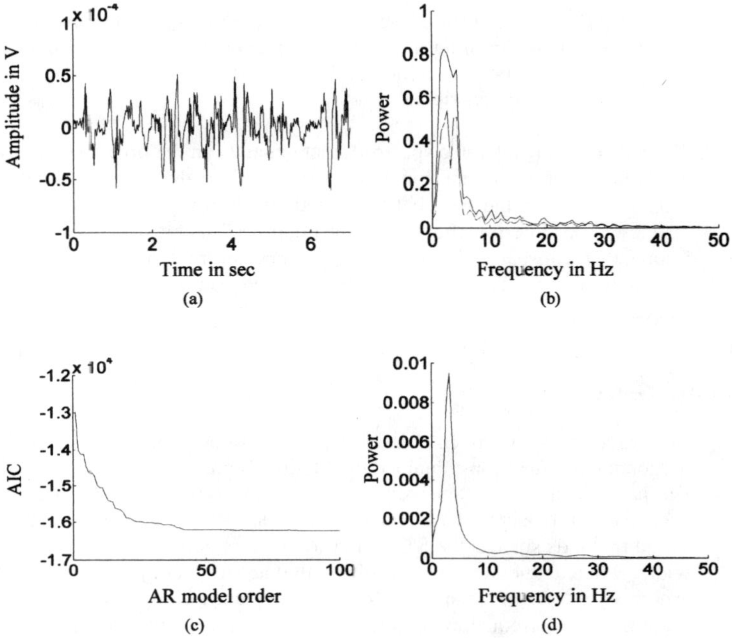

FIGURE 18.3 Comparison of a periodogram-based approach and a parametric approach to spectral analysis of stochastic signals. (*a*) A typical segment of EEG. (*b*) The power spectrum estimated using the periodogram approach (lower curve) along with the 95 percent confidence interval (upper curve). There is a higher uncertainty in the estimates in the low-frequency regime. (*c*) The autoregressive (AR) model order is estimated using an AIC criteria. The value of AIC seems to reach a minimum at approximately a model order of 50. (*d*) The power spectrum of the EEG signal estimated using an AR model of order 50. The power spectrum from the AR model has a higher resolution than (*b*).

averaged estimate in these cases and will fail to capture the dynamics of the underlying generators. Alternative analysis techniques that determine time-localized spectral estimates can be used when the user suspects that the signal sequence under consideration is not under steady-state physiological conditions. These methods work well with both deterministic and random signals in noise.

Some of the commonly used algorithms for time-frequency representations of spectral estimates include the short-time Fourier transform (STFT), the Wigner-Ville transform, wavelet transforms, etc. A good summary of the mathematical basis of each of these techniques can be found in Ref. 11.

18.4.1 Short-Time Fourier Transform

The STFT as introduced by Gabor involves multiplying the signal by a short-duration time window that is centered around the time instant of interest. The Fourier transform of the product then gives an estimate of the spectral content of the signal at that time instant. The short-time-duration window is subsequently slid along the time axis to cover the entire duration of the signal and to obtain an estimate of the spectral content of the signal at every time instant. The signal is assumed to be stationary within the short-duration window. Mathematically,

$$P(m, n) = \int_{-\infty}^{\infty} e^{imt\omega_0} h(t - n\tau_0) x(t) dt \tag{18.5}$$

where $P(m, n)$ = short time Fourier transform (STFT) at time instant $n\tau_0$ and frequency $m\omega_0$
τ_0 and ω_0 = sampling interval and fundamental angular frequency, respectively
$h(t)$ = short-duration window
$x(t)$ = given signal

The STFT thus results in a spectrum that depends on the time instant to which the window is shifted. The choice of Gaussian functions for the short-duration window gives excellent localization properties despite the fact that the functions are not limited in time. Alternatively, STFT can also be viewed as filtering the signal "at all times" using a bandpass filter centered around a given frequency f whose impulse response is the Fourier transform of the short-duration window modulated to that frequency. However, the duration and bandwidth of the window remain the same for all frequencies.

18.4.2 Wavelet Transforms

The wavelet transform, in contrast, is a signal-decomposition (or analysis) method on a set of orthogonal basis functions obtained by dilations, contractions, and shifts of a prototype wavelet. Wavelets have been used extensively for processing biomedical images[12–15] and for processing almost every kind of biosignal with nonstationarities.[16–25] The main distinction between Fourier transform–based methods such as the STFT and wavelet transforms is that the former use windows of constant width, whereas the latter use windows that are frequency-dependent.[7,26] Wavelet transforms enable arbitrarily good time resolution for high-frequency components and arbitrarily good frequency resolution for low-frequency components by using windows that are narrow for high-frequency components and broad for low-frequency components. In general, the continuous wavelet transform can be represented mathematically as

$$S(a, b) = \int x(t)\psi^*_{a,b}(t)\, dt \tag{18.6}$$

where $x(t)$ = given signal
a = scale factor
b = time
$*$ = complex conjugate

The orthogonal basis functions denoted by $\psi_{a,b}(t)$ are obtained by scaling and shifting a prototype wavelet function $\psi(t)$ (also sometimes called a *mother wavelet*) by scale a and time b, respectively, as shown below:

$$\psi_{a,b}(t) = \frac{1}{\sqrt{|a|}}\psi\left(\frac{t-b}{a}\right) \tag{18.7}$$

By adjusting the scale factor, the window duration can be arbitrarily changed for different frequencies. By choosing a and b appropriately, a discrete time version of the orthogonal basis functions can be represented as follows:

$$\psi(m, n) = a_0^{-m/2}\psi(a_0^{-m}t - nb_0) \tag{18.8}$$

where a_0 and b_0 are fundamental scale factor and time shift, respectively (m and n represent multiples of scale factor and the time shift).

Wavelet analysis works especially well with signals that have short durations of high-frequency components and long durations of low-frequency components, e.g., EEG signals or signals or variations in interbeat (R-R) intervals, etc. I describe in this subsection a simple procedure for obtaining time-frequency representation of signals using wavelet transforms as outlined in Refs. 7, 27, and 28. A MATLAB function that implements the same procedure is detailed next.

1. Using FFT, find the Fourier transform $X(\omega)$ of the discrete time signal N-point signal $x(n)$.
2. For every equally spaced discrete frequency of interest f_i (>0). ($f_{min} < f_i < f_{max}$),
   a. Determine the scale factor $a_i = 1/f_i$.
   b. Denoting the analyzing wavelet by $\psi(n)$ (being implicitly dependent on the scale factor a) and its DFT by $\Psi(\omega)$, which is centered around ω_0, evaluate the product vector $Z(\omega) = \Psi(a\omega)X(\omega)$ for $\omega = -N/2$ to $N/2 - 1$.
   c. Determine the inverse Fourier transform of $Z(\omega)$ and scale the resulting time series by $\sqrt{a}$.

$$S(n, \omega) = \sqrt{a}\int_{\omega_{min}/a}^{\omega_{max}/a} \Psi(a\omega)X(\omega)e^{jn\omega}\,d\omega \tag{18.9}$$

   d. Repeat steps a to c for every discrete frequency within the range f_{min} and f_{max}.
   e. The resulting output $S(n, \omega)$ is the time-frequency representation of the signal $x(n)$.

The Fourier transform of one of the prototype wavelets (called a *Morlet wavelet*) is given by

$$\Psi(\omega) = \exp[-(\omega - \omega_0)^2/2] + \text{small correction terms}$$

The correction terms are theoretically necessary to make the negative-frequency components of the term $\exp[-(\omega - \omega_0)^2/2]$ zero (an additional constraint imposed by Kronland-Martinet et al.[27]). In order to make the correction terms negligible, the center frequency ω_0 is chosen to be anywhere between 5.0 and 6.0. The corresponding time-domain form of the analyzing wavelet is a modulated Gaussian of the form (up to a constant normalization factor)

$$\psi(n) = e^{i\omega_0 n}\exp(-n^2/2)$$

The square of $S(n, \omega)$ is often referred to as a *scalogram*, and it gives an estimate of the time-localized power spectrum. A MATLAB function that can be used to implement this procedure is detailed below[28]:

```
function S=wt_morl(x,fs,nfft,nout,f0,fmin,fmax,pas,pass,sigm);
%function S=wt_morl(x,fs,nfft,nout,f0,fmin,fmax,pas,pass,sigm);
% time frequency analysis using Morlet wavelet
% x - signal vector
%fs: sampling frequency
%nfft: number of points in the signal
%nout: number of points not use in the beginning of signal (=0)
%f0: center frequency of the wavelet=1
%fmin: minimum frequency of the representation
%fmax: maximum frequency
%pas: increment in the scale between fmin and fmax
%pass: increment in time of the result (steps in time resolution)
%sigm: time frequency resolution general case =0.7;
%S: result (it is better to plot abs(S))

y=fft(x(1+nout:nfft+nout));
for f=fmin:pas:fmax
a=1/f;
wavelet=exp(-2/sigm*(pi)^2*(([0:nfft/2-1,-nfft/2:-1]*(fs*a/nfft))-(f0)).^2);
temp=wavelet.'.*y;
t=ifft(temp).';
S(fix(1/pas*(f-fmin))+1,:)=(a^0.5).*t(1:pass:nfft);
end;
```

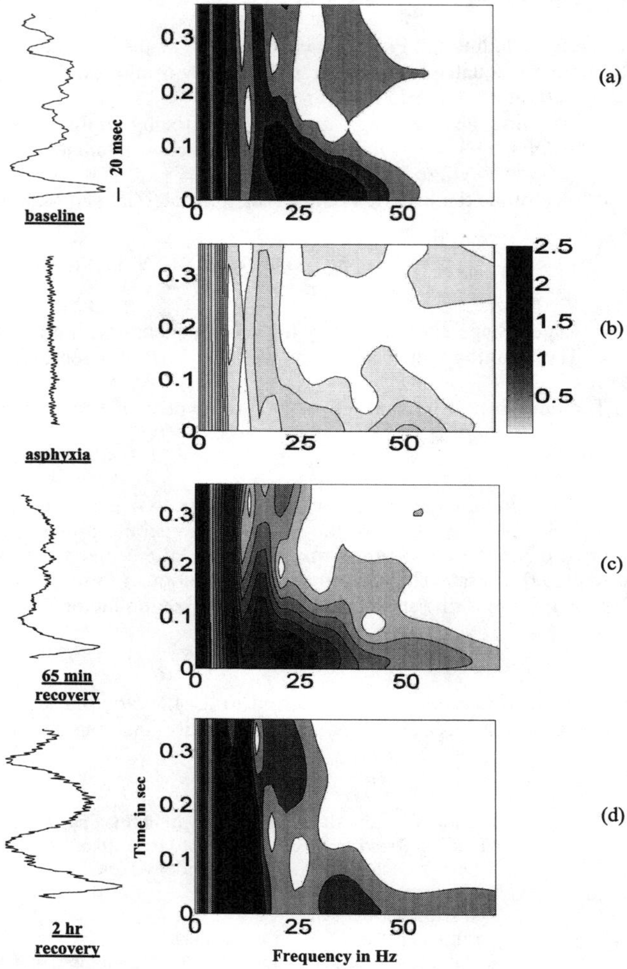

FIGURE 18.4 Analysis of a *nonstationary stochastic biosignal* using a wavelet-based approach for time-frequency localization of signal power. Somatosensory evoked potentials in response to an electrical stimulation to the contralateral forelimb recorded at different stages of brain injury are shown to the left of the vertical axis of the colored contour plots. The signals are recorded (*a*) at baseline—before the rat was subject to a controlled duration of asphyxia or oxygen deprivation; (*b*) during asphyxia—showing electrical silence or isoelectric behavior; (*c*) after approximately 65 minutes into recovery from the injury—showing partial recovery of low frequency amplitudes; (*d*) after approximately 2 hours into recovery—showing greater recovery of the low-frequency and high-frequency generators in the evoked potentials. The time-frequency distribution of evoked potentials seems to indicate a greater vulnerability of the high-frequency generators in the brain to oxygen deprivation.

APPLICATION *I have used wavelet transforms to track the changes in the spectral components of the somatosensory evoked potentials with an ischemic injury (induced by asphyxia) to the brain in a rodent model, as illustrated in Fig. 18.4. The high-frequency components disappear first at the induction of injury and recover long after recovery of the low-frequency components after resuscitation of the animal. The results seem to suggest differential vulnerabilities in the generators of brain potentials. The somatosensory responses were obtained by averaging 100 stimulus-response pairs.*

18.5 PRINCIPAL COMPONENTS ANALYSIS

Signal Types—Deterministic or Stochastic Signals in Colored Noise. So far we have represented (by projection) biosignals in terms of Fourier basis functions, autoregressive parameters, or wavelet-basis functions depending on application. In general, the signal can be represented in terms of any convenient set of orthogonal basis functions. One such orthogonal basis, called the *principal components* of the given signal, has been widely used for analysis of biosignals. It is also called *Karhoenen-Loeve analysis* for signals with zero mean.

Broadly speaking, principal component analysis has been used for at least two different applications where signal representation using other basis functions does not provide the optimal solution. First, when the presence of a known signal waveform (hence deterministic) has to be detected in colored (implying correlation among noise terms) Gaussian noise, projecting the noisy signal along its principal components allows for optimal detection of signal. When the noise is white (independent, identically distributed) Gaussian, a matched filter (or a correlator, to be discussed in a subsequent subsection on cross-correlation) provides the most optimal detection. The second broad application of principal components analysis has been in pattern classification, where it has been used for identifying the feature vectors that are most discriminating. Selecting the most discriminating features leads to a reduction in the dimensionality of input feature vectors. Indeed, one could reasonably argue that the first application is a special case of the second. Nevertheless, I will treat them distinctly in this section and outline methods for both applications.

18.5.1 Detecting Deterministic or Known Signals in Colored Gaussian Noise

The observation vectors (successive time samples could also be used to construct the observation vector) are the sum of a known signal and noise terms that are correlated. The key idea in this application is to emphasize the projection of the signal along directions where the noise energy is minimal. The assumption is that we know the covariance matrix of the noise a priori. Even if it is not known, it can be easily estimated from the covariance of a *training set* (trial observations) of observation vectors. The element K_{ij} of the covariance matrix K is the estimated covariance between ith and jth components of the observation vector. Having estimated K, we find the eigenvalues and the corresponding eigenvectors of the matrix. If the matrix is positive definite, all the eigenvalues will be distinct, and the corresponding eigenvectors will be orthogonal. By suitably scaling the eigenvectors, we obtain a set of orthonormal basis functions ϕ_i, $i = 1, 2, \ldots, N$, where N is the dimensionality of the observation vector. The known signal vector **S** and the observation vector **X** are projected onto the set of orthonormal eigenvectors. Mathematically, the $(N \times N)$ matrix

$$\mathbf{\Phi} = \begin{bmatrix} \phi_1^T \\ \phi_2^T \\ \vdots \\ \phi_N^T \end{bmatrix}$$

represents the set of N orthonormal eigenvectors (each of dimension $N \times 1$). $\mathbf{S}' = \mathbf{\Phi S}$ represents the projection of the known signal vector along each eigenvector. Similarly $\mathbf{X}'$ represents the projection of the received observation vectors (known signal plus colored noise) along the orthonormal eigenvectors. The optimal detector is designed by evaluating

$$\sum_{i=1}^{N} \frac{S_i'}{\lambda_i} X_i'$$

for every observation vector, where λ_i are the corresponding eigenvalues. This ratio (also called the *likelihood ratio* after normalization by the norm of $\mathbf{S}'$; it is really a cross-correlation of lag zero between projections of the known signal and projections of the observation vector) is compared against a threshold that is calculated from a priori known probabilities to detect the presence of the known signal. As the ratio indicates, the signal projection is emphasized (amplified) in directions where the eigenvalues are small, implying smaller noise energy in that direction. The threshold Λ_0 for an optimal Bayes detector is given by Ref. 29,

$$\Lambda_0 = \frac{p_0(C_{10} - C_{00})}{p_1(C_{01} - C_{11})} \tag{18.10}$$

where p_1 = a known a priori probability of the presence of the known signal (the frequency of occurrence of the known signal can be estimated by prior observation or knowledge of the generating process)

$p_0 = (1 - p_1)$

$C_{ij}(0 \leq C_{ij} \leq 1)$ = cost of deciding i when j actually occurred (where 0 indicates absence and 1 indicates presence of known signal in the noise)

Typically, C_{00} and C_{11} can be assumed to be zero, whereas C_{01} and C_{10} can each be assumed to be unity. However, sometimes the cost of false-positive C_{10} (detecting a signal when there is none) may be different from the cost of false-negative C_{01} (missing the known signal when it was actually present) depending on the application. Therefore, when the likelihood ratio exceeds the Bayes threshold, the detector indicates the presence of the known signal, and vice versa.

In the statistical analysis toolbox of MATLAB, functions *princomp* and *pcacov* enable the computation of principal components. The same method can be extended to detecting a finite number M of known signals. For further information on M-ary detection, the reader is referred to Ref. 29.

18.5.2 Principal Components Analysis for General Pattern Classification

Biosignals are often examined for recognizable patterns that indicate change in pathology or underlying physiology and hence aid diagnosis. In the absence of any signal analysis, the sample points in the measured biosignals have to be directly used to discriminate changes in underlying physiology. Analysis of the signal along spectral or wavelet components may not be the most discriminating since the components are not optimized for *discrimination* but are optimized for *spectral decomposition* and localization of energy. However, principal components analysis offers an elegant way to choose a set of orthonormal basis functions that are optimized specifically for maximum discrimination between two or more known classes of biosignals. In most cases, this method results in a drastic reduction in the number of signal components that one has to work with and also improves the classification or diagnosis as well.

Given a set of observation vectors (containing samples of the biosignals) from two different physiological or pathological states, we form a scatter matrix, which is really a scaled version of the sample covariance matrix. The scatter matrix $\mathbf{K}$ is defined as[30]

$$\mathbf{K} = \sum_{k=1}^{L} (\mathbf{X}_k - \mathbf{M})(\mathbf{X}_k - \mathbf{M})^t \tag{18.11}$$

where L = number of N-dimensional observation vectors $\mathbf{X}_i$
$\mathbf{M}$ = sample mean of the observation vectors
t = transpose operator

All the observations from different physiological states are pooled together to form the scatter matrix. Since the scatter matrix is real and positive, all the eigenvectors of $\mathbf{K}$ are orthogonal. The eigenvectors are then normalized as in the preceding application to yield an orthonormal set of basis vectors. The magnitude of the corresponding eigenvalues indicates the degree of scatter along those eigenvectors. We then choose the eigenvectors corresponding to the largest eigenvalues as the preferred directions of projections and determine the projections of the subsequent observations along those directions.

APPLICATION *Principal component analysis is commonly used for sorting neuronal spikes (action potentials). Extracellular data collected using microelectrodes implanted in the brain typically record action potentials or spikes from more than one neuron. Different neurons typically register action potentials of different shapes and sizes at the microelectrode. A trace of multineuronal action potential data (or multiunit data) is shown in Fig. 18.5b. The three different action potentials found in the trace are shown separately in an enlarged fashion in Fig. 18.5a. The result of principal components analysis of the three action potential waveforms is shown in Fig. 18.5c. When plotted in the space spanned by the two most dominant principal components, the corresponding projections of the three different action potentials show three distinct clusters that are easily separable. Instead of using all the time samples that constitute the waveform of the action potentials, principal components analysis identifies just two features (projections along two dominant principal components or eigenvectors) that are sufficient for discrimination.*

18.5.3 Independent Components Analysis

Signal Types—Stochastic Signals in Noise and Mixtures of Several Stochastic Signals. A recently developed extension of principal components analysis called *independent components analysis* (ICA) was originally used for *blind-source separation.*[31] In other words, when the measured signal vector (the number of channels of measured signal is given by the dimension of the vector) is an unknown linear mixture or combination of an equal number of independent, non-Gaussian sources, then this technique is useful in arriving at an estimate of the unknown original sources of the signal. For instance, when multiple channels of EEG are collected from several locations on the scalp, the potentials are presumably generated by mixing some underlying components of brain activity. While principal components analysis is useful in identifying the orthogonal directions (eigenvectors) that contain significant signal energy, independent components analysis is useful in identifying the independent, non-Gaussian components of the signal. This technique will not work if there is a strong indication that the original independent sources are Gaussian (if just one of the independent components is Gaussian, the ICA model may still be estimated[31]).

If the independent signal sources are represented by $\mathbf{s} = \{s_1, s_2, \ldots, s_N\}$ (could be N different kinds of sources) and an unknown *mixing* matrix $\mathbf{A}$ (inherent to the generating source or the medium between the source and the measurement point) generates the measured or received signals $\mathbf{r} = \{r_1, r_2, \ldots, r_n\}$, then the ICA signal model can be represented as $\mathbf{r} = \mathbf{As}$. This technique has been used successfully on EEG signals,[32,33] to decompose evoked potentials,[34] and to remove artifacts in magnetoencephalography (MEG).[35]

The goal of ICA is to design a matrix $\mathbf{F}$ specifying linear spatial filters that inverts the mixing process due to the $\mathbf{A}$ matrix. The matrix $\mathbf{F}$ is therefore often referred to as the *separation matrix* that helps to generate a scaled and permuted version $\mathbf{y} = \{y_1, y_2, \ldots, y_N\} = \mathbf{Fr}$ of the original sources $\mathbf{s}$. The basic principle of ICA is centered on the central limit theorem in statistics, which states that the distribution of a sum of independent random variables tends toward a Gaussian distribution under certain conditions. Therefore, the signal $\mathbf{r}$ (mixture of several independent non-Gaussian sources) will be more Gaussian. The idea then is to formulate a measure of non-Gaussianity and then project

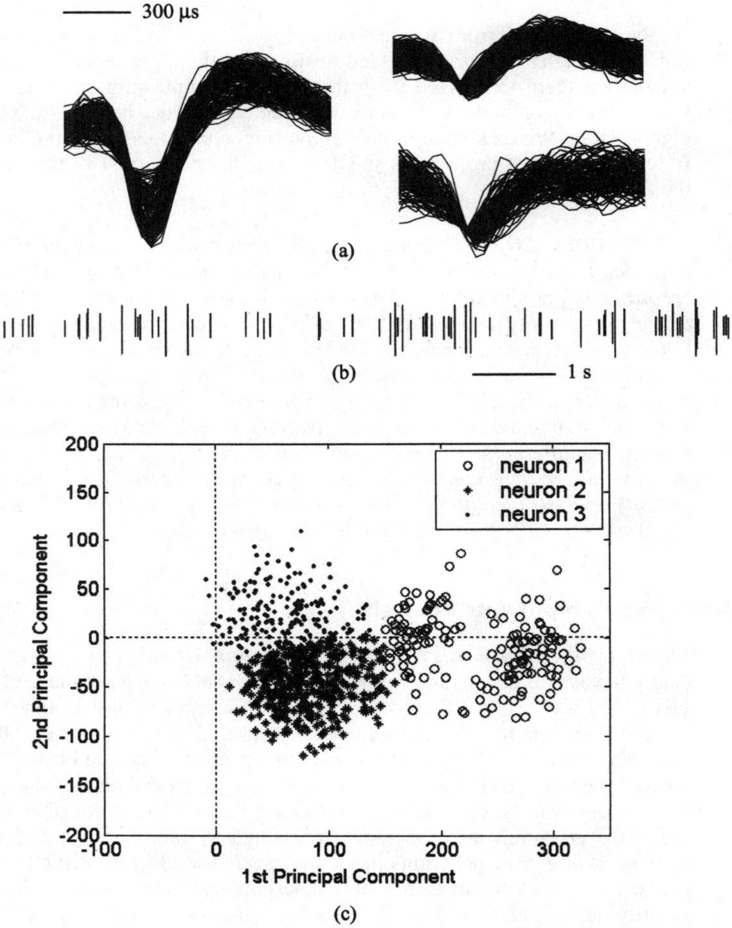

FIGURE 18.5 The measured biosignal (multiunit activity from a single microelectrode) is projected along an alternate (as opposed to Fourier basis functions shown earlier) set of orthogonal basis vectors (eigenvectors or principal components) for the purpose of *sorting* action potentials belonging to different neurons. The three different action potentials recorded by the microelectrode are shown in (*a*). Multiple traces within each action potential indicate multiple occurrences of similar action potentials superimposed to demonstrate variability. The three action potentials differ from each other in duration and peak amplitudes. On a more compressed scale in (*b*) the three action potentials appear as spikes of different heights. Plotting the action potential waveforms along two of their principal components in (*c*) gives a clear separation between the three different action potentials. Thus, instead of using all the points of the action potential waveform for sorting or classification, the use of projections along the principal components reduces the number of features required for sorting to just two.

the measured or received signal along specific directions in an iterative manner (as specified by the **F** matrix), which will maximize the measure of non-Gaussianity. Different measures of non-Gaussianity have been used to determine the optimal matrix **F**, such as the information-maximization approach[36] or methods based on cumulants[37] and negentropy.[38]

Using kurtosis (or the fourth-order cumulant) of a signal as a measure of its non-Gaussianity, an implementation of ICA originally derived by Hyvarinen and Oja[39] is discussed here. For the kth source signal s_k, the kurtosis is defined as

$$\text{Kurt}(s_k) = E\{s_k^4\} - 3[E\{s_k^2\}]^2 \tag{18.12}$$

where $E\{\cdot\}$ denotes the mathematical expectation. The kurtosis is negative for source signals whose amplitude has sub-Gaussian probability densities (probability distributions flatter than Gaussian), positive for super-Gaussian (probability distributions sharper than Gaussian), and zero for Gaussian densities. Maximizing the norm of the kurtosis leads to identification of the independent sources. This method was used in Ref. 33 to eliminate ECG interference in EEG signals.

A brief outline is presented here. For a more detailed presentation, the reader is referred to an excellent review.[31] The first step in ICA processing is to center **r** by subtracting its mean vector **m** so as to make **r** a zero-mean variable. After estimating the mixing matrix with centered data, the procedure is completed by adding the mean vector **m** of **r** back to the centered estimate that was subtracted in the first step. Subsequent to centering, the measured vector is sometimes whitened as part of preprocessing the signal before applying the ICA algorithm. Whitening the observed vector **r** (resulting in a flat power spectrum) removes the second-order relationships and produces a new vector $\mathbf{u} = \{u_1, u_2, \ldots, u_n\}$ whose components are uncorrelated. The variance of the individual components of **u** is unity. In matrix notation, this is

$$\mathbf{u} = \mathbf{Ur} \tag{18.13}$$

where **U** is the whitening matrix. Expressing **r** in terms of **s** and **A**, the model **u** becomes

$$\mathbf{u} = \mathbf{UAs} \tag{18.14}$$

The solution is

$$\mathbf{y} = \mathbf{Wu} \tag{18.15}$$

where **W** is the separating matrix for the measured signals after whitening (Fig. 18.6).

Maximizing the absolute value of the kurtosis of the components of **y** (Eq. 18.12), one of the columns of the separating matrix **W** is found, and so one independent component at a time is identified. The other columns are estimated subsequently. The algorithm has a cubic convergence and typically convergence by 20 iterations.[33] From Eqs. (18.13), (18.14), and (18.15), the output matrix of independent components **y** can be written as

$$\mathbf{y} = \mathbf{Wu} = \mathbf{WUr} = \mathbf{Fr} \tag{18.16}$$

The rows of this matrix are the time course of activation of the individual ICA components. A MATLAB implementation of the algorithm (FastICA) is available at http://www.cis.hut.fi/projects/ica/fastica/.[31]

APPLICATION *ICA has been used to separate ECG interference in EEG signals recorded from adult rats undergoing controlled ischemic brain injury and subsequent recovery.[33] The measured signal vector* **r** *consisted of two EEG signals recorded from right and left parietal cortex areas and one channel of ECG, as illustrated in Fig. 18.7. The signals shown at the top in Fig. 18.7 were recorded right after an asphyxic injury to the brain during which the EEGs became isoelectric. The early recovery from such injury is characterized by low-amplitude waveforms in the EEG that carry information about the early recovery mechanisms in the cortex. However, they are prone to corruption from the larger-amplitude ECG signals. The reconstructed EEG is free from any interference*

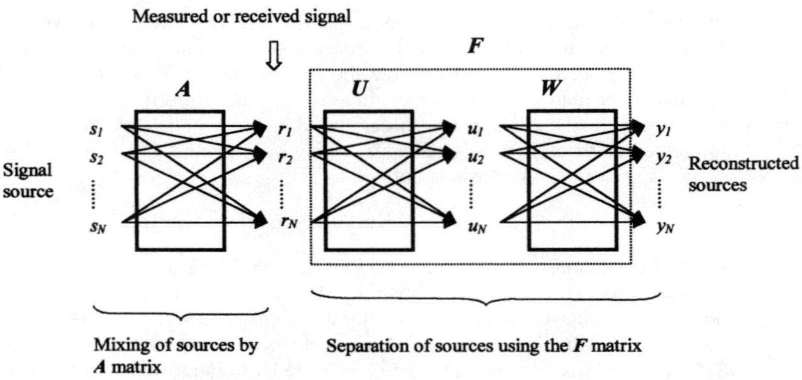

FIGURE 18.6 Schematic of independent components analysis technique for a blind source separation application. The signal from the sources gets mixed by an unknown mixing matrix **A** before being measured. The goal of ICA is to design a separating matrix **F** of linear spatial filters that will invert the measured signal vector **r** back into the independent components represented by the vector **y**. The components of **y** should be approximations of the source vector **s** although they (the reconstructed sources) do not have to match the order in which they appear in **s**.

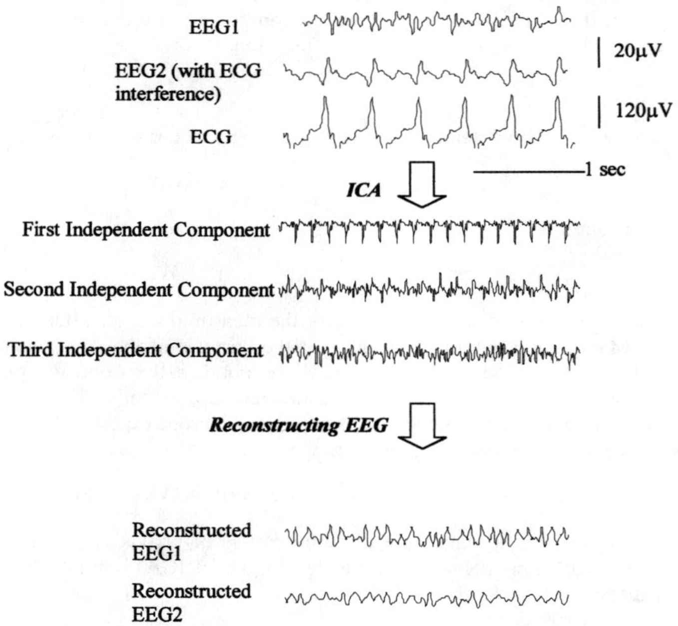

FIGURE 18.7 Independent components analysis has been used to separate ECG interference in the EEG.[33] Two channels of EEG and an interfering ECG waveform are shown at the top. The EEG data were recorded from an adult rat just after an episode of brain injury due to oxygen deprivation resulting in electrical silence in the brain. EEG during the early recovery stages shows very low-amplitude activity resulting in pronounced interference from the ECG waveforms. The independent components are used for reconstructing the two EEG waveforms shown at the bottom.

from ECG. The EEG was reconstructed from the independent components by setting the independent component in **y** *that corresponds to ECG to zero and evaluating* $\mathbf{r} = \mathbf{F}^{-1}\mathbf{y}$ *[from Eq. (18.16)].*

18.6 CROSS-CORRELATION AND COHERENCE ANALYSIS

Signal Types—Pairs of Stochastic Signals, Pairs of Stochastic and a Deterministic Signal, and Multichannel Signal Analysis. The signal-analysis methods discussed so far represented the measured signals along a set of orthogonal basis functions (Fourier, wavelet, or principal components) for noise removal (thus enhancing the signal-to-noise ratio), feature extraction, signal-source separation, etc. A bioengineer will also find situations where he or she has to deal with pairs of signals. Cross-correlation and coherence analysis are often simple tools that are extremely effective in quantifying relationships among pairs of biosignals.[40–46] Some of the typical application situations are as follows:

1. Detecting a known (deterministic) signal in a noisy environment
2. Estimating time delay or propagation delay between two biosignals (deterministic or stochastic signals in noise)
3. Estimating the transfer function of the signal-generating system (deconvolution, estimating the degree of linearity, etc.)

The cross-correlation between two measured biosignals $x(n)$ and $y(n)$ is defined statistically as $R_{yx}(k) = E[y(n)x(n + k)]$, where the operator E represents the statistical expectation or mean and k is the amount of time signal $x(n)$ is delayed with respect to $y(n)$. Given two time sequences of $x(n)$ and $y(n)$ each of N points, the commonly used estimate of cross-covariance $c_{yx}(k)$ is as follows[4]:

$$c_{yx}(k) = \begin{cases} \dfrac{1}{N}\displaystyle\sum_{n=0}^{N-k-1} [y(n) - \hat{m}_y][x(n+k) - \hat{m}_x] & k \geq 0 \\ \dfrac{1}{N}\displaystyle\sum_{n=0}^{N-k-1} [y(n+k) - \hat{m}_y][x(n) - \hat{m}_x] & k \leq 0 \end{cases} \tag{18.17}$$

where $\hat{m}_y$ and $\hat{m}_x$ are the sample means of the signal sequences $y(n)$ and $x(n)$, respectively. The sample means can be omitted in this expressions if the signals have been detrended using a MATLAB function such as *detrend.*

Statistics of the Estimate. The cross-correlation estimator is however biased[4] because

$$E[c_{yx}(k)] = \left(1 - \frac{|k|}{N}\right) C_{yx}(k) \tag{18.18}$$

where $C_{yx}(k)$ is the true value of the cross-covariance or cross-correlation at a lag of k. The bias reduces for increasing length N of the signal sequence. However, the variance of this estimate is complicated. If the two sequences are Gaussian and completely uncorrelated, then the variance in the estimate of the cross-correlation is $\sigma_x^2\sigma_y^2/N$, where σ_x^2 and σ_y^2 are variances of $x(n)$ and $y(n)$ (the white noise sequences), respectively. In general, estimation of correlation between two sequences has to be done cautiously, as outlined in Ref. 4. Each individual signal sequence is modeled using a parametric model outlined in Sec. 18.3.3. The cross-correlation between the two error sequences (should be white Gaussian noise if the model is accurate) that were generated by modeling each signal must be estimated. The 95 percent confidence limits on the cross-correlation between the error sequences must be computed using the variance expression given earlier. Finally, the cross-correlation between the two signal sequences $x(n)$ and $y(n)$ can be estimated. Values that exceed the 95 percent confidence limits of the cross-correlation between the two error sequences indicate statistically significant correlation.

18.6.1 Detection of Known Signals (Deterministic) in White Gaussian Noise

The simple cross-correlation estimator is used extensively in the form of a matched filter implementation to detect a finite number of known signals (in other words, simultaneous acquisition of multiple channels of known signals). When these deterministic signals are embedded in white Gaussian noise, the matched filter (obtained from cross-correlation estimate at zero lag, $k = 0$, between the known signal sequence and the observed noisy signal sequence) gives the optimum detection performance (in the Bayes' sense[29]).

If the known N-dimensional signal vector is $\mathbf{S}$ (could be any signal of fixed morphology, such as the ECG, evoked potential, action potentials, etc.) and the measured N-dimensional observation vector with white Gaussian noise is $\mathbf{X}$, then we evaluate a likelihood ratio that is defined as

$$\sum_{i=1}^{N} \frac{S_i X_i}{\|\mathbf{S}\|^2}$$

and compare it against a Bayes threshold Λ_0 identical to the one described in the section on principal component analysis. When the likelihood ratio exceeds the threshold, the detector flags the presence of the known signal in the measurement.

APPLICATION *Matched filters can be used to detect and sort action potentials in noise. The multiunit data in Fig. 18.8b were collected from the ventral posteriolateral thalamic nuclei using a microelectrode. The segment shows two different types of action potentials (corresponding to two*

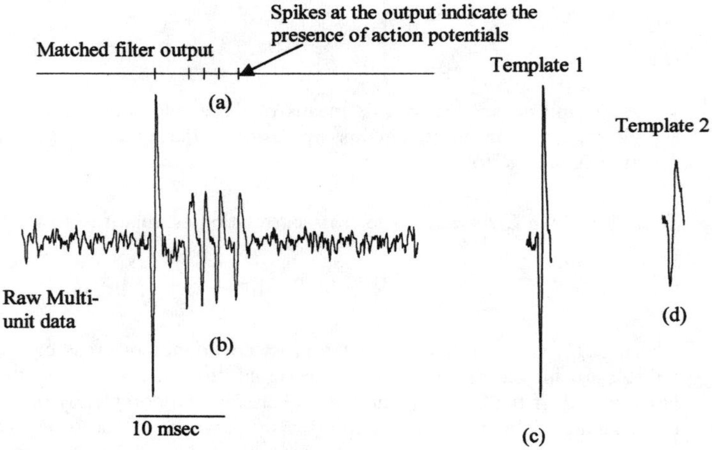

FIGURE 18.8 Cross-correlation-based method to detect the presence of spikes or action potentials in multiunit activity. The continuous waveform with the action potentials is shown in (*b*). The shapes of the action potentials to be detected are known a priori (deterministic signal). Using a cross-correlation of the known signal [templates (*c*) and (*d*)] with the measured signal, the presence of action potentials is indicated by a vertical line in (*a*). This approach is optimal in the presence of white Gaussian noise.

different neurons at different locations relative to the microelectrode). Action potential shapes from trial data were recognized and stored as templates. Two of the templates presumably corresponding to two different neurons are shown in Fig. 18.8c and d. Action potentials (or spikes) in the subsequent data set are detected (indicated by vertical lines in the detector output in Fig. 18.8a) and sorted based on these templates (detecting a known signal in noise). The noise is assumed to be white Gaussian, and the matched filter works reasonably well in detecting these spikes. However, the detector often fails when two action potentials overlap.

In contrast to the principal component (PCA)–based approach for detection and discrimination, the cross-correlation or matched-filter approach is based on a priori knowledge of the shape of the deterministic signal to be detected. However, the PCA-based method also requires some initial data (although the data could be noisy, and the detector does not need to know a priori the label or the class of the different signals) to evaluate the sample covariance matrix **K** and its eigenvectors. In this sense, the PCA-based detector operates in an unsupervised mode. Further, the matched-filter approach is optimal only when the interfering noise is white Gaussian. When the noise is colored, the PCA-based approach will be preferred.

18.6.2 Coherence Estimation

The estimation of coherence is useful to determine if *spectral* components between two signal sequences are significantly correlated. It is also useful in determining the linearity of biological systems. Although most biological systems can be expected to be nonlinear, changes in the degree of nonlinearity often are correlated with changes in underlying physiology or pathology. For instance, Kong et al.[47] used coherence to monitor changes in blood oxygenation level in the brain by calculating the linearity of the somatosensory evoked responses. Assuming that $x(n)$ and $y(n)$ are the input and output of a linear system, the cross-spectral density $S_{yx}(f)$ is given by $S_{yx}(f) = S_x(f)H(f)$, where $S_x(f)$ is the amplitude spectral density of the input signal at f and $H(f)$ is the transfer function of the linear system. A coherence function defined as

$$C\hat{o}h_{xy}(f) = \frac{|S_{yx}(f)|^2}{S_y(f)S_x(f)} \tag{18.19}$$

is therefore unity (maximum value) for linear systems. Consequently, any deviation from unity can be used as an indicator of the presence of nonlinearity in the system. The coherence function is estimated using the values of $S_x(f)$, $S_y(f)$, and $S_{yx}(f)$ obtained through FFTs. In MATLAB, the *spectrum* and *cohere* functions can be used to evaluate the cross-spectral density and the coherence.

Statistics of the Estimates. The mean and the variance of the sample estimates of the coherence function are derived in Ref. 4, and I only reproduce the final results here. In general, the cross-spectral density is evaluated by doing the Fourier transform of a windowed (using a lag window) sequence of cross-correlation estimates. The choice of the smoothing window therefore determines the variance in the estimates of cross-spectral density (numerator in the expression of coherence). The variance of the smoothed coherence estimator is given by[4]

$$\frac{1}{2V}[1 - C\hat{o}h_{xy}(f)]^2 \tag{18.20}$$

where V is a variance reduction factor that depends on the smoothing window used. The different types of windows and their respective variance reduction factors can be found in Ref. 4. In general, the user has to be careful in choosing a large enough lag window that it includes all the largest cross-correlation estimates. An insufficient length in the lag window can give unreasonably large bias in the estimate of cross-spectral density. The coherence estimate is also biased because

$$E[C\hat{o}h_{xy}(f)] \approx \left(1 - \frac{\tau}{NT}\right) Coh_{xy}(f) \tag{18.21}$$

where τ is the time shift between the two sequences[4,48] that can be found by locating the lag corresponding to the peak estimate of cross-correlation. The source of bias can then be minimized by aligning the signals so that τ becomes negligibly small before estimating the cross-spectral density.

APPLICATION *An adaptive implementation of coherence estimation has been used in Ref. 47 to track hypoxic injury related changes in somatosensory evoked potential signals, as shown in Fig. 18.9. Fig. 18.9a shows the change in amplitude of one of the peaks in the somatosensory evoked response (SEP amplitude) from a cat at different levels of oxygen concentration in the inhaled air. After the administration of 9 percent oxygen, the SEP amplitude shows a gradual decrease until the oxygen is restored to 100 percent. A linearity index derived using the coherence measure gives a more sensitive indication of injury, as shown in Fig. 18.9b.*

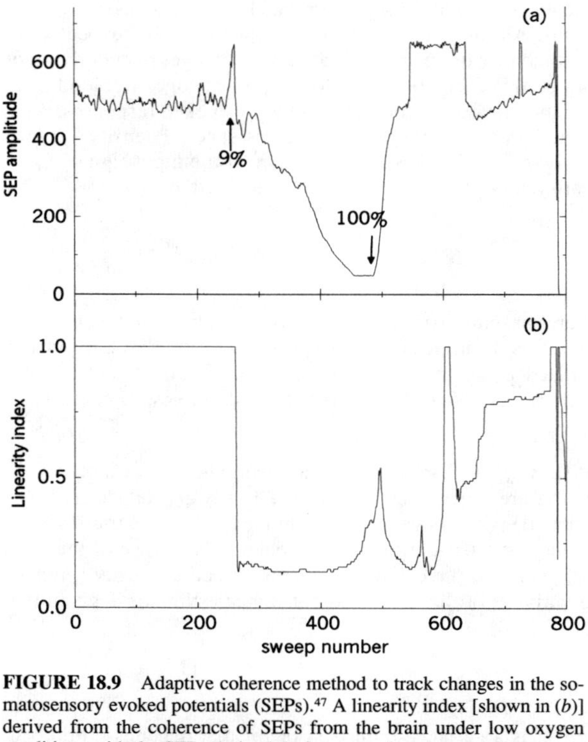

FIGURE 18.9 Adaptive coherence method to track changes in the somatosensory evoked potentials (SEPs).[47] A linearity index [shown in (*b*)] derived from the coherence of SEPs from the brain under low oxygen conditions with the SEPs derived during normal oxygen conditions was shown to be a very sensitive indicator of changes in brain oxygen content. The conventional amplitude of the SEP [shown in (*a*)] was not found to be that sensitive.

18.7 CHAOTIC SIGNALS AND FRACTAL PROCESSES

Signal Types—Random Signals That Are Not Stochastic But Are Generated by a Nonlinear Dynamic System. Fractal signals are scale-invariant, meaning that they look very similar at all levels of magnification. There is very good evidence to indicate that the beat-to-beat interval in heart rate is a fractal. Chaotic signals, on the other hand, are deterministic signals that cannot be predicted beyond a short time in the future.[2] They are extremely sensitive to initial conditions. Further, their behavior does not depend on any random inputs. The nonlinear time-series analysis techniques discussed briefly here have been used to analyze heart rate,[49–51] nerve activity,[52] renal flow,[53] arterial pressure, EEG,[54,55] and respiratory waveforms.[56]

The randomness of a chaotic time series is not due to noise but rather due to deterministic dynamics of a small number of dynamical variables in the nonlinear generating system. Therefore, they need to be distinguished from stochastic processes discussed earlier and analyzed appropriately. The randomness in a chaotic signal is therefore not revealed by statistical analysis but by dynamical analysis based on phase-space reconstruction. The phase space of a dynamical system (both nonlinear and linear) is the space spanned by its dynamical variables, and the phase plot is the plot of the time variation in the dynamical variables within the phase space. The phase plot of a dynamical system generating a chaotic time series is a strange attractor whose dimensionality (dimension of the set of points comprising the phase plot) is not an integer but a fraction. Hence they are said to have a *fractal* dimension. Standard objects in Euclidean geometry have integer dimensions. For instance, the dimension of a point in space is zero, that of a line is one, and that of an area is two. The fractal dimension D of a strange attractor (name of the phase plot of a nonlinear system generating a chaotic signal) is related to the minimum number of dynamical variables needed to model the dynamics of the strange attractor. Analysis of chaotic signals typically is geared toward (1) understanding how complex the nonlinear system is, (2) determining if it is chaotic or not, (3) determining the number of dynamic variables that dominate the system, and (4) assessing the changes in dynamic behavior of a system with different rhythms.

The following subsection details some methods to find out if a given random time series from a biological source is chaotic. Further, methods to determine the dimensionality and draw the phase plot or portrait of the chaotic signal are outlined briefly.

18.7.1 Analysis of Chaotic Signals

Frequency Analysis. The statistical analysis of chaotic signals includes spectral analysis to confirm the absence of any spectral lines, since chaotic signals do not have any periodic deterministic component. Absence of spectral lines would indicate that the signal is either chaotic or stochastic. However, chaos is a complicated nonperiodic motion distinct from stochastic processes in that the amplitude of the high-frequency spectrum shows an exponential decline. The frequency spectrum can be evaluated using FFT-based methods outlined the earlier sections.

Estimating Correlation Dimension of a Chaotic Signal. One of the ways to measure the dimensionality of the phase portrait of a chaotic signal is through what is known as the *correlation dimension* D_2. It is defined as

$$D_2 = \frac{\log C(r)}{\log r}$$

as $r \rightarrow 0$, where r is the length of the side of the hypercubes that are needed to cover the phase plot and $C(r)$ is correlation sum. The correlation sum is defined as $C(r) = \sum_{i=1}^{M(r)} p_i^2$, where p_i is the probability of a finding a single point belonging to the phase plot within the hypercube. $M(r)$ is the number of m-dimensional cells or hypercubes of side r needed to cover the entire phase plot.

One of the key steps in the analysis is therefore to reconstruct the phase plot or phase space of the dynamical system. The embedding theorems of Takens[57] and Sauer et al.[58,59] help us in reconstructing the phase space using a time series from the dynamical system rather than using the dynamical variables of the system. Given a discrete time series $x(n)$ from a biological source, we construct a k-dimensional point in the phase space using $k - 1$ time-delayed samples of the same series as represented by

$$\mathbf{x}(n) = \{x(n), x(n + \tau), x(n + 2\tau), \ldots, x[n + (k - 1)\tau]\}$$

The selection of the time delay τ is done in such a way that it makes every component of phase space uncorrelated. Therefore, τ is determined from estimate of the autocorrelation function of the time series. The time lag that corresponds to the first zero in the autocorrelation is often chosen as a good approximation for τ.[60,61] The determination of D_2 in practice can be done using the Grassberger-Procaccia algorithm[62] outlined below. Consider a pair of points in space with m dimensions ($m < k$) at time instants i and j:

$$\mathbf{x}_m(i){:}\{x(i), x(i + \tau), x(i + 2\tau), \ldots, x[i + (m - 1)\tau]\}$$

$$\mathbf{x}_m(j){:}\{x(j), x(j + \tau), x(j + 2\tau), \ldots, x[j + (m - 1)\tau]\}$$

where m is called the *embedding dimension* of the phase plot. However, m is not known a priori. Therefore, we determine the correlation dimension D_2 for different embedding dimensions of the attractor in the phase space. The minimum embedding dimension is then given by $m + 1$, where m is the embedding dimension above which the measured value of the correlation dimension D_2 for the corresponding phase plot remains constant. The Euclidean distance between the two points is given by $r_{ij}(m) = \|\mathbf{x}_m(i) - \mathbf{x}_m(j)\|$. For a critical distance r, a correlation integral (an approximation of the correlation sum defined earlier) is evaluated that gives the probability of the distance between the two given points being less than r:

$$C_2(r, m) = \frac{1}{N^2} \sum_{\substack{i,j=1 \\ i \neq j}}^{N} \theta[r - \|\mathbf{x}_m(i) - \mathbf{x}_m(j)\|]$$

where $N = k - (m - 1)\tau$
θ = Heaviside function
$C_2(r, m)$ = correlation integral

An estimate of the correlation dimension D_2 is given by

$$D_2(m) = \frac{\log[C_2(r, m)]}{\log(r)}$$

The log-log plot of $C_2(r, m)$ versus r corresponding to the given m has a linear region called the *scaling region*, the slope of which gives an estimate of the correlation dimension. The reliability of the estimated slope in the linear scaling region can be a major source of error in the Grassberger-Procaccia algorithm. If N_c point pairs $\{\log[C_2(r_i, m)], \log(r_i) | i = 1, 2, \ldots, N_c\}$ exist in the scaling region, then $D_2(m)$ is given by[63]

$$D_2(m) = \frac{N_c \sum_{i=1}^{N_c} \log(r_i) \log[C_2(r_i, m)] - \sum_{i=1}^{N_c} \log(r_i) \sum_{i=1}^{N_c} \log[C_2(r_i, m)]}{N_c \sum_{i=1}^{N_c} [\log(r_i)]^2 - \left[\sum_{i=1}^{N_c} \log(r_i)\right]^2}$$

The value of m beyond which $D_2(m)$ gradually saturates determines the embedding dimension m_c. That is, $D_2(m_c) = D_2(m_c + 1) = D_2(m_c + 2) = \cdots = D_2(m_c)$ gives an estimate of the correlation dimension.

Phase-Space Analysis. Having determined the time lag τ for decorrelating the components of the phase space and the embedding dimension, a line joining the m_c dimensional points given by

$$\mathbf{x}_{m_c}(j):\{x(j), x(j+\tau), x(j+2\tau), \ldots, x[j+(m_c-1)\tau]\}$$

(where $j = 1, 2, \ldots, n$) gives a portrait of the evolution of dynamical variables of the system. According to the Takens theorem, for an attractor having an integer-dimension manifold, the phase plot obtained from the time series preserves the topological properties (such as dimension) of the attractor. Phase-plane analysis of chaotic processes enables quantitative description subsequently through calculation of Lyapunov exponents, Poincare mapping, etc. The goal of this characterization is to obtain a portrait of evolution of the state variables of the system. Additional parameters such as the Lyapunov exponents and complexity measures can be determined as well once the biological process begins to exhibit chaotic behavior. The user is referred to Refs. 64 to 66 for further information.

There are several issues that a user has to be aware of while doing a nonlinear time-series analysis. If the biosignal in question is indeed generated by a deterministic dynamical system, then the sampling frequency of the time series should be sufficiently large to capture the deterministic rule governing the evolution of the series. The length of the sequence should also be sufficiently large. Fortunately, there is a rule of thumb given by Ruelle[67] for the minimum length of time series that is quite helpful. It states that estimates of the dimension of the phase plot $D \geqslant 2 \log_{10} N$ should be regarded as unreliable where N is the length of the time series. Finally, noise contamination during measurement could obscure detection of deterministic dynamics and hence degrade forecasting of biologically significant events. The user is referred to Ref. 66 for MATLAB codes for the Grassberger-Procaccia algorithm, phase-space reconstruction, and forecasting.

Surrogate Data to Compare Chaotic and Stochastic Processes. Surrogate data are used in nonlinear time-series analysis to make useful comparisons between the given biosignal, which could be chaotic, and a set of artificially generated stochastic signals that share the same essential statistical properties (mean, variance, and the power spectrum) with the biosignal.[66,68] If the measured topological properties of the biosignal lie within the standard deviation of the topological properties measured from the artificially generated surrogate data, then the null hypothesis that the biosignal is just random noise cannot be ruled out. A systematic method for generating surrogate data with the same mean, variance, and power spectrum as a given biosignal is detailed in Ref. 68, and a MATLAB code is available in Ref. 66. The main steps in the procedure are outlined below.

The DFT $X(k)$ of the observed signal sequence $x(n)$ is evaluated using Eq. (18.1). Keeping the magnitudes of the Fourier coefficients intact, the phases of the Fourier coefficients are randomized (maintaining symmetry about the midpoint so that we still get a real time series after inverse Fourier transform). The Fourier sequence is now inverted to produce a real time series that is a surrogate of the original data. With different random assignments for the phase values of Fourier coefficients, a different time series can be generated every iteration. All the different time series generated using this procedure would be part of the surrogate data to the biosignal. The random phase assignments are done as follows: If the length of the measured sequence is N (assumed to be even), a set of random phase values $\phi_m \epsilon [0, \pi]$, for $m = 2, 3, \ldots, N/2$, is generated using a function such as *rand* in MATLAB. The phase of the Fourier coefficients are randomized to generate a new sequence of Fourier coefficients $X_s(k)$ as shown below:

$$X_s(k) = \begin{cases} X(k) & \text{for } k = 1 \text{ and } \dfrac{N}{2} + 1 \\ |X(k)|e^{j\phi_m} & \text{for } k = 2, 3, \ldots, \dfrac{N}{2} \\ |X(N-k+2)|e^{j\phi_{N-k+2}} & \text{for } k = \dfrac{N}{2} + 2, \dfrac{N}{2} + 3, \ldots, N \end{cases}$$

For every new set of $\phi_m \epsilon [0, \pi]$, we can generate a new surrogate time series $x_s(n)$ by performing the inverse Fourier transform of $X_s(k)$.

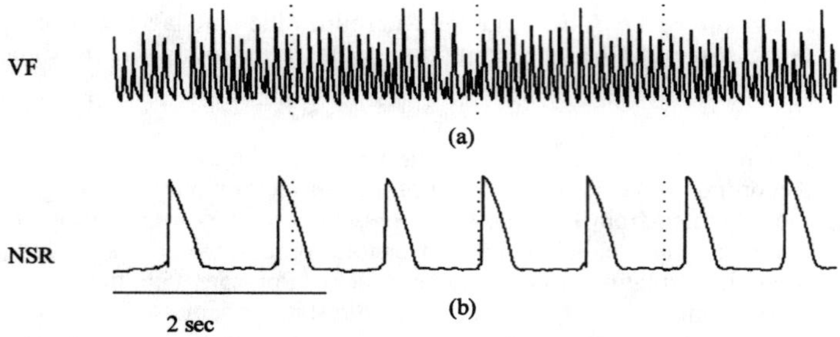

FIGURE 18.10 Action potential waveforms from an isolated heart (Langendorff preparation) are shown[63] (*a*) during ventricular fibrillation (VF) and (*b*) during normal sinus rhythm (NSR).

APPLICATION *Myocardial cell action potentials collected from a Langendorff setup of an isolated rabbit heart experiment have been found to exhibit chaotic behavior.*[63] *Two different cardiac cell action potential waveforms are illustrated in Fig. 18.10 corresponding to ventricular fibrillation (VF) in (a) and normal sinus rhythm (NSR) in (b). The corresponding frequency spectrum is shown in Fig. 18.11. VF certainly shows chaotic characteristics in Fig. 18.11a, whereas NSR shows distinct spectral lines in Fig. 18.11b, indicating the presence of deterministic components. Using the calculated correlation dimensions for the two different action potentials (VF = 5.629, NSR = 2.704), the embedding dimension is m = 6 for VF. The projections of the trajectories for VF in a three-*

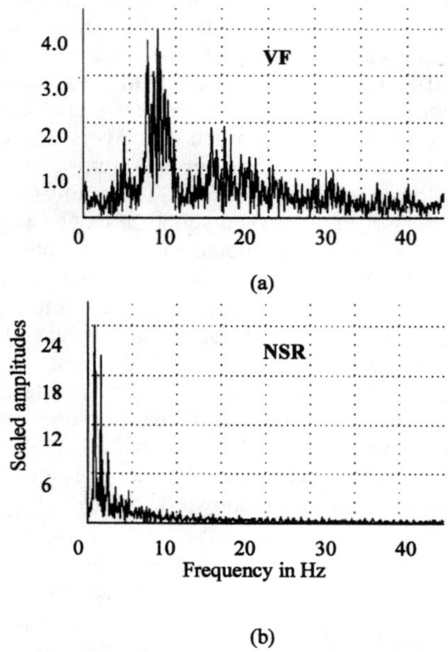

FIGURE 18.11 Power spectrum of the action potential waveform during[63] (*a*) VF and (*b*) NSR. The power spectrum of VF has no spectral components or peaks in contrast with the power spectrum of NSR, which has distinct spectral lines indicating that it is a periodic and not a stochastic or chaotic waveform.

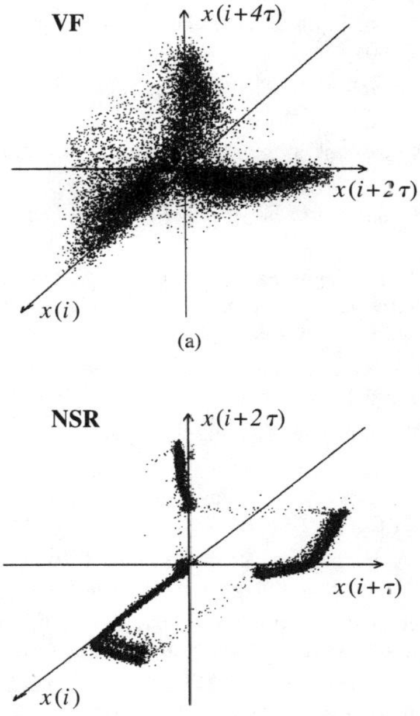

FIGURE 18.12 Phase plots of the action potential waveform during[63] (*a*) VF and (*b*) NSR. The time delay required to make the individual components of the phase plot uncorrelated was determined to be $6\delta t$, and $41\delta t$, respectively, where $\delta t = 5$ ms is the sampling interval. The embedding dimension of the VF waveform was determined to be 6, and that of the NSR waveform was 3. The phase plot of the VF waveform is shown in (*a*) spanned by $\{x(i), x(i+2\tau), x(i+4\tau)\}$, demonstrating a strange attractor, whereas the phase plot of the NSR waveform is shown in (*b*) spanned by $\{x(i), x(i+\tau), x(i+2\tau)\}$.

dimensional phase space spanned by $\{x(i), x(i+2\tau), x(i+4\tau)\}$ *is shown in Fig. 18.12a, whereas for NSR the embedding dimension* $m = 3$. *The corresponding three-dimensional phase portrait spanned by* $\{x(i), x(i+\tau), x(i+2\tau)\}$ *is shown in Fig. 18.12b. For VF and NSR, the decorrelating time* τ *was determined to be* $6\delta t$, *and* $41\delta t$ *respectively, where* $\delta t = 5$ *ms is the sampling interval. Obviously, VF demonstrates chaotic character, whereas NSR demonstrates a periodic or quasi-periodic nature.*

REFERENCES

1. M. Akay, *Biomedical Signal Processing*, Academic Press, San Diego, 1994.
2. E. N. Bruce, *Biomedical Signal Processing and Signal Modeling*, Wiley, New York, 2001.
3. S. M. Kay and S. L. Marple, Spectrum analysis: A modern perspective, *Proc. IEEE* **69**(11):1380–1419 (1981).
4. R. Shiavi, *Introduction to Applied Statistical Signal Analysis*, 2d ed. Academic Press, San Diego, 1999.
5. P. M. T. Broersen. Facts and fiction in spectral analysis, *IEEE Trans. Instum. Measur.* **49**:766–772 (2000).

6. P. M. T. Broersen, Finite sample criteria for autoregressive order selection, *IEEE Trans Signal Proc.* **48**: 3550–3558 (2000).
7. J. Muthuswamy and N. V. Thakor, Spectral analysis of neurological signals, *J. Neurosci. Methods* **83**:1–14 (1998).
8. H. Akaike, A new look at the statistical model identification, *IEEE Trans. Automatic Control* **AC-19**:716–723 (1974).
9. J. P. Burg, *Maximum entropy spectral analysis*, Dept. of Geophysics, Stanford University, Stanford, Calif., 1975.
10. K. A. Wear, R. F. Wagner, and B. S. Garra, A comparison of autoregressive spectral estimation algorithms and order determination methods in ultrasonic tissue characterization, *IEEE Trans Ultrason. Ferroelect. Frequency Control* **42**:709–716 (1995).
11. F. Hlawatsch and G. F. Boudreaux-Bartels, Linear and quadratic time-frequency signal representations, *IEEE Signal Processing Magazine*, pp. 21–67, 1992.
12. J. Z. Wang, Wavelets and imaging informatics: A review of the literature, *J. Biomed. Inform.* **34**:129–141 (2001).
13. V. von Tscharner and K. R. Thulborn, Specified-resolution wavelet analysis of activation patterns from BOLD contrast fMRI, *IEEE Trans. Med. Imaging* **20**:704–714 (2001).
14. M. Desco, J. A. Hernandez, A. Santos, and M. Brammer, Multiresolution analysis in fMRI: Sensitivity and specificity in the detection of brain activation, *Hum. Brain Mapp.* **14**:16–27 (2001).
15. D. F. Schomer, A. A. Elekes, J. D. Hazle, J. C. Huffman, S. K. Thompson, C. K. Chui, and W. A. Murphy, Jr., Introduction to wavelet-based compression of medical images, *Radiographics* **18**:469–481 (1998).
16. K. Wang, H. Begleiter, and B. Porjesz, Spatial enhancement of event-related potentials using multiresolution analysis, *Brain Topogr.* **10**:191–200 (1998).
17. P. J. Phillips, The design of matching pursuit filters, *Network* **9**:1–17 (1998).
18. K. L. Park, K. J. Lee, and H. R. Yoon, Application of a wavelet adaptive filter to minimise distortion of the ST-segment, *Med. Biol. Eng. Comput.* **36**:581–586 (1998).
19. N. D. Panagiotacopulos, J. S. Lee, M. H. Pope, and K. Friesen, Evaluation of EMG signals from rehabilitated patients with lower back pain using wavelets, *J. Electromyogr. Kinesiol.* **8**:269–278 (1998).
20. J. Jalife, O. Berenfeld, A. Skanes, and R. Mandapati, Mechanisms of atrial fibrillation: Mother rotors or multiple daughter wavelets, or both? *J. Cardiovasc. Electrophysiol.* **9**:S2–12 (1998).
21. R. Gobbele, H. Buchner, and G. Curio, High-frequency (600 Hz) SEP activities originating in the subcortical and cortical human somatosensory system, *Electroencephalogr. Clin. Neurophysiol.* **108**:182–189 (1998).
22. Q. Fang and I. Cosic, Protein structure analysis using the resonant recognition model and wavelet transforms, *Australas. Phys. Eng. Sci. Med.* **21**:179–185 (1998).
23. P. Carre, H. Leman, C. Fernandez, and C. Marque, Denoising of the uterine EHG by an undecimated wavelet transform, *IEEE Trans. Biomed. Eng.* **45**:1104–1113 (1998).
24. N. Aydin, S. Padayachee, and H. S. Markus, The use of the wavelet transform to describe embolic signals, *Ultrasound Med. Biol.* **25**:953–958 (1999).
25. V. J. Samar, A. Bopardikar, R. Rao, and K. Swartz, Wavelet analysis of neuroelectric waveforms: A conceptual tutorial, *Brain Lang.* **66**:7–60 (1999).
26. M. Akay and C. Mello, Wavelets for biomedical signal processing, presented at 19th International Conference of IEEE/EMBS, Chicago, 1997.
27. R. Kronland-Martinet, J. Morlet, and A. Grossmann, Analysis of sound patterns through wavelet transforms, *Int. J. Pattern Recog. Artif. Intell.* **1**:273–301 (1987).
28. B. Gramatikov, S. Yi-chun, P. Caminal, H. Rix, and N. V. Thakor, Multiresolution wavelet analysis of the body surface ECG before and after angioplasty, *Ann. Biomed. Eng.* **23**:553–561 (1995).
29. H. L. van Trees, *Detection, Estimation and Modulation Theory*, Vol. **I**, Wiley, New York, 1969.
30. R. O. Duda, P. E. Hart, and D. G. Stork, *Pattern Classification*, Wiley, New York, 2001.
31. A. Hyvarinen and E. Oja, Independent component analysis: Algorithms and applications, *Neural Networks* **13**:411–430 (2000).

32. S. Makeig, A. Bell, T. P. Jung, and T. J. Sejnowski, Independent component analysis of electroencephalographic data, in *Advances in Neural Information Processing Systems*, Vol. **8**, pp. 145–151, MIT Press, Cambridge, Mass., 1996.

33. S. Tong, A. Bezerianos, J. Paul, Y. Zhu, and N. Thakor, Removal of ECG interference from the EEG recordings in animals using independent component analysis, *J. Neurosci. Methods* **108**:11–17 (2001).

34. R. Vigario, J. Sarela, and E. Oja, Independent component analysis in wave decomposition of auditory evoked fields, presented at International Conference on Artificial Neural Networks (ICANN '98), Skovde, Sweden, 1998.

35. R. Vigario, V. Jousmaki, M. Hamalainen, R. Hari, and E. Oja, Independent component analysis for identification of artifacts in magnetoencephalographic recordings, in *Advances in Neural Information Processing Systems*, pp. 229–235, MIT Press, Cambridge, Mass., 1998.

36. J. P. Nadal and N. Parga, Nonlinear neurons in the low noise limit: a factorial code maximized information transfer, *Network* **5**:565–581 (1994).

37. P. Comon, Independent component analysis: A new concept? *Signal Processing* **36**:287–314 (1994).

38. M. Girolami and C. Fyfe, Extraction of independent signal sources using a deflationary exploratory projection pursuit network with lateral inhibition, *IEEE Proc. Vision Image Signal Proc.* **14**:299–306 (1997).

39. A. Hyvarinen and E. Oja, A fast fixed-point algorithm for independent component analysis, *Neural Comput.* **9**:1483–1492 (1997).

40. P. H. Boeijinga and F. H. Lopes da Silva, A new method to estimate time delays between EEG signals applied to beta activity of the olfactory cortical areas, *Electroencephalogr. Clin. Neurophysiol.* **73**:198–205 (1989).

41. Z. Bohdanecky, P. Lansky, and T. Radil, An integral measure of the coherence function between pairs of EEG recordings, *Electroencephalogr. Clin. Neurophysiol.* **54**:587–590 (1982).

42. P. Brown, A. Oliviero, P. Mazzone, A. Insola, P. Tonali, and V. Di Lazzaro, Dopamine dependency of oscillations between subthalamic nucleus and pallidum in Parkinson's disease, *J. Neurosci.* **21**:1033–1038 (2001).

43. S. Cerutti, M. Alberti, G. Baselli, O. Rimoldi, A. Malliani, M. Merri, and M. Pagani, Automatic assessment of the interaction between respiration and heart rate variability signal, *Med. Prog. Technol.* **14**:7–19 (1988).

44. M. P. Davey, J. D Victor, and N. D. Schiff, Power spectra and coherence in the EEG of a vegetative patient with severe asymmetric brain damage, *Clin. Neurophysiol.* **111**:1949–1954 (2000).

45. L. Fendelander, P. W. Hsia, and R. J. Damiano, Jr., Spatial coherence: A new method of quantifying myocardial electrical organization using multichannel epicardial electrograms, *J. Electrocardiol.* **30**:9–19 (1997).

46. T. B. Kuo, C. C. Yang, and S. H. Chan, Transfer function analysis of ventilatory influence on systemic arterial pressure in the rat, *Am. J. Physiol.* **271**:H2108–2115 (1996).

47. X. Kong and N. Thakor, Adaptive coherence estimation reveals nonlinear processes in injured brain, in *IEEE Proc. Intl. Conf. Acoustics, Speech, and Signal Processing, ICASSP-93*, **1**:89–90 (1993).

48. G. Carter, Coherence and time-delay estimation, in C. Chen (ed.), *Signal Processing Handbook*, Dekker, New York, 1988.

49. M. R. Guevara, L. Glass, and A. Shrier, Phase locking, period doubling bifurcations, and irregular dynamics in periodically stimulated cardiac cells, *Science* **214**:1350–1352 (1981).

50. D. R. Chialvo, D. C. Michaels, and J. Jalife, Supernormal excitability as a mechanism of chaotic dynamics of activation in cardiac purkinje fibers, *Circ. Res.* **66**:525–545 (1990).

51. K. M. Stein, N. Lippman, and P. Kligfield, Fractal rhythms of the heart, *J. Electrocardiol.* **24**(suppl.):72–76 (1992).

52. Z. S. Huang, G. L. Gebber, S. Zhong, and S. Barman, Forced oscillations in sympathetic nerve discharge, *Am. J. Physiol.* **263**:R564–571 (1992).

53. K. P. Yip, N. H. Holstein-Rathlou, and D. J. Marsh, Chaos in blood flow control in genetic and renovascular hypertensive rats, *Am. J. Physiol.* **261**:F400–408 (1991).

54. L. D. Iasemidis, L. D. Olson, R. S. Savit, and J. C. Sackellares, Time dependencies in the occurrences of epileptic seizures, *Epilepsy Res.* **17**:81–94 (1994).

55. C. A. Skarda and W. J. Freeman, How brains make chaos in order to make sense of the world, *Behav. Brain Sci.* **10**:161–195 (1987).

56. D. Hoyer, K. Schmidt, R. Bauer, U. Zwiener, M. Kohler, B. Luthke, and M. Eiselt, Nonlinear analysis of heart rate and respiratory dynamics, *IEEE Eng. Med. Biol. Mag.* **16**:31–39 (1997).

57. F. Takens, Detecting strange attractors in turbulence, D. A. Rand and L. S. Young (eds.), in *Dynamical Systems and Turbulence*, Springer, Berlin, 1981.

58. T. Sauer and J. A. Yorke, Rigorous verification of trajectories for computer simulation of dynamical systems, *Nonlinearity* **4**:961–979 (1991).

59. T. Sauer, J. Yorke, and M. Casdagli, Embedology, *J. Stat. Phys.* **65**:579–616 (1994).

60. D. Holton and R. M. May, Distinguishing chaos from noise, in *The Nature of Chaos*, Chap. 7, Oxford University Press, Oxford, England, 1993.

61. A. M. Fraser and H. L. Swinney, Using mutual information to find independent coordinates of strange attractors, *Phys. Rev. [A]* **33**:1134–1139 (1986).

62. P. Grassberger and I. Procaccia, Characterization of strange attractors, *Phys. Rev. Lett.* **50**:346–349 (1983).

63. X. S. Zhang and Y. S. Zhu, The extraction of dynamical information of myocardial cell electrophysiology by fractal and chaotic analysis methods, *Automedica* **18**:9–26 (1999).

64. F. Kasper and H. G. Schuster, Easily calculable measure for the complexity of spatiotemporal patterns, *Phys. Rev. [A]* **36**:842–848 (1987).

65. A. Lempel and J. Ziv, On the complexity of finite sequences, *IEEE Trans. Inform. Technol.* **22**:75–81 (1976).

66. B. Henry, N. Lovell, and F. Camacho, Nonlinear dynamics time series analysis, M. Akay (ed.), in *Nonlinear Biomedical Signal Processing*, Vol. II, pp. 1–39, IEEE Press, New York, 2000.

67. D. Ruelle, Deterministic chaos: The science and the fiction, *Proc. R. Soc. Lond. [A]* **427**:241–248 (1990).

68. J. Theiler, B. Galdrikian, A. Longtin, S. Eubank, and J. D. Farmer, Using surrogate data to detect nonlinearity in time series, M. Casdagli and S. Eubank (eds.), in *Nonlinear Modeling and Forecasting*, Addison-Wesley, Reading, Mass., 1992.

P • A • R • T • 5

DESIGN OF MEDICAL DEVICES AND DIAGNOSTIC INSTRUMENTATION

CHAPTER 19
MEDICAL PRODUCT DESIGN

James P. O'Leary
Tufts University, Medford, Massachusetts

19.1 INTRODUCTION/OVERVIEW

The design of a medical product is a complex task. All design activities involve the resolution of conflicts and compromise among the desired features, but in medical products the conflicts tend to be more intense, the stakes are often higher, and the background information is often more uncertain. This section describes a process that has been found to be useful in bringing successful products to market. It is based on an approach to design that has recently been described as *concurrent engineering*.

This section opens with some groundwork on getting a program started, follows that with a somewhat specific set of steps to be carried out as the design is developed (Fig. 19.1), and then examines some issues that pervade the entire process, reviewing how these concerns might affect a design and in particular the design of a medical device. Figure 19.1 shows the steps in the process to be described.

In order to be more specific about some of the details, an example is sometimes necessary. In this section that product is an improved system of exterior fixation of long bone fractures. In exterior fixation, pins of some type are attached through the skin to the bones above and below the fracture and these pins are in turn attached to an external structure that maintains the position of the bone segments during healing. This is a problem that is easy to picture and understand, which makes it a good example for this document. It is a difficult area in which to make progress, however, and none will be made here. It will only serve as an illustration.

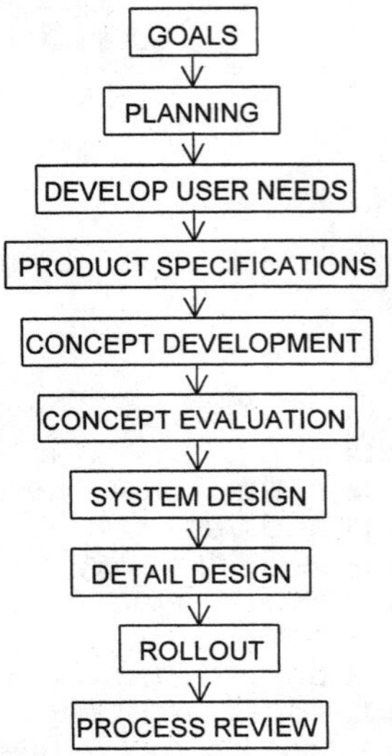

FIGURE 19.1 Elements of the medical device design process.

Everything has to start somewhere, and a medical product may emanate from a variety of originating events, but at some point a decision is made to pursue a certain need, question, request, or idea, and to devote some resources toward that effort. At that point there is an embryonic project, and the device development effort has begun. A preliminary examination is made, a goal is defined, perhaps a report is prepared, some estimates are made of the market, the cost of development, the time to market, the fit with the existing organization plans, and the competition. If the aggregate of these estimates is attractive, a decision is made to develop a new product, and a planning stage is funded. The process here will start with that planning activity.

Sometime before, after, or during the planning process, the development team is assembled. Even though it may not follow in that sequence, the process of building the team will be discussed even before the planning phase. The steps in building the team are probably as important as any in a serious development task. Even if it is to occur later, it must be considered carefully in laying out the plan.

With a plan and a team in place, the next step is to carefully define the characteristics that the product is to have. A medical product fulfills a set of needs, which are referred to here as the user needs. It is most important that the product does indeed fill a legitimate and clearly defined need. A process is described here for arriving at and documenting the need, through contact with potential users, medical professionals, patients, and others who will be affected by the device.

With a clear vision of the desired outcome, the team can then proceed to converting the desires of the various parties into a set of specifications for the design. These requirements are to be stated in measurable terms, allowing for careful evaluation of ideas in the early stages and progress as the design is solidified.

The next step in the process is to generate concepts. Although it is frequently the case that a program starts with a solution, all too often the starting idea is not the best one. A concerted effort must be made to seek the best possible idea or concept to exploit. The approach taken is to generate as long and broad a set of alternatives as possible. The logic and some processes for doing this will be discussed.

With a large set of alternatives in hand, the list must be winnowed down to a number that can be pursued. This may be a single idea, but it is preferred that several are taken through the next few steps. There are selection tools or methodologies that can help in this process. A great deal of judgment is required, and great care is advised here.

With the concept selection completed, a design must be executed. This is accomplished in the system and detail design phases. Elements here are highly dependent on the nature of the product, so that the process here is less explicitly described than the previous steps. This activity is greatly influenced by several of the issues mentioned previously, so a framework will be laid here and then these ancillary issues will be discussed along with ways they might interact with the design itself.

With a medical product designed, it must be tested thoroughly, not just to verify its efficacy and safety, but to assure that it has those characteristics that embody a successful product. There are strategies for managing the prototyping and testing process in ways that enhance the product's likelihood of success.

The last activity in the process is the start-up of manufacturing and the product rollout. This event is the result of careful planning done though the duration of the project, often starting on the very first day! Some of the special considerations discussed here cannot be allowed to wait, but are integrated into all the activities. Nonetheless, it will be discussed as a phase near the end of the section.

The last part of this discussion will be devoted to some overarching issues: documentation, the roll of design tools and other resources, and regulatory issues and their impact on the design activity.

19.2 SCOPE

The term *medical product* can describe things that vary in size, scope, complexity, importance, and cost by several orders of magnitude. All of the material covered here is applicable over virtually all of that range, but the relative importance of and effort spent in various elements will vary in the same way. A stage that requires a year in the development of one product may be completed as part of a single meeting on another. The reader is cautioned to exercise judgment in this regard. Judgment will be found to play a major role throughout any product development effort.

19.3 QUALITIES OF SUCCESSFUL PRODUCT DESIGN

Before proceeding it may be useful to define the objectives of this section more specifically. Eppinger and Ulrich[1] suggest that there are five aspects to the quality of a product design effort:

- Product quality
- Product cost
- Development cost
- Development time
- Enhanced development capability

Product quality in this context is the overall accumulation of desired characteristics—ease of use, precision, attractiveness, and all of the other things that one finds attractive in artifacts. Cost is, of course, the cost to manufacture. In general, the market determines a product's selling price, while the cost is determined by the design. There are differing opinions about where in the design process cost becomes locked in—in the conception, the details, or the manufacturing start-up—and it no doubt differs for various products, but it is always within the scope of the material discussed here. The profitability of the project depends on the difference between the two, the gross margin. That difference must cover the other expenses of the enterprise with something left over. In the medical device business, those other expenses are much higher than in most sectors. Costs of selling, tracking, providing customer service, etc., are higher for medical devices than for other products of equivalent function. The development cost of the product is another of those expenses that must be covered by the margin. (It is also one that is higher in the medical device sector.) It represents an investment and is apportioned over the life of the product. Development time is often tightly related to development cost, as projects tend to have a spending rate. In addition, the investment in development cost must be recovered in a timely way, which only begins when the product begins to sell. Furthermore, it is generally accepted that the product will eventually be replaced by another in the market, and that the "window" of opportunity opens when the product is introduced and begins to close when its successor appears. The design team can widen the window by an earlier introduction.

The last of these qualities needs some explanation. The process used here results in good documentation of the design activities. This results in an opportunity for those engaged to re-evaluate the work done and to learn from that effort. Coming out of a design effort with an organization better equipped to undertake the next task is an important benefit of doing it well. Design of medical products is always done in a rapidly changing milieu where the rate of learning is very important.

19.4 CONCURRENT ENGINEERING

Although a recently coined term, concurrent engineering embodies an approach that has been around for a very long time. Simply stated, it asks that a design be developed from the outset by a team that is capable of respecting all aspects of the problems faced. This has been contrasted with the style of design organization sometimes described as the waterfall scheme, where a group does part of the design task and then passes their results on to another group, which makes its contribution and in turn passes the result on again. The most popular image of this is someone throwing a document package "over a wall" to the next phase. Concurrent engineering is based on assembling a complete design team at the time the project is begun, and having it work as a unit throughout the design cycle. A great deal of the success of this technique depends on the composition of the team.

19.5 GOALS

The preliminary efforts should have defined a goal. In the corporate world, this must usually be well defined in order to be approved, but it's a good idea to step back at the beginning and be sure that the goal is clearly defined and agreed on by all parties. There are times when the goal of the project starts out as an assignment or charge from a higher point in the organization or it might originate within the group. Regardless of origin, it's a good idea to critique it, edit it, and turn it into a single sentence, often with a few important adjectives. When sufficiently polished, it should be displayed in some way that assists the team in remaining focused.

In the case of the external fixation system, the goal might be: *Develop an external fixation system, that can be quickly installed at all of the common long bone fracture sites, using as few as possible distinct elements.*

19.6 TEAM/TALENT

There are two important aspects to building or recruiting a product design team: covering all of the specialties required to accomplish the task, and assembling a group of individuals who can work together effectively and efficiently. It is much easier to discuss the first of these. Although every function within a corporation will have some impact on how successful the effort is, the product design team is usually built around a core of people representing marketing, engineering, and production. In the case of a medical device, an additional specialty in regulatory affairs is important. The size of this group, the core team, will depend on the nature of the project and could be from four people, each at one-quarter time, to several dozen full-time key people.

The team will have a working relationship with one or more clinicians who are expected to be heavy users of the product. Very often this will be an individual who suggested the product opportunity. This person can be categorized as an expert user, but plays a different role than the users that the team will deal with later. In almost every case, this is a small part of this person's professional activity. He or she has an active practice and is involved here for reasons of intellectual pride or curiosity. The help of these individuals is vital in formulating the problem and sometimes in maintaining a correct focus. Finding high-caliber practitioners who are effective in this role can be very difficult. Many find the give and take of the product design and evolution process frustrating, or do not find the design team environment very comfortable. Recruiting a knowledgeable clinician who is helpful in this regard should be an early priority.

The team leader, who might carry the title product manager or some variation of that, may come from any of the four specialties mentioned earlier, but must be capable of providing insight in each of them and leadership to the entire team. If possible, the leader should have some control over the membership of the core team. The effectiveness of communication among the core team will have much to do with the success of the program, so the leader must feel that the team will be candid, both with the leader and with each other as problems and challenges arise.

Another position often mentioned in the product design literature is the "champion." This is an individual who is high enough in the organization to influence the spending of resources at the level required and who is personally committed to the project. In some cases the champion may be the team leader, but more often is the person the leader reports to. Regardless of the titles and terminology, the champion is a vital part of the team. If this person doesn't exist, the project will not weather the first storm.

The team leader will carry the primary project management responsibilities. This includes drafting the original plan, defining schedule and resource requirements, assessing the various risks that things will not proceed as planned, and watching the progress of the plan in all of its details. Much has been written about the role of the leader, and most professionals with some experience recognize the importance of the leadership, management, and technical skills required for a successful program.

If there are team members new to the medical device field, they should bear in mind that many of the rules of thumb that are useful in managing a product design effort need to be recalibrated in this sector. Things take longer, cost more, and are more difficult to pin down. The same items are important but the parameters are often different.

19.7 PLANNING/RESOURCES

Before the project gets underway, a fairly detailed plan must be put into place. A good plan is the best possible start for the project. The team leader along with the core team should compile a document that spells out the schedule, the people and other assets required to accomplish the various tasks.

The plan should identify gating events. These are the points in the project where an evaluation of progress and potential outcomes is made. The plan is held to, changed, or abandoned on the basis of the gating events. These points should be well defined at the planning stage. The level of formality at the gate points depends on the nature of the product and the corporate culture. The need for meeting regulatory requirements will guide the choice of some of these gates, but other issues will also play a role. The points where the rate of spending increases or when other commitments must be made are good choices.

The planning process also involves the setting of targets. If the five measures of success mentioned earlier are considered, there is at least one target imbedded in each of them. First there is a product of high quality, with quality measured in market acceptance, performance, and perhaps other attributes. Target values should be spelled out and serve as a focus throughout the project. Likewise there are manufacturing cost, development cost, and development time targets that should be delineated. These should be aggressive without being absurd. In the current business environment, there is little room for any but the best of products. To be successful, the product must have a strong advantage over its competition. It will have to offer something unattainable elsewhere. A strong focus on the advantage to be attained is very important. If the thing is worth doing, it is worth doing well.

It is difficult to set measures for the enhancement of capabilities that is the final measure of success. One set of objectives could be the establishment and documentation of procedures if the group is doing things in a really new way. A second approach is to challenge the time and cost schedules. In a larger, more mature organization, a formal benchmarking process can be used to measure the strength and effectiveness of the development team. See Tucker et al.[2] for an overview of this approach.

19.8 DEVELOPING USER NEEDS

Before proceeding to design a product, it is extremely important to establish clearly what it is that the product will do. It is important that this sequence be understood and respected. First the function required of the device will be defined, and only then are the means to accomplish the function sought. Many design efforts have resulted in failure because the object was defined first and the major effort was aimed at proving the device was needed, trying to make the market fit the design.

It doesn't work that way unless a team is extremely lucky. The place to start is to define what is referred to here as the need, with no regard to the means of meeting the need.

Before digging into the need definition process, the term *stakeholder* must be introduced. In this context, the stakeholders are all of the individuals who are going to be affected by the introduction of this product. There are often a large number of individuals represented here. They range from the those who work in the manufacturing facility where the device is produced, to the patient who benefits most directly, to the organization responsible for disposing of it when it is no longer functional, with many people in between. All of these people have some interaction with the product, will influence its effectiveness, and will in some way contribute to the success or failure of the venture. The definition of need begins with the identification of the stakeholders and an evaluation of the contribution each can make to the process.

The actual user of the device obviously heads the list of those who must be consulted, and this discussion will focus on this individual first. This might be a surgeon, a therapist, a nurse, an operating room technician, or some other member of the health care delivery system. Regardless of the role, the purpose of the device under development is to in some way assist that individual, to do something that could not be done before, to make something easier to do, to make it safer, less invasive, less traumatic, or in some other form more desirable. A chronic problem with designers is that they try to make things better from their own point of view, rather than that of the user.

The knowledge that is sought from these stakeholders concerns the desired characteristics of the product. What would make it a good or attractive device from their point of view? It is important to maintain a separation here from any particular design or solution, and not to ask the individual to "design" the device. (Although often they will wish to do just that.) As an example, a useful piece of knowledge is that "the device should weigh as little as possible." Some stakeholders will want to suggest making the product out of aluminum or foamed plastic, in essence jumping over the issue, toward a design solution. We know that choosing a low-density material is not always the best way to attain a lighter weight in the finished product. By exploring the rationale for the suggestion, the developer can get a clearer understanding of the user's desires. Try to understand the attitude and try and get it into functional requirements in the user's own words.

In medical products the stakeholder group also includes the patient, and sometimes the patient's family. If we consider the external fixation device, the person most interested in keeping the weight down is the patient who may be encumbered by it for a period of months. That is probably not the only problem, however. In this instance, an effort should be made to locate a number of patients who have had external fixations and gather the available information.

The patients never see many medical devices, so that the potential contribution from this source must be evaluated for each product. Examine the circumstances carefully. The error that is often made is to rely on the clinician to present the patient's point of view. If the patient is conscious of interacting with the device, he or she should be spoken to directly.

There are a variety of techniques for soliciting the opinion of the stakeholder in this context; questionnaires, interviews, focus groups, and direct observation have all been used effectively. The best choice in a particular situation will depend on many circumstances. It is important that the choice not be based on convenience, cost, or time constraints, but on the nature of the product, the state of development of products of this type, the user group itself, and a myriad of other issues.

There are some concerns in selecting the individual users for this process. Von Hipple[3] has identified a group of individuals he calls "lead users." These are the pacesetters in the area in which they work. They are always looking for new ideas and new ways to do things. Lead users should be part of the user group studied for two reasons. They are often critical of ideas and are very helpful in identifying new directions. In addition, these individuals are often the first to be willing to try out new products and their peers will look to them for guidance in this regard. It is also important that those queried include representatives of average users of a product. It is possible to design a product that can only be used by or is only useful to the a few practitioners at the very top of their field. Producing a product of this kind can be prestigious, but it is seldom a good business decision. (There are many examples of prototype and custom-built devices, particularly surgical instruments, made for these individuals, that were successful in the hands of the target user but were not commercialized because there was no expectation of a sizable market.)

An additional concern that should be recognized here is that many aspects of medical practice vary, often geographically and sometimes in more complex patterns. A procedure may be done frequently in one facility and rarely in another, which would seem to have a similar patient base. At the international scale, the variations can be striking. Rutter and Donaldson[4] provide an interesting examination of some of these issues. In assessing the needs for a device, patterns of treatment that are relative to the problem should be examined carefully.

The material collected from this exercise should be sorted, characterized, and summarized into a document that is the first real step toward the design. Individual statements from stakeholders should be preserved verbatim. Condensation should only remove redundancy. If we have seven people telling us that the design must be such that it can be operated with one hand, at least one of them should be quoted in the needs document. At a later point in the process, it will be important to examine whether the request was met, and it should be tested against the user's words, not the developer's interpretation of the user's need.

When a list of the needs has been established, the team should assign weights to each item. This can be done on a scale of 1 to 5 or 1 to 10, or even a three-level scale of "highly," "moderately," and "minimally" important.

This is the first time that the team doing numerical evaluations has been mentioned, and it brings up several potential problems. In this instance, there may be some tendency to bias the scores according to what is perceived to be obtainable. If we know that keeping the weight of the product down is going to be difficult, we know that we will feel better if somehow it is not so important. Unfortunately, how we feel has little to do with how important it might be, so the team must gather itself up and evaluate things from the point of view of the user, regardless of the difficult position into which it may put itself.

The second risk in team scoring is team members who strategize. Typically, this takes the form of stating a position that is different and more extreme than that which is held, in order to offset the position of another team member. If I think the item is a 6, and I think you want to score it an 8, I would give it a 4 to offset your opinion. Some individuals have difficulty resisting the desire to do this, and some may even do it unconsciously. It is one of the responsibilities of the team leader to recognize that this is happening and to deal with it. One measure of an effective team is the absence of this effect. If it were to happen frequently, the team could be considered dysfunctional.

The most highly developed and detailed processes for obtaining user input have been described under the title *quality function deployment* (QFD). For a more detailed description of this process, see Hauser and Clausing.[5]

19.9 PRODUCT SPECIFICATIONS

The document that clarifies the needs is expressed in the language of the user. The next step in our process is to transform these needs into specifications that can be used to guide the design decisions. These product specifications become the targets for the product development and as such they need to be measurable. In general there should be a specification for each identified need, but there are exceptions to this that run in both directions.

The important characteristic of specifications is that they must be expressed in measurable terms, e.g., "The device must weigh less than 2.3 pounds," or "The power consumption must be less than 350 watts." While the user needs were often qualitative, the purpose now is to write requirements that can be used to evaluate the ideas and designs produced. Before proceeding to define the specifications, the metric or unit of measure for each of the needs must be identified. These are often obvious, but sometimes require some thought. Most requirements will state maximum or minimum values, like weight and power consumption. Sometimes there are actual target values that are desired, such as "Must fit in size B front end."

The metrics selected for each requirement should be considered carefully. They should be the common measures for simple characteristics. The difficulty of making the measurements must be considered. These specifications could easily turn into quality assurance requirements in the product, necessitating numerous repetitions of the measurement. It will be much more convenient to have easily determined objectives. It is almost always possible to reduce the issue to easily examined variables. One exception would be a reliability specification where many units would have to be tested to arrive at an acceptable failure rate.

Once the list of metrics has been set, the values to be required must be determined. Here we have several leads to go on. We have the results of our user studies, which might have set some specific guidelines. We have the existing products that our competitors or we market and products that will be perceived to fall into the same category. These must be examined for similar user characteristics, sizes, operating forces, etc. What are the current accepted values? What are the values we hope to achieve? It is a good idea to obtain samples of all these products if it has not yet been done, even those that are in no way competitive, if they are going to be thought of by the user as in the same class. If there is a range of quality here, the level that is being aimed for must be determined and the values set accordingly. Having reviewed all the pertinent information available, specification values should be set. (See Table 19.1.)

TABLE 19.1 Idealized Example of Some User Needs Translated to Specifications for the External Fixation System

User comment	Specification	Units	Value
Make it from aluminum or carbon fiber.	The system mass will be no more than *X*	Kilograms	0.75
It must be easy to install.	Installation time will be less than *X*	Minutes	20
It must not weaken the bone too much.	Holes in bones must have diameters no more than *X*	Millimeters	4.0

The documentation of this process is important because in most cases the target specifications arrived at here will be unobtainable. (You find that you cannot make the laproscopic tool strong enough and still fit it through a 7-millimeter port!) Later in the project there will be an effort to reach some compromises and having at hand the details of how the existing values were chosen will save more than half of the discussion time and, more often, allow the discussion to proceed in a more civilized manner, preserving the team morale.

19.10 CONCEPT DEVELOPMENT

With the specifications in place, it is now time to seriously look for solutions. The concept development phase is the time to search out ideas that will meet the need. Much has been written about ways to generate new ideas, and some of them will be touched on here, but more important than method is motivation. The objective is to generate an absolutely superb product, one that meets all the requirements in a cost-effective, safe, elegant way, and to do it quickly and efficiently. Unfortunately, at this stage of the process, quickly and efficiently often dominate the first part of that sentence in a manner that is very shortsighted. There is a tendency to proceed with the first idea that appears to have merit. In the general scheme of things, a little extra time at this stage might very well produce a much better idea, or perhaps even a scheme that is only slightly better. Consider the trade-off.

Time spent on concept generation is relatively inexpensive. Prototypes are not being built, clinical trials aren't being conducted. A great deal of paper may get expended and even some computer time, but if one looks at the rate at which resources are consumed, the concept phase of a development project is on the lower end of the scale. If one thinks of the concept phase as a search for the best idea, it's not like searching for a needle in a haystack, but perhaps something slightly larger than a needle. If we search for a while, we may find an attractive idea. If we search further, we may find a better idea. If we walk away from the haystack, someone else may come along and find a better idea than the best one that we found and bring it to market shortly after we bring ours—and our product will be obsolete.

The primary argument for shortening the development time, or the "time to market" as it is thought of, is that eventually another product will displace ours. If we think of that product's introduction as an event over which we have no control, we lengthen our market window by an early introduction. If, however, we pass up the best product to get to market a short time earlier, we are inviting the early arrival of product termination.

Generating a long list of alternative design ideas is a very difficult task, requiring creativity, resourcefulness, energy and, most importantly, stamina. It is for this reason that so much emphasis has been placed on the quantity of concepts. After a short time and a few seemingly credible ideas, it is hard to continue to dig for alternatives. Different teams and leaders advocate different modes of idea generation, each having worked well for them. A starting point is to ask each team member to spend some time alone making a concept list. A comfortable place to work undisturbed, a pad of paper, and a few hours should produce many more ideas than each had at the outset. These lists can be brought together in a team meeting or brainstorming session. (Brainstorming has two working definitions. One is a formal process developed to produce ideas from a group. It has a clear procedure and set of rules. See Osborn.[6] The word is also used to describe any group working together to find a problem solution. Either definition could be applied here.)

At this point in our project we do not want to discard ideas. It is widely believed that some of the really good ideas are stimulated by some less sane proposals, so the objective is to lengthen the list, and hold off on criticism or selectivity.

There are some techniques that can assist when things begin to slow down here. One is to break out some characteristic of the device and make a list of all of the ways one could supply that function. For example, an active device requires energy. Ways to supply energy would include electric power from the utility, batteries, springs, pneumatics, hydraulics, hand power, and foot power. Trying to think up design ideas that use each of these functions will often lead to some different kinds of solutions. Other types of characteristics are specific material classes, cross-section shapes, etc.

It is also fruitful to look at the ways that analogous medical devices function. Is there an equivalent functionality that has been used in a different medical problem? One example of this is access ports that might function for feeding in one instance, as a drug delivery mode in another, and as a monitoring channel in a third. All have similar problems to overcome. Surgical instruments can provide many examples of devices that are variations of previous designs.

Another means of expanding the list is to combine ideas with features from other ideas. Look especially at the ideas that seem novel and find as many permutations on them as possible.

Some means of compiling the concept list is necessary. Index cards, spreadsheets, or even Post-it Notes® may be used. It helps if the scheme allows the ideas to be sorted and rearranged, but the importance of this will vary a great deal with the situation. Ultimately, you should emerge from this phase with a list of ideas to be considered. The list should be long, have many ideas that are of little value, and hopefully have several that have the potential to turn into world class products.

19.11 CONCEPT EVALUATION

Having spent a great deal of time and effort developing an extended set of concepts, the team must now select those that will be developed further. There are several important points to keep in mind as this process begins. The first is to have an even-handed approach. It is unwise to compare an idea that has been sketched out in 30 seconds to one that has had many hours of development, and discard the first because it isn't understood. As much as is reasonably possible, concepts should be compared at an equal state of development. That having been said, there will be ideas brought up that can be recognized immediately as inappropriate. (This is particularly true if the team is diligent about seeking ideas.) The notably bad ideas should be quickly but carefully (and sometimes even gently) culled out and discarded.

The second issue revolves around selection criteria. The process will (or should) pick the idea that provides to the team and the organization the best opportunity to develop a successful product. In making that selection, the capabilities of the organization come into consideration. What kind of products is the group experienced with? What kind of manufacturing facilities does it have at hand? What are the core competencies of the corporation? The concern here is the good idea that doesn't fit. As a simplistic example, if the organization builds mechanical things, and the best concept for the product under discussion is an electronic solution, there is an impasse. The kind of design available is second rate, and will fail in the marketplace. (Someone will market the electronic version!) The options are (1) abandon the product plan, (2) acquire the needed skills and competency, or (3) partner with one or more organizations that can provide the talent. Contracting with a design firm for development and an outside manufacturing organization might accomplish the latter. This decision will probably be made at a higher level in the organization, on the basis of the evaluation of concept potential done within the team.

Having reduced the list to ideas that have real possibilities, a selection process should be used that rates the concepts on all-important criteria. The specification document will provide not only the list of criteria but also some guidance as to the importance of each item. This list should be used in a two-step process; the first step will screen the concepts, and the second will select those to be developed further.

For the screening we will use a method called Pugh concept selection. See Pugh.[7] Place a list of the criteria to be used in the first column of a matrix, use the list of the remaining concepts as the headings for the adjacent columns. A spreadsheet program will be very useful for this purpose. Choose one of the concepts as a reference. This should be a well-understood idea, perhaps embodied in a current product, yours or a competitor's. Now compare each concept against the reference concept for each criterion and score it with a plus sign if it does better and a minus sign if it is inferior. Use a zero if there is no clear choice. When all the cells are full, add the number of pluses and subtract the number of minuses to get the score for each idea. Many of the scores will be close to zero, and of course the reference idea will get exactly zero. (See Table 19.2.)

TABLE 19.2 Partial Scoring of Concepts for External Fixation

Criteria	Concept A	Concept B	Concept C	Baseline concept
Weight	+	−	+	0
Installation time	−	+	+	0
Weakening	+	0	−	0
Totals	+#	−#	+#	0

At this point the number of concepts under consideration should be cut back to about 10 or 15. The first criterion for the reduction is the score on the requirements sheet. Before discarding any of the ideas, however, examine each one to determine why it has done so poorly. See if the idea may be modified to increase its potential. If so, retain it for the next cycle.

With the field narrowed to a reasonable number of candidates, it is now possible to devote a little effort to refining each of them, get to understand them better, and then construct a new matrix. This time a weighting factor should be agreed to for each of the criteria. A range of 5 for very important and 1 for minimally important will suffice. Now score each of the concepts on a basis of 1 to 10 on each criterion and compute the sums of the weighted scores. This will allow the ideas to be ranked and a selection made of the ideas to be seriously developed.

The major decision remaining here is how many concepts should be pursued in depth. This is another judgment decision, to be guided by a number of factors: How much confidence is there in the first two or three ideas? How long will it take to prove out the uncertainties? What resources are available? It is highly probable that the idea that emerges from the selection process with the highest score will have some yet-to-be-discovered flaw. This stage is much too early to commit all of the resources to a single concept. It is a time when options should be kept open.

Conclude this process by documenting the selection. Include all of the scoring sheets, concept descriptions, and related data. This will become part of the medical device master file, and perhaps play a role in getting product approval and it will be useful when concepts need to be reconsidered later in order to make final decisions.

19.12 ARCHITECTURE/SYSTEM DESIGN

It was mentioned early that the term *medical device* describes a wide variety of products. The process of proceeding from concept through system design and detail design will vary greatly thoughout the spectrum of products. Design development should in general follow a path similar to that of physically similar nonmedical items. The major accommodation is in documentation. It is important to maintain a detailed record of the decisions made and their basis. This will be important through the product's manufacturing start-up and afterward, when alterations are proposed for manufacturing reasons, to accommodate later product upgrades, etc. The ability to return to well-kept documentation and follow the original logic will provide guidance, sometimes supporting the change, often indicating its inappropriateness. There is a tendency in these situations to repeat the same mistakes. A change in a medical device is a much more expensive undertaking than it would be in other objects of equivalent complexity because of the qualification, verification, and testing that is so often required. Good documentation can often prevent the initiation of some misguided efforts.

The objective in the system-level design is to deconstruct the product into separate elements that can be considered independently. Once this has been done, the various team members can proceed to define the elements of the product. The system design must define the interfaces where the separate elements meet. This includes shapes, dimensions, and connection characteristics, such as currents, digital signals, fluid pressures, and forces. Decisions made at this stage have a profound effect on the complete design. The team and the leader must exercise judgment here. In a sense these decisions are allocating a budget. In certain examples this is obvious, as in space allocated or power consumption limits. Sometimes it may be subtler than this, but much of the budgetary sense will remain.

The actual relationships defined in the system design are called the architecture. There are two extremes that are considered in system architecture, integrated and modular. Modular systems are made up of components or modules that have clear interfaces with each other, and are easily visualized as systems. Integrated products, on the other hand, appear to be a single element, with what interfaces exist being so soft and blurred as to be hard to identify. Each of these styles has its advantages. Integration allows the device to be technically more efficient. With the overall device optimized, there is no loss due to inefficiencies of connections, etc.

Modularity provides flexibility at several levels, which is often extremely desirable. As mentioned earlier, it can make the design effort simpler. It also allows for modifications and upgrades. If a system component is changed, one can test the new component for performance more efficiently than testing the entire system. All of the engineering issues tend to be less challenging in this modular environment. In addition, a modular design enables element replacement. This is important in several ways. An element of the system can be designed with the expectation that its life will be shorter than that of the system, and it will be replaced when needed. The clearest case of this in medical devices is a single-use or disposable element such as a blade. An appropriate architecture allows a very efficient design, replacing the elements that cannot be reused without wasting those that can.

Modularity also enables variety in design by permitting variation in one or more components. This is a way to provide devices in various sizes, or sometimes at various levels of performance. In some cases, this is accomplished by selling the components of the system separately, in others the unit can be thought of as a platform product with several different products, all based on the same system and sharing a large number of elements. See Meyer.[8]

In addition to the economic and technical reasons that support modularity, it can assist in a medical product gaining acceptance. If the product can be viewed by the clinician as having a number of modules that are familiar, along with some new functionality, it may be easier to introduce than a system that is viewed as an entirely new approach. Most medical professionals are somewhat conservative and like to be on ground that is at least somewhat familiar. This could be a consideration in choice of architecture and even in the details of the deconstruction process.

In many designs, the layout of the system is obvious. Similar products have been broken out the same way to everyone's satisfaction, and no other arrangement seems possible. On the other hand, it is good practice to ask if this is indeed the optimum architecture for the given product, and explore the advantages and disadvantages of altering the architecture.

19.13 DETAIL DESIGN

The detail design phase is where the individual components of the system are fully defined. Again, the issues here vary greatly across the product spectrum, but primarily the issue is function against the various penalties of weight, space, cost, etc. None of these are unique to the medical device area, but the costs tend to be higher, the penalty for error much higher, and therefore the need for care very intense.

It is good practice to use commercially available components as much as possible as long as they do not compromise the design functionality in any way. Each and every part that is unique to the product will require careful specification, manufacturing study, shape, and manufacturing documentation, manufacturing qualification, etc. It is difficult to estimate the cost of adding a part to a medical product production system, but Galsworthy[9] quotes a source from 1994 stating that the average in the commercial world is no less than $4000. The medical product equivalent must be an order of magnitude larger.

As the parts are designed, consideration must be given to not only the manufacturing process to be used, but also the method of testing and verifying functionality. Some foresight at the detail level can provide "hooks" that enable the initial testing of the part, the system, and even the ongoing quality assurance testing that good manufacturing practice mandates. Providing an electrical contact point or flat on which to locate a displacement probe can make what would otherwise be a project into a routine measurement.

19.14 DESIGN FOR MANUFACTURE

The need for design for manufacture goes without saying, since one of the primary measures of success in product development is cost to manufacture. There is a strong belief among developers that the cost to manufacture a product is largely determined at the concept selection stage, but there is a great deal of evidence that indicates details of design based on selection of manufacturing methods are the real determining factors. The major stumbling block here is that so many manufacturing processes are volume sensitive. The design of parts for the same physical function and annual volumes of 1000, 100,000, and 10 million would call for three totally different manufacturing processes. In detailing the part, it is important to know the production targets. Things can be designed to transition from low volume to high volume as the product gains market, but it needs careful planning and good information. This is one of the places that the team's skills in marketing, design, and manufacture can pay large dividends.

19.15 ROLLOUT

Bringing the product out of the trial phase and into manufacturing and the marketplace is the last responsibility of the team. The initial product planning should have defined target sales volumes, and the marketing function of the team should have been updating these estimates throughout the development and trials. Products are often introduced at shows, resulting in a national or even international introduction. This means that there should be sufficient quantities of product "on the shelf" to meet the initial round of orders. It is difficult to recover from a product scarcity that develops in the initial launch. All of the stakeholders become disenchanted.

The arrangements for product sales and distribution are not within the scope of this book, but the plan for this should be understood by the development team well in advance of the introduction. Issues like training the sales staff, providing printed materials, and samples should have been worked out well in advance.

19.16 PROCESS REVIEW

This is also the time for the team to evaluate the job it has done, the lessons it has learned, and specifically how it performed against the five objectives set forth at the start of this section. Quality, cost to manufacture, time to market, and cost to develop will be well recognized. The fifth objective of improving the capability of the organization will require some thought. An effort should be made to determine what aspects of the process did not work as well as they should have. A team meeting may be the best way to do this, or the team leader may choose to meet members individually.

The basis for assessment should be the plan and the process. Were the plan and the process followed closely? If not why? Should the plan have been different on the basis of the knowledge at the time it was drawn? Is the process, as laid out, the best approach for this group? How should it be altered? Were the resources and external support adequate? If not, in what way?

The answers to these and related questions should be compiled into a report and that document circulated in a way that maximizes the learning function of the project. It is important to review this document again as the next product development project is being proposed.

19.17 PROTOTYPING

Making prototypes is so much a part of device development that it is often taken for granted. Prototyping plays a very important part in the process but it must be thought out and planned, if the maximum benefit is to be derived. Clearly the size and scope of the product and the effort will control to some extent this activity. In some cases, many prototypes will be built before the design is frozen while in other projects one or two may be an efficient plan. Of course, a clinical trial will require many copies of the prototype design in most cases.

In planning the project, the goals of each prototyping effort should be defined. They are often associated with milestones in the plan. A prototype having particular working characteristics is to be functional by a particular date. In setting that goal, the purpose of that prototype should be understood and stated. We generally think of the prototype as proving the feasibility of the design, but close examination often shows that not to be the case. Clausing[10] describes situations where the effort that should be focused on the design is going into debugging prototypes that have already become obsolete. We need to have focus.

In many situations a partial prototype, one that might be more properly termed a test bed, is in order. This is a device that mimics some part of the proposed design, and allows critical characteristics of the design to be altered and tested. This system may bear little resemblance to the final product; it really only needs to have the significant physical characteristics of the segment under study. This might be the working parts of a transducer or the linkage of some actuator. It may be constructed to evaluate accuracy, repeatability, durability, or some other critical feature. It should be thought of as an instrument. If it is worth the trouble, time, and effort to build it, then an experiment should be designed to garner as much information as it has to offer.

With partial medical device experiments, usually called bench testing, an often-encountered problem is the replication of the biological materials that the device interacts with. If we return momentarily to the external fixation device, it requires that the pins be affixed to the fractured bone. In considering the need, it is recognized that the fracture might represent an athletic injury to a young healthy person with very strong bones, or it may have resulted from a mild impact to an individual with relatively weak bones. We would like our device to function over this range, and must choose test media accordingly. We would start with commercial bone emulation material (e.g., Sawbones®), but then probably move to some animal models. Careful analysis of this situation is important. Why is the test carried out? Are we interested in failure mode to better understand the device or are we proving that this design is superior? Caution is called for.

19.18 TESTING

The testing and evaluation of the final device is a most important activity. This is dealt with here separately from the prototyping activity because it reaches beyond the design team in many respects and because the process and attitude required are more distinct from nonmedical products here than in the prototyping activities that are really internal to the design process.

Once the design is finalized, all of the evaluation and test data become important in the regulatory approval process, so each step should be carefully planned, executed, and documented. Each test should be done with the clear objective of validating some aspect of the design. The level of performance required should be established in advance. The test conditions should be carefully spelled out.

Bench or laboratory testing should measure the integrity, precision, and other "engineering" aspects of the device. They can also verify the production techniques, as at this stage the units being tested should come from the manufacturing processes that will be used in full production. Besides testing for normal performance, it is wise to test devices to failure at this stage, to ascertain that there are no unforeseen failure modes or collateral effects.

Many products move next to cadaver testing, where the biologically driven uncertainties cause complications. Sometimes it is possible to use animal models first in this phase, which allows for much more consistent conditions. Human cadaver tests require careful notation of the history and condition of the cadavaric materials. Results may be correlated with this data if there are anomalies that require explanation.

Clinical trials with human subjects require extensive documentation. They are most often done in institutions that have in place extensive procedures to ensure the subjects are informed about any experimental activities that they participate in. A committee reviews the forms and procedures for doing this along with the details of the study before institutional approval is granted. These things require long lead times in many cases, so the planning of the experiments ought to be done well in advance.

It is common practice to run clinical trials in several settings, often distant from each other. The point was made earlier that the device should be usable by an average practitioner, not just the ultrahighly skilled. The trials should include the opportunity to evaluate this aspect, as well as to evaluate training materials and other ancillary items that contribute to product function. Care and forethought continue to be the watchwords.

19.19 DOCUMENTATION

Documentation of a medical device design is critical. This applies not only to the usual drawings and manufacturing documents that accompany commercial products, but also to the needs and requirements documents that have been discussed earlier, as well as test data, evaluation reports, results of clinical trials, and any other pertinent information about the device, its history, or experience.

The Food and Drug Administration requires that for each device there be a device master file. The length and breadth of that rule becomes clearer when it is explained that the "file" need not be in one place! It is the collection of all the material just mentioned and it could be in several states, or even on several continents. The need to provide all of this documentation and to have it accessible requires that the project have someone in the role of librarian. If the project gets too far along and too much material is generated before a document control system is in place, it may be a situation from which there is no recovery. Document control should be part of the initial project planning. If the organization has experience in device development, it probably has already established formats and templates to handle the materials and data, and a new master file can probably be built around existing standards. If the current project is a first, an experienced person will be needed to take charge of these matters and will need to have the power to demand that procedures are followed.

19.20 TOOLS

There are a number of software products and related items that can be very useful in various aspects of product development. These include computer-aided design (CAD), computer-assisted manufacturing (CAM), various engineering analysis tools for things like finite-element stress analysis or circuit analysis, planning tools, scheduling products, systems that generate rapid prototypes, and a vast number of other things that have the potential to save effort, and more important, time. Most if not all of these tools have associated with them a learning process that requires time and effort. It is wise to develop at the outset of the project a clear picture of what tools of this kind will be needed and to be sure that the team includes members who have experience in their use.

As alluded to above, the choice to make use of a particular tool is an investment decision, and should be dealt with in that way. Software and hardware will have to be acquired, and individuals may require training in their use. The selection of these tools from the outset allows training to be done early when it will have the longest-term benefit. (It also requires the expenditure of the corresponding funds at a time that the project may appear to be overspending!) In most categories, there are competing products from which to choose. Once the requirements are met, the most desirable characteristic is familiarity. If the people on the team have used it (that might include venders and others you need to exchange information with), you are well ahead selecting that product. The purchase price can't be disregarded, but the cost in time spent learning to use a computer tool is usually higher than its price.

With most tools and particularly with CAD and similar products, there is a more or less continuous version upgrade problem. If at all possible, this is to be avoided. If you select a product that fills your needs at the outset, you should be able to get along without the three new "hotkeys" and the six new plotter interfaces that have been added to version 37.3. Installing a new release will cost time that you do not have, increase the likelihood of losing data, and will not pay off in a productivity increase that is noticeable.

The previous paragraphs presume an intense design effort on a single product. If a group is spread out continuously over a large number of projects, upgrades are a fact of life. One can, and many do, skip some of them. With some products, installing every third or fourth version will keep the group's tools sufficiently up-to-date.

19.21 REGULATORY ISSUES

Manufacturing and marketing a medical product in the United States must be done under the regulations of the Food and Drug Administration (FDA). Medical devices are handled by the agency's Center for Devices and Radiological Health (CDRH). Complying with these requirements constitutes one of the major resource requirements in the development process. As with so many other topics in this section, the extent of effort will vary greatly, depending on the nature of the product, the potential of it causing harm, and the history of similar products and devices in similar areas. Having available a person knowledgeable about the agency's dealings with similar products is a must.

Under the law granting the FDA jurisdiction over devices, products already on the market were permitted to continue. Products "equivalent to" products on the market require only a formal notification of the intent to market, but the agency is the arbiter of what is "equivalent." (Note also that the required notification documents can constitute a large package!) Devices without a predicate product are held to much more stringent controls, usually requiring a series of clinical trials, showing statistically that the device is safe and effective.

The agency has a responsibility to monitor organizations producing and distributing devices, to see that they do business in ways that do not endanger the public. This includes assuring that quality is adequately monitored, that complaints about device performance are promptly investigated, that steps are taken to correct any potential problems, etc.

The FDA has a very large and difficult responsibility and, given the resources it has to work with, carries it very well. Those who carefully design products so that they are safe and effective tend to resent having to explain the way they have gone about it, as they view the effort as unproductive. Nonetheless, this will continue to be a problem and the team must prepare for it and accept the burden.

In keeping with its responsibility, the FDA publishes documents that can be very helpful to those developing products. Most of the useful documents are in the category of "guidance." Available on many topics, including specific categories of products, these publications aim to assist people to understand what the agency expects. It is important to understand that these documents do not carry the force of law as do the regulations, but they are much more readable and can be very helpful. A suggested example is the Design Control Guidance for Medical Device Manufacturers.[11] These documents are indexed on the Web site www.accessdata.fda.gov/scripts/cdrh/cfdocs/cfTopic/topicindex/topindx.cfm.

19.22 CLOSURE

This is clearly a very short introduction to the topic of medical device design. It has been only recently that design process activities have been written on extensively, and with a few exceptions where the process is easily codified, most of what has been written is very general. All of the books in the references address the broader questions, but they provide strategy and advice that is useful to the medical product designer. All can be read with great benefit. In addition to those cited previously, attention can be paid to Cooper[12] on planning, to Leonard and Swap[13] for concept generation and teamwork. The remaining three: Otto and Wood,[14] Patterson,[15] and Wheelwright and Clark[16]; all provide deep insight into the product design world.

REFERENCES

1. Ulrich, Karl T., and Eppinger, Steven D., *Product Design and Development*, 2d ed., Irwin/McGraw-Hill, 2000.
2. Tucker, Frances G., Zivan, Seymour M., and Camp, Robert C., "How to Measure Yourself Against the Best," *Harvard Business Review*, January–February 1987, p. 8.
3. Von Hipple, Eric, *The Sources of Innovation*, Oxford University Press, 1988.
4. Rutter, Bryce G., and Donelson, Tammy Humm, "Measuring the Impact of Cultural Variances on Product Design," *Medical Device and Diagnostic Industry*, October 2000.
5. Hauser, John, and Clausing, Don, "The House of Quality," *Harvard Business Review*, vol. 66, no. 3, May–June 1988, pp. 63–73.
6. Osborn, Alex F., *Applied Imagination*, 3d ed., Scribners, 1963.
7. Pugh, Stuart, *Total Design*, Addison Wesley, 1991.
8. Meyer, Mark H., and Lehnerd, Alvin P., *The Power of Product Platforms*, The Free Press, 1997.

9. Galsworthy, G. D., *Smart Simple Design*, Oliver Wight Publications, Wiley, 1994.
10. Clausing, Don, *Total Quality Development*, ASME Press, 1994.
11. "Design Control Guidance for Medical Device Manufacturers," FDA Center for Devices and Radiological Health.
12. Cooper, Robert G., *Product Leadership*, Perseus Books, 2000.
13. Leonard, Dorothy, and Swap, Walter, *When Sparks Fly*, Harvard Business School Press, 1999.
14. Otto, Kevin, and Wood, Kristin, *Product Design*, Prentice Hall, 2001.
15. Patterson, Marvin L., with Lightman, Sam, *Accelerating Innovation*, Van Nostrand Reinhold.
16. Wheelwright, Steven C., and Clark, Kim B., "*Revolutionizing Product Development*," Free Press, 1992.

CHAPTER 20
CARDIOVASCULAR DEVICES

Kenneth L. Gage and William R. Wagner
University of Pittsburgh, Pittsburgh, Pennsylvania

20.1 INTRODUCTION

Cardiovascular disease remains the leading cause of mortality among men and women in Western countries. Many biomedical engineers have focused their careers on the study of cardiovascular disease and the development of devices to augment or replace function lost to the disease process. The application of engineering principles to device design has improved device function, while minimizing some of the detrimental side effects. Progress to date has allowed complex, challenging cardiovascular surgical procedures (e.g., open-heart surgery) and medical therapies (e.g., dialysis) to become routine, although limitations remain. In this chapter, eight major categories of cardiovascular devices are addressed, including cardiac valves, stents and stent grafts, pacemakers and implantable defibrillators, vascular grafts, hemodialyzers, indwelling catheters, circulatory support devices, and blood oxygenators. For each topic, the market size, indications for device use, device design, complications and patient management, and future trends are covered. The intent is to provide a brief introduction to the current status of cardiovascular device development and application and to identify challenges that remain in the field.

20.2 ARTIFICIAL HEART VALVES

20.2.1 Market Size

There were at least 60,000 valve replacement operations performed in the United States during 1996 (Vongpatanasin et al., 1996). About two-thirds of the artificial heart valve market in the United

States consists of various mechanical valves, with the remaining one-third being distributed among the bioprosthetic, or tissue-based, models. Worldwide, the mechanical valve market is slightly less, approximately 60% of market share (Akins, 1995).

20.2.2 Indications

The medical indications for valve replacement are thoroughly described in a report from the American College of Cardiology/American Heart Association Task Force on Practice Guidelines, which address the management of patients with valve disease (Bonow et al., 1998). The etiology of valve disease differs, depending on the patient group and the valve location, as does the preferred corrective treatment. In general, the young suffer from congenital valve defects, while older adults exhibit acquired valve disease. Valve replacement can be performed upon all valves of the heart but most cases involve the aortic or mitral valves. Common reasons for native valve replacement are severe stenosis and regurgitation with or without symptoms, which may include chest pain, shortness of breath, and loss of consciousness. The reduction in effective orifice size associated with a stenotic lesion results in a large transvalvular pressure gradient that may exceed 50 mmHg in the aortic position for severe cases (Bonow et al., 1998). In regurgitation, the blood pumped forward into the recipient vessel or ventricle spills back into the adjacent pumping chamber through an incompetent valve, minimizing forward movement. The end effects of chronic stenosis or regurgitation are compensatory anatomic changes that accommodate, for a limited time, the reduced pumping efficiency due to restricted blood movement. In general, heart valve replacement is performed when repair is not possible, as the implantation of an artificial heart valve brings with it another set of problems. Total replacement and removal of native valve components in the mitral position is particularly limited, as the mitral valve is anatomically and functionally integral to the left ventricle (David et al., 1983; Yun et al., 1999). Concomitant illnesses such as congestive heart failure, atrial fibrillation, and coronary artery disease can alter the indication for valve replacement, as can the surgical need to correct other cardiac disease.

20.2.3 Current and Historical Device Design

Artificial heart valve design has a long and colorful history, with more than 80 different versions of valves being introduced since the 1950s (Vongpatanasin et al., 1996). The two general types of replacement valves, mechanical and biologic, each have their own set of indications, complications, and performance factors. The mechanical valve can be further categorized into three major design lines: caged ball, single tilting disk, and bileaflet (Vongpatanasin et al., 1996). Caged-ball valves have been largely supplanted by the more modern single-tilting-disk and bileaflet valves. Biologic valves are divided according to the source of the tissue material, with the term *bioprosthetic* reserved for valves constructed from nonliving, animal-source tissue. Homograft biologic valves are preserved human aortic valves or pulmonary valves surgically placed within a recipient patient (Bonow et al., 1998). Heterograft bioprosthetic valves consist of porcine heart valves or bovine pericardial tissue formed into a valve over a support structure (Vongpatanasin et al., 1996). Because mechanical and bioprosthetic valves have different design considerations, the categories are discussed separately.

Mechanical Valves. The assorted mechanical valve designs use different approaches to achieve the same functional goal. Caged-ball valves use a free-floating polymeric sphere constrained by a metal cage to periodically occlude the valve orifice. Single-disk valves possess a central disk occluder that is held in place by struts projecting from the housing ring. The disk opens through a combination of tilting and sliding over the struts to reveal primary and secondary orifices. Bileaflet valves feature leaflets that are hinged into the housing ring. The opened valve presents three orifices, two along the housing ring edge and one central orifice between the leaflet mount points. Figure 20.1 shows both orifice and profile views of representative tilting-disk and bileaflet mechanical valves.

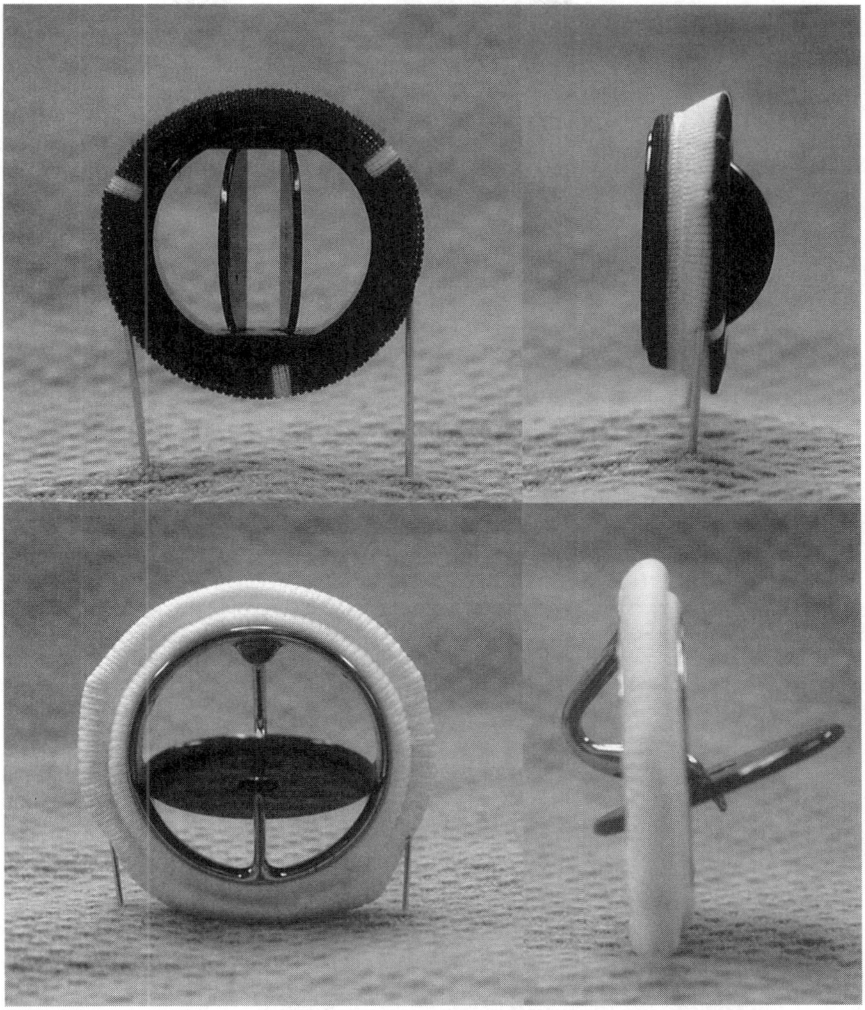

FIGURE 20.1 Orifice and profile views of representative bileaflet and tilting-disk valves are provided in the upper and lower halves of the figure. Note the larger percentage of open orifice area present in the bileaflet valve. The tilting disk projects farther into the downstream ejection zone than the leaflets on the bileaflet valve.

Mechanical valves are expected to perform flawlessly for decades with minimal patient burden. Criteria used to evaluate designs during development and clinical use can be divided into *structural* and *hemodynamic* groups, although there is considerable overlap. Structural considerations involve fatigue and device integrity, valve profile, rotatability, and occluder interference (Akins, 1995). To accommodate the wear associated with operating hundreds of millions of times, current mechanical valves are manufactured with durable metal and carbon alloys (Vongpatanasin et al., 1996; Helmus and Hubbell, 1993), and include a polymer fabric sewing ring for surgical placement. Rotatability of the valve is desirable, as evidence suggests that optimum orientations minimizing turbulence and microembolic signals exist for mechanical heart valves in vivo (Kleine et al., 2000; Laas et al.,

1999). Concerns regarding valve profile and occluder interference focus on possible negative interactions between the valve, adjacent ventricular structures, native valve remnants, and surgical suture material. The impingement of these structures into the valve could prevent complete closure of the valve or cause binding of the occluder. Although not a structural requirement, devices tend to be radio-opaque to aid in visualization during radiographic procedures.

Hemodynamic performance factors that should be considered during functional evaluation of a valve design are the transvalvular pressure gradient, rate and duration of valve opening, dynamic regurgitant fraction, and static leak rate (Akins, 1995). The transvalvular pressure gradient is a function of the effective orifice area of the valve and the flow regime (turbulent or laminar) encountered. The mass of the occluder and mechanism of action play a significant role in the rate and duration of valve actuation, and similarly in the dynamic regurgitant fraction, which represents the percentage of blood that flows back into the originating chamber prior to valve closure. Finally, some leak is expected in the closed valve position. Consistent convective flow over these surfaces is believed to aid in the minimization of thrombotic deposition.

Bioprosthetic Valves. Engineering design concerns have little influence on homograft valves because of their anatomic origin, thereby limiting the focus to heterograft bioprostheses. The bioprosthetic tissue found in heterografts is treated with glutaraldehyde to cross-link the proteins that make up the tissue structure. The treatment is cytotoxic, disrupts the antigenic proteins that can cause an immune response, and improves the toughness of the tissue by cross-linking the structural collagen (Bonow et al., 1998). Some bioprosthetic valves are further treated with surfactants, diphosphonates, ethanol, or trivalent metal cations to limit the rate of calcification and associated structural deterioration (Schoen and Levy, 1999).

Porcine heterograft valves can be mounted on a support scaffold with a sewing ring, although unmounted designs have been introduced to improve flow characteristics and structural endurance. Heterograft valves constructed of bovine pericardium are formed over a scaffold with a sewing ring to mimic an anatomic valve shape. Because the pericardial valves are constructed to design criteria rather than harvested, the orifice size can be made larger to improve flow characteristics, while the higher collagen content may allow improved graft resilience when cross-linked (Bonow, 1998). Figure 20.2 shows representative porcine and bovine pericardial heterograft valves.

Design Performance Evaluation. The design of artificial heart valves has benefited from the advent of computational fluid dynamics and other computationally intensive modeling techniques. Simulations have been used to predict the performance of both bioprosthetic (Makhijani et al., 1997) and mechanical (Krafczyk et al., 1998) valve designs. Results from computer modeling can be

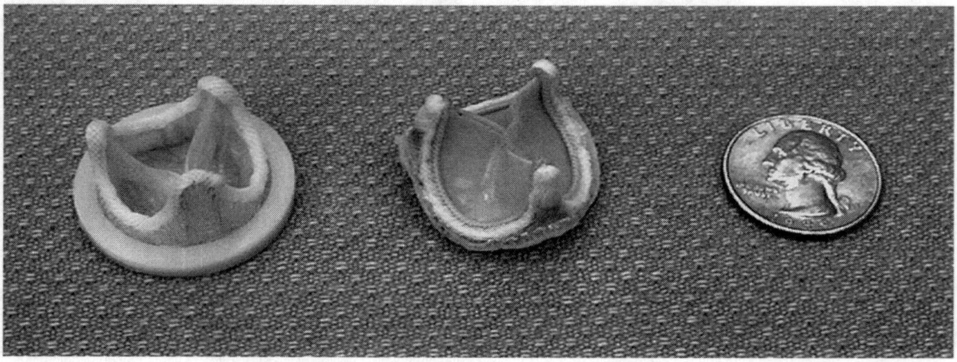

FIGURE 20.2 Two tissue-based artificial heart valves are shown above with a U.S. quarter dollar for size comparison. The valve on the far left is a porcine heart valve, while the other valve is constructed of bovine pericardium. Both valves are intended for aortic placement.

compared with findings from experimental studies using such methods as particle image velocimetry (PIV) (Lim et al., 1998) and high-speed photography of valve structure motion (De Hart et al., 1998). Such comparisons provide necessary model validation, revealing inadequacies in the numerical model and capturing phenomena not predicted using existing model assumptions.

20.2.4 Complications and Patient Management

Mechanical and bioprosthetic valves suffer from complications that dictate follow-up care and preventive measures. Possible complications facing heart valve recipients include thromboembolism, hemolysis, paravalvular regurgitation, endocarditis, and structural failure of the valve (Vongpatanasin et al., 1996). Some preventive measures are indicated for both mechanical and biologic valve recipients, such as the use of antibiotics during dental surgery and invasive procedures to avoid infective endocarditis (Dajani et al., 1997). Other preventive treatments, such as long-term anticoagulation, are administered differently, depending on the type of valve implanted.

Because of the high incidence of thromboembolic complications associated with mechanical artificial heart valves, chronic anticoagulation is required. Anticoagulation with warfarin and an antiplatelet agent such as aspirin is indicated for both mechanical and heterograft bioprosthetic valves for the first 3 months after implantation (Stein et al., 2001; Bonow et al., 1998; Heras et al., 1995). After that time warfarin is discontinued for heterograft bioprosthetic valves unless the patient possesses a risk factor that increases susceptibility to thromboembolic complications (Bonow et al., 1998). Despite the use of chronic anticoagulant therapy, a recent review reported that between 0.4 and 6.5 percent of mechanical heart valve recipients will experience a thromboembolic event per year, a range that is dependent upon valve type, number, placement, and other risk factors, as well as the level of anticoagulation (Stein et al., 2001). The long-term risk of thromboembolism in bioprosthetic heart valve recipients is comparatively low, ranging from 0.2 to 2.6 percent per year (Stein et al., 2001).

In contrast to the anticoagulation requirement associated with mechanical valves, the largest problem facing patients with bioprosthetic valves is progressive structural deterioration due to calcification, which can eventually require valve replacement and the risks of a reoperation (Hammermeister et al., 2000; Schoen and Levy, 1999). Heterograft bioprosthetic valves exhibit accelerated deterioration in younger patients, which promotes the use of mechanical valves in this age group when homograft valves are unavailable (Bonow et al., 1998).

Although the literature is rife with comparisons between valve designs in regard to their complication rates, the general lack of large randomized trials using standardized methods makes valid comparisons problematic (Horstkotte, 1996). To reduce some of the confusion surrounding outcome reporting in heart valve studies, the Society of Thoracic Surgeons and the American Association of Thoracic Surgery have updated guidelines for reporting common surgical and nonsurgical artificial valve complications (Edmunds et al., 1996). The guidelines distinguish between structural and nonstructural valve dysfunction, thrombosis, embolism, bleeding events, and infection. In addition, various types of neurologic events are graded, and methods of statistical data analysis are suggested on the basis of the type of data being collected and analyzed. Adherence to such guidelines should allow valid comparisons to be made between manuscripts reporting on outcomes with different valve types and clinical approaches (Edmunds et al., 1996).

20.2.5 Future Trends

Increasing efforts in the area of tissue engineering hold great promise for the development of replacement valves with improved biocompatibility. As reviewed in a recent article (Jankowski and Wagner, 1999), researchers have attempted to stimulate the growth of endothelial cells on existing bioprosthetic valves to limit valve degradation and thromboembolic complications, while others have endeavored to grow valves and individual leaflets de novo using cell-seeded polymeric scaffolds. Endothelialization of commercial bioprosthetic valve tissue is hampered by the cytotoxic aftereffects of glutaraldehyde fixation (Eybl et al., 1992), directing efforts toward alternative fixation techniques and coatings to improve cell adhesion and growth. Although these efforts have reached in vivo experimentation, results for endothelialized bioprosthetic valves have been mixed (Lehner et al., 1997). Efforts to produce artificial valves de novo using harvested endothelial cells and fibroblasts grown on a biodegradable scaffold have resulted in functional pulmonary valve leaflets in a lamb model (Shinoka et al., 1996). More recently, a biohybrid trileaflet pulmonary valve has been successfully tested in lambs for a limited period (Sodian et al., 2000a). In vitro (Sodian et al., 2000b) and recent in vivo (Sodian et al., 2000a) work has shown much promise although challenges in creating valves with ideal performance characteristics remain. Successful development of a viable autologous biologic valve capable of long-term stability and growth would likely revolutionize the management of heart valve disease through a reduction of implant-related morbidity and mortality, with particular applicability to pediatric patients.

20.3 STENTS AND STENT-GRAFTS: PERCUTANEOUS VASCULAR THERAPIES

20.3.1 Market Size

According to *Taber's Cyclopedic Medical Dictionary*, the term *stent* refers to "any material used to hold tissue in place or to provide a support for a graft or anastomosis while healing is taking place" (Thomas, 1989). As this broad definition would suggest, stents are indicated for a wide range of disease states in the genitourinary, hepatobiliary, gastrointestinal, reconstructive, and vascular fields. The focus of the current section is limited to those devices used in the vasculature, primarily the peripheral and coronary arteries. It is worth noting that stents approved by the FDA for use in a particular intervention are often used for other tasks, such as the use of biliary stents in coronary bypass grafts (Holmes et al., 1998).

Stents and stent-grafts are widely used to treat vascular lesions and are being employed in a growing number of cases. Many patients suffering from atherosclerotic vascular disease possess focal narrowings inside their blood vessels. These narrowed regions, termed stenoses, are made up of fatty plaques and can reach a size where they restrict the movement of blood to the tissue, resulting in sequelae ranging from pain upon physical exertion to tissue breakdown in severe cases. One dangerous scenario is thrombus formation at the lesion site with the potential for embolization distally. Thrombosis and thromboembolization can directly lead to tissue ischemia (oxygen starvation) and possible tissue death. Percutaneous transluminal coronary angioplasty (PTCA), otherwise known as balloon angioplasty, is a procedure wherein a catheter-mounted balloon is moved to the lesion site and inflated, displacing the tissue and creating a wider lumen in the vessel. Stents are designed to keep the treated lesion open through forceful opposition with the vessel wall; in essence, the stent braces the disrupted lesion in an expanded position. In 1998 it was estimated that each year more than a million patients worldwide received some type of nonsurgical cardiac revascularization procedure (Holmes et al., 1998). Stents are placed during most of these coronary interventions either as a primary or adjunct therapy (Al Suwaidi et al., 2000), and recent guidelines report that less than than 30 percent of interventions are composed of PTCA alone (Smith et al., 2001). Estimates for the number of interventional procedures in the peripheral vasculature exceeded 200,000 per year in 1997, with the expectation that up to 50 percent of traditional vascular procedures would be replaced with an endovascular analogue in the near future (Krajcer and Howell, 2000). One such endovascular intervention is the abdominal aortic stent-graft and its use in abdominal aortic aneurysm repair, which is traditionally a high-risk surgical procedure with perioperative mortalities ranging from 3 to 60 percent, depending upon patient characteristics (Krajcer and Howell, 2000). Stent-grafts consist of a polymer fabric or sheet coating mounted on a stenting device that excludes the vessel wall from the flowing blood, thus reducing its exposure to the pulse pressure wave. More than 4000 abdominal aortic stent-grafts were placed in 1999 (Krajcer and Howell, 2000).

20.3.2 Indications

Consensus documents from the American College of Cardiology and the American Heart Association detail coronary stenting indications currently supported by literature findings (Holmes et al., 1998; Smith et al., 2001). It is worth noting that few patients qualify for stent placement under the FDA guidelines associated with device approval (Holmes et al., 1998), yet stent utilization has increased to encompass pathologies not within the approved sphere of use despite the lack of strong supporting clinical evidence (Smith et al., 2001). Stents are most often placed in nonemergent settings to optimize outcome and improve patency of vessels that have undergone balloon angioplasty (Al Suwaidi et al., 2000). An ideal acute PTCA outcome is variously defined, but a residual vessel diameter stenosis of <35 percent along with a ratiometric increase in coronary blood flow of 2.5 under adenosine stimulation has been predictive of improved patient outcomes (Serruys et al., 1997). Unfortunately, an ideal PTCA result occurs in only 25–50 percent of patients, thus stenting is an important adjunctive therapy (Holmes et al., 1998). Stents are also placed emergently to treat or prevent acute closure of an artery following balloon angioplasty. Other less well-established indications include stent placement for the treatment of chronic total vessel occlusion, saphenous vein bypass graft lesions, acute myocardial infarction, gradual tissue overgrowth in an existing stent, and lesions with difficult morphology, such as long, diffuse stenoses, stenoses in small vessels, and lesions at a bifurcation or vessel opening (Holmes et al., 1998). Randomized trials currently underway or in the planning stages are expected to clarify these and other potential indications.

A recent review article listed the current indications for peripheral, noncoronary vascular stent placement (Mattos et al., 1999). The reported indications include immediate treatment of balloon angioplasty complications such as intimal dissection and flap formation; the correction of unsatisfactory angioplasty outcomes such as residual stenosis, spasm, recoil or occlusion; treatment of complicated, chronic lesions or occlusions; and as a routine combination treatment with angioplasty. The most common reason for placement of a stent in the peripheral vasculature is an unsatisfactory result from angioplasty.

20.3.3 Vascular Stent Design

The ideal stent would possess a number of characteristics designed to ease handling and permit stable, long-term function. Desired handling characteristics include a simple, effective deployment method, high radiopacity for visualization under fluoroscopy, flexibility to ease passage through tortuous vessels, limited shortening during expansion so placement is optimal, high expansion ratio to allow smaller profiles, and ease of retrievability or removal if misplaced (Henry et al., 2000; Mattos et al., 1999). Preferred functional characteristics include a high hoop strength to counteract arterial spasm, biocompatibility that minimizes short-term and long-term complications, plus durability in the stressful, corrosive environment of the human body (Henry et al., 2000; Mattos et al., 1999). Despite the large number of stent designs, with additional models under development, no one stent possesses all these particular characteristics (Henry et al., 2000).

Stents can be divided into two main groups on the basis of the method of expansion. *Balloon-expandable* stents either arrive premounted on a balloon angioplasty catheter or are mounted by the doctor prior to the procedure. A balloon catheter with inflation apparatus is shown in Fig. 20.3. While mounted, the stent is moved into place and the balloon inflated to expand the stent to the desired diameter. Figure 20.4 illustrates the placement and inflation procedure for balloon-expandable stents. In contrast, *self-expanding* stents come premounted or sheathed. Once deployed to the treatment area, the sheath is pulled back, allowing the stent to expand to its predetermined diameter. Balloon-expandable stents can be further subdivided into slotted-tube and coil-based designs (Oesterle et al., 1998).

Each stent type possesses particular advantages and disadvantages. Self-expanding stents can experience shortening during deployment, which may complicate placement, although more recent stent designs are able to mitigate this effect to a significant degree (Oesterle et al., 1998). Balloon-expandable stents exhibit a greater stiffness than the self-expanding models, which can cause diffi-

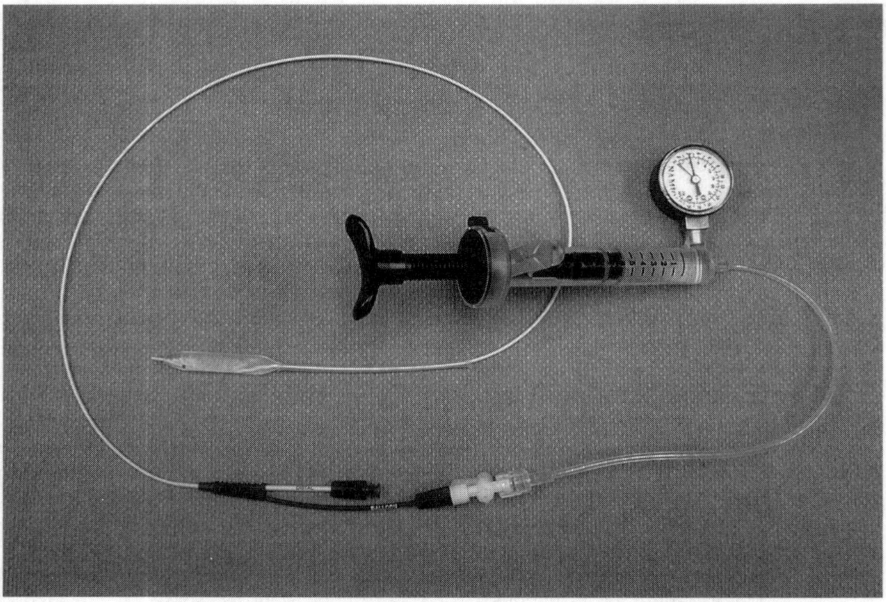

FIGURE 20.3 A balloon catheter and inflation pump are shown. Balloon catheters are used to widen stenosed or narrowed vessels and to expand stents. The balloon is often inflated with a mixture of saline and contrast agent to aide radiographic visualization.

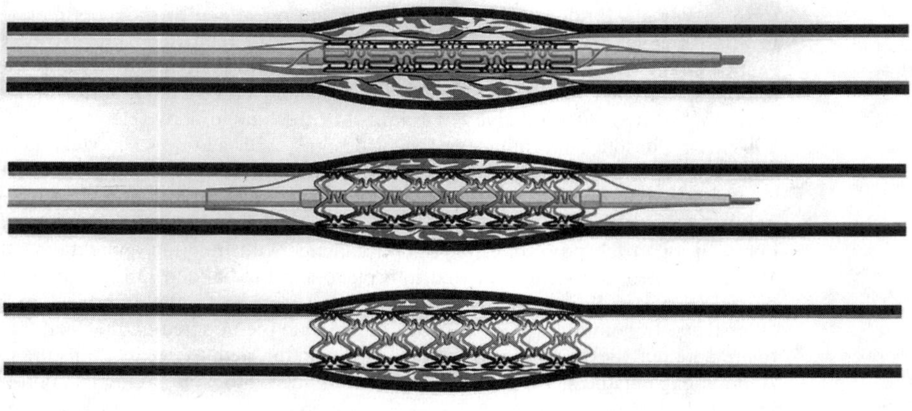

FIGURE 20.4 The stent implantation procedure for a balloon-expandable stent is demonstrated in the above figure. The balloon-mounted stent is first guided into place inside a damaged vessel under fluoroscopy. The balloon is then inflated, expanding the stent to lie in opposition to the vessel wall. The balloon is then deflated and withdrawn, leaving the stent in place to prevent constriction of the vessel. (*Cordis Corporation, Miami Lakes, Florida.*)

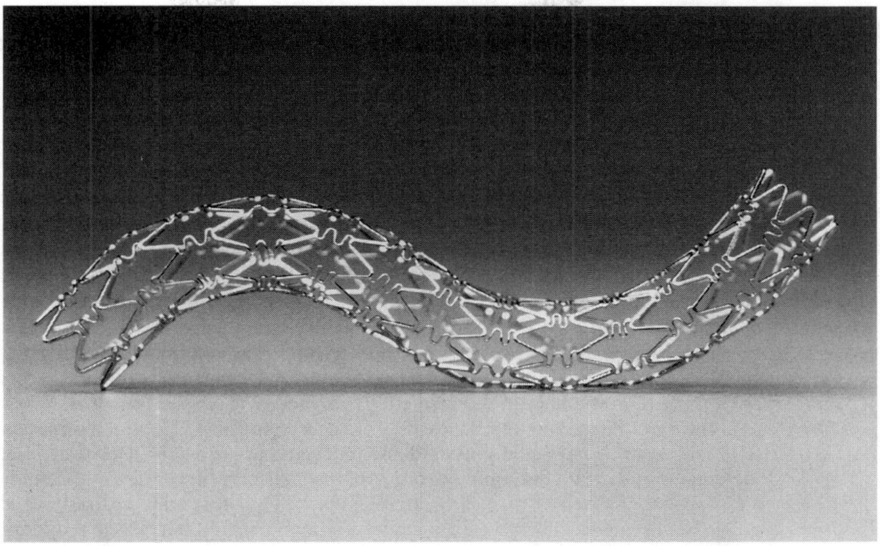

FIGURE 20.5 The flexibility of a modern stent design (Cordis Corporation 7-cell BX Velocity) is demonstrated. (*Cordis Corporation, Miami Lakes, Florida.*)

culty navigating long or tortuous lesions and anatomy (Mattos et al., 1999). In general, coil design stents have greater flexibility than the slotted tube models, making them attractive for more complex and difficult to reach lesions such as those found at bifurcations and in side branches (Oesterle et al., 1998). Some more recent stent designs combine features of slotted tube and coil models. Both stent types usually undergo further balloon expansion to optimize the resulting placement (Oesterle et al., 1998). Figure 20.5 illustrates the flexibility that can be achieved by a modern stent.

Most stent designs use metal as a construction material. Traditional alloys include tantalum and certain stainless steels (304 and 316L) (Mattos et al., 1999). Nitinol, a nickel-titanium alloy, has been used in self-expanding stent designs for its shape memory properties (Mattos et al., 1999). Both biodegradable and nondegradable polymeric stents have been developed, but the positive results of metallic stents over the long term, coupled with the technical challenges of producing a mechanically viable polymeric stent, have limited efforts in this area (Bertrand et al., 1998).

To accommodate the variety of arterial pathologies encountered, stents come in an ever-increasing array of sizes. Coronary stent diameters span from 2.5 to 4 mm, with lengths ranging from 8 to 38 mm (Al Suwaidi et al., 2000). Stents for the peripheral vasculature are of a considerably greater size because of the much larger vessels in the thorax, abdomen, and proximal extremities. The various stent designs appear to differ in their ability to maintain postexpansion lumen size (Okabe et al., 1999), which could affect clinical outcomes such as long-term patency (Fischman et al., 1994).

20.3.4 Management and Complications

As of 1997, incidence rates for complications following coronary stenting were under 1 percent for thrombosis when the patient was treated with antiplatelet agents, less than 25 percent for repeated arterial narrowing, and less than 15 percent of patients required an additional procedure on the stented lesion during the follow-up period (Oesterle et al., 1998). *Restenosis* is a term referring to the repeated narrowing or closure experienced in stented lesions typically due to an overgrowth of smooth muscle cells. Affecting up to a full quarter of patients, restenosis is a common complication following coronary stent placement and remains the major stumbling block in the long-term success of stenting (Virmani et al., 1999). Evidence from animal studies suggest that the amount of restenosis is related to the amount of damage incurred at stent implantation (Schwartz et al., 1992), while other evidence implicates the stent design as affecting the rate of restenosis and thrombosis (Rogers and Edelman, 1995). Additional causes implicated in restenosis include excessive thrombosis, inflammation, poor stent apposition, and the presence of large amounts of necrotic tissue or ruptured plaque (Oesterle et al., 1998). Restenosis is not limited to stented lesions, but actually occurs at a higher rate in lesions receiving PTCA only (Serruys et al., 1994) and is a significant contributor to stent placement in previously treated lesions.

A large assortment of potential stent coatings and materials have been evaluated in an effort to reduce the number of complications associated with stent implantation. Approaches have included both degradable and nondegradable polymeric coatings for surface passivation and drug elution, coating with the anticoagulant heparin to limit platelet activation, endothelial cell seeding to create a "natural" surface, and the use of radioactive sources to inhibit cellular proliferation (Bertrand et al., 1998). As detailed in a recent review, some nondegradable polymers under investigation include polyethylene terephthalate, poly(dimethyl)siloxane (van der Giessen et al., 1996), poly(organo) phosphazene (De Scheerder et al., 1995), methacrylphosphorylcholine/laurylmethacrylate copolymer (Chronos et al., 1995), and polyurethane (Bertrand et al., 1998). Degradable polymer systems have included polyhydroxybutyrate valerate, polyorthoester, polyethylene oxide/polybutylene terephthalate (van der Giessen et al., 1996), polyglycolic/polylactic acid, and polycaprolactone (Bertrand et al., 1998). Nonpolymeric coatings such as silicon carbide (Ozbek et al., 1997) and the hemostatic protein fibrin have also been investigated (Bertrand et al., 1998). The spectrum of results reported in humans and various animal models suggests that the overall direct effect of polymer coatings on stent biocompatibility is unclear at this time (Bertrand et al., 1998).

Inclusion of various drugs into the polymeric coatings further complicates analysis, but recent animal studies have shown a reduction of neointimal proliferation using such an approach with polylactic acid and related copolymer coated stents containing a common steroid (Strecker et al., 1998). Promising results from studies using other drugs with polylactic acid acting as a degradable stabilizer (Alt et al., 2000) indicate that the polymer-based delivery system shows future potential. Heparin coating of stents has been shown to limit thrombosis without an apparent long-term effect on neointimal growth (Serruys et al., 1998), but it is speculated that improvements in deployment technique and antithrombotic drug regimens might have contributed to improved outcomes (Bertrand et al., 1998). Radiation treatment of the stented area has been attempted with both gamma- and beta-type radiation sources being delivered via catheter or radioactive stent to reduce cell proliferation (Waksman, 1999). Recent results with a gamma radiation source have shown improved rates of clinical and angiographic restenosis in patients treated for instent restenosis; unfortunately, the treatment was associated with a higher rate of late thrombosis and subsequent heart attack (Leon et al., 2001). Beta irradiation of lesions treated with PTCA without concomitant stent placement has been shown to reduce restenosis in a dose-dependent manner (Verin et al., 2001), and could eliminate the need for eventual stenting in some patients.

20.3.5 Future Developments

Progress in stent technology will undoubtedly focus on reducing the incidence of restenosis and on the improvement of stent outcomes in challenging vascular lesions. Recent developments include PTFE-coated coronary stents designed primarily for placement in saphenous vein grafts (Baldus et al., 2000), which usually exhibit diffuse atherosclerotic disease and high complication rates when stented (Oesterle et al., 1998). Stents coated with native arterial and venous tissue are also under testing (Stefanadis et al., 2000a; 2000b). Drug-eluting stents show promise for mitigating neointimal formation and late-phase restenosis, as high local concentrations of therapeutic and preventative agents may be achieved. It is not difficult to envision a stent capable of minimizing vessel injury during deployment, sealing the injured site when expanded, and releasing radiation, drugs, or other factors in a manner responsive to the particular characteristics of the lesion involved. Wound-responsive "smart" stents could result in improved patency rates even in difficult lesions where enhanced thrombotic deposition or neointimal proliferation is likely to occur. Improved knowledge regarding the pathobiologic causes of stent complications is required, as is additional insight into technical details such as elution rates, radiation dosing, and other responsive features.

20.4 PACEMAKERS AND IMPLANTABLE DEFIBRILLATORS

20.4.1 Market Size

The market for pacemakers has grown over the past 4 decades as a result of improved device function and an expanding set of clinical indications. Approximately 1 million patients in the United States had permanent pacemakers in 1996 (Kusumoto and Goldschlager, 1996), many of whom will need either a lead or generator replacement sometime in the future. Recent incidence data from nonfederal, short-stay hospitals suggests at least 135,000 permanent cardiac pacemakers were implanted in the United States in 1997, with a minimal estimate of 26,000 automatic implantable cardioverter-defibrillator (AICD) units being placed during the same year (Owings and Lawrence, 1999).

20.4.2 Indications

The American College of Cardiology and American Heart Association consensus practice guideline lists current indications for artificial pacemaker and implanted cardioverter-defibrillator use (Gregaratos et al., 1998). The purpose of a pacemaker is to deliver an electrical impulse of sufficient magnitude to depolarize the heart chamber in a spreading, coordinated fashion, as occurs in a normal heart beat. In contrast, defibrillators are used to depolarize the entire heart at once in an effort to terminate uncoordinated contractions. The natural refractory period of the myocardial tissue usually prevents erratic residual electrical activity from propagating for a short period of time, restoring coordinated muscular contraction.

In general, a pacemaker is warranted in certain cases where electric impulse conduction or initiation in the heart is blocked, slowed, or triggered in an irregular, variable fashion. Specific diseases for which pacemaker therapy is employed include certain forms of atrioventricular and fascicular conduction block, sinus node dysfunction, and some forms of neurocardiogenic syncope (Gregaratos et al., 1998). The most popular indications for first-time implantation of cardiac pacemakers have changed over time. In 1993, over half of all new pacemakers were placed to treat sinus node dysfunction, the most common indication for pacemaker implantation (Bernstein and Parsonnet, 1996a). Congenital or acquired atrioventricular (AV) conduction block was second, accounting for approximately 31 percent of primary implantations, followed by drug-induced bradycardia, AV block secondary to ablation, and tachyarrhythmia (Bernstein and Parsonnet, 1996a). Emerging indications, particularly pacemaker treatment of congestive heart failure, could significantly alter the percentages indicated above (Cazeau et al., 2001; Gerber et al., 2001).

The indications for implantation of a cardioverter-defibrillator are based primarily on the past presence of a potentially fatal ventricular arrhythmia due to nontransient causes, regardless of the specific illness (Gregaratos et al., 1998). Recent evidence has suggested that the contributing illness might alter therapy effectiveness, leading to changes in the recommendations in the near future (Gregaratos et al., 1998). Two indications, ventricular tachycardia or fibrillation and aborted sudden death, dominated the list of indications for implantable cardioverter-defibrillator (ICD) placement in a 1993 survey, accounting for 97 percent of the devices implanted for which an indication was reported (Bernstein and Parsonnet, 1996a).

20.4.3 Device Design

The wide range of electrophysiologic disorders treated with pacemakers and defibrillators requires devices with various capabilities and settings. Generic classification codes have been developed to

TABLE 20.1 The five-position NASPE/BPEG Generic Pacemaker Code. The purpose of the code is to allow medical practioners to recognize the basic features and capabilities of a given pacemaker. The code alone does not provide a complete description of device features, but is clear and simple to use

Positions and description	I. Chamber(s) paced	II. Chamber(s) sensed	III. Pacemaker response	IV. Rate adaptive and programming features	V. Antiarrhythmic pacing, cardioversion, and defibrillator functions
	Atrium	**A**trium	**T**riggered	**R**ate Adaptive	**P**acing
	Ventricle	**V**entricle	**I**nhibited	**S**imple Programmability	**S**hock
	Dual	**D**ual	**D**ual	**M**ultiple Programmability	**D**ual
	NOne	**NO**ne	**NO**ne	**C**ommunicating	**NO**ne
				NOne	

Source: *Adapted from Bernstein et al., (1987.)*

ease the identification of the different pacemakers, defibrillators, and associated leads presented both in the literature and medical practice. One such code is the North American Society for Pacing and Electrophysiology/British Pacing and Electrophysiology Group Generic Pacemaker Code, which contains five positions and defines the chambers that are paced and sensed, the potential electrical response to sensed rhythms, rate modulation functions and the degree of programmability, and antitachyarrhythmia functions, if present (Bernstein et al., 1987). The code represents an expansion of previous generic classification schemes to incorporate rate-adaptive functions and cardioversion-defibrillation tasks (Bernstein et al., 1987). Table 20.1 summarizes the possible features for each location. More than 60 percent of pacemakers implanted in 1993 were of the DDD type, possessing dual chamber sensing and pacing with both excitatory and inhibitory functions, with half of that number having additional rate adaptive features (DDDR) (Bernstein and Parsonnet, 1996a). Similar classification codes exist for defibrillators (Bernstein et al., 1993) and pacing leads (Bernstein and Parsonnet 1996b).

Pacemaker and defibrillator systems consist of two implanted and one external component: the generator, cardiac leads, and programmer, respectively (Morley-Davies and Cobbe, 1997). Although the clinical functions of pacemakers and defibrillators differ, the desired component characteristics are similar and include low complication rates coupled with small size, durability, and longevity.

Generators. Despite enhanced functional performance and an ever-increasing array of features, current ICD generators displace less than half the volume of those devices implanted a short time ago as a result of improved circuit, packaging, and power storage technology (Morris et al., 1999). Similarly, modern pacemaker generators are small, thin, and weigh only 20–30 g (Kusumoto and Goldschlager, 1996). Generator and battery technology allows pacemakers to last approximately 5 to 9 years in the body (Morley-Davies and Cobbe, 1997), with the more complex dual-chamber designs having a shorter lifespan than the simpler, single-chamber units (Kusumoto and Goldschlager, 1996). An example of state-of-the-art pacemaker and implantable defibrillator generators is shown in Fig. 20.6.

Power technology has steadily improved since the first mercury-zinc batteries of the 1960s, with modifications in other generator components and usage algorithms further increasing battery endurance (Jeffrey and Parsonnet, 1998). Novel power supplies have included the nuclear slug and biokinetic sources, but the advent of the lithium-based power source allowed for increased generator longevity over mercury-zinc types and was amenable to being hermetically sealed (Jeffrey and Parsonnet, 1998). The current batteries used in pacemakers and defibrillators differ in formulation. Pacemakers utilize lithium-iodine batteries (Kusumoto and Goldschlager, 1996), while ICDs utilize lithium-silver-vanadium batteries and a capacitor network for discharges (Morris et al., 1999).

Electrical Leads. The pacing lead consists of five components: the connector, conductor, insulating material, electrode(s), and fixation mechanism (Crossley, 2000). Electrical leads have to fulfill a number of conflicting requirements, although reliable performance remains the dominant criterion

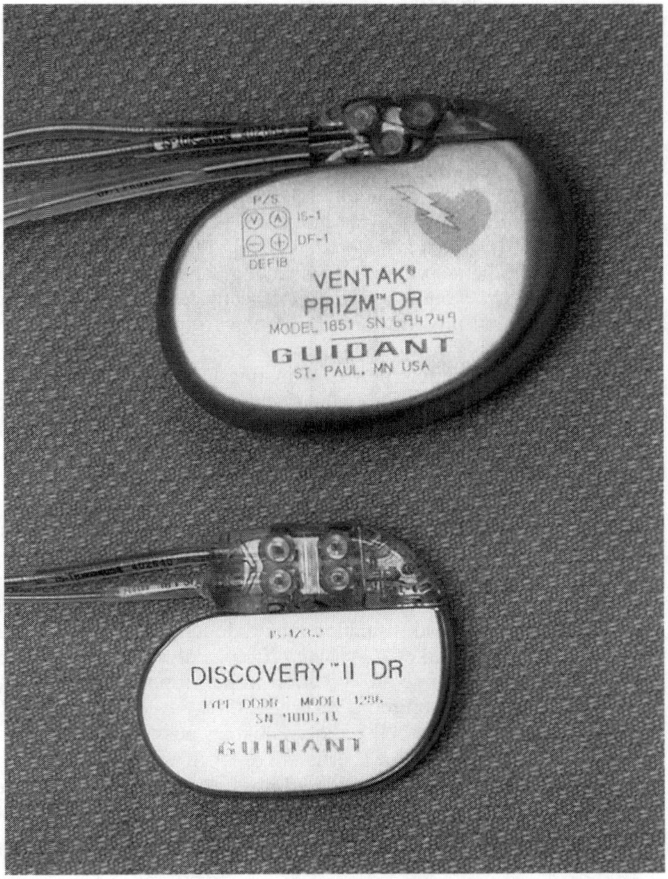

FIGURE 20.6 The generators of a modern cardioverter-defibrillator (Ventak Prizm DR, Guidant Corporation, Minneapolis) and pacemaker (Discovery II DR, Guidant Corporation, Minneapolis) are shown. Both are dual-chamber devices, with the cardioverter-defibrillator possessing pacing features in addition to the core cardioversion hardware. The added hardware and power requirements demand a larger housing for this device in comparison to the pacemaker.

(de Voogt, 1999). Leads should possess a small diameter, be flexible enough to place but durable enough to resist wear, possess favorable power consumption characteristics, anchor in place to prevent migration, and enjoy good biocompatibility.

Pacing leads can be subdivided into a number of groups based on the area of placement and method of stimulation. Leads can be placed either on the epicardium (external surface) of the heart or, by a transvenous route, onto the endocardium (internal surface) of the right heart atrium or ventricle. Epicardial leads are used for permanent pacing in pediatric cases, where size considerations or congenital defects prevent transvenous placement, and in patients who have undergone tricuspid valve replacement (Mitrani et al., 1999). Transvenous placement of the lead is the preferred route in most patients.

The depolarization shock can be delivered through either a *unipolar* or *bipolar* lead. Most older leads utilize the unipolar design (Morley-Davies and Cobbe, et al., 1997), in which a single insulated electrode is placed near the myocardium of the heart and acts as a cathode (Tyers et al., 1997;

Morley-Davies and Cobbe, 1997). The generator shell acts as the anode of the resulting circuit. Modern leads use a bipolar design where the cathode and anode are both in the lead and spaced a short distance apart (Morley-Davies and Cobbe, 1997). Three general approaches have been developed for placing two insulated conducting wires within the lead body for the bipolar design. The original bipolar leads were fabricated with two conducting wires alongside one another, enclosed in a large silicone rubber sheath (Tyers et al., 1997). These designs gave way to coaxial systems where a layered approach was used. Here, a conducting wire at the center is sheathed in an insulator and surrounded by another conducting layer and a final layer of insulation. Coaxial pacing leads tend to be smaller than the side-by-side models (Tyers et al., 1997). The most recent approach to bipolar lead design is described as a coradial pacing lead. The two conducting wires are coated with a thin layer of insulation and wound in a helical manner along the length of the pacing lead (Schmidt and Stotts, 1998). The wound helix is also sheathed in another layer of insulation to improve handling characteristics and provide some insulation redundancy. The compact nature of the coradial inner structure results in a very small lead.

To avoid the incompatibilities in mating technologies that arise from having a large number of electrophysiology (EP) device manufacturers, internationally accepted standards have been developed to define the mating connection between pacemakers and leads, thus allowing leads from one manufacturer to be mated to the generator of another. Originally developed by the British Standards Institution as the IS-1 specification for low-profile connectors for implantable pacemakers, the standard has been revised and rereleased in 2000 as the ISO 5841-3 standard from the International Organization for Standardization as part of its publications governing the design of cardiac pacemakers. A similar standard (DF-1) exists for ICD shock leads (Morris et al., 1999).

A number of materials have been used for insulation in cardiac pacing leads, including polyethylene, polysiloxanes, polyurethanes, and poly(ethylene-co-tetraflouroethylene) (ETFE). Polyethylene was the original lead insulation material, but poor long-term performance has eliminated its use in the pacing lead market (Crossley, 2000). Polysiloxanes are used in modern electrophysiology leads but have undergone an evolution in performance and formulation over the years. A limited resistance to tearing required a relatively thick layer until newer, high-performance polysiloxanes were released (Crossley, 2000). Polyurethanes possess a high tear strength and, in contrast to early polysiloxane formulas, a low coefficient of friction, both of which served to popularize polyurethane use; leads could be made smaller and possessed improved placement characteristics, especially in multiple-lead applications (Crossley, 2000; Schmidt and Stotts, 1998). The newest coradial bipolar leads utilize a thin layer of ETFE as an insulation coating for the interior helically wound leads, with a redundant layer of exterior polyurethane insulation (Crossley, 2000; Schmidt and Stotts, 1998). The coradial leads are small and appear to have a reduced complication rate compared to coaxial designs (Tyers et al., 1997).

To prevent migration, lead tips are fitted with either passive or active fixation mechanisms. Passive devices use hooks or tines to secure the line into place, while active fixation methods rely upon a screwlike tip to burrow the lead into the heart tissue. Although active fixation leads have been considered less likely to dislodge or suffer from complications than passive leads, few major studies have been performed (Crossley, 2000), and some doubt that differences in performance exist in the hands of an experienced implanter (Mitrani et al., 1999). Some situations do warrant active fixation leads, however. Congenital heart disease may require lead placement in an odd area, which can make active fixation useful (Mitrani et al., 1999). Active fixation leads appear to be preferred for atrial placement, while passive fixation leads are the overwhelming choice when leads are placed in the ventricle (Bernstein and Parsonnet, 1996a). Figure 20.7 provides a close-up view of a variety of lead tips, revealing the fixation equipment used for both passive and active methods.

Rate-adaptive pacing requires some method of rate modulation, preferably without patient intervention, that mimics as closely as possible the normal behavior of the intact sinus node during exertion (Leung and Lau, 2000). A very large number of sensors, control techniques, and algorithms have been proposed in an effort to provide optimal rate responses to exercise and activity in those patients unable to generate an adequate heart rate increase (Leung and Lau, 2000). Because no single sensor and associated circuitry can perfectly replace the sinus node, dual sensor systems have been introduced to accommodate for particular deficiencies in either sensor (Mitrani et al., 1999). The

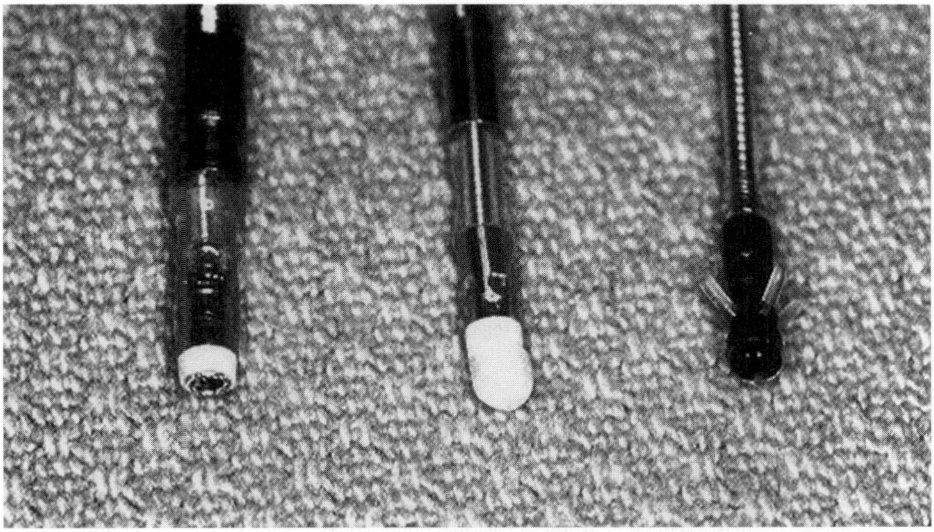

FIGURE 20.7 Close-up views of three pacemaker and/or cardioverter-defibrillator leads. The lead on the far left is an active fixation lead with a retractable screw embedded in the lead tip. It can be extended after implantation to attach the lead to the heart wall. The middle lead possesses a soluble cap that dissolves within a few minutes inside the body to reveal a hook or screw for tip fixation. A tined passive fixation lead is shown on the right. The soft tines become lodged in the irregular inside surface of the heart, preventing lead migration.

most popular sensors are mechanical devices that measure vibration, which roughly indicates that body movement is taking place (Morley-Davies and Cobbe, 1997; Leung and Lau, 2000). These units have the benefit of being compatible with existing lead technology (Leung and Lau, 2000).

The current standard pacemaker lead utilizes many of the innovations described above and possesses a low coil resistance coupled with a high electrode impedance and a steriod-eluting tip (de Voogt, 1999). The steroid reduces the inflammatory reaction to the implanted lead that can increase the stimulation threshold over time (Mitrani et al., 1999; Crossley, 2000).

Programmer. The programmer allows the physician to adjust pacemaker and defibrillator settings to meet the particular needs of the patient. Modern microcomputer-based systems use radio-frequency waves or magnetic fields to communicate with the EP device noninvasively (Kusumoto and Goldschlager, 1996; Morley-Davies and Cobbe, 1997). The programmer can retrieve device settings and performance data, including failure episodes, electrocardiograms, and battery function, allowing the physician to optimize device performance or analyze an existing problem (Kusumoto and Goldschlager, 1996). A recent review divided the most common programmable features into six categories, including pacing mode selection, energy output and characteristics, electrical sensitivity, rate limits, refractory periods and their duration, and the various rate-adaptive features and functions (Kusumoto and Goldschlager, 1996). The ever-growing range of features and functions on the modern EP device complicates patient management but does allow tailored therapy.

20.4.4 Complications

Complications associated with electrophysiology devices can be divided into those that are a consequence of device failure or malfunction and those that are secondary to device implantation, extraction, or patient care. In general, complications are reported to occur at a greater rate with those physicians who implant pacemakers less frequently (Bernstein and Parsonnet, 1996a). Clinically significant perioperative complications such as hemothorax (blood in the chest cavity), pneumothorax (air in the chest cavity), infection, and hematoma (blood collection around insertion site) are relatively rare at 1 to 2 percent (Morley-Davies and Cobbe, 1997).

Generators. Although there are a number of problems that could necessitate generator replacement, the most common reason is a depleted battery at the end of its service life, which indicates that proper generator function and longevity is the norm rather than the exception. However, a relatively common emergent indication for generator replacement is apparent electrical component failure (Bernstein and Parsonnet, 1996a).

Electrical Leads. Pacing (de Voogt, 1999) and defibrillator electrical leads are the most problematic components in their respective electrophysiology systems. As a common site of malfunction and a significant factor in device power consumption, the electrical pacemaker lead has been subject to a variety of design and usage improvements (de Voogt, 1999). A recent survey of cardiac pacing revealed that the most common reason for lead replacement was insulation failure, followed by a high stimulation threshold for cardiac depolarization and displacement of the lead electrode itself (Bernstein and Parsonnet, 1996a). Bipolar leads suffered from a higher complication rate in both the atrial and ventricular positions, and ventricular leads secured with passive fixation devices also experienced a higher reported complication rate (Bernstein and Parsonnet, 1996a). Results from the Danish Pacemaker Register (Moller and Arnsbo, 1996) revealed significantly lower reliability in bipolar leads versus unipolar leads, but unipolar leads do have shortcomings. Unipolar sensing units suffer from electrical interference due to extracardiac muscle activity or electromagnetic interference (EMI), which can cause pacemaker malfunction (Morley-Davies and Cobbe, 1997; Mitrani et al., 1999). Furthermore, unipolar pacemaker leads are incompatible with implantable defibrillator devices (Mitrani et al., 1999).

Among common lead insulation materials, polyurethane 80A, a polyetherurethane (PEU), is known to suffer from higher rates of degradation and failure and is responsible for a large proportion of lead malfunctions (Crossley, 2000; Tyers et al., 1997). The PEU 80A degrades via three mechanisms: environmental stress cracking, metal ion catalyzed oxidation, and calcification (Schmidt and Stotts, 1998; Crossley, 2000). Alternative polymers such as ETFE, polycarbonateurethanes, and more durable formulations of PEU are being used to overcome the limitations of PEU 80A (Schmidt and Stotts, 1998). These materials do not suffer from the same mechanisms of failure, or at least exhibit increased resilience to degradation via those pathways (Schmidt and Stotts, 1998).

20.4.5 Future Developments

Up to a fifth of all patients with implanted cardioverter-defibrillators have also demonstrated a pacemaker requirement (Pinski and Trohman, 2000). The dual requirement for pacemakers and ICDs has led to the development of units combining the two technologies on an advanced level, complete with dual chamber activity (Morris et al., 1999; Pinski and Trohman, 2000). The benefit of a combined EP device is the elimination of separate costs for each system and potential harmful interactions between pacemakers and ICDs, including a reduction in the number of leads (Morris et al., 1999; Pinski and Trohman, 2000).

Steady improvements in microelectronics and software will further expand the capabilities of EP devices toward the analysis and treatment of other pathologic heart conditions currently on the fringe of electrophysiology, such as atrial arrhythmias (Morris et al., 1999). These enhanced units will monitor a wide variety of patient and system data to optimize performance to a greater extent than current systems. There is room for improvement in power technology and lead design, as well as improved algorithms able to perform advanced rhythm discrimination (Morris et al., 1999). As suggested in the literature, the development of a universal programmer capable of interfacing with any EP device would be a significant advance both in cost and ease of use for providers now faced with a multitude of different programming units (Kusumoto and Goldschlager, 1996). Improved and simplified programmer interfaces would also benefit the implanting physicians, who are now overwhelmed with feature-heavy and highly adjustable systems (Morris et al., 1999).

20.5 ARTIFICIAL VASCULAR GRAFTS

20.5.1 Market Size

According to a recent discharge survey of nonfederal short-stay hospitals, at least 66,000 blood vessel resections with graft replacement were performed in 1997, with the majority of those reported involving the abdominal aorta (Owings and Lawrence, 1999). Because of the limited availability of biologic tissues with the proper size and length for the replacement of large vessels, artificial vascular prostheses are often used in the clinical setting (Brewster, 2000; Greisler, 1991). The placement of vascular access grafts for dialysis treatment represents a significant use of artificial vessels, with 52 percent of dialysis patients receiving therapy through this route (Tokars et al., 2001). Treatment guidelines for renal failure from the National Kidney Foundation advocate the use and placement of an artificial graft for permanent vascular access when a native fistula cannot be created (NKF-DOQI, 1997).

20.5.2 Indications

A number of situations can necessitate the placement of an artificial vascular graft. Vascular grafts can be placed for primary diseases of the vasculature, such as aneurysms or severe atherosclerotic narrowing, for secondary causes such as trauma, or for chronic vascular access issues such as hemodialysis. In the case of primary disease or injury, the indications for placement are dependent upon the level of the vascular tree in which the lesion or damage is located, as well as the etiology of the illness. The choice of interventional therapy also varies, depending on whether the target vascular lesion is cardiac, thoracic, carotid, or peripheral. Replacement or bypass of a native vascular conduit is a procedure that lies on one end of the interventional spectrum for treatment of vascular insufficiency. The initial approach to addressing a case of vascular insufficiency (stenosis, occlusion, etc.) is usually a percutaneous intervention such as balloon angioplasty and stenting, followed by surgical repair or replacement of the native vessel if necessary. Some vascular patients with severe coexisting conditions are unable to tolerate the profound sedation required to directly excise and replace a damaged blood vessel. For these patients, long-tunneled extra-anatomic grafts placed outside the path of the existing diseased vessel are the preferred therapy, despite poorer long-term outcomes and patency rates compared to traditional excisional grafting (Biancari and Lepantalo, 1998; Foster et al., 1986; Connolly, 1984).

20.5.3 Current Graft Designs

Vascular grafts or prostheses can be classified as originating primarily either from biologic or synthetic sources. Biologic grafts can be harvested from other species (xenograft), from other humans (allograft), or the vasculature of the patient (autograft). As the focus of the present section is on the engineering aspects of vascular grafts, the following discussion will be limited to conduits of synthetic origin. Tissue-engineered biologic grafts will be discussed in the section covering future trends.

The ideal synthetic vascular graft would possess a number of characteristics designed to mimic native conduits, ease surgical placement and limit manufacturing complexity and cost (Brewster, 2000). A number of features supporting these design goals are listed in Table 20.2. Initially, vascular graft research focused upon the development of completely passive conduits that would not elicit a biological response when exposed to blood. More recent research has focused on the development of grafts that generate a favorable biological response from the body, as it has been recognized that a perfectly inert surface may be an unattainable goal (Greisler, 1991).

TABLE 20.2 Desirable Characteristics of Prosthetic Vascular Grafts

Biologic features	Handling and implant features
Biocompatible	Range of sizes
Infection resistant	Ease of sterilization
Antithrombotic	Flexible without kinking
Compliant	No blood leakage
Durable	Good suturability

Source: *Adapted from Brewster (2000)*

Current commercial polymeric graft designs can be divided into both *textile* and *nontextile* forms (Brewster, 2000). Textile-based grafts are generally composed of woven or knitted polyethylene terephthalate (Dacron), while the nontextile grafts are usually fashioned from expanded polytetrafluoroethylene (ePTFE). Other polymeric materials investigated or used in the manufacture of vascular grafts include polyamides or nylons, nonexpanded PTFE, polyvinyl alcohol (Ivalon), vinyl chloride–vinyl acetate copolymers (Vinyon-N), polyacrylonitrile (Orlon), and polyurethanes (Brewster, 2000; Greisler, 1991; Eberhart et al., 1999]. Although the majority of these materials has been abandoned, polyurethanes have enjoyed continued interest as a source material, especially for small-diameter vascular grafts (Eberhart et al., 1999), despite their mixed showing in the few human studies that have been attempted (Zhang et al., 1997; Allen et al., 1996; Bull et al., 1992) and increased concerns over their long-term degradation performance (Eberhart et al., 1999).

The preferred graft material differs for both implantation site and use, and selection is based upon patency rates, absence of complications, convenience, or handling characteristics, and cost (Brewster, 2000). Large-diameter grafts for use in the aorta and major arteries have generally been made of Dacron, while medium-size grafts are primarily constructed of ePTFE. Smaller vessels located in the infrainguinal region (below the area where the legs connect with the torso) represent a significant challenge because of the traditionally poor patency rates with artificial grafts in these vessels (Londrey et al., 1991; "Comparative . . . ," 1988; Veith et al., 1986). Current evidence suggests replacement of the damaged vessel with an autogenous vein or artery graft is the procedure of choice rather than implantation of an artificial graft (Faries et al., 2000). Unfortunately, preexisting disease, anatomic or size limitations, and other factors may rule out an autogenic source for vessel replacement, thereby forcing the use of an artificial (usually ePTFE) vascular graft. It is in these smaller vessels that the promise of tissue-engineered prosthesis is expected to have the greatest impact.

The different construction techniques used to manufacture textile grafts have an effect on the final device properties. Graft porosity is considered to be a prime factor determining a number of handling, performance, and outcome (Wesolowski et al., 1961) characteristics for textile implants. Knitted textile grafts possess a high porosity and generally require a special yet simple procedure called *preclotting* prior to implantation to prevent excessive leakage of blood through the graft wall. Traditionally, preclotting is performed with a sample of the patient's blood; coagulation is initiated to fill the open pores and interstices with fibrin. As of 1997, most (>80 percent) implanted grafts came presealed with collagen, gelatin, or other biological compounds (e.g., albumin) directly from the manufacturer in an effort to reduce time spent on graft preparation (Brewster, 2000) and limit surface thrombogenicity (Greisler, 1991). In contrast, woven textile grafts do not generally require preclotting but possess less desirable handling properties such as increased stiffness (Brewster, 2000) and a tendency to fray when cut (Greisler, 1991). Examples of woven and knitted polyester aortic grafts are shown in Fig. 20.8. Nontextile PTFE grafts can possess differing levels of porosity,

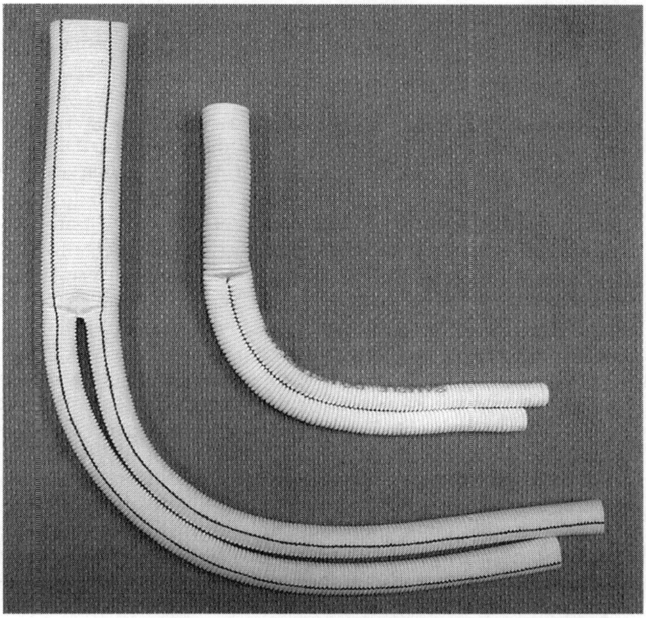

FIGURE 20.8 Two polyester aortobifemoral grafts (Meadox Division, Boston Scientific Corporation, Natick, Massachusetts) are shown. These large grafts are used to repair abdominal aortic aneurysms and other lesions affecting the distal aortic segment. The larger graft on the left is made of woven polyester, while the smaller graft is of knit construction. The knitted graft is significantly more elastic and can be stretched to a greater extent than the corresponding woven graft.

depending on the processing technique employed. This feature may influence the extent of the healing process (Golden et al., 1990; Kohler et al., 1992; Contreras et al., 2000).

20.5.4 Complications and Management

Graft "healing" or incorporation into the vasculature is a complex process that is the subject of numerous studies and reviews (Greisler, 1991; Davids et al., 1999). Graft healing is affected by the graft material, the graft microstructure and surface characteristics, hemodynamic and biomechanical factors, and eventually the interplay between these characteristics and the cellular and humoral components involved in the ongoing incorporation of the graft (Greisler, 1991). Incorporation of the graft into the vasculature can occur through ingrowth of tissue from the anastomotic ends, or from tissue growth through the graft wall (Davids, 1999). Humans, in contrast to many animal models, usually fail to fully endothelialize the inner surface of a vascular graft, possessing instead some near-anastomic endothelialization and scattered islands of endothelial cells on a surface rich in fibrin and extracellular matrix (Pasquinelli et al., 1990; Sauvage et al., 1975; Sauvage et al., 1974; Berger et al., 1972). The lack of full healing has prompted extensive research into the mechanism of endothelialization and methods to improve it in the clinical setting.

Graft complications can be split into those that occur with high or low frequency. High-frequency complications include graft thrombosis and anastomotic pseudointimal hyperplasia, a condition of cellular overgrowth that occurs at the site where the artificial graft meets the remaining native vessel.

Less frequent complications include prosthesis infection and structural changes such as graft dilatation, a problem more prevalent in knitted Dacron prostheses than other graft types (Robinson et al., 1999).

Because of their grave consequences, graft thrombosis and infection are of particular concern to the implanting surgeon. Thrombosis in vascular grafts is a function of surface characteristics, hemodynamic properties, and the patient's hemostatic system, and represents a common cause of early graft failure. In hemodialysis patients, for whom vascular access is critical, the thrombosis rate of ePTFE access grafts averages 45 percent per year (Kaufman, 2000) and is often secondary to venous stenosis (Palder 1985). Although treatment differs by graft location, results from a randomized clinical trial involving thrombosed lower-extremity grafts revealed that catheter-based thrombolytic therapy with urokinase or t-PA (tissue plasminogen activator) restores patency in most patients for whom catheter access is successful. This treatment can also reduce the extent of surgical revision is such revision is subsequently required (Comerota et al., 1996).

Graft infection occurs in less than 1 percent of cases as an early (<30 days) complication, but may afflict as many as 5 percent of graft recipients over the longer term (Bandyk and Esses, 1994). The usual culprit is a staphylococcal species, but many other species have been known to infect vascular grafts (Bunt, 1983), with different complications and rates of progression (Bandyk and Esses, 1994). Although there is some leeway in regard to the level of therapy, infections can often be fatal unless aggressive surgical means are coupled with antibiotic treatment. Because residual infection is a common cause of therapeutic failure, extensive debridement and graft excision with in situ or extra-anatomic prosthetic replacement is warranted in most cases (Calligaro and Veith, 1991; Seeger, 2000). Even with aggressive therapy and placement of an extra-anatomic bypass, graft infections result in amputation rates up to 30 percent and a mortality rate approaching 40 percent (Bandyk and Esses, 1994).

20.5.5 Future Trends

An excellent text on efforts to produce improved vascular grafts through tissue engineering has been recently released (Zilla and Greisler, 1999). Current research is focused on improving the performance of existing vascular graft designs, although attempts have been made to create entire tissue-engineered vessels as well (L'Heureux et al., 1998; Niklason et al., 1999). Novel coatings have been developed to steer the biological response to an implanted vascular graft toward endothelialization and improved biocompatibility, while moving away from the proliferation of cellular components linked to vessel narrowing and failure (Greisler, 1996). Various approaches to improving graft performance, healing, and long-term patency include endothelial cell seeding (Park et al., 1990; Williams et al., 1992) and the use of antithrombotic and antibiotic coatings (Devine and McCollum, 2001). Clinical experiences with some of these approaches (endothelialization) have shown promising results (Deutsch et al., 1999) compared to traditional grafts, while others, such as antibiotic coatings, lack definitive outcomes (Earnshaw, 2000).

20.6 ARTIFICIAL KIDNEYS

20.6.1 Market Size

According to the United States Renal Data System, the number of Americans in end-stage renal disease (ESRD) is expected to rise from around 370,000 current patients to greater than 660,000 in 2010 (USRDS, 2000). Treatment modalities used to replace the failing kidney are few, including hemodialysis with an artificial kidney or hemodialyzer, peritoneal dialysis, or kidney transplantation. Hemodialysis remains the dominant therapy for ESRD in the United States, doubling in use in the last decade with about 280,000 patients supported in 2000 (USRDS, 2000). Hemodialysis is also the dominant therapy worldwide, with 71 percent of ESRD patients receiving dialysis in 1991

(Klinkmann and Vienken, 1995). In the United States, kidney transplantation follows hemodialysis in the number of patients treated, but a limited donor pool prevents dramatic expansion of this option. Peritoneal dialysis trails hemodialysis and transplantation with only 13.5 percent of U.S. ESRD patients supported (USRDS, 2000). The combined cost for ESRD is staggering, with approximately $11 billion expended in the Medicare ESRD program alone in 1998, most of which is applied to hemodialysis treatment (USRDS, 2000). The present section focuses on the use of hemodialyzers as a cardiovascular device addressing kidney failure.

20.6.2 Indications

Two physical processes, diffusion and convection, are in widespread use as methods to mimic the excretion functions of the native kidney. Therapies utilizing predominantly convective transport are accurately termed hemofiltration, while diffusion-dependent methods are grouped under hemodialysis. Both hemodialysis and hemofiltration are used in the inpatient and outpatient setting to treat renal and nonrenal diseases. In general, the purpose of these therapies is to either remove toxins circulating in the blood or reduce the blood volume by removal of water.

The indications to initiate hemodialysis for end stage renal disease are variable but involve clinical and symptomatic manifestations related to uremia, a buildup of metabolic toxins in the blood. Indications requiring immediate initiation of therapy include encephalopathy or a change in mental status, disturbances in either the sensory or motor pathways, effusions in the pericardial space or pleura due to uremic irritation, uncontrollable metabolic derangements such as high serum potassium and low pH, and excessive water retention (Denker et al., 2000). Because of evidence suggesting improved outcomes in patients receiving early dialysis intervention, patient symptoms resulting in a reduced quality of life (fatigue, cognitive changes, itching, and malnutrition) can be considered as indications for the initiation of dialysis (Hakim and Lazarus, 1995).

Indications for renal replacement therapy in the acute setting and for other disease processes are different from those for ESRD. A common mode of ESRD therapy in the outpatient setting is intermittent hemodialysis (IHD) where a patient receives intense treatment over the course of a few hours several times a week. Acute renal failure in the inpatient setting is often treated with continuous renal replacement therapy (CRRT), which is applied for the entire duration of the patient's clinical need and relies upon hemofiltration to a higher degree than IHD (Meyer, 2000). Other nonrenal indications for CRRT are based on the theoretical removal of inflammatory mediators or toxins and elimination of excess fluid (Schetz, 1999). These illnesses include sepsis and systemic inflammatory response syndrome, acute respiratory distress syndrome, congestive heart failure with volume overload, tumor lysis syndrome, crush injury, and genetic metabolic disturbances (Schetz, 1999).

20.6.3 Device Design

Three physical processes determine the removal rate for uremic toxins through membrane-based devices. *Convection* results in toxin removal through a semipermeable membrane that separates blood from dialysate and can be used to remove excess fluid. A pressure gradient across the membrane is responsible for the solvent flow, and toxins are removed as a function of their concentration in solution, the ultrafiltration rate or rate of fluid removal, and the sieving coefficient of the particular toxin across the membrane barrier. Membranes for convection-based therapies exclude molecules larger than their pore size but permit improved removal of the middle molecules (500 to 5000 daltons) implicated in uremia (Meyer, 2000). Depending on the amount of fluid removed, replacement electrolyte solution may be required to maintain adequate hemodynamic volume. *Diffusion*-based solute removal primarily affects smaller molecules with high diffusion coefficients and possessing a favorable concentration gradient from the blood to the dialysate. *Adsorption* is the third and least characterized method of solute removal in renal replacement therapies (Klinkmann and Vienken, 1995). Controlled by electrostatic and Van der Waals forces between solute and membrane, adsorption-based removal can be beneficial or harmful depending on the compound involved, such as removal of proinflammatory cytokines versus a needed anticoagulant (Klinkmann and Vienken, 1995). Convection and diffusion remain the dominant physical processes by which membranes and devices are designed.

Although hemodialysis and hemofiltration are often considered separate therapies, some clinical treatments rely on a combination of the two and therefore can be classified as hemodiafiltration procedures. Treatment techniques can be further stratified as to whether they are intermittent or continuous in nature, and whether the vessels accessed are both venous, or arterial and venous. Due to a lack of definitive prospective randomized trials, the relative advantage of continuous versus intermittent treatment is unknown, although continuous administration is felt to be more "gentle" in some circles, allowing greater time for toxin equilibration and removal (Meyer, 2000).

Unit and Membrane Design. Hemodialysis units have undergone a variety of changes since the first practical design, a rotating drum dialyzer, was introduced in the 1940s (Kolff et al., 1944). As recently reviewed by Clark, subsequent unit designs have progressed through coil and parallel flow dialyzers to the current dominant design of the hollow fiber dialyzer, which was introduced to address performance and use limitations inherent in the earlier devices (Clark, 2000). Subsequent to the introduction of the hollow fiber dialyzer, much of the improvement and development in artificial kidney devices has focused on the membrane barrier materials and device operating conditions. An example of a standard hollow fiber dialyzer is shown in Fig. 20.9, and there have been few major changes in the design since its introduction decades ago (Ronco et al., 2000).

(a)

(b)

FIGURE 20.9 Hemodialysis exchangers are disposable units similar to membrane oxygenators in construction. (*a*) Note the simple design of the device as shown in the upper portion of the figure (Artificial Organs Division, Travenol Laboratories, Inc., Deerfield, Illinois). (*b*) Close-up view of hollow fibers used to separate the patient's blood from the dialysate fluid. Toxins in the blood diffuse through the fiber walls to the dialysate fluid.

As of 1995, more than 30 different polymer blends were being used in the manufacture of membranes for hemodialysis and hemofiltration (Klinkmann and Vienken, 1995). The various membrane types used for renal replacement therapy can be divided into membranes derived from cellulose (83 percent of 1991 worldwide total) and from synthetic materials (the remaining 17 percent) (Klinkmann and Vienken, 1995). Synthetic membranes have been constructed from such materials as polyacrylonitrile (PAN), polysulfone, polyamide, polymethylmethacrylate, polycarbonate, and ethylvinylalchohol copolymer (Klinkmann and Vienken, 1995). In the United States, use of cellulosic materials for membrane construction predominates at around 95 percent of the total number of membranes used (Klinkmann and Vienken, 1995).

Membrane material selection is dependent upon the mode of therapy employed. Convective therapies such as hemofiltration require a high hydraulic permeability and a large pore size, which might permit large molecules such as cytokines to pass through the fiber wall. Synthetic membranes are well suited for this role and are desired for most continuous, convective techniques (Jones, 1998).

Design Performance Evaluation. Removal of uremic toxins and excess fluid is the central purpose of an artificial kidney. A proposed artificial kidney design should undergo transport testing that encompasses the spectrum of molecules encountered in a uremic state at the appropriate flow conditions. Mock circulatory circuits can approximate hemofiltration and hemodialysis flow environments, generating ultrafiltration rates, sieving coefficients (larger molecules) and clearances (smaller molecules) for sample fluids as simple as salt solutions up to uremic human blood (Leypoldt and Cheung, 1996). Many of the toxins responsible for uremia are unknown and are approximated with marker molecules spanning a large size spectrum, including small solutes, middle molecules, and albumin (Leypoldt and Cheung, 1996). Small solutes used for in vitro studies include the clinically relevant compounds urea and creatinine, while middle molecules such as beta2-microglobulin can be approximated with inulin and dextrans (Leypoldt and Cheung, 1996). Albumin is used to approximate high-molecular-weight oncotic substances in the blood, and its clearance is believed to be negligible (Leypoldt and Cheung, 1996). The transport properties derived using mock circulatory loops may not reflect clinical performance because of complex solute—carrier protein interactions and membrane surface fouling by plasma proteins (Leypoldt and Cheung, 1996).

20.6.4 Complications and Patient Management

Complications that occur during hemodialysis and hemofiltration can be divided into problems related to vascular access and those due to exposure of the blood to the exchange circuit. Depending upon the method used, vascular access problems associated with renal replacement therapy are similar to those experienced in patients with vascular grafts or catheters and are covered in those respective sections.

The complications associated with blood exposure to the dialysis or filtration membrane are related to the activation and upregulation of various homeostatic systems. Inflammatory and hemostatic cascades can become activated by protein adsorption onto the significant surface area that these devices necessarily present to the blood. As part of the inflammatory response to dialyzer perfusion, activation of the complement system results in the generation of anaphylatoxins C3a and C5a and can potentiate activation of polymorphonuclear leukocytes and monocytes (Johnson, 1994; Grooteman and Nube, 1998). These activated leukocytes release cytotoxic compounds and inflammatory mediators that normally would be reserved for the killing of bacteria and infectious agents. The release of these inflammatory mediators is implicated in sequelae such as fever, cardiovascular instability, and increased catabolism of muscle protein (Pertosa et al., 2000). Since the nature and extent of protein adsorption impacts the inflammatory response, the surface properties of dialyzer membranes are of interest as a means to limit this response. In studying surface properties and complement activation, the presence of surface hydroxl groups has been implicated as a potential trigger for pathway activation (Chenoweth, 1984).

Activation of the coagulation cascade and platelet deposition in the artificial kidney circuit is obviously undesirable, leading to reduced device performance or failure. To minimize this phenomenon, patients on hemodialysis are usually anticoagulated with heparin during the dialysis session on either a systemic or regional (extracorporeal circuit only) level (Denker et al., 2000). Patients with a sensitivity or contraindication to heparin therapy can be anticoagulated regionally with citrate as well (Denker et al., 2000). Minimizing the potential of membrane surfaces to activate both the hemostatic and inflammatory pathways is of interest to device manufacturers as they seek to reduce complications associated with artificial kidney use.

20.6.5 Future Trends

Artificial kidney designs will likely continue to experience incremental improvements in the materials and hemodynamic areas. New developments in biocompatible materials, superior transport methods for toxin removal, and improved patient management techniques will allow further maturation of hemodialysis and hemofiltration therapy. For example, considerable benefits could be realized from selective toxin removal without concomitant elimination of beneficial proteins. It has been suggested that future devices might utilize the absorption removal pathway with affinity methods as a primary technique to eliminate specific uremic toxins (Klinkmann and Vienken, 1995).

The promise of a true revolution in artificial kidney design comes from the area of tissue engineering. The living kidney performs a number of important metabolic, endocrine, and active transport functions that are not replaced with current hemofiltration and hemodialysis therapy. An artificial kidney that successfully replaces these functions could be a significant improvement when used in conjuction with existing therapies. Researchers have developed a bioartificial renal tubule assist device that successfully reproduces many of the homeostatic functions of the native kidney during in vitro studies and that responds in the proper manner to known physiologic regulators of the various homeostatic functions (Humes et al., 1999).

20.7 INDWELLING VASCULAR CATHETERS AND PORTS

20.7.1 Market Size

Catheters, in their simplest form, are merely tubes inserted into a body cavity for the purpose of fluid removal, injection, or both (Thomas, 1989). The term *catheter* has been expanded to include a number of tubing-based percutaneous interventional devices used for tasks such as stent delivery and deployment, clot removal, radio-frequency ablation, and intra-aortic balloon cardiac support. Because of their prevalence and representative uses, the present section will be limited to vascular infusion catheters and access ports. Stenting and cardiac support applications utilizing catheter-based techniques are discussed elsewhere in this chapter.

In 1991 it was estimated that more than 150 million intravascular catheters were being procured in the United States each year (Maki and Mermel, 1998). Of this number, more than 5 million were central venous catheters (Maki and Mermel, 1998). Catheters have a critical role in modern health care and are used in increasing numbers for central access of the major arteries and veins, as well as for an ever-expanding array of invasive procedures (Crump and Collignon, 2000).

20.7.2 Indications

Catheters are placed when there is a clinical need for repeated sampling, injection, or vascular access, usually on a temporary basis. In kidney failure, catheters allow emergent blood access for hemodialysis and hemofiltration (Canaud, 2000), and provide temporary access as more permanent sites such as arteriovenous fistulas or grafts mature (Trerotola, 2000). Placement of a catheter or access port is routine for the administration of chemotherapeutic agents and intravenous nutritional supplements. Catheters are often placed when frequent, repeated doses of medication are to be injected, blood samples are to be taken, and hemodynamic performance is to be monitored in critically ill patients (Pearson, 1996).

The anatomic location for temporary central venous catheter (CVC) insertion and placement can be dictated by certain patient or disease restrictions, but the most common sites are the internal jugular vein (neck), the femoral vein (groin), and the subclavian position (upper chest). The internal jugular approach is the first choice for placement of a hemodialysis CVC, while femoral placement is favored when rapid insertion is essential (Canaud et al., 2000). Subclavian vein access has fallen from favor because of a higher incidence of thrombosis and stenosis associated with this site, which can ultimately prevent use of the veins in the downstream vascular tree for high-flow applications such as dialysis (Cimochowski et al., 1990; Schillinger et al., 1991).

20.7.3 Device Design

Design considerations for vascular access devices include ease of handling, insertion, and use; minimal thrombotic and other biocompatibility-related complications; structural and operational reliability over time; and optimization for application-specific performance issues (Canaud et al., 2000). Three different catheter tips are shown in Fig. 20.10 to illustrate these variations in design and structure. Because of the distinct characteristics of the different treatments and agents deployed through catheters, it is not practical to provide specific values for flow rates, pressure drops, viscosities, and other important transport properties.

Catheter device selection is based on a number of factors, including the planned application and placement site, duration of implantation, composition of fluids infused, and frequency of access (Namyslowski and Patel, 1999). Vascular catheters can be divided into two general groups: short-term, temporary catheters that are placed percutaneously, and long-term, indwelling vascular catheters that usually require a surgical insertion. Temporary catheters include short peripheral venous and arterial catheters, nontunneled central venous and arterial catheters, and peripherally inserted central catheters (Pearson, 1996). Tunneled central venous catheters and totally implantable intra-

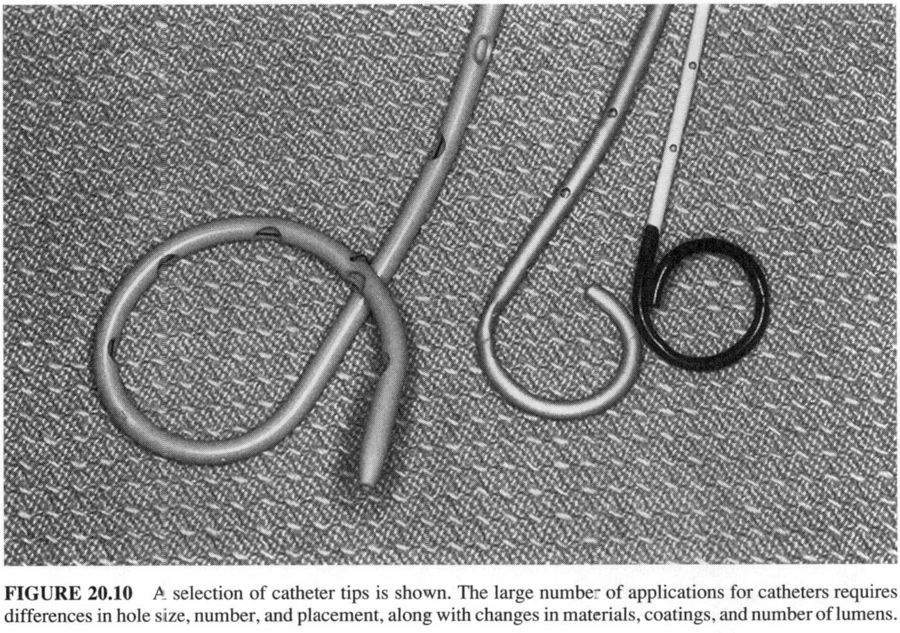

FIGURE 20.10 A selection of catheter tips is shown. The large number of applications for catheters requires differences in hole size, number, and placement, along with changes in materials, coatings, and number of lumens.

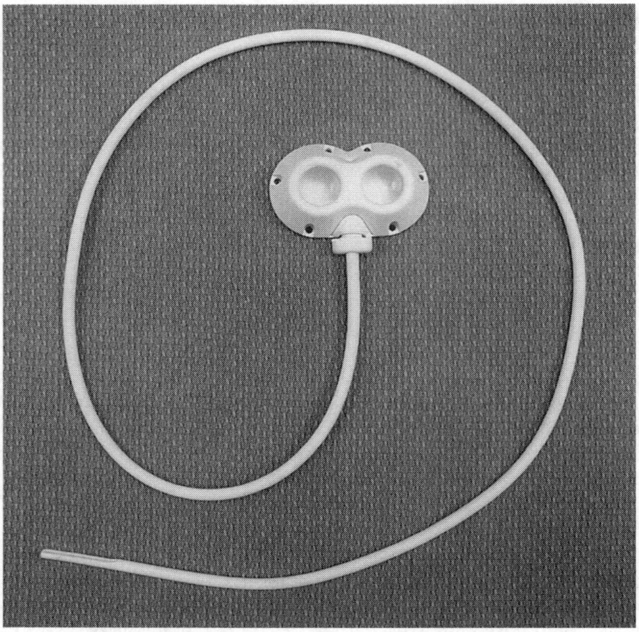

FIGURE 20.11 An implantable vascular access port is shown. The port is accessed through the skin via needle and is resilient enough to allow hundreds of punctures and infusions before needing to be replaced.

vascular devices (i.e, ports) are used for therapies requiring long-term vascular access (Pearson, 1996). The term *tunneled* refers to the placement of the catheter exit site at a location away from the area where the vasculature is penetrated, with the portion of the catheter between these two locations lying in a subcutaneous position. Peripheral venous catheters are the most common devices used for intravascular access, while the nontunneled central venous catheter is the most common central catheter (Pearson, 1996). Subcutaneous ported catheters are the preferred long-term route given an infrequent need for vascular access (Namyslowski and Patel, 1999). Figure 20.11 contains an example of a double-ported fully implantable catheter used for longer-term access.

As mentioned previously, hemodialysis is a common indication for catheter placement. Polysiloxane tunneled central venous catheters and, more recently, totally implantable intravascular port devices are utilized in hemodialysis for long-term vascular access (Schwab and Beathard, 1999). Temporary central venous catheters for hemodialysis access are further subdivided into different groups based on catheter flexibility and the length of time the device will be in use (Canaud et al., 2000). The longer the catheter will be in place, the more supple the material used for construction. Short-term use catheters possess a high stiffness and are fabricated from polytetrafluoroethylene, polyethylene, polyvinyl chloride, and rarely polyurethane (Schwab and Beathard, 1999; Canaud et al., 2000). Medium-term (8 to 30 days) catheters are primarily contructed of polyurethane, while catheters implanted for longer periods of time are usually based on polysiloxane, although polyurethane can be used (Canaud et al., 2000).

Catheters have been permeated, coated, or surface-modified with a variety of compounds in an effort to minimize thrombosis, infection, and friction (Marin et al., 2000; Triolo and Andrade, 1983). The ultimate goal is an improvement in catheter handling characteristics and long-term performance. Some of the more common strategies for imparting microbial resistance include saturating the catheter material with silver sulfadiazine and chlorhexidine (Maki et al., 1997), coating the surface with antibiotics (Raad et al., 1997), or bonding heparin to the surface of the catheter (Appelgren et al.,

1996). Results with antibacterial and antiseptic coatings have been mixed, but a recent meta-analysis involving several randomized controlled trials has shown a significant reduction in hospital-acquired infections when catheters modified for bacterial resistance are used (Marin et al., 2000). The meta-analysis revealed a significant decrease in the number of catheter-related infections for all experimental catheters versus standard devices, plus a substantial reduction in infections for catheters employing antimicrobial systems *other* than silver sulfadiazine and chlorhexadine when compared to those systems (Marin et al., 2000). A randomized trial comparing bacterial colonization and catheter-related bloodstream infection rates associated with two antimicrobial catheters came to a similar conclusion, with minocycline- and rifampin-coated catheters being linked with significantly fewer events than chlorhexadine- and silver sulfadiazine–coated catheters (Darouiche et al., 1999). In an effort to improve handling characteristics, radio-frequency glow discharge has been used to alter the surface properties of common catheter materials to increase hydrophilicity (Triolo and Andrade, 1983) and reduce friction. This latter property is important in the double catheter systems used for some interventional procedures (Triolo and Andrade, 1983).

20.7.4 Management and Complications

Infection and thrombosis are common across a variety of catheter designs and applications, while other complications arise as a result of the particular nature of the therapy being administered, such as recirculation in hemodialysis. Catheter malfunction and related morbidity and mortality represent an area where significant strides are currently being made.

Infection is a common complication associated with intravenous catheters and represents the primary cause of hospital-acquired bloodsteam infection (Valles et al., 1997), resulting in significant morbidity and mortality. Proper care and maintenance is essential for the continued functioning of a catheter, and a number of strategies have been implemented in an effort to minimize infection risk. The use of special coatings and antimicrobial-saturated devices was discussed above in the section on device design. Careful exit site management with a dedicated catheter care team, antibiotic flushes, and possibly catheter tunneling can lower the risk of infection (Raad, 1998), as can the use of a totally implantable intravascular device, which has the lowest infection rate among standard vascular access devices (Maki and Mermel, 1998). Tunneled catheters are fitted with a Dacron cuff that stimulates accelerated healing and ingrowth of tissue distal to the insertion site, thereby providing a host barrier to pathogen migration. In addition, antiseptic cuffs have been placed on catheters to inhibit bacterial migration (Maki et al., 1988; Hasaniya, 1996) but overall results are mixed, suggesting that infecting organisms often migrate through the luminal route or that the cuff loses its antibiotic function over time (Sitges-Serra, 1999). Once a catheter-related infection is detected, the approach to care is dependent upon the patient condition, the number of remaining access sites, and other factors such as the suspect organism, but the catheter is usually removed and antibiotic therapy initiated (Mermel et al., 2001).

Thrombotic complications are common with catheter use. The development of a fibrin sheath is a near universal occurrence on intravascular devices such as central venous (Hoshal et al., 1971) and hemodialysis catheters, and can have a profound effect upon blood flow in the device (Trerotola, 2000). This sheath can be removed either by stripping (Rockall et al., 1997; Crain et al., 1996) or fibrinolytic medical therapy (Twardowski et al., 1998), or the catheter can be replaced to restore adequate flow performance. Recent randomized controlled clinical trials have revealed improved long-term outcomes with catheter exchange versus fibrin sheath stripping (Merport, 2000), while no outcome differences were realized in patients randomized to either fibrin sheath stripping or thrombolytic therapy (Gray, 2000).

20.7.5 Future Trends

Improvements in catheter design are likely to be evolutionary, representing progress in materials, surface treatments, and functional monitoring as catheters are expected to perform for increasingly extended periods. Surfaces with active antithrombotic and antimicrobial activity will continue to be developed and evaluated, as will surfaces designed to minimize protein and cellular adhesion.

20.8 CIRCULATORY SUPPORT DEVICES

20.8.1 Market Size

Heart disease remains the leading cause of death in the United States (AHA, 2000). The devices discussed in the current section provide cardiac support for a spectrum of indications and durations, spanning damage due to an acute myocardial infarction to the long-term decline that accompanies chronic congestive heart failure. The section excludes therapies such as extracorporeal membrane oxygenation and the use of cardiopulmonary bypass pumps as a support method since these are described in the following section on artificial lungs.

The most commonly used means of mechanical circulatory support is the intra-aortic balloon pump (IABP). In 1990, it was estimated that IABP therapy was provided to 70,000 patients annually (Kantrowitz, 1990). As described below, the IABP can provide only limited cardiovascular support, as its effects are limited to pressure unloading of the ventricle, in contrast to artificial hearts and ventricular assist devices, which provide volume unloading (Mehlhorn et al., 1999). To be effective, the IABP requires that the patient maintain some native pumping capacity, as the movement of blood due to the balloon is minimal.

Unlike the IABP, ventricular assist devices (VADs) and total artificial hearts (TAHs) aid or replace the function of the native organ for an extended period of time. A cardiac transplant is the last resort for many patients who fail other medical and surgical therapy. Unfortunately, donor organs remain limited, creating the market for both temporary and extended cardiac support. It has been estimated that up to 100,000 patients in the United States alone would benefit from the implantation of a long-term cardiac support device, with between 5000 and 10,000 requiring biventricular support (Willman et al., 1999). However, the need for mechanical support has not been met by existing designs because of mechanical and biocompatibility limitations. The ongoing need continues to spur development of novel approaches to chronic cardiac support.

20.8.2 Indications

As discussed in recent articles (Mehlhorn et al., 1999; Torchiana et al., 1997), the indications for IABP placement have changed over the years, as has the insertion method. Current indications can be divided into *hemodynamic indications*, due to cardiogenic shock, congestive heart failure, and hypotension, which are characterized by low cardiac output and systemic perfusion, or *ischemic indications* such as those caused by coronary artery disease, which result in poor cardiac perfusion and dysfunction (Torchiana et al., 1997). Current trends indicate an increased use of IABP support for ischemia in patients undergoing percutaneous cardiac therapies such as balloon angioplasty, with a dramatic shift from almost 100 percent surgical implantation to near universal percutaneous implantation (Mehlhorn et al., 1999; Torchiana et al., 1997). The number of IABPs placed for ischemic indications now exceeds those placed for hemodynamic concerns, with the trend for hemodynamic causes staying relatively flat (Torchiana et al., 1997).

The indications for implantation of TAHs and VADs are similar to those for the IABP, but are usually reserved for patients who have failed balloon pump support and/or maximal medical therapy. Current FDA-approved VADs are placed for postcardiotomy support or as a bridge to either transplantation or recovery (Willman et al., 1999). Investigators are also evaluating chronic mechanical circulatory support as an alternative to transplantation for patients who do not meet the criteria to become a donor heart recipient (Rose et al., 1999).

20.8.3 Current Device Design

Intra-aortic Balloon Pump (IABP). The first clinical use of the intra-aortic balloon pump was reported in 1968 (Kantrowitz et al., 1968). Although updated with electronics and computer control, the basic equipment of the modern IABP system remains similar to units introduced decades ago. An IABP system consists of an external pump control console that monitors physiologic patient variables (electrocardiogram and blood pressure) and delivers a bolus of gas to a catheter-mounted balloon located within the patient's aorta (Bolooki, 1998a). Figure 20.12 demonstrates the approximate location of the balloon inside the patient along with the exit site in the femoral artery. Gas delivery is controlled via a solenoid valve and is timed to correspond with the onset of diastole, during which the left ventricle is filling with blood and the aortic valve is closed (Bolooki, 1998a). Inflation of the balloon at this time, as demonstrated in Fig. 20.13*a*, results in blood being pushed back toward the heart and forward to the systemic vasculature, allowing improved perfusion of the target tissues. Figure 20.13*b* demonstrates active collapse (via vacuum) of the balloon during systole or ventricular contraction, which results in a reduction of the pressure the ventricle must work against and eases blood ejection. The reduced workload lowers myocardial oxygen consumption, reducing angina and other more serious consequences of a heart oxygen deficit (Bolooki, 1998b).

The intra-aortic balloon consists of a single, long (approximately 20 cm) inflatable polyurethane sac mounted circumferentially upon a polyurethane catheter (Bolooki, 1998c). Multichambered balloons have been investigated (Bai et al., 1994) but failed to enter clinical use despite potential theoretical advantages. Because of its lower viscosity and better transport speeds, helium is used as the shuttle gas to inflate modern balloons, although carbon dioxide and even air were used in older models (Bolooki, 1998a).

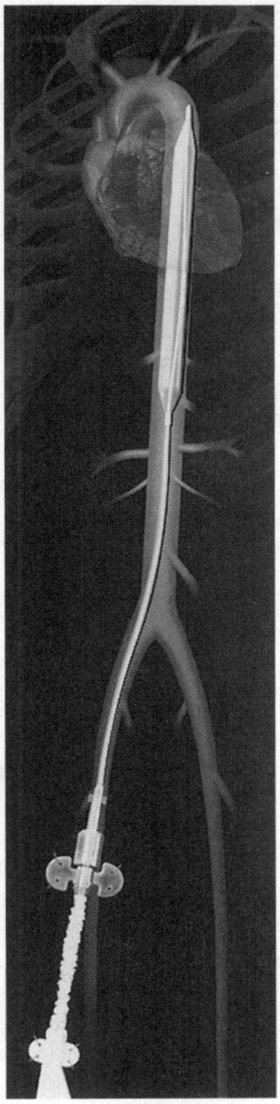

FIGURE 20.12 The anatomical placement of an intra-aortic balloon is shown. The balloon portion of the catheter is located in the aorta distal to the main vessels supplying the head and upper extremities. The catheter providing gas to the balloon is threaded through the iliac artery and aorta, emerging from a puncture site in the femoral artery in the groin. (*Datascope Corporation, Cardiac Assist Division, Fairfield, New Jersey.*)

Ventricular Assist Device (VAD) and Total Artificial Heart (TAH). Ventricular assist devices can be divided into pulsatile and nonpulsatile designs. Available pulsatile VADs can be further classified as *intracorporeal* or *extracorporeal*, depending on whether the pump is placed internally or externally. There are three intracorporeal pulsatile LVADs available for commercial use in the United States: the Novacor electric pump (World Heart Corp., Ottawa, Ontario) and the pneumatic and vented electric Heartmate pumps (Thoratec Corp., Pleasanton, California) (McCarthy and Hoercher, 2000). These devices all utilize pusher-plate technology, with the Novacor using opposing, electri-

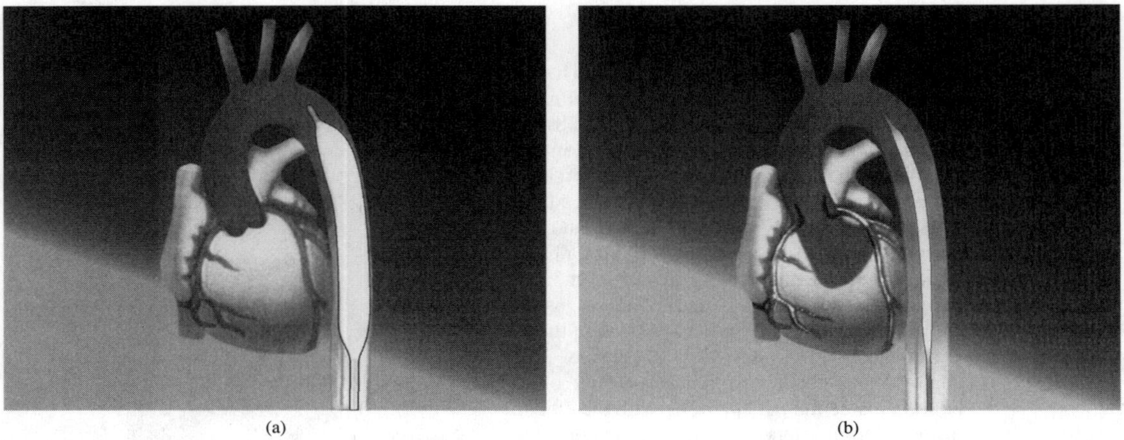

FIGURE 20.13 The balloon inflation-deflation cycle is demonstrated. (*a*) Balloon inflation occurs during diastole, or filling of the ventricle. At this point in the cardiac cycle, the aortic valve is closed. Inflation of the balloon forces blood back toward the heart and into the systemic circulation. The increased pressure developed by the balloon allows better perfusion of the heart muscle during the filling phase. (*b*) Balloon deflation coincides with systole, or ejection of blood from the heart. Deflating the balloon at this time decreases the pressure in the aorta, reducing the effort required by the heart to eject blood. (*Datascope Corporation, Cardiac Assist Division, Fairfield, New Jersey.*)

cally activated plates to compress a smooth polyurethane sac, and the Heartmate using one moving plate to compress a polyurethane sheet into a housing coated with sintered titanium. The force comes from compressed air in the case of the pneumatic Heartmate device and from an electrically driven cam mechanism in the vented-electric model. Each device requires a percutaneous lead for power, venting, and device monitoring (McCarthy and Hoercher, 2000). Figure 20.14 demonstrates the anatomic placement of the Heartmate VE and other current implantable VAD designs.

Available extracorporeal VAD designs include the ABIOMED BVS-5000 (ABIOMED, Inc., Danvers, Massachusetts) and Thoratec (Thoratec Corp., Pleasanton, California) VADs (Dowling and Etoch, 2000). The potential advantages of extracorporeal systems include the ability to be implanted in smaller patients (<1.5 m^2 body surface area), multiple sites for cannulation of the heart, reduced surgical time, and most importantly, biventricular support if needed by allowing the simultaneous use of two devices (Dowling and Etoch, 2000). Drawbacks include large percutaneous cannulation sites with a potential for infection, plus limited patient mobility because of the larger console size that is required for the pneumatic drivers of both devices (Dowling and Etoch, 2000). Figure 20.15 presents a sampling of current FDA-approved intra- and extracorporeal VAD designs and demonstrates the relative size differences between the intracorporeal (Heartmate VE and Novacor) and extracorporeal (Thoratec) designs. Both the Thoratec and BVS-5000 consist of polyurethane blood sacs housed in rigid plastic shells that are compressed with air provided by a driver console. For all practical purposes, the BVS-5000 does not allow patient mobility because of the large driver console and the need for gravity filling for the artificial ventricles. The Thoratec dual driver console is a wheeled, dishwasher-sized unit that is not intended for extensive mobile use. Of note, the recently FDA-approved Thoratec II portable pneumatic driver console promises to allow much more patient freedom.

A number of new device designs will be reaching maturity in the near future, including pulsatile and nonpulsatile VADs and pulsatile TAHs. Of the nonpulsatile or continuous-flow designs, there are two general categories: axial-flow pumps and centrifugal pumps. The axial-flow devices that have recently entered or will soon enter clinical trials in the United States include the Jarvik 2000 (Jarvik Heart Inc., New York, New York), the MicroMed DeBakey (MicroMed Technology, Inc., Houston, Texas), and the Heartmate II (Thoratec Corp., Pleasanton, California) VADs. An example

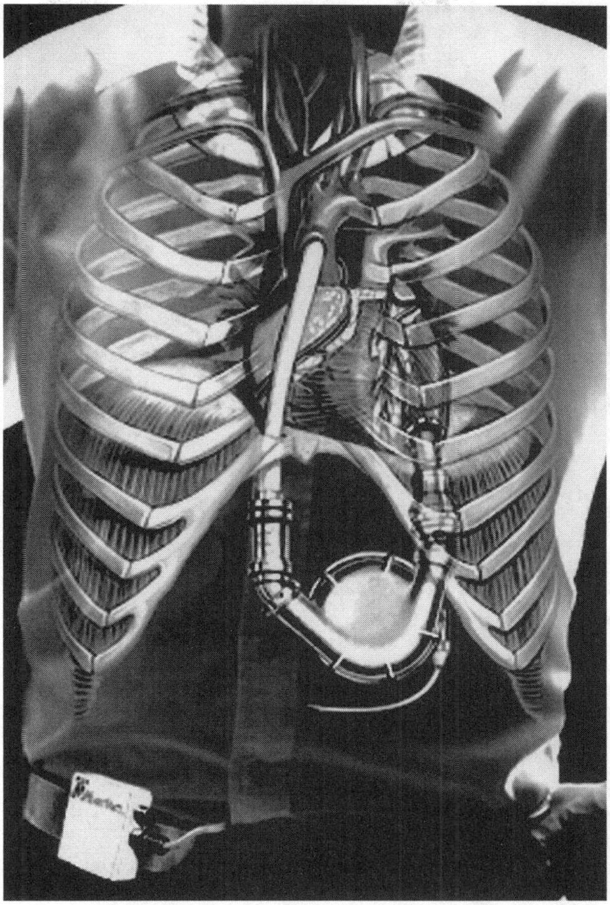

FIGURE 20.14 The anatomic placement of a Heartmate VE left ventricular assist device (LVAD) is shown. The left ventricular apex is cannulated to allow the blood to drain from the ventricular chamber into the assist device during filling. The outflow graft is connected to the aorta, where the pump contents are discharged during the pump ejection phase. The pump itself lies in an abdominal pocket, with the inflow and outflow grafts penetrating into the thoracic cavity. A number of implantable LVADs use the anatomic placement and cannulation sites described here. *(Thoratec Corporation, Pleasanton, California.)*

of the latter device is displayed in Fig. 20.16. Centrifugal pumps at this stage of development include the Heartmate III (Thoratec Corp.) and the TandemHeart percutaneous VAD (CardiacAssist, Inc., Pittsburgh, Pennsylvania) shown in Fig. 20.17. Other centrifugal pumps are under development at various universities and companies throughout the world (Clark and Zafirelis, 2000). The potential advantages of the nonpulsatile VADs include smaller size (with correspondingly smaller patient size requirements), fewer moving parts (increased reliability and lifespan), and reduced power requirements. The lack of valves could reduce the likelihood of thrombotic complications involving these sites, which have been shown to be problem areas in previous studies (Wagner et al., 1993). However,

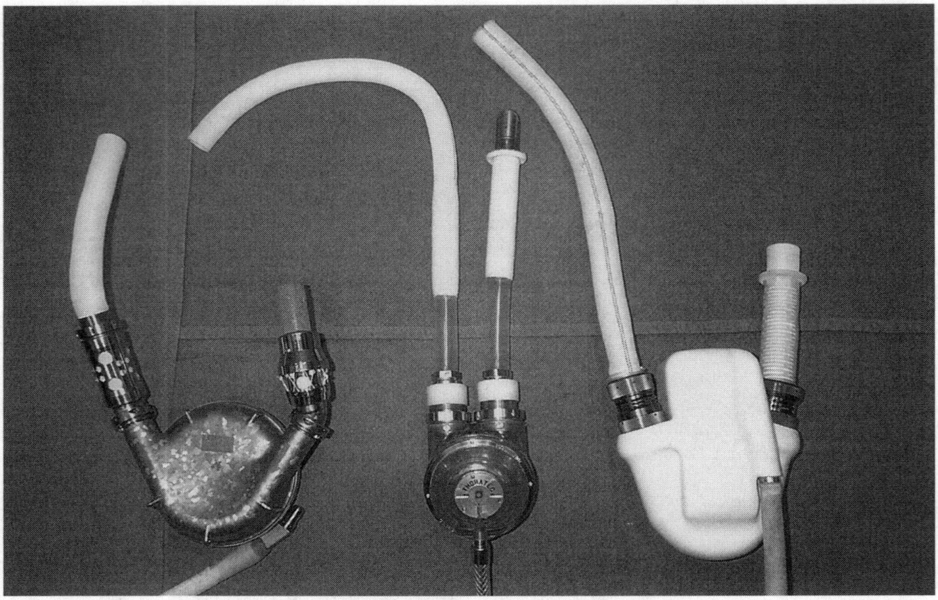

FIGURE 20.15 Some of the FDA-approved ventricular assist devices are displayed. On the left is the implantable Heartmate vented electric VAD (Thoratec Corporation, Pleasanton, California). It uses a single moving plate to push a polyurethane disk into the rigid titanium pumping chamber. The middle device is the paracorporeal Thoratec VAD (Thoratec Corporation), which is capable of providing left or right heart support. The rigid pumping chamber contains a polyurethane sac that is compressed pneumatically. The final VAD on the right is the Novacor left ventricular assist system (Baxter Healthcare Corporation, Berkeley, California). The Novacor device uses opposing plates actuated with a solenoid to compress a polyurethane blood sac.

the bearings used to support the rotors in most continuous flow devices could act as a nidus for thrombus formation (Stevenson et al., 2001). Hemolysis due to the high shearing forces is of some concern, and the effect of chronic nonpulsatile flow upon the human body is unknown.

20.8.4 Management and Complications

IABP support is an invasive procedure possessing an overall complication rate ranging from 12 to 30 percent in most reports (Cohen et al., 2000). The most common complications seen during balloon pump support are vascular in nature, with limb ischemia being the dominant problem (Busch et al., 1997). Other vascular complications include bleeding and hematoma formation (Busch et al., 1997), aortic dissection (Busch et al., 1997), embolism (Kumbasar et al., 1999), and rare events such as paraplegia due to spinal infarction (Hurle et al., 1997). Nonvascular complications include infection and mechanical or device-related failures such as balloon rupture (Scholz et al., 1998; Stavarski, 1996). Factors increasing the risk of complications vary among different studies, but generally include peripheral vascular disease, diabetes, and female sex (Cohen et al., 2000; Arafa et al., 1999; Busch et al., 1997). Results have been mixed as to whether smaller catheter sizes and shorter periods of support can limit the complication rates, despite the theoretical advantages (Cohen et al., 2000; Scholz et al., 1998; Arafa et al., 1999).

Complications associated with TAH and VAD use can be divided into those experienced shortly after implantation and events occurring later in the implant period. Postoperatively, the major con-

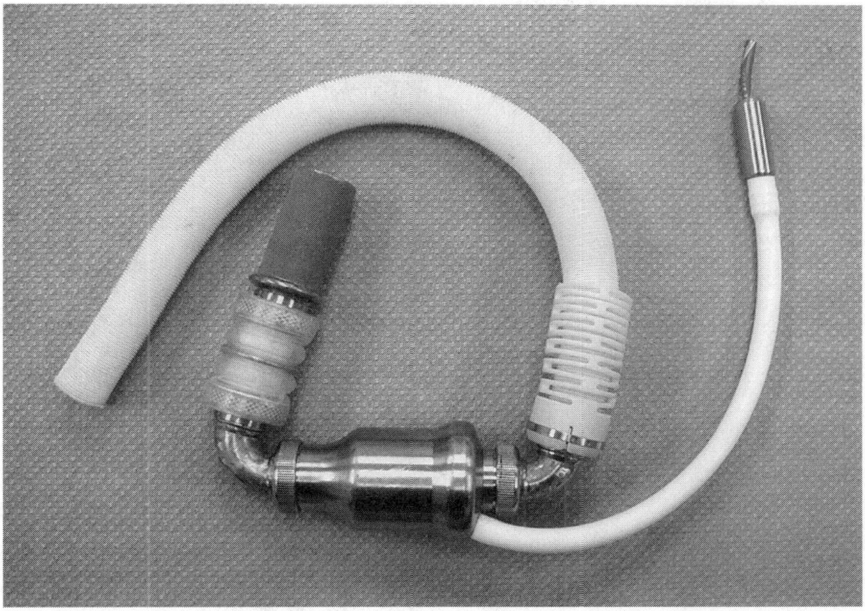

FIGURE 20.16 The Heartmate II (Thoratec Corporation, Pleasanton, California) is an axial-flow ventricular assist device currently under development. Unlike the pusher plate and pneumatic sac designs shown in Fig. 20.15, the Heartmate II provides continuous flow support without pulsatile pressure changes. The Heartmate II is much smaller than the Novacor, Thoratec, and Heartmate VE VADs.

cerns include hemorrhage, and in the case of isolated left ventricular VAD support, right heart failure or dysfunction resulting in low pump outputs (Heath and Dickstein, 2000). Bleeding management can include infusion of blood products, but in some cases surgical intervention may be required. Right heart performance can often be improved through reduction of the pulmonary vascular resistance with various drug infusions and inhalants (Heath and Dickstein, 2000). Long-term VAD complications include infection, thromboembolism, and rarely, device failure (Kasirajan et al., 2000). Infection can involve (1) the percutaneous connections of the pump or (2) the pump interior or exterior surfaces, or can be systemic in nature (Holman et al., 1999). Management includes antibiotics, debridement, and in severe cases involving the pump pocket, removal or replacement of the pump (Holman et al., 1999; Kasirajan et al., 2000). Thromboembolism is limited through the use of anticoagulant medications such as heparin perioperatively and warfarin in the long term, although not all pumps require such therapy (Kasirajan et al., 2000). Antiplatelet regimens such as aspirin are also used to prevent emboli (Kasirajan et al., 2000). Pump failure is a rare occurrence (Kasirajan et al., 2000), but the dire consequences require careful monitoring of pump performance.

20.8.5 Future Trends

Improvements in electronics and software development will undoubtedly be incorporated into future IABP designs. Closed-loop designs requiring minimal operator intervention have been investigated (Kantrowitz et al., 1992), and results suggest that independent device control could be achieved. Catheters coated with antithrombotic agents have been shown to reduce thrombosis and thrombotic deposition even in immobile (noninflating) balloons when compared to similar uncoated designs

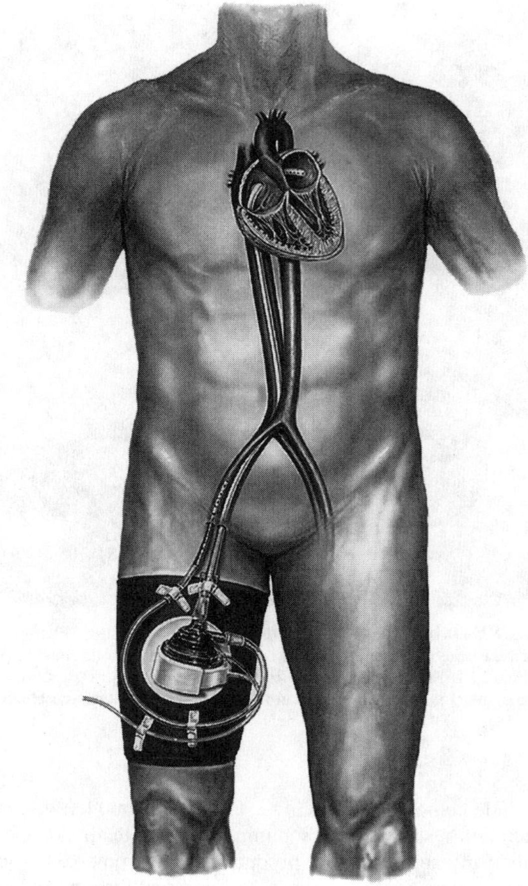

FIGURE 20.17 The TandemHeart percutaneous VAD (CardiacAssist, Inc., Pittsburgh) is shown. Intended as a short-term support device, the TandemHeart has the advantage of not requiring a major surgical procedure for implantation, and can be placed by interventional cardiologists in a catheterization laboratory. It is a centrifugal pump with novel integrated anticoagulation system that discharges into the pump chamber. It has been used successfully in Europe and is undergoing clinical trials in the United States. (*CardiacAssist, Inc., Pittsburgh*).

(Lazar et al., 1999; Mueller et al., 1999). Hydrophilic coatings have also been placed on balloon catheters, with recent results showing a 72 percent reduction in ischemic vascular complications when modified devices are used (Winters et al., 1999). Antibiotic coatings for indwelling catheters could be used to coat intra-aortic balloons if such treatment is proven to be efficacious.

In contrast to the IABP, VADs are undergoing a revolution in design and indications for use. The current rush of devices entering clinical trials shows improvements in a variety of areas, including power transmission (transcutaneous versus percutaneous), method of propulsion (electric centrifugal and axial flow versus pneumatic pulsatile), and size reduction. The largest obstacle to the widespread use of these devices on a scale commensurate with heart disease are the complications that face these patients over the long term. The most critical future developments will focus on patient management issues with a significant impact on morbidity and mortality, such as infection control and thromboembolism, as well as device issues affecting quality of life and durability.

20.9 ARTIFICIAL LUNGS

20.9.1 Market Size

According to the National Hospital Discharge Survey (Owings and Lawrence, 1999), approximately 300,000 membrane oxygenators were used in the United States in 1997 for acute surgical cardiopulmonary bypass. The number of oxygenators required for acute cardiopulmonary bypass use dwarfs the number used for extended support of the failing lung, a modality termed extracorporeal membrane oxygenation (ECMO). Changing ECMO demographics suggest new developments in alternative medical therapies might be causing a reduction in the number of neonatal patients supported via ECMO therapy, a standard of care for respiratory support for over a decade (Roy et al., 2000).

20.9.2 Indications

The indications for cardiopulmonary bypass are surgical in nature, and are based on whether the procedure requires the heart to be stopped. Currently, most cardiac surgical procedures fall into this category (McGiffin and Kirklin, 1995). Cardiopulmonary bypass provides the surgeon with a stable, blood-free field to perform intracardiac repairs and procedures such as coronary artery bypass grafting. Recent trends in minimally invasive surgery have led to surgical systems that allow some procedures such as coronary artery bypass grafting to be performed in the absence of oxygenator support (Svennevig, 2000).

In contrast to cardiopulmonary bypass, medical criteria are the primary indicators for ECMO support. Conventional treatment of acute respiratory failure calls for high-pressure mechanical ventilation with an elevated percentage of oxygen in the ventilation gas. Unfortunately, the high oxygen concentration can result in oxidative damage to lung tissue (oxygen toxicity) and, in the case of the newborn, proliferation of blood vessels in the retina leading to visual damage (retinoproliferative disorder) (Anderson and Bartlett, 2000). The high pressures used to achieve maximum ventilation area also cause lung damage through a process known as barotrauma. In essence, the lungs are being subjected to further damage by the therapy employed, preventing the healing necessary to restore proper lung function. The purpose of ECMO is to take over the burden of gas exchange and allow the native lung tissue time to heal.

ECMO is considered a standard therapy for the treatment of respiratory failure in neonatal patients (Anderson and Bartlett, 2000). In adult and pediatric patients, it is a treatment of last resort for individuals who would otherwise die despite maximal therapy (Anderson and Bartlett, 2000; Bartlett et al., 2000). Even in neonatal cases, ECMO is a therapy reserved for those patients with severe respiratory compromise and a high risk of death who are failing traditional ventilator-based interventions. Common causes of respiratory failure in the neonatal population that are treatable with ECMO support include pneumonia or sepsis, meconium aspiration syndrome, respiratory distress syndrome, persistent fetal circulation, and congenital diaphragmatic hernia (Anderson and Bartlett, 2000). Contraindications to ECMO support include root causes that are unresolvable, such as a major birth defect or genetic abnormality, and comorbid conditions such as intracranial hemorrhage or fetal underdevelopment that suggest a poor outcome (Anderson and Bartlett, 2000). Indications for ECMO use in the pediatric and adult populations are not dissimilar from those of the neonate, but the causes for respiratory or cardiopulmonary failure are different, and many individuals suffer from comorbid conditions. Indications include pneumonia, aspiration pneumonitis, acute respiratory distress syndrome, and recoverable heart failure as caused by infection or postsurgical complications (Anderson and Bartlett, 2000).

20.9.3 Device Design

Although the first oxygenators were described in the late 1800s, it would not be until 1951 that total cardiopulmonary bypass would be performed on a human patient (Stammers, 1997). Oxygenators, or artificial lungs, have undergone a dramatic evolution in the 5 decades since the first total CPB operation. The initial clinical units are described as *film oxygenators* because a rotating cylinder was used to generate a large, thin film of blood on the cylinder surface where it contacted the exchange gas (Wegner, 1997). Although effective, these early film oxygenators suffered from a number of failings that eventually led to their replacement. The direct gas-blood interface allowed for adequate gas exchange but extensive cellular damage and protein denaturation resulted from the blood-gas interface (Wegner, 1997). The large blood priming volume and time-consuming, complicated maintenance and use procedures characteristic of film oxygenators were addressed through the advent of *bubble oxygenators* (Stammers, 1997). Direct gas-blood contact remained in bubble oxygenators, but the large surface area of the dispersed oxygen bubbles resulted in greater mass transfer and a reduction in priming volume (Wegner, 1997). In addition, these devices were simple and disposable, consisting of a bubbling chamber, defoaming unit and a return arterial reservoir (Wegner, 1997; Stammers, 1997). However, the blood damage seen with film oxygenators was not corrected with the new bubbling technology, and concerns regarding blood trauma during longer perfusions contributed to the movement toward *membrane oxygenators* (Wegner, 1997; Stammers, 1997). The use of a semipermeable membrane to separate the blood and gas phases characterizes all membrane oxygenator designs. Membrane oxygenators can be further divided into flat-sheet/spiral-wound and hollow-fiber models. The flat-sheet designs restrict blood flow to a conduit formed between two membranes, with gas flowing on the membrane exterior; these systems were the first membrane oxygenators to enter use (Wegner, 1997). Spiral-wound oxygenators use membrane sheets as well, but are arranged in a roll rather than the sandwich formation of the original flat-sheet assemblies. Polymers such as polyethylene, cellulose (Clowes and Hopkins, 1956), and polytetrafluoroethylene (Clowes and Neville, 1957) were used for membranes in these early designs as investigators searched for a material with high permeability to oxygen and carbon dioxide but that elicited mild responses when in contact with blood. The introduction of polysiloxane as an artificial lung membrane material in the 1960s provided a significant leap in gas transfer efficiency, particularly for carbon dioxide (Galletti and Mora, 1995). These membranes remain in use today for long-term neonatal ECMO support.

Development of the microporous hollow fiber led to the next evolution in lung design, the hollow-fiber membrane oxygenator (Stammers, 1997). Increased carbon dioxide permeability compared to solid membranes, coupled with improved structural stability, has secured the standing of these devices as the market leader (Wegner, 1997; Stammers, 1997). The current, standard artificial lung is constructed of hollow microporous polypropylene fibers housed in a plastic shell. An extraluminal cross-flow design is used for most models and is characterized by blood flow on the exterior of the fibers with gas constrained to the fiber interior. Intraluminal flow designs utilize the reverse blood-gas arrangement, with blood constrained to the fiber interior. The laminar conditions experienced by blood flowing inside the fibers result in the development of a relatively thick boundary layer that limits gas transfer. Extraluminal-flow devices are less susceptible to this phenomenon, and investigators have used barriers and geometric arrangements to passively disrupt the boundary layer, resulting in large gains in mass transfer efficiency (Drinker, 1978; Galletti, 1993). Extraluminal-flow hollow fiber membrane oxygenators have come to dominate the market because of their improved mass transfer rates and decreased flow resistance, the latter of which minimizes blood damage (Wegner, 1997). Figure 20.18 presents a collection of commercial extraluminal-flow membrane oxygenators demonstrating a diversity of design arrangements.

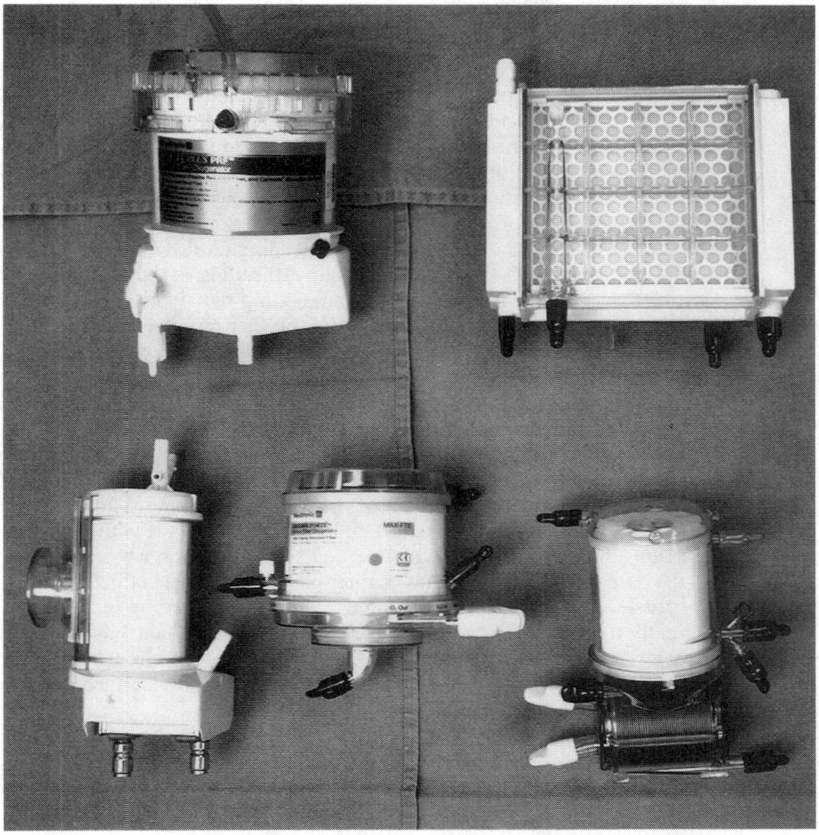

FIGURE 20.18 An assortment of commercial membrane oxygenators are shown to provide an indication of the possible arrangements and designs that can be used. Beginning with the upper left, the Maxima Plus (Medtronic, Inc., Minneapolis) membrane oxygenator is shown. The bottom portion of the device is a heat exchanger for regulating blood temperature, while the upper portion contains a helically wound gas exchange fiber bundle. On the upper right is the rectangular William Harvey HF-5000 (C. R. Bard, Haverhill, Massachusetts) oxygenator. The bottom row displays the Cobe Optima (Cobe Cardiovascular, Arvada, Colorado) oxygenator on the left, with the Maxima Forte and Affinity oxygenators (Medtronic, Inc., Minneapolis) following on the right. Note the smaller size of these designs compared to the oxygenators at the top.

20.9.4 Management and Complications

Despite the major advances in surgical and supportive therapy engendered by the introduction of the artificial lung, substantial limitations remain. The complications faced by patients on oxygenator support are not infrequent and are often life-threatening. The limited duration of the average open heart procedure provides less time for major complications to occur, although a few deserve mention.

A relatively common complication suffered by patients who undergo cardiopulmonary bypass (CPB) is called *postpump syndrome* and is characterized by complement activation, cytokine production, activation of various white blood cell types, and depressed lung function. The depressed lung function infrequently progresses to acute respiratory distress syndrome or ARDS (0.5 to 1.7 percent of CPB patients developed ARDS), but tends to be fatal if such progression occurs (50 to 91.6 percent mortality) (Asimakopoulos et al., 1999). The inflammatory response observed is not attributed solely to blood exposure to the extracorporeal circuit but is likely also due to a combination of operative trauma, reperfusion injury, and exposure to endotoxin during the procedure (Asimakopoulos et al., 1999). Hemorrhagic and thrombotic problems during CPB are of grave concern, and traditionally, CPB support has required the use of high doses of anticoagulant to prevent clotting in the extracorporeal circuit at the expense of a possible increase in bleeding. This bleeding risk is further magnified by the consumption of coagulation factors and platelets by extracorporeal surfaces and activation of fibrinolytic pathways. Investigations into minimizing this phenomenon have led to the development of anticoagulant (heparin) fiber coatings for the oxygenators (Wendel and Ziemer, 1999). The presence of a heparin coating has led some to reduce the amount of systemic heparin provided during CPB in an effort to limit bleeding (Aldea et al., 1996; Aldea et al., 1998; Kumano et al., 1999). Although there is evidence that bleeding is reduced, the approach is controversial in many circles, as clinical markers of blood clotting remain elevated, resulting in lingering thromboembolism concerns (Kuitunen et al., 1997). Further research is required to elucidate what and how much protective effect is provided by heparin-coated fibers.

The more chronic support provided by extracorporeal membrane oxygenation (ECMO) is plagued by complications and poor outcomes. Survival rates for ECMO support range from 88 percent for neonates in respiratory failure to 33 percent for adults in cardiac failure (Bartlett et al., 2000). As in CPB, hemorrhagic and thrombotic complications occur and have led to intense investigation into the long-term neurologic outcomes of these patients (Graziani et al., 1997; Nield et al., 2000), although the exact relationship between ECMO and outcomes remains unclear (Vaucher et al., 1996; Rais-Bahrami et al., 2000). In addition, longer-term support with hollow-fiber membrane oxygenators results in the manifestation of a particular phenomenon where plasma from the blood side seeps into the pores of the fibers in a process termed *weeping*. The effect of this phenomenon is to increase the diffusion distance for oxygen and carbon dioxide, thereby reducing mass transfer performance. Originally thought to be due to temperature changes and condensation of water in the fiber interior (Mottaghy et al., 1989), evidence suggests deposition of circulating phospholipid at the fluid-gas interface results in a change in hydrophobicity on the pore surface that mediates penetration of the plasma (Montoya et al., 1992).

20.9.5 Future Trends

Future developments in cardiopulmonary bypass will focus on improving the biocompatibility of the device through minimization of hematologic alterations. One current approach involves coating the oxygenator fibers with a layer of polysiloxane in an effort to limit the postperfusion syndrome (Shimamoto et al., 2000). ECMO, with its higher longevity requirements, will similarly benefit from new coatings and materials. General measures to improve the biocompatibility of the fibers include coatings and additives to limit plasma infiltration into the device (Shimono et al., 1996), limit platelet deposition on the surface (Gu et al., 1998) and platelet activation (Defraigne et al., 2000), and minimize leukocyte and complement activation (Watanabe et al., 1999; Saito et al., 2000). Although not a new concept, novel methods to improve mass transfer through the thinning or disruption of the boundary layer are currently being developed. Some approaches include the use of fibers mounted on a rapidly spinning disk (Borovetz et al., 2000; Reeder et al., 2001) or cone (Makarewicz et al., 1994), the pulsing and distention of silicone sheets (Fiore et al., 2000), and the pulsation of a balloon mounted inside circumferentially organized fibers designed for placement in the vena cava (Federspiel et al., 1997). Some of these devices offer the opportunity to combine the features of a blood pump and artificial lung and therefore represent a significant leap over current bypass systems.

ACKNOWLEDGMENTS

The support of the McGowan Institute for Regenerative Medicine at the University of Pittsburgh is acknowledged. Fellowship support was provided to Kenneth Gage during the writing of this chapter from the Department of Energy under the Krell Institute's Computational Science Graduate Fellowship (CSGF).

REFERENCES

Akins, C. W., "Results with mechanical cardiac valvular prostheses," *Ann. Thorac. Surg.*, 1995; **60**:1836–1844.

Al Suwaidi, J., Berger, P. B., and Holmes, Jr., D. R., "Coronary artery stents," *JAMA*, 2000; **284**(14):1828–1836.

Aldea, G. S., Doursounian, M., O'Gara, P., Treanor, P., Shapira, O. M., Lazar, H. L., and Shemin, R. J., "Heparin-bonded circuits with a reduced anticoagulation protocol in primary CABG: A prospective, randomized study," *Ann. Thorac. Surg.*, 1996; **62**(2):410–418.

Aldea, G. S., O'Gara, P., Shapira, O. M., Treanor, P., Osman, A., Patalis, E., Arkin, C., Diamond, R, Babikian, V., Lazar, H. L., et al., "Effect of anticoagulation protocol on outcome in patients undergoing CABG with heparin-bonded cardiopulmonary bypass circuits," *Ann. Thorac. Surg.*, 1998; **65**(2):425–433.

Allen, R. D., Yuill, E., Nankivell, B. J., and Francis, D. M., "Australian multicentre evaluation of a new polyurethane vascular access graft," *Aust. N. Z. J. Surg.*, 1996; **66**(11):738–742.

Alt, E., Haehnel, I., Beilharz, C., Prietzel, K., Preter, D., Stemberger, A., Fliedner, T., Erhardt, W., and Schomig, A., "Inhibition of neointima formation after experimental coronary artery stenting: A new biodegradable stent coating releasing hirudin and the prostacyclin analogue iloprost," *Circulation*, 2000; **101**(12):1453–1458.

AHA, "2001 heart and stroke statistical update," American Heart Association, Dallas, 2000.

Anderson III, H. L., and Bartlett, R. H., "Extracorporeal life support for respiratory failure and multiple organ failure," Chap. 123 in Grenvik, A., Ayres, S. M., Holbrook, P. R., and Shoemaker, W. C. (eds.), *Textbook of Critical Care*, 4th ed., W. B. Saunders, Philadelphia, 2000; pp. 1356–1362.

Appelgren, P., Ransjo, U., Bindslev, L., Espersen, F., and Larm, O., "Surface heparinization of central venous catheters reduces microbial colonization in vitro and in vivo: Results from a prospective, randomized trial," *Crit. Care. Med.*, 1996; **24**(9):1482–1489.

Arafa, O. E., Pedersen, T. H., Svennevig, J. L., Fosse, E., and Geiran, O. R., "Vascular complications of the intraaortic balloon pump in patients undergoing open heart operations: 15-year experience," *Ann. Thorac. Surg.*, 1999; **67**(3):645–651.

Asimakopoulos, G., Smith, P. L. C., Ratnatunga, C. P., and Taylor, K. M., "Lung injury and acute respiratory distress syndrome after cardiopulmonary bypass," *Ann. Thorac. Surg.*, 1999; **68**(3):1107–1115.

Bai, J., Lin, H., Yang, Z., and Zhou, X., "A study of optimal configuration and control of a multichamber balloon for intraaortic balloon pumping," *Ann. Biomed. Eng.*, 1994; **22**(5):524–531.

Baldus, S., Koster, R., Elsner, M., Walter, D. H., Arnold, R., Auch-Schwelk, W., Berger, J., Rau, M., Meinertz, T., Zeiher, A. M., et al., "Treatment of aortocoronary vein graft lesions with membrane-covered stents: A multicenter surveillance trial," *Circulation*, 2000; **102**(17):2024–2027.

Bandyk, D. F., and Esses, G. E., "Prosthetic graft infection," *Surg. Clin. North. Am.*, 1994; **74**(3):571–590.

Bartlett, R. H., Roloff D. W., Custer, J. R., Younger, J. G., and Hirschl, R. B., "Extracorporeal life support: The University of Michigan experience," *JAMA*, 2000; **283**(7):904–908.

Berger, K., Sauvage, L. R., Rao, A. M., and Wood, S. J., "Healing of arterial prostheses in man: Its incompleteness," *Ann. Surg.*, 1972; **175**(1):118–127.

Bernstein, A. D., Camm, A. J., Fisher, J. D., Fletcher, R. D., Mead, R. H., Nathan, A. W., Parsonnet, V., Rickards, A. F., Smyth, N. P. D., Sutton, R., et al., "North American Society of Pacing and Electrophysiology Policy Statement. The NASPE/BPEG Defibrillator Code," *Pacing Clin. Electrophysiol.*, 1993; **16**(9):1776–1780.

Bernstein, A. D., Camm, A. J., Fletcher, R. D., Gold, F. D., Rickards, A. F., Smyth, N. P. D., Spielman, S. R., and Sutton, R., "The NASPE/BPEG Generic Pacemaker Code for antibradyarrhythmia and adaptive-rate pacing and antitachyarrhythmia devices," *Pacing Clin. Electrophysiol.*, 1987; **10**(4 pt 1):794–799.

Bernstein, A. D., and Parsonnet, V., "Survey of cardiac pacing and defibrillation in the United States in 1993," *Am. J. Cardiol.*, 1996a; **78**(2):187–196.

Bernstein, A. D., and Parsonnet, V., "The NASPE/BPEG Pacemaker-Lead Code (NBL Code)," *Pacing Clin. Electrophysiol.*, 1996b;**19**(11 pt 1):1535–1536.

Bertrand, O. F., Sipehia, R., Mongrain, R., Rodes, J., Tardif, J.-C., Bilodeau, L., Cote, G., and Bourassa, M. G., "Biocompatibility aspects of new stent technology," *J. Am. Coll. Cardiol.*, 1998; **32**(3):562–571.

Biancari, F., and Lepantalo, M., "Extra-anatomic bypass surgery for critical leg ischemia. A review," *J. Cardiovasc. Surg.* (Torino), 1998; **39**:(3):295–301.

Bolooki, H., "Balloon Pump Equipment," Chap. 6 in Bolooki, H., (ed), *Clinical Application of the Intra-Aortic Balloon Pump*, 3d ed., Futura Publishing, Armonk, N.Y., 1998a; pp. 73–85.

Bolooki, H., "Physiology of Balloon Pumping," Chap. 8 in Bolooki, H. (ed.), *Clinical Application of the Intra-Aortic Balloon Pump*, 3d ed., Futura Publishing, Armonk, N.Y., 1998b: pp. 109–161.

Bolooki, H., "Balloon Pump Consoles and Catheters," Chap. 7 in Bolooki, H. (ed.), *Clinical Application of the Intra-Aortic Balloon Pump*, 3d ed., Futura Publishing, Armonk, N.Y., 1998c; pp. 87–107.

Bonow, R. O., Carabello, B., De Leon, Jr., A. C., Edmunds, Jr., L. H., Fedderly, B. J., Freed, M. D., Gaasch, W. H., McKay, C. R., Nishimura, R. A., O'Gara, P. T., et al., "ACC/AHA Guidelines for the management of patients with valvular heart disease: A report of the American College of Cardiology/American Heart Association Task Force on Practice Guidelines (Committee on Management of Patients with Valvular Heart Disease)," *J. Am. Coll. Cardiol.*, 1998; **32**(5):1486–1588.

Borovetz, H. S., et al., inventors, University of Pittsburgh, assignee, "Membrane apparatus with enhanced mass transfer via active mixing," patent number 6,106,776, 2000.

Brewster, D. C., "Prosthetic grafts," Chap. 37 in Rutherford, R. B. (ed.), *Vascular Surgery*, 5th ed., W. B. Saunders Company, Philadelphia, 2000, pp. 559–584.

Bull, P. G., Denck, H., Guidoin, R., and Gruber, H., "Preliminary clinical experience with polyurethane vascular prostheses in femoro-popliteal reconstruction," *Eur. J. Vasc. Surg.*, 1992; **6**(2):217–224.

Bunt, T. J., "Synthetic vascular graft infections. I. Graft infections," *Surgery*, 1983; **93**(6):733–746.

Busch, T., Sirbu, H., Zenker, D., and Dalichau, H., "Vascular complications related to intraaortic balloon counterpulsation: An analysis of ten years experience," *Thorac. Cardiovasc. Surg.*, 1997; **45**(2):55–59.

Calligaro, K. D., and Veith, F. J., "Diagnosis and management of infected prosthetic aortic grafts," *Surgery*, 1991; **110**(5):805–813.

Canaud, B., Leray-Moragues, H., Kamoun, K., and Garrigue, V., "Temporary vascular access for extracorporeal technologies," *Therapeutic Apheresis*, 2000; **4**(3):249–255.

Cazeau, S., Leclercq, C., Lavergne, T., Walker, S., Varma, C., Linde, C., Garrigue, S., Kappenberger, L., Haywood, G. A., Santini, M., et al., "Effects of multisite biventricular pacing in patients with heart failure and intraventricular conduction delay," *N. Engl. J. Med.*, 2001; **344**(12):873–880.

Chenoweth, D. E., "Complement activation during hemodialysis: Clinical observations, proposed mechanisms and theoretical implications," *Artif. Organs*, 1984; **8**(3):281–290.

Chronos, N. A. F., Robinson, K. A., Kelly, A. B., et al., "Thromboresistant phosphorylcholine coating for coronary stents" (abstract), *Circulation*, 1995; **92**(suppl I):1685.

Cimochowski, G. E., Norley, E., Rutherford, W. E., Sartain, J., Blondin, J., and Harter, H., "Superiority of the internal jugular over the subclavian access for temporary dialysis," *Nephron*, 1990; **54**(2):154–161.

Clark, R. E., and Zafirelis, Z., "Future devices and directions," *Prog. Cardiovasc. Dis.*, 2000; **43**(1):95–100.

Clark, W. R., "Hemodialyzer membranes and configurations: A historical perspective," *Semin. Dial.*, 2000; **13**(5): 309–311.

Clowes, Jr., G. H. A., Hopkins, A. L., and Neville, W. E., "An artificial lung dependent upon diffusion of oxygen and carbon dioxide through plastic membranes," *J. Thorac. Surg.*, 1956; **32**(5):630–637.

Clowes, Jr., G. H. A., and Neville, W. E., "Further development of a blood oxygenator dependent upon the diffusion of gases through plastic membranes," *Trans. Am. Soc. Artif. Intern. Organs.*, 1957; **3**:52–58.

Cohen, M., Dawson, M. S., Kopistansky, C., and McBride, R., "Sex and other predictors of intra-aortic balloon counterpulsation-related complications: Prospective study of 1119 consecutive patients," *Am. Heart. J.*, 2000; **139**(2 pt 1):282–287.

Comerota, A. J., Neaver, F. A., Kosking, J. D., Froehlich, J., Folander, H., Sussman, B., and Rosenfield, K., "Results of a prospective, randomized trial of surgery versus thrombolysis for occluded lower extremity bypass grafts," *Am. J. Surg.*, 1996; **172**(2):105–112.

"Comparative evaluation of prosthetic, reversed, and in situ vein bypass grafts in distal popliteal and tibial-peroneal revascularization," Veterans Administration Cooperative Study Group 141, *Arch. Surg.* 1988; **123**(4): 434–438.

Connolly, J. E., Kwaan, J. H., Brownell, D., McCart, P. M., and Levine, E., "Newer developments of extra-anatomic bypass," *Surg. Gynecol. Obstet.*, 1984; **158**(5):415–418.

Contreras, M. A., Quist, W. C., and LoGerfo, F. W., "Effect of porosity on small-diameter vascular graft healing," *Microsurgery*, 2000; **20**(1):15–21.

Crain, M. R., Mewissen, M. W., Ostrowski, G. J., Paz-Fumagalli, R., Beres, R. A., and Wertz, R. A., "Fibrin sleeve stripping for salvage of failing hemodialysis catheters: technique and initial results," *Radiology*, 1996; **198**(1):41–44.

Crossley, G. H., "Cardiac pacing leads," *Cardiol. Clin.*, 2000; **18**(1):95–112.

Crump, J. A., and Collignon, P. J., "Intravascular catheter-associated infections," *Eur. J. Clin. Microbiol. Infect. Dis.*, 2000; **19**(1):1–8.

Dajani, A. S., Taubert, K. A., Wilson, W., Bolger, A. F., Bayer, A., Ferrieri, P., Gewitz, M. H., Shulman, S. T., Nouri, S., Newburger, J. W., et al., "Prevention of bacterial endocarditis. Recommendations by the American Heart Association," *JAMA*, 1997; **277**(22):1794–1801.

Darouiche, R. O., Raad, I. I., Heard, S. O., Thornby, J. I., Wenker, O. C., Gabrielli, A., Berg, J., Khardori, N., Hanna, H., Hachem, R., et al., "A comparison of two antimicrobial-impregnated central venous catheters," *N. Engl. J. Med.*, 1999; **340**(1):1–8.

David, T. E., Vden, D. E., and Strauss, H. D., "The importance of the mitral apparatus in left ventricular function after correction of mitral regurgitation," *Circulation*, 1983; **68**(suppl. II):II76–II82.

Davids, L., Dower, T., and Zilla, P., "The lack of healing in conventional vascular grafts," Chap. 1 in Zilla, P., and Greisler, H. P., (eds.), *Tissue engineering of vascular prosthetic grafts*, R. G. Landes Co., Austin, Tex., 1999; pp. 3–44.

De Hart, J., Cacciola, G., Schreurs, P. J. G., and Peters, G. W. M., "A three-dimensional analysis of a fibre-reinforced aortic valve prosthesis," *J. Biomech.*, 1998; **31**(7):629–638.

De Scheerder, I. K., Wilczek, K. L., Verbeken, E. V., Vandorpe, J., Lan, P. N., Schacht, E., De Geest, H., and Piessens, J., "Biocompatibility of polymer-coated oversized metallic stents implanted in normal porcine coronary arteries," *Atherosclerosis*, 1995; **114**(1):105–114.

de Voogt, W. G., "Pacemaker leads: Performance and progress," *Am. J. Cardiol.*, 1999; **83**(5B):187D–191D.

Defraigne J., Pincemail, J., Dekoster, G., Larbuisson, R., Dujardin, M., Blaffart, F., David, J., and Limet, R., "SMA circuits reduce platelet consumption and platelet factor release during cardiac surgery," *Ann. Thorac. Surg.*, 2000; **70**(6):2075–2081.

Denker, B. M., Chertow, G. M., and Owen, Jr., W. F., "Hemodialysis," Chap. 57 in Brenner, B. M., (ed.), *The Kidney*, 6th ed., W. B. Saunders, Philadelphia, 2000; pp. 2373–2453.

Deutsch, M., Meinhart, J., Fischlein, T., Preiss, P., and Zilla, P., "Clinical autologous in vitro endothelialization of infrainguinal ePTFE grafts in 100 patients: A 9-year experience," *Surgery*, 1999; **126**(5):847–855.

Devine, C., and McCollum, C., "Heparin-bonded Dacron or polytetrafluoroethylene for femoropopliteal bypass grafting: A multicenter trial," *J. Vasc. Surg.*, 2001; **33**(3):533–539.

Dowling, R. D., and Etoch, S. W., "Clinically available extracorporeal assist devices," *Prog. Cardiovasc. Dis.*, 2000; **43**(1):27–36.

Drinker, P. A., and Lehr, J. L., "Engineering aspects of ECMO technology," *Artif. Organs*, 1978; **2**(1):6–11.

Earnshaw, J. J., "The current role of rifampicin-impregnated grafts: pragmatism versus science," *Eur. J. Vasc. Endovasc. Surg.*, 2000; **20**(5):409–412.

Eberhart, A., Zhang, Z., Guidoin, R., Laroche, G., Guay, L., De La Faye, D., Batt, M., and King, M. W., "A new generation of polyurethane vascular prostheses: rara avis or ignis fatuus?," *Biomed. Mater. Res.*, 1999; **48**(4):546–558.

Edmunds, Jr., L. H., Clark, R. E., Cohn, L. H., Grunkemeier, G. L., Miller, D. C., and Weisel, R. D., "Guidelines for reporting morbidity and mortality after cardiac valvular operations," *Ann. Thorac. Surg.*, 1996; **62**(3):932–935.

Eybl, E., Grimm, M., Grabenwoger, M., Bock, P., Muller, M. M., and Wolner, E., "Endothelial cell lining of bioprosthetic heart valve materials," *J. Thorac. Cardiovasc. Surg.*, 1992; **104**(3):763–769.

Faries, P. L., LoGerfo, F. W., Arora, S., Hook, S., Pulling, M. C., Akbari, C. M., Campbell, D. R., and Pomposelli, Jr., F. B., "A comparative study of alternative conduits for lower extremity revascularization: All-autogenous conduit versus prosthetic grafts," *J. Vasc. Surg.*, 2000; **32**(6):1080–1090.

Federspiel, W. J., Hout, M. S., Hewitt, T. J., Lund, L. W., Heinrich, S. A., Litwak, P., Walters, F. R., Reeder, G. D., Borovetz, H. S., and Hattler, B. G., "Development of a low flow resistance intravenous oxygenator," *ASAIO J.*, 1997; **43**(5):M725–M730.

Fiore, G. B., Constantino, M. L., Fumero, R., and Montevecchi, F. M., "The pumping oxygenator: Design criteria and first in vitro results," *Artif. Organs*, 2000; **24**(10):797–807.

Fischman, D. L., Leon, M. B., Baim, D. S., Schatz, R. A., Savage, M. F., Penn, I., Detre, K., Veltri, L., Ricci, D., Nobuyoshi, M., et al., "A randomized comparison of coronary-stent placement and balloon angioplasty in the treatment of coronary artery disease," *N. Engl. J. Med.*, 1994; **331**(8):496–501.

Foster, M. C., Mikulin, T., Hopkinson, B. R., and Makin, G. S., "A review of 155 extra-anatomic bypass grafts," *Ann. R. Coll. Surg. Engl.*, 1986; **68**(4):216–218.

Galletti, P. M., "Cardiopulmonary bypass: A historical perspective," *Artif. Organs*, 1993; **17**(8):675–686.

Galletti, P. M., Mora, C. T., "Cardiopulmonary bypass: The historical foundation, the future promise," Chap. 1 in Mora C. T. (ed.), *Cardiopulmonary bypass: Principles and techniques of extracorporeal circulation*, Springer-Verlag, New York, pp. 3–18.

Gerber, T. C., Nishimura, R. A., Holmes, Jr., D. R., Lloyd, M. A., Zehr, K. J., Tajik, A. J., and Hayes, D. L., "Left ventricular and biventricular pacing in congestive heart failure," *Mayo Clin. Proc.*, 2001; **76**(8):803–812.

Golden, M. A., Hanson, S. R., Kirkman, T. R., Schneider, P. A., and Clowes, A. W., "Healing of polytetrafluoroethylene arterial grafts is influenced by graft porosity," *J. Vasc. Surg.*, 1990; **11**(6):838–845.

Gray, R. J., Levitin, A., Buck, D., Brown, L. C., Sparling, Y. H., Jablonski, K. A., Fessahaye, A., and Gupta, A. K., "Percutaneous fibrin sheath stripping versus transcatheter urokinase infusion for malfunctioning well-positioned tunneled central venous dialysis catheters: A prospective, randomized trial," *J. Vasc. Interv. Radiol.*, 2000; **11**(9):1121–1129.

Graziani, L. J., Gringlas, M., and Baumgart, S., "Cerebrovascular complications and neurodevelopmental sequelae of neonatal ECMO," *Clin. Perinatol.*, 1997; **24**(3):655–675.

Gregoratos, G., Cheitlin, M. D., Conill, A., Epstein, A. E., Fellows, C., Ferguson, T. B., Jr., Freedman, R. A., Hlatky, M. A., Naccarelli, G. V., Saksena, S., et al., "ACC/AHA guidelines for implantation of cardiac pacemakers and antiarrhythmia devices: A report of the American College of Cardiology/American Heart Association Task Force on Practice Guidelines (Committee on Pacemaker Implantation)," *J. Am. Coll. Cardiol.*, 1998; **31**(5):1175–1209.

Greisler, H. P., *New Biologic and Synthetic Vascular Prostheses*, R. G., Landes Co., Austin, Tex, 1991.

Greisler, H. P., "Growth factor release from vascular grafts," *J. Control. Release.*, 1996; **39**(2–3):267–280.

Grooteman, M. P. C., and Nube, M. J., "Hemodialysis-related bioincompatibility: Fundamental aspects and clinical relevance," *Neth. J. Med.*, 1998; **52**(5):169–178.

Gu, Y. J., Boonstra, P. W., Rijnsburger, A. A., Haan, J., and van Oeveren, W., "Cardiopulmonary bypass circuit treated with surface-modifying additives: Clinical evaluation of blood bicompatibility," *Ann Thorac Surg.*, 1998; **65**(5):1342–1347.

Hakim, R. M., and Lazarus, J. M., "Initiation of dialysis," *J. Am. Soc. Nephrol.*, 1995; **6**(5):1319–1328.

Hammermeister, K., Sethi, G. K., Henderson, W. G., Grover, F. L., Oprian, C., and Rahimtoola, S. H., "Outcomes 15 years after valve replacement with a mechanical versus a bioprosthetic valve: Final report of the Veterans Affairs Randomized Trial," *J. Am. Coll. Cardiol.*, 2000; **36**(4):1152–1158.

Hasaniya, N. W., Angelis, M., Brown, M. R., and Yu, M., "Efficacy of subcutaneous silver-impregnated cuffs in preventing central venous catheter infections," *Chest*, 1996; **109**(4):1030–1032.

Heath, M. J. S., and Dickstein, M. L., "Perioperative management of the left ventricular assist device recipient," *Prog. Cardiovasc. Dis.*, 2000; **43**(1):47–54.

Helmus, M. N., and Hubbell, J. A., "Materials selection," *Cardiovasc. Pathol.*, 1993; **2**(3 suppl):53S–71S.

Henry, M., Klonaris, C., Amor, M., Henry, I., and Tzvetanov, K., "Which stent for which lesion in peripheral interventions?," *Tex. Heart Inst. J.*, 2000; **27**(2):119–126.

Heras, M., Chesebro, J. H., Fuster, V., Penny, W. J., Grill, D. E., Bailey, K. R., Danielson, G. K., Orszulak, T. A., Pluth, J. R., Puga, F. J., et al., "High risk of thromboemboli early after bioprosthetic cardiac valve replacement," *J. Am. Coll. Cardiol.*, 1995; **25**(5):1111–1119.

Holman, W. L., Skinner, J. L., Waites, K. B., Benza, R. L., McGiffin, D. C., and Kirklin, J. K., "Infection during circulatory support with ventricular assist devices," *Ann. Thorac. Surg.*, 1999; **68**(2):711–716.

Holmes, D. R., Hirshfeld, Jr., J., Faxon, D., Vlietstra, R. E., Jacobs, A., and King, III, S. B., "ACC expert consensus document on coronary artery stents," *J. Am. Coll. Cardiol.*, 1998; **32**(5):1471–1482.

Horstkotte, D., "Results with mechanical cardiac valvular prostheses" (letter), *Ann. Thorac. Surg.*, 1996; **62**(5): 1565–1567.

Hoshal, Jr., V. L., Ause, R. G., and Hoskins, P. A., "Fibrin sleeve formation on indwelling subclavian central venous catheters," *Arch. Surg.*, 1971; **102**(4):353–358.

Humes, H. D., MacKay, S. M., Funke, A. J., and Buffington, D. A., "Tissue engineering of a bioartificial renal tubule assist device: In vitro transport and metabolic characteristics," *Kidney Int.*, 1999; **55**(6):2502–2514.

Hurle, A., Llamas, P., Meseguer, J., Casillas, J. A., "Paraplegia complicating intraaortic balloon pumping" (letter), *Ann. Thorac. Surg.*, 1997; **63**(4):1217–1218.

Jankowski, R. J., and Wagner, W. R., "Directions in cardiovascular tissue engineering," *Clin. Plast. Surg.*, 1999; **26**(4):605–616.

Jeffrey, K., and Parsonnet, V., "Cardiac pacing, 1960–1985: A quarter century of medical and industrial innovation," *Circulation*, 1998; **97**(19):1978–1991.

Johnson, R. J., "Complement activation during extracorporeal therapy: Biochemistry, cell biology and clinical relevance," *Nephrology Dialysis Transplantation*, 1994; **9**(suppl 2):36–45.

Jones, C. H., "Continuous renal replacement therapy in acute renal failure: membranes for CRRT," *Artif. Organs*, 1998; **22**(1):2–7.

Kantrowitz, A., "Origins of intraaortic balloon pumping," *Ann. Thorac. Surg.*, 1990; **50**(4):672–674.

Kantrowitz, A., Freed, P. S., Cardona, R. R., Gage, K., Marinescu, G. N., Westveld, A. H., Litch, B., Suzuki, A., Hayakawa, H., Takano, T., "Initial clinical trial of a closed loop, fully automatic intra-aortic balloon pump," *ASAIO J.*, 1992; **38**(3):M617–M621.

Kantrowitz, A., Tjonneland, S., Freed, P. S., Phillips, S. J., Butner, A. N., and Sherman, J. L., "Initial clinical experience with intraaortic balloon pumping in cardiogenic shock," *JAMA*, 1968; **203**(2):113–118.

Kasirajan, V., McCarthy, P. M., Hoercher, K. J., Starling, R. C., Young, J. B., Banbury, M. K., and Smedira, N. G., "Clinical experience with long-term use of implantable left ventricular assist devices: indications, implantation, and outcomes," *Semin. Thorac. Cardiovasc. Surg.*, 2000; **12**(3):229–237.

Kaufman, J. S., "Antithrombotic agents and the prevention of access thrombosis," *Semin. Dial.*, 2000; **13**(1): 40–46.

Kleine, P., Perthel, M., Hasenkam, J. M., Nygaard, H., Hansen, S. B., and Laas, J., "Downstream turbulence and high intensity transient signals (HITS) following aortic valve replacement with Medtronic Hall or St. Jude Medical valve substitutes," *Eur. J. Cardiothorac. Surg.*, 2000; **17**(1):20–24.

Klinkmann, H., and Vienken, J., "Membranes for dialysis," *Nephrology Dialysis Transplantation*, 1995; **10**(suppl 3):39–45.

Kohler, T. R., Stratton, J. R., Kirkman, T. R., Johansen, K. H., Zierler, B. K., and Clowes, A. W., "Conventional versus high-porosity polytetrafluoroethylene grafts: Clinical evaluation," *Surgery*, 1992; **112**(5):901–907.

Kolff, W. J., Berk, H. T. J., Ter Welle, M., van der Leg, J. W., van Dijk, E. C., and van Noorwijk, J., "The artificial kidney: A dialyzer with a great area," *Acta Med. Scand.*, 1944 **117**:121–134.

Krafczyk, M., Cerrolaza, M., Schulz, M., and Rank, E., "Analysis of 3D transient blood flow passing through an artificial aortic valve by Lattice-Boltzmann methods," *J. Biomech.*, 1998; **31**(5):453–462.

Krajcer, Z., and Howell, M. H., "Update on endovascular treatment of peripheral vascular disease: new tools, techniques, and indications," *Tex. Heart. Inst. J.*, 2000; **27**(4):369–385.

Kuitunen, A. H., Heikkila, L. J., and Salmenpera, M. T., "Cardiopulmonary bypass with heparin-coated circuits and reduced systemic anticoagulation," *Ann. Thorac. Surg.*, 1997; **63**(2):438–444.

Kumano, H., Suehiro, S., Hattori, K., Shibata, T., Sasaki, Y., Hosono, M., and Kinoshita, H., "Coagulofibrinolysis during heparin-coated cardiopulmonary bypass with reduced heparinization," *Ann. Thorac. Surg.*, 1999; **68**(4): 1252–1256.

Kumbasar, S. D., Semiz, E., Sancaktar, O., Yalcinkaya, S., and Deger, N., "Mechanical complications of intra-aortic balloon counterpulsation," *Int. J. Cardiol.*, 1999; **70**(1):69–73.

Kusumoto, F. M., and Goldschlager, N., "Cardiac pacing," *N. Engl. J. Med.*, 1996; **334**(2):89–97.

L'Heureux, N., Paquet, S., Labbe, R., Germain, L., and Auger, F. A., "A completely biological tissue-engineered human blood vessel," *FASEB J.*, 1998; **12**(1):47–56.

Laas, J., Kleine, P., Hasenkam, J. M., and Nygaard, H., "Orientation of tilting disc and bileaflet aortic valve substitutes for optimal hemodynamics," *Ann. Thorac. Surg.*, 1999; **68**(3):1096–1099.

Lazar, H. L., Bao, Y., Rivers, S., Treanor, P. R., and Shemin, R. J. "Decreased incidence of arterial thrombosis using heparin-bonded intraaortic balloons," *Ann. Thorac. Surg.*, 1999; **67**(2):446–449.

Lehner, G., Fischlein, T., Baretton, G., Murphy, J. G., and Reichart, B., "Endothelialized biological heart valve prostheses in the non-human primate model," *Eur. J. Cardiothorac. Surg.*, 1997; **11**(3):498–504.

Leon, M. B., Teirstein, P. S., Moses, J. W., Tripuraneni, P., Lansky, A. J., Tani, S., Wong, S. C., Fish, D., Ellis, S., Holmes, D. R., et al., "Localized intracoronary gamma-radiation therapy to inhibit the recurrence of restenosis after stenting," *N. Engl. J. Med.*, 2001; **344**(4):250–256.

Leung, S.-K., and Lau, C.-P., "Developments in sensor-driven pacing," *Cardiol. Clin.*, 2000; **18**(1):113–155.

Leypoldt, J. K., and Cheung, A. K., "Characterization of molecular transport in artificial kidneys," *Artif. Organs*, 1996; **20**(5):381–389.

Lim, W. L., Chew, Y. T., Chew, T. C., and Low, H. T., "Steady flow dynamics of prosthetic aortic heart valves: A comparative evaluation with PIV techniques," *J. Biomech.*, 1998; **31**(5):411–421.

Londrey, G. L., Ramsey, D. G., Hodgson, K. J., Barkmeier, L. D., and Sumner, D. S., "Infrapopliteal bypass for severe ischemia: Comparison of autogenous vein, composite and prosthetic grafts," *J. Vasc. Surg.*, 1991; **13**(5): 631–636.

Makarewicz, A. J., Mockros, L. F., and Anderson, R. W., "A pumping artificial lung," *ASAIO J.*, 1994; **40**(3): M518–M521.

Makhijani, V. B., Hang, H. Q., Dionne, P. J., and Thubrikar, M. J., "Three-dimensional coupled fluid-structure simulation of pericardial bioprosthetic aortic valve function," *ASAIO J.*, 1997; **43**(5):M387–M392.

Maki, D. G., Cobb, L., Garman, J. K., Shapiro, J. M., Ringer, M., and Helgerson, R. B., "An attachable silver-impregnated cuff for prevention of infection with central venous catheters: A prospective randomized multi-center trial," *Am. J. Med.*, 1988; **85**(3):307–314.

Maki, D. G., and Mermel, L. A., "Infections due to infusion therapy," Chap. 44, in Bennett, J. V., Brachman, P. S. (eds.) *Hospital Infections*, 4th ed., Lippincott-Raven Publishers, Philadelphia, 1998; pp. 689–724.

Maki, D. G., Stolz, S. M., Wheeler, S., and Mermel, L. A., "Prevention of central venous catheter-related bloodstream infection by use of an antiseptic-impregnated catheter. A randomized, controlled trial," *Ann. Intern. Med.*, 1997; **127**(4):257–266.

Marin, M. G., Lee, J. C., and Skurnick, J. H., "Prevention of nosocomial bloodstream infections: Effectiveness of antimicrobial-impregnated and heparin-bonded central venous catheters," *Crit. Care Med.*, 2000; **28**(9): 3332–3338.

Mattos, M. A., Hodgson, K. J., Hurlbert, S. N., Henretta, J. P., Sternbach, Y., Douglas, M. G., Mansour, M. A., Hood, D. B., and Sumner, D. S., "Vascular stents," *Curr. Prob. Surg.*, 1999; **36**(12):909–1053.

McCarthy, P. M., and Hoercher, K., "Clinically available intracorporeal left ventricular assist devices," *Prog. Cardiovasc. Dis.*, 2000; **43**(1):37–46.

McGiffin D. C., and Kirklin, J. K., "Cardiopulmonary bypass for cardiac surgery," Chap. 32 in Sabiston J., and Spencer, F. C., (eds.), *Surgery of the chest*, 6th ed., W. B. Saunders, Philadelphia, 1995; p. 1256–1271.

Mehlhorn, U., Kroner, A., and de Vivie, E. R., "30 years clinical intra-aortic balloon pumping: Facts and figures," *Thorac. Cardiovasc. Surg.*, 1999; **47**(suppl 2):298–303.

Mermel, L. A., Farr, B. M., Sherertz, R. J., Raad, I. I., O'Grady, N., Harris, J. S., and Craven, D. E., "Guidelines for the management of intravascular catheter-related infections," *Clin. Infect. Dis.*, 2001; **32**(9):1249–1272.

Merport, M., Murphy, T. P., Egglin, T. K., and Dubel, G. J., "Fibrin sheath stripping versus catheter exchange for the treatment of failed tunneled hemodialysis catheters: Randomized clinical trial," *J. Vasc. Interv. Radiol.*, 2000; **11**(9):1115–1120.

Meyer, M. M., "Renal replacement therapies," *Crit. Care. Clin.*, 2000; **16**(1):29–58.

Mitrani, R. D., Simmons, J. D., Interian, Jr., A., Castellanos, A., and Myerburg, R. J., "Cardiac pacemakers: Current and future status," *Curr. Prob. Cardiol.*, 1999; **24**(6):341–420.

Moller, M., and Arnsbo, P., "Appraisal of pacing lead performance from the Danish Pacemaker Register," *Pacing Clin. Electrophysiol.*, 1996; **19**(9):1327–1336.

Montoya, J. P., Shanley, C. J., Merz, S. I., and Bartlett, R. H., "Plasma leakage through microporous membranes. Role of phospholipids," *ASAIO J.*, 1992; **38**(3):M399–M405.

Morley-Davies, A., and Cobbe, S. M., "Cardiac pacing," *Lancet*, 1997; **349**(9044):41–46.

Morris, M. M., KenKnight, B. H., Narren, J. A., and Lang, D. J., "A preview of implantable cardioverter defibrillator systems in the next millennium: An integrative cardiac rhythm management approach," *Am. J. Cardiol.*, 1999; **83**(5B):48D–54D.

Mottaghy, K., Oedekoven, B., Starmans, H., Muller, B., Kashefi, A., Hoffmann, B., and Bohm, S., "Technical aspects of plasma leakage prevention in microporous capillary membrane oxygenators," *ASAIO Trans.*, 1989; **35**(3):640–643.

Mueller, X. M., Tevaearai, H. T., Hayoz, D., and von Segesser, L. K., "Thrombogenicity of deflated intraaortic balloon: Impact of heparin coating," *Artif. Organs*, 1999; **23**(2):195–198.

Namyslowski, J., and Patel, N. H., "Central venous access: A new task for interventional radiologists," *Cardiovasc. Intervent. Radiol.*, 1999; **22**(5):355–368.

Nield, T. A., Langenbacher, D., Poulsen, M. K., and Platzker, A. C., "Neurodevelopmental outcome at 3.5 years of age in children treated with extracorporeal life support: Relationship to primary diagnosis," *J. Pediatr.*, 2000; **136**(3):338–344.

Niklason, L. E., Gao, J., Abbott, W. M., Hirschi, K. K., Houser, S., Marini, R., and Langer, R., "Functional arteries grown in vitro," *Science*, 1999; **284**(5413):489–493.

"NKF-DOQI clinical practice guidelines for vascular access. National Kidney Foundation-Dialysis Outcomes Quality Initiative," *Am. J. Kidney Dis.*, 1997; **30**(4 suppl 3):S150–S191.

Oesterle, S. N., Whitbourn, R., Fitzgerald, P. J., Yeung, A. C., Stertzer, S. H., Dake, M. D., Yock, P. G., and Virmani, R., "The stent decade: 1987 to 1997. Stanford Stent Summit Faculty," *Am. Heart J.*, 1998; **136**(4 pt 1):578–599.

Okabe, T., Asakura, Y., Ishikawa, S., Asakura, K., Mitamura, H., and Ogawa, S., "Evaluation of scaffolding effects of five different types of stents by intravascular ultrasound analysis," *Am. J. Cardiol.*, 1999; **84**(9):981–986.

Owings, M. F., and Lawrence, L., "Detailed diagnoses and procedures," National Hospital Discharge Survey, 1997. National Center for Health Statistics. *Vital Health Stat.*, **13**(145). 1999.

Ozbek, C., Heisel, A., Gross, B., Bay, W., and Schieffer, H., "Coronary implantation of silicone-carbide-coated Palmaz-Schatz stents in patients with high risk of stent thrombosis without oral anticoagulation," *Cathet. Cardiovasc. Diagn.*, 1997; **41**(1):71–78.

Palder, S. B., Kirkman, R. L., Whittemore, A. D., Hakim, R. M., Lazarus, J. M., and Tilney, N. L., "Vascular access for hemodialysis. Patency rates and results of revision," *Ann. Surg.*, 1985; **202**(2):235–239.

Park, P. K., Jarrell, B. E., Williams, S. K., Carter, T. L., Rose, D. G., Martinez-Hernandez, A., and Carabasi III, R. A., "Thrombus-free, human endothelial surface in the midregion of a Dacron vascular graft in the splanchnic venous circuit—Observations after nine months of implantation," *J. Vasc. Surg.*, 1990; **11**(3):468–475.

Pasquinelli, G., Freyrie, A., Preda, P., Curti, T., D'Addato, M., and Laschi, R., "Healing of prosthetic arterial grafts," *Scanning Microsc.*, 1990; **4**(2):351–362.

Pearson, M. L., "Guideline for prevention of intravascular device-related infections. Part I. Intravascular device-related infections: An overview. Part II. Recommendations for the prevention of nosocomial intravascular device-related infections," *Am. J. Infect. Control*, 1996; **24**(4):262–293.

Pertosa, G., Grandaliano, G., Gesualdo, L., and Schena, F. P., "Clinical relevance of cytokine production in hemodialysis," *Kidney Int.*, 2000; **58**(suppl 76):S104–S111.

Pinski, S. L., and Trohman, R. G., "Permanent pacing via implantable defibrillators," *Pacing Clin. Electrophysiol.*, 2000; **23**(11 part 1):1667–1682.

Raad, I., "Intravascular-catheter-related infections," *Lancet*, 1998; **351**(9106):893–898.

Raad, I. I., Darouiche, R. O., Dupuis, J., Abi-Said, D., Gabrielli, A., Hachem, R., Wall, M., Harris, R., Jones, J., Buzaid, A., et al., "Central venous catheters coated with minocycline and rifampin for the prevention of catheter-related colonization and bloodstream infections. A randomized, double-blind trial. The Texas Medical Center Catheter Study Group," *Ann. Intern. Med.*, 1997; **127**(4):267–274.

Rais-Bahrami, K., Wagner, A. E., Coffman, C., Glass, P., and Short, B. L., "Neurodevelopmental outcome in ECMO and near-miss ECMO patients at 5 years of age," *Clin. Pediatr.*, 2000; **39**(3):145–152.

Reeder, G. D., et al., inventors, University of Pittsburgh, assignee, "Membrane apparatus with enhanced mass transfer, heat transfer, and pumping capabilities via active mixing," patent number 6,217,826, 2001.

Robinson, D. A., Lennox, A., Englund, R., and Hanel, K. C., "Graft dilatation following abdominal aortic aneurysm resection and grafting," *Aust. N. Z. J. Surg.*, 1999; **69**(12):849–851.

Rockall, A. G., Harris, A., Wetton, C. W. N., Taube, D., Gedroyc, W., and Al-Kutoubi, M. A., "Stripping of failing haemodialysis catheters using the Amplatz Gooseneck snare," *Clin. Radiol.*, 1997; **52**(8):616–620.

Rogers, C., and Edelman, E. R., "Endovascular stent design dictates experimental restenosis and thrombosis," *Circulation*, 1995; **91**(12):2995–3001.

Ronco, C., Ballestri, M., and Brendolan, A., "New developments in hemodialyzers," *Blood Purif.*, 2000; **18**(4): 267–275.

Rose, E. A., Moskowitz, A. J., Packer, M., Sollano, J. A., Williams, D. L., Tierney, A. R., Heitjan, D. F., Meier, P., Ascheim, D. D., Levitan, R. G., et al., "The REMATCH trial: Rationale, design, and endpoints. Randomized Evaluation of Mechanical Assistance for the Treatment of Congestive Heart Failure," *Ann. Thorac. Surg.*, 1999; **67**(3):723–730.

Roy, B. J., Rycus, P., Conrad, S. A., and Clark, R. H., "The changing demographics of neonatal extracorporeal membrane oxygenation patients reported to the Extracorporeal Life Support Organization (ELSO) Registry," *Pediatrics*, 2000; **106**(6):1334–1338.

Saito, N., Motoyama, S., and Sawamoto, J., "Effects of new polymer-coated extracorporeal circuits on biocompatibility during cardiopulmonary bypass," *Artif. Organs*, 2000; **24**(7):547–554.

Sauvage, L. R., Berger, K., Beilin, L. B., Smith, J. C., Wood, S. J., and Mansfield, P. B., "Presence of endothelium in an axillary-femoral graft of knitted dacron with an external velour surface," *Ann. Surg.*, 1975; **182**(6):749–753.

Sauvage, L. R., Berger, K. E., Wood, S. J., Yates II, S. G., Smith, J. C., and Mansfield, P. B., "Interspecies healing of porous arterial prostheses: Observations, 1960 to 1974," *Arch. Surg.*, 1974; **109**(5):698–705.

Schetz, M., "Non-renal indications for continuous renal replacement therapy," *Kidney Int.*, 1999; **56**(S72):S88–S94.

Schillinger, F., Schillinger, D., Montagnac, R., and Milcent, T., "Post catheterisation vein stenosis in haemodialysis: Comparative angiographic study of 50 subclavian and 50 internal jugular accesses," *Nephrol. Dial. Transplant.*, 1991; **6**(10):722–744.

Schmidt, J. A., and Stotts, L. J., "Bipolar pacemaker leads: New materials, new technology," *J. Invest. Surg.*, 1998; **11**(1):75–81.

Schoen, F. J., and Levy, R. J., Founder's Award, 25th Annual Meeting of the Society for Biomaterials, Providence, R. I., April 26–May 2, 1999, "Tissue heart valves: Current challenges and future research perspectives," *J. Biomed. Mater. Res.*, 1999; **47**(4):439–465.

Scholz, K. H., Ragab, S., von zur Muhlen, F., Schroder, T., Werner, G. S., Mindel, L., and Kreuzer, H., "Complications of intra-aortic balloon counterpulsation. The role of catheter size and duration of support in a multivariate analysis of risk," *Eur. Heart. J.*, 1998; **19**(3):458–465.

Schwab, S. J., and Beathard, G., "The hemodialysis catheter conundrum: Hate living with them, but can't live without them," *Kidney Int.*, 1999; **56**(1):1–17.

Schwartz, R. S., Huber, K. C., Murphy, J. G., Edwards, W. D., Camrud, A. R., Vlietstra, R. E., and Holmes, D. R., "Restenosis and the proportional neointimal response to coronary artery injury: Results in a porcine model," *J. Am. Coll. Cardiol.*, 1992; **19**(2):267–274.

Seeger, J. M., "Management of patients with prosthetic vascular graft infection," *Am. Surg.*, 2000; **66**(2):166–177.

Serruys, P. W., de Jaegere, P., Kiemeneij, F., Macaya, C., Rutsch, W., Heyndrickx, G., Emanuelsson, H., Marco, J., Legrand, V., Materne, P., et al., "A comparison of balloon-expandable-stent implantation with balloon angioplasty in patients with coronary artery disease," *N. Engl. J. Med.*, 1994; **331**(8):489–495.

Serruys, P. W., di Mario, C., Piek, J., Schroeder, E., Vrints, C., Probst, P., de Bruyne, B., Hanet, C., Fleck, E., Haude, M., et al., "Prognostic value of intracoronary flow velocity and diameter stenosis in assessing the short- and long-term outcomes of coronary balloon angioplasty: The DEBATE study (Doppler Endpoints Balloon Angioplasty Trial Europe)," *Circulation*, 1997; **96**(10):3369–3377.

Serruys, P. W., van Hout, B., Bonnier, H., Legrand, V., Garcia, E., Macaya, C., Sousa, E., van der Geissen, W., Colombo, A., Seabra-Gomes, R., et al., "Randomised comparison of implantation of heparin-coated stents with balloon angioplasty in selected patients with coronary artery disease (Benestent II)," *Lancet*, 1998; **352**(9129): 673–681.

Shimamoto, A., Kanemitsu, S., Fujinaga, K., Takao, M., Onoda, K., Shimono, T., Tanaka, K., Shimpo, H., and Yada, I., "Biocompatibility of silicone-coated oxygenator in cardiopulmonary bypass," *Ann. Thorac. Surg.*, 2000; **69**(1):115–120.

Shimono, T., Shomura, Y., Tahara, K., Hioki, I., Tenpaku, H., Maze, Y., Hirano, R., Shimpo, H., Shionoya, Y., Yokoyama, A., et al., "Experimental evaluation of a newly developed ultrathin silicone layer coated hollow fiber oxygenator," *ASAIO J.*, 1996; **42**(5):M451–M454.

Shinoka, T., Ma, P. X., Shum-Tim, D., Breuer, C. K., Cusick, R. A., Zund, G., Langer, R., Vacanti, J. P., and Mayer, Jr., J. E., "Tissue-engineered heart valves: Autologous valve leaflet replacement study in a lamb model," *Circulation*, 1996; **94**(suppl II):II164–II168.

Sitges-Serra, A., "Strategies for prevention of catheter-related bloodstream infections," *Support Care Cancer*, 1999; **7**(6):391–395.

Smith, Jr., S. C., Dove, J. T., Jacobs, A. K., Kennedy, J. W., Kereiakes, D., Kern, M. J., Kuntz, R. E., Popma, J. J., Schaff, H. V., and Williams, D. O., "ACC/AHA guidelines for percutaneous coronary intervention: Executive summary and recommendation: A report of the American College of Cardiology/American Heart Association Task Force on Practice Guidelines (Committee to Revise the 1993 Guidelines for Percutaneous Transluminal Coronary Angioplasty)," *Circulation*, 2001; **103**(24):3019–3041.

Sodian, R., Hoerstrup, S. P., Sperling, J. S., Daebritz, S., Martin, D. P., Moran, A. M., Kim, B. S., Schoen, F. J., Vacanti, J. P., and Mayer, Jr., J. E., "Early in vivo experience with tissue-engineered trileaflet heart valves," *Circulation*, 2000a; **102**(suppl III):III22–III29.

Sodian, R., Hoerstrup, S. P., Sperling, J. S., Daebritz, S. H., Martin, D. P., Schoen, F. J., Vacanti, J. P., and Mayer, Jr., J. E., "Tissue engineering of heart valves: In vitro experiences," *Ann. Thorac. Surg.*, 2000b; **70**(1): 140–144.

Stammers, A. H., "Historical aspects of cardiopulmonary bypass: From antiquity to acceptance," *J. Cardiothorac. Vasc. Anesth.*, 1997; **11**(3):266–274.

Stavarski, D. H., "Complications of intra-aortic balloon pumping. Preventable or not preventable?," *Crit. Care. Nurs. Clin. North. Am.*, 1996; **8**(4):409–421.

Stefanadis, C., Toutouzas, K., Tsiamis, E., Kallikazaros, I., Stratos, C., Vlachopoulos, C., Vavuranakis, M., and Toutouzas, P., "Implantation of stents covered by autologous arterial grafts in human coronary arteries: A new technique," *J. Invasive Cardiol.*, 2000a; **12**(1):7–12.

Stefanadis, C., Toutouzas, K., Tsiamis, E., Vlachopoulos, C., Kallikazaros, I., Stratos, C., Vavuranakis, M., and Toutouzas, P., "Stents covered by autologous venous grafts: Feasibility and immediate and long-term results," *Am. Heart. J.*, 2000b; **139**(3):437–445.

Stein P. D., Alpert, J. S., Bussey, H. I., Dalen, J. E., and Turpie, A. G., "Antithrombotic therapy in patients with mechanical and biologic prosthetic heart valves," *Chest*, 2001; **119**(suppl 1):220S–227S.

Stevenson, L. W., Kormos, R. L., Bourge, R. C., Gelijns, A. C., Griffith, B. P., Hershberger, R. E., Hunt, S., Kirklin, J., Miller, E. W., Pae, Jr., W. E., et al., "Mechanical cardiac support 2000: Current applications and future trial design," *J. Am. Coll. Cardiol.*, 2001; **37**(1):340–370.

Strecker, E.-P., Gabelmann, A., Boos, I., Lucas, C., Xu, Z., Haberstroh, J., Freudenberg, N., Stricker, H., Langer, M., and Betz, E., "Effect on intimal hyperplasia of dexamethasone released from coated metal stents compared with non-coated stents in canine femoral arteries," *Cardiovasc. Intervent. Radiol.*, 1998; **21**(6):487–496.

Svennevig, J. L., "Off-pump vs. on-pump surgery," *Scand. Cardiovasc. J.*, 2000; **34**(1):7–11.

Thomas, C. L., (ed.), *Taber's Cyclopedic Medical Dictionary*, 16th ed., F. A. Davis Co.: Philadelphia, 1989.

Tokars, J. I., Miller, E. R., Alter, M. J., et al., "National surveillance of dialysis-associated diseases in the United States, 1999," National Center for Infectious Diseases, Centers for Disease Control and Prevention, Altanta, 2001.

Torchiana, D. F., Hirsch, G., Buckley, M. J., Hahn, C., Allyn, J. W., Akins, C. W., Drake, J. F., Newell, J. B., and Austen, W. G., "Intraaortic balloon pumping for cardiac support: Trends in practice and outcome, 1968 to 1995," *J. Thorac. Cardiovasc. Surg.*, 1997; **113**(4):758–769.

Trerotola, S. O., "Hemodialysis catheter placement and management," *Radiology*, 2000; **215**(3):651–658.

Triolo, P. M., and Andrade, J. D., "Surface modification and evaluation of some commonly used catheter materials, I. Surface properties," *J. Biomed. Mater. Res.*, 1983; **17**(1):129–147.

Twardowski, Z. J. "High-dose intradialytic urokinase to restore the patency of permanent central vein hemodialysis catheters," *Am. J. Kidney Dis.*, 1998; **31**(5):841–847.

Tyers, G. F. O., Mills, P., Clark, J., Cheesman, M., Yeung-Lai-Wah, J. A., and Brownlee, R. R., "Bipolar leads for use with permanently implantable cardiac pacing systems: A review of limitations of traditional and coaxial configurations and the development and testing of new conductor, insulation and electrode designs," *J. Invest. Surg.*, 1997; **10**(1–2):1–15.

U.S. Renal Data System, *USRDS 2000 Annual Data Report*, National Institutes of Health, National Institute of Diabetes and Digestive and Kidney Diseases, Bethesda, Md., 2000.

Valles, J., Leon, C., and Alvarez-Lerma, F., "Nosocomial bacteremia in critically ill patients: A multicenter study evaluating epidemiology and prognosis," *Clin. Infect. Dis.* 1997; **24**(3):387–395.

van der Giessen, W. J., Lincoff, A. M., Schwartz, R. S., van Beusekom, H. M. M., Serruys, P. W., Holmes, Jr., D. R., Ellis, S. G., and Topol, E. J., "Marked inflammatory sequelae to implantation of biodegradable and nonbiodegradable polymers in porcine coronary arteries," *Circulation*, 1996; **94**(7):1690–1697.

Vaucher, Y. E., Dudell, G. G., Bejar, R., and Gist, K., "Predictors of early childhood outcome in candidates for extracorporeal membrane oxygenation," *J. Pediatr.*, 1996; **128**(1):109–117.

Veith, F. J., Gupta, S. K., Ascer, E., White-Flores, S., Samson, R. H., Scher, L. A., Towne, J. B., Bernhard, V. M., Bonier, P., Flinn, W. R., et al., "Six-year prospective multicenter randomized comparison of autologous saphenous vein and expanded polytetrafluoroethylene grafts in infrainguinal arterial reconstructions," *J. Vasc. Surg.*, 1986; **3**(1):104–114.

Verin, V., Popowski, Y., de Bruyne, B., Baumgart, D., Sauerwein, W., Lins, M., Kovacs, G., Thomas, M., Calman, F., Disco, C., et al., "Endoluminal beta-radiation therapy for the prevention of coronary restenosis after balloon angioplasty," *N. Engl. J. Med.*, 2001; **344**(4):243–249.

Virmani, R., and Farb, A., "Pathology of in-stent restenosis," *Curr. Opin. Lipidol.* 1999; **10**(6):499–506.

Vongpatanasin, W., Hillis L. D., and Lange, R. A., "Prosthetic heart valves," *N. Engl. J. Med.*, 1996; **335**(6): 407–416.

Wagner, W. R., Johnson, P. C., Kormos, R. L., and Griffith, B. P., "Evaluation of bioprosthetic valve-associated thrombus in ventricular assist device patients," *Circulation*, 1993; **88**(5 pt 1):2023–2029.

Waksman, R., "Intracoronary radiation therapy for restenosis prevention: Status of the clinical trials," *Cardiovasc. Radiat. Med.*, 1999; **1**(1):20–29.

Watanabe, H., Hayashi, J., Ohzeki, H., Moro, H., Sugawara, M., and Eguchi, S., "Biocompatibility of a silicone-coated polypropylene hollow fiber oxygenator in an in vitro model," *Ann. Thorac. Surg.*, 1999; **67**(5):1315–1319.

Wegner, J. A., "Oxygenator anatomy and function," *J. Cardiothorac. Vasc. Anesth.*, 1997; **11**(3):275–281.

Wendel, H. P., and Ziemer, G., "Coating-techniques to improve the hemocompatibility of artificial devices used for extracorporeal circulation," *European Journal of Cardiothoracic Surgery*, 1999; **16**(3):342–350.

Wesolowski, S. A., Fries, C. C., Karlson, K. E., DeBakey, M., and Sawyer, P. N., "Porosity: Primary determinant of ultimate fate of synthetic vascular grafts," *Surgery*, 1961; **50**:91–96.

Williams, S. K., Carter, T., Park, P. K., Rose, D. G., Schneider, T., and Jarrell, B. E., "Formation of a multilayer cellular lining on a polyurethane vascular graft following endothelial cell sodding," *J. Biomed. Mater. Res.*, 1992; **26**(1):103–117.

Willman, V. L., Anderson, J. M., Klesges, R., et al., "Expert panel review of the NHLBI Total Artificial Heart (TAH) Program June 1998—November 1999," National Institutes of Health, National Heart, Lung and Blood Institute, Bethesda, Md., 1999.

Winters, K. J., Smith, S. C., Cohen, M., Kopistansky, C., and McBride, R., "Reduction in ischemic vascular complications with a hydrophilic-coated intra-aortic balloon catheter," *Cathet. Cardiovasc. Intervent.*, 1999; **46**(3):357–362.

Yun, K. L., Sintek, C. F., Miller, D. C., Schuyler, G. T., Fletcher, A. D., Pfeffer, T. A., Kochamba, G. S., Khonsari, S., and Zile, M. R., "Randomized trial of partial versus complete chordal preservation methods of mitral valve replacement: A preliminary report," *Circulation*, 1999; **100**(suppl II):II90–II94.

Zhang, Z., Marois, Y., Guidoin, R. G., Bull, P., Marois, M., How, T., Laroche, G., and King, M. W., "Vascugraft polyurethane arterial prosthesis as femoro-popliteal and femoro-peroneal bypasses in humans: Pathological, structural and chemical analysis of four excised grafts," *Biomaterials*, 1997; **18**(2):113–124.

Zilla, P., and Greisler, H. P. (eds.), *Tissue engineering of vascular prosthetic grafts*, R. G. Landes Co., Austin, Tex., 1999.

CHAPTER 21
DESIGN OF RESPIRATORY DEVICES

David M. Shade
Johns Hopkins School of Medicine, Baltimore, Maryland

Arthur T. Johnson
University of Maryland, College Park, Maryland

21.1 INTRODUCTION

Respiratory medical devices generally fall into categories designed to measure volume, flow, pressure, or gas concentration. Many of these devices have been used in some form for many years, so design is both sophisticated and incremental. A thorough knowledge of pulmonary physiology and existing devices can be very helpful when proposing improvements. Most respiratory devices are composed of one or more simpler components, often linked to a processing unit of some type, such as a stand-alone personal computer (PC). Remarkably varied and sophisticated diagnostic instruments can be constructed from these basic building blocks combined with valves, tubing, pumps, and mouthpieces.

21.2 PULMONARY PHYSIOLOGY

Before delving into the instrumentation, it will be helpful to review briefly the function of the respiratory system and the parameters most often measured in pulmonary medicine.

The human respiratory system is composed of two lungs contained within the thorax, or chest cavity. The primary function of the lungs is gas exchange with the blood, providing a continuous source of oxygen for transport to body tissues, and eliminating carbon dioxide produced as a waste product of cellular metabolism. Gas exchange occurs in the alveoli, tiny thin-walled air-filled sacs numbering approximately 300 million in normal adult lungs. The alveoli are connected to the outside environment through a system of conducting airways that ends with the oral and nasal cavities. Alveoli are surrounded by and in close proximity to pulmonary capillaries, the tiny vessels containing blood to participate in gas exchange. The network of alveoli and airways is interdependent, so that the walls of the alveoli are shared among neighboring lung units.

Inspiration of fresh air occurs when respiratory muscles, chiefly the diaphragm, contract, expanding the chest wall and decreasing slightly the pressure surrounding the lungs but within the chest cavity (the *pleural pressure*). This drop in pleural pressure tends to pull outward on the outer lung surfaces, which in turn pull outward on more central lung units because of interdependence, and so on, with the result that the pressure within the alveolar air spaces falls. When alveolar pressure falls below the surrounding atmospheric pressure, air flows down the pressure gradient, filling the alveoli with inspired gas. Because lung tissue is elastic, it will tend to deflate when inspiratory muscle forces are released. In a normal person at rest, expiration is a passive phase and requires no muscular activity.

If all of a person's muscles were paralyzed and the lung were then permitted to empty, the lung volume would decrease to the *functional residual capacity* (FRC). At this volume, lung elasticity tends to favor further emptying, but the chest wall favors expansion. At higher volumes, both structures favor emptying (see Fig. 21.1). FRC, then, represents an equilibrium between the tendency of the lungs to empty further and the tendency of the chest wall to spring outward. In order to decrease the volume of the lung below FRC, expiratory muscles must be used to contract the chest wall and increase the pleural pressure. When the lung is emptied as much as is possible, there remains trapped within the alveoli and airways a volume of gas termed the *residual volume* (RV). The gas becomes trapped when the airways through which gas must travel collapse because of the high pleural pressure surrounding them. The amount of additional gas that may be expired with effort after the lung has reached FRC is termed the *expiratory reserve volume* (ERV). Thus, FRC is the sum of RV and ERV (FRC = RV + ERV).

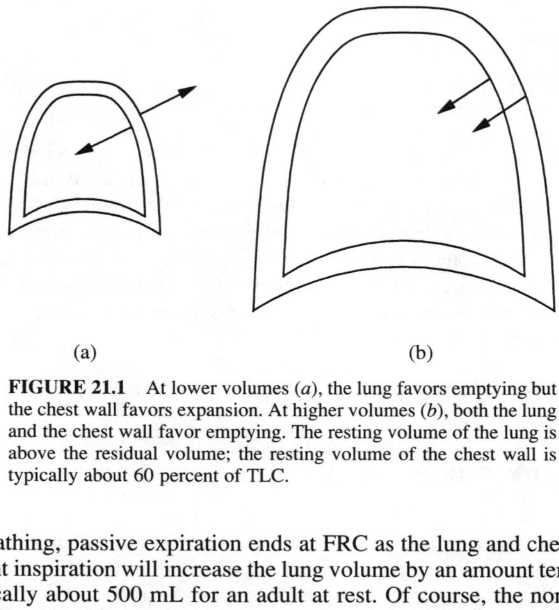

FIGURE 21.1 At lower volumes (*a*), the lung favors emptying but the chest wall favors expansion. At higher volumes (*b*), both the lung and the chest wall favor emptying. The resting volume of the lung is above the residual volume; the resting volume of the chest wall is typically about 60 percent of TLC.

During quiet breathing, passive expiration ends at FRC as the lung and chest wall reach equilibrium. The subsequent inspiration will increase the lung volume by an amount termed the *tidal volume* (TV), which is typically about 500 mL for an adult at rest. Of course, the normal resting breath is far smaller than a deep breath; maximum inspiration occurs when the lung is inflated to the *total lung capacity* (TLC), the largest volume that can be contained within the lung. TLC is determined by the balance between the increasing recoil (or "pull") of the lung and chest wall as they are inflated and the decreasing strength of the muscles as they are stretched. The difference between the TLC and the RV is termed the *vital capacity* (VC) and represents the largest volume of gas that can be exhaled, starting from full inflation, in a single breath. Thus, TLC = RV + VC. Figure 21.2 shows a diagram of the various lung volumes and capacities.

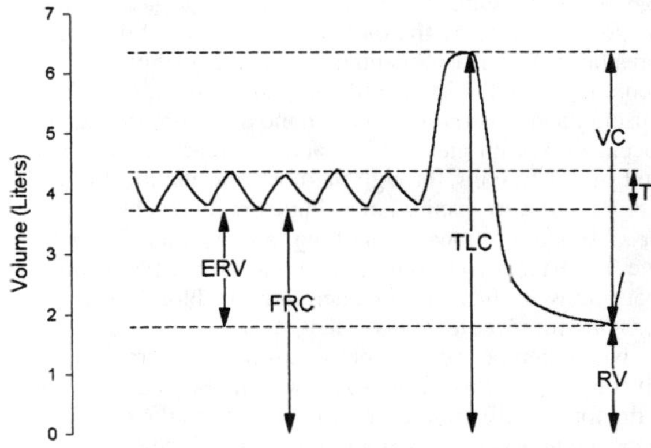

FIGURE 21.2 Spirometer tracing showing the lung volumes. Note that since the zero point of the spirometer may be set arbitrarily, only volumes not depending on zero (VC, TV, and ERV on this figure) may be measured directly.

Unlike TV, which is determined in large part by effort and ventilatory drive, TLC, RV, and VC are parameters that reflect certain properties of the pulmonary physiology. For example, a condition tending to increase the propensity of the airways to collapse would tend to increase the RV, and thereby decrease the VC, without changing the TLC. Asthmatics, in fact, may exhibit just this propensity during an asthma attack. Likewise, a condition increasing the elasticity (or "stiffness") of the lung would tend to decrease the TLC, and perhaps the VC, as the muscles became unable to supply sufficient force for continued inspiration. One condition causing such a change is asbestosis. Measurement of the various static lung volumes can be helpful both in classifying a lung disease and in tracking its progress or responsiveness to treatment.

During the process of ventilation, gases must travel through the airways linking the alveoli with the nose and mouth. These airways impose a pneumatic resistance to the flow of air, which, like an electrical resistance that limits current, reduces the flow for a given pressure gradient. During quiet breathing, most of the resistance to airflow occurs in the upper airway and not in the smaller airways within the lung. The resistance of the airways themselves is affected by the volume of the lung at the time of measurement because, as the lung approaches full inflation, the airways are stretched radially, achieving their greatest diameter and hence their lowest resistance. Another lung parameter affecting both static and dynamic changes is the lung compliance, a measure of tissue elasticity. The compliance is defined as the change in volume divided by the change in pressure across the wall of the lung, has units of volume over pressure, and is typically measured as the slope of a plot of lung volume against transmural pressure. Even in a person having normal muscle strength, it is possible to have marked reductions in the static lung volumes (such as TLC, VC, and RV) because of increases in the stiffness of the lung tissue. Like the resistance, compliance depends on the volume of gas contained within the lung. As the lung volume approaches TLC, it takes considerably greater pressure to continue inflation when compared with similar inflations taken nearer to FRC, representing a reduction in lung compliance at higher lung volumes.

Like electrical resistance and capacitance, airway resistance and lung compliance together impose a frequency-dependent impedance to ventilation. Thus, the normal lung emptying passively follows an exponential decay with a single time constant equal to the product of the resistance and the compliance. Because the lung is composed of millions of lung units, potentially having regional differences in impedance, dynamic characteristics of ventilation affect how quickly the lung may inflate or deflate, and may also result in portions of the lung being underventilated, while other portions are either normally ventilated or even overventilated.

In addition to ventilating the alveoli, another important function of the respiratory system is gas exchange, the process of moving oxygen into and carbon dioxide from the blood supply. Gas exchange occurs through the passive diffusion of gases across the alveolar membrane. Diffusive gas exchange is affected by the thickness and permeability of the alveolar membrane, the total surface area available for diffusion, and the availability of blood flow to receive or deliver the diffused gases. One common pulmonary test, the *diffusion capacity*, indirectly assesses the alveolar membrane by measuring the diffusion of carbon monoxide from the lungs into the blood. Of course, gas exchange can occur only with gas that has actually reached the alveoli, which is considerably less than the amount of gas moving through the nose and mouth. In fact, in a normal person at rest, only about two-thirds of the air inspired in a single breath reaches the alveoli. The remaining one-third occupies the *dead space*, portions of the lung, airways, and nasopharynx that do not participate in gas exchange. In certain conditions, alveoli that are ventilated may not participate in gas exchange because of aberrations in diffusion or alterations in blood flow that reduce or eliminate perfusion. These nonexchanging alveoli increase the *physiologic dead space* (as distinguished from *anatomic dead space*, which comprises the mouth, nasopharynx, trachea, and other conducting airways) and reduce the alveolar ventilation. Dead space ventilation, then, is "wasted" ventilation and must be subtracted from the total ventilation in order to determine the effective alveolar ventilation.

Once the pulmonary circulation has undergone gas exchange with ventilated alveoli, the oxygen-enriched blood circulates to body tissues, which consume the oxygen and replace it with carbon dioxide. The net consumption of oxygen ($\dot{V}_{O_2}$) is not equal to the net respiratory production of carbon dioxide ($\dot{V}_{CO_2}$), because cellular metabolic pathways do not maintain a 1:1 balance between the two, and also because carbon dioxide may be excreted in the urine as well as through the respiratory system. At rest, the ratio of $\dot{V}_{CO_2}$ to $\dot{V}_{O_2}$, the *respiratory quotient* (RQ), is typically 0.7. During maximum exercise, this value may rise to well above 1.0. Measurements of $\dot{V}_{O_2}$, $\dot{V}_{CO_2}$, and RQ are important in assessing the adequacy of nutrition of a critically ill patient, the diagnosis and treatment of patients with various pulmonary and cardiovascular diseases, and the training of elite athletes.

Most of the physiological parameters related to the respiratory system vary across age groups, gender, and ethnicity, and most are affected significantly by the size of the individual being assessed (large people tend to have large lungs). Therefore, interpreting the meaning of measured parameters often relies on comparing results with those obtained in large population studies of "normal" (disease-free) people of similar characteristics. There are many sources for the equations available to predict these normal values. While many devices attempt to include information about the predicted values, the most useful devices will allow the user to choose from among many sets.

21.3 IMPORTANT PRINCIPLES OF GAS PHYSICS

Most measurements made in the context of the respiratory system assume that the gas or gas mixture involved obeys the *ideal gas law*:

$$PV = nRT \tag{21.1}$$

where P = pressure, atmospheres (atm)
V = volume, liters (L)
n = moles of gas, mol
T = temperature, K
R = gas constant = 0.082057 atm · L/(mol · K)

The amount of each gas present in a mixture of gases may be represented conveniently in several ways: by its percentage (or fraction) of the whole, by its absolute volume in a known total volume,

TABLE 21.1 Constituents of Air

Species	Ambient		Inspired		Alveolar		Expired	
	mmHg	%	mmHg	%	mmHg	%	mmHg	%
Nitrogen (N_2)	584.1	76.86	556.8	73.26	566.2	74.50	583.7	76.80
Oxygen (O_2)	156.6	20.60	149.2	19.64	100.0	13.16	103.1	13.56
Carbon dioxide (CO_2)	0.2	0.03	0.2	0.03	40.0	5.26	41.2	5.43
Water (H_2O)	12.0	1.58	47.0	6.18	47.0	6.18	25.0	3.29
Other	7.1	0.93	6.8	0.89	6.8	0.89	7.0	0.92

Assume ambient temperature of 24°C, ambient relative humidity of 50%, barometric pressure of 760 mmHg, and body temperature of 37°C.

or by its partial pressure. The partial pressure of a gas in a mixture is the pressure that would be exerted by that gas on the walls of its container in the absence of all other gases. According to *Dalton's law*, the sum of the partial pressures of all gases present in a mixture is equal to the total pressure exerted by the mixture. For a gas mixture at atmospheric pressure, then,

$$P_B = \sum_{i=1}^{N} P_i \tag{21.2}$$

where P_B = the ambient atmospheric pressure (typically ~760 mmHg at sea level)
P_i = the partial pressure of the ith species
N = the total number of gases present in the mixture

The fractional concentration of a species is simply its partial pressure divided by the total pressure. Typical components of inspired and expired air are shown in Table 21.1.

Special note must be made of the presence of water vapor, as it presents an often-overlooked factor. The amount of water a gas mixture is *capable* of containing is limited by the availability of water and the temperature of the mixture. The relative humidity measures the percentage of this upper limit actually achieved. The relationship between temperature and the maximum possible partial pressure of water has been determined empirically; one suitable equation over common temperature ranges is

$$P_{H_2O} = 14.47 - 0.705T + 0.0428T^2 \tag{21.3}$$

where P_{H_2O} = partial pressure of water vapor (mmHg) and T = temperature (degrees Celsius).

At room temperature, the maximum possible P_{H_2O} is approximately 25 mmHg. The maximum possible P_{H_2O} rises to about 47 mmHg as the temperature rises to normal body temperature (37°C). Thus, upon inspiration, a sample of air is warmed and humidified, while expiration leads to cooling and condensation of excess water vapor. This is important for several reasons. First, the concentration of a gas as it exists within the lung differs from that in the air prior to inspiration or following expiration, as it is diluted or concentrated by changes in water vapor concentration. Second, even if a known fixed volume of gas is injected into the lung, such as by a ventilator, the volume of the lung does not increase by that same volume. Lastly, a warm moist gas that is exhaled undergoes cooling and condensation, and even if rewarmed will not reach 100% relative humidity and will therefore contain less water than initially, unless a source of water is available as the gas is rewarmed.

The combined effects of changes in temperature, pressure, and water vapor concentration are frequently accounted for by applying an appropriate correction factor, calculated as shown in Table 21.2.

Under ideal conditions, seldom encountered, gas flowing through a straight tube would exhibit only *laminar flow* (the absence of all turbulence). In laminar flow, every molecule in the gas stream has only axial velocity and moves only in the direction of the bulk flow. Under conditions of complete turbulence, the gas molecules move in all directions, with no net flow. Flows encountered in practice exhibit some characteristics of both laminarity and turbulence (Fig. 21.3). Factors tending to favor

TABLE 21.2 Gas Conversion Factors (Volume and Flow)

To convert from	To	Multiply by: Formula	Multiply by: Typical value
ATPS	BTPS	$\frac{BP - pH_2O}{BP - 47} \cdot \frac{310}{273 + T}$	1.080
	STPD	$\frac{BP - pH_2O}{760} \cdot \frac{273}{273 + T}$	0.892
	ATPD	$\frac{BP - pH_2O}{BP}$	0.971
ATPD	BTPS	$\frac{BP}{BP - 47} \cdot \frac{310}{273 + T}$	1.113
	STPD	$\frac{BP}{760} \cdot \frac{273}{273 + T}$	0.919
	ATPS	$\frac{BP}{BP - pH_2O}$	1.030
BTPS	STPD	$\frac{BP - 47}{760} \cdot \frac{273}{310}$	0.826
	ATPS	$\frac{BP - 47}{BP - pH_2O} \cdot \frac{273 + T}{310}$	0.926
	ATPD	$\frac{BP - 47}{BP} \cdot \frac{273 + T}{310}$	0.899
STPD	BTPS	$\frac{760}{BP - 47} \cdot \frac{310}{273}$	1.210
	ATPS	$\frac{760}{BP - pH_2O} \cdot \frac{273 + T}{273}$	1.121
	ATPD	$\frac{760}{BP} \cdot \frac{273 + T}{273}$	1.088

BP = barometric pressure
T = temperature (°C)
ATPS = ambient temperature and pressure, saturated
ATPD = ambient temperature and pressure, dry
STPD = standard temperature and pressure, dry
BTPS = body temperature and pressure, saturated

These typical values assume a barometric pressure of 760 mmHg and a temperature of 24°C. Standard pressure is 760 mmHg and standard temperature is 0°C (= 273 K). Body temperature is 37°C (= 310 K).

increased turbulence include decreasing tube radius, increasing gas density, and increasing bulk flow rate. Turbulence is increased by sharp angles and geometric obstructions in the gas stream. Laminar flow is increased by having a straight, smooth-bore tube with a length at least 6 times its diameter prior to flow measurement. The pressure required to move gas along a straight tube under conditions of laminar flow is given by the Poiseuille equation:

$$\Delta P = \frac{\dot{V} 8 \eta l}{\pi r^4} \tag{21.4}$$

where ΔP = pressure drop
$\dot{V}$ = flow
η = gas viscosity
l = length of tube
r = radius of tube

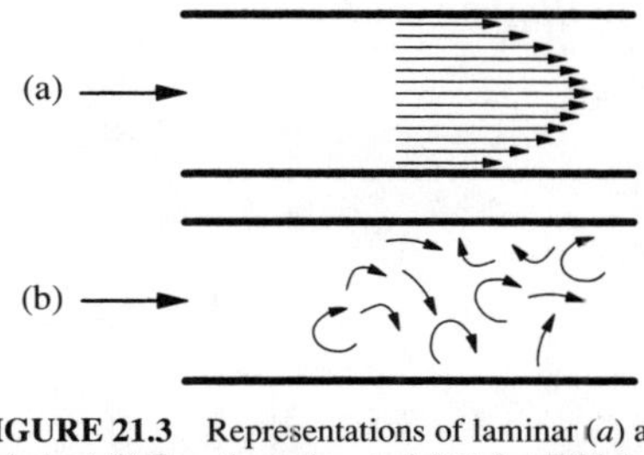

FIGURE 21.3 Representations of laminar (*a*) and turbulent (*b*) flow through a straight tube. With laminar flow, the pressure drop across a length of tube is proportional to the flow. With turbulence, the pressure approaches proportionality with the square of flow. The velocity profile under laminar flow conditions is parabolic, with higher flows near the center of the tube than near the boundary.

Thus, for a given straight tube, Poiseuille's equation predicts a linear relationship between pressure and flow and gives rise to the pneumatic analog of Ohm's law:

$$\Delta P = \dot{V}R \qquad \text{(compare with } \Delta V = IR\text{)} \tag{21.5}$$

where R = resistance, given by $8\eta l/\pi r^4$.

This Ohm's law representation of the pressure-flow relationship forms the basis for measurements of airway resistance and is the principle on which several flow-measuring devices are based. The pressure required to move gas under conditions of turbulence is always greater than that required to achieve the same flow under laminar conditions, as additional pressure (or energy) is required to accelerate molecules in directions other than the direction of bulk flow. This will be manifested as an upward curve and deviation from linearity on a plot of flow (x axis) versus pressure drop (y axis).

Turbulence in flow through a straight tube may be estimated by calculating the *Reynold's number*:

$$N_R = \frac{2sr\rho}{\eta} \tag{21.6}$$

where N_R = Reynold's number
s = linear velocity of flow
ρ = gas density

and other variables are as shown above.

Reynold's numbers greater than 2000 are predictive of significant turbulence. It should be noted that, while turbulent flow depends on both gas density and viscosity, laminar flow depends only on viscosity.

21.4 DEVICE COMPONENTS

21.4.1 Volume

Gas volumes may be measured directly by using one of several volume displacement *spirometers*. The simplest and oldest design, the water-sealed spirometer, uses a hollow cylinder, or bell, which is inverted and lowered into a bucket of water containing a tube for the movement of gas (Fig. 21.4). The bell rises or lowers as gas moves into or out of the gas space trapped between the bell and the

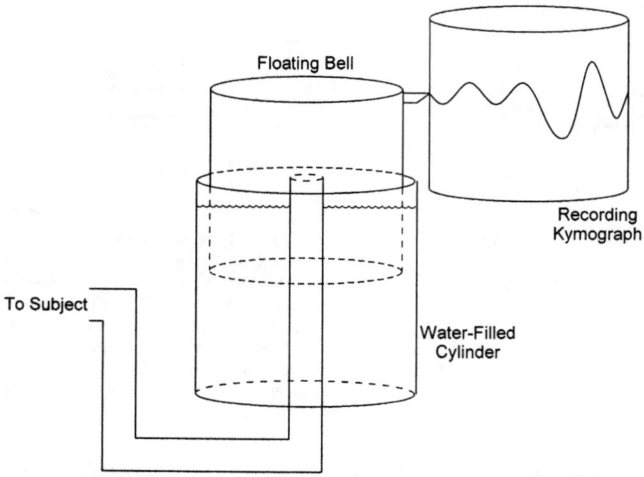

FIGURE 21.4 Schematic drawing of the water-sealed spirometer. The kymograph rotates at a constant speed, allowing for the inscription of a volume (y axis) versus time (x axis) plot as the bell moves up and down in response to gas movement at the outlet.

water. In order to prevent excess compression of the gas within the bell, earlier models used a chain and counterweight, although newer models are designed with lightweight bells in which gas compression is not significant. A pen is often attached to the bell, which may graph the volume-time tracing on an attached *kymograph*, a rotating drum with recording paper. Many spirometers also incorporate a linear potentiometer attached to the bell for easy electrical recording via a simple amplifier. The basic principle of operation is that the height of the bell is related to its volume by the formula for the volume of a cylinder:

$$V = \pi r^2 h \tag{21.7}$$

A similar approach is used in the dry-seal spirometer. In this case, the bell is sealed to its base with a thin layer of latex (or some other thin and flexible material). As gas is introduced into the bell, the latex prevents its escape and forces the bell to move, as with the water-sealed spirometer. Dry-seal spirometers may be mounted horizontally, and may employ a moving piston instead of a moving bell. Manual and electrical recording are achieved as in the water-sealed spirometer. A third type of volume displacement spirometer, somewhat less common recently, is the bellows, or wedge, spirometer. In this device, the gas to be measured is contained within a bellows whose expansion is recorded via a pen or a rotational potentiometer.

Volume displacement spirometers offer the advantage of simple construction and use. They do not require computers or processors for simple volume and time measurements. Perhaps most important, they are easy to calibrate and do not depend on the composition of the gases they are used to measure. However, they do suffer from some disadvantages. First, they are bulky. Water-sealed spirometers, in particular, can be heavy when they are filled with water, and they are prone to spillage when tipped. Second, owing to their mechanical action, they have a limited frequency response and are not well suited to rapidly changing signals (although they do have satisfactory frequency response for most common measurements). Last, the maximum volume they can measure is limited by their size. Thus, for an experiment in which tidal volume is measured over a period of 5 minutes, the volume displacement spirometer would be difficult to employ, without a series of complicated valves, as it would be filled before the experiment was over. Nevertheless, the volume displacement spirometer remains popular for simple measurements of volume and time.

A completely different approach, applicable only to volumes of gas as inspired or expired from the body, relies on the changes in chest (and sometimes abdominal) geometry that accompany breathing. One design uses two elastic bands, one placed around the chest, the other around the abdomen. The bands contain strain gages that measure the relative expansion of each of the compartments. This device requires calibration with each patient or subject on whom it is to be used. It is affected by movement artifact, changes in patient position, or changes in the relative movements of chest and abdomen during breathing. It is best employed on patients at rest, during quiet breathing, such as during sleep. Its accuracy seldom exceeds ±20 percent. A similar device uses electrodes placed on the chest wall to measure transthoracic impedance as a means to estimate changes in chest geometry. These devices offer the advantage of easy long-term measurements without the need for mouthpieces or other cumbersome connections, but their relative inaccuracy limits their usefulness.

21.4.2 Flow

Flow is the time derivative of volume, or

$$\dot{V} = \frac{dV}{dt} \tag{21.8}$$

Thus, any device capable of measuring either volume or flow can also report the other, given an appropriate time measurement and the needed processing. For this reason, and others given below, flow measuring devices have become popular methods for measuring both volumes and flows (although volume displacement spirometers also can easily calculate and report flow measurements).

There are three common groups of flow-measuring sensors. The pressure-drop pneumotachometers rely on the Poiseuille equation by imposing a fixed resistance on the gas flow and measuring the pressure drop across the resistance. Assuming laminar flow, the pressure drop will be linearly related to the flow rate. The resistive element may come in several forms, but two of the most common are the Lilly and the Fleisch (Fig. 21.5). The Lilly type uses a fine screen (similar to a window screen) to provide resistance. Often, three screens are used, with the outer screens meant to maintain laminar flow and to prevent debris and exhaled droplets from damaging the middle screen across which the pressure drop is measured. The Fleisch type uses a group of small parallel tubes as its resistive element.

In their simplest designs, both types rely on laminar flow to maintain a linear output. Thus, they may be rated for a maximum flow rate, above which linearity is compromised, as predicted by increases in the Reynolds number. In practice, deviations from linearity may be seen even below the maximum rated flow. Sometimes this may be due to suboptimal geometry between the device and the connecting tubes and valves. Another cause of inaccuracy may be condensation of water from the warm moist exhaled gas as it contacts the colder surfaces of the pneumotachometer. In addition to causing computational difficulties because of changes in gas volume as water vapor is lost, the condensed water that is deposited on the resistive element may change its resistance and diminish its accuracy. This is often countered by heating the device to body temperature, thus preventing condensation. The reliance on linearity minimized the usefulness of this class of device until sufficient portable processing power was available that linearity no longer was required. Most devices now used no longer assume linearity, but instead characterize the flow versus pressure relationship over the entire useful range. This may be done by measuring the response to a series of known flows, but is more often calculated using an algorithm described by Yeh and colleagues (1982) using a calibration syringe of known volume. Under conditions of laminar flow, where only gas viscosity is important, the specific composition of the gas being measured is usually of little concern, since most gases present in common respiratory measurements are of similar viscosity (see Table 21.3). However, if gases with a different viscosity will be used, the device must be carefully calibrated with gases having that same composition.

A different variety of pressure-drop flow sensor relies on the Pitot effect instead of Ohm's law. In one design of this type of device, two pressure-sensing taps are placed within the gas stream, one

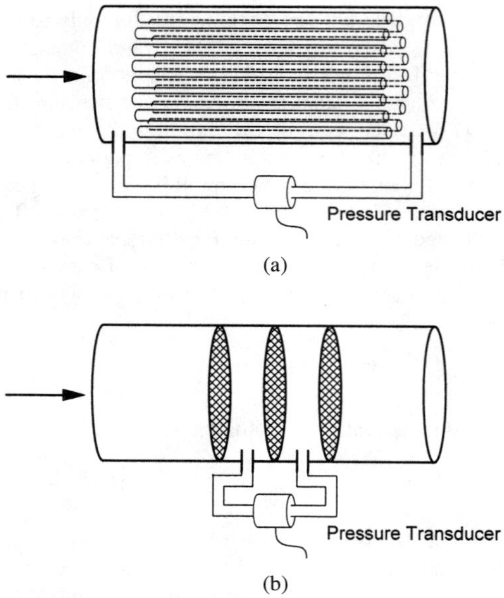

FIGURE 21.5 Schematic representations of the Fleisch (*a*) and Lilly (*b*) types of pneumotachographs for measuring flow. In both types, the small open ports are connected to opposite sides of a differential pressure transducer. The pressure difference between the open ports is related to the flow through the device by an equation that depends in part on the degree of turbulence present.

pointed directly upstream and the other pointed directly downstream (Fig. 21.6). The pressure difference is no longer described by the Poiseuille equation, but rather is described by Bernoulli's law, which states that the pressure difference between the taps is a function of the density of the gas measured and the square of its velocity. Velocity is related to volumetric flow by the cross-sectional area of the apparatus. Because this device is affected by gas density, it is usually employed in conjunction with analyzers providing precise measurements of the types and concentrations of gases present, allowing for calculation of the mixture's density, but limiting the device's usefulness as a stand-alone flow sensor.

TABLE 21.3 Physical Constants of Common Respiratory Gases

Species	Viscosity, μP	Density, g/L	Thermal conductivity, 10^{-6} cal/(s · cm² · °C/cm)	Magnetic susceptibility, 10^{-6} cgs units
N_2	178.1	0.625	56.20	−12.0
O_2	201.8	0.714	57.24	3449.0
CO_2	148.0	0.982	33.68	−21.0
He	194.1	0.089	333.50	−1.88
H_2	87.6	0.089	405.00	−3.98
CO	175.3	0.625	53.85	−9.8
Ne	311.1	0.446	107.03	−6.74
CH_4	108.7	0.446	71.08	

All values taken near room temperature (19 to 30°C) from tables in *CRC Handbook of Chemistry and Physics*, Robert C. Weast, ed., 61st edition, 1981. Magnetic susceptibility is dimensionless, but is defined differently in cgs and mks units.

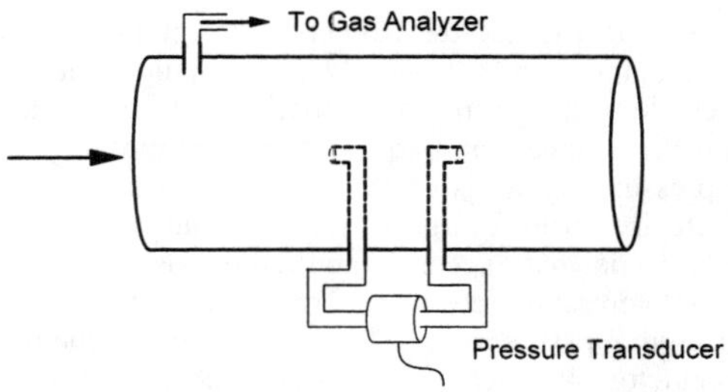

FIGURE 21.6 Schematic representation of a pitot tube. Some designs incorporate elements to better control the degree of turbulence. The gas analyzer is used to measure the density of the gas.

The second group of flow-measuring sensors, commonly referred to as hot-wire anemometers, is based on the principle that flowing gas will cool a heat source and that greater flows will cause greater cooling of the heat source. The devices contain a heated wire (some designs incorporate a second heated wire, at a different temperature, as a reference), which is maintained at a constant temperature via feedback circuitry (Fig. 21.7). The current required to maintain the stable temperature is related to the bulk flow through the device. The hot-wire anemometer is affected slightly by the thermal conductivity of the gases present, but is generally less sensitive to changes in gas composition than pressure-drop flow sensors. The accuracy of this type of flow sensor, too, is affected by changing amounts of turbulence.

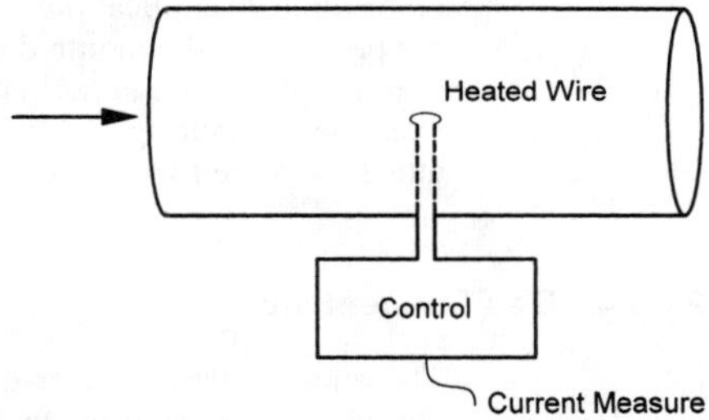

FIGURE 21.7 Schematic representation of a hot-wire anemometer. The flow is related to the current required by the feedback control circuitry to maintain the heated wire at a constant temperature.

The third group of flow sensors is the turbines, devices that are based on a rotating vane placed in the gas stream, much like a wind speed indicator on a simple weather station. The rotation of the vane is proportional to the velocity and density of the gas stream. As with the Pitot effect devices, gas velocity is related to volumetric flow by the cross-sectional area of the apparatus. Turbine flow sensors can be purely mechanical, with precision gearings and a display dial, making for a very portable device requiring no electricity. Battery-powered models using a light-emitting diode (LED) and a light sensor to count the rotations of the vane are also common. Turbine sensors must overcome several challenges to remain accurate, however. First, the moving parts must be lightweight to reduce the effects of inertia, which tend to impair the frequency response of the device. At the same time, however, the vanes must be made strong enough to withstand the forces of high flow rates.

The chief advantages of flow sensors are their small size and their ability to measure continuously in flow-through designs. They all offer good dynamic range, with frequency responses that exceed that needed for most or all respiratory measurements. They suffer from the need for more complex calibrations and generally require a computer or processor unit. The Pitot effect device requires simultaneous measurement of gas concentrations. The hot-wire anemometer requires external circuitry capable of heating the wires sufficiently and somewhat sophisticated feedback circuitry. The turbine sensors can be fragile and susceptible to damage.

21.4.3 Pressure

Although there are a variety of pressure transducer types available, for most respiratory applications the newer solid-state pressure transducers offer the most attractive alternative. Variable-capacitance transducers generally use the deflection of conductive plates by the input pressure differential to generate an output signal. Similarly, variable-inductance transducers use the input pressure differential to create a change in inductance. In both cases, the transducer requires an input ac excitation signal in order to create an output signal. The circuitry used to create the excitation, and to output a usable pressure signal, is often referred to as a *carrier-demodulator*. These types of transducers are very precise, offer a large variety of input ranges, and maintain stable calibrations. However, they can be somewhat expensive and require dedicated external circuitry.

Solid-state pressure transducers come in a single package, often quite small, and also offer a wide choice of input pressure ranges. Typically, they use the input pressure differential to deform slightly a semiconductor plate separating two chambers exposed to each of the two input pressure levels. Any difference between the two levels will deform the separating plate, changing its conductive properties. On-board circuitry usually allows the transducer to accept a dc excitation voltage, and provides a linear, temperature-corrected dc output voltage. These transducers are offered by a large number of vendors at affordable pricing.

All of these pressure transducers offer frequency response characteristics acceptable for most or all respiratory applications. In situations requiring the measurement of a single pressure (for example, the pressure at the mouth during breathing), one of the two input ports to the transducer is left open to atmospheric pressure. In other situations requiring the measurement of a pressure differential (for example, measuring the output of a pressure-drop flow sensor), the two ports of the transducer are attached to the two taps of the sensor.

21.4.4 Gas Concentration

There are a large number of gas analysis technologies available to measure the concentrations of various species present in a gas mixture. Some of the more common types used in respiratory applications are based on principles of *thermal conductivity, infrared absorption, zirconium fuel cell technology, paramagnetism, emission spectroscopy, gas chromatography*, and *mass spectrometry*. In many cases, some gases in a mixture interfere with the analysis for other gases and must be removed. Most often, interfering gases to be removed are water vapor and carbon dioxide. Water vapor may be removed by passing the gas stream through a canister of calcium sulfate ($CaSO_4$). Alternatively, water vapor may be removed, or equilibrated to a known level, by passing the gas mixture through a length of nafion, tubing with walls that are able to remove water because they are impregnated with a material having a high affinity for water. Placing the nafion within the lumen of larger tubing through which is run dry nitrogen (running in the direction opposite to the gas mixture to be analyzed) will remove all the water vapor from the mixture, whereas leaving the nafion tubing exposed to room air will equilibrate the water vapor concentration with ambient levels. Carbon dioxide is most often removed by passing the gas stream through a canister containing either barium hydroxide [$Ba(OH)_2$] or sodium hydroxide (NaOH). In both cases, the chemical combines with the gaseous CO_2, removing it from the gas mixture. Also, both chemical CO_2 scrubbers produce water as a by-product of the reaction and therefore should be placed upstream to a water removal system if needed. Usually, any chemical scrubber will be impregnated with an indicator that changes color as the chemical becomes consumed, alerting the user to the need for replacement.

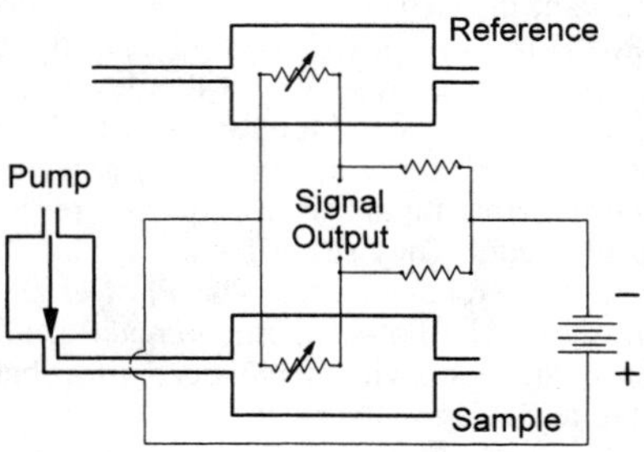

FIGURE 21.8 Schematic drawing of a thermal conductivity meter. The electrical elements are configured in a typical Wheatstone bridge circuit. The variable resistors are usually thermistors.

Thermal conductivity meters measure the ability of a gas mixture to conduct heat away from a heat source (Fig. 21.8). They usually employ two thermistors, one exposed to sample gas and the other to a reference sample containing none of the gas to be measured, arranged in a conventional Wheatstone bridge. Thermal conductivity is most commonly used to measure helium concentration, although it can be used to measure carbon dioxide concentration if used carefully. Both carbon dioxide and water vapor interfere with helium analysis and must be removed from the gas mixture prior to helium analysis. Water vapor, like helium, has a higher thermal conductivity than other gases found in air and thus will cause false elevation in the helium reading. Carbon dioxide, on the other hand, has a lower thermal conductivity and will cause a lowered helium reading. Thermal conductivity meters tend to be quite linear, making them easy to use, but they have relatively slow response times, on the order of 20 to 30 seconds.

Infrared (IR) absorption is a common method of gas analysis and can be used for measurements of carbon dioxide, carbon monoxide, and methane, among others (Fig. 21.9). IR analyzers run the gas mixture through a sample chamber illuminated at one end with an IR light source. At the other end of the chamber is an IR detector. Parallel to the sample chamber is a reference chamber, which is usually filled with room air and uses the same IR source and detector as the sample chamber. The gases to be analyzed absorb some of the IR radiation, decreasing the amount reaching the detector. A motor rotates a "chopper blade," which alternately blocks one of the two chambers from the source. By synchronizing the chopper rotation with the signal at the detector, the analyzer can determine the relative absorption in the reference chamber. This synchronization can be achieved by using a standard LED and light detector to determine the position of the chopper opening. The IR analyzer is sensitive to the pressure within the sample chamber, and thus requires either a constant flow through the chamber using a stable pump, or no gas flow after a pump has filled the chamber with the gas to be analyzed. IR analyzers can be designed to have rapid response times. When used as a CO analyzer, CO_2 is an interfering gas and must be removed or subtracted mathematically. H_2O is an interfering gas for almost all other gases and should be removed or equilibrated prior to analysis.

The zirconium fuel cell employs a zirconium substrate coated by platinum to create an electrode sensitive only to oxygen. Oxygen diffuses through the platinum coating on one side, passes through the zirconium, and completes the circuit on the other side of the fuel cell. So long as there is an oxygen concentration difference on the two sides, the movement of oxygen ions creates a current and induces a voltage, which is logarithmically related to the oxygen concentration difference. The zirconium is heated to high temperature (>700°C), requiring adequate warm-up time (20 to 30 minutes) and power requirements not suited to highly portable applications. The analyzer is sensitive

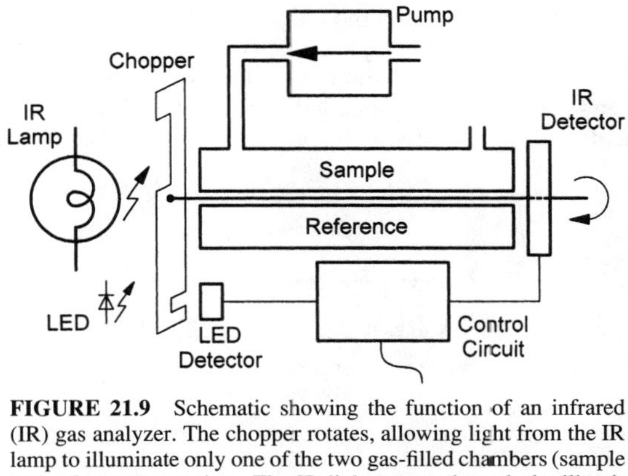

FIGURE 21.9 Schematic showing the function of an infrared (IR) gas analyzer. The chopper rotates, allowing light from the IR lamp to illuminate only one of the two gas-filled chambers (sample and reference) at a time. The IR light passes through the illuminated chamber, where it is differentially absorbed by the gases present. The IR detector measures the amount of IR light transmitted through the whole chamber. The LED and LED detector allow the control circuit to determine which of the two gas-filled chambers is illuminated at any time. The control circuit measures and linearizes the differential IR absorbance of the two chambers, and outputs the usable signal.

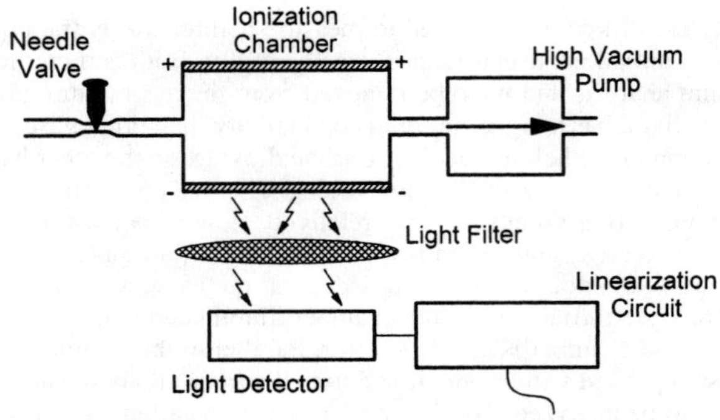

FIGURE 21.10 Schematic drawing of an emission spectroscopy analyzer, typically used to measure nitrogen concentrations. The output varies with the pressure within the ionization chamber, so the needle valve and vacuum pump must carefully regulate that pressure.

to the pressures within it and thus requires a stable sample flow. It offers a rapid response time and is relatively insensitive to interfering gases.

Paramagnetism refers to the propensity for certain molecules to align themselves when exposed to a magnetic field. Oxygen exhibits very high magnetic susceptibility compared to other common respiratory gases. The paramagenetic oxygen analyzer introduces a gas mixture into a hollow dumbbell-shaped structure suspended in a magnetic field. The greater the oxygen content, the greater the force tending to move the dumbbell toward alignment in the field. The oxygen concentration may be measured either as the deflection of the dumbbell or as the force required to prevent the dumbbell from moving. The analyzer is sensitive to pressure, and thus flow rates must be held steady during measurement. Response times on modern analyzers are quite fast and may be suitable for rapid sampling.

Emission spectroscopy is used to measure nitrogen concentrations. Nitrogen is drawn via a strong vacuum through a needle valve into a *Giesler tube*, a very low pressure chamber where it is exposed to a strong electric field (on the order of 25 kV) (Fig. 21.10). The N_2 ionizes and emits light, which is filtered and picked up by a photodetector, where the signal is amplified and linearized, and output to a recorder. The output is quite sensitive to the pressure within the chamber and thus the vacuum pump and needle valve must both be quite stable. This type of nitrogen analyzer has a rapid response time and is relatively insensitive to interfering gases. However, achieving and maintaining the re-

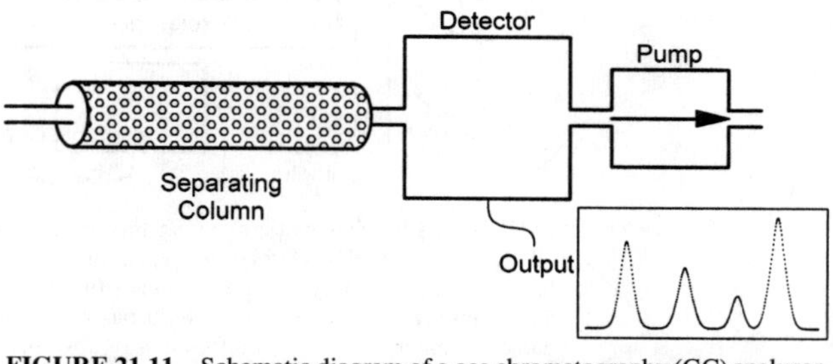

FIGURE 21.11 Schematic diagram of a gas chromatography (GC) analyzer. The gas mixture to be analyzed is drawn through the column where it is separated into its constituent species. The separated gases are then drawn through another gas detector (often a thermal conductivity analzyer), where concentrations are measured.

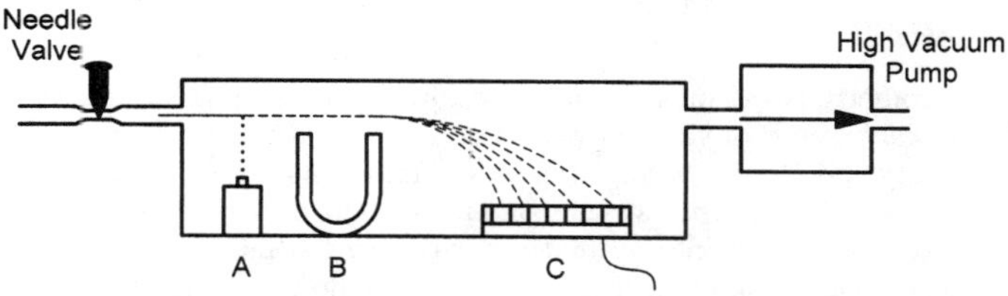

FIGURE 21.12 Schematic drawing of a mass spectrometer. The gas is ionized by an ion gun (A) and then passes through a magnetic field (B). The charged particles follow a circular path with a radius dependent on the mass-to-charge ratio. The collector (C) is composed of a series of particle detectors at various distances along its length. Each detector's output indicates the concentration of a different gas species. The spectrometer collector does not distinguish between ionized particles having the same mass-to-charge ratio, and thus there are certain gases which cannot be analyzed separately.

quired vacuum can be challenging to the designer. An alternative approach to the analysis of nitrogen concentration is to measure O_2 and CO_2 concentrations and, if water vapor pressure has been removed, calculate the remainder to be nitrogen (assuming no other gases are present).

Gas chromatography separates a mixture of gases by passing the sample through a column containing a material that selectively impedes the progress of different species along its length (Fig. 21.11). The gas sample is mixed with a carrier gas (usually helium, which thereby renders the analyzer unable to measure helium concentrations) and run through the column. Species that are less impeded by the column material exit the column first, followed sequentially by other species. The concentrations of the now separated species are measured using another type of detector, frequently a thermal conductivity meter. Water vapor is usually removed prior to passing the gas through the column. The gas chromatograph offers the advantage of being able to measure several different gases with one analyzer, but has relatively slow response times and is unsuitable for continuous sampling with changing inputs.

The mass spectrometer, like the gas chromatograph, is capable of measuring many or all constituents in a gas mixture within a single analyzer. The gas mixture is drawn via a vacuum into a low-pressure chamber where its molecules are ionized by an electron gun (Fig. 21.12). The charged ions are accelerated by an electric field down a chamber, and are then exposed to a magnetic field. The ions are deflected along an arc by the magnetic field according to their mass and charge: larger ions exhibit a greater radius of curvature than smaller ions with the same charge. Detectors count the number of ions at various locations within the analyzer. Because the analyzer distinguishes species on the basis of molecular mass, different ions with the same mass and charge cannot be separated. The mass spectrometer has a rapid response time and is well suited to continuous sampling and measurement. It is, however, quite large and expensive. Water vapor is usually removed from the gas stream prior to analysis.

All of the gas analyzers described above have been used successfully in clinical practice, and although they have different advantages and disadvantages, all are well suited to certain types of applications.

21.5 COMMON RESPIRATORY MEASUREMENTS

Putting the preceding discussions into practice is the everyday mission of the *pulmonary function laboratory*, a common clinical unit designed to make respiratory measurements on hospital patients and research subjects. Some of the most common measurements and the devices used to make them will be described below.

21.5.1 Spirometry

Spirometry is the most common respiratory measurement and reports, in a single maneuver, a remarkable range of valuable parameters. In this test, the patient inspires fully to TLC and then expires rapidly and forcefully, blowing hard until the lung volume reaches RV. A tracing of volume versus time is obtained, from which many parameters may be measured, including FVC, FEV_1 (the volume of air exhaled during the first second), and the ratio of FEV_1/FVC (Fig. 21.13). Spirometry also reports the maximum, or peak, expiratory flow, as well as instantaneous flows at specified lung volumes. Spirometry may be performed using a volume displacement spirometer, with electrical or mathematical differentiation to measure flows, or using a flow sensor with mathematical integration to obtain volumes. Standards for spirometry have been published by numerous medical societies (see references at the end of this chapter). For such a simple-to-perform test, spirometry is remarkably powerful. It is exceedingly reproducible when compared with other medical tests. It is commonly used for a wide variety of medical purposes, including the detection and classification of various forms of lung disease, the responsiveness to various therapeutic interventions, and the progression of pulmonary and nonpulmonary conditions. Indeed, the FEV_1 has been shown in numerous studies to correlate more reliably with mortality than many other common medical tests, for reasons not completely understood. All spirometric parameters are reported at BTPS conditions.

21.5.2 CO Diffusing Capacity

The CO diffusing capacity, DLCO, is calculated by measuring the difference in alveolar CO concentrations at the beginning and end of a period of breath holding. The test begins by having the patient exhale completely to RV and then inspiring rapidly to TLC a breath of gas with a known CO concentration. After a 10-second breath-hold, the patient exhales rapidly (Fig. 21.14). The initial portion of this exhalation is discarded, as it contains gas from the dead space, and a portion of the subsequently exhaled gas, assumed to be well-mixed alveolar gas, is analyzed for CO content. The initial alveolar concentration of CO is not the inspired concentration, as the inspired gas is diluted

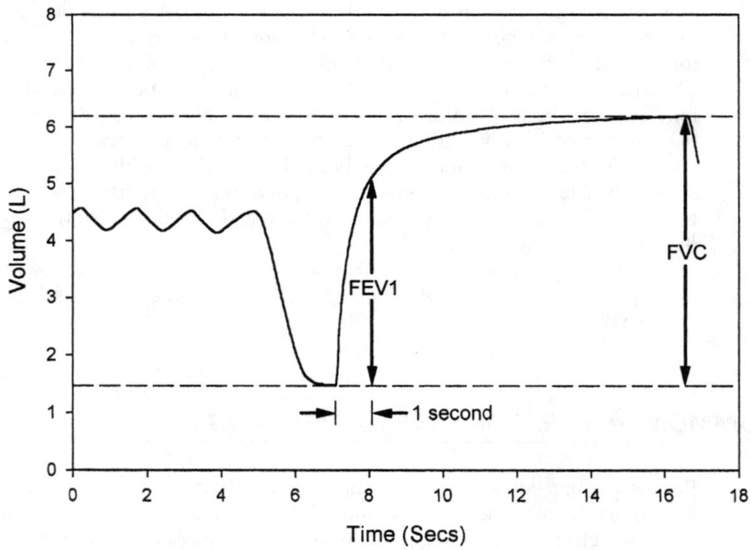

FIGURE 21.13 Spirometer volume-time tracing typical of spirometry during forced expiration. In this representative test, the FEV1 is 3.64 L and the FVC is 4.73 L.

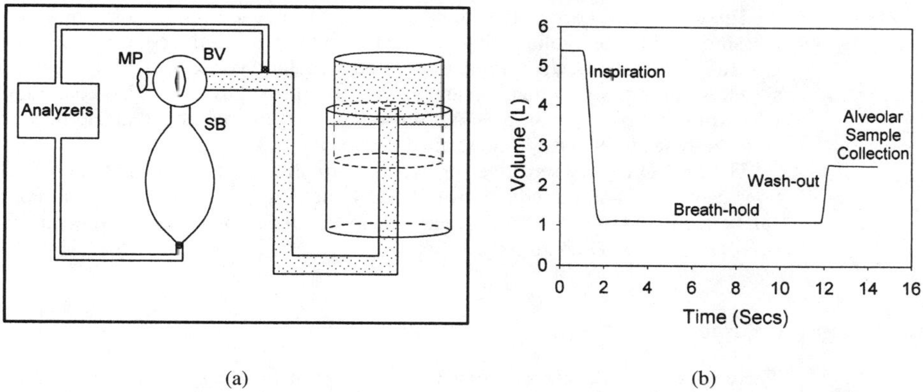

FIGURE 21.14 Illustration of equipment for performing a single-breath diffusing capacity test (*a*). The patient breathes through the mouthpiece (MP). Initially, the breathing valve (BV) is turned so that all breathing is in from and out to the room. After the patient exhales to RV, the breathing valve is turned to attach the patient to the spirometer filled with test gas containing carbon monoxide and helium. The patient inspires rapidly to TLC, breath-holds for 10 seconds, and exhales rapidly. The breathing valve is left connected to the spirometer for the initial portion of this expiration, allowing the spirometer to record the amount of gas washing out the dead space. After the dead space has been flushed, the breathing valve is turned so that an alveolar sample is collected in the sample bag (SB). Gas analyzers measure the inspired gas concentrations from the spirometer, and the expired gas concentrations from the sample bag. A representative spirometer tracing is shown on the right (*b*).

with gas remaining in the lung prior to inspiration (the RV). In order to assess this dilutional reduction in CO concentration (as contrasted with the later reduction due to diffusion), an inert gas that is readily diluted but does not diffuse or dissolve is added to the inspired gas. Suitable tracer gases for this purpose include helium, methane, and neon. The concentration drop of the tracer gas from beginning to end of breath holding is used to calculate the initial CO alveolar concentration as follows:

$$F_{ACO} = F_{ICO}\frac{F_{ETR}}{F_{ITR}} \tag{21.9}$$

where F_A = fractional concentration in alveolar gas
F_I = fractional concentration in inspired gas
F_E = fractional concentration in expired gas
CO = carbon monoxide
TR = tracer gas

Over the period of breath holding, the alveolar concentration of CO falls exponentially according to its partial pressure gradient between the gas and blood sides of the alveolar membrane (it is assumed that the blood concentration of CO is zero throughout the short breath-hold). Then, the DLCO is calculated as

$$\text{DLCO} = V_I\frac{F_{ITR}}{F_{ETR}}\frac{1}{T}\frac{1}{\text{BP} - 47}\ln\left(\frac{F_{ACO}}{F_{ECO}}\right) \tag{21.10}$$

where V_I = volume of inspired gas (usually adjusted by an estimate of dead space)
T = duration of breath-hold
BP = ambient barometric pressure

and other parameters are as defined above.

The equipment needed to calculate the DLCO includes either a spirometer or a flow sensor to measure the inspired volume and gas analyzers to measure the concentrations of CO and the tracer gas. In some systems, the sample of expired alveolar gas is collected in a bag for analysis; in other systems with rapidly responding analyzers, the expired gas is sampled continuously for calculation. It is worth noting that using a flow sensor for this test requires that the flow sensor be calibrated with the gases to be used, as described above.

The DLCO is very sensitive for lung abnormalities but is also quite nonspecific. Several conditions can cause marked reductions in DLCO, including emphysema and pulmonary fibrosis. DLCO can be increased in some cases of early heart failure and in cases of pulmonary hemorrhage. DLCO is reported at STPD conditions.

21.5.3 Lung Volumes

There are several different methods available for measuring the TLC, RV, and FRC. These parameters, regardless of how they are measured, are reported at BTPS conditions.

In the *helium dilution* method, a two-port spirometer is connected with tubes to a breathing valve and a blower motor. The spirometer is filled with a known concentration of helium and the patient, with no helium in the lungs, is attached at FRC by turning the breathing valve. At the beginning of the test, the total amount of helium contained within the system is equal to the initial concentration times the system volume. After 3 to 7 minutes of rebreathing, when the helium has been distributed evenly between the patient's lungs and the system, the final helium concentration is recorded (Fig. 21.15). Then, owing to the fact that helium is inert and does not leave the lung via solution in tissues nor diffusion into the blood,

$$\text{FRC} = V_S \frac{F_{I\text{HE}} - F_{F\text{HE}}}{F_{F\text{HE}}} \tag{21.11}$$

where V_S = system volume
$F_{I\text{HE}}$ = initial helium concentration
$F_{F\text{HE}}$ = final helium concentration

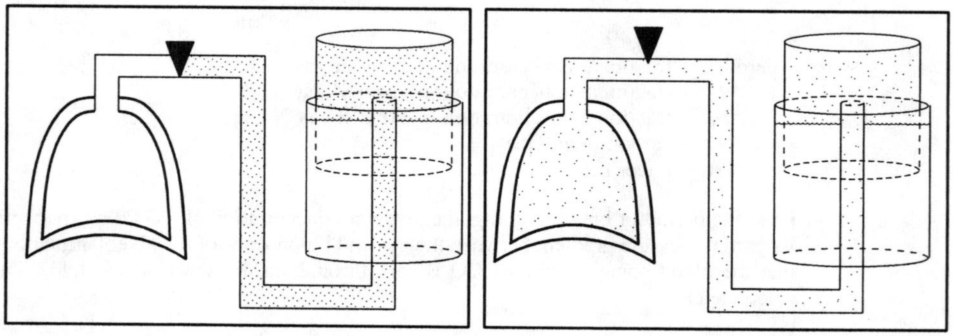

(a) (b)

FIGURE 21.15 Diagram illustrating the helium-dilution lung volume test. Prior to the test (*a*), the spirometer system is filled with a known concentration of helium. When the test is completed (*b*) after the helium has distributed and reached equilibrium with the gas in the lungs, the ratio of final to initial concentration of helium is equal to the ratio of spirometer volume to spirometer plus lung volume. In practice, the spirometer used generally has two ports, not only one as pictured, and allows for better mixing of the gases.

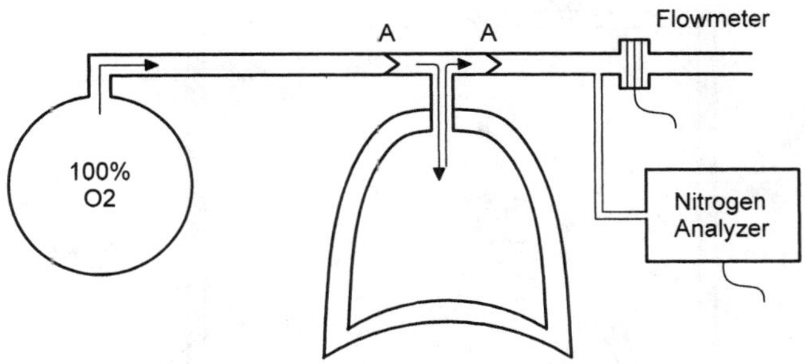

FIGURE 21.16 Equipment setup for a nitrogen-washout lung volume test. The one-way valves (A) keep separate the inspired and expired gas streams, as shown by the directional arrows. The total volume of nitrogen exhaled is calculated by integrating the continuous expired nitrogen concentration over the volume signal, which is in turn generated by integrating the flow signal over time. The nitrogen analyzer is also used to determine when the test is over, as indicated by an expired nitrogen concentration close to zero following the replacement of all lung nitrogen by other gases (oxygen and carbon dioxide).

This test requires that some of the gas be circulated through a CO_2 removal canister to prevent exhaled CO_2 from rising to uncomfortable or dangerous levels during the rebreathing period. Oxygen is added periodically to maintain a constant total system volume at end expiration. Once FRC is obtained, TLC and RV may be calculated, after the patient performs one or more VC maneuvers, as described earlier. The helium dilution system, since it is a closed-system test, is almost always performed using a volume displacement spirometer. It also requires the use of a helium analyzer.

The *nitrogen washout* method attempts to "wash out" all of the nitrogen from the lung during a period of 100 percent oxygen breathing (Fig. 21.16). All of the exhaled gas is analyzed for volume and N_2 concentration, which is integrated to give the total volume of nitrogen washed from the lung. By assuming a known initial concentration of nitrogen within the lung, and having measured its volume, the total volume of gas contained within the lung may be calculated as follows:

$$\mathrm{FRC} = \frac{V_{EN_2}}{F_{AN_2}} \tag{21.12}$$

where V_{EN_2} = the total volume of exhaled N_2 and F_{AN_2} = the initial alveolar N_2 concentration (typically taken to be ~0.79).

The total volume of exhaled N_2 may be calculated by collecting all the expired gas in a very large spirometer and making a single measurement of its nitrogen concentration, or the expired gas may be analyzed continuously for N_2 content with simultaneous use of a flow sensor. Even when the expired gas is collected separately, a continuous N_2 signal is helpful to detect leaks in the system that will interfere with accuracy and to determine when the test is completed.

Both the helium dilution and nitrogen washout tests also give information about the distribution of gas within the lung, as both will yield a curve closely following exponential decay when the helium concentration or the end-expiratory nitrogen concentration is plotted against time. Although this information may be useful in describing gas mixing within the lung, it has not been widely used in clinical practice.

A third method for measuring lung volumes uses a device known as a *body plethysmograph*, also referred to as a *body box*. The body box is a large rigid-walled structure in which the patient is seated, after which the door is closed and sealed completely. In one variety, a small hole in the wall of the cabinet leads to a spirometer or flow sensor. Respiratory efforts within the box cause changes in volume to be recorded on this spirometer as chest wall movement displaces air within the box. In a second variety of body box, there is no hole in the wall of the box and respiratory efforts instead

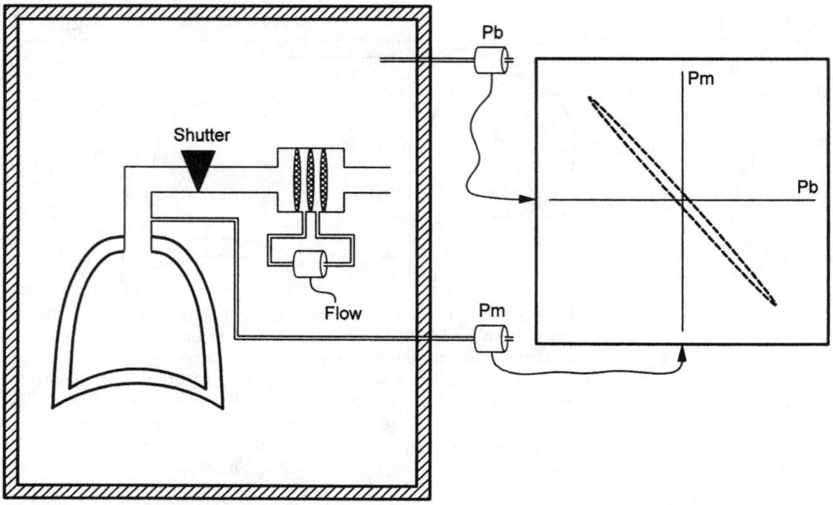

FIGURE 21.17 Illustration of a body plethysmograph, or body box. The patient sits within the sealed rigid-walled box. For measurements of lung volumes, the patient pants against the closed shutter while box pressure (P_b) and mouth pressure (P_m) are recorded as shown. The slope of the relationship between P_b and P_m determines the volume of gas being compressed within the lung. For measurements of airways resistance, the shutter is then opened and an additional recording is made, this time with flow rather than P_m on the y axis.

cause pressure swings within the tightly sealed box. At FRC, a valve at the patient's mouth is closed and the patient is instructed to pant. The rhythmic respiratory efforts cause rises and falls in alveolar pressure (measured at the mouth) and opposite pressure or volume swings in the box around the patient (Fig. 21.17). According to Boyle's law, the product of the volume being compressed and its pressure remains constant. Thus

$$\text{FRC} \times P_m = (\text{FRC} - \Delta V)(P_m + \Delta P) \tag{21.13}$$

where P_m = pressure measured at the mouth at the beginning of a pant
ΔV = change in lung volume due to compression during a single pant
ΔP = change in pressure at the mouth during a single pant

The ΔV is measured either from the change in volume through the port in the side of the box, or by calibrating changes in box pressure to volume changes. The body plethysmograph allows for rapid multiple determinations of lung volume. It requires the large rigid box, one or two pressure transducers, and usually a flow sensor or spirometer.

21.5.4 Airways Resistance

Measurements of airways resistance generally assume that Ohm's law applies and therefore that

$$P_A = \dot{V} R_{\text{aw}} \tag{21.14}$$

where P_A = alveolar pressure (measured relative to atmospheric pressure)
$\dot{V}$ = flow
R_{aw} = airways resistance

One method uses a body plethysmograph with a pneumotachograph at the mouth (Fig. 21.17). It repeats the measurements described above in the discussion of lung volumes, and adds a series of pants with the mouthpiece valve *open*, allowing for airflow. During open-valve panting, the slope of the relationship between airflow at the mouth and changes in box pressure is recorded. During closed-valve panting, the slope of the relationship between changes in alveolar pressure and changes in box pressure is recorded. The ratio of these two measurements yields the airways resistance as follows:

$$\frac{\text{Closed valve}}{\text{Open valve}} = \frac{P_A/P_{\text{Box}}}{\dot{V}/P_{\text{Box}}} = \frac{P_A}{\dot{V}} = R_{\text{aw}} \tag{21.15}$$

where P_{Box} = pressure within the box and other symbols are as shown above.

A second, nonplethysmographic, technique for measuring airways resistance makes use of an airflow perturbation device (APD) (Fig. 21.18). In this approach, the patient breathes through a mouthpiece connected to a flow sensor. A pressure transducer measures the pressure at the mouth. At the outlet of the flow sensor is a rapidly rotating wheel that intermittently creates a *partial* obstruction to airflow. During the time that there is no obstruction, the airflow is described by

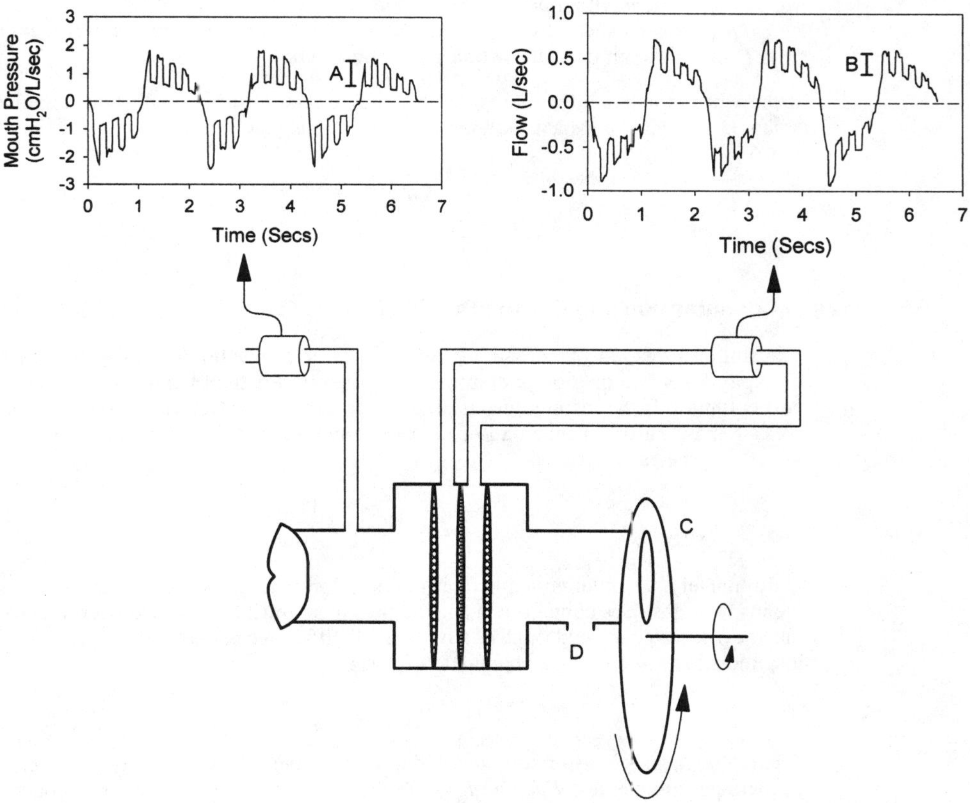

FIGURE 21.18 Diagram of the airflow perturbation device (APD). The rotating disk (C) creates changes in resistance through which the patient breathes. The ratio of delta pressure (A) to delta flow (B) with the partial obstructions is equal to the patient's airways resistance. The side opening (D) provides a flow outlet for periods when the rotating disk obstructs the outlet of the flow sensor.

$$P_A = \dot{V}_{open}(R_{aw} + R_{open}) \tag{21.16}$$

$$R_{open} = \frac{P_{m,open}}{\dot{V}_{open}} \tag{21.17}$$

where P_A = alveolar pressure (relative to atmospheric)
$\dot{V}_{open}$ = flow with no obstruction
R_{aw} = airways resistance
R_{open} = resistance of the flow sensor and the device with no obstruction
$P_{m,open}$ = mouth pressure with no obstruction

When there is a partial obstruction, airflow is described by

$$P_A = \dot{V}_{obs}(R_{aw} + R_{obs}) \tag{21.18}$$

$$R_{obs} = \frac{P_{m,obs}}{\dot{V}_{obs}} \tag{21.19}$$

where $\dot{V}_{obs}$ = flow with a partial obstruction
R_{obs} = resistance of the flow sensor and the device with a partial obstruction
$P_{m,obs}$ = mouth pressure with a partial obstruction

If the wheel spins rapidly enough so that each partial obstruction is short, the alveolar pressure is assumed to remain constant. Solving these equations yields

$$R_{aw} = \frac{P_{m,obs} - P_{m,open}}{\dot{V}_{open} - \dot{V}_{obs}} = \frac{\Delta P}{-\Delta \dot{V}} \tag{21.20}$$

21.5.5 Oxygen Consumption and Carbon Dioxide Production

The simplest method for measuring carbon dioxide production ($\dot{V}_{CO_2}$) is to collect a timed sample of expired gas in a balloon or spirometer, measuring the total volume of gas collected and its CO_2 concentration. If the inspired gas is assumed to contain no CO_2 (a reasonable assumption for most measurement purposes), then all of the CO_2 contained within the expired gas came from the $\dot{V}_{CO_2}$, which may be calculated as

$$\dot{V}_{CO_2} = \frac{VF_{ECO_2}}{\text{time}} \tag{21.21}$$

Unfortunately, a similar equation does not hold for measurements of $\dot{V}_{O_2}$. The reason for this is that the inspired gas does contain oxygen, and thus there must be included a term to account for the fact that the oxygen consumed is the relatively small difference between the large amounts of total oxygen inspired and expired. Thus, an equation for $\dot{V}_{O_2}$ is

$$\dot{V}_{O_2} = \dot{V}_i F_{IO_2} - \dot{V}_e F_{EO_2} \tag{21.22}$$

where $\dot{V}_e$ = average inspiratory ventilation in liters per minute, $\dot{V}_i$ = average expiratory ventilation in liters per minute, and F_{IO_2} and F_{EO_2} are the inspired and expired oxygen concentrations, respectively.

Note that, as described earlier, the $\dot{V}_{O_2}$ does not ordinarily equal the $\dot{V}_{CO_2}$, which means as a consequence that $\dot{V}_e$ does not equal $\dot{V}_i$. Some devices do measure separately the $\dot{V}_e$ and the $\dot{V}_i$, either with two flow sensors or a single flow sensor measuring in both directions (inspiration and expiration). However, it is possible to obtain a reasonably accurate calculation of $\dot{V}_{O_2}$ by noting that, in the steady state, there is no net consumption nor production of N_2. Thus, the following equation for N_2 consumption may be set to zero:

$$\dot{V}_{N_2} = \dot{V}_i F_{IN_2} - \dot{V}_e F_{EN_2} = 0 \tag{21.23}$$

Solving this for $\dot{V}_i$ yields

$$\dot{V}_i = \dot{V}_e \frac{F_{EN_2}}{F_{IN_2}} \tag{21.24}$$

Both F_{EN_2} and F_{IN_2} are known, so long as inspired concentrations of oxygen and carbon dioxide are known and no other gases (besides water vapor) are present, as follows:

$$F_{IN_2} = 1 - F_{IO_2} - F_{ICO_2} \tag{21.25}$$

$$F_{EN_2} = 1 - F_{EO_2} - F_{ECO_2} \tag{21.26}$$

Thus, combining these equations, $\dot{V}_{O_2}$ may be calculated as follows:

$$\dot{V}_{O_2} = \dot{V}_e \left[\frac{1 - F_{EO_2} - F_{ECO_2}}{1 - F_{IO_2} - F_{ICO_2}} \right] F_{IO_2} - \dot{V}_e F_{EO_2} \tag{21.27}$$

Measurements of $\dot{V}_{O_2}$ and $\dot{V}_{CO_2}$, while possible with balloons ("Douglas bags") or large spirometers, is more commonly performed with a flow sensor and continuous O_2 and CO_2 sampling.

21.6 OTHER DEVICES

It is not possible in this space to describe all respiratory measurements and devices. However, those that have been described do represent some of those most commonly encountered in routine clinical and research applications. Commercially available systems, complete with software and printers, make easy work of performing these common measurements in the usual manner, and are constructed with components as described above. Often, however, these systems are ill suited to making measurements that differ, even only slightly, from common practice. Thus, an understanding of their function may benefit both the researcher and the clinician.

21.7 DESIGN OF RESPIRATORY DEVICES

The designer of respiratory devices is, in many ways, faced with the same challenges as designers of other medical devices. There are many different aspects to consider, with device function being but one of these. In addition to device performance, there are safety, user interface, legal, biocompatibility, marketing, cost, and adaptability issues to face.

21.7.1 Concurrent Engineering

Modern engineering design methods often employ concurrent engineering, wherein designs are the result of teams of people representing various specialties working together toward a common goal. Especially for larger projects, teams may be formed from design engineers, marketing specialists, manufacturing engineers, packaging specialists, legal and regulatory specialists, servicing specialists, and others. Their common goal is to design an acceptable product in the shortest possible time for the smallest cost.

Before the adoption of concurrent engineering practices, each of these functions occurred in sequence: the marketing specialist surveyed past and potential users to determine what people liked and didn't like about prior models from the firm and from the competition; design engineers used computer model methods to create a new basic design; packaging specialists worked to create an attractive instrument that functioned according to user expectations; manufacturing people took this instrument and developed the means to mass-produce it and the quality assurance tests to be sure that each device met or exceeded minimum standards; the legal and regulatory specialists developed the data to meet government requirements in the country of use; sales personnel found potential users, compared the new device with the competition, and adapted the device to specific user needs; then the servicing specialists responded to customer concerns by developing quick field tests, parts kits, and field manuals to repair defective devices on site.

This procedure was time-consuming and expensive. If the manufacturing engineer found that the device could not be produced effectively as designed, then the whole design process had to be revisited. The same holds true for other steps in the process. There are legends about instruments that would not function without failure and that could not be serviced economically. The results were very expensive calamities.

Simultaneously considering all of these design aspects shortens the design process and allows industry to respond much quicker to customer needs or technological breakthroughs. What once took 5 to 10 years to develop can now be done in 1 year or less. Perhaps because of this, attention has turned to regulatory delays for device approval, which now take an unproportionately large amount of time.

Small projects by small companies may not use formal concurrent engineering teams in the way that large projects in large companies do. However, the same functions need to be represented in the same parallel way. Because many medical devices are produced by small companies with few employees, the design engineer working for one of these companies must be able to wear many hats.

21.7.2 Technical Design

Design of medical devices employs a combination of fundamental engineering principles and empirical data. Almost all respiratory devices incorporate the need to move air from one place to another. At the least, it is usually desired to reduce airflow resistance and dead volume of the air circuit. There is no substitute for a thorough knowledge of fluid mechanics in order to realize these goals. Minimizing the number of sharp bends, sudden constrictions, and obstructions can reduce resistance; reducing turbulence, tubing length, and compliant members can reduce dead volume. This knowledge comes from engineering principles and can be predicted beforehand.

A field as mature as the field of respiratory devices develops around a great deal of empirical knowledge. This information could, perhaps, be predicted from first principles, but often is determined by experimental measurement on previous devices or on prototypes of newly developed devices. Two devices where empirical knowledge is important are the body plethysmograph and the hospital ventilator. Each of these is a complex device that has undergone many embodiments over the years; each has been required to become more accurate or more versatile; each has been improved through knowledge gained by use.

The design process, then, is to begin a new device from basic engineering considerations and to make improvements based more and more on empirical information. Computer-aided design (CAD) and computer-aided manufacturing (CAM) programs are often constructed to incorporate the state of knowledge in compact and easily used form. This has the advantage of allowing the information base to be constantly accessible without dependence on the memories of any particular individual. This makes these proprietary computer programs some of the most valuable assets of any company, and gives companies that have been manufacturing the same kind of medical device for many iterations an almost insurmountable advantage over younger companies that attempt to improve the same kind of device. Thus, newer companies are usually found developing newer kinds of devices that do not require as much empirical knowledge to produce. When these newer companies develop substantial amounts of their own empirical technical knowledge, they become valuable for their technical information bases, and they become acquisition targets by larger companies wishing to develop their own devices of that type.

21.7.3 Safety

Most respiratory devices are noninvasive, which reduces the safety burden somewhat because there is no direct route for introduction of microbes into the interior of the body as with other types of devices. One of the main causes for safety concern is due to saliva: this fluid could be the pathway for stray electrical currents to enter the body. Even more important is the transmission of respiratory diseases.

Saliva contains a mixture of ionic substances and can conduct electricity. For this reason, there can be no direct electrical pathway from the ac supply to the patient. Thus use of isolation transformers should be seriously considered, especially when young children are to be measured. An alternative is the use of battery-powered units, which are especially attractive because there cannot be a stray path to ac electrical ground as long as the unit is not hooked to an electrical outlet. Although most hospitals have grounded ac outlets and ground fault interrupters, respiratory medical devices are being used in homes, clinics, and schools where such safety precautions may not be installed.

Saliva can carry disease organisms. To minimize the transmission of disease, disposable cardboard mouthpieces are used with many respiratory measurement devices. Respiratory devices should, if possible, be designed to include sterilizable parts in the airflow path. This would reduce the possibility that microbes hitchhiking on aerosol particles would be passed from patient to patient. When appropriate, disposable filters may also help prevent microbe transmission.

Some ventilators and pulmonary function equipment connects to cylinders of gas, for example, to supplemental oxygen, carbon monoxide, helium, and others. Connectors for gas cylinders of different gases are different to prevent accidentally connecting to the wrong kind of gas. If a mistake is made, asphyxiation and death may be the result. There have been hospital deaths attributed to this simple mistake. Be sure to use the correct connector.

21.7.4 Costs

The range of costs for pulmonary function equipment is from less than $300 for simple spirometers to $50,000 for hospital body plethysmographs. The cost must be appropriate for the use. In general, home-use devices are much less costly then hospital devices. Home devices are usually much simpler and may not be as accurate or reliable as hospital devices. The design engineer must know the market for a new device before investing inordinate amounts of time or before including too many options that lead to inappropriate purchase prices.

Much of the cost of a new medical device is incurred because of governmental regulatory requirements. There is no practical solution to this because approval to manufacture requires amounts of device details and human trials to assure that the device is safe and effective.

21.7.5 Materials

Many of the materials incorporated in respiratory medical devices do not directly contact the person using the device. Therefore, materials are not subject to the same biocompatibility constraints as are implantable or surface contact devices. There is a need, however, to be sure that materials are rugged enough to stand up to repeated use under sometimes frantic circumstances. An endotracheal tube, for instance, must not fail during insertion into the trachea. In addition, materials in the air passage cannot give off toxic gases or undesirable odors or flavors. Finally, parts that are to be disinfected periodically must be able to withstand the disinfection process. This is especially important given the prevalence of chemical disinfection with fairly corrosive agents.

21.7.6 Legal Liability

All medical devices are possible sources for legal action. Misuse or malfunctioning of the device can bring tort claims against the hospital, the owner of rented equipment, and the company of manufacture. The designer of respiratory medical devices must design the device to minimize accidental misuse by the user by making the device as foolproof as possible, especially during emergency situations where the attention is on the patient and not the device. If an injury can be attributed to a device, either because of malfunction or faulty design, the manufacturer can face severe penalties.

21.7.7 Optional Functions

There is a tendency on the part of a designer to incorporate many additional functions to give the device additional capabilities. There are cases where this is counterproductive, especially if the complexity of device use increases to the point where mistakes can be made. Increasing the number of options often shifts the burden for a measurement from the device to the nurse or technician. Especially if these options are rarely used, they may not be useful even if they are appealing to the designer and marketing specialists. If options are to be included, make them hierarchical: the most important functions should be obtained with little effort by the user. Additional functions can be accessed with extra effort. For instance, the display on a computerized pulmonary function device should not show all possible measurements made with the device; only the one to three most important measurements should be displayed. Additional values, tables, or graphs can be displayed with additional switches, knobs, or touching the screen. Keep it simple.

21.7.8 Calibration

All hospital equipment requires periodic calibration to assure accurate measurements. Automatic calibration features allow for quick calibration at the point of use. If the instrument can self-calibrate, even for just the most important points, then it will be much more useful than if it must be moved to a shop for calibration. Some instruments undergo self-calibration when they are turned on. Other instruments are normally always powered, and can be calibrated either by pressing a button or by a timer (although this may interfere with a measurement). More thorough calibrations still need to be completed in a biomedical maintenance laboratory.

If the device has a linear input-output response, then there are normally only two calibration points, one at the device zero (null input), and the other at the span value (maximum input). It is not uncommon that significant drift occurs in the zero value alone; it is not so common that the span value drifts independent of the zero value. That is fortunate, because a null input is easier to calibrate than is a span input. The instrument can be made to measure the output for null input and correct all readings accordingly.

Calibration of the device should be an important consideration when the product is being designed.

21.7.9 Human Factors Issues

Human factors considerations can be easily overlooked in the initial design of a medical instrument, which can result in the need for costly redesigns and delays in testing and approval. In these days when concurrent engineering practices are being applied to designs of everything from light bulbs to automobiles, it is important for the biomedical engineer to understand the medical environment in which the device is to be used. Especially important is to understand the various expectations for the device from personnel involved in its use.

The Patient. There is not one stereotypical patient, but several. Patients may be normal, healthy individuals who are undergoing a routine physical examination. The examination might be for school, for personal reasons, or for work. There may be some apprehension exhibited by such patients when confronted by medical surroundings. Especially with a new medical device or test, patients will wonder what it will do to them. These patients will likely exhibit slight hyperventilation, their respiratory systems may not be entirely normal, and the operation of the device can be viewed with suspicion.

Other patients may enter the hospital with a severe pulmonary condition. They may suffer from extreme dyspnea, can be nearly unconscious, and may exhibit symptoms of panic. Movement to standard pulmonary function testing equipment may require too much exertion, and they may be too preoccupied to fully cooperate. Special breathing maneuvers may not be within their capabilities. The operation of the device must be fast and able to occur where they are located, and, unlike the first group of patients, they see hope, not a threat in the instrument.

The Technician. The technician is highly trained, but not in the way an engineer is trained. The technician is oriented toward the patient, and can coax and cajole the best effort and the best measurements from the patient. The technician can also add consistency to the measurement, often adding a great deal of discrimination when it comes to deciding whether a measurement should be kept or repeated.

The technician does not coax and cajole instruments very well. Operation of the instrument is ideally automatic up to the point where a decision is made whether or not to repeat a test. Too many buttons and choices take too much attention away from the patient, which the technician may not like.

The Physician. Physicians shoulder the ultimate responsibility for diagnosis and treatment. They are busy, and have little time to devote to undependable instrumentation. Physicians, like technicians, feel that the medical device must do what they want instead of the other way around. Physicians must have confidence in both the technician and the device in order that they may be able to rely on measurement results and make a determination of the problem without wasting unnecessary time. The device should give unequivocal results whenever possible so as not to create confusion or uncertainty. Physicians generally prefer new medical devices that are familiar and relate to other devices and methods they have used in the past. Once accepted by a physician, the assumption is made that the new medical device gives accurate results. In reality, the device does not always need to be extremely accurate, but it must be consistent.

The Engineer. It is the engineers who know each flaw in the measurement. They want to be proud of the instrument, and want it to be perfect. Engineers wish to be appreciated for all the bits of creativity and insight that have been incorporated into the machine and are disappointed when these are not appreciated. Their tendency, in the face of a compromise between machine and patient, is to require more of the patient. This tendency must be resisted. Engineers must have faith in their abilities, patience with those in the medical profession, and care in their approach. Above all, they must realize that the medical world is different from their own, and that, while nothing can stand between medicine and a technique that it craves, there is no path more strewn with obstacles than the path toward acceptance of a new medical device.

21.7.10 Physical Characteristics

Aesthetic design is important for device acceptance. Many respiratory devices are relatively small, and compactness is usually a positive attribute. There was a time when the very size and massive appearance of a device connoted robustness, but styles change, and now the appearance of elegance is in vogue. The device must look clean, small, lightweight, and sanitary. Computer displays should have an appearance similar to popular computer programs of the day. The color of the device should be rather neutral instead of gaudy.

Most medical devices are accepted better if they are lightweight. Respiratory devices may be used in the home, and home may be located up several flights of stairs. Even hospital equipment must be moved for cleaning or storage, so lighter devices are appreciated. Portability is beneficial.

Medical devices should be quiet. They should not appear to be contraptions to nurses and technical staff, and they should not cause loss of confidence by patients. They should not add to the din of a hospital environment, especially during emergencies, when communications among health care professionals are most critical.

Devices must be rugged. They should be designed not to break if they are to fall on the floor during a medical emergency. They must be able to withstand fluids that are sometimes all around in a medical setting. They must be able to tolerate electrical surges, and they should operate even when placed in unconventional orientations.

If they can be made to work correctly in dusty or dirty environments, or at extreme temperatures, then they can be used under severe field conditions. However, most medical devices are expected to be used in clean conditions at moderate temperatures. Such conditions prevail in most modern hospitals in the developed world. In third-world countries or during combat, however, medical environments are not as well controlled.

21.7.11 Governmental Regulatory Requirements

The U.S. Food and Drug Administration (FDA) is the main regulatory body for medical device approval in the United States. It is the job of the FDA to determine that a new medical device is both safe and effective. Each medical device must meet criteria of both safety (it can do no harm) and effectiveness (it must do what it purports to do).

Most new medical devices must undergo a process of premarket notification. Certain Class I devices (see the FDA Web site at www.fda.gov) are exempt from this requirement. There are certain respiratory-related devices in the list of Class I devices, but, in general, they are ancillary to respiratory diagnostic measurement and health care.

If the medical device is intended to be used in a new way or is based on a fundamental scientific technology different from that of other devices, then the approval process is extremely thorough and is called premarket approval (PMA). If the device can be justified as a minor modification of a device that has received prior FDA approval for manufacture, then it undergoes an abbreviated 501(k) approval process. Neither of these processes is a trivial step, and specialists are often employed just to guide the device through to approval.

Medical device approval processes in other countries may or may not be similar to the process in the United States. Mutual recognition agreements may be negotiated between different governments to allow judgments by national conformity assessment bodies to be accepted in other countries.

REFERENCES

American Association for Respiratory Care, 1994, "Clinical Practice Guideline: Body Plethysmography," *Respir. Care*, **39**(12):1184–1190.

American Association for Respiratory Care, 1994, "Clinical Practice Guideline: Static Lung Volumes," *Respir. Care*, **39**(8):830–836.

American Association for Respiratory Care, 1996, "Clinical Practice Guideline: Spirometry," *Respir. Care*, **41**(7): 629–636.

American Association for Respiratory Care, 1999, "Clinical Practice Guideline: Single-breath Carbon Monoxide Diffusing Capacity," *Respir. Care*, **44**(5):539–546.

American Thoracic Society, 1991, "Lung Function Testing: Selection of Reference Values and Interpretive Strategies," *Am. Rev. Respir. Dis.*, **144**(5):1202–1218.

American Thoracic Society, 1995, "Single Breath Carbon Monoxide Diffusing Capacity (transfer factor)," *Am. Rev. Respir. Dis.*, **152**:2185–2198.

American Thoracic Society, 1995, "Standardization of Spirometry—1994 Update," *Am. Rev. Respir. Dis.*, **152**: 1107–1136.

Clausen, J. L. (ed.), 1982, *Pulmonary Function Testing Guidelines and Controversies: Equipment, Methods, and Normal Values*, Grune & Stratton, Orlando.

CNN, 2000, "*Oxygen tank mix-up blamed in deaths of Ohio nursing home residents*," Dec. 14, 2000; http://www.cnn.com/2000/US/12/14/nursinghome.deaths.ap.

Coates, A. L., R. Peslin, D. Rodenstein, and J. Stocks, 1997, "Measurement of Lung Volumes by Plethysmography," *Eur. Respir. J.*, **10**:1415–1427.

Dubois, A. B., S. Y. Botelho, and J. H. Comroe, 1956, "A New Method for Measuring Airway Resistance in Man Using a Body Plethysmograph: Values in Normal Subjects and in Patients with Respiratory Disease," *J. Clin. Invest.*, **35**:327–335.

Fessler, H. E., and D. M. Shade, 1998, "Measurement of Vascular Pressure," in Tobin, M. J. (ed.), *Principles and Practice of Intensive Care Monitoring*, McGraw-Hill, New York.

Gardner, R. M., J. L. Hankinson, and B. J. West, 1980, "Evaluating Commercially Available Spirometers," *Am. Rev. Respir. Dis.*, **121**(1):73–82.

Geddes, L. A., and L. E. Baker, 1989, *Principles of Applied Biomedical Instrumentation*, 3d ed., J. Wiley, New York.

Johnson, A. T., and J. D. Bronzino, 1995, "Respiratory System," in Bronzino, J. D. (ed.), *The Biomedical Engineering Handbook*, CRC Press and IEEE Press, New York.

Johnson, A. T., 1986, "Conversion Between Plethysmograph and Perturbational Airways Resistance Measurements," *IEEE Trans. Biomed*, BME **33**:803–806.

Johnson, A. T., and C.-S. Lin, 1983A, "Airflow Perturbation Device for Measuring Airway Resistance of Animals," *Trans. ASAE*, **26**:503–506.

Johnson, A. T., and C.-S. Lin, 1983B, "Airflow Resistance of Conscious Boars," *Trans. ASAE*, **26**:1150–1152.

Johnson, A. T., and J. M. Milano, 1987, "Relation Between Limiting Exhalation Flow Rate and Lung Volume," *IEEE Trans. Biomed.*, BME **34**:257–258.

Johnson, A. T., C.-S. Lin, and J. N. Hochheimer, 1984, "Airflow Perturbation Device for Measuring Airways Resistance of Humans and Animals," *IEEE Trans. Biomed.*, BME **31**:622–626.

Johnson, A. T., H. M. Berlin, and S. A. Purnell, 1974, "Perturbation Device for Noninvasive Measurement of Airway Resistance," *Med. Instr.* **8**:141.

Johnson, A. T., J. N. Hochheimer, and J. Windle, 1984, "Airflow Perturbation Device for Respirator Research and Medical Screening," *J. ISRP*, **2**:338–346.

Lausted, C. G., and A. T. Johnson, 1999, "Respiratory Resistance Measured by an Airflow Perturbation Device," *Physiol. Meas.*, **20**:21–35.

Leff, A. R., and P. T. Schumacker, 1993, *Respiratory Physiology: Basics and Applications*, W. B. Saunders, Philadelphia.

Lin, C.-S., A. T. Johnson, and N. Yaramanoglu, 1985, "Model Analysis of the Airflow Perturbation Device," *Inn. Tech. Biol. Med.*, **6**:461–472.

Meneely, G. R., and N. L. Kaltreider, 1949, "The Volume of the Lung Determined by Helium Dilution," *J. Clin. Invest.*, **28**:129–139.

Miller, M. R., and A. C. Pincock, 1986, "Linearity and Temperature Control of the Fleisch Pneumotachograph," *J. Appl. Physiol.*, **60**(2):710–715.

Morris, A. H., R. E. Kanner, R. O. Crapo, and R. M. Gardner, 1984, "Clinical Pulmonary Function Testing," *A Manual of Uniform Laboratory Procedures*, 2d ed., Intermountain Thoracic Society, Salt Lake City.

Neas, L. M., and J. Schwartz, 1998, "Pulmonary Function Levels as Predictors of Mortality in a National Sample of U.S. Adults," *Am. J. Epidemiol.*, **147**(11):1011–1018.

Nelson, S. B., R. M. Gardner, R. O. Crapo, and R. L. Jensen, 1990, "Performance Evaluation of Contemporary Spirometers," *Chest*, **97**:288–297.

Ogilvie, C. M., R. E. Forster, W. S. Blakemore, and J. W. Morton, 1957, "A Standardized Breath Holding Technique for the Clinical Measurement of the Diffusing Capacity of the Lung for Carbon Monoxide," *J. Clin. Invest.*, **36**:1–17.

Porszasz, J., T. J. Barstow, and K. Wasserman, 1984, "Evaluation of a Symmetrically Disposed Pitot Tube Flowmeter for Measuring Gas Flow During Exercise," *J. Appl. Physiol.*, **77**(6):2659–2665.

Punjabi, N. M., D. Shade, and R. A. Wise, 1998, "Correction of Single-Breath Helium Lung Volumes in Patients with Airflow Obstruction," *Chest*, **114**:907–918.

Roussos, C., and P. T. Macklem, 1982, "The Respiratory Muscles," *New Eng. J. Med.*, **307**:786–797.

Ruppel, G. L., 1997, *Manual of Pulmonary Function Testing*, 7th ed., Mosby Year Book, St. Louis.

Rutala, D. R., W. A. Rutala, D. J. Weber, and C. A. Thomann, 1991, "Infection Risks Associated with Spirometry," *Infect. Control. Hosp. Epidemiol.*, **12**:89–92.

Turner, M. J., I. M. MacLeod, and A. D. Rothberg, 1989, "Effects of Temperature and Composition on the Viscosity of Respiratory Gases," *J. Appl. Physiol.*, **67**(1):472–477.

Wanger, J. (ed.), 1998, *Pulmonary Function Laboratory Management and Procedure Manual*, American Thoracic Society, New York.

Wanger, J., 1996, *Pulmonary Function Testing: A Practical Approach*, 2d ed., Williams & Wilkins, Baltimore.

Wilson, A. F. (ed), 1985, *Pulmonary Function Testing Indications and Interpretations*, Grune & Stratton, Orlando.

Yeh, M. P., T. D. Adams, R. M. Gardner, and F. G. Yanowitz, 1984, "Effect of O_2, N_2, and CO_2 Composition on Nonlinearity of Fleisch Pneumotachograph Characteristics," *J. Appl. Physiol.*, **56**(5):1423–1425.

Yeh, M. P., R. M. Gardner, T. D. Adams, and F. G. Yanowitz, 1982, "Computerized Determination of Pneumotachometer Characteristics Using a Calibrated Syringe," *J. Appl. Physiol.*, **53**(1):280–285.

CHAPTER 22
DESIGN OF CONTROLLED-RELEASE DRUG DELIVERY SYSTEMS

Steve I. Shen, Bhaskara R. Jasti, and Xiaoling Li
University of the Pacific, Stockton, California

With the advances in science and technology, many new chemical molecules are being created and tested for medical use. The United States Food and Drug Administration (FDA) approved 22 to 53 new molecular entities each year between 1993 and 1999.[1] Creation of these active ingredients is only part of the arsenal against diseases. Every drug molecule needs a delivery system to carry the drug to the site of action upon administration to the patient. Delivery of the drugs can be achieved using various types of dosage forms including tablets, capsules, creams, ointments, liquids, aerosols, injections, and suppositories. Most of these conventional drug delivery systems are known to provide immediate release of the drug with little or no control over delivery rate. To achieve and maintain therapeutically effective plasma concentrations, several doses are needed daily, which may cause significant fluctuations in plasma levels (Fig. 22.1). Because of these fluctuations in drug plasma levels, the drug level could fall below the minimum effective concentration (MEC) or exceed the minimum toxic concentration (MTC). Such fluctuations result in unwanted side effects or lack of intended therapeutic benefit to the patient.

Sustained-release and controlled-release drug delivery systems can reduce the undesired fluctuations of drug levels, thus diminishing side effects while improving the therapeutic outcome of the drug (Fig. 22.1). The terms *sustained release* and *controlled release* refer to two different types of drug delivery systems, although they are often used interchangeably. Sustained-release dosage forms are systems that prolong the duration of the action by slowing the release of the drug, usually at the

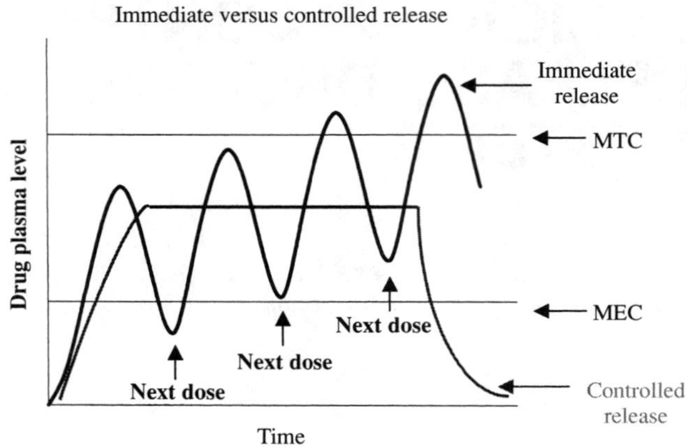

FIGURE 22.1 Therapeutic or toxic levels and immediate- versus controlled-release dosage form.

cost of delayed onset and its pharmacological action. Controlled-release drug systems are more sophisticated than just simply delaying the release rate and are designed to deliver the drug at specific release rates within a predetermined time period. Targeted delivery systems are also considered as a controlled delivery system, since they provide spatial control of drug release to a specific site of the body.

Advantages of controlled release drug delivery systems include delivery of a drug to the required site, maintenance of drug levels within a desired range, reduced side effects, fewer administrations, and improved patient compliance. However, there are potential disadvantages that should not be overlooked. Disadvantages of using such delivery systems include possible toxicity of the materials used, dose dumping, requirement of surgical procedures to implant or remove the system, and higher manufacturing costs. In the pharmaceutical industry, design and development of controlled/sustained-release delivery systems have been used as a strategic means to prolong the proprietary status of drug products that are reaching the end of their patent life. A typical example is modifying an existing drug product that requires several doses a day to a single daily dosing to maintain the dominance over generic competition. For some drugs, controlled delivery is necessary, since immediate release dosage forms cannot achieve the desired pharmacological action. These include highly water soluble drugs that need slower release and longer duration of action, highly lipophilic drugs that require enhancement of solubility to achieve therapeutic level, short half-life drugs that require repeated administration, and drugs with nonspecific action that require the delivery to target sites.

An ideal drug delivery system should deliver precise amounts of a drug at a preprogrammed rate to achieve a drug level necessary for treatment of the disease. For most drugs that show a clear relationship between concentration and response, the drug concentration will be maintained within the therapeutic range, when the drug is released by zero-order rate. In order to design a controlled-release delivery system, many factors such as physicochemical properties of the drug, route of drug administration, and pharmacological and biological effects must be considered.

22.1 PHYSICOCHEMICAL PROPERTIES OF DRUG

Physicochemical properties such as solubility, stability, lipophilicity, and molecular interactions play a major role in biological effectiveness of a drug. Solubility is a measure of the amount of solute that can be dissolved in the solvent. For a drug to be absorbed, it must first dissolve in the physiological fluids of the body at a reasonably fast dissolution rate. Drug molecules with very low aqueous solubility often have lower bioavailability because of the limited amount of dissolved drug at the site of absorption. In general, drugs with lower than 10 mg/mL in aqueous solutions are expected to exhibit low and erratic oral bioavailability.

Once the drug is administered, biological fluids that are in direct contact with a drug molecule may influence the stability of the drug. Drugs may be susceptible to both chemical and enzymatic degradation, which results in a loss of activity of the drug. Drugs with poor acidic stability, when coated with enteric coating materials, will bypass the acidic stomach and release the drug at a lower portion of the gastrointestinal (GI) tract. Drugs can also be protected from enzymatic cleavage by modifying the chemical structure to form prodrugs.

The ability of a drug to partition into a lipid phase can be evaluated by the distribution of drug between lipid and water phase at equilibrium. A distribution constant, the partition coefficient K, is commonly used to describe the equilibrium of drug concentrations in two phases.

$$K = \frac{\text{Drug}_{\text{lipid}}}{\text{Drug}_{\text{water}}} \tag{21.1}$$

The partition coefficient of a drug reflects the permeability of the drug through the biological membrane and/or the polymer membrane. Commonly, the partition coefficient is determined by equilibrating the drug in a saturated mixture of octanol (lipid phase) and water. Drugs with a high partition coefficient can easily penetrate biological membranes, as they are made of lipid bilayers, but are unable to proceed further because of a higher affinity to the membrane than the aqueous surroundings. Drugs with a low partition coefficient can easily move around the aqueous areas of the body, but will not cross the biological membranes easily.

In addition to the inherent properties of drug molecules, molecular interactions such as drug-drug, drug-protein, and drug–metal ion binding are important factors that can significantly change the pharmacokinetic parameters of a drug. These factors should also be taken into consideration in designing controlled drug delivery systems.

22.2 ROUTES OF DRUG ADMINISTRATION

Various routes of administration pose different challenges for product design. As a result of the different barriers and pathways involved, selection of an administration route is an important factor for design of drug delivery system. For example, the oral route is most widely utilized route because of its ease of administration and the large surface area of the GI tract (200 m^2). The presence of microvilli makes this the largest absorptive surface of the body (4500 m^2).[2] The challenges of oral administration are short GI transit time, extreme acidic pH, abundant presence of digestive enzymes, and first-pass metabolism in the liver. Several products were designed to prolong the retention time of a drug in the gastrointestinal tract. A hydrodynamically balanced drug-delivery system (HBS) is designed to achieve bulk density of less than 1 when contacted with gastric fluids rendering the drug formulation buoyant. This dosage form is also called *floating* capsules or tablets because of this characteristic.[3]

Another commonly used route for drug delivery is parenteral administration. The routes used for parenteral therapy include intradermal, subcutaneous, intravenous, intracardiac, intramuscular, intra-arterial, and intrasynovial. Parenteral administrations offer immediate response, in such situations as cardiac arrest or shock, and good bioavailability for drugs that undergo degradation by digestive enzymes in the GI tract. The disadvantages of parenteral administrations are difficulty of administration, requirement of sterile conditions, and cost of manufacturing.

In addition, skin, with surface area of 2 m^2, is a commonly used route for drug delivery. Advantages of the transdermal route include avoidance of the first-pass effect, potential of multiday therapy with a single application, rapid termination of drug effects, and easy identification of medication in an emergency. The limitations are skin irritation and/or sensitization, variation of intra- and interindividual percutaneous absorption efficiency, the limited time that the patch can remain affixed, and higher cost.[4]

Most of the controlled-release delivery systems available in the market for systemic delivery of drugs utilize oral, parenteral, and transdermal route for their administration. Advances in biotechnology produced many gene, peptide, and protein drugs with specific demands on route of delivery. Thus, other routes such as buccal, nasal, ocular, pulmonary, rectal, and vaginal are gaining more attention.

22.3 PHARMACOLOGICAL AND BIOLOGICAL EFFECTS

It is important to consider the human dimension in the design of the drug delivery systems. Biological factors, such as age, weight, gender, ethnicity, physiological processes, and disease state, will change the pharmacokinetics and pharmacodynamics of a drug. For example, dosing newborn infants requires caution because of their immature hepatic function and higher water content in the body. Geriatric patients may suffer from reduced sensitivity of certain receptors that may lead to insensitivity to certain drugs. It has been found that different ethnic groups respond to drugs differently. Diuretics and calcium channel blockers are recommended as first-line therapy in hypertensive Black patients, while beta blockers work better for Caucasian patients. Pathological changes may influence the distribution and bioavailability of the drug by altering the physiological process. Decreased kidney and/or liver functions will affect the clearance of many drugs.

In this chapter, the discussion of the design of drug delivery systems is based on various approaches: prodrug approach, diffusion-controlled reservoir and matrix systems, dissolution/coating-controlled systems, osmotically controlled systems, ion-exchange resin systems, and approaches for macromolecular drug delivery. The aim of this chapter is to introduce the basic concepts for the designs of various drug delivery systems. Readers can refer to the references for further details.[2–9]

22.4 PRODRUG

The molecule with the most potent form does not always have the desired physicochemical properties needed for drug dissolution and/or absorption. If fact, of all the pharmaceutically active ingredients, 43 percent are sparingly water soluble or insoluble in water. In the prodrug approach for drug delivery, active ingredients are chemically modified by connecting specialized functional groups that will be removed in the body after administration, releasing the parent molecule.[8] These latent groups are used in a transient manner to change the properties of the parent drug to achieve a specific function, e.g., alter permeability, solubility, or stability. After the prodrug has achieved its goal, the functional group is removed in the body (enzymatic cleavage or hydrolysis) and the parent compound is released to elicit its pharmacological action (Fig. 22.2).

The prodrug approach has been used for one or more of the following reasons:

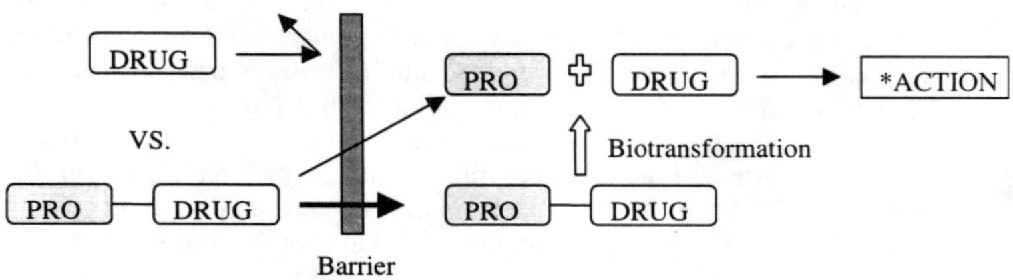

FIGURE 22.2 Schematic of prodrug.

To change half-life. Half-life is defined as the time required by the biological system for removing 50 percent of administered drug. Drugs with very short half-life may not be therapeutically beneficial unless this characteristic is improved. Attaching the drug to a polymer as part of a pendent will enhance its half-life. Modification of the drug to protect the site of degradation or metabolism is another method to achieve longer half-life.

To cross a biological barrier. Drugs with unbalanced hydrophilic or hydrophobic properties will not effectively cross the biological barriers. Attachment of functional groups can change the properties of the parent drug and allow the prodrug to cross the barrier.

To increase retention time. When intended for a part of the body with high tissue turnover rate, such as intestinal mucosa, a drug linked to a mucoadhesive polymer can increase adhesion to the site and increase bioavailability of a drug that has low residence time.

To target a specific site. Connecting specialized functional groups that have site-specific affinity (peptide, antibody, etc.) can allow the parent drug to be delivered to the targeted area of the body to produce site specific therapeutic action.

22.5 DIFFUSION-CONTROLLED DELIVERY SYSTEMS

Diffusion process has been utilized in design of controlled release drug delivery systems for several decades. This process is a consequence of constant thermal motion of molecules, which results in net movement of molecules from a high concentration region to a low concentration region. The rate of diffusion is dependent on temperature, size, mass, and viscosity of the environment.

Molecular motion increases as temperature is raised as a result of the higher average kinetic energy in the system.

$$E = \frac{kT}{2} = \frac{mv^2}{2} \tag{22.2}$$

where E = kinetic energy
k = Boltzmann's constant
T = temperature
m = mass
v = velocity

This equation shows that an increase in temperature is exponentially correlated to velocity (v^2). Size and mass are also significant factors in the diffusion process. At a given temperature, the mass of molecule is inversely proportional to velocity [Eq. 22.2]. Larger molecules interact more with the surrounding environment, causing them to have slower velocity. Accordingly, large molecules diffuse much slower than light and small particles. The viscosity of the environment is another important parameter in diffusion, since the rate of molecular movement is associated with the viscosity of the environment. Diffusion is fastest in the gas phase, slower in the liquid phase, and slowest in the solid phase.

Mathematically, the rate of drug delivery in diffusion-controlled delivery systems can be described by Fick's laws. Fick's first law of diffusion is expressed as[9]:

$$J = -D\frac{dC}{dx} \tag{22.3}$$

where J = flux of diffusion
D = diffusivity of drug molecule
$\frac{dC}{dx}$ = concentration gradient of the drug molecule across diffusional barrier with thickness dx

According to the diffusion principle, controlled-release drug delivery systems can be designed as a reservoir system or a matrix system. Drugs released from both reservoir and matrix type devices follow the principle of diffusion, but they show two different release patterns as shown in Fig. 22.3.

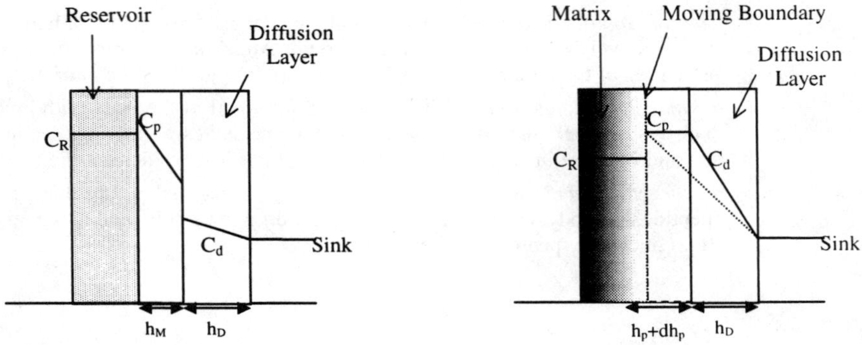

FIGURE 22.3 Schematic illustrations of reservoir versus matrix systems.

In Fig. 22.3, C_R is drug concentration in the reservoir or matrix compartment, C_P is solubility of drug in the polymer phase, C_D is the concentration in the diffusion layer, h_m is the thickness of the membrane, h_d is thickness of the diffusion layer, and $h_P + dh_P$ indicates the changing thickness of the depletion zone of matrix.

In a reservoir system, if the active agent is in a saturated state, the driving force is kept constant until it is no longer saturated. For matrix systems, because of the changing thickness of the depletion zone, release kinetics is a function of the square root of time.[10] A typical reservoir system for transdermal delivery consists of a backing layer, a rate-limiting membrane, a protective liner, and a reservoir compartment. The drug is enclosed within the reservoir compartment and released through a rate-controlling polymer membrane (Fig. 22.4).

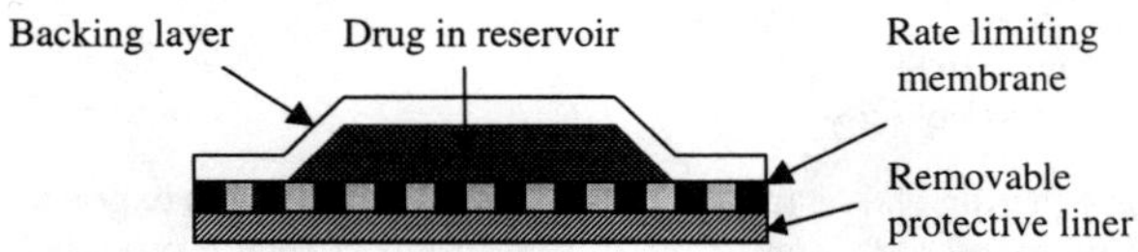

FIGURE 22.4 An example of a reservoir-type transdermal drug delivery system.

Membranes used to enclose the device can be made from various types of polymers. The rate of release can be varied by selecting the polymer and varying the thickness of the rate-controlling membrane. The drug in the reservoir can be in solid, suspension, or liquid form.

Analysis of diffusion-controlled reservoir or matrix drug delivery systems requires a few assumptions:

1. The diffusion coefficient of a drug molecule in a medium must be constant.
2. The controlled drug release must have a pseudo-steady state.
3. Dissolution of solid drug must occur prior to the drug release process.
4. The interfacial partitioning of the drug is related to its solubility in polymer and in solution as defined by

$$K = \frac{C_s}{C_p} \tag{22.4}$$

where K = partition coefficient of the drug molecule from polymer to solution
C_s = solubility of drug in the solution phase
C_p = solubility of drug in polymer phase

With the above assumptions, the cumulative amount Q of drug released from a diffusion-controlled reservoir-type drug delivery device with a unit surface area can be described as follows[2]:

$$Q = \frac{C_p K D_d D_m}{K D_d h_m + D_m h_d} t - \frac{D_d D_m}{K D_d h_m + D_m h_d} \int_0^t C_{b(t)}\, dt \tag{22.5}$$

where D_m = diffusivity of the drug in a polymer membrane with thickness h_m
D_d = diffusivity of hydrodynamic diffusion layer with thickness h_d
C_b = concentration of drug in reservoir side
t = time

Under a sink condition, where $C_{b(t)} \approx 0$ or $C_s \gg C_{b(t)}$, Eq. (22.5) is reduced to

$$Q = \frac{C_p K D_d D_m}{K D_d h_m + D_m h_d} t \tag{22.6}$$

This relationship shows that release of drug can be a constant, with the rate of drug release being

$$\frac{Q}{t} = \frac{C_p K D_d D_m}{K D_d h_m + D_m h_d} \tag{22.7}$$

In extreme cases, the rate of release may depend mainly on one of the layers, either the polymer membrane layer or the hydrodynamic diffusion layer. If the polymer membrane is the rate-controlling layer, $K D_d h_m \gg D_m h_d$, the equation can be simplified to:

$$\frac{Q}{t} = \frac{C_p D_m}{h_m} \tag{22.8}$$

which shows that the release rate is directly proportional to the solubility of the drug in polymer and inversely proportional to thickness of the polymer membrane.

Delivery systems designed on this principle can be administered by different routes: intrauterine such as Progestasert, implants such as Norplant, transdermal such as Transderm-Nitro, and ocular such as Ocusert.

A matrix system, often described as monolithic device, is designed to uniformly distribute the drug within a polymer as a solid block. Matrix devices are favored over other design for their simplicity, low manufacturing costs, and lack of accidental dose dumping, which may occur with reservoir systems when the rate controlling membrane ruptures.

The release properties of the device depend highly upon the structure of the matrix: whether it is porous or nonporous. The rate of drug release is controlled by the solubility of the drug in the polymer and the diffusivity of the drug through the polymer for nonporous system. For a porous matrix, the solubility of the drug in the network and the tortuosity of the network add another dimension to affect the rate of release. In addition, drug loading influences the release, since high loading can complicate the release mechanism because of formation of cavities as the drug is leaving the device. These cavities will fill with fluids and increase the rate of release.

The cumulative amount released from a matrix-controlled device is described by[2]

$$Q = \left(C_A - \frac{C_p}{2}\right) h_p \tag{22.9}$$

where C_A is initial amount of drug, C_P is solubility of drug in polymer, and h_p is a time dependent variable defined by

$$h_p^2 + \frac{2(C_A - C_p)D_p h_d h_p}{\left(C_A - \frac{C_p}{2}\right)D_d \bar{k} K} = \frac{2C_p D_p}{C_A - \frac{C_p}{2}}\,t \tag{22.10}$$

where $\bar{k}$ is a constant for relative magnitude of the concentration in the diffusion layer and depletion zone, D_p is the diffusivity of drug in the polymer devices, and other parameters are the same as described for Eqs. (22.4) to (22.9). At a very early stage of the release process, when there is a very thin depletion zone, the following will be true:

$$h_p^2 \ll \frac{2(C_A - C_p)D_p h_d h_p}{\left(C_A - \frac{C_p}{2}\right)D_d \bar{k} K}$$

Equation (22.10) can be reduced to

$$h_p \approx \frac{C_p D_d \bar{k} K}{(C_A - C_p)h_d} \tag{22.11}$$

and placing Eq. (22.11) into Eq. (22.9) gives

$$\frac{Q}{t} = \frac{C_p D_d \bar{k} K}{h_d} \qquad \left(\text{if } C_A - C_p \approx C_A - \frac{C_p}{2}\right) \tag{22.12}$$

Since $KC_p = C_s$, Eq. (22.12) becomes

$$\frac{Q}{t} = \frac{C_s D_d \bar{k}}{h_d} \tag{22.13}$$

The $\bar{k}$ term implies that the matrix system is more sensitive to the magnitude of concentration difference between depletion and diffusion layers.

If

$$h_p^2 \gg \frac{2(C_A - C_p)D_p h_d h_p}{\left(C_A - \frac{C_p}{2}\right)D_d \bar{k} K}$$

where the depletion zone is much larger and the system has a very thin diffusion layer, Eq. (22.10) becomes

$$h_p \approx \left(\frac{2C_p D_p}{\left(C_A - \frac{C_p}{2}\right)}\,t\right)^{1/2} \tag{22.14}$$

and placing Eq. (22.14) into Eq. (22.9) makes

$$\frac{Q}{t^{1/2}} = [(2C_A - C_p)C_p D_p]^{1/2} \tag{22.15}$$

Equation (22.15) indicates that after the depletion zone is large enough, the cumulative amount of drug released (Q) is proportional to the square root of time ($t^{1/2}$).

22.6 DISSOLUTION/COATING-CONTROLLED DELIVERY SYSTEMS

Controlled release of drug can be achieved by utilizing the rate-limiting step in the dissolution process of a solid drug with relatively low aqueous solubility. The dissolution rate can be quantitatively described by the Noyes-Whitney equation as follows.

$$\frac{dC}{dt} = \frac{DA}{h}(C_0 - C_t) \tag{22.16}$$

where $\frac{dC}{dt}$ = rate of drug dissolution
D = diffusion coefficient of drug in diffusion layer
h = thickness of diffusion layer
A = surface area of drug particles
C_0 = saturation concentration of the drug in diffusion layer
C_t = concentration of drug in bulk fluids at time t

The surface area A of the drug particle is directly proportional to the rate of dissolution. For a given amount of drug, reducing the particle size results in a higher surface area and faster dissolution rate. However, small particles tend to agglomerate and form aggregates. Using a specialized milling technique with stabilizer and other excipients, aggregation can be prevented to make microparticles smaller than 400 nm in diameter to improve the dissolution of the drug in the body.

The saturation solubility C_0 can also be manipulated to change the rate of dissolution. Both the physical and chemical properties of a drug can be modified to alter the saturation solubility. For example, salt forms of a drug are much more soluble in an aqueous environment than the parent drug. The solubility of a drug can also be modified when the drug forms a complex with excipients, resulting in a complex with solubility different from the drug itself.

Controlled or sustained release of drug from delivery systems can also be designed by enclosing the drug in a polymer shell or coating. After the dissolution or erosion of the coating, drug molecules become available for absorption. Release of drug at a predetermined time is accomplished by controlling the thickness of coating. In Spansule® systems, drug molecules are enclosed in beads of varying thickness to control the time and amount of drug release. The encapsulated particles with thin coatings will dissolve and release the drug first, while a thicker coating will take longer to dissolve and will release the drug at later time. Coating-controlled delivery systems can also be designed to prevent the degradation of the drug in the acidic environment of the stomach, which can reach as low as pH 1.0. Such systems are generally referred as *enteric-coated systems*. In addition, enteric coating also protects the stomach from ulceration caused by drug agents. Release of the drug from coating-controlled delivery systems may depend upon the polymer used. A combination of diffusion and dissolution mechanisms may be required to define the drug release from such systems.

22.7 BIODEGRADABLE/ERODIBLE DELIVERY SYSTEMS

Biologically degradable systems contain polymers that degrade into smaller fragments inside the body to release the drug in a controlled manner. Zero-order release can be achieved in these systems as long as the surface area or activity of the labile linkage between the drug and the polymeric backbone are kept constant during drug release. Another advantage of biodegradable systems is that, when formulated for depot injection, surgical removal can be avoided. These new delivery systems can protect and stabilize bioactive agents, enable long-term administration, and have potential for delivery of macromolecules.

22.8 OSMOTIC PUMP

This type of delivery device has a semipermeable membrane that allows a controlled amount of water to diffuse into the core of the device filled with a hydrophilic component.[11] A water-sensitive component in the core can either dissolve or expand to create osmotic pressure and push the drug out of the device through a small delivery orifice, which is drilled to a diameter that correlates to a specific rate. In an elementary osmotic pump, the drug molecule is mixed with an osmotic agent in the core of the device (Fig. 22.5*a*). For drugs that are highly or poorly water soluble, a two-compartment push-pull bilayer system has been developed, in which the drug core is separated from the push compartment (Fig. 22.5*b*). The main advantage of the osmotic pump system is that constant release rate can be achieved, since it relies simply on the passage of water into the system, and the human body is made up of 70 percent water. The release rate of the device can be modified by changing the amount of osmotic agent, surface area and thickness of semipermeable membrane, and/or the size of the hole.

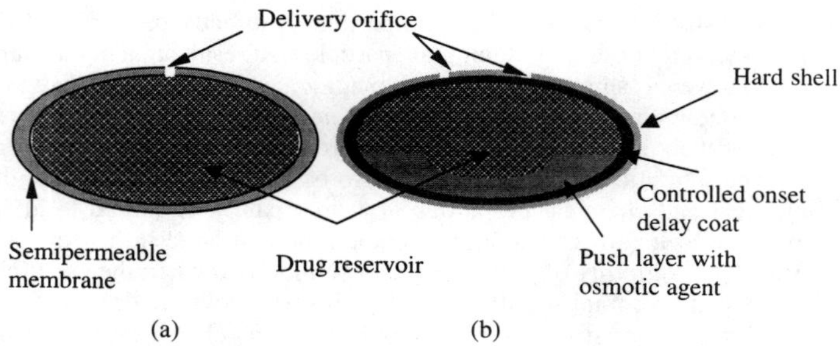

FIGURE 22.5 Schematic illustration of an elementary osmotic pump (*a*) and a push-pull osmotic pump device (*b*).

The rate of water diffusing into the osmotic device is expressed as[12]

$$\frac{dV}{dt} = \left(\frac{AK}{h}\right)(\Delta\pi - \Delta P) \tag{22.17}$$

where $\frac{dV}{dt}$ = change of volume overchange in time
A, K, h = area, permeability, and thickness of membrane, respectively
$\Delta\pi$ = difference in osmotic pressure between drug device and release environment
ΔP = difference in hydrostatic pressure

If the osmotic pressure difference is much larger than the hydrostatic pressure difference ($\Delta\pi \gg \Delta P$), the equation can be simplified to

$$\frac{dV}{dt} = \left(\frac{AK}{h}\right)(\Delta\pi) \tag{22.18}$$

The rate at which the drug is pumped out the device dM/dt, cab be expressed as

$$\frac{dM}{dt} = \left(\frac{dV}{dt}\right)C \tag{22.19}$$

where C is the drug concentration. As long as the osmotically active agent provides the constant osmotic pressure, the delivery system will release the drug at a zero-order rate. The zero-order delivery rate can be expressed as

$$\left(\frac{dM}{dt}\right)_z = \frac{AK}{h}\pi_s C \tag{22.20}$$

where π_s is osmotic pressure generated by saturated solution and all other symbols are the same as described earlier.

22.9 ION EXCHANGE RESINS

The ion exchange resin system can be designed by binding drug to the resin. After the formation of a drug/resin complex, a drug can be released by an ion exchange reaction with the presence of counterions. In this type of delivery system, the nature of the ionizable groups attached determines the chemical behavior of an ion exchange resin (Fig. 22.6).

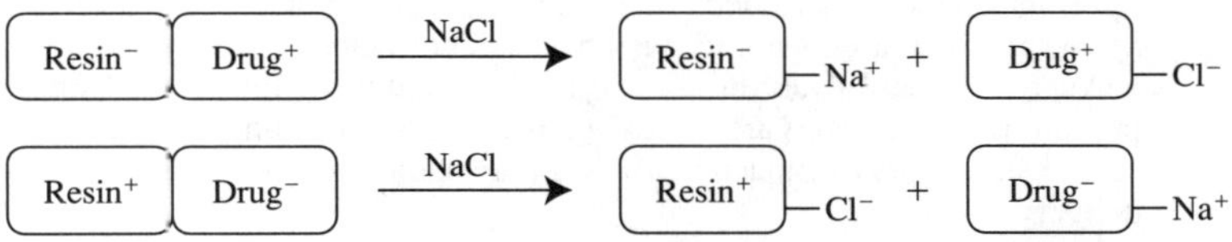

An ion exchange reaction can be expressed as

$$[A^+] + [R^-][B^+] \leftrightarrow [B^+] + [R^-][A^+]$$

and the selectivity coefficient K_B^A is defined as

$$K_B^A = \frac{[A_R^+][B^+]}{[A^+][B_R^+]} \tag{22.21}$$

where $[A^+]$ = concentration of free counterion
$[B_R^+]$ = concentration of drug bound of the resin
$[B^+]$ = concentration of drug freed from resin
$[A_R^+]$ = concentration of counterion bound to the resin

Factors that affect the selectivity coefficient include type of functional groups, valence and nature of exchanging ions, and nature of nonexchanging ions. Although it is known that ionic strength of GI fluid is maintained at a relatively constant level, first-generation ion-exchange drug delivery systems had difficulty controlling the drug release rate because of a lack of control of exchange ion concentration (Fig. 22.6*a*). The second-generation ion-exchange drug delivery system (Pennkinetic system) made an improvement by treating the drug-resin complex further with an impregnating agent such as polyethylene glycol 4000 to retard the swelling in water (Fig. 22.6*b*). These particles are then coated with a water-permeable polymer such as ethyl cellulose to act as a rate-controlling barrier to regulate the drug release.

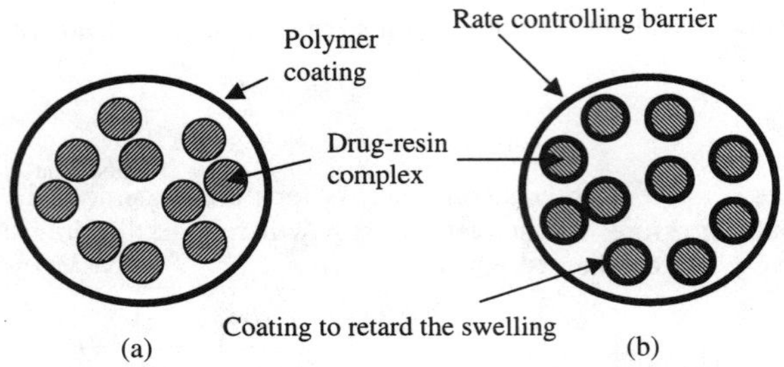

FIGURE 22.6 Schematic illustration of first generation (*a*) and second generation (*b*) ion exchange drug delivery system.

22.10 NEW MACROMOLECULAR DELIVERY APPROACHES

The advances in biotechnology have introduced many proteins and other macromolecules that have potential therapeutic applications. These macromolecules bring new challenges to formulation scientists, since the digestive system is highly effective in metabolizing these molecules, making oral delivery almost impossible, while parenteral routes are painful and difficult to administer. A potential carrier for oral delivery of macromolecules is polymerized liposomes.[14] Liposomes are lipid vesicles that target the drug to selected tissues by either passive or active mechanisms.[15] Advantages of liposomes include increased efficacy and therapeutic index, reduction in toxicity of the encapsulated agent, and increased stability via encapsulation. One major weakness of liposomes is the potential leakage of encapsulated drugs due to the stability of liposome. Unlike traditional liposomes, the polymerized liposomes are more rigid because of cross-linking and allow the polymerized liposomes to withstand harsh stomach acids and phospholipase. This carrier is currently being tested for oral delivery of vaccines.

Pulmonary route is also being utilized as route for delivery of macromolecules. The lung's large absorptive surface area of around 100 m^2 makes this route a promising alternative route for protein administration. Drug particle size is a key parameter to pulmonary drug delivery. To reduce the particle size, a special drying process called *glass stabilization technology* was developed. By using this technology, dried powder particles can be designed at an optimum size of 1 to 5 μm for deep lung delivery. Advantages of powder formulation include higher stability of peptide and protein for longer shelf life, lower risk of microbial growth, and higher drug loading compared to liquid formulation.[16] Liquid formulations for accurate and reproducible pulmonary delivery are now made possible by technology which converts large or small molecules into fine-particle aerosols and deposits them deep into the lungs. The device has a drug chamber that holds the liquid formulation

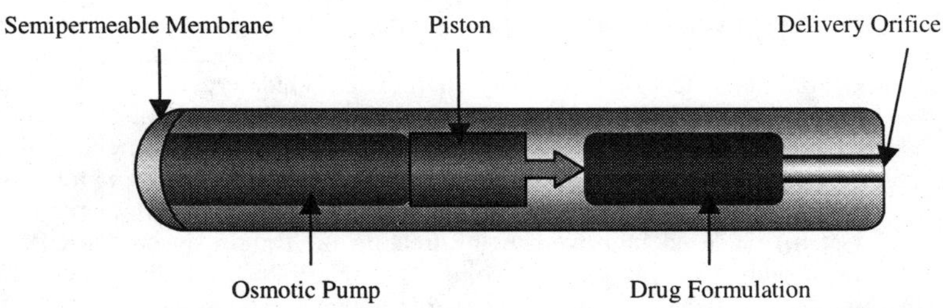

FIGURE 22.7 Illustration of an implantable osmotic pump.

TABLE 22.1 Examples of Various Delivery Approaches

Type of device	Product name	Active ingredient	Route	Developer/manufacturer
Diffusion (reservoir)	Estraderm	Estradiol	Transdermal	Alza/Novartis
	Norplant	Levonorgestrel	Subdermal implant	Wyeth-Ayerst Laboratory
	Ocusert	Pilocarpine	Ocular	Alza
	Progestasert	Progesterone	Intrauterine	Alza
	Transderm Scop	Scopolamine	Transdermal	Alza/Novartis
Diffusion (matrix)	Nitro-Dur	Nitroglycerine	Transdermal	Key Pharmaceutical
	Nitrodisc	Nitroglycerine	Transdermal	Searle
Mixed (matrix-reservoir)	Catapress-TTS	Clonidine	Transdermal	Alza/Boehinger Ingelheim
	Deponit	Nitroglycerine	Transdermal	Pharma-Schwarz
Hydrodynamically balanced system	Medopar CR	Levodopa and benserazide	Oral tablet	Roche
	Valrelease	Diazepam	Oral tablet	Roche
Ion exchange	Colestid	Colestipol	Oral tablet or granules	Upjohn
	Questran	Cholestyramine	Oral tablet or powder	Bristol Labs
	Tussionex Pennkinetic	Chlorpheniramine and hydrocodone	Oral suspension	Fisons
Coating	Compazine Spansule	Prochlorperazine	Oral capsules	Smith Kline Beecham
	Dexedrine CR	Dextroamphetamine	Oral capsules	Smith Kline Beecham
	Fefol-Vit	Ferrous sulfate and vitamins	Oral capsules	Smith Kline Beecham
	Ecotrin	Aspirin	Oral tablet	Smith Kline Beecham
	Voltaren	Diclofenac	Oral tablet	Geigy
Nanocrystal Technology	Rapamune	Sirolimus	Oral tablet	Elan/Wyeth-Ayerst Laboratory
Osmotic pumps	Calan SR	Verapamil	Oral tablet	Alza/G. D. Searle
	Cardura XL	Doxazosin	Oral tablet	Alza/Pfizer
	Covera HS	Verapamil	Oral tablet	Alza/G. D. Searle
	DUROS	Potential carrier for macromolecules	Implant	Alza
	Ditropan	Oxybutynin	Oral tablet	Alza/UCB Pharma
	Minipress XL	Prazosin	Oral tablet	Alza/Pfizer
	Teczem	Enalapril and diltiazem	Oral tablet	Merck/Aventis
	Volmax	Albuterol	Oral tablet	Alza/Muro Pharmaceuticals
Liposome	AmBisome	Amphotericin B	Parenteral	NeXstar Pharmaceuticals
	Amphotec	Amphotericin B	Parenteral	Sequus Pharmaceuticals
	Doxil	Doxorubicin	Parenteral	Sequus Pharmaceuticals
	DaunoXome	Daunorubicin	Parenteral	NeXstar Pharmaceuticals

and upon activation, the pressure will drive the liquid through fine pores creating the microsize mist for pulmonary delivery.

Transdermal needleless injection devices are another candidate for protein delivery.[17] The device propels the drug with a supersonic stream of helium gas. When the helium ampule is activated, the gas stream breaks the membranes that hold the drug. The drug particles are picked up by a stream of gas and propelled fast enough to penetrate the stratum corneum (the rate-limiting barrier of the skin). This delivery device is ideal for painless delivery of vaccine through the skin to higher drug loading. Limitations to this device are the upper threshold of approximately 3 mg and temporary permeability change of skin at the site of administration. An alternative way to penetrate the skin barrier has been developed utilizing thin titanium screens with precision microprojections to physically create pathways through the skin and allow for transportation of macromolecules. Another example of macromolecular delivery is an implantable osmotic pump designed to deliver protein drugs in a precise manner for up to 1 year (Fig. 22.7). This implantable device uses osmotic pressure to push the drug formulation out of the device through the delivery orifice.

22.11 CONCLUSION

Controlled-release delivery devices have been developed for more than 30 years. Most of the devices utilize the fundamental principles of diffusion, dissolution, ion exchange, and osmosis (Table 22.1). Optimal design of a drug delivery system will require a detailed understanding of release mechanisms, properties of drugs and carrier materials, barrier characteristics, pharmacological effect of drugs, and pharmacokinetics. With development in the field of biotechnology, there is an increase in the number of protein and other macromolecular drugs. These drugs introduce new challenges and opportunities for design of drug delivery systems.

REFERENCES

1. CDER 1999 Report to the Nation, *Improving Public Health Through Human Drugs. US Department of Human Services*, Food and Drug Administration and Center for Drug Evaluation and Research.
2. Y. E. Chien, *Novel Drug Delivery Systems*, 2d ed., Marcel Dekker, New York, 1992.
3. V. S. Gerogiannis, D. M. Rekkas, and P. P. Dallas, "Floating and Swelling Characteristics of Various Excipients Used in Controlled-Release Technology," *Drug Dev. Ind. Pharm.*, **19**:1061–1081, 1993.
4. R. O. Potts and G. W. Cleary, "Transdermal Drug Delivery: Useful Paradigms," *J. Drug Target*, **3** (4):247–251, 1995.
5. J. R. Robinson and V. H. Lee, (eds.), *Controlled Drug Delivery: Fundamentals and Applications*, 2d ed., Marcel Dekker, New York, 1987.
6. S. D. Bruck (ed.), *Controlled Drug Delivery*, vols. 1 and 2, CRC Press, Boca Raton, Florida, 1984.
7. S. Cohen and H. Bernstein (eds.), *Microparticulate Systems for Delivery of Proteins and Vaccines*, Marcel Dekker, New York, 1996.
8. H. Bundgaard (ed.), *Design of Prodrugs*, Elsevier Science, New York, 1985.
9. J. Crank, *The Mathematics of Diffusion*, 2d ed., Oxford Press, New York, 1975.
10. R. A. Lipper and W. I. Higuchi, "Analysis of Theoretical Behavior of a Proposed Zero-Order Drug Delivery System," *J. Pharm. Sci.*, **66**(2):163–164, 1977.
11. F. Theeuwes, "Elementary Osmotic Pump," *J. Pharm. Sci.*, **64**:1987–1991, 1975.
12. C. Kim, *Controlled Release Dosage Form Design*, Technomic, Lancaster, Pa., 2000.
13. S. Motycha and J. G. Naira, "Influence of wax coatings on release rate of anions from ion-exchange resin beads," *J. Pharm. Sci.*, **67**(4):500–503, 1978.
14. J. Okada, S. Cohen, and R. Langer, "In vitro Evaluation of Polymerized Liposomes as an Oral Drug Delivery System," *Pharm. Res.*, **12**(4):576–582, 1995.
15. A. D. Bangham, "Diffusion of Univalent Ions Across the Lamellae of Swollen Phospholipids," *J. Mol. Biol.*, **13**:238–252, 1965.
16. J. R. White and R. K. Campbell, "Inhaled Insulin: An Overview," *Clinical Diabetes*, **19**(1):13–16, 2001.
17. T. L. Brukoth, B. J. Bellhouse, G. Hewson, D. J. Longridge, A. G. Muddle, and D. F. Saphie, "Transdermal and Transmucosal Powdered Drug Delivery," *Crit. Rev. Ther. Drug Carrier Sys.*, **16**(4):331–384, 1999.

CHAPTER 23
STERILE MEDICAL DEVICE PACKAGE DEVELOPMENT

Patrick J. Nolan
DDL Inc., Eden Prairie, Minnesota

This chapter provides an overview of the process of designing and developing a package system for a medical device. A comprehensive discussion of this subject is beyond the scope of this chapter; however, numerous references are provided for further detail and study of the subject. The information provided in this chapter is a compilation from the references cited as well as from the experiences of the author in developing package systems for medical devices.

Introduction

Implementation of a standard for the process of designing and developing a package for terminally sterilized medical devices is essential to the overall endeavor of marketing a sterile device in the international and domestic communities. It is incumbent upon the medical device manufacturer to ensure that a safe, reliable, and fully functional device arrives at the point of end use. This assurance is complicated by the fact that the package must maintain its barrier integrity throughout its intended shelf life and through the rigors of manufacture and shipping and handling. The total product development effort must include the packaging design process and encompasses the package design, manufacturing process, sterilization process, and distribution environment effects. The intended sterilization method and the intended use, shelf life, transport, and storage all influence the package design and choice of packaging materials.

The issue of developing a package system seems uncomplicated and elementary. In actuality, the package process is complicated by the fact that device packages must allow effective sterilization of their contents by a wide array of methods; therefore the materials must be compatible with the sterilization method. Consequently, the package must provide a consistent and continuous barrier to environmental microorganisms and bacteria so as to maintain product sterility. The package must be designed to prevent product damage and loss of functionality from the dynamic hazards of shock and vibration inherent in the distribution environment. In addition, the manufacturer must have documented evidence that the performance of the package system is not adversely affected over time. The interactions of the materials and product, combined with the processes required to bring the product to its end use, influence the package design and manufacturing of the finished product.

The importance of packaging to the implementation of a medical device is illustrated in a speech by George Brdlik in 1982 that is no less true today. Brdlik stated:

> Packaging is too often neglected as an important characteristic of a medical device. When sterile medical devices are involved, deficient packaging can cause the following problems:
>
> - Increased risk of patient infection if product sterility is compromised by defective seals, pinholes fragile packaging materials, or packaging which shreds, delaminates, or tears upon opening.
> - Hampering of a surgical procedure because of difficulties in product identification or aseptic transfer, or if a product selected for use must be replaced because the package is either initially defective or damaged upon opening.
> - Increased hospital costs due to discarded products or excessive storage space requirements
> - Increased manufacturing costs for refund/replacement of damaged products and recall of products with potentially compromised sterility or integrity.
>
> In essence, poor packaging can transform the best, safest, and most economical medical device into an inferior, unsafe, and expensive product.

This chapter will provide a systematic and standard approach to developing a comprehensive package system that meets regulatory hurdles and ensures a high degree of confidence that the sterile medical device product will meet its performance specifications at the point of end use. These elements include:

- Selection of materials
- Design of the package
- Process validation
- Final package design validation

All of these elements must be combined to produce a package system that meets regulatory, industry, and consumer's requirements.

23.1 REGULATORY HISTORY

The regulatory burden for validating the manufacturing process and package system has become significant and considerable. It was started in 1938 with the amended Food and Drug Act of 1906 in which medical devices were first regulated, and then progressed to the Quality System Regulation (QSR), which specifies the requirements for components, device master record, and environmental controls, to name a few.

It is appropriate to present a brief history of how the medical device industry became regulated and how eventually the FDA recognized the importance of the package as an integral part, and in fact a component, of the medical device. As mentioned earlier, the Food and Drug Administration began regulating medical devices in 1938. At that time, the Federal Food, Drug, and Cosmetic Act extended the FDA's legal authority to control foods and drugs and bestowed the agency with new legal powers over cosmetics and medical devices. However, the act was limited in scope in that the regulatory actions could only be taken after a device was introduced into interstate commerce, and only after the device was found to be adulterated or misbranded. Surprisingly, the burden was on the government to provide evidence of violation of the act. In addition, the 1938 act could not prevent the introduction and marketing of "quack" medical devices. However, there was also an explosion of legitimate and sophisticated new devices using postwar biotechnology. These devices not only presented huge potential benefits to patient healthcare, but also caused an increased risk for harm. It became apparent that additional legal controls were necessary in order to harness the potential good from the new technologies.

A government committee studied the best approach to new comprehensive device legislation, and, as a result, in 1976 a new law amended the 1938 Act and provided the FDA with significant additional authority concerning the regulation of medical devices. The amendments included classification of all devices with graded regulatory requirements, medical device manufacturer registration, device listing, premarket approval (PMA), investigational device exemption (IDE), good manufacturing practices (GMP), records and reporting requirements; pre-emption of state and local regulations, and performance standards. Two years later, in 1978, the FDA published the GMP regulations that provided a series of requirements that prescribed the facilities, methods, and controls to be used in the manufacture, packaging, and storage of medical devices. The law has since been modified, with the most substantive action occurring in 1990 with the passage of the Safe Medical Devices Act (SMDA). It broadly expanded FDA's enforcement powers by authorizing the levying of civil penalties and creating a series of postapproval controls for monitoring the performance of medical devices.

Over the past 18 years, the FDA has significantly changed the way medical devices are regulated. The issuance of guidance documents effectively circumvented rulemaking and public comment. Publishing FDA's interpretation of the GMP effectively causes the manufacturer to comply with that interpretation. Legally, guidances are not binding on the public, whereas certain rules are. But for all practical purposes, there is little difference between the two. For example, two of these guidance documents are

- Guideline on General Principles of Process Validation—1987
- Preproduction Quality Assurance Planning: Recommendations for Medical Device Manufacturers

FDA issues dozens of guidances each year on specific products and processes. The last significant piece of legislation for medical devices came with the signing of the Food and Drug Administration Modernization Act of 1997 (FDAMA). This legislation essentially changed FDA's approach to standards-based enforcement and adds a system to recognize national and international standards in product reviews. The FDA will publish the recognized standards in the *Federal Register,* and these standards will then serve as guidance, enabling companies to use them to satisfy premarket submission requirements through a declaration of conformity. The list of recognized standards is provided in the FDA guidance document entitled "Guidance for Recognition and Use of Consensus Standards: Availability." This legislative change enables companies to submit a one-page 510(k) that simply states that the device complies with a stated list of recognized standards.

Significant legislation affecting medical devices and packaging includes:

- 1938—Federal Food, Drug, and Cosmetic Act enacted
- 1976—Medical Device Amendments (MDA) passed
- 1978—GMP Regulations published in *Federal Register*
- 1990—Safe Medical Device Act (SMDA) enacted
- 1997—Food and Drug Administration Modernization Act (FDAMA) passed

So, medical device manufacturers are subject to the provisions of the FDAMA and the GMP when doing business in the United States. The regulations are published in 21CFR (Code of Federal Regulations), part 820. The section dealing specifically with the requirements for packaging is contained in section 820.130. FDA approves products for use through the investigational device exemption (IDE), premarket application (PMA), and 510(k) processes. There may be additional regulatory burdens when the device is being marketed outside the United States. Some of these regulations are discussed in Sec. 23.1.1.

23.1.1 International Regulations

International regulations play a significant role in the marketing and use of medical devices. The European Union "Council Directive concerning medical devices" is the international equivalent of the FDA regulations. The directive (93/42/EEC), also known as the Medical Device Directive (MDD), as published in the EEC in 1993, lists the essential requirements for devices and packages, and all medical devices sold on the European free market must meet the specifics of this directive, which overrides all national regulations. The "Essential Requirements" section for medical devices and packages are found in Annex I of the MDD. General requirements for ensuring that the characteristics of the medical device are not altered or adversely affected during their intended use as a result of transport and storage are found in Sec. 5. More specific requirements for infection and microbial contamination as it relates to packaging is found in Sec. 8. It is incumbent upon the medical device manufacturer to conform to all of the sections of the "Essential Requirements," not just the packaging requirements.

ISO 11607. An ISO standard approved in 1997 by the international community has become essential to the development and validation of a package system for a terminally sterilized medical device. The ISO 11607 standard, entitled *Packaging for terminally sterilized medical devices,* provides the guidance for selecting materials (Clause 4), designing and validating the manufacturing process (Clause 5), and validating the final package design (Clause 6). Compliance with this standard and other European standards ensures compliance with the packaging provisions of the MDD, "Essential Requirements." The FDA has recognized this standard for product reviews in its 1997 *Guidance for Recognition and Use of Consensus Standards: Availability.* This standard specifies the basic attributes that materials must have for use in packaging for terminally sterile medical devices. In addition, it provides the producer or manufacturer the guidance to conduct a formal qualification and validation of the packaging operations. There must be a documented process validation program that demonstrates the efficacy and reproducibility of all sterilization and packaging processes, to ensure the package integrity at the point of end use.

Finally, the ISO standard provides a series of recommended tests and criteria to evaluate the performance of the complete package under all of the stresses and hazards inherent in the packaging and distribution process. These tests include the following types:

- Internal pressure
- Dye penetration
- Gas sensing
- Vacuum leak
- Seal strength
- Transportation simulation
- Accelerated aging
- Microbial barrier

EN 868 Part 1. This European standard, entitled *Packaging Materials and Systems for Medical Devices Which Are to Be Sterilized: General Requirements and Test Methods,* provides detailed guidance on meeting the requirements of the MDD. It includes more detail on the selection and validation of packaging materials than does Clause 4 of the ISO 11607 standard. However, there are some differences, and both standards must be considered in evaluating the packaging system for compliance to the FDA and MDD regulations.

Standards within the EN 868 series fall into two distinct categories—horizontal and vertical. EN 868 Part 1 is a horizontal standard, since it specifies the requirements for a broad range of packaging materials, types, and designs. The requirements of Part 1 are essentially the same as ISO 11607, Clause 4; however, the ISO document includes requirements for package forming and final package validation. Vertical standards within the 868 Series include detailed requirements for individual materials or specific package types or medical device products. These standards are designated Parts 2 through 10. They specify limits for material properties for:

- Sterilization wraps (Part 2)
- Paper for use in the manufacture of paper bags, pouches, and reels (Parts 3, 4, 5)

- Paper for the manufacture of packs for medical use for sterilization by ethylene oxide (ETO) or irradiation (Part 6)
- Adhesive-coated paper for the manufacture of heat-sealable packs for medical use for sterilization by ETO or irradiation (Part 7)
- Reusable sterilization containers for steam sterilization conforming to EN 285 (Part 8)
- Uncoated nonwoven polyolefin materials for use in the manufacture of heat-sealable pouches, reels, and lids (Part 9)
- Adhesive-coated nonwoven polyolefin materials for use in the manufacture of heat-sealable pouches, reels, and lids (Part 10).

The "Essential Requirements" of the MDD can be effectively met by complying with the requirements of the ISO 11607 and EN 868, Part 1 standards.

CE Mark. A CE Mark can be affixed to the medical device when all of the essential requirements of the MDD and other directives, as appropriate, are met. The declaration of conformity that contains the documented evidence that all requirements have been met achieves this.

23.2 FUNCTIONS OF A PACKAGE

The first step in designing and developing a package system for a medical device is the selection of materials appropriate for and compatible with the device. Packages are intended to contain the product. However, for medical devices, there are other functions the package serves; it provides protection, identification, processability, ease of use, and special applications. A basic knowledge of the product's use, dimensions, shape, and special characteristics (sharp edges, points, fragility, etc.); distribution environment, application, and barrier requirements are essential to selecting appropriate materials and designing the final package.

23.2.1 Protection

Protection of the device by the package may be provided in several different ways. Obviously, the sterile medical device must be protected from the bacteria and microorganisms natural to the environment. The package must provide a protective barrier from the environment but must also allow the product to be terminally sterilized, be opened easily by medical professionals, and maintain integrity until the point of end use. Materials must allow for the most efficient and effective sterilization method but not be degraded by that method. The package must also provide protection to the product from the rigors of the distribution and handling environment. In addition, there cannot be any damage to the package itself from shock or impacts associated with handling in shipment, or loss of seal integrity. Materials should be resistant to impacts and abrasion. The package must be designed so as to prevent sharp objects from piercing the materials or damaging seals, by eliminating movement of the device inside the package. In some applications, the product may be so fragile that the package must have cushioning characteristics that prevent excess shock energy from being transmitted to the device. Protection of the device over an extended shelf life is another function of the package design requiring material stability over time.

In summary, the protective features a package for a sterile medical device the package must have are

- *Sterilizability:* provide the ability to terminally sterilize the device by one or more methods without degrading the material.
- *Stability:* provide a barrier throughout the intended shelf life of the product.

- *Environmental resistance:* provide a barrier to moisture, air, bacteria, oxygen, light.
- *Physical:* provide dynamic protection; resist impacts and abrasion, provide structural support.

The materials most commonly used for medical device packages today incorporate the characteristics required for a protective package.

23.2.2 Identification

Packages must not only provide protection to the product, but they must also communicate what the product is, instructions, warnings and safety information, and other pertinent information such as; lot number, sterilization method, and expiration date. Space must be provided on the package for conveying this information either by printing directly on the package or by applying a label. Often, there must be adequate space for the information in two or more languages. The information must be legible at the point of end use; therefore abrasion, water, and the sterilization process must not damage the printing and labels.

23.2.3 Processability

By *processability* we mean the ability of the packaging material along with its product to be processed through a series of manufacturing operations that includes mechanical stresses in filling and sealing, chemical or physical actions during sterilization, stresses of shipping and handling, and the environmental effects of aging before the product is finally used. This processability requirement is clearly stated in 21 CFR Part 820.130, "Device Packaging":

> The device package and any shipping container for a device shall be designed and constructed to protect the device from alteration or damage during the customary conditions of processing, storage, handling and distribution.

23.2.4 Ease of Use

Parallel with the increase in the use of disposable medical devices is the requirement for easy-to-open packages for the sterile field. The package has to be designed such that the materials are strong enough to withstand the rigors of processing but can be opened without tearing or excessive stress on the package or product.

23.2.5 Special Applications

In some cases the package may serve as a tool in the procedure. The obvious application is as a delivery mechanism for the product to the sterile field; for example, heart valve holders for asceptic transfer to the surgery site. However, even more complex applications may be designed into the package to aid in the procedure.

23.3 PACKAGE TYPES

The form the package takes is fundamentally dependent upon the characteristics of the device such as size, shape, profile, weight, physical state, irregularities, density, application of the device, and other considerations.

The medical device industry uses a limited variety of types but many different materials in each form. Over the past 20 years, the industry has standardized on the following basic medical device package types.

23.3.1 Thermoform Trays (Semirigid)

Thermoform trays are made from a variety of plastics by molding them to the desired shape through the thermoforming process. Trays are particularly suited for devices with irregular shapes and high profiles, since the tray can be designed to accommodate these device characteristics. Trays are ideal for procedure kits that contain several devices, as they can be molded with special flanges, undercuts, and snap fits or ridges for securing the components. Semirigid trays are self-supporting.

In designing a tray for a medical device, several criteria must be considered in the selection of the material:

- Tensile strength
- Stiffness
- Impact resistance
- Clarity
- Ease of forming and cutting
- Heat stability
- Compatibility with sterilization processes
- Cost, on a yield basis, versus performance
- Product compatibility
- Ability to be sealed with lidding material

In using a tray for a sterile medical device, in order to perform the most basic protection function, the tray must be sealed to prevent loss of sterility. This is accomplished by using a lidding material that is sealed around the flange area of the tray. Until the advent of Tyvek™ into the market, it was difficult to provide lidding material that would provide a means for terminal sterilization using the common sterilization methods of the day. However, Tyvek has allowed widespread use of thermoform trays for many applications.

23.3.2 Flexible Formed Pouches

This type of package is one in which a flexible material is drawn by the thermoform process into a flexible "tray." The package essentially is a formed pouch that allows containment of high profile devices. These packages are generally not self-supporting.

Characteristics of the formed flexible packages are:

- Relatively low cost, suitable to high-volume, low-cost devices
- May be made to be heat sealable
- Ease of forming
- Available for form-fill-seal operations
- Suited to relatively simple tray configurations
- Can conform to product
- Good visibility of product
- Cannot be preformed
- Offer little structural protection

- Limited variety of materials available
- Relatively lower heat resistance

Like the semirigid tray, this package type must also be sealed using a lidding material or top web. The top web material must be designed with the particular barrier properties needed to be compatible with bottom web material and the chosen sterilization process. The selection of the top web material must therefore be based on the following three factors:

- *Type of device.* Environmental or product barrier requirements
- *Method of sterilization.* Gas, radiation, steam,
- *Method of dispensing.* Peel, cut, tear, puncture

23.3.3 Flexible Nonformed Pouches

The most common package in this category, and probably for most single-use medical devices, is the two-web peel pouch. The package form is very common for a variety of medical devices including gloves, catheters, tubing, adhesive bandages, dressings, sutures, and other low-profile and lightweight products. This flat pouch design is suitable for high-volume, low-cost devices, as it provides the basic protection for devices (e.g., sterile barrier, sterilizable by one or more methods, efficient manufacturing, easily opened for dispensing). The most popular form of flat pouch is known as the *chevron* pouch. It gets its name from the peak-shaped end of the package where the initial peeling of the seal begins. This design provides ease of peeling as the peel forces are distributed angularly along the seal line rather than across the entire seal end. Other forms of the flat pouch can be achieved by forming seals across the corner of the package, leaving a tab to initiate the peel.

The typical peel pouch used for devices is made from two separate web materials that are heat sealable or adhesive coated. Since these packages usually contain sterile disposable devices that are terminally sterilized inside the primary package, a porous material is required for one of the webs. Either paper or Tyvek is used as one of the web materials along with a plastic film such as a laminated polyester and polyethylene.

Some of the benefits of the peel pouch are

- Relatively low cost
- Suitable for small-run or high-volume uses
- Can be fabricated from a wide variety of materials
- Can be prefabricated or formed in-line
- Provides a sterile delivery capability
- Product visibility
- Easy opening
- Printable with product information and instructions

Some of the disadvantages are

- Not useful for high-profile devices
- Low dynamic protection capabilities
- Not suitable for irregularly shaped devices
- Not suitable for kits or multicomponent devices

Another type of peelable pouch is known as the *header* bag. This package is essentially a two-web pouch in which a portion of one web is a peelable Tyvek or paper vent. The header bag provides high permeability, ease of opening, and convenient product dispensing. An advantage of this package type is that a basic flexible bag can contain a high-profile device.

23.4 PACKAGING MATERIALS

This part provides a basic overview of some of the more common packaging materials used for medical device packages. Since entire books are published describing the chemical characteristics, applications, and performance properties of packaging materials, it is beyond the scope of this chapter to provide all of the necessary information for the selection of materials for the specific medical device. Consult the references for additional information.

23.4.1 Primary Materials

Tyvek. Tyvek, a spun-bonded olefin, is used in almost every major form of sterile package including peelable pouches, header bags, and lid stock of thermoform trays and kits. Tyvek is a fibrous web material composed entirely of extremely fine, continuous strands of high-density polyethylene. This material has exceptional characteristics that distinguish it from other materials. It has superior dry and wet strength, chemical inertness, and dimension stability. Its excellent puncture and tear resistance and toughness allow for a wide range of applications for medical devices, particularly irregularly shaped and bulky products. This material has an unusual fiber structure that allows for rapid gas and vapor transmission but at the same time provides a barrier to the passage of microorganisms. Tyvek is used most often with ethylene oxide (ETO) sterilization methods because of its unique combination of properties of high porosity and microbial impermeability. Tyvek provides several other attributes useful to package integrity and aesthetics:

- *Water repellency.* Repels water but is porous to moisture vapor.
- *Chemical resistance.* Resists usual agents of age degradation (e.g., moisture, oxidation, rot, mildew, and many organic chemicals).
- *Radiation stability.* Unaffected by common levels of radiation used for sterilization.
- *Low-temperature stability.* Retains strength and flexibility at subzero temperatures.
- *High-temperature stability.* Can be used in steam sterilization methods.
- *Aesthetic qualities.* Bright, white, clean appearance for printing.

Since Tyvek does not readily adhere to other plastics, except other polyolefins, through the application of heat and pressure, it has been made a more versatile packaging material by applying coatings that enable it to bond with a wide variety of plastics. There are several grades of Tyvek used for medical packaging applications including 1059B, 1073B, and 2FS.

Paper. For many years, paper was the only choice for package types until the introduction of Tyvek as a medical packaging material. However, paper still plays a significant role in the medical device industry. Over the years before the introduction of Tyvek, paper materials compiled a significant performance record of product protection and patient safety. Although Tyvek has taken a majority share of the medical device package market, the industry is finding ways to utilize paper in combination with plastics and foils to provide the needed performance characteristics with favorable economics.

The significant features of paper materials that enable it to continue as a feasible packaging material alternative are:

- Cost
- Disposability
- Sterilization
- Combination with other materials
- Versatility

- Peelability
- Range of grades

Some of the limitations of paper as a medical device packaging material are:

Strength—low tear and puncture resistance
Dimensional stability
Moisture sensitivity
Aging—limited under certain environmental conditions

Paper can be used as lidding material for semirigid and flexible trays, and for peelable pouches. Adhesive coatings are required to allow sealing.

Films, Laminates, and Coextrusions. Many films are used in medical device packaging applications. Both flexible formed and nonformed pouches, as well as bags, use films for their manufacture. These materials offer a high degree of versatility and are available in a countless variety of forms in monofilms, laminations, and coextrusions. The specific material to be used for a medical device is dependent upon the performance properties required for the device application. For example:

- Sterilization method (e.g., the material must tolerate high temperature)
- Protection requirements (e.g., high puncture resistance)
- Peel requirements (e.g., easily peelable)
- Package style (e.g., formable versus nonformable pouch)
- Barrier properties (e.g., moisture or oxygen barrier)
- Packaging process (e.g., in-line sealing versus form-fill-seal)
- Packaging aesthetics (e.g., visibility of product)

The flexible materials used for medical device packages include a plastic film that is usually a lamination or extrusion-coated material. The material most commonly used for flexible packaging applications is oriented polyester (e.g., Mylar™), which is used as a base for properties such as dimensional stability, heat resistance, and strength with an adhesively laminated seal layer such as low-density polyethylene, which provides the film structure with heat sealability. The variety of film combinations is virtually unlimited and the performance properties of the film can be customized to meet the requirements of the package specifications and the medical device. Other examples of film constructions are

- Polyster/Pe/EVA
- Polyester/Surlyn
- Polyester/nylon/Pe
- Polyester/nylon/PP
- Polyester/PVDV/Pe
- Metallized polyester/Pe
- Polyester/foil/Pe
- Polyester/foil/Polyester/Surlyn
- Oriented PP/Pe
- Polycarbonate/Pe/EVA

The thermoplastic films used in flexible applications are suited only for sealing to themselves or to chemically related materials. The sealing of like materials produces fused bonds that may not be peelable and thus applicable for single use medical devices. To overcome the limitations of sealing like materials, adhesives specifically tailored for seal-peel functionality are applied to the film surface allowing films to remain unaltered and to retain their performance characteristics. Uncoated or coextruded materials for medical device packages are limited in their application because they allow only a narrow sealing range, provide limited sealability on high-speed equipment, allowing sealing of only chemically similar materials, and commonly overseal to paper and Tyvek materials. On the other hand, materials coated with an adhesive provide versatility and greater benefits such as a wider sealing range, easy and consistent sealability to porous materials such as Tyvek and paper, barrier properties, lower cost, and versatility in adhesive properties dependent upon the application (e.g., pouch or tray application).

Foils. Foil laminate materials are used in applications where high moisture, gas, and light barriers are essential. Foil can be used in all forms of packaging and for both medical devices and pharmaceuticals. The lamination of the foil with plastic films is required to provide sealability. Foil materials are being used for lidding of thermoform tray packages where high moisture and gas barriers are required and where the sterilization method allows it (e.g., gamma, e-beam, steam, etc.).

Wet devices such as dressings, solutions, sponges, swabs, and other saturated products requiring a high moisture barrier are particularly suited to foil packages. Foil laminations with high-density polyethylene or polypropylene are common constructions for these package types. For solutions, a form-fill-seal application would be ideal, as the pouch is formed and filled in a multiphase operation on a single machine.

The trend in medical device packaging over the past 10 years has been to flexible packages, as they are less costly, more resistant to shipping damage, easier to handle, and produce less packaging waste. A foil-laminated package offers many benefits such as strength, high environmental barrier, peelability, easy opening, temperature resistance, opacity for light-sensitive products, sterilizer resistance, ease of formability, compatibility with many products, and tamper evidence.

Thermoformable Plastics. Thermoformed plastics are among the most widely used package materials because of aesthetic appeal, medical device applications, and versatility for customized designs to fit contours of medical devices or several components of procedure kits. The selection of a material for a specific medical device is dependent upon several factors such as barrier requirements, sterilization method, and cost. There are many thermoformable plastics; however, not all have the ideal properties that lend themselves to medical device packaging applications. For example, an acrylic-based plastic has very low structural flexibility, low impact resistance, and poor clarity, but has a very high radiation resistance. The polyethylene terephthalate (PET) plastics have excellent clarity, structural flexibility, impact resistance, sealability, and radiation resistance, but only marginal water vapor barrier and heat resistance. So each material has its favorable and unfavorable properties, and the material that most closely fits the desired packaging application must be selected. The most common packaging materials for thermoform tray applications are discussed in some detail.

Polyethylene Terephthalate (PET). The generic material called PET or polyethylene terephthalate is probably the most widely used material for medical packaging applications because of its favorable characteristics, as mentioned previously. This material forms easily in thermoforming operations and provides good barrier performance and sealability with various lidding materials. The material imparts excellent clarity, flexibility, and radiation resistance—all important characteristics for packaging medical devices. It is produced in forms for injection or blow molding of rigid containers such as bottles and jars, and in sheet form for thermoforming trays, and blisters. When PET is coextruded with other materials such as glycol to make PETG, the barrier performance characteristics of the material are improved. PETG is not heat sealable, so the lidding stock must be adhesive coated to facilitate a functional seal for the package.

Polycarbonate (PC). Polycarbonate is used for high-performance package applications where high strength and toughness are required because of the size, density, or shape of the product. In some applications PC is used because of its superior clarity and the aesthetic appeal of the product. PC has the most impact resistant of all the plastics but has only average moisture and gas barrier properties. The cost of PC is somewhat prohibitive in a high-volume product application. However, for low-volume, high-priced devices such as pacemakers, defibrillators, and other implantable devices, it is an excellent material for thermoform trays. Most of the common sterilization methods such as autoclave, steam, ethylene oxide, gamma, and e-beam can be used on packages made from polycarbonate. Typically, PC film for thermoform applications is coextruded with a polyolefin heat-seal layer.

Polyvinyl Chloride (PVC) and Polyvinylidene Chloride (PVdC). The material known as PVC or polyvinyl chloride is one vinyl-based polymer used commonly in packaging applications. Another material in the same family is polyvinylidene chloride or PVdC. These materials differ from polyethylene in having a chlorine atom that replaces one hydrogen atom in its chemical structure. This is important, since it is this chlorine atom that has caused the material to lose favor for packaging applications because of environmental concerns. The environmental concern is that when incinerated, the material generates a hydrogen chloride gas. Several European countries have banned the use of vinyl-based materials. The criticism is controversial. The perceived environmental threat has caused many PVC applications to be replaced by PET. PVC is used most frequently in packaging application for blow-molded bottles, blisters, and thermoform trays. PVC is tough and clear and has excellent barrier properties as well as high impact resistance.

Polystyrene (PS). Polystyrene (PS) is one of the most versatile, easily fabricated, and cost-effective plastic used in the packaging industry. It can be molded, extruded, and foamed. It is probably best known for its use as cushioning materials for electronic products. There are two types of polystyrene available for packaging applications: general purpose and high impact (HIPS). It is the high-impact type that is used for medical device packaging applications. High-impact polystyrene contains a small amount of rubberlike polybutadiene blended in to overcome the brittleness of the general-purpose material. This makes the HIPS tougher but less clear, usually translucent or opaque. The material may be acceptable for applications where visibility of the device is not required. The advantages of the material are its cost, heat resistance, and ease of formability. However, it may be susceptible to impact damage during shipping and handling. Another styrene-based material is styrene butadiene copolymer (SBC) and is commonly processed into containers, sheet, and film. It is used extensively in medical packaging applications because of its ability to be sterilized by both gamma irradiation and ethylene oxide.

The styrene materials are commonly recycled in communities where economics or legislation is favorable. However, where these materials are incinerated, PS, like PVC, causes unacceptable gaseous emissions and thus have come under intense environmental pressure and outright banning in some communities.

Other Materials and Properties. There is a host of other materials used in thermoform packaging applications. Some are specifically engineered for high barrier applications while others are resistant to high temperature. Although these materials have their greater use for medical device components, some materials are finding use for various types of packages such as tubes, blown containers, molded closures, and in some cases thermoform sheet material. Table 23.1 shows barrier and mechanical strength properties for the most common thermoformable plastics.

23.4.2 Secondary Materials

Secondary packaging is often used with primary packages to provide several functions in the overall distribution system for a medical device. Secondary packages are defined as containers that enclose one or more primary packages. One function the secondary package provides is the communication of information about the device. Protection of the device through the rigors of distribution and handling is another function a secondary package provides. In addition, the secondary package allows for storage of primary packages in a neat and orderly manner before use.

Paperboard Cartons. The most common form of secondary package used for primary medical device packages is the paperboard carton. This package is used for all types of primary packages including the semirigid tray, flexible formed pouch, chevron pouch, and header bag. It is used most often when the primary package requires some additional protection and as a "shelf box" for storage at the point of end use. A paperboard carton is usually inserted into a shipping container (i.e., shipper)

TABLE 23.1 Barrier and Mechanical Strength Properties

Plastic	WVTR, nmol/m · s	Oxygen, nmol/m · GPa	Tensile strength, psi	Elongation, %	Impact strength, kg/cm	Tear strength, g/0.001 in
PET	0.45	6–8	25,000–30,000	120–140	25–30	13–80
Polycarbonate	2.8	N.A.	10,000	92–115	100	16–25
PVdC	0.005–0.05	0.02–0.03	8,000	60–100	5–15	4–6
HDPE	0.095	200–400	3,000–7,500	10–500	1–3	15–300
LDPE	0.35	500–700	1,000–3,500	225–600	4–6	50–300
PVC	0.55	10–40	2,000–16,000	5–500	12–20	Varies
PP	0.16	300–500	9,000–25,000	60–100	5–15	4–6
Polystyrene	1.8	500–800	N.A	N.A.	N.A.	N.A.

Source: Diana Tweede and Ron Goddard, *Packaging Materials,* Pira International, Leatherhead, Surrey, U.K.

that provides more substantial protection for transport. Many paperboard cartons may be consolidated into a single shipper.

Materials used to fabricate paperboard cartons may also be variously known as boxboard, cartonboard, chipboard, containerboard, and solid fiberboard. They are made in the same manner as paper and allow semirigid formability as well as surface strength and printability. Solid bleached boxboard is the highest quality, as it is made from the purest virgin bleached chemical pulp. This grade of paperboard is most often used for medical devices for its aesthetic good looks and excellent printability for graphics and product information. Various styles of paperboard carton are available to suit a particular product or primary package type or application.

Corrugated Fiberboard Boxes. A corrugated fiberboard box is used to transport the medical device through the distribution environment and to its ultimate user. This package may be known as the shipping container, shipper, shipping box, transport package, or other name that denotes its purpose as the final package to be offered for shipment. This package may contain only primary package types, or single or multiple secondary packages containing primary packages. In this case the package system may be considered to have a primary, secondary, and tertiary package. Most shippers are made from a material known as *corrugated fiberboard.* The paper-based material consists of a corrugated medium sandwiched between two kraft paper faces. It is characterized by the thickness and spacing of the medium (fluting), the weight of the facing layers, and the quality of the paper used. Most medical devices are transported in a grade and style of shipper known as a single wall, C flute, or regular slotted container (RSC).

23.5 COMMON TESTING METHODS

This section details some of the testing methods used and accepted within the medical device industry for characterizing the performance of the package. These methods will be used to validate the packaging processes and to establish performance specifications for continuous monitoring of quality.

23.5.1 Introduction

The package for a medical device plays a key role in safely delivering specialized treatment to the patient for which the device was designed and developed. It must ensure the efficacy of the device from the point of manufacture to the point of final use. Most single-use terminally sterilized medical devices must be delivered with a very high confidence that the device has remained in a sterile condition throughout its storage, handling, and transport environment. In addition, packaging may have a direct function in the application of the treatment, as it may act as a fixture or dispenser for the physician. Thus mechanical damage to the package may not be tolerated. The design and development of the packaging system has come under closer and closer scrutiny by both the international and domestic regulatory agencies. This scrutiny has placed a great deal of emphasis on standardizing the package development process. Some standardization of the packaging process has come in the form of the international standard entitled ISO 11607, "Packaging for terminally sterilized medical devices." This section will specifically discuss the current consensus (ASTM) and industry test methods available for evaluating the integrity and strength of packages.

23.5.2 Package Integrity Versus Package Strength

First, there seems to be some confusion within the medical device industry regarding the strength of a package versus the integrity of a package. Package strength concerns the force required to separate two components of the package. It could be the force to separate two flexible components of a pouch, or a flexible lid and a thermoform tray. These forces may be measured in pounds per inch width, as in the seal/peel test, or in pounds per square inch, as in the burst test method. Alone, these tests of package strength values do not necessarily prove the integrity of the entire package. For example, since the seal/peel test per ASTM F-88 evaluates only a 1-inch segment of the package, there may be other areas of the package that are not sealed adequately to prevent contamination of the product. In fact, the seal width that was actually measured may be within the strength specification but may have a channel leak that could breach the package and negate integrity.

Likewise, the ASTM F-1140 burst test method as referenced by ISO 11607 also has its pitfalls. This method evaluates the whole package by applying pressure to all areas of the package, however the pressure is not applied equally at all points as a result of package irregularities and distortions. This can lead to a relatively high degree of variability between tests. Further, the burst test may not detect breaches in the package, such as pinholes and channel leaks, even though the burst test values have met the performance specification.

Even though the package strength specifications are confirmed, the package integrity is not necessarily proved. Package integrity is defined by ISO 11607 as *unimpaired physical condition of a final package*. Seal integrity is defined as *condition of the seal, which ensures that it presents a microbial barrier to at least the same extent as the rest of the packaging*. Neither definition refers to the *strength* of the seal. Package integrity is independent of package strength, although a strong package seal is a convincing indicator of a safe package. Further, if the entire seal area is proved to be homogeneous and continuous, then one could say that the package seals provide integrity. However, this says nothing about the package surfaces that may have pinholes or leaks not detected by seal strength tests. Other mechanical tests may be appropriate for determining package seal homogeneity.

Seal strength is important in the overall scheme of developing the package process, but the seal strength performance specification is used most effectively to monitor the process, not to determine ultimate acceptance. Seal strength is also an important determinant for establishing package process parameters. In fact, the ISO 11607 standard requires that the *seal strength shall be determined at the upper and lower limits of the defined critical sealing process variables and shall be demonstrated to be suitable for the intended purpose*. To restate, seal strength is an important performance attribute for the package and provides suitable guidance in establishing statistical process control limits, but is not the absolute determinant of the acceptability of the package for its intended use. Package integrity at the point of final use is the principal acceptance criterion for a sterile medical device package. However, both performance attributes are essential to the package design and development process.

23.5.3 Determining Package Strength

The performance specification of the package may be based on the seal and burst test values of packages produced from a specific validated production line. These tests are performed using standardized test methods developed by American Society for Testing and Materials (ASTM). The seal strength test procedure is described in ASTM F-88, "Seal Strength of Flexible Barrier Materials." This test covers the measurement of the strength of a seal of a given width at a specific point of the package. It does not measure the seal continuity. Other methods such as the 180° peel test may be used to determine the seal continuity or peeling characteristics. The seal strength test is performed by cutting a 1-in-wide strip from the seal of the package. The strip is placed in the tensile test machine by clamping each leg of the sample in the grips, aligning the specimen so that the seal is perpendicular to the direction of pull. The seal is pulled apart at a rate of 10 to 12 in/min. The peak force required to pull the seal completely apart is recorded. It would be appropriate to perform the test at several points of the package, including the manufacturer's seals (seals produced by the vendor of the package), and the production seals (seals produced by the manufacturer of the product). Typical seal strength values lie in the range between 1 and 4 pounds. The optimum seal strength varies according to the type of package being tested and its specific applications.

The burst test procedure is described in ASTM Standard D-1140, "Failure Resistance of Unrestrained and Nonrigid Packages for Medical Applications." This method covers the determination of the ability of package materials or seals to withstand internal pressurization. Since packages may be produced from substandard materials or with inadequate seals, or both, package integrity may be compromised during production, distribution, or storage. Burst testing may provide a rapid means of evaluating overall package quality during production, and overall package integrity after dynamic events associated with shipping and handling.

Two methods of burst testing are provided in the standard: the open-package test and the closed-package test. The open-package test is performed in a fixture that clamps the open end but provides a means for pressurizing the package. The pressure is increased in the package at a rate greater then the permeability of the porous package component, until a failure occurs. The type and location of the failure is recorded as well as the maximum pressure at which failure occurred. The open-package test is most useful as a quality assurance procedure on incoming materials to ensure that the supplier of the material is meeting pre-established specifications for seal strength.

The closed-package test is performed on production samples as an internal quality assurance procedure. This method is performed by inserting the pressure source through a component of the package and then increasing the pressure until a failure occurs. The pressure at failure and location and type of failure are recorded. Burst test values typically fall in the range between 0.5 and 3 psi. No correlation has been made between the burst test value and seal strength values. A recent study has shown that unrestrained pressure testing may lead to inconsistencies in test results while more consistent test results are achieved by restraining the test specimen between parallel plates (Hackett, 1996).

23.5.4 Determining Package Integrity

One category of package integrity test methods has been available for over 10 years and involves microbial challenge and product sterility. As you will read later in this chapter, these are not the only means of determining package integrity and these methods are coming under tighter examination, as alternate test procedures are developed. In fact, the FDA has recognized ISO 11607 as a consensus standard, which states, "The manufacturer shall demonstrate the integrity of the package by testing the package. This can be accomplished by physical tests." Examples of physical tests as described in the ISO 11607 standard include: internal pressure test, dye penetration test, gas sensing test, vacuum leak test. All of these methods have their advantages and disadvantages.

Microbial Challenge/Product Sterility Test Methods. There are really two types of microbial barrier test: those performed on materials and those performed on whole packages. Microbial barrier tests on materials are performed by packaging manufacturers to ensure that their materials are impervious to microorganisms while allowing sterilant gases to permeate for product sterilization purposes. These tests are typically performed using ASTM F-1608, "Microbial Ranking of Porous Packaging Materials (Exposure Chamber Method)." Microbial barrier testing of materials is significantly less controversial than microbial testing of whole packages, since this methodology lends itself to some level of standardization and control. Determining the microbial barrier characteristics of materials is very different from the methods required for a whole package. A whole package is significantly more complex than a single material.

Aerosol Challenge. At the risk of oversimplifying the procedural demands of microbial testing, here is a summary of how a microbial challenge/product sterility test is performed. There are two types of whole-package microbial barrier tests currently in use.

One method takes a sterile finished primary package containing an actual device and fixtures it into a vacuum chamber. The chamber is loaded using a specific configuration and then performance is qualified to establish a homogeneous distribution of the indicator organism prior to the actual test runs. After the performance qualification, the test packages are subjected to a microbial challenge of a high concentration of bacteria, which is nebulized into an aerosol and circulated in the chamber for a specified period of time. Next, the outer surfaces of the package are decontaminated and the product is aseptically extracted from the package. The product may even need to be manipulated further at this point to facilitate the sterility test. The product sterility test determines if any of the indicator micro-organisms were able to breach the package and contaminate the product.

Although this method would appear to be the best at determining package integrity since it is a direct indicator of product sterility or nonsterility, it has several deficiencies:

1. It is very expensive to perform the test using an adequate sample size while providing statistical significance.
2. It is prone to false positive results due to the high precision necessary for lab technicians to aseptically handle and manipulate packages and products.
3. Each package type, configuration, and size must be prequalified in the chamber.
4. Several well known studies by HIMA and member companies have indicated that it may not even detect obvious breaches in the package integrity (Jones et al., 1995).
5. There is no standardized method to ensure the reliability and repeatability of the test.

Dust/Talc Challenge. The other whole-package microbial method involves a similar concept of challenging the package with a high concentration of micro-organisms and then performing a product sterility test. This method uses talc or dust mixed with a specific micro-organism. The package is exposed to the dust by shaking in a chamber. Similarly, the outer package surfaces are decontaminated prior to product removal and sterility testing. Likewise, this method has deficiencies similar to the aerosol method.

The microbial methods are still in use to evaluate package integrity essentially because the regulatory agencies are still asking manufacturers for data using these methods, or medical device manufacturers have always evaluated their packages for integrity and they are hesitant to change their protocols. However, there are alternative methods.

Physical Test Methods. Some of the physical test methods have been available for many years as published ASTM standards. Recently, industry has taken a closer look at the validity and effectiveness of these methods and has developed new methods for evaluating package integrity.

Visual Inspection. ISO 11607 handles visual inspection for package integrity in Sec. 6.2, which is very detailed in the requirements and procedures. ASTM Subcommittee F2.60 on Medical Packaging recently published standard F-1886-98. "Standard Test Method for Determining Integrity of Seals for Medical Packaging by Visual Inspection," to help further detail a method for visual inspection. This standard describes a method to visually detect channel defects in package seals down to a width of 0.003 in with a 60 to 100 percent probability, depending upon the package type and size of the channel. It provides attribute data (accept/reject) for package integrity of finished, unopened packages. It is generally not effective in detecting pinholes and minute tears in package substrates.

In addition, visual inspection cannot be used for packages with two opaque substrates, as transparency of the seal surfaces is essential to the inspection. Its most applicable attribute is for monitoring package quality in production to detect any significant changes in heat-sealing process parameters, which may provide the first indication that the process is out of control. Additional testing using more sensitive methods for leak detection of packages under suspicion of having defects may be warranted to confirm whether the channel or void is in fact an unsealed area. Visual inspection is not considered to be the only means by which the manufacturer should evaluate for package integrity.

Internal Pressure Test. ISO 11607 describes the internal pressure test as applying an internal pressure to the sterile package while it is submerged in water and noting any escaping air bubbles. A Flexible Packaging Association (FPA) committee, Sterilization Packaging Manufacturers Council (SPMC), has published several standards for testing packaging. One of those standards, FPA/SPMC Standard 005–96, "Standard Test Method for Detection of Leaks in Heat Seal Packages-Internal Pressurization Method," details the internal pressure test method.

The advantages of using this method for determining package integrity are that it is very easy to perform the test. In addition, it is inexpensive to test a large sample size and obtain statistical significance in the test sample set. The equipment costs are low, since all that is required is a pressure source and a water bath.

This method has not been validated by round robin testing, and no precision and bias statement has been made as to its repeatability and reproducibility. Nor is its sensitivity for detecting leak size known. However, independent verification has proved its usefulness in detecting gross leaks in packages. Gross leaks in packages occur most often as a result of handling and distribution hazards that cause tears, gouges, and punctures. Package validations most often fail as a result of the rigors of shipping and distribution. This test is sufficiently sensitive to detect those types of defects caused by the hazards of distribution. Leaks in seals and in material surfaces can be detected using this method.

The method can be used for both porous and nonporous packaging materials. For packages with porous materials, the porous material substrate is sealed using a label or coating to reduce the porosity of the material. This facilitates the pressurization of the package and reduces the interpretation of what constitutes a leak and where a leak is occurring in the package. The porous material is not evaluated for leakage, as the coating may mask or block leaks. However, pinholes, tears, gouges, and channel leaks are readily apparent under an internal pressure that does not begin the separate the seals.

Validation of the method for the package under investigation must be performed to determine the proper internal pressure, and to evaluate the ability to detect channel and pinhole leaks over the permeation of air through the porous substrate.

Vacuum Leak Test. The vacuum leak test is similar in concept to the internal pressure leak test in that the result is a pass/fail for the detection of bubbles emanating from the package while submersed in a water bath. The method is described in ASTM D3078, "Standard Test Method for Leaks in Heat-Sealed Flexible Packages." The pressure differential is obtained by evacuating the chamber, causing the package to expand.

The difficulty in using this method for porous packages is that the pressure differential may not reach a point at which air passes through a channel or material leak before air passes readily through the porous material. Lowering the porosity of the material by coating it with a lacquer or other means could reduce this problem. This test is more suitable for nonporous packages that will expand under vacuum and create an internal pressure adequate to force air through leaks.

Dye Penetration Test. The ASTM F2 Committee has recently approved a dye penetration test method. The new standard, designated F-1929, "Standard Test Method for Detecting Seal Leaks in Porous Medical Packaging by Dye Penetration," finally provides a standardized method for conducting leak testing of package seals using a low-surface-tension solution and dye indicator. The basis of the test is that, when the test solution comes in contact with a channel or breach in the package seal, it will flow through the channel by capillary action. The leak will be indicated by a blue streak visible in the seal and/or a profuse and consistent flow of the dye through the channel.

This test method is generally considered to be more sensitive than the whole-package microbial challenge methods discussed earlier in this chapter. It is reported in a study on Tyvek-to-plastic pouches that seal defects down to 0.0015 in were readily detected with a blue dye solution (Nolan, 1996). The published test standard has verified by round robin testing that the smallest channel that can be reliably detected is on the order of 0.002 in. In fact, the detection rate for breathable pouches and trays with breathable lids was found to be 98 to 99 percent. It was discovered during the testing that significant reductions in test performance can be observed when indicator dyes other than toluidine blue were used. Also, the round robin results are specific for the wetting agent (Triton X-100) used for the solution.

The most effective application for the dye penetration test method is for detecting breaches in the seals of transparent packages, since seal defects must be observed easily. It is possible to use this method for opaque packages; however, observation of the seal leak must be made at the seal's outside edge and the exact location of the leak may be difficult to ascertain. One characteristic of this test methodology is that it is difficult to use for detecting leaks in the surfaces of package components. That is, pinholes, gouges, or abrasions of the materials cannot be detected, since the dye cannot be easily contacted with all of the package surfaces. So, although the dye penetration test is a sensitive leak indicator for seals, it is not a good whole-package integrity test. Other means must be used to detect material leaks, such as the bubble emission leak test. Other characteristics of this test method must be considered before incorporating it into a package validation protocol. The method is difficult to use for packages having a paper component, as the dye solution can destroy the material in a very short time—maybe even faster than the dye would travel through a channel. Other porous packages may allow the dye solution to wick through, causing difficulty in distinguishing a true leak from the permeation or wicking of the solution through the material. Since the dye solution is injected into the package, the method is destructive to the package and, in many instances, also to the product.

Gas Sensing Test Method. Up until now, there has never been a cost-effective means of performing this type of test. The introduction of a new technology that allows a trace gas (helium) to permeate through the porous component of a package has made nondestructive package integrity testing possible. The test is performed by first placing the test package into a specially designed housing.

This system is ideal for thermoformed trays with porous lids and flexible pouches with one porous side. The test has been shown to detect leaks as small as 0.002 in. Guidant and Medtronic have demonstrated the reliability of detecting leaks in blind tests and have quantified 100 percent of the purposely manufactured leaks in thermoformed trays (Hackett, 1996). In addition, there were no false positive readings in any of the unaltered packages.

The test method will be suitable for testing packages for package validation in which the package system is being designed and developed. In the short term, it is thought that this test methodology could replace the whole-package microbial challenge test methods, as it provides greater reliability, reduces the risks of false positives, and is similar in cost. In the long term, since the test housing is designed and manufactured for the package, this test methodology could be incorporated into a quality assurance program to validate the integrity of each and every package being manufactured. An ongoing 100 percent inspection or lot-to-lot sampling program would ensure the efficacy of the package process. The risk of a nonsterile package finding its way into the operating room would be virtually eliminated.

23.5.5 Conclusion

Package seal strength does not necessarily equate to package integrity. These two attributes of a finished medical device package are separate considerations in proving the efficacy of the package. Industry has developed methods for seal strength testing that are used to validate the package process. Although package seal strength is an important performance attribute, the ultimate acceptance of the package is based on its complete integrity. There are several means available for evaluating the integrity of sterile medical device packages. The application of a particular integrity test depends upon may factors, including the type of package, materials of construction, size, desired sensitivity, and objective of the test.

23.6 PACKAGE PROCESS VALIDATION

This section provides an overview of the package manufacturing and the elements that must be considered for validating the process.

23.6.1 Introduction

The product engineering team has developed an exciting new medical device that will improve the quality of life for many patients. The product has been tested and retested. Regulatory questions concerning the product have been defined and answered. Clinical trials to show that the product performs as intended have been completed. The manufacturing process has proved to be consistent and is fully documented. However, the challenge of bringing the device to the market is just beginning. Many more questions must be answered before the product can be safely distributed and used by the patient's caregiver. The most basic one is "How will I get the product to the caregiver in the condition required for safe and proper use?" The most basic answer is "By designing a package system that will combine with the device to create a total product that performs efficiently, safely, and effectively in the hands of the user.

At first glance, the issue of developing a package system seems uncomplicated and elementary. After all, what could be difficult about placing the device into a thermoform tray, covering it with a Tyvek lid, inserting it into a paperboard carton, and consolidating the cartons into a shipping unit? In actuality, the process of designing and developing a package for terminally sterilized medical devices is complex and complicated. This is because of all of the interactions of various processes, equipment, materials, and environments that combine to influence the package design and manufacture of the finished product.

For example, the product engineering team has developed the product as a disposable sterile device that must remain sterile at the point of end use. Therefore, the microbial barrier properties of the packaging materials, along with the suitability of forming and sealing, are crucial for assuring package integrity and product safety. So, the product and package materials must be compatible with the chosen sterilization process. In addition, the product will need to survive the rigors of transportation with its intrinsic hazards of shock, vibration, and environmental conditions. Finally, the manufacturer must have documented evidence that the performance of the package is not adversely affected over time.

Unfortunately, the product engineering team was unaware that there are regulations within the FDA and International community that require a formal package system qualification process and a documented validation program demonstrating the efficacy and reproducibility of all sterilization and packaging processes (i.e., forming, sealing, capping, cutting, and handling). At this point, the engineering staff has realized that the package design and development process should have been an integral part of the product development program and should not have been left to the very end of the development process. Serious delays in distribution of the product have resulted, since the package validation process requires significant time and effort to complete. The engineering team now turns to the Regulatory Affairs (RA) department for help in identifying the regulatory requirements for packaging.

Investigation by the Regulatory Affairs (RA) Department for the requirements imposed on packaging reveals an array of documents on the subject. Foremost is the Quality Systems Regulation (QSR), found in Title 21 CFR, Part 820. The requirements for components, device master records, and environmental controls that affect the selection and use of packaging appear throughout the QSR. However, the specific requirements for packaging are in Sec. 820.130. Further investigation discloses two international documents regulating the design and development of packaging include the International Standards Organization (ISO) 11607 "Packaging for terminally sterilized medical devices" and European Norm (EN) 868-1, "Packaging materials systems for medical devices which are to be sterilized—Part 1: General requirements and test methods." Both of these documents provide an outline of general requirements and test methods for validating the complete package system. RA has reviewed the two international documents and has found that they are very similar, but with a few significant differences. "What standard do we follow?" becomes the next basic question to answer.

FDA has helped answer this question by acknowledging the importance of international consensus standards. The FDA stated in the FDA Modernization Act of 1997: *Guidance for the Recognition and Use of Consensus Standards:*

> . . . conformance with applicable consensus standards can provide a reasonable assurance of safety and/or effectiveness. Therefore, information submitted on conformance with such standards will have a direct bearing on determination of safety and effectiveness made during the review of IDE's and PMA's. Furthermore, if a premarket submission contains a declaration of conformity to recognized consensus standards, this will in most cases, eliminate the need to review actual test data for those aspects of the device addressed by the standard.

Consequently, FDA has recognized the ISO 11607 standard as the consensus standard for manufacturing and quality control of packaging processes, materials, product package and design, and sterilization processes. Confusion about the existence of two packaging standards to conform to is a concern for medical device companies. However, conformance to the EN 868-1 standard will become a moot issue as the ISO 11607 standard undergoes a revision by the ISO TC 198 Working Group 7 to harmonize the differences. So now we know what needs to be accomplished in regards to packaging, right? We just need to perform a package process validation. That's simply a matter of following the ISO 11607 standard. Yes, but unfortunately it's not a cookbook recipe to success.

23.6.2 What Is Process Validation (PV)?

The FDA defines validation as "establishing by objective evidence that the process, under anticipated conditions, including worst case conditions, consistently produces a product which meets all predetermined requirements (and specifications)" Likewise, the ISO 11607 standard, "Packaging for terminally sterilized medical devices," defines validation as a "documented procedure for obtaining and interpreting the results required to establish that a process will consistently yield product complying with predetermined specifications." What these definitions are really saying in a practical sense is that a process validation must address:

1. The requirements or application of the package
2. The interaction of people and equipment used in the manufacture of the package
3. The consistency with which a package can be made
4. The effects of processing (e.g., sterilization) on the performance of the package
5. The storage and handling of the package

A manufacturer must become intimately involved with how the product is packaged and how to maintain consistency and uniformity. It must have proof that a process performs as it was intended.

The process validation (PV) consists of a series of qualifications of the processes making up the complete package system. These processes include the installation qualification (IQ), operational qualification (OQ), and performance qualification (PQ). Each facet of the packaging system must be challenged and qualified in order to claim validation of the entire system.

ISO 11607 addresses the package system validation in three phases, or clauses. Clause 4 specifies the basic attributes required for a wide range of materials as they combine and interact with various medical devices, packaging designs, sterilization methods, and distribution modes. Clause 5 defines the framework of activities to qualify the processes used to make and assemble the final package configuration. Clause 6 is intended to assist in the selection of tests and to provide criteria that can be used to evaluate the performance of the final package.

23.6.3 Why Is Package Validation Important?

The primary objective of a package process validation should be to provide the medical device manufacturer with a high degree of assurance that the product will reach the user in a condition suitable for optimum functionality and for its intended purpose.

The specific benefits of the package process validation include not only reducing the manufacturer's risk of product malfunction or the potential of a nonsterile operating condition but also improved customer satisfaction, improved manufacturing efficiencies, reduced costs, reduced development time, and compliance with regulatory requirements. The "Guideline on General Principles of Process Validation" provides valuable understanding of quality systems requirements and may be relied upon with the assurance of its acceptability to FDA.

23.6.4 The Total Validation Process

Prior to beginning any work on a validation, it is essential to write a protocol. The protocol provides a blueprint stating how testing is to be conducted, including the purpose, scope, responsibilities, test parameters, production equipment and settings, and the acceptance criteria for the test. Validation requires careful planning and preparation and it begins with a well-conceived and -written protocol. As was mentioned earlier, the validation process consists of a series of qualifications of unique processes that make up the complete package process system. This total package process system includes the final package design, the materials chosen for the package, and the ability to sterilize the product inside its package. The design of the package and its dynamic interactions with the product, the machinery used to assemble the package, the setup and maintenance of the machine, and consistency of production are other important considerations. If one of these processes is not right, the entire system breaks down and the manufacturer is at risk of malfeasance.

23.6.5 Package Forming and Sealing

While working with a packaging vendor, the package design has been completed. Vendors are usually responsible for validating that the materials are compatible with the sterilization process, and that compliance qualification tests are conducted. Appropriate materials have been selected and validated for compatibility with the intended product by the manufacturer. But how will we assemble the product into the package using the most efficient and consistent process? The package-sealing equipment for the job is identified and purchased; however, it must be properly installed before producing finished packages. ISO 11607 states that "before starting final process development, it shall be demonstrated that the process equipment and ancillary systems are capable of consistently operating within the established design and operating limits and tolerances." Clause 5 of the ISO standard addresses all of the issues in validation, such as equipment qualification, process development, process performance qualification, process control, and process certification and revalidation.

An equipment installation qualification is important because all production facilities have specific requirements for utilities, cleanliness, temperature and humidity, and other variables. For this reason, the equipment should be installed in its intended production location before qualification. The equipment is run at its operating limits to determine its performance capabilities relative to the manufacturer's specifications. In addition, all equipment change parts and accessories are assembled and checked out for proper fit. This phase of the validation includes verifying that the equipment will perform its intended function, establishing calibration and maintenance procedures, and identifying

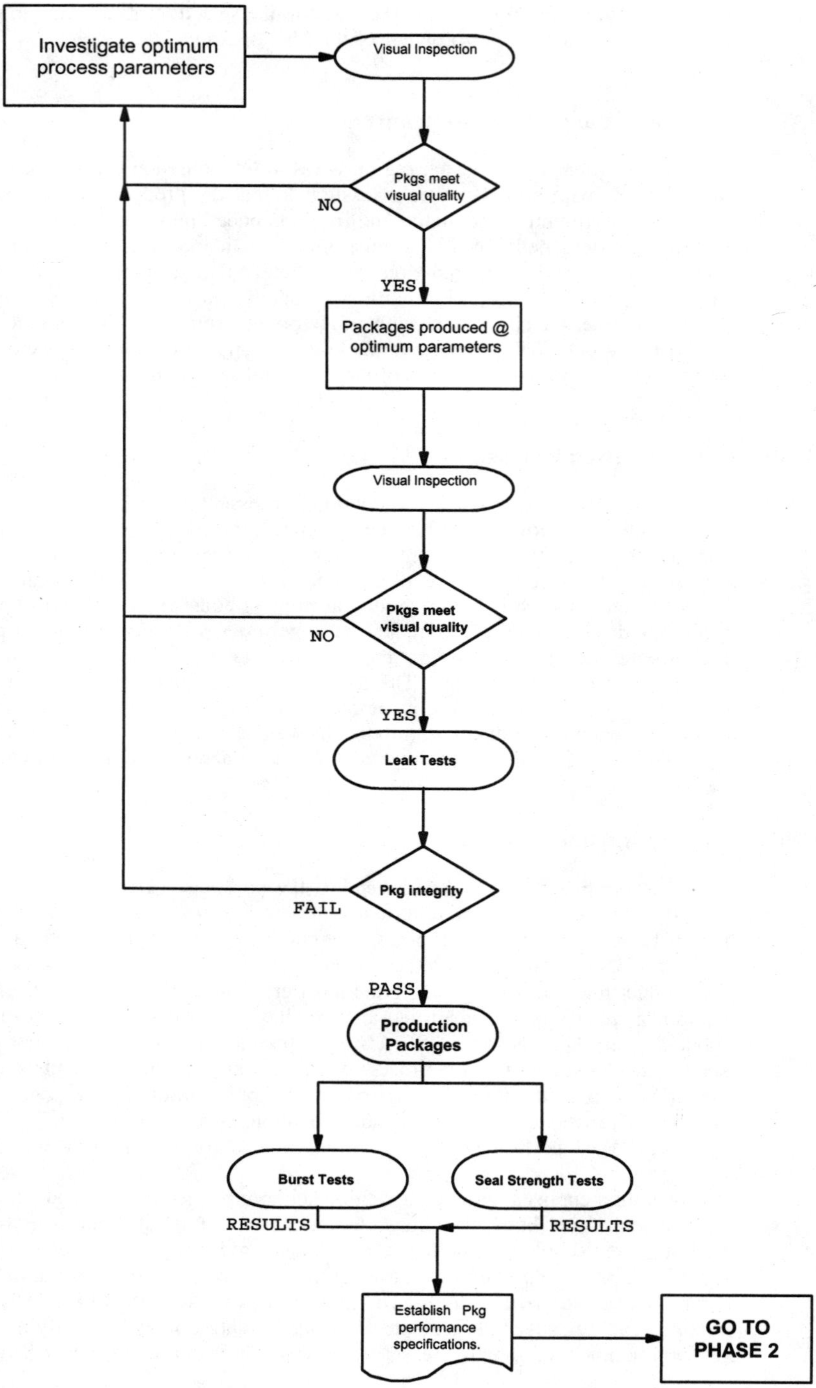

FIGURE 23.1 Flowchart of package manufacturing process qualification steps, phase 1.

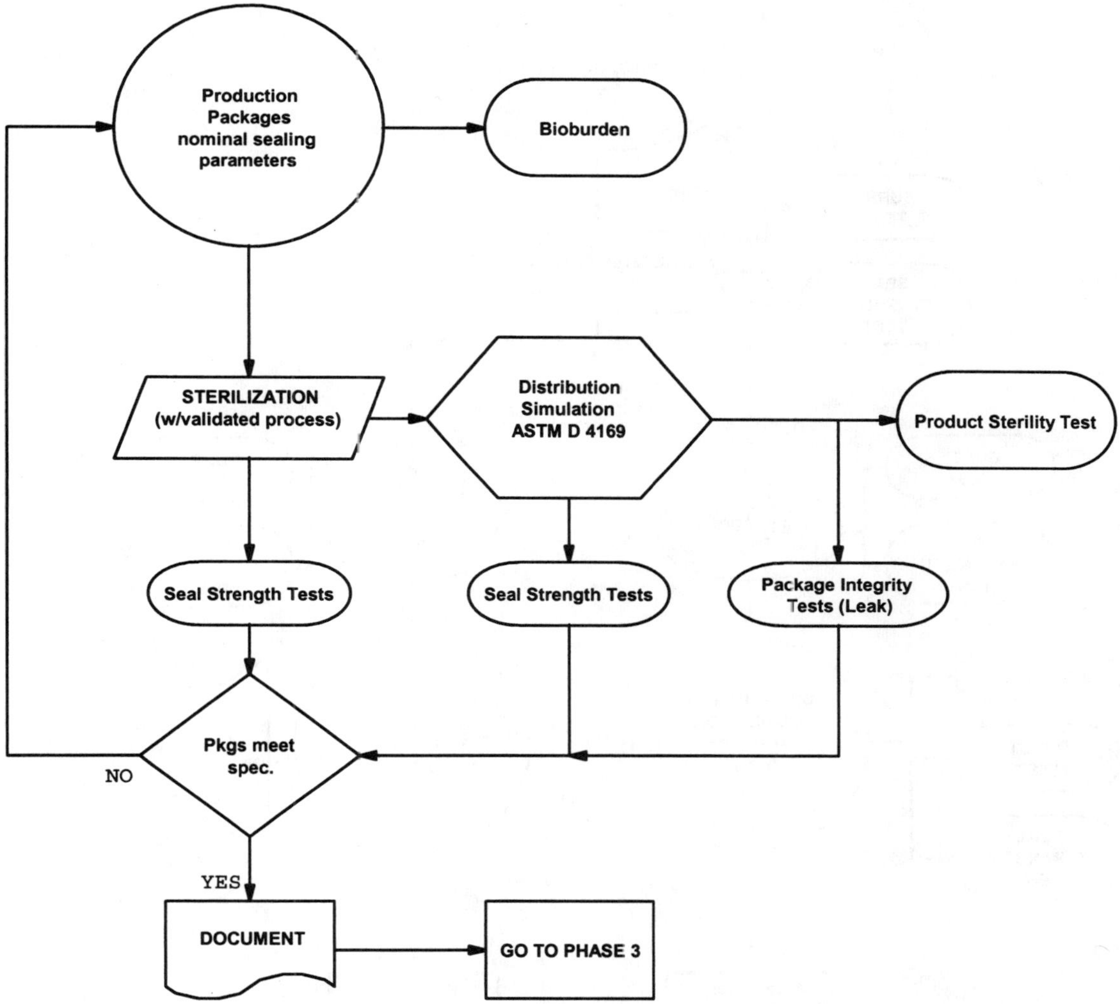

FIGURE 23.2 Flowchart of package process performance qualification steps, phase 2.

monitoring and control issues. Standard operating procedures (SOPs) must be written for calibration, maintenance, and repair.

Facilities management has confirmed that the equipment has been properly installed and that it performs in accordance with the manufacturer's specification. We can now begin to produce finished packages on the new equipment. The operational or process performance qualification is designed to provide a rigorous test to the effectiveness and reproducibility of the process. This is the most critical and time-consuming phase of the validation process. It requires a large degree of performance testing and evaluation. Tests must be chosen that measure relevant performance characteristics of the package for important attributes such as seal strength and integrity. ISO provides examples of tests that may be used for measuring these attributes, including ASTM F-88, ASTM D-903, ASTM F-1140 for seal strength, and several unpublished methods such as internal pressure, dye penetration,

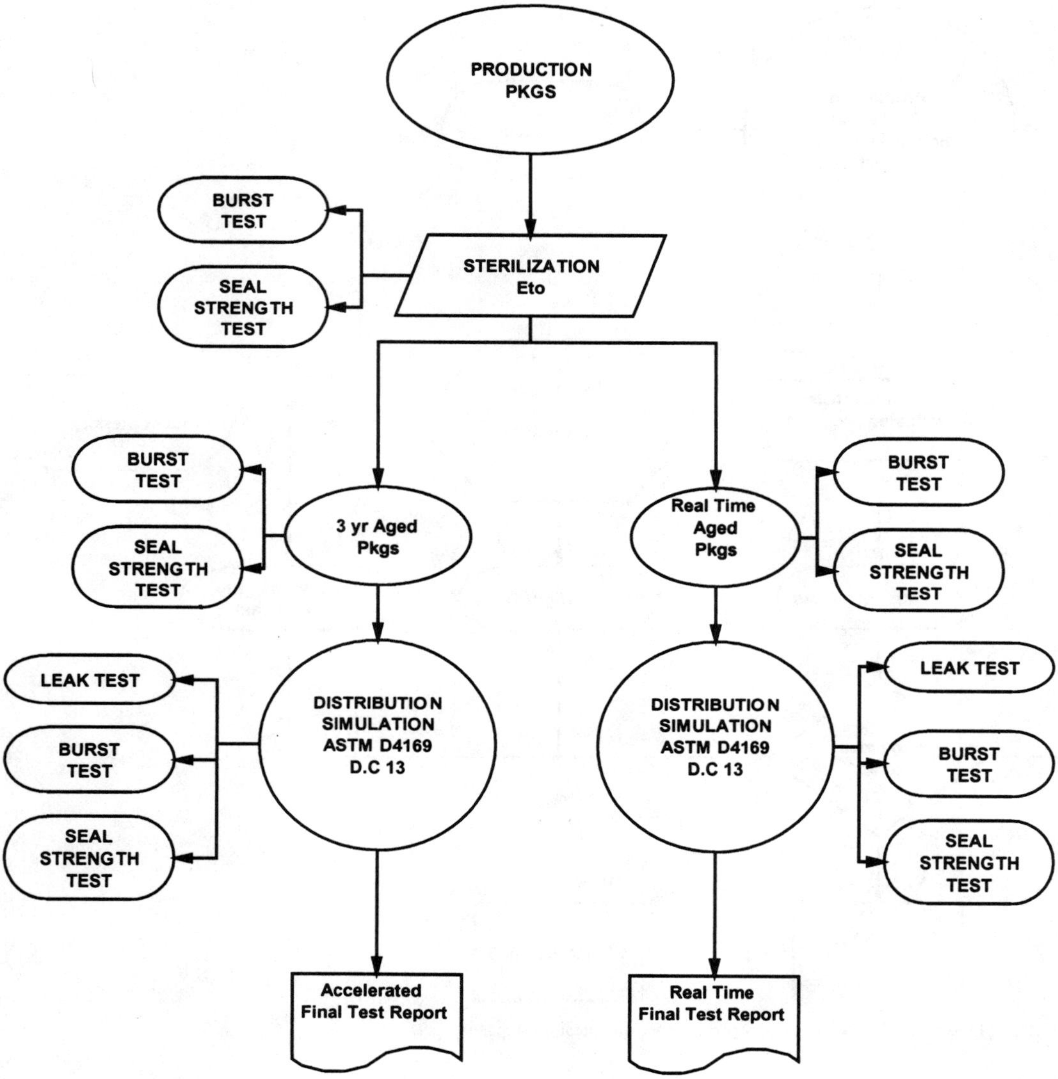

FIGURE 23.3 Flowchart of final package performance qualification, phase 3.

gas sensing, and vacuum leak tests for package integrity. These are physical test methods that ISO acknowledges can be used for demonstrating the integrity of the sterile package.

The first step in this phase of the validation is to establish the process parameters that produce acceptable package performance. This may be accomplished by testing packages produced from a matrix of process parameter variable combinations, or by a design of experiments (DOE) which will test only packages produced at the extreme range of the process parameters. Where limited quantities of packages are available, one combination of process parameters may be used to produce packages on the basis of historical experience and then test them for strength and integrity. If these process parameters do not produce packages meeting the prescribed performance specifications, then the process parameters are adjusted until acceptable packages are produced. The flowchart in Fig. 23.1 depicts one process for establishing the machine process parameters.

The packages are produced under normal operating conditions and on process equipment that has completed an installation qualification. When the optimum machine process parameters have been established, it is essential to determine the effects of sterilization, storage, and shipping and handling on the performance of the critical package attributes. This can be accomplished by measuring the seal strength and integrity after each of the individual processes and after the cumulative effects of all the processes. The flowchart in Fig. 23.2 depicts how this can be accomplished in a comprehensive protocol.

Through the rigorous step of process performance qualification, the manufacturing department has now documented the machine process parameters and established the package performance specifications for the package system being developed. The department is satisfied that the process is in control after measuring the consistency with which packages meet the performance specifications.

One final phase of the process validation should be to demonstrate that the combined effects of manufacturing, sterilization, storage, and handling do not have an adverse effect on the performance of the package produced under standard operating procedures. Clause 6 of the ISO standard address the issues associated with final package qualification. The flowchart in Fig. 23.3 depicts one protocol for assessing the integrity of the package after exposure to simulated, but realistic, events that the package will encounter during its useful life. These events may include, but are not be limited to, the manufacturing process itself, the sterilization process, storage or aging, and handling and shipping hazards.

23.6.6 Conclusion

Medical devices are developed using engineering principles and process qualification techniques to ensure that they perform as intended. So too must the package design and development process be qualified and validated. The complete validation requires a series of qualifications of the entire system, which ensures that the package will perform in harmony with the product in a consistent and safe manner. This is accomplished by developing a comprehensive plan that cannot be simplified or short-circuited. Product engineering has realized that, to accomplish the task of package process validation, the package system must be developed in tandem with the product development. Otherwise, delays of 6 to 12 months could result while the package system is being validated. The ISO 11607 standard provides guidance to assist medical device companies in developing a sterile medical device package system that performs efficiently, safely, and effectively in the hands of the caregiver. Since the standard provides designers and manufacturers of medical devices with a *framework* of laboratory tests and evaluations that can be used to qualify the overall performance of the package, there are many means within this framework to achieve the end result.

23.7 SHELF LIFE STUDIES

This section provides guidance for conducting accelerated aging studies for medical device packages. Developers of medical device packaging have struggled for years to justify shelf life claims and establish expiration dating for packaged medical devices. Much has been published over the past decade describing techniques for conducting accelerated aging programs. However, the theory of accelerated aging is complex enough for homogeneous materials, let alone device systems involving several different materials, such as complete medical device packages. The rapidly changing marketplace, technological developments, and regulations that govern them, demand that the manufacturer be responsive, which places a priority on the ability of the manufacturer to develop products meeting all of the regulatory burdens in a timely and expeditious manner. Establishing shelf life claims can be a significant bottleneck in the product development timeline. Real-time aging protocols would significantly hamper the product development cycle, as well as its marketability, and are impracticable in today's fast-paced environment.

The adoption of the European Medical Device Directive (MDD) in June of 1998 and the mandatory implementation of the CE label on all sterile medical devices marketed in the European Community have resulted in the compulsory use of expiration dates on all medical device packages. In order to obtain the CE label, *all* the "Essential Requirements" of the directive must be met. The MDD states that "the label must bear . . . where appropriate, an indication of the date by which the device should be used, in safety, expressed as the year and month."

The MDD's "Essential Requirements" are complied with by using harmonized standards. These standards may be European Norm (EN) or International Standards Organization (ISO) standards that meet the essential requirements of the Directive. For the development of medical device package systems, ISO 11607 has been developed and is used to meet essential *packaging* requirements of the Directive. Specifically, for meeting the Directive requirement as stated above, the ISO 11607 provision states "for medical devices with a defined shelf life, the manufacturer shall have documented evidence that the performance of the packaging is not adversely affected by storage under specified conditions for a period not less than the shelf life of the medical device." The net result is that manufacturers must supply documented evidence to support product-expiration claims. This is accomplished by monitoring measurable characteristics before, during, and after the test to determine the effects of time on package performance.

Expiration claims could be documented by real-time shelf life testing, however, the timelines for product development would be adversely affected. The developers of the ISO 11607 standard recognized this hindrance and therefore have allowed that "accelerated aging tests may be undertaken in addition to real-time aging tests by storage under conditions of increased severity." This provision is beneficial; however, no guidance is provided as to what conditions of adverse severity are permissible or technically reliable. It therefore has become crucial that guidance and standards be provided to help manufacturers establish product shelf life and expiration claims.

23.7.1 10 Degree Rule

There are no published standards for performing an accelerated aging study. Some guidance on accelerated aging of packages has been provided in the past. A landmark technical paper by Robert Reich (1988) introduced the Von't Hoff theory as an appropriate rationale for the accelerated aging of packaging. This theory, based on the Arrhenius rate kinetics theory of materials, states simply that a rise in temperature of 10°C will double the rate of a chemical reaction. The rule is commonly expressed as a Q_{10} value. So, for example, a doubling of the chemical reaction rate makes the Q_{10} value 2.0. The aging factor (AF) is derived from the following equation:

$$\mathrm{AF} = Q_{10}^{(T_H - T_L)/10}$$

where Q_{10} = rate of chemical reaction (usually 2.0)
T_H = high temperature (test temperature)
T_L = low temperature (ambient)

Figure 23.4 indicates the relationship between the aging temperatures versus equivalency to a 1-year room temperature aging using various Q_{10} values. Other authors such as Geoffrey Clark (1991) of the Food and Drug Administration (FDA) have used the Q_{10} rule as rationale for accelerated aging protocols. Clark's guidance document, *Shelf Life of Medical Devices* uses a test temperature of 40°C and a Q_{10} value of 1.8 for intraocular and contact lenses. This guidance has been applied by industry to other medical devices and package systems and represents a very conservative estimate for real-time aging equivalents.

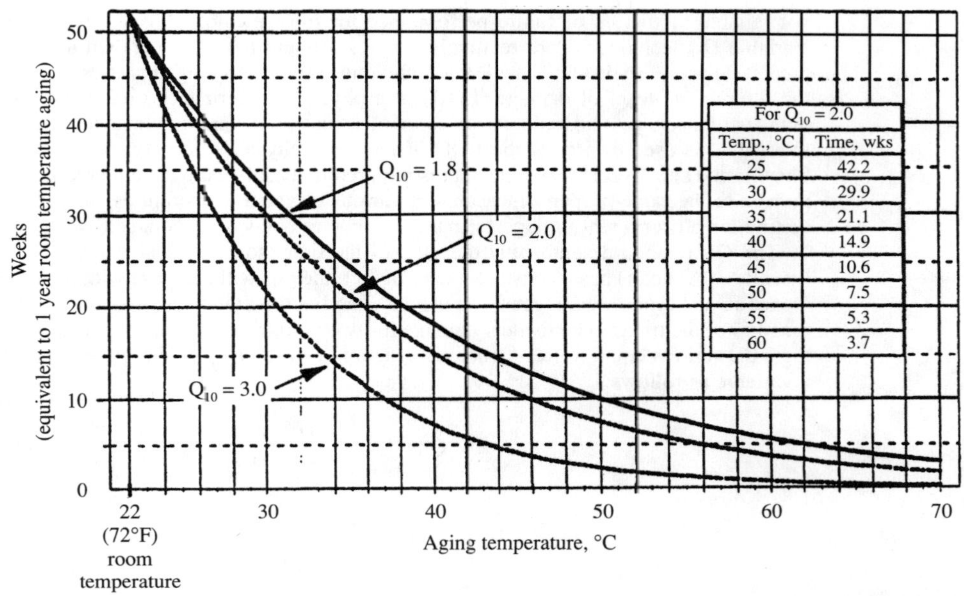

FIGURE 23.4 Accelerated aging of polymers: time versus temperature. Time (weeks) equivalent to one year room temperature aging when a polymer is heat aged at a selected temperature (°C). *(From Hemmerich, K., "General Aging Theory and Simplified Protocol for Accelerated Aging of Medical Devices," Medical Plastics and Biomaterials, July/August 1998,* pp. 16–23.) Q_{10} = Δ10°C reaction rate constant; Q_{10} = 1.8, conservative rate as suggested by G. Clark (FDA, 1991); Q_{10} = 2.0, conventionally accepted rate for first order chemical reaction; Q_{10} = 3.0, more aggressive rate; 60°C is the suggested upper temperature for most medical polymers.

Karl Hemmerich (1998) described the 10-degree rule (Q_{10}) in his article "General Aging Theory and Simplified Protocol for Accelerated Aging of Medical Devices." In it, he concludes, "the 10-degree rule will likely be conservative in the prediction of shelf life. However, the technique depends on numerous assumptions that must be verified by real-time validation testing conducted at room temperature." Reich suggested that using this approach for accelerated aging of medical grade packaging *should be used with some reservations, since the rate kinetics of the (packaging) systems are not fully understood*. Further, the Q_{10} values are based on the rate kinetics of a single chemical reaction, but the concept of accelerated aging of packages involves *assumptions regarding the uniform aging rates of one or more packaging materials, plus any adhesive reactions*. In addition, caution should be exercised that the aging temperatures do not produce unrealistic failure conditions that would never occur under real-time, ambient conditions. A temperature of 60°C is the suggested upper temperature limit for most medical polymers.

Reich concludes, however, that the *concept can be useful (as a rationale) for the accelerated aging of packages*. Hemmerich concurs that "this type of conservative relationship is appropriate for a wide range of medical polymers that have been previously characterized." Nevertheless, "the simplified protocol for accelerated shelf-life testing is not a replacement for more complex and advanced accelerated aging (techniques)."

23.7.2 Advanced Aging Techniques

John Donohue and Spiro Apostolou (1998) offer more complex and advanced techniques for predicting shelf life of medical devices. Their contention is that the Arrhenius and Q_{10} techniques are not reliable predictors of future performance for most medical devices. However, the D&A and Variable Q_{10} techniques "are relatively easy to use and have been shown to be more accurate in predicting actual shelf life." The D&A technique assumes nothing, and uses only the data to predict the future. The level of damage (LOD) of a physical performance property—such as brittleness, number of package seal failures, or color of a plastic at various elevated temperatures and time intervals—is used to predict the LOD of the same physical property of real-time aged materials. Short-term (i.e., 1 year) real-time data is required to establish the benchmark performance for comparison to the same property measured at various elevated temperatures, and for subsequently predicting longer-term real-time performance or time to equivalent damage (TED).

The Q_{10} method assumes that the ratio of the times to equivalent damage at low temperatures (usually 10°C apart) has a constant value. In fact, the Q_{10} will decrease with increasing temperature. Donohue and Apostolou suggest the use of a modified method, the variable Q_{10} method, in which the ratio of the times to equivalent damage for two temperatures is used as a variable. In this method the TED ratio is equal to the Arrhenius equation, and the Q_{10} is determined with the TED ratio as a variable as follows:

$$Q_{10}^{(T_H-T_L)/10} = \frac{\text{TED at } T_L}{\text{TED at } T_H}$$

So

$$Q_{10} = \left(\frac{\text{TED at } T_L}{\text{TED at } T_H}\right)^{1/[(T_H-T_L)/10]}$$

Again, it is necessary to acquire performance data for ambient storage conditions as well as for elevated conditions in order to determine the TED ratio, and before this method can be employed for predicting a variable Q_{10}.

Lambert and Tang (1997) describe a method of aging using an iterative process that provides an opportunity to refine and validate the initial, conservative aging factor (Q_{10}). The basic concept is to collect a number of parallel real-time aged and accelerated aged data points at early time points such that a correlation between the two can be developed, thereby defining the actual aging factor of the system under investigation. One limitation of this method is that real-time aged package performance data is required in which to compare accelerated aged data and make iteration on the conservative Q_{10}.

The American Society for Testing and Materials (ASTM) Committee F-2 on Flexible Barrier Materials published ASTM F-1980, "Standard Guide for Accelerated Aging of Sterile Medical Device Packages." The scope of the Guide is *to provide information for developing accelerated aging protocols to rapidly determine the effects due to the passage of time and environmental effects on the sterile integrity of packages and the physical properties of their component packaging materials.* The information obtained from utilizing this Guide may be used to support expiration date claims for medical device packages. It is hoped that it will provide the necessary rationale for accelerated aging protocols that satisfies both the FDA's Quality System Regulations (QSR) and the essential requirements for packaging in the MDD.

The Guide provides referenced documents (many of which are cited in this article) that give credibility to the current suggested methodology for aging medical device packages. The Guide condones the simplified Q_{10} method as a rationale for using accelerated aging for medical device packages. The basic eight-step concept was flowcharted by Lambert and Tang as shown in Fig. 23.5. The Guide states, "Conservative accelerated aging factors must be used if little information is known about the package under investigation." Although the method provides conservative estimates of product/package shelf life, resulting in longer test durations than would be necessary using more complex aging methods, it does not require benchmark real-time data up-front in the development process that could further delay introduction of new products. In addition, it requires fewer samples and conditioning resources. Still, it may be advantageous to refine the aging process in subsequent studies using the more complex techniques summarized in this article. With more information about the system under investigation and with information demonstrating the correlation between real-

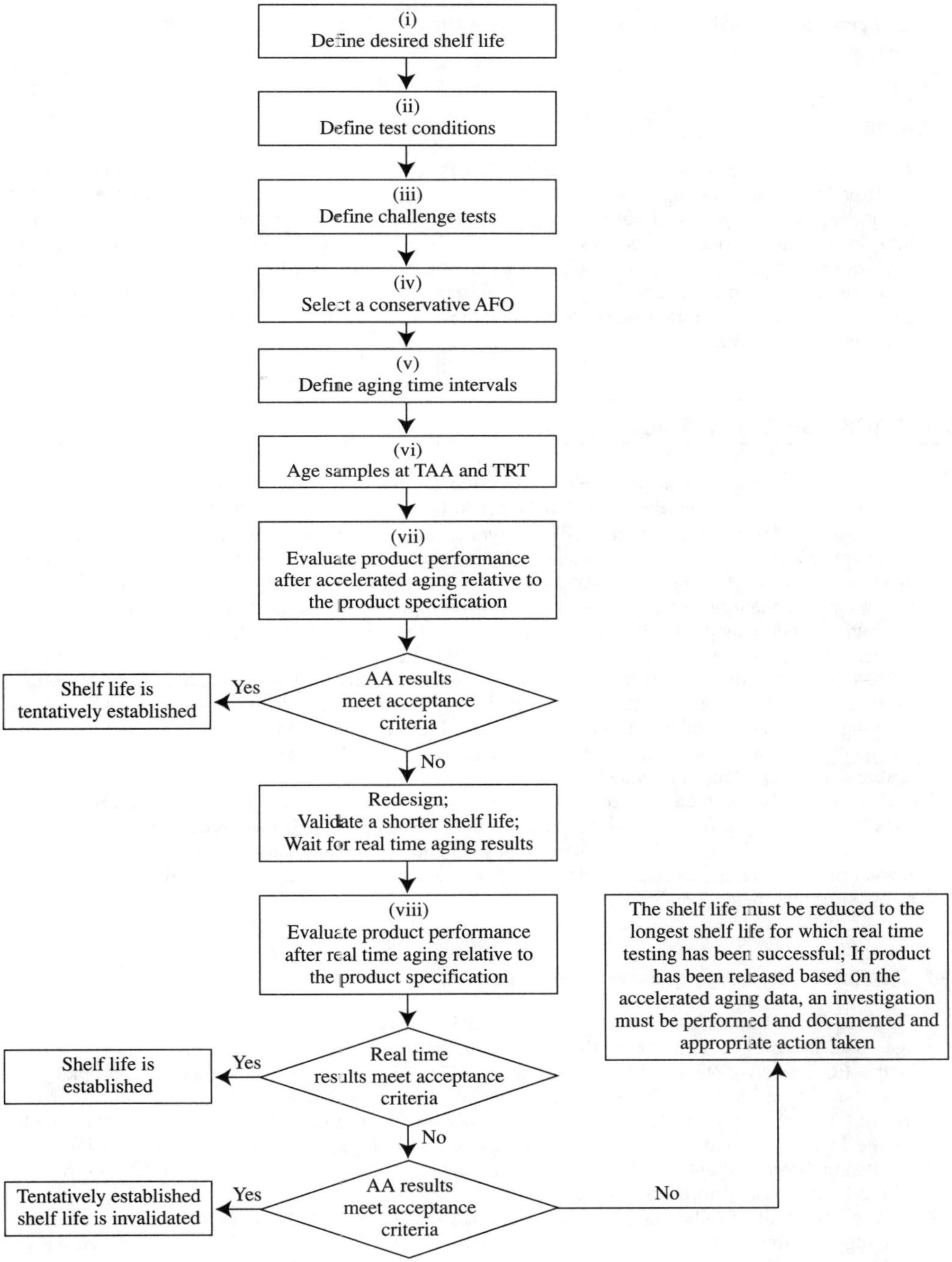

FIGURE 23.5 Eight-step process for accelerated aging.

time performance and accelerated aging performance, more aggressive and accurate aging factors may be defined.

23.7.3 Conclusion

There is no shortage of rationale to support accelerated aging protocols as demonstrated by the published literature. Any manufacturer using techniques described in the literature will be successful in meeting the provisions of national and international regulations. Some techniques require very little information about the system under investigation and make assumptions about material rate kinetics resulting in conservative estimates, while others require real-time performance data in order to define material rate kinetics and predict long-term performance. Which technique to choose for an accelerated aging program will depend upon the manufacturer's resources, expertise, and product development time lines.

23.8 FINAL PACKAGE VALIDATION

The efficacy of sterile medical device packages at the point of end use is of great concern to not only the producer of the product, but also the general public, and foremost the regulatory community. The Food and Drug Administration (FDA) has the regulatory responsibility to ensure that medical devices perform their intended function and pose no undo risk to the patient. Not only must the product itself meet stringent regulatory requirements, but the package must also perform consistently under variable manufacturing conditions, sterilization procedures, distribution hazards; and perhaps over an extended shelf life. It is apparent that over the years, the FDA has become increasingly concerned over the number of packaging-related field failures, and has issued a growing number of FDA-483 observations (Spitzley, 1991). Nonetheless, there have been few guidelines for determining the effectiveness of sterile medical packages. Industry and medical device organizations have been working for a number of years to develop standard test methods and guidelines for validating the integrity of packages, and to answer the questions of what constitutes the best package and how to ensure that it performs as intended.

It is generally accepted industry practice to evaluate the integrity of sterile medical device packages by subjecting a fully processed package to extremes in sterilization processes, performing a simulated shelf life or accelerated aging study, conducting a simulated distribution and handling stress test, and then evaluating the efficacy of the package for sterility through microbial challenge or physical test methods.

23.8.1 Shelf Life/Accelerated Aging/Expiration Dating

FDA requires documented evidence to support published expiration dates on medical device packages. The European Union has required expiration dates on all medical device packages as specified in the EC Directive 93/42/EEC, which states, "The label must bear . . . where appropiate, an indication of the date by which the device should be used, in safety, expressed as the year and month." Manufacturers are being forced to comply with European directives based on ISO Standards. As the FDA is moving toward harmonization with the European standards through revision of its GMP, and through adoption of ISO and CEN Standards, the need for guidance on the performance of accelerated aging protocols is crucial.

The net result of publishing expiration dates is that there must be some documented evidence that supports the product expiration claims, thus the need to perform shelf life studies. However, real-time shelf life studies are not an alternative in a fast-changing industry that can see two to three generations of products developed over the time it would take to document a 2-year shelf life claim. So, the need for accelerated aging protocols as an alternative in developing a product and introducing it into the marketplace in a timely fashion is essential, although concurrent real-time studies should be performed to substantiate results of accelerated studies.

Ideally, accelerated aging involves a single measurable characteristic under extreme conditions to simulate, in a short time, the conditions the package would likely be subjected to during its designated shelf life (Henke and Reich, 1992). The protocol shown in the flowchart rotates the packages through three environments designed to simulate the aging process—high temperature and high humidity, high temperature and low humidity, and freezing conditions. Low temperature is included since it has been implicated in package failure through cold creep and material embrittlement, and packages may, in fact, be exposed to these temperatures in wintertime distribution systems or in the cargo areas of aircraft.

There are no published standards for performing an accelerated aging study. Recently, Geoffrey Clark of CDRH's Division of Small Manufacturers Assistance stated FDA's position that companies develop their own packaging test protocols based on the regulations appropriate for a given device (Henke and Reich, 1992). These regulations are found mainly in the 21 CFR and guidance documents published by CDHR. The information in these documents can be reduced to four basic principles for determining shelf life and consequent expiration dating:

- Determine an acceptable target expiration date based on R&D data, on the likely distribution and storage conditions that the product will encounter prior to its use, and on the company's marketing strategies.
- Select the parameters that will be tested.
- Conduct the testing under consistent procedures.
- Once all the testing has been completed, validate the test data (Henke and Reich, 1992).

Notice that these principles do not explicitly define the test parameters. However, the guidance documents developed by CDHR do provide accelerated aging protocols for specific devices within their jurisdiction based on the Q_{10} theory for chemical reactions. So, the theory postulated by Von't Hof using the Q_{10} value (which states that a rise in temperature of 10°C will double the rate of chemical reaction) is the most convenient method of estimating the approximate ambient storage time equivalent at a selected accelerated aging temperature, despite the known limitations and concerns for use on complex and dissimilar material structures.

For the example shown in the flowchart, using an accelerated aging temperature of 55°C, the equivalent ambient storage time for 1 year is 26 days. Caution must be taken not to accelerate the aging too much, since elevating the temperature of packaging materials could result in a mode of failure that might never be observed in real life (material/product interaction, creep, deformation, etc.) (Reich et al., 1988).

It appears that the protocol offered in this example, using the Q_{10} theory as a basis, is becoming a default model for establishing data for expiration date claims. Until industry devotes the time and effort involved in correlating real-time aging performance to simulated aging performance, it appears this is the best methodology available.

23.8.2 Distribution Simulation Stress Testing

The second phase of the package validation protocol is based on the accepted fact that sterile medical device packages do not typically lose their sterility simply by being stored on a shelf. Package failures are a result of a dynamic event that may have occurred during the manufacturing process, during shipping and handling to the sterilization facility, or during distribution to the point of end use. All of these processes may subject the finished package to dynamic events involving handling shocks, vibration, and, high and low temperature extremes. The GMP for Medical Devices Part 820.130 states, "the device package and any shipping container for a device shall be designed and constructed to protect the device from alteration or damage during the customary conditions of processing, storage, handling, and distribution."

There are optional methods available to satisfy this segment of the package validation process. First, the package could be tested by simply shipping it to a destination using the anticipated shipping mode (overnight parcel, common carrier, etc.). This method, although economical, does not lend itself to a high degree of control and repeatability. Alternatively, laboratory simulations provide a means of subjecting packages to the anticipated distribution hazards of shock, vibration, and dynamic compression in a controlled and repeatable manner. Observations of the package performance, as it is subjected to various hazards, can be accomplished in the laboratory and corrective action can be taken to alleviate any anticipated problems in a timely fashion.

ISO/DIS 11607, "Packaging for Terminally Sterilized Medical Devices," references the International Safe Transit Association (ISTA) Project 1A preshipment test procedure and ASTM D-4169, "Performance Testing of Shipping Containers and Systems." This author believes that the ASTM procedure provides a better simulation of the distribution environment since it sequences a number of distribution "elements" or tests that use realistic test intensity levels. The ASTM method also allows users who have significant knowledge of their distribution system to design a test sequence that more closely matches their own environment.

The most typical distribution simulation used for medical device package validation is Distribution Cycle 13, which is designed for packages weighing less than 100 pounds that are being transported by air and motor freight (by UPS, Fed Ex, etc.). This test "provides a uniform basis of evaluating in the laboratory, the ability of shipping units to withstand the distribution environment. This is accomplished by subjecting the packages to a test plan consisting of a sequence of anticipated hazard elements encountered in the chosen distribution environment" (ASTM D-4169). The preferred method for performing the vibration test in Element G, "Vehicle Vibration," is the random option, since it provides the most realistic simulation of actual transport vibration environments.

23.8.3 Package Integrity Evaluation

Of course, to simply subject a medical device package to extremes in temperature and humidity conditions for an extended period of time, and then to "shake, rattle and roll" them during transportation simulation does not indicate the package's ability to maintain its sterile barrier. Herein lies one of the most controversial and difficult tasks facing the packaging engineer. What tests to use to determine the sterility of the package—microbial challenge or physical test methods?

FDA inspectors, reviewers, and compliance staff are increasingly asking for (microbial barrier) data during routine facility inspections, and as part of product submissions. Also, draft international standards contain provisions for specific microbial-challenge test protocols (Henke and Reich, 1992). However, microbial challenge testing is time-consuming, labor intensive, and costly. Also the method most commonly used is difficult to adapt to the myriad of package sizes currently available in the medical device industry (Freiherr, 1994). An alternative validation method would involve the use of microbial challenge to assess the barrier properties of various packaging materials, but would use various forms of physical testing to evaluate the integrity of the package as a whole (Freiherr, 1994). The premise for this method is that if the materials provide an adequate barrier and the seals and the material components of the package are undamaged and intact, then the package as a whole will provide a barrier to infectious agents. Regulators have some difficulty accepting this premise, however, since physical testing can indicate sterility of the package only indirectly, as opposed to the certainty of a direct microbial indicator inside the package.

The packaging task force assembled by the Health Industry Manufacturers Association (HIMA) is working to evaluate whether physical test methods are, in fact, more sensitive and better indicators of package sterility. Critics (of microbial challenge testing) note that physical testing methods (e.g., visual seal examination, burst and creep testing, seal strength tests, dye penetration tests, and black light fluorescence examinations) are more repeatable, reliable, and controllable (Spitzley, 1993). ASTM has developed test methods for determining tensile strength of materials, peel test methods for evaluating the peelability of lid stock from trays or pouches, burst resistance tests, and leak tests. Attempts to correlate these test methods with microbial challenge results are ongoing; however, no scientific evidence has been generated to date to substantiate the efficacy of one method over another. Some combination of physical and microbial test methods for both materials and whole packages appears to be the most advantageous program for meeting the expectations of regulators and fulfilling the corporate responsibility of ensuring the safety and effectiveness of its products.

23.8.4 Summary

It is now generally recognized that manufacturers must conduct all three types of tests-physical, transportation simulation, and microbial challenge-to validate packaging materials and processes (Henke and Reich, 1992). The protocol presented here offers the most comprehensive and justifiable methodologies, based on the published literature, for determining the effectiveness of a medical device package design to maintain its sterile condition from the point of sterilization to point of end use.

REFERENCES

AAMI Technical Information Report (TIR) No. 22-1998, "Guidance for ANSI/AAMI/ISO 11607-1997, Packaging for terminally sterilized medical devices," Association for the Advancement of Medical Devices, Arlington, VA, July 1998.

American Society for Testing and Materials, ASTM D-4169, "Performance Testing of Shipping Containers and Systems," ASTM Book of Standards, vol. 15.09.

ANSI/AAMI/ISO 11607:1997, "Packaging for terminally sterilized medical devices," Association for the Advancement of Medical Instrumentation, Arlington, VA, February 1997.

Clark, G., Shelf Life of Medical Devices, FDA Division of Small Manufacturers Assistance, Rockville, MD, April 1991.

The Council of the European Communities, Directive 93/42/EEC, *Medical Device Directive (MDD)*, June 1993.

Code of Federal Regulations 21 CFR Part 820, "Good Manufacturing Practices for Medical Devices: General."

Donohue, J., and Apostolou, S., "Predicting Shelf Life from Accelerated Aging Data: The D&A and Variable Q10 Techniques," *Medical Device and Diagnostic Industry,* June 1998, pp. 68–72.

Dyke, Denis G., "Medical Packaging Validation: Complying with the Quality System Regulation and ISO 11607," *Medical Device and Diagnostics Industry*, August 1998.

EN 868-1:1997, "Packaging materials and systems for medical devices which are to be sterilized—Part 1: General requirements and test methods." February 1997.

FDA Modernization Act of 1997, "Guidance for the Recognition and Use of Consensus Standards; Availability," *Federal Register*, February 25, 1998, vol. 63, no. 37, pp. 9561–9569 (www.fda.gov/cdrh/modact/fro225af.html).

Fielding, Paul, "Medical Packaging Legislation in Europe," *Medical Device and Diagnostics Industry*, November 1999.

Freiherr, G., "Issues in Medical Packaging: Cost-Consciousness Leads the Way in New World Order," *Medical Device and Diagnostics Industry,* August 1994, pp. 51–57.

Hackett, Earl T., "Dye Penetration Effective for Detecting Package Seal Defects," *Packaging Technology and Engineering,* August 1996.

Hemmerich, K., "General Aging Theory and Simplified Protocol for Accelerated Aging of Medical Devices," *Medical Plastics and Biomaterials,* July/August 1998, pp. 16–23.

Henke, C., and Reich, R., "The Current Status of Microbial-Barrier Testing of Medical Device Packaging," *MD&DI,* August 1992, pp. 46–49, 94.

Hooten, Fred W., "A Brief History of FDA Good Manufacturing Practices," *Medical Device and Diagnostics Industry*, May 1996.

Hudson, B., and Simmons, L., "Streamlining Package Seal Validation," *Medical Device and Diagnostics Industry*, October 1992, pp. 49–52, 89.

International Safe Transit Association, ISTA Project 1A, "Pre-Shipment Test Procedures."

ISO 11607, "Packaging for terminally sterilized medical devices," 1997.

Jones, Lois, et al., "In Quest of Sterile Packaging: Part 1, Approaches to Package Testing," *Medical Device and Diagnostics Industry,* August 1995.

Jones, Lois, et al., "In Quest of Sterile Packaging: Part 2, Physical Package Integrity Test Methods," *Medical Device and Diagnostics Industry*, September 1995.

Lambert, B., and Tang, F., "Overview of ANSI/AAMI Material Qualification Guidance; Iterative Accelerated Aging Method," Proceedings 1, Session 108 (55–64), *Medical Design and Manufacturing*, West, Anaheim, CA, 1997.

Medical Device Directive (MDD), "Council Directive 93/42/EEC," *Official Journal European Communities*, **14**(10), 1992.

Nolan, Patrick J., "Physical Test Methods for Validating Package Integrity," *The Validation Consultant*, July 1996, vol. 3, no. 6.

Nolan, Patrick J., "Medical Device Package Design: A Protocol for Sterile Package Integrity Validation," *Medical Device and Diagnostics Industry*, November 1995.

O'Brien, Joseph D., *Medical Device Packaging Handbook*, Marcel Dekker, New York, 1990.

Process Validation Field Book, Smartware, 1997

Reich, R., Sharpe, D., and Anderson, H., "Accelerated Aging of Packages: Consideration, Suggestions, and Use in Expiration Date Verification," *Medical Device and Diagnostics Industry,* March 1988, pp. 34–38.

Spitzley, J., "A Preview of the HIMA Sterile Packaging Guidance Document," *MD&DI*, December 1991, pp. 59–61.

Spitzley, J., "How Effective Is Microbial Challenge Testing for Intact Sterile Packaging?," *Medical Device and Diagnostics Industry*, August 1993, pp. 44–46.

Tweede, Diana, and Goddard, Ron, *Packaging Materials*, Pira International, Leatherhead, Surrey, U.K., 1998.

21 CFR, Part 820, "Good Manufacturing Practices for Medical Devices: General."

CHAPTER 24
DESIGN OF MAGNETIC RESONANCE SYSTEMS

Daniel J. Schaefer
General Electric Medical Systems, Milwaukee, Wisconsin

24.1 INTRODUCTION

Atomic nuclei containing odd numbers of nucleons (i.e., protons and neutrons) have magnetic moments. Hydrogen (^1H) nuclei (protons) have the highest magnetic moment of any nuclei and are the most abundant nuclei in biological materials. To obtain high signal-to-noise ratios, hydrogen nuclei are typically used in magnetic resonance imaging and spectroscopy. Note that many other nuclei (e.g., ^2H, ^{13}C, ^{19}F, ^{23}Na, ^{31}P, and ^{39}K) may also be studied using magnetic resonance.

In the absence of an external static magnetic field, magnetic moments of the various nuclei point in random directions. So, without a static magnetic field, there is no net magnetization vector from the ensemble of all the nuclei. However, in the presence of a static magnetic field, the magnetic moments tend to align. For ^1H nuclei, some nuclei align parallel with the static magnetic field, which is the lowest energy state (and so the most populated state). Other ^1H nuclei align antiparallel with the static magnetic field. The energy of nuclei with a magnetic moment, m. in a static magnetic field, B_0, may be expressed as[1]:

$$W_m = mB_0 \tag{24.1}$$

The difference in energy between protons aligned with the static magnetic field and those aligned antiparallel is the energy available in magnetic resonance (MR) experiments. This energy is twice that given in Eq. (24.1). Recall that the kinetic energy of the same nuclei at temperature T may be expressed as [2]:

$$W_T = KT \tag{24.2}$$

where K is Boltzmann's constant. The fraction of nuclei aligned with B_0 may be expressed as:

$$\eta = 1 - e^{-2mB_0/KT} \approx \frac{2mB_0}{KT} \tag{24.3}$$

For protons at 1.5 T at body temperature (37°C), about one proton in 100,000 is aligned with the

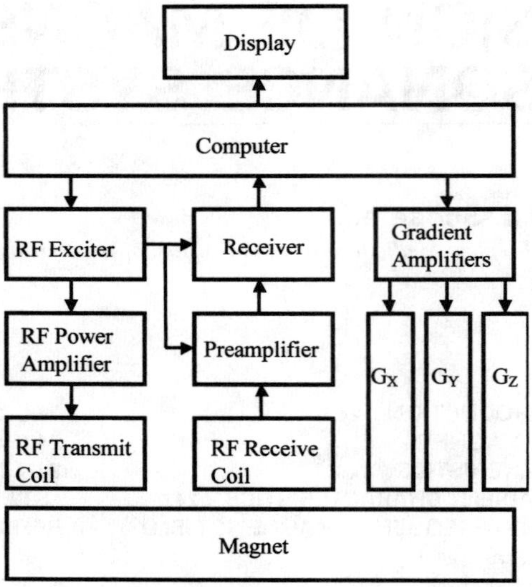

FIGURE 24.1 Components composing a MR system.

static magnetic field. Aligned protons provide the MR signal. So, assuming all other parameters equal, higher static magnetic fields provide higher signal levels.

Nuclei with magnetic moments precess in static magnetic fields at frequencies proportional to the local static magnetic field strength. Let B_0 represent the static magnetic field strength, let γ represent a proportionality constant called the magnetogyric ratio, and let the radian precession frequency be ω (= $2\pi f$, where f is the linear frequency of precession). Then the relationships between these quantities may be expressed mathematically as the Larmor[3] equation:

$$\omega = \gamma B_0 \tag{24.4}$$

Properly designed coils may receive signals induced by the time-varying magnetic flux. Ideally, magnetic resonance scanners would produce perfectly homogeneous magnetic fields. In magnetic resonance spectroscopy (MRS), nearby nuclei with magnetic moments may alter the local static magnetic field and the precession frequency so that various chemical components may be identified by the received spectrum.

If small, linear "gradient" magnetic fields are added to the static magnetic field, then received frequency would correlate to physical location. Magnetic resonance imaging uses magnetic field gradients to spatially encode all three dimensions. Note that the most widely used nucleus in MR is the hydrogen nucleus or proton.

For diagnostic purposes, signals from various tissues should differ sufficiently to provide contrast to distinguish them. There are two relaxation processes in magnetic resonance.[4] One mechanism is called spin-lattice or T_1 relaxation. In the absence of a static magnetic field, a collection of nuclei with magnetic moments are randomly oriented and the net macroscopic magnetic moment vector is zero. In the presence of a static magnetic field, the collection of nuclei with magnetic moments has a net macroscopic magnetic moment vector aligned with the static magnetic field. Consider a static magnetic field in which there are nuclei with magnetic moments. When resonant RF pulses excite the nuclei, the macroscopic magnetic moment vector tips by some angle related to the RF waveform. Gradually, the nuclei lose energy to the lattice and the macroscopic magnetic moment vector relaxes back to alignment with the static magnetic field. This type of relaxation is called spin-lattice or longitudinal or T_1 relaxation. Biologically relevant T_1 values are typically in the 100- to 2000-ms range.[5]

The other relaxation mechanism is called spin-spin or transverse or T_2 relaxation. The presence of other nuclei with magnetic moments causes changes in the local magnetic field. These changes lead to slightly different precession frequencies for the spins. As the spins get out of phase, signal is lost. Note that $T_2 \leq T_1$, because T_1 depends on T_2 loss mechanisms as well as others. Typical T_2 values of biological interest are in the 20 to 300 ms range.[5]

Fortunately, various tissues differ in their T_1 and T_2 properties. Different imaging sequences and pulse parameters can be used to optimize contrast between tissues. So, MR pulse sequences are analogous to histological stains; different sequences and parameters can be used to highlight (or obscure) differences.

Magnetic resonance scanners use static magnetic fields to produce conditions for magnetic resonance (see Fig. 24.1). In addition, three coil sets (along with amplifiers and eddy-current correction devices) are needed to spatially encode the patient by producing time-varying gradient magnetic fields. Radio-frequency (RF) transmit and receive coils, amplifiers, and receivers are used to excite the nuclei and to receive signals. Computers are useful to control the scanner and to process and display results (i.e., images, spectra, or flow velocities). Other equipment includes patient tables, patient gating systems, patient monitoring equipment, and safety systems.

24.2 MR MAGNET CHARACTERISTICS

Static magnetic fields of MR scanners are generated either by resistive electromagnets, permanent magnets, or (more commonly) by superconducting magnets. Superconducting magnets are usually the least massive. Superconducting magnets use cryogens. When superconducting magnets quench (i.e., when they warm up and are no longer superconducting), proper venting must prevent asphyxiation hazards from developing. In addition, mechanical design must prevent magnet damage from quenches.

Typically, the static magnetic field is parallel to the floor and aligned with the long (superior/inferior) axis of the patient. However, there are systems where the static magnetic field is along the anterior/posterior axis of the patient and some where the static field is along the left/right patient axis. While patients are typically horizontal, there are some magnets that allow the patient may be vertical. Most superconducting magnets have horizontal openings for the patient and are typically at fields strengths of 0.5 to 3 T. Most vertical magnets are permanent or resistive, though there are resistive superconducting magnets as well. Vertical magnets currently have field strengths up to 0.7 T.

Magnetic fringe fields from large, strong, unshielded magnets used in MR could require large areas to accommodate siting. To alleviate this problem, passive shielding can be achieved using ferromagnetic materials arranged as numerically determined. Often, magnets are actively shielded (sometimes in addition to some passive shielding). Bucking coils that oppose the static magnetic field are added to increase the rate the static magnetic field diminishes with distance. Actively shielded magnets decrease siting costs.

Many superconducting magnets employ recirculation devices to prevent loss of cryogens. Such systems have lower operating costs.

24.2.1 Field Strength and Signal-to-Noise Ratio (SNR)

As discussed above, the fraction of nuclei that are available for MR interactions increases with static magnetic field strength B_0. Noise in MR scans depends on the square root of the product of 4 times the bandwidth, temperature, Boltzmann's constant, and the resistance of the object to be imaged.

Note that increased bandwidth (leading to a higher noise floor) may be needed as B_0 inhomogeneity becomes worse. As discussed below, B_0 inhomogeneities may also lead to some signal loss.

In magnetic resonance imaging raw data are acquired over some period of time. The raw data are then converted into image data through the use of Fourier transforms.[4] Any temporal instability in B_0 will result in "ghosts," typically displaced and propagating from the desired image. Energy (and signal) in the desired image is diminished by the energy used to form the ghosts. So, image signal is lost and the apparent noise floor increases with B_0 temporal instabilities. B_0 fields of resistive magnets change with power line current fluctuations and with temperature. Resistive magnets are not used for imaging until some warm-up period has passed. Permanent-magnet B_0 will drift with temperature variations. Superconducting magnets have the highest temporal B_0 stability a few hours after ramping to field.

Another source of B_0 instability is the movement of nearby objects with large magnetic moments such as trucks and forklifts. Such objects may vary the static magnet field during imaging, resulting in image ghost artifacts propagating from the intended image. This effect depends on the size of the magnetic moment, its orientation, and on its distance from the magnet isocenter. Siting specifications typically are designed to prevent such problems. Note that a common misperception is that actively shielded magnets reduce susceptibility to B_0 instabilities from nearby magnetic moments. Unfortunately, this is not the case.

24.2.2 B_0 Homogeneity

Inhomogeneous static magnetic fields can result in apparent T_2 values called T_2^*, which are shorter than T_2. Let the inhomogeneity be ΔB_0, then T_2^* may be expressed as[6]:

$$\frac{1}{T_2^*} = \frac{1}{T_2} + \frac{\gamma \, \Delta B_0}{2} \tag{24.5}$$

Spin echo pulse sequences use a 90° RF pulse followed after half an echo time (TE) by a 180° RF pulse. Spin-echo signals s decay as[7]

$$s \propto e^{-T_E/T_2} \tag{24.6}$$

Shorter T_2 results in less signal. The static field of MR scanners must be very uniform to prevent signal loss and image artifacts. Typically, B_0 inhomogeneity of MR scanners is about 10 parts per million (ppm) over perhaps a 40-cm-diameter spherical volume (dsv) for imaging.[8] In spectroscopy, measurements of small frequency shifts must be accurately made and B_0 inhomogeneity is typically limited to perhaps 0.1 ppm over a 10-cm dsv.[8]

Clearly, MR magnets demand high homogeneity of the static magnetic field. In solenoidal magnets, geometry and relative coil currents determine homogeneity. A perfectly uniform current density flowing orthogonal to the desired B_0 vector on a spherical surface will produce a perfectly uniform B_0 field in the sphere.[8] Patient access needs render such a design impractical. A Helmholtz pair (two coils of the same radius spaced half a radius apart with the same current flowing in the same direction) is a first approximation to the uniform spherical current density. Typically four or six primary (not counting active shield coils) coils are used. Higher degrees of homogeneity are possible as more coils are used.

No matter how clever the magnet design, the local environment may perturb the desired static magnetic field. Field "shims," in the form of either coils (which may be resistive or superconducting) or well-placed bits of ferromagnetic material, or both, are used to achieve the desired magnet homogeneity. The static magnetic field is sampled at numerous points on a spherical or cylindrical surface. Then field errors are expanded in terms of, for example, spherical harmonics. Shim coils typically are designed[9] to produce fields that approximate the desired harmonic (or other expansion). The current appropriate for correcting each term is then set for each coil shim. Alternatively the correct size and position of each shim is calculated. The lowest-order shims can usually be achieved by placing constant currents on the three gradient coil axes.

24.2.3 Forces

Forces on ferrous objects near magnets may be of concern. The acceleration a (normalized to that of gravity g) of objects experiencing magnetic forces depends on the permeability of free space, μ_0, susceptibility χ, density ρ, and magnetic field B and its spatial gradient[10]:

$$a = \frac{\chi}{\mu_0 \rho g} B \frac{\partial B}{\partial z} \tag{24.7}$$

From Eq. (24.7) it is clear that the greatest forces on ferromagnetic objects occur where the product of field strength and spatial gradient is the largest. For superconducting magnets, this position is normally close to the coil windings.

24.3 GRADIENT CHARACTERISTICS

A computer commonly generates digital waveforms for the three gradient axes and for the radio-frequency coils. These waveforms (which for gradients may include corrections for eddy currents) are converted into analog signals, amplified, and sent to the appropriate coils. Received signals are converted into digital signals and reconstructed into images using Fourier transforms.[4] The reconstructed images are then electronically displayed. The computer system may also monitor MR scanner subsystems, including those associated with patient safety.

In MR, the static magnetic field is usually taken as the z direction. Linear variations of the static magnetic field (i.e., $\partial B_z/\partial x$, $\partial B_z/\partial y$, and $\partial B_z/\partial z$) are produced by separate gradient coil sets for the three (x, y, and z) coordinates. Only gradient field components in the z direction matter for MR imaging physics. However, magnetic fields form closed loops. So, other non-z components are produced as well. These other components may produce unwanted imaging artifacts and may influence patient physiological responses to switched gradients.

Considerations in gradient coil design include gradient linearity, gradient slew rate (i.e., how quickly the gradient amplitude can change), gradient power dissipation, eddy currents, and gradient-induced nerve stimulation. For simplicity in discussing these issues, consider the Maxwell pair (see Fig. 24.2).[11] If two filamentary circular loops carry current I in opposing directions, have radii a, and are spaced a distance $2d$ apart, then the magnetic induction B may be expressed as

$$\begin{aligned} B &= \frac{\mu_0 I a^2}{2(a^2 + (z - d)^2)^{3/2}} - \frac{\mu_0 I a^2}{2(a^2 + (z + d)^2)^{3/2}} \\ &\approx \left(\frac{3\mu I a^2 d}{(a^2 + d^2)^{5/2}}\right) z - \left(\frac{5\mu I a^2 d(3a^2 - 4d^2)}{(a^2 + d^2)^{9/2}}\right) z^3 + \left(\frac{21\mu I a^2 d(5a^4 - 20a^2 d^2 + 8d^4)}{8(a^2 + d^2)^{13/2}}\right) z^5 + \text{higher orders} \end{aligned} \tag{24.8}$$

The portion of Eq. (24.8) following the approximation sign is a Taylor series expansion about $z = 0$. Note that if the factor of z^3 were zero, gradient error would be reduced to terms dependent on z^5. Selecting $d = 0.5\sqrt{3}$ makes the z^3 factor vanish. The remaining relative error from the ideal gradient may be approximated by dividing the fifth-order z factor by the first-order z factor:

$$\text{Error} \approx \frac{176 z^4}{343 a^4} \tag{24.9}$$

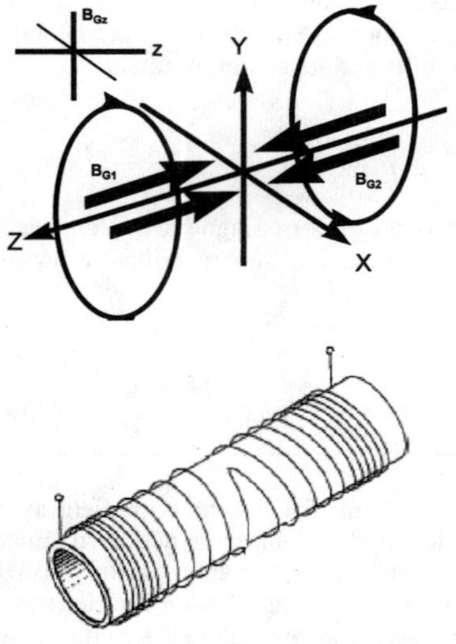

FIGURE 24.2 A filamentary, single-turn Maxwell coil pair z gradient coil and a typical unshielded z gradient coil.

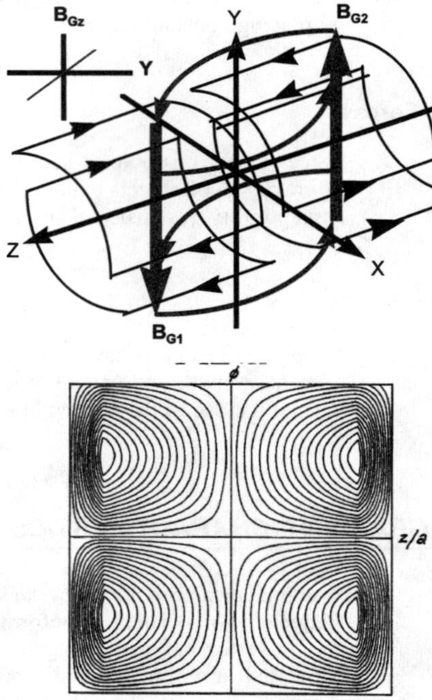

FIGURE 24.3 A filamentary, single-turn, saddle transverse gradient coil set and patterns for a typical unshielded transverse gradient coil is shown.

Equation (24.9) predicts that, for Maxwell coils, the gradient deviates 10 percent from the ideal linearity at $z/a = 0.66$. For $a = 0.3$ m, this deviation takes place at $z = 0.2$ m (e.g., the field of view would be 40 cm if 10 percent deviation from linearity was the maximum desired).

The first term in z of the expansion of Eq. (24.8) is the gradient amplitude G for a Maxwell pair. So from Eq. (24.7), the current needed to produce a gradient G may be expressed as:

$$I = \frac{49Ga^2\sqrt{21}}{144\mu_0} \tag{24.10}$$

Let R_L be the resistance per unit length of the coil. The total resistance of a Maxwell pair then is $4\pi aR_L$. The power P dissipated in the gradient coil depends on the product of the total resistance and the square of the current:

$$P = 4\pi aR_L\left(\frac{49Ga^2\sqrt{21}}{144\mu}\right)^2 = \frac{16{,}807\pi R_L G^2 a^5}{1728\mu^2} \tag{24.11}$$

Equation (24.11) illustrates that gradient dissipation goes with the fifth power of coil diameter and with the square of gradient strength. So, a Maxwell head gradient coil with half the diameter of a whole-body Maxwell gradient coil dissipates only 3 percent as much power (assuming gradient strength is unchanged).

Axial gradient coil sets of commercial MR scanners typically use more than the two coils that make up a Maxwell pair. Normally, gradient field series expansions are made and coil location, radius, and current are selected to obtain high linearity or low inductance.[12,13] One such axial coil[12] is shown in Fig. 24.1. Often an outer bucking gradient coil is combined with the inner coil to cancel gradient-induced fields on nearby conductive structures.[14] Undesirable magnetic fields (and resulting image artifacts) associated with eddy currents can then be considerably reduced.

Typical transverse (x and y) gradient coils for superconducting systems include four planar coil sets bent around a cylinder. One very simplified configuration is shown in Fig. 24.3. The pair of coils at one end of the cylinder produces a magnetic vector that for a y gradient, for example, points up. The pair at the other end of the cylinder produces a magnetic field that points down. Near isocenter, magnetic vectors of transverse gradient coils point along z and produce the desired gradient field. Magnetic induction from these coils will be highest near coil conductors where fields point up or down. For transverse coils, the largest gradient magnetic field components near patients is not along z. A "thumbprint" transverse coil[13] is also shown in Fig. 24.3.

Switched gradient coils with inductance L will experience a voltage V that depends on the time (t) rate of change of gradient current:

$$V = L\frac{dI}{dt} \tag{24.12}$$

Gradient coils must be designed to avoid electrical breakdown for the highest desired dI/dt levels.

24.3.1 Gradient-Induced Eddy Currents

Gradient-induced eddy currents produce unwanted distortions to the desired magnetic field. Eddy currents can cause signal loss, ghosting, and incomplete cancellation of static material in angiographic imaging. Eddy currents may be considerably reduced by using actively shielded gradient coils to null fields on conductive surfaces.[14] The addition of such "bucking" coils reduces gradient strength per unit current for the coils.

It is also possible to reduce eddy current effects by predistorting gradient waveforms. Eddy currents combined with predistorted waveforms result in intended gradient waveforms.[15–17] Predistortion can not correct for spatially dependent eddy current effects.

24.3.2 Gradient-Induced Stimulation

As MR imaging has evolved, so has the demand for higher gradient slew rates. Higher slew rates translate to shorter echo times, higher signal, less distortion artifact, and the possibility of imaging faster biological events. Lossy inductances have associated time constants of inductance/resistance. So, higher gradient slew rates imply lower gradient coil inductances and typically larger, faster gradient amplifiers. High gradient slew rates will induce electric fields in patients. It is imperative that these gradient-induced electric fields be limited to values incapable of harming patients. Safety standards[18–20] are designed to protect patients from cardiac stimulation by a significant margin through avoiding gradient-induced patient discomfort from peripheral nerve stimulation.[21–47]

24.3.3 Acoustic Noise

Time-varying magnetic field gradients spatially encode the anatomy of the patient in MR imaging. Switched gradients may also produce acoustic noise. There are two sources of gradient-generated acoustic noise. A conductor of length l carrying current I in a static magnetic field B_0, will experience a force F_m due to the magnet[48,49]:

$$\mathbf{F}_m = I\mathbf{l} \times \mathbf{B}_0 \tag{24.13}$$

There is also a force on the coil that is independent of the static magnetic field. Consider the potential energy U_b of a gradient coil with inductance L[49]:

$$U_b = \tfrac{1}{2}LI^2 = \int F_c \, dq \tag{24.14}$$

F_c is the force on the coil due to a displacement and q is an arbitrary coordinate. Note that the force on the coil will be in a direction that increases inductance. Hence, the force will try to increase the radius or compress the turns axially. From the derivative of the above equation, an expression for the force may be obtained[49]:

$$F_c = \frac{I}{2\left(\dfrac{dL}{dq}\right)} \tag{24.15}$$

The acoustic noise produced by forces on gradient coils must be limited to appropriate levels to avoid hearing loss.[18–20,50–55] Various schemes to reduce gradient-produced acoustic noise have been reported.[56–58]

24.4 RADIO-FREQUENCY MAGNETIC FIELD AND COILS

Resonant radio-frequency (RF) magnetic fields, orthogonal to the static magnetic field, are used in magnetic resonance to interrogate (excite) a region of interest for imaging or for spectroscopy.[3,4] The patient may absorb some portion of the transmitted RF energy.[59–63] Heating is the potential safety concern with absorbed RF energy.[64] It is essential for patient safety to limit whole-body and localized heating to appropriate levels.[18–20,59–84]

Resonant frequency scales with static field strength and nuclei of interest. For protons, the resonant RF frequency is 42.57 MHz/T[3]. Adjusting tip angle maximizes received signals in MR. Tip angles are proportional to area under the envelope of RF waveforms. For a given waveform, RF energy is proportional to the square of tip angle. Only the magnetic component of the RF field is useful in MR. Efforts are made by manufacturers to reduce electric field coupling to patients. The distribution of RF power deposition in MR tends to be peripheral, because of magnetic induction.[59–61] Note that plane wave exposures (in non-MR applications) may lead to greater heating at depth.[63,64]

RF pulses are typically transmitted by resonant RF coils. Transmit RF coils may be whole-body coils or local coils. Safety precautions with whole-body RF transmit coils are primarily to limit whole-body temperature elevation to appropriate levels. As shall be explored later, elevation of core body temperatures to sufficiently high levels may be life-threatening.[59–84] With local transmit coils, the primary safety concern is to limit local heating to prevent localized burns.[85–87]

Average RF power is proportional to the number of images per unit time. Patient geometry, RF waveform, tip angle, and whether the system is quadrature during transmit determine peak power. Quadrature excitation lowers RF power requirements by a factor of 2 and stirs any field inhomogeneities.[60,63] Both mechanisms lower the local specific absorption rate (SAR).

24.4.1 Transmit Birdcage Coils

One means of achieving rather homogeneous radio-frequency magnetic fields in MR is through the use of birdcage transmit coils[88] (see Fig. 24.4). Birdcage coils ideally would produce uniform B_1 fields. Let **A** be the magnetic vector potential. Components of **A** (and thus the electric field as well) must be parallel to the current density on the conductors that produced them. Perfectly uniform B_1 requires an infinitely long birdcage coil (or a spherical current density). The B_1 field is related to magnetic vector potential:

$$\mathbf{B}_1 = \nabla \times \mathbf{A} = \hat{a}_x\left[\frac{\partial A_z}{\partial y} - \frac{\partial A_y}{\partial z}\right] + \hat{a}_y\left[\frac{\partial A_x}{\partial z} - \frac{\partial A_z}{\partial x}\right] + \hat{a}_z\left[\frac{\partial A_y}{\partial x} - \frac{\partial A_x}{\partial y}\right] \tag{24.16}$$

Let a be the radius of the birdcage coil. Assume that (at the moment of time we look) $B_1 = B_{1x}$ ($B_{1y} = B_{1z} = 0$). Assume conductors of the RF coil lie only parallel to the z direction. Then $A_y = A_x = 0$. Let θ be the angle between the conductor, the center of the cylinder, and the x axis. Then it is possible to find B_{1x} and A_z (remember that B_1 is constant, independent of z):

$$B_{1x} = \frac{\partial A_z}{\partial z} \quad \rightarrow A_z = B_1 y = B_1 a \sin\theta \qquad (24.17)$$

So, an infinitely long coil with infinite conductors parallel to z, lying on the surface of a cylinder, will produce a uniform B_1 field, provided current varies with sin θ.

Of course, real birdcage coils are not infinitely long. Current is returned through end rings at the ends of the finite, cylindrical coil. End rings sum (or integrate) current from the straight coil conductors (which I will call *legs*). So, if coil leg current varies as sin θ, then current in the end rings varies as cos θ. Let N be the number of legs in the birdcage coil. Schenck[89,90] showed that peak end ring current amplitude is a factor $1/[2 \sin(\pi/N)]$ larger than the peak current in the coil legs. Let D be the length-to-diameter ratio of the birdcage coil. Let a be the coil radius, and let I be the maximum current in the legs of the birdcage coil. Schenck also showed that the radio-frequency magnetic induction B_1 at the center of a birdcage coil is given by:

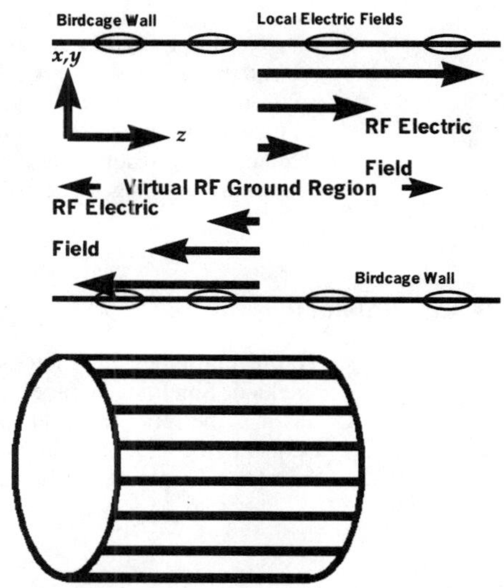

FIGURE 24.4 Electric fields inside a low-pass RF birdcage coil. Note that electric fields reach their maximum magnitude at the coil and fall to zero along the coil axis. Capacitors along the coil wall may also give rise to locally high electric fields. Any conductors should be routed along regions of low electric field or orthogonal to the electric field. At the bottom is an illustration of a birdcage coil.

$$B_1 = \frac{\mu_0 I N \sin\left(\dfrac{\pi}{N}\right)}{2\pi a} \frac{D(D^2 + 2)}{(1 + D^2)^{3/2}} \qquad (24.18)$$

B_1 as expressed in Eq. (24.18) is maximum when $D = \sqrt{2}$. However, Eq. (24.18) is within 95 percent of its maximum value for $D > 0.9$. Longer birdcage coils result in higher radio-frequency deposition in patients. Shorter birdcage coils are desirable. Typically, birdcage coils are designed with $D \approx 1$.

In MR, the nuclear spins of interest precess about the static magnetic field. Consider a transmit B_1 field in which the B_1 vector constantly points parallel or antiparallel to, for example, the y axis. This linear polarization with respect to the B_1 component may be considered to consist of two counterrotating components of equal magnitude (see Fig. 24.5). One of the rotating components will be in the same direction as the nuclear precession and so will contribute to MR

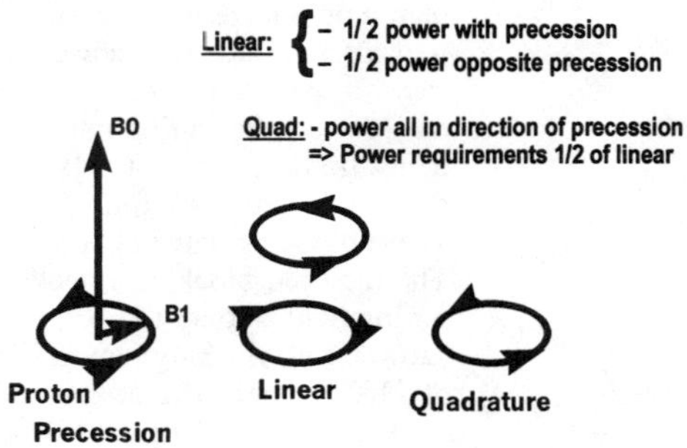

FIGURE 24.5 Comparisons of quadrature and linear RF excitation of spins are shown. Note that linear excitation wastes half the applied power.

physics. The other component will not contribute to MR physics; instead, that component will only waste energy by unnecessarily depositing RF energy in patients. Quadrature coils are driven electrically and physically 90° apart. Quadrature transmit coils are designed to excite only the B_1 component that rotates in the same direction as the precessing nuclei. Peak power requirements are reduced by a factor of two using quadrature coils.

Quadrature receive coils receive correlated signals but uncorrelated noise from the two receive channels. The result is a $\sqrt{2}$ improvement in signal-to-noise ratio over linear coils. In addition, quadrature imaging reduces diagonal shading in images.

24.4.2 Shielding

To prevent undesirable interactions with surrounding conductive structures, transmit coils are often shielded. Shielding reduces coil losses and in receive coils reduces noise as well. The coil quality factor is the ratio of the inductive reactance of the coil to the resistance of the coil. Let the radian frequency be ω, let coil inductance be L, let coil losses be R, let BW be the bandwidth, and let f_c be the center frequency of the coil; then the coil quality factor Q may be expressed as[89]:

$$Q = \frac{f_c}{\mathrm{BW}} = \frac{\omega L}{R} \tag{24.19}$$

High quality factors are improve signal to noise ratios for receive coils. Note that extremely high Q may result in make precise coil tuning more critical.

24.4.3 Receive Coils

RF receive coils receive signal and noise from the matter in the coil. If only a small region were to be imaged, then signal may be generated only from the region of interest while noise is received from the entire sample in the coil.[86,87,91] To reduce noise in the image, it is sensible to receive with the smallest coil capable of spanning the region of interest. This concept is referred to as *fill factor*.

Transmit coils may also double as receive coils. Frequently, a larger, relatively homogeneous coil such as a birdcage body coil will be used to transmit the excitation pulses. Then, a smaller, less homogeneous receive-only coil called a *surface coil*[86,87,91] will be used to receive the signal. The smaller coil generally has a better fill factor and so produces higher signal-to-noise ratios (SNR) than would have been possible with the larger coil.

Large surface coil currents could result if receive-only surface coils were resonant while RF transmit pulses are generated on the body coil. Such current can produce opposing B_1 fields which may destroy transmit homogeneity. In addition, these large currents could result in locally high RF deposition near the coils. Boesiger[85] has shown conditions where the surface coil amplified normal body coil heating 47-fold. To prevent such problems, blocking networks[86,87] are used (see Fig. 24.6). These blocking networks present high impedances to surface coil currents during body coil transmit. The required blocking impedance depends on coil area, magnet frequency, and how large an opposing field is to be allowed. Typical blocking impedances are a few hundred ohms. Poorly designed surface coil blocking networks may become warm. IEC 60601-1[92] sets a surface temperature limit of 41°C for objects that may touch people.

The high blocking impedance is switched out during receive. During receive, the surface coil is made resonant. The transmit body coil normally would couple noise from the rest of the body into surface coils during receive, degrading images. To prevent this sort of coupling, body coils are detuned during the receive phase.

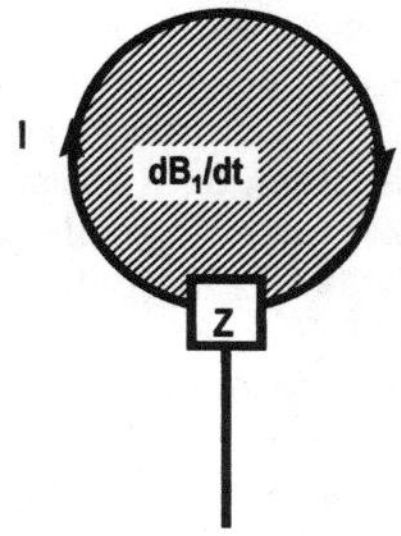

FIGURE 24.6 A receive-only surface coil with a blocking network is shown. During body coil transmit the blocking network, *Z* becomes a high impedance to prevent high induced currents from flowing on the coil. Such currents could lead to very high local SAR levels. During surface coil receive, the blocking network becomes a very low impedance to improve the image signal-to-noise ratio.

Further increases in SNR might be obtained using phased-array surface coils.[93] Phased-array coils are designed to be orthogonal (received noise is uncorrelated). If data from each coil are separately received and reconstructed, then SNR can be significantly increased over levels possible with individual coils that are not orthogonal. The coils are made orthogonal by slightly overlapping them until their mutual inductance approaches zero.

At higher field strengths, most of the RF losses are due to patients. However, for low field systems, patient losses are much smaller than coil losses. One approach for reducing the SNR impact of such low field coil losses, is to use superconducting surface coils.[94] These coils are typically limited in size and require attached cryostats. In addition, very high Q translates to very narrow bandwidths and tight tolerances on tuning.

24.4.4 Preamplifiers

Low noise figure preamplifiers with noise impedances matched and relatively near the receiver coils are required to avoid degrading SNR. Quadrature systems with preamplifiers on each channel have small SNR advantages over quadrature systems employing low-loss combiners and a single preamplifier.

24.4.5 SAR

During MR scans below 3 T or so, RF power deposition in patients can be approximated from quasistatic analysis, assuming electric field coupling to patients can be neglected.[10,34] Let R be the radius, σ the conductivity, and ρ the density of a homogeneous, sphere of tissue (Fig. 24.1). Assume that this sphere is placed in a uniform RF magnetic field of strength B_1 and frequency ω. Let the radiofrequency duty cycle be η. Then average specific absorption rate (SAR), $\mathrm{SAR}_{\mathrm{ave}}$, may be expressed as:[63]

$$\mathrm{SAR}_{\mathrm{ave}} = \frac{\sigma\eta\omega^2 B_1^2 R^2}{20\rho} \tag{24.20}$$

For homogeneous spheres, it turns out that the maximum peak SAR at a point is located on the outer radius of the sphere and is 2.5 times the average for the sphere.

RF heating during MR is by magnetic induction. Power deposition in homogeneous spheres immersed in uniform RF magnetic fields increases with the fifth power of the radius. Heating is largely peripheral with little deep body heating.[63]

As discussed above, RF body coils induce electric fields in the body of patients. The induced electric fields are largest near the RF coil conductors (Fig. 24.4). RF coils may have high electric fields near capacitors on the coil as well. Ensuring patients are kept well away from coil conductors (using pads, for example), especially during high SAR exams, may reduce local heating concerns. Note that in low-pass birdcage coils, the centerline of the coil is nearly a virtual ground. Any conductors that must be introduced into the bore will minimally affect local SAR if they are placed along this virtual ground.

If conductive loops (e.g., monitoring equipment or even coiled transmission line) are introduced into the scanner, high local SAR levels may result (Fig. 24.6). Even straight conductors may increase local SAR significantly (Fig. 24.7). For patient safety, fiber optic devices should be used instead of conductors, when possible.

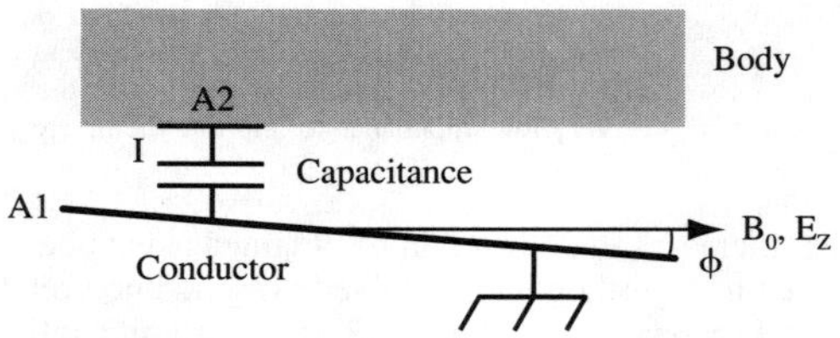

FIGURE 24.7 The effect of conductors on local power deposition is illustrated. A straight conductor experiences a gradient-induced electrical potential related to the vector component of the electric field over the conductor length. The conductor contacts a patient over some cross-sectional area (A2). Local SAR is 40 W/Kg (well beyond the 8 W/Kg guideline), if the local current density is as little as 18 mA/cm^2. Local SAR may be limited by increasing conductor impedance, by increasing contact area, by orienting the conductor orthogonal to the electric field, or by making the conductor shorter.

24.5 OTHER MR SYSTEMS

24.5.1 Computer Systems

Typically, MR pulse sequences (including radio-frequency and gradient waveforms) are computer generated and computer controlled. The computer also generates eddy-current compensation for the gradients and perhaps other types of compensation for imperfections. Patient tables are typically under computer control. RF receivers interface with computers that convert the raw received data into images. Computers typically display and possibly monitor patient comfort and patient safety parameters. Computers typically monitor system hardware as well.

24.5.2 Gating

Often MR scans need to be synchronized with (gated by) a physiological input such as the electrocardiogram or peripheral pulse or possibly respiration. Many MR systems provide the ability to gate the scanner from these waveforms. It is imperative that the gating waveforms be sufficiently free of artifacts induced by the MR system (gradient or RF interference or B_0 enhanced "T" waves appearing to be "R" waves) to permit accurate gating. It is also imperative that gating hardware should not increase the chances of local power deposition in patients. Elimination of conductors (for example by using fiber-optic devices) in the scanner greatly reduces local power deposition concerns. Finally, the gating equipment must not introduce signals that interfere with the scanner and show up as image artifacts.

24.5.3 Other MR Hardware

Other essential MR hardware may include patient tables, patient comfort systems (pads, lights, airflow, or headphones), and venting systems for magnets with cryogens. Patient tables may be used to transport patients, to move the region of interest into the center of the magnet for the examination, and to permit rapid removal of patients from scanners during emergencies. Pads may add to patient comfort and may be essential in reducing concerns of localized RF power deposition. Lighting and airflow may reduce patient anxiety.

24.6 SAFETY STANDARDS

MR is a rapidly evolving technology. It is imperative that MR safety standards protect patient safety during exams, while not preventing safe development of future diagnostic techniques. While many

TABLE 24.1 MR Safety Standards

Safety parameter	FDA significant risk criteria (http://www.fda.gov/cdrh/ode/magdev.html)	IEC 60601-2-33 normal mode	IEC 60601-2-33 first controlled mode (operator takes deliberate action to enter)	IEC 60601-2-33 second controlled mode (needs IRB approval)
B_0	$B_0 > 4$ T	$B_0 \leqslant 2.0$ T	$2\text{ T} < B_0 \leqslant 4$ T	$B_0 > 4$ T
WB SAR	Whole-body ave. SAR $\geqslant 4$ W/kg with 15 min ave.	Whole body ave. SAR $\leqslant 2$ W/kg with 6 min ave.	Whole body ave. SAR $\leqslant 4$ W/kg with 6 min ave.	Whole body ave SAR > 4 W/kg 6 min ave.
Head SAR	Ave. head SAR $>$ 3 W/kg with 10 min ave.	Ave. head SAR $\leqslant$ 3.2 W/kg with 6 min ave.	Ave. head SAR $\leqslant$ 3.2 W/kg with 6 min ave.	Ave. head SAR $>$ 3.2 W/kg with 6 min ave.
Local SAR	Local SAR > 8 W/kg in any gram (head or torso)	Local SAR $\leqslant 10$ W/kg in head or trunk, $\leqslant 20$ W/kg in extremities with 6 min ave.	Local SAR $\leqslant 10$ W/kg in head or trunk, $\leqslant 20$ W/kg in extremities with 6 min ave.	Local SAR > 10 W/kg in head or trunk, > 20 W/kg in extremities with 6 min ave.
Partial body SAR	N/A	SAR $\leqslant$ 10–8 W/kg (exposed mass/total mass limits between 2 and 10 W/kg): 6 min ave.	Partial SAR $\leqslant$ 10–6 W/kg (exposed mass/total mass limits between 4 and 10 W/kg): 6 min ave.	Partial SAR $>$ first controlled mode
Short-term SAR	N/A	SAR over any 10 s period may be up to 3 times appropriate long-term SAR level	SAR over any 10 s period may be up to 3 times appropriate long-term SAR level	SAR $>$ first controlled mode
dB/dt	Sufficient for severe discomfort	$dB/dt \leqslant 80\%$ of peripheral nerve mean	$dB/dt \leqslant 100\%$ of peripheral nerve mean	$dB/dt >$ first controlled mode

TABLE 24.2 NEMA Standards*

Standard (as listed in NEMA catalog)	Title
MS 1-2001	Determination of Signal-to-Noise Ratio (SNR) in Diagnostic Magnetic Resonance Images
MS 2-1989 (R-1996)	Determination of Two-Dimensional Geometric Distortion in Diagnostic Magnetic Resonance Images
MS 3-1989 (R-1994)	Determination of Image Uniformity in Diagnostic Magnetic Resonance Images
MS 4-1989 (Revision 1998)	Acoustic Noise Measurement Procedure for Diagnostic Magnetic Resonance Images
MS 5-1991 (R-1996)	Determination of Slice Thickness in Diagnostic Magnetic Resonance Imaging
MS 6-1991 (Revision 2000)	Characterization of Special Purpose Coils for Diagnostic Magnetic Resonance Images
MS 7-1993 (Revision 1998)	Measurement Procedure for Time-Varying Gradient Fields (*dB/dt*) for Diagnostic Magnetic Resonance Imaging Devices
MS 8-1993 (R-2000)	Characterization of the Specific Absorption Rate for Magnetic Resonance Imaging Systems
MS 9-2001	Characterization of Phased Array Coils for Diagnostic Magnetic Resonance Imaging Systems
MS 10-2001*	Determination of Local Specific Absorption Rate for Magnetic Resonance Imaging Systems
MS 11-2001*	Alternate Measurement Procedure for Time-Varying Gradient Fields (*dB/dt*) by Measuring Electric Field Gradients for Magnetic Resonance Imaging Systems

* Not finished at publication time.

safety standards govern various aspects of MR hardware development, two that are unique to MR are listed in Table 24.1. The United States Food and Drug Administration (FDA) published "Non-Significant Risk Criteria" for magnetic resonance devices.[19] These criteria state the conditions under which MR patient studies need investigational device exemption (IDE). The International Electrotechnical Commission (IEC)[20] developed a widely used MR safety standard. The IEC MR safety standard is three-tiered. The *normal operating mode* is for routine scanning of patients. The operator must take a deliberate action (usually an ACCEPT button) to enter the *first controlled operating mode*. This mode provides higher scanner performance, but requires more operator monitoring of the patient. Finally, there is a *second controlled operating mode* used only for research purposes under limits controlled by an Investigational Review Board (IRB). In Table 24.1, values from the new, recently approved, second edition of the IEC MR Safety Standard are presented along with FDA criteria.

Another IEC safety standard [92] also establishes safety criteria for electrical safety and limits surface contact temperatures to 41°C. Note that during high SAR scans, skin temperature approaches 37°C (a 4°C margin for temperature rise). During very low SAR scans, skin temperature is typically 33°C (an 8°C margin).

24.7 NEMA MR MEASUREMENT STANDARDS

The National Electrical Manufacturers Association (NEMA) has developed a number of useful standards for measurement of MR parameters. The current list of NEMA MR standards is given in Table 24.2. For more information see http://www.nema.org/

REFERENCES

1. Halliday, D., and R. Resnick, 1966, *Physics*, John Wiley, New York, p. 826.
2. Moore, W. J., 1972, *Physical Chemistry*, Prentice-Hall, Englewood Cliffs, New Jersey, pp. 140–143.
3. Mansfield, P., and P. G. Morris, 1982, "NMR Imaging in Biomedicine," in *Advances in Magnetic Resonance*, suppl. 2, J. S. Waugh, ed., New York, Academic Press, p. 32.
4. Keller, P. J., 1990, *Basic Principles of Magnetic Resonance Imaging*, General Electric Company, pp. 16–37.
5. Bottomley, P. A., T. H. Foster, R. E. Argersinger, and L. M. Pfiefer, 1984, "A review of normal tissue hydrogen NMR relaxation times and relaxation mechanisms from 1–100 MHz: Dependence on tissue type, NMR frequency, temperature, species, excision, and age," *Med. Phys.*, **11**:425.
6. Glover, G. H., 1993, "Gradient Echo Imaging," in the American Association of Physicists in Medicine (AAPM) Monograph No. 21: *The Physics of MRI*, P. Sprawls and M. Bronskill, eds., American Institute of Physics, New York, pp. 188–205.
7. McVeigh, E., and E. Atalar, 1993, "Balancing Contrast, Resolution, and Signal-to-Noise Ratio in Magnetic Resonance Imaging," in the American Association of Physicists in Medicine (AAPM) Monograph No. 21: *The Physics of MRI*, P. Sprawls and M. Bronskill, eds., American Institute of Physics, New York, pp. 234–267.
8. Thomas, S. R., 1993, "Magnets and Gradient Coils: Types and Characteristics," in the American Association of Physicists in Medicine (AAPM) Monograph No. 21: *The Physics of MRI*, P. Sprawls and M. Bronskill, eds., American Institute of Physics, New York, pp. 56–97.
9. Golay, M. J., 1968, "Field Homogenizing Coils for Nuclear Spin Resonance Instrumentation," *Rev. Sci. Inst.*, **29**:313–315.
10. Schenck, J. F., 2000, "Safety of Strong, Static Magnetic Fields," *JMRI*, **12**:2–19.
11. Mansfield, P., and P. G. Morris, 1982, "NMR Imaging in Biomedicine," in *Advances in Magnetic Resonance*, suppl. 2, J. S. Waugh, ed., Academic Press, New York, p. 271.
12. Schenck, J. F., 1986, "Axial magnetic field gradient coil suitable for use with NMR apparatus," U.S. Patent number 4617516.
13. Schenck, J. F., M. A. Hussain, and W. A. Edelstein, 1987, "Transverse gradient coils for nuclear magnetic resonance imaging," U.S. Patent number 4646024.
14. Roemer, P. B., and J. S. Hickey, 1986, "Self-shielded gradient coils for nuclear magnetic resonance imaging," U.S. Patent number 4737716.
15. Hughes, D. G., S. Robertson, and P. S. Allen, 1992, "Intensity Artifacts in MRI Caused by Gradient-Switching in an Animal-Size NMR Magnet," *Magnetic Resonance in Medicine*, **25**:167–179.
16. Henkelman, R. M., and M. J. Bronskill, 1987, "Artifacts in Magnetic Resonance Imaging," *Reviews of Magnetic Resonance in Medicine*, **2**(1):1–126.
17. Jehenson, P., M. Westphal, and N. Schuff, 1990, "Analytical Method for the Compensation of Eddy-Current Effects Induced by Pulsed Magnetic Field Gradients in NMR Systems," *J. Magnetic Resonance*, **90**:264–278, 1990.
18. FDA, 1988 (August 2), "*Guidance for content and review of a magnetic resonance diagnostic device 510(k) application: safety parameter action levels*," Center for Devices and Radiological Health Report, Rockville, Maryland.
19. FDA, 1997, "Magnetic Resonance Diagnostic Devices Criteria for Significant Risk Investigations," at http://www.fda.gov/cdrh/ode/magdev.html.
20. IEC 60601-2-33, 2002 (2d ed.), Medical Electrical Equipment, part 2: "*Particular Requirements for The Safety of Magnetic Resonance Equipment for Medical Diagnosis*," International Electrotechnical Commission (IEC); 3, rue de Varembé, P.O. Box 131, CH1211 Geneva 20, Switzerland. [In the United States, copies of this standard can be obtained from the American National Standards Institute (ANSI), 11 West 42nd Street, New York, NY 10036.]
21. Nyenhuis, J. A., J. D. Bourland, A. V. Kildishev, and D. J. Schaefer, 2001, "Health Effects and Safety of Intense Gradient Fields," in *Magnetic Resonance Procedures: Health Effects and Safety*, F. G. Shellock (ed.), CRC Press, New York, pp 31–54.

22. Schaefer, D. J., J. D. Bourland, and J. A. Nyenhuis, 2000, "Review of Patient Safety in Time-Varying Gradient Fields," *JMRI*, **12**(1):20–29.

23. Bourland, J. D., J. A. Nyenhuis, and D. J. Schaefer, 1999, "Physiologic Effects of Intense MRI Gradient Fields," *Neuroimaging Clinics of North America*, **9**(2):363–377.

24. Schaefer, D. J., 1998, "Safety Aspects of Switched Gradient Fields," E. Kanal (ed.), *MRI Clinics of North America*, **6**(4):731–747.

25. Reilly, J. P., 1989, *Cardiac Sensitivity to Electrical Stimulation*, U.S. Food and Drug Administration Report MT 89–101.

26. Cohen, M. S., R. Weisskoff, and H. Kantor, 1990, "Sensory stimulation by time varying magnetic fields," *Magn. Reson.*, **14**:409–414.

27. Bourland, J. D., J. A. Nyenhuis, G. A. Mouchawar, L. A. Geddes, D. J. Schaefer, and M. E. Riehl, 1990, "Human peripheral nerve stimulation from z-gradients," *Abstracts of the SMRM, Works in Progress* p. 1157.

28. Budinger, T. F., H. Fischer, D. Hentshel, H. E. Reinfider, and F. Schmitt, 1991. "Physiological effects of fast oscillating magnetic field gradients," *J. Comput. Assist. Tomogr.*, **15**:609–614.

29. Bourland, J. D., J. A. Nyenhuis, G. A. Mouchawar, T. Z. Elabbady, L. A. Geddes, D. J. Schaefer, and M. E. Riehl, 1991, "Physiologic Indicators of High MRI Gradient-Induced Fields," *Book of Abstracts, Works in Progress*, Society of Magnetic Resonance in Medicine, Tenth Annual Meeting, San Francisco, p. 1276.

30. Nyenhuis, J. A., J. D. Bourland, G. A. Mouchawar, T. Z. Elabbady, L. A. Geddes, D. J. Schaefer, and M. E. Riehl, 1991, "Comparison of Stimulation Effects of Longitudal and Transverse MRI Gradient Coils," abstract in *Works in Progress, SMRM*, p. 1275.

31. Bourland, J. D., J. A. Nyenhuis, G. A. Mouchawar, L. A. Geddes, D. J. Schaefer, and M. E. Riehl, 1991, "Z-Gradient Coil Eddy-Current Stimulation of Skeletal and Cardiac Muscle in the Dog," *Book of Abstracts, Society of Magnetic Resonance in Medicine*, Tenth Annual Meeting, San Francisco, p. 969.

32. Reilly, J. P., 1992, "Principles of Nerve and Heart Excitation by Time-Varying Magnetic Fields," *Annals of the New York Academy of Sciences*, **649**:96–117.

33. Nyenhuis, J. A., J. D. Bourland, D. J. Schaefer, K. S. Foster, W. E. Schoelein, G. A. Mouchawar, T. Z. Elabbady, L. A. Geddes, and M. E. Riehl, 1992, "Measurement of Cardiac Stimulation Thresholds for Pulsed z-Gradient Fields in a 1.5 T Magnet," abstract in *Eleventh Annual SMRM*, Berlin, p. 586.

34. J. D. Bourland, J. A. Nyenhuis, D. J. Schaefer, K. S. Foster, W. E. Schoelein, G. A. Mouchawar, T. Z. Elabbady, L. A. Geddes, and M. E. Riehl, 1992, "Gated, Gradient-Induced Cardiac Stimulation in the Dog: Absence of Ventricular Fibrillation," abstract in *Works in Progress, Eleventh Annual SMRM*, Berlin, p. 4808.

35. Mansfield, P., and P. R. Harvey, 1993, "Limits to Neural Stimulation in Echo-Planar Imaging," *Magnetic Resonance in Medicine*, **29**:746–758.

36. Irnich, W., and F. Schmitt, 1995, "Magnetostimulation in MRI," *Magn. Reson. Med.*, **33**:619–623.

37. Mouchawar, G. A., J. A. Nyenhuis, J. D. Bourland, L. A. Geddes, D. J. Schaefer, and M. E. Riehl, 1993, "Magnetic Stimulation of Excitable Tissue: Calculation of Induced Eddy-Currents With a Three-Dimensional Finite Element Model," *IEEE Trans. Magn.*, **29**(6):3355–3357.

38. D. J. Schaefer, J. D. Bourland, J. A. Nyenhuis, K. S. Foster, W. F. Wirth, L. A. Geddes, and M. E. Riehl, 1994, "Determination of Gradient-Induced, Human Peripheral Nerve Stimulation Thresholds for Trapezoidal Pulse Trains," abstract, *Society of Magnetic Resonance, Second Meeting*, San Francisco, p. 101.

39. Ehrhardt, J. C., C. S. Lin, V. A. Magnotta, S. M. Baker, D. J. Fisher, and W. T. C. Yuh, 1993, "Neural Stimulation in a Whole-body Echo-planar Imaging System," abstract, *Twelfth Annual SMRM*, New York, vol. **3**, p. 1372.

40. Rohan, M. L., and R. R. Rzedzian, 1992, "Stimulation by Time-Varying Magnetic Fields," *Annals of the New York Academy of Sciences*, **649**:118–128.

41. Schaefer, D. J., J. D. Bourland, J. A. Nyenhuis, K. S. Foster, P. E. Licato, and L. A. Geddes, 1995, "Effects of Simultaneous Gradient Combinations on Human Peripheral Nerve Stimulation Thresholds," *Society of Magnetic Resonance, Third Meeting*, Nice, p. 1220.

42. Bourland, J. D., J. A. Nyenhuis, W. A. Noe, D. J. Schaefer, K. S. Foster, and L. A. Geddes, 1996, "Motor and Sensory Strength-Duration Curves for MRI Gradient Fields," *Abstracts of the Society of Magnetic Resonance, Fourth Meeting*, New York p. 1724.

43. Nyenhuis, J. A., J. D. Bourland, and D. J. Schaefer, 1996, "Analysis from a Stimulation Perspective of Magnetic Field Patterns of MR Gradient Coils," *Abstracts of the Magnetism and Magnetic Materials Conference*.

44. Bourland, J. D., J. A. Nyenhuis, K. S. Foster, G. P. Graber, D. J. Schaefer, and L. A. Geddes, 1997, "Threshold and Pain Strength-Duration Curves for MRI Gradient Fields," *Abstracts of the International Society of Magnetic Resonance in Medicine*, Vancouver, p. 1974.

45. Havel, W., J. Nyenhuis, J. Bourland, K. Foster, L. Geddes, G. Graber, M. Waninger, and D. Schaefer, 1998, "Comparison of Rectangular and damped sinusoidal dB/dt waveforms in magnetic stimulation," *Abstracts of the 7th Joint MMM-INTERMAG Conference*, San Francisco.

46. Abart, J., K. Eberhardt, H. Fischer, W. Huk, E. Richer, F. Schmitt, T. Storch, and E. Zeitler, 1997, "Peripheral Nerve Stimulation by Time-Varying Magnetic Fields," *Journal of Computer Assisted Tomography*, **21**(4): 532–538.

47. Ham, C. L. G., J. M. L. Engels, G. T. van de Weil, and A. Machielsen, 1997, "Peripheral nerve Stimulation during MRI: Effects of High Gradient Amplitudes and Switching Rates," *JMRI*, **7**(5):933–937.

48. Halliday, D., and R. Resnick, 1966, *Physics*, John Wiley, New York, pp. 819–820.

49. D. J. Schaefer, 1993, "Bioeffects of MRI and Patient Safety," in the American Association of Physicists in Medicine (AAPM) Monograph No. 21: *The Physics of MR Imaging*, American Institute of Physics, New York, pp. 607–646.

50. Hedeen R. A., Edelstein W. A., 1997, "Characteristics and prediction of gradient acoustic noise in MR imagers," *Magn. Reson. Med.*, **37**:7–10.

51. McJury, M., 2001, "Acoustic Noise And Magnetic Resonance Procedures," in *Magnetic Resonance Procedures: Health Effects and Safety*, F. G. Shellock (ed.), CRC Press, New York, pp. 115–138.

52. McJury, M., and F. G. Shellock, 2000, "Auditory Noise Associated with MR Procedures: A Review," *JMRI*, **12**:37–45.

53. Melnick, W., 1979, "Hearing loss from noise exposure," in C. M. Harris (ed.), *Handbook of Noise Control*, McGraw-Hill New York, p. 2.

54. Robinson, D. W., 1976 "Characteristics of occupational noise-induced hearing loss," in D. Henderson, R. P. Hamernik, D. S. Dosjanjh, and J. H. Mills (eds.), *Effects of noise on hearing*. Raven Press, New York, pp. 383–405.

55. Brummett, R. E., J. M. Talbot, and P. Charuhas, 1988, "Potential hearing loss resulting from MR imaging," *Radiology*, **169**:539–540.

56. Mansfield, P. M., P. M. Glover and R. W. Bowtell, 1994, "Active acoustic screening: design principles for quiet gradient coils in MRI," *Meas. Sci. Technol.*, **5**:1021–1025.

57. Cho, Z. H., S. T. Chung, J. Y. Chung, et al., 1998, "A new silent magnetic resonance imaging using a rotating DC gradient," *Magn. Reson. Med.*, **39**:317–321.

58. Chen, C. K., T. D. Chiueh and J. H. Chen, 1999, "Active cancellation system of acoustic noise in MR imaging," *IEEE Trans. Biomed. Eng.*, **46**:186–190.

59. Shellock, F. G., 2000, "Radiofrequency Energy-Induced Heating During MR Procedures: A Review," *JMRI*, **12**:30–36.

60. D. J. Schaefer, "Health Effects and Safety of Radiofrequency Power Deposition Associated with Magnetic Resonance Procedures," 2001, in *Magnetic Resonance Procedures: Health Effects and Safety*, F. G. Shellock (ed.), CRC Press, New York, pp. 55–74.

61. Shellock, F. G., and D. J. Schaefer, 2001, "Radiofrequency Energy-Induced Current Density Distributions for Radiofrequency and Gradient Magnetic Fields Used for Magnetic Resonance Procedures," in *Magnetic Resonance Procedures: Health Effects and Safety*, F. G. Shellock (ed.), CRC Press, New York, pp. 75–97.

62. Smith, C. D., J. A. Nyenhuis, and A. V. Kildishev, 2001, "Health Effects of Induced Electric Fields: Implications for Metallic Implants," in *Magnetic Resonance Procedures: Health Effects and Safety*, F. G. Shellock (ed.), CRC Press, New York, pp. 393–414.

63. Schaefer, D. J., 1998, "Safety Aspects of Radio Frequency Power Deposition in Magnetic Resonance," E. Kanal (ed.), *MRI Clinics of North America*, **6**(4):775–789.

64. Elder, J. E., 1984, "Special senses," in *Biological Effects of Radio-frequency Radiation*, J. E. Elder and D. F. Cahill (eds.), EPA-600/8-83-026F, U.S. Environmental Protection Agency, Research Triangle Park, North Carolina, sec. 5, pp 64–78.

65. Bottomley, P. A. and E. R. Andrew, 1978, "RF Penetration, Phase-Shift, and Power Dissipation in Biological Tissue: Implications for NMR Imaging," *Phys. Med. Biol.*, **23**:630–643.

66. Abart, J., G. Brinker, W. Irlbacher, and J. Grebmeir, 1989, "Temperature and heart rate changes in MRI at SAR levels up to 3 W/kg," poster presentation, *SMRM Book of Abstracts*, p. 998.

67. Adair, E. R., and L. G. Berglund, 1986, "On the thermoregulatory consequences of NMR imaging," *Magnet. Reson. Imag.*, **4**(4):321–333.

68. Adair, E. R., and L. G. Berglund, 1989, "Thermoregulatory consequences of cardiovascular impairment during NMR imaging in warm/humid environments," *Magnet. Reson. Imag.*, **7**(1):25–37.

69. Athey, T. W., 1989, "A model of the temperature rise in the head due to magnetic resonance imaging procedures," *Magnetic Resonance in Medicine*, **9**(2):177–184.

70. Athey, T. W., 1992, "Current FDA guidance for MR patient exposure and considerations for the future," *Annals of the New York Academy of Sciences*, **649**:242–257.

71. Barber, B. J., D. J. Schaefer, C. J. Gordon, D. C. Zawieja, and J. Hecker, 1990, "Thermal effects of MR imaging: worst-case studies on sheep," *AJR*, **155**:1105–1110.

72. Benjamin, F. B., 1952, "Pain reaction to locally applied heat," *J. Appl. Physiol.*, (52):250–263.

73. Bernhardt, J. H., 1992, "Non-ionizing radiation safety: radio-frequency radiation, electric and magnetic fields," *Phys. Med. Biol.*, **4**:807–844.

74. Budinger, T. F., and C. Cullander, 1984, "Health effects of in-vivo nuclear magnetic resonance," chap.7 in *Clinical Magnetic Resonance Imaging*, A. R. Margulis, C. B. Higgins, L. Kaufman, and L. E. Crooks, Radiology Research and Education Foundation, San Francisco.

75. Carlson, L. D., and A. C. L. Hsieh, 1982, *Control of energy exchange*, Macmillan, London, pp. 56, 73, 85.

76. Goldman, R. F., E. B. Green, and P. F. Iampietro, 1965, "Tolerance of hot wet environments by resting men," *J. Appl. Physiol.*, **20**(2):271–277.

77. Guy, A. W., J. C. Lin, P. O. Kramer, and A. F. Emery, 1975, "Effect of 2450 MHz Radiation on the Rabbit Eye," *IEEE Trans. Microwave Theory Tech.*, **MTT-23**:492–498.

78. Kanal, E., 1992, "An overview of electromagnetic safety considerations associated with magnetic resonance imaging," *Annals of the New York Academy of Sciences*, **649**:204–224.

79. Saunders, R. D., and H. Smith, 1984, "Safety aspects of NMR clinical imaging," *British Medical Bulletin*, **40**(2):148–154.

80. Shellock, F. G., D. J. Schaefer, W. Grunfest, and J. V. Crues, 1987, "Thermal effects of high-field (1.5 Tesla) magnetic resonance imaging of the spine: clinical experience above a specific absorption rate of 0.4 W/kg," *Acta Radiologica*, suppl. **369**:514–516.

81. Shellock, F. G., and J. V. Crues, 1987, "Temperature, heart rate, and blood pressure changes associated with clinical magnetic resonance imaging at 1.5 tesla," *Radiology*, **163**:259–262.

82. Shellock, F. G., 1989, "Biological effects and safety aspects of magnetic resonance imaging," *Magnetic Resonance Quarterly*, **5**(4):243–261.

83. Shellock, F. G., D. J. Schaefer, and J. V. Crues, 1990, "Alterations in body and skin temperatures caused by MR imaging: is the recommended exposure for radiofrequency radiation too conservative," *British Journal of Radiology*, **62**:904–909.

84. Smith, D. A., S. K. Clarren, and M. A. S. Harvey, 1978, "Hyperthermia as a possible teratogenic agent," *Pediatrics*, **92**(6):878–883.

85. Boesigner, P., R. Buchli, M. Saner, and D. Meier, 1992, "An overview of electromagnetic safety considerations associated with magnetic resonance imaging," *Annals of the New York Academy of Sciences*, **649**: 160–165.

86. Edelstein, W. A., C. J. Hardy, and O. M. Mueller, 1986, "Electronic Decoupling of Surface-Coil Receivers for NMR Imaging and Spectroscopy," *J. Magn. Reson.*, **67**:156–161.

87. Haase, A., 1985, "A New Method for the Decoupling of Multiple NMR Probes," *J. Magn. Reson.*, **61**:130–136.

88. Hayes, C. E., W. A. Edelstein, J. F. Schenck, O. M. Mueller, and M. Eash, 1985, "An Efficient, Highly Homogeneous Radiofrequency Coil for Whole-Body NMR Imaging at 1.5 T," *J. Magn. Reson.*, **63**:622–628.

89. Schenck, J. F., "Radiofrequency Coils: Types and Characteristics," 1993, in American Association of Physicists in Medicine (AAPM) Monograph No. 21, *The Physics of MRI*, P. Sprawls and M. Bronskill (eds.), American Institute of Physics, New York, pp 98–134.

90. Schenck, J. F., E. B. Boskamp, D. J. Schaefer, W. D. Barber, and R. H. Vander Heiden, 1998, "Estimating Local SAR Produced by RF Transmitter Coils: Examples Using the Birdcage Coil." *Abstracts of the International Society of Magnetic Resonance in Medicine, Sixth Meeting*, Sydney, p. 649.

91. Edelstein, W. A., T. H. Foster, and J. F. Schenck, 1985, "The Relative Sensitivity of Surface Coils to Deep Lying Tissues," *Abstracts of the Society of Magnetic Resonance in Medicine*, Berkeley, 964–965.

92. IEC 60601-1-1, *Medical Electrical Equipment—Part 1: General Requirements for Safety; Safety Requirements for Medical Electrical Systems, 1992-06, Amendment 1, 1995-11 (General), International Electrotechnical Commission (IEC)*; 3, rue de Varembé, P.O.Box 131, CH-1211 Geneva 20, Switzerland, [In the United States, copies of this standard can be obtained from the American National Standards Institute (ANSI), 11 West 42nd St., New York, NY 10036.]

93. Roemer, P. B., W. A. Edelstein, C. E. Hayes, S. P. Souza, and O. M. Muller, 1990, "The NMR Phased-Array," *Magn. Reson. Med.*, **16**:192–225.

94. Ginefri, J. C., L. Darrasse, P. Crozat, 2001, "High-temperature superconducting surface coil for in vivo microimaging of the human skin," *Magn. Reson. Med.*, **45**(3):376–382.

CHAPTER 25
INSTRUMENTATION DESIGN FOR ULTRASONIC IMAGING

Kai E. Thomenius
GE Corporate Research and Development, Schenectady, New York

25.1 INTRODUCTION

The purpose of this chapter is to show how piezoelectric transduction, sound wave propagation, and interaction with scattering targets are taken advantage of in image formation with an ultrasound instrument. These instruments have evolved over the last 40 years from relatively simple hand-moved scanners built around an off-the-shelf oscilloscope to rather sophisticated imaging computers. Much technology has been perfected during this evolution. For example, transducers have grown from circular single-element probes to precision arrays with more than 1000 elements. With better front-end electronics, the operating frequencies have increased as weaker echoes can be handled. As the gate counts of VLSI ASICs* have increased, the numbers of processing channels in array-based systems have risen. With the introduction of reasonably low cost high-speed (20 to 40 MHz) 8- to 12-bit analog-to-digital converters, digital beam formation has become the standard.

The organization of this chapter is based on the discussion of a block diagram of a generalized ultrasound system. Each component of the block diagram will be reviewed in considerable detail. Different design approaches for the various blocks will be reviewed and their advantages and disadvantages discussed. Finally those areas of the block diagram that are targets of significant current research are summarized.

25.2 BASIC CONCEPTS

25.2.1 Image Formation

Image formation in medical ultrasound is accomplished by a pulse-echo mechanism in which a thin ultrasound beam is transmitted and the echoes generated by the interaction of that beam with scat-

* Very large-scale integration application-specific integrated circuits.

tering targets are received by a transducer or a set of transducer elements. The transmit and receive processing used to create this beam is referred to as *beam formation*. Because of its central role in ultrasonic imaging, beam formation will be discussed in detail later on.

The strength of the received echoes is usually displayed as increased brightness on the screen (hence the name for the basic ultrasonic imaging mode, B-mode, with B for brightness). A two-dimensional data set is acquired as the transmitted beam is steered or its point of origin is moved to different locations on the transducer face. The data set that is acquired in this manner will have some set of orientations of the acoustic rays. The process of interpolating this data set to form a TV raster image is usually referred to as scan conversion. With Doppler signal processing, mean Doppler shifts at each position in the image can be determined from as few as 4 to 12 repeated transmissions. The magnitudes of these mean frequencies can be displayed in color superimposed on the B-mode image and can be used to show areas with significant blood flow.

25.2.2 Physical Constants and Typical System Operating Parameters

It may be useful to consider typical system operating parameters. The following table lists some physical constants that help define the operating parameters of today's systems:

Typical attenuation in tissue	0.5 dB/cm · MHz for one-way travel
Speed of sound in tissue	1540 m/s (or approximately 13 μs/cm for round-trip travel)

One of the challenges of ultrasonic imaging relates to that very high attenuation. To put this in numerical terms, a typical 5-MHz transducer is expected to penetrate about 10 cm. Thus, the signal leaving the transmitting transducer will undergo attenuation in the order of 25 dB before it reaches a scattering site. At that point, a small fraction of the energy will be reflected; let us say the reflected echo will be another 30 dB down, and the return will bring about another 25 dB. Thus the entire attenuation has been about 80 dB. Needless to say, there is a strong need for careful low-noise designs for ultrasound front ends.

The following table gives some of the typical system design parameters commonly used for B-mode imaging:

Transducer frequencies	2–15 MHz
Transducer bandwidth	50–90% fractional bandwidth
Typical depths of penetration	18 cm (abdominal imaging) 16 cm (cardiac imaging) 4 cm (small parts and peripheral vascular imaging)
Time to acquire one 20-cm acoustic ray	~ 260 μs
Pulse repetition frequency (PRF)	4 kHz
Typical number of rays in an image	128–400
Data acquisition frame rates	10–80 frames/s

Because of frequency-dependent attenuation, applications with greater penetration requirements use the lower transducer frequencies. The instrument parameters have been selected or have evolved to their current values as the manufacturers have optimized their instruments to the needs of the various clinical applications. Given the compromises that have to be made, resolution of scanners is been limited, ranging from roughly 0.5 mm with 7- to 10-MHz transducers to about 2 to 4 mm with 2.5- to 3.5-MHz transducers. The penetration at which these resolutions can be achieved is about 3 cm for the higher frequencies and 15 or more cm for the lower frequencies. Whether or not this performance can be achieved on any given patient is dependent on factors such as uniformity of speed of sound, which is highly patient dependent (O'Donnell, 1988). The degree of and correction for sound speed variations in ultrasound systems continues to receive much attention (for example, Fink, 1992; Flax, 1988; Li, 1995; Nock, 1988; O'Donnell, 1988; Trahey, 1988; Zhu, 1993a and 1993b; Krishnan et al., 1996; Rigby, 2000; Silverstein, 2001).

25.2.3 Clinical Applications

B-mode imaging has found numerous uses in today's clinic (Goldberg, 1990; Sarti, 1987). Some of these are:

Abdominal imaging	Identification of tumors, cysts in liver, kidneys
Cardiology	Valvular insufficiency (flail leaflet), myocardial dyskinesis, septal defects, congenital malformations
Obstetrics	Fetal growth, congenital malformations
Peripheral vasculature	Extent of plaque, blood vessel tortuosity

Many of these diagnoses are based on relatively gross anatomical information that is available from the ultrasound image. In addition there is a large amount of additional information imparted to the echoes by the scattering process. Some of this information is displayed on B-mode images and is of major value to the clinician. This information is critical to diagnoses such as diffuse disease processes, the identification of tumors, quality of myocardial dynamics, and so forth. It is for these reasons the signal integrity and retention of maximum dynamic range is of key value in the image formation process.

Sales of ultrasound instruments are divided among the four areas listed above roughly as follows (source: Klein Biomedical Consultants):

Radiology	39%
Cardiology	35%
Obstetrics/gynecology	16%
Peripheral vasculature	5%

This gives a rough idea of the level of utilization in the several areas; however, it also should be noted that the marketplace is currently undergoing significant changes with the reform activity in health care. Also, there is much overlap between the segments. For example, many radiologists do perform obstetric or peripheral vascular examinations and echocardiologists perform increasing amounts of peripheral vascular work. In terms of instrumentation sales, these are believed to be approximately $2.5 billion in year 2000.

25.2.4 Classifications of Ultrasound Instruments

Ultrasonic instruments can be classified in many of different ways (Christensen, 1988). Among these are:

Types of electronic beam steering:	Phased arrays versus steering by element selection
Clinical applications:	See Sec. 25.2.3
Nature of beam formation:	Analog versus digital
Portability	Console-based systems versus hand-helds

FIGURE 25.1 Range of ultrasound scanners. As can be seen the ultrasound scanners come in a range of physical sizes. (*Courtesy of GE Medical Systems.*)

With respect to steering methods, the great majority of instruments (perhaps more than 95 percent) sold today are electronically (as opposed to mechanically) steered. Phased-array systems are dominant in echocardiographic applications where aperture size is limited by rib spacings, while the other beam steering methods are more often used in radiologic and obstetric and gynecological examinations. The latter grouping is sometimes referred to as *general imaging ultrasound*. The shift to digital beam formation is accelerating, and most of the instruments sold have digital beam formers. This is true even for the lower-price points. Figure 25.1 shows a range of ultrasound scanners.

25.3 TYPICAL SYSTEM BLOCK DIAGRAM

The block diagram in Fig. 25.2 shows the signal processing steps required for B-mode image formation. The actual implementations vary considerably among manufacturers and the types of systems. For example, the grouping of functions might be different from one system to the next, depending on the choices made by the system designers; however, the basic functionality shown by each block has to be there.

One point that the block diagram may not convey adequately is the degree of duplication of functions in today's systems For example, in systems with 128 processing channels, there will usually be 128 pulsers, 128 transmit/receive switches (T/R switches), and so forth. In such systems the use of large-scale integration and application specific integrated circuits (ASIC's) is highly important for cost and space reduction. For the most part, the block diagrams for digital and analog beam formers are quite similar, although there will usually be a large number of additional support circuitry required for synchronization, interpolation, and decimation of the sampled waveforms, and so forth.

Depending on the particular system implementation, the full RF bandwidth will be retained through the top part of the block diagram on a typical analog design. The class of heterodyned systems performs frequency mixing at the beam former level, and the signal spectrum is shifted to either an intermediate frequency or all the way to the baseband. With digital beam formers, A/D conversion occurs after the variable gain stages. The digital beam former systems can be designed with similar heterodyning approaches, although there are many different approaches to delay generation. In addition to the digital and analog forms of beam formers, it is also possible to use charge-coupled devices for this purpose and these, in fact, are receiving considerable attention at the present time.

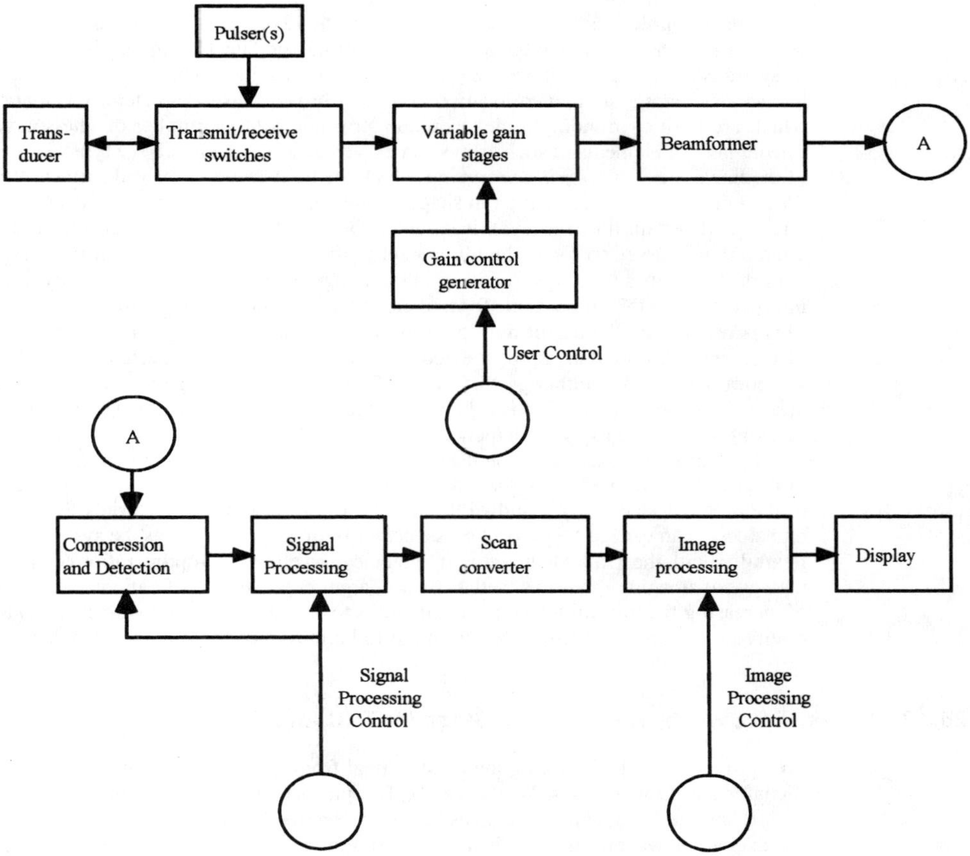

FIGURE 25.2 Block diagram of a typical ultrasound system. The blank circles represent points at which user control is introduced.

The following paragraphs introduce and describe briefly the functions of the most important blocks in Fig. 25.2. The most important of these will receive greater attention in the remainder of the chapter.

25.3.1 B-Mode Transducers

The mode of transduction in ultrasound systems takes advantage of the piezoelectric characteristic of certain ceramics. There are several types of transducers currently in use, and the nature of processing of acoustic data is different among them. These differences are highlighted in later sections as appropriate.

The piezoceramics can be fabricated in a number of ways to perform B-mode imaging. In its simplest form, the B-mode transducer is a circular *single-element transducer* with a fixed geometric focus. To a large degree this type of a transducer has been replaced today by more sophisticated multielement transducers; however, there are still commercial systems where single-element transducers are used. The next more complicated designs are *annular arrays*, which are also circular but, as the name implies, are composed of several (4 to 12) rings. These are the transducers most often used with mechanically steered systems (to be discussed later). Both the single-element and annular array transducers usually have a concave curved transmitting surface or an acoustic lens to focus the acoustic energy at a given location. The next broad transducer category is that of *linear arrays*, which are built by dividing a piezoceramic strip into a large number of line-source-like elements. The number of elements in such arrays can exceed 200, although 128 is a typical number. A variation of the linear arrays is *curvilinear arrays* which are built around a curved surface. With linear arrays, the acquired image is rectangular in shape, while with curvilinear arrays (and single element, annular arrays, and certain linear arrays), it is sector shaped. Both linear and curvilinear arrays either have a lens or are curved to focus the ultrasound in the plane perpendicular to the imaging plane. With both of these types of arrays, focusing in the image plane is done electronically. Finally, in order to improve on slice thickness, some form of elevation focusing is being introduced into today's systems. This is realized by the use of multiple rows of elements which form a two-dimensional (2-D) array of elements. Arrays that are connected to form a symmetrical aperture about the center line are sometimes referred to either as 1.25-D or 1.5-D arrays depending on whether improved slice thickness in the elevation direction is achieved by increasing the aperture size or by performing electronic focusing with those rows of elements.

The transducer or transduction itself has received considerable attention from researchers because of its critical position in the signal processing sequence (Hunt, 1983). Much progress has been made in the areas of sensitivity, bandwidth, and the use of composite materials. Bandwidths of today's transducers can exceed 80 percent of the center frequency. This gives the system designer additional flexibility with the available imaging frequencies and allows the optimization of image quality among the various imaging modes as well as over a larger cross section of patients. There is some potential of increasing the transmitted signal bandwidths to above 100 percent with a new class of transducers referred to as *single-crystal relaxors*, about to be introduced into the marketplace.

25.3.2 Pulser, T/R Switch, Variable Gain Stage (or TGC Amplification)

This group of blocks is among the most critical from the analog signal-to-noise ratio point of view (Schafer and Lewin, 1984; Wells, 1977). The piezoceramic array elements are energized with the pulser, and the transduction occurs as has been described. With the newer generation of instruments, the excitation waveform is typically a short square-wave burst with one to three cycles. Earlier generations (pre-Doppler) of instruments tended to use shock excitation with very wide bandwidths. The transducer element, with its limited bandwidth, filters this signal during both transmission and reception to a typical burst. The pulser voltages used vary considerably but values around 150 volts are common. Use of a short burst (say with a bandwidth in the 30 to 50 percent range) can give the system designer the ability to move the frequency centroid within the limits of the transducer bandwidth. In some imaging modes such as B mode, the spatial-peak temporal-average intensity (I_{spta}, an FDA-regulated acoustic power output parameter) value tends to be low; however, the peak pressures tend to be high. This situation has suggested the use of coded excitation, or transmission of longer codes that can be detected without loss of axial resolution. In this manner the average acoustic power output can be increased and greater penetration depth be realized.

The T/R switches are used to isolate the high voltages associated with pulsing from the very sensitive amplification stage(s) associated with the variable gain stage [or time gain compensation (TGC) amplifier]. Given the bandwidths available from today's transducers (80 percent and more in some cases), the noise floor assuming a 50-Ω source impedance is in the area of few microvolts rms. With narrower bandwidths this can be lowered but some imaging performance will be lost. If the T/R switch can handle signals in the order of 1 volt, the dynamic range in the neighborhood of 100+ dB may be achieved. It is a significant implementation challenge to have the noise floor reach the thermal noise levels associated with a source impedance; in practice there are a good number of interfering sources that compromise this. In addition, some processing steps such as the T/R switching cause additional losses in the SNR.

The TGC stages supply the gain required to compensate for the attenuation brought about by the propagation of sound in tissue. During the echo reception time, which ranges from 40 to 240 μs, the gain of these amplifiers is swept over a range approaching 60 to 70 dB, depending on the clinical examination. The value of this gain at any depth is under user control with a set of slide pots often referred to as the TGC slide pots. The dynamic range available from typical TGC amplifiers is in the order of 60 dB. One can think of the TGC amplifiers as providing a dynamic range window into the total range available at the transducer. This is illustrated in Fig. 25.3.

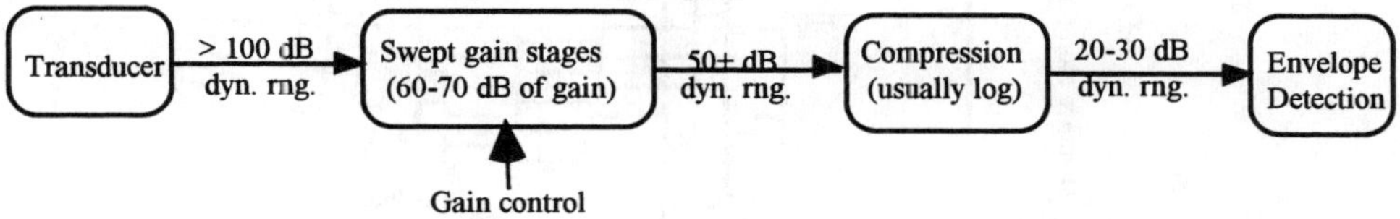

FIGURE 25.3 Front-end block diagram with signal processing steps and corresponding dynamic ranges.

It is interesting to note that the commercial impact of multichannel ultrasound instruments is such that a special purpose TGC amplifier IC has been developed for this function by a major integrated circuit manufacturer (Analog Devices in 1990).

25.4 BEAM FORMATION

Beam formation can be considered to be composed of two separate processes: beam steering and focusing (Macovski, 1983). The implementation of these two functions may or may not be separated, depending on the system design. Focusing will be discussed first.

25.4.1 Focusing

Analogously to optics, the spatial variation in system sensitivity can be modified by the action of focusing on the transmitted acoustic beam and, during reception, on its echoes. One can view focusing as the modification of the localized phases (or, more correctly for wideband systems, time shifts) of the acoustic beam so as to cause constructive interference at desired locations. One simple way to accomplish focusing is by curving the transducer element so as to form a phase front that, after traveling a defined distance, will cause the beam to add constructively at a desired focal point. With transducer arrays, the formation of the desired phase front during transmission can be accomplished by electronically altering the sequence of excitation of the individual elements. Similarly, during reception, the signals from the array elements can be routed through delay lines of appropriate lengths so that echoes from specific locations will have constructive interference (Thurstone, 1973). These processes are shown schematically in Fig. 25.4.

As suggested by Fig. 25.4, the echoes from a point source will have a spherical wavefront. The center elements of the array will receive these echoes at first while the outer elements will receive them last. To compensate for this and to achieve constructive interference at the summer, the center elements will be given the longest delays, as suggested by the length dimension of the delay lines. The calculation to determine the differential delays among the received echoes is straightforward.

An attractive formalism for expressing the array-based focusing in mathematical terms is due to Trahey (1988). The total transmitted pressure wave $T(t)$ at a point p can be expressed as a sum of the contributions from N array elements as follows:

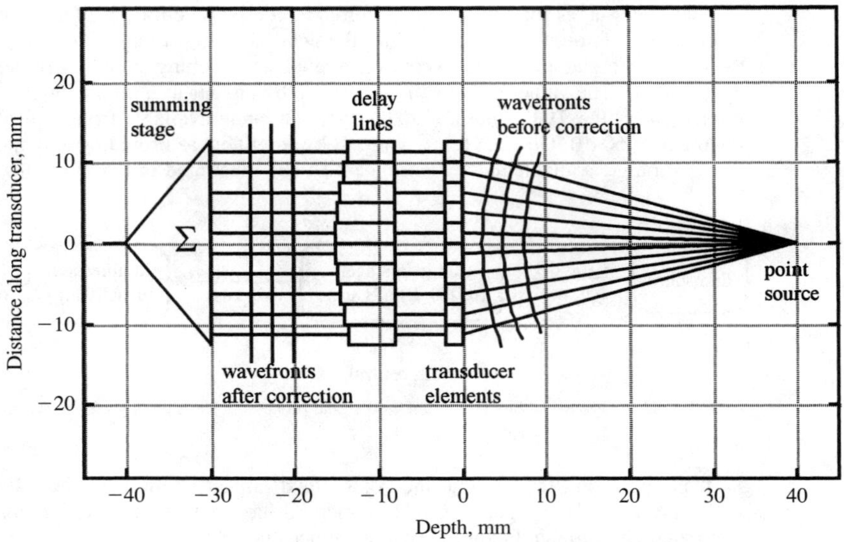

FIGURE 25.4 Schematic of focusing during reception. The echoes from a point source (at 40 mm) are shown impinging on a transducer array. The difference in the reception times is corrected by the delay lines. As an example, the echo will be received first by the center elements. Hence, their delays are the longest.

$$T(t) = \sum_{n=1}^{N} \frac{A_T(n)}{r(n)} S\left[t - t_T(n) + \frac{r(n)}{c}\right] \tag{25.1}$$

where $A_T(n)$ = pressure amplitude contribution of the nth element of the array at point p
$r(n)$ = distance from the nth element to the focus
$S(t)$ = waveshape of the pressure pulse generated by any given element of the array
$t_T(n)$ = focusing time delay for element n shown as the length of delay lines in Fig. 25.4
c = speed of sound in the medium

Assuming that at location p there is a point target with reflectivity W_p, then the signal after the summer in Fig. 25.4 can be described by

$$R(t) = W_p \sum_{n=1}^{N} \frac{A_R(n)}{r(n)} T\left[t - t_R(n) + \frac{r(n)}{c}\right] \tag{25.2}$$

where $A_T(n)$ = pressure amplitude contribution of the nth element to echoes from point p
$t_R(t)$ = receive focusing delay for element n

and $T(t)$ is given by Eq. (25.1). The remaining parameters of Eq. (25.2) were defined in Eq. (25.1). It should be noted that the $A_T(n)$ and $A_R(n)$ terms in Eq. (25.1) and (25.2) will, in general, be different since the transmit and receive operation need not be symmetric. The terms include tissue attenuation, element sensitivity variation, and transmit or receive apodizations.

It might be useful at this point to discuss several methods by which the receive delays for either focusing or beam steering are implemented. The previous paragraph refers to the use of delay lines for this purpose. Analog delay lines are an older method, albeit a very cost-effective one. However, lumped-constant delay lines do suffer from several limitations. Among these is the limited bandwidth associated with longer delay lines. Delays needed for focusing for most apertures are less than

0.5 μs; however, for phased-array beam steering (see below) they may be as long as 8 μs for larger apertures required for 2.5- to 3.5-MHz operation or up to 5 μs required for 5- to 7-MHz operation. Delay lines suitable for the latter case are relatively expensive. In addition, there are concerns about the amplitude variations with tapped delay lines as different taps are selected, delay uniformity over a production lot, and delay variations with temperature. In response to these difficulties, there has been a major migration to digital beam formation over the last 10 years (Thomenius, 1996).

An alternative method of introducing focusing delays for both analog and digital beam formers is by heterodyning (for example, Maslak, 1979). This is usually done in conjunction with mixing the received signal with a lower local oscillator frequency with the goal of moving the received energy to a different location on the frequency spectrum. If the phase of the local oscillator is varied appropriately for each of the array signals, the location of constructive interference can be placed at a desired location. The limitations of this are associated with the reduced bandwidth over which the delay correction will be accurate and the reduced range of phase correction that is possible. Finally, as noted above, focusing (and beam steering) can be accomplished by relatively straightforward digital techniques in a digital beam former. A number of different methods of digital beam former implementation have been published in the sonar and ultrasound literature (Mucci, 1984; Steinberg, 1992).

Figure 25.5 shows the formation of the focal region for a 20-mm-aperture circular transducer with a geometric focal distance of 100 mm and excited with a CW signal of 3.0 MHz. At the left-hand side of the pressure profile, the rectangular section from −10 to 10 mm corresponds to the pressure at the transducer aperture. In the near field, there are numerous peaks and valleys corre-

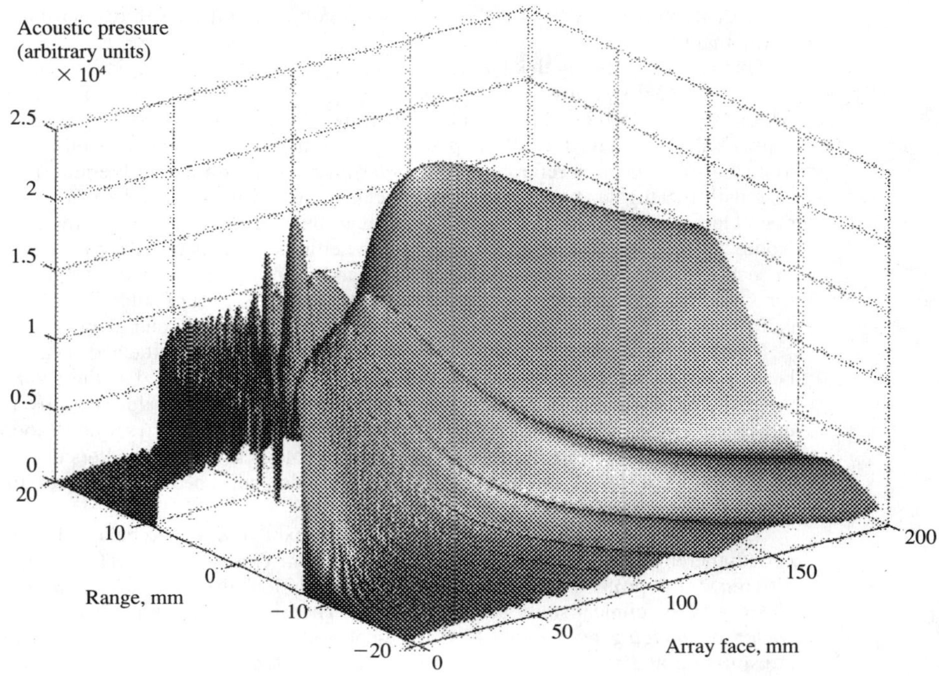

FIGURE 25.5 Spatial cross-sectional pressure distribution due to a circular transducer with a diameter of 20 mm, frequency of 3.0 MHz, and a radius of curvature of 100 mm. The left-hand side corresponds to the transducer aperture. All spatial dimensions are in millimeters; the *x* and *y* axes are not to scale. This pressure distribution was determined by the use of angular spectrum techniques. (*Schafer, 1989.*)

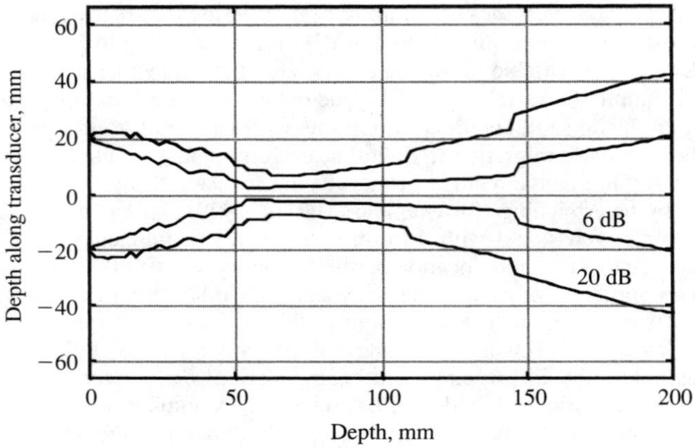

FIGURE 25.6 The 6- and 20-dB beam contours for the beam generated by a 3.0-MHz, 19-mm-aperture, 100-mm-focus transducer undergoing CW excitation. The *x* and *y* axes are to scale so that one can get a sense of the beam dimensions with respect to the depth of penetration.

sponding to areas where there is partial constructive and destructive interference. As one looks closer to the focal region, these peaks and valleys grow in size as the areas of constructive and destructive interference become larger. Finally, at the focal point the entire aperture contributes to the formation of the main beam.

One way of assessing the quality of a beam is to look at its beamwidth along the transducer axis. The 6- and 20-dB beamwidths are plotted on Fig. 25.6. It is important to recognize that the beamwidths shown are those for a circular aperture. Because of the axial symmetry, the beamwidths shown will be achieved in all the planes, i.e., in the imaging plane as well as plane perpendicular to it (this plane is often referred to as the *elevation plane*, from radar literature). This will not be the case with rectangular transducers. With rectangular transducers, the focusing in the image plane is done electronically, i.e., in a manner similar to that for annular arrays. However, in the elevation plane, the focusing in today's systems is done either by a lens or by the curvature of the elements. In such cases the focal location will be fixed and cannot be changed electronically. Remedying this limitation of rectangular transducers is currently an active area of study. The introduction of the so-called *elevation focusing* will be discussed in greater detail in a later chapter.

There is considerable diagnostic importance that has to be attached to the 20-dB and higher beamwidths. Sometimes the performance at such levels is discussed as the *contrast resolution* of a system. The wider the beam is, say at 40 dB below the peak value at a given depth, the more unwanted echoes will be brought in by these sidelobes. Small cysts and blood vessels may be completely filled in by such echoes. Also, if a pathological condition alters the backscatter strength of a small region by a modest amount, this variation may become imperceptible because of the acoustic energy introduced by the sidelobes.

With array-type transducers, the timing of the excitation or the delays during reception can be varied, thereby causing a change in the focal location. This is demonstrated in Fig. 25.7, where three different 6-dB profiles are shown. During transmission, the user can select the focal location as dictated by the clinical study being performed. There are operating modes sometimes referred to as *composite imaging modes* in which the final image is a composite of the data acquired during transmission and reception from several distributed focal locations. Not only can one change the transmit focal location but also the aperture size and transmit frequency between the transmissions. With this approach, one can achieve superior image uniformity at the expense of frame rate, which will be decreased by the number of transmissions along a single look angle.

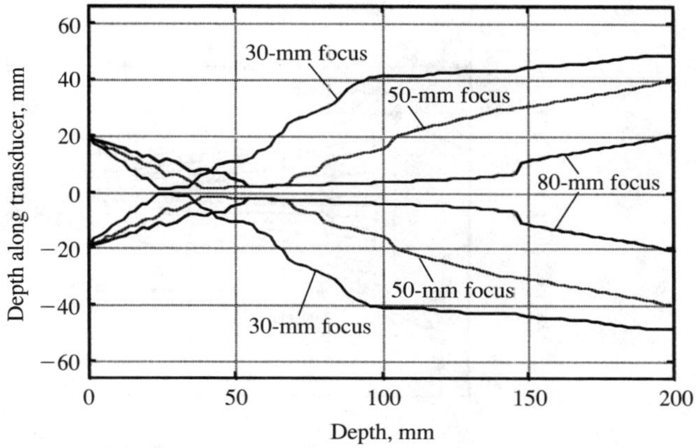

FIGURE 25.7 Three 6-dB beam profiles for an array-type transducer. The radii of curvature for the three cases are 30, 50, and 80 mm.

During reception there is yet another possibility to improve the performance of the system. As the transmitted wavefront travels away from the transducer, the delays introduced by the focusing can be varied, thereby changing the receive focus with the location of the wavefront. This approach is referred to as *dynamic focusing* and is now a standard feature of most systems. With analog delay lines, dynamic focusing is often implemented with tapped delay lines by changing the tap selection as the required focal delays change. Implementation of dynamic focusing by this method does add a number of challenges to the system designer, among which are the introduction of switching noise in the case of analog delay lines.

25.4.2 Beam Steering

There are a number of different methods of beam steering currently in use. These can be grouped into three categories:

1. Mechanical
2. Element selection
3. Phased array

The major implication of the selection of the beam steering approach is in the cost of the instrument. Mechanically steered systems tend to be the simplest and hence the least expensive while the phased arrays are the most expensive. Some imaging systems incorporate all three types; the great majority of recent vintage scanners have the latter two types of beam steering. The following paragraphs will discuss the relevant features of each of the three types.

Mechanical Steering. The simplest method of beam steering is to use a mechanism to reorient a transducer (usually a circular aperture) to a predetermined set of orientations so as to capture the required two-dimensional data set. This approach was dominant at first; however, in the last 15 years, electronically steered systems have become, by far, the most popular. Mechanical systems usually use either a single-element transducer or an annular array transducer (Fig. 25.8). The former will have a fixed focus while the latter does allow the focal point to be moved electronically. This will be discussed more fully later.

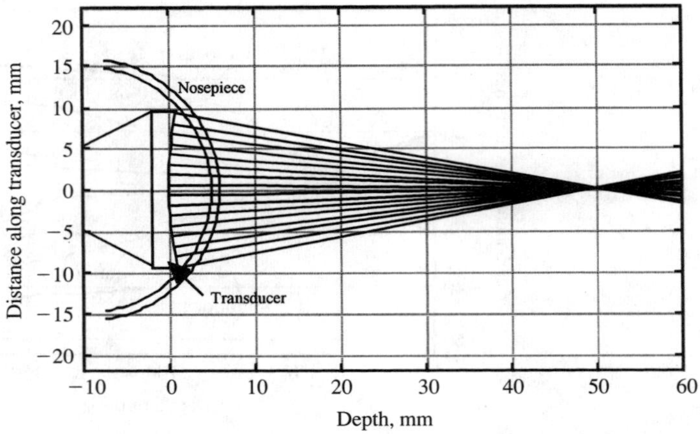

FIGURE 25.8 Sketch of a mechanically steered transducer. The transducer is isolated by a nosepiece and bathes in a coupling fluid. Included in the sketch are the ray paths due to the radius of curvature. A motor or a motor drive (not shown) moves the transducer back and forth. Because of that motion, this design is sometimes referred to as a *wobbler*.

There are a number of very attractive aspects to mechanical systems. Among these are low cost and the ability to focus the sound beam electronically in all planes. The low cost arises from the relatively low cost associated with the mechanisms used to move the transducer in comparison to the multielement transducer arrays and supporting electronics needed with electronic beam steering. The ability to focus the acoustic energy in all planes is a unique advantage, since most mechanically steered systems use either single-element or annular array type transducers. With the annular arrays, one has the capability to move the focus electronically in all planes, as opposed to the electronically steered arrays, which are usually rectangular in shape and will have electronic focusing in only one plane. The number of transducer elements in an annular array is usually less than 12, typically 6 or 8. With electronically steered arrays, the element count can go as high as 192 or more. As a consequence, the costs tend to be higher.

With mechanical steering, one can vary the acoustic line density by changing the speed of rotation of the transducer. The system designer has the choice of maintaining a constant pulse repetition frequency but winding up with a variable line density or, by making the pulse repetition timing a function of the location of the transducer at any point of the sweep and thereby achieving uniform line density. The best alternative may be to try to achieve a linear sweep with the transducer across the area to be imaged at the same time that the transducer firing is kept as a function of the transducer angle. The last choice makes the challenges of scan conversion more tolerable.

Some of the drawbacks associated with mechanical steering involve the inertia associated with the transducer, the mechanism, and the fluid within the nosepiece of the transducer. The inertia introduces limitations to the frame rate and clearly does not permit random access to look angles as needed (the electronically steered approaches supply this capability). The ability to steer the beam at will is important in several situations but most importantly in Doppler applications. Further, electronic beam formation affords numerous advanced features to be implemented such as the acquisition of multiple lines simultaneously and elimination of the effects due to variations in speed of sound in tissue.

Steering by Element Selection. Another relatively low cost beam-steering approach involves steering of the beam by element selection. In this approach one doesn't strictly steer the beam but rather changes the location of its origin, thereby achieving coverage over a two-dimensional tomographic

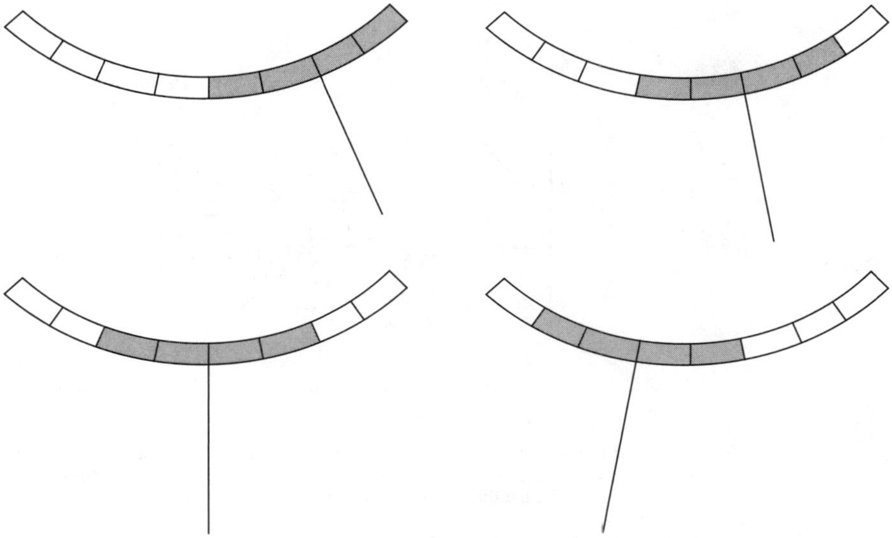

FIGURE 25.9 Steering by element selection for a curvilinear array. The beam will shift to a new location as the center of the active aperture is shifted over the entire array.

slice. This method is applied with both linear and curvilinear arrays. Figure 25.9 shows the application in the case of curvilinear arrays. For this particular case, the two-dimensional image will be sector shaped; with linear arrays, it will, of course, be rectangular. This is a relatively low cost approach since aside from the multiplexing required for element selection, the electronics required to accomplish beam formation are merely the focusing circuitry.

The line densities achievable with this mode of beam steering are not as variable as with mechanical steering, since it they will be dependent on element center-to-center spacing. There are methods by which one can increase the achieved line density. Figure 25.9 shows an acquisition approach sometimes referred to as *full stepping*. The line density with full stepping will be equal to the element density since the beam center will always be at the junction between two elements. It is possible to change the sizes of the transmit and receive apertures and thereby change the transmit and receive beam centers. This changes the effective location of the resultant beam and introduces the possibility of an increased line density. Half- and even quarter-stepping schemes exist, although care has to be taken that the resulting beam travels along the expected path.

Steering with Phased Arrays. The most complicated form of beam steering involves the use of phased-array concepts derived from radar (for example, Steinberg, 1976; Thurstone, 1973; Thomenius, 1996). Most ultrasonic phased-array transducers have between 64 and 128 elements. Transmit beam steering in phased-array systems is achieved by adding an incremental delay to the firing time of each of the array elements that is linearly related to the position of that element in the array. Similarly, during reception the delay that is applied to each of the echoes received by the array elements is incremented or decremented by a position-dependent factor. This differential time delay Δt is given by

$$\Delta t = \frac{x_n}{c} \tan \theta \tag{25.3}$$

where x_n is the location of the array element n and θ is the desired beam steering angle. The application of such a delay increment during reception is illustrated in Figure 25.10. Since the beam

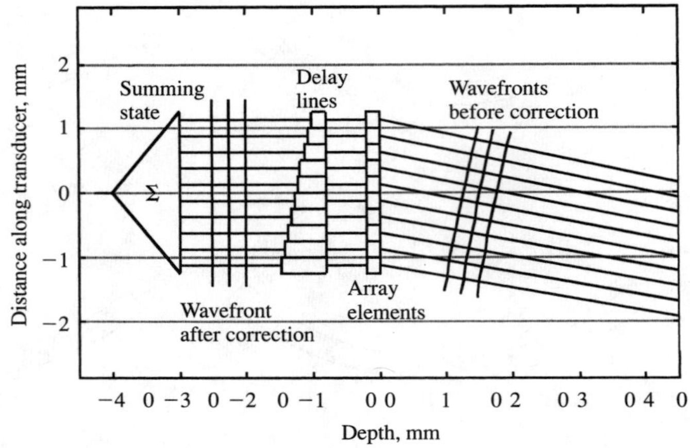

FIGURE 25.10 Beam steering in a phased-array system during reception. A linearly increasing time delay differential is introduced for each of the delay lines to correct for the linear time difference in the arrival times.

steering angle is such that the echoes will reach the array elements toward the bottom of the figure first, the longest delays will be imposed on the echoes from those elements. Since the wavefront is linear, the arrival times of the remaining echoes have a linear relationship, hence the linear decrement on the delays from one element to the next.

In addition to beam steering, one can combine focusing and beam steering delays in the same process. This is illustrated in Figure 25.11. Echoes from a near-field point source (at 40 mm) are shown arriving at the array elements. The arrival times of the echoes have a nonlinear component

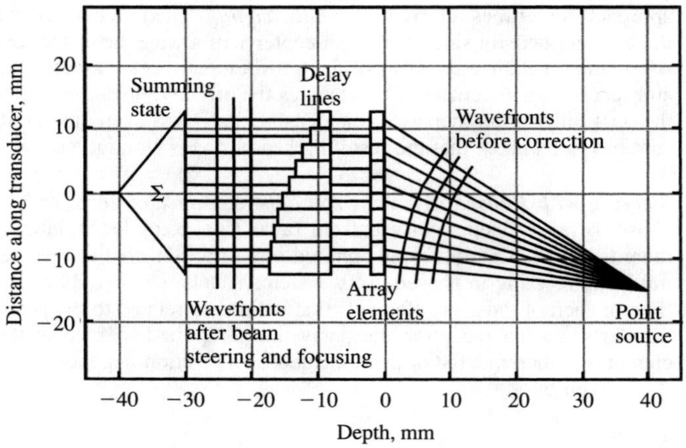

FIGURE 25.11 Schematic of beam steering and focusing during reception in a phased array type system. In addition to the focusing correction (also shown in Fig. 25.4), phased-array systems add a linearly increasing time shift to the receive delay lines to achieve constructive interference in the desired direction.

so the delay lines cannot compensate with a simple linear increment as in Figure 25.10. It can be shown easily that the differential time delay between channels can be determined from the law of cosines.

As the process of steering and focusing is repeated for a sequence of look angles, a sector-shaped image data set is acquired. The line density available from phased-array scanners is not as restricted as with curvilinear arrays, but some limitations exist in certain systems, due to the relatively large size of delay increments available in tapped delay lines. Heterodyned systems and digital beamformers have less limitations in this area.

All linear and curvilinear array systems have limitations or design constraints associated with the existence of *grating lobes*, which are due to leakage of acoustic energy in unwanted angles. It turns out that for certain larger center-to-center array element spacings, there will be constructive interference at look angles other than the main beam. This difficulty is particularly serious for the case of phased-array systems because of the need for beam steering. It turns out that the grating lobes move with the steering angle and can be brought into the visible region by the simple act of beam steering.

Grating lobes can be completely avoided by keeping the center-to-center spacing at one-half the wavelength at the highest contemplated operating frequency. [It turns out this is completely analogous to the familiar sampling theorem, which states the temporal sampling has to occur at a frequency that is twice that of the highest spectral component of the signal being processed (Steinberg, 1976).] This has the drawback of forcing the use of a larger number of array elements and their processing channels. This, and the expensive processing required for each channel, causes the phased-array systems to be more expensive than the other types.

25.4.3 Harmonic Imaging

A recent development in the area of B-mode imaging is that of imaging of the harmonics generated during propagation of acoustic waves in tissue (Averkiou, 1997; Jiang, 1998). While all the discussion so far has assumed that the propagation of these waves is linear, this is actually not the case. There is a difference in the speed of sound in the compressional and rarefactional parts of the acoustic pressure wave. As a consequence, the positive half of a propagating sine wave will move faster than the negative half; this results in the formation of harmonic energy. An image formed from such harmonics will be superior to that from the fundamental part of the spectrum because of reduced reverberant energy and narrower main beam. The acceptance of this form of imaging has been so rapid that in certain clinical applications (e.g., echocardiology) harmonic imaging is the default operating mode. From the point of view of beam former design, there is relatively little that needs to be done differently other than developing the ability to transmit at a lower frequency while receiving at twice the transmit frequency.

25.4.4 Compression, Detection, and Signal Processing Steps

The sequence of the processing steps between the beam former and the scan conversion is different among the various commercial systems but the goals of the steps remain the same. The beam former output will be a wideband RF, an IF, or a complex baseband signal, which will usually be bandpass filtered to reduce out-of-band noise contributions. In systems with very wideband processing, frequency diversity techniques (e.g., split spectrum processing) can be brought into play to try to reduce the impact of coherent interference or speckle.

With most of today's systems, there is a logarithmic compression of the amplified signal after beam formation amplification. The goal of this is to emphasize the subtle gray level differences between the scatterers from the various types of tissues and from diffuse disease conditions.

There are a number of ways that envelope detection has been implemented. In purely analog approaches, simple full-wave rectification followed by a low-pass filtering has been shown to work quite well. It is also possible to digitize the RF signals earlier in the processing chain, perform the compression and detection processes digitally, and use quadrature detection to determine the signal envelope.

25.5 SIGNAL PROCESSING AND SCAN CONVERSION

As has been noted earlier, the image data are acquired on a polar coordinate grid for sector scanners (e.g., mechanically steered, curvilinear arrays, and phased-array systems) and in a rectangular grid for linear arrays. It is clearly necessary to convert this image data into one of the standard TV raster formats for easier viewing, recording, computer capture, etc. This is performed in a module in most systems referred to as the *scan converter*. The major function of the scan converter is that of interpolation from, say, a polar data grid to that of the video pixel space. Given the data rates required, it is a challenging task but one that most commercial systems appear to have well under control.

The early scan converters used several clever schemes to accomplish the required function. For example, with sector scanners, certain systems would modulate the A/D converter sampling rate by the look angle of the beam former in such a way that every sample would fall onto a raster line. Once the acoustic frame was completed, and the data along each horizontal raster line was read out, simple one-dimensional interpolation was performed (Park, 1984). The need for more sophisticated processing brought about two-dimensional interpolation methods. Among the first in this area were Larsen et al. (Larsen, 1980), whose approach involved the use of bilinear interpolation. With the Larsen et al. approach, the sampling rate along each scan line was held at a uniform value which was high enough meet the sampling theorem requirements. With two axial samples along two adjacent acoustic lines, the echo strength values at each of the pixels enclosed by the acoustic samples were determined. This could be done by either (1) interpolating angularly new axial sample values along a synthetic acoustic ray that traversed through the pixel in question and then performing an axial interpolation along the synthetic ray or (2) interpolating a new axial sample along both real acoustic rays and then performing the angular interpolation. This basic approach has become the most widely used scan conversion method among the various manufacturers and seems to have stood the test of time.

More recently, researchers have continued to study the scan conversion problem with different approaches. For example, Berkhoff et al. have evaluated "fast algorithms" for scan conversion, i.e. algorithms which might be executed by software as opposed to dedicated hardware. Given the rapid trend to faster, more powerful, and cost-effective computers and their integration with ultrasound systems, it is likely that more of the scan conversion function will be done in software. Berkhoff et al. recommend two new algorithms which they compare with several conventional interpolators (Berkhoff, 1994). With the speed of computers improving at a steady pace, these approaches are increasingly attractive (Chiang, 1997). In other work, with the cost of some of the hardware components such as A/D converters coming down, oversampling the acoustic line data may permit the replacement of bilinear interpolation with simple linear interpolation (Richard and Arthur, 1994). Oversampling by a factor of two along with linear interpolation was found to be superior to bilinear interpolation under certain specific circumstances. It is clear there is additional work in this area yet.

25.6 SUMMARY

This chapter has reviewed the fundamentals of the design of ultrasound scanners with a particular focus on the beam formation process. Different types of image data acquisition methods are described.

REFERENCES

Averkiou, M. A., Roundhill, D. N., and Powers, J. E., (1997), "A new imaging technique based on the nonlinear properties of tissues," *IEEE Ultrasonics Symposium Proceedings*, pp. 1561–1566.

Berkhoff, A. P., Huisman, H. J., Thijssen, J. M., Jacobs, E. M. G. P., and Homan, R. J. F. (1994), "Fast scan conversion algorithms for displaying ultrasound sector images," *Ultrasonic Imaging*, **16**:87–108.

Chiang, A. M., and Broadstone, S. R. (1997), "Portable Ultrasound Imaging System," U.S. Patent 5,590,658, issued January 7.

Christensen, D. A. (1988), *Ultrasonic Bioinstrumentation*, Chap. 6, John Wiley & Sons, New York.

Fink, M. (1992), "Time reversal of ultrasonic fields: Part I—Basic Principles," *IEEE Trans. Ultrason. Ferroelec. Freq. Contr.*, **39**:555–566.

Flax, S. W., and O'Donnell, M. (1988), "Phase aberration correction using signals from point reflectors and diffuse scatterers: Basic principles," *IEEE Trans. Ultrason. Ferroelec. Freq. Contr.*, **35**:758–767.

Goldberg, B. B., and Kurtz, A. B. (1990), *Atlas of Ultrasound Measurements*, Year Book Medical Publishers, Chicago.

Hunt, J. W., Arditi, M., and Foster, F. S. (1983), "Ultrasound transducers for pulse-echo medical imaging," *IEEE Trans. Biomed. Eng.*, **30**:453–481.

Jiang, P., Mao, Z., and Lazenby, J. (1998), "A New Tissue Harmonic Imaging Scheme with Better Fundamental Frequency Cancellation and Higher Signal-To-Noise Ratio," *Proceedings, IEEE Ultrasonics Symposium.*

Krishnan, S., Li, P.-C., O'Donnell, M. (1996), "Adaptive compensation of phase and magnitude aberrations," *IEEE Trans. Ultrason. Ferroelect. Freq. Control*, vol. 43, pp. 44–55, January.

Larsen, H. G., and Leavitt, S. C. (1980), "An image display algorithm for use in real-time sector scanners with digital scan conversion," *IEEE Ultrasonics Symposium Proceedings*, pp. 763–765.

Li, P.-C., and O'Donnell, M. (1995), "Phase aberration correction on two-dimensional conformal arrays," *IEEE Trans. Ultrason. Ferroelec. Freq. Contr.*, **42**:73–82.

Macovski, A. (1983), *Medical Imaging Systems*, chap. 10, Prentice-Hall, Englewood Cliffs, N.J.

Maslak, S. H., (1979), "Acoustic Imaging Apparatus," U.S. Patent 4,140,022, issued February 20.

Mucci, R. A. (1984), "A comparison of efficient beamforming algorithms," *IEEE Trans. Ac., Speech, Sig. Proc.*, **32**:548–558.

Nock, L., Trahey, G. E., and Smith, S. W. (1988), "Phase aberration correction in medical ultrasound using speckle brightness as a quality factor," *J. Acoust. Soc. Am.*, **85**:1819–1833.

O'Donnell, M. O., and Flax, S. W. (1988), "Phase aberration measurements in medical ultrasound: human studies," *Ultrasonic Imaging*, **10**:1–11.

Park, S. B., and Lee, M. H. (1984), "A new scan conversion algorithm for real-time sector scanner," *IEEE Ultrasonics Symposium Proceedings*, pp: 723–727.

Richard, W. D., and Arthur, R. M. (1994) "Real-time ultrasonic scan conversion via linear interpolation of oversampled vectors," *Ultrasonic Imaging*, **16**:109–123.

Rigby, K. W. (2000), "Real-time correction of beamforming time delay errors in abdominal ultrasound imaging," *Proc. SPIE*, **3982**:342–353.

Sarti, D. A. (1987), *Diagnostic Ultrasound—Text and Cases*, Year Book Medical Publishers, Chicago.

Schafer, M. E., Lewin, P. A. (1984), "The influence of front-end hardware on digital ultrasonic imaging," *IEEE Trans. Ultrason. Ferroelec. Freq. Contr.*, **31**:295–306.

Schafer, M. E., and Lewin, P. A. (1989), "Transducer characterization using the angular spectrum method," *J. Acoust. Soc. Amer.*, **85**:2202–2214.

Silverstein, S. D. (2001), "A Robust Auto-focusing Algorithm for Medical Ultrasound: Consistent Phase References from Scaled Cross-correlation Functions," *IEEE Signal Processing Letters*, **8**(6):177–179.

Steinberg, B. D. (1976), *Principles of Aperture and Array System Design*, Wiley, New York.

Steinberg, B. D. (1992), "Digital Beamforming in Ultrasound," *IEEE Trans. on Ultrasonics, Ferro., and Freq. Cont.*, **39**(6):716–721, November.

Thomenius, K. E. (1996), "Evolution of Ultrasound Beamforming," *Proc. IEEE Ultrasonics Symposium*, IEEE Cat. No. 96CH35993, pp. 1615–1622.

Thurstone, F. L., and von Ramm, O. T. (1973), "A new ultrasound imaging technique employing two-dimensional electronic beam steering," vol. 5 in *Acoustical Holography*, Booth Newell (ed.), pp. 249–259, Plenum Press, New York.

Trahey, G. E., and Smith, S. W. (1988), "Properties of acoustical speckle in the presence of phase aberration. Part I: first-order statistics," *Ultrasonic Imaging*, **10**:12–28.

Wells, P. N. T. (1977), *Biomedical Ultrasonics*, Academic Press, San Diego.

Zhu, Q., and Steinberg, B. D. (1993a), "Wavefront amplitude distortion and image sidelobe levels: Part I—Theory and computer simulations," *IEEE Trans. Ultrason. Ferroelec. Freq. Contr.*, **40**:747–753.

Zhu, Q., and Steinberg, B. D. (1993b), "Wavefront amplitude distortion and image sidelobe levels: Part II—In vivo experiments," *IEEE Trans. Ultrason. Ferroelec. Freq. Contr.*, **40**:754–762.

CHAPTER 26
THE PRINCIPLES OF X-RAY COMPUTED TOMOGRAPHY

Peter Rockett
Department of Engineering Science, Oxford University, Oxford, United Kingdom

Ge Wang
Department of Radiology, University of Iowa, Iowa City, Iowa, USA

26.1 INTRODUCTION

The object of x-ray computed tomography is to reproduce, in the form of digital data, the internal structure of three-dimensional bodies with as little distortion as possible. This is achieved by the mathematical reconstruction of a series of two-dimensional images produced by the projection of an x-ray beam through the body in question. In this respect the form and functioning of x-ray computed tomography systems are closely linked to the physics of the interaction of x-ray photons with the material body and to the principles that govern the mathematical process of image reconstruction. In depth these are complex matters, but the fundamentals can be simply explained to provide a sound basis for further study.[1,2,3,4]

The relative values of absorption and transmission of adjacent elemental ray paths in an x-ray beam provide the contrast in a projected image. The associated interaction of the x-rays with the material body can be defined in terms of collision cross sections that describe the probability of particular events occurring along the ray path.[5] Here, the principal effects, for x-ray energies below 1 MeV, are identified in Sec. 26.2 as the photoelectron event and the Compton scattering event.[6,7] These effects are combined to define a linear attenuation coefficient $\mu(x)$ that describes the variation of photon intensity Φ along the ray path x according to the exponential relation $\Phi = \Phi_0 \exp\left(-\int_l \mu(x)\, dl\right)$, where Φ_0 is the initial intensity of the ray and l is the path length.

Taking account of the role of attenuation in the formation of image contrast, the method by which the material features along the ray path are reconstructed can be mathematically formulated in detail. In this process it is recognized that the two-dimensional object function $f(x, y)$ is represented by the attenuation coefficient $\mu(x, y)$. A key assumption in this process is that x-ray photons travel in straight lines and that diffraction effects may be ignored. Given the energetic nature of x-ray photons, this is a reasonable approach and enables considerable simplification of the underlying principles that are discussed in Sec. 26.3. It also permits the observation that from a mathematical view, a perfect representation of the body is theoretically possible, provided the requisite type and number of projected images are available.

An important feature in the formulation of the tomographic reconstruction process is the assumption of linearity attached to the various operations involved. This leads to the concept of a spatially invariant point-spread function that is a measure of the performance of a given operation. In practice the transformation associated with an operation involves the convolution of the input with the point-spread function in the spatial domain to provide the output. This is recognised as a cumbersome mathematical process and leads to an alternative representation that describes the input in terms of sinusoidal functions. The associated transformation is now conducted in the frequency domain and with the transfer function described by the Fourier integral. In discussing these principles, the functions in the spatial domain and the frequency domain are considered to be continuous in their respective independent variables. However, for practical applications the relevant processes involve discrete and finite data sampling. This has a significant effect on the accuracy of the reconstruction, and in this respect certain conditions are imposed on the type and amount of data in order to improve the result.

In Sec. 26.3 the mathematical principles of computed tomography are developed by considering the formation of images by the relative attenuation of a series of parallel ray paths. The fundamental quantity in this process is the line integral $\int_{AB} f(x, y)\, ds$ of the object function $f(x, y)$, which is the radon transform, or projection, of the object along the ray path AB. This leads to the concept that a parallel projection of an object, taken at an angle θ in physical space, is equivalent to a slice of the two-dimensional Fourier transform of the object function $f(x, y)$, inclined at an angle θ in frequency space. According to a process known as the filtered back-projection, the object can be reconstructed by taking the Fourier transform of the radon transform, applying a weighting to represent the effect of discrete steps in projection angle θ, and back-projecting the result through the reconstruction volume.

A natural progression is to consider the advantages that divergent ray-path scanning provides and the increased complexity that this introduces. It should be appreciated that the cardinal measure for developments in computed tomography is, according to the mathematical view, increased computational efficiency with enhanced modeling accuracy, and from the systems view is faster data capture with improved image contrast and higher spatial resolution. These two aspects are inextricably linked, though it is reasonable to conclude that the extensive and continuous development in reconstruction algorithms has tended to be the catalyst for advancement.

The question of absolute accuracy will depend on considerations of what is practically achievable and directly relates to the type and extent of measured data and the configuration of the associated system. The components of a basic laboratory system will comprise a source of x-rays, a specimen stage, one or more imaging detector/camera devices, and a data acquisition system. The type of x-ray source employed will depend on the permissible exposure period and on the degree of resolution required. Large-area sources, measured in square millimeters, with high x-ray flux will be used where the object is liable to movement and the length of exposure is important, as is usually the case in medical imaging. However, small-area sources, measured in square micrometers, with relatively low x-ray flux will be used where resolution is important. This is usually the case for biological research where the application is generally referred to as microtomography, as discussed in Sec. 26.4. This activity is properly classed as microscopy, where the physical layout of the related system can provide high geometric magnification at the detector input plane of about 100 times and high spatial resolution of about 1 μm or better.

With provision for projected high magnification, the beam-line area is relatively extensive and can accommodate a range of ancillary equipment. This attribute makes the laboratory microtomography facility a very versatile instrument. It can operate in a variety of modes and support scanning strategies that are tailored to particular subject shapes, sizes, and material content. With large-area detectors, magnified images can be recorded with divergent cone-beam projection and provide information of the internal structure with a combination of specimen rotation and translation motion. With small-area detectors, images can be recorded with a collimated pencil-beam and transverse raster scanning of the specimen in the manner of a conventional light microscope. These detectors can be energy sensitive and provide information of the chemical composition in addition to structural morphology. Further, by optically collimating the primary x-ray beam to a secondary focus, it is possible to form images by the x-ray fluorescence emission process and hence provide a map of the atomic composition.

A further indication of the versatility of the projection x-ray microscope is the possibility of splitting the beam path in various ways to provide enhanced imaging detail by phase contrast. Although the assumption that diffraction effects may be neglected has proved to be an acceptable approximation for most attenuation contrast imaging applications, they can nevertheless be manipulated to reveal additional information. A medium of varying density will, by virtue of the associated changes in refractive index, create phase differences in the propagating waves of adjacent ray-paths.[8] Under normal circumstances these effects are small and would not be seen in the image. However, if the difference in phase is converted into spatial displacement, very fine detail emerge as phase contrast in the image. This method is currently receiving a great deal of attention as means of providing improved contrast.[9]

26.2 THE INTERACTION OF X-RAYS WITH MATTER

26.2.1 The Collision Model

X-rays are located at the high-energy end of the electromagnetic spectrum and obey the same laws and possess the same wave-particle duality characteristics attributed to visible light (Fig. 26.1). Their location in the energy E (J) distribution is measured by either the wavelength λ (m) value or the frequency ν (Hz) value of the radiation. As the energy of the radiation increases, the wavelength decreases, frequency increases, and the particle description becomes more appropriate. Since x-rays are to be found at the high-energy end, roughly covering the region 10^{18} to 10^{17} Hz, or 0.1 to 10 nm, we may usefully consider them to be photon particles with energy given by $E = h\nu = hc/\lambda$, where Plank's constant $h = 6.6 \times 10^{-34}$ J · s and the speed of light $c = 3.0 \times 10^8$ m/s. Our description of the interaction of x-rays with matter may thus follow the concept of photon particle encounters with material particles such as atoms and free electrons.

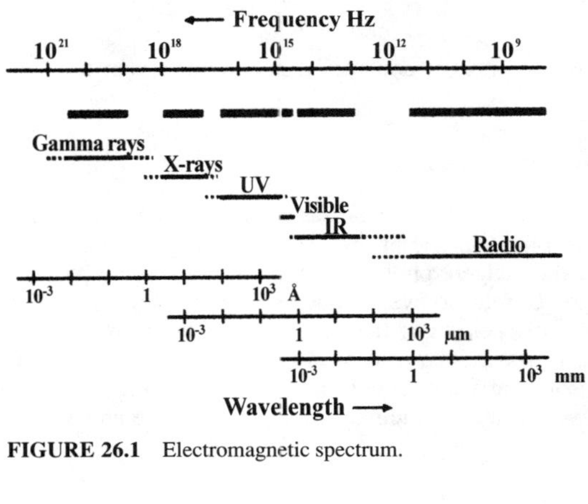

FIGURE 26.1 Electromagnetic spectrum.

We may regard each encounter as a single identifiable process, involving an isolated absorbing or scattering center that acts independently of the surroundings. This is a classical approach to collision processes and leads to the perception of a measured cross section σ^τ (m^2) as representing the area, with reference to a particular collision event τ, occluded by a single absorbing or scattering center. With this representation, the probability P_τ of absorption, or scattering, for a monoenergetic beam of intensity Φ, passing through a thin homogeneous layer of thickness dx and unit cross section, can be written as $P_\tau = d\Phi/\Phi = n_p\sigma^\tau\, dx$, where n_p is the number density of material atomic or electron particles. Hence, the number of photons dN_s/dV removed per unit volume for event τ is given by

$$\frac{dN_s}{dV} = -\frac{d\Phi}{dx} = \Phi n_p \sigma^\tau \tag{26.1}$$

The angular distribution of photons for scattering events is determined by considering the number of photons scattered at an angle θ into an elemental solid angle $d\Omega$. This leads to the concept of the differential collision cross section $(d\sigma/d\Omega)_\theta$, which can be defined by differentiating Eq. (26.1) with respect to $d\Omega$ to give

$$\frac{d^2N_s}{dV\, d\Omega} = \Phi n_p \left(\frac{d\sigma^\tau}{d\Omega}\right)_\theta \tag{26.2}$$

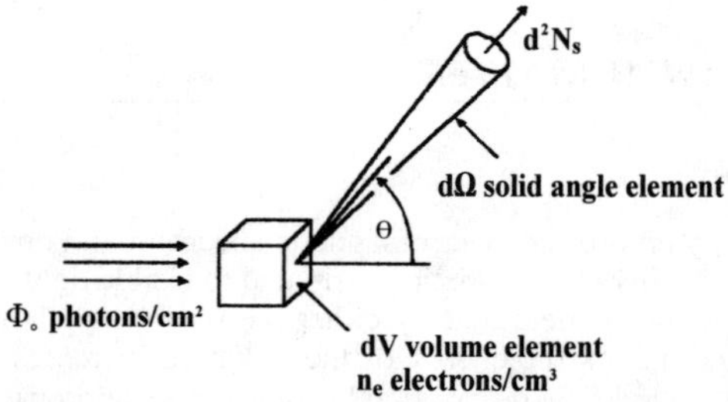

FIGURE 26.2 Photons scattered Into solid angle $d\Omega$.

Here, $d\sigma$ is the probability that the incident photon will be deflected into the elemental solid angle $d\Omega$, so that d^2N_s is the number of photons scattered into $d\Omega$ from volume element dV, for incidence flux Φ (Fig. 26.2).

If the radiation is unpolarized, the differential collision cross section depends only on the angle of deflection θ, so the cross section σ can be determined by integrating $(d\sigma/d\Omega)_\theta$ over a hemisphere, according to

$$\sigma = \int_0^\pi \left(\frac{d\sigma}{d\Omega}\right)_\theta 2\pi \sin\theta\, d\theta \tag{26.3}$$

Since we may interpret the cross section σ^τ as the probability that a photon will be removed from the primary beam, for a single τ event collision site per unit area, the total cross section σ^T for all events can be expressed as

$$\sigma^T = \Sigma_\tau \sigma^\tau \tag{26.4}$$

where the σ^τ are now considered as components to the overall attenuation process.

26.2.2 Principal Photon Interactions

The principal collision processes observed for x-rays, in the low- to medium-energy range 0 to 200 keV, are photoelectric absorption and Compton scattering. Absorption losses refer to events that remove the incident photon from the primary beam by the complete conversion of its energy. Scattering losses refer to events that remove the incident photon from the primary beam by a fractional exchange of energy and the redirection of the path of propagation. The relative proportion of either process to the primary beam attenuation varies with x-ray energy. Another scattering process is Thompson scattering, which occurs at low energy and imparts very little energy to the absorber. This involves the interaction of the photon wave and an electron, with the electron emitting electromagnetic radiation. Encounters of this type cause phase shifts in the incident photon wave and are the basis of phase contrast imaging. However, for attenuation contrast imaging, the effect of phase shift is assumed to be negligible.

For the photoelectric effect, a photon with energy $h\nu_0$ encounters an atom, is completely annihilated, and in the process ejects a bound electron from the atom with kinetic energy E_k. The vacant atomic bound state is repopulated from an adjacent level with the emission of a fluorescent x-ray photon. For Compton scattering, a photon with energy $h\nu_0$ encounters a free electron, is redirected though an angle θ with changed energy $h\nu'$ and the free electron recoils though an angle ϕ with received energy E_k. Both collision processes continuously remove photons from the x-ray beam as it progresses through a material body.

The Photoelectric Effect. The K, L, and M shell electrons of the colliding atom are the principle participants in the photoelectric interaction. When the incident photon possesses an energy $h\nu_0$ greater than the binding, or ionisation energy I, the shell electron will be ejected with a kinetic energy E_k according to the conservation of energy relation

$$h\nu_0 = E_k + I \tag{26.5}$$

The value of the absorption cross section increases with increasing binding energy I, so that a K-shell interaction is roughly 5 times more probable than an L-shell interaction. The event will most readily occur if the photon energy is comparable to the binding energy, in a manner somewhat like a resonance process. This behavior is borne out in practice where the photoelectric cross section σ^{pe} is seen to vary with the inverse cube of the photon energy $h\nu_0$, according to

$$\sigma^{pe} = k\frac{Z^4}{(h\nu_0)^3} \tag{26.6}$$

where the variation with atomic weight Z is a function of the penetration of the incident photon into the Coulomb field presented by the electron shell and k is a constant for that particular shell.

The momentum of the incident photons is mostly imparted to the atomic nucleus with the direction ϕ of the emitted photoelectrons predominantly toward larger angles (Fig. 26.3). The angle is determined by the direction of the electric field associated with the incident radiation. If it is linearly polarized, the photoelectron has a large component of velocity parallel to the electric field. However, at higher energies the photoelectron acquires a forward component of velocity from the incident radiation, as required from conservation of momentum.

The Compton Effect. The Compton effect refers to the scattering of incident photons by the interaction with free electrons. Here, an incident photon of energy $h\nu_0$ is scattered into the angle θ, with reduced energy $h\nu'$, and the electron recoils through the angle ϕ, with increased kinetic energy E_k (Fig. 26.4). The fact that the electrons are free and not bound to an atom reduces the complexity

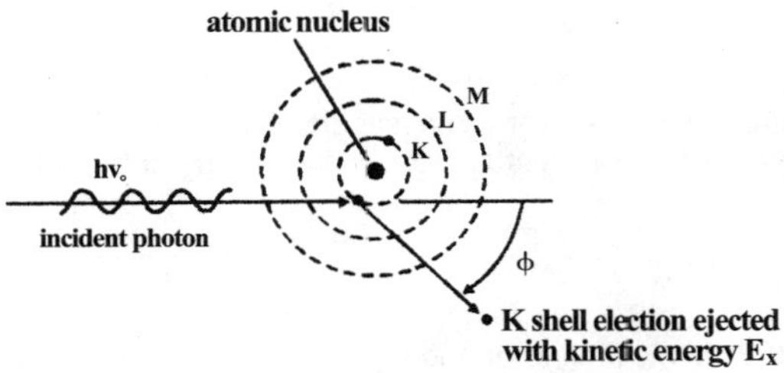

FIGURE 26.3 Photoelectric absorption with energy transfer to ionization and photoelectron motion.

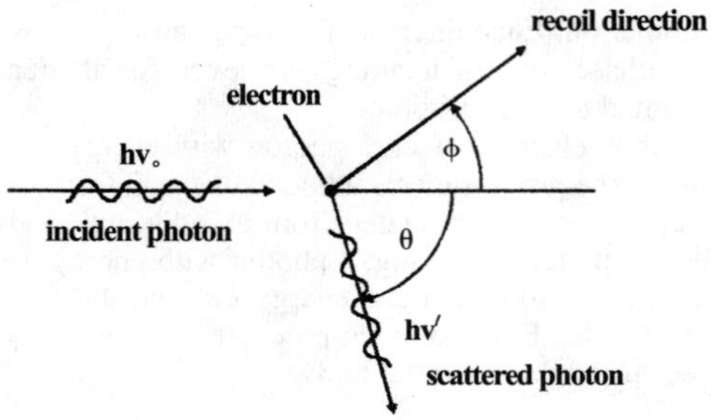

FIGURE 26.4 Geometry of Compton scattering.

of the problem. As a consequence, the principles of conservation of momentum can be invoked to determine the division of energy between the colliding particles. For the conservation of energy we have $h\nu_0 - h\nu' = E_k$. For conservation of momentum we have in the horizontal direction

$$\frac{h\nu_0}{c} = \frac{h\nu'}{c}\cos\phi + \sqrt{2m_0E_k}\cos\theta \qquad (26.7a)$$

and in the vertical direction

$$0 = \frac{h\nu'}{c}\sin\phi + \sqrt{2m_0E_k}\sin\theta \qquad (26.7b)$$

where m_0 is the rest mass of the electron. Combining these relations, we obtain an expression for the energy of the scattered photon in terms of the scattering angle:

$$h\nu' = \frac{h\nu_0}{1 + \alpha(1 - \cos\theta)} \qquad (26.8)$$

and the angle of the scattered photon in terms of the electron recoil angle:

$$\cot\phi = (1 + \alpha)\tan\left(\frac{\theta}{2}\right) \qquad (26.9)$$

where $\alpha = h\nu_0/m_0c^2$ and $m_0c^2 = 0.511$ MeV. It is convenient to rewrite Eq. (26.8) in terms of wavelength, so that

$$\frac{c}{\nu'} - \frac{c}{\nu_0} = \lambda' - \lambda = \frac{h}{m_0c^2}(1 - \cos\theta) \qquad (26.10)$$

where $h/m_0c^2 = 2.246 \times 10^{-12}$ m is the Compton wavelength. This expression indicates that the change in wavelength, for a given angle of scattering θ, is independent of the incident photon energy.

The collision differential cross section is given by the Klein-Nishina relation

$$d\sigma = \frac{r_e^2}{2}\left[\left\{1 + \frac{\alpha^2(1 - \cos\theta)^2}{(1 + \cos^2\theta)[1 + \alpha(1 - \cos\theta)]}\right\}\frac{1 + \cos^2\theta}{[1 + \alpha(1 - \cos\theta)]^2}\right]d\Omega \qquad (26.11)$$

where the classical electron radius $r_e = (1/4\pi\epsilon_0)(e^2/mc^2)$. The scattered radiation is strongly peaked in the forward direction, where the cross section falls with increasing angle θ and falls off more rapidly for higher values of α (Fig. 26.5). The cross section σ^c for the Compton interaction is found by integrating (2.11) over a hemisphere, to give

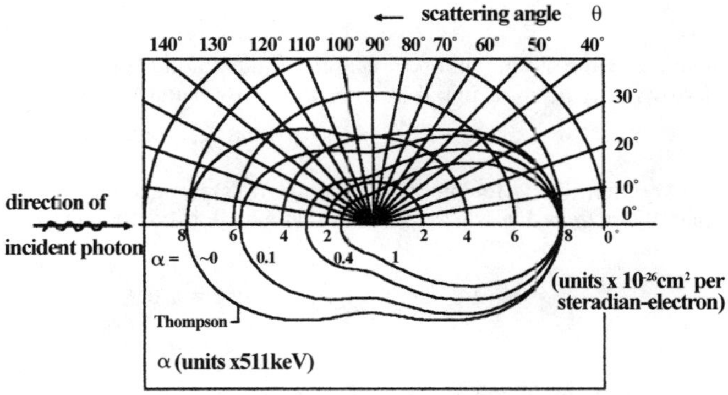

FIGURE 26.5 Compton collision differential cross section.

$$\sigma^c = \frac{3}{4}\sigma_o\left[\frac{2(1+\alpha)^2}{\alpha^2(1+2\alpha)} + \frac{\ln(1+2\alpha)}{\alpha}\left(\frac{1}{2} - \frac{1+\alpha}{\alpha^2}\right) - \frac{1+3\alpha}{(1+2\alpha)^2}\right] \tag{26.12}$$

where $\sigma_0 = 8\pi r_e^2/3$ is the cross section for Thomson scattering $\alpha \approx 0$.

With reference to Eq. (26.1), the energy ϵ lost from the primary beam per unit volume is given by

$$\epsilon = -\frac{d(\Phi h\nu_0)}{dx} = \Phi h\nu_0 n_e \sigma^c \tag{26.13}$$

This can be resolved into the energy that appears as scattered radiation ϵ_s and the energy imparted to the recoil electron ϵ_e, so that $\epsilon = \epsilon_s + \epsilon_e$. With this interpretation we can define a scattering cross section according to $\epsilon_s = \Phi h\nu_0 n_e \sigma_s^c$ and where

$$\sigma_s^c = \frac{3}{8}\sigma_0\left[\frac{2(2\alpha^3 - 3\alpha - 1)}{\alpha(1+2\alpha)^2} + \frac{8\alpha^2}{3(1+2\alpha)^3} + \frac{\ln(1+2\alpha)}{\alpha^3}\right] \tag{26.14}$$

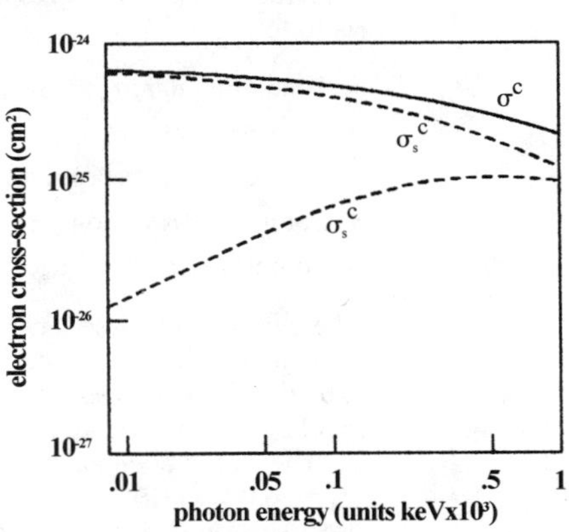

FIGURE 26.6 Compton interaction cross sections.

Similarly, we can define an absorption cross section $\epsilon_e = \Phi h\nu_0 n_e \sigma_e^c$, that is related to the total cross section according to

$$\sigma_e^c = \sigma^c - \sigma_s^c \tag{26.15}$$

The energy dependencies σ^c, σ_s^c, and σ_e^c can be evaluated from Eqs. (26.12) to (26.15) (Fig. 26.6).

26.2.3 X-Ray Path Attenuation

For x-rays with energies in the range 0 to 200 keV, the principal mechanisms for attenuating the photon flux are those due to photoelectric absorption with n_a atoms/cm^3 and cross section σ^{pe} cm^2 and Compton collisions with n_e electrons/cm^3 and cross section σ^c cm^2. Therefore, according to Eq. (26.4), we can write the total atomic cross section as

$$\sigma^T = \sigma^{pe} + Z\sigma^c \tag{26.16}$$

where Z is the atomic number. Therefore, integrating Eq. (26.1) for the removal of photons from a narrow x-ray beam of initial flux/m^2 Φ_0, over a path of length x, gives

$$\Phi = \Phi_0 \exp(-\sigma^T n_a x) \tag{26.17}$$

This relation holds for monochromatic x-rays propagating through a homogeneous medium. The quantity $\sigma^T n_a$ is known as the linear attenuation coefficient μ, with photoelectric and Compton components

$$\begin{aligned} \mu^{pe} &= n_a \sigma^{pe} \\ \mu^c &= n_e \sigma^c \end{aligned} \tag{26.18}$$

so that the overall attenuation coefficient is given by

$$\mu = \sigma^T n_a = \mu^{pe} + \mu^c \tag{26.19}$$

According to Eq. (26.15), the linear attenuation coefficient for the Compton event can be separated into that due to absorption and that due to scattered radiation, so that

$$\mu^c = \mu_e^c + \mu_s^c \tag{26.20}$$

This means that for a given material substance we have the linear attenuation coefficients grouped as follows.

- Total attenuation $\mu^{pe} + \mu^c$
- Total absorption $\mu^{pe} + \mu_e^c$
- Photoelectric absorption μ^{pe}
- Compton absorption μ_e^c
- Compton scattering μ_s^c

The variation of these terms with x-ray energy is usefully demonstrated for water (Fig. 26.7). With reference to the overall attenuation coefficient [Eq. (26.19)], and to the associated cross section [Eq. (26.16)], for each element there is a photon energy for which $\mu^{pe} = \mu^c$. Values of these energies can be used to indicate the region in which either process is dominant (Fig. 26.8).

The linear attenuation coefficient can be written as $\mu = \sigma^T N_A \rho / A$, where Avagadro's number $N_A = 6.02 \times 10^{23}$ mol^{-1}, A is the atomic weight, and ρ is the density. Given that the interaction cross section σ^T is not a function of the density of the medium, a mass absorption coefficient $\mu/\rho = \sigma^T N_A / A$ can be defined. This is related to the mass of a material required to attenuate an x-ray beam by a given amount. It is the form most often quoted for x-ray attenuation in tables of physical constants.

According to Eq. (26.17), the beam photon flux Φ can be written as

$$\Phi = \Phi_0 \exp(-\mu x) \tag{26.21}$$

which is the familiar Beer-Lambert law for the attenuation of an x-ray beam passing through matter. When the medium is nonhomogeneous, the attenuation must be integrated along the ray-path length l according to $\Phi = \Phi_0 \exp(-\int_r \mu(x)\, dl)$. For the location $\mathbf{r}$ in two or three-dimensional space, this becomes

$$\Phi = \Phi_0 \exp\left(-\int_l \mu(\mathbf{r})\, dl\right) \tag{26.22}$$

If the x-ray source is polychromatic, the attenuation must be integrated over the wavelength λ as well, to give

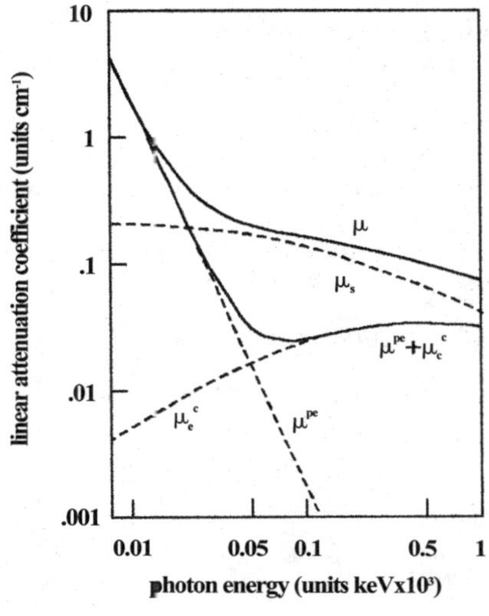

FIGURE 26.7 Linear attenuation coefficients for H_2O.

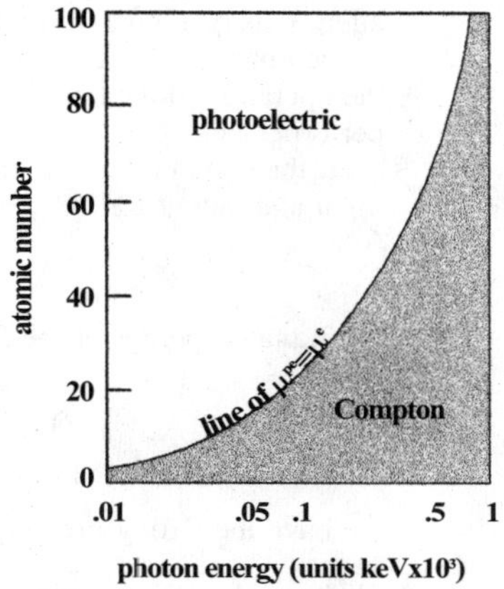

FIGURE 26.8 Region of dominant attenuation process.

$$\Phi = \Phi_0 \exp\left(- \int_\lambda \int_l \mu(\mathbf{r}, \lambda)\, dl\, d\lambda \right) \tag{26.23}$$

The presence of water in biological substances, such as soft tissue, has lead to the creation of a unit of attenuation coefficient for tomographic imaging. This is the Hounsfield unit (HU), defined as

$$\text{HU} = \frac{\mu(\text{substance}) - \mu(\text{water})}{\mu(\text{water})} \times 1000 \tag{26.24}$$

The usefulness of this unit for biological imaging is evident by comparing the values in HU for the widely disparate attenuating substances, such as air and bone. Here, air has a value of about -1000 HU and bone about $+1000$ HU.

26.3 THE MATHEMATICAL MODEL

26.3.1 Linear Systems and the Point-Spread Function

In dealing with the processes that are involved in tomographic reconstruction, the fundamental assumption is made that the individual systems perform linear operations. This ensures that simple relations exist between the input signals and the responding outputs. These systems may be treated schematically as black boxes, with an input $g_{\text{in}}(u)$, producing a corresponding output $g_{\text{out}}(u)$, where the independent variable u may be a one-dimensional time t, or displacement x, a two-dimensional position upon an x, y plane, or a three-dimensional position within an x, y, z volume (Fig. 26.9).

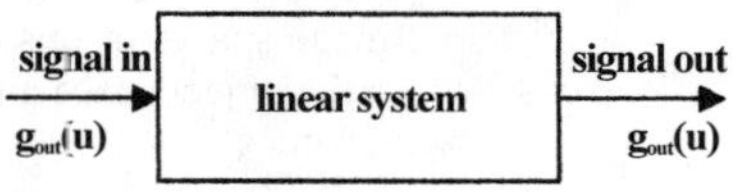

FIGURE 26.9 Black box representation of a linear system.

The principle of linearity prescribes that the weighted input $ag_{\text{in}}(u)$ will produce the output $ag_{\text{out}}(u)$, and the sum of two weighted inputs $ag_{\text{in}}^{(1)}(u) + bg_{\text{in}}^{(2)}(u)$ will produce the output $ag_{\text{out}}^{(1)}(u) + bg_{\text{out}}^{(2)}(u)$, for any real numbers a and b. We can further propose that the linear system be space invariant so that changing the position of the input merely changes the location of the output without altering its functional form. Hence, the image forming system can be treated as a linear superposition of the outputs arising from each of the individual points on the object. An estimate of the effect of the operation of the linear system on the contribution from each point will provide a measure of the performance.

If the operation of the linear system is symbolically represented as $L\{\ \}$, the relation between the input and output for independent variable u can be written as

$$g_{\text{out}}(u) = L\{g_{\text{in}}(u)\} \tag{26.25}$$

From the properties of the Dirac δ function, which represents a unit area impulse, defined as

$$\delta(u) = \begin{cases} \infty & u = 0 \\ 0 & \text{otherwise} \end{cases} \qquad \text{and} \qquad \int_{-\infty}^{\infty} \delta(u)\, du = 1$$

we have the sifting property $\int_{-\infty}^{\infty} g(u)\delta(u - u_0)\, du = g(u_0)$. Applying this property to Eq. (26.25) gives

$$g_{\text{out}}(u) = L\left\{\int_{-\infty}^{\infty} g_{\text{in}}(u')\ \delta(u' - u)\ du'\right\} \tag{26.26}$$

which expresses $g_{\text{out}}(u)$ as a linear combination of elementary δ functions, each weighted by a number $g_{\text{in}}(u')$. From the linearity principle, the system operator $L\{\ \}$ can be applied to each of the elementary functions so that Eq. (26.26) can be written as

$$g_{\text{out}}(u) = \int_{-\infty}^{\infty} g_{\text{in}}(u')L\{\delta(u' - u)\}\ du' \tag{26.27}$$

The quantity $L\{\delta(u' - u)\}$ is the response of the system, measured at point u in the output space, to a delta function located at the point u' in the input space, and is called the *impulse response*.

For image forming systems the impulse response is usually referred to as the *point-spread function* (PSF). In this case we consider the response at vector point $\mathbf{r}$ in the image space, to an irradiance distribution $\Phi_0(\mathbf{r}')$ located in object space. An element $d\mathbf{r}'$ located at $\mathbf{r}'$ will emit a radiant flux of $\Phi_0(\mathbf{r}')\ d\mathbf{r}'$, which will be spread by the linear operation $L\{\ \}$ over a blurred spot defined by the function $p(\mathbf{r}'; \mathbf{r})$. The flux density at the image point $\mathbf{r}$, from the object point $\mathbf{r}'$, will be given by $d\Phi_i(\mathbf{r}) = p(\mathbf{r}'; \mathbf{r})\Phi_0(\mathbf{r})\ dr'$, so that the contribution from all points in the object space becomes

$$\Phi_i(\mathbf{r}) = \int_{-\infty}^{\infty} \Phi_0(\mathbf{r}')p(\mathbf{r}'; \mathbf{r})d\mathbf{r}' \tag{26.28}$$

where $p(\mathbf{r}'; \mathbf{r})$ is the point-spread function. If the system is space invariant, a point source moved over the object space $\mathbf{r}'$ will produce the same blurred spot at the corresponding locations $\vec{r}$ in the image space. In this case the value of $p(\mathbf{r}'; \mathbf{r})$ depends only on the difference between the location in the object space and the image space, namely $(\mathbf{r}' - \mathbf{r})$, so that $p(\mathbf{r}'; \mathbf{r}) = p(\mathbf{r}' - \mathbf{r})$ and Eq. (26.28) becomes

$$\Phi_i(\mathbf{r}) = \int_{-\infty}^{\infty} \Phi_0(\mathbf{r}')p(\mathbf{r}' - \mathbf{r})\ d\mathbf{r}' \tag{26.29}$$

The expression (26.29) is known as the two-dimensional [i.e., $\vec{r} = (x, y)$] *convolution integral*, written as

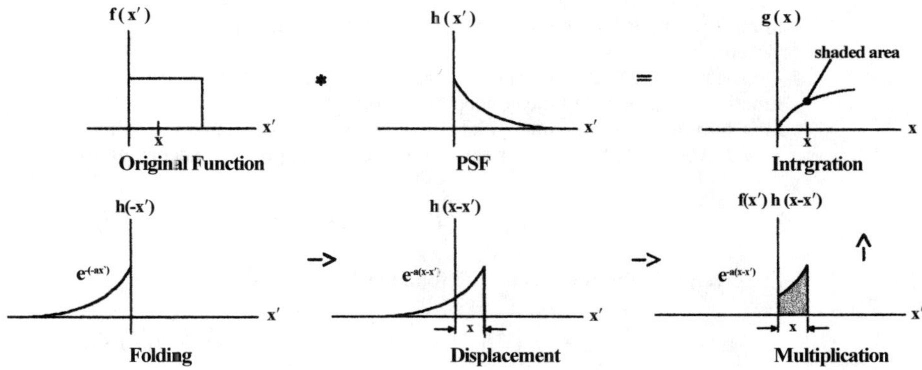

FIGURE 26.10 The convolution process.

$$g(x, y) = \int_{-\infty}^{\infty}\int_{-\infty}^{\infty} f(x', y')h(x - x', y - y')\, dx'dy' \tag{26.30}$$

This has the shorthand notation $f^{**}h = h^{**}f$, where ** denotes convolution with double integration. Similarly, $f^*h = h^*f$ denotes convolution with single integration.

A physical description of convolution can be expressed in terms of graphical manipulation of the one-dimensional function $f(x')$ (Fig. 26.10). Here, the function $h(x')$ is *folded* by taking the mirror image about the ordinate axis, a *displacement* $h(-x')$ is applied along the abscissa axis, followed by a *multiplication* of the shifted function $h(x - x')$ by $f(x')$. The convolution at location x is then determined by *integration* along the abscissa axis, by finding the area under the product of $h(x - x')$ and $f(x')$. The result of convolution will be the same if the role of the two functions is reversed, i.e., whether $h(x')$ or $f(x')$ is folded and shifted.

26.3.2 Frequency Domain

The method of resolving the input into a linear combination of δ functions presents a framework for the analysis of linear shift-invariant systems in cartesian space. However, solutions involve the multiple integration of the product of functions and are cumbersome to carry out in practice. Since δ functions are not the only form of input signal decomposition it is expedient to examine the superposition of complex exponential functions as a possible alternative. In this case the input $g_{in}(\mathbf{u}) = \exp(i2\pi\mathbf{k} \cdot \mathbf{u})$, so that Eq. (26.29) becomes

$$g_{out}(\mathbf{u}) = \int_{-\infty}^{\infty} p(\mathbf{u}' - \mathbf{u}) \exp(+i2\pi\mathbf{k} \cdot \mathbf{u}')\, d\mathbf{u}' \tag{26.31}$$

where $\mathbf{u}$ is the independent variable vector and $\mathbf{k}$ is a given real number vector.
Changing the variables according to $\mathbf{u}'' = \mathbf{u} - \mathbf{u}'$, we have

$$g_{out}(\mathbf{u}) = \exp(i2\pi\mathbf{k} \cdot \mathbf{u}) \int_{-\infty}^{\infty} p(\mathbf{u}'') \exp(-i2\pi\mathbf{k} \cdot \mathbf{u}'')\, d\mathbf{u}'' \tag{26.32}$$

Since the integrand is independent of $\mathbf{u}$, this reduces to

$$g_{out}(\mathbf{u}) = \text{constant} \times g_{in}(\mathbf{u}) \tag{26.33}$$

This indicates that passing a complex exponential through a linear shift-invariant system is equivalent to multiplying it by a complex factor depending on **k**, which in effect is the *transfer function* of the system.

The result (26.32) allows us to consider each image to be constructed from a series of sinusoidal functions. This leads to the concept that a function $f(x)$, having a spatial period λ, can be synthesized by a sum of harmonic (sinusoidal) functions whose wavelengths are integral submultiples of λ (i.e., λ, $\lambda/2$, $\lambda/3$, etc.). The Fourier series represents this process, according to

$$f(x) = \frac{A_0}{2} + \sum_{m=1}^{\infty} A_m \cos mkx + \sum_{m=1}^{\infty} B_m \sin mkx \tag{26.34}$$

For given $f(x)$, the coefficients are found as

$$A_m = \frac{2}{\lambda}\int_0^{\lambda} f(x') \cos mkx' \, dx' \qquad B_m = \frac{2}{\lambda}\int_0^{\lambda} f(x') \sin mkx' dx'$$

where the angular spatial frequency $k = 2\pi/\lambda$.[10]

We make the observation that purely sinusoidal waves have no actual physical reality and that we are invariably dealing with anharmonic features in practice. The synthesis of a periodic square wave into harmonic components provides a good demonstration of the principle (Fig. 26.11).

The effect of reducing the width of the wave is to introduce higher-order harmonics with smaller wavelengths. Consequently, the dimensions of the smallest feature being reproduced determine the total number of terms in the series. If we let the wavelength increase without limit and keep the width of the square-wave peaks constant, they become isolated pulses. This is in the character of nonperiodic functions, where it is no longer meaningful to speak of a fundamental frequency and associated harmonics. We are in effect now considering a series of terms of size approaching zero as the number of the terms approaches infinity. Hence, as we allow $\lambda \to \infty$, the Fourier series is replaced by the Fourier integral, according to

$$f(x) = \frac{1}{\pi}\left[\int_0^{\infty} A(k) \cos kx \, dx + \int_0^{\infty} B(k) \sin kx \, dx\right] \tag{26.35}$$

with the coefficients derived from the given function $f(x)$ written as

$$A(k) = \int_0^{\infty} f(x') \cos kx' \, dx' \qquad B(k) = \int_0^{\infty} f(x') \sin kx' \, dx'$$

The quantities $A(k)$ and $B(k)$ are interpreted as the amplitudes of the *sine* and *cosine* contributions in the range of angular spatial frequency between k and $k + dk$, and are referred to as the *Fourier cosine* and *sine* transforms. If we consolidate the *sine* and *cosine* transforms into a single complex exponential expression, we arrive at the complex form of the Fourier integral. This is the integral in Eq. (26.32), known as the *Fourier transform*, which for the one-dimensional function $f(x)$ is

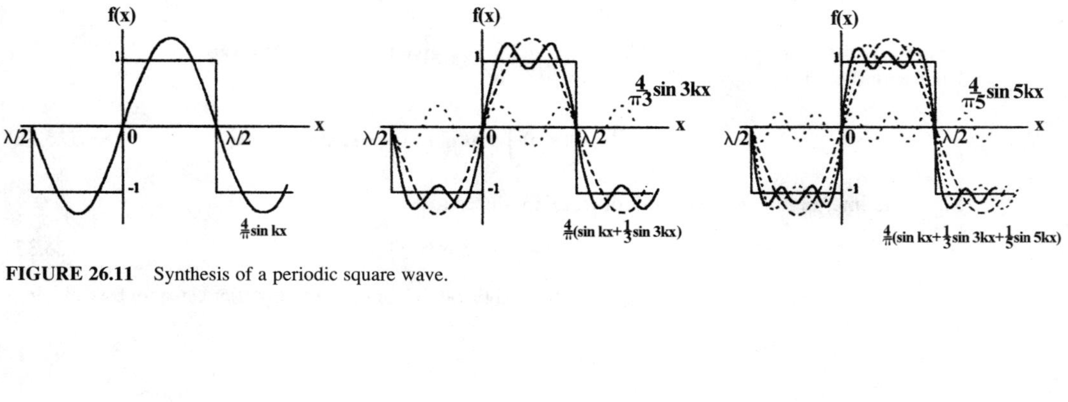

FIGURE 26.11 Synthesis of a periodic square wave.

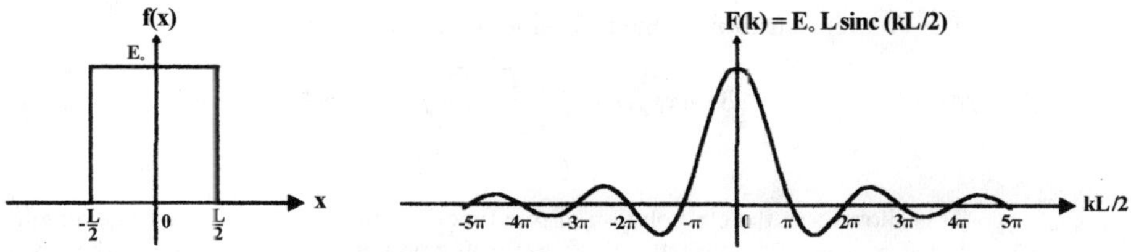

FIGURE 26.12 Fourier transform of a square pulse.

$$F(\nu) = \Im\{f(x)\} = \int_{-\infty}^{\infty} \exp(-i2\pi\nu x)\, f(x)\, dx \tag{26.36}$$

where the function $\exp(-i2\pi\nu x)$ is periodic in x with frequency $\nu = k/2\pi$.[11] Hence, to continue with the square-wave example, using Eq. (26.36) we can synthesize an isolated pulse in the *frequency domain* (Fig. 26.12).

For the two-dimensional or three-dimensional function $f(\mathbf{r})$, the Fourier transform is

$$F(\boldsymbol{\rho}) = \Im\{f(\mathbf{r})\} = \int_{-\infty}^{\infty} \exp(-i2\pi\boldsymbol{\rho}\cdot\mathbf{r})\, f(\mathbf{r})\, d\mathbf{r} \tag{26.37}$$

We are particularly interested in two-dimensional space where the function $f(\mathbf{r}) = f(x, y)$. The exponential term $\exp(-i2\pi\boldsymbol{\rho}\cdot\mathbf{r}) = \exp\{-i2\pi(ux + vy)\}$ is periodic in $\mathbf{r} = (x, y)$ with frequency $\boldsymbol{\rho} = (u, v)$. Thus the function $F(\boldsymbol{\rho})$ resides in the spatial *frequency domain*, whereas the function $f(\mathbf{r})$ resides in the physical *space domain*. The usual form of the inverse of the one-dimensional transform is written as

$$f(x) = \Im^{-1}\{F(\nu)\} = \int_{-\infty}^{\infty} \exp(+i2\pi\nu x)\, F(\nu)\, d\nu \tag{26.38}$$

and that for the two- or three-dimensional transform is written as

$$f(\mathbf{r}) = \Im^{-1}\{F(\boldsymbol{\rho})\} = \int_{-\infty}^{\infty} \exp(+i2\pi\boldsymbol{\rho}\cdot\mathbf{r})\, F(\boldsymbol{\rho})\, d\boldsymbol{\rho} \tag{26.39}$$

The two-dimensional Fourier transform $\Im\{f(x, y)\} = F(u, v)$ has the linearity property

$$\begin{aligned} F\{af_1(x, y) + bf_2(x, y)\} &= aF\{f_1(x, y)\} + bF\{f_2(x, y)\} \\ &= aF_1(u, v) + bF_2(u, v) \end{aligned} \tag{26.40}$$

the scaling property

$$F\{f(\alpha x, \beta y)\} = \frac{1}{|\alpha\beta|} F\left(\frac{u}{\alpha}, \frac{v}{\beta}\right) \tag{26.41}$$

and the shift property

$$F\{f(x - \alpha, y - \beta)\} = F(u, v) \exp[-i2\pi(u\alpha + v\beta)] \tag{26.42}$$

The Fourier transform of the two-dimensional convolution (26.30) is

$$\begin{aligned}\Im\{g(x, y)\} &= F\left\{\int_{-\infty}^{\infty}\int_{-\infty}^{\infty} f_1(x', y')f_2(x - x', y - y')\,dx'dy'\right\}\\ &= F\{f_1(x, y)\}F\{f_2(x, y)\} = F_1(u, v)F_2(u, v)\end{aligned} \tag{26.43}$$

This shows that the convolution of two functions in the space domain is equivalent to multiplication in the frequency domain. It is this simplifying property that demonstrates the advantage of conducting signal processing in the frequency domain rather than the space domain.

26.3.3 Discrete and Finite Data Sampling

The data contained in a digitally recorded image is an ordered *finite* array of *discrete* values of intensity (grayscale). To manipulate this data, the continuous integrals defining the Fourier transform and convolution must be expressed as approximating summations. For a series of discrete samples $x(n\tau)$ of a continuous function $x(t)$, a representation in the frequency domain can be written as

$$X(w) = \sum_{-\infty}^{\infty} x(n\tau)\exp(-iwn\tau) \tag{26.44}$$

where τ is the sample interval. For a discrete function $x(0), x(2\tau), \ldots, x((N-1)\tau)$ of N elements, Eq. (26.44) becomes

$$X\left(u\frac{1}{N\tau}\right) = \frac{1}{N}\sum_{n=0}^{N-1} x(n\tau)\exp\left\{-i2\pi\left[u\left(\frac{1}{N\tau}\right)\right](n\tau)\right\} \tag{26.45}$$

where the substitution $w = u(1/N\tau)$ has been made, with $u = 0, 1, 2, \ldots, N-1$.

This in effect provides samples $X_0, X_1, \ldots, X_{N-1}$ of the continuous function $X(w)$ at intervals $1/N\tau$ of the frequency w. Hence, for the discrete function $x_0, x_1, \ldots, x_{N-1}$, we can express the Fourier transform as

$$X_u = \frac{1}{N}\sum_{n=0}^{N-1} x_n \exp\left\{-i\left(\frac{2\pi}{N}\right)un\right\} \tag{26.46}$$

for $u = 0, 1, 2, \ldots, N-1$. Hence, a sampling interval of τ in the t domain is equivalent to a sampling interval of $1/N\tau$ in the frequency domain. The inverse of Eq. (26.46) is given by

$$x_n = \sum_{u=0}^{N-1} X_u \exp\left\{i\left(\frac{2\pi}{N}\right)un\right\} \tag{26.47}$$

for $n = 0, 1, 2, \ldots, N-1$. The finite Fourier transform Eq. (26.46) and its inverse Eq. (26.47) define sequences that are periodically replicated, according to the expressions $X_{N,m+k} = X_k$, $x_{N,m+k} = x_k$, for all integer values of m.

To determine the convolution for a discrete sample we follow Eq. (26.43) and find the product of two finite Fourier transforms $Z_u = X_u Y_u$ and take the inverse of the result, according to

$$z_n = \sum_{u=0}^{N-1} \exp\left\{i\left(\frac{2\pi}{N}\right)un\right\}X_u Y_u \tag{26.48}$$

Substituting from Eq. (26.46) and rearranging the order of summation gives

$$z_n = \frac{1}{N^2}\sum_{p=0}^{N-1}\sum_{q=0}^{N-1} x_p y_q \sum_{u=0}^{N-1} \exp\left\{i\left(\frac{2\pi}{N}\right)un - up - uq\right\} \tag{26.49}$$

Because of the replication of sequences, the convolution (26.49) can be simplified to (26.50)

$$z_n = \frac{1}{N} \sum_{q=0}^{N-1} x_{n-q} y_q \tag{26.50}$$

for $n = 0, 1, 2, \ldots, N-1$.

With discrete data sampling, the interval τ between data points determines how well the original signal is modeled. The choice of τ is referred to the Nyquist criterion, which states that a signal must be sampled at least twice during each cycle of the highest signal frequency. This means that if a signal $x(t)$ has a Fourier transform such that

$$X(w) = 0 \qquad \text{for} \qquad w \geqslant \frac{w_N}{2} \tag{26.51}$$

samples of x must be taken at a rate greater than w_N, so that we require $2\pi/\tau \geqslant w_N$.

We can illustrate the effect of the sampling interval τ by considering the band-limited transform $X(w)$ of the continuous function $x(t)$. Here, $X(w)$ is zero for values of w outside the interval $(-W, W)$ (Fig. 26.13*a*). If we multiply $x(t)$ by a sampling function $y(t)$, which is a train of impulses with interval τ (Fig. 26.13*b*), we obtain the sampled version $x(n\tau) = y(t)x(t)$ (Fig. 26.13*c*). The transform $F\{y(t)x(t)\} = F\{y(t)\}*F\{x(t)\} = Y(w)*X(w)$ is periodic with interval $1/\tau$. The repetitions of $X(w)$ can overlap, with the nearest centers of the overlapped region occurring at $w = \pm 1/2\tau$, providing $1/2\tau < W$. Therefore, to avoid overlapping we require

$$\tau \leqslant \frac{1}{2W} \tag{26.52}$$

Hence, by decreasing the sampling interval τ, we can separate the transformed functions (Fig. 26.13*d*). This means that we can multiply the transform by the function

$$H(w) = \begin{cases} 1 & -W \leqslant w \leqslant W \\ 0 & \text{elsewhere} \end{cases}, \tag{26.53}$$

to isolate $X(w)$ (Fig. 26.13*e*). The inverse Fourier transform $F^{-1}\{X(w)\}$ of the isolated function will yield the original *continuous* function $x(t)$ (Fig. 26.13*f*). The complete recovery of a band-limited function, from samples whose spacing satisfies Eq. (26.51) has been formulated as the *Whittaker-Shannon sampling theorem*. If the condition of Eq. (26.51) is not met, the transform in the interval $(-W, W)$ is corrupted by contributions from adjacent periods. This leads to a phenomenon known as *aliasing* and prevents the complete recovery of an undersampled function.[12]

Shannon's sampling theorem is closely related to the Nyquist criterion. With respect to telegraph transmission theory, Nyquist proposed the basic idea that the minimum signal rate must contain the bandwidth of the message. For sampled signals, the bandwidth really ranges from $-f_s/2$ to $f_s/2$, where f_s is the sampling rate. So the minimum sampling rate f_s (known as the Nyquist rate), does contain the signal bandwidth, whose highest frequency is limited to the Nyquist frequency f_s. Shannon gives credit to Nyquist for pointing out the fundamental importance of the time interval $1/2W$ in telegraphy. However, he further established that if a function $f(t)$ contains no frequencies higher than W, it is completely determined by samples at a series of points spaced $W/2$ apart.

26.3.4 Line Integrals and Projections

To describe the process of tomographic reconstruction, we need to formulate a description of the geometry of x-ray paths and the process of absorption along the paths that result in the projected two-dimensional image. This will provide the basic framework from which to develop the method for building up a digital representation of the solid object. We define a single ray path as the straight

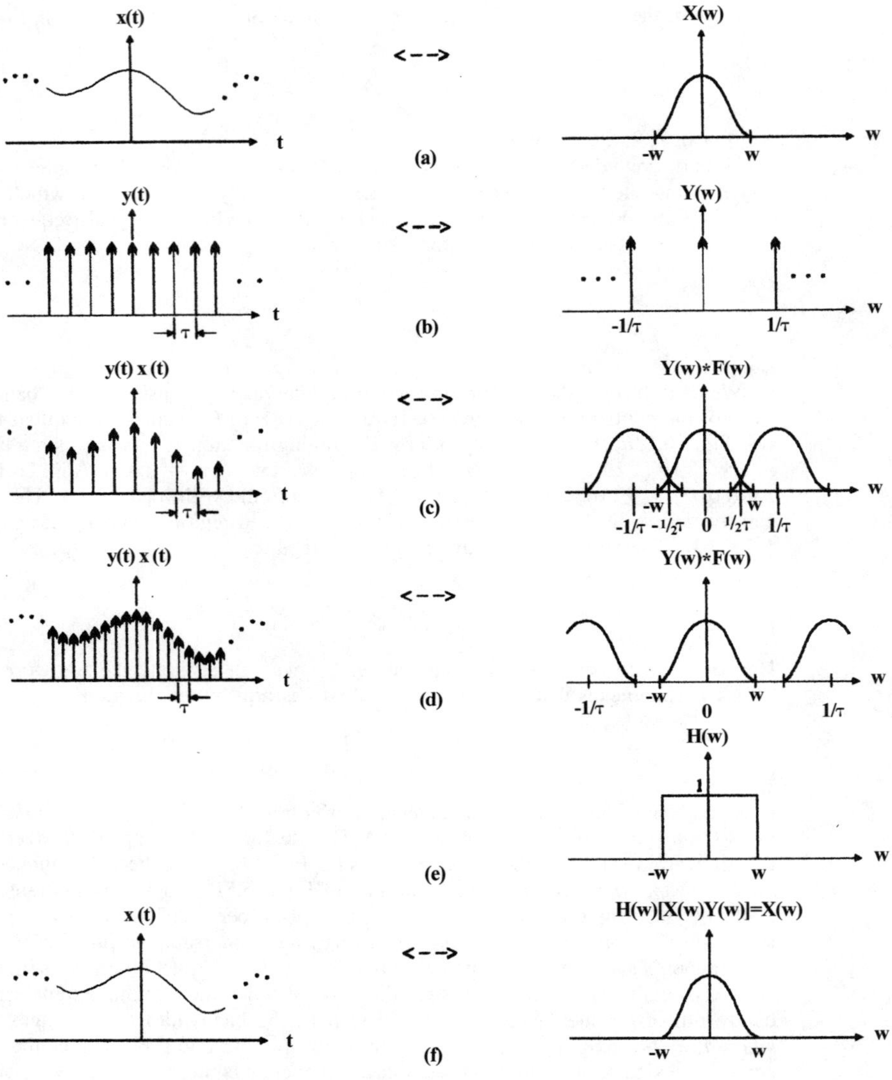

FIGURE 26.13 The effects of sampling interval on a band-limited function.

line joining the x-ray source to a detector cell and passing through the absorbing object (Fig. 26.14). The object is represented by the function $f(x, y)$. The location r and angle θ of ray AB with respect to the (x, y) coordinates is defined by

$$x \cos \theta + y \sin \theta = r \tag{26.54}$$

The level of input signal to the detector cell is determined by the integrated absorption of photons

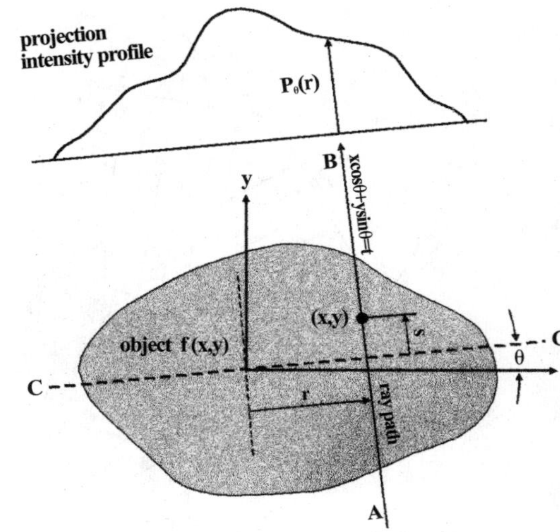

FIGURE 26.14 Projection of object $f(x, y)$ for angle θ.

along the ray path according to Eq. (26.22). For the path defined by the (θ, r) coordinates, the photon attenuation, over length l of ray AB in the s direction, is written as

$$\Phi_\theta(r) = \Phi_0 \exp\left(-\int_{AB} f(x, y)\, ds\right) \tag{26.55}$$

This indicates that the object function $f(x, y)$ is the linear attenuation coefficient $\mu(x, y)$ in the plane of the projected rays. Taking the natural logarithm of Eq. (26.55) gives the linear relation

$$P_\theta(r) = -\ln \frac{\Phi_\theta(r)}{\Phi_0} = \int_{AB} f(x, y)\, ds \tag{26.56}$$

This is a line integral of $f(x, y)$ and the quantity $P_\theta(r)$ is called the *radon transform* or the *projection* of $f(x, y)$ along the ray path. It is important to note that the process of reconstructing slices of objects from their projections is simply that of inverting Eq. (26.56). A projection is formed by combining a set of line integrals $P_\theta(r)$, the simplest one being a collection of parallel ray paths with constant θ. This is known as a parallel projection (Fig. 26.15).

The Fourier Slice Theorem. The two-dimensional Fourier transform of the object function, according to Eq. (26.37), is given by

$$F(u, v) = \int_{-\infty}^{\infty} f(x, y) \exp[-i2\pi(ux + vy)]\, dx\, dy \tag{26.57}$$

where (u, v) is the Fourier space, or frequency domain variables, conjugate to the physical space variables (x, y). For a parallel projection at an angle θ, we have the Fourier transform of the line integral (26.56), written as

$$S_\theta(w) = \int_{-\infty}^{\infty} P_\theta(r) \exp(-i2\pi wr)\, dr \tag{26.58}$$

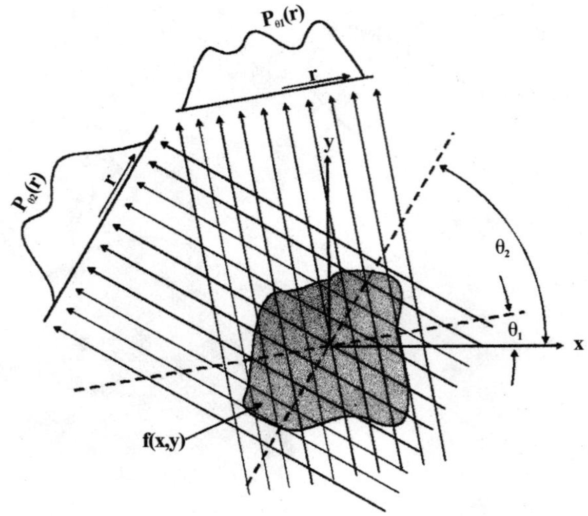

FIGURE 26.15 Parallel projections with varying angle.

The relationship between the rotated (r, s) coordinate system and the fixed (x, y) coordinate system is given by the matrix equation

$$\begin{bmatrix} r \\ s \end{bmatrix} = \begin{bmatrix} \cos\theta & \sin\theta \\ -\sin\theta & \cos\theta \end{bmatrix} \begin{bmatrix} x \\ y \end{bmatrix} \tag{26.59}$$

Defining the object function in terms of the rotated coordinate system, for the projection along lines of constant r, we have

$$P_\theta(r) = \int_{-\infty}^{\infty} f(r, s)\, ds \tag{26.60}$$

Substituting this into Eq. (26.58) gives

$$S_\theta(w) = \int_{-\infty}^{\infty} \left[\int_{-\infty}^{\infty} f(r, s) \right] \exp(-i2\pi wr)\, dr \tag{26.61}$$

This can be transformed into the (x, y) coordinate system using Eq. (26.59), to give

$$S_\theta(w) = \int_{-\infty}^{\infty}\int_{-\infty}^{\infty} f(x, y) \exp[-i2\pi w(x\cos\theta + y\sin\theta)]dx\, dy \tag{26.62}$$

The right-hand side of this expression is the two-dimensional Fourier transform at the spatial frequency ($u = w\cos\theta$, $v = w\sin\theta$). Hence, we have the relationship between the projection at angle θ and the two-dimensional transform of the object function, written as

$$S_\theta(w) = F(w,\theta) = F(w\cos\theta, w\sin\theta) \tag{26.63}$$

This is the Fourier slice theorem, which states that the Fourier transform of a parallel projection of an object taken at angle θ to the x axis in physical space is equivalent to a slice of the two-dimensional transform $F(u, v)$ of the object function $f(x, y)$, inclined at an angle θ to the u axis in frequency space (Fig. 26.16).

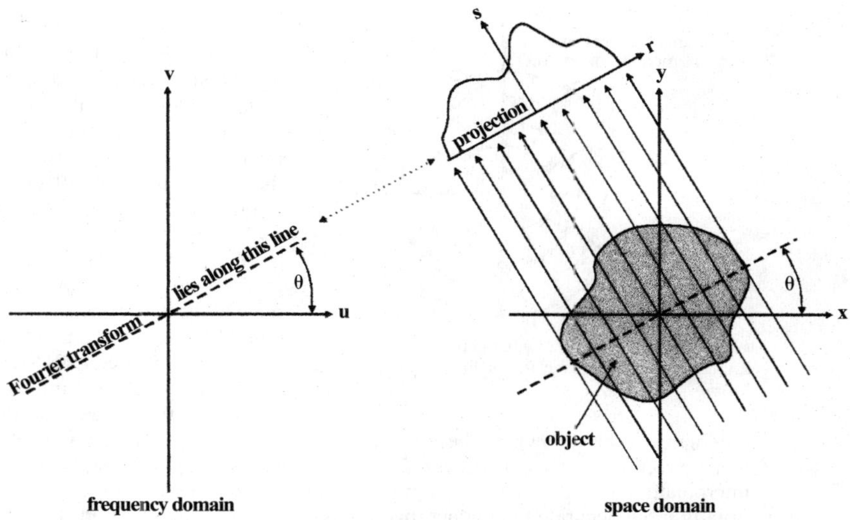

FIGURE 26.16 The Fourier transform of a projection relates to the Fourier transform of the object along a radial line.

If an infinite number of projections were taken, $F(u, v)$ would be known at all points in the u, v plane. This means that the object function could be recovered by using the two-dimensional inverse Fourier transform (26.39), written as

$$f(x, y) = \int_{-\infty}^{\infty}\int_{-\infty}^{\infty} F(u, v) \exp[+i2\pi(ux + vy)]du\, dv \tag{26.64}$$

In practice, the sampling in each projection is limited, so that $f(x, y)$ is bounded according to $-A/2 < x < A/2$ and $-A/2 < y < A/2$. Also, physical limitations ensure that only a finite number of Fourier components are known, so that $F(u, v)$ is bounded according to $-N/2 < u < N/2$ and $-N/2 < v < N/2$. With reference to the discrete inverse Fourier transform (26.47), and in view of the limiting bounds, Eq. (26.64) becomes

$$f(x, y) = \frac{1}{A^2} \sum_{m=-N/2}^{N/2} \sum_{n=-N/2}^{N/2} F\left(\frac{m}{A}, \frac{n}{A}\right) \exp\left[+i2\pi\left(\frac{m}{A}x + \frac{n}{A}y\right)\right] \tag{26.65}$$

Similarly for the discrete form of the Fourier transform (26.46), we have

$$F(u, v) = \frac{1}{M} \sum_{m=0}^{M-1} \left\{\frac{1}{N} \sum_{n=0}^{N-1} f(m, n) \exp\left[-i\frac{2\pi}{N}nv\right]\right\} \exp\left[-i\frac{2\pi}{M}mu\right] \tag{26.66}$$

for $u = 0, \ldots, M-1$; $v = 0, \ldots, N-1$. The expression within the curly brackets { } is the one-dimensional finite Fourier transform of the mth row of the projected image and can be computed using the fast Fourier transform (FFT) algorithm. To compute $F(u, v)$, each row in the image is replaced by its one-dimensional FFT, followed by the one-dimensional FFT of each column. The spatial resolution of the reconstructed $f(x, y)$, in the plane of the projection, is determined by the range N of Fourier components used.

The proposition of an infinite number of projections is not realistic in practice so that the function

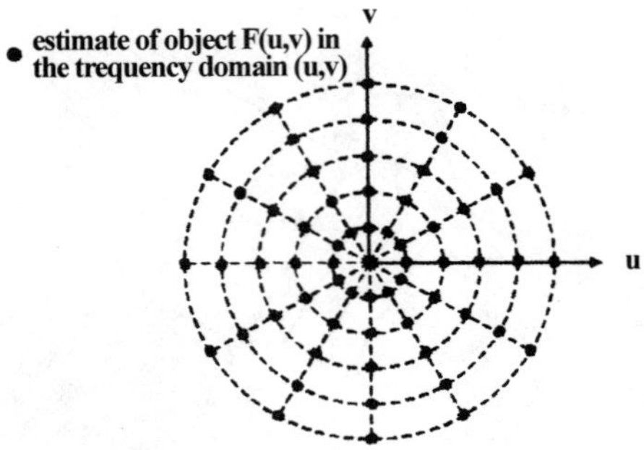

FIGURE 26.17 Finite number of projections provide estimates of Fourier transforms $F(u, v)$ along radial lines.

$F(u, v)$ is known only along a finite number of radial lines (Fig. 26.17). To implement Eq. (26.65), points on a square grid are interpolated from the values of $F(u, v)$ at the given radial points. However, because the density of the radial points becomes progressively sparser the farther away from the center, the interpolation error increases. This implies that there is greater error in the calculation for higher-frequency components in an image.

While the Fourier slice theorem implies that given a sufficient number of projections, an estimate of the two-dimensional transform of the object could be assembled and by inversion an estimate of the object obtained, this simple conceptual model of tomography is not implemented in practice. The approach that is usually adopted for straight ray tomography is that known as the *filtered back-projection algorithm.* This method has the advantage that the reconstruction can be started as soon as the first projection has been taken. Also, if numerical interpolation is necessary to compute the contribution of each projection to an image point, it is usually more accurate to conduct this in physical space rather than in frequency space.

The Filtered Back-Projection. The Fourier slice theorem indicates that each projection is almost an independent operation, a fact that is not immediately obvious from considerations in the space domain. The cautionary "almost" is because the only common information in the Fourier transforms of projections at different angles is the constant [also referred to as "direct current" (dc)] term. Using the argument of independence in the frequency domain, we consider the inverse Fourier transform of each radial line as representing the object with contributions from all other projections reduced to zero. The conception is that the single-projection reconstruction so formed is equivalent to the Fourier transform of the original object multiplied by a simple narrow bandwidth filter (Fig. 26.18*a*). However, because of the angular segmentation, the desired contribution to the summation of projections is that of a pie-wedge filter (Fig. 26.18*c*).

To achieve an estimation of the pie wedge, a simple weighting is taken in the frequency domain, such as multiplying the Fourier transform of the projection $S_\theta(w)$ by the width of the wedge at that frequency. Thus, if there are K projections over 180°, for frequency w, each wedge has width $|w|(2\pi/K)$ and the estimate is a vertical wedge (Fig. 26.18*b*) with the same mass as the pie wedge. The approximation can be made increasingly accurate the greater the number of projections taken.

A formal description of the filtered back-projection process is best served by expressing the object function $f(x, y)$ defined in Eq. (26.64), in an alternative coordinate system. Here, the rectangular coordinate system in the frequency domain (u, v) is exchanged for the polar coordinate system (w, θ), so that

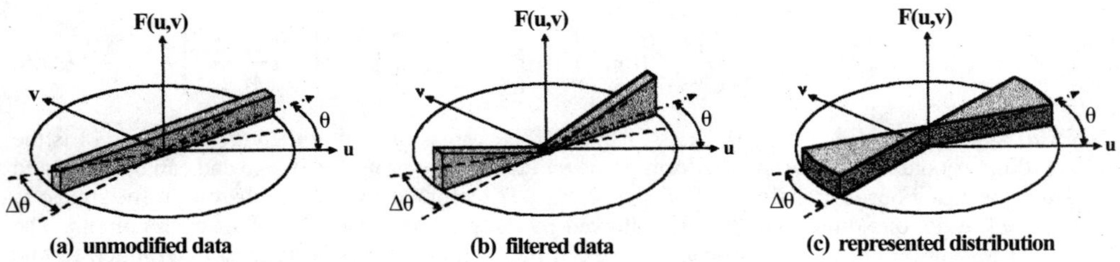

FIGURE 26.18 Frequency domain data from one projection.

$$f(x, y) = \int_0^{2\pi}\int_0^{\infty} F(w, \theta) \exp[+i2\pi w(x \cos \theta + y \sin \theta)]w \, dw \, d\theta \tag{26.67}$$

given that $u = \cos \theta$ and $v = w \sin \theta$. Splitting the integral into two parts, in the ranges 0 to π and π to 2π, using the property $F(w, \theta + \pi) = F(-w, \theta)$, and substituting from Eqs. (26.59) and (26.62), we get

$$f(x, y) = \int_0^{\pi}\left[\int_{-\infty}^{\infty} S_{\theta}(w)|w| \exp(+i2\pi wr) \, dw\right] d\theta \tag{26.68}$$

If we write the integral inside the straight brackets [] as

$$Q_{\theta}(r) = \int_{-\infty}^{\infty} S_{\theta}(w)|w| \exp(+i2\pi wr) \, dw \tag{26.69}$$

the object function, for projections over 180°, becomes

$$f(x, y) = \int_0^{\pi} Q_{\theta}(x \cos \theta + y \sin \theta) \, d\theta \tag{26.70}$$

The function $Q_{\theta}(r)$ is a filtered projection so that Eq. (26.70) represents a summation of back-projections from the filtered projections. This means that a particular $Q_{\theta}(r)$ will contribute the same value at every point (x, y) in the reconstructed image plane that lies on the line defined by (r, θ) according to Eq. (26.59). This is equivalent to smearing the filtered back-projection along the line *GH* in the reconstructed image plane (Fig. 26.19).

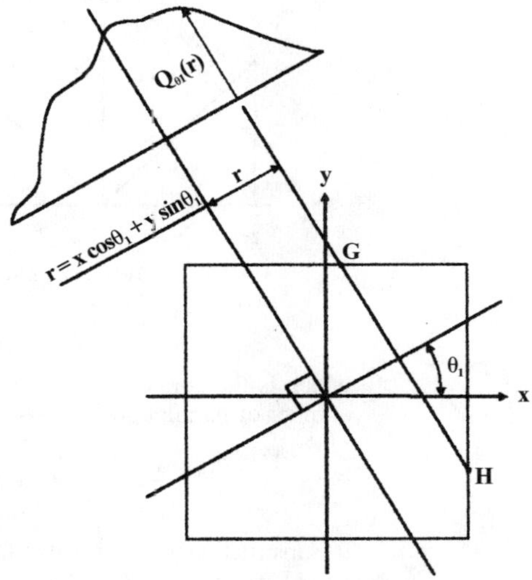

FIGURE 26.19 Filtered back-projection along line *GH*.

When the highest frequency in the projection is finite, we can express the filtered projection (26.69) as

$$Q_{\theta}(r) = \int_{-\infty}^{\infty} S_{\theta}(w)H(w) \exp(+i2\pi wr) \, du \tag{26.71}$$

where

$$H(w) = |w|b_w(w) \qquad \text{and} \qquad b_w(w) = \begin{cases} 1 < |w| < W \\ 0 \quad \text{otherwise} \end{cases}$$

The filter transfer function $H(w)$ is a ramp in frequency space with a high-frequency cutoff $W = 1/2\tau$ Hz (Fig. 26.20). In the physical space $H(w)$ has the impulse response $h(r) = \int_{-\infty}^{\infty} H(w) \exp(+i2\pi wr) \, dw$, so that, according to Eq. (26.43), the product $S_{\theta}(w)H(w)$ can be written as $F\{\int_{-\infty}^{\infty} P_{\theta}(r')h(r - r') \, dr'\}$. Hence the filtered projection (26.71) becomes

$$Q_{\theta}(r) = F^{-1}\{S_{\theta}(w)H(w)\} = \int_{-\infty}^{\infty} P_{\theta}(r')h(r - r') \, dr' \tag{26.72}$$

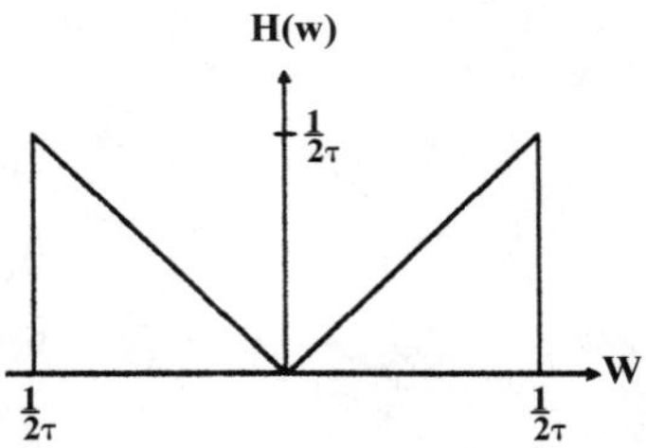

FIGURE 26.20 Ramp filter with high-frequency cutoff.

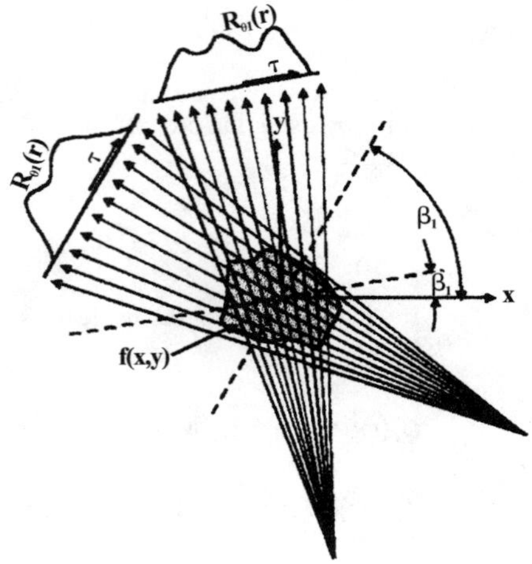

FIGURE 26.21 Fan beam projection with varying angle.

which is the convolution of the line integral with the filter function in physical space. For a finite number of parallel projections, where $P_\theta = 0$ for $|r| > r_m$, Eq. (26.72) becomes

$$Q_\theta (r) = \int_{-\infty}^{\infty} P_\theta(r) h(x \cos \theta + y \sin \theta - r)\, dr \tag{26.73}$$

It is instructive to observe that for an unfiltered back-projection, the response to a point object, where $f(x, y) = f(r, s) = (\delta r)\ \delta(s)$, gives the point-spread function $PSF \approx 1/\pi R$, where R is the radial direction from the center. This is generally considered an unfavorable response, and the aim of the filtering is to make the PSF as close to a two-dimensional delta function as possible.

The process of producing a series of parallel ray paths in order to form a one-dimensional projection image involves a linear scanning action across the object. This slow process has to be repeated at each angular step in the rotation of the system. This results in lengthy data collection periods that have proved unacceptable in many cases. These shortcomings can be overcome by generating the line integrals simultaneously for a single projection by using a fan beam source of x-rays in conjunction with a one-dimensional array of detector cells (Fig. 26.21).

The Fan Beam Reconstruction. With a slit collimated fan beam of x-rays, a projection is formed by the illumination of a fixed line of detector cells. A common detector structure in this respect is the equally spaced collinear array. The projection data for this geometry is represented by the function $R_\beta(p)$, where β is the projection angle of a typical ray path SB and p is the distance along the detector line D_1D_2 to a point at B (Fig. 26.22). To simplify the algebra, the fan beam projection $R_\beta(p)$ is referred to the detector plane moved to D_1'/D_2'. The ray path integral along SB is now associated with the point A on this imaginary detector line at $p = OA$ (Fig. 26.23).

If a parallel projection were impressed upon this geometry, the ray SA defined by (r, θ) would belong to the one-dimensional image $P_\theta(r)$. The relationship between the parallel-beam and fan-beam cases is given by

$$r = p \cos \gamma \qquad \theta = \beta + \gamma \tag{26.74}$$

$$r = \frac{pD}{\sqrt{D^2 + s^2}} \qquad \theta = \beta + \tan^{-1}\frac{p}{D}$$

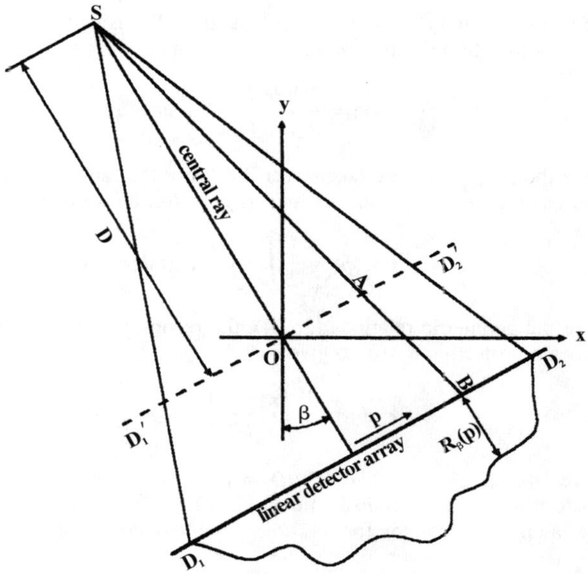

FIGURE 26.22 Linear array of evenly spaced detector cells.

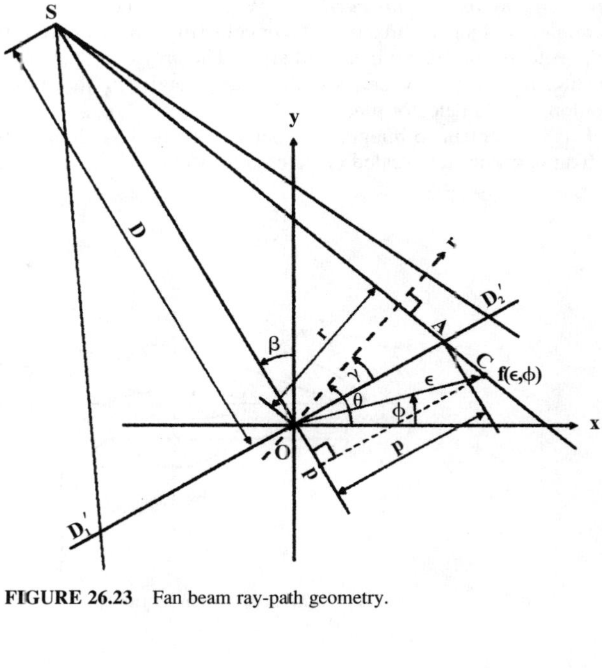

FIGURE 26.23 Fan beam ray-path geometry.

The reconstruction $f(x, y)$ at a point C is given by the substitution of the filtered projection (26.73) into the projection summation (26.70), written as

$$f(x, y) = \frac{1}{2}\int_0^{2\pi}\int_{-t_m}^{t_m} P_\theta(r)h(x \cos \theta + y \sin \theta - r)\, dr\, d\theta \tag{26.75}$$

where the projections are taken over 360°. For the fan beam geometry it is convenient to work in polar coordinates (ϵ, ϕ), so that for $f(x, y) = f(\epsilon, \phi)$ we have

$$f(\epsilon, \phi) = \frac{1}{2}\int_0^{2\pi}\int_{-t_m}^{t_m} P_\theta(r)h[\epsilon \cos (\theta - \phi) - r]\, dr\, d\theta \tag{26.76}$$

Using the geometric relations (26.74), the reconstruction (26.76) can be expressed in terms of the fan beam projection $R_\beta(p)$, to give

$$f(\epsilon, \phi) = \frac{1}{2}\int_0^{2\pi}\frac{1}{U^2}\int_{-\infty}^{\infty} R_\beta(p)h(p' - p)\frac{D}{\sqrt{D^2 + p^2}}\, dp\, d\beta \tag{26.77}$$

where $U(\epsilon, \phi, \beta) = (SO + OP)/D = [D + \epsilon \sin (\beta - \phi)]/D$. Here, $h(p)$ is the inverse Fourier transform of the filter transfer function in Eq. (26.71) and the variable p' is the location p of the pixel along the detector for the object point (ϵ,ϕ) given by $p' = D\{\epsilon \cos (\beta - \phi)/[D + \epsilon \sin (\beta - \phi)]\}$

Although the fan beam geometry has definite advantages, it is nevertheless a two-dimensional reconstruction method. Like the parallel beam method, it relies on the stacking of sections, with interpolation, to reconstruct the three-dimensional object. Given the advent of large-area-format x-ray detectors, a more efficient technique is to completely illuminate the object with a cone beam and perform the reconstruction as a volume operation rather than an independent slice operation. Consequently, the ray integrals are measured through every point in the object in a comparable time to that taken in measuring a single slice.

The Cone Beam Reconstruction. With a cone beam of x-rays, a projection is formed by the illumination of a fixed area of detector cells (Fig. 26.24). A common detector structure in this respect is the equally spaced collinear cell array. The projection data for this geometry is represented by the function $R_\beta(p_D, q_D)$, where β is the source angle, p_D the horizontal position, and q_D the vertical position, on the detector plane.

It is convenient to imagine the detector be moved along the detector-source axis to the origin, with an appropriately scaled detector cell location (p, q), according to

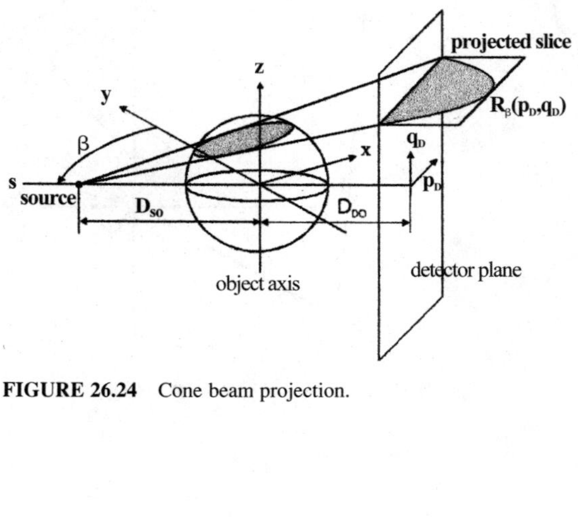

FIGURE 26.24 Cone beam projection.

$$p = \frac{p_D D_{SO}}{D_{SO} + D_{DO}} \qquad q = \frac{q_D D_{SO}}{D_{SO} + D_{DO}} \tag{26.78}$$

where D_{SO} is the distance from the source to the origin and D_{DO} the distance from the origin to the detector. Each cone beam ray terminating at the relocated detector cell (p, q) is contained in a tilted fan specified by the angle of tilt $\bar{\gamma}$ of the central ray and the value of the normal $t = \bar{t}$ to the central ray, given by

$$\bar{t} = q\frac{D_{SO}}{\sqrt{D_{SO}^2 + q^2}} \qquad \bar{\gamma} = \tan^{-1}\frac{q}{D_{SO}} \tag{26.79}$$

The idea is that an object function can be approximately reconstructed by summing the contributions from all the tilted fans. This means that the back-projection is applied within a volume rather than across a plane. The volume elemental cell is a voxel and has the same implication for resolution that the pixel has in the planar representation.

To develop the related analysis, first consider a two-dimensional fan beam rotated about the z axis by β and lying in the x, y plane. If the location of a point (ϵ, ϕ) in polar coordinates for the x, y plane is defined in terms of the rotated coordinate system (r, s), we have the coordinate conversion in the r, s plane, given by

$$\begin{aligned} r &= x\cos\beta + y\sin\beta \qquad & s &= -x\sin\beta + y\cos\beta \\ x &= \epsilon\cos\phi & y &= \epsilon\sin\phi \end{aligned} \tag{26.80}$$

so that

$$p' = \frac{D_{SO} r}{D_{SO} - s} \qquad U(x, y, \beta) = \frac{D_{SO} - s}{D_{SO}} \tag{26.81}$$

Hence, the reconstructed object function according to (26.77) may be written as

$$f(r, s) = \frac{1}{2}\int_0^{2\pi} \frac{D_{SO}^2}{(D_{SO} - s)^2}\int_{-\infty}^{\infty} R_\beta(p) h\left(\frac{D_{SO} r}{D_{SO} - s} - p\right)\frac{D_{SO}}{\sqrt{D_{SO}^2 + p^2}}\, dp\, d\beta. \tag{26.82}$$

To contribute to a voxel (r, s, z) for $z \neq 0$ in the cone beam geometry, the fan beams must be tilted out of the r, s plane to intersect the particular voxel (r, s, z) from various x-ray source orientations. As a result, the location of the reconstruction point in the tilted system is now determined by a new coordinate system $(\bar{r}, \bar{s})$ (Fig. 26.25). Consequently, the fan beam geometry in these new coordinates will change. Specifically, the new source distance is defined by

$$\bar{D}_{SO} = \sqrt{D_{SO}^2 + q^2} \tag{26.83}$$

where q is a detector cell row and represents the height of the z axis intersection of the plane of the fan beam. The incremental angular rotation $d\beta$ will also change according to

$$\bar{D}_{SO}\overline{d\beta} = D_{SO}\overline{d\beta} \qquad \overline{d\beta} = \frac{d\beta D_{SO}}{\sqrt{D_{SO}^2 + q^2}}. \tag{26.84}$$

Substituting these changes in Eq. (26.82), we have

$$f(\bar{r}, \bar{s}) = \frac{1}{2}\int_0^{2\pi} \frac{\bar{D}_{SO}^2}{(\bar{D}_{SO} - \bar{s})^2}\int_{-p_m}^{p_m} R_\beta(p, q) h\left(\frac{\bar{D}_{SO}\bar{r}}{\bar{D}_{SO} - \bar{s}} - p\right)\frac{\bar{D}_{SO}}{\sqrt{\bar{D}_{SO}^2 + p^2}}\, dp\, \overline{d\beta} \tag{26.85}$$

In order to work in the original (r, s, z) coordinate system we make the following substitutions in Eq. (26.85):

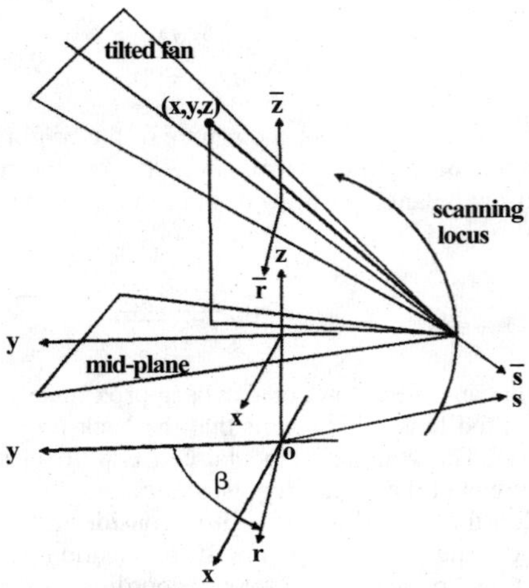

FIGURE 26.25 Tilted fan coordinate geometry.

$$\bar{r} = r \qquad \frac{\bar{s}}{D_{SO}} = \frac{s}{D_{SO}} \qquad \frac{q}{D_{SO}} = \frac{z}{D_{SO} - s} \tag{26.86}$$

to give the well-known Feldkamp reconstruction formula[13]

$$f(r, s) = \frac{1}{2}\int_0^{2\pi} \frac{D_{SO}^2}{(D_{SO} - s)^2} \int_{-p_m}^{p_m} R_\beta(p, q) h\left(\frac{D_{SO} r}{D_{SO} - s} - p\right) \frac{D_{SO}}{\sqrt{D_{SO}^2 + p^2}} \, dp \, d\beta \tag{26.87}$$

To apply these relations in practice the cone beam reconstruction algorithm would involve the following arithmetic operations.

1. Multiplication of the projection data $R_\beta(p, q)$ by the ratio of D_{SO} to the source-detector cell distance:

$$\overline{R}_\beta(p, q) = \frac{D_{SO}}{\sqrt{D_{SO}^2 + q^2 + p^2}} R_\beta(p, q)$$

2. Convolution of the weighted projection $R_\beta(p, q)$ with 1/2 $h(p)$ by multiplying their Fourier transforms with respect to p for each elevation q:

$$Q_\beta(p, q) = \overline{R}_\beta(p, q) * \tfrac{1}{2} h(p)$$

3. Back-projection of each weighted projection over the three-dimensional reconstruction grid:

$$f(r, s, z) = \int_0^{2\pi} \frac{D_{SO}}{(D_{SO} - s)^2} Q_\beta\left(\frac{D_{SO} r}{D_{SO} - s}, \frac{D_{SO} z}{D_{SO} - s}\right) d\beta$$

The inaccuracies resulting from the approximation associated with this type of reconstruction can be made diminishingly small with decreasing cone angle, which is typically ~10° for microtomography.

Algorithms that are based upon the Feldkamp principles have found favor in practice because of their good image quality and fast computational speed. However, the circular path that the source follows lies in a single plane and consequently projects a limited view of the object. Hence, intuitively it is observed that the greater the number of planes containing a source point, the more accurate the reconstruction of the object. To this effect it is possible to state a sufficient condition on the nature of source paths for exact cone-beam reconstruction after the formulation of Smith[14]: "If on every plane that intersects the object there exists at least one cone-beam source point, then the object can be reconstructed precisely." The equivalent statement for a fan-beam reconstruction is: "If there exists at least a fan-beam source on any straight line intersecting an object, an exact reconstruction is possible."

The scanning schemes that have been adapted to the above principle may involve the spiral/helical motion that requires lateral displacement with rotation.[15] Extensions of the Feldkamp algorithm, for quite general three-dimensional scanning loci, have been developed by Wang et al.[16] For further discussions on the approximate and accurate cone-beam reconstruction the reader is referred to Ref. 17.

26.4 THE MICROTOMOGRAPHY SYSTEM

26.4.1 The X-Ray Source

The spatial resolution is the principal factor that distinguishes microtomography from conventional medical tomography. In practice, medical systems must provide very high x-ray fluence in order to minimize the exposure times. This means that the x-ray source has a relatively large extended emission area of ~5 mm × 5 mm. The source of x-rays for medical/laboratory systems is created by the bombardment of a solid metal target (tungsten-rhenium) with a directed high-energy electron beam. The conventional design produces electrons from a tungsten spiral wire filament (cathode) held at high voltage. This is heated to a very high temperature, and electrons are extracted in a high voltage field ΔV_{CA} formed between the filament housing and the grounded x-ray target (anode) housing (Fig. 26.26). The associated geometry acts as an electrostatic lens and focuses the electron beam onto the target. There is provision for a reduced focal spot size of ~2.5 mm × 2.5 mm from a smaller auxiliary filament with a corresponding reduction in x-ray emission.

The result of the extended source is a geometric unsharpness in the image that can be interpreted as that due to a continuous distribution of point sources representing the x-ray emission area. The extent of the unsharpness U, or blurring, at the edges of the image depends on the extent of the source, according to

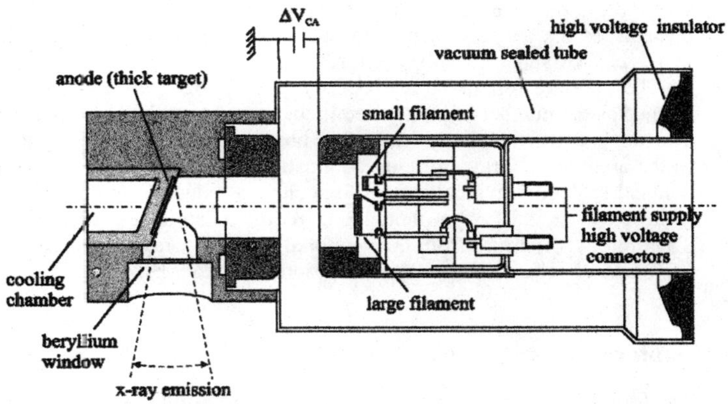

FIGURE 26.26 Medical x-ray tube.

$$U = F\left(\frac{D_{SD}}{D_{SO} - 1}\right) \tag{26.88}$$

where F = source size
D_{SO} = source to object distance
D_{SD} = source-to-detector (or image) distance

Hence, if the source size is reduced indefinitely, we have as $F \rightarrow 0$ the unsharpness $U \rightarrow 0$, and the magnification M is given by

$$M = \frac{D_{SD}}{D_{SO}} \tag{26.89}$$

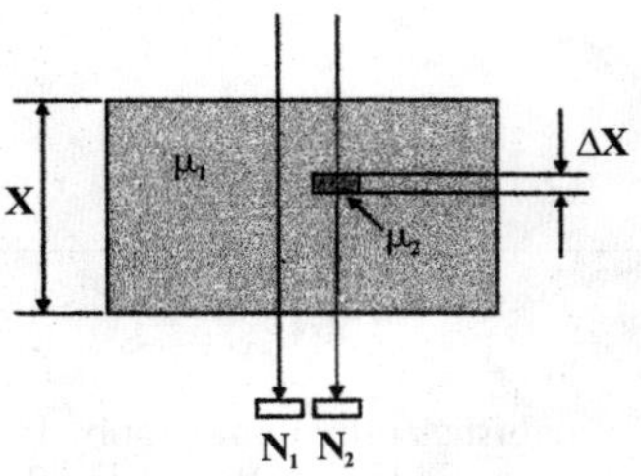

FIGURE 26.27 Threshold of detection for small contrasting object.

To achieve high-resolution $\Delta x \approx 1\ \mu m$, it is necessary to produce a source size $F \approx 1\ \mu m$. However, the penalty for a reduction in source size is a corresponding reduction in photon emission. In order to provide a level of contrast that will support the spatial resolution, consideration must be given to the photon flux. In this respect, a signal-to-noise ratio (SNR) of ~5 is usually chosen as a suitable threshold, for a contrasting feature of relative scale $\Delta x/x$ along a ray path of length x (Fig. 26.27). Here we consider a small contrasting object of thickness Δx and linear attenuation coefficient μ_2, within a larger object of thickness x and attenuation coefficient μ_1. The signal-to-noise ratio for the photon count in adjacent detector cells can be expressed approximately as

$$SNR = \frac{S}{\sigma_s} = \frac{N_1 - N_2}{\sqrt{N_1 + N_2}} \tag{26.90}$$

where the signal S is the difference $N_1 - N_2$ in the number of primary photons counted in the two identical detectors and σ_s is the standard deviation in S.[18] The presence of scattered radiation has been neglected in this derivation. For small contrast, where $|\mu_2 - \mu_1|\Delta x \ll 1$, Eq. (26.90) can be written as

$$SNR = \sqrt{\frac{N_0}{2}\exp(-\mu_1 x)}(\mu_2 - \mu_1)\,\Delta x \tag{26.91}$$

where N_0 is the unattenuated primary photon count for either detector cell. Hence, to observe the presence of Δx, the x-ray source must provide a sufficiently intense photon flux $\geqslant N_0$ as a function of μ_1, μ_2, Δx, and x, to satisfy a minimum $SNR \approx 5$.

The measure for the strength of x-ray emission from a point-like source is the brightness β, defined as the number of photons emitted per unit steradian (solid angle), per unit area of the source. The aim is to provide a value as high as possible while recognizing that there are practical limitations to the amount of x-ray emission obtainable with reduction in x-ray source size. For example, if a medical system is overcollimated to achieve a reduction in effective source size, the brightness is insufficient to satisfy the photon flux requirements at the high resolutions in the microdomain. Therefore, a departure from conventional medical x-ray source design is necessary to achieve the performance required for microtomography applications.

26.4.2 The Microfocal X-Ray Source

The formation of the electron beam is the key to resolution and to the level of photon flux from the x-ray source.[19] In this respect, the electron source brightness β_E and focal quality are functions of

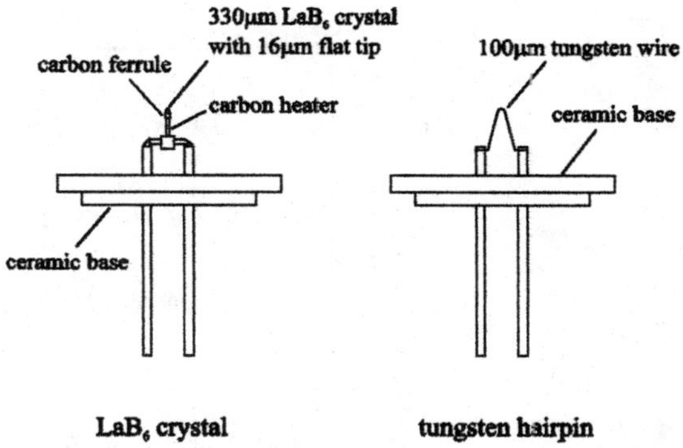

FIGURE 26.28 Electron emission cathodes.

the electron energy and the electron current.[20] The principal components that perform this task are the electron gun and the electron lens. The electron gun consists of an electron emitter, or cathode, and two beam-forming electrodes. The electron lens can be either electrostatic or electromagnetic. The latter is usually preferred because the associated focal aberrations are less.

Cathode Condition and Electron Optics. The principal conditions sought for an electron emitter are high brightness, stable geometry, and long life. To achieve a high-quality focus it is important to produce a well-defined virtual source at the electron gun. Here, efficient use is made of the peak cathode brightness of the emitted beam if a single round crossover disk, of effective diameter d_{CO}, is produced by the electron trajectories close to the cathode. Following electron microscopy practice, the choice of cathode structure generally relates to those based on a refractory metal, usually tungsten, or a hexaboride compound, usually lanthanum or cerium. In this respect, a 100-μm wire tungsten (W) hairpin and a lanthanum hexaboride (LaB_6) crystal with 16-μm flat tip, both mounted on a standard Philips base, offer alternative commercial solutions (Fig. 26.28). The tungsten hairpin is low cost, is structurally robust, has a high brightness, and can operate in fairly poor vacuum. However, it suffers from limited lifetime and poor mechanical stability at its high working temperature. The lanthanum hexaboride crystal has a very high brightness, good mechanical stability at its relatively low working temperature, and a very long lifetime (Fig. 26.29). However, it is high in cost, prone to poison unless operated in ultrahigh vacuum, and is structurally delicate.

The shape of the crossover diameter d_{CO}, which is the virtual electron source, is controlled by the geometry of the electrodes that compose the electron gun and the associated voltages (Fig. 26.30). The size, shape, and location of the electrodes, namely the Wehnelt (grid) and the first anode (extractor), relate to the diameter d_W of the Wehnelt aperture, the diameter d_A of the first anode aperture, and to the spacing D_{WA} between them. The brightness of the electron source is also determined by the ratio d_W/h, where h is the

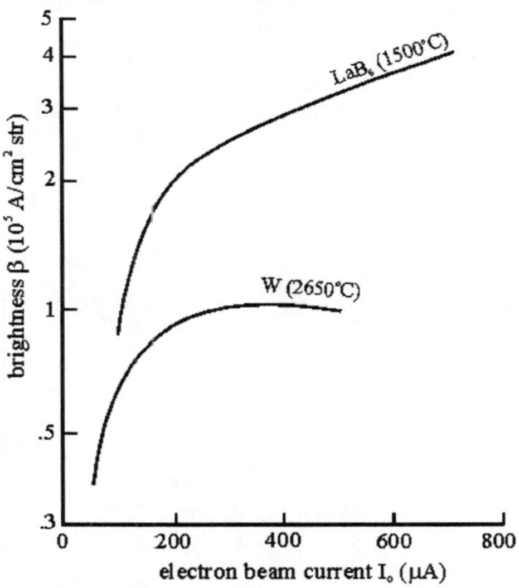

FIGURE 26.29 Electron emission LaB_6 and W cathodes.

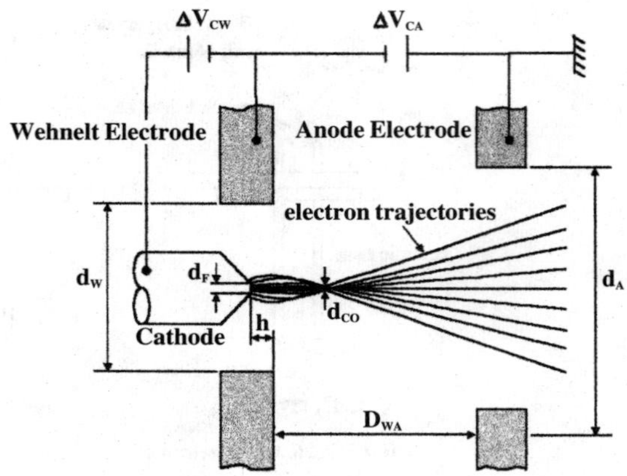

FIGURE 26.30 Electron gun cathode-electrode geometry.

distance of the cathode tip from the Wehnelt face. The voltage difference ΔV_{WA} between the Wehnelt and first anode produces the electron extraction-accelerating field, and the voltage difference ΔV_{CW} between the cathode and Wehnelt produces the electron emission control–retarding field. The size of the crossover diameter d_{CO} is dependent on the type of cathode and varies with the beam current I_0 extracted (Fig. 26.31). The electric fields provided by ΔV_{WA} and ΔV_{CW} are prone to interact at high electron beam currents I_0, causing distortion of the crossover d_{CO}.

With the location of the virtual electron source at the crossover confirmed, the image formation at the x-ray generating target (second anode) is determined according to the principles of electron optics. To achieve high x-ray flux Φ, the electron beam current I_0 must be commensurately high. Therefore, the electron beam shaping apertures are made as large as possible without compromising the focal spot size. A significant source of focal spot aberration arises from off-axis astigmatism caused by axial misalignment of the various optical elements. To minimize this effect, a combination of a single electromagnetic lens with a single defining electron beam aperture is a suitable configuration for high electron beam current and small focal spot.

According to Eq. (26.89) the object must be placed close to the x-ray source in order to work at high x-ray magnification M. Consequently the lens must provide a focus at z_i that is well clear of

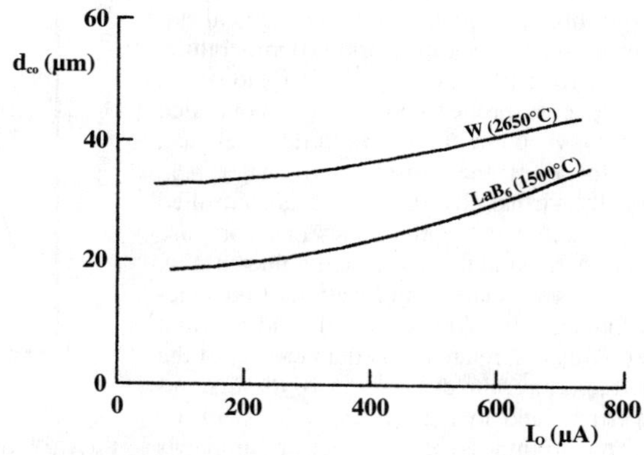

FIGURE 26.31 Crossover diameter, LaB_6 and W.

the base of the electron beam column. In this respect, a suitable shape for the lens can be found by numerically computing the magnetic vector potential **A** distribution and the flux density **B** distribution throughout the magnetic circuit and coil windings.[21] The calculation involves minimizing the functional

$$F = \iiint_{\text{volume}} \left\{ \frac{1}{2\mu} (\nabla \times \mathbf{A}) \cdot (\nabla \times \mathbf{A}) - \mathbf{J} \cdot \mathbf{A} \right\} dv \tag{26.92}$$

where $\mathbf{A}$ = vector potential, defined by $\mathbf{B} = \nabla \times \mathbf{A}$,
∇ = gradient del operator
μ = permeability
$\mathbf{J}$ = current density at any point

The optical properties of the lens are computed numerically from the paraxial electron trajectories $r(z)$, given by

$$\frac{d^2r}{dz^2} + \frac{\eta}{8V_r} B^2(z)r = 0 \tag{26.93}$$

where $B(z)$ = axial magnetic flux density distribution
η = electron charge/rest-mass ratio
V_r = relativistically corrected beam voltage (Fig. 26.32)

There are three components to the focal spot size d_f that strongly influence the design of the electron column, namely, the geometric focus d_m of the crossover d_{CO}, the spherical aberration d_s and the chromatic aberration d_c. The geometric focus of the crossover is controlled by the demagnification M_L of the electron lens and is given by

$$d_m = M_L d_{CO} \tag{26.94}$$

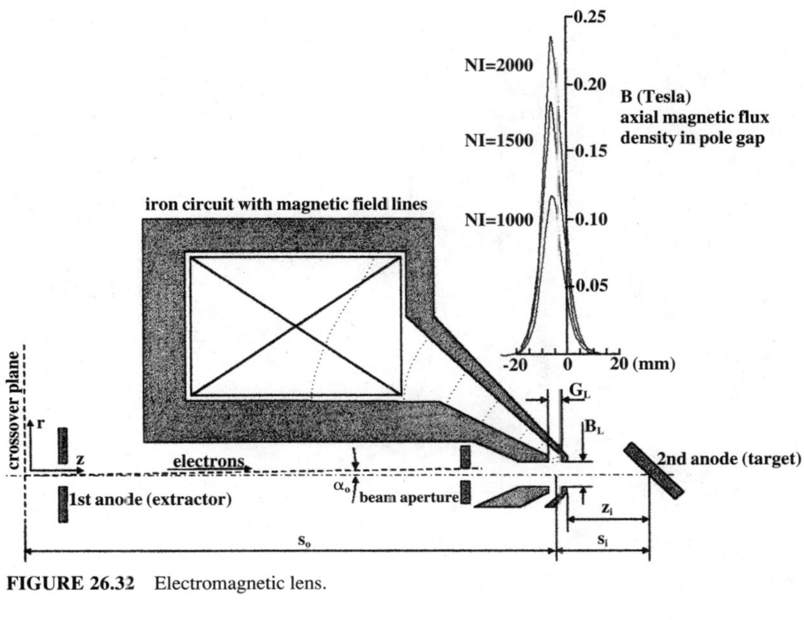

FIGURE 26.32 Electromagnetic lens.

An approximate value of M_L can be calculated from the simple optical relation

$$M_L \approx \frac{s_i}{s_o} \tag{26.95}$$

where s_O and s_i are the respective distances from the center of the lens pole gap to the crossover and the focus. Hence, the choice of s_O and s_i, to achieve a desired value of d_m, is the first step in an iterative process that will evaluate M_L from the numerical analysis, Eqs. (26.92) and (26.93). The spherical aberration caustic is given by

$$d_s = \frac{1}{2}C_{so}(\alpha_0)^3 = \frac{1}{2}C_{si}(\alpha_i)^3 \tag{26.96}$$

The spherical aberration coefficients C_{so} and C_{si} are referred to the object side and to the image side, respectively, and the beam semiangles α_o and α_i are referred to in a similar manner. The chromatic aberration caustic is given by

$$d_c = 2C_{co}\alpha_o\frac{\delta V_0}{V_0} = 2C_{ci}\alpha_i\frac{\delta V_0}{V_0} \tag{26.97}$$

The chromatic aberration coefficients C_{co} and C_{ci} are referred to the object side and to the image side respectively and $\pm\delta V_O$ is the variation about the beam voltage V_0. The focal spot size d_f is found by adding the components d_m, d_s, and d_c in quadrature according to

$$d_f = \sqrt{d_m^2 + d_s^2 + d_c^2} \tag{26.98}$$

For good focal quality, low values of C_{si}, C_{ci}, and M_L are sought. For weak lenses, these parameters are dependent on the excitation parameter $NI/\sqrt{V_r}$, where NI is the ampere-turns (excitation) of the lens. In this respect, they are seen to decrease with increasing excitation parameter (Fig. 26.33*b* and *c*). However, to seek a high value of z_i will result in a low value of excitation parameter, which is

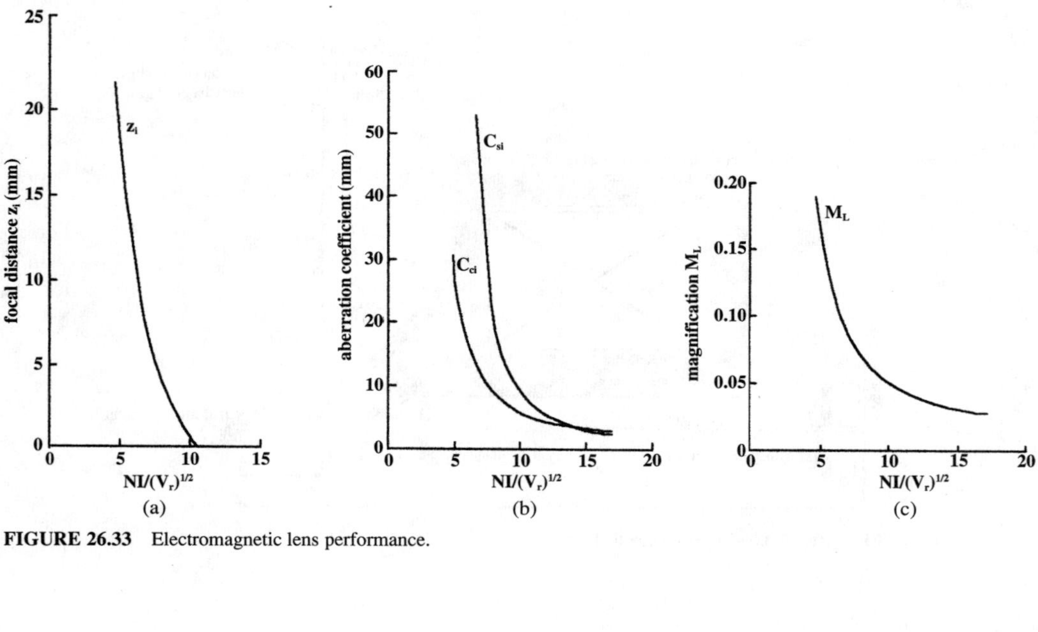

FIGURE 26.33 Electromagnetic lens performance.

EC - Electron Gun Chamber
EG - Electron Gun
EI - Electron Gun Ceramic Insulator
CA - Cathode Assembly
WE - Wehnelt Electrode
FA - First Anode
BA - Electron Beam Defining Aperture

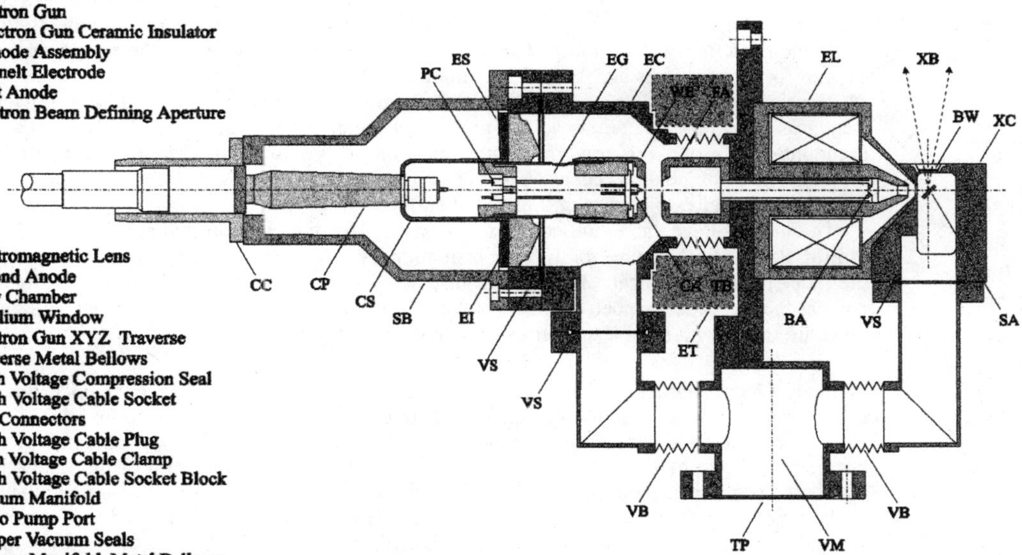

EL - Electromagnetic Lens
SA - Second Anode
XC - X-ray Chamber
BW - Beryllium Window
ET - Electron Gun XYZ Traverse
TB - Traverse Metal Bellows
ES - High Voltage Compression Seal
CS - High Voltage Cable Socket
PC - Pin Connectors
CP - High Voltage Cable Plug
CC - High Voltage Cable Clamp
SB - High Voltage Cable Socket Block
VM - Vacuum Manifold
TP - Turbo Pump Port
VS - Copper Vacuum Seals
VB - Vacuum Manifold Metal Bellows
XB - X-ray Beam

FIGURE 26.34 Microfocal x-ray source.

not the desired outcome (Fig. 26.33*a*). Consequently, a balance must be struck between the need for close access to the x-ray source and the need for a small focal spot size. An important lens parameter in this respect is the pole bore-to-gap ratio B_L/G_L. If this is chosen for optimum performance, a large value of B_L will allow an increase in z_i, without a corresponding increase in aberrations. However, for a given excitation NI, a larger lens bore leads to a reduction in the value of the axial flux density. To compensate, the size of the lens must be increased. A further compromise must be made in the choice of the diameter d_B of the beam-defining aperture. This determines the value of beam current I_0 and also fixes the beam semiangle α_0 referred to the object side, where $I_0 \propto d_B^2$ and $\alpha_o \propto d_B$. An increase in d_B produces a corresponding increase in x-ray flux Φ at the cost of an increase in focal spot size d_f.

A Practical Design. The technology involved in the development of the electron beam column conforms to the principles and practices associated with electron microscopes (Fig. 26.34). To provide versatility with optimum performance, the electron gun accommodates lanthanum hexaboride cathodes, as well as the more conventional tungsten hairpin cathodes. Therefore, the construction complies with standard ultrahigh vacuum (UHV) practice to avoid cathode poisoning. In this respect, the interior vacuum envelope is fabricated from UHV-grade stainless steel, the vacuum seals are flat copper rings (conflat), moving components are sealed with stainless steel bellows, and electrical insulators are metal-ceramic. To maintain the required vacuum $< 10^{-9}$ mbar, the system is turbo-pumped and all resins, rubber o rings, greases, etc., are rigorously excluded.

The electron chamber has a demountable triode electron gun with a very compact metal-ceramic insulator profile that can support voltages up to 100 kV. Cathode replacement is a routine bench operation with adjustments for concentricity and axial spacing relative to the Wehnelt electrode for optimum ΔV_{CW}. The electron gun is mounted on a three-axis traverse for alignment with the first anode and for adjustment in the extraction gap to provide optimum ΔV_{WA} over the energy range 0 to 100 to keV.

The electromagnetic lens iron circuit is external to the vacuum envelope and provides continuous focus according to $NI \propto \sqrt{V_r}$. The coil windings are cast with high-thermal-conductivity resin to avoid the need for forced cooling. The special geometry permits access to the x-ray source to within 17 mm for large objects and 10 mm for small objects.

The x-ray chamber has a relatively thin (~0.5 mm) beryllium window and houses magnification standards within the vacuum envelope to monitor the electron beam focus. The second anode stage is mounted on a two-axis traverse with a selection of five different target materials for x-ray generation. The individual targets are 6 mm × 8 mm in extent to provide fresh material in the case of damage due to electron beam pitting of the surface.

The high voltage cable and connector geometry, for the electron gun supply, are based upon a Philips standard cable plug for 100-kV insulation. A silicon seal prevents high-voltage tracking on the atmosphere side of the electron gun insulator. A resin filled cable block has a slightly raised conical face to expel trapped air pockets while compressing the silicon seal. The electrical connections are made with a cable socket in the core of the cable block.

Measured Performance. Under the conditions of space invariance and incoherence, an image can be expressed as the convolution of the object irradiance and the point-spread function, Eq. (26.15). The corresponding statement in the spatial frequency domain, Eq.(26.28), is obtained by taking the Fourier transform of Eq. (26.15). This states that the frequency spectrum of the image irradiance equals the product of the frequency spectrum of the object irradiance distribution and the transform of the point-spread function. In this manner, optical elements functioning as linear operators transform a sinusoidal input into an undistorted sinusoidal output [Eq. (26.33)]. Hence the function that performs this service is the transform of the point-spread function $\Im\{h(x, y)\}$, known as the *optical transfer function* $O[u, v]$ (OTF).[22] This is a spatial frequency-dependent complex function with a modulus component called the *modulation transfer function* $M[u, v]$ (MTF) and a phase component called the *phase transfer function* $\Phi[u, v]$ (PTF). The *MTF* is the ratio of the image-to-object modulation, while the *PTF* is a measure of the relative positional shift from object to image.

Taking a one-dimensional distribution for simplicity, we have

$$\Im\{h(x)\} \equiv O[v] = M[v]e^{i\Phi(v)} \tag{26.99}$$

where $M[v]$ and $\Phi[v]$ are *MTF* and *PTF* respectively. It is customary to define a set of normalized transfer functions by dividing by its zero spatial frequency value

$$O[0] = \int_{-\infty}^{+\infty} h(x)\, dx$$

according to

$$O_n[v] = \frac{\Im\,\{h(x)\}}{\int_{-\infty}^{+\infty} h(x)\, dx} = \Im\{h_n(x)\} = M_n[v]e^{i\Phi(v)} \tag{26.100}$$

where

$$h_n(x) = \frac{h(x)}{\int_{-\infty}^{+\infty} h(x)dx}$$

is the normalized line-spread function and

$$M_n[v] = \frac{M[v]}{O[0]}$$

is the normalized modulation transfer function. If the line-spread function is symmetrical, there is no phase shift so that Im $O_n[v] = 0$, $M_n[v] = \text{Re } O_n[v]$, and $\Phi(v) = 0$. A useful parameter in evaluating the performance of a system is the contrast or modulation, defined by

$$\Gamma\,[v] = \frac{I(v)_{\max} - I(v)_{\min}}{I(v)_{\max} + I(v)_{\min}} \tag{26.101}$$

where $I(v)_{\max}$ is the peak of the profile and $I(v)_{\min}$ is the trough of the profile. Since the *MTF* is the ratio of the image-to-object modulation, we may write, for sinusoids of period λ and corresponding frequency $v = 1/\lambda$ the expression

$$M_n[v] = \frac{\Gamma(v)_{\text{image}}}{\Gamma(v)_{\text{object}}} \tag{26.102}$$

It is common practice to use the *MTF* to quantify the resolution capability of linear systems associated with image processing. This is because if the individual *MTF*s are known for the components of a system, the overall *MTF* is often simply their product. A natural *resolution standard* would comprise a continuous series of parallel lines with sine wave intensity profiles of increasing frequency. However, this is difficult to produce in practice and it is usual to adopt a simpler pattern known as the *Sayce chart*, which has a parallel line, or rectangular, wave grating with a decreasing spatial period δ and hence increasing spatial frequency $\sigma = 1/\delta$.[23] The rectangular pattern profile $L(x)$ may be expressed in terms of its Fourier series components [Eq. (26.34), according to

$$L(x) = L_0\left\{1 + \frac{4}{\pi}\Gamma_0(\sigma)_r\sum_{k=1}^{\infty}\frac{(-1)^{k+1}}{2k-1}\cos 2\pi(2k-1)\sigma\right\} \tag{26.103}$$

where the object contrast [Eq. (26.101), or modulation, is

$$\Gamma_0(\sigma)_r = \frac{L(\sigma)_{\max} - L(\sigma)_{\min}}{L(\sigma)_{\max} + L(\sigma)_{\min}} \tag{26.104}$$

and

$$L_0 = \frac{L(\sigma)_{\max} + L(\sigma)_{\min}}{2}$$

For the object contrast we have the condition $\Gamma_0(\sigma)_r = 1$ for $L(\sigma)_{\min} = 0$.

In seeking a relationship for the value of the *MTF* for a system, we consider the contribution from a single component and offer the output as the input for the next component in the system chain. Hence, the incoherent image $I(x)$ of the rectangular wave grating, formed by convolution of the object $L(x)$ with the line-spread function $h(x)$ of the first component, is given by

$$I(x) = \int_{-\infty}^{\infty} h(x')L(x - x')\,dx' \tag{26.105}$$

If $h(x)$ possesses symmetry, the modulation transfer function value $M_n[(2k-1)\sigma]$, for the spatial frequencies $(2k-1)\sigma$, according to Eq. (26.100) may be written as

$$M_n[(2k-1)\sigma] = \frac{\int_{-\infty}^{+\infty} h(x')\cos\,[2\pi(2k-1)\sigma x]\,dx'}{\int_{-\infty}^{+\infty} h(x')\,dx'}$$

Substituting the expression for the object function (26.103) into Eq. (26.105), we obtain

$$I(x) = L_0 H_0 \times \left\{1 + \frac{4}{\pi}\Gamma_0(\sigma)_r \sum_{k=1}^{\infty} \frac{(-1)^{k+1} M_n[(2k-1)\sigma]}{2k-1} \cos\left[2\pi(2k-1)\sigma x\right]\right\} \tag{26.106}$$

where

$$H_0 = \int_{-\infty}^{+\infty} h(x')\, dx' = O[0]$$

The image contrast $\Gamma_i(v)_r$ can be written in a similar manner to the object contrast [Eq. (26.104), so that an overall rectangular-wave response, at the spatial frequency $\sigma = 1/\delta$ may be written as

$$M(\sigma)_r = \frac{\Gamma_i(\sigma)_r}{\Gamma_0(\sigma)_r} \tag{26.107}$$

where

$$\Gamma_i(\sigma)_r = \frac{I(\sigma)_{\max} - I(\sigma)_{\min}}{I(\sigma)_{\max} + I(\sigma)_{\min}}$$

Substituting for $L(x)_{\min}$, $L(x)_{\max}$ from Eq. (26.103) and $I(x)_{\min}$, $I(x)_{\max}$ from Eq. (26.106), we find the overall response [Eq. (26.107)] becomes

$$\begin{aligned} M(\sigma)_r &= \frac{4}{\pi}\sum_{k=1}^{\infty} \frac{(-1)^{k+1}}{2k-1} M_n[(2k-1)\sigma] \\ &= \frac{4}{\pi}\left[M_n[\sigma] - \frac{1}{3}M_n[3\sigma] + \frac{1}{5}M_n[5\sigma] - \cdots\right] \end{aligned} \tag{26.108}$$

Solving for the $M_n[\sigma]$ by successively subtracting series for $M_n[k\sigma]/k$, with k chosen to eliminate progressively higher terms in $M_n[k\sigma]$, gives the expression

$$\begin{aligned} M_n[\sigma] = \frac{\pi}{4}\Big[&M(\sigma)_r + \frac{1}{3}M(3\sigma)_r - \frac{1}{5}M(5\sigma)_r + \frac{1}{7}M(7\sigma)_r \\ &+ B_k \frac{1}{k} M(k\sigma)_r \cdots\Big], \end{aligned} \tag{26.109}$$

where k takes on the odd values 1, 3, 5, etc., and B_k is 1, 0, or -1 according to

$$B_k = (-1)^m(-1)\frac{k-1}{2} \qquad \text{if } r = m$$

$$B_k = 0 \qquad \text{if } r < m$$

Here, m is the total number of primes into which k can be factored and r is the number of different prime factors in k. According to Eq. (26.109) the modulation transfer function can be evaluated in terms of the measured modulation, or contrast values $M(\sigma)_r$. However, since a value of $M(0)_r = 1$ from Eq. (26.108) gives a corresponding value $M_n[\sigma] = 0.9538$ from Eq. (26.109), we must apply a normalizing factor of 1.0484 to the calculated result. We note that the values of M_n are compounded according to the product of the individual MTFs of the imaging components. For the simple x-ray source and detector combination, we have $M_n = M_n^s M_n^D$, where M_n^s is the x-ray source *MTF* and M_n^D is the detector MTF. If the detector is a film emulsion, the *MTF* for the digital scanning device should be included in the product.

It is difficult to construct a suitable resolution standard of the Sayce type for x-ray imaging. Typically the lines would be etched in a gold layer to provide good absorption, but as the line space

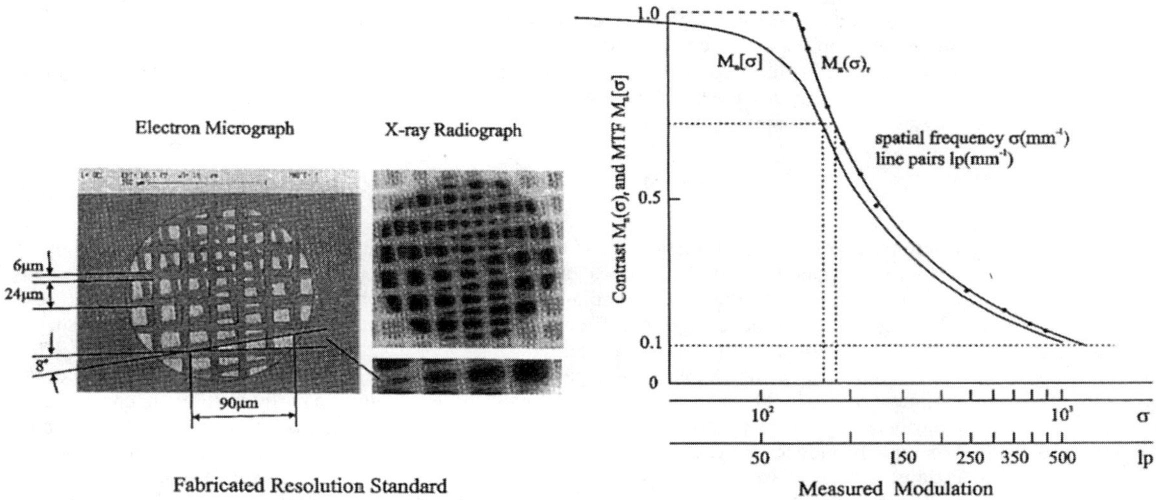

FIGURE 26.35 X-ray source measured resolution.

decreases the lithography process would require the layer to become progressively thinner to maintain the line discrimination. A limited alternative can be constructed from electron microscopy standard 8-μm-thick gold square grid structures. If two of these are superposed and skewed ~8°, the tapered grid intersections will create a continuously varying spatial frequency. A suitable scale is provided by a standard electron microscopy grid of spatial period of $\delta = 24$ μm with a 6-μm bar and 18-μm gap. To stabilize the structure, the grid assembly is sandwiched between standard 8-μm-thick gold disks with suitably sized apertures. This arrangement provides useful contrast values for varying spatial frequency σ but it is not a simple one-dimensional function as discussed in the foregoing analysis. Also, the x-ray source is not a line source with a line-spread function as assumed. Hence, the calculated values of the *MTF* will have a limited accuracy.

The values of $M(\sigma)_r$ and $M_n[\sigma]$ for the microfocal x-ray source are determined from a radiographic film image of the fabricated resolution standard at ×240 magnification Fig. (26.35). Here, the modulation of intensity profiles is measured at steps along the grid taper, to provide the variation of the contrast with frequency σ. A spatial calibration and modulation reference is provided by an electron micrograph of the resolution standard. The measurements indicate that the resolving power of the x-ray source has a bandwidth of 90 line-pairs/mm (i.e., for $M_r = 0.7071$). At a frequency of 1000 cycles/mm the modulation $M_r = 0.1$, which indicates that features of ~1 μm are resolvable.

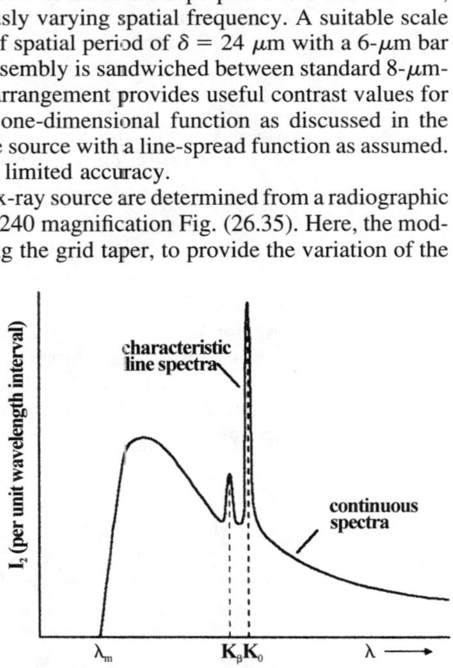

FIGURE 26.36 Spectral distribution of x-ray intensity.

X-Ray Generation. We are principally concerned with the deposition of electron beam energy in thick, or electron-opaque, targets. For this type of target the energy is mainly converted internally with the remainder backscattered externally. The x-ray emission is created by deceleration of the incident electrons through the Coulomb interaction with the nuclei of the target and by the removal of bound atomic electrons with the subsequent repopulation of the vacant orbits from adjacent-electron orbits. The former process relates to a broadband continuous spectrum and the latter to a process that produces narrowband discrete line spectra (Fig. 26.36).

In general, the electrons undergo several collisions before coming to rest so that the continuous spectrum is fairly broadband.[24] The shape of the distribution is dependent on the electron beam energy V_0, so that the envelope $\Phi(\lambda)$ is given by

$$\Phi(\lambda) = \frac{C_1 c Z}{\lambda^2}\left\{\frac{c(\lambda - \lambda_m)}{\lambda\lambda_m} + C_2 Z\right\} \tag{26.110}$$

where the short-wavelength cutoff $\lambda_m = hc/V_0$, C_1 and C_2 are constants, and Z is the atomic number of the target material.[25] Hence, the wavelength of maximum intensity is roughly $3\lambda_m/2$. To reduce the effects of beam hardening, the radiation distribution can be altered by varying the value of V_0 to change the short-wavelength cutoff and by inserting low atomic number material filters such as aluminum, to suppress the long-wave tail.

The intensity of the characteristic spectrum is proportional to Z and the energy depends on the atomic level K, L, M, of the interaction. Usually only the higher-energy K series is considered to contribute significantly to the overall radiation. Below 70 keV this is considered negligible and is less than 10 percent for energies in the range 80 to 150 keV. However, for monochromatic applications, such as phase contrast imaging, the presence of characteristic spectra is an important factor.

Historically, the angular distribution of emitted radiation from a thin target has been thoroughly researched. Similar information for electron opaque targets is very limited. From measurements of radiation in the forward hemisphere, the degree of anisotropy for electron opaque targets, expressed as the angle of deviation from the incident electron beam, was found to be ~60°.[26] However, a measurement in the backward hemisphere of a conventional x-ray tube indicated that the radiation distribution was isotropic.[27]

The relation for efficiency η_p of x-ray production of the continuous spectrum, defined as the ratio of x-ray power to that of incident electron beam power, is expressed as

$$\eta_p = KZV_0 \tag{26.111}$$

where $K \approx 10^{-9}$ for V_0 in electron-volts. Since the beam power $W_0 = V_0 I_0$ for beam current I_0, the continuous spectrum x-ray power W_c is given by

$$W_c = KZV_0^2 I_0 \tag{26.112}$$

This is a very inefficient process with most of the energy appearing as heat. As a consequence, the temperature of the target material can reach melting point. The liquid metal is subsequently removed by electron beam pressure, resulting in the excavation of a pit or groove. This degradation of the target material creates an enlarged focal spot. Therefore, it is important to determine the maximum power density that a target material can safely accommodate in order to produce maximum x-ray flux to satisfy the threshold of detection condition [Eq. (26.91)].

For the large-focal-diameter electron beams associated with medical x-ray sources, the heat dissipation is considered to occur at the surface and the analytical geometry is that of a disk-heating model.[28] However, this model is inaccurate for focal diameters of small extent where the volume dissipation of heat is an important factor. In this respect the penetration of the beam electrons into the target material must be taken into account (Fig. 26.37). Furthermore, the extent of the electron distribution will contribute to the enlargement of the focal spot, since it represents the volume of x-ray generation within the bulk of the material. For example, at 50 keV the electron penetration depth is ~10 μm which is a significant amount in comparison with a typical microfocal electron beam focal diameter of ~5 μm.

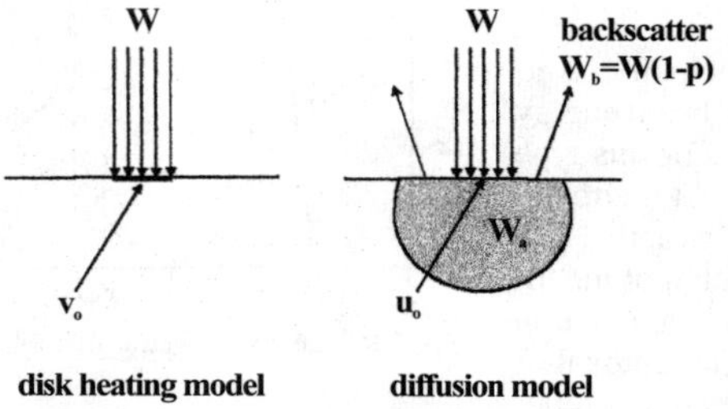

FIGURE 26.37 Heat transfer models.

A simple model for the electron motion in the target is to assume that the electrons travel in straight lines to a *depth of complete diffusion*, after which they diffuse evenly in all directions, to cover a total distance called the *range*.[29] Along the *range* an electron is assumed to loose energy according to the Thomson-Whiddington law, written as

$$V_0^2 - V^2 = kx \tag{26.113}$$

where V is the energy after the electron travels a distance x and the constant k is given by

$$k = \frac{b\rho Z}{A} \tag{26.114}$$

where ρ, Z, and A are density, atomic number, and atomic weight, respectively. The parameter b varies slightly with the type of target material but may be assumed to have the constant value $b = 7.75 \times 10^{10}\ \mathrm{eV^2 \cdot m^2/kg}$. From Eq. (26.113), the electron *range* x_0 is given by

$$x_0 = \frac{V_0^2}{k} \tag{26.115}$$

and the depth of diffusion[29] may be taken to be

$$z_d = \frac{40x_0}{7Z} \tag{26.116}$$

Hence, the diffusion center is nearer to the surface the larger the atomic number of the target material (Fig. 26.38).

The volume in which heat is dissipated is the sphere of radius $x_0 - z_d$ centered at D, which is the center of complete diffusion. The shaded region contains the backscattered beam power. For a small element of the surface, the initial electron beam current dI will be redistributed to a current density dJ at $x(r_1, z_1)$. The corresponding volume density of power dissipated, dW, will be given by

$$dW = -dJ\frac{dV}{dx} \tag{26.117}$$

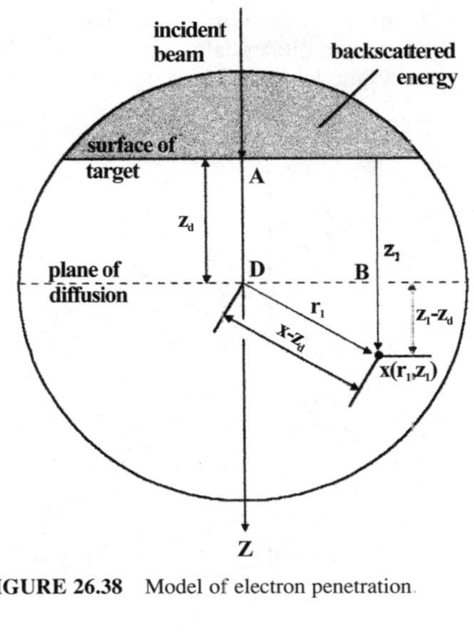

FIGURE 26.38 Model of electron penetration.

For uniform diffusion from D, the current density dJ is given by

$$dJ = \frac{dI}{4\pi(x - z_d)^2} \tag{26.118}$$

where it is assumed that $z_1 > z_d$. Substituting Eq. (26.118) into Eq. (26.117) gives

$$dW = \frac{k\, dI}{8\pi(x - z_d)^2(V_0^2 - kx)^{1/2}} \tag{26.119}$$

The total power density dissipation at $x(r_1, z_1)$ is determined by integrating over the beam elements dI. To do this, we consider a gaussian beam current density profile written as

$$J(r) = \frac{I}{1.44\pi a^2} \exp\left(-\frac{r^2}{1.44a^2}\right) \tag{26.120}$$

where I = total beam current
r = radial distance at the surface
a = beam radius at which the current density is $J_{(r=0)}/2$

Hence, the current element dI is given by $J\, dS$. The region of beam cross section that is within the range of $x(r_1, z_1)$ is bounded by a circle on the surface with radius $[(x_0 - z_d)^2 - (z_1 - z_d)^2]^{1/2}$ and center at $x(r_1, 0)$. If $z_1 < z_d$ then $x(r_1, z_1)$ lies between the surface and the depth of diffusion. In this case there is an additional contribution to dW from the incoming beam at $r = r_1$. This is found by replacing dJ in Eq. (26.117) by J at $r = r_1$ from Eq. (26.120), to give

$$dW_0 = \frac{I}{1.44\pi a^2} \exp\left(-\frac{r_1^2}{1.44a^2}\right) \frac{k}{2(V_0^2 - kx)^{1/2}} \tag{26.121}$$

The steady-state temperature u, in the presence of a volume distributed heat input $W(r, z)$, is determined by the Poisson equation, which for axisymmetric geometry is written as

$$\frac{\partial^2 u}{\partial r^2} + \frac{1}{r}\frac{\partial u}{\partial r} + \frac{\partial^2 u}{\partial z^2} = -\frac{W(r, z)}{\kappa} \tag{26.122}$$

where κ is the thermal conductivity. With the appropriate boundary conditions, solutions to Eq. (26.122) can be computed numerically. In this process it is convenient to normalize the temperature u by the disk-heating model temperature v_0 at the beam center, so that

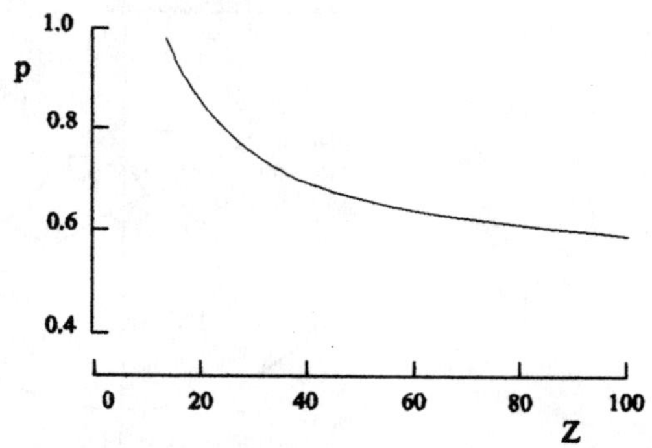

FIGURE 26.39 Fraction of power absorbed by the target.

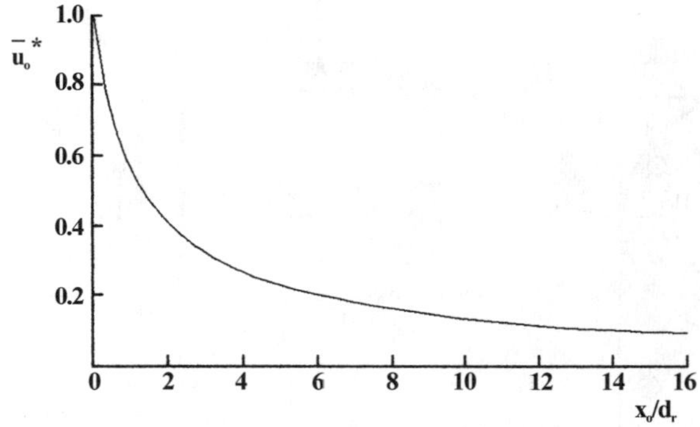

FIGURE 26.40 Temperature rise at the focal spot center.

$$\bar{u} = \frac{u}{v_0} \qquad and \qquad v_0 = \frac{0.099W_0}{\pi^{1/2}\kappa a} \tag{26.123}$$

If the power that is absorbed in the target is W_a and the backscattered power in the shaded area (Fig. 26.37) is W_b, the total beam power $W_0 = W_a + W_b = V_0 I_0$. The power retention factor $p = W_a/W_0 = 1 - W_b/W_0$, which is a function of atomic number Z and is independent of voltage V (Fig. 26.39). According to Eqs. (26.115) and (26.116), scaling the electron beam voltage by f will increase the *range* and the *depth of diffusion* by f^2, thus altering the power distribution by $1/f^2$, which in effect is equivalent to changing the beam diameter d_f by the same amount. Hence, changes in beam voltage and beam diameter result in the same outcome. Therefore, it is sensible to measure *range* in beam diameters as x_0/d_f, noting that for large diameter sources $x_0/d_f \rightarrow 0$ and $\bar{u} \rightarrow p$. This suggests an additional normalization by the power retention factor p to remove the effect of backscatter according to $\bar{u}^* = \bar{u}/p = u/(pv_0)$, which results in a temperature variation with electron range in beam diameters x_0/d_f, based upon absorbed power and independent of the material Z (Fig. 26.40).

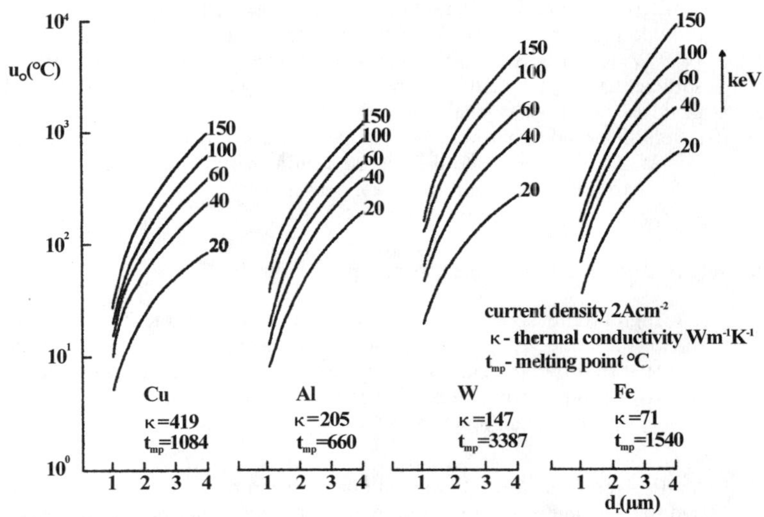

FIGURE 26.41 Heating of a semi-infinite block by a gaussian beam.

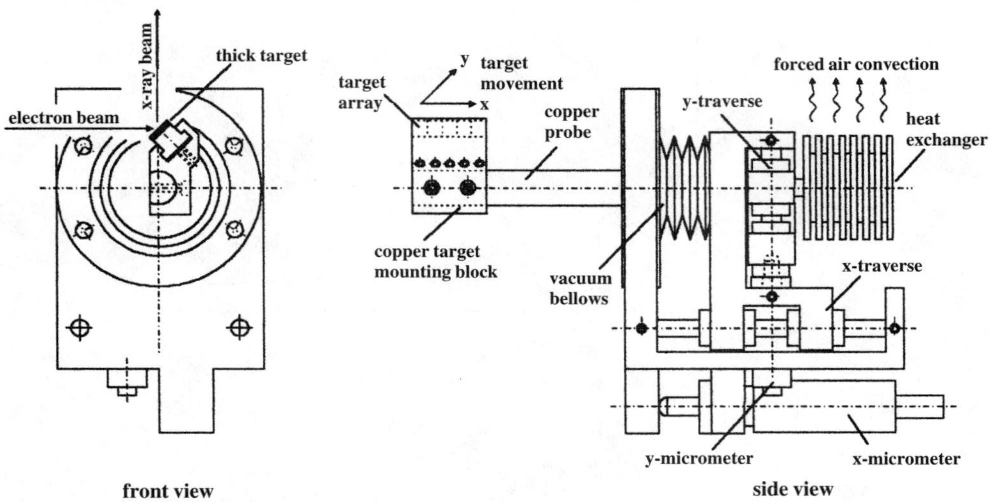

FIGURE 26.42 Microfocal target assembly.

If the beam power W_0 is absorbed at the target surface, the temperature rise v_0 is given by Eq. (26.123). This must be corrected for the effects of backscatter and source dispersal to give u_0. In this process, the power retention factor p is taken from Fig. 26.39 for a given material Z, and the temperature rise $\bar{u}^*_0$ is taken from Fig. 26.40 for x_0 calculated from Eqs. (26.114) and (26.115). With the values of v_0, p, and $\bar{u}^*_0$, the true temperature rise u_0 at the beam center is found from $u_0 = \bar{u}_0{}^* p v_0$ (Fig. 26.41).

A high-speed rotating anode is used in conventional x-ray tubes to overcome the heat loading.[28] However, the choice of cooling method is problematical for microfocal applications because of the need to place the specimen close to the x-ray source to achieve high magnification, according to Eq. (26.89). In this respect, rotating anodes are bulky and are difficult to install in a confined space. Furthermore, a range of target (anode) materials should be available within the vacuum envelope to provide the versatility required for research purposes. Since the power dissipation is quite small, $\sim$10 W, it can be argued that thermal conduction in copper, plus forced air convection, should be sufficient to maintain the temperature of the target at an acceptable level. The target assembly installed in the x-ray chamber (Fig. 26.34) adopts this principle with a copper probe supporting several material targets and an external cylindrical heat exchanger receiving forced cooling air (Fig. 26.42). The copper probe is supported on a two-dimensional traverse to select different materials and move fresh material under the electron beam.

26.4.3 Detectors and the Mode of Operation

The high brightness of the microfocal x-ray source enables the microtomography system to operate in two basic modes. These two modes are distinguished by whether the beam is collimated or not and lead to different data acquisition procedures. Also, with the installation of a monochromator, it is possible to perform phase contrast imaging that has a greater sensitivity to density variation than conventional contrast imaging.

Small-Area Detector. In the collimated beam mode the detector is a small-area single-cell device and the specimen is pointwise scanned, in the manner of conventional light microscopy, to record a single frame for each projection (Fig. 26.43). The advantage is that single-cell Si(Li), or HPGe,

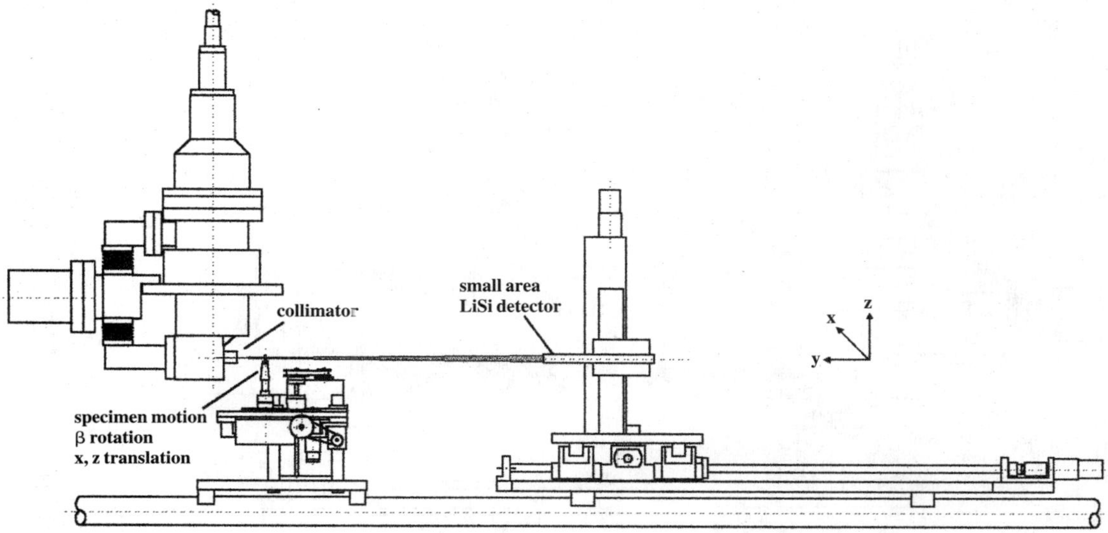

FIGURE 26.43 Raster scan imaging with small-area detector.

crystal detectors are available as energy-discriminating devices.[30] These are highly sensitive and possess good energy resolution. They respond to the characteristic spectra and are used in a photon pulse counting mode. This function can provide information on atomic composition of the specimen. A disadvantage is that the data acquisition is a relatively slow process. Also, the magnification is a function of the area raster scanned. The accuracy depends on the specimen scanning stage and detector collimation.

Large-Area Detector. In the uncollimated beam mode, the detector is a large-area multicell device that can record a single frame for each projection without pixel scanning (Fig. 26.44). Consequently data acquisition can be based on the rapid techniques associated with commercially available frame-grabbing hardware/software. However, the device is energy integrating and cannot determine atomic composition. Also, scattering may compromise the image quality, though at the higher magnifications the effect would be reduced because of the larger source to detector distance D_{SD}.

The relative separation of the source, specimen and detector provides direct magnification according to Eq. (26.39). Here, the magnification can be chosen so that the resolution limit of the detector does not compromise the resolution of the x-ray source. For a given detector pixel size d_p at the input plane, the corresponding voxel element size $d_v = d_p/M$. This also determines the size of the specimen to be imaged, since the frame size (height $H_F = d_pN_H$, width $W_F = d_pN_W$) relates to the size of the reconstruction volume (height $H_V = H_F/M$, width $W_V = W_F/M$, depth $D_V = W_F/M$), where $N_H \times N_W$ is the number of frame pixels.

If a wide range of specimen sizes is to be accommodated, the detector/camera input plane needs be of adequate extent, >50 mm. However, the charged coupled device (CCD) that is the actual detector is small in size, ~7 mm × 7 mm. Also, this device is principally sensitive to light and not to x-rays. Hence, the x-ray photons must be down shifted in frequency at the input plane and the image must be reduced to match the CCD. If the image demagnification m is provided by a simple lens system, with an f-number $f/\#$, the light-gathering power will be $\propto 1/(f/\#)^2$. Therefore, $f/\#$ needs to be kept as small as possible which can be achieved by a reduction in m. If a demagnifying intensifier is used, then a portion of the demagnification is provided without loss (Fig. 26.45). Furthermore, if the photon conversion scintillator is coated directly onto the intensifier, rather than onto a conventional fiber-optic faceplate, additional losses are avoided. The aim is to provide as

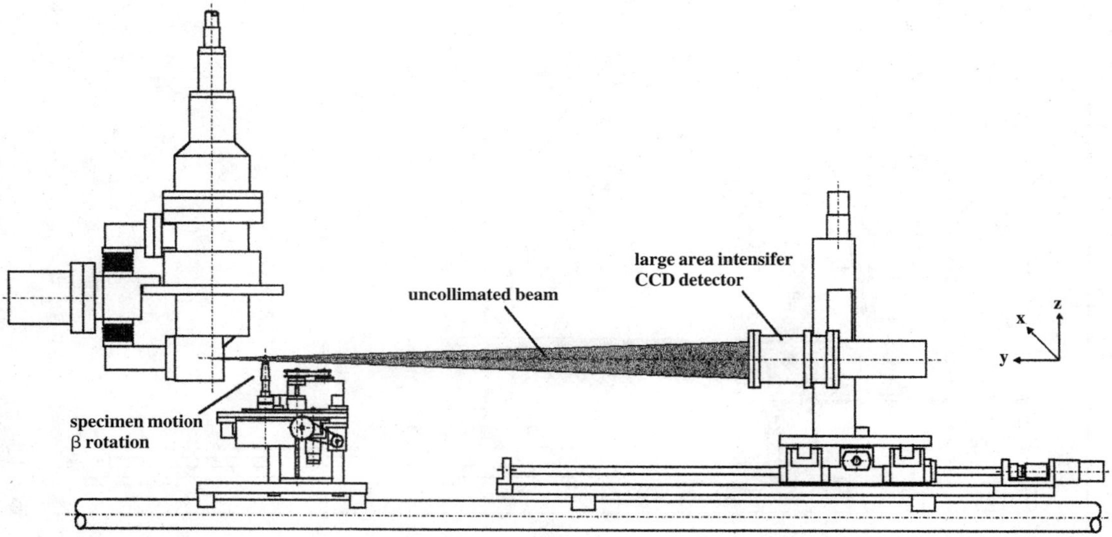

FIGURE 26.44 Direct magnification imaging with large-area detector.

high a *detective quantum efficiency* (DQE) as possible, where DQE = $(\text{snr}/\text{SNR})^2$ and *snr*, *SNR* are the signal-to-noise ratios at input and output respectively.[31]

Phase Contrast Imaging. In conventional x-ray imaging, the contrast is dependent on the photoelectric and Compton scattering processes with sensitivity to density variations $\Delta\rho/\rho$ depending on the relative absorption of adjacent rays. In phase contrast imaging, the contrast is dependent on the elastic Thompson scattering process with sensitivity to density variation $\Delta\rho/\rho$ determined by the refraction of adjacent rays at the boundaries of the structured density variation. The differences in refraction create a slight deviation in the x-ray wavefront so that a phase delay occurs. The small phase differences are separated in a crystal analyzer to form the image. The advantage over absorption contrast is that the phase delay remains significant even when the detail becomes very small.

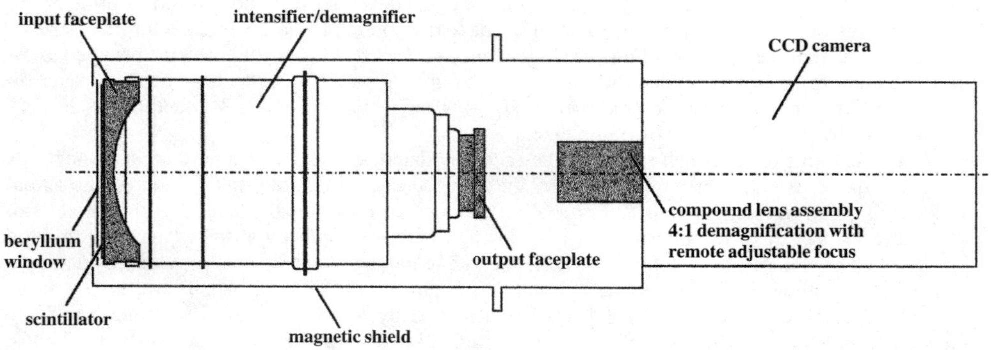

FIGURE 26.45 80-mm intensifier/demagnifier CCD digital x-ray camera.

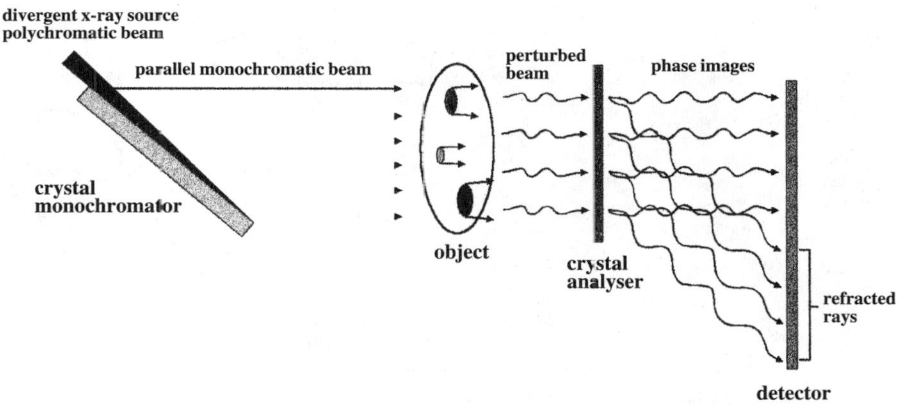

FIGURE 26.46 Phase contrast imaging.

To distinguish the phase delay, the x-rays must be monochromatic and parallel. With a sufficiently coherent source this can be achieved with crystal monochromator (Fig. 26.46).

26.4.4 Motion Systems

The specimen motion required for cone-beam reconstruction is a lateral axes (x, z) translation and a vertical axis β rotation. The scanning cycle needs to be under computer control in order to synchronize mechanical movement with data acquisition. Further, the control must provide a level of accuracy that, at the very least, matches the measured resolution of the x-ray source. In practice, an encoded accuracy in lateral translation of 10,000 counts per mm and a rotational accuracy of 2000 counts per degree of revolution can be achieved with commercial components.

The system shown in Fig. 26.47 has a vertical z movement of ~100 mm to facilitate specimen mounting and spiral scanning and a lateral x movement of ~40 mm for specimen alignment. It has

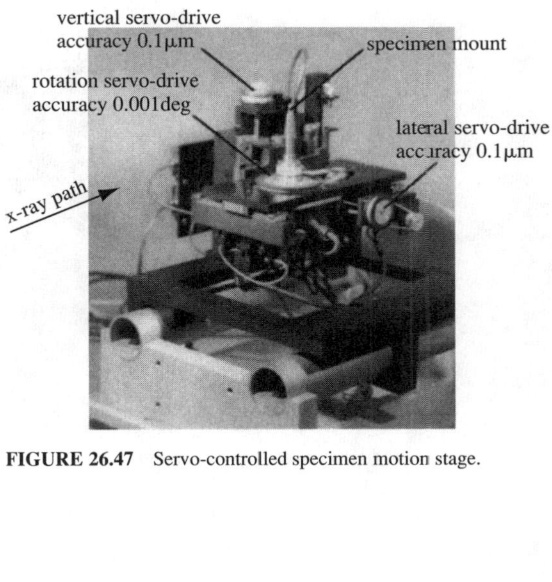

FIGURE 26.47 Servo-controlled specimen motion stage.

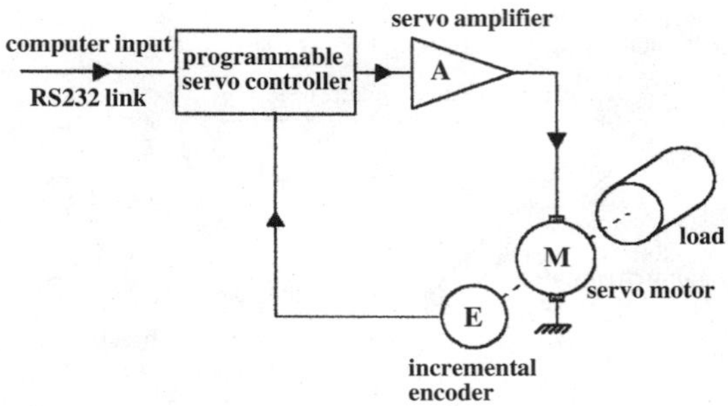

FIGURE 26.48 Servo controller RS232 interface.

an RS232 interface with a PC through a three-axis servo control that uses incremental encoder feedback to monitor position, speed, and direction. The level of proportional, integral, and differential (PID) feedback can be set for the optimum response and accuracy. Motion commands in ASCII format can be executed in immediate real time or loaded for programmed operation (Fig. 26.48). The camera three-axis traverse is servo controlled with a similar RS232 interface (Fig. 26.49). The motion control interfaces are daisy-chain-linked to the controlling PC (Fig. 26.50).

System Alignment. In a basic setting where the detector quarter-shift scheme is not considered, the back-projection function of the cone-beam reconstruction algorithm assumes that the central ray, which passes through the origin of the coordinate system, intersects the center point of the camera input plane. It also assumes that the z axis, which is the rotation axis, is aligned with the detector cell columns of the camera input plane. Hence, the camera must be correctly located with respect to the x-ray source central ray and the specimen stage rotation axis must be correctly located and aligned with the camera.

In order to center the camera on the central ray, a double-aperture collimator is located over the x-ray chamber window. The aperture size and spacing is selected to provide a 2° cone beam at the first aperture, which is subsequently stopped down to a 1° cone beam by the second aperture. The double aperture ensures that the central ray is orthogonal to the electron beam column and parallel

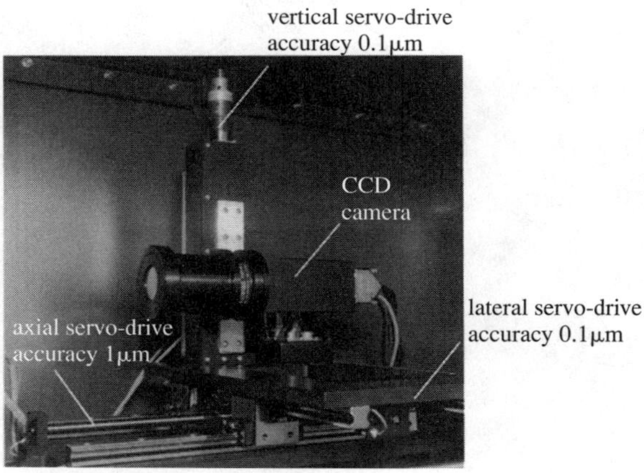

FIGURE 26.49 Servo-controlled camera traverse.

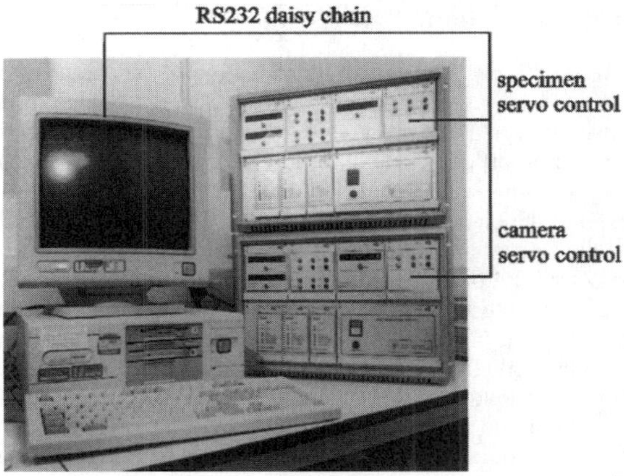

FIGURE 26.50 Motion systems PC interface racks.

to the instrumentation support rails. The location of the aperture axis can be adjusted to coincide with the x-ray focal spot. The geometry of this arrangement provides a projected image of the collimator aperture for alignment with the camera center point (Fig. 26.51). The camera is translated in x, z until the center of the image, given by $(a + b)/2$ and $(c + d)/2$, is located at the pixel coordinates (511.5, 511.5) that define the frame center.

A small 1.5-mm-diameter precision-machined spindle is mounted on the specimen motion stage to provide an image for alignment of the camera vertical axis with the specimen rotation axis. The camera can be rotated about its longitudinal axis for this purpose. Two images are recorded, at 0° and 180°, to take account of any eccentricity, or tilt of the spindle with respect to the rotational center (Fig. 26.52). The camera is rotated in θ until the average of the coordinates of two locations along the spindle images have the same lateral pixel location.

To align the specimen rotation axis with the camera center point, two images are recorded at 0° and 180° to take account of any eccentricity of the spindle with respect to the rotational center. The spindle is translated in x until the average of the coordinates of the location a, b or c, d is equal to the pixel value 511.5 at the center of the frame (Fig. 26.53).

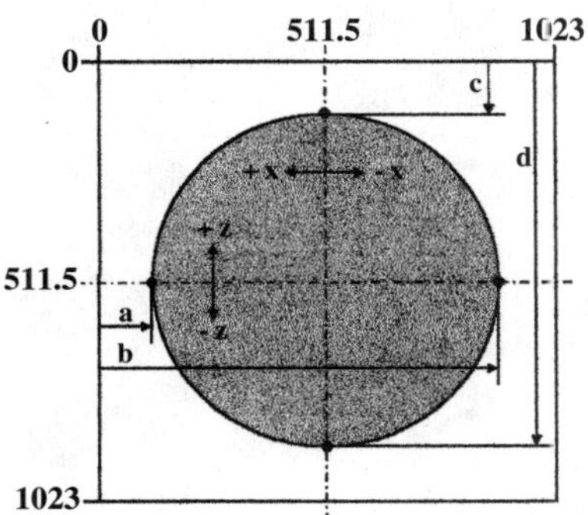

FIGURE 26.51 Camera center point alignment.

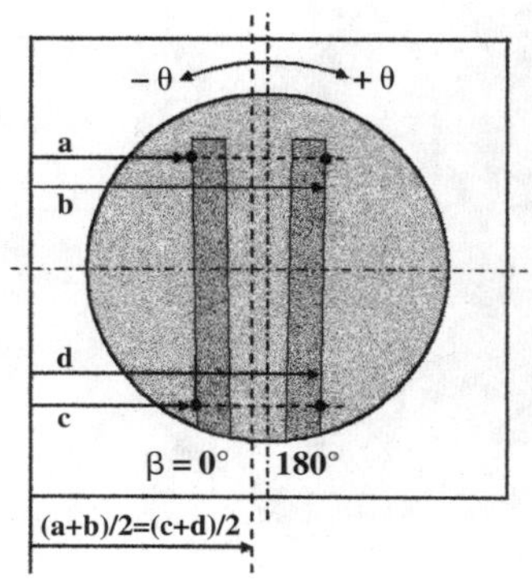

FIGURE 26.52 Camera θ rotation alignment.

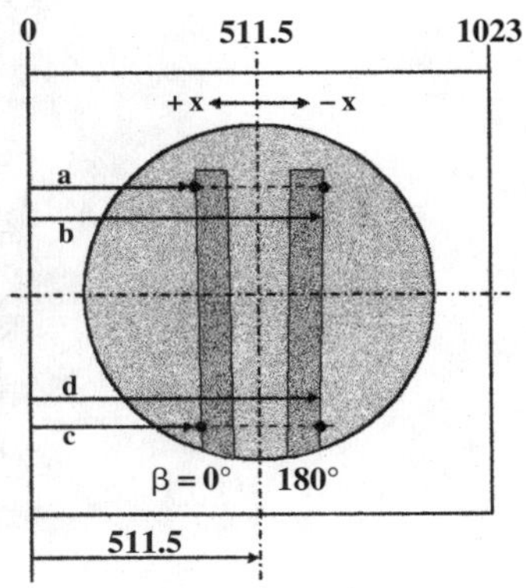

FIGURE 26.53 Specimen rotation axis alignment.

26.4.5 Ancillary Instrumentation

Data Acquisition. The broad interpretation of data acquisition relates to the various operations that are executed in a sequential fashion to secure a complete data set for subsequent computational reconstruction. In a narrower sense, this principally involves specimen motion, frame grabbing, and storing. For an efficient operation, these functions must be carried out automatically with provision for programming operational parameters such as the motion step value and total number of steps/frames, and the camera integration period. The system described in this section has an NT PC with an installed programmable PCI board and a fiber-optic link to control the camera and also communicates with a programmable servo controller for specimen motion via the RS232 port.

The organization of frame grabbing and specimen motion is performed by a programmable interface written in Python, which is an interpreted, interactive, object-oriented language. This provides the handshaking function that is essential for error-free sequential operation. The interface will pick up configuration files that set the camera parameters and servo control parameters. Run-specific details are entered in proffered dialog boxes at the outset of a data acquisition sequence. The image frames are temporarily stored on the PC hard disk and subsequently downloaded onto a Sun Micro Systems SCSI multiple-disk drive.

To provide an example of data quality, a sample of hollow glass spheres (ballotini), size ranging roughly from 10 to 100 μm, is reconstructed using a 2-mm-bore beryllium tube specimen holder (Fig. 26.54). The rotation step interval is 0.5° over 360° to give 720 projected frames of 1024 $\times$ 1024 pixel size. The volume is reconstructed using the cone-beam algorithm with a voxel size of 2.5 μm for a projection magnification of $\times$18. The illustrated tomograph is quarter-size child volume extracted from the full reconstruction. A reference standard was provided by two-dimensional electron micrograph images of a similar sample.

Specimen Stage. The delicate labile nature of biological materials presents a particularly stringent set of conditions for the design of the specimen stage, especially since controlled variations in temperature are often required. The difficulties are compounded by the need to move the specimen in a precisely controlled and accurate manner during the acquisition of projection data for tomographic reconstruction. To illustrate these points, a particular example of a design solution, one that has a proved successful performance, is described as follows.

The requirement is for (1) very accurate linear motion along the lateral x, z axes and very accurate β rotation motion about the vertical z axis, (2) very accurate temperature control, and (3) versatile

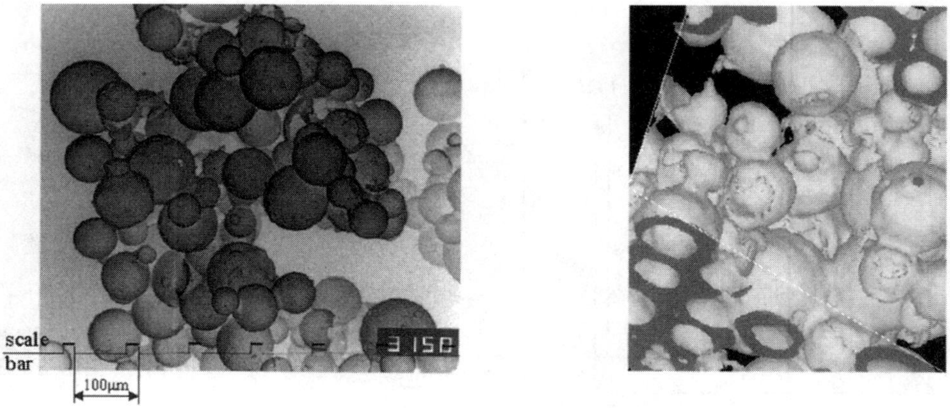

(a) (b)

FIGURE 26.54 Tomographic reconstruction of hollow glass spheres (ballotini). (*a*) Electron micrograph (two-dimensional); (*b*) x-ray tomograph (three-dimensional).

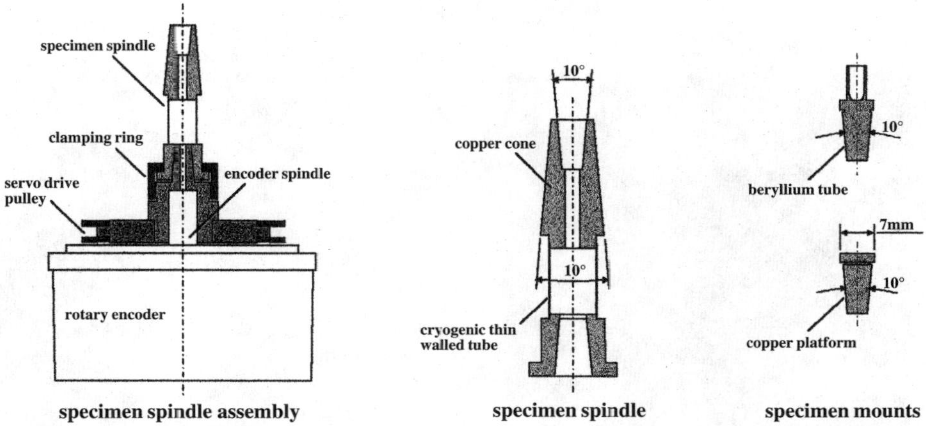

FIGURE 26.55 Specimen rotary stage.

specimen mounting. An important step is to uncouple the motion control system from the thermal control system so that requirements 1 and 2 can function independently. The solution is to let the motion system engage mechanically with the thermal system but allow the latter to passively follow the movement of the specimen. This means that the thermal system must apply only the weakest of constraint to three-dimensional displacement and offer minimal resistance to rotation in the vertical axis.

A cone-on-cone contact is a suitable geometry for mechanical engagement, provided that the cone angle and material are correctly chosen. Since good thermal contact is required, copper is a natural choice, and a 10° cone will provide solid location while keeping the mechanical contact stiction and friction to within acceptable bounds. To minimize shaft bearing inaccuracies, the specimen spindle is built directly onto the highly accurate rotary encoder shaft (Fig. 26.55). Here, a thin-walled stainless steel tube thermally isolates the specimen mount receptacle from the spindle clamping assembly. The specimen mount also has a conical surface for good thermal contact and alignment. Typical mounting arrangements consist of a copper platform to which stable specimens can be lightly bonded and a thin-walled beryllium tube into which unstable specimens can be inserted.

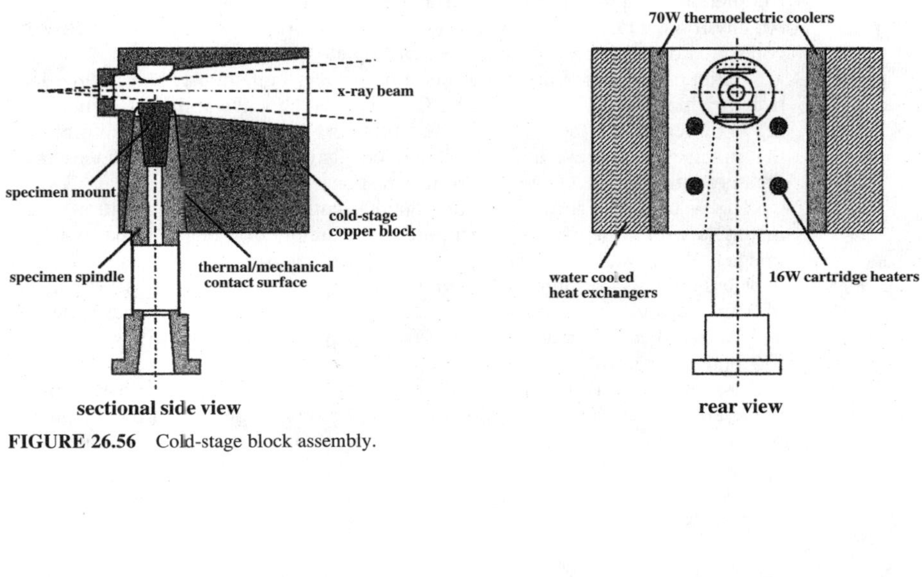

FIGURE 26.56 Cold-stage block assembly.

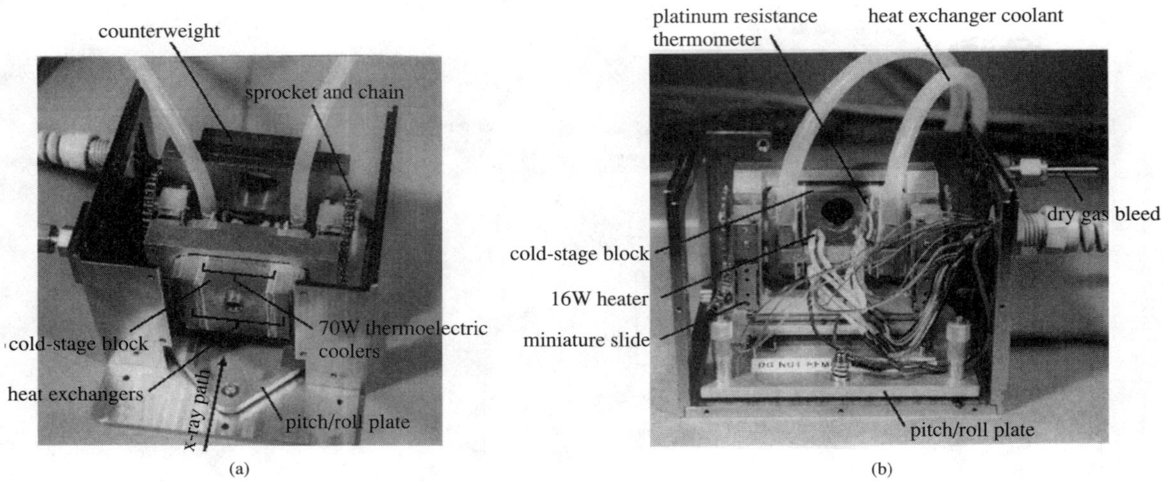

FIGURE 26.57 Thermal control stage assembly. (*a*) Top view; (*b*) rear view with counter weight removed.

The contacting component of the thermal system, namely, the copper cold-stage block, is required to float with a small downward force in order to engage positively with the specimen spindle of the motion system. The cold-stage block has a conical bore to receive the specimen spindle and controls the temperature of the specimen mount through thermal conduction across the conical contact surface. Primary heat transfer through thermoelectric coolers maintains the base temperature variation, and secondary heat transfer from buried cartridge heaters maintains temperature control. Solutions to Eq. (26.122) were computed numerically to confirm that a uniform temperature distribution could be maintained with the specific shape and size of the cold-stage block for given values of heat input and heat output. On the basis of these calculations, two 70-W thermoelectric coolers were located at the outer surfaces of the block and four 16-W cartridge heaters were inserted in one surface of the block (Figure 26.56).

The cold-stage block is counterbalanced through a sprocket-and-chain mechanism to offset the weight and is attached to an unrestrained three-axis miniature slide to permit motion within a 40-mm cubic space (Fig. 26.57). The complete assembly is supported on a micrometer-adjustable pitch/roll plate for alignment of the conical engagement bore with the rotational axis of the motion stage. An environmental chamber, with thin beryllium x-ray path windows, encloses the cold stage and a controlled dry gas bleed prevents water droplet condensation.

Platinum resistance thermometers monitor the temperature, and a three-term tunable PID controller, which is RS232-linked to a PC, stabilizes the value at the set point to an accuracy of at least 0.1°C by regulating the secondary heat transfer. The set point temperature can be programmed, so that precise thermal cycling is available, or manually entered for step variation. The working temperature range is ±50°C with a thermal response ~0.2°C/s.

The system is ideal for examining the rearrangement of microstructural composition of soft-solid materials with variation in temperature. An example of a material that is practically impossible to image in the natural state by conventional optical microscopy is shown in (Fig. 26.58*a*). The volumes depicted are child volumes, containing a region of interest (ROI), extracted from the reconstruction of a frozen four-phase soft-solid structure. The reconstructed volume is derived from 720 filtered and back-projected frames using the Feldkamp cone-beam algorithm, with a resolution of 512 × 512 pixels per frame.

The material was initially chilled to −20°C and subsequently warmed to −10°C. The alteration in the air cell network (black) and the structural rearrangement of the ice plus matrix phase (white) with temperature rise is clearly defined. To provide quantitative measurement of microstructural

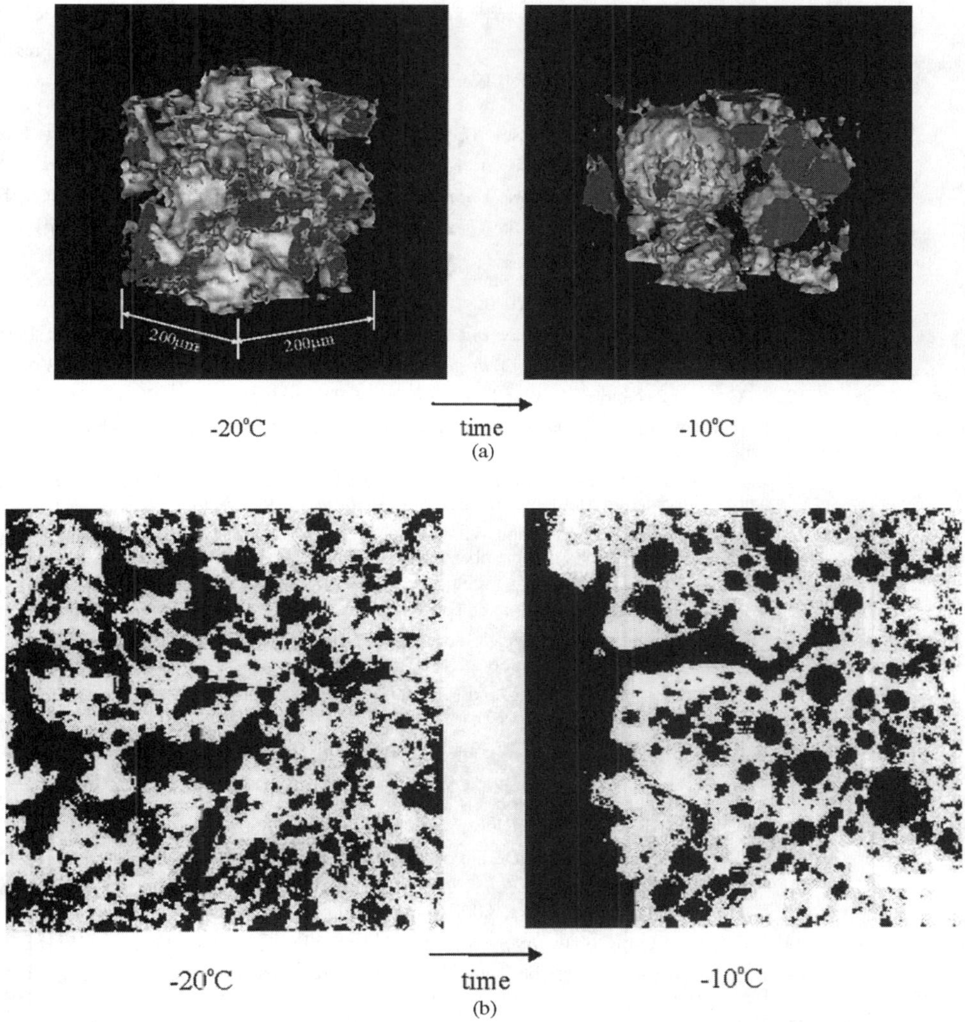

FIGURE 26.58 Soft-solid four-phase composite material. (*a*) Three-dimensional Reconstruction; (*b*) two-dimensional slice.

detail, the volume is sliced across a region of interest (Fig. 26.58*b*). Here, the slice pixel size is 5 μm for a projection magnification of $\times 18$.

REFERENCES

1. G. T. Herman, *Image Reconstruction from Projections—The Fundamentals of Computerized Tomography*, Academic Press, New York, 1980.

2. F. Natterer, *The Mathematics of Computerized Tomography*, John Wiley and Sons, New York, 1986.
3. A. C. Kak and M. Slaney, *Principles of Computerized Tomographic Imaging*, IEEE Press, New York, 1988.
4. H. H. Barrett and W. Swindell, *Radiological Imaging—The Theory of Image Formation, Detection, and Processing*, Academic Press, New York, 1981.
5. N. F. Mott and H. S. W. Massey, *The Theory of Atomic Collisions*, Oxford University Press, London, 1965.
6. A. H. Compton and S. K. Allison, *X-Rays in Theory and Experiment*, D. Van Nostrand, New York, 1935.
7. N. A. Dyson, *X-Rays in Atomic and Nuclear Physics*, Cambridge University Press, Cambridge, 1973.
8. V. Kohn, *Selected Topics in the Theory of Coherent Scattering of X-Rays—Hamburg Lectures*, Kurchatov Institute, Moscow, 1998.
9. J. Baruchel, J.-Y. Buffiere, E. Maire, P. Merl, and G. Peix, *X-Ray Tomography in Material Science*, Hermes Science Publications, Paris, 2000.
10. E. Kreyszig, *Advanced Engineering Mathematics*, John Wiley and Sons, New York, 1988.
11. R. N. Bracewell, *The Fourier Transform and Its Applications*, McGraw-Hill, New York, 1978.
12. R. C. Gonzalez and R. E. Woods, *Digital Image Processing*, Addison-Wesley, New York, 1992.
13. L. A. Feldkamp, L. C. Davis, and J. W. Kress, "Practical Cone-Beam Algorithm," *J. Opt. Soc. Am.*, A **1**(6), June 1984.
14. B. D. Smith, "Image Reconstruction From Cone-Beam Projections—Necessary and Sufficient Conditions and Reconstruction Methods," *IEEE Trans. Med. Imag.*, **4**(1), March 1985.
15. Ge Wang, C. R. Crawford, and W. A. Kalender, "Multirow Detector and Cone-Beam Spiral/Helical CT," *IEEE. Med. Imag.*, **19**(9), September 2000.
16. Ge Wang, S. Zhao, and P. Cheng, "Exact and Approximate Cone-Beam X-ray Microtomography, Modern Microscopies (I)—Instrumentation and Image Processing," *World Scientific*, Singapore, 1998.
17. Ge Wang and M. W. Vannier, "Computerized Tomography," *Encyclopedia of Electrical and Electronics Engineering*, John Wiley and Sons, New York, 1998.
18. C. A. Carlsson, G. Matscheko, and P. Spanne, "Prospects for Microcomputerized-Tomography Using Synchrotron Radiation," *Biological Trace Element Research*, **13**, 1987.
19. P. Rockett and R. W. Parish, "A Wide-Energy Range, High-Resolution Microfocal X-Ray Source," *British Journal of NDT*, March 1986.
20. V. E. Cosslett and W. C. Nixon, *X-Ray Microscopy*, Cambridge Monographs in Physics, N. Feather and D. Shoenberg (eds.), Cambridge University Press, London, 1960.
21. E. Munro, "Computer-aided Design of Electron Lenses by the Finite Element Method," *Image Processing and Computer Aided Design in Electron Optics*, P. W. Hawkes (ed.), Academic Press, London, 1973.
22. E. Hecht and A. Zajac, *Optics*, Addison-Wesley, London, 1979.
23. J. C. Dainty and R. Shaw, *Image Science*, Academic Press, London, 1974.
24. S. T. Stephenson, "The Continuous X-Ray Spectrum," *Handbuch der Physik*, S. Flugge (ed.), Springer, Berlin, 1957.
25. B. K. Agarwal, *X-ray Spectroscopy*, Springer Series in Optical Sciences, vol. **15**, Berlin, 1979.
26. N. A. Dyson, "The Continuous Spectrum from Electron-Opaque Targets," *Proc. Phys. Soc*, **73**(6), December 1958.
27. A. Bouwers and P. Diepenhorst, *X-Ray Research and Development*, Philips, Eindhoven, 1933.
28. W. J. Oosterkamp, "The Heat Dissipation in the Anode of an X-Ray Tube," *Philips Res. Rep. 3*, 1948.
29. J. Vine and P. A. Einstein, "Heating Effect of an Electron Beam Impinging on a Solid Surface, Allowing for Penetration," *Proc. IEE*, 1964.
30. J. C. Elliott, P. Anderson, G. R. Davis, F. S. L. Wong, and S. D. Dover, "Computed Tomography, Part II: The Practical Use of a Single Source and Detector," *Journal of Microscopy*, March 1994.
31. J. D. Vincent, *Fundamentals of Infrared Detector Operation and Testing*, John Wiley and Sons, 1990.

CHAPTER 27
NUCLEAR MEDICINE IMAGING INSTRUMENTATION

Mark T. Madsen
University of Iowa, Iowa City, Iowa

27.1 INTRODUCTION

Nuclear medicine is a diagnostic imaging modality that is used to obtain clinical information about most of the major tissues and organs of the body. Diagnostic information is obtained from the way the tissues and organs process radiolabeled compounds (radiopharmaceuticals). The radiopharmaceutical is typically administered to the patient though an intravenous injection. The radiopharmaceutical is carried throughout the body by the circulation where it localizes in tissues and organs. Images of these distributions are acquired with a scintillation camera. Ideally, the radiopharmaceutical would go only to abnormal areas. Unfortunately, this is never the case and the abnormal concentration of the radiotracer is often obscured by normal uptake of the radiopharmaceutical in the surrounding tissues. Images of higher contrast and better localization can be obtained with tomographic systems designed for nuclear medicine studies (SPECT systems). These are described in detail below.

The imaging of radiotracers in the body presents special challenges that are unique. The flux of gamma rays available for imaging is orders of magnitude less than that used in x-ray radiography or x-ray computed tomography (CT). In addition, the high energy of the gamma rays makes detection more difficult. As a result, the images produced in nuclear medicine studies are much noisier and have worse spatial resolution. In order to appreciate these problems and how they affect the design of nuclear medicine imaging devices, we will briefly review the physics of gamma ray interactions.[1]

The intensity of gamma rays traveling through material is gradually reduced by absorption or scattering. This loss of gamma rays is referred to as *attenuation* and is described by the exponential equation

$$I(x) = I_0 \exp(-\mu x) \tag{27.1}$$

where I_0 = initial intensity
$I(x)$ = intensity of rays after traveling a distance x through the material
μ = linear attenuation coefficient of the material

Over the range of gamma ray energies used in radionuclide imaging, the two primary interactions that contribute to the attenuation coefficient are photoelectric absorption and Compton scattering. Photoelectric absorption refers to the total absorption of the gamma ray by an inner shell atomic electron and is the primary interaction in high-*Z* materials such as sodium iodide (the detector material used in the scintillation camera) and lead. In low-*Z* materials such as body tissues, its contribution to attenuation is relatively small. Compton scattering occurs when the incoming gamma ray interacts with a loosely bound outer shell electron. A portion of the gamma ray energy is imparted to the electron and the remaining energy is left with the scattered photon. The amount of energy lost in the event depends on the angle between the gamma ray and scattered photon. Compton scattering is the dominant interaction in body tissues.

High attenuation is desirable in detecting and shielding materials. Ideally materials used for these purposes would absorb every gamma ray. In the body, attenuation is very undesirable, but unfortunately, unavoidable. Attenuation reduces the intensity of gamma rays available for detection and scattered radiation that reaches the detector causes a significant loss of contrast.

27.2 SCINTILLATION CAMERAS

The scintillation camera is the primary imaging instrument used in nuclear medicine and is often referred to as a *gamma camera*.[2] The scintillation camera is a position-sensitive gamma ray imager. Although the entire field of view is available for detection, it processes one event at a time. The spatial resolution is approximately 10 mm and it yields a count rate of 200 to 300 cpm/μCi in the field of view (cpm = counts per minute). The field of view covers a large portion of the body and is typically 40 $\times$ 50 cm, although other sizes are available.

The rectilinear scanner was the first practical nuclear medicine imaging device and it was still in use through the 1970s. It was invented by Benedict Cassen in 1950. The rectilinear scanner used the detected count rate of a radiation detector to control the brightness of a small light bulb masked to expose a small area of a film. The movement of the radiation detector and the bulb were linked so that, as the detector moved in a raster pattern over the patient, the bulb tracked a corresponding path over the film. The developed film revealed the internal distribution of the radiopharmaceutical sampled by the scanning probe. While the invention of the rectilinear scanner was a major step forward, it had several shortcomings that were inherent in its design. Because it sampled only one small area at a time, image acquisitions took a long time (10 to 20 minutes). In addition, only static imaging was possible. Dynamic studies, such as those that followed the progression of a tracer through the body, required an imaging system with a large field of view where all areas are equally sampled, i.e., a scintillation camera.

The first scintillation camera was developed by Hal O. Anger in 1958.[3] Although this system was very crude, it contained the fundamental components of all future designs: NaI(Tl) as the primary detector and weighted signals from an array of photomultiplier tubes to determine the location of detected events. Table 27.1 gives typical performance values for a modern scintillation camera.

The gamma ray–sensitive element of the scintillation camera is a large, thin piece of NaI(Tl). Although the crystals originally had a circular cross section, most scintillation cameras now use a rectangular crystal with dimensions as large as 40 $\times$ 50 cm. The thickness of NaI(Tl) in most conventional cameras is 9.5 mm, but in systems that are used for coincidence detection, the crystal may be twice as thick. NaI(Tl) is a scintillator; It converts gamma ray energy into visible light. The amount of light generated is directly proportional to the absorbed energy. NaI(Tl) is very efficient at this and the absorption of one 140-keV gamma ray will yield 5000 visible light photons. There are a number of advantages associated with NaI(Tl) in addition to its high light output. It efficiently absorbs 140-keV gamma rays (with a photopeak efficiency of 85%) and it has a moderate energy resolution. Energy resolution is an important property since it provides the means to discriminate against scattered radiation. Gamma rays that undergo scattering within the patient degrade the quality of images. However, scattered gamma rays necessarily have less energy than unscattered gamma rays and can be selectively eliminated on that basis. Another positive feature of NaI(Tl) is that it

TABLE 27.1 Scintillation Camera Specifications

Parameter	Specification
Crystal size	40 × 50 cm
Crystal thickness	9.5 mm
Efficiency at 140 keV	0.86 (photopeak); 0.99(total)
Efficiency at 511 keV	0.05 (photopeak); 0.27(total)
Energy resolution	10%
Intrinsic spatial resolution	3.5 mm
System count sensitivity	325 counts/min/μCi
System spatial resolution at 10 cm	9.5 mm
Maximum count rate (SPECT)	250,000 counts/s
Maximum count rate (coincidence)	>1,000,000 counts/s

can be manufactured in many shapes and sizes. There are disadvantages though. NaI(Tl) actively absorbs water vapor from the air and loses its transparency. It must be hermetically sealed, and loss of this seal results in irreparable damage. Another disadvantage is that the persistence of the scintillation is long enough that it limits the count rate that the crystal can accurately handle. Most nuclear medicine imaging is performed far below this limit. However, some first pass studies do result in significant count rate losses from this limit. The biggest problem is encountered in coincidence imaging.

Converting the gamma ray energy to visible light is only part of the battle. In order for the information from the scintillation to be useful, it has to be converted into an electronic signal. This is accomplished with a photomultiplier tube (Fig. 27.1). The photomultiplier tube is a vacuum tube with a photoemissive surface called the *photocathode*. Visible light hitting this surface knocks off electrons. These electrons are accelerated to an electric terminal called a *dynode*. The first dynode

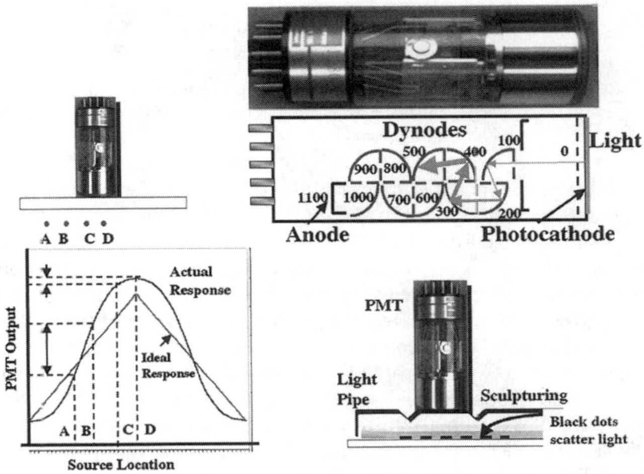

FIGURE 27.1 Photomultiplier tube. The photomultiplier tube converts the scintillation into an electronic pulse preserving the linear relationship between the magnitude of the scintillation and the energy of the interaction. The location of a source can be inferred from the magnitude of the signal change. The relationship between the signal magnitude and source position is nonlinear, and positioning errors occur both when the source is far from the PMT and when it is directly under it. The ideal response can be approximated with the use of a light pipe.

has a potential approximately 100 volts higher than the photocathode, and the electrons hit it with enough force to knock off about 4 more new electrons. The next dynode is another 100 volts higher, so the process is repeated. The same process occurs over a total series of nine dynodes resulting in a signal amplification of 1,000,000. Proportionality is maintained throughout this amplification so that the size of the electron pulse is directly proportional to the energy deposited by the gamma ray.

A scintillation camera needs to be able to record gamma ray events over a large area. This requires uniform sampling by an array of photomultiplier tubes (PMTs). The PMTs are arranged in a close packed array that covers the entire surface of the NaI(Tl) crystal (Fig. 27.2). The PMTs used in scintillation cameras are usually 2 or 3 in across, so that as many as 120 PMTs may be used. PMTs have been manufactured in variety of different cross sections in order to maximize their areal coverage. Circular, hexagonal, and square tubes have all been used. The signals obtained from the PMTs will be used to determine two important properties about the gamma ray interaction: where did it occur and how much energy was deposited? At first blush, it may seen that even 2-in PMTs are too coarse to determine the event location. However, we will see that magnitude of the PMT output is fairly sensitive to the location of a source.

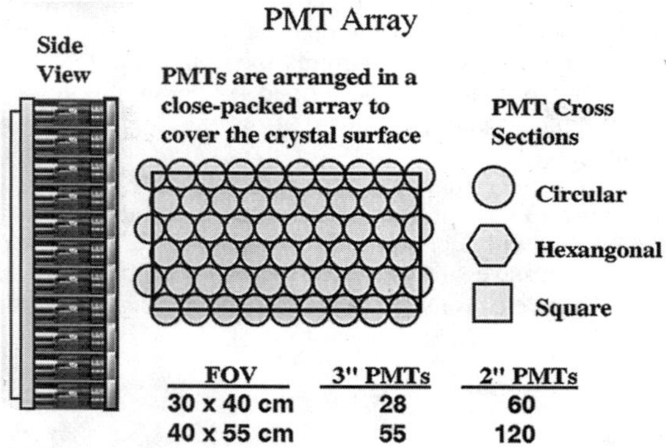

FIGURE 27.2 Photomultiplier tube array. The photomultiplier tubes are arranged in a close-packed array to cover the back surface of the NaI(Tl) crystal.

If a PMT is mounted to an NaI(Tl) crystal and the signal output is plotted as a source is moved from left to right, the result shown by the solid line (Fig. 27.1) will be obtained. When the source is positioned far from the PMT, the signal is weak and the location of the source is not certain. When the source is directly under the PMT, the signal is strong, but the dependence on the position is modest. However, when the source is just to the left or right of the PMT, the signal change with source location is large. Over this region, the location of the source can be accurately tracked. With only a single PMT, we could not tell if the source was on the left or right (or front or back), but by considering the signals from other surrounding PMTs, that can be determined. The main problem is what to do about the poor response near the center of PMT. What can be done to transform the measured PMT output into the ideal signal shown in Fig. 27.1? The early solution to this was to displace the PMTs from the NaI(Tl) crystal with a light pipe.

A light pipe is a transparent material such as lucite or quartz that is interspersed between the crystal and PMT array. The displacement of the PMTs causes the light from the scintillations to spread out, yielding a more favorable signal output. To achieve even better results, some manufacturers have used special mask patterns on the front surface of the light pipe along with sculptured grooves in the back. These efforts pay off in more accurate positioning of the event locations; however, there is a price to pay. The effort to scatter the light also results in higher light losses, leading to increases in statistical fluctuations and ultimately degradations in spatial resolutions. In the early gamma cameras, there was no way around this dilemma. However, digital electronics provides the capability of nonlinear mapping, which greatly reduces the demands put on the light pipe. This has allowed light pipes to be made very thin and even eliminated.

Figure 27.3 shows a schematic of an analog scintillation camera. The scintillation light from an absorbed gamma ray is transmitted through to the photomultiplier tube array. The energy of the event is determined by summing all the photomultiplier tube signals. As will be seen, the energy signal is used both for scatter discrimination and for normalizing the position signals. The position of the event is determined by summing weighted outputs from each photomultiplier tube. These weighting factors are determined by the location of the photomultiplier tube in the array. Separate weighting factors are used for determining the x and y signals. This process is referred to as Anger logic, since it is the scheme developed by Hal Anger in the first scintillation camera. In the initial designs, literally all the photomultiplier tubes participated in the energy and position signal summations. It was subsequently found that the signals from photomultiplier tubes located far from the event contributed mostly noise. In modern designs, the photomultiplier tube signal must exceed a threshold before it is included in the sum. Another point that should be made is that all the processing is performed on each detected event. The decision to include the event as a valid count is not made until the end of the processing when the pulse height analysis is done. If the event falls within the selected energy window, the normalized x and y signals are available for either an analog display or digital storage.

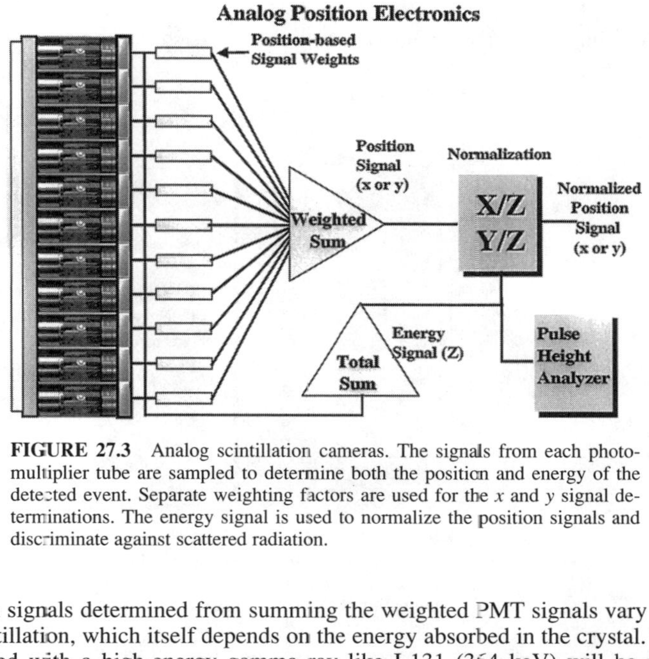

FIGURE 27.3 Analog scintillation cameras. The signals from each photomultiplier tube are sampled to determine both the position and energy of the detected event. Separate weighting factors are used for the x and y signal determinations. The energy signal is used to normalize the position signals and discriminate against scattered radiation.

The position signals determined from summing the weighted PMT signals vary with the brightness of the scintillation, which itself depends on the energy absorbed in the crystal. This means that an object imaged with a high-energy gamma ray like I-131 (364 keV) will be magnified when compared to the same object imaged with Tc-99m (140 keV). This magnification is a concern even when only one gamma ray energy is imaged because of the finite energy resolution of the scintillation camera system. The pulse heights from the absorptions of identical gamma rays vary enough to cause slight minifications and magnifications, ultimately degrading spatial resolution. The solution to this problem is to normalize the position signals with the measured energy signal. This removes the image size dependence with energy, thereby improving spatial resolution and allowing the simultaneous imaging of more than one radionuclide without distortion. This feature is the primary component for guaranteeing good multiwindow spatial registration.

As has been previously noted, gamma rays that are scattered within the patient have distorted spatial information and degrade image contrast. Because scattered gamma rays necessarily lose energy, they can be selectively avoided by accepting only events that have pulse heights corresponding to the primary gamma ray energy. The pulse height analyzer provides this capability. A "window" is centered to cover 15 to 20 percent of the photopeak. All energy pulses that meet this criterion generate a logic pulse that indicates to the system that a valid event has occurred. This output enables the recording of the x and y position information in a computer or display.

The precision of locating gamma ray events by a scintillation camera is referred to as *intrinsic spatial resolution.* The original Anger camera had very poor intrinsic spatial resolution (~12 mm full-width-half-maximum). With the improvement in electronics and pulse processing methods, the spatial resolution of the scintillation camera has improved steadily, approaching 3 mm in modern systems.[4] These improvements include: better-quality, low-noise PMTs; improvements in the Anger logic electronics including signal thresholding; smaller PMTs, and improved PMT quality control. The most recent improvement has been the replacement of most of the analog processing with digital electronics.

With improvements in the speed of digitization electronics and decreases in component costs, the trend in scintillation cameras has been to digitize the PMT signals (Fig. 27.4).[5] The analog-to-digital converters assign a numeric magnitude to the PMT signals. All subsequent determinations of energy and positions can then be done by computer algorithms that can accurately model the nonlinear behavior of the PMT signals with source position. The success of this approach has allowed the reduction and even total elimination of the light pipe. This moves the PMTs closer to the scintillation, thereby improving both the precision of the position determination (i.e., spatial resolution) and the energy resolution.

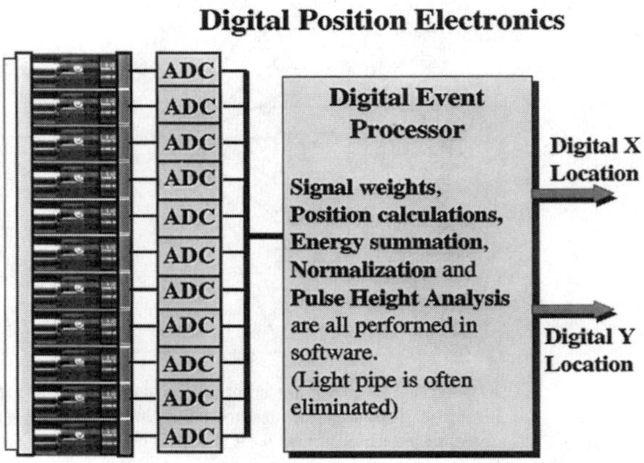

FIGURE 27.4 Digital scintillation cameras. By digitizing the output of the PMTs, the analog weighting electronics can be replaced by a nonlinear positioning algorithm. This allows a more accurate correction, culminating with the elimination of the light pipe.

Once there is an x and y coordinate that locates a valid event, this information has to be stored as image data. Although it is possible on some scintillation camera systems to store the individual coordinates sequentially (referred to as *list mode acquisition*), most systems store the information directly in histogram or matrix mode. With this method, an array of computer memory, typically 128 × 128 or 256 × 256, is reserved for each image frame. The matrix elements or pixels are initially set to 0. The coordinates for each event point to one of the pixels and this pixel is incremented by 1. When the acquisition-stopping criteria are met, the image is complete. The information in the matrix is either gray-scale or color encoded to display the image data. The entire process is shown schematically in Fig. 27.5. A gamma ray originating in the patient is absorbed in the NaI(Tl) crystal.

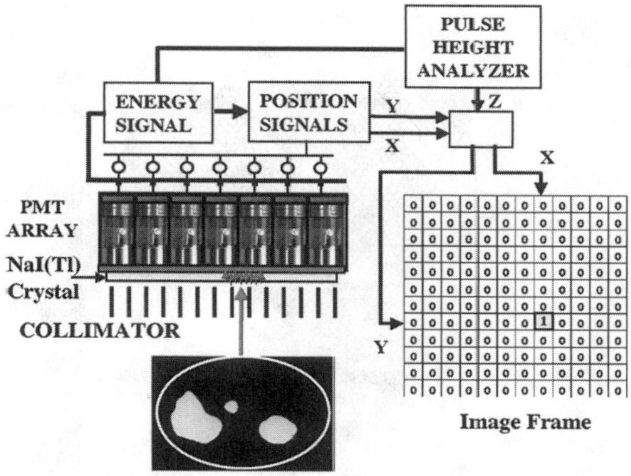

FIGURE 27.5 Scintillation camera. The scintillation camera processes each detected event to determine the x, y, and energy. If an event falls within the selected energy range, the memory location pointed to by the x and y coordinates is incremented. This process continues until the stopping criteria (number of counts or acquisition time) are met.

The light from the scintillation is sampled by the PMT array, which determines both the x and y coordinates of the event and its energy. If the energy signal falls within the window of the pulse height analyzer, the x and y coordinates are used to increment the appropriate pixel. This process is repeated for every detected event.

In order to form images with a scintillation camera, a collimator must be placed in front of the NaI(Tl) crystal. The collimator (Fig. 27.6) is the image-forming aperture of the camera system, and it is necessary for the imaging process. The collimator projects the gamma ray pattern originating in the patient onto the NaI(Tl) crystal. It does this by selectively absorbing gamma rays. The collimator is a close-packed array of holes in a lead plate. Most often the holes are parallel, but fan-beam converging and diverging collimators are available. Gamma rays whose trajectory takes them through a hole get to interact with the NaI(Tl). All the others are absorbed. The design of collimators depends on the gamma ray energy and the ever-present trade-off between count sensitivity and spatial resolution.[6,7] Collimators used for imaging Tc-99m typically have holes that are 1 to 1.5 mm across and are 20 to 40 mm thick. Typical collimator design parameters are given in Table 27.2.

Although a collimator is necessary for the formation of images, it represents the limiting factor in the both count sensitivity and spatial resolution of the scintillation camera. Because of the brute-force absorption approach to forming images with collimators, they are very inefficient. Less than 1 in 5000 gamma rays that hit the front surface of the collimator get through to the crystal. To improve the count sensitivity, the collimator hole size could be increased and the hole length shortened. Unfortunately, these changes degrade the spatial resolution. The spatial resolution of the collimator is constrained by the geometry of the holes and is typically in the range of 6 to 8 mm at 10 cm when used with Tc-99m. This is the dominant factor in determining the overall system resolution, since the intrinsic spatial resolution is in the range of 3 to 4 mm.

One very important property to remember about collimators is that the spatial resolution gets worse as the source-to-collimator distance increases. This is illustrated in the set of phantom images that were acquired from 5 to 30 cm from the collimator surface. To obtain the best-quality images, spatial resolution comes at the price of count sensitivity; therefore it is crucial to keep the collimator as close to the patient as possible.

The modern scintillation camera has improved performance because of improvements in the components and electronics. The availability of digital electronics has allowed the elimination of

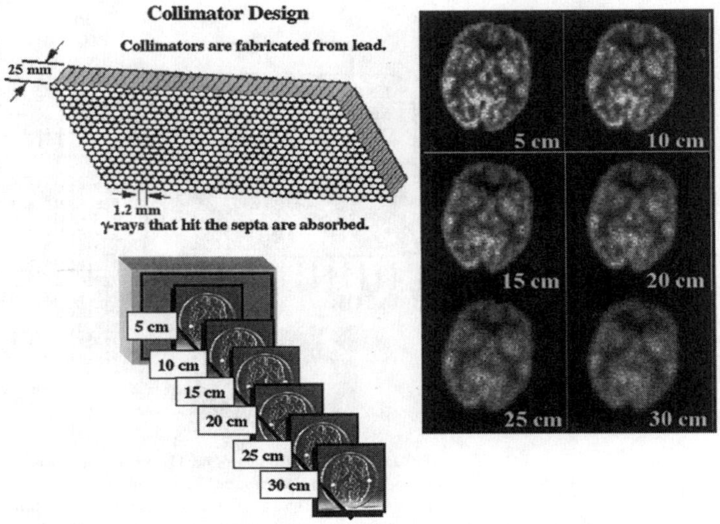

FIGURE 27.6 Collimation. The collimator is the image-forming aperture of the scintillation camera. It projects an image of the radionuclide distribution onto the NaI(Tl) crystal by brute force absorption of all gamma rays except those whose trajectory takes them through the holes. The collimator is also the limiting factor of both spatial resolution and count sensitivity. The spatial resolution significantly degrades with source-to-collimator distance.

the light pipe, which improves both energy and spatial resolution. However, this requires additional corrections because of the nonlinear response of the PMT array to the scintillations. If a collimated point source were focused on a portion of the NaI(Tl) crystal located exactly on a photomultiplier tube center, the energy spectrum would be distinctly different than one that was acquired from a point between two tubes (Fig. 27.7). This difference reflects the efficiency for collecting all the

TABLE 27.2 Collimator Specifications

Collimator	Energy, keV	Hole size, mm	Hole length, mm	Septal thickness, mm	Relative sensitivity	R_{col} at 10 cm, mm
Parallel general purpose	140	1.4	24	0.20	1.0	8.3
Parallel high sensitivity	140	2.0	24	0.20	2.0	11.9
Parallel high resolution	140	1.1	24	0.15	0.7	6.3
Parallel ultrahigh resolution	140	1.1	36	0.15	0.3	4.5
Ultrahigh fan beam	140	1.4	35	0.20	0.8	7.1
Cone beam	140	1.9	41	0.25	1.2	7.1
Parallel medium energy	300	3.5	50	1.30	0.8	11.6
Parallel high energy	364	4.2	63	1.30	0.5	13.4
Parallel ultrahigh energy	511	4.2	80	2.40	0.3	12

scintillation light. When the source is located directly on a photomultiplier tube, more of the scintillation is sampled by the photomultiplier tubes, and the pulses are therefore somewhat larger on the average than at other points. This position-dependent shift in the energy spectrum causes an overall loss in energy resolution. It also means that portions of the crystal will be accepting proportionately more scattered radiation. The solution to this problem is to locally sample the energy spectra and regionally adjust the energy window for each area. Typically the camera field of view is divided into a 64 × 64 matrix and energy window adjustments are made for each of the 4096 regions.

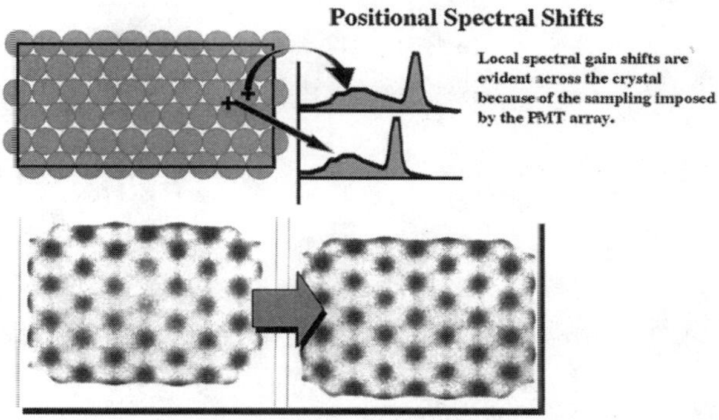

FIGURE 27.7 Energy correction. Because of the spatial arrangement of the PMTs, the magnitude of the energy signal varies with location, degrading the overall energy resolution. This problem is overcome by setting multiple local energy windows across the field of view. This energy correction does not improve uniformity, but it does remove the dependence of the scintillation camera on scatter conditions.

Figure 27.7 shows the effect of energy correction when the scintillation camera is exposed to a uniform flux of gamma rays. First, it should be noted that both the corrected and uncorrected images are highly nonuniform and are not adequate for imaging. The energy correction simply makes sure that each region of the crystal is contributing valid photopeak events to the image. This results in only a subtle improvement in uniformity at this stage. However, it makes the subsequent corrections more robust, since there will be much less dependence on the effects of scattered radiation, which can vary over a large range, depending on the imaging situation.

Because of the nonlinear response of the photomultiplier tubes, detected events are not correctly positioned using Anger logic alone. This mispositioning of events has a profound effect on field uniformity.[8] The parameter that quantifies how well-detected events are positioned is called *spatial linearity*. The optimization of spatial linearity requires the acquisition of an image from a well-defined distribution (Fig. 27.8). Typically this is accomplished with a highly precise rectangular hole pattern that is placed directly on the NaI(Tl) crystal. A distant point source of radioactivity is used to project an image of the hole pattern onto the scintillation camera. The image of this pattern appears similar to the image on the left with distortions caused by the mispositioning of events. Because the actual and measured location of the holes is known, regional displacements to the x and y coordinates can be calculated for each hole. Displacements for regions in between holes that are not directly sampled are interpolated at a very high sampling frequency (1024 × 1024). This information is stored and is available as a lookup table. This measurement is usually done by the vendor at the factory and may be repeated several times a year. When a valid event is detected, the initial x and y coordinates are modified by the appropriate displacements that are read from the lookup table. Using this approach, events can be accurately positioned to better than 0.5 mm. The improvement in spatial linearity has a profound effect on field uniformity. Both images show the response of the scintillation camera to a uniform flux of gamma rays. With spatial linearity correction, the field becomes uniform to within ±10 percent of the mean image counts. This level of uniformity is adequate for most conventional imaging.

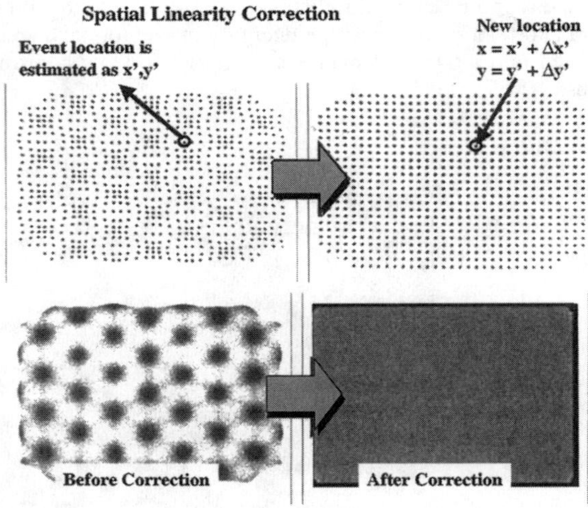

FIGURE 27.8 Spatial linearity correction. Residual positioning errors are corrected by imaging a precision hole phantom. A correction factor table is generated with the appropriate *x* and *y* offsets to reposition events to their correct location. The application of spatial linearity correction has a profound effect on image uniformity.

There are still some residual nonuniformities that exist in the scintillation camera even after energy and spatial linearity correction have been applied. These can be further reduced by applying uniformity correction (Fig. 27.9). Typically, a high count flood is acquired and a map of the nonuniformities is stored in a memory buffer. During acquisition, the number of valid events that is acquired is modulated by this map to ensure uniformity. With this additional correction, the field uniformity can be reduced to within ±3 percent of the mean image counts. It should be noted that field uniformity can be degraded by a number of factors, including the energy of the gamma ray. The most crucial factor for a system that is operating properly is the setting of the energy window. Figure 27.9 illustrates the dependence of uniformity with the energy window setting. Some scintillation cameras are more forgiving than others, but all show more nonuniformity when the energy window is displaced from the center of the photopeak.[9] There is often a gamma ray energy dependence as well. Most scintillation cameras are optimized for the best performance for the 140-keV gamma rays of Tc-99m. In some systems, uniformity significantly degrades at other gamma ray energies.

Photomultiplier tubes are relatively unstable components. Their performance changes as they age and is also sensitive to variations in temperature and humidity. In order for the energy, spatial linearity, and uniformity corrections to remain valid, there must be some way of maintaining the photomultiplier tubes at a constant operating point. Most scintillation camera systems have photomultiplier tube stabilization firmware that dynamically adjusts the photomultiplier tubes in response to a known reference signal. Some vendors use a constant-output light-emitting diode inside the electronics housing that flashes 10 times per second. The individual photomultiplier tube signals from these calibration flashes are sensed by electronics that can compare the measured output to the desired value, and then make appropriate adjustments to maintain the operating point. Another approach uses the ratio between the count rates in a photopeak and scatter window to maintain constant photomultiplier tube response. Photomultiplier tubes are also very sensitive to magnetic

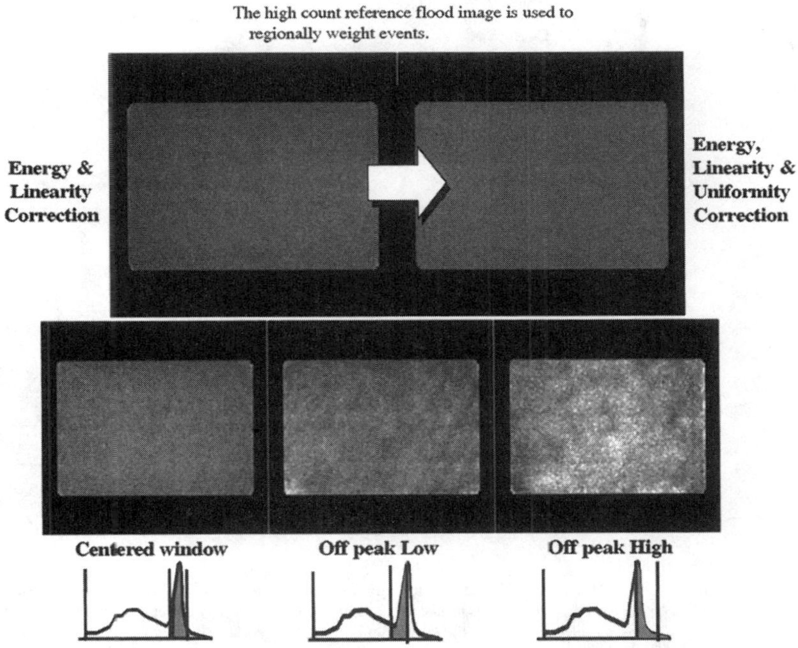

FIGURE 27.9 Uniformity correction. Nonuniformities in the field that remain after spatial linearity correction are corrected from the acquisition of a high count reference flood image. In many cameras, the field uniformity degrades when the energy window is not centered on the photopeak.

fields, and changes in the orientation of the camera with respect to the earth's magnetic field are enough to cause measurable changes in field uniformity.[10] To reduce this effect, each photomultiplier tube is individually surrounded with mu-metal shielding.

The emission of gamma rays from a radioactive source has an exponential distribution. This means that for any particular event rate, short intervals between events occur much more often than long intervals. Because the scintillation light persists for a finite time, there will eventually be light emitted from more than one event as the count rate increases. In a scintillation camera, pulse pile-up becomes evident at count rates as low as 20,000 counts per second (cps), and it gets increasingly worse as the count rate increases. Since information about pulse height becomes compromised, the performance of the scintillation camera degrades at high count rates. In most conventional and SPECT imaging, the count rate is low enough that pulse pileup is not a major concern. However, high count rates are encountered in some first-pass studies, and it is the primary problem in coincidence imaging. Corrections can be made for pulse pileup, since the physics of scintillations is well known (Fig. 27.10). Pileup can be detected on the basis of pulse height analysis. If the pileup is the result of only two events, the contribution to the second pulse from the first can be accurately estimated and subtracted, thus preserving both events. Multiple pileups can be identified and discarded. The count rate performance of scintillation cameras has improved dramatically in recent years because of the demands of coincidence imaging. When a conventional scintillation camera is recording count data at 120,000 cps, it is losing about 20 percent of the valid events. This loss increases with increasing count rate. In addition to the loss of sensitivity, both uniformity and spatial resolution get progressively worse as the count rate increases. Finally, it should be recognized that Anger logic will produce artifacts at very high count rates. "Virtual" sources will appear midway between real sources because of the signal averaging.[11] This problem has been addressed in coincidence systems by using maximum likelihood estimation instead of the conventional Anger logic.

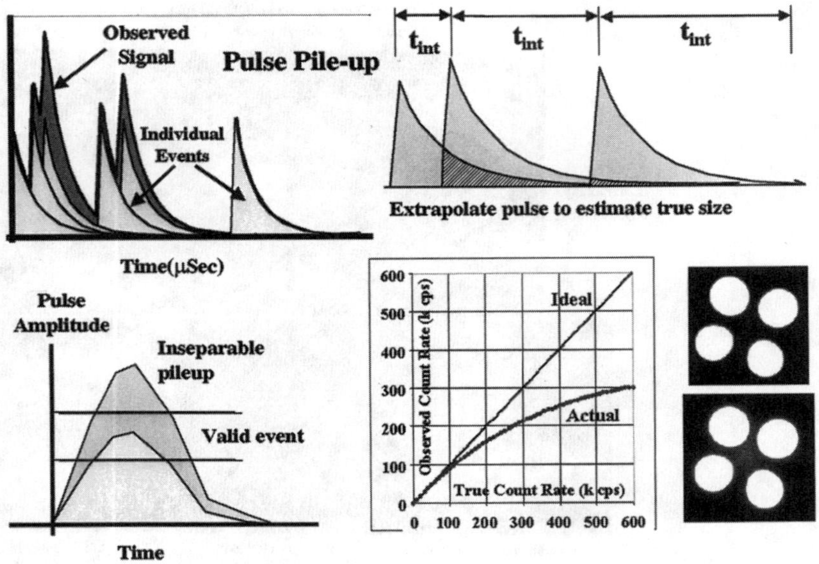

FIGURE 27.10 Pulse pileup. The persistence of the scintillation causes pulse pileup when the event rate exceeds 50,000 counts/s. Techniques can be applied to recover some of the lost information. This is especially important for coincidence imaging. Note also that the Anger positioning algorithm causes "ghost" images when the count rate becomes high.

The performance of scintillation cameras is often specified in terms of the spatial resolution. Spatial resolution is a measure of image blur and it is often specified by the width of the point-spread function. On a perfect imaging system, the image of a point has no dispersion so that all the counts fall at the same point. On a real imaging system the point is blurred and a count profile through it has a gaussian shape. The width of this count profile at the half-maximum level (FWHM) is a commonly used method for specifying the spatial resolution. The imaging system will more or less blur every point in a similar way, leading to a loss of contrast. There are two spatial resolution parameters that are of interest with scintillation cameras. The intrinsic spatial resolution describes how precisely the event location is determined when a gamma ray interacts with the crystal. The extrinsic or system spatial resolution combines the effect of the collimation with the intrinsic resolution. The intrinsic spatial resolution varies from about 3 mm FWHM on systems with thin crystals and 2-in PMTs to about 4.5 mm on systems with thicker crystals and 3-in PMTs.

The system spatial resolution depends on a variety of factors including the gamma ray energy, collimation, the source to collimator distance, and the intrinsic spatial resolution. Higher-energy gamma rays require thicker septal walls, which limits the resolving power of the collimator. As indicated above, there is a strong dependence on the distance that the source is from the collimator. The collimator and intrinsic resolution combine in quadrature like the sides of a right-angle triangle to yield the system resolution. The collimator resolution is generally more than 50 percent larger than the intrinsic resolution and therefore is the dominant factor. The system resolution at 10 cm ranges from 6.5 mm for an ultrahigh-resolution collimator to about 9.5 mm for a general-purpose collimator. It should be noted that there is nearly a factor of 3 loss of count sensitivity with the ultrahigh-resolution collimator compared to the general-purpose collimator.

Scintillation cameras have the capability of simultaneously imaging different-energy gamma rays. Most scintillation cameras handle at least three and many can handle six or more energy windows. It is important that there is no significant distortion of the images obtained at the different energies. The parameter that monitors the correspondence between images acquired at different energies is referred to as multiwindow spatial registration. The multiwindow spatial registration should be less than 3 mm on a modern scintillation system.

In the future there will be further improvements in scintillation camera technology.[12] These include photomultiplier tube replacements, new scintillators, and solid-state alternatives. Photomultiplier tubes are expensive, bulky, and susceptible to drifting. Avalanche photodiodes are a possible replacement for photomultiplier tubes (Fig. 27.11). The photodiodes use solid-state components that are very compact and rugged. Photomultiplier tubes still have superior performance specifications and are better matched for NaI(Tl). But considerable progress has been made with the photodiode approach and special-purpose devices that use this technology are beginning to appear. Figure 27.11 shows a schematic for a miniature camera module that uses avalanche photodiodes. The scintillator used in this application is CsI(Tl). The scintillation light emitted for CsI(Tl) is better matched to the properties of the photodiode. Because of the compactness offered by such designs, scintillation cameras can be designed in novel ways for specific applications. One potential application is breast cancer imaging, where several of these devices could be arranged around the breast to collect SPECT data more efficiently than bulky conventional scintillation cameras.

Solid-state detectors directly convert the absorbed gamma ray energy into collection of electric charge and do not need photomultiplier tubes. Since the PMT is a bulky and expensive component, this represents a significant breakthrough. Cadmium zinc telluride is an attractive solid-state detec-

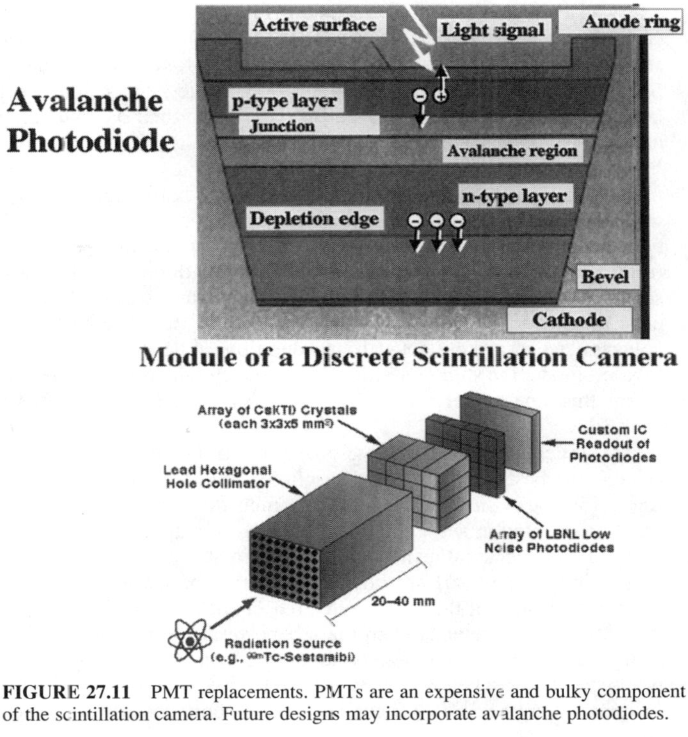

FIGURE 27.11 PMT replacements. PMTs are an expensive and bulky component of the scintillation camera. Future designs may incorporate avalanche photodiodes.

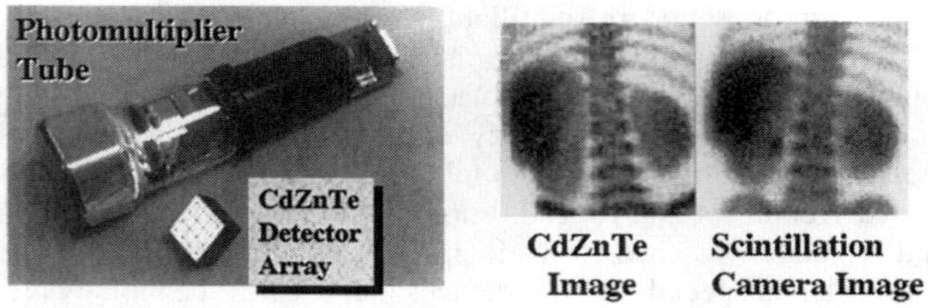

FIGURE 27.12 Solid state imagers. Although NaI(Tl)-based scintillation cameras will be the choice for the near future, improvements in solid-state detectors such as CdZnTe may lead to competitive imaging systems.

tor[13] (Fig. 27.12). It can be manufactured in a pixelated array and has comparable gamma ray detection efficiency to NaI(Tl) at 140 keV. Another advantage is that the energy resolution with CdZnTe is nearly a factor of 2 better than that of NaI(Tl). However, several production problems need to be solved before CdZnTe becomes a viable replacement for NaI(Tl).

27.3 SPECT SYSTEMS

Single-photon-emission computed tomography (SPECT) produces tomographic images of the internal distribution of radiopharmaceuticals.[14,15] It is most commonly used in the diagnosis of coronary artery disease and in tumor detection. Projection images collected by one or more scintillation cameras are mathematically reconstructed to obtain the tomographic slices. Most clinical SPECT studies are qualitative with simplistic corrections for attenuation and scattered radiation. Quantitative SPECT requires corrections for attenuation, scatter, and spatial resolution, although these have not been routinely implemented in the past because of their computational load. SPECT instrumentation has evolved to include coincidence imaging of positron-emitting radiopharmaceuticals, specifically ^{18}F fluorodeoxyglucose.

A SPECT system consists of one or more scintillation cameras mounted to a gantry that can revolve about a fixed horizontal axis (the axis of rotation)[16–19] (Fig. 27.13). SPECT studies are usually acquired over a full 360° arc, although myocardial perfusion studies typically use only data from the 180° arc that minimizes tissue attenuation. SPECT acquisitions are performed with the scintillation camera located at preselected angular locations (step-and-shoot mode), or in a continuous rotation mode. In the step-and-shoot mode, the detector rotates to each angular position and collects data in a preselected frame duration while the detector is motionless. In the continuous-rotation mode, the study duration is selected and the rotation speed is adjusted to complete the orbit during this time. Projections are collected as the detector rotates and are binned into 60 to 120 frames over 360°.

It is crucial to maintain close proximity to the body as the detector rotates about the patient to achieve the best possible spatial resolution. Although a number of different approaches have been used to accomplish this, the most common method moves the detectors radially in and out as a function of rotation angle. Myocardial perfusion studies are the most commonly performed SPECT procedures. Because the heart is located in the left anterior portion of the thorax, gamma rays originating in the heart are highly attenuated for views collected from the right lateral and right posterior portions of the arc. For this reason, SPECT studies of the heart are usually collected using the 180° arc that extends from the left posterior oblique to the right anterior oblique view.[20] This results in reconstructed images with the best contrast, although distortions are often somewhat more pronounced than when 360° data are used.[21] Because of the widespread use of myocardial perfusion imaging, many SPECT systems have been optimized for 180° acquisition by using two detectors

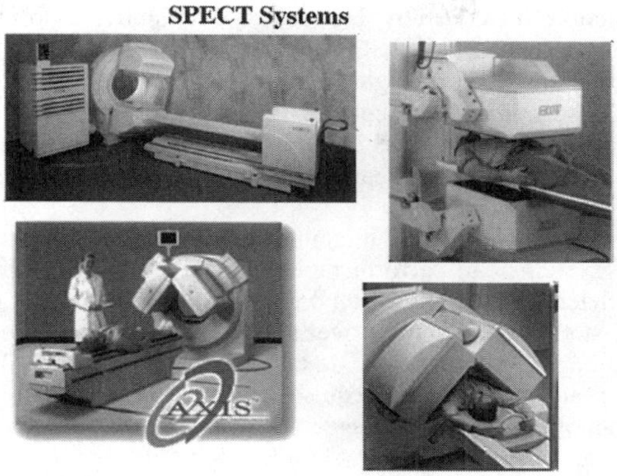

FIGURE 27.13 SPECT acquisition. One or more scintillation cameras collect images at typically 60 to 120 angles around a 360° orbit. The scintillation camera acquires projection images from a large volume simultaneously.

arranged at ~90° (Fig. 27.14). This reduces the acquisition time by a factor of 2 over single detectors and is approximately 30 percent more efficient than triple detector SPECT systems. Positioning the detectors at 90° poses some challenges for maintaining close proximity. Most systems rely on the motion of both the detectors and the SPECT table to accomplish this.

The heart is continually moving during the SPECT acquisition, and this further compromises spatial resolution. Because the heart beats many times per minute, it is impossible to directly acquire

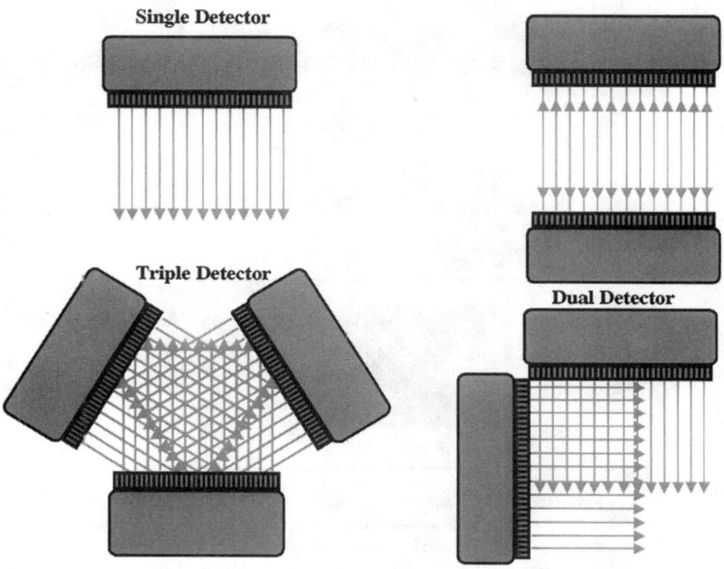

FIGURE 27.14 SPECT system configurations. Although a single scintillation camera can be used to acquire SPECT data, multiple detectors improve the overall sensitivity. Two detectors arranged at either 180 or 90° are the most common configuration.

a stop-action SPECT study. However, since the heart motion is periodic, it is possible to obtain this information by gating the SPECT acquisition.[22] In a gated SPECT acquisition, the cardiac cycle is subdivided and a set of eight images spanning the ECG R-R interval is acquired for each angular view. These images are acquired into predetermined time bins based on the patient's heart rate, which is monitored by the ECG R wave interfaced to the SPECT system. As added benefits of gating, the motion of the heart walls can be observed and ventricular volumes and ejection fractions can be determined.[22,23]

Although most SPECT imaging samples more or less static distribution of radionuclides, some SPECT systems can perform rapid sequential studies to monitor tracer clearance. An example of this is determination of regional cerebral blood from the clearance of ^{133}Xe.[24] Multiple 1-minute SPECT studies are acquired over a 10-minute interval. When one acquisition sample is completed, the next begins automatically. In order to minimize time, SPECT systems that perform these studies can alternately reverse the acquisition direction, although at least one SPECT system utilizes slip-ring technology so that the detectors can rotate continuously in the same direction.

In order to produce accurate tomographic images, projection data representing the line integrals of activity in the internal distribution have to be acquired. This information is not directly available because of tissue attenuation. Simple attenuation correction methods can be used in regions of the body such as the abdomen or head where the tissue density is more or less uniform. However, compensation for attenuation in the thorax requires an accurate attenuation map for each tomographic plane. This is especially important for myocardial perfusion studies, since the artifacts resulting from tissue attenuation mimic the patterns caused by coronary artery disease. In recent years, all the SPECT manufacturers have offered systems that can perform transmission measurements along with the emission studies. These systems use the scintillation camera to detect the transmission of gamma rays from an external source.[17,25,26] Several different configurations are available. Most use a line source that is translated across the camera field of view at each angular stop in much the same way as a first-generation CT scanner (Fig. 27.15). Typically a source such as Gd-153 or Ba-133 is used

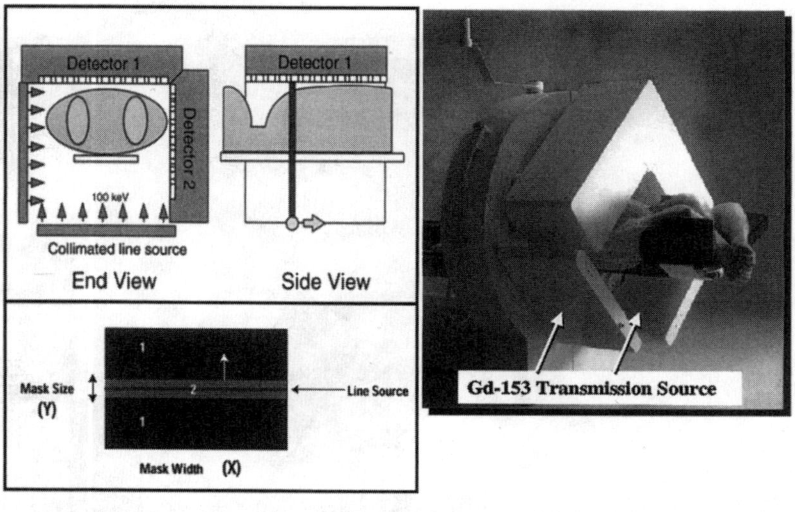

FIGURE 27.15 Attenuation correction. To obtain accurate SPECT results, corrections must be made for tissue attenuation. In regions where there are large variations in tissue density such as the thorax, this requires an independent transmission study. This shows one possible configuration where line sources of Gd-153 are translated across the field of view to collect the transmission data. This information is reconstructed to obtain a crude CT image of the thorax to correct myocardial perfusion studies. (*Courtesy of GE Medical Systems.*)

that has a different gamma ray energy than those used in the emission study so that the studies can be acquired simultaneously (with appropriate correction for cross talk). Other approaches use multiple stationary line sources to collect this information. At least one vendor has an x-ray tube and separate detectors to obtain a moderate-quality CT scan.[27] In each configuration, the transmission data are collected and reconstructed to yield the attenuation maps. This information can be incorporated into an iterative algorithm to effect the correction.

In spite of the energy discrimination available on all SPECT systems, Compton scattered radiation still accounts for about 30 to 40 percent of the acquired counts in SPECT imaging. Scattered radiation decreases contrast and can impact other corrections. For example, when attenuation correction is applied without also correcting for scattered radiation, the count density in the heart walls near the liver may be overenhanced. SPECT systems in the future may resort to other detectors that have substantially better energy resolution than that of NaI(Tl), but for now, scatter compensation routines must be employed. Scatter correction has been performed in several different ways.[15,28–33] The easiest to implement is the subtraction method, where information is simultaneously acquired into a second energy window centered below the photopeak in the Compton scatter region of the energy spectrum. After establishing an appropriate normalization factor, the counts from the scatter window are subtracted from the photopeak window. The corrected projections are then used in the reconstruction algorithm. The disadvantage of this approach is that it increases noise and it is difficult to establish an accurate normalization factor. To accommodate this type of correction (and also to image different-energy gamma rays), SPECT systems allow the simultaneous acquisition from multiple energy windows. The number of energy windows available varies for each manufacturer, although every system is capable of imaging from at least four energy windows.

27.3.1 SPECT Image Reconstruction

The details of SPECT image reconstruction are beyond the scope of this article. However, the demands of image reconstruction do impact the features required by the computer. The CPU must be fast enough and have enough memory to accommodate the entire SPECT data set. This is well within the capability of home PCs. A typical SPECT study is less than 5 Mbyte and the current processor speeds approaching 1 GHz are fast enough to render reconstructions using either filter backprojection or optimized iterative algorithms in an acceptable time (less than 10 minutes).

27.3.2 SPECT System Performance

Typical performance specifications for SPECT imaging systems are summarized in Table 27.3. As with conventional planar imaging, the scintillation cameras, and the associated collimation are the primary factors affecting the performance. SPECT spatial resolution is nearly isotropic with an FWHM of 8 to 10 mm for brain imaging where the detectors can get close to the radioactive source.

TABLE 27.3 SPECT System Performance

Parameter	Specification
Number of scintillation cameras	One, two, or three
Count sensitivity per camera	250 cpm/μCi per detector
Matrix size	64 × 64; 128 × 128
Pixel size	6 mm; 3 mm
Spatial resolution (brain studies)	8 mm
Spatial resolution (heart studies)	14 mm
SPECT uniformity	15%
Contrast of 25.4-mm sphere	0.45

The spatial resolution degrades to 12 to 18 mm for body imaging because the detectors can not be positioned as close. The components of SPECT spatial resolution and their relative importance can be identified from the equation shown below:

$$R_{\text{SPECT}} = \sqrt{R_{\text{col}}^2 + R_{\text{filter}}^2 + R_{\text{int}}^2}$$

As before, R_{int} and R_{col} represent the intrinsic and collimator resolution components. R_{filter} is the FWHM of the smoothing kernel required to yield an acceptable reconstruction. The intrinsic spatial resolution is the least important factor in this calculation, since it is usually a factor of 2 or more smaller than the other components. The trade-off between spatial resolution and count sensitivity is explicit in this equation. Decreasing R_{col} to improve spatial resolution will often require R_{filter} to become larger to compensate for increased noise.

27.3.3 SPECT/PET Hybrid Systems

The primary reason for the success of nuclear medicine imaging is the availability of radiopharmaceuticals that provide crucial diagnostic information. For cancer diagnosis and followup, ^{18}F flourodeoxyglucose (^{18}F FDG) is an exquisite imaging agent for a wide variety of malignancies including lung, colon, breast, melanoma, and lymphoma. Because ^{18}F is a positron emitter that yields very high energy x-rays (511 keV) when the positron combines with a free electron, it can not be imaged well on conventional SPECT systems. The thin NaI(Tl) crystals have a low efficiency for detection at this energy (less than 10 percent). Also, the collimators designed for the 511-keV photons have low count sensitivity and poor spatial resolution. The collimation can be eliminated if coincidence detection is used. Annihilation photons from positrons are always colinear. This feature can be exploited to count only events that are simultaneously detected by opposed detectors. Two opposed scintillation cameras with their collimators removed and the addition of coincidence electronics will turn a SPECT system into a PET tomograph (Fig. 27.16).[34,35] The efficiency for coincidence detection equals the product of the individual efficiencies, so that the coincidence efficiency is about 1 percent for conventional scintillation cameras. This is actually much higher than the efficiency with collimators. Coincidence efficiency can be improved by using thicker NaI(Tl) crystals, and all the vendors have done this. However, thicker crystals degrade the intrinsic spatial resolution when the scintillation cameras are used for conventional (i.e., collimated) studies. A solution to this prob-

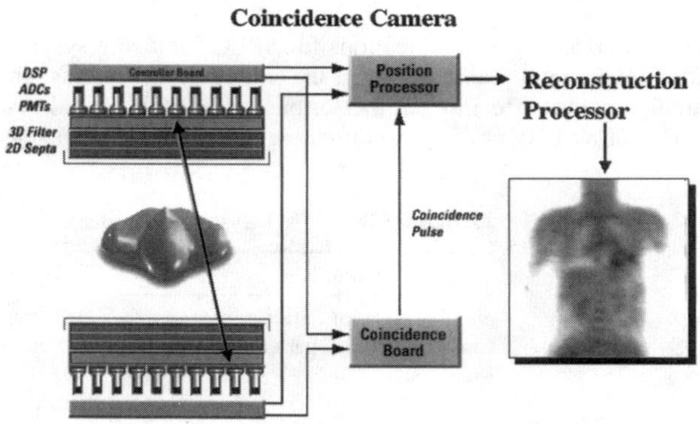

FIGURE 27.16 Two or more opposed scintillation cameras can be used as coincidence detection systems for PET imaging. Valid events are established when the two detectors record events within the 10- to 15-ns timing window. Graded filters and lead septa are placed in front of the detectors to limit scattered radiation. *(Courtesy of GE Medical Systems.)*

lem, called StarBrite, has been recently introduced. Thick NaI(Tl) crystals (25 mm) have slots machined in the PMT side of the detector to prevent the scintillation light from scattering throughout the crystal to maintain good spatial resolution. There are other concerns, primarily the count rate capability. With most SPECT imaging studies, there are essentially no count rate losses resulting from the time it takes to process each event. The amount of radioactivity that can be safely administered and the low sensitivity of the collimation keep the count rate in the range that the electronics can easily handle. However, when the collimators are removed for coincidence imaging, the NaI(Tl) crystals are exposed to count rates above the capacity of the conventional electronics. Much effort over the past decade has been devoted to increasing the count rate capability of the scintillation cameras. In the early 1990s, the maximum observed counting rate for a scintillation camera was in the 200,000 to 400,000 count/second range. This rate has been extended to over 1,000,000 counts/second by shortening the integration time on the pulses and implementing active baseline restoration. Because the light is proportional to the energy deposited in the crystal, one can shorten the pulse integration time without extreme degradation. Even with this substantial improvement in count rate, the maximum activity the system can handle is about 3 mCi. Typical performance values for a coincidence scintillation camera system are given in Table 27.4.

TABLE 27.4 SPECT/PET Hybrid System Performance

Parameter	Specification
Number of scintillation cameras	2
NaI(Tl) thickness, mm	15.9–19
Matrix size	128 × 128
Pixel size	3 mm
Maximum singles rate, counts/sec	1,000,000–2,000,0000
Maximum coincidence rate, counts/sec	10,000
Spatial resolution, mm	5

In addition to specialized electronics, other measures have been taken to help reduce the count rate burden. One example of these is a graded absorber placed in front of each detector to help reduce the scattered radiation component.[36] Because scattered radiation has lower energy than the unscattered photons, low-energy scatter will be preferentially absorbed by lead, since the photoelectric cross section is inversely proportional to the cube of the gamma ray energy. If only lead is used, the lead characteristic x-ray resulting from the absorption will be detected. Additional layers of tin, copper, and aluminum will absorb the respective characteristic x-rays. This graded filter causes a significant reduction in the low-energy scattered radiation and, since the scintillation cameras process every event, also reduces the overall count rate burden.

Even though the thin NaI(Tl) crystals have low intrinsic efficiency, the uncollimated detectors present a large solid angle to the annihilation photons. Maximum sensitivity is achieved when all coincidences, even those at large angles, are accepted. This makes the camera sensitivity extremely sensitive to the source location. Sources located near the central axis of the detectors have a large solid angle, while those at the periphery can interact only with a very small portion of the detectors. In addition, the scatter component increases to well over 50 percent when large coincidence acceptance angles are used. This problem has been addressed by using lead slits aligned perpendicular to the axial direction of the system to restrict the angular extent of the coincidences. While this reduces the sensitivity of the imaging system, it also reduces the scatter component to less than 30 percent and limits the solid angle variability.

The intrinsic spatial resolution of the hybrid systems is comparable to that of the dedicated PET systems with a FWHM of 4 to 5 mm. However, the count sensitivity is at least an order of magnitude lower. This, along with the maximum count rate constraint, guarantees that the coincidence camera data will be very count poor and therefore require substantial low-pass filtering when reconstructed. As a result the quality of the reconstructed images is perceptibly worse than the dedicated PET images. In head-to-head comparisons, it has been found that the hybrid systems perform well on tumors greater than 2 cm in diameter located in the lung.[37–39] Tumors smaller than these and those located in high-background areas are detected with a much lower sensitivity. These results are important since they provide a guide where the application of the coincidence camera will be useful.

In a conventional scintillation camera, Anger logic is used to determine the location of an interaction. The tacit assumption in this approach is that only one event is occurring at a time. At the high count rate encountered in coincidence imaging, multiple interactions are likely, and when this occurs, the events are improperly located somewhere between the two true locations. Improved algorithms have been developed that can identify multiple hits and that use a maximum likelihood calculation to correctly determine event locations.

The projection data collected by the coincidence cameras require correction for random coincidences, scatter, and attenuation if accurate tomographic images are to be obtained. Random or accidental coincidences occur when two unrelated photons are detected. These random coincidences increase rapidly with the count rate and give rise to a nondescript background that has to be subtracted from the projections. Typically randoms are either monitored in a separate time window or are calculated from the singles count rate and are subtracted. Scatter correction is sometimes ignored in coincidence PET, or maybe monitored by a separate energy window and subtracted as discussed for SPECT imaging.

Accurate reconstruction of PET data requires correction for attenuation since the degree of attenuation for coincidence imaging is very high, approaching values of 100 or more. Attenuation correction requires information about the transmission of the annihilation radiation through the body at the coincidence lines of response. When attenuation correction is ignored, severe artifacts are seen in the reconstructed images. As with SPECT imaging, a separate transmission study with an external source, typically Cs-137, must be acquired to provide the attenuation map used in the correction.

27.4 SUMMARY

SPECT imaging is expected to play a continuing important role in medical imaging. Future improvements in SPECT instrumentation are likely to include new detectors and collimation schemes. The coincidence scintillation cameras will also continue their evolution with the addition of more cameras and multidetector levels optimized for SPECT and coincidence imaging. Reconstruction algorithms will evolve as new techniques are developed and as the performance of the computer expands. In spite of the importance of instrumentation, the primary motivating factor in SPECT imaging will continue to be the creation and implementation of new radiopharmaceuticals. While SPECT will continue to be highly utilized for myocardial perfusion imaging, SPECT use in tumor imaging will probably experience the largest growth. Applications will include treatment planning for internal radiation therapy as well as diagnostic studies.

REFERENCES

1. Hubbell J. H., "Review of photon interaction cross section data in the medical and biological context," *Phys. Med. Biol.*, 1999; **44**:R1–R22.
2. Graham L. S., et al., "Nuclear medicine from Becquerel to the present," *Radiographics*, 1989; **9**:1189–1202.
3. Anger H. O., et al., "Recent applications of the scintillation camera," *Strahlentherapie* [Sonderb] 1967; **65**: 70–93.
4. White W., "Resolution, sensitivity, and contrast in gamma-camera design: A critical review," *Radiology*, 1979; **132**:179–187.
5. Genna S., S. C. Pang, and A. Smith, "Digital scintigraphy: Principles, design, and performance," *J. Nucl. Med.*, 1981; **22**:365–371.

6. Gunter D., *Collimator Characteristics and Design*, 1st ed., *Nuclear Medicine*, R. Henkin, et al., (eds.), vol. 1., 1996, Mosby, St. Louis, pp. 96–124.
7. Moore S. C., K. Kouris, and I. Cullum, "Collimator design for single photon emission tomography," *European Journal of Nuclear Medicine*, 1992; **19**:138–150.
8. Muehllehner G., J. G. Colsher, and E. W. Stoub, "Correction for field nonuniformity in scintillation cameras through removal of spastial distortion," *J. Nucl. Med.*, 1980; **21**:771–776.
9. Graham L. S., R. L. LaFontaine, and M. A. Stein, "Effects of asymmetric photopeak windows on flood field uniformity and spatial resolution of scintillation cameras," *J. Nucl. Med.*, 1986; **27**:706–713.
10. Bieszk J. A., "Performance changes of an Anger camera in magnetic fields up to 10 G," *J. Nucl. Med.*, 1986; **27**:1902–1907.
11. Strand S. E., and I Larsson, "Image artifacts at high photon fluence rates in single-crystal NaI(Tl) scintillation cameras," *J. Nucl Med.*, 1978; **19**:407–413.
12. Kuikka J. T., et al., "Future developments in nuclear medicine instrumentation: A review," *Nucl. Med. Commun.*, 1998; **19**:3–12.
13. Yaffe, M. J., and J. A. Rowlands, "X-ray detectors for digital radiography," *Phys. Med. Biol.*, 1997; **42**: 1–39.
14. Madsen M. T., "The AAPM/RSNA physics tutorial for residents. Introduction to emission CT," *Radiographics*, 1995; **15**:975–991.
15. Tsui B. M., et al., "Quantitative single-photon emission computed tomography: Basics and clinical considerations," *Semin. Nucl. Med.*, 1994; **24**:38–65.
16. Jaszczak R. J., and R. E. Coleman, "Single photon emission computed tomography (SPECT). Principles and instrumentation," *Invest. Radiol.*, 1985; **20**:897–910.
17. Fahey F. H., "State of the art in emission tomography equipment," *Radiographics*, 1996; **16**:409–420.
18. Links J. M., "Advances in nuclear medicine instrumentation: Considerations in the design and selection of an imaging system," *Eur. J. Nucl. Med.*, 1998; **25**:1453–1466.
19. Rogers W. L., and R. J. Ackermann, "SPECT instrumentation," *Am. J. Physiol. Imaging*, 1992; **7**:105–120.
20. Tsui B. M., et al., "Quantitative myocardial perfusion SPECT," *J. Nucl. Cardiol.*, 1998; **5**:507–522.
21. LaCroix K. J., B. M. Tsui, and B. H. Hasegawa, "A comparison of 180 degrees and 360 degrees acquisition for attenuation-compensated thallium-201 SPECT images," *J. Nucl. Med.*, 1998; **39**:562–574.
22. White M. P., A. Mann, and M. A. Saari, "Gated SPECT imaging 101," *J. Nucl. Card.*, 1998; **5**:523–526.
23. Berman D. S., and G. Germano, "Evaluation of ventricular ejection fraction, wall motion, wall thickening, and other parameters with gated myocardial perfusion single-photon emission computed tomography," *J. Nucl. Card.*, 1997, **4**:S169–S171.
24. Bruyant P., et al., "Regional cerebral blood flow determination using 133Xe and a standard rotating gamma-camera," *Comput. Biol. Med.*, 1998; **28**:27–45.
25. King M. A., B. M. Tsui, and T. S. Pan, "Attenuation compensation for cardiac single-photon emission computed tomographic imaging: Part 1. Impact of attenuation and methods of estimating attenuation maps," *J. Nucl. Cardiol.*, 1995; **2**:513–524.
26. Celler A., et al., "Multiple line source array for SPECT transmission scans: Simulation, phantom and patient studies," *J. Nucl. Med.*, 1998; **39**:2183–2189.
27. Patton J. A., "Instrumentation for coincidence imaging with multihead scintillation cameras," *Semin. Nucl. Med.*, October 2000, **30**:239–254.
28. Buvat I., et al., "Comparative assessment of nine scatter correction methods based on spectral analysis using Monte Carlo simulations," *J. Nucl. Med.*, 1995; **36**:1476–1488.
29. Beekman F. J., C. Kamphuis, and E. C. Frey, "Scatter compensation methods in 3D iterative SPECT reconstruction: A simulation study," *Phys. Med. Biol.*, 1997; **42**:1619–1632.
30. Beekman F. J., H. W. de Jong, and E. T. Slijpen, "Efficient SPECT scatter calculation in non-uniform media using correlated Monte Carlo simulation," *Phys. Med. Biol.*, 1999; **44**:N183–N192.
31. Riauka T. A., and Z. W. Gortel, "Photon propagation and detection in single-photon emission computed tomography—An analytical approach," *Medical Physics*, 1994; **21**:1311–1321.
32. Haynor D. R., et al., "Multiwindow scatter correction techniques in single-photon imaging," *Medical Physics*, 1995; **22**:2015–2024.

33. Rosenthal M. S., et al., "Quantitative SPECT imaging: A review and recommendations by the Focus Committee of the Society of Nuclear Medicine Computer and Instrumentation Council," *J. Nucl. Med.*, 1995; **36**:1489–1513.
34. Jarritt P. H., and P. D. Acton, "PET imaging using gamma camera systems: A review," *Nucl. Med. Commun.*, 1996; **17**:758–766.
35. Lewellen T. K., R. S. Miyaoka, and W. L. Swan, "PET imaging using dual-headed gamma cameras: an update," *Nucl. Med. Commun.*, 1999; **20**:5–12.
36. Muehllehner G., R. J. Jaszczak, and R. N. Beck, "The reduction of coincidence loss in radionuclide imaging cameras through the use of composite filters," *Phys. Med. Biol.*, 1974; **19**:504–510.
37. Shreve P. D., et al., "Oncologic diagnosis with 2-[fluorine-18]fluoro-2-deoxy-D-glucose imaging: Dual-head coincidence gamma camera versus positron emission tomographic scanner," *Radiology*, 1998; **207**:431–437.
38. Shreve P. D., R. S. Steventon, and M. D. Gross, "Diagnosis of spine metastases by FDG imaging using a gamma camera in the coincidence mode," *Clin. Nucl. Med.*, 1998; **23**:799–802.
39. Coleman R. E., C. M. Laymon, and T. G. Turkington, "FDG imaging of lung nodules: A phantom study comparing SPECT, camera-based PET, and dedicated PET," *Radiology*, 1999; **210**:823–828.

CHAPTER 28
BREAST IMAGING SYSTEMS: DESIGN CHALLENGES FOR ENGINEERS

Mark B. Williams
Department of Radiology, University of Virginia, Charlottesville

Laurie L. Fajardo
The Russell Morgan Department of Radiology and Radiological Sciences, Johns Hopkins Medical Institutions, Baltimore, Maryland

28.1 INTRODUCTION

Breast cancer is the second greatest cause (after lung cancer) of cancer-related death among American women, accounting for approximately 40,000 deaths each year. At the present time, early detection and characterization of breast cancers is our most effective weapon, since local disease is in most cases curable. Breast imaging systems can thus be potentially useful if they either (1) are useful for *detection* of cancers or (2) are useful for *characterization* of suspicious lesions that may or may not be cancerous. Similarly, from a clinical perspective, methodologies used for breast cancer diagnosis (as opposed to therapy) fall into one of two broad categories: screening or diagnostic. Screening pertains to the population of women exhibiting no symptoms. Diagnostic imaging (otherwise known as problem-solving imaging) is used when there is some suspicion of disease, as a result of the manifestation of some physical symptom, of a physical exam, or of a screening study. The relative effectiveness of a given imaging modality at the tasks of detection and characterization determines whether it will be employed primarily in a screening or diagnostic context. At the present time, x-ray mammography is the only FDA-approved modality for screening, and is by far the most effective modality because of its ability to detect small cancers, when they are most treatable (i.e., prior to metastasis). The sensitivity (fraction of all cancers that are detected) by screen-film mammography is approximately 85 percent. Ultrasound, MRI, scintimammography, and electrical impedance scanning are FDA approved as diagnostic procedures following detection of an abnormality via x-ray mammography.

In design of a system for either screening or diagnostic imaging of the breast, several characteristics of the breast itself present unique engineering challenges. First, unlike parts of the body supported by bone, the breast is a malleable organ. Thus obtaining the exact same configuration for successive imaging studies is difficult, if not impossible. This complicates correlation between follow-up images of a given modality, or between concurrently obtained images from two different modalities. A second challenge arises from the fact that cancers can arise in areas of the breast very close to the chest wall, which presents special difficulties. For example, the focal spot in x-ray mammography must be positioned directly above the chest wall edge of the image receptor in order to assure that x-rays passing through tissue adjacent to the chest wall are imaged. The proximity of the chest and shoulders also presents geometric hindrance affecting MRI coil design and nuclear medicine scanning. A third challenging aspect of breast cancer imaging is the similarity of many of the physical attributes of cancerous material and normal breast tissue. For example, the x-ray attenuation and acoustic impedance of cancerous masses are very similar to those of healthy fibroglandular breast tissue. Thus the imaging process must result in a sufficiently high signal-to-noise ratio that such subtle differences can be ascertained.

28.2 BREAST ANATOMY

The breast is composed primarily of fat and glandular tissue. The glandular tissue is sandwiched between layers of fat and lies above the pectoralis muscle and chest wall. The combination of the

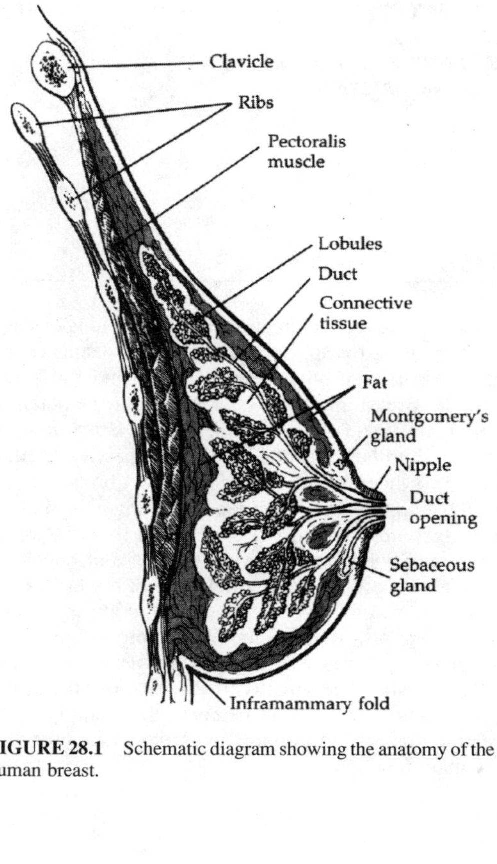

FIGURE 28.1 Schematic diagram showing the anatomy of the human breast.

adipose and glandular tissue provides the radiographic contrast detected on x-ray mammography. When the ratio of adipose tissue to glandular tissue in the breast is greater, greater radiographic contrast is achieved. Breast tissue composed of only fibroglandular tissue results in a mammogram with lesser radiographic contrast that is more difficult for radiologists to evaluate.

The breast is considered a modified sweat gland and the milk it produces is a modification of sweat. Breast lobules, which produce milk during lactation, are connected by the breast ducts. The breast lobules and ducts are supported by the surrounding connective tissue. Deep and superficial facial layers envelop the stromal, epithelial, and glandular breast elements. Cooper's ligaments, a criss-crossing network of fibrous supporting structures, course between the deep and superficial layers of fascia. Surrounding the cone of glandular tissue is a layer of subcutaneous fat.

The nipple and areola contain erectile smooth muscle as well as sebaceous glands. Between five and nine separate ductal systems intertwine throughout the breast and have separate openings at the nipple. Each major lactiferous duct extends from the nipple-areolar complex into the breast in a branching network of smaller ducts. The area drained by each duct network is called a *lobe*, or *segment*, of the breast. The duct network is lined by two types of cells, an inner epithelial layer surrounded by a thinner layer of myoepithelial cells. The final branch from a segmental duct is called the *extralobular terminal duct* and terminates in several acini. The anatomic unit comprising the extralobular duct and its lobule of acini is histologically designated as the terminal ductal lobular unit. It is postulated that most cancers arrive in the extralobular terminal ducts, just proximal to the lobule.

Figure 28.1 is a schematic drawing depicting the anatomy of a healthy breast.

28.3 CURRENT CLINICAL BREAST IMAGING

28.3.1 Screening and Diagnostic Mammography

X-ray mammography is a projection onto two dimensions of the three-dimensional x-ray attenuation distribution of the breast. By federal regulation (the Mammography Quality Standards Act), dedicated and specialized radiographic equipment must be used for mammography. Mammographic systems utilize x-ray tubes with small focal spot size (0.1 mm and 0.3 mm nominal diameters), high-resolution x-ray receptors, and scatter reduction grids between breast and receptor. The x-ray focal spot is positioned directly above the edge of the image receptor closest to the patient, so that structures immediately adjacent to the chest wall may be visualized. A typical screening mammogram consists of two views of each breast. The two views are separated by approximately 45 to 60°, in order to help resolve ambiguities produced by overlapping breast structure in a given projection. In addition, one view maximizes visualization of the structures near the lateral portion of the breast, such as the axiallary lymph nodes. Both are obtained with the breast under compression, using a flat acrylic paddle with low x-ray attenuation. Compression reduces the amount of scatter radiation reaching the detector by reducing the breast thickness, and also spreads out the tissue, thereby reducing the amount of structural overlap in the projection image.

Follow-up diagnostic procedures include additional x-ray views (magnification views, spot views, or unusual projections such as lateral views), ultrasound, and, more recently, MRI and nuclear medicine imaging (positron emission mammography or single gamma emission scintigraphy). We first present an overview of some of the technical aspects of screening mammography, then describe challenges associated with current and upcoming diagnostic imaging techniques.

X-ray Mammography Image Considerations. X-ray mammography is perhaps the most exacting of all x-ray based imaging tasks. The main reasons for this are (1) the small difference in x-ray attenuation properties between various breast structures, and between normal and cancerous tissue, and (2) the requirement that physically small objects such as microcalcifications be imaged with enough clarity to be detected by the radiologist (microcalcifications are calcium-containing deposits that are associated with early breast cancers, although many calcifications are merely benign growths). Clinically significant microcalcifications may be 0.2 mm or less in size. They often appear in clusters, and their individual shapes and relative orientations can be a clue as to the likelihood of an associated malignancy. The simultaneous requirements of high-contrast resolution and high spatial resolution, along with the desire to minimize radiation dose to the breast, dictate that the image sensor have high sensitivity, low noise, and a narrow point response function. In addition, depending on the size and composition of the breast, the range of x-ray fluence incident on the sensor surface in a given mammogram can be 400:1 or more (Johns and Yaffe, 1987; Nishikawa et al., 1987). Thus the sensitivity and resolution must be maintained over an appreciable dynamic range.

Film-Based Imaging. Most mammography is currently performed using screen-film systems; that is systems that employ image receptors consisting of film placed in direct contact with an x-ray-to-light converting screen. The screen-film combination, housed in a light-tight cassette, is placed below the breast, with a rotating anode x-ray tube placed 60 to 65 cm above the cassette. A major technical challenge in screen-film mammography arises from the fact that the film is used both as an acquisition and display medium. That means that system parameters such as film speed, detection and light conversion efficiency of the screen, and x-ray spectrum must be chosen not only to maximize the efficiency of the image acquisition process, but also to result in the optimum film darkening to maximize the visual contrast between structures when the film is viewed on a light box. Furthermore, as with other film-based imaging procedures, there is an inescapable trade-off between image contrast and dynamic range. One motivation for the development of digital detectors to replace screen-film receptors in mammography is the desire to increase dynamic range without a concomitant loss of image contrast.

Digital Mammography *Current Technologies for Image Acquisition.* The principal motivations for the development of electronic receptors to replace film-based receptors are (1) the limited dynamic range of x-ray intensities over which the sensitivity of film-based systems is appreciable and (2) the inherent noise properties of film due to the random distribution of the grains on the film substrate (Nishikawa and Yaffe, 1985). The desire to overcome these limitations, coupled with the advantages inherent in having images in digital form (e.g., computer-aided diagnosis, easier image storage, remote transmission of images), has led to the development of several technical solutions to digital mammographic image acquisition during the past decade. Leading large area technologies for full-field digital mammography currently include: (1) CsI(Tl) converters coupled via demagnifying fiber-optic tapers to arrays of charge coupled devices (CCDs), (2) CsI(Tl) converters coupled to arrays of amorphous silicon (a-Si) photodiodes that are read out using a-Si thin-film-transistor (flat panel) technologies, and (3) storage phosphor plates (typically europium-doped barium fluorobromide) that retain a metastable latent image during x-ray exposure that can be de-excited and quantified using a raster-scanned laser beam. On the horizon is a second type of flat panel detector, which uses a layer of amorphous selenium instead of the CsI(Tl)-photodiode combination. This is a so-called direct conversion detector, in which absorbed x-rays generate electron-hole pairs in the selenium, rather than using an intermediate x-ray to visible-photon conversion stage and a photodetector to generate charge. In each of the above digital mammography detectors, the detector area is equal to the imaged area. A fourth approach uses a slot shaped CCD-based detector that spans the imaged area in one dimension (the chest wall to nipple direction), and is scanned from left to right during image acquisition. The x-ray beam, collimated to a similar slot shape, is scanned in synchrony with the detector. With the first three technologies, x-ray targets (molybdenum or rhodium) and filtration (molybdenum or rhodium) combinations similar to those used in screen-film mammography are employed. For the fourth technology, a tungsten target is used, and the tube voltage is somewhat higher than that used with Mo or Rh targets.

Figure 28.2 compares analog (screen-film) and digital cranial-caudal images of the right breast of a woman with moderately radiodense breast tissue and an irregularly-shaped mass with spiculated margins in the lateral aspect of the breast. The digital image (Fig. 28.2*b*) shows improved contrast, enhanced depiction of the suspicious mass and better visualization of peripheral tissue and skin line than the analog image (Fig. 28.2*a*).

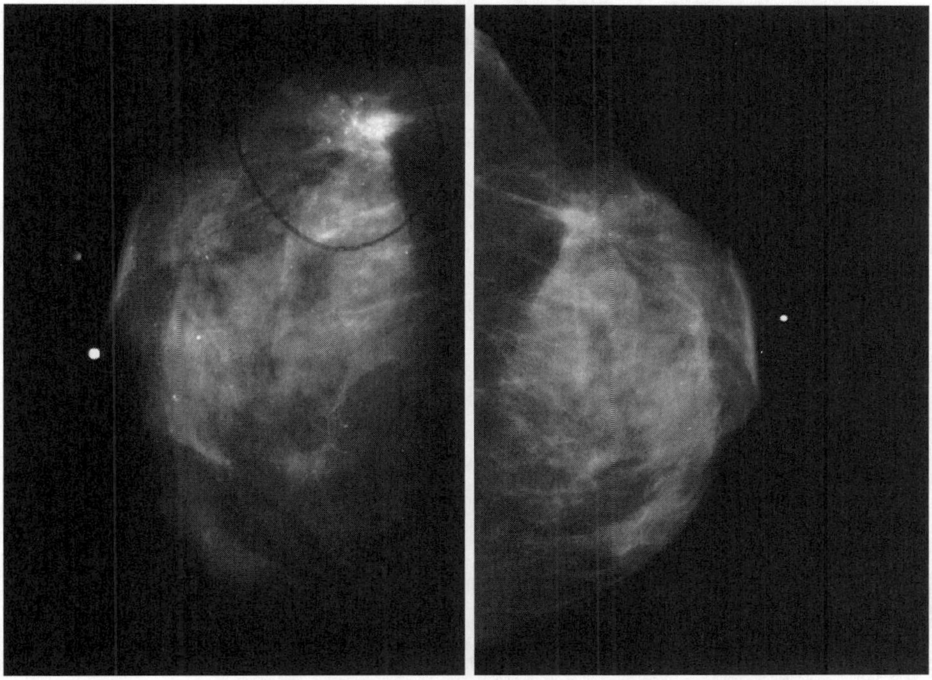

FIGURE 28.2 Analog (left) and digital cranial-caudal images of the right breast of a woman with moderately radiodense breast tissue and an irregularly-shaped mass with spiculated margins. The improved contrast, enhanced depiction of the suspicious mass and better visualization of peripheral tissue and skin line in the right image are a result of the larger dynamic range of the digital receptor.

As of this writing, two manufacturers have received approval from the Food and Drug Administration (FDA) to market full-field digital mammography (FFDM) systems (one is an x-ray phosphor/flat panel system, the other is a scanned CCD-based system), and at least two other manufacturers are in the process of gathering data for submission to the FDA for approval.

Challenges Facing Digital Mammography. Although detector development is well under way, there are major challenges that must be addressed before digital mammography becomes a clinical mainstay. Perhaps the most immediate obstacles have to do with image display. The very large size of the digital images (up to 30 megapixels, with 12 to 16 bits per pixel) far exceeds the matrix size (up to 2000 × 2500 pixels) and grayscale depth (typically 8 bits) of current commercially available high-resolution displays. Furthermore, a typical screening study involves simultaneous viewing of four current views (two per breast) alongside the corresponding four views from the previous screening study. Partially for this reason, many early practitioners of FFDM use laser film printers to produce hard copies of the digitally obtained mammograms, so that they can be viewed on a conventional light box, alongside the previous (analog) study.

Another area of ongoing research is the development of image processing techniques for both laser film and grayscale monitor viewing. In the former case (hard copy display), the characteristics of the laser film automatically apply a nonlinear dynamic range compression to the digital pixel values. In the case of workstation viewing (soft copy display), it is desirable to limit the amount of time that the radiologist must spend adjusting the image display (i.e., adjusting the range of pixel values mapped into the available grayscale levels of the display), so some means of reducing the dynamic range in the mammogram without compromising image quality is needed. One approach being tested is *thickness compensation* (Bick et al., 1996; Byng et al., 1997). This is a software technique that compensates for the decrease in attenuating thickness near the breast periphery by locating the region between the area of uniform thickness (where the breast is in contact with the flat compression paddle) and the skin line. Pixel values corresponding to that region are scaled downward to make their values more similar to those in the region of uniform thickness.

Diagnostic Ultrasound. Ultrasound (US) is used as a diagnostic (as opposed to screening) breast imaging modality, in part because it lacks the sensitivity of x-ray mammography for small masses and for microcalcifications. However, US is recommended by the American College of Radiology as the initial imaging technique for evaluation of masses in women under 30 and in lactating and pregnant women. In the majority of breast-imaging cases, the primary diagnostic use of ultrasound is the differentiation of solid masses from fluid-filled cysts. There has been a general lack of confidence, however, in the ability of US to characterize solid masses as benign versus malignant. Although at least one study has reported that benign and malignant solid breast masses could be differentiated based on US alone (Stavros et al., 1995), subsequent studies have not confirmed this hypothesis, and it is now generally believed that at the present time there are no ultrasound features that, by themselves, are sufficient evidence to forgo biopsy. Newer techniques being explored to improve the ability of US to differentiate benign and malignant masses include intensity histogram analysis (Kitaoka et al., 2001) and disparity processing, in which the sonographer slightly varies the pressure of the probe on the breast surface, and the apparent displacement of the tissue is measured by analysis of the correlation between images obtained at different parts of this compression cycle (Steinberg et al., 2001). This measurement of the elastic properties of the lesion is similar to that employed in breast elastography (briefly described below). In addition to these diagnostic tasks, US also plays a major role in biopsy guidance.

One technical issue affecting US is that its results tend to be more operator-dependent than the other modalities because of variations in positioning of handheld transducers. Automated transducers are much less operator-dependent than handheld transducers. However, handheld transducers permit a more rapid exam, and are better suited for biopsy guidance.

28.4 NEW AND DEVELOPING BREAST IMAGING MODALITIES

28.4.1 Introduction

While x-ray mammography is unquestionably the leading currently available modality for early detection of small cancers, it suffers from a relatively low positive predictive value (the fraction of lesions identified as positive that ultimately turn out to be positive). As a result, 65 to 85 percent of all breast biopsies are negative (Kerlikowske et al., 1993; Kopans, 1992). Therefore, adjunct modalities that can differentiate benign and malignant lesions detected by mammography are considered highly desirable. In a recent report issued by the National Research Council, the Committee on Technologies for the Early Detection of Breast Cancer identified the breast imaging technologies and their current status, as shown in Table 28.1 (*Mammography and Beyond*, 2001).

The x-ray imaging modalities have been discussed above. Below, we discuss some of the more advanced (i.e., they have been written up in the scientific literature, and have undergone at least preliminary clinical evaluation) and most promising adjunct modalities: breast MRI, scintimammography, breast PET, and electrical impedance imaging. We also briefly discuss optical imaging and elastography.

28.4.2 MRI

At present, mammography is the primary imaging modality used to detect early clinically occult breast cancer. Despite advances in mammographic technique, there are limitations in sensitivity and

TABLE 28.1 Breast-Imaging Technologies

Technology	Screening*	Diagnosis*	FDA approved?
Film-screen mammography	+++	+++	Yes
Full field digital mammography	++	++	Yes
Computer-assisted detection	++	0	Yes
Ultrasound (US)	+	+++	Yes
Novel US methods	0	0	No
Elastography (MR and US)	0	0	No
Magnetic resonance imaging (MRI)	+	++	Yes
Magnetic resonance spectroscopy (MRS)	−/0	+/0	No
Scintimammography	0	+	Yes
Positron emission tomography	0	0	Yes
Optical imaging	0	+	No
Optical spectroscopy	−	0	No
Thermography	0	+	Yes
Electrical potential measurements	0	+	No
Electrical impedance imaging	0	+	Yes
Electronic palpation	0	NA	No
Thermoacoustic computed tomography, microwave imaging, Hall effect imaging, magnetomammography	NA	NA	No

*Status description:

− Technology is not useful for the given application.

NA No data are available regarding use of the technology for given application.

0 Preclinical data are suggestive that the technology might be useful for breast cancer detection, but clinical data are absent or very sparse for the given application.

+ Clincial data suggest that the technology could play a role in breast cancer detection, but more study is needed to define a role in relation to existing technologies.

++ Data suggest that the technology could be useful in selected situations because it adds (or is equivalent) to existing technologies, but not currently recommended for routine use.

+++ The technology is routinely used to make clinical decisions for the given application.

Source: From *Mammography and Beyond* (2001).

specificity that remain. These limitations have stimulated exploration into alternative or adjunctive imaging techniques. MRI of the breast provides higher soft tissue contrast than conventional mammography. This provides the potential for improved lesion detection and characterization. Studies have already demonstrated the potential for breast MRI to distinguish benign from malignant breast lesions and to detect mammographically and clinically occult cancer (Dash et al., 1986; El Yousef et al., 1984; Stelling et al., 1985). The first studies using MRI to detect both benign and malignant breast lesions concluded it was not possible to detect and characterize lesions on the basis of signal intensities on T1 and T2 weighted images (Dash et al., 1986; El Yousef et al., 1984; Stelling et al., 1985). However, reports on the use of gadolinium enhanced breast MRI were more encouraging. Cancers enhance relative to other breast tissues following the administration of intravenous Gd-DTPA (Kaiser and Zeitler, 1989). Indeed, in one study, 20 percent of cancers were seen only after the administration of Gd-DTPA (Heywang et al., 1989). In addition, 2 studies reported on MRI detection of breast cancer not visible on mammography (Harms et al., 1993; Heywang et al., 1989). The detection of mammographically occult multifocal cancer in up to 30% of patients has led to the recommendations that MRI can be successfully used to stage patients who are potential candidates for breast conservation therapy (Harms et al., 1993).

However, the presence of contrast enhancement alone is not specific for distinguishing malignant from benign breast lesions. Benign lesions frequently enhance after Gd injection on MRI, including fibroadenomas, benign proliferative change, and inflammatory change. To improve specificity, some investigators recommend dynamic imaging of breast lesions to evaluate the kinetics of enhancement (Heywang et al., 1989; Kaiser and Zeitler, 1989; Stack et al., 1990). Using this technique, cancers have been found to enhance intensely, in the early phases of the contrast injection. Benign lesions enhance variably but in a more delayed fashion than malignant lesions.

Because x-ray mammography is limited by its relatively low specificity, MRI has been suggested as an adjunct imaging study to evaluate patients with abnormal mammograms. One application of breast MRI may be to reduce the number of benign biopsies performed as a result of screening and diagnostic mammography workup. To distinguish benign from malignant breast lesions using MRI scanning, most investigators rely on studying the time course of signal intensity changes of a breast lesion after contrast injection. However, among reported studies, the scan protocols varied widely and differ with respect to emphasis on information acquired. For example, Boetes et al. (Boetes et al., 1994) used a protocol consisting of a single-slice, nonfat, suppressed gradient echo imaging sequence with 2.6 × 1.3-mm in-plane spatial resolution (10-mm slice) at 2.3-s time intervals. They used the criterion that any lesion with visible enhancement in less than 11.5 s after arterial enhancement was considered suspicious for cancer. These criteria resulted in 95 percent sensitivity and 86 percent specificity for the diagnosis of cancer. Similar sequences have been reported by others using protocols with the time resolution varying from 6 to 60 s. However, problems locating a lesion on the precontrast images in order to perform a single-slice dynamic enhanced examination and the need to detect and evaluate other lesions within the breast, have resulted in recommendations that a multislice technique that captures dynamic data from the entire breast after injection of contrast is necessary (Gilles et al., 1994; Hickman et al., 1994; Perman et al., 1994). These investigators advocate using multislice two-dimensional (2D) gradient echo, 3D gradient echo, and echo planar techniques with the time resolution varying from 12 s to 1 min, varying special resolution and varying section thickness. Keyhole imaging techniques that dynamically sample the center of *k*-space after contrast administration have been suggested as a technique to obtain dynamic high-resolution 3D images of the entire breast (Van Vaals et al., 1993). However, the spatial resolution of enhanced tissue is limited with keyhole techniques because only part of the breast is sampled after contrast is injected. Keyhole imaging is criticized as being suboptimal for assessing lesion architecture.

Also widely varying among investigators are their criteria used for differentiating benign from malignant lesions. Criteria vary from simplistic models that report the percent of lesion enhancement at 2 min following contrast injection to more sophisticated physiologic models that take into account the initial T1 characteristics of a lesion and calculate Gd concentration as a function of time in order to extract pharmacokinetic parameters. Thus, wide variability in the accuracy cited by these investigators for differentiating benign from malignant lesions has been reported (66 to 93 percent) (Daniel et al., 1998; Esserman et al., 1999; Gilles et al., 1994; Hickman et al., 1994; Hylton, 1999; Orel et al., 1994; Perman et al., 1994; Van Vaals et al., 1993). Despite the many differing techniques, it is clear that there is a tendency for cancer to enhance more rapidly than benign lesions after bolus injection of Gd chelate. However, it is also clear that overlap exists in dynamic curves between cancerous and no cancerous lesions, resulting in false negative diagnosis in all reported series and false positive diagnosis in many.

Alternative approaches to characterizing enhancing lesions on breast MRI include extracting architectural features that describe breast lesions. The superior soft tissue contrast of MRI and use of higher spatial resolution techniques have prompted investigations in this area and the development of a potential lexicon that might be widely applicable to the reporting and interpretation of breast MRI scans (Gilles et al., 1994; Nunes et al., 1997; Orel et al., 1994; Schnall et al., 2001). Such an advancement would improve the widespread general reliability and comparability among breast MRI examinations performed from one institution to another. Clearly, the relative importance of spatial and temporal resolution in this regard requires further evaluation.

Other reported investigations of breast MRI suggest that MRI can demonstrate more extensive cancer than indicated by mammography or predicted by clinical breast examination. Several investigators have now demonstrated that MRI can detect breast cancer that is mammographically occult (Harms et al., 1993; Heywang et al., 1989), and suggest that MRI may have a role as a screening examination for patients with a high genetic predisposition to breast cancer and in those populations of women having extremely radiodense breast tissue on x-ray mammography.

28.4.3 Nuclear Imaging Methods

Nuclear imaging involves the injection of pharmaceutical compounds that have been labeled with radioisotopes. The compounds are selected such that they couple to some sort of biological process such as blood flow, metabolic activity, or enzyme production, or such that they tend to accumulate at specific locations in the body, e.g., binding to certain cell receptor sites. Thus the relative concentrations of these radiotracers in various areas of the body gives information about the relative degree to which these biological activities are occurring. Measurement of this concentration distribution therefore provides *functional* information very different from the *structural* information supplied by modalities such as x-ray mammography and ultrasound. For this reason, nuclear medicine techniques are being explored as adjunct imaging approaches to the structurally oriented x-ray mammography.

Nuclear medicine tracers being considered for breast imaging can conveniently be divided into two groups: those emitting single gamma rays and those emitting positrons. In both cases, the concentration distribution in the breast is mapped by one or more imaging gamma detectors placed in proximity to the patient. In the case of single gamma emitters, the gamma detectors are equipped with some type of physical collimators whose function is to create a unique correlation between a point on the detector surface and a line through the spatially distributed radioactivity source (e.g., the breast). Physical collimation is not necessary in the case of positron emitting isotopes because of the near colinearity of the gamma ray pair produced by annihilation of the emitted positron with a nearby electron. However, opposing gamma detectors and timing circuitry must be used to detect the two gamma rays of a pair in coincidence. Below we discuss some of the unique challenges presented by nuclear imaging of the breast, and describe some technical solutions currently being explored.

Scintimammography *Conventional scintimammography.* Breast scintigraphy (or scintimammography) is nuclear medicine imaging of the breast using a single gamma emitting tracer and an imaging gamma detector fitted with a collimator. Recent studies suggest that scintimammography can provide diagnostic information complementary to that of x-ray mammography (Allen et al., 2000; Buscombe et al., 2001; Palmedo et al., 1998; Scopinaro et al., 1997). Comparable performance has been reported between scintimammography and contrast-enhanced MRI as adjuncts to x-ray mammography (Imbriaco et al., 2001). Like contrast-enhanced MRI, scintimammography is particularly useful for women with radiodense breasts, for whom mammographic interpretation can be difficult.

Technical challenges associated with scintimammography are: (1) positioning the camera close to the breast, (2) dealing with the significant scatter radiation arising from gamma rays emitted from regions of the heart and liver, and (3) correcting for contrast degradation due to partial volume averaging and attenuation of gamma rays emitted from the lesion. The first of these issues is driven by the fact that the spatial resolution of cameras with parallel hole collimators is approximately a linear function of the distance between source and collimator. To date, scintimammography has been performed primarily using conventional gamma cameras, with the patient lying prone on a table with holes for the breasts. Prone positioning is favored over supine positioning because gravity acts to pull the breast tissue away from the chest wall. Typically two lateral views are obtained, and one anterior-posterior view to aid in medial-lateral tumor localization. The majority of clinical trials to date have employed one of two ^{99m}Tc-labeled pharmaceuticals; sestamibi or tetrafosmin. Reported values for sensitivity and specificity for planar scintimammography performed under these conditions vary according to several factors, a principle one being the distribution of lesion sizes represented in the particular study. In a three-center European trial, sensitivities of 26, 56, 95, and 97 percent were reported for category pT1a (<0.5 cm), pT1b (0.5 to 1.0 cm), pT1c (1.0 to 2.0 cm), and pT2 (>2 cm) cancers, respectively (Scopinaro et al., 1997). A recent clinical trial (134 women scheduled for open breast biopsy were enrolled) investigating the use of prone ^{99m}Tc-sestamibi scintimammography for pT1 tumors (4.7% pT1a, 46.7% pT1b, and 48.6% pT1c) reported sensitivity, positive predictive value, negative predictive value and accuracy of 81.3, 97.6, 55.6 and 83.6 percent, respectively (Lumachi et al., 2001). The corresponding values for x-ray mammography were 83.2, 89.9, 48.6, and 79.1 percent. A European multicenter trial evaluating palpable and nonpalpable breast lesions demonstrated an overall sensitivity and specificity of 80 and 73 percent, respectively (Bender et al., 1997). These and other early studies reveal that while scintimammography has excellent sensitivity for tumors larger than about 1 cm, sensitivity is generally poor for smaller, nonpalpable, or medially located lesions.

Dedicated Cameras. One reason for the low sensitivity of conventional scintimammography for small lesions is that it is difficult to position the conventional Anger cameras close to the breast. This is due to their large size, and their appreciable inactive borders. The result is that the lesion-to-collimator distance can often exceed 20 cm. At this distance, the spatial resolution of conventional gamma cameras with high-resolution parallel-hole collimators can be 15 to 20 mm. Thus counts originating from the lesion are smeared out over a large area of the image, and small lesions, providing few counts, are lost. One potential method for improving sensitivity is the development of smaller gamma cameras with narrow inactive borders, permitting camera placement adjacent to the chest wall, and near the breast. Several groups have developed such dedicated breast gamma cameras (Majewski et al., 2001; Pani et al., 1998). Early clinical evaluation of these dedicated systems has shown promising results (Scopinaro et al., 1999).

Positron Emission Mammography. In positron emission mammography (PEM), positron-emitting radiotracers are utilized, rather than single-gamma-ray emitters. Radiotracer location within the breast is determined by detection of the pair of simultaneously emitted 511-keV gamma rays resulting when the positron annihilates with an electron in the breast. Timing coincidence circuitry is used to identify gamma rays originating from a single annihilation event. To date, breast PET has been based primarily on assessment of either glucose metabolic rate via 2-[^{18}F]fluoro-2-deoxyglucose (FDG) or of estrogen or progestin receptor density using 16α-[^{18}F]fluoroestradiol (FES). In a prospective study of 144 patients (185 breast lesions), FDG-PET showed a sensitivity of 68.2 and 91.9 percent for pT1 and pT2 tumors, respectively. The overall specificity was only 75.5 percent. However, the positive predictive value (fraction of cases positive on PET that were shown by histology to be malignant) was 96.6 percent. A recent Japanese study compared FDG PET and sestamibi SPECT in the detection of breast cancer and of axillary node metastases (Yutani et al., 2000). The study concluded that MIBI-SPECT is comparable to FDG-PET in detecting breast cancer (overall sensitivities of 76.3 and 78.9 percent, respectively) but that neither FDG-PET nor MIBI-SPECT is sufficiently sensitive to rule out axillary lymph node metastasis.

As in the case of single-gamma breast imaging, dedicated small-field-of-view gamma cameras are being developed (Doshi et al., 2000; Raylman et al., 2000; Thompson et al., 1994). The emphasis to date has been on two-head systems with planar, opposing detectors, rather than the typical ring structure used in whole-body PET scanners. Technical challenges for breast PET are similar to those cited above for single-gamma imaging. In addition, because scattered gamma rays and random coincidences (arising from two different annihilation events) add to the true coincidences to give the total number of counted coincidences, the counting rates encountered in PET are significantly higher than those in single-gamma breast imaging, placing more stringent requirements on the readout electronics.

28.4.4 Electrical Impedance Scanning

Electrical impedance scanning (EIS) relies on the differences in conductance and capacitance (dielectric constant) between healthy tissue and malignant tissue. Studies have reported both conductivity and capacitance values 10 to 50 times higher in malignant tissue than in the surrounding healthy tissue (Jossinet, 1996; Morimoto et al., 1993). A commercial EIS scanner called *T-scan*, developed by Siemens Medical, was approved by the FDA in 1999 to be used as an adjunct tool to mammography (Assenheimer et al., 2001). That device uses a patient-held metallic wand and a scanning probe that is placed against the skin of the breast to complete the electrical circuit. 1.0 to 2.5 volts ac is used. Conductive gel is used to improve conductivity between the skin of the breast and the scanning probe. The scanning probe is moved over the breast and its sensors (256 sensors in high-resolution mode and 64 sensors in normal-resolution mode) measure the current signal at the skin level.

In a study to determine the accuracy of EIS in the differentiation of benign and malignant breast lesions (MRI), 100 mammographically suspicious lesions were examined using EIS, and histology was acquired through either lesion biopsy or surgical excision. EIS correctly identified 50 of 62 malignant lesions (81 percent overall sensitivity), and 24 of 38 benign lesions (63 percent specificity). Negative predictive value and positive predictive value of 67 and 78 percent were observed, respectively. A three-dimensional electrical impedance scanner has also been constructed and tested clinically (Cherepenin et al., 2001). The primary technical challenges for EIS center around the reduction of the high false-positive rate due to artifacts produced in imaging superficial skin lesions, and resulting from poor contact or air bubbles (Malich et al., 2001).

28.4.5 Other Techniques

Techniques for using optical radiation for breast imaging (sometimes referred to as optical mammography) have been under development for more than a decade. The principle obstacles to both structural and functional optical breast imaging are the significant levels of attenuation and scattering of optical photons in biological tissue. Because of its relatively low attenuation in human tissue, near infared (NIR) radiation is the wavelength band most frequently employed in optical breast imaging. NIR breast imaging can be described as either endogenous or exogenous. The former relies on the IR absorption of hemoglobin and deoxyhemoglobin and on the differences in vasculature between normal and malignant tissues (angiogenesis). The latter utilizes injected NIR-excitable fluorescent dyes to increase contrast between tumors and surrounding healthy tissue.

Elastography is a technique whose goal is to characterize breast masses by measuring their elastic properties under compression. Studies of excised breast specimens have demonstrated that while fat tissue has an elastic modulus that is essentially independent of the strain level (the amount of compression), normal fibroglandular tissue has a modulus that increases by 1 to 2 orders of magnitude with increasing strain (Krouskop et al., 1998). Furthermore, carcinomas are stiffer than normal breast tissue at high strain level, with infiltrating ductal carcinomas being the stiffest type of carcinoma tested (Krouskop et al., 1998).

28.5 FUTURE DIRECTIONS—MULTIMODALITY IMAGING

There is a general consensus in the breast-imaging community that no single imaging modality is likely to be able to detect and classify early breast cancers, and that the most complete solution for diagnostic breast imaging is likely to be some combination of complementary modalities. However, again the unique properties of the breast create challenges for successfully merging the information. In particular, the mechanically pliant nature of the breast permit optimization of breast shape for the particular modality used (compressed for x-ray mammography, coil-shaped for breast MRI, pendulant for breast scintigraphy, etc.). The result is that multimodality image fusion is extremely difficult. One approach to overcoming this problem is to engineer systems permitting multimodality imaging of the breast in a single configuration. Toward this end, a system combining digital x-ray mammography and planar gamma emission scintigraphy has been developed (Williams et al., 2002). The system employs an upright mammography gantry that is fitted with both a CCD-based x-ray detector and a high-resolution gamma camera. The latter is mounted on a retractable arm. The digital mammogram is first obtained with the gamma camera retracted, and the breast under mild compression. Then, without removing the compression, the gamma camera is positioned immediately above the paddle and the scintigram is obtained. The x-ray and gamma ray images are then fused to obtain the dual modality image. Figure 28.3 shows an x-ray and gamma ray image, respectively, obtained on the system. The dual modality scanner is currently being evaluated for its ability to

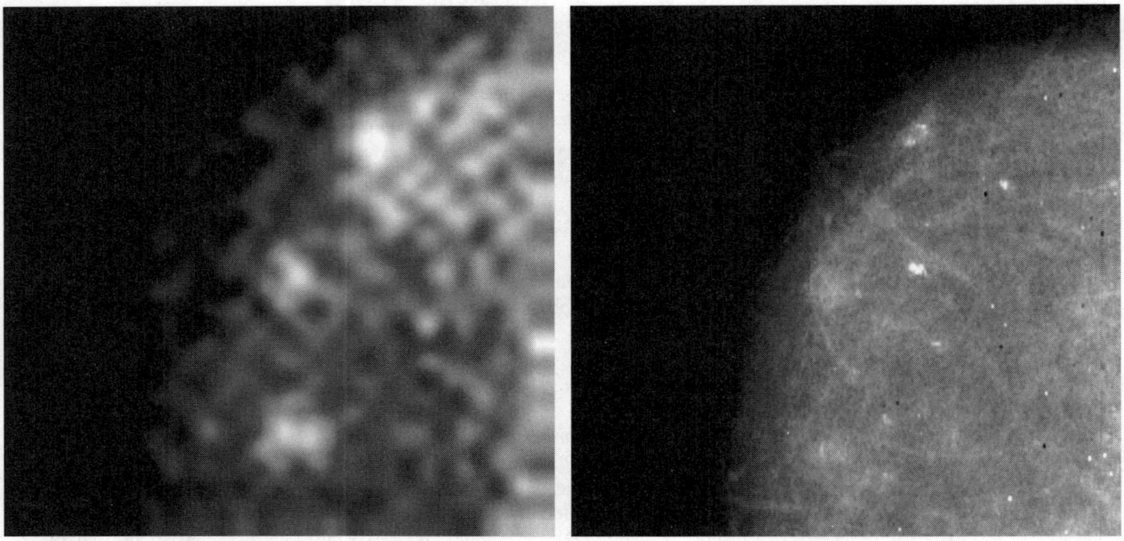

FIGURE 28.3 Scintigram (left) and digital mammogram (right) from the dual modality scanner at the University of Virginia. Note the correlation between the locations of the three areas of increased radiotracer uptake on the scintigram and the three areas of increased radiodensity on the mammogram.

distinguish benign and malignant lesions in patients scheduled for biopsy at the University of Virginia. The system is only one example of many other possible multimodality breast imaging approaches that are sure to be developed in the upcoming years.

REFERENCES

Allen, M. W., Hendi, P., Schwimmer, J., Bassett, L., and Gambhir, S. S.(2000). "Decision analysis for the cost effectiveness of sestamibi scintimammography in minimizing unnecessary biopsies," *Quarterly Journal of Nuclear Medicine*, **44**(2):168–185.

Assenheimer, M., Laver-Moskovitz, O., Malonek, D., Manor, D., Nahaliel, U., Nitzan, R., and Saad, A. (2001). "The T-SCAN technology: electrical impedance as a diagnostic tool for breast cancer detection," *Physiological Measurement*, **22**:1–8.

Bender, H., Kerst, J., Palmedo, H., Schomburg, A., Wagner, U., Ruhlmann, J., and Biersack, H. J. (1997). "Value of (18)fluoro-deoxyglucose positron emission tomography in the staging of recurrent breast carcinoma," *Anticancer Research*, **17**:1687–1692.

Bick, U., Giger, M., Schmidt, R., Nishikawa, R. M., and Doi, K. (1996). "Density correction of peripheral breast tissue on digital mammograms," *Radiographics*, **16**:1403–1411.

Boetes, C., Barentsz, J., Mus, R., et al. (1994). "MRI characterization of suspicious breast lesions with a gadolinium-enhanced turbo FLASH subtraction technique." *Radiology*, **193**:777–781.

Buscombe, J. R., Cwikla, J. B., Holloway, B., and Hilson, A. J. (2001). "Prediction of the usefulness of combined mammography and scintimammography in suspected primary breast cancer using ROC curves," *Journal of Nuclear Medicine*, **42**(1):3–8.

Byng, J., Critten, J., and Yaffe, M. (1997). "Thickness-equalization processing for mamographic images," *Radiology*, **203**:568.

Cherepenin, V., Karpov, A., Korjenevsky, A., Kornienko, V., Mazaletskaya, A., Mazourov, D., and Meister, D. (2001). "A 3D electrical impedance tomography (EIT) system for breast cancer detection," *Physiological Measurement* **22**:9–18.

Daniel, B., Yen, Y., Glover, G., Ikeda, D., Birdwell, R., Sawyer-Glover, A., Black, J., Plevritis, S., Jeffrey, S., and Herfkens, R. (1998). "Breast disease: dynamic spiral MR imaging," *Radiology*, **209**:499–509.

Dash, N., Lupetin, A., and Daffner, R. (1986). "Magnetic resonance imaging in the diagnosis of breast disease," *AJR*, **146**:119–125.

Doshi, N. K., Shao, Y., Silverman, R. W., and Cherry, S. R. (2000). "Design and evaluation of an LSO PET detector for breast cancer imaging," *Medical Physics*, **27** (7):1535–1543.

El Yousef, S., Duchesneau, R., and Alfidi, R. (1984). "Magnetic resonance imaging of the breast," *Radiology*, **150**:761–766.

Esserman, L., Hylton, N., George, T., and Weidner, N. (1999). "Contrast-enhanced magnetic resonance imaging to assess tumor histopathology and angiogenesis in breast carcinoma," *The Breast Journal*, **5**:13–21.

Gilles, R., Guinebretiere, J., Lucidarme, O., et al. (1994). "Nonpalpable breast tumors: diagnosis with contrast-enhanced subtraction dynamic MRI imaging," *Radiology*, **191**:625–631.

Harms, S., Flaming, D., Hesley, K. L., et al. (1993). "MRI imaging of the breast with rotating delivery of excitation off resonance," *Radiology*, **187**:493–501.

Heywang, S., Wolf, A., and Pruss, E. (1989). "MRI imaging of the breast with Gd-DTPA: use and limitations," *Radiology*, **171**:95–103.

Hickman, P., Moore, N., and Shepstone, B. (1994). "The indeterminate breast mass: assessment using contrast enhanced magnetic resonance imaging," *The British Journal of Radiology*, **67**:14–20.

Hylton, N. (1999). "Vascularity assessment of breast lesions with gadolinium-enhanced MR imaging," *MRI Clinics of North America*, **7**:411–420.

Imbriaco, M., Del Vecchio, S., Riccardi, A., Pace, L., Di Salle, F., Di Gennaro, F., Salvatore, M., and Sodano, A. (2001). "Scintimammography with 99mTc-MIBI versus dynamic MRI for non-invasive characterization of breast masses," *European Journal of Nuclear Medicine*, **28**(1):56–63.

Johns, P. C., and Yaffe, M. J. (1987). "X-ray characterization of normal and neoplastic breast tissue," *Phys. Med. Biol.*, **32**:675–695.

Jossinet, J. (1996). "Variability of impedivity in normal and pathological breast tissue." *Medical & Biological Engineering & Computing*, **34**:346–350.

Kaiser, W., and Zeitler, E. (1989). "MRI imaging of the breast: Fast imaging sequences with and without Gd-DTPA," *Radiology*, **170**:681–686.

Kerlikowske, K., Grady, D., Barclay, J., Sickles, E., Eaton, A., and Ernster, V. (1993). "Positive predictive value of screening mammography by age and family history of breast cancer," *JAMA*, **270**:2444–2450.

Kitaoka, F., Sakai, H., Kuroda, Y., Kurata, S., Nakayasu, K., Hongo, H., Iwata, T., and Kanematsu, T. (2001). "Internal echo histogram examination has a role in distinguishing malignant tumors from benign masses in the breast," *Clinical Imaging*, **25**:151–153.

Kopans, D. B. (1992). "The positive predictive value of mammography," *Am. J. Roentgenol.*, **158**:521–526.

Krouskop, T., Wheeler, T., Kallel, F., Garra, B., and Hall, T. (1998). "Elastic moduli of breast and prostate tissues under compression," *Ultrasonic Imaging*, **20**:260–274.

Lumachi, F., Ferretti, G., Povolato, M., Marzola, M. C., Zucchetta, P., Geatti, O., Bui, F., and Brandes, A. A. (2001). "Sestamibi scintimammography in pT1 breast cancer: alternative or complementary to X-ray mammography?" *Anticancer Research*, **21**:2201–2205.

Majewski, S., Kieper, D., Curran, E., Keppel, C., Kross, B., Palumbo, A., Popov, V., Weisenberger, A. G., Wojcik, R., Williams, M., Goode, A., More, M., and Zhang, G. (2001). "Optimization of dedicated scintimammography procedure using detector prototypes and compressible phantoms," *IEEE Trans. Nucl. Sci.* **48**(3):822–829.

Malich, A., Boehm, T., Facius, M., Freesmeyer, M. G., Fleck, M., Anderson, R., and Kaiser, W. A. (2001). "Differentiation of mammographically suspicious lesions: evaluation of breast ultrasound, MRI mammography and electrical impedance scanning as adjunctive technologies in breast cancer detection." *Clinical Radiology* **56**:278–283.

Mammography and Beyond: Developing Technologies for the Early Detection of Breast Cancer (2001). S. Nass, I. Henderson, and J. Lashof (eds.), National Academy Press, Washington, D.C.

Morimoto, T., Kimura, S., Konishi, Y., Komaki, K., Uyama, T., Monden, Y., Kinouchi, Y., and Iritani, T. (1993). "A study of the electrical bio-impedance of tumors," *Journal of Investigative Surgery*, **6**:25–32.

Nishikawa, R. M., Mawdsley, G. E., Fenster, A., and Yaffe, M. J. (1987). "Scanned-projection digital mammography," *Med. Phys.*, **14**:717–727.

Nishikawa, R. M. and Yaffe, M. J. (1985). "Signal-to-noise properties of mammographic film-screen systems," *Med. Phys.*, **12**:32–39.

Nunes, L., Schmidt, M., Orel, S., et al. (1997). "Breast MR imaging: interpretation model," *Radiology*, **202**: 833–841.

Orel, S., Schnall, M., LiVolsi, V., and Troupin, R. (1994). "Suspicious breast lesions: MR imaging with radiologic-pathologic correlation," *Radiology*, **190**:485–493.

Palmedo, H., Biersack, H. J., Lastoria, S., Maublant, J., Prats, E., Stegner, H. E., Bourgeois, P., Hustinx, R., Hilson, A. J. W., and Bischof-Delaloye, A. (1998). "Scintimammography with technetium-99m methoxyisobutylisonitrile: results of a prospective European multicentre trial," *European Journal of Nuclear Medicine*, **25**: 375–385.

Pani, R., De Vincentis, G., Scopinaro, F., Pellegrini, R., Soluri, A., Weinberg, I. N., Pergola, A., Scafe, A., and Trotta, G. (1998). "Dedicated gamma camera for single photon emission mammograpy (SPEM)," *IEEE Trans. Nucl. Sci.*, **45**:3127–3133.

Perman, W., Heiberg, E., Grunz, J., et al. (1994). "A fast 3D imaging technique for performing dynamic Gd-enhanced MRI of breast lesions," *Mag. Reson. Imag.*, **12**:545–551.

Raylman, R. R., Majewski, S., Wojcik, R., Weisenberger, A. G., Kross, B., Popov, V., and Bishop, H. A. (2000). "The potential role of positron emission mammography for detection of breast cancer. A phantom study," *Medical Physics*, **27**:1943–1954.

Schnall, M., Rosten, S., Englander, S., Orel, S., and Nunes, L. (2001). "A combined architectural and kinetic interpretation model for breast MR images," *Academic Radiology*, **8**:591–597.

Scopinaro, F., Schillaci, O., Ussof, W., Nordling, K., Capoferro, R., De Vincentis, G., Danieli, R., Ierardi, M., Picardi, V., Tavolaro, R., Colella, A. C. (1997). "A three center study on the diagnostic accuracy of 99m Tc-MIBI scintimammography," *Anticancer Research*, **17**:1631–1634.

Scopinaro, F., Pani, R., De Vincentis, G., Soluri, A., Pellegrini, R., and Porfiri, L. M. (1999). "High-resolution scintimammography improves the accuracy of technetium-99m methoxyisobutylisonitrile scintimammography: use of a new dedicated gamma camera," *European Journal of Nuclear Medicine*, **26**:1279–1288.

Stack, J., Redmond, A., Codd, M., et al. (1990). "Breast disease: tissue characterization with Gd-DTPA enhancement profiles," *Radiology*, **174**:491–494.

Stavros, A. T., Thickman, D., Rapp, C. L., Dennis, M. A., Parker, S. H., and Sisney, G. A. (1995). "Solid breast nodules: use of sonography to distinguish between benign and malignant lesions" (see comments), *Radiology*, **196**:123–134.

Steinberg, B. D., Carlson, D. L., and Birnbaum, J. A. (2001). "Sonographic discrimination between benign and malignant breast lesions with use of disparity processing," *Academic Radiology*, **8**:705–712.

Stelling, C., Wang, P., and Lieber, A. (1985). "Prototype coil for magnetic resonance imaging of the female breast," *Radiology*, **154**:457–463.

Thompson, C. J., Murthy, K., Weinberg, I. N., and Mako, F. (1994). "Feasibility study for positron emission mammography," *Medical Physics* **21**:529–538.

Van Vaals, J., Brummer, M., Dixon, W., Tuithof, H., Engels, H., Nelson, R. G. B., Chezmas, J., and den Boer, J. (1993). "Keyhole method for accelerating imaging of contrast agent uptake," *J. Mag. Reson. Imaging*, **3**:671.

Williams, M., More, M., Narayanan, D., Weisenberger, A. G., Wojcik, R., Stanton, M., Phillips, W., and Stewart, A. (2002). "Combined structural and functional Imaging of the breast," *Technology in Cancer Research and Treatment* **1**:39–42.

Yutani, K., Shiba, E., Kusuoka, H., Tatsumi, M., Uehara, T., Taguchi, T., Takai, S. I., and Nishimura, T. (2000). "Comparison of FDG-PET with MIBI-SPECT in the detection of breast cancer and axillary lymph node metastasis," *Journal of Computer Assisted Tomography*, **24**:274–280.

P · A · R · T · 6

ENGINEERING ASPECTS OF SURGERY

CHAPTER 29
COMPUTER-INTEGRATED SURGERY AND MEDICAL ROBOTICS

Russell Taylor
Department of Computer Science Center for Computer-Integrated Surgical Systems and Technology, The Johns Hopkins University, Baltimore, Maryland U.S.A.

Leo Joskowicz
School of Computer Science and Engineering Computer-Aided Surgery and Medical Image Processing Laboratory, The Hebrew University of Jerusalem, Israel

29.1 INTRODUCTION: COUPLING INFORMATION TO SURGICAL ACTION

The growing demand for complex and minimally invasive surgical interventions is driving the search for ways to use computer-based information technology as a link between the preoperative plan and the tools utilized by the surgeon. Computers, used in conjunction with advanced surgical assist devices, will fundamentally alter the way that procedures are carried out in twenty-first century operating rooms.

Computer-integrated surgery (CIS) systems make it possible to carry out surgical interventions that are more precise and less invasive than conventional procedures, while judiciously tracking and logging all relevant data. This data logging, coupled with appropriate tracking of patient outcomes, will make possible a totally new level of quantitative patient outcome assessment and treatment improvement analogous to "total quality management" in manufacturing.

The goals of CIS systems are to enhance the dexterity, visual feedback, and information integration of the surgeon. While medical equipment is currently available to assist the surgeons in specific tasks, it is the synergy between these capabilities that gives rise to a new paradigm. The goal is to complement and enhance surgeons' skills and always leave them in control, never to replace them.

CIS systems are instances of an emerging paradigm of human-computer cooperation to accomplish delicate and difficult tasks. In some cases, the surgeon will supervise a CIS system that carries out a specific treatment step such as inserting a needle or machining bone. In other cases, the CIS system will provide information to assist the surgeon's manual execution of a task, for example, through the use of computer graphic overlays on the surgeon's field of view. In some cases, these modes will be combined.

From an engineering systems perspective, the objective can be defined in terms of two interrelated concepts:

- *Surgical CAD/CAM systems* transform preoperative images and other information into models of individual patients, assist clinicians in developing an optimized interventional plan, register this preoperative data to the patient in the operating room, and then use a variety of appropriate means, such as robots and image overlay displays, to assist in the accurate execution of the planned interventions.
- *Surgical assistant systems* work interactively with surgeons to extend human capabilities in carrying out a variety of surgical tasks. They have many of the same components as surgical CAD/CAM systems, but the emphasis is on intraoperative decision support and skill enhancement, rather than careful preplanning and accurate execution.

Two other concepts related to CIS are *surgical total information management* (STIM) and *surgical total quality management* (STQM), which are analogous to total information management and total quality management in manufacturing enterprises.

Table 29.1 summarizes some of the factors that must be considered in assessing the value of CIS systems with respect to their potential application. Although the main focus of this chapter is the technology of such systems, an appreciation of these factors is very important both in the development of practical systems and in assessing the relative importance of possible research topics.

The CIS paradigm started to emerge from research laboratories in the mid 1980s, with the introduction of the first commercial navigation and robotic systems in the mid-1990s. Since then, a few hundreds of CIS systems have been installed in hospitals and are in routine clinical use, and a few tens of thousands of patients have been treated with CIS technology, with their number rapidly growing. The main clinical areas for which these systems have been developed are neurosurgery, orthopedics, radiation therapy, and laparoscopy. Preliminary evaluation and short-term clinical studies indicate improved planning and execution precision, which results in a reduction of complications and shorter hospital stays. However, some of these systems have in some cases a steep learning curve and longer intraoperative times than traditional procedures, indicating the need to carry out preoperative analysis and elaborate a surgical plan of action. This plan can range from simple tasks such as determining the access point of a biopsy needle, to complex gait simulations, implant stress analysis, or radiation dosage planning. Because the analysis and planning is specific to each surgical procedure and anatomy, preoperative planning and analysis software is usually customized to each clinical application. These systems can be viewed as medical CAD systems, which allow the user to manipulate and visualize medical images, models of anatomy, implants, and surgical tools, perform simulations, and elaborate plans. To give the reader an idea of the current scope of these systems, we will briefly describe two planning systems, one for orthopedics and one for radiation therapy.

In orthopedics, planning systems are generally used to select implants and find their optimal placement with respect to anatomy. For example, a planning system for spinal pedicle screw insertion shows the surgeon three orthogonal cross sections of the acquired CT image (the original *xy* slice and interpolated *xz* and *yz* slices) and a three-dimensional image of the vertebrae surfaces. The surgeon selects a screw type and its dimensions, and positions it with respect to the anatomy in the three cross-sectional views. A projection of the screw CAD model is superimposed on the images, and its position and orientation with respect to the viewing plane can be modified, with the result displayed in the other windows. Once a satisfactory placement has been obtained, the system stores it with the screw information for use in the operating room. Similar systems exist for total hip and total knee replacement, which, in addition, automatically generate in some cases machining plans

TABLE 29.1 Key advantages of CIS systems

Advantage	Important to whom	How quantify	Summary of key leverage
New treatment options	Clinical researchers Patients	Clinical and preclinical trials	Transcend human sensory-motor limits (e.g., in microsurgery). Enable less invasive procedures with real-time image feedback (e.g., fluoroscopic or MRI-guided liver or prostate therapy). Speed clinical research through greater consistency and data gathering.
Quality	Surgeons Patients	Clinician judgment; revision rates	Significantly improve the quality of surgical technique (e.g., in microvascular anastomosis), thus improving results and reducing the need for revision surgery
Time and cost	Surgeons Hospitals Insurers	Hours, hospital charges	Speed or time for some interventions. Reduce costs from healing time and revision surgery. Provide effective intervention to treat patient condition.
Less invasive	Surgeons Patients	Qualitative judgment; recovery times	Provide crucial information and feedback needed to reduce the invasiveness of surgical procedures, thus reducing infection risk, recovery times, and costs (e.g., percutaneous spine surgery).
Safety	Surgeons Patients	Complication and revision surgery rates	Reduce surgical complications and errors, again lowering costs, improving outcomes and shortening hospital stays (e.g., robotic total hip replacement, steady-hand brain surgery).
Real-time feedback	Surgeons	Qualitative assessment Quantitative comparison of plan to observation Revision surgery rates	Integrate preoperative models and intraoperative images to give surgeon timely and accurate information about the patient and intervention (e.g., fluoroscopic x-rays without surgeon exposure, percutaneous therapy in conventional MRI scanners). Assure that the planned intervention has in fact been accomplished.
Accuracy or precision	Surgeons	Quantitative comparison of plan to actual	Significantly improve the accuracy of therapy dose pattern delivery and tissue manipulation tasks (e.g., solid organ therapy, microsurgery, robotic bone machining).
Documentation and follow-up	Surgeons Clinical researchers	Databases, anatomical atlases, images, and clinical observations	CIS systems inherently have the ability to log more varied and detailed information about each surgical case than is practical in conventional manual surgery. Over time, this ability, coupled with CIS systems' consistency, has the potential to significantly improve surgical practice and shorten research trials.

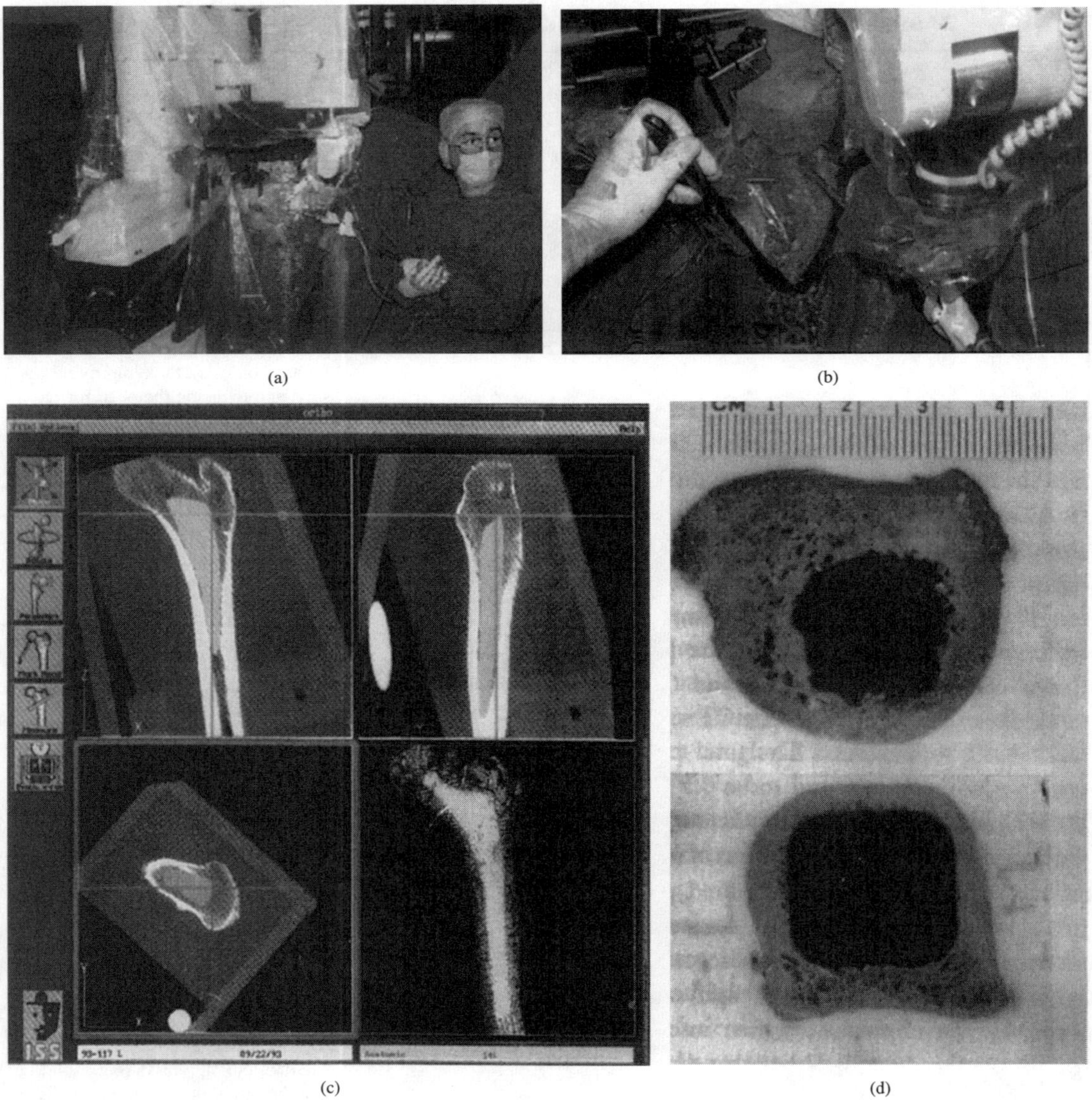

FIGURE 29.1 (*a, b*) ROBODOC system for orthopedic replacement surgery (photos by author). (*c*) Early version ROBODOC planning screen and (*d*) comparative cross sections showing conventionally broached (top) and robotically machined (bottom) cadaver femurs.[1,156,157]

(cut files) for intraoperative surgical robots. Other systems also extract kinematic or fine-element models and perform gait and stress analysis that help surgeons estimate the effectiveness of the proposed solution.

Another example of a complex planning system is in the field of radiation therapy. The goal of radiation therapy is to kill tumor cells by exposing them to a radiation beam while affecting as little as possible the surrounding healthy cells. One way of achieving this is to expose the tumor cells to radiation beams from different directions so that the cumulative radiation effect on the tumor cells destroys them while preserving the surrounding healthy cells. The planning task consists of identifying the tumor and the critical areas where no radiation should be present from MRI images, and then selecting the number of beams, their radius, intensity, duration, and placement that maximizes the radiation to the tumor cells while minimizing the radiation to other cells, especially to improve them.

To make our discussion more concrete, we briefly present two examples of deployed CIS systems: ROBODOC® (Integrated Surgical Systems, Davis, California) an active medical robotics system, and the StealthStation® (Medtronic Surgical Navigation Technology, Boulder, Colorado), an intraoperative navigation system used in neurosurgery and orthopedics.

The ROBODOC system[1–7] is an active medical robot developed clinically by Integrated Surgical Systems from a prototype developed at the IBM T. J. Watson Research Center in the late 1980s (Fig. 29.1). ROBODOC is a computer-integrated system for cementless primary total hip replacement. In primary total hip replacement procedures, a damaged joint connecting the hip and the femur is replaced by a metallic implant inserted into a canal broached in the femur. ROBODOC allows surgeons to plan preoperatively the procedure by selecting and positioning an implant with respect

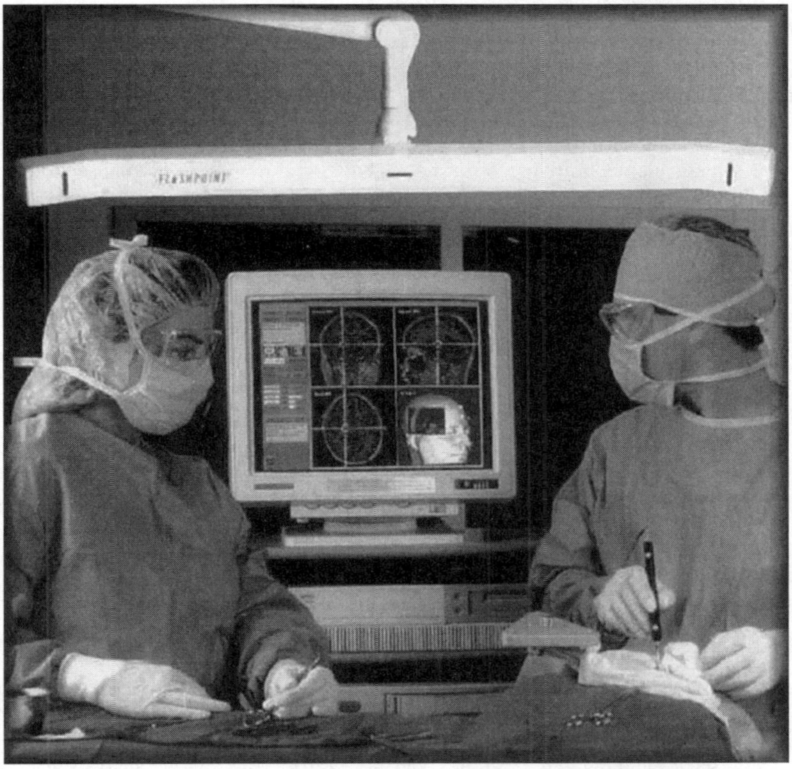

FIGURE 29.2 A CIS navigation system in action. The surgeon (left) is performing a brain tumor biopsy with the help of a navigation system. He is holding a pointer with light-emitting diodes whose position and orientation is precisely determined in real time by a stereo tracking camera (top). The computer display (center) shows the preoperative MRI and the current position of the pointer. The image shows three orthogonal cross sections and a three-dimensional reconstruction of the MRI data. The cross hair in each view shows the position of the pointer. The surgeon moves the pointer toward the desired position by watching the pointer location move on the screen.

to a computer tomography (CT) study and intraoperatively mill the corresponding canal in the femur with a high-speed tool controlled by a robotic arm. It consists of an interactive preoperative planning software, and an active robotic system for intraoperative execution. Preclinical testing showed an order-of-magnitude improvement in precision and repeatability in preparing the implant cavity. As of 2001, about 40 systems were in clinical use, having performed an estimated 8000 procedures, with very positive results documented in follow-up studies.

The StealthStation[8] is representative of current surgical navigation systems (Fig. 29.2). It allows surgeons to intraoperatively visualize the relative locations of surgical tools and anatomy in real time and perform surgical actions accordingly. The anatomical model used for navigation is constructed from preoperative CT or MRI data. The locations of instruments and rigid anatomy are obtained in real time by attaching to them frames with light-emitting diodes that are accurately tracked with a stereoscopic optical tracking camera. The preoperative model is registered to the intraoperative situation by touching with a tracked probe predefined landmarks or points on the anatomy surface and matching them to their corresponding location on the model. Intraoperative navigation allows for less invasive surgery and more precise localization without the need of repeated intraoperative x-ray or ultrasound two-dimensional imaging. For example, to perform a biopsy of a tumor on the brain, the surgeon directs the instrumented drill on the patient's skull with the help of the images, and drills directly toward the tumor instead of making an incision on the skull and visually looking for the tumor.

The key technical enabling factors that led the development of CIS systems were the increasing availability of powerful imaging modalities, such as CT, MRI, NMT, and live video, powerful computers with graphics capabilities, novel algorithms for model construction and navigation, and integrative systems and protocol development. This article reviews the main technical issues of CIS systems. It is organized as follows: The next section presents an overview of CIS systems, their main elements architecture, and information flow. The following section describes the main enabling technologies of CIS systems: imaging devices, image processing, visualization and modeling, preoperative analysis and planning, registration, tracking and sensing, robotics, human-machine interfaces, and systems integration technology. Then, we describe in detail examples of CIS systems including navigation systems, augmented reality navigation systems, and virtual reality systems. We conclude with perspectives and possible directions for future development.

29.2 AN OVERVIEW OF CIS SYSTEMS

Figure 29.3 shows a generic block diagram of a CIS system. At the core is a computer (or network of computers) running a variety of modeling and analysis processes, including image and sensor processing, creation and manipulation of patient-specific anatomical models, surgical planning, visualization, and monitoring and control of surgical processes. These processes receive information about the patient from medical imaging devices and may directly act on the patient through the use of specialized robots or other computer-controlled therapy devices. The processes also communicate with the surgeon through a variety of visualization modules, haptic devices, or other human-machine interfaces. The surgeon remains at all times in overall control of the procedure and, indeed, may do all of the actual manipulation of the patient using hand tools with information and decision support from the computer. The modeling and analysis processes within the computer will often rely upon databases of prior information, such as anatomical atlases, implant design data, or descriptions of common surgical tasks. The computer can also retain nearly all information developed during surgical planning and execution, and store it for postoperative analysis and comparison with long-term outcomes.

Essential elements of CIS systems are devices and techniques to provide the interfaces between the "virtual reality" of computer models and surgical plans to the "actual reality" of the operating room, patients, and surgeons. Broadly speaking, we identify three interrelated categories of interface technology: (1) imaging and sensory devices, (2) robotic devices and systems, and (3) human-machine interfaces. Research in these areas draws on a broad spectrum of core engineering research disciplines, such as materials science, mechanical engineering, control theory, and device physics.

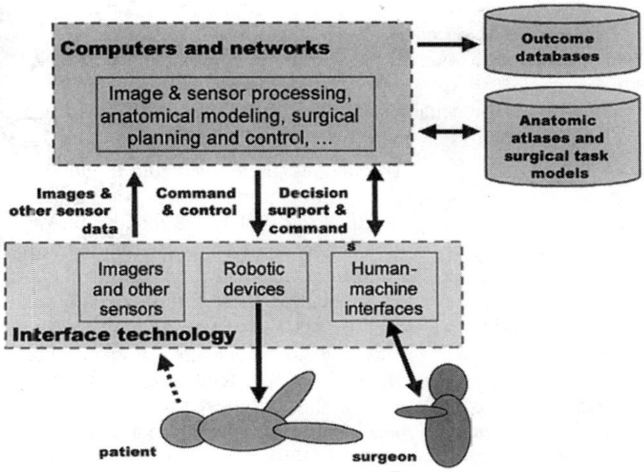

FIGURE 29.3 The architecture of CIS systems: elements and interfaces.

The fundamental challenge is to extend the sensory, motor, and human-adaptation abilities of computer-based systems in a demanding and constrained environment. Particular needs include compactness, precision, biocompatibility, imager compatibility, dexterity, sterility, and human factors.

Figure 29.4 illustrates the overall information flow of CIS systems from the surgical CAD/CAM paradigm perspective. The CIS systems combine preoperative and intraoperative modeling and planning with computer-assisted execution and assessment. The structure of the surgical assistant systems is similar, except that many more decisions are made intraoperatively, since preoperative models and plans may sometimes be relatively less important. Broadly speaking, surgery with a CIS system comprises three phases, all drawing upon a common technology base.

- *Preoperative phase.* A surgical plan is developed from a patient-specific model generated from preoperative images and a priori information about human anatomy contained in an anatomical

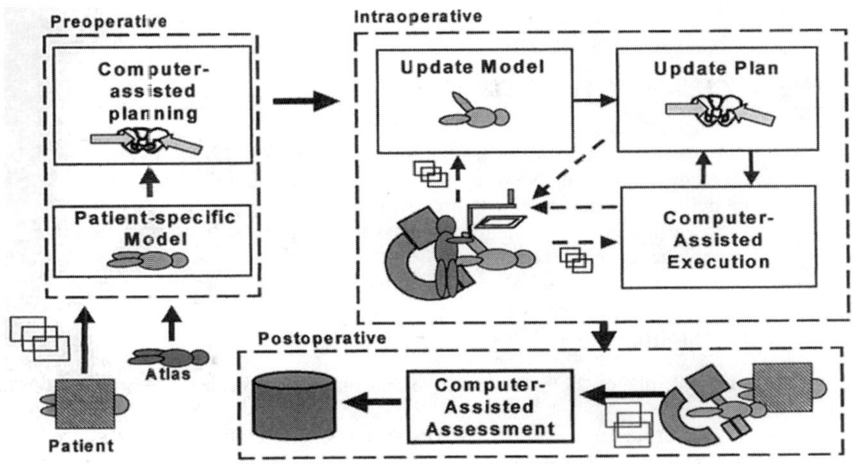

FIGURE 29.4 Major information flow in CIS systems.

atlas or database. Planning is highly application dependent since the surgical procedures are greatly different. In some cases, it may be simple interactive simulations or the selection of some key target positions, such as performing a tumor biopsy in neurosurgery. In other cases, such as in craneofacial surgery, planning can require sophisticated optimizations incorporating tissue characteristics, biomechanics, or other information contained in the atlas and adapted to the patient-specific model.

- *Intraoperative phase*. The images, patient-specific model, and plan information are brought into the operating room and registered to the patient, on the basis of information from a variety of sensors, such as a spatial tracking system and/or intraoperative imaging device. In some cases, the model and plan may be further updated, based on the images. The computer then uses a variety of interface devices to assist the surgeon in executing the surgical plan. Depending on what is most appropriate for the application these interfaces may include active devices such as robots, "smart" hand tools, and information displays. As the surgery proceeds, additional images or other measurements may be taken to assess progress and provide feedback for controlling tools and therapy delivery. On the basis of this feedback, the patient model may be updated during the procedure. This updated model may be used to refine or update the surgical plan to ensure that the desired goals are met. Ideally, intraoperative imaging and other feedback can ensure that the technical goals of the surgical intervention have been achieved before the patient leaves the operating room. Further, the computer can identify and record a complete record of pertinent information about the procedure without significant additional cost or overhead.
- *Postoperative phase*. The preoperative and intraoperative information are combined with additional images and tests, both to verify the technical results of the procedure and to assess the longer-term clinical results for the patient. Further, the results of many procedures may be registered back to an anatomical atlas to facilitate statistical studies relating surgical technique to clinical outcomes.

Note that the above description is of an idealized CIS system: specific systems do not necessarily require all these capabilities, and some of them are beyond the current state of the art. However, we will use this generic description to organize the technical discussion in the following section.

From a surgeon's perspective, the key difference between advanced medical equipment and a CIS system is the information integration, both between phases and within each phase. This new capability requires in most cases modifications to existing surgical protocols, and in a few cases radically new protocols. It could also enable more surgeons to perform certain difficult procedures that require much coordination and knowledge available to only a few experienced specialists, or perform procedures that are currently not feasible.

29.3 THE TECHNOLOGY OF CIS SYSTEMS

This section describes the main technical elements of CIS systems. We begin with a brief summary of medical imaging devices, and then present methods for image processing, visualization, and modeling. We describe next preoperative planning and analysis, followed by registration of data from various sources. Then we discuss tracking and sensing, robotics, man-machine interfaces, and systems integration technology.

29.3.1 Medical Imaging

Medical images, both preoperative and intraoperative, are the main source of information for CIS systems.

Since they are used in all CIS systems, we briefly discuss their technical characteristics and typical uses.

We distinguish between preoperative and intraoperative imaging devices. Preoperative imaging devices, such as film and digital x-rays, computed tomography (CT), magnetic resonance imaging (MRI), and nuclear magnetic tomography (NMT), in various forms, are used to obtain images for diagnosis and surgical planning. In most cases, the imaging devices are large and are located outside the surgical suite. Two-dimensional film x-ray images are the most common, with superb spatial resolution, gray-value range, and contrast, and negligible noise and geometric distortion. However, they are two-dimensional projections of spatial structures, and are not amenable to processing for further use unless scanned. CT and MRI images are used to visualize anatomical structures, with CT best suited for bony structures and MRI best suited for soft tissue. They consist of a series of two-dimensional parallel cross-sectional images with high spatial resolution, little geometric distortion and intensity bias, good signal-to-noise ratio, and a wide field of view. Typical data sets consist of 80 to 150 images of size 512 × 512 12-bit gray-level pixel images with pixel size of 0.4 × 0.4mm at 1- 2-mm intervals. They can be used to visualize anatomical structures, perform spatial measurements, and extract three-dimensional anatomical models. NMT images show functional anatomy, such as nerve activity, and are mostly used in the brain. They also consist of a series of two-dimensional parallel slices, although their quality is lower. They are usually viewed in conjunction with MRI images. The main drawback of preoperative images is that they are static and don't always reflect the position and orientation of anatomical structures which have moved between the time the images were taken and the surgery is performed.

Intraoperative imaging devices include fluoroscopic x-ray, ultrasound, and video image streams from endoscopes, laparoscopes, and surgical microscopes. Fluoroscopic x-ray is widely used in orthopedics to visualize and adjust the position of surgical instruments with respect to bones, or to locate kidney stones. The images are obtained from a mobile C-arm unit, which allows capturing two-dimensional projection images from different viewpoints while the patient lies on the table. The circular images are usually displayed on a video monitor. They have a narrow field of view (6 to 12 in, 400 pixels in diameter), limited spatial resolution and contrast, and present varying, position-dependent intensity and geometric distortions. They are mostly used for qualitative evaluation, and have cumulative radiation as a side effect. Ultrasound images (both static and as sequences) are used to obtain images of anatomy close to the skin. Unlike x-ray images, they have no ionizing radiation, but present significant imaging artifacts, such as speckling, noise, and spatial distortion. They also have a narrow field of view and have the resolution and image quality of the standard video monitor where they are displayed. Video image streams became commonplace with the introduction of minimally invasive surgery in the 1980s. They are used to support tumor biopsies, gall bladder removals, and colon explorations, among many others. They allow the surgeon to visualize in real time anatomy and surgical instruments inserted in a body cavity. The limitations of these imaging devices are that they have a narrow field of view (about 3 in), have no depth perception, uneven illumination, distortion due to the use of wide-angle lenses, and require direct line of sight. Surgeons must learn how to move and point the camera while respecting various point-of-entry and location constraints. The main advantage of intraoperative images is that they provide an up-to-date image of the surgical situation. However, the field of view and image quality are far inferior to preoperative images. More recent intraoperative imaging devices include surgical open MRI, surgical CT, and three-dimensional (3D) ultrasound, which overcome some of the limitations of the more common imaging devices.

The main limitation of current practice is that there is no quantitative correlation between high-quality preoperative images and intraoperative images. The surgeon must mentally establish the spatial correlation between the images and base decisions on this correlation.

29.3.2 Image Processing, Visualization, and Modeling

After image acquisition, the first task is usually visualization for diagnosis, evaluation, and planning. The visualization can take place on displays other than those of the devices where they were acquired, and can require various image processing techniques for better evaluation. These include image balancing and enhancement, distortion and contrast correction, denoising, and spatial aggregation.

For example, individual two-dimensional x-ray and ultrasound images can be processed using an array of standard image processing techniques to improve their clinical value. They can be visualized using zooming, cropping, and other imaging techniques. They can also be combined to create new, multimodal images.

Visualization of CT, MRI, and nuclear medicine images can greatly benefit from specialized visualization techniques, since they are series of two-dimensional cross sections. Instead of having the surgeon mentally correlate consecutive slices and create a mental three-dimensional view, it is desirable to directly reconstruct the three-dimensional information and show it as a new computed image. There are two families of visualization algorithms: volume visualization and surface visualization. We describe them briefly next.

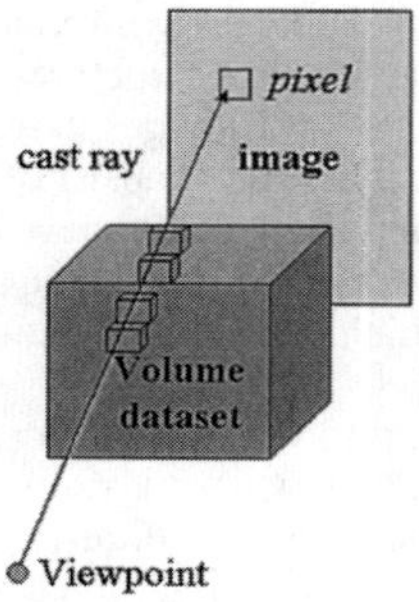

FIGURE 29.5 Volumetric rendering by ray casting. (*Adapted from [Ref. 9.]*)

Volume visualization algorithms[9] take as input slices and produce a three-dimensional image from any desired viewpoint. The most common method of generating the three-dimensional images is ray casting (Fig. 29.5). The data set is viewed as a volumetric data set, in which the space is divided into small volume units, called *voxels*. The voxels are rectangular blocks whose upper and lower faces are consecutive slice pixels in the vertical direction, and whose height is the slice interval distance. To each voxel is associated an intensity value, which is interpolated from the nearby pixel intensity values. To obtain the three-dimensional image, rays emanating from the viewpoint's location toward the image plane are cast on the volume. The pixel intensities in the new image are computed according to an attenuation function, which indicates how to compose the voxel intensity values that the ray traverses. Different choices of attenuation function produce various effects, such as opaque bodies, semitransparency, or anatomy isolation according to predefined intensity ranges. For example, if only bony surfaces are to be shown, only voxels whose intensity values are within the range of bone intensity are considered in the attenuation function. The advantage of this method is its simplicity, as no previous segmentation or surface extraction is necessary. However, it is computationally expensive, as hundreds of thousands of voxels need to be examined for each new image. Various hardware (Z buffering) and software techniques (precomputed views, ray arrays) have been developed to speed up the rendering process. Another disadvantage is that no model of the anatomy is created, restricting the types of analysis that can be performed on it. Volume visualization is best suited for complex anatomy with fine details, such as the brain gray matter.

Surface-based visualization algorithms rely on geometric surface models of the anatomy to be visualized. The inputs are usually objects described as triangular meshes extracted from the original data representing the surface of the anatomical structures of interest, such as the skull, femur, kidneys, and colon. The objects are then displayed as CAD models on viewers that can take advantage of standard graphics hardware. The main advantage of surface-based visualization is that it has to handle smaller data sets and is thus computationally much more efficient than volume visualization, allowing for near-real-time positioning and manipulation on standard computers. Another advantage is that CAD models of implants and surgical instruments can be readily incorporated into the image. However, surface-based visualization requires extracting the surface models, which can be difficult for complex anatomical structures with many fine details and complex geometry. Surface-based algorithms are best suited for anatomy with relatively large and clearly defined surfaces, such as bones and intestinal conduits.

Model construction algorithms are a prerequisite to surface-based visualization and for all tasks that require a geometric model of the anatomy: preoperative planning, contour-based registration, anatomical atlas construction, and matching. Their input is a series of slices, and a predefined intensity threshold interval that defines the image intensity ranges of the anatomy of interest. The output is one or more triangular meshes describing the geometry of the surfaces. Mesh extraction algorithms can be divided into two families: 2D contour extraction algorithms and 3D surface reconstruction algorithms. Contour extraction algorithms work by segmenting (manually or automatically) the contour of the anatomy of interest in each slice, and then connecting the resulting successive 2D contours to form a 3D surface. A point p_1 on the contour extracted in slice i is connected to the next point p_2 on the same contour at a predefined distance, and both are connected to the closest point p_3 in slice $i + 1$ to form a triangle $p_1p_2p_3$ which represents a surface element. By alternating between consecutive slices, a triangulated ribbon is created between the boundary contours. The drawback of this method is that ambiguities can arise as to how points should be selected to create triangles, resulting in topologically inconsistent surfaces (holes, self-intersections, etc).

To alleviate this problem, surface reconstruction algorithms work directly on the volumetric data to identify the voxels, which are intersected by the object surface and determine its geometry. The most commonly used algorithm in this category is the so-called marching cubes algorithm.[10] The algorithm proceeds as follows: A moving cube whose vertices are the pixels of two subsequent slices is formed. The eight vertices of the cube have associated with them a binary number (0 or 1), which indicates if the corresponding pixel intensity value is above or below a prespecified threshold (Fig. 29.6). When all eight vertices have a value of 0 (1), the voxel is entirely outside (inside) the anatomical object of interest. Cubes with mixed values (one or more 0 and 1) are at the boundary of the object. Depending on which vertex values are zero or one, one or more triangular surfaces cutting the cube can be constructed. There are $2^8 = 256$ cases, which can be reduced by symmetry to 14 cases and stored in a lookup table for reference. The algorithm proceeds by moving the cube from the topmost, upper corner of the first two slices to the lowest, bottom corner of the last two slices in sequence. Depending on the vertex values, the table is accessed and the corresponding triangles are constructed. The advantage of this algorithm is its locality, and that the surfaces constructed are topologically consistent (ambiguities in surface construction can be resolved locally). Variants of this algorithm include a tetrahedron instead of a cube, for which there are only two cases with no ambiguities, but which produce 2 to 5 times more triangles. The resulting meshes are typically several tens to several hundreds of thousands of triangles, depending on the slice spacing on the original data set. Mesh simplification algorithms can then be applied to the resulting models to reduce their complexity with minimal loss of accuracy.

While surface models are the most commonly used in CIS systems, they are by no means the only types of models. Functional models, containing relevant information specific to an anatomical structure or procedure can also be extracted with custom techniques. For example, a kinematic model of the leg bones and joints is of interest when planning a total knee replacement. To construct this

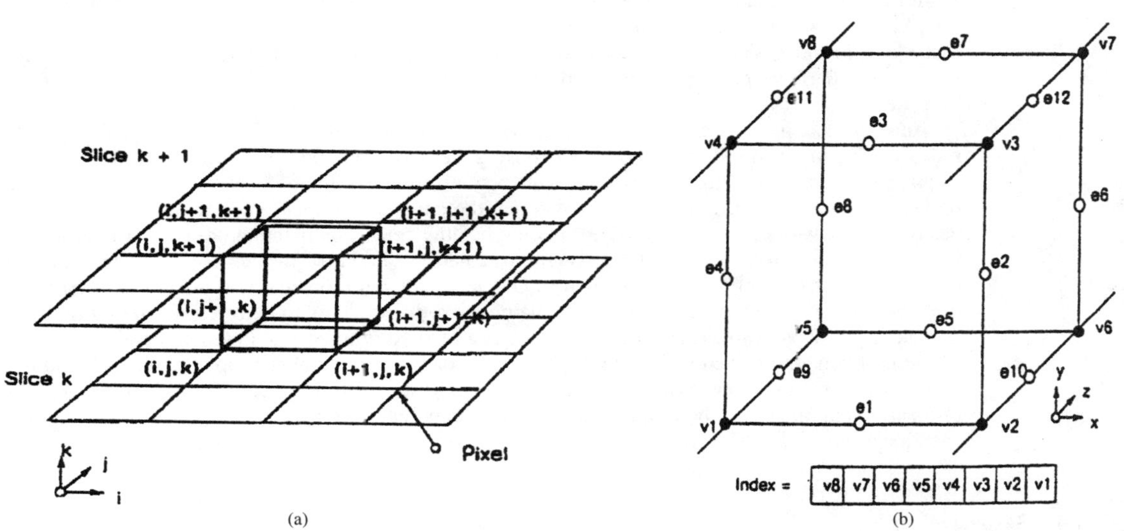

FIGURE 29.6 (*a*) Principle of the marching cubes algorithm: indexes for the marching cube and (*b*) coding scheme. (*Reproduced with permission from Ref. 10.*)

model, geometric entities such as the mechanical axis of the femur, the center of the femoral head, and other anatomical landmarks should be extracted. Each surgical application requires the construction of its model and the simulation associated with it.

Another type of model used in CIS is a digital atlas. Digital atlases are constructed from detailed imaging data of a person and are used for visualization, planning, and educational purposes. An example of this type of data is the Visible Human Project, which has detailed CT, MRI, and photographic data of a male and a female. The data are carefully segmented and labeled, and a database of organs is constructed from the data. The model can then be inspected, for example, using the VOXEL-MAN software,[11] or used to match to other patient data.

29.3.3 Preoperative Analysis and Planning

Once the diagnosis has been made and it has been decided that surgery is necessary, the next step is to carry preoperative analysis and elaborate a surgical plan of action. This plan can range from simple tasks such as determining the access point of a biopsy needle, to complex gait simulations, implant stress analysis, or radiation dosage planning. Because the analysis and planning is specific to each surgical procedure and anatomy, preoperative planning and analysis software is usually custom to each clinical application. These systems can be viewed as medical CAD systems, which allow the user to manipulate and visualize medical images, models of anatomy, implants, and surgical tools, perform simulations, and elaborate plans. To give the reader an idea of the current scope of these systems, we will briefly describe two planning systems, one for orthopedics and one for radiation therapy.

In orthopedics, planning systems are generally used to select implants and find their optimal placement with respect to anatomy. For example, a planning system for spinal pedicle screw insertion shows the surgeon three orthogonal cross sections of the acquired CT image (the original *xy* slice and interpolated *xz* and *yz* slices) and a three-dimensional image of the vertebrae surfaces. The surgeon selects a screw type and its dimensions, and positions it with respect to the anatomy in the three cross-sectional views. A projection of the screw CAD model is superimposed on the images, and its position and orientation with respect to the viewing plane can be modified, with the result displayed in the other windows. Once a satisfactory placement has been obtained, the system stores it with the screw information for use in the operating room. Similar systems exist for total hip and total knee replacement, which, in addition, automatically generate in some cases machining plans (cut files) for intraoperative surgical robots. Other systems also extract kinematic or finite-element models and perform gait and stress analysis that help surgeons estimate the effectiveness of the proposed solution.

Another example of a complex planning system is in the field of radiation therapy. The goal of radiation therapy is to kill tumor cells by exposing them to a radiation beam while affecting as little as possible the surrounding healthy cells. One way of achieving this is to expose the tumor cells to radiation beams from different directions so that the cumulative radiation effect on the tumor cells destroys them while preserving the surrounding healthy cells. The planning task consists of identifying the tumor and the critical areas where no radiation should be present from MRI images, and then selecting the number of beams, their radius, intensity, duration, and placement that maximizes the radiation to the tumor cells while minimizing the radiation to other cells, especially those in the critical areas. This problem is formulated as a geometric minimum-maximum constrained optimization problem, and solved with a combination of geometric and nonlinear optimization techniques. The planning system includes a data visualization and volume definition module, and outputs a series of location commands to the robotic arm carrying the radiation source, and the beam information at each location.

29.3.4 Registration

Multimodal registration is one of the key steps for information integration in CIS systems. The goal of the registration process is to allow the combination of data from several modalities, possibly taken at different times, so that they can be viewed and analyzed jointly. Registering two data sets consists of finding a transformation that aligns common features in two modalities, so that their spatial locations coincide. Registration is necessary for many tasks such as:

- Combine information of the same patient taken with different modalities, such as CT and MRI or MRI and PET
- Combine information of the same patient before, during, and after surgery, such as preoperative CT and intraoperative x-ray fluoroscopy, preoperative MRI and intraoperative video from a microscope or an endoscope, or CT and x-rays from before and after surgery
- Create real-time virtual reality views of moving anatomy and surgical tools by matching preoperative models from CFT or MRI with intraoperative tracking data
- Perform a statistical study of patient data

Most CIS applications require more than one transformation to link two data sets, and thus have more than one registration problem. For example, in the ROBODOC system, the preoperative plan has to be registered to the intraoperative position of the bone so that the robot tip can machine the desired canal shape in the planed position. To obtain this transformation, we must compute the transformation from the bone coordinate system to the implanted fiducials, then from the fiducials to the robot tip, to the robot coordinate system, and then to the cut volume. The series of mathematical transformations that align one data set with another is called the *registration chain*.

The registration task is in fact not one but many different problems. There are great differences on technical approaches depending on the type of data to be matched, the anatomy involved, and the clinical and technical requirements of the procedure. There is a vast body of literature on registration, which is comprehensively surveyed in Refs. 12 and 13 and can be classified according to the following characteristics:

- *Modalities*. Refers to the sources from which data is acquired, e.g., x-ray, CT, MRI, PET, video, tracker. The combinations can be unimodal (same data source) or multimodal (different data sources), which can be two images, an image to a model, or an image to a patient (tracker data).
- *Dimensionality*. Refers to the spatial and temporal dimensions of the two data sets to be matched (two or three dimensional, static or time varying). The registration dimensionality can be static 2D/2D (x-ray images), 2D/3D (ultrasound to MRI), 3D/3D (PET to MRI) or time varying, such as digital subtraction angiography (DSA).
- *Registration basis*. Refers to the image features that will be used to establish the alignment. These can be extrinsic registration objects, such as a stereotactic frame or fiducial markers, or intrinsic, such as anatomical landmarks, anatomical contours, or pixel intensity values.
- *Nature and domain of mathematical transformation*. Refers to the type of mathematical transformation that is used to perform the alignment. The transformation can be rigid, affine, projective, or generally curved (deformable registration), and can be applied to parts of the image (local) or to the entire image (global).
- *Solution method*. Refers to how the transformation is computed. This can include direct solutions when an analytic solution or an appropriate approximation is found, or iterative solutions, where there is a search and numerical optimization methods are used.
- *Type of interaction*. Refers to the type of input that the user has to supply. The registration is interactive when it is performed entirely by the user, automatic when no user intervention is required, or semiautomatic when the user supplies an initial estimate, helps in the data segmentation, or steers the algorithm by accepting or rejecting possible solutions.
- *Subject*. Refers to the patient source from which the images are taken; it can be the same patient (intrasubject), two different patients (intersubject), or a patient and an atlas.
- *Anatomy*. Refers to the anatomy being imaged. This can be the head (brain, skull, teeth, nasal cavities), the thorax (heart, breast, ribs), the abdomen (kidney, liver, intestines), the pelvis and the perineum, or the limbs (femur, tibia, humerus, hand).

TABLE 29.2 Basic Steps of a Registration Algorithm

Input: Two data sets to be matched, image acquisition parameters
Process each data set separately to correct for geometric and intensity distortion and to reduce noise.
While *there is dissimilarity* **and** *there is improvement* **do**:
Extract features or regions of interest from both images
Pair features from each data set
Formulate a similarity measure based on pairing
Find the transformation T that most reduces the dissimilarity
Transform one of the data sets by T
Output: Transformation T

The main steps of registration algorithms are summarized in Table 29.2. Before attempting to match the datasets, each data set should be corrected for distortions so that the errors resulting from imaging artifacts do not affect the accuracy of the registration process. Next, what should be matched is identified in each image. This can be point landmarks, contours, surfaces, pixel values and their gradients, or regions of interest. The pairwise correspondence between these is established so that a measure of similarity between the data sets can be established. The more the features are apart, the larger the dissimilarity is. The similarity is usually formulated as a constrained minimization problem whose minimum is the transformation T that reduces the dissimilarity the most. If no closed form solution exists, the local minimum is found by numerical optimization. One of the data sets is moved by the transformation, and the process is repeated until the match is sufficiently good or no further improvement is possible.

Technically, registration techniques can be classified as rigid or deformable, and geometry or intensity based. Rigid registration computes a transformation of position and orientation between two data sets. It is applicable to rigid structures that change their position but not their shape, such as bones, implanted fiducials, and stereotactic frames, as an approximation to quasi-rigid structures, such as tumors or brain white matter. It is also used as the first step of deformable registration, which computes a general global or local curved map. Deformable registration is necessary for matching soft tissue organs (e.g., brain images before and after brain shift) for time-dependent comparisons (e.g., tumor growth evaluation), and for cross-patient and atlas comparisons. The main difficulties of deformable registration are that the problem is ill posed, since there are usually infinitely many transformations that match the data, and error measurements and comparisons are difficult. The geometric approach uses the spatial disparity (usually the distance) between geometric entities, such as points, contours, or surfaces. The intensity-based approach uses the pixel intensity values and the intensity gradient between pixels to maximize the image correlation.

Examples of common registration tasks are:

- Rigid geometric registration between a surface model obtained from preoperative CT and intraoperative surface data on the same anatomy obtained by touching landmarks or collecting sample points with a tracker. This method is widely used in CIS orthopedic systems, such as pedicle screw fixation, total hip and knee replacement, and trauma.
- Deformable intensity-based registration between brain MRI data sets before and after brain shift.

29.3.5 Positional Tracking and Other Sensing

An important feature of many CIS systems is the ability to accurately determine in real time the location of selected anatomical structures, surgical instruments, and implants during surgery. This information is necessary for visualization, navigation, and guidance. The component that delivers this information to the CIS system is called a *tracker* or a *localizer*.

There are many technologies available for positional tracking, including encoded mechanical linkages, acoustic tracking, electromagnetic tracking, optical tracking using specialized devices, and optical tracking using conventional computer vision methods. Typically, these systems measure the motion relative to some base device of individual elements (which we will call *markers*) attached to the objects to be tracked. Several excellent surveys are available on this subject.[14,15] Each method has advantages and drawbacks. The main comparison parameters include setup requirements, work volume characteristics, number of objects that can be tracked simultaneously, the update frequency, the static and dynamic accuracy, the variability and repeatability of the readings, and cost.

Currently, the most commonly used position tracking approaches are based on specialized optical devices such at the Optotrak® and Polaris® systems (Northern Digital, Waterloo, Ontario) and Pixsys® and FlashPoint® systems (Image Guided Technologies, Boulder, Colorado). These devices use two or more optical cameras to identify light-emitting diodes or reflective markers in the camera image and compute their location by stereo triangulation. They can be quite accurate, providing 3D localization accuracies ranging from 0.1 to about 0.5 mm in typical applications. Their drawbacks include cost and the necessity of maintaining a clear line of sight between the sensors and the markers. Magnetic tracking systems such as the Polhemus® (Rockwell International, Milwaukee, Wisconsin), Flock-of-Birds® (Ascension Technology, Burlington, Vermont) and Aurora® (Northern Digital, Waterloo, Canada) systems are also widely used. These systems do not have line-of-sight constraints, but may be subject to field distortion from materials in the operating room.

Force sensors are commonly used in medical robotic systems to measure and monitor tool-to-tissue and tool-to-surgeon interaction forces.[16–21] Generally speaking, the technology used in these sensors is the same as that used in other applications, although specific issues of sterility and compactness often present unusual design strategies.

More broadly, a very wide variety of sensors may be used to determine any number of local tissue properties. Examples include electrical conductivity, optical coherence tomography, near-infrared sensing, and temperature sensing, to name a few.

29.3.6 Robotics

Medical robot systems have the same basic components as any other robot system: a controller, manipulators, end effectors, communications interfaces, etc. Many of the design challenges are familiar to anyone who has developed an industrial system. However, the unique demands of the surgical environment, together with the emphasis on cooperative execution of surgical tasks, rather than unattended automation, do create some unusual challenges. Table 29.3 compares the strengths and weaknesses of humans and robots in surgical applications.

TABLE 29.3 Complementary Strengths of Human Surgeons and Robots

	Strengths	Limitations
Humans	Excellent judgment. Excellent hand-eye coordination. Excellent dexterity (at natural "human" scale). Able to integrate and act on multiple information sources. Easily trained. Versatile and able to improvise.	Prone to fatigue and inattention. Tremor limits fine motion. Limited manipulation ability and dexterity outside natural scale. Bulky end effectors (hands) Limited geometric accuracy. Hard to keep sterile. Affected by radiation, infection.
Robots	Excellent geometric accuracy. Untiring and stable. Immune to ionizing radiation. Can be designed to operate at many different scales of motion and payload. Able to integrate multiple sources of numerical and sensor data.	Poor judgment. Hard to adapt to new situations. Limited dexterity. Limited hand-eye coordination. Limited ability to integrate and interpret complex information.

Safety is paramount in any surgical robot, and must be given careful attention at all phases of system design. Each element of the hardware and software should be subjected to rigorous validation at all phases, ranging from design through implementation and manufacturing to actual deployment in the operating room. Redundant sensing and consistency checks are essential for all safety-critical functions. Reliability experience gained with a particular design or component adapted from industrial applications is useful but not sufficient or even always particularly relevant, since designs must often be adapted for operating room conditions. It is important to guard against both the effects of electrical, electronic, or mechanical component failure and the more insidious effects of a perfectly functioning robot system correctly executing an improper motion command caused by improper registration between the computer's model of the patient and the actual patient. Further excellent discussion may be found in Refs. 22 and 23 and in a number of papers on specific systems.

Sterility is also a crucial concern. Usually, covering most of the robot with a sterile bag or drape and then separately sterilizing the instruments or end effectors provides sterility. Autoclaving, which is the most universal and popular sterilization method, can unfortunately be very destructive for electromechanical components, force sensors, and other components. Other common methods include gas (slow, but usually kindest to equipment) and soaking.

Manipulator design is very important in medical robots. Several early systems (e.g., Ref. 24) used essentially unmodified industrial robots. Although this is perhaps marginally acceptable in a research system that will simply position a guide and then be turned off before any contact is made with a patient, any use of an unmodified robot capable of high speeds is inherently suspect. Great care needs to be taken to protect both the patient and operating room personnel from runaway conditions. It is generally better to make several crucial modifications to any industrial robot that will be used in surgery. These include:

- Installation of redundant position sensing
- Changes in gear ratios to slow down maximum end-effector speed
- Thorough evaluation and possible redesign for electrical safety and sterility

Because the speed/work volume design points for industrial and surgical applications are very different, a more recent trend has emphasized design of custom manipulator kinematic structures for specific classes of applications.[25–30]

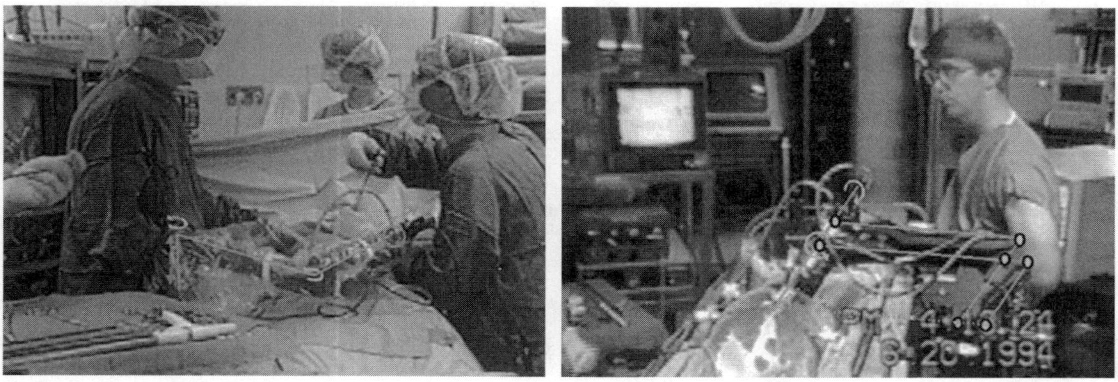

(a) (b)

FIGURE 29.7 Two robotic assistants for laparoscopic surgery. (*a*) The AESOP® system is a widely deployed commercial system for manipulating a laparoscopic camera. The robot combines active joints with a 2 degree of freedom wrist to achieve four controlled motions of the endoscope while preventing lateral forces from being exerted at the entry point into the patient's body. (*b*) The experimental IBM/JHU LARS system uses a five-bar linkage to decouple rotational and translational motions at the entry point. Both approaches have been used in a number of experimental and clinically deployed surgical robots. (*AESOP photo courtesy Computer Motion, Inc.*)

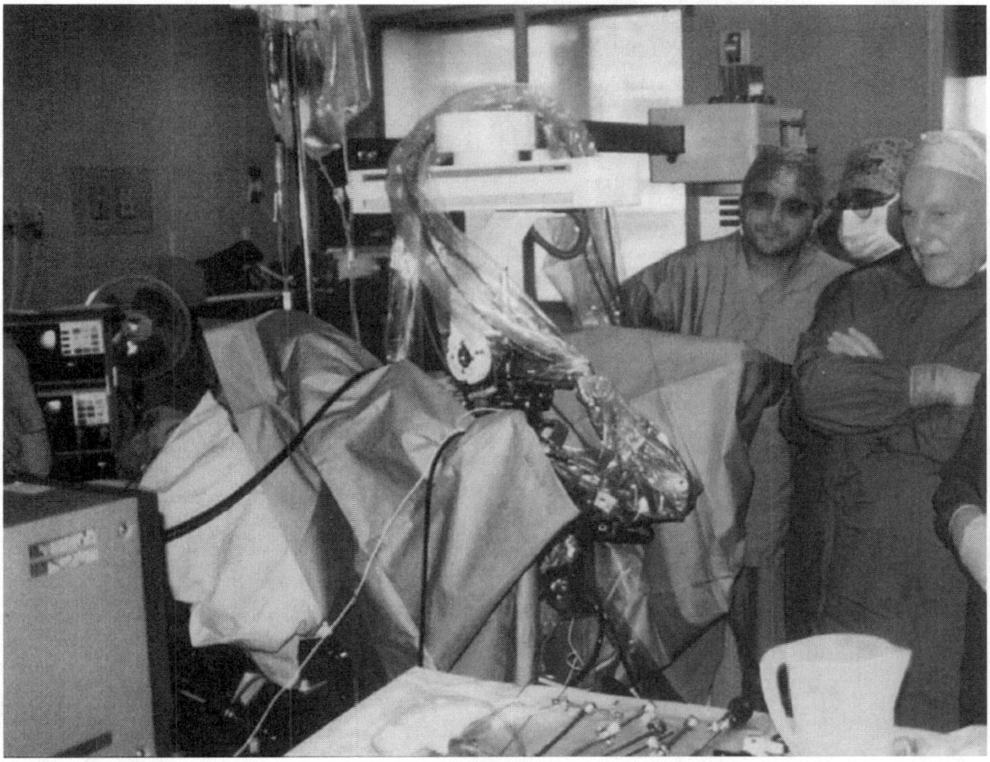

FIGURE 29.8 Robot system for transurethral prostate surgery.[25] This system uses goniometer arcs to provide conical motions about an apex point remote from the mechanism. (*Photo courtesy Brian Davies.*)

Many surgical applications (e.g., in laparoscopy or neuroendoscopy) require surgical instruments to pass through a narrow opening into the patient's body. This constraint has led a number of groups to consider two rather different approaches in designing robots for such applications. The first approach (e.g., Figs. 29.7*b*, 29.8, 29.9, and 29.10)[25,26,31,32] uses goniometers, chain drives, parallel five-bar linkages, or other means to decouple instrument motions about an "isocenter" which is placed at the entry portal. The second approach (e.g., Fig. 29.7*a*)[33–35] relies on passive compliance to cause the surgical instrument to comply with the entry portal constraint. In this case, the robot's "wrist" typically has two unactuated, but encoded, rotary axes proximal to the surgical instrument holder. Both approaches have merit, and they can be combined fruitfully.[36] The first approach is usually more precise and provides a more stable platform for stereotactic procedures. The second approach has the advantages of being simple and of automatically accommodating patient motions. A fuller discussion of the trade-off can be found in Ref. 36.

Surgical manipulators are not always active devices. Often, the human surgeon provides some or all of the motive power, while the computer provides real-time navigational or other assistance.[25,27,32,37–39]

Because medical robots are often used together with imaging, materials are also an important concern in surgical manipulator design equipment.[27,40] Figure 29.9 shows one example of a simple 1-degree-of-freedom radiolucent mechanism that can be used to drive needles into soft tissue.[27] This device is designed for use with fluoroscopic x-rays or CT scanners, and it can be employed either with a simple support clamp or as the end effector of an active robot. Fiducial geometry can be

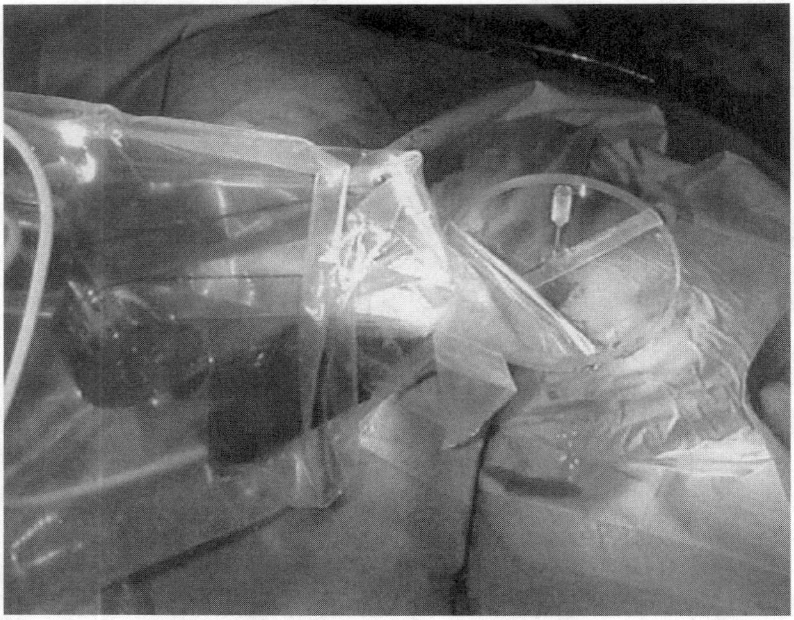

FIGURE 29.9 RCM robot with radiolucent PAKY needle driver used in percutaneous kidney stone removal.[127,158–162] (*Photo courtesy Dan Stoianovici.*)

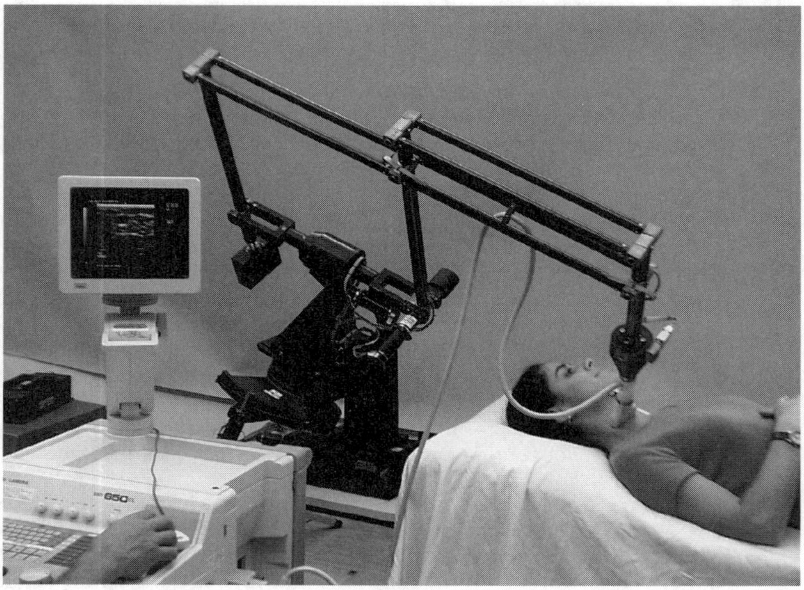

FIGURE 29.10 Robotic system for diagnostic ultrasound.[148] (*Photo courtesy S. Salcudean.*)

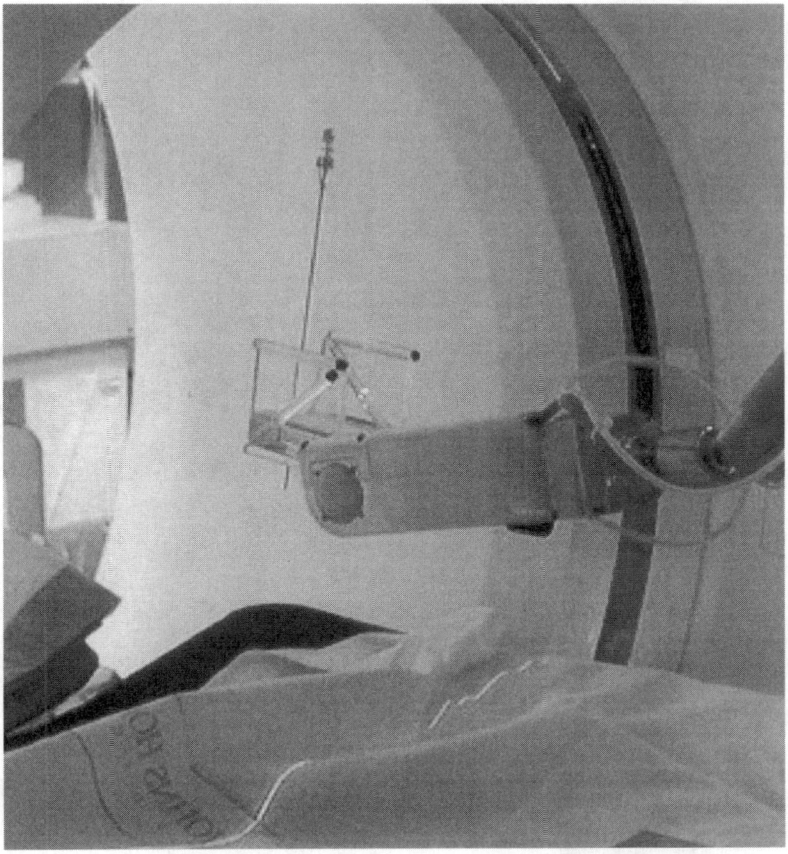

FIGURE 29.11 RCM robot with radiolucent CT-compatible needle driver.[42,44,45]

added easily to the robot or end effectors to assist in registration of the robot to the images (Fig. 29.11).[41–45]

Development of robotic devices for use with magnetic resonance imaging (MRI) poses special challenges, because of the strong magnetic fields and RF signals involved. Figures 29.12 and 29.13 show two typical systems.[40,46]

29.3.7 Human-Machine Interfaces

Computer-based systems that work cooperatively with humans must communicate with them, both to provide information and to receive commands and guidance. As with surgical robots, surgical human-machine interfaces (HMIs) have much in common with those for other application domains, and they draw upon essentially the same technologies (speech, computer vision and graphics, haptics, etc.) that have found use elsewhere. In many cases, HMI subsystems that have been developed for other uses may be adapted with little change for surgical use. However, attention must be given to the unusual requirements of surgical applications.[47] Surgeons tend to have very high expectations about system responsiveness and transparency but have very low tolerance for interfaces that impede

FIGURE 29.12 MRI-compatible robot for breast biopsy.[120] (*Photo courtesy Harald Fischer.*)

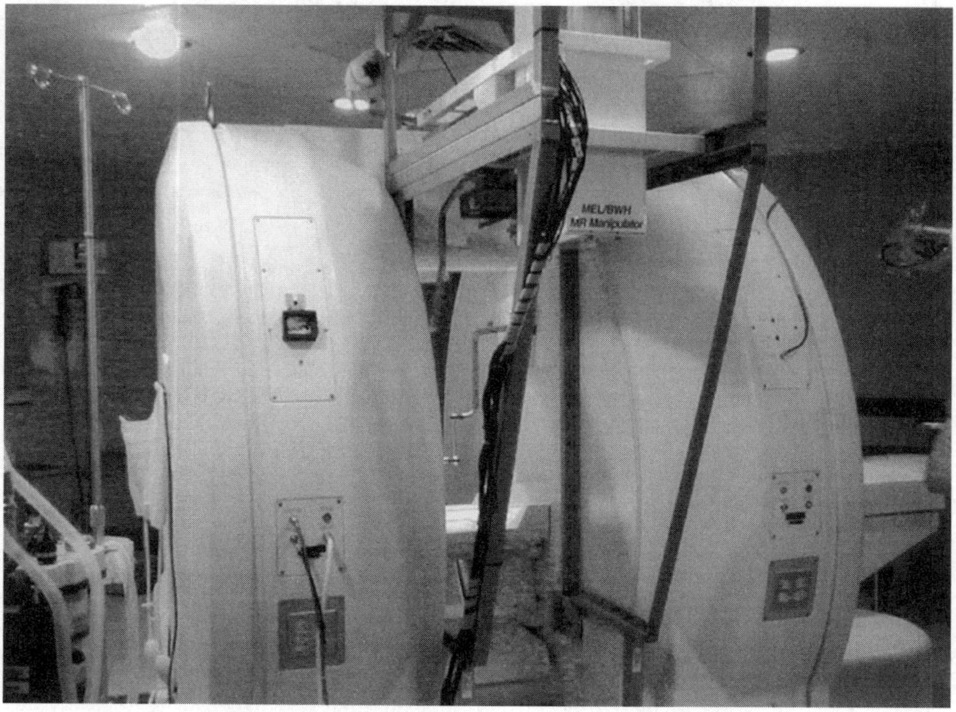

FIGURE 29.13 Robot system for use in open-magnet MRI system.[117]

their work. On the other hand, they can also be quite willing to put up with great inconvenience if the system is really performing a useful function that truly extends their capabilities.

Surgeons overwhelmingly rely on vision as their dominant source of feedback during surgery. Indeed, the explosion in minimal access surgery over the past decade has very largely been the result of the availability of compact, high-resolution video cameras attached to endoscopic optics. In these cases, the surgeon's attention is naturally focused on a television monitor. In such cases, it is often possible for the computer to add computer graphics, text, and other information to the video stream.[48,49] Similarly, surgical navigation systems [8,9,45–52] provide computer graphic renderings and feedback based on tracked surgical instrument positions and preoperative images. The so-called "virtual fluoroscopy" systems [58–61] show predicted x-ray projections based on intraoperative fluoroscopic images and tracked instrument positions. One very important challenge in the design of such systems is providing useful information about the *imprecision* of the system's information, so that the surgeon does not make decisions based on a false determination of the relative position of a surgical instrument and target anatomy. One common approach is to display a circle or ellipse representing likely registration uncertainty, but significant advances are needed both in the modeling of such errors and in the human factors associated with their presentation.

One limitation of video overlay systems is the limited resolution of current-generation video cameras. This is especially important in microsurgical applications, where the structures being operated on are very small, or in applications requiring very good color discrimination. Consequently, there is also interest in so-called optical overlay methods in which graphic information is projected into the optical path of a microscope[62] or presented on a half-silvered mirror[63,64] so that it appears to be superimposed on the surgeon's field of view in appropriate alignment. The design considerations for these systems are generally similar to those for systems using video displays, but the registration problems tend to be even more demanding and the brightness of the display can also be a problem.

All of the common interfaces (mice, joysticks, touch screens, push buttons, foot switches, etc.) used for interactive computer applications are used to provide input for surgical systems as well. For preoperative planning applications, these devices are identical to those used elsewhere. For intraoperative use, sterility, electrical safety, and ergonomic considerations may require some design modifications. For example, the LARS robot[48] repackaged the pointing device from an IBM Thinkpad® computer into a three-button "mouse" clipped onto the surgeon's instruments. As another example, a tracked stereotactic wand has been used to provide a configurable "push button" interface in which functions are selected by tapping the tip of the pointer onto a sterilized template.[65]

Surgeons routinely use voice to communicate with operating room personnel. Further, their hands (and feet) are frequently rather busy. Accordingly, there has long been interest in using voice as a two-way command and control system for surgical applications.[35,48,66–68]

Force and haptic feedback is often important for surgical simulation[68,69] and telesurgery applications.[19,21,70–73] Again, the technical issues involved are similar to those for other virtual reality and telerobotics applications, with the added requirement of maintaining sterility and electrical safety.

29.3.8 Systems

Computer-integrated surgery is highly systems oriented. Well-engineered systems are crucial both for use in the operating room and to provide context for the development of new capabilities. Safety, usability and maintainability, and interoperability are the most important considerations. We discuss them briefly next.

Safety is very important. Surgical system designs must be both safe, in the sense that system failures will not result in significant harm to the patient and the system will be perceived to be safe. Good system design typically will require careful analysis of potential failure sources and the likely consequences of failures. This analysis is application dependent, and it is important to remember that care must be taken to ensure that system component failures will not go undetected and that the system will remain under control at all times. Wherever possible, redundant hardware and software subsystems should be provided and cross-checked against each other. Rigorous software engineering practices must be maintained at all stages. Discussion of general safety issues for surgical robots may be found in Refs. 22 and 74–77. An excellent case study of what can happen when good practices are ignored may be found in Ref. 78, which discusses a series of accidents involving a radiation therapy machine.

Many discussions of safety in CIS systems tend to focus on the potential of active devices such as robots or radiation therapy machines to do great harm if they operate in an uncontrolled manner. This is a valid concern, but it should not be forgotten that such "runaway" situations are not usually the main safety challenge in CIS systems. For example, both robotic and navigation assistance systems rely on the accuracy of registration methods and the ability to detect and/or compensate for patient motion to ensure that the surgical instruments do not stray from the targeted anatomy. A human surgeon acting on incorrect information can place a screw into the spinal cord just as easily as a robot can. This means that analysis software and sensing must be analyzed just as carefully as motion control. Surgeons must be fully aware of the limitations as well as the capabilities of their systems and system design should include appropriate means for surgeons' "sanity checking" of surgical actions.

System usability and maintainability are also important design considerations. Clearly, the ergonomic design of the system from the surgeon's perspective is important.[79,80] However, the interfaces provided for the operating room staff that must set up the equipment, help operate it, and provide routine maintenance are also crucial both for safety and economic reasons. Similarly, CIS systems should include interfaces to help field engineers troubleshoot and service equipment. In this regard, the ability of computer-based systems to log data during use can be especially useful in postfailure analysis and in scheduling preventive maintenance, as well as in providing data for improvement in surgical outcomes and techniques. Although most systems make some use of such facilities, they are probably underused in present-day commercial systems.

System interoperability is currently a major challenge. Commonly accepted open standards permitting different equipment to work together in a variety of settings are badly needed. Several companies have proposed proprietary standards for use by alliances of vendors, and there has been some academic and government-supported work to provide tool kits, especially in software. However, these efforts are still very fragmented.

29.4 EXAMPLES OF CIS SYSTEMS

There are already a few dozen CIS systems available commercially or as prototypes in research laboratories worldwide. Although it is not practical to present an exhaustive survey, this section describes a few examples of integrated systems that use parts of the technology described above. For the purposes of this overview, we distinguish between four types of systems:

1. Information enhancement systems
2. Robotic systems for precise preoperative plan execution
3. Robotic systems for human augmentation
4. Other robotic systems

Note that many real systems could logically fit in several of these categories.

29.4.1 Information Enhancement Systems

The purpose of information enhancement systems is to provide the surgeon and surgical team with accurate, up-to-date, and useful data and images during the surgery so that they can best develop and update their plan of action and perform surgical actions. To achieve this goal, information enhancement systems usually combine information from different modalities, such as preoperative CT and MRI data, real-time tracking data of tools and anatomy, intraoperative images such as ultrasound and fluoroscopic x-ray images, video sequences from endoscopic cameras, and more. In some cases, such as virtual diagnostic endoscopy, a simulated environment replaces the actual procedure. Information enhancement systems are by far the most commonly used CIS systems. What distinguishes them from other CIS systems is that it is the surgeon who performs all surgical gestures without any physical assistance from mechanical devices.

We classify information enhancement systems into three categories:

1. Navigation systems
2. Augmented reality navigation systems
3. Virtual reality systems

We describe them briefly next.

Navigation Systems. The purpose of intraoperative navigation systems is to provide surgeons with up-to-date, real-time information about the location of surgical tools and selected anatomy during surgery. The goal is to improve the surgeon's hand/eye coordination and spatial perception, thereby improving the accuracy of the surgical gestures. They support less invasive procedures, can shorten surgery time, and can improve outcomes.

The basic elements of a navigation system are

1. A real-time tracking system to follow one or more moving objects (anatomy, surgical tools, or implants)
2. Tracking-enabled tools and reference frames
3. A display showing the intraoperative situation
4. A computer to integrate the information (Fig. 29.2)

Since the patient is usually not immobilized, a dynamic reference frame is attached to the anatomy to correct the relative position of the tools to the images.

What is displayed depends on the type of images that are available. The navigation systems can be based on:

- Preoperative images, such as CT or MRI augmented with CAD models of tools and implants.
- Intraoperative images, such as fluoroscopic x-ray, ultrasound, or open MR images augmented with projections of tool CAD models and implant axes.
- Intraoperative video streams from an endoscopic camera or a surgical microscope, shown alongside or fused with preoperative CT or MRI images.

Navigation systems based on preoperative CT or MRI images are typically used as follows: Shortly before surgery, a preoperative CT or MRI study of the anatomy of interest is acquired. In some cases, fiducial markers that will be used for registration are attached to the patient skin or implanted to the anatomy so that they appear in the images. The data are downloaded to a computer, and a model of the anatomy is created. When there are fiducials, they are identified and their precise relative spatial location is computed. The surgeon can visualize the data and elaborate the surgical plan. Before the surgery starts, the preoperative data, model, and plan are downloaded to the computer in the operating room. A dynamic reference frame is attached to the patient, and the intraoperative situation is registered with the preoperative data by either touching the fiducials with a tracked tool, or by acquiring a cloud of points on the surface of the anatomy. Once the registration has taken place, a display showing the preoperative images and model with the CAD models of the tools superimposed is created on the basis of the current tool and anatomy position obtained from the tracker (Fig. 29.14). Several commercial systems are currently available for a variety of procedures.

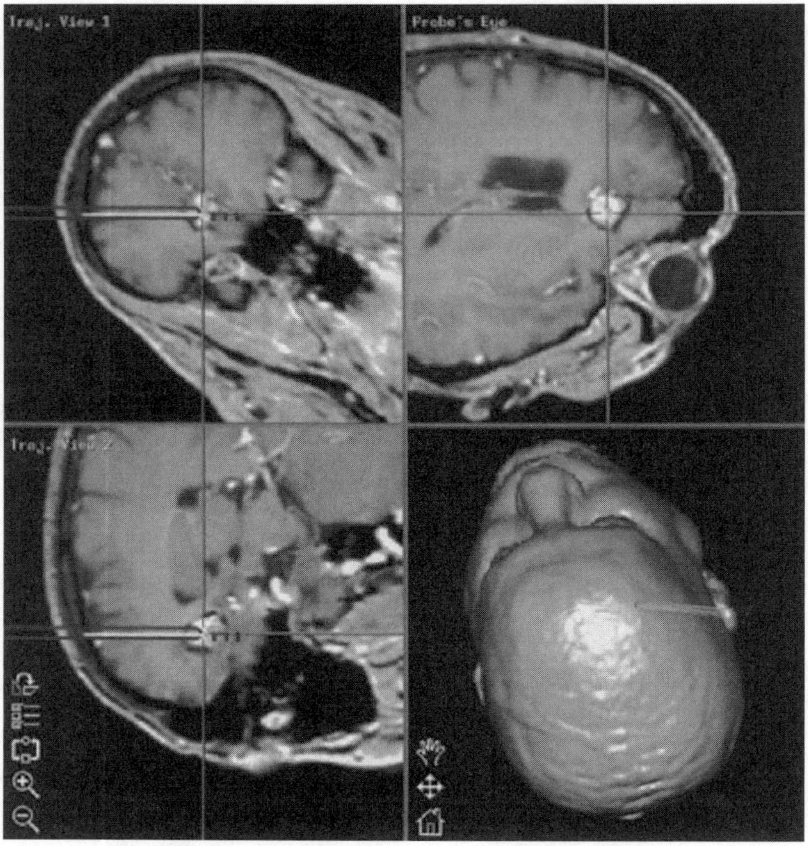

FIGURE 29.14 Typical screen display of a CIS navigation system in action. The top left, right, and bottom windows show orthogonal cross-sectional views of the MRI data set. The cross hair in each shows the position of the tool tip at the center of the tumor. The bottom right window shows the spatial view, with the tool shown in light gray. (*Photo courtesy of BrainLab.*)

Clinical studies report millimetric accuracy on tool and implant positioning. These types of systems have been applied extensively in orthopedics, (the spine,[58,81,82]) pelvis, [83,84] fractures,[85–89] hip,[56,90,91] and knee[57,92–94] neurosurgery, and craneofacial and maxillofacial surgery.

Navigation systems based on intraoperative images combine intraoperative images with position data from surgical tools and implants to create augmented intraoperative views. An example of such system is the FluoroNav system, which uses fluoroscopic x-ray images.[61] During surgery, a tracking device is attached to the fluoroscopic C arm, and one or more images are acquired with it. Projections of the tools are then superimposed on the original images and updated in real time as the tools move (Fig. 29.15). Since the camera and the tools are tracked simultaneously, there is no need for registration. The advantages of these systems are that they do not require a preoperative study and that no registration is necessary. However, the views remain two dimensional, requiring the surgeon to mentally recreate the spatial intraoperative situation. Recent clinical studies show that these systems are having excellent acceptance, since they are closest to current practice, and beginning to be used successfully.[95]

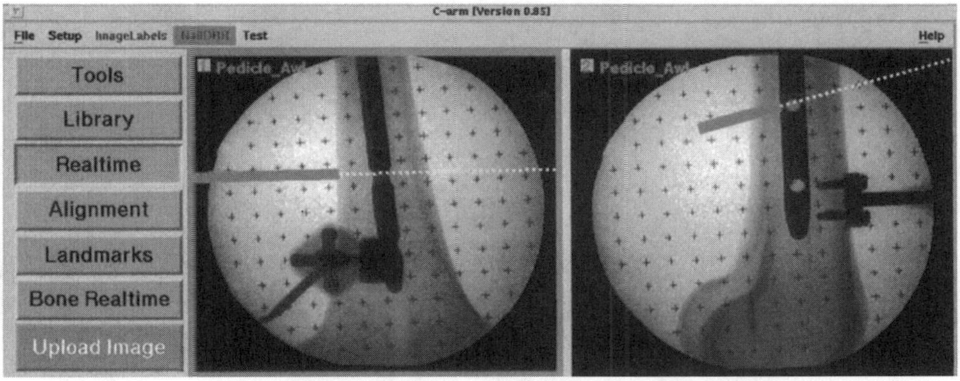

FIGURE 29.15 Screen display of a fluoroscopy-based navigation system during intramedullary nail distal locking. The windows show AP and lateral fluoroscopic x-ray images with the tool (solid line) and its extension (dotted line). The goal is to align the tool with the nail's hole axis.[59,61] (*Photo Courtesy of L. P. Nolte.*)

Other navigation systems combine video stream data obtained from endoscopic cameras or surgical microscopes, with data from preoperative studies, such as CT or MRI. The camera is tracked, so its position and orientation during surgery are known and can be shown, after registration, together with the preoperative images. The video stream can be shown side by side with a preoperative study,

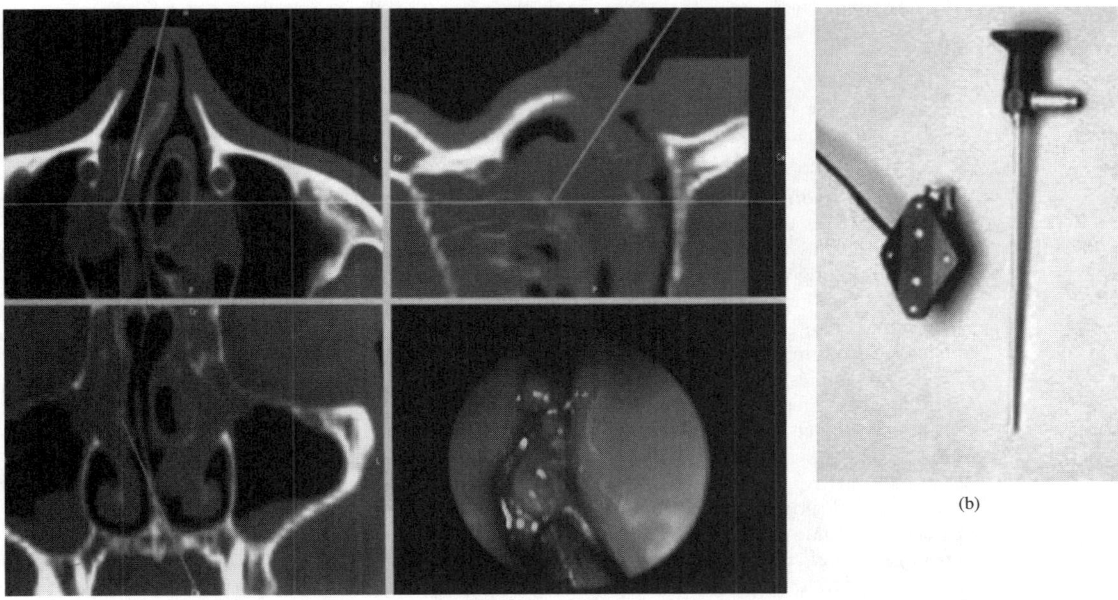

(a)

(b)

FIGURE 29.16 (*a*) Typical screen display of a CIS navigation system for endoscopic surgery. The left windows show orthogonal cross-sectional views of the CT data set and the endoscope image. The diagonal line (which is green in the original image) shows the position of the endoscope. (*b*) The endoscope and the tracking plate. (*Reproduced from Ref. 98.*)

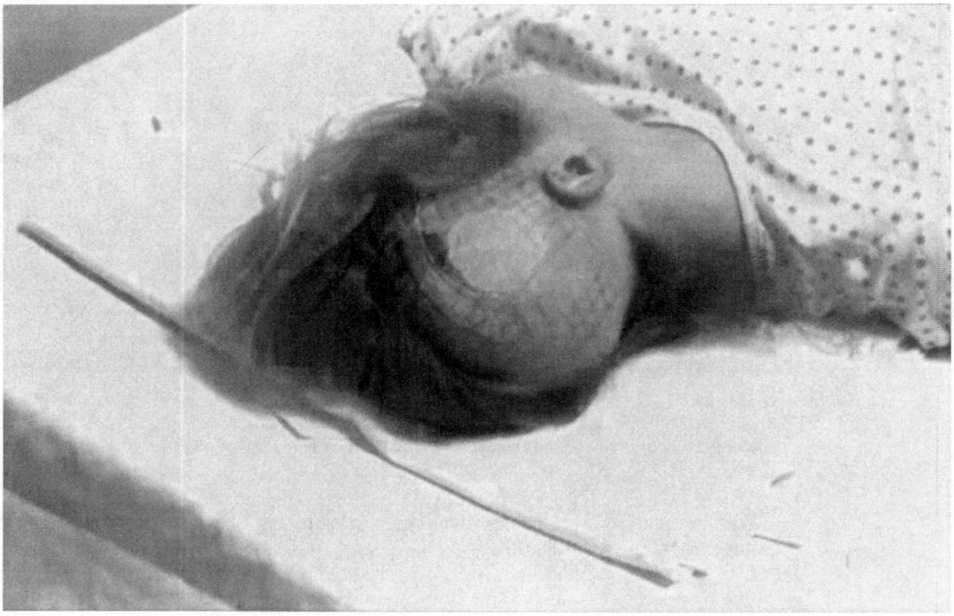

FIGURE 29.17 Augmented reality CIS system for brain tumor surgery. The tumor image (round area) and brain structure (below tumor image) are projected directly on the patient's skull. (*Photo courtesy of Ron Kikinis.*)

as in Fig. 29.16, or selected information from it can be inserted in the video stream. The main advantage of these systems is that they allow surgeons to see beyond the surfaces shown in the video and to obtain spatial location information that overcomes the narrow field of view of the cameras.[96]

Augmented Reality Navigation Systems. One of the drawbacks of the navigation systems described above is that they require the surgeon to constantly shift attention from the patient to the computer display and back. Augmented reality navigation systems attempt to overcome this drawback by bringing the display right where the surgeon needs it. The data is viewed through glasses worn by the surgeon, projected directly on the patient, or displayed on a transparent screen standing between the surgeon and the patient. The surgeon's head is usually tracked, so that the data can be displayed from the correct viewpoint.

Two examples of this type of system are the augmented reality CIS system for neurosurgery[97] and the CMU image overlay system[63] for orthopedics. The augmented reality CIS system for neurosurgery projects colored segmented volumetric data of brain tumors and other neural structures directly on the patient's skull (Fig. 29.17). This allows the surgeon to directly see where to start the minimally invasive procedure. The HipNav navigation system was developed to assist orthopedic surgeons in positioning the acetabular cup in total hip replacement surgery. In this system, a transparent glass that serves as a projection screen is placed between the surgeon and the patient on the operating table. After registration, the hip and pelvis models extracted from the preoperative CT data are projected on the glass screen, thereby providing the surgeon with an x-ray-like view of what lies beneath.

Virtual Reality Diagnosis Systems. The third type of information-enhancing system is virtual reality diagnosis systems. These systems, typically used in diagnostic endoscopy and colonoscopy, replace an actual exploration on the patient with a virtual exploration on MRI images. A three-

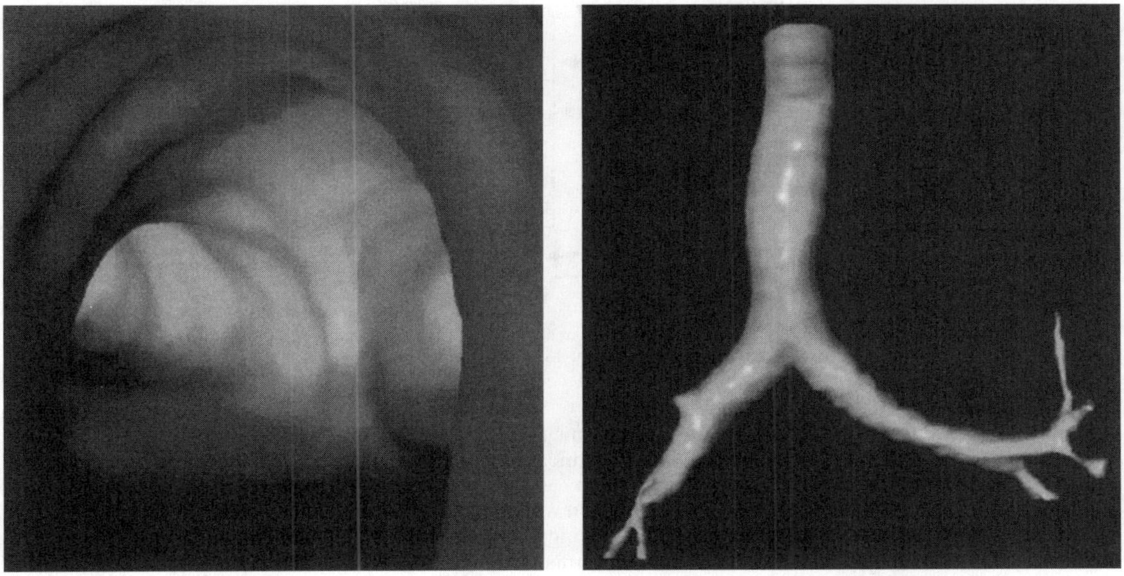

(a) (b)

FIGURE 29.18 Virtual endoscopy: (*a*) three-dimensional reconstruction of structure to be viewed and (*b*) view from the inside fly-through. (*Reproduced from Ref. 98.*)

dimensional reconstruction of the anatomical structures of interest, typically tubelike, is constructed from the data set, and a fly-through inspection path inside the structure is computed. The clinician is then presented with a virtual movie that simulates the actual endoscopic exploration (Fig. 29.18). On the basis of these images, the clinician can look for and identify certain pathologies, such as tumors, and then determine if an actual examination or surgery is necessary. Several algorithms have been developed for model construction, fast visualization, and computation of fly-through path.[98]

29.4.2 Robotic Systems for Precise Preoperative Plan Execution

One of the drawbacks of navigation systems is that they cannot guarantee that a planned surgical gesture, such as screw placement or needle insertion, will be executed precisely as planned. To ensure not only precise positioning but also precise execution, surgical robots have been developed. We describe next two examples of the most common types of active surgical robots: the ROBODOC system discussed earlier and the LARS robot for percutaneous therapy.

Robotic Orthopedic Surgery. Because bone is rigid and relatively easy to image in CT, and because geometric precision is often an important consideration in orthopedic surgical procedures, orthopedic surgery has been an important domain for the development of CIS systems. For example, the ROBODOC system has been in clinical use since 1992 and combines CT-based preoperative planning with robotic machining of bone. Both ROBODOC and a very similar subsequently introduced system called CASPAR®[99] have been applied to knee surgery,[100–102] as well as hip surgery. Other robotic systems have been proposed or (in a few cases) applied for hip or knee surgery.[38,39,103–107]

These applications fit naturally within the context of surgical CAD/CAM systems. For example, Fig. 29.19 shows the information flow for the current ROBODOC implementation. The information flow in the CASPAR system is very similar. CT images of the patient's bones are read into a planning

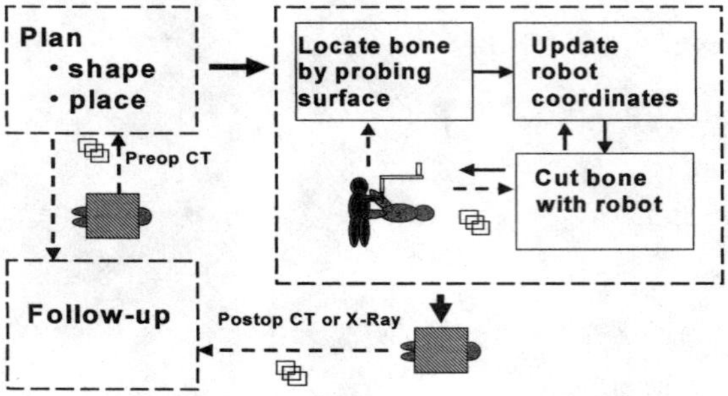

FIGURE 29.19 Information flow in a typical orthopedic robotic system (in this case, the ISS ROBODOC system).

workstation and a simple segmentation method is used to produce an accurate surface model of key anatomical areas. After some key anatomical measurements are made from the images, the surgeon selects an implant design from a library and determines its desired placement in the patient by manipulating a CAD model of the implant with respect to selected mutually orthogonal cross sections through the CT data volume. The planning workstation computes a cutter trajectory relative to CT coordinates and all of the planning information is written to a magnetic tape along with the patient images and model.

In the operating room, robotic hip replacement surgery proceeds much as manual surgery until after the head of the femur (for the case of primary hip surgery) or failing implant (for revision surgery) is removed. Then the femur is fixed to the base of the robot and a redundant position sensor is attached to the bone to detect any slipping of the bone relative to the fixation device. Then a 3D digitizer is used to locate a number of points on the bone surface. These points are used to compute the coordinate transformation between the robot and CT images used for planning and (thus) to the patient's bone. The surgeon then hand-guides the robot to an approximate initial position using a force sensor mounted between the robot's tool holder and the surgical cutter. The robot then cuts the desired shape while monitoring cutting forces, bone motion, and other safety sensors. The surgeon also monitors progress and can interrupt the robot at any time. If the procedure is paused for any reason, there are a number of error recovery procedures available to permit the procedure to be resumed or restarted at one of several defined checkpoints. Once the desired shape has been cut, surgery proceeds manually in the normal manner. The procedural flow for robotic knee replacement surgery is quite similar.

Robotically Assisted Percutaneous Therapy. One of the first uses of robots in surgery was positioning of needle guides in stereotactic neurosurgery.[24,108,109] This is a natural application, since the skull provides a rigid frame of reference. However, the potential application of localized therapy is much broader. Percutaneous therapy fits naturally within the broader paradigm of surgical CAD/CAM systems. The basic process involves planning a patient-specific therapy pattern, delivering the therapy through a series of percutaneous access steps, assessing what was done, and using this feedback to control therapy at several time scales. The ultimate goal of current research is to develop systems that execute this process with robotic assistance under a variety of widely available and deployable image modalities, including ultrasound, x-ray fluoroscopy, and conventional MRI and CT scanners.

Current work at Johns Hopkins University is typical of this activity. Our approach has emphasized the use of "remote center-of-motion" (RCM) manipulators to position needle guides under real-time image feedback. One early experimental system,[110,111] shown in Fig. 29.20, was used to establish the feasibility of inserting radiation therapy seeds into the liver under biplane x-ray guidance. In this work, small pellets were implanted preoperatively and located in CT images used to plan the pattern of therapy seeds. The fiducial pellets were relocated in the biplane x-rays and used to register the preoperative plan to a modified LARS robot[112,113] used to implant the treatment seeds. Although this

FIGURE 29.20 Early percutaneous therapy experiments at Johns Hopkins University using the LARS robot.[110,111]

experiment and related work directed at placing needles into the kidney[114,115] established the basic feasibility of our approach, we concluded that significant improvements in the robot would be needed.

Subsequent work has focused on development of a modular family of very compact component subsystems and end effectors that could be configured for use in a variety of imaging and surgical environments. Figure 29.9 shows a novel RCM linkage with a radiolucent needle driver ("PAKY") developed by Stoianovici et al. that forms a key component in this next generation system. Figure 27.11 shows the RCM device with a novel end-effector developed by Susil and Masamune that permits the computer to determine the needle pose to be computed with respect to a CT or MRI scanner using a single image slice.[42,44,45] This arrangement can have significant advantages in reducing setup costs and time for in-scanner procedures and also eliminates many sources of geometric error. Figure 29.21 shows another variation of the RCM used as a high dexterity wrist in a system designed for manipulating ultrasound probes for diagnosis and ultrasound-guided biopsies.[116]

Related work at Brigham and Women's Hospital in Boston is illustrated in Fig. 29.13. This system[117] is designed to operate in an open-magnet MRI system and uses a common control architecture developed jointly by MIT, Brigham and Women's Hospital, and Johns Hopkins.[118,119] One early application will be MRI-guided prostate therapy. Figure 29.12 shows another MRI-compatible robot system, this one designed for breast biopsy.[120]

29.4.3 Robotic Systems for Human Augmentation

The emphasis in surgical assistant robots in the use of these systems cooperatively to enhance human performance or efficiency in surgery. Much of the past and current work on surgical augmenta-

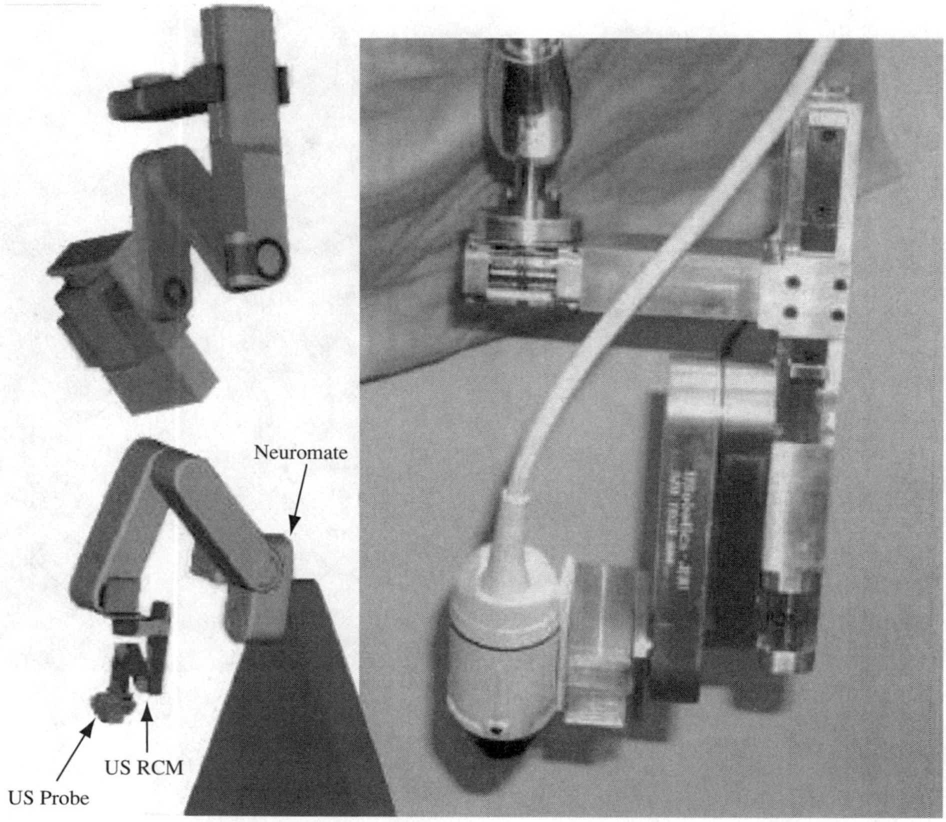

FIGURE 29.21 Dexterous RCM end effector for ultrasound and similar applications[116] mounted on an Integrated Surgical Systems Neuromate robot. (*Photo courtesy Randy Goldberg.*)

tion[67,70,73,121–125] has focused on teleoperation. There is considerable interest in the use of master-slave manipulator systems to improve the ergonomic aspects of laparoscopic surgery. Figure 29.22 shows a typical example (the Da Vinci® system[122] marketed by Intuitive Surgical). In this case, three slave robots are used. One holds an endoscopic camera and two others manipulate surgical instruments. In the case of the Da Vinci, the surgical instruments have high dexterity wrists, as shown in Fig. 29.22*b* Other systems with varying degrees of complexity (e.g., the Zeus® system marketed by Computer Motion) are also in use, and this area of application may be expected to grow in the future.

Although the primary impact of teleoperated robots in surgical applications over the next years will probably be in applications in which the surgeon remains close to the patient, there has also been considerable interest in remote telesurgery.[70,126,127] In addition to the design issues associated with local telesurgery, these systems must cope with the effects of communication delays and possible interruptions on overall performance.

The manipulation limitations imposed by human hand tremor and limited ability to feel and control very small forces, together with the limitations of operating microscopes have led a number of groups to investigate robotic augmentation of microsurgery. Several systems have been developed for teleoperated microsurgery using a passive input device for operator control. Guerrouad and Vidal[128] describe a system designed for ocular vitrectomy in which a mechanical manipulator was

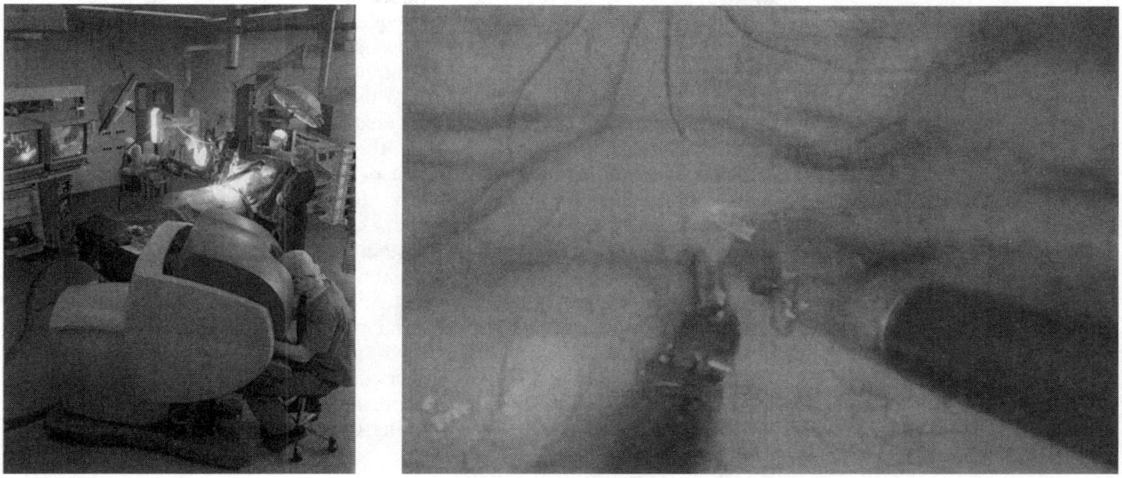

(a) (b)

FIGURE 29.22 Telesurgical augmentation system: In this "telepresence" system the surgeon sits at a control console [(*a*) foreground] and manipulates a pair of "master" robot arms while "slave" robots [(*a*) background and (*b*)] mimic his motions inside the patient's body.[122] (*Photos courtesy Intuitive Surgical Systems.*)

constructed of curved tracks to maintain a fixed center of rotation. A similar micromanipulator[129] was used for acquiring physiological measurements in the eye using an electrode. While rigid mechanical constraints were suitable for the particular applications in which they were used, the design is not flexible enough for general-purpose microsurgery and the tracks take up a great deal of space around the head. An ophthalmic surgery manipulator built by Jensen et al.[130] was designed for retinal vascular microsurgery and was capable of positioning instruments at the surface of the retina with submicrometer precision. While a useful experimental device, this system did not have sufficient range of motion to be useful for general-purpose microsurgery. Also, the lack of force sensing prevented the investigation of force/haptic interfaces in the performance of microsurgical tasks.

Many microsurgical robots[70,124,131–133] are based on force-reflecting master-slave configurations. This paradigm allows an operator to grasp the master manipulator and apply forces. Forces measured on the master are scaled and reproduced at the slave and, if unobstructed, will cause the slave to move accordingly. Likewise, forces encountered by the slave are scaled and reflected back to the master. This configuration allows position commands from the master to result in a reduced motion of the slave and for forces encountered by the slave to be amplified at the master.

While a force-reflecting master-slave microsurgical system provides the surgeon with increased precision and enhanced perception, there are some drawbacks to such a design. The primary disadvantage is the complexity and cost associated with the requirement of providing two mechanical systems, one for the master and one for the slave. Another problem with telesurgery in general is that the surgeon is not allowed to directly manipulate the instrument used for the microsurgical procedure. While physical separation is necessary for systems designed to perform remote surgery, it is not required during microsurgical procedures. In fact, surgeons are more likely to accept assistance devices if they are still allowed to directly manipulate the instruments.

The performance augmentation approach pursued by the CIS group at Johns Hopkins University, which has also been explored independently by Davies et al.,[37–39] and which has some resemblances to the work of Kazerooni,[134] emphasizes cooperative manipulation, in which the surgeon and robot both hold the surgical tool. The robot senses forces exerted on the tool by the surgeon and moves to comply. Our initial experiences with this mode in ROBODOC indicated that it was very popular with surgeons and offered means to augment human performance while maximizing the surgeon's natural hand-eye coordination within a surgical task. Subsequently, we incorporated this mode into the IBM/JHU LARS system.[26,36,49,135–139] Figure 29.23 shows one early experiment using LARS to evacuate simulated hematomas with a neuroendoscopic instrument.[54,140–142] We found that the surgeon took slightly longer (6 versus 4 minutes) to perform the evacuation using the guiding, but evacuated much less surplus material (1.5 percent excess versus 15 percent).

More recently, we have been exploring the extension of these ideas into microsurgery and other precise manipulation tasks. We have extended our model of cooperative control, which we call "steady hand" guiding, to permit the compliance loop to be closed on the basis of a scaled combination of forces exerted by the surgeon and tissue interaction forces, as well as on other sensors such as visual processing. The result is a manipulation system with the precision and sensitivity of a machine, but with the manipulative transparency and immediacy of hand-held tools for tasks characterized by compliant or semirigid contacts with the environment.[18] We have also begun to develop higher levels of control for this system, incorporating more complex behaviors with multiple sensing modalities,[72,143–146] using microsurgical tasks drawn from the fields of ophthalmology and otology. Figure 29.24 shows a typical experiment using our current robot to evaluate robot-assisted stapedotomies. Figure 29.25 shows a comparison of instrument tremor and drift with and without robotic assistance. We have also demonstrated 30:1 scaling of forces in compliant manipulation tasks.

29.4.4 Other Robotic Assistants

The use of robotic systems to assist surgeons by performing routine tasks such as laparoscopic camera manipulation is becoming commonplace.[33,35,135,147] Some of the manipulator design issues

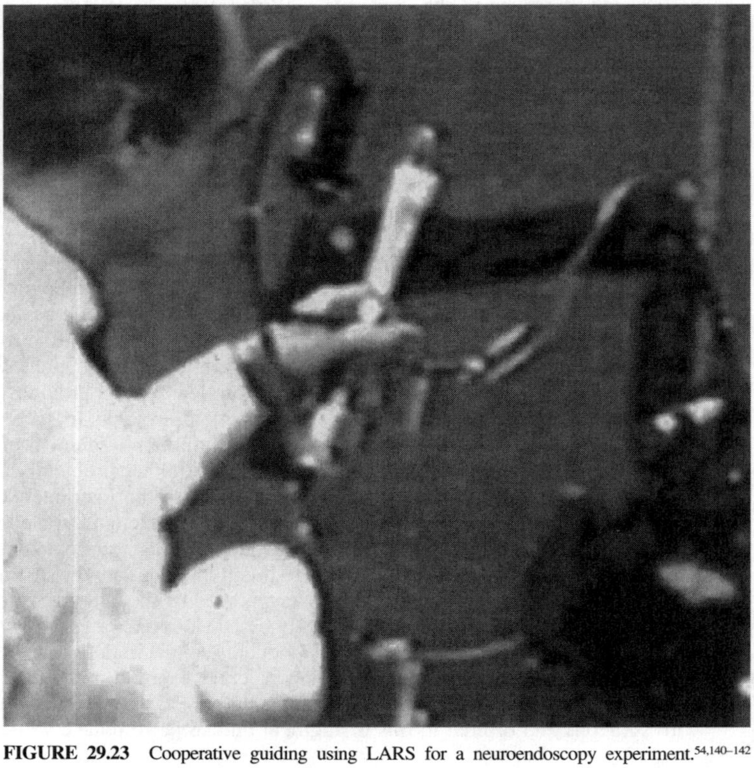

FIGURE 29.23 Cooperative guiding using LARS for a neuroendoscopy experiment.[54,140–142]

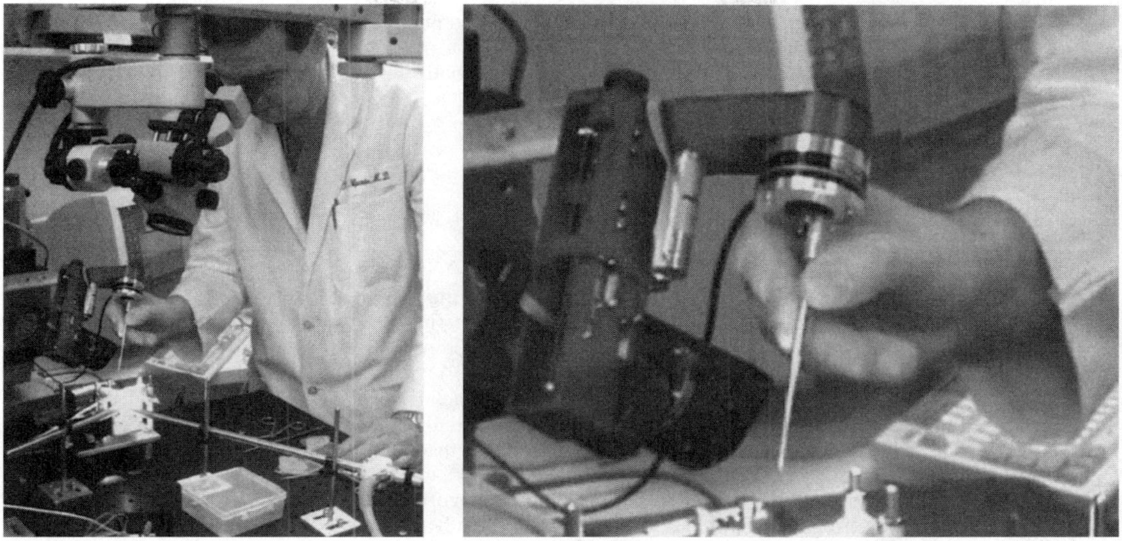

FIGURE 29.24 Microsurgical augmentation experiments with the Johns Hopkins steady-hand robot.[163] Shows the current generation of the robot being used to evaluate robotically assisted stapedotomy. The current generation comprises an RCM linkage,[164] a custom end-effector assembly,[21,165,166] and off-the-shelf components. (*Photo courtesy Dan Rothbaum.*)

associated with such systems were discussed in Sec. 29.3.6. For human-machine interfaces, these systems provide a joystick or foot pedal to permit the surgeon to control the motion of the endoscope. However, other interfaces include voice, tracking of surgeon head movements, computer vision tracking of surgical instruments, indication of desired gaze points by manipulating a cursor on the computer screen, etc. Figure 29.7*a* shows a typical installation of a voice-controlled commercial system (the AESOP™, developed by Computer Motion, Inc.).

More recently, there has been interest in robotic systems for manipulating ultrasound probes.[116,148–151] Figures 29.10 and 29.21 show typical current research efforts in development of such robotic systems. Most of this activity has targeted diagnostic procedures such as systematic examination of carotid arteries for occlusions. However, these systems have the potential to become as ubiquitous as the robotic endoscope holders discussed above. Our research group at Johns Hopkins

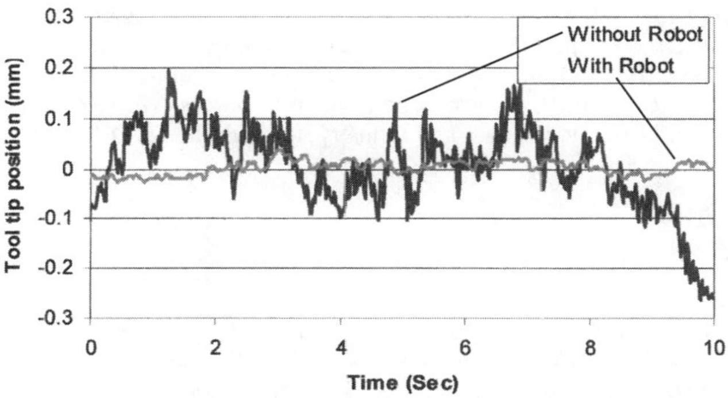

FIGURE 29.25 Comparative performance of human tremor and drift without a robot and with steady-hand manipulation augmentation.[67]

University has begun to explore applications such as precise ultrasound-guided biopsies and other interventional procedures.

There has also been work in the use of flexible robotic devices for intralumenal applications such as colonoscopy and angioplasty.[152–155] Generally, these devices are snakelike, though there have been a few efforts[155] to develop autonomous crawlers.

29.5 PERSPECTIVES

Computer-integrated surgery is a new, rapidly evolving paradigm that will change the way surgery will be performed in the future. Technical advances in medical imaging, tracking, robotics, and integration are paving the way to a new generation of systems for minimally invasive surgery.

We believe that computer-integrated surgery (CIS) will have the same impact on health care in the coming decades that computer-integrated manufacturing has had on industrial production in the recent past. Achieving this vision will require both significant advances in basic engineering knowledge and the development of robust, flexible systems that make this knowledge usable in real clinical applications.

It is important to remember that the ultimate payoff for CIS systems will be in improved and more cost-effective health care. Quantifying these advantages in practice can be problematic, and sometimes the final answer may take years to be demonstrated. The consistency, enhanced data logging, and analysis made possible by CIS systems may help in this process. It will not be easy to figure out how to apply these capabilities. However, we believe that the CIS paradigm is here to stay.

ACKNOWLEDGMENTS

Any survey or critical summary must necessarily draw upon the work of many people. We would like especially to acknowledge the contributions of many colleagues over the past decade who have helped develop an evolving shared understanding of medical robotics and computer-integrated surgery. We are especially grateful to those individuals who generously provided photographs and other information about the specific systems that we have used as examples. In some cases, these colleagues have also worked with us in developing some of these systems.

We also gratefully acknowledge the many agencies and organizations that have contributed financial and other support to the development of much of the work that we have reported. The National Science Foundation supports the Engineering Research Center for Computer-Integrated Surgical Systems and Technology under cooperative agreement number EEC9731478. Further support has been provided by other NSF grants, the National Institutes of Health, the Whitaker Foundation, IBM Corporation, Integrated Surgical Systems, The Johns Hopkins University, and numerous other government, industry, and academic sources. We also acknowledge the partial support of the Israeli Ministry of Trade and Industry, through the IZMEL Consortium for Image-Guided Therapy.

REFERENCES

1. Taylor, R. H., et al., "An Image-directed Robotic System for Precise Orthopedic Surgery," *IEEE Transactions on Robotics and Automation*, 1994, **10**(3):261–275.
2. Mittelstadt, B. D., et al., "The Evolution of a Surgical Robot from Prototype to Human Clinical Trial," *Proc. Medical Robotics and Computer Assisted Surgery*, 1994, Pittsburgh.

3. Paul, H., et al., "Accuracy of Implant Interface Preparation: Hand-held Broach vs. Robot Machine Tool," *Proc. Orthopedic Research Society*, 1992, Washington, D.C.
4. Joskowicz, L., and R. H. Taylor, "Preoperative Insertability Analysis and Visualization of Custom Hip Implants," *1st International Symposium on Medical Robotics and Computer Assisted Surgery*, 1994, Pittsburgh.
5. Bauer, A., "Primary THR Using the ROBODOC System," *CAOS/USA*, 1999, Pittsburgh.
6. Joskowicz, L., et al., "Computer Integrated Revision Total Hip Replacement Surgery: Preliminary Report," *Second Annual International Symposium on Medical Robotics and Computer Assisted Surgery*, 1995, Baltimore.
7. Taylor, R. H., et al., "Computer-Integrated Revision Total Hip Replacement Surgery: Concept and Preliminary Results," *Medical Image Analysis*, 1999, **3**(3):301–319.
8. Smith, K. R., K. J. Frank, and R. D. Bucholz, "The Neurostation—a highly accurate minimally invasive solution to frameless stereotactic neurosurgery," *Comput. Med. Imaging Graph.*, 1994, **18**:247–256.
9. Levoy, M., "Efficient ray tracing of volume data," *ACM Transactions on Graphics*, 1990, **9**:245–261.
10. Lorensen, W. E., and H. E. Cline, "Marching Cubes: a high resolution 3D surface reconstruction algorithm," *Computer Graphics*, 1987, **21**:163–169.
11. Hohne, K. H., M. Bomans, and M. Reimer, "A 3D anatomical atlas based on a volume model," *IEEE Computer Graphics and Applications*, 1992, **2**(1):72–78.
12. Lavallee, S., "Registration for Computer-Integrated Surgery: Methodology, State of the Art," in *Computer-Integrated Surgery*, R. H. Taylor et al. (eds.), 1996, MIT Press, Cambridge, Mass., pp. 77–98.
13. Maintz, J. B., and M. A. Viergever, "A survey of medical image registration," *Medical Image Analysis*, 1998, **2**(1):1–37.
14. Reinhardt, H. F., "Neuronagivation: A ten years review," in *Computer-Integrated Surgery*, R. Taylor et al., (eds.), 1996, MIT Press, Cambridge, Mass., pp. 329–342.
15. Maciunas, R. J., *Interactive Image-Guided Neurosurgery*, American Association of Neurological Surgeons.
16. Taylor, R. H., et al. (eds.), "An Image-Directed Robotic System For Precise Orthopedic Surgery," *Computer-Integrated Surgery, Technology and Clinical Applications*, 1995, pp. 379–396.
17. Taylor, R. H., et al. (eds.), *Computer-Integrated Surgery, Technology and Clinical Applications*, 1995, MIT Press, Cambridge, Mass.
18. Taylor, R., et al., "A Steady-Hand Robotic System for Microsurgical Augmentation," *International Journal of Robotics Research*, 1999, **18**(12).
19. Howe, R. D., et al., "Remote Palpation Technology," *IEEE Engineering in Medicine and Biology*, 1995, pp. 318–323.
20. Glucksberg, M. R., and R. Dunn, "Direct measurement of retinal microvascular pressures in the live, anesthetized cat," *Microvascular Research*, 1993, **45**(2):158–165.
21. Berkelman, P. J., et al., "A Miniature Instrument Tip Force Sensor for Robot/Human Cooperative Microsurgical Manipulation with Enhanced Force Feedback." in *Medical Image Computing and Computer-Assisted Interventions*, 2000, Springer, Pittsburgh.
22. Taylor, R. H., "Safety," in *Computer-Integrated Surgery*, R. H. Taylor, et al. (eds.), 1996, MIT Press, Cambridge, Mass., pp. 283–286.
23. Davies, B. L., "A discussion of safety issues for medical robots," in *Computer-Integrated Surgery*, R. H. Taylor, et al. (eds.), 1996, MIT Press, Cambridge, Mass., pp. 287–298.
24. Kwoh, Y. S., J. Hou, E. A. Jonckheere, et al., "A robot with improved absolute positioning accuracy for CT guided stereotactic brain surgery," *IEEE Trans. Biomed. Eng.*, 1988, **35**(2):153–161.
25. Nathan, M. S., et al., "Devices for Automated Resection of the Prostate," in *Proc. 1st International Symposium on Medical Robotics and Computer Assisted Surgery*, 1994. Pittsburgh.
26. Eldridge, B., et al., "A Remote Center of Motion Robotic Arm for Computer Assisted Surgery," *Robotica*, 1996, **14**(1)103–109.
27. Stoianovici, D., et al., "An efficient needle injection technique and radiological guidance method for percutaneous procedures," in *First Joint Conference: CRVMed II & MRCAS III*, March 1997, Grenoble, France.

28. Erbse, S., et al., "Development of an automated surgical holding system based on ergonomic analysis," in *Proc. First Joint Conference of CVRMed and MRCAS*, 1997, Grenoble, France, Springer.
29. Grace, K. W., et al., "Six degrees of freedom micromanipulator for ophthalmic surgery," in *IEEE Int. Conf. Robotics and Automation*, 1993, Atlanta, IEEE.
30. Brandt, G., et al., "A compact robot for image-guided orthopedic surgery: concept and preliminary results," in *Proc. First Joint Conference of CVRMed and MRCAS*, 1997, Grenoble, France, Springer.
31. Neisius, B., P. Dautzenberg, and R. Trapp, "Robotic Manipulator for Endoscopic Handling of Surgical Effectors and Cameras," in *Proc. Medical Robotics and Computer Assisted Surgery*, 1994, Pittsburgh.
32. Taylor, R. H., et al., "A Model-Based Optimal Planning and Execution System with Active Sensing and Passive Manipulation for Augmentation of Human Precision in Computer-Integrated Surgery," in *Proc. 1991 Int. Symposium on Experimental Robotics*, 1991. Toulouse, France, Springer-Verlag.
33. Sackier, J. M., and Y. Wang, "Robotically Assisted Laparoscopic Surgery: from Concept to Development," in *Computer-Integrated Surgery*, R. Taylor et al. (eds.), 1996, MIT Press, Cambridge, Mass., pp. 577–580.
34. Wang, Y., "Robotically Enhanced Surgery," in *Medicine Meets Virtual Reality II*, 1994. San Diego.
35. Hurteau, R., et al., "Laparoscopic Surgery Assisted by a Robotic Cameraman: Concept and Experimental Results," in *IEEE Conference on Robotics and Automation*, 1994, San Diego.
36. Funda, J., et al., "Comparison of two manipulator designs for laparoscopic surgery," in *1994 SPIE Int. Symposium on Optical Tools for Manufacturing and Advanced Automation*, 1994, Boston.
37. Troccaz, J., M. Peshkin, and B. L. Davies, "The use of localizers, robots, and synergistic devices in CAS," in *Proc. First Joint Conference of CVRMed and MRCAS*, 1997, Grenoble, France, Springer.
38. Ho, S. C., R. D. Hibberd, and B. L. Davies, "Robot Assisted Knee Surgery," *IEEE EMBS Magazine*, Sp. Issue on Robotics in Surgery, April–May 1995, pp. 292–300.
39. Harris, S. J., et al., "Experiences with robotic systems for knee surgery," in *Proc. First Joint Conference of CVRMed and MRCAS*, 1997, Grenoble, France, Springer.
40. Masamune, K., et al., "A newly developed stereotactic robot with detachable driver for neurosurgery," in *Proc. 2nd Int. Symp. on Medical Robotics and Computer Assisted Surgery (MRCAS)*, 1995, Baltimore, Center for Orthop. Res. Shadyside Hospital, Pittsburgh.
41. Taylor, R. H., et al., "Computer-Integrated Revision Total Hip Replacement Surgery: Concept and Preliminary Results," *Medical Image Analysis*, 1999, **3**(3):301–319.
42. Susil, R. C., J. H. Anderson, and R. H. Taylor, "A Single Image Registration Method for CT Guided Interventions," in *2nd Int. Symposium on Medical Image Computing and Computer-Assisted Interventions (MICCAI99)*, 1999, Cambridge, England, Springer.
43. Yao, J., et al., "A C-arm fluoroscopy-guided progressive cut refinement strategy using a surgical robot," *Computer Aided Surgery*, 2000, **5**(6):373–390.
44. Masamune, K., et al., "Development of CT-PAKY frame system—CT image guided needle puncturing manipulator and a single slice registration for urological surgery," in *Proc. 8th annual meeting of JSCAS*, 1999, Kyoto.
45. Masamune, K., et al., "Guidance system for robotically assisted percutaneous procedures with computed tomography," *Journal of Computer Aided Surgery*, 2000. 2001, **6**(6):370–375.
46. Masutani, Y., et al., "Computer Aided Surgery (CAS) System for Stereotactic Neurosurgery," in *Proc. Computer Vision and Robotics in Medicine (CVRMED)*, 1995. Nice, France, Springer.
47. "Human-Machine Interfaces," in *Computer-Integrated Surgery*, R. H. Taylor et al. (eds.), 1996, MIT Press, Cambridge, Mass, pp. 201–254.
48. Taylor, R. H., et al., "A Telerobotic Assistant for Laparoscopic Surgery," in *Computer-Integrated Surgery*, R. Taylor et al. (eds.), 1996, MIT Press, Cambridge, Mass., pp. 581–592.
49. Funda, J., et al., "Image Guided Command and Control of a Surgical Robot," in *Proc. Medicine Meets Virtual Reality II*, 1994, San Diego.
50. Adams, L., et al., "CAS—A Navigation Support for Surgery," in *3D Imaging in Medicine*, 1990, Springer-Verlag, Berlin, pp. 411–423.
51. Kosugi, Y., et al., "An Articulated Neurosurgical Navigation System Using MRI and CT Images," *IEEE Transactions on Biomedical Engineering*, February 1988, pp. 147–152.

52. Watanabe, E., T. Watanabe, and S. Manka, et al., "Three-dimensional digitizer (neuronavigator): New equipment for computed tomography-guided stereotaxic surgery," *Surg. Neurol.*, 1987, **27**:543–547.

53. Cutting, C. B., F. L. Bookstein, and R. H. Taylor, "Applications of Simulation, Morphometrics and Robotics in Craniofacial Surgery," in *Computer-Integrated Surgery*, R. H. Taylor, et al. (eds.), 1996, MIT Press, Cambridge, Mass., pp. 641–662.

54. Auer, L. M. "Virtual Endoscopy for Planning and Simulation of Minimally Invasive Neruosurgery," in *Proc. First Joint Conference of CVRMed and MRCAS*, 1997, Grenoble, France, Springer.

55. Bucholz, R. D., et al., "Intraoperative Localization Using a Three Dimensional Optical Digitizer," in *Proceedings Medical Robotics and Computer Assisted Surgery*, 1994, Shadyside Hospital, Pittsburgh.

56. DiGioia, A. M., et al., "HipNav: Pre-operative Planning and Intra-operative Navigational Guidance for Acetabular Implant Placement in Total Hip Replacement Surgery," *Computer Assisted Orthopedic Surgery*, 1996.

57. Picard, F., et al., "Computer-assisted navigation for knee arthroplasty: Intra-operative measurements of alignment and soft tissue balancing," in *First Annual Meeting of CAOS International*, 2001, Davos, Switzerland.

58. Nolte, L. P., et al., "Use of C-arm for Surgical Navigation in the Spine," in *CAOS/USA '98*, 1998, Pittsburgh.

59. Hofstetter, R., et al., "Principles of Precise Fluoroscopy Based Surgical Navigation," in *4th International Symposium on CAOS*, 1999, Davos, Switzerland.

60. Hofstetter, R., et al., "Fluoroscopy based surgical navigation-concept and clinical applications," in *Proceedings of Computer Assisted Radiology and Surgery, CAR '97*, 1997, Elsevier, Berlin.

61. Hofstetter, R., et al., "Fluoroscopy as an imaging means for computer-assisted surgical navigation," *Computer-Aided Surgery*, 1999, **4**(2):65–76.

62. N. Hata, W. M. Wells, M. Halle, S. Nakajima, P. Viola, R. Kikinis, and F. A. Jolesz, "Image guided microscopic surgery system using mutual information based registration," in *VBC*, 1996, Hamburg.

63. Blackwell, M., et al., "An Image Overlay System for Medical Data Visualization," *Medical Image Analysis*, 2000, **4**(1):67–72.

64. Masamune, T., et al., "Three dimensional slice image overlay system with accurate depth perception for surgery," in *Medical Image Computing and Computer-Assisted Intervention*, MICCAI 2000, 2000, Pittsburgh, Springer.

65. Nolte, L. P., H. Visarius, et al., *Computer Assisted Orthopedic Surgery*, 1996. Hofgrefe & Huber.

66. Uecker, D. R., et al., "A Speech-Directed Multi-Modal Man-Machine Interface for Robotically Enhanced Surgery," in *First Int. Symp. on Medical Robotics and Computer Assisted Surgery (MRCAS '94)*, 1994, Shadyside Hospital, Pittsburgh.

67. Reichenspurner, H., et al., "Use of the voice controlled and computer-assisted surgical system zeus for endoscopic coronary artery surgery bypass grafting," *J. Thoracic and Cardiovascular Surgery*, 1999, **118**(1).

68. Confer, R. G., and R. C. Bainbridge, "Voice control in the microsurgical suite," in *Proc. of the Voice I/O Systems Applications Conference '84*, 1984, American Voice I/O Society, Arlington, Va.

69. d'Aulignac, D., R. Balaniuk, and C. Laugier, "A haptic interface for a virtual exam of the human thigh," in *IEEE Conf. on Robotics and Automation*, 2000, San Francisco.

70. Mitsuishi, M., et al., "A Telemicrosurgery System with Colocated View and Operation Points and Rotational-force-feedback-free Master Manipulator," in *Proc. 2nd Int. Symp. on Medical Robotics and Computer Assisted Surgery, MRCAS '95*, 1995, Baltimore, Center for Orthop. Res, Shadyside Hospital, Pittsburgh.

71. Howe, R. D., and M. R. Cutkosky, "Dynamic Tactile Sensing: Perception of Fine Surface Features with Stress Rate Sensing," *IEEE Trans. Robotics & Automation*, 1993, **9**(2):140–151.

72. Kumar, R., et al., "Preliminary Experiments in Cooperative Human/Robot Force Control for Robot Assisted Microsurgical Manipulation," in *IEEE Conference on Robotics and Automation*, 2000, San Francisco.

73. Green, P., "Telepresence Surgery," in *NSF Workshop on Computer Assisted Surgery*, 1993, Washington, D.C.

74. Taylor, R., et al., "Redundant Consistency Checking in a Precise Surgical Robot," in *12th Annual Conference on Engineering in Medicine and Biology*, 1990, Philadelphia, IEEE Press.

75. Taylor, R., et al., "Taming the Bull: Safety in a Precise Surgical Robot," in *Intl. Conf. on Advanced Robotics (ICAR)*, 1991, Pisa, Italy.

76. B. Davies, "A Discussion of Safety Issues for Medical Robots," in *Computer-Integrated Surgery*, R. Taylor, et al., (eds.), 1996, MIT Press, Cambridge, Mass., pp. 287–296.
77. Varley, P., "Techniques of development of safety-related software in surgical robots," *IEEE Trans. on Information Technology in Biomedicine*, 1999, **3**(4):261–267.
78. Levensen, N. G., and C. S. Turner, "An investigation of the Therac-25 accidents," *Computer*, 1993, **26**(7): 18–41.
79. Rau, G., et al., "Aspects of Ergonomic System Design Applied to Medical Work Stations," in *Computer-Integrated Surgery*, R. H. Taylor et al. (eds.), 1996, MIT Press, Cambridge, Mass. pp. 203–222.
80. Sheridan, T., "Human Factors in Telesurgery," in *Computer-Integrated Surgery*, R. H. Taylor et al. (eds.), 1996, MIT Press, Cambridge, Mass., pp. 223–230.
81. Merloz, P., et al., "Computer-assisted versus manual spine surgery: Clinical report," in *Proc. First Joint Conference of CVRMed and MRCAS*, 1997, Grenoble, France, Springer.
82. Nolte, L. P., et al., "A Novel Approach to Computer Assisted Spine Surgery," in *First Int. Symp. on Medical Robotics and Computer Assisted Surgery (MRCAS 94)*, 1994, Shadyside Hospital, Pittsburgh.
83. vanHellenMondt, G., M. deKleuver, and P. Pavlov, "Computer assisted pelvic osteotomies; clinical experience in 25 cases," in *First Annual Meeting of CAOS International*, 2001, Davos, Switzerland.
84. Arand, M., L. Kinzl, and F. Gebhard, "CT-based navigation in minimally invasive screw stabilization of the iliosacral joint," in *First Annual Meeting of CAOS International*, 2001, Davos, Switzerland.
85. Joskowicz, L., et al., "FRACAS: A System for Computer-Aided Image-Guided Long Bone Fracture Surgery," *Journal of Computer Assisted Surgery*, 1999.
86. Tockus, L., et al. "Computer-Aided Image-Guided Bone Fracture Surgery: Modeling, Visualization, and Preoperative Planning," in *MICCAI '98*, 1998, Cambridge, Mass.
87. Verheyden, A., et al., "Percutaneous stabilization of dorsal pelvic ring fractures—transiliosacral screw placement in the open MRI," in *First Annual Meeting of CAOS International*, 2001, Davos, Switzerland.
88. Grutzner, P., et al., "Computer-aided reduction and fixation of long bone fractures," in *First Annual Meetings of CAOS International*, 2001, Davos, Switzerland.
89. Suhm, N., et al., "Computer assisted distal locking of intramedullary implants: A controlled clinical study including 84 patients," in *First Annual Meeting of CAOS International*, 2001, Davos, Switzerland.
90. DiGioia, A. M., B. Jaramaz, and R. V. O'Toole, "An Integrated Approach to Medical Robotics and Computer Assisted Surgery in Orthopedics," in *Proc. 1st Int. Symposium on Medical Robotics and Computer Assisted Surgery*, 1994. Pittsburgh.
91. Digioia, A., et al., "Clinical Measurements of Acetabular Component Orientation Using Surgical Navigation Technologies," in *First Annual Meeting of CAOS International*, 2001, Davos, Switzerland.
92. Kunz, M., et al., "Development and verification of a non-CT based total knee arthroplasty system for the LCS prosthesis," in *First Annual Meeting of CAOS International*, 2001, Davos, Switzerland.
93. Stulberg, S. D., P. Loan, and V. Sarin, "Computer-Assisted Total Knee Replacement Surgery: An Analysis of an Initial Experience with the Orthopilot (TM) System," in *First Annual Meeting of CAOS International*, 2001, Davos, Switzerland.
94. Saragaglia, D., et al., "Computer-Assisted Total Knee Replacement Arthroplasty: comparison with a conventional procedure. Results of a 50 cases prospective randomized trial," in *First Annual Meeting of CAOS International*, 2001, Davos, Switzerland.
95. Wirth, S., et al., "C-arm based computed tomography: A comparative study," in *Proc. of the 15th Conf. on Computer-Aided Radiology and Surgery*, 2001, Elsevier, Berlin.
96. Shahidi, R., "Advances in video laparoscopic surgery using three-dimensional image enhanced endoscopy," *MDVista Journal*, 2000, May, pp. 56–65.
97. Grimson, W. E. L., et al., "An automatic registration method for frameless stereotaxy, image guided surgery and enhanced reality visualization," *IEEE Trans. on Medical Imaging*, 1996, **15**(2):129–140.
98. Ecke, U., et al., "Virtual Reality: preparation and execution of sinus surgery," *Computer-Aided Surgery*, 1998, **4**(2):45–50.
99. Peterman, J., et al., "Implementation of the CASPAR System in the reconstruction of the ACL," in *CAOS/USA*, 2000, Shadyside Hospital, Pittsburgh.
100. Wiesel, U., et al., "Total Knee Replacement Using the Robodoc System," in *Proc. First Annual Meeting of CAOS International*, 2001, Davos, Switzerland.

101. Tenbusch, M., et al., "First results using the Robodoc system for total knee replacement," in *First Annual Meeting of CAOS International*, 2001, Davos, Switzerland.

102. Mai, S., C. Lorke, and W. Siebert, "Motivation, Realization, and First Results of Robot Assisted Total Knee Arthroplasty," in *Proc. 1st Annual Meeting of CAOS International*. 2001, Davos, Switzerland.

103. Garbini, J. L., et al., "Robotic Instrumentation in Total Knee Arthroplasty," in *Proc. 33rd Annual Meeting*, orthopedic Research Society, 1987, San Francisco.

104. Fadda, M., et al., "Computer-Assisted Knee Arthroplasty at Rizzoli Institutes," in *Proc. 1st International Symposium on Medical Robotics and Computer Assisted Surgery*, 1994, Pittsburgh.

105. Kienzle, T. C., et al., "An integrated CAD-robotics system for total knee replacement surgery," in *Proc. IEEE Int. Conf. on Robotics and Automation*, 1993, Atlanta.

106. Leitner, F., et al., "Computer-assisted knee surgical total replacement," in *Proc. First Joint Conference of CVRMed and MRCAS*, 1997, Grenoble, France, Springer.

107. Marcacci, S., et al., "Computer-Assisted Knee Arthroplasty," in *Computer-Integrated Surgery*, R. H. Taylor et al. (eds.), 1996, MIT Press, Cambridge, Mass. pp. 417–423.

108. Cinquin, P., et al., "IGOR: Image Guided Operating Robot," *Innovation et Technonogie en Biologie et Medicine*, 1992, pp. 374–394.

109. Lavallee, S., et al., "Image-Guided Operating Robot: A Clinical Application in Stereotactic Neurosurgery," in *Computer-Integrated Surgery*, R. H. Taylor et al. (eds.), 1996, MIT Press, Cambridge, Mass., pp. 343–352.

110. Bzostek, A., et al., "A Testbed System for Robotically Assisted Percutaneous Pattern Therapy," in *Medical Image Computing and Computer-Assisted Surgery*, 1999. Springer, Cambridge, England.

111. Schreiner, S., et al., "A system for percutaneous delivery of treatment with a fluoroscopically-guided robot," in *Joint Conf. of Computer Vision, Virtual Reality, and Robotics in Medicine and Medical Robotics and Computer Surgery*, 1997. Grenoble, France.

112. Taylor, R., et al., "An Experimental System for Computer Assisted Endoscopic Surgery," in *IEEE Satellite Symposium on Neuroscience and Technoloy*, 1992, Lyons, IEEE Press.

113. Taylor, R. H., et al., "A Telerobotic Assistant for Laparoscopic Surgery," in *IEEE EMBS Magazine*, Special Issue on Robotics in Surgery, 1995, pp. 279–291.

114. Bzostek, A., et al., "An automated system for precise percutaneous access of the renal collecting system," in *Proc. First Joint Conference of CVRMed and MRCAS*, 1997, Grenoble, France, Springer.

115. Caddedu, J. A., et al., "A Robotic System for Percutaneous Renal Access," *Urology*, 1997.

116. Goldberg, R., *A Robotic System for Ultrasound Image Acquisition*, 1999, Johns Hopkins University, Baltimore.

117. Chinzei, K., et al., "MR Compatible Surgical Assist Robot: System Integration and Preliminary Feasibility Study," in *Proceedings of Third International Conference on Medical Robotics, Imaging and Computer Assisted Surgery*, 2000, Pittsburgh.

118. Bzostek, A., et al., "Distributed Modular Computer-Integrated Robotic Systems: Implementation Using Modular Software and Networked Systems," in *Medical Image Computing and Computer-Assisted Interventions*, 2000, Pittsburgh, Springer.

119. Schorr, O., et al., "Distributed Modular Computer-Integrated Robotic Systems: Architecture for Intelligent Object Distribution," in *Medical Image Computing and Computer-Assisted Interventions*, 2000, Pittsburgh, Springer.

120. Kaiser, W. A., et al., "Robotic system for biopsy and therapy of breast lesions in a high-field whole-body magnetic resonance tomography unit," *J. Investigative Radiology*, 2000, **35**(8):513–519.

121. Green, P., et al., "Mobile Telepresence Surgery," in *Proc. 2nd Int. Symp. on Medical Robotics and Computer Assisted Surgery, MRCAS '95*, 1995, Baltimore. Center for Orthop. Res., Shadyside Hospital, Pittsburgh.

122. Guthart, G. S., and J. K. Salisbury, "The Intuitive Telesurgery System: Overview and Application," in *Proc. of the IEEE International Conference on Robotics and Automation (ICRA2000)*, 2000, San Francisco.

123. Charles, S., R. E. Williams, and B. Hamel, "Design of a Surgeon-Machine Interface for Teleoperated Microsurgery," *Proc. of the Annual Int'l Conf. of the IEEE Engineering in Medicine and Biology Society*, 1989, **11**:883–884.

124. Salcudean, S. E., S. Ku, and G. Bell. "Performance measurement in scaled teleoperation for microsurgery" in *First Joint Conference Computer Vision, Virtual Realtiy and Robotics in Medicine and Medical Robotics and Computer-Assisted Surgery*, 1997, Grenoble, France, Springer.

125. Ku, S., and S. E. Salcudean, "Dexterity enhancement in microsurgery using a motion-scaling system and microgripper," in *IEEE Int. Conf. on Systems, Man and Cybernetics*, 1995, Vancouver, B.C.

126. Satava, R., "Robotics, telepresence, and virtual reality: A critical analysis fo the future of surgery," *Minimally Invasive Therapy*, 1992, **1**:357–363.

127. Lee, B. R., J. T. Bishoff, S. Micali, L. L. Whitcomb, R. H. Taylor, and L. R. Kavoussi, "Robotic Telemanipulation for Percutaneous Renal Access," in *16th World Congress on Endourology*, 1998, New York.

128. Guerrouad, A., and P. Vidal, "S.M.O.S.: Stereotaxical Microtelemanipulator for Ocular Surgery," *Proc. of the Annual Int'l Conf. of the IEEE Engineering in Medicine and Biology Society*, 1989, **11**:879–880.

129. Pournaras, C. J., et al., "New ocular micromanipulator for measurements of retinal and vitreous physiologic parameters in the mammalian eye," *Exp. Eye Res.*, 1991, **52**:723–727.

130. Jensen, P. S., et al., "Toward robot assisted vascular microsurgery in the retina," *Graefes Arch. Clin. Exp. Ophthalmol.*, 1997, **235**(11):696–701.

131. Charles, S., "Dexterity enhancement for surgery," *Proc. First Int.'l Symp. Medical Robotics and Computer Assisted Surgery*, 1994, **2**:145–160.

132. Hunter, I. W., et al., "Ophthalmic microsurgical robot and associated virtual environment," *Computers in Biology and Medicine*, 1995, **25**(2):173–182.

133. Schenker, P. S., H. O. Das, and R. Timothy, "Development of a new high-dexterity manipulator for robot-assisted microsurgery," in *Proceedings of SPIE—The International Society for Optical Engineering: Telemanipulator and Telepresence Technologies*, 1995, Boston.

134. Kazerooni, H., and G. Jenhwa, "Human extenders," *Transaction of the ASME: Journal of Dynamic Systems, Measurement and Control*, 1993, **115**(2B):218–290, June.

135. Taylor, R. H., et al., "Telerobotic assistant for laparoscopic surgery," *IEEE Eng. Med. Biol.*, 1995, **14**(3): 279–288.

136. Funda, J., et al., "Constrained Cartesian motion control for teleoperated surgical robots," *IEEE Transactions on Robotics and Automation*, 1996.

137. Funda, J., et al., "Control and evaluation of a 7-axis surgical robot for laparoscopy," in *Proc. 1995 IEEE Int. Conf. on Robotics and Automation*, 1995, Nagoya, Japan, IEEE Press.

138. Funda, J., et al., "An experimental user interface for an interactive surgical robot," in *1st International Symposium on Medical Robotics and Computer Assisted Surgery*, 1994, Pittsburgh.

139. Funda, J., et al., "Optimal Motion Control for Teleoperated Surgical Robots," in *1993 SPIE Intl. Symp. on Optical Tools for Manuf. & Adv. Autom.*, 1993, Boston.

140. Kumar, R., et al., "Performance of Robotic Augmentation in Microsurgery-Scale Motions," in *2nd Int. Symposium on Medical Image Computing and Computer-Assisted Surgery*, 1999, Cambridge, England, Springer.

141. Goradia, T. M., R. H. Taylor, and L. M. Auer, "Robot-assisted minimally invasive neurosurgical procedures: First experimental experience," in *Proc. First Joint Conference of CVRMed and MRCAS*, 1997, Grenoble, France, Springer.

142. Kumar, R., et al., "Robot-assisted microneurosurgical procedures, comparative dexterity experiments," in *Society for Minimally Invasive Therapy 9th Annual Meeting, Abstract Book*, vol 6, supplement 1, 1997, Tokyo.

143. Kumar, R., An Augmented Steady Hand System for Precise Micromanipulation, Ph.D. thesis, Computer Science, The Johns Hopkins University, Baltimore, 2001.

144. Kumar, R., and R. Taylor, "Task-Level Augmentation for Cooperative Fine Manipulation Tasks in Surgery," in *MICCAI 2001*, 2001.

145. Kumar, R., et al., "An Augmentation System for Fine Manipulation," in *Medical Image Computing and Computer-Assisted Interventions*, 2000, Pittsburgh, Springer.

146. Kumar, R., P. Jensen, and R. H. Taylor, "Experiments with a Steady Hand Robot in Constrained Compliant Motion and Path Following," in *8th IEEE International Workshop on Robot and Human Interaction (RO-MAN)*, 1999, Pisa, Italy.

147. Faraz, A., and S. Payandeh, "A robotic case study: Optimal design for laparoscopic positioning stands," *J. Robotics Research*, 1998, **17**(9):986–995.

148. Abolmaesumi, P., et al., "A User Interface for Robot-Assisted Diagnostic Ultrasound," *IEEE Robotics and Automation Conference*, 2001, Seoul.

149. Goldberg, R., A Modular Robotic System for Ultrasound Image Acquisition, M.S. thesis, Mechanical Engineering, Johns Hopkins University, Baltimore, 2001.

150. Degoulange, E., et al., "HIPPOCRATE: an intrinsically safe robot for medical applicaions," in *IEE/RSH International Conference on Intelligent Robots and Systems*, 1998, Victoria, B.C.

151. Mitsuishi, M., et al., "Remote Ultrasound Diagnostic System," in *Proc. IEEE Conf. on Robotics and Automation*, 2001, Seoul.

152. Carrozza, M., et al., "The development of a microrobot system for colonoscopy," in *Proc. CVRMed and MRCAS—1205*, 1997, Grenoble, Springer Verlag.

153. Ikuta, K., M. Tsukamoto, and S. Hirose, "Shape Memory Alloy Servo Actuator System with Electric Resistance Feedback and Application for Active Endoscope," in *Computer-Integrated Surgery*, R. H. Taylor et al. (eds.), 1996, MIT Press, Cambridge, Mass, pp. 277–282.

154. Sturges, R., and S. Laowattana, "A voice-actuated, tendon-controlled device for endoscopy," in *Computer-Integrated Surgery*, R. H. Taylor, et al. (eds.), 1996, MIT Press, Cambridge, Mass.

155. Asari, V. K., S. Kumar, and I. M. Kassim, "A fully autonomous microrobotic endoscopy system," *Journal of Intelligent and Robotic Systems: Theory and Applications*, 2000. **28**(4):325–342.

156. Mittelstadt, B., et al., "Accuracy of Surgical Technique of Femoral Canal Preparation in Cementless Total Hip Replacement," in *Annual Meeting of American Acadamy of Orthopedic Surgeons*, 1990, New Orleans.

157. Mittelstadt, B., et al., "The Evolution of a Surgical Robot from Prototype to Human Clinical Use," in *Computer-Integrated Surgery*, R. H. Taylor et al., (eds.), 1996, MIT Press, Cambridge, Mass., pp. 397–407.

158. Bishoff, J. T., et al., "RCM-PAKY: Clinical application of a new robotics system for precise needle placement," *Journal of Endourology*, 1998, **12**:S82.

159. Cadeddu, J. A., et al., "A Robotic System for Percutaneous Renal Access Incorporating a Remote Center of Motion Design," *Journal of Endourology*, 1998, **12**:S237.

160. Stoianovici, D., J. A. Cadeddu, L. L. Whitcomb, R. H. Taylor, and L. R. Kavoussi, "A Robotic System for Precise Percutaneous Needle Insertion," *Thirteenth Annual Meeting of the Society for Urology and Engineering*, 1988, San Diego.

161. Stoianovici, D., et al., "Friction Transmission with Axial Loading and a Radiolucent Surgical Needle Drive," 1997, Johns Hopkins University (provisional patent application filed 17 February 1997).

162. Stoianovici, D., et al., "A Modular Surgical Robotic System for Image-Guided Percutaneous Procedures," in *Medical Image Computing and Computer-Assisted Interventions (MICCAI-98)*, 1998, Cambridge, Mass., Springer.

163. Berkelmann, P. J., et al., "Performance Evaluation of a Cooperative Manipulation Microsurgical Assistant Robot Applied to Stapedotomy," in *Medical Image Computing and Computer-Assisted Interventions (MICCAI 2001)*, 2001.

164. Stoianovici, D., et al., "A modular surgical robotic system for image guided percutaneous procedures," in *International Conference on Medical Image Computing and Computer-Assisted Intervention*. 1998, Cambridge, Mass.

165. Barnes, A., H. Su, and W. Tam, "An End-Effector for Minimally Invasive Robot Assisted Surgery," 1996, Johns Hopkins University, Dept. of Mechanical Engineering, Baltimore.

166. Barnes, A., A modular robotic system for precise minimally invasive surgery, M.S. thesis, Mechanical Engineering, The Johns Hopkins University, Baltimore, 1999.

167. Gomez-Blanco, M. A., C. N. Riviere, and P. K. Khosla, "Intraoperative instrument motion sensing for microsurgery," in *Proc. 20th Annu. Conf. IEEE Eng. Med. Biol. Soc.*, 1999, Atlanta.

P • A • R • T • 7

REHABILITATION ENGINEERING

CHAPTER 30
TECHNOLOGY AND DISABILITIES

Albert M. Cook
University of Alberta, Edmonton, Alberta, Canada

30.1 INTRODUCTION

This chapter is about the design of electronic assistive devices and the roles they play in the lives of people who have disabilities. It is also a chapter about the role that rehabilitation engineers play in the design and application of these devices and the unique design constraints placed on them in meeting the needs of people who have disabilities. Electronic assistive technologies have a relatively brief history that parallels the development of electronics in general. However, the key to successful application of electronic assistive devices for persons who have disabilities lies in the development of a thorough understanding of the needs that the person has, the context in which the technology will be used, and the skills that the person brings to the task.

30.2 THE CONTEXT FOR REHABILITATION ENGINEERING

Rehabilitation engineering can be described as the design, development, and application of engineering methods and devices to amelioration of the problems faced by persons who have disabilities. Future developments in assistive technologies and the successful application of these technologies to meet the needs of people who have disabilities will be driven by the combination of rapid technological advances and the requirement to meet the needs of persons with disabilities. We must design devices that are accessible to people who have disabilities, if possible. When this is not possible, then we must provide adaptations that ensure accessibility in a timely manner. If we do not meet these objectives, then individuals with disabilities may be left behind as technological advances occur.

30.3 A WORKING DEFINITION OF ASSISTIVE TECHNOLOGIES

One widely used definition of *assistive technologies* is that provided in Public Law 100–407, the Technical Assistance to the States Act in the United States. The definition of an assistive technology device is

> Any item, piece of equipment or product system whether acquired commercially off the shelf, modified, or customized that is used to increase or improve functional capabilities of individuals with disabilities.

This definition also has been incorporated into other legislation in the United States and is used as a working definition in other countries as well (Cook and Hussey, 2002). Note that the definition includes commercial, modified, and customized devices. This allows us to include products made for the general population in our working definition of assistive technologies. This definition also emphasizes functional capabilities of individuals who have disabilities. The inclusion of customized devices emphasizes the role of the rehabilitation engineer in designing or modifying devices to meet the unique needs of an individual person who has a disability.

30.4 CONTROL INTERFACES FOR ELECTRONIC ASSISTIVE TECHNOLOGIES

There are three elements that make up the user interface for assistive technologies: the control interface, the selection set, and the selection method (Cook and Hussey, 2002). The *control interface* is the boundary between the user and an electronic or mechanical assistive technology device. This is what allows the individual to operate or control the device. For electronic assistive technology systems, control interfaces include joysticks for powered wheelchairs, keyboards and mouse input for computers, and communication devices and single switches used to control household devices such as lights or radios.

The *selection set* is a presentation of the items from which the user makes choices. The elements in the selection set correspond to the elements of a specific activity output. Selection sets may have written letters, words and sentences, symbols used to represent ideas, computer icons, or line drawings/pictures. They may be presented in visual (e.g., letters on keys), tactile (e.g., Braille), or auditory (e.g., voice synthesis) form. We can define two *selection methods* through which the user makes selections using the control interface: *direct selection* and *indirect selection* (Cook and Hussey, 2002). For any particular application, the three elements of the human-technology interface will be chosen based on the best match to the consumer's skills (motor, sensory, linguistic, and cognitive) (Cook and Hussey, 2002).

The fastest and easiest selection method to understand and use is *direct selection*. In this method, each possible choice in the selection set is available at all times, and the user merely chooses the one that he or she wants. *Indirect selection* methods were developed to provide access for individuals who lacked the motor skills to use direct selection. Indirect selection methods are scanning, directed scanning, and coded access. Each of the indirect selection methods involves one or more intermediate steps between the user's action and entry of the choice into the device. One of the most commonly used methods is *scanning*. Although there are a variety of implementations of scanning, they all rely on the basic principle of presenting the selection-set choices to the user sequentially and having the user indicate when his or her choice is presented. The indication may be by a single movement of any body part. Since scanning is inherently slow, there have been a number of approaches that increase the rate of selection (Cook and Hussey, 2002). These involve selecting groups of characters first to narrow the choices and then selecting the desired item from the selected group.

In a combined approach, called *directed scanning*, the user first activates the control interface to select the direction (vertically or horizontally) in which the selection set is scanned by the device. The user then sends a signal (either by waiting or by hitting an additional switch) to the *processor* to make the selection when the desired choice is reached. Joysticks or other arrays of switches (two to eight switches, including diagonal directions) are the control interfaces that typically are used with directed scanning.

Coded access requires the individual to use a unique sequence of movements to select a code corresponding to each item in the selection set. These movements constitute a set of intermediate steps that are required to make the selection using either a single switch or an array of switches as the control interface. One example of coded access used in electronic assistive devices is Morse code, in which the selection set is the alphabet. An intermediate step [e.g., holding longer (dash) or shorter (dot)] is necessary to make a selection. Two-switch Morse code is also used, in which one switch sends dots and the other sends dashes. As long as either switch is held down, the dots or dashes are sent. A major goal in the development of Morse code was its efficiency, and this can be very useful when speed of entry is the goal. Codes are usually memorized, and this means that the selection set need not be visually displayed. This allows use by individuals who have visual limitations. Coded access also requires limited physical skill but significant cognitive skill, especially memory and sequencing.

30.4.1 Methods of Activation Used for Control Interfaces

Control interfaces may be characterized by the way in which the consumer activates them (Cook and Hussey, 2002). Three types of actions by the user can result in activation of the control interface: movement, respiration, and phonation. These are shown in Table 30.1. Movements can be detected in three basic ways. First, a force may be generated by a body movement and detected by the control interface. These mechanical control interfaces (e.g., switches, keyboard keys, joysticks, mouse, and trackball) represent the largest category. Many mechanically activated switches are available for control of assistive devices. Most often movement of the hand, head, arm, leg, or foot activates these switches. Hand, foot, or head movement (e.g., chin) is generally used to activate multiple-switch

TABLE 30.1 Methods of Activation for Control Interfaces

Signal sent, user action (what the body does)	Signal detected	Examples
1. Movement (eye, head, tongue, arms, legs)	a. Mechanical control interface activation by the application of a force	a. Joystick, keyboard, tread switch
	b. Electromagnetic control interface: activation by the receipt of electromagnetic energy such as light or radio waves	b. Light pointer, light detector, remote radio transmitter
	c. Electrical control interface: activation by detection of electrical signals from the surface of the body	c. EMG, EOG, capacitive or contact switch
	d. Proximity control interface: activation is by a movement close to the detector, but without contact	d. Heat-sensitive switches
2. Respiration (inhalation/expiration)	2. Pneumatic control interface: activation by the detection of respiratory air flow or pressure	2. Puff and sip
3. Phonation	3. Sound or speech control interface: activation by the detection of articulated sounds or speech	3. Sound switch, whistle switch, automatic speech recognition

Source: From: Cook, A. M., and Hussey, S. M. (2002), *Assistive Technologies: Principles and Practice*, 2d ed., Mosby Year Book Publishers, St. Louis, with permission.

arrays, including some joysticks. These vary in the amount of force required, the sensory feedback provided, and the degree to which the consumer can mount them for easy access.

Electromagnetic control interfaces can also be used to detect movement at a distance through either light (visible or infrared) or radiofrequency (rf) energy. These interfaces include head-mounted light sources or detectors and transmitters used with environmental control systems for remote control.

The third type of movement detection is electrical. These control interfaces sense bioelectric signals. Switches of this type require no force, and a common example of this type of interface is capacitive switches used on some elevator buttons. Bioelectrical switches detect muscle electrical activity (EMGs) or eye movement (EOG) by attaching electrodes to the skin.

Another type of body-generated signal is respiration, which is detected by measuring either airflow or air pressure using what is called a *sip-and-puff switch* activated by blowing air into the switch or sucking air out of it. Arrays of single puff-and-sip switches are often used for multiple functions. The most common use has been two puff-and-sip switches mounted side by side for wheelchair control. Puffing on both causes forward motion, sipping on both leads to reverse, and puffing on only one results in a turn in that direction.

30.4.2 Control Interfaces for Direct Selection

The most common control interface for direct selection is the keyboard. However, there are many different types of keyboards, and each requires unique user skills. Some consumers have limited range of movement but good fine motor control. In this case, a contracted keyboard (closely spaced keys over a small range) can be more effective than the standard keyboard. Other individuals have difficulty accessing small targets, but they have good range of movement. In this case, an expanded keyboard, in which the size of each key can be up to several inches, may allow direct selection. Different keyboard layouts can also be used (e.g., for left-hand- or right-hand-only typing). It is also possible to make the keys in different sizes and different shapes on the same keyboard. Touch screens allow a user choice from the selection set by pointing directly to the item on the screen. Because the item chosen is the one that is pointed at, this method can be easier for some users to understand.

Automatic speech recognition (ASR), in which the individual uses sounds, letters, or words as a selection method, is another alternative to keyboard input. In most such systems, the speech recognition is *speaker-dependent*, and the user "trains" the system to recognize his or her voice by producing several samples of the same element (Comerford et al., 1997). ASR system use is increasing in the mainstream commercial market for use on the Internet, dictation, general telephone use, and most other computer activities. Persons with disabilities will be the beneficiaries of this expanded use of ASR systems. These systems all use continuous ASR techniques in which the user can speak in almost normal patterns with slight pauses between words.

Speaker-independent systems recognize speech patterns of different individuals without training (Gallant, 1989). These systems are developed using samples of speech from hundreds of people and information provided by phonologists on the various pronunciations of words (Baker, 1981). The total recognition vocabulary is generally small. Discrete ASR systems are used in assistive technology applications for wheelchair control and electronic aids to daily living for appliance control. These systems require the user to pause between words and result in very abnormal speech patterns. They are only used in limited applications requiring a few commands to be recognized.

Often the microphones supplied with ASR systems are not adequate when the user has limited breath support, special positioning requirements, or low-volume speech (Anson, 1997). Many individuals who have disabilities may not be able to independently don and doff headset microphones that are normally supplied with commercial ASR systems. In these cases, desk-mounted types are often used. Current ASR systems utilize commonly available sound cards rather than separate hardware installed in the computer (Anson, 1999).

30.4.3 Control Interfaces for Indirect Access

Indirect methods of selection use a single switch or an array of switches. Cook and Hussey (2002) describe a number of interfaces that are used for indirect selection.

30.5 COMPUTER ACCESS BY PERSONS WITH DISABILITIES

Computer use by persons who have disabilities has opened up new opportunities for education, employment, and recreation. The computer offers (1) flexibility (multiple options with the same hardware), (2) adaptability (e.g., as user's skills change over time), (3) customization to a specific user and need (e.g., settings of scanning rate for indirect selection), and (4) specific applications and/or upgrades that can be based on software rather than hardware (e.g., specific user profile for Internet browsing in Braille) (Cook and Hussey, 2002). Computer use is often difficult for individuals who have motor and/or sensory impairments. Successful computer use requires sensory and perceptual abilities for processing computer outputs, motor control for generating input to the computer, and cognitive skills (e.g., problem solving, decision making, memory, and language) for understanding the computer functions. When a person with one or more disabilities has difficulty carrying out these functions, engineers are called on to adapt the computer to make it accessible.

30.5.1 Adapted Computer Inputs

The most common user computer input is provided via either the keyboard or a mouse. We can provide adapted computer input in many ways depending on the needs of the consumer. Keyboard alternatives were discussed earlier in this chapter. There are also software adaptations for the most common problems experienced by people with disabilities when using a standard keyboard. These are shown in Table 30.2. These software adaptations are included in Accessibility Options in Windows* and Universal Access in Macintosh† operating systems. They are accessed and adjusted for

*Microsoft Corporation, Seattle, Wash.
†Microsoft Corporation, Seattle, Wash.

TABLE 30.2 Minimal Adaptations to the Standard Keyboard and Mouse

Need addressed	Software approach*
Modifier key cannot be used at same time as another key	StickeyKeys (M,W)
User cannot release key before it starts to repeat	RepeatKeys(M)
User accidentally hits wrong keys	SlowKeys(M)
	BounceKeys(M)
	FilterKeys (W)
User cannot manipulate mouse	MouseKeys (M,W)
User wants to use augmentative communication device as input	SerialKeys (M,W)

**StickeyKeys:* User can press modifier key and then press second key without holding both down simultaneously; *RepeatKeys:* User can adjust how long key must be held before it begins to repeat; *SlowKeys:* A delay can be added before the character selected by hitting a key is entered into the computer; this means that the user can release an incorrect key before it is entered; *BounceKeys:* Prevents double characters from being entered if the user bounces on the key when pressing and releasing; *FilterKeys:* The combination of SlowKeys, BounceKeys, and RepeatKeys in Microsoft Windows; *MouseKeys:* Substitutes arrow keys for mouse movements; *SerialKeys:* Allows any serial input to replace mouse and keyboard.

Source: From: Cook, A. M., and Hussey, S. M. (2002), *Assistive Technologies: Principles and Practice*, 2d ed., Mosby Year Book Publishers, St. Louis, with permission.

an individual user through the control panel. Universal Access for the Macintosh includes Easy Access and Closeview. Easy Access features are those shown in Table 30.2. Closeview enables enlarged characters on the screen. It is described later in this chapter.

For persons who have cognitive difficulties, we can increase access by using *concept keyboards*. These keyboards replace the letters and numbers of the keyboard with pictures, symbols, or words that represent the concepts required by the software. For example, a program designed to teach monetary concepts might use a concept keyboard in which each key is a coin rather than a number or letter. The user can push on the coin and have that amount entered into the program. Such keyboards have been used in point-of-sale applications to allow individuals who have intellectual disabilities to work as cashiers. The Intellikeys keyboard* is often used as a concept keyboard.

Alternatives to the use of a mouse for computer input that are often used by persons with disabilities include trackballs, a head-controlled mouse, a continuous joystick, eye pointing, and the use of the arrow keys on the numeric keypad (called *mouse keys*; see Table 30.2.).† Head control for mouse emulation employs an infrared system that detects a reflected beam to measure head position relative to a fixed reference point for the cursor (the center of the screen). As the user moves his or her head away from this point in any direction, the cursor is moved on the screen. Commercial systems‡ use a wireless approach in which a reflective dot is place on the user's forehead and serves as the reflective target. This allows the user to move around more freely. These head-controlled systems are intended for individuals who have limited arm and hand control and who can accurately control head movement. For example, persons with high-level spinal cord injuries who cannot use any limb movement often find these head pointers to be rapid and easy to use. On the other hand, individuals who have random head movement (e.g., due to cerebral palsy) or who do not have trunk alignment with the vertical axis of the video screen because of poor sitting posture often have significantly more trouble using this type of input device.

In some severe physical disabilities, the user may only be able to move his or her eyes. In this case, we can use a system that detects the user's eye movements to control mouse pointing. Two basic approaches are used in the design of eye-controlled systems. One of these uses an infrared video camera mounted below the computer monitor. An infrared beam is aimed at the eye and reflected back into the camera. As the user directs his or her eye gaze at different points on the computer monitor screen, signal-processing software is used to analyze the camera images and determine where and for how long the person is looking at the screen. The user makes a selection by looking at it for a specified period of time, which can be adjusted according to the user's needs. The EyeGaze System§ and Quick Glance¶ are examples of two eye-controlled systems that emulate mouse pointing. The second approach uses a head-mounted viewer attached to one side of the frame of a standard pair of glasses in front of the eye. Eye movements are tracked and converted into keyboard input by a separate control unit. One example of this type of system is VisionKey.** An eye-controlled system is beneficial for individuals who have little or no movement in their limbs and may also have limited speech, e.g., someone who has had a brain stem stroke, has amyotrophic lateral sclerosis (ALS), or has high-level quadriplegia.

In each of these alternative pointing devices, input to the pointing device moves a pointer on the screen. Once the pointer is moved to the desired item, the user can make a selection by either pausing for a preset time (called *acceptance time selection*) or by pressing a switch (called *manual selection*).

The *graphic user interface* (GUI) is used as the selection set for mouse entry. There are limitations to use of a GUI for people with disabilities. The first of these is that the GUI requires significant eye-hand coordination. Pointing devices rely on a significant amount of coordination between the body site (hand, foot, or head for most individuals with disabilities) executing the movement of the screen pointer and the eyes following the pointer on the screen and locating the targets. Second, it

*Intellitools, Richmond, Calif.; *www.intellitools.com.*

†Included with Windows and Macintosh operating systems.

‡Head Mouse, Orign Instruments; *www.orin.com/*; Tracker, Madentec Limited; *http:\\www.madentec.com/.*

§LC Technologies, Inc., Fairfax, Virginia; *www.lctinc.com/.*

¶EyeTech Digital Systems, Mesa, Arizona; *www.eyetechds.com/.*

**H. K. EyeCan, Ltd., Ottawa, Canada; *www.cyberbus.ca/~eyecan/.*

relies on a visual image of the screen for operation. If either of these is not present (e.g., muscle weakness or blindness), then the GUI is difficult to use and must be adapted or an alternative found.

The sensory feedback provided by the particular type of pointing device can vary widely. For individuals who have some vision, a major form of feedback is vision, i.e., following the cursor on the screen. Devices controlled by the hand (e.g., mouse, trackball, or joystick) also provide rich tactile and proprioceptive feedback. Head- and eye-controlled pointing devices provide much less sensory feedback, and other means, such as a light-emitting diode (LED) that lights when the signal is being received, are included to aid the user. The type of sensory feedback affects the user's performance, and the more feedback available, the more likely the use will be successful.

30.5.2 Adapted Computer Outputs

User output from the computer is typically provided by either a video display terminal (VDT), a printer, or speech or sounds. Use of any of these requires an intact sensory system. Alternatives that can be provided may substitute auditory (e.g., speech) or tactile (e.g., Braille) output for the VDT output or visual cues for sounds or speech. In the case of low vision, the standard size, contrast, and spacing of the displayed information are inadequate. For individuals who are blind, alternative computer outputs based on either auditory (hearing) or tactile (feeling) modes are used. Persons who are deaf or hard of hearing may also experience difficulties in recognizing auditory computer outputs. Adaptations that facilitate some of these functions are included in Accessibility Options* in Windows. These include ShowSounds, which displays captions for speech and sounds, and SoundSentry, which generates a visual warning when the system generates a sound.

30.5.3 Alternatives to Visual Input for Individuals Who Have Low Vision

The major problem with visual computer displays for individuals with low vision is that the text characters and icons are not easily readable. The three factors that affect the readability of text characters are (1) size (vertical height), (2) spacing (horizontal distance between letters and width of letters), and (3) contrast (the relationship of background and foreground color). Commercial screen magnifiers are used to compensate for several of these problems. In most cases these are software approaches only, but some include a hardware component as well. Screen-magnifying software enlarges a portion of the screen. The unmagnified screen is called the *physical screen*. Screen magnifiers enlarge one section of the physical screen to fit the whole display screen with a magnification of 2 to 32 times or more. The user has access to only the portion of the physical screen that appears in the magnified viewing window. In order to help the user navigate, the viewing window must track any changes that occur on the physical screen by moving the viewing window to show the portion of the physical screen in which changes are occurring or where input is required. For example, if a navigation or control box is active, then the viewing window highlights that box. If mouse movement occurs, then the viewing window can track the mouse cursor movement. If text is being typed in, then the viewing window will highlight that portion of the physical screen. Scrolling allows the user to slowly move the viewing window over the physical screen image. This may be automatic or manual.

Adaptations that allow persons with low vision to access the computer screen are available in several commercial forms. Lazzaro (1999) describes several potential methods of achieving computer access for people who have low vision. The simplest and least costly are built-in screen-enlargement software programs provided by the computer manufacturer. One of these is Close View.† This program allows for magnification from 2 to 16 times and has fast and easy text handling and graphics capabilities. Magnifier‡ displays an enlarged portion of the screen (from 2 to 9 times magnification)

*Microsoft Corporation, Seattle, Wash.

†Macintosh Operating System, Apple Computer, Cupertino, Calif.

‡Windows, Microsoft, Corporation, Seattle, Wash.

in a separate window. Software programs that are purchased separately rather than being built in offer wider ranges of magnification and have more features than built-in screen magnifiers. Hardware and software combinations have other features such as multiple enlarged windows, smoother scrolling, and a wider range of magnification. Cook and Hussey (2002) and Lazzarro (1999) describe commercial approaches to screen-magnification utilities. Most of these approaches also allow adjustment of background and foreground colors to address difficulties with contrast.

30.5.4 Alternatives to Visual Input for Individuals Who Are Blind

For individuals who are blind, computer outputs must be provided in either auditory or tactile form or both. Auditory output is typically provided through systems that use voice synthesis and tactile output as in Braille. These adaptations are generally referred to as *screen readers*. Commercial screen readers have two modes: application and review. In the application mode, the user can utilize all the screen features as well as the features of the application program. This mode is normally used for reading the screen contents. In review mode, the user can utilize all the screen-reading features but cannot access the features of the application program (e.g., editing of a word processor text file). Current systems eliminate the use of the two modes for most applications (e.g., word processing, spread sheets, Web browsing) and provide full review capability while in an application program. For some applications (e.g., a complex document with multiple columns or a complicated Web page), the review mode is used.

Cursor routing allows a user to mark a location in an application so that the screen-reading cursor can return to it when the user exits the review mode and goes into application mode. Macros in screen-reader programs are useful in moving between windows or locating key areas (e.g., a dialog box or a window requiring input from the user). Screen readers are ideally suited for applications that consist of text only.

However, the power of any particular screen reader is in the degree to which the other capabilities dictated by the use of the GUI are achieved. The fundamental differences in the ways that a text-only command-line interface (CLI) and a GUI provide output to the video screen present access problems for persons who are blind. These are related to both the ways in which internal control of the computer display is accomplished and the ways in which the GUI is employed by the computer user (Boyd et al., 1990). The CLI-type interfaces use a memory buffer to store text characters for display. Since all the displayed text can be represented by ASCII code, it is relatively easy to use a software program and to divert text from the screen to a speech synthesizer or Braille display. However, this type of screen reader is unable to provide access to charts, tables, or plots because of their graphic features. This type of system is also limited in the features that can be used with text. For example, features such as size, shape, and font or alternative graphic forms are not captured in standard ASCII text code.

GUI design uses a totally different approach that creates many more options for the portrayal of graphic information to video display control. Since each character or other graphical figure is created as a combination of dots, letters may be of any size, shape, or color, and many different graphical symbols can be created. This is very useful to sighted computer users because they can rely on the use of visual metaphors (Boyd et al., 1990) to control a program. *Visual metaphors* use familiar objects to represent computer actions. For example, a trash can may be used for files that are to be deleted, and a file cabinet may represent a disk drive. The graphic labels used to portray these functions are referred to as *icons*. Screen location is important in using a GUI, and this information is not easily conveyed via alternative means such as screen readers. Visual information is spatially organized, and auditory information (including speech) is temporal (time-based). It is difficult to convey screen location of a pointer by speech alone. An exception to this is screen locations that never change (e.g., the edges of the screen, such as "right border" and "top of screen"). Another major problem is that mouse pointer location on the screen is relative, and the only information available is the direction of the movement and how far the mouse has moved. One approach to this problem is the Microsoft Screen Access Model.* This is a set of technologies designed to facilitate the development of screen readers and other accessibility utilities for Windows that provide alternative ways to store and access information about the contents of the computer screen. The Screen Access Model also includes software driver interfaces that provide a standard mechanism for accessibility utilities to send information to speech devices or refreshable Braille displays. GUI access also requires the ability to locate open windows, monitor them for changes, and output information to the user if changes occur. Screen-reader programs also provide assistance in this navigation function by using keyboard commands such as movement to a particular point in the text, finding the mouse cursor position, providing a spoken description of an onscreen graphic or special function key, or accessing help information. Screen readers also monitor the screen and take action when a particular block of text or a menu appears (Lazzaro, 1999). This feature allows the user to automatically have pop-up window and dialog boxes read to him or her. Screen readers can typically be set to speak by line, sentence, or paragraph.

For Windows-based computers, the majority of commercial products include both a voice synthesizer and a software-based screen reader to provide access. Many of these bundled software programs also work with computer soundboards to generate high-quality synthetic speech. Tactile (Braille) display of screen information is the other major alternative for screen readers. This requires the use of a translator program to convert from text characters to Braille cell-dot patterns. Computer output systems use either a refreshable Braille display consisting of raised pins or hard copy via Braille printers. Refreshable Braille displays consist of 20, 40, or 80 separate cells. Rather than the standard 6-cell Braille used for print materials, a unique 8-dot cell format is available in which the seventh and eighth dots are used for indicating the location of the cursor and to provide single-cell presentation of higher-level ASCII characters. The latter feature is necessary because the normal 6-cell Braille display can only generate 64 permutations and full ASCII has 132 characters. Braille embossers produce hard-copy (printed) output. Cook and Hussey (2002) describe a number of commercial approaches to screen readers with speech and Braille output as well as embossers for hard copy.

30.6 AUGMENTATIVE AND ALTERNATIVE COMMUNICATION

The term *augmentative and alternative communication* (AAC) is used to describe any communication that requires something other than the person's own body. This includes things such as a pen or pencil, a letter or picture communication board, a typewriter, or an electronic communication device. There are two basic communication needs that lead to the use of augmentative and alternative communication systems: *conversation* and *graphics* (Cook, 1988).

Conversational needs are those that would typically be accomplished using speech if it were available. Conversational use typically focuses on interaction between two or more people. Light (1988) describes four types of communicative interaction: (1) expression of needs and wants, (2) information transfer, (3) social closeness, and (4) social etiquette. Each of these has unique features that dictate the AAC characteristics to be included. Since speech allows communication at a rapid rate, between 150 and 175 words per minute (Miller, 1981), an individual using an AAC system must be as rapid as possible. In all AAC systems, some form of letter or symbol selection is required, and in many cases persons who are unable to speak use a keyboard to type their messages that are then spoken by an AAC device. If the person has limited motor skills, this can result in significantly lower rates of communication than for speech (as low as a few word per minute). Cook and Hussey (2002) describe other relationships between AAC characteristics and effective conversational skills.

Graphic communication describes all the things that we normally do using a pencil and paper, typewriter, word processor, calculator, and other similar tools, and it includes writing, mathematics, and drawing and/or plotting. Each of these serves a different need, and therefore, AAC devices

**microsoft.com/enable/products/microsoft.*

designed to meet each type of need have different characteristics (Cook and Hussey, 2002). For example, all AAC systems for writing must be capable of providing output in a hard-copy format. This can either be generation of normal text (letters, numbers, and special characters) or special symbols. If spelling is difficult for the user, some devices allow the selection of whole words that are then output to a printer. While it is possible to learn basic arithmetic without being able to write the numbers down, it is much more difficult. AAC systems used for math allow the cursor to move from left to right as the user enters numbers to be added, but once the user has a column of numbers, the cursor moves right to left as he or she enters the sum. Special symbols (e.g., Greek letters) and the use of superscripts and subscripts are also required for algebra. AAC device characteristics that are required for drawing include cursor movement in all four directions; choice of colors, line widths, and other features; and the ability to save a drawing for later editing.

AAC systems may be thought of as having three major components: (1) user interface, (2) processor, and (3) output. The user interface was described earlier in this chapter. It includes the user control interface, selection method and selection set, and an optional user display to provide feedback for self-correction. AAC devices often use special symbols in the selection set. These include miniatures of objects, color or black and white pictures, line drawings, pictographic symbols, text characters, and multiple meaning icons* (Beukelman and Mirenda, 1998). For AAC systems, the processor has several specific functions: (1) selection technique, (2) rate enhancement and vocabulary expansion, (3) vocabulary storage, and (4) output control. The output is conversational and/or graphic communication. Communication takes place in many different settings.

In order to maximize the production of output, AAC devices use techniques to increase the rate of entry by the user. Any approach that results in the number of characters generated being greater than the number of selections that the individual makes will increase rate. Rate-enhancement techniques can be grouped into two broad categories: (1) encoding techniques and (2) prediction techniques. There are several types of codes that are currently used in AAC devices. Numerical codes can be related to words or complete phrases or sentences. When the user enters one or more numbers, the device outputs the complete stored vocabulary item. Abbreviation expansion is a technique in which a shortened form of a word or phrase (the abbreviation) stands for the entire word or phrase (the expansion). When an abbreviation is entered, it is automatically expanded by the device into the desired word or phrase. Vanderheiden and Kelso (1987) discuss the major features of abbreviation-expansion systems and the strategies for developing stored vocabulary using this approach. An alternative approach that is based on coding of words, sentences, and phrases on the basis of their meaning is called *semantic encoding* (Baker, 1982). In this approach, pictorial representations, called *icons*, are used in place of numerical or letter codes. For example, using a picture of an apple for "food" and a sun rising for "morning" and then selecting "apple" and "sunrise" as a code for "What's for breakfast" is easier to remember than an arbitrary numerical or letter code for the same phrase. The apple can also represent the color red, eating, fruit, etc.

It is also possible to realize significant increases in rate by using word-prediction or word-completion techniques with any selection method (Swiffin et al., 1987). Devices that use these techniques typically have a list on the screen that displays the most likely words based on the letters that have previously been entered. The user selects the desired word, if it is listed, by entering a number listed next to the word. If the desired word is not displayed, the user continues to enter letters, and the listed words change to correspond to the entered letters. In adaptive word completion, the ranking of the presented list is changed based on the user's frequency of use.

A selection set for AAC consists of a maximum of 128 selections on most devices, and many have far fewer available at any one time. However, even a child has a vocabulary of thousands of words. Thus it is necessary to have methods to allow easy access to vocabulary items that are not displayed. The rate-enhancement techniques are one way to do this, but they all require memory and cognitive skills such as sequencing. One approach is to present the selection sets on a dynamic display accessed through a touch screen. The displayed information is changed based on previous entries. For example, a general selection set might consist of categories such as work, home, food,

---

*Minsymbols, Prentke Romich Co., Wooster, Ohio.

clothing, greetings, or similar classifications. If one of these is chosen, either by touching the display surface directly or by using scanning, then a new selection set is displayed. For example, a variety of food-related items and activities (eat, drink, ice cream, pasta, etc.) would follow the choice of "foods" from the general selection set. Thus the user does not have to remember what is on each level. Cook and Hussey (2002) describe typical approaches used in AAC system design.

30.7 AIDS FOR MANIPULATION

Many individuals who have limited function of their arms and hands experience difficulty in manipulating objects, controlling appliances (e.g., television, telephone, etc.), reading a book, feeding themselves, or doing self-care. Assistive technologies that aid manipulation may be simple mechanical aids (e.g., reachers or enlarged-handle eating utensils), special-purpose electromechanical devices (e.g., page turners or feeders), or general-purpose devices (e.g., electronic aids to daily living and robotics). In this section I will discuss only electronic aids to daily living (EADLs). The others are described in Cook and Hussey (2002).

Many objects that need to be manipulated are electrically powered devices such as appliances (e.g., television, room lights, fans, and kitchen appliances such as blenders or food processors), which use standard house wiring (110 volts ac in North America). EADLs are designed to allow people with limited upper extremity function to control these appliances. *They were formerly known as environmental control units (ECUs).* A typical EADL for turning appliances on and off is shown in Fig. 30.1. The user control interface may be a keypad, as shown, or a single switch with an indirect selection method. The appliances are plugged into modules that are controlled by the EADL. The most common type of on/off module is the X-10.* Three types of wireless transmission are used in EADL design. The most common is *infrared* (IR) transmission like that used in most TV remote units. A second method, also sometimes used for TV control, is *ultrasound* transmission. The third method, often used in garage door openers, is *radiofrequency* (rf) transmission. IR and ultrasound require line-of-sight transmission. Rf transmission does not have this requirement.

The system in Fig. 30.1 may be modified to include remote control over TV or VCR functions such as volume control, channel selection, play, fast forward, and reverse. In this case, the level (for volume) or number (for channel) is under the control of the human user. Often these functions are incorporated into EADLs by modifying standard TV or VCR remote controls. This may be done merely by adding a single switch to the remote control or by more elaborate adaptations that allow indirect selection. Sometimes universal remote controls that can "learn" the signal for a particular TV or VCR are used. This allows several appliances to be controlled from the same EADL. Cook and Hussey (2002) describe several commercial approaches to this type of EADL.

Persons with physical disabilities of the upper extremities often have difficulty in carrying out the tasks associated with telephone use. These include lifting the handset, dialing, holding the handset while talking, and replacing the handset in its cradle. There are several options for accomplishing these tasks. Nonelectronic methods such as mouth sticks or head pointers can be used to press a button to open a line on a speakerphone, dial, and hang up. EADLs perform these same telephone tasks electronically. For persons who require single-switch access to the system, the control interface is connected to a control unit that also interfaces with a display and with standard telephone electronics. A typical approach is for the device to present numbers sequentially on the display, and the user presses a switch when the desired number to be dialed is on the display. By repeating this process any phone number can be entered and then sent through the standard telephone electronics for automatic dialing. Since many persons with disabilities respond slowly, all practical systems use stored numbers and automatic dialing. Another unique feature is the inclusion of a Help or Emergency phone number that can be dialed quickly. Most systems have a capacity of 50 to 100 stored numbers. Some telephones are IR-controlled, and they can be included with EADLs that learn device codes.

* X-10 Powerhouse System, Northvale, N.J.

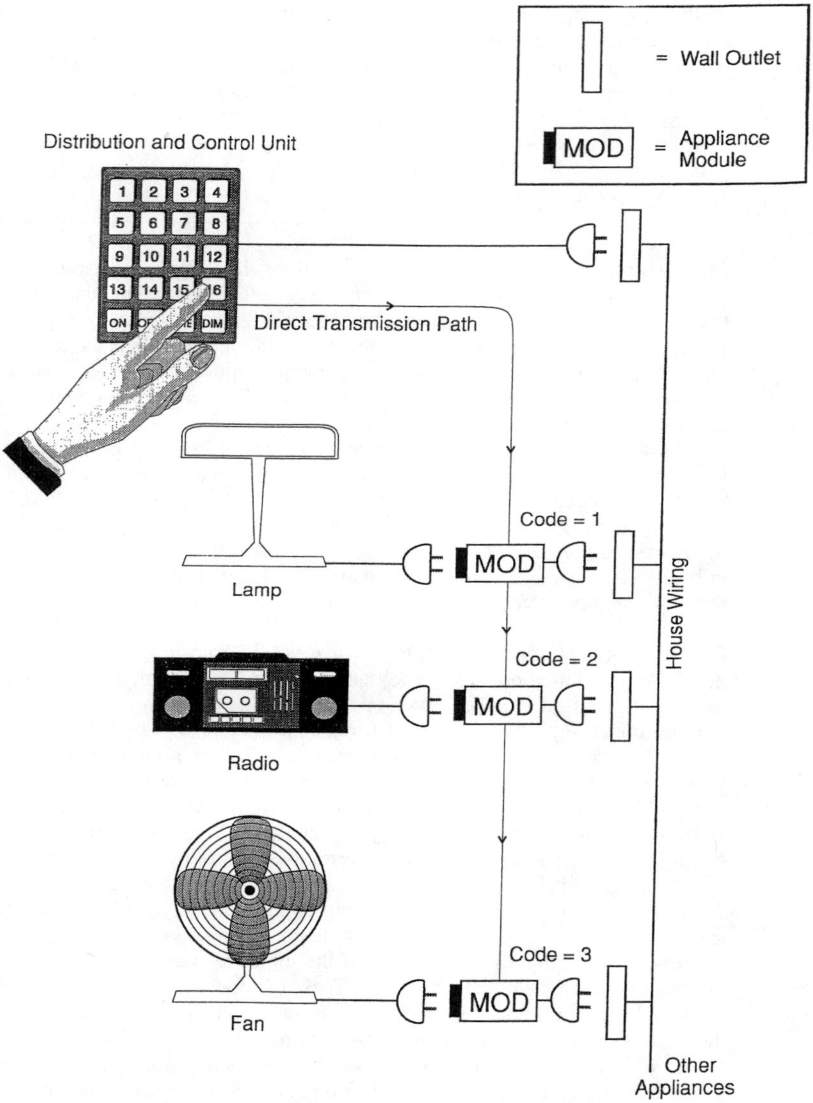

FIGURE 30.1 A direct-selection ECU. Each appliance has a numerical code, and the keypad is used to select the appropriate module. Control functions such as "on," "off," and "dim" are also activated by pressing the proper key on the keypad. This figure also illustrates the use of house wiring for distribution of the control signals to the appliance modules. (From Cook, A. M., and Hussey, S. M., *Assistive Technologies: Principles and Practice, 2d ed., Mosby Year Book Publishers, St. Louis, 2002, with permission.*)

30.8 FUTURE TRENDS

The editors of *PC Magazine* identified 10 trends that are likely to have a profound impact on assistive technology design and application over the first 10 years of the twenty-first century (Miller et al.,

1999.). One prediction is that computers will have more human attributes such as reacting to spoken words (using ASR) or hand-written instructions. In general, the emphasis is on making the user interface more "friendly." Telephone access to the Internet via ASR is already available. ASR will continue to improve, but a major challenge will continue to exist for people who have unintelligible (dysarthric) speech. Current ASR systems do not provide good results for this population, and this is clearly an area in which assistive technologies must be developed to allow persons with speech disabilities to access the new user interfaces. Similar problems could occur for individuals who have poor fine motor control if user interfaces require recognition of handwriting.

The *chaterbot*, a virtual character that responds to questions using voice synthesis, is another feature designed to make the user interface more human-like. With this user interface, the user asks questions in a particular subject area using ASR. Successful use of these interfaces depends on expected questions and answers in a given subject area. This is how the interface appears "smart." However, the chaterbot cannot reason or solve problems. A natural-language interface of this type has great appeal for use by people who have intellectual abilities. The question is how forgiving these interfaces will be if the user becomes confused or uses nonstandard forms of language and speech. Rehabilitation engineers can ensure that these general-purpose technologies are inclusive of all individuals by employing universal design principles and by adapting the user interface so that it is accessible to individuals with motor, sensory, and cognitive limitations.

Another example of a human-like user interface is the inclusion of emotion (Miller et al., 1999). Current work in this area includes cameras that monitor facial expressions and sensors that detect physiological changes. These signals provide the information necessary for an *affective tutor* to adjust a program to react to emotions. For example, if the tutor detects confusion, an alternative explanation could be offered. Researchers are also working on user interfaces that express emotion to the user as well. These are based on animated characters that change facial expressions to react to input questions. These advances could be of benefit to individuals who have intellectual disabilities and require more concrete interactions.

Changes in the nature of networks will occur over the next few years. These changes have the potential to benefit those who have disabilities. Existing networks will be expanded into the home, work, and community environments to provide the capability for unique and powerful connections. For example, automobile networks will be connected to toll booths and automated fuel pumps. They will also make assisted driving possible, with a potential benefit to persons with disabilities. Because networks will become wireless in the future, connectivity will be a function of local resources, not existing hard-wired communications providers. This offers the opportunity for people with disabilities to use their assistive technologies to connect to the network. If this is to be realized, then rehabilitation engineers must design assistive technologies that keep pace with constant changes in the design of network configurations. A second major trend affecting networks is an increase in the "intelligence" of the Internet. Web browsers will be able to track sites by the characteristics the user typically looks for. These might include styles, colors, and sizes of clothing and particular items of interest. Software will also determine a profile for the user based on these preferences and then compare the profiles with those of other users who have similar tastes and interests. The intelligent agent can then recommend products and services matching the profile. These changes require XML (for Extensible Markup Language), an enhanced programming language. Accessibility guidelines for XML are being developed by the World Wide Web Consortium Web Access Initiative (*www.w3.org/WAI*).

In the future, appliances from watches to dishwashers will have embedded microcomputers, making it possible to network them within a home and to control them remotely. For example, a meal can be started from work by turning on the stove remotely. Some appliances such as cellular phones will also include access to the Internet and other computing functions. One of the keys to these applications is the use of hardware and software that allows digitally controlled appliances to self-organize into a community without a server to manage the network. This type of interconnectivity and remote control are ideal for individuals who have EADLs or who cannot see appliance controls. To ensure accessibility to persons with disabilities, the software and networking capabilities in these appliances must meet the accessibility guidelines described earlier. They must also be compatible with assistive devices (e.g., desktop computers, personal digital assistants, control interfaces, AAC devices) used for input or output from the network. There will also be major changes in e-mail and chat rooms. They will proceed beyond pure text to text-linked and three-dimensional graphics. The Internet (e-mail and chat rooms) has the advantage of anonymity, and this can be a major benefit to individuals who have disabilities (Blackstone, 1996). Some of these advantages are

1. Composition at a slower speed since the partner reads it at a later time.
2. User can communicate with another person without someone else being present.
3. Because the person's disability is not immediately visible, people with disabilities report that they enjoy establishing relationships with people who experience them first as a person and then learn of their disability.

As the ability of the Internet to connect people grows, the benefits to people who have disabilities will also increase. The engineering to make this a reality is that these individuals must continue to have access to both input to and output from the Internet as this growth progresses.

Other trends, cited by Miller et al. (1999), include smarter and more powerful software with faster computing hardware, expanded memory and functionality, and decreasing or stable cost. This means that the size of information-processing electronics will continue to shrink, with greater and greater capability in smaller and smaller packages. This will have implications for assistive technologies that depend on large amounts of signal processing with severe size and weight limitations such as hearing aids and AAC devices.

30.9 SUMMARY

Electronic assistive devices provide expanded opportunities for people who have disabilities. Those discussed in this chapter include access to computers for both input and output functions, Internet access, electronic aids for manipulation, and communication devices (AAC). As advances are made in electronics, computing, and network applications, there is a constant challenge to ensure that these systems remain accessible to persons who have disabilities. Rapidly changing communication technologies may result in the need to redesign accessible interfaces. Developments that broaden the scope, applicability, and usability of the user interface will be driven, at least in part, by the needs of people who have disabilities. Engineers who focus their efforts on rehabilitation applications can have a significant influence on these developments and on the lives of people who have disabilities.

REFERENCES

Anson, D. (1999), Speech recognition technology, *OT Prac.* (January-February):59–62.

Anson, D. K. (1997), *Alternative Computer Access: A Guide to Selection*, F. A. Davis Company, Philadelphia.

Baker, J. M. (1981), How to achieve recognition: A tutorial/status report on automatic speech recognition, *Speech Technol.* (Fall):30–31, 36–43.

Beukelman, D. R., and Mirenda, P. (1998), *Augmentative and Alternative Communication, Management of Severe Communication Disorders in Children and Adults*, 2d ed., Paul H. Brooks, Baltimore.

Boyd, L. H., Boyd, W. L., and Vanderheiden, G. C. (1990), The graphical user interface: Crisis, danger, and opportunity, *J. Vis. Impair. Blindness* **84**(10):496–502.

Comerford, R., Makhoul, J., and Schwartz, R. (1997), The voice of the computer is heard in the land (and it listens too), *IEEE Spectrum* **34**(12):39–47.

Cook, A. M., and Hussey, S. M. (2002), *Assistive Technologies: Principles and Practice*, 2d ed., Mosby Year Book Publishers, St Louis.

Gallant, J. A. (1989), Speech-recognition products, *EDN* (Jan):112–122.

Lazzaro, J. L. (1999), Helping the Web help the disabled, *IEEE Spectrum* **36**(3):54–59.

Miller, M. J., Kirchner, J., Derfler, F. J., Grottesman, B. Z., Stam, C., Metz, C., Oxer, J., Rupley, S., Willmott, D., and Hirsch, N. (1999), Future technology, *PC Magazine* **18**(12):100–148.

CHAPTER 31
APPLIED UNIVERSAL DESIGN

Ronald S. Adrezin
University of Hartford, West Hartford, Connecticut

31.1 MOTIVATION

Engineers are expert problem solvers. They love challenges and regularly handle the stress and excitement of balancing cost with safety and reliability. However, many wonderful devices on the market were not designed by engineers. So what makes engineers so valuable? They can apply their skills across industries. An engineer can safely strip every ounce out of a weight-critical design. And when a problem is properly defined, an engineer can solve it in a cost-effective manner.

Designing a product that is accessible to persons with disabilities will increase your market share and produce a better product for all. If this accessibility is designed into the product, there is a negligible effect on the price. This is the heart of what we call *universal design*. By applying one's creativity, countless products can be manufactured that do not neglect people below the fifth or above the ninety-fifth percentiles. These individuals may be found among your family, friends, neighbors, and colleagues. Do not exclude them!

It is my sincere hope that this chapter will assist you in designing for persons with disabilities. The consumers of these products have been divided into three broad categories. The first includes custom devices designed for a sole user. Here, the design team knows the needs, desires, and anthropometrics of the final user, who is generally part of the team. A second category covers devices to be marketed to persons with a common disability. Examples include hearing aids, wheelchairs, and environmental control units. Their design may include the ability to be customized by the end user. The last category describes products to be used by the general population. These mass-produced devices include television remote controls, appliances, toys, and computers. Whether for a sole user, a group of persons with a common disability, or a mass-produced product where universal design principles are applied, I have attempted to provide engineers with a practical starting point. This chapter stresses new designs of devices for persons with disabilities. It is not focused on the selection or customization of existing assistive devices.

Engineers sometimes forget the importance of form while focusing on function. Persons with disabilities are often subjected to stares and the stigmas attached to the use of many assistive devices. In the design process, therefore, you must consider the aesthetics of the device. There is often an assumption that everyone loves technology as much as engineers. For instance, there was a time when many counselors would push a person with a spinal chord injury into the field of computer programming. Yes, the interface to the computer was available but not necessarily the person's interest. Can everyone in the general population become a programmer? So why should everyone with a disability? It is our role to utilize technology to expand the options available. As a result of the Americans with Disabilities Act,* companies must make reasonable accommodations to allow those with disabilities to succeed in the workplace.

It is my hope that readers will consider lending their talents to this challenging but satisfying field. I also encourage you to fully incorporate universal design into all your products.

31.2 DESIGN METHODOLOGY

This section has been developed to supplement the many fine design textbooks available. The order of the subsections from problem definition to embodiment design is similar to the organization of many of these texts. The conceptual generation stage includes techniques such as brainstorming and functional decomposition. This section will simply present functional decomposition as an example. Embodiment design, where the design concept begins its physical form, combines product architecture, configuration, and parametric design.[1] Here, the focus is on human factors and universal design. Evaluation methods and detail design are not included, but your concepts should satisfy universal design principles in the evaluation stage.

31.2.1 Problem Definition

This subsection addresses special issues in the problem definition stage of the design process. It then demonstrates how techniques such as quality function deployment can be applied to this field.

31.2.2 Avoid an Ill-Defined Problem

A properly defined engineering problem is essential to a successful design. You must understand the desires and needs of the device's user, as well as those who will interact with, prescribe, service, or pay for this device. (Read the Sec. 31.2.5 for more details.) Many articles and texts treat this topic, and it is there that I refer you.[1–3,7] As you study this chapter, pay careful attention to the person's age, prognosis, and the many social-psychological issues as the problem is defined. If the user is a child, his or her size will change, as will his or her physical and cognitive abilities. A person with a disease may have his or her condition improve or worsen. The prognosis is the best indication of what might happen. Finally, if the device has any negative connotations associated with it, it might not be used in public or at all.

31.2.3 Striving to Increase Independence

Assistive devices for persons with disabilities are not designed to "fix" the person but rather to improve the person's surroundings.[4] Devices that can improve and enhance a person's independence are of great value. A device that helps a person bathe his or her elderly parent is useful. Imagine a product that allows this parent to bathe himself or herself; now that is wonderful! Whether applying universal design principles to develop a product to satisfy an aging population, an adult to succeed

*The Americans with Disabilities Act was signed into law in 1990 and went into effect in January 1992. *Reasonable accommodations* refer to efforts that may include, among other adjustments, making the workplace structurally accessible to disabled individuals, restructuring jobs to make best use of an individual's skills, modifying work hours, reassigning an employee with a disability to an equivalent position as soon as one becomes vacant, acquiring or modifying equipment or devices, appropriately adjusting or modifying examinations, training materials, or policies, and providing qualified readers for the blind or interpreters for the deaf.

in the workplace, a teen to attend school, or a child to participate in social activities with both able-bodied and disabled friends, an engineer has the ability to increase a person's independence and improve the quality of his or her life. In addition, for often a negligible differential in per-unit cost when designed into the product, a company will increase its market share by tapping into this large population.

31.2.4 Reality Check

Commitment to a design refers to the point at which you have invested too many resources to make substantial revisions. At this point, your options may be to continue with your current design or scrap it entirely. Best practice suggests that you keep your design on paper (or computer) as long as possible before investing funds on the product's manufacture.[7] As part of good design practice, before you are committed to a design, do a reality check. This is where you ensure that the features that you have selected to satisfy your customers are appropriate. Simply selecting to meet the requirements calculated to be of greatest importance as a result of a focus group may lead to a design that no one wants. For example, you might miss the features that are commonplace and therefore expected. This may result from a focus group's failure to list important features because they assume that all products will have them. Examples include a remote control for a television, the automatic shutoff of a coffee pot, the ability to fold a manual wheelchair, and the light weight of a walker. When designing products for a broad population, the statistics from the quality function deployment, if improperly interpreted, may lead to the production of a two-door minivan with seven sporty bucket seats that cannot accommodate infant seats. However, when applied correctly, techniques such as quality function deployment will aid the design team in fully understanding the customers' needs and wants. It is one of many important tools that will lead to a successful, safe, and cost-effective design.

31.2.5 Quality Function Deployment

Brief Overview. Quality function deployment (QFD) is a powerful and widely used method to define your customers, determine their needs, benchmark the competition, and define engineering parameters and targets that, when met, will lead to a successful product. The QFD diagram, referred to as the *house of quality* because of its shape (Fig. 31.1), provides an important view of the customers' needs. Understanding your customers and the environment in which you must compete is crucial to your problem definition.

There are variations to the house of quality.[1,3,5,7] Some people try to quantify the relative importance of each requirement[1]; others are more qualitative. Figure 31.1 shows a house of quality with each "room" numbered. Figure 31.2 illustrates an example of the following hypothetical product: A manufacturer of toaster ovens would like its next generation of products to satisfy universal design requirements. In this brief overview of QFD, select issues will be highlighted in each room. Follow both Figs. 31.1 and 31.2 in the house tour below.

Room 1: Customer requirements. These are generated by focus groups, surveys, interviews, etc. The requirements include expecters, spokens, unspokens, and exciters.[1] *Expecters* are standard features that one would expect in a product. *Spokens* are features that the customer will tell you are desired. *Unspokens* are important features that the customer has forgotten, was unwilling to mention, or did not realize at the time. These first three categories must be addressed for consumer satisfaction. The last category, *exciters*, are typically unique features of your product that will boost sales. Since technology advances quickly and competition is stiff, today's exciters will become next quarter's spokens and next year's expecters. This is evidenced in both the computer and consumer electronics industry. Some requirements are directed toward function and others toward form. Requirements also may include price, serviceability, etc.

Room 2: Identification of the customers and their ranking of the customer requirements. When identifying your customers, consider internal customers as well. Are you developing a product

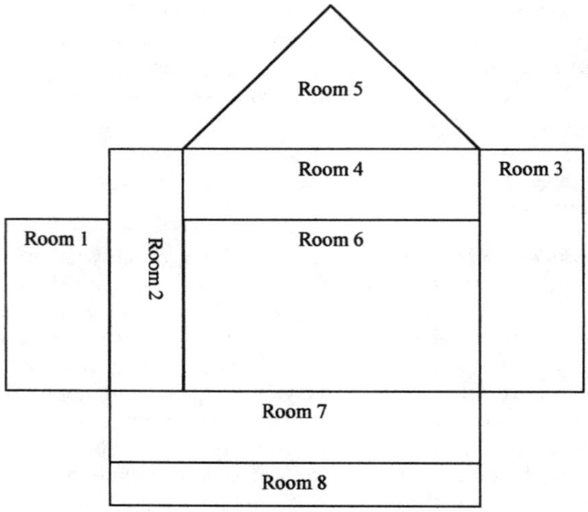

FIGURE 31.1 House of quality.

that your manufacturing personnel can produce? Does your sales force need a claim of cheapest, fastest, or lightest? List salespersons, assemblers, etc., as customers in the house of quality. Table 31.1 primarily illustrates external customers that an engineer new to this field may be unfamiliar with. Multiple house-of-quality diagrams may need to be generated to handle all the customers on manageable-sized diagrams. In addition to the primary customer, the device user, other customers include all who must interact with it. For example, will a family member or health professional need to customize or maintain this device. How about the insurance agency or organization that may be paying for it? Have their requirements been addressed? Typically, each class of customer ranks the importance of each requirement. One common ranking scheme is to have the customer list all requirements in the order of importance beginning with one. Although this forces the customer to distinguish between requirements, it does not show the relative weight of each. A second approach is to mark each requirement that is extremely important with a 5, very important with a 4, and continue down to 1 for minimally important. (Blank can be reserved for not important at all.)

Room 3: Identification and customer evaluation of the competition or environmental scan. The competition must be identified. Once again, look internal as well. Will this new product compete with your existing product? If so, list it here. An important competitor to list when developing any assistive technology product is "no product at all." For example, how many of you have had a grandparent who could clearly benefit by a hearing aid but refused it, stating, "I have heard all that I need to hear already!"? In this room, the customers assess how satisfied they are with the competition for each requirement. A scale from 1 to 5 may be used, with 5 representing completely satisfied and 1 not satisfied at all. Although the competitor "no product at all" may score a 1 on its ability to amplify in this hearing aid example, it would score straight 5s on great comfort, good looks, low cost, and serviceability.

Room 4: Engineering parameters. How do we ensure that the entire design team is visualizing the same magnitude when the members hear customer requirements such as low cost or very powerful? If customers indicate that they want a powerful automobile, one engineer may be envisioning 200 horsepower, and another, 400. Clearly, each would lead to very different design concepts in the conceptual design phase. Furthermore, what is a low-cost automobile, \$6000, \$16,000, or \$26,000? The engineering parameters would include the sales price and the vehicle's horsepower. Engineering parameters (also called *engineering characteristics*) are measures that

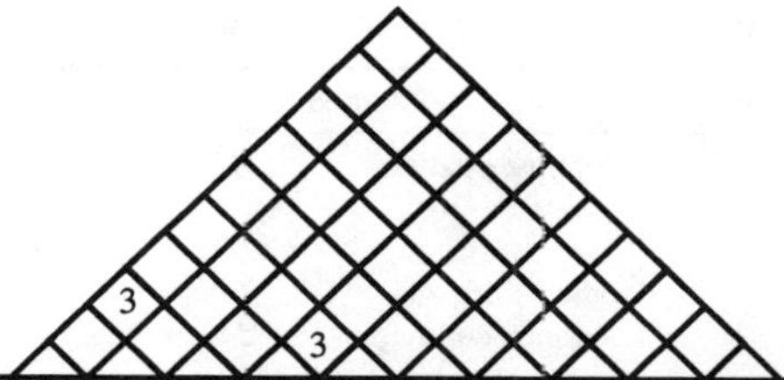

	User with Low Vision	User with Arthritis	User with No Disabilities	Hold Temperature within X degrees [°C]	Force to Open Door [N]	% of Food Burned	# of Color Options	% of Focus Groups Finding Design Appealing	% of Input/Output Devices Compatible with	Size of Smallest Font [point]	Size of Smallest Input Device [cm^2]	Maximum Exterior Case Temperature [°C]	Sales Price [$]	Competitor A	Competitor B	Competitor C	No Product
Cooks Well	5	5	5	9		9								4	4	4	1
Easy to Open	2	5	4		9									3	3	3	
Cannot Burn Food	5	4	4			9								1	1	1	
Looks Nice	3	4	3				3	9		1	1			3	2	4	
Compatible with I/O	4	5	1						9					1	1	1	
Easy to Read Settings	5	3	3				1	3	3	9				2	2	3	
Easy to Adj. Settings	2	5	3						3		9			3	4	2	
Low Price	2	3	5	3					3				9	4	5	4	5
Case Not Too Hot	3	5	3			1						9		1	1	1	5
Easy to Clean	5	5	5		1	3								3	4	4	5
Can Hear When Done	4	3	3											2	1	1	
Competitor A				5	0.1	NA	2	40	0	6	3	200	30				
Competitor B				4	0.2	NA	1	25	0	6	2	210	15				
Competitor C				4	0.2	NA	3	60	0	8	3	190	65				
No Product													0				
Target				4	0.1	0	1	80	100	18	9	30	60				

FIGURE 31.2 QFD, toaster oven example.

TABLE 31.1 Potential customers to satisfy with your design

Category[a]	Member	Description
A,C	Client/user/patient	Person who will use the product.
B,C	Client's family	They have valuable insight. They also may interact with the device.
B	Home health aide	Assists client with activities of daily living (ADL).
C	Insurance company (public or private)	May fund the purchase of an assistive device.
C	Vocational rehabilitation	May fund the purchase of an assistive device.
D	Assembler	
D	Repairperson	
D	Engineering technologist	Often the assistive technology provider that installs or adapts a device for client.
E	Audiologist	Concerned with alternative and augmented communication (AAC) devices; *www.asha.org*
E	Speech pathologist	Concerned with alternative and augmented communication (AAC) devices; *www.asha.org*
E	Physical therapist	Works with mobility aids; *www.apta.org*
E	Occupational therapist	Works with devices that assist in the activities of daily living (bathing, toileting, cooking, etc.). Concerned with devices necessary to perform at an occupation; *www.aota.org*
E	Recreational therapist	Concerned with devices that allow a client to enjoy avocations, dine out, and play sports.
E	Physician	In addition to traditional medical duties, prescribes devices. Coordinates the care of the client. A physiatrist specializes in rehabilitation/physical medicine.
E	Nurse	Cares for the client. Interacts with the devices.
E	Certified orthotist/prosthetist	Fits and produces orthotic/prosthetic device.
F	Salesperson	May need certain claims (e.g., fastest, lightest, least expensive) to sell the device. May be a therapist, etc.

[a] Category classification: A, primary user of the device; B, may interact with the device while in use; C, pays for the device; D, maintains, builds, or modifies the device; E, medical personnel who may train the user or prescribe the device.

allow us to quantify the customer requirements. If property selected and our room 8 target is met, we will have increased confidence in the future success of our product.

Room 5: Correlation matrix. Is there a strong interdependence between our engineering parameters? In the room 4 automotive example, is there a relationship between the horsepower of the vehicle and its cost? The following scale is often used: 9 (strong relationship), 3 (moderate), and 1 (weak), and the space is left blank if there is no relationship at all. This room will result in an increased awareness of trade-offs at the beginning of the design process.

Room 6: Relationship matrix. Applying the scale from room 5, consider the following questions: Have I developed engineering parameters that will clearly show if I have satisfied each customer requirement? In other words, do I have enough 3s and 9s in each row? Are my engineering parameters repetitive, with unnecessary measures of a requirement? Restated, do I have too many 3s and 9s in each row? Each engineering parameter may have a substantial cost associated with it. It may require significant testing or analysis.

Room 7: Benchmarking. For each competitor listed in room 3, apply the engineering parameters to their products. In the automotive example in room 4, what is the horsepower and sales price of each of your competitors? Some benchmarking (measuring your product against the competition) may only require reading published specifications; others might require expensive testing.

Room 8: Target. You have made it to the last room in the house of quality. Now we understand who the customers are and what they want. We have sized up our competition, in both our customers' eyes and our own tests. Engineering parameters have been developed to allow us to measure success throughout the design process. With this understanding, it is now time to select the product's targets. We will develop our target for each engineering parameter. This is often selected with the help of experts: marketing people who understand what price the market will bear; senior engineers who can assess technical feasibility in the selection of weight, power, and speed requirements; and strategic planners who can ensure that the product is consistent with the company's mission.

One model for selecting an appropriate target is illustrated in Fig. 31.3. It is a graph of price versus differentiation. *Differentiation* refers to the uniqueness of the product or service. Most successful businesses operate in one of the four corners. For example, a successful business in the low-price/low-differentiation corner (3) is Wal-Mart. Products found here include many remote controls and walkers. The high-price/low-differentiation corner (4) would include many sneaker manufactures whose products are similar, but they distinguish themselves through extensive advertising. The Ferrari is in the high-price/high-differentiation corner (2) with its impressive power, design, and sticker price. Many custom assistive devices would fit here. They are designed specifically for the consumer, but this may result in a high price. The final corner (1), low price/high differentiation includes consumer electronics manufactures that can add exciters to their products quickly and cheaply. When a business that is successful in operating in a particular corner adds a product line that pulls it toward the center (danger zone), that business often has problems as it begins to lose its market identity, and thus eventual profitability. Does this shift follow the corporation's strategic plan? This model is used to define a company's targeted strategic market segment based on its competitive advantages and manufacturing strategies.[6]

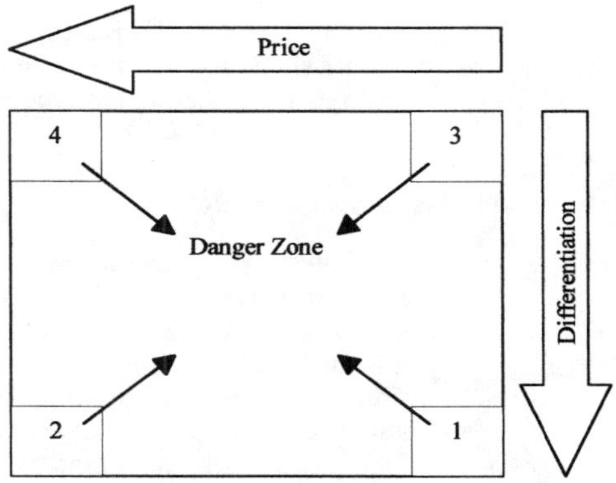

FIGURE 31.3 Target strategy.

31.2.6 The Team

The balance of the team and their involvement will primarily depend on whether the product is targeted for one specific person (denoted here as *sole user*), a class of persons with specific types of disabilities, or a general consumer product applying universal design principles to maximize its customer base. The members of the team include those drawn from the QFD customers list, as shown in Table 31.1. The specific backgrounds of the engineering team are left for other sources to detail.[1,3] This section will focus on the consumer and medical team.

Sole User. The most critical member of the design team is the user of the product. Methods, such as quality function deployment, stress understanding the customers' requirements, so why not involve the user of this custom device in the process? Just as a product that is high-tech but almost works is of little value to the customer, a device that works flawlessly but is too large and visually unappealing (function without form) may also be put aside. An example, most would agree that a device that improves its wearer's vision by 25 percent is significant, but if it were built into a helmet, most would not trade in their eyeglasses. Section 31.3.4 provides further details.

Other team members include the user's family. They are particularly important when the user is a child or a cognitively impaired adult. If the device is an alternative or augmentative communication aid, a speech pathologist is needed. The speech pathologist would perform the evaluation of the user and often would provide training with the device. Similarly, a physical therapist might be involved when the device is a walking aid such as a walker, cane, or crutch. Activities of daily living (bathing, cooking, toileting, etc.), vehicle modifications, specialized seating to minimize the formation of pressure sores, and workplace accommodations are examples of situations where an occupational therapist may be critical to the team. Nurses and physicians are just two more professions that might be required to design an appropriate device. Note that the team leader is often not an engineer but a clinician (a medically trained person such as a physician or therapist).

Multiple Users. The issues discussed earlier still apply in the design of mass-produced products, but the single user on the team may be replaced by focus groups, surveys, etc. Examples of other methods for gathering required information can be found in both marketing and design texts.[7]

31.2.7 Necessary Research

Engineers are experts at gathering and interpreting technical information. The medical world, however, introduces an entirely foreign vocabulary. Those trained as biomedical engineers have begun to learn medicine as a second language, but for the rest, it is just one more challenge. Traditionally, it is the biomedical engineer's job to speak the language of the clinician, not the clinician's responsibility to learn our engineering jargon. The medical team members will lead you through the medical issues briefly described below. For example, they will determine the user's diagnosis and prognosis. Together you will determine the product's efficacy and safety. Hybrid areas such as ergonomics are concerned with the interaction of a person and his or her surroundings. An ergonomist (typically with a clinical background) will be aware of the proper positioning of a person while performing a specified task. An engineer, however, is of great value in the measurement of specific forces on the body, as well as the estimation of joint forces that may be created by the use of products under development. Both anthropomorphic and traditional analysis software is useful in these activities.

Pathology → Impairment → Disability → Handicap. *Pathology* is the underlying cause of the disability. In this context, it refers to diseases and traumatic events such as amputations and spinal chord injuries. *Impairment,* the effect of the pathology on the body, describes the loss of specific functions. For example, the individual cannot move his or her right leg. *Disability* often refers to one or more impairments that interfere with specific activities. Examples include an inability to walk or feed oneself. *Handicap* generally implies the loss of one or more of life's major functions.

EXAMPLE *A dancer, Tom, suffers a traumatic head injury as the result of a drunk driver. He can walk but becomes dizzy when making any quick movements while standing. Pathology: traumatic head injury; Impairment: cannot move quickly while standing; disability: cannot dance; handicap: ??? Tom might consider himself handicapped due to his inability to dance, since that was a major life function for him. But what if he performs in a wheelchair? He might decide that he has overcome his handicap. A second person with a balance disorder might not consider it a handicap at all if he or she had no interest in dancing to begin with.*

Handicap is a term left to the person with disabilities to apply. A great percentage of those who society views as having handicaps would not describe themselves that way.[8] Despite this definition, the term *handicap* has specific legal meanings that are often not politically correct.

Diagnosis Versus Prognosis. In the preceding section, the diagnosis by the clinician would tell us the pathology. But the engineer generally focuses on the impairment and disability, so how does the pathology affect our design? The answer is the *prognosis.* Will the user be regaining function? Is the level of impairment increasing with time? What is the projected timetable? Is the person medically stable? Are there important secondary and tertiary effects of the pathology?

EXAMPLE *Amy has been diagnosed with amyotrophic lateral sclerosis (ALS), also known as Lou Gehrig's disease. The design team is assembled to create a device or modify the environment to allow her to continue to teach college physics. At the time the design team is assembled, Amy has begun to lose function in her legs and uses a self-propelled wheelchair. The team developed some initial concepts. These included building a ramp and a platform in front of the chalkboard so that she may write as she had before. A second idea is to reposition an overhead projector so that she may comfortably write on transparencies while seated in her wheelchair. The second concept was chosen due to cost and safety concerns. Shortly after it was adapted, Amy lost control of her arms. If the team had considered the prognosis, it would have realized the path of the progression of ALS. Eventually, Amy will only be able to communicate through her eyes. Locked-in syndrome, where*

TABLE 31.2 Sources on the Web

Description	URL
ABLEDATA: Links to manufacturers of assistive technology	*www.abledata.com*
American Foundation for the Blind: Reviews products for persons who are blind or have low vision. Produces fact sheets on many topics.	*www.afb.org*
Canadian Institutes of Health Research: Begun in June 2000, it is a federal agency that supports medical research.	*www.cihr.ca*
Canadian Standards Association (CSA): works with standards and the certification and testing of products	*www.csa.ca*
Eastem Paralyzed Veterans Association (EPVA): Provides materials on the Americans with Disabilities Act, workplace accommodation, and more.	*www.epva.org*
FDA: Check the codes and standards that affect durable medical equipment and other medical devices.	*www.fda.gov*
Health Canada: The federal department responsible for helping the people of Canada maintain and improve their health.	*www.hc-sc.gc.ca*
Medical Device Register: Directory of medical products manufacturers.	*www.devicelink.com/mdr*
Medline: Online database of medical and biomedical engineering articles; searches and abstracts are free, but there is a fee for ordering the full article.	*www.ncbi.nlm.nih.gov/PubMed/ gateway.nlm.nih.gov.gw.cmd*
National Institute on Disability and Rehabilitation Research, U.S. Department of Education: Supports research in assistive technology	*www.ed.gov/offices/OSERS/NIDRR/*
National Institutes of Health (NIH): A first stop to leam about many disorders including clinical trials.	*www.nih.gov*
Patent site: Conduct free patent searches and view recent patents online.	*www.uspto.gov*
Rehabilitation Engineering and Assistive Technology Society of North America (RESNA): Professional society with a strong focus on assistive technology; members are engineers and clinicians; sponsors a licensing exam for assistive technology providers.	*www.resna.org*
Sites for designing accessible software and Web sites.	*www.w3.org/TR/WAI-AUTOOLS* *www.microsoft.com/enable/* *www.artictech.com/howprog.htm* *www.access-board.gov* *ocfo.ed.gov/coninfo/clibrary/ software.htm*
Thomas Register: Directory of manufacturers.	*www.thomasregister.com*
Whitaker Foundation: Supports biomedical engineering education and research.	*www.whitaker.org*

she understands everything as always but cannot easily communicate, is one of the final stages. An appropriate device might be a computer and projector to display her notes for her class. As the disease progresses, the input device can be repeatedly replaced. Ultimately, Amy may require a blink switch or eye-tracking device.

Sources. Engineers are familiar with such databases for articles as the Engineering Index[9] and sources for manufacturers, including the *Thomas Register*.[10] Medical articles can be searched online with the database Medline.[11] The medical members of the team can aid in this search. What about existing medical devices? The *Medical Device Register*[12] lists medical device manufacturers by product and location. The National Institute on Disability and Rehabilitation Research supports the database ABLEDATA,[13] which lists assistive technology. Table 31.2 lists a few traditional engineering Web sites but focuses on assistive technology sites.

31.2.8 Conceptual Design

After completion of techniques such as QFD, we have a strong understanding of the customers' product requirements. The problem is defined in a written problem statement or product design specification, and the design team is finally ready to brainstorm its initial concepts. It is important to consider all options for each of the product's functions. What are my options for turning my device on? Should it use a lever, button, sensor, or timer? What about voice activation? How should information be displayed? Does it require a gauge, light-emitting diode (LED), buzzer, or synthetic speech?

Functional Decomposition. Functional decomposition is a method of exploring many design options for each product function. Figure 31.4 shows the overall system diagram for the toaster oven example. Inside the box is the overall function of the product. At the top of the box, there are the objects with which the toaster oven will interface. It must accept pans and trays and sit on a countertop or be installed below an upper cabinet.

The legend shows lines with arrowheads for energy, material, and information flow. Anything that does not remain in the toaster oven must be shown to leave. For example, food is considered a material flow. It enters cold and then leaves hot. The design team can then brainstorm ways to go from cold to hot. An example of information flow is "No fire?" and "No fire!" The question mark designates that we need an answer. The exclamation point tells us that we have an answer. Once again, the team will brainstorm, this time searching for methods of preventing fires due to grease, etc.

The overall system diagram is reduced to subfunction diagrams. Figure 31.5 shows four of the subfunctions. These shall be further subdivided as necessary. These diagrams allow us to separate each function of the product for individual and thorough consideration. Subfunction 3, for example, notifies the cook when the oven is preheated. How will this be accomplished? Tone, light, speech synthesizer, flag, or LED display? Table 31.3 is designed to heighten awareness of sample input and output devices used by persons with disabilities. Ideally, the product should suit all cooks, including those with disabilities. At times, appropriate jacks need to be designed into the product so that specialized input-output (I/O) devices can be used. For example, a person with a hearing impairment would simply connect his or her own amplified bell. Of course, try to build such features into your product. Requiring an interface would lead to its own subfunctions, such as "Connect external devices." Brainstorming might result in concepts that include quick connectors and infrared sensors (control the appliance from your wheelchair or recliner).

While brainstorming subfunction 2, questions that may arise include: Do I need any human force at all, or can it be voice-activated? Is it better to turn a knob, push a slider, or use an electronic keypad? Can it be controlled by remote? Why not have all controls and feedback designed into a remote unit?

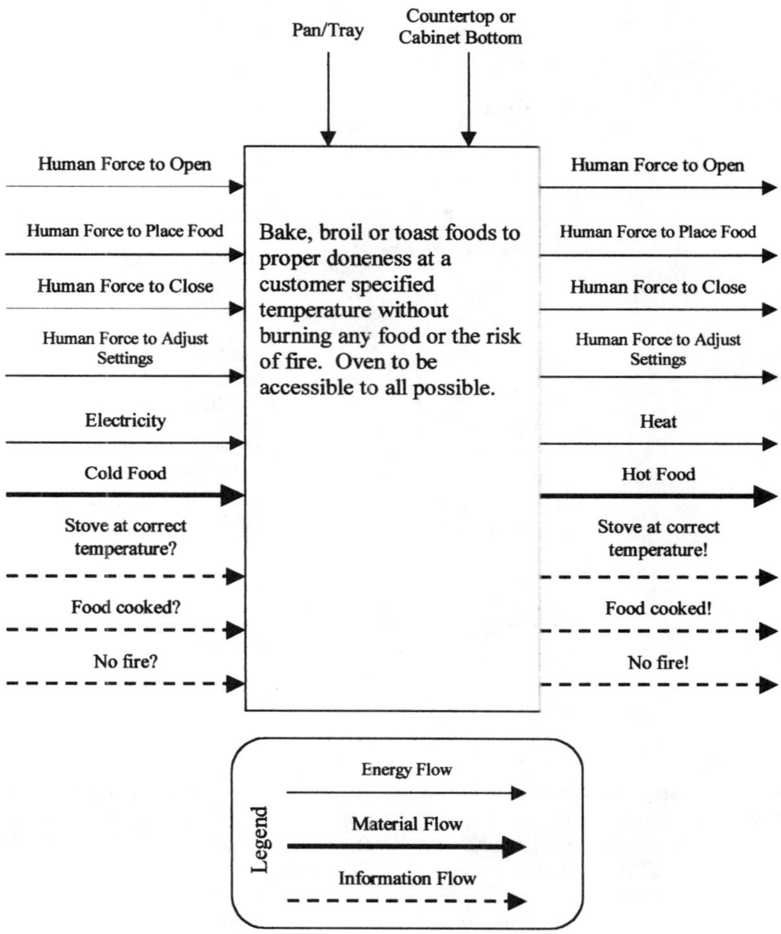

FIGURE 31.4 Overall system diagram, toaster oven example.

The resources in Table 31.2, especially ABLEDATA,[13] will help you customize and update Table 31.3 as technology changes. During the evaluation phase, remember to consider universal design principles in the assessment of all concepts.

31.2.9 Embodiment Design

Once again, I refer you to your design texts for a review of embodiment design. Since this is the time when the product takes shape, I introduce the topics of human factors and universal design. Because of the equal importance of both form and function, the design team should include an industrial designer from the initial design stages. In addition, consultants with expertise in the area of accessibility for persons with various disabilities must also be included. It is far less expensive and more effective to apply universal design principles in the earliest stages. In addition, both the industrial designer and the consultants will be invaluable in the development of the QFD.

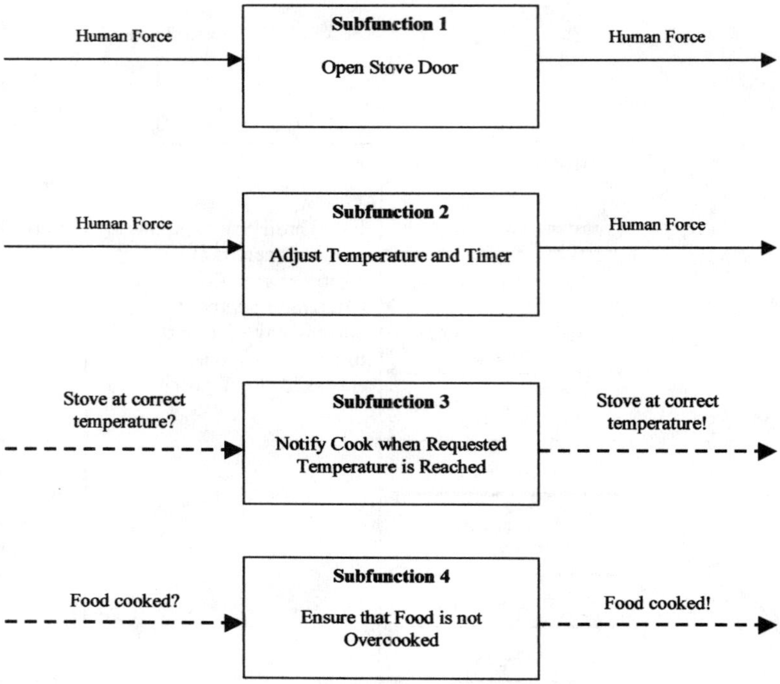

FIGURE 31.5 Subfunctions, toaster oven example.

Human Factors. Human factors or ergonomics are critical in any design with which a person will interface. Are the controls, warning lights, etc., laid out in a logical way? Does the color red mean hot and blue cold? Brake on the left and accelerator on the right? Rewind before the fast-forward button? Is it simple to operate? We also desire to minimize the energy that must be expended. Are the controls positioned appropriately? Is there a concern for repetitive-motion injuries? A person might have limited neuromuscular control or may be wearing gloves while operating the device outdoors. The person may have low vision on be in a dimly lit room. Is the user color-blind?

When designing for a specific person, certain anthropometric measurements (mass, lengths, radii of gyrations, etc., of the human body) can easily be made. These include height, weight, and reach. What if you are creating a product for the mass market? One approach is to use anthropometric tables. You can estimate, for example, the height of the fifth, fiftieth, and ninety-fifth percentiles of American women. However, what do you design for? If you design for even the ninety-fifth percentile, you are still excluding the tallest 5 percent. What if you are attempting to estimate the mass of someone's leg to be used in a gait analysis (how someone walks)? The tables are based on cadaveric studies with a sample of generally fewer than a dozen. Did any of the cadavers fit the build of your target consumer?[14]

Disability categories include vision, hearing, speech, limb loss, neuromuscular impairment, other, and multiple impairments.[8] Categories may easily be subdivided. For example, vision may be divided into low vision and blind. Each may lead to different solutions. Similarly, is a person hard of hearing or profoundly deaf? The following is a list of suggestions or questions to consider in your design. It is a starting point, not a comprehensive list. Some are geared more to the design of a device for a sole user and others to a mass market:

1. What are a child's personality traits, intelligence, developmental and functional levels, physical size, and growth factors? Do we understand their physical, financial, and social environment?[8]

TABLE 31.3 Sample Input/Output Devices

Name	Description
Amplification/magnification	Sound may need to be amplified or a display magnified.
Braille display	Multiple-character display in which pins extend to represent Braille characters and are continuously refreshed by the computer.
EEG switch	Receiver worn on the forehead can detect EEG signals to activate a switch.
Electromyographic (EMG) switch	A surface electrode detects a muscle contraction and triggers this switch; it can use small, controlled muscle movements.
Environmental control unit (ECU)	A black box that accepts most input devices (e.g., sip and puff and touch switches) and may be connected to multiple devices (e.g., telephones, bed controls, lamps, and appliances); it has the ability to scan through different options, such as "call home," "turn lamp on," and "increase the volume of the television," with a minimum of one switch.
Eye/eyebrow switch and eye trackers	Detect blinks, eyebrow motion, or the wrinkling of the forehead; eye trackers are more sophisticated systems that can detect the coordinates of the eye to replace a computer mouse, etc.
Flashers	A smoke detector, door bell, or a baby's cry can alert a person who is deaf with a flashing light.
Functional electrical stimulation (FES)	Electrodes inserted into muscles to activate (stimulate) selected muscles for those with paralysis.
Optical pointer	Laser that can be mounted on a user to allow for precise pointing.
Proximity and temperature sensors	Technologies include ultrasound and can detect the presence of a person through his or her motion or body temperature.
Remote I/O devices	These include remote keyboards that allow you to roll up in a wheelchair near your personal computer without having to physically connect or carry a laptop; television, lamp, and appliance remotes can be purchased at local electronics stores; these include infrared and FM-based systems.
Scanning device	Using as little as a single input device, the device switches between multiple output options.
Sip-and-puff switch	Ultrasensitive pneumatic switch actuated by slight mouth pressure.
Skin surface stimulation/subdermal stimulation[8]	Electrodes are placed on the surface of the skin or implanted to provide sensory feedback.
Speech output	These include computer-generated speech synthesizers and systems that replay human speech.
Tactile	Includes textured surfaces, scales, thermometers, and watch faces that can be felt (raised and Braille lettering).
Touch pads	Adhesive-backed flexible plastic pad; it is activated by touch and can be washed.
Touch switch	Single switch activated by very light pressure.
Vibrators	Includes smaller vibrators used in pagers and larger units that may be placed underneath a mattress to wake a person who is deaf.
Voice- and sound-activated systems	From systems that respond to a clap to word processors that can type what you speak, the sophistication and price cover a broad spectrum.

2. What is the diagnosis and prognosis?
3. Is the product compatible with the user's existing I/O devices? If computer-based, can it be operated without a pointing device such as trackball or mouse? Will it interface with a voice dictation system? What about speech output? Can keystroke commands be entered sequentially? For example, pressing the Control, Alternate, and Delete buttons simultaneously is impossible for many people.[15,16]
4. Is documentation in Braille, electronic format, large print, or on audiotape, etc.?[15,17]
5. How can print be made more readable for someone with low vision? Use an 18-point font (but no less than 16 points) in standard roman or sans serif. Avoid decorative fonts, italics, and all

capital letters. Bold type increases the thickness of the letters and is preferable. Avoid the use of color fonts for headings. However, if used, consider dark blues and greens. Maximize contrast. Some people prefer white or light-yellow letters on a black background over the reverse. Avoid glossy paper due to the glare. Use a line spacing of 1.5 rather than single space. Spacing between letters should be wide. Since low-vision devices such as closed-circuit televisions and magnifiers often require a flat surface to read, use a margin of 1.5 in (but no less than 1 in).[17]

6. How can one design a product to be easier to use by someone with low vision? Maximize contrast between components. For example, use a dark switch on a light device or a light handle on a dark door. Avoid glares.[18]
7. How can my software application be improved to help those with visual impairments? Use standard user-interface elements compatible with screen readers. Is the application compatible with screen-magnification and speech-synthesis software? Do not require a pointing device and allow the user to modify the interface. For example, can the user select the font style, color, and size? What about the background color? Do not place time limits on input activities or messages. Do not rely on color alone to convey information. For example, a green-filled circle that becomes red when done may not be seen.[16]
8. Have I considered the ergonomics of the design? Is the device easy to grip? What does it weigh? How must the user be positioned to operate the device? Will its use lead to back or neck pain? Are repetitive-motion injuries a concern? There is commercial software available to determine forces at the back and extremities for various anatomic positions and loading cases. Parameters such as the user's height may be specified.

Universal Design. Over the years, industry has moved away from telling consumers that they know what is best for them to seeking their input early on in the design process. Designing quality into the product has also become the norm. There is a great need for universal design to become the industry standard as well. Universal design results in a more usable product for all consumers, improves sales by increasing the customer base, and is good for society as a whole. There will likely be great resistance and some problems in the transition phase as industry learns these new requirements. Ultimately, it will be profitable, as corporations have seen with total quality management.

Universal Design Principles. Universal design, as in all other design requirements, depends on sufficient and correct background information. Many codes and standards actually meet only minimum requirements. For example, the standard recommended pitch for a wheelchair ramp requires significant upper body strength. A less steep incline is actually preferred. Simply observing the letter of the law does not necessarily lead to a satisfactory design. Check with the experts.

An accommodation that benefits one type of disability may cause new concerns for others. A curb cut, for example, is an absolute necessity for persons in wheelchairs but will make it difficult for a person who is blind to use a cane to detect where the sidewalk ends and the roadway begins. As a result, the ramp portion of the curb cut is often textured to provide tactile feedback. Elimination of the need to step up onto a sidewalk has also benefited countless persons pushing baby carriages or hand trucks.

Increasing the size and contrast of the letters on a remote control will not only aid users with low vision but also will aid the general population in low light. Choose color combinations that can be seen by those with color blindness.[19] Also, larger switches arranged in an expected, logical order will make for a better product. Those with limited mobility, including the millions with arthritis, will be able to operate such a remote with ease.

Devices that increase a person's independence are essential. But do not forget to apply universal design to fun products too! Work, school, and activities of daily living are not enough. Persons with disabilities do not lose their desire to participate in leisure activities.

In addition to the universal design principles that follow, Table 31.4 shows a set of principles developed by the Center for Universal Design at North Carolina State University.

1. Maximum inclusion will increase your market share and result in a better product for all consumers.

TABLE 31.4 Universal Design Principles

1. *Equitable use*. The design is useful and marketable to people with diverse abilities.

   Guidelines

   a. Provide the same means of use for all users: identical whenever possible; equivalent when not.
   b. Avoid segregating or stigmatizing any users.
   c. Provisions for privacy, security, and safety should be equally available to all users.
   d. Make the design appealing to all users.

2. *Flexibility in use*. The design accommodates a wide range of individual preferences and abilities.

   Guidelines:

   a. Provide choice in methods of use.
   b. Accommodate right- or left-handed access and use.
   c. Facilitate the user's accuracy and precision.
   d. Provide adaptability to the user's pace.

3. *Simple and intuitive use*. Use of the design is easy to understand, regardless of the user's experience, knowledge, language skills, or current concentration level.

   Guidelines:

   a. Eliminate unnecessary complexity.
   b. Be consistent with user expectations and intuition.
   c. Accommodate a wide range of literacy and language skills.
   d. Arrange information consistent with its importance.
   e. Provide effective prompting and feedback during and after task completion.

4. *Perceptible information*. The design communicates necessary information effectively to the user, regardless of ambient conditions or the user's sensory abilities.

   Guidelines:

   a. Use different modes (pictorial, verbal, tactile) for redundant presentation of essential information.
   b. Provide adequate contrast between essential information and its surroundings.
   c. Maximize "legibility" of essential information.
   d. Differentiate elements in ways that can be described (i.e., make it easy to give instructions or directions).
   e. Provide compatibility with a variety of techniques or devices used by people with sensory limitations.

5. *Tolerance for error*. The design minimizes hazards and the adverse consequences of accidental or unintended actions.

   Guidelines:

   a. Arrange elements to minimize hazards and errors: most used elements, most accessible; hazardous elements eliminated, isolated, or shielded.
   b. Provide warnings of hazards and errors.
   c. Provide fail-safe features.
   d. Discourage unconscious action in tasks that require vigilance.

6. *Low physical effort*. The design can be used efficiently and comfortably and with a minimum of fatigue.

   Guidelines:

   a. Allow user to maintain a neutral body position.
   b. Use reasonable operating forces.
   c. Minimize repetitive actions.
   d. Minimize sustained physical effort.

7. *Size and space for approach and use*. Appropriate size and space are provided for approach, reach, manipulation, and use, regardless of user's body size, posture, or mobility.

   Guidelines:

   a. Provide a clear line of sight to important elements for any seated or standing user.
   b. Make reach to all components comfortable for any seated or standing user.
   c. Accommodate variations in hand and grip size.
   d. Provide adequate space for the use of assistive devices or personal assistance.

Note: The principles of universal design address only universally usable design, whereas the practice of design involves more than consideration for usability. Designers must also incorporate other considerations, such as economic, engineering, cultural, gender, and environmental concerns, in their design processes. These principles offer designers guidance to better integrate features that meet the needs of as many users as possible.

Source: Reprinted with permission of North Carolina State University, The Center for Universal Design, 1997.

2. Provide appropriate interfaces (jacks, plugs, etc.) to allow a user to connect his or her own specialized I/O devices.
3. Designing for inclusion has a significant positive societal impact.

31.3 SPECIAL TOPICS

31.3.1 Stochastic Processes

The human body is subjected to many environmental forces that may be treated as stochastic (random) processes. For example, how would you estimate the forces on an individual during an automobile crash? How about the stress on a hip as a person transverses a rocky terrain? A second area of interest focuses on the treatment of uncertainties in the measurement of anthropometric data or human performance. Refer to Sec. 31.2.9 ("Human Factors") for more details. A clinician estimates the mass of a man's arm as 0.4 kg from anthropometric tables, but how do we quantify the effects that might occur due to an error in the estimate? One method to quantify the effect of stochastic processes is the Monte Carlo simulation.[20–23]

The Monte Carlo simulation allows an engineer to methodically study the effect of a parameter on the results. The parameter's range is identified, and then a random sample is selected from within this range. Imagine using a roulette wheel in Monte Carlo to select these values. In practice, programs such as MS Excel and MathWorks Matlab or random-number tables are used to select the values. The Monte Carlo simulation will be demonstrated by example.

EXAMPLE *A medical device manufacture is developing a hearing aid that will reduce the chances of injury to its user during a car accident. The manufacturer would like to know the forces on an individual's head during an accident. An automobile manufacturer has software to simulate the forces on the head for any size driver during a simulated car accident. What height should the manufacturer choose? Can we just estimate the worst case? Is the shortest driver less safe because of his or her proximity to the steering column and air bag? Would a very tall driver experience added forces if he or she is too tall for the headrest? Would an average size driver be more likely to strike the top of the steering wheel?*

What heights do we analyze? How many do we choose? Have we selected appropriately?

For simplification of this example, let us assume that the peak force on the head during a crash is a function of height only and follows the following hypothetical equation:

$$
\begin{aligned}
F &= 2h^4 + 7 && \text{when } h < 60 \text{ in} \\
&= 2h^3 + 6h && \text{when } 60 \leq h < 72 \text{ in} \\
&= 2h^5 && \text{when } 72 \leq h \text{ in}
\end{aligned}
$$

where F is the peak force on the head and h is the subject's height when standing.

Steps

1. *Determine the range of heights.*
2. *Select initial sample size n.*
3. *Using a random-number generator that can produce a uniform random distribution, select n values of h.*
4. *Calculate a total of n peak forces by calculating the force for each height.*
5. *Apply all relevant statistical measures to these peak forces. They might include mean, standard deviation, correlation coefficients, etc.*
6. *Repeat steps 2 to 5 with a larger sample size.*

7. *Compare the statistical measures from each sample size. Repeat steps 2 to 5 with an ever-increasing sample size until the results of these statistical measures converge. Use these results in your design.*

TABLE 31.5 Random Heights

Height, in
58.25513474
69.04791406
71.38007752
65.59837642
76.76573992
59.89321574
75.82985931
76.9483932
63.6593524
74.27295755

Suppose that h is in the range from 58 to 77 in and n is initially selected to be 10. A random-number generator has generated the sample heights using the random number seed 123 shown in Table 31.5.

Figure 31.6 shows the mean, standard deviation, and maximum for the peak force for n = 10 above, where n = 15 and the random seed is arbitrarily chosen as 7,638, n = 20 with seed 418, n = 50 with seed 6,185, n = 100 with seed 331, and n = 200 with seeds 43 and 3,339. Note the range of values at n = 200 when two different seeds were selected. The engineer would then decide if he or she is satisfied with the results. Remember, in an actual computer simulation, the CPU time may be significant, and a sample of even n = 10 may be expensive.

31.3.2 Weight Is Critical

Many products that a person with disabilities may use are carried, worn, or a mounted on a wheelchair. The heavier the product, the more energy that a person must exert to carry it. When a device

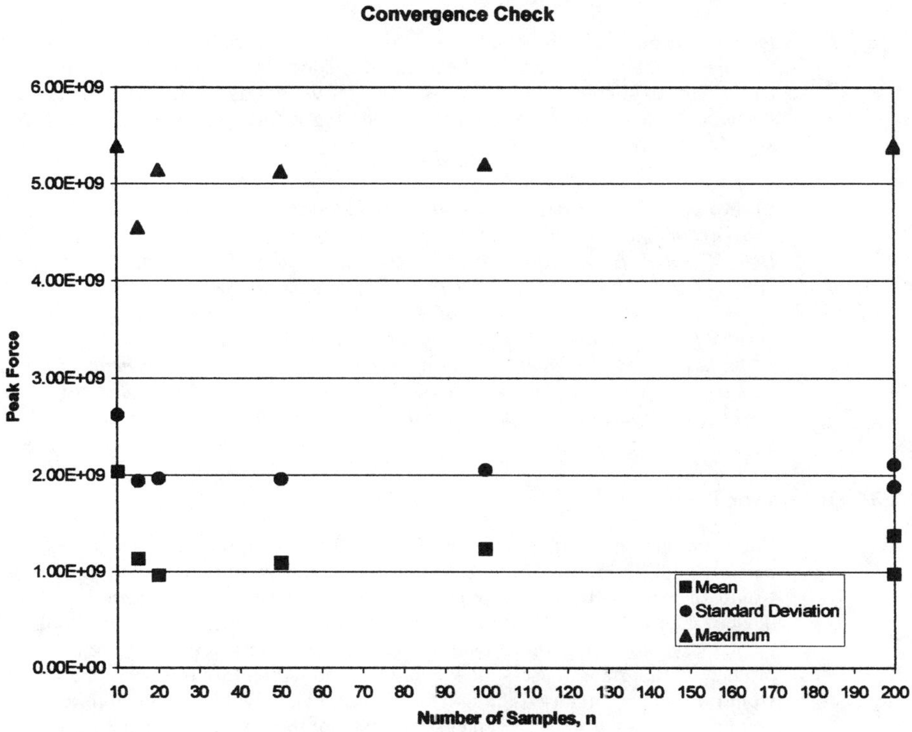

FIGURE 31.6 Monte Carlo simulation sample results.

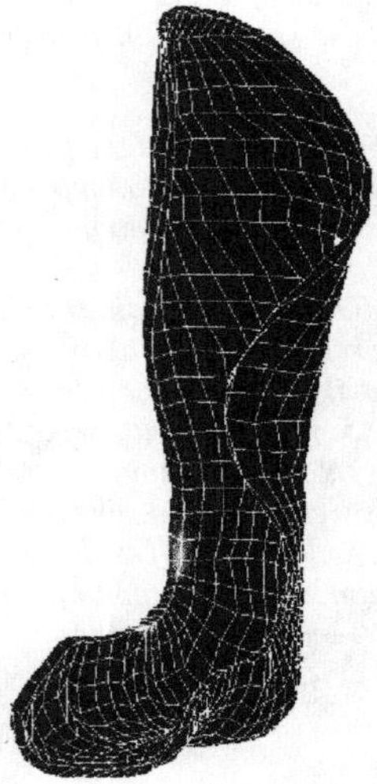

FIGURE 31.7 FEM of a standard ankle-foot orthotic. (*Reprinted with permission of Khamis Abu-Hasaballah and the University of Hartford.*)

FIGURE 31.8 FEM of an ankle-foot orthotic after weight reduction. (*Reprinted with permission of Khamis Abu-Hasaballah and the University of Hartford.*)

is mounted on a self-propelled wheelchair, the user must also expend additional energy to account for this weight. On an electric wheelchair, the added load will shorten the interval between recharges and reduce available power. The reduction of weight as a critical factor is one that is shared with the aerospace industry. Additional engineering efforts to reduce the product's weight will yield significant benefits to the consumer.

EXAMPLE *Lisa, a 15-year-old high school student, wears an angle-foot orthosis (AFO). This AFO, worn around her calf and under her foot, supports her ankle. To the casual observer, the AFO seems light. Through Lisa's daily activities, the extra weight is noticed. View Fig. 31.7 of her AFO. An engineer may initially check that the peak stresses do not exceed their limit. But as an aerospace engineer would ask, why do we allow all these areas of low stress? If the material is thinned or removed from the low-stress regions, the AFO will not fail, and its weight will be reduced. Figure 31.8 shows a finite-element model of the AFO after non-loading-bearing material is reduced. An added benefit is the reduction of perspiration experienced due to wearing an AFO in the summer. Technology transfer from other industries would certainly benefit Lisa.*

31.3.3 Design for Failure

Engineers are constantly considering failure modes in their designs. When designing a device for persons with disabilities, great care must continue. What will happen to its user if a failure occurs? Could the person become injured? Trapped until someone happens by? Who would repair this device? Can it be completed onsite? Would it require special training or tools? From the instant the device fails, how long will it take before the user has the repaired or replaced device in use? If this product is important to users, how could they do without it? If it is not that important, they probably would not have purchased the unit in the first place. If this is a custom design for a sole user who you know personally, it is even easier to picture the possible impact of a failure.

31.3.4 Social-Psychological Perspective

A team of talented engineers is asked to design a device that will automatically open doors for a teenage girl who has no use of her arms. The team, after 3 months, produces a device that can open standard doors, sliding doors, light doors, heavy doors, and even bifold doors with ease. It is lightweight and works off a single AA battery that lasts for a year. The team proudly delivers the invention to the client, but she refuses to wear it. You see, the device is built into a helmet with a robotic arm jutting out. Most teens will not leave their homes if they are wearing the "wrong" clothes. Why would a teen with a physical disability be any different? Many people, young or old, will not use the appropriate assistive technology because of the stigma attached. They might avoid using a wheelchair that is too ungainly, even if they are in a wheelchair already. Many of us have friends and family members who would benefit from a hearing aid but refuse one because of what they fear people might say. Others may be concerned that it would make them feel old.

These are complex issues that must be considered in the design of a product. The following are questions to consider: Will my client use the device as intended if it is designed and fabricated? What if the device could be hidden? Can it be made aesthetically pleasing? What if through universal design the product was desired by the general population, thereby eliminating the stigma?[8]

31.4 WEB SITES

This is the first edition of this chapter. Although already used in a classroom setting, it will undoubtedly change between editions. Please submit suggestions, comments, and corrections to *adrezin@hartford.edu* or *http://uhaweb.hartford.edu/adrezin/handbook*. Responses may be posted and web hyperlinks updated.

31.5 WHAT'S NEXT?

Toward the final stages of the embodiment design phase, there are often two or three "most promising" designs that the team may still be considering. There are still many factors to consider. These include technical feasibility, safety, cost, and many manufacturing issues. Is the device at risk for failure due to fatigue or corrosion? Are all applicable codes and standards satisfied? Must the product be registered with the Food and Drug Administration (FDA) or other regulatory agency? Will it require clinical trials? Once the final design is selected, it is time for the detail design phase.

The detail design phase will include the preparation of detail drawings and product specifications. A bill of materials must be prepared along with a detailed cost estimate. A final design review is conducted before releasing all drawings to be manufactured.

REFERENCES

1. Dieter, G. E., *Engineering Design: A Materials and Processing Approach*, 3d ed., McGraw-Hill, New York, 2000.
2. Fogler, H. S., and LeBlanc, S. E., *Strategies for Creative Problem Solving*, Prentice-Hall, Englewood Cliffs, N.J., 1995.
3. Shetty, D., *Design for product success*, Society of Manufacturing Engineers, Dearborn, Mich., 2002.
4. *www.resna.org*.
5. Ulrich, K. T., and Eppinger, S. D., *Product Design and Development*, 2d ed., McGraw-Hill, New York, 2000.

6. Birch, John, *Strategic Management Process*, The Birch Group, Farmington, Conn., 1997.
7. Ullman, D. G., *The Mechanical Design Process*, 2d ed., McGraw-Hill, New York, 1997.
8. Smith, R. V., and Leslie, J. H., Jr., *Rehabilitation Engineering*, CRC Press, Boca Raton, Fla., 1990.
9. *Engineering Index*, *www.ei.org*.
10. *Thomas Register*, *www.thomasregister.com*.
11. *Medline*, *www.ncbi.nlm.nih.gov/PubMed/*.
12. *Medical Device Register*, Medical Economics Co., Montvale, N.J., 2000.
13. ABLEDATA, Silver Spring, Md., *www.abledata.com*.
14. Chaffin, D. B., Andersson, G. B. J., and Martin, B. J. (1999), *Occupational Biomechanics*, 3d ed., Wiley, New York.
15. Missouri Assistive Technology Project, Basic Questions to Ask When Purchasing Technology, Missouri Technology Center for Special Education, Kansas City, Mo., September 1996.
16. National Technology Access Program, *Creating Applications Accessible to People Who Are Visually Impaired*, American Foundation for the Blind, New York, June 1998.
17. Orr, A. L., *Tips for Making Print More Readable*, American Foundation for the Blind, New York, June 1999.
18. AFB Aging Program, *A Checklist for Environmental Safety and Access*, American Foundation for the Blind, New York, August 1998.
19. Guyton, A. C., *Textbook of Medical Physiology*, 7th ed., Saunders, Philadelphia, 1986.
20. Adrezin, R., and Benaroya, H., Monte Carlo simulation of a partially submerged compliant structure, *ASME IMECE*, November 1998, pp. 191–198.
21. Benaroya, H., *Mechanical Vibration: Analysis, Uncertainties, and Control*, Prentice-Hall, Upper Saddle River, N.J., 1998.
22. Adrezin, R., and Benaroya, H., Nonlinear stochastic dynamics of tension leg platforms, *Journal of Sound and Vibration*, **220**(1):27–65 (1999).
23. Adrezin, R., and Benaroya, H., Dynamic modelling of tension leg platforms, in *Stochastically Excited Nonlinear Ocean Structures*, World Scientific, Singapore, 1998.
24. Nowak, M. D., Abu-Hasaballah, K. S., and Cooper, P. S., Design enhancement of a solid ankle-foot orthosis: Real-time contact pressures evaluation, *J. Rehabil. Res. Dev.*, **37**(3):273–281 (2000).

CHAPTER 32
DESIGN OF ARTIFICIAL ARMS AND HANDS FOR PROSTHETIC APPLICATIONS

Richard F. ff. Weir, Ph.D.,
Northwestern University Prosthetics Research Laboratory, and Rehabilitation Engineering Research Center, Chicago, Illinois

32.1 INTRODUCTION

The design of fully functioning artificial arms and hand replacements with physiological speeds-of-response and strength (or better) that can be controlled almost without thought is the goal of upper-extremity prosthetics research. Unfortunately, current prosthetic components and interface techniques are still a long way from realizing this goal. The current state-of-the-art prosthesis can be considered to be a tool rather than a limb replacement.

The prosthesis as a tool makes no pretense of trying to replace the lost limb physiologically but is there as an aid to help provide some of the functions that were lost. The prosthesis as a tool is an interchangeable device that is worn and used as needed, and then ignored. Much effort in the field of upper-extremity prosthesis research is directed toward the creation of prostheses as true limb replacements; however, in current practice we are mostly limited to prostheses as tools.

The major factors limiting prostheses to tools are practical ones due to the severe weight, power, and size constraints of hand/arm systems as well as the difficulty in finding a sufficient number of appropriate control sources to control the requisite number of degrees of freedom. Of these, it is the lack of independent control sources that imposes the most severe impediment to the development of today's prosthetic hand/arm systems. As a result, upper-limb prosthetics research is somewhat dominated by considerations of control. Still, the importance of better actuators and better multifunctional mechanisms must not be ignored. Control is useless if effective hand and arm mechanisms are not available.

The problems associated with the design of artificial hand and arm replacements are far more challenging than those associated with the design of robotic arms or terminal devices. In fact, robotics and prosthetics design has much less in common than one might expect. Robotics concepts have had little impact on commercial prosthetics because of the significantly different physical constraints required for a prosthetic device to be successful. Although some size, weight, and power constraints must be placed on robots and manipulators, robotic actuators can often be as large and as heavy as required to achieve a specific result. Power is usually not an issue since it can be obtained from the power mains. Prosthetic arm and hand design can be viewed as a subset of the greater field of robot and manipulator arm and end-effector design.

Robot arms look impressive. However, announcements by robot arm designers who, when searching for an application for their new mechanical arms, claim their new robot will be a boon to the field of prosthetics but should be treated with skepticism. The issue has never been about an inability to build mechanical arms and hands. The MIT/Utah dexterous hand (Jacobsen et al., 1984) is an example of a mechanical hand that mimics the function of a hand. This hand was designed for use in research studying robot dexterity. This device could never be used in prosthetics because the actuators and computer system required to control this hand occupy the space of two small filing cabinets, and power is supplied externally from electrical mains. The real issue in upper-limb prosthetics, of which most robot arm designers seem to be unaware, is "How does one interface this arm to the person?" and "How is the arm to be controlled?"

Current state-of-the-art prosthetic hands are single-degree-of-freedom (opening and closing) devices often implemented with myoelectric control (a form of control that uses the electric fields generated as a by-product of normal muscle contraction). Prosthetic arms requiring multi-degree-of-freedom control most often use sequential control with locking mechanisms to switch control from one degree of freedom to another. Generally, since vision is the primary source of feedback, the greatest number of functions that are controlled in parallel is two. Otherwise, the mental loading becomes excessive. Switch, myoelectric, or harness and cable are the primary modes of control for today's upper-limb prostheses.

Eugene F. Murphy, Ph.D., (1955) a former chief of the Research & Development Division, Prosthetic & Sensory Aids Service, Veterans Administration (Hays, 2001), probably articulated best the awe and frustration associated with the task of trying to replicate the function of the natural hand by mechanical means, when he wrote in "Engineering—Hope of the Handless":

> The human hand, with its elaborate control system in the brain, is doubtless the most widely versatile machine that has ever existed anywhere. Its notorious deficiency lies in its persistent inability to create a similar machine as versatile as itself. This circumstance accounts for the fact that, while there has been from earliest times a great need for hand replacements, all attempts to produce successful hand substitutes have thus far ended in only a rather crude imitation of a very few of the many attributes of the living counterpart. For want of complete knowledge of the natural hand-brain complex, and of the ingenuity requisite even to the most modest simulation of the normal hand, artificial hands have always resembled the natural model in a superficial way only. Voltaire is said to have remarked that Newton, with all his science, did not know how his own hand functioned.

The design of artificial arms and hands is a multidisciplinary endeavor. The design team needs an understanding of the mechanics of mechanisms, such as gears, levers, and points of mechanical advantage, and electromechanical design, such as switches, dc motors, and electronics. *Mechatronics* is the new word used to describe this marriage of mechanical and electronic engineering. In addition to mechatronics, the prosthetics designer must also have knowledge of muscloskeletal anatomy and muscular- as well as neurophysiology.

It is the goal of this chapter to serve as a resource for designers of artificial limbs who come to the problem with different areas of expertise. As such, I am compelled to refer the reader to the *Atlas of Limb Prosthetics* (Bowker and Michael, 1992). This text is a comprehensive guide to the surgical, prosthetic, and rehabilitation techniques employed in current clinical practice. From a historical perspective, *Human Limbs and Their Substitutes* (Klopsteg and Wilson, 1954) is invaluable and contains much biomechanical data on persons with amputation that is still valid. Consequently, both these texts are referenced throughout this chapter. Also, following the philosophy of "Why reinvent the wheel?" wherever possible, examples are provided of commercially available devices that use the mechanisms, algorithms, or control approaches mentioned in the text.

It is hoped that the reader will find the information in this chapter to be practical and useful and an aid in eliminating much of the startup time usually associated with familiarizing oneself with a new topic. Ultimately, it is hoped that the information imparted will aid and facilitate in the design of better artificial arm and hand replacements for people with upper-limb amputations and deficiencies.

32.2 THE NATURE OF THE PROBLEM

There are over 30 muscles acting on the forearm and hand. The human hand has 27 major bones, and at least 18 joint articulations with 27 or more degrees of freedom (DOF). The arm contributes another 7 degrees of freedom. The primary role of the arm is to position the hand in space. The primary role of the hand is to enable a person to interact with the environment. Control of a person's arm is directed at controlling the position of the arm's hand. Even though people control their arms with great facility, this is a highly complex and demanding task.

A backhoe is essentially a mechanical arm that is under the control of an operator. To control this mechanical arm the operator uses both arms, both feet, both eyes, and all his or her concentration (Fig. 32.1). The driver uses both arms to pull levers, both feet to press pedals to operate the arm, and both eyes to monitor the task being performed by the digger—all this to control a single mechanical arm. Now consider a person with amputations of both arms above the elbow and one begins to have some appreciation of the task such a limbless person faces in controlling an artificial (prosthetic) arm or arms.

As for performance, the anatomical hand is capable of speeds in excess of 40 rad/s (2290 degrees/s), and grasps involving all fingers of the hand can exert up to about 400 N (90 lbf) of force. Average physiological speeds for every day pick-and-place tasks have been found to be in the range of 3 to 4 rad/s (172 to 200 degrees/s), while most activities of daily living (ADL) require prehension forces in the range 0 to 67 N (0 to 15 lbf) [these forces are dependent on the coefficient of friction between the gripping surface and the object held (Heckathorne, 1992)].

For the wrist and forearm normal ranges of motion (ROM) (Fig. 32.2, top) are 85 to 90 degrees of pronation, 85 to 90 degrees of supination, 15 degrees of radial deviation, 30 to 45 degrees of ulnar deviation; 80 to 90 degrees of wrist flexion, and 70 to 90 degrees of wrist extension (Magee, 1987). It has been found that for ADLs, 100 degrees of forearm rotation, 80 degrees of wrist flexion-extension, and 60 degrees of radial-ulnar deviation are sufficient (Heckathorne, 1992). The forearm can achieve maximum rotational velocities in excess of 14 rad/s (800 degrees/s) for pronations (inward rotations) and 20 rad/s (1150 degrees/s) for supinations (outward rotations).

For the elbow, normal range of motion (Fig. 32.2, bottom, left) is 140 to 150 degrees of flexion and 10–15 degrees of hyperextension. Peak anatomic elbow speeds of 261 degrees/s (4.5 rad/s) have been found (Doubler, 1982) for 90-degree movements, and the average male can generate an elbow torque of 138 N·m (102 ft·lbf) in flexion with the elbow at 90 degrees. In extension the average male can generate 75 percent of the maximum flexion torque.

For the upper arm, normal ranges of motion (Fig. 32.2, center) are 90 degrees of medial (inward, toward the midline) humeral rotation and 40 degrees of lateral (outward away from the midline) humeral rotation, 180 degrees of flexion (forward rotation of the arm about the shoulder) and 45 degrees of extension (backward rotation of the upper arm about the shoulder), and 180 degrees of elevation (abduction, outward rotation about the shoulder) and 20 degrees of depression (adduction, inward rotation of the upper arm about the shoulder).

For the shoulder complex, the normal ranges of motion (Fig. 32.2, bottom, center and right) are 40 degrees of elevation and 10 degrees of depression, 15 degrees of extension (scapular adduction), and 20 degrees of flexion (scapular abduction). Motion at the shoulder is not due to a single joint but is instead a combination of the motion associated with the scapulothoracic interface, the sternoclavicular joint, and the glenohumeral joint. The glenoid cavity in which the head of the humerus sits is made up of the acromion of the scapula, the spine of the scapula, and the clavicle. As such there is no single shoulder joint but rather a shoulder complex or girdle. The primary power source for body-powered prostheses uses a combination of glenohumeral flexion [forward rotation of the upper arm about the shoulder (glenohumeral joint)] and biscapular abduction (rounding of the shoul-

FIGURE 32.1 A digger, or backhoe, is essentially a large mechanical arm. To control this arm, an operator needs to use both hands (to operate the joysticks, see inset), both feet (to operate multiple pedals), and both eyes and almost all of their concentration (to watch what they are doing). These are the requirements for a "whole" individual to control just one artificial arm. Consider now the control problems experienced by persons who have lost both arms at the shoulder. These individuals do not have hands to help in the control of their artificial arms, their feet are needed for walking, and the mental burden placed on them to control their prosthetic arms cannot be so great that controlling their prostheses takes more effort than it is worth. (*Many thanks to Mr. John Marbes of American Demolition for taking me on to the job site and allowing me to take these pictures.*)

ders). This motion can result in excursions of up to 10 cm (4 in) with a force of 178 to 266 N (40 to 60 lbf). Knowledge of the normal ranges of motions for the arm and hand is important, particularly when designing body-powered systems. For more in-depth information, Sarrafian (1992) provides extensive anatomical range-of-motion data relating individual muscle contributions to different arm motions.

Another property of the physiologic arm is that it has "give," that is, it is compliant or springlike. This compliance is not a fixed quantity but can be varied depending on the task requirements: a stiff arm for bracing oneself against an expected blow or a relaxed arm for playing the piano. This inherent compliance of the human arm also provides protection for the joints and musculo-skeletal system. Because the musculoskeletal system is compliant, it can withstand external shock loads far better than can a stiff-jointed equivalent.

Interaction with the real world is something current robotic and prosthetic actuators (dc electric motors with gear trains) do not do well. When a stiff robot arm comes into contact with a hard

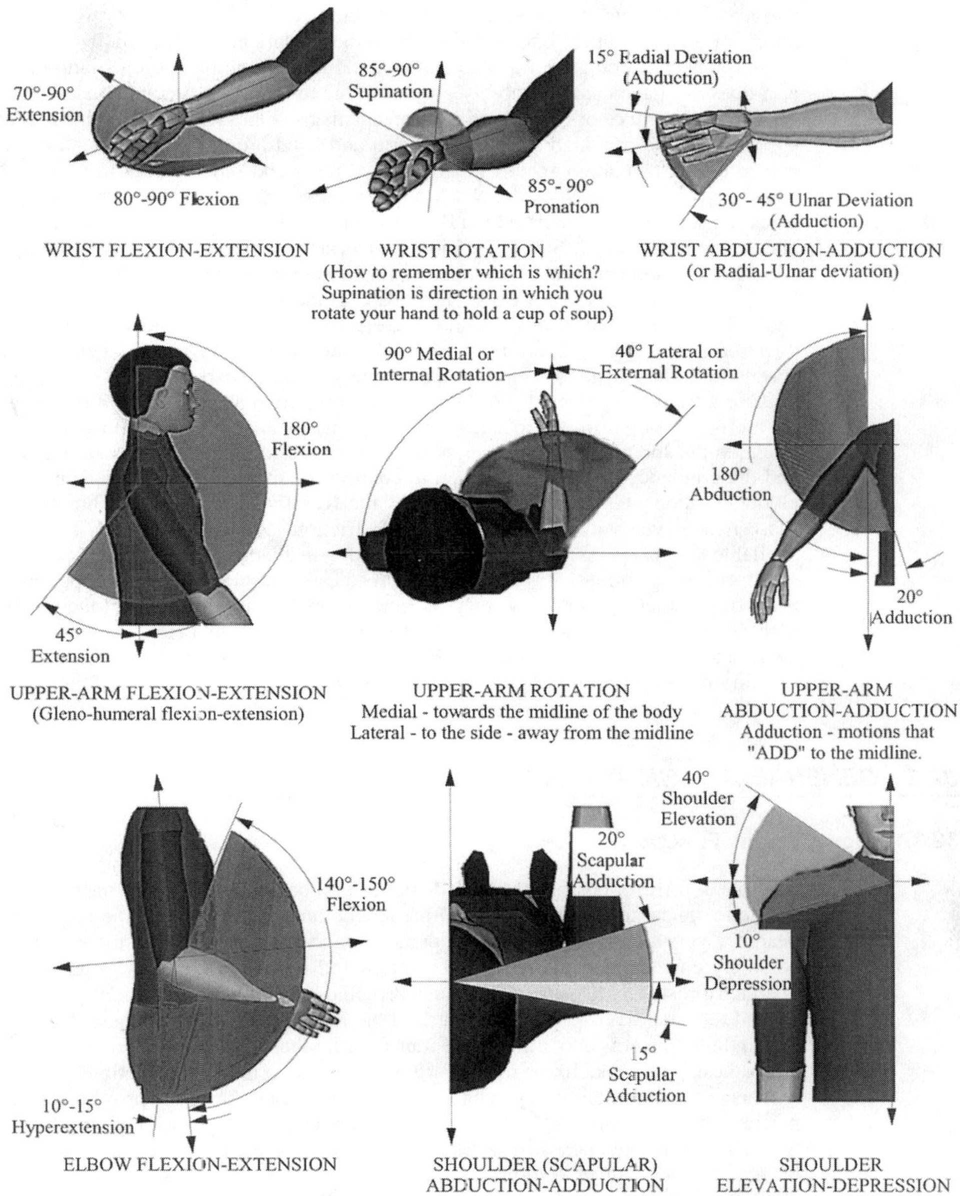

FIGURE 32.2 Ranges of motion for the wrist (top), upper-arm (center), elbow (bottom-left), and shoulder (bottom, center and right). Normal ranges of motion are for the wrist and forearm, −85° to 90° of pronation, 85° to 90° of supination, 15° of radial deviation, 30° to 45° of ulnar deviation, 80° to 90° of wrist flexion, and 70° to 90° of wrist extension; for the elbow, 140° to 150° of flexion and 10° to 15° of hyperextension; for the upper arm, 90° of medial humeral rotation and 40° of lateral humeral rotation, 180° of flexion and 45° of extension, and 180° of abduction and 20° of adduction; for the shoulder, 40° of elevation and 10° of depression, 15° of scapular adduction and 20° of flexion. (*Many thanks to Ms. Pinata Hungspreugs for her help in making this image*).

surface, a phenomenon known as contact instability can arise. Unless robot, environment, and the nature of the mechanical interaction between the two are precisely modeled mathematically, contact instabilities can occur, even if the mathematical representation is only slightly in error. The human arm does not have contact stability problems due to its inherent compliance.

The performance of current artificial mechanisms comes nowhere close to meeting the maximum speed and force of which the anatomic arm and hand are capable, although hand mechanisms are available that can attain speeds in excess of 3 rad/s and pinch forces in excess of 110 N (25 lbf). These mechanisms, such as the NU-VA Synergetic Prehensor, (Hosmer-Dorrance, Calif.), the Otto Bock Sensor and DMC hands (Otto Bock Healthcare, Duderstadt, Germany), and the Motion Control Hand (Motion Control, Salt Lake City, Utah) can achieve this performance when overvoltaged as is commonly done in clinical practice. As for wrist components, with the exception of the Otto Bock Electric Wrist Rotator (Otto Bock Healthcare, Duderstadt, Germany) and the VASI (Varity Ability Systems, Inc, Toronto, Canada) child-size electric wrist, all prosthetic wrist components are body-powered and, when used, are used for positioning purposes. As such, these devices operate at low speeds (less than 1 rad/s) with minimal torque-generating capability.

Current electric powered prosthetic elbows can attain about 12.2 N·m (9 ft·lbf) of "live-lift" (lift by the elbows own motor mechanism) at speeds of about 2 rad/s (Boston Elbow III, Liberating Technology, Inc., Mass.). Body-powered elbows are limited by the speed and strength of the user and the efficiency of the linkage used to connect the user and the component. Humeral rotation for elbow components, with the exception of the RIMJET body-powered humeral rotator (RIMJET, Fla.), is achieved with manually positioned friction joints or turntables. The only shoulder joints available are also passive, manually-positioned units that use friction or a lock to hold their position.

Thus it is apparent that while the user-prosthesis interface is a major impediment to the advancement of prosthetic technology, there is much room for improvement in the prosthetic components themselves. The limitations of current systems are not due to a lack of innovative design but rather are due to the very severe nature of the physical constraints that are placed on the designer and the inability of current technology to match the power and energy density of natural muscle.

32.3 GENERAL DESIGN CONSIDERATIONS

32.3.1 Form versus Function

The role of form, or cosmesis, in prosthetics cannot be overstated. Often the design team will sacrifice cosmetic appeal to achieve increased prehensile function. However, the relative importance of appearance versus function is highly dependent on the person with the amputation. Some people may be solely concerned with visual presentation and reject a highly functional body-powered, cable-operated prosthesis because of the unacceptable appearance of the control harness or of a hook-shaped terminal device. Others might find the function provided by these devices sufficient to outweigh their concerns about their appearance. Still others prefer a more toollike appearance believing a prosthesis that looks like a natural arm or hand but is not is a deception.

It is not uncommon for a person with an amputation to have two prostheses, one that emphasizes mechanical function, one for work, and an interchangeable one with a more humanlike appearance for social occasions, i.e., different tools for different jobs. Choice of prosthesis is ultimately based on many psychological, cultural, and practical factors. Other factors affecting the issue are age, gender, occupation, degree of physical activity, the amputee's attitude toward training, the type of amputation involved, and whether it is unilateral or bilateral limb loss. (Beasley and de Bese, 1990; Pillet and Mackin, 1992).

Cosmesis is the term used in prosthetics to describe how a particular device looks. A device is considered to be cosmetic in appearance if it is aesthetically pleasing and looks like the limb it is designed to replace in both its lines and color. However, a device may be statically "correct" in appearance but it can look "wrong" or lifeless when it is in motion. In this instance, the device has good *static cosmesis* but poor *dynamic cosmesis*.

People see what they expect to see, so if a person with an artificial limb replacement looks and moves in an expected manner the fact of the artificial limb will often go unnoticed by casual observers. Often a hand replacement can have only a vague resemblance to the natural hand but because it is moved in a natural-looking manner, it can pass undetected unless closely scrutinized. Here the device has poor *static cosmesis* but good *dynamic cosmesis*.

Dynamic cosmesis is frequently the more important of the two forms of cosmesis, but it is frequently overlooked because it is difficult to achieve. Dynamic cosmesis can be enhanced by preserving as much of the person's residual motion as possible. For example, a partial hand prosthesis should not interfere with residual wrist motion because the wrist is used extensively in the positioning of the hand in space. Finally, a device can be considered to be *functionally cosmetic* if at a glance it is not immediately recognizable as an artificial hand regardless of whether it is in motion or not or whether it is or is not handlike when stationary.

Cosmetic gloves, made of polyvinyl chloride (PVC) or silicone, are generally employed to cover artificial hand mechanisms to give them good static cosmesis (Fig. 32.3). These coverings serve to increase the passive adaptability of a prosthetic hand to the shape of a grasped object and to increase the coefficient of friction of the hand. The price paid for this increased cosmesis and grasping function is a restricted range of motion and hindered performance (speed/force output) of the hand. This is so because the hand motors must overcome the elastic forces inherent in the glove. The ideal glove should enhance the cosmesis of the prosthesis without interfering with its performance.

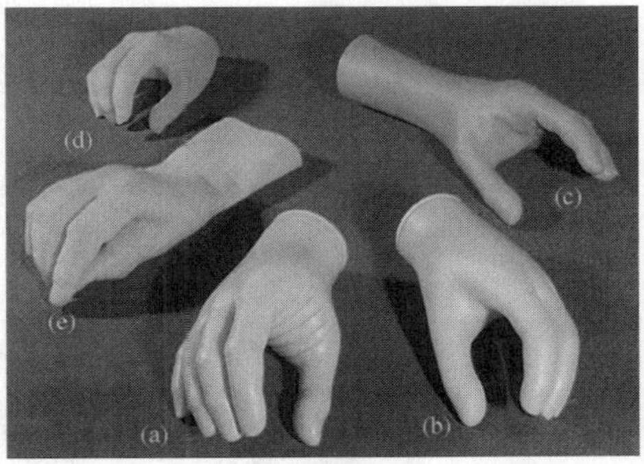

FIGURE 32.3 Photograph of various hand mechanism liners (*a*, *b*) and their associated cosmetic glove covers (*c*, *d*, *e*). Hand liner (*b*) is an older version of liner (*a*). Liner (*a*) has "grooves" or serrations added (in the thumb finger web space) in an attempt to reduce the elastic forces the liner imposes on the hand mechanism when it is trying to open or close the hand. Cosmetic glove (*c*) is made from polyvinyl chloride (PVC), whereas gloves (*d*) and (*e*) are made from silicone rubber. Silicone cosmetic gloves can be made with much more lifelike detail but tend not to be as sturdy as their PVC counterparts.

Limb replacements should be anthropomorphic in general shape, size, and outline. This does not mean that an artificial hand or arm needs to look exactly like its human counterpart. However, there should be a joint that operates like an elbow joint where one would expect to see an elbow joint, and the various limb segments should be of a size consistent with a normal human being, i.e., any replacement should have similar kinematics and kinetics. With regard to the issue of hand size, artificial hands are usually selected to be smaller than their physiological counterparts. This is so because artificial hands are perceived to be larger than they really are, probably due to their rigid structure and essentially static appearance.

32.3.2 Weight

Final weight of a prosthesis is critical to the success of any prosthetic fitting. Contrary to what one might think, one should not make an artificial limb replacement the same weight as the limb it replaces. The weight of an adult male arm is about 10 kg (20 lb). Total arm replacements that exceed 3.5 kg (~7.5 lb) cannot be expected to be worn and used for a full day because of the discomfort associated with suspending that much weight from the body. Artificial arms need to be as light as possible or else they will end up in a closet. The lack of an intimate connection between amputee and limb replacement means that the prosthesis is perceived as an external load and therefore as something that must be carried. To be effective, artificial arms should be worn by their users for periods in excess of 8 to 12 hours a day. To gain some insight into how this might feel, consider carrying a 6-lb laptop computer slung from your shoulder for a day.

32.3.3 Power Sources

As is the case for all portable devices, power is scarce. Choice of power source defines a prosthesis, in that it determines the choice of actuator. If the power source is to be the person, i.e., body power, then the actuator is the person's own musculature and the prosthesis should not require excessive effort to use. Mechanical mechanisms need to be efficient and frictional losses need to be minimized to avoid tiring the user over the course of a day. If the artificial limb is externally powered (i.e., uses a power source other than the body, usually electric storage), the limb should be able to run for a day from the same power source without needing to be replaced or recharged. In addition, it is desirable for the power source to be contained within the prosthesis.

Electrochemical batteries are the main source of energy for modern externally powered prosthetic arms, although pneumatic gas cylinders have been used in the past. There are a number of other technologies that could replace batteries as portable sources of electricity in the future. These include electromechanical flywheel systems that store energy in a rotating disk, and miniature Wankel type rotary combustion engines. However the most promising technology is that of ethanol- or methanol-based fuel cells. These devices are already moving into production for interim cell phone products. All are heavy and occupy excessive space. If electricity is the power source, then for the foreseeable future, dc electric motors will be the actuators. The problem of portable prosthesis power is analogous to the power issues in the laptop computer and cellular phone industry, where a major contributor to the weight and space of these portable devices is the power source, which for the moment is the battery. In addition, prostheses need high power density as well as high energy density.

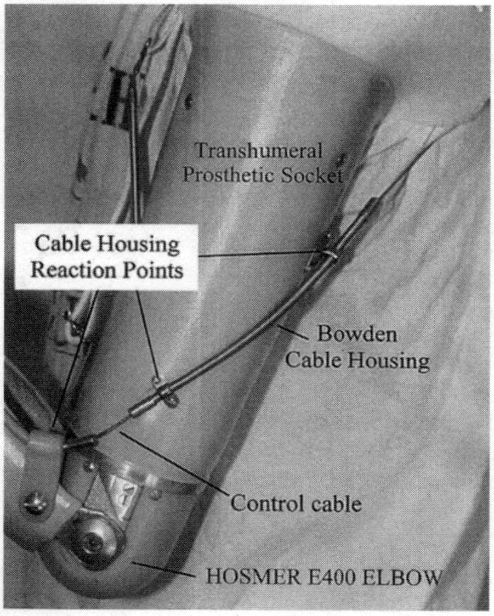

FIGURE 32.4 Photograph of how a Bowden cable is set up for use as the control cable of a transhumeral (above-the-elbow) body-powered prosthesis. The housing acts as a guide or channel for the control cable, which transmits forces developed by a harnessed body part, in this case the contralateral shoulder, to the prosthesis. Retainers on the housing fasten it to the prosthesis and serve as reaction points in the transmission of force by the cable (*Photograph courtesy of Mr. C. Heckathorne of the Northwestern University Prosthetics Research Laboratory.*)

32.3.4 Body-Powered Power Sources

In a body-powered device, the person uses his or her own muscular power to operate the prosthesis, usually via a cable link called a *Bowden cable* (Fig. 32.4). A Bowden cable consists of two parts, an outer housing and an inner tension cable. The housing is fixed at both ends and serves as a flexible bridge between two points, maintaining a constant length regardless of any motion. The cable is free to slide within the housing.

The Raleigh Bicycle Company first introduced Bowden cables as bicycle brake actuators in the later part of the nineteenth century. They were then adopted by the fledgling aircraft industry of the early twentieth century for the control of aircraft flight surfaces. At the end of World War II, many former aircraft designers, in particular engineers at Northrop-Grumman, were set to the task of designing better prostheses for returning U.S. veterans. One of their more important contributions was the use of Bowden cables to operate upper-limb prostheses.

The typical prosthesis control system consists of a Bowden cable with appropriate terminal fittings. The terminal fittings are used to anchor one end of the cable to a body-harness, and the other end to the prosthetic component to be controlled. Between the two end points, the cable crosses the prosthetic joint to be controlled and the physiological joint that is moved to actuate this prosthetic joint. The housing through which the cable slides acts as a guide or channel for the transmission of force by the cable. Retainers on the housing fasten it to the prosthesis, and serve as reaction points in the transmission of force by the cable.

The basic configuration of Bowden cables in prostheses has changed little over the intervening years and is still in use today. In fact, if prehensile function is the primary goal of the prosthetic fitting, the device of choice for most persons with amputations is a body-powered, Bowden-cable-operated prosthesis with a split hook-shaped terminal device. This is in spite of all the technological advances in electronics, computers, and dc motor technology that have occurred since the end of World War II.

Low technology does not imply wrong or bad technology. In fact, ease of maintenance, ease of repair in the event of failures in the field (one can use a piece of heavy cord to get by if the control cable should break), and the intuitive understanding of pulling on one end of the cable to effect motion at the other are probably major reasons for the success of Bowden cables in prosthetics. This is in addition to the ability of users to sense prosthesis state by the pull or feel of the control cable and harness on their skin.

For transradial (below-elbow) prostheses, only one Bowden cable is needed to open and close the terminal device (Fig. 32.5). In transhumeral (above-elbow) prostheses, two Bowden cables are needed, one to lock and unlock the elbow and another to flex and extend the elbow when the elbow

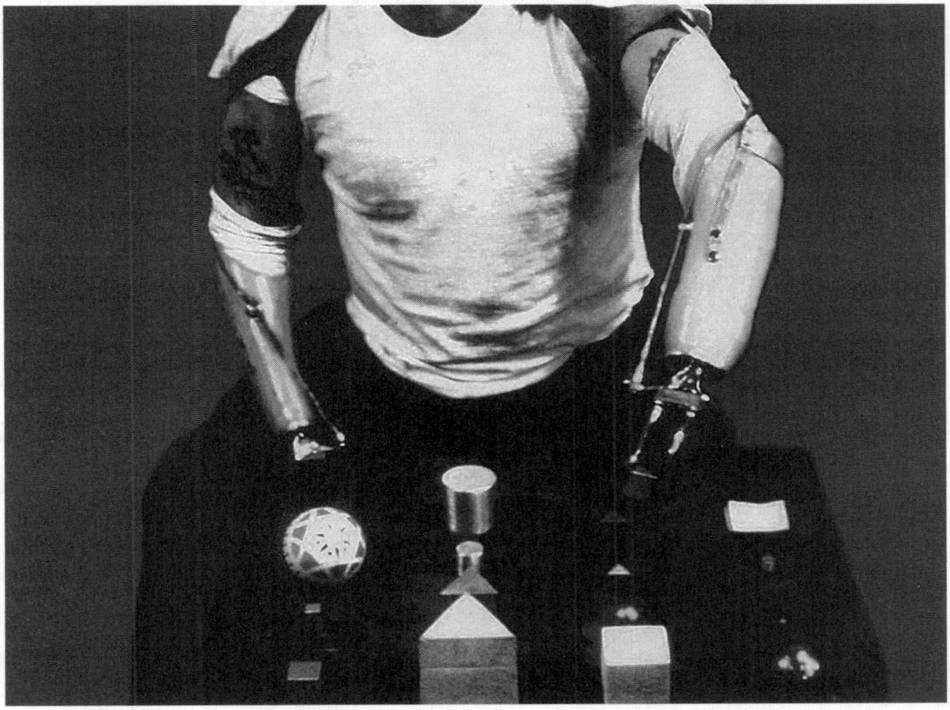

FIGURE 32.5 Photograph of a person with bilateral transradial (below-the-elbow) amputations using body-powered prostheses to perform standard object manipulation tasks during an occupational therapy session. Tasks of this nature usually take the form of pick and place trials. These trials require the subject to move objects of various sizes and shapes as quickly as possible between two different locations. This teaches the subject how to coordinate the prosthesis control while moving the terminal device. In another set of trials the subject moves objects of various sizes and shapes as quickly as possible from one hand to the other, for the purpose of teaching bimanual dexterity. Such tasks are used for training and evaluation purposes.

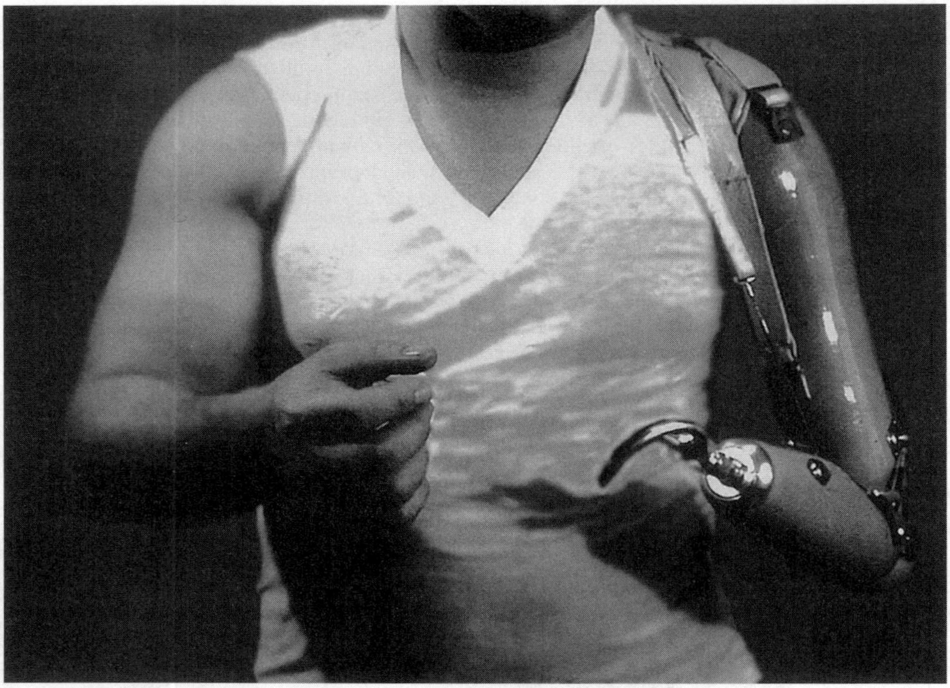

FIGURE 32.6 Photograph of a person with a unilateral transhumeral (above-the-elbow) amputation using a body-powered prosthesis. This picture clearly shows the elbow lock control cable (strap and cable on the anterior of the humeral section of the prosthesis). In a transhumeral prosthesis this cable is needed to switch control from the elbow to the terminal device. When the elbow is locked, the terminal device is under the control of the control cable; when the elbow is unlocked, then the elbow is under the control of the control cable.

is unlocked or to open and close the terminal device when the elbow is locked (Fig. 32.6). Both the transradial and transhumeral prostheses use a harness worn about the shoulders to suspend the prosthesis and for attachment of the control cable.

A combination of glenohumeral flexion and shoulder abduction is the primary body-control motion used to affect terminal device opening and closing or elbow flexion and extension. Typically, a total excursion of 10 cm (4 in) and upward of 222 N (50 lbf) of force are possible using these motions. Elbow lock control is affected by a complex shoulder motion which involves downward rotation of the scapula combined with simultaneous abduction and slight extension of the shoulder

TABLE 32.1 Force and Excursions for Different Body-Powered Control Sites for Average-Strength Males with Arm Amputations

	Force available		Excursion available	
Control source	N	lbf	mm	in
Arm flexion	280	63	53	2.10
Shoulder flexion	271	61	57	2.25
Arm extension	247	56	58	2.30

Source: From Taylor (1954).

joint. Figure 32.7 shows the basic harnessing of these cables for each case. There are many different variations on these basic harnessing schemes depending on specific needs of the prosthesis user. These variations and their application are described in Fryer (1992). Fryer and Michael (1992) provide an excellent overview of available body-powered components.

Typical forces and excursions for different body-powered control sources for an average adult male with an arm amputation can be found in Table 32.1. As can be seen, the average person has fairly high magnitude sources of force and excursion. By varying the locations of the reaction points of the Bowden cable or through the use of a pulley mechanism arranged to amplify force, or excursion, force can be interchanged with excursion or vice versa. However, the total power (force times excursion) remains constant.

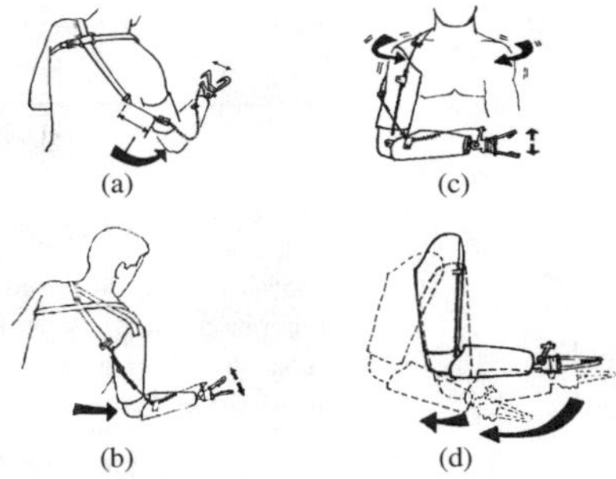

FIGURE 32.7 Schematics showing the harnessing and control motions used in both transradial and transhumeral body-powered prostheses. (*a*) Glenohumeral flexion—forward motion of the upper arm about the shoulder. This schematic shows the harnessing for a transradial prosthesis. (*b*) Glenohumeral flexion for a person harnessed for a transhumeral prosthesis. (*c*) Biscapular abduction (rounding of the shoulders) used by person with either transradial or transhumeral amputations. A combination of glenohumeral flexion and biscapular abduction is the most common mode of body-powered prosthesis control. (*d*) Shoulder depression followed by glenohumeral extension is the control motion used by persons with transhumeral amputations to actuate the elbow lock.

In prosthetics, stainless steel Bowden cables come in three sizes: Light Duty, Standard Duty, and Heavy Duty. The basic configuration is a Standard size multistranded steel cable in a Standard size housing. A heavy-duty user will use a Heavy-Duty multistranded steel cable in a Heavy-Duty housing. Friction is the primary cause of transmission losses in a Bowden cable. If friction is an issue then a Heavy duty housing with a Teflon liner and a Standard size cable can be used.

Composed of ultra-high-molecular-weight polyethylene fibers, Spectra cable is another option that is available to body-powered prosthesis designers and users. Spectra cable is now widely used at almost all major upper-limb fitting centers and could be considered the standard for contemporary fittings. Spectra's is used extensively as fishing line for deep-sea sport fishing because one of its properties is that it does not stretch. In addition, Spectra is very lightweight, quiet, strong, and it has a low coefficient of friction, making it an ideal alternative for numerous prosthetic applications, especially for children. Spectra is almost certainly always used when friction losses need to be avoided.

In Europe it is common to use a Perlon cable in a standard housing for light to medium users. While Spectra, like Perlon, can be used with standard cable housings, lower friction can be achieved by using Spectra or Perlon with a Heavy Duty housing with a Teflon insert. Note, the Heavy Duty housing is not needed for strength but for its wider inside diameter to accommodate the liner. This way one has plastic on plastic, reducing wear and friction and increasing lifetime. It should be noted that standard swage fittings will not work with Spectra and that tie-off fittings are required. Spectra is a registered trademark of Allied-Signal, Inc.

Tip. Bowden housings are essentially tightly wound steel spirals (tightly wound springs) and as such can be screwed into an appropriately sized hole. For a Standard housing use a ⅛-in drill and a No. 8–32 Tap (note that this is a nonstandard Drill and Tap combination).

32.3.5 Voluntary Opening Versus Voluntary Closing Devices

Body-powered prehension devices can be either voluntary opening or voluntary closing. A voluntary opening device requires a closing force, usually exerted by rubber bands, to be overcome before it will open (default to close position). The maximum pinch force in a voluntary opening device is limited to the closing force exerted by the rubber bands. Typically, a standard prosthesis rubber band results in 7 to 9 Newtons (~1½ to 2 lbf) of prehension force and a healthy consistent user will have anywhere from 3 to 8 rubber bands depending on strength, amputation level, and personal preference.

TABLE 32.2 Force and Excursions Needed to Actuate a Number of Standard Body-Powered Prosthetic Components

	Force needed		Excursion needed	
Component/operation	N	lbf	mm	inches
Elbow Flexion (No Load on Hook)	40	9	51	2
Prehension, APRL Voluntary Closing Hook	40–155	9–35	38	1.5
Prehension, Sierra Two-load* Voluntary Opening Hook	44.4, 89	10, 20	38	1.5
Elbow Lock Actuation	9–18	2–4	15	0.6

*The Sierra Two-Load Hook, formerly the Northrop Two-Load Hook, was a voluntary opening hook with two closing force settings. A manually controlled switch on the side of the hook moved the hook between force settings.

Source: Taylor (1954).

Voluntary closing devices require an opening force to be overcome before the terminal device will close (default to open position). Pinch force in a voluntary closing terminal device is directly proportional to the force applied to the device through the control cable.

The voluntary closing principle is the closest to natural hand prehension (Fletcher, 1954). However, tension must be maintained in the control cable to hold a constant gripping force or position, just as must be applied by the human hand to maintain grasp. The natural neuromuscular mechanism has the ability to maintain grasp over long periods of time without undue strain. Automatic or manual prehension locking mechanisms have been used for this function in voluntary closing devices. A voluntary opening device does not have this problem as long as the object held can withstand the maximum closing force.

The Therapeutic Recreation Systems (TRS, Boulder, Colo.) range of terminal devices and the Army Prosthetic Research Laboratory (APRL) hook, which was used predominantly by muscle tunnel cineplasty amputees, are examples of successful voluntary closing devices that are available. The APRL hook has automatic locking, whereas the TRS devices do not.

Operating requirements (force and displacement) for a number of standard body-powered prosthetic components are given in Table 32.2. Neither the Army Prosthetics Research Laboratory Hook nor the Sierra Two-Load Hook are used much today. It is most common to find a voluntary opening split hook on current body-powered systems. However, if one were to compare new contemporary fittings with refitting of existing prostheses, the percentage of voluntary opening split hooks is declining among new contemporary fittings in the United States. At the transradial level, the TRS devices are becoming the prehensor of choice with sales of TRS devices far exceeding sales of APRL hooks and hands. Properly harnessed, the TRS devices are nearly closed in the relaxed, or neutral, position.

32.3.6 Electric Power Sources (Batteries)

Battery technology, specifically rechargeable battery technology, is vital to portable electronic equipment and is driven by the billions of dollars spent by the laptop computer and cellular phone industries. The field of prosthetics and orthotics (P&O) sits on the sidelines and picks up anything that looks like it could be of use. In an electrically powered prosthesis, the main current draw comes from the dc motor(s) used to actuate the device. In a dc motor, the output torque is directly proportional to the amount of current drawn. Motor use in prostheses is not continuous but is intermittent. Consequently, it is important not only to know how much energy a battery can provide but also how fast the battery can provide it.

The maximum amount of current drawn by a dc motor is the *stall current*. This is the current drawn by the motor when it is fully "on" but unable to move, such as occurs when a hand has grasped an object and the fingers can close no further. This is also the point of maximum torque output of the motor and is known as the *stall torque*. Running a dc motor in stall for extended periods of time may damage the motor. However, the stall current is the upper-limit on the amount of current required by a dc motor driven mechanism. As such, the stall current determines the size of the MOSFET's (field-effect transistor) and/or H-bridges used to deliver current to the motor.

Batteries are a chemical means of producing electricity, in which the choice of electrode materials is dictated by their relative location in the electrochemical series. Compounds that are at the extremes of the series are desirable. Many battery chemistries exist. Batteries are usually packages of a number of cells stacked in series to achieve a desired voltage. The voltage of the individual battery cells depends on the chemistry of the cell. The nonrechargeable off-the-shelf batteries that one buys in the store are referred to as *alkaline batteries*. In P&O we are most interested in rechargeable batteries with solid chemistries, i.e., nonliquid batteries.

Size, weight, capacity, and power density are the primary selection considerations for batteries in externally powered prosthetic design applications. The most popular types of rechargeable batteries in use in prosthetics today are nickel-cadmium (NiCd), nickel–metal-hydride (NiMH), and lithium-ion (Li-ion). Li-ion is fast becoming the chemistry of choice because of its high capacity-to-size (weight) ratio and low self-discharge characteristic.

The amount of time it takes to discharge a battery depends upon the battery capacity C expressed in milliamp hours (mAh) (more is better) and the amount of current drawn by the load. Battery charge and discharge currents are normalized with respect to battery capacity and are expressed in terms of C-rate. C-rate $= C/1$ hour, e.g., to a 150-mAh battery has a C-Rate of 150 mA. The current corresponding to 1C is 150 mA and 0.1C, 15 mA. For a given cell type, the behavior of cells with varying capacity is similar at the same C-rate. More in depth information about batteries can be found in the *Handbook of Batteries* (Linden, 1995).

Battery capacity is the important measure of performance in the cell phone and laptop industries because once the device is switched "on," the load current remains fairly constant. The *discharge rate* is an important measure for prosthetics applications because it is a measure of the rate at which energy can be delivered to the load. It is the maximum allowable load or discharge current, expressed in units of C-rate. The higher the discharge rate, the faster a battery can meet the demand for energy.

The *self-discharge rate* is the rate at which a battery discharges with no load. Li-ion batteries are a factor of two better than NiCd or NiMH batteries. The number of charge and discharge cycles is the average number of times a battery can be discharged and then recharged and is a measure of the service life. For prosthetics applications, a battery ought to provide at least a day of use before needing to be recharged; thus the average life expectancy for a rechargeable battery in prosthetics use should be about 3 years. Table 32.3 tabulates these quantities for the chemistries of interest.

The problem of *memory* that one hears about in association with NiCd batteries is relatively rare. It can occur during cyclic discharging to a definite fixed level and subsequent recharging. Upon discharging, the cell potential drops several tenths of a volt below normal and remains there for the rest of the discharge. The total ampere-hour capacity of the cell is not significantly affected. Memory usually disappears if the cell is almost fully discharged (*deep discharge*) and then recharged a couple of times. In practice, memory is often not a problem because NiCd battery packs are rarely discharged to the same level before recharging. Environmental concerns exist regarding the proper disposal of NiCd batteries because of the hazardous metal content. NiMH and Li-ion batteries do not contain significant amounts of pollutants, but nevertheless, some caution should be used in their disposal. Discharge profiles for these three popular types of batteries are shown in Fig. 32.8. Note that NiCd batteries have a greater discharge rate than LiM batteries (making them suitable for prosthetic applications) however their energy density is lower.

Slow charging (charging times greater than 12 hours) is straightforward for all battery chemistries and can be accomplished using a current source. Charge termination is not critical, but a timer is sometimes used to end slow charging of NiMH batteries. Li-ion slow-chargers should have a voltage limit to terminate charging of Li-ion batteries. *Trickle charging* is the charging current a cell can accept continually without affecting service life. A safe trickle charge for NiMH batteries is typically $0.03C$. *Fast charging* (charge time of less than 3 hours) is more involved. Fast-charging circuits must be tailored to the battery chemistry and provide both reliable charging and charge termination. Overcharging can damage the battery by causing overheating and catastrophic outgassing of the

TABLE 32.3 Figures of Merit Used to Characterize the Common Rechargeable Battery Chemistries in Use Today.

	Nickel cadmium (NiCd)	Nickel metal hydride (NiMH)	Lithium ion (Li-ion)	Lithium metal (LiM)
Cell voltage (V)	1.2	1.25	3.6	3.0
Energy density (watt-hours/liter [Wh/L])	45	55	100	140
Energy density (watt-hours/kilogram [Wh/Kg])	150	180	225	300
Cost ($/watt-hour [$/Wh])	0.75–1.5	1.5–3.0	2.5–3.5	1.4–3.0
Self-discharge rate (%/month)	25	20–25	8	1–2
Discharge rate	>10C	<3C	<2C	<2C
Charge/discharge cycles	1000	800	1000	1000
Temperature range (°C)	−10 – +50	−10 – +50	−10 – +50	−30 – +55
Memory effect	Yes	No	No	No
Environmental concerns	Yes	No	No	No

Note: Lithium metal chemistry is a new chemistry that is not yet widely available. The values shown here are based on an AA size cell.

Source: Kester and Buxton (1998, p. 5.3).

electrolyte. The gas released from any outgassing may be dangerous and corrosive. In some instances overcharging may cause the battery to explode (Kester and Buxton, 1998).

Standard operating voltages used in prosthetics are 4.8, 6, 9, and 12 V depending on the manufacturer and component to be driven. All Otto Bock System 2000 children's hands (Fig. 32.14) (Otto

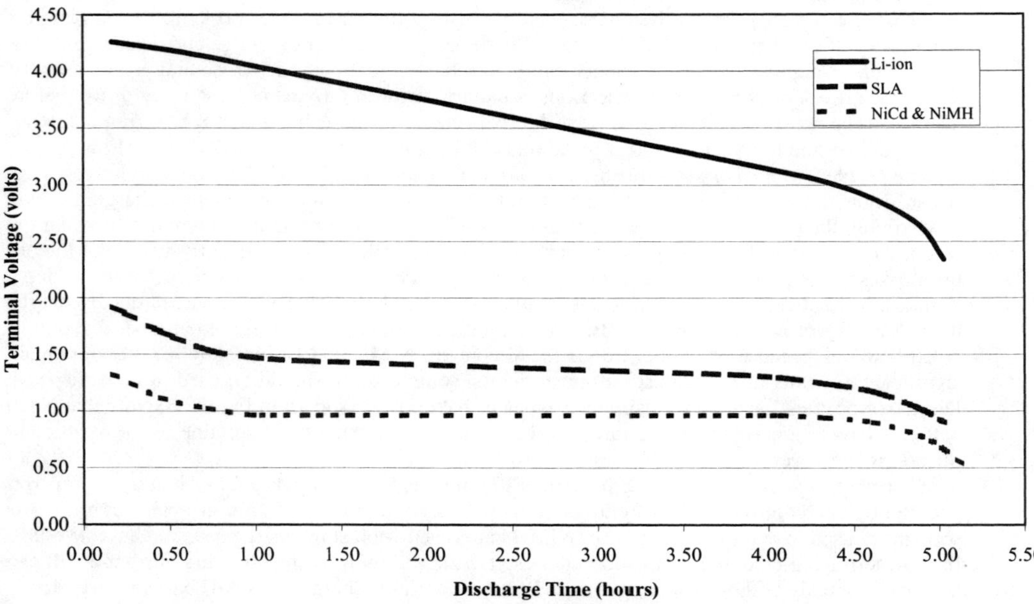

FIGURE 32.8 Discharge profiles of different kinds of batteries. A discharge current of 0.2 C was used in each case. NiCd and NiMH batteries have a relatively flat profile, whereas Li-ion batteries have a nearly linear discharge profile. This feature is useful when designing "fuel gauges" for Li-ion batteries (*Generated using data supplied in*: *Kester & Buxton 1998*).

Bock Orthopedic Industry Inc., Duderstadt, Germany) run off 4.8 V and will not run at higher voltages. Otto Bock adult hands (Fig. 32.12), Steeper hands/prehensors (Fig. 32.20*c*) (RSLSteeper, Rochester, U.K.), VASI hands and elbows [Variety Ability System, Inc. (VASI), Toronto, Canada] and Hosmer NYU elbow (Hosmer-Dorrance Corp., Campbell, Calif.) are specified to run off 6 V but can sometimes be run at higher voltages to boost performance. The Hosmer NU-VA Synergetic Prehensor (Fig. 32.20*b*) and Motion Control ProControl System (Motion Control, Inc., Salt lake City, Utah) use 9 V. The Motion Control Utah Arm and LTI Boston Elbow [Liberating Technology, Inc. (LTI), Holliston, Mass.] run off 12 V, while Motion Control's electric hand is specified to run on any voltage between 6 and 18 V. In general, adult-size electric hands use 6 V while adult-size electric elbows use 12 V. Elbows need the extra voltage so that they can have a useful live-lift capacity (torque to actively lift a load). Typical currents are 1 to 3 A (~12 to 36 W) for hand mechanisms and 9 to 12 A (~120 W) for some of the brushless dc motors used in elbow prostheses.

The preponderance of 6-V electric hands is due to the dominance and history of Otto Bock in the field of prosthetics. Otto Bock was one of the first companies to produce an electric hand with its associated electronics and power source in the early 1970s. Their standard battery was the 757B8 five-cell NiCd battery pack, where each cell has a voltage of 1.2 V. Otto Bock's current standard battery is now a Li-ion battery. This is a custom-designed battery pack and battery holder. The battery holder is designed to be laminated into the prosthetic socket so that batteries can be easily interchanged should one discharge completely during use. Even if the battery chemistry has changed, this battery holder's form has been around for many years. The disadvantage of a custom form is that if the battery dies in the field and the user does not have any spares then they are stuck without a working component. Liberating Technology, Inc. and Motion Control elbows use custom battery backs, whereas the Hosmer-Dorrance elbows use packs made up of off-the-shelf AA rechargeable batteries (Fig. 32.9).

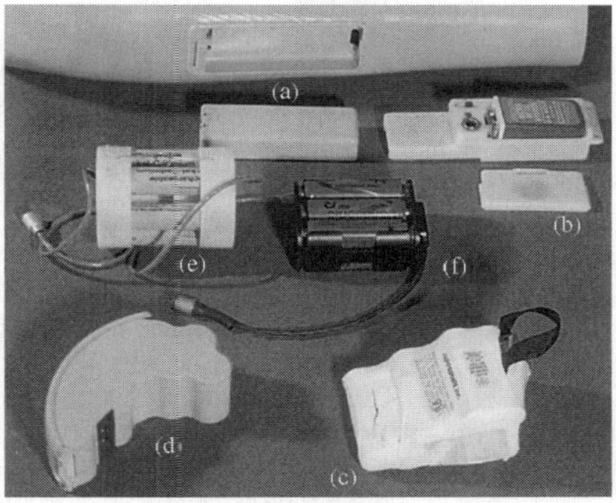

FIGURE 32.9 Battery packs in prosthetics come in all different shapes and sizes. (*a*) Standard Otto Bock (Otto Bock Orthopedic Industry, Inc., Duderstadt, Germany) NiCd Battery pack and holder. (*b*) The Hosmer-Dorrance (Hosmer-Dorrance Corp., Campbell, Calif.) battery holder for a standard 9-V transistor battery (NiCd or NiMH). (*c*) Liberating Technology (Liberating Technology, Inc., Holliston, Mass.) NiCd battery pack for the Boston Elbow II. (*d*) Motion Control's (Motion Control, Salt Lake City, Utah) custom battery pack for the Utah Arm, and (*e*) and (*f*) Hosmer Dorrance (Hosmer-Dorrance Corp., Campbell, Calif.) NiCd battery packs for their NY-Hosmer Elbow. Battery packs c, e, & f consist of multiple AA rechargeable batteries packaged together to create a battery of the desired output voltage and capacity.

Our laboratory's preference is to use rechargeable batteries in a 9-V transistor battery form. This enables the prosthesis to be recharged overnight, while a standard 9 V transistor battery form allows commercially available off-the-shelf 9-V alkaline batteries to be easily purchased and used should the rechargeable battery run out of charge unexpectedly. Motion Control sells an Otto Bock battery case that has been modified to accommodate a standard 9-V transistor battery.

32.3.7 DC Electric Motors

By far the most common actuator for electrically powered prostheses is the permanent magnet dc electric motor with some form of transmission (Fig. 32.10). While there is much research into other electrically powered actuator technologies, such as shape memory alloys and electroactive polymers, none is to the point where it can compete against the dc electric motor. A review of the available and developing actuator technologies with their associated advantages and disadvantages as well as their power and force densities can be found in Hannaford and Winters (1990) and Hollerbach et al.

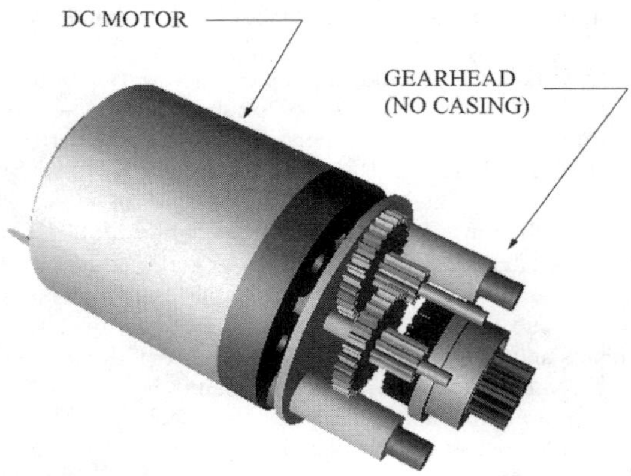

FIGURE 32.10 Drawing of a dc motor and gear head. The high output speed of small dc motors must be converted into a usable torque in order to build functional components. This transformation of speed into torque is performed by a gear train (shown in the figure without its casing).

(1991). In both these reviews, biological muscle is used as the benchmark for comparison of these technologies.

For electrically powered prosthetic hand mechanisms, a coreless, or ironless, dc motor with a fitted gear head transmission is the actuator of choice. For elbows, coreless dc motors, or brushless dc motors and a transmission are used. Brushless dc motors are becoming more common now that both the motor and its electronics are small enough to be accommodated in a prosthetic elbow (they are still too large for hand mechanisms, but this is likely to change). Although brushless motors require much more complicated control electronics, their use is justified because they have substantially higher performance than their coreless counterparts. In addition, recent advances in surface mounted integrated circuit (IC) technology greatly facilitate the design of controllers for these motors. A broad range of driver and controller ICs are available in surface-mount forms from companies like Texas Instruments, International Rectifier, ST Electronics, Vishay-Siliconix, Zetex, among others, and application notes explaining the use of these chips are readily available on company Web pages.

32.3.8 Sizing a DC Motor for Maximum Performance

In general, because there is no actuator technology that can match muscle tissue's speed and force capabilities for a given weight and volume, one chooses the largest motor-gearhead combination that will fit within the physical constraints of the mechanism being designed. When one discusses maximizing performance, one is talking about maximizing output speed and output torque to attain physiological speeds and torques. Almost without exception, powered prosthetics components use MicroMo-Faulhaber (Faulhaber Group, Clearwater, Fla.) motors. This company manufacturers coreless motors, brushless motors, and their associated gearheads. Their motors range in diameter from 6 up to 35 mm.

The key issue to keep in mind when choosing a motor is that one trades no-load speed for stall torque and vice versa (Fig. 32.11). A prosthetic hand usually operates at high speed and low torque during hand closing and opening and then at high torque and low speed while gripping an object. To meet both these requirements, assuming an automatic transmission is not used, a single motor must be sized to provide both the requisite stall torque as well as the requisite no-load speed. Thus,

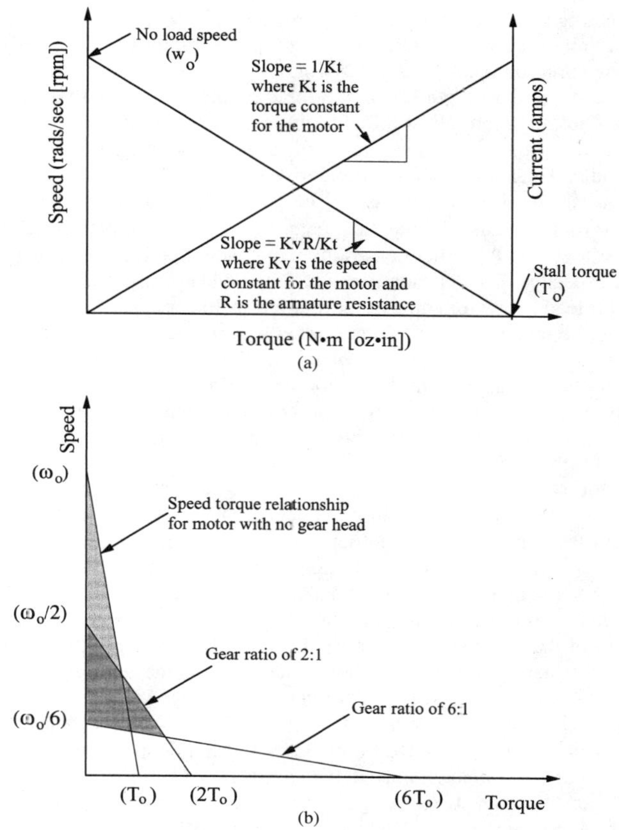

FIGURE 32.11 (*a*) Typical speed-torque and current-torque relationships for a dc motor. The key issue is that stall torque times no-load speed is constant (stall torque × no-load speed = const.) regardless of the ratio of the gear train driven by the motor. The main points of interest on this curve are no-load speed (output speed of the motor when running without load), the stall torque (torque generated by the motor when fully "ON" but unable to move), and the area under the curve (which is the total power available to the motor). Typical units are included in brackets—imperial equivalents are in the square brackets (careful use of the correct conversion factors is required when working with imperial units). (*b*) Adding a gear head to the motor output changes the no-load speed and stall torque of the drive train, but the area under the new speed-torque relationships remains the same as for the motor without the gearhead (assuming a lossless gearhead, i.e., 100 percent efficient). One trades no-load speed for stall torque and vice versa, but the total available power remains the same.

the motor must be capable of providing the mechanical power output given by the product of the no-load speed times the stall torque.

Choice of gear ratio depends on the no-load speed and stall torque of the motor selected, as well as on the desired output speed and torque of the mechanism. Remember speed times torque into and out of the gear train is constant (assuming an ideal gear train). Usually, when using a dc motor, speed is traded for torque. One must also allow for the overall efficiency of the gearhead and the maximum allowable intermittent output torque for a given gearhead. The MicroMo catalog (www.micromo.com) provides worked examples of how to choose the correct motor and gearhead for a particular output speed and force requirement (Faulhaber MicroMo Application Notes).

The maximum allowable intermittent output torque is a function of the tooth strength (this is a material property) of the gears and, as such, is an upper limit on the torque a particular gear can handle. Thus, although a particular gear ratio might suggest that the output torque is attainable, the gears in the drive train may not be physically able to handle the loads placed on them. Planetary gear trains can usually handle greater torques than simple spur gear trains, but usually at the expense of efficiency. A further constraint is to ensure that the motor-gearhead maximum axial load will not be exceeded. This is usually not an issue when using spur gears on the output. However, if a worm gear, lead screw, or some other rotation-to-linear conversion mechanism is used then this parameter must be taken into consideration and a thrust bearing used to protect the gearhead if necessary. A good reference is the *Handbook of Small Electric Motors* (Yeadon and Yeadon, 2001).

Often, for a given size of motor, a manufacturer will offer motors with different magnet materials. To maximize the performance (force-speed output) of a particular size of motor, choose the "rare earth" version (neodymium-iron/neodymium-iron-boron magnets). (Note: rare-earth magnet motors tend to pull higher currents in stall.) In addition, when specifying the nominal drive voltage of the motor, consider deliberately overvoltaging the motor to boost overall performance. This can be done because prosthesis use is intermittent in nature. A nominal 6-V motor can be run at 9 V yielding a ~50 percent increase in output speed and torque. Care must be taken, however, because the stall current will also increase. Prolonged exposure to increased stall current can damage the motor unless the rise in current as the motor stalls is detected and the supply voltage cut off, or the current limited. A final word of caution, ultimate drive system torque performance is limited by the amount of current that the battery can provide. If the battery cannot provide the stall current needed then the drive system will not reach the torque levels suggested by your calculations.

Current flowing in the motor can be monitored using a number of different ICs or power FETs available from any number of suppliers. These devices include current shunt ICs (Maxim, Texas Instruments Zetex), sense FETs (Fairchild Semiconductor), HEX sense power FETs (International Rectifier), and fault-protected switches (fancy MOSFETs, Micrel) to name but a few. Current-shunt ICs are common because they find use in dc-to-dc converters, consequently, a large number of different manufacturers make them in many different forms and flavors (see Fig. 32.17).

Figure 32.17 shows an implementation of high-side current monitoring on a *n*-channel H-bridge. (Note: this circuit is not limited to use with just *n*-channel bridges.) In this circuit a shunt resistor measures current flowing to the bridge, and an amplifier scales the voltage from the shunt resistor to logic levels, where a 5-V signal corresponds to the maximum current anticipated. This voltage level is compared against some threshold level to ensure that too much current is not being drawn. If it is, the comparator output goes high or low depending on your logic of choice (logic-low will fail-safe if the power goes out) and another supervisory circuit (microprocessor, etc.) takes action to switch off the motor. This same circuit can be implemented as a low-side current monitor just by placing it between ground and the bridge. Since the bridge is under your control, you should know the motor's direction. If current sensing for each motor direction is required, then this circuit needs to be duplicated and a sense resistor placed in each side of the bridge at the location of the switches shown in the *pn* complementary H-bridge.

An implementation of limit switch protection of the motor is shown in Fig. 32.17 on a *pn* complimentary H-bridge. In this case electromechanical switches are physically located at the ends of the range-of-motion of the component (full flexion or full extension of an elbow, for example). When the elbow extends fully it physically hits/bumps/actuates the switch, causing it to open. This has the effect of "dropping out" the ground line for the extension direction motor supply, causing the motor to stop extending. Further extension drive signals will not extend the component. A drive signal of the reverse polarity travels through the other side of the H-bridge enabling the motor to run in the opposite direction (flex in this example). Again this example is not limited to *pn* bridges and could also be implemented on the high side.

While the output power of a motor may remain constant, how that power is used, i.e., whether all the power is used to generate high force at low speed or high speed at low force, is up to the designer. Since there is no one motor small enough to be placed in an artificial hand that can meet both the speed and force requirements (even with overvoltaging), other techniques must be employed to increase hand mechanism speed of opening and closing and grip force. Two techniques are currently employed in commercially available prosthetic prehensors to increase the performance of prosthetic hands to levels approaching those of the physiological hand.

The first technique is to use a single large motor and an automatic transmission. In this case, the motor opens and closes the hand in a high speed, low force gear ratio. When the hand closes against an object, the rise in force (as detected by an internal spring) automatically triggers the transmission to switch into a high-force, low-speed mode that allows the hand mechanism to build up prehension/pinch force. The main disadvantage of these mechanisms is associated with tightly grasped objects. In this instance, during opening, a perceptible amount of time appears to elapse before the hand visibly opens. This is so because time must be allowed for the force built up during the grasping cycle to drop below the force threshold where the hand automatically switches from high force, low speed to low force, high speed. Thus, although the hand is opening, the gear ratio is such that little motion occurs until the transmission switches into the high-speed mode. To the user, this time spent "unwinding the hand" appears as if nothing is happening in response to the open command. A further disadvantage of automatic transmissions is that they are mechanically complex and are often beyond the means of the individual to design in a form compact enough for use in hand prostheses. The Otto Bock range of System Electric Hands (Fig. 32.12) for adults has used and refined its automatic

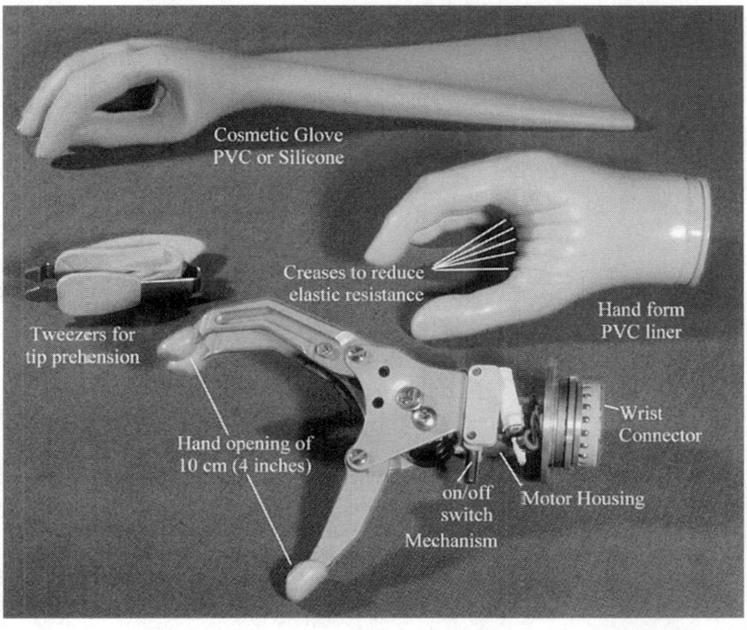

FIGURE 32.12 Otto Bock System Electrohand (Otto Bock Orthopedic Industry, Inc., Duderstadt, Germany). The hand consists of a mechanism over which a liner is placed. A cosmetic glove is then pulled over the liner. (Note: Otto Bock does not provide silicone gloves.) Liner creases have been placed in the finger thumb web space in an effort to reduce its elastic resistance. Also shown is a pair of tweezers that are provided with each hand to make tip prehension possible. Prosthetic hands have very poor tip prehension without the aid of some sort of extra tool such as these tweezers.

transmission for many years, to the point where their design is a marvel of compactness and robustness: Why reinvent the wheel? The Motion Control hand, which owes much of its mechanical design to the Otto Bock hands, also uses an automatic transmission.

The second technique is to use multiple smaller motors configured for what Childress (1973) called *synergetic prehension*. In synergetic prehension, one motor is geared for high speed and low force, and another is geared for high force at low speed. A simple synergetic prehensor consists of two motors that open and close a split hook (Fig. 32.13*a*). One motor gives one tine of the hook high speed and excursion but little force (fast side), the other motor gives the other tine of the hook high force but little speed and excursion (force side). In this way the motors work in synergy to boost overall prehension performance (Fig. 32.13*b*).

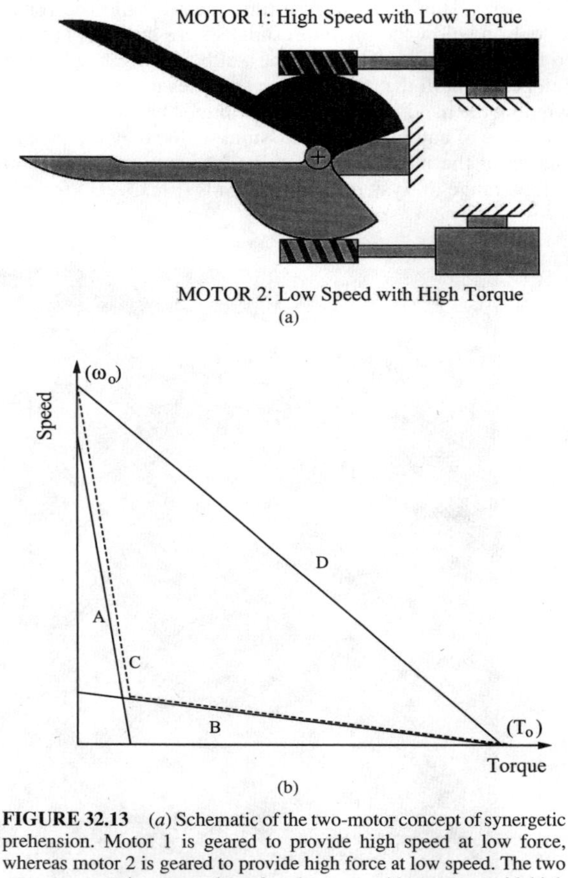

FIGURE 32.13 (*a*) Schematic of the two-motor concept of synergetic prehension. Motor 1 is geared to provide high speed at low force, whereas motor 2 is geared to provide high force at low speed. The two motors operate in synergetic prehension to provide a system with high speed and high force. (*b*) The effective speed torque relationship (*C*) of two drive systems having characteristics *A* and *B* operating in synergy. *A* is the characteristic for a high-speed, low-torque motor, whereas *B* is the characteristic for a low-speed, high-torque motor. *D* is the required speed-torque relationship for a single dc motor and gearhead to achieve the same no-load speed and stall torque as the synergetic system.

32.3.9 Synergetic Prehension

The principle of synergetic prehension stems from Childress' (1973) observation that the act of grasping an object usually requires little real work. When an object is grasped, a force is exerted with very little excursion, where as excursion of the grasping fingers when approaching the object usually occurs in space and requires very little force. In both cases the work involved is minimal. The exception is grasping a compliant object, where both force and excursion are required (e.g., squeezing a lemon). This condition, however, is not the usual case in prosthetics. Synergetic prehension can be readily implemented using multiple motors that operate together, or in synergy. Each synergetic motor pair controls a single degree of freedom and has its own synergetic motor controller. The combination of the synergetic controller and synergetic motor pair allows it to be treated like a single motor. Placing a voltage across the input lines of the synergetic controller will drive the motors in the synergetic motor pair. The polarity of this voltage determines the direction the motors in the pair will rotate. Drive signals in the form of either pulse-width-modulated (PWM) signals for proportional speed and force control or on/off signals for switch or single speed control can be used.

The Hosmer-Dorrance NU-VA Synergetic Prehensor was developed using this theory and achieved angular velocities in excess of 3 rad/s and prehension forces greater than 20 lbf (89 N) using a small 9-V transistor battery (Fig. 32.20*b*). The Otto Bock System 2000 children's hands (except for the smallest size) (Fig. 32.14) and the RSLSteeper Powered Gripper (RSLSteeper, U.K.), also use the principle of synergetic prehension (Fig. 32.20*c*). An advantage of synergetic systems is that they can potentially develop high force and high speed at low power. The main disadvantage associated with synergetic systems is the use of multiple motors. More motors mean more parts that can fail. In addition, the motor and its gearhead tend to be some of the most expensive components in a hand mechanism.

Not all commercially available hands use automatic transmissions or multiple motors operating in synergy. The Variety, Ability System Inc. (VASI, Canada) range of children's hands, Centri AB's

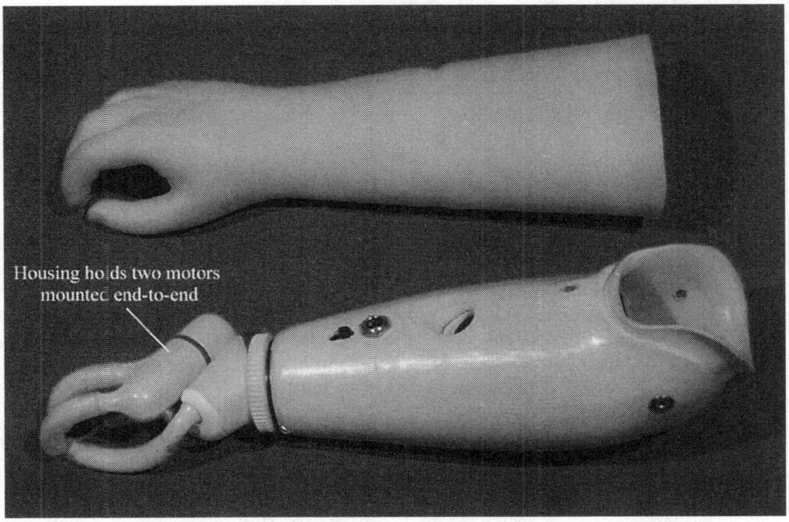

FIGURE 32.14 Otto Bock System 2000 children's hand (Otto Bock Orthopedic Industry, Inc., Duderstadt, Germany). This mechanism uses two gear motors attached end to end that operate using the principle of synergetic prehension. Notice that this hand does not use a liner. The cosmetic glove is pulled directly over the mechanism. This is so because for a child's hand, the forces are not as high as for an adult mechanism and the tips will not punch through the glove at the lower grip force.

(Sweden) hands, and RSLSteeper's Adult Electric Hands (RSLSteeper, U.K.) are all examples of single-motor designs with a fixed gear ratio. Typical performance for these devices is correspondingly lower [35 N (8 lbf) pinch, 10 cm/s (4 in/s) speed] than typical performance for those prehensors that utilize synergy or automatic transmissions [80 N (18 lbf) pinch, 11 cm/s (4.3 in/s) speed].

All hands need some form of backlock mechanism to prevent the fingers from being forced open or backdriven once power is removed. This is so that a hand mechanism will hold its position, or applied force, once the control signal is removed. While this is nonphysiological, (i.e., muscles must keep contracting to maintain a position or an applied force), this preserves power. Synergetic mechanisms need a backlock mechanism to prevent the force side from backdriving the speed side, in addition to maintaining applied force in the absence of power.

The RSLSteeper Electric Hands use a worm gear for this purpose. A worm gear is inherently nonbackdrivable. The VASI hands, the NU-VA Synergetic prehensor, and the LTI Boston Elbow all use variations on a roller clutch (backlock) first used in ViennaTone hands of the 1960s (Fig. 32.15). Locking occurs when a cam inside the mechanism wedges a roller(s) between itself and the outside cylindrical surface in response to an external torque. Driving the input shaft from the motor side causes the cam to move away from the roller(s), allowing the roller(s) to move away from the outside surface, thus freeing up the mechanism. These roller backlocks can resist external torques in either direction or only one direction depending on the number and position of the roller(s) and shape of the cam. Roller backlocks should be placed as high up in the gear train (close to the motor) as possible to minimize backlash (play) and to minimize the force exerted on the cylindrical surface as well as the force needed to unlock the mechanism. *Machinery's Handbook* (Oberg et al., 2000) is a good reference on different types of mechanism and all issues pertaining to the design and machining of mechanical components, including material properties, tooling, machining, manufacturing, gearing, threading, fasteners, etc.

Backlocks are also found in some prosthetic elbows (Boston Elbow III, LTI; Otto Bock Body-Powered Elbow) as a means of holding elbow position in the absence of a drive torque. This ability allows users to park their prosthetic arm at any desired position and then remove (decouple) themselves from the arm's control.

The Otto Bock Body-Powered Elbow mechanism uses a springlike clamp that locks down on a shaft when twisted in one direction but opens to release a shaft when twisted in the other. This is a unidirectional mechanism whose main advantage is simplicity. A similar mechanism can be found in some children's bicycles, where the child backpedals to brake the bike. A disadvantage of this type of mechanism is that over time the lock can slip because of wear and tear on the spring on the elbow shaft.

32.3.10 Electrical Power Distribution

The torque output of a dc electric motor is linearly related to the amount of current the motor can handle. The limiting factor is usually heat. Current flow in the windings of the motor generates heat. If the heat in a motor can be dissipated fast enough, then the motor is able to handle more current. Thus heat sinks can be used to dissipate heat generated and boost motor performance. The same issue applies to the electronic components used to switch the direction of current used to control the motor. In order to drive the motor, an H-bridge, generally made up of four power MOSFETs, is usually employed to direct current to the motor under the direction of a controller. As a rule, the higher the current, the larger the MOSFETs in the H-bridge must be to handle the current and to dissipate the heat generated, although heat sinks can once again be employed to enable smaller MOSFETs handle larger currents.

The important issues from a prosthetics standpoint for the selection and/or design of bridge circuits is that they be as small as possible and consume as little power as possible. In general, one wants as much of the battery's power as possible to be directed to drive the mechanism motors rather than expending it running the prosthesis/bridge controller. *An empirical rule of thumb based on experience in our laboratory is that no more than 5 percent of the available energy should be used to power the controller circuitry.*

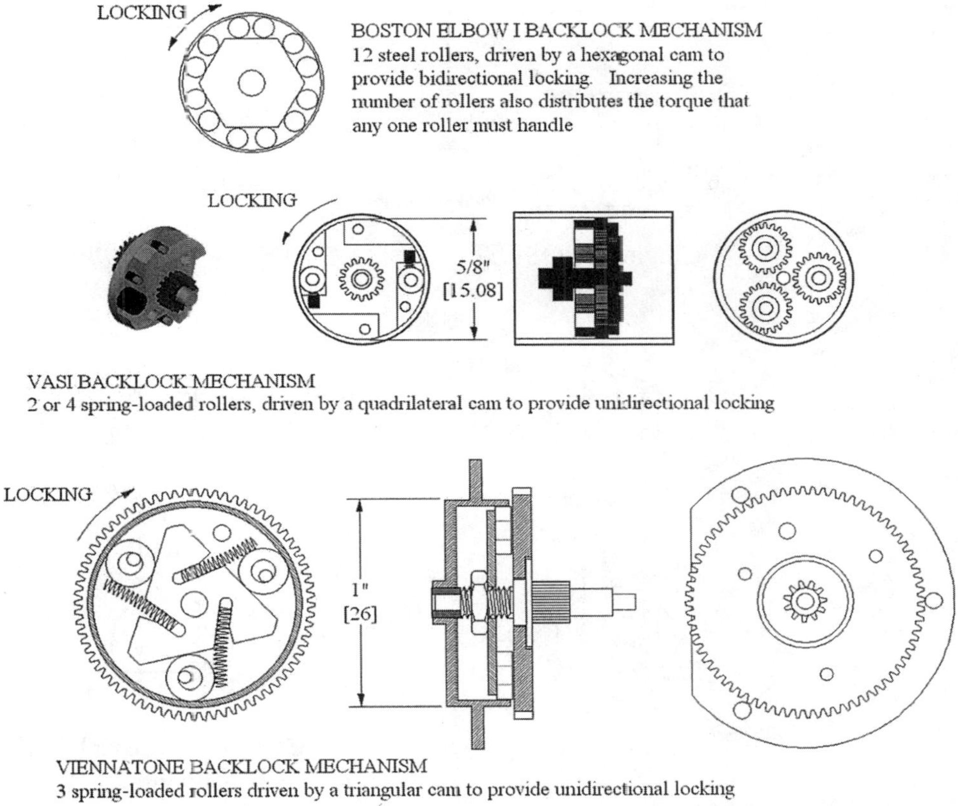

FIGURE 32.15 Backlock mechanisms or roller clutches/brakes. The drawing shows three variations on a backlock (roller clutch/brake) mechanism first used in the Viennatone Hands of the 1960s. Backlocks find extensive use in prosthetic components because the components must often hold their position in the absence of power even in the presence of externally applied forces. Locking occurs when a cam inside the mechanism wedges a roller(s) between itself and the outside cylindrical surface in response to an external torque. Depending on the component, this ability to resist external loads can be in one or both directions. In synergetic mechanisms, the backlock prevents the high-speed, low-force side from being back driven, or forced open, by the high-force, low-speed side. The Boston Elbow I (Liberating Technology, Inc., Holliston, Mass.) has a hexagon-shaped roller cam to provide bidirectional locking. VASI hands (Variety Ability Systems, Inc., Toronto, Canada) have a quadrilateral cam that provides unidirectional locking, and the NU-VA Synergetic Prehensor (Hosmer-Dorrance Corp., Campbell, Calif.), uses a triangular cam to provide unidirectional locking. The NU-VA Synergetic Prehensor backlock is very similar to the original Vienna Tone design.

MOSFET H-bridges come in two basic configurations: *n*-channel half-bridge and *p*- and *n*- channel (complementary) half-bridge (Fig. 32.16). Simplicity of the MOSFET's gate drive is the main advantage of the *p*- and *n*- complementary half-bridge. When an *n*-channel MOSFET is used for the high-side (or upper) switch, the gate drive signal requires level shifting, resulting in increased complexity and cost. However, most power IC processes are optimized for *n*-channel devices. Also, *n*-channel power MOSFETs (*n*-FETs) are more efficient then *p*-channel power MOSFETs (*p*-FETs) in terms of die size for a given current and voltage. For a given breakdown voltage rating, a *p*-FET can be 2.5 to 4 times the area of a comparable *n*-channel device.

The H-bridge and/or MOSFETs are some of the larger components required in the electronics used to control a prosthetic mechanism (Fig. 32.17). Surface-mount technology is employed in

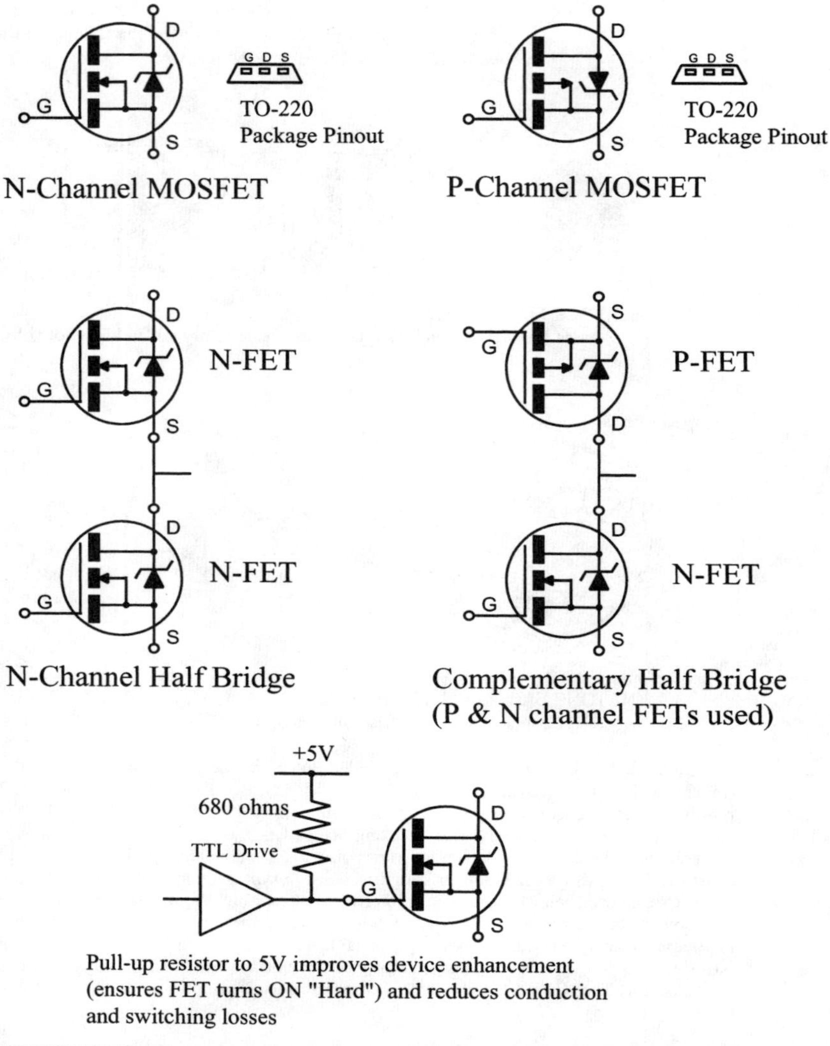

FIGURE 32.16 There are two types of power MOSFETs:-*n* (top left) and *p* (top right). When used to drive a motor or a speaker, these MOSFETs are usually configured as either an *n*-channel half-bridge (bottom left) or a complementary (*pn*) half-bridge (lower right). One talks about half-bridges because when driving multiphase motors a half-bridge is needed to drive each phase. Brushless motors need three half-bridges, whereas dc motors need two. Often a logic circuit with logic circuit voltage levels (5 V) is used to drive the bridge FETs. To ensure that the FET turns on fully, it is common to use a pull-up resistor, as shown in the bottom diagram, to improve FET performance. Also shown are the associated pinouts for a standard TO-220 package. Note: FETs do not normally have logic-level gate drive signals, and consequently, this must be specified when ordering them.

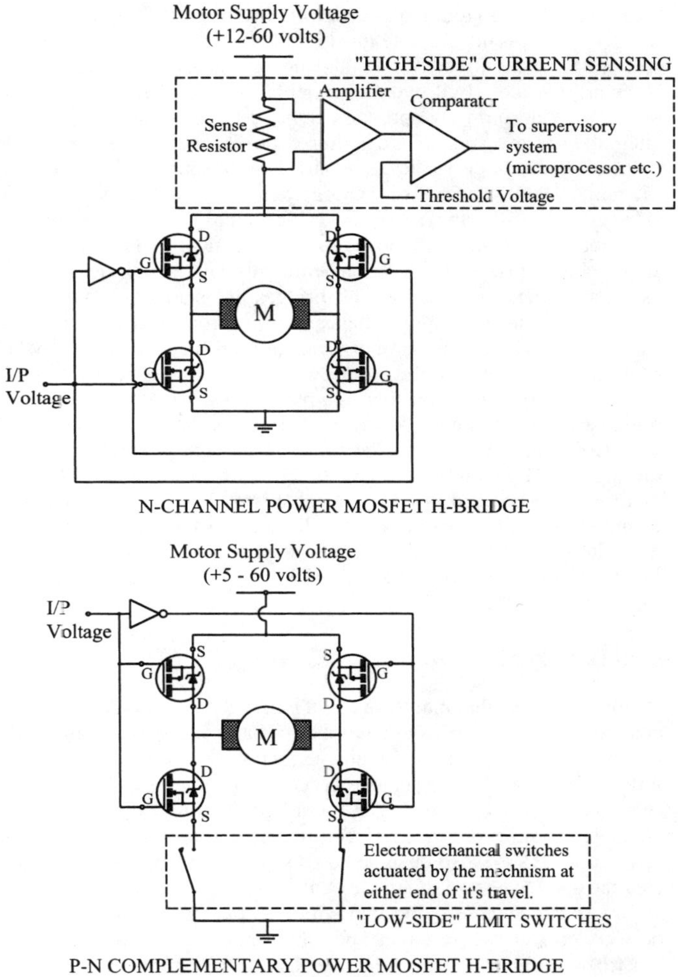

FIGURE 32.17 Diagram shows a full *n*-channel and *pn* (complementary) channel H-bridges with their associated control. It is important to notice that for the *n*-channel H-bridge *diagonally* opposite *n*-FETs are controlled as a pair, whereas for the *pn* H-Bridge, FETs on the *same side* of the bridge are controlled as a pair. Also notice how the *p*-FET is wired as compared with the *n*-FET in terms of their drain and source connections. The use of the inverter symbol is to highlight that both sides of the bridge are never on at the same time, and in fact, break-before-make switching ought to be used to prevent current transients during the switching process. An implementation of high-side current monitoring is shown on the *n*-channel H-bridge while an implementation of limit switch protection of the motor is shown on the *pn* complimentary H-bridge. (See text for further discussion.)

prosthetics control electronics packages to reduce overall circuit size. Surface-mount IC bridges and bridge drivers are available, but there is a problem. In prosthetics the supply voltage is usually 6 to 12 V. In the field of industrial motor control, the standard operating voltage, for which all the commercially available high-current IC bridges and controllers are designed, is anywhere from 12 to 60 V. This means that one must build one's own bridge using four discrete MOSFETs and some sort of MOSFET driver. While this appears straightforward, H-bridge design can be something of a black art, particularly if two *n*-channel half-bridges are used, and the resulting bridge can take time to get working properly and can occupy a large amount of printed circuit-board (PCB) space. An alternative is to use standard motor control components and a charge pump to boost the supply voltage to a voltage sufficient to drive both upper and lower MOSFET gates directly. (Vishay-Siliconix, 1992). The use of a charge pump resolves the issue of how to drive the gate of the high side (upper) *n*-FET above the half-bridge supply to fully turn the device "on."

Another possibility would be to use a number of low-current (1 A maximum), surface-mount, single-chip H-bridges (e.g., Vishay-Siliconix Si9986 or Zetex) and wire them in parallel. These chips use complementary *p*- and *n*- half-bridges in a standard 8-pin SOIC surface-mount package and as such can be run on supply voltages ranging from 4 to 13 V and have a smaller footprint than the standard TO-220 surface-mount package used for some power MOSFETs. Also, using single-chip bridges, rather than building your own, enables a prosthetics designer to benefit from the protection features incorporated into these chips. One such protection feature ensures that both sides of the bridge are never turned on together. Turning on both sides of the bridge at the same time will create a short and burn out the bridge FETs very quickly. Another common cause of FET failure is not turning the FET "on" hard enough (i.e., not turning the FET all the way on or off). If multiple motors are to be used in a hand mechanism these chips are a nice solution since a multiple motor configuration will generally use smaller motors that individually do not draw high currents. A very good introduction to, and general reference on, electronics and electronics design is the *Art of Electronics* by Horowitz and Hill (1995).

32.3.11 Notes on Pneumatic Power

Historically, the other major source of external power in prosthetics was pneumatic power. Pneumatic actuators were employed in a number of hand and arm designs in the 1970s. Cool and Van Hooreweder (1971) developed a pneumatically powered hand prosthesis with adaptive, internally powered fingers. This hand could achieve good grasping with low forces because it was able to adapt to the shape of the object grasped. The Edinburgh Arm was a pneumatically powered arm that saw limited clinical use (Simpson, 1973). Unfortunately, this arm was mechanically complex and prone to failure. Kenworthy (1974) also designed a CO_2-powered hand for use with the Edinburgh arm. Otto Bock Healthcare (Duderstadt, Germany) also sold a number of CO_2-powered systems up until the mid-1970s. Disadvantages associated with pneumatics are that a cylinder of gas (carbon dioxide) has to be secreted somewhere either upon the user or in the device and a ready supply of these cylinders must be available. Because the user has to have tubes running from the gas supply to his or her hand, self-containment of the whole prosthesis becomes an issue.

Currently, there are no pneumatically powered prostheses commercially available, but there are a number of groups working on pneumatically powered devices for prosthetic applications. The WILMER group at the Technical University of Delft in the Netherlands is working on a hand mechanism that uses small disposable CO_2 cartridges called "Sparklets" as a power source (Plettenburg, 2002). "Sparklets" are available in retail stores in Europe for making soda water.

What are claimed to be a new class of pneumatic actuator drive a humanoid hand under development at the Karlsruhe Research Center (Schulz et al., 2002). The actuators consist of cavities that change their size when inflated with a pressurized gas or liquid. Forces of up to 10 N can be obtained by driving these flexible pneumatic actuators with air at a pressure of 3 to 5 bar, allowing motion frequencies of up to 10 Hz. The five-fingered hand mimics the functions and resembles the anatomy of the human hand. The thumb consists of two finger segments and an opposition mechanism; all other fingers consist of three finger segments. All fingers and the wrist can be positioned independently. The compact size of the actuators allows for their full integration into the finger. A safe grasp is accomplished through the flexible self-adaptability of the actuators. The maximized contact surface allows for the use of relatively low pressures.

The BioRobotics Lab in Seattle is experimenting with pneumatic McKibben muscle actuators to mimic the musculature of the natural arm. These actuators, which were first conceived of in the 1950s, consist of an inflatable elastic tube covered with a flexible braided mesh. When pressurized, the elastic tube inside expands but is constrained by the mesh. The flexible mesh shortens or contracts like a muscle due to the expanding tube. It has been found that McKibben muscles can exhibit passive behavior very similar to biological muscle since both have series and parallel elasticity (Klute et al., 1999). From a robotics perspective McKibben muscles can have a high force-to-weight ratio. However this ratio ignores the weight of the compressed air source and associated tubing, which is an issue for a self-contained prosthetic device. In addition McKibben muscles have a shorter range of motion than human muscle and so far have not proven themselves to be reliable long-term mechanisms.

The major attraction of pneumatic systems for prosthetics applications is their inherent compliance, which tends to give these systems a very natural look and feel. While pneumatic systems did find some measure of success in prosthetics, hydraulic systems did not. Hydraulics tended to be messy, with hydraulic units leaking hydraulic fluid. In addition, a fluid reservoir and fluid are required, adding to the total weight of the mechanism.

32.3.12 Note on Hybrid Systems

When body-powered and externally powered systems are linked together, they are called *hybrid systems*. Hybrid systems are used most frequently with persons who have high-level amputations, that is, amputations above the elbow, or who have bilateral arm amputations. Such systems can provide the user with the high gripping and/or high lifting capacities of powered systems and the fine control of body power. For example, providing a person with bilateral limb loss at the shoulder level with a body-powered limb on one side and with an electrically powered limb on the other side decouples the limbs (Fig. 32.23). The prosthetic fitting shown in Fig. 32.23 highlights a number of important issues.

First, by providing the person with a body-powered limb on one side and an electric-powered limb on the other side, the wearer is able to use the limb that is most appropriate for a specific task. This method also decouples the limbs so that body motions used to operate the body-powered side do not influence the state of the electrically powered limb and vice versa, control of each side is independent of the other.

Second, this fitting demonstrates the highly modular nature of upper-limb prosthetics today—components from many different manufacturers were used in a mix-and-match approach to obtain the most functional set of prostheses for the user.

Third, this fitting highlights the severity of the control interface problem when faced with amputations at this level. Everything is used—conventional harness control and biomechanical switches, using the chin, are extensively used in the control of the body-powered prosthetic components. These same chin motions are used to control externally powered components through the use of electromechanical switches and linear transducers.

Fourth, this fitting demonstrates the high degree of creativity required on the part of the prosthetists in being able to customize and modify prosthetic components meant for other applications and the high degree of customization that is required in general, in the fabrication of these high-level prosthetic fittings. This particular set of arms did not "just happen" but is the result of iterative and incremental improvements that have occurred, as new components and devices have become available, over a period of some ten or more years.

Finally, the use of locking mechanisms to switch control from one component to the next allows for a kind of two-state impedance control. The prosthesis is either in a state of high impedance—rigid or locked, or in a state of low impedance—free or unlocked. The prosthesis, when locked, has very high impedance and becomes a rigid extension of the user. The user can use this rigid extension to interact with the real world by pulling and pushing on objects. Because the prosthesis is rigid the user "feels" the forces exerted on the prosthesis and has a sense of feedback about what the prosthesis is doing. The user has extended his or her proprioception into the prosthesis, similar to how a person extends their proprioception into a hammer when hammering a nail. This is the embodiment of Simpson's (1973) extended physiological proprioception (EPP). The other state is when the device is unlocked, now it has very nearly zero impedance. The device in an unlocked state also embodies EPP control. The cable linking the user's physiological joint to the selected prosthetic joint enables the user to control the position and speed of the prosthetic joint directly through the movement of the physiological joint. Forces exerted on the prosthesis are reflected back through the control cable to the physiological joint. The coupling of the control cable allows the user to extend proprioception into the prosthesis even during movement of the prosthesis.

32.3.13 Notes on Unpowered or Passive Prostheses

There is a class of terminal devices that do not offer prehensile function. Devices in this class, usually hands or adapted tools, are regarded as passive or passively functional prostheses. They have no moving parts and require no cables or batteries for operation. They are typically lightweight and reliable. Generic (standard) passive hand prostheses may consist of a cosmetic outer glove over a soft plastic hand with wire reinforcements in the fingers. Traditionally, cosmetic gloves have been made of polyvinyl chloride (PVC), although silicone is becoming the material of choice (Fig. 32.3). Individualized hands, when expertly done, have a preferable appearance to generic hand replacements. Highly realistic hands, fingers, and finger parts can be custom sculpted and painted to an individual's size and skin coloring. Such prostheses confer to persons what Beasley has called the highly important function of "social presentation" (Beasley & de Bese, 1990).

Passive work prostheses may be a simple post to provide opposition, or they may incorporate specialized features to aid in certain occupations. A custom-designed system that serves only one function may aid the wearer more than one that is supposed to be multifunctional. In such cases, the prosthetic device is worn on those occasions when it is needed. These devices range from tool adapters to sports mitts.

In the past, a prosthetic fitting was deemed to have been unsuccessful if the patient did not wear an actively controlled prosthesis for prehensile function. In a recent study, Fraser (1998) showed that amputees used their prostheses most frequently for nonprehensile tasks, such as stabilizing, pushing, or pulling an object. In addition, Fraser (1998) showed that amputees with passive or cosmetic prostheses used their devices for nonprehensile tasks on average just as frequently as amputees with active prostheses. These results show that just because a prosthetic device is passive or cosmetic does not imply that it is not functional.

32.4 ANATOMICAL DESIGN CONSIDERATIONS

32.4.1 Prehension or Grasp

Hand function is generally limited to modes of prehension that are used the most often. Numerous studies of how the hands grasp objects have been performed (Schlesinger et al., 1919; Keller et al., 1947; Napier, 1956; Kamakura et al., 1980, to name but a few). Broadly speaking, hand tasks can be subdivided into nonprehensile functions and prehensile functions. Nonprehensile functions of the hand are functions where grasping is not required; for example, pushing an object, holding an object between the body and forearm, flicking, brushing, percussive motions such as playing the piano, etc. Prehensile hand functions are those cases where an object is grasped and held partly or wholly within the hand. The 6 grasping patterns adapted by Keller et al. (1947) from Schlesinger et al. (1919) 12 patterns, are the most widely accepted in the field of prosthetics (Fig. 32.18) and have endured the test of time. These patterns are

1. Tip prehension
2. Palmar prehension
3. Lateral prehension
4. Hook prehension
5. Spherical prehension
6. Cylindrical prehension

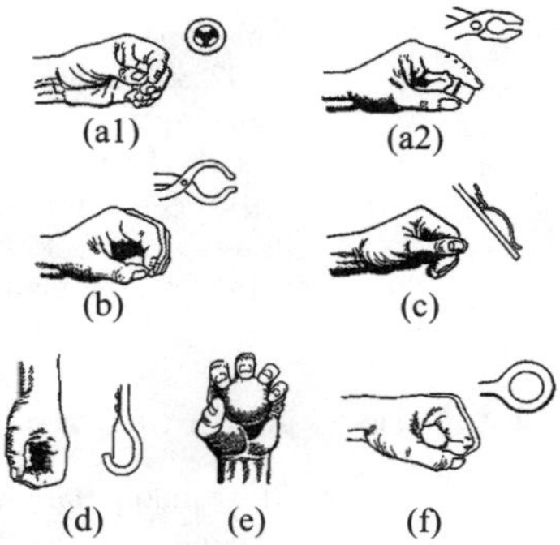

FIGURE 32.18 Schematic of the prehension patterns of the hand as defined by Keller, Taylor, and Zahn (1947): (*a*1) palmar prehension (three-jaw chuck), (*a*2) palmar prehension (two finger), (*b*) tip prehension, (*c*) lateral prehension, (*d*) hook prehension, (*e*) spherical prehension, (*f*) cylindrical prehension. In a handlike prosthesis, it takes two to four independently controlled degrees of freedom to implement these prehension patterns. In a non-hand-like device, a single-degree-of-freedom device such as a split hook can be used.

Napier (1956) described tip prehension, palmar prehension, and lateral prehension as precision grips and spherical and cylindrical prehension as power grasp, whereas hook prehension falls outside both these categories. Precision grips primarily involve the thumb working in opposition with the index and middle fingers. Tip prehension, or fingernail pinch, is used mainly to grasp small objects. In lateral prehension, the thumb holds an object against the side of the index finger, as is the case when using a key. In palmar prehension (sometimes referred to as tridigital pinch or three-jaw chuck), the thumb opposes either a single finger or two or more fingers.

Power grasps use all the fingers of the hand to provide an encompassing grasp that firmly stabilizes the object being held. Hook prehension is achieved by flexing the fingers into a hook; the thumb is either alongside the index finger or opposes the index and middle fingers to lock the object held. Carrying a brief case is a good illustration of this kind of prehension. Keller et al. found that palmar prehension or tridigital pinch was the most frequently used prehensile pattern for static grasping whereas lateral prehension is used most often for dynamic grasping.

The finding by Keller et al. (1947) that palmar prehension was the most frequently used pattern and reduction of most prosthetic terminal devices to a single DOF has meant that most prosthetic hands incorporate palmar prehension as the dominant grasp pattern. The persistence of this pattern, combined with a wide width of opening in prosthetic hand designs and its general acceptance over the years tends to support this compromise (Heckathorne, 1992).

A study done at the University of California at Los Angeles (UCLA) (Taylor, 1954) on human prehension force indicated that adult males could produce maximum mean forces of 95.6 N (21.5 lbf) of palmar prehension, 103 N (23.2 lbf) of lateral prehension, and 400 N (90 lbf) of cylindrical grasp. In light of another (unpublished) UCLA study that showed that forces up to 68 N (15 lbf) were needed for carrying out activities of daily living, Peizer et al. (1969) proposed that 68 N (15 lbf) be a minimum standard for the maximum prehension force for electric prehensors (Heckathorne, 1992).

32.4.2 Dominant and Nondominant Hands

The issue of hand dominance, whether one is right or left-handed, must also be considered. People use their dominant hand differently from their nondominant hand. The role of the nondominant hand is to hold things while the dominant hand is working or waiting to work on them. A unilateral amputee will always use his or her prosthesis in a nondominant role even if the prosthesis is a replacement for what was once the amputee's dominant hand. The unilateral amputee will also tend to pick things up with his or her dominant hand and then place them into the nondominant hand. Consequently, van Lunteren et al. (1983) suggested that there is a difference in the type of grasps that ought to be incorporated in devices for unilateral as opposed to bilateral amputees. However,

Toth (1991) and Heckathorne et al. (1995) showed that the findings of Keller et al. (1947) held regardless of whether the dominant or nondominant hand was used.

The Rancho Los Amigos Easy-Feed Hand (Lansberger et al., 1998) is an example of a new mechanical hand design that is based on this observation that prosthetic hands for persons with unilateral amputations tend to be used to hold objects placed into them with the sound hand. The Easy-Feed Hand is a mechanical children's hand that is easy to push objects into but difficult to pull things out of, making it good for grasping or hanging onto objects. A version of this concept is being commercialized as the Live Touch hand from TRS (Boulder, Colo.).

32.4.3 Hand Width-of-Opening

Most manipulations of the hand are precision manipulations of the palmar prehension kind where the thumb directly opposes the index finger and/or the middle finger. In this mode, most of the hand's actions are performed with a hand opening of about 5 cm (2 in) (Keller et al., 1947). When designing for dominant hand function, palmar prehension is the desirable pattern limited emphasis on wide opening. For nondominant hand function, where the hand is used essentially as a portable vice with objects being placed into it, a wide opening becomes more important. From a design perspective, allowing the hand mechanism to open 10 cm (3.5 to 4 in), instead of 5 cm (2 in) enables the mechanism to perform the cylindrical prehension power grasp with minimal extra design effort. In general an artificial hand should be able to open at least 10 cm (3.5 to 4 in), or enough to grasp a beverage can or a Mason jar.

32.4.4 Passive Adaptation During Grasping

The grip of the hand is improved by the ability of the hand to passively adapt to the shape of an object grasped. A grasped object depresses, or indents, the skin and underlying soft tissues of the hand, at first meeting little reaction force. Consequently, the soft tissue adapts easily to the shape of the object grasped. However, the mechanical properties of the soft tissue are nonlinear, and the conforming tissue becomes more rigid as pressure is increased. The rise in tissue stiffness after conformation to shape enables objects to be grasped securely. This feature of the human hand would seem to be useful for robotic and prosthetic systems. In prosthetics, the passive adaptability afforded by the soft tissue of the hand is mimicked, to some extent, by lining the prosthesis mechanism with a soft plastic and covering it with a cosmetic glove. In robotics, it is common to use a compliant coating on an end effector to stabilize a robot arm during contact with hard surfaces.

32.4.5 Non-Hand-Like Prehensors

The reduction of most prosthetic terminal devices to a single degree of freedom was a compromise to make the best use of the available control sources. A common transhumeral (above-elbow) body-powered prosthesis has two control cables (two active DOFs), one for an elbow lock to switch the control of the other control cable from the elbow to the terminal device. The terminal device is generally a split hook. A split hook is used when maximum function is desired (Fig. 32.19). Although a split hook is a single DOF device, a split hook can reproduce tip, lateral, cylindrical, or hook prehension depending on which part of the hook is used, making it a very simple and versatile device. This is a contributing factor to the success of body-powered prostheses over externally powered prostheses. Another contributing factor is the "thin" nature of the hook fingers, which only minimally restrict the user's vision of the object being manipulated.

The use of split hooks highlights the trade-off made between form and function. The hook bears little resemblance to the natural hand but is widely used because of the function it affords if only one DOF is available for terminal device control. Split-hooks are available in many variations from

Hosmer-Dorrance Corp. and Otto Bock Healthcare. Fryer & Michael (1992) provide a thorough review of the various types of hooks currently available.

In an effort to capitalize on the function of the split hook the Hosmer-Dorrance NU-VA Synergetic Prehensor uses a split hook in an externally powered configuration. Other commercially available, externally powered, non-hand-like prehensors include the Otto Bock Greifer and the RSLSteeper Powered Gripper. The NU-VA Synergetic Prehensor, the Otto Bock Greifer, and the Steeper Powered Gripper incorporate many of the previously mentioned prehension patterns (Fig. 32.20).

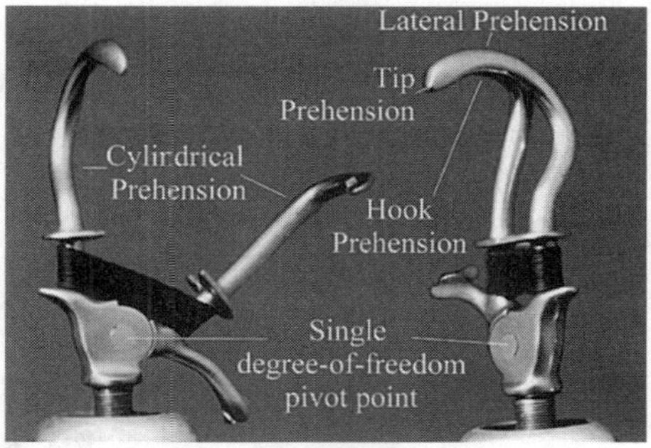

FIGURE 32.19 A split hook with the prehension surfaces pointed out of the page. Here a single-degree-of-freedom device can recreate four of the prehension patterns of Keller, Taylor, and Zahn (1947): tip prehension; lateral prehension; cylindrical prehension; and hook prehension. This is one of the reasons why the hook, though often thought unsightly, has been so successful in the field of prosthetics.

32.4.6 Hand-Like Prehensors

The de facto standard externally powered hand-like prosthesis is the single-DOF Otto Bock System Electrohand (Fig. 32.12). When used in a prosthetic fitting, a plastic hand-form liner is pulled over the mechanism and a PVC or silicone-rubber cosmetic glove is then pulled over the liner. This gives the hand good overall *static* cosmesis at the expense of reduced overall mechanism performance. The liner and cosmetic glove act as springs to oppose the opening of the hand by the mechanism, thus degrading the overall performance of the hand. De Visser and Herder (2000) advocated the use of compensatory mechanisms to reduce the effect of the liner and glove on the mechanism's performance.

RSLSteeper Ltd. (Roehampton, England), and Centri (Sweden) also manufacture single-DOF devices for the adult. Single-DOF child-size hands are also available from Variety Village, Otto Bock Orthopaedic, Inc., and RSLSteeper Ltd., among others. Michael (1986) provides a nice overview of the commercially available powered-hand mechanisms of the day, whereas Heckathorne (1992) provides in-depth descriptions and technical specifications of all the externally-powered components available at the time of writing.

To be clinically viable, the fingers and thumb of most prosthetic hands are nonarticulated and have a single axis of rotation. This minimizes the number of moving parts, reduces complexity and increases robustness. In these handlike prehensors palmar prehension is achieved by single-joint fingers that are fixed in slight flexion at a position approximating the interphalangeal joint. The resulting finger shape also creates a concave inner prehension surface that can be used to provide cylindrical prehension (Heckathorne, 1992). A

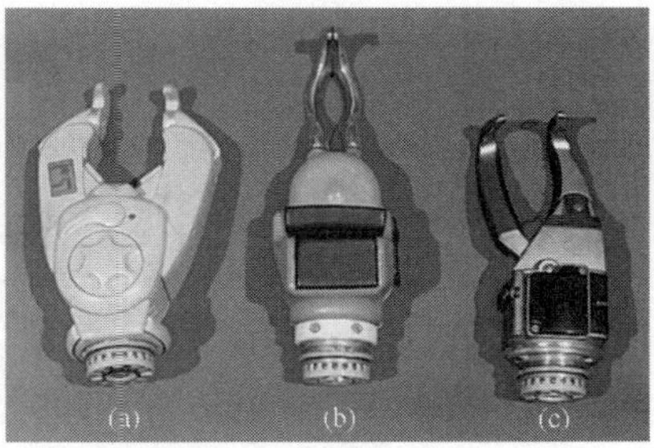

FIGURE 32.20 Examples of powered non-hand-like prehension devices: (*a*) Otto Bock Greifer (Otto Bock Orthopedic Industry, Inc., Duderstadt, Germany), (*b*) NU-VA Synergetic Prehensor (Hosmer-Dorrance Corp., Campbell, Calif.), (*c*) RSLSteeper Powered Gripper (RSLSteeper, Ltd., England). Both the NU-VA Synergetic Prehensor and the RSLSteeper Powered Gripper use the principle of synergetic prehension to maximize performance, whereas the Greifer uses Otto Bock's automatic transmission. The NU-VA Synergetic Prehensor sought to capitalize on the success of the split hook by basically creating a powered version of it. All the mechanisms are shown here with an Otto Bock quick-wrist disconnect, a de facto standard type of wrist interface for powered components. The NU-VA Synergetic Prehensor and the Otto Bock Griefer are capable of high pinch forces, high speeds, and large widths of opening, making them very functional at the expense of cosmesis.

single joint thumb opposes these fingers by rotating in the same plane as, but in the opposite direction to, the fingers. All these mechanisms are typically used in a prosthesis that has no wrist flexion or extension. This can be a problem when trying to pick up small objects from a surface. The fixed wrist combined with poor line of sight of the object to be grasped can lead to nonphysiological movements resulting in poor *dynamic* cosmesis.

32.4.7 Planes of Motion of the Thumb

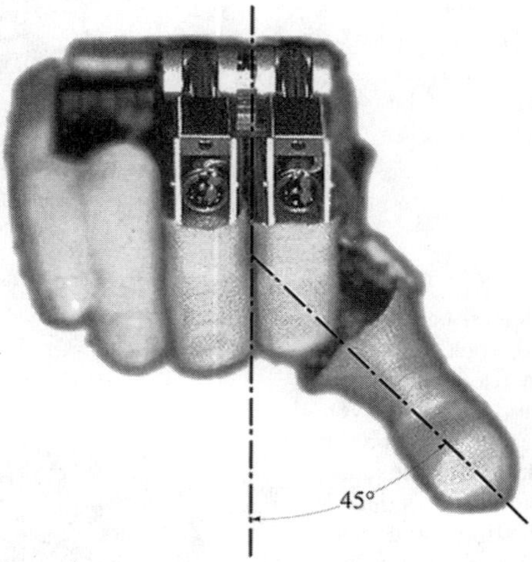

FIGURE 32.21 Preferred operating plane of the thumb in an artificial hand. The physiological thumb has two axes of rotation, the metacarpophalangeal (MP) joint and the carpometacarpal (CP) joint. A practical prosthetic hand can have only one axis of rotation for the thumb. Lozac'h (1984; Vinet et al., 1995) determined that the preferred working plane of the thumb lay between 45 and 55 degrees.

The physiological thumb has two axes of rotation, the metacarpophalangeal (MP) joint and the carpometacarpal (CP) joint giving it many degrees of freedom. Blair and Kramer (1981) claim that the thumb accounts for up to 40 percent of the function of the hand. However, prosthetic hands only have a single axis thumb. Lozac'h (1984; Vinet et al., 1995) performed a series of experiments to find if there was a preferred working plane for a single-DOF active thumb. From these experiments he determined that the preferred working plane of the thumb lay between 45 and 55 degrees (Fig. 32.21). These findings conform to those reported by Taylor (1954) who, from a statistical analysis on natural unrestricted prehension, concluded that: "in palmar prehension, the thumb approaches the fingers in a plane inclined approximately 45 degrees to the palmar plane."

This finding was implemented in a single-degree-of-freedom, multifunctional hand design (Lozac'h et al., 1992; Vinet et al., 1995). Because the hand had articulated fingers that could move independently of each other it was capable of forming an adaptive grip with which to grasp objects. An adaptive grip meant that it required much lower forces than conventional prosthetic hands to hold objects. Lozac'h et al. (1992) also found that the hand reduced the number of arm and body compensatory movements during both the approach and utilization phases of prehension, as well as greatly improving object visibility, prehension cosmesis (dynamic cosmesis) and grip stability, particularly for large objects. Unfortunately, they also found that they had to further improve the design for greater durability and long-term reliability.

The finding that the preferred working plane of the thumb lay between 45 and 55 degrees is not reflected in many other prosthetic or orthotic devices. Kenworthy (1974) designed a hand in which the thumb moved at 90 degrees to the fingers, i.e., the hand used a side-pinch type of grip (lateral prehension). This hand was developed for use with the CO_2-powered arms (Simpson, 1973) of the Orthopaedic Bio-Engineering Unit (OBEU) of Princess Margaret Rose Orthopaedic Hospital, Edinburgh, Scotland.

The motivation for placing the thumb at this angle came from trying to improve the visibility of the object to be grasped during the final stages of the grasping process and to improve flat-surface operation. It was an attempt to overcome the poor dynamic cosmesis that results from the use of a traditional pincer-grip prostheses (such as the Otto Bock System Electrohand) with a rigid wrist.

Traditional pincer-grip prostheses are easy to use when picking up tall objects (greater than about 25 mm above the surface), but the only way that low objects (less than 25 mm) can be grasped is by approaching the object from above. This requires the amputee to maintain an unsightly line of attack (poor dynamic cosmesis) in order to see the object being grasped and to permit flat surface operation. This unusual motion, in turn, draws attention to the user.

Kenworthy's hand had the fingers fixed with only the thumb being able to move. While flat-surface operation is important, it is perhaps not the most important feature of prosthesis design. This hand was designed primarily for use in bilateral arm systems in which one hand was a conventional pincer-type hand and the other was of Kenworthy's design. The 90 degrees of the Kenworthy hand was chosen so that at least one hand of the bilateral arm system could have flat-surface operation.

Another hand, developed by Davies et al. (1977) also at the OBEU, was a monofunctional body-powered device in which the thumb moved about an axis inclined at an angle of 60 degrees to the middle and index fingers. This hand, called the *OBEU Hand*, was a compromise between the more traditional pincer-type of hand and the Kenworthy hand. The motivation for this hand was to retain the pincer-type function while improving the overall visibility of the object to be grasped. Additionally flat-surface operation could be achieved by allowing the fingers and thumb to sweep the surface. In this design both the thumb and fingers moved simultaneously. The 60 degrees for the OBEU hand was chosen because it allowed the loci of the tips of the thumb and index and middle fingers to move approximately in a horizontal plane when the wrist axis was at about 25 degrees to the horizontal so that they could sweep the surface. More recently the hand under development by the WILMER group (Plettenburg, 2002) also has its thumb at an angle of 45 degrees to the fingers. All these designs were an attempt to improve the visibility of the object being grasped and consequently the dynamic cosmesis of the prosthesis while using single-DOF handlike mechanisms.

32.5 MULTIFUNCTIONAL MECHANISMS

There have always been attempts to design fully articulated arms and hands in effort to recreate the full function of the hand. In prosthetics, it was not until after the Second World War and in particular during the 1960s and 1970s that much time, effort, and money was invested in the development of externally powered multifunctional hand-arm systems. Much of the impetus for the research of the 1960s was as a result of the drug thalidomide that was prescribed to a large number of pregnant women in Europe, Canada, and Australia. The drug acted on the fetus in such a way as to inhibit development of the limbs, causing the child to be born with one or more fetal-sized limbs that never developed further.

Worldwide, many new government initiatives were established to help in the development of externally powered, multifunctional, complete arm systems for these children. Prime among these being the Edinburgh Arm (Simpson, 1969), the Boston Arm (Mann, 1968; Mann & Reimers, 1970), the Philadelphia Arm (Taylor & Wirta 1970; Taylor & Finley, 1974), the Wasada Hand (Kato et al., 1970), the Belgrade Hand (Razic, 1973; Stojiljkovic & Saletic, 1975), the Sven Hand (Herberts et al., 1978), the Utah Arm (Jacobsen et al., 1982).

While many ideas were tried and tested during this period, only a few devices ever made it from the laboratory into everyday clinical practice. The Edinburgh Arm, which was pneumatically powered, saw some clinical use, but it was too complex and as a result lacked robustness making it susceptible to breaking down. While this arm is not available today, it is important because it was an implementation of Simpson's ideas on extended physiological proprioception (EPP). The Boston Arm, developed at MIT, was the first myoelectrically controlled elbow. This elbow was extensively redesigned (Williams, 1989) for commercial production. The current version is the Boston Elbow III (Liberating Technology, Inc.). The Utah Arm is commercially available through Motion Control, Inc. (Sears et al., 1989).

The Sven Hand never found widespread clinical use, even though it was extensively used in multifunction control research using pattern recognition of myoelectric signals (Lawrence and Kadefors, 1974). Henry Lymark, the director of the Handikappinstitutet of Stockholm, Sweden, later created a simplified version of the Sven hand called the *ES hand*. This hand was an attempt to produce a more robust and hence more clinically viable version of the Sven hand. The ES hand possessed an adaptive grip and a passive two-position thumb. It was powered by a single motor with differential drives to the fingers. Unfortunately, Lymark died soon after the initial development of this hand and the project was never continued.

The Philadelphia Arm of Taylor and Wirta (1970) and Taylor and Finley (1974) also never found clinical use, but like the Sven Hand found use as a research tool for multifunction control using weighted filters for the pattern recognition problem. The Belgrade hand too was never used clinically but has ended up in the robotics field in the form of the Belgrade/USC robotic hand (Beattie et al., 1994).

In the end, most multifunctional prosthesis designs are doomed by practicality, even before the control interface becomes an issue. Prosthesis users are not gentle with their devices; they expect them to work in all sorts of situations never dreamed of by their designers. Most mechanisms fail because of poor durability, lack of performance, and complicated control. No device will be clinically successful if it breaks down frequently. A multifunctional design is by its nature more complex than a single-DOF counterpart. From a maintenance standpoint this means the device will have more components that are likely to fail. Articulated joints on fingers are more likely to fail than monocoque, or solid finger, designs. However, a compromise must be reached if increased function is to be achieved. Some of the robustness and simplicity of a single-DOF device must be traded to achieve the increase in performance possible with a multi-DOF hand.

Another practical consideration is performance. The hand must be able to generate enough torque and speed and have a sufficient width of opening to be useful to the user. Many devices have been designed in the laboratory that have insufficient performance once a cosmetic glove is added. A cosmetic glove is standard for prosthetic hands, and unless a mechanism is specifically designed not to require a glove, the effect of the glove on performance must be taken into consideration. The ES Hand was designed to work in its own cover. The Belgrade Hand needed an additional cover. The pinch force of a multifunctional hand does not have to be as high as that possible with current commercially available single-DOF hands because the adaptive nature of the grip enables objects to be encompassed within the hand. But they should still be capable of high speeds of opening and have a pinch force of at least 68 (15 lbf) in accordance with Peizer et al. (1969).

Most of the early artificial arms for the high-level amputee were designed as complete arm systems. But as these systems failed to provide the expected function, there was a move away from designing complete arm prostheses to the design of specific or modular components, such as externally or body-powered elbow joints, powered or passive wrist rotators, passive humeral rotators, and whole ranges of different hooks, hands, and prehensors. The current trend is to use a "mix and match" approach to optimize the function available. This modular approach has the advantage of providing great flexibility and practicality for system design. However, it will probably never be able to attain the high functional goals that may be possible from a more integrated standpoint. Many of today's commercially available externally powered elbow systems (Boston Elbow, Liberating Technology, Inc; NYU Elbow, Hosmer-Dorrance; and Utah Elbow, Motion Control) owe their origins to this era of upper-limb research.

32.5.1 Is There an Optimum Number of Degrees of Freedom for Multifunctional Hands?

To be clinically viable a multifunctional hand must be robust. This was one of the reasons why the Sven Hand was simplified to the ES Hand—so that it might find some clinical application. A possible compromise to the dilemma of robustness versus increased function possible with a multi-DOF hand is to limit the device to those degrees of freedom necessary to replicate Keller et al. (1947) grasp patterns. This idea of providing sufficient DOFs to recreate Keller et al. grasp patterns turns up in many unrelated fields when compromise must be made between function and some other variable.

Professional scuba diver gloves trade function for warmth in order to extend dive times (Fig. 32.22). A mitten is warmest, whereas a glove with individual fingers is the most functional. Professional scuba diver gloves are a compromise, having the thumb and index fingers free and the middle, ring, and little fingers together. This configuration affords the diver the basic prehension patterns of the hand while at the same time keeping the bulk of the hand warm.

In the area of remote manipulation, the SARCOS system (Jacobsen et al., 1990) uses a three-

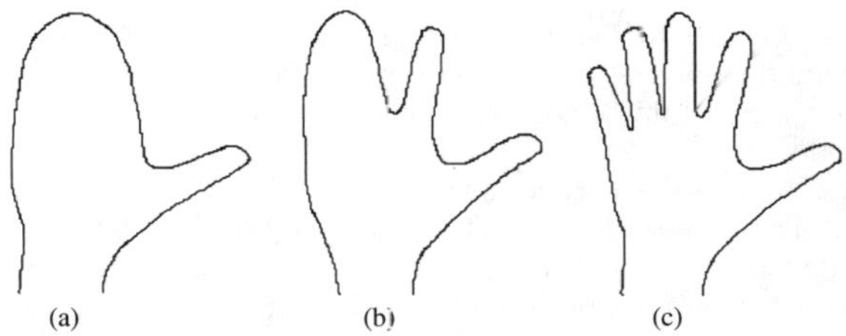

FIGURE 32.22 Sketches of the three types of gloves used by SCUBA divers: (*a*) a mitten is warmest but has the poorest function; (*b*) compromise glove provides warmth and allows diver to achieve the prehension patterns of Keller et al. (1947); (*c*) standard glove with individual fingers provides the least warmth but the most function.

DOF hand for the slave manipulator terminal device and limits the hand of the operator to the same three DOFs when controlling the master arm. Constraining the operator's hand to the same DOFs as the slave, and vice versa, enables the operator to extend his or her proprioception into the remotely controlled terminal device. Forces experienced by the slave are reflected back to the master and experienced by the operator. In this mechanism, the thumb has two DOFs while the three fingers (middle, ring, and little) have the third DOF. The index finger was kept rigid, providing a stable platform against which to operate.

For space suit gloves, the Direct-Link Prehensor ({xe "Direct-Link Prehensor, (1991a&b) limits the motions of the operator's hand to three DOFs. Space suit gloves are bulky and stiff due to the suit's pressurization. This stiffness results in limited external dexterity and excessive hand fatigue. Also, as in the case with diver's gloves, tactile sensation and manual dexterity are lost because the hand is gloved. The Direct-Link Prehensor is a two finger device with the thumb mounted at 45 degrees to the fingers similar to the work of Lozac'h (1988).

In the area of surgery, Beasley (1983) described a surgical procedure to provide a functional four-DOF hand for persons with C5–C6 quadriplegia. This procedure makes possible precision prehension with careful positioning of the stabilized thumb to oppose the actively flexed index and middle fingers. The result is a functional hand that retains some of its sense of touch.

In the field of functional electrical stimulation (FES), surgical techniques similar to those of Beasley (1983) have been combined with implantable four-channel FES electrodes to enable patients with flail arms to reproduce the palmar and lateral grasp patterns of Keller et al. (Triolo et al., 1996).

In each of these examples either tactile sensation (feedback) was compromised [by gloves or remote nature of the terminal device] or muscular function (control) was impaired. In all cases enabling the hand to recreate Keller et al.'s prehension patterns optimized hand function vs. the number of available control sources. This observation suggests that a prosthetic hand capable of implementing Keller et al.'s patterns would also optimize function versus available control sources (assuming sufficient control sites can be found).

Initially it would appear that such an artificial handlike prehensor would require three or four DOFs to adequately reproduce all six prehension patterns: at least one, more usually two, for the thumb; one for the index finger; and one for the middle, ring and little (MRL) fingers which are combined to move as a single unit. Limiting the thumb to a single degree of freedom and orienting it such that it operates along its preferred plane, 45° (Lozac'h, 1988, Vinet et al., 1995), reduces the number of DOFs to be controlled to three.

Furthermore, if one considers the role of the thumb during grasping it becomes apparent that the prehension pattern adopted by the hand can be made to be a function of the timing (or speed) of thumb closure with respect to the timing (or speed) of finger closure. If a "close" command to the thumb and fingers results in tridigital palmar prehension, in which the thumb engages both the index finger and MRL finger unit, then delaying closure of the thumb a fraction (or reducing thumb speed) will result in tip prehension, in which the thumb only engages the index finger, because the MRL finger unit will pass and miss the thumb (due to the delay) to close on themselves. Delaying thumb closure more will result in lateral prehension, in which the thumb closes on the side of the index finger, because the thumb will engage neither the index finger nor MRL finger unit and both will close on themselves. Power grasps result from digit shape and a wide width of opening.

A system of this kind, where the timing, or speed, of thumb is controlled, can be configured so a single "open" signal drives all digits (fingers and thumb) back to their start positions. But two "close" signals are needed, one for the index and MRL finger drives and a second for the thumb drive. This implies a one and one-half degree-of-freedom system. If a wiffle tree structure, or a differential drive, is used to drive the fingers, then a single motor could be used to drive all the fingers. It should be noted that prehension (grasping) should not be confused with manipulation. For dexterous manipulation many more degrees of freedom of control are required.

The new Gung-Ho Grip™ for G.I. Joe comprises a 2-DOF hand in which the index finger (trigger finger) is free to move by itself while the middle, ring, & little (MRL) fingers move together. The thumb is rigid and at an angle to the palmar surface of the hand (Lee Publications, 2001).

32.6 SAFETY

Safety considerations are an integral part of any design that has a person in the loop. Limit switches, or a suitable substitute (current sensing), should be used to shut off the motor at the limits of a component's range of motion (Fig. 32.17). These limit switches should be backed up by mechanical stops capable of physically stopping the component's drive should the electrical limit switches fail. In this way, while the drive may burn-out, should it run up against the mechanical stops due to a limit switch failure, the user remains safe. Manual overrides for mechanisms should also be included so that if a prehensor fails while it is gripping an object the user can still open it manually. Motion Control has a manual override built into the on-off switch of its new hand. Otto Bock has a break away mechanism that consists of a friction brake. In the Otto Bock mechanism there are multiple interdigitated plates that are preloaded by way of a set screw. The problem with the Otto Bock mechanism is that it will protect the drive train from high external loads by slipping at a preset friction, but it must be screwed tight enough that it cannot be manually unloaded.

32.7 CONTROL

Although the physical design constraints of weight, volume, and power are severe, they are not so severe that multifunctional arms and hands cannot be built that would be of acceptable weight and size. The real problem is, as has been alluded to before, the issue of how to interface a multifunctional arm or hand to an amputee in a meaningful way. It is for this reason that upper-limb prosthetics is often dominated by consideration of control. That is, how can the prosthesis be controlled in such a fashion that it will be an aid rather than a burden to the user?

Childress (1992) presented the following attributes of prosthesis control as desirable. While some of these attributes may be difficult, if not impossible, to achieve in practice, they are still seen as desirable goals.

1. *Low mental loading or subconscious control.* The prosthesis should be able to be used without undue mental involvement. The prosthesis should serve the user; the user should not be the servant of the prosthesis.
2. *User friendly or simple to learn to use.* Any device should be intuitive and natural. An amputee should be able to learn to use the prosthesis quickly and easily.
3. *Independence in multifunctional control.* Control of any function or degree of freedom, should be able to be executed without interfering with the other control functions of a multifunctional prosthesis.
4. *Simultaneous, coordinated control of multiple functions (parallel control).* User should have the ability to coordinate multiple functions simultaneously in effective and meaningful ways without violating the first and third attributes.

5. *Direct access and instantaneous response (speed of response).* All functions, if possible, should be directly accessible to the user, and these functions should respond immediately to input commands.
6. *No sacrifice of human functional ability.* The prosthesis should be used to supplement, not subtract, from available function. The control system should not encumber any natural movement that an amputee can apply to useful proposes.
7. *Natural appearance.* Movements that appear mechanical in nature attract unwanted attention in social situations and may not be pleasing to the eye.

As one might imagine, the level of amputation has very important consequences for the control of a prosthetic device. The level of amputation determines the number of control sites available versus the number of control sites that are needed. This is an inverse relationship. That is, the higher the level of amputation, the fewer are the available control sources but the greater is the amount of function that must be replaced. A control source is the means used by the amputee to control a specific function, or degree of freedom, of the prosthesis, e.g., opening and closing of an artificial hand. The fewer the number of independent control sources, the greater are the number of compromises that must be made in order to achieve a clinically practical device.

Practical inputs typically come from muscular activity, (1) directly, (2) indirectly through joints, and (3) indirectly from by-products of muscular contraction (myoelectricity, myoacoustics, muscle bulge, and mechanical/electrical impedance). Although signals can be obtained from brain waves, voice, feet, eyes, and other places, these sources of control have not been shown to be practical for artificial limb control (Childress 1992).

The primary sources of control for body-powered devices are biomechanical in nature. Movement, or force, from a body joint or multiple joints is used to change position, or develop a force/pressure that can be transduced by a harness and Bowden cable and/or mechanical switches. Typically, inputs such as chin and head force/movement, glenohumeral flexion/extension or abduction/adduction, biscapular and scapular abduction, shoulder elevation and depression, chest expansion, and elbow or wrist movements are used. However, direct force/motion from muscle(s) has also been used by way of surgical procedures such as muscle tunnel cineplasty (Sauerbruch, 1916) and the Krukenberg cineplasty (Krukenberg, 1917).

For externally powered devices, electromechanical switches and myoelectric control are the main sources of control. The usual input to most externally powered prosthetic components is the plus-minus drive wires of the dc motor used to drive the component. As long as the motor receives a voltage of either polarity across these wires, it will run in one direction or the other. This voltage can be an on/off type voltage in which the motor is fully "on" or fully "off" (switch control). Or it can take the form of a variable dc level voltage, in which case motor speed is proportional to the dc voltage level (proportional control). Or it can be a stream of pulses in which the motor drive voltage level is proportional to the amount of "on" time of the pulses (pulse width modulation or PWM). PWM also is a type of proportional control. In all instances, a control source is needed to transduce some body motion or biological artifact into a voltage to drive the motor. While biomechanical inputs are used most extensively in the control of body-powered prosthetic components, the same inputs can be used, through the use of appropriate electromechanical switches or force transducers, to control externally powered components. Fig. 32.23 shows a high level bilateral hybrid fitting that highlights many of these issues.

32.7.1 Comments on Proportional Control and Pulse Width Modulation

In proportional control, the amount/intensity of a controlled output variable is directly related (proportional) to the amount of the input signal. For example, the output speed of a dc motor is proportional to the amount of voltage applied to its terminals. This is why dc motors are said to be *speed controlled.* This is also the reason why most of today's commercially available prosthetic components are speed controlled—it is simple. Output speed is proportional to the amount of input signal. Proportional control is used where a graded response to a graded input is sought.

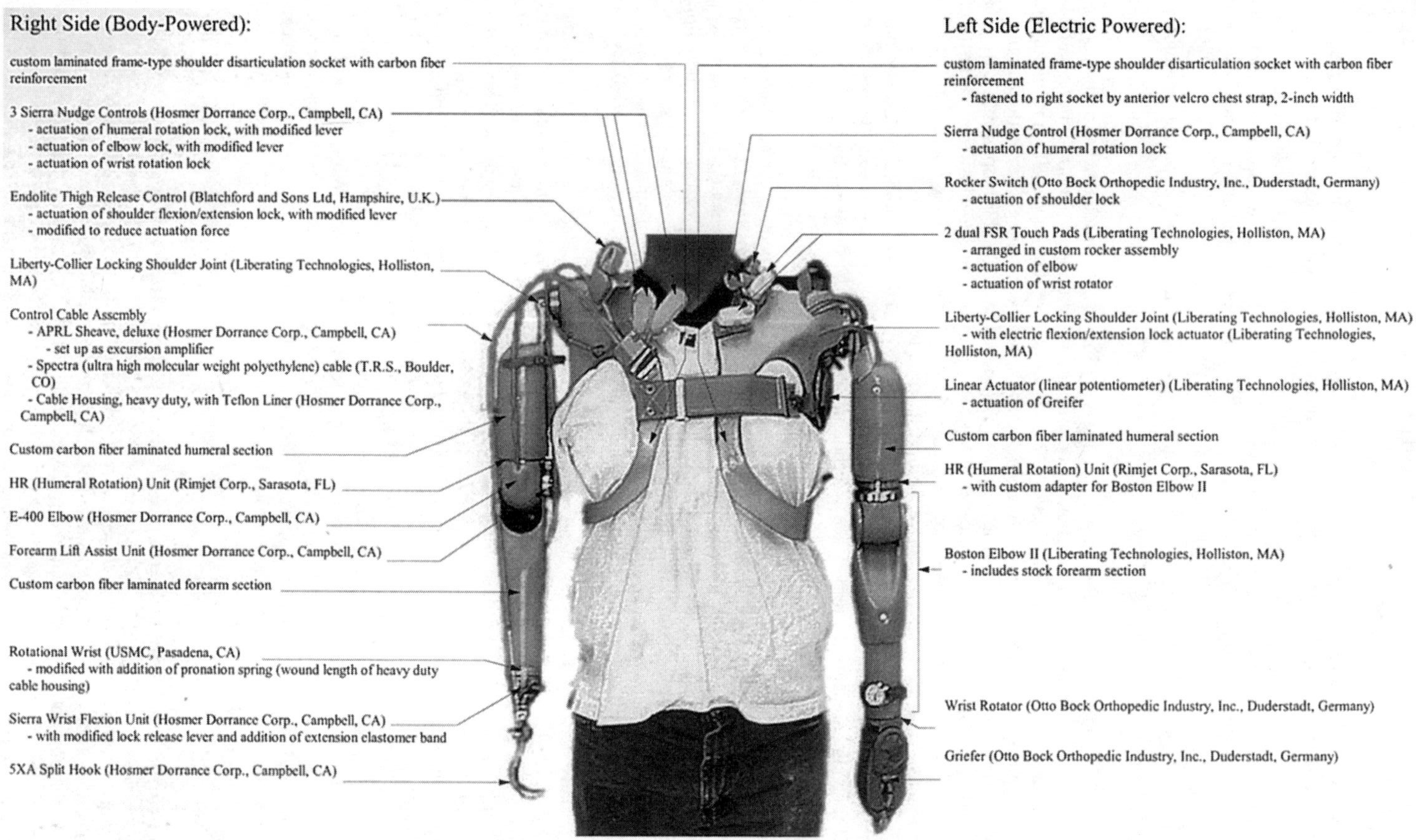

FIGURE 32.23 This photograph shows a person with bilateral shoulder disarticulations who has been fitted bilaterally using hybrid prostheses. The right side is fitted with a body-powered, single-control cable, four-function system. This is a sequential control system in which the control cable controls elbow flexion-extension, wrist rotation, wrist flexion or terminal device opening and closing depending on which components are locked or unlocked by the chin-actuated mechanical switches. Chin-actuated levers lock and unlock a shoulder flexion-extension unit, a humeral rotator, an elbow, and a wrist rotator. A lever on the wrist locks and unlocks a wrist flexion unit. The terminal device is a voluntary-opening split hook. Shoulder flexion-extension and humeral rotation are under gravity control once they are unlocked. The left side, or electric-powered side, still uses mechanical chin actuated levers to lock and unlock a shoulder flexion-extension joint and a humeral rotator, otherwise the remaining components are externally powered. The arm has an electric-powered elbow and a wrist rotator that are controlled by chin movement against two rocker switches. Pulling on a linear potentiometer actuates the powered prehensor. Output speed, or force, of the prehensor is proportional to how much the transducer is pulled. For this system, the body-powered side is used as the dominant arm, with the electric-powered side assisting in a nondominant role. The prostheses are used for activities of daily living. This fitting highlights a number of important issues that are discussed further in the text.

In position control the position of the prosthetic joint is proportional to the input amount/intensity. The input amount/intensity might be the position of another physiological joint or a force level. If the position of another joint is used as the input then the system is known as a *position actuated, position servomechanism.* If the amount of force applied by some body part is the input, then the system is a *force actuated, position servomechanism.* An example is the power steering of a car. Here, the position of the steering wheel is related directly (proportional) to the position of the front wheels. Such a system is an example of a position follower (the position of the wheels follows the position of the steering wheel) or a position servomechanism.

If a further constraint is added whereby the input and output are physically (mechanically) linked such that change in position of the output cannot occur without a change in position of the input and vice versa, then as is the case in the power steering example, the system becomes what Simpson (1974) called an *unbeatable* position servomechanism or a system that has inherent extended physiological proprioception (EPP). Unbeatable servomechanisms are a subset of position servomechanisms as a whole where the input and output must move together. One cannot beat the other.

With position control the amputee's ability to perceive and control prosthesis position is directly determined by his or her ability to perceive and control the input signal. A major disadvantage of position control is that, unlike velocity control, it must maintain an input signal to hold an output level other than zero. This means that power must be continuously supplied to the component to maintain a commanded position other than zero. This is one of the reasons why speed or velocity control is the dominant mode of control in externally-powered prosthetics today, despite the fact that it has been shown that position control for positioning of the terminal device in space is superior to velocity control (Doubler & Childress, 1984a&b). Equally, it has been observed that velocity control may be better suited to the control of prehension (Carlson & Primmer, 1978; McKenzie, 1970) but this observation may be due to the poor performance of the prosthetic hand mechanisms of the day and consequently the manner in which amputees tended to use them.

Although proportional control for externally powered prostheses has been around since the 1970s it is only comparatively recently that proportional controllers have become more widely accepted. This is so because of the poor speed and speed of response of earlier systems. If a mechanism is slow, a user will tend to drive it using either a full open signal or full close signal regardless of the type of controller, i.e., slow devices will be used in essentially a switch or "bang bang" mode. It is the bandwidth of the mechanism and controller together that determines speed of response. If the mechanism is fast but the controller introduces delays due to processing, then the speed of response of the overall system will be limited by the controller. However, it is only recently that any prosthetic components have had sufficiently high mechanical bandwidth that users could perceive or have a use for a proportional-type control. The Hosmer-Dorrance NU-VA Synergetic Prehensor, Otto Bock System Electrohands, the Motion Control Hand, The Utah Elbow, and the Boston Elbow III all benefit from the use of proportional controllers. For some users, these devices need a proportional controller so that they can be controlled in a meaningful way because they are too fast to be operated effectively in a switch control mode.

In this day of digital circuits and microprocessor based controllers pulse width modulation is the preferred method of supplying a graded (proportional) control signal to a component. A PWM stream only requires a single digital output line and a counter on the microprocessor to be implemented, whereas a conventional analog signal (linear dc voltage level) requires a full digital-to-analog (D/A) converter (Fig. 32.24). PWM techniques are used extensively in switched-mode power supply design and audio amplifiers (Israelsohn, 2001) and as such, there is a large array of resources available to the designer to choose from.

As implemented on most microprocessors, a period register is loaded with a base period (1/PWM frequency; PWM frequency is usually in the range of 30 kHz for switched mode power supplies but can be as low as 300 Hz in some motor control applications) that is some fraction of the microprocessor's clock frequency (D'Souza and Mitra, 1997). A duty-cycle register is loaded with a percentage of the base period that corresponds to the desired proportional output voltage level as a percentage of the maximum possible output voltage. At the beginning of each period, the output at the PWM port is high, and the microprocessor counts down until the duty cycle register is zero, at which time the PWM port goes low. The microprocessor continues to count down until the base

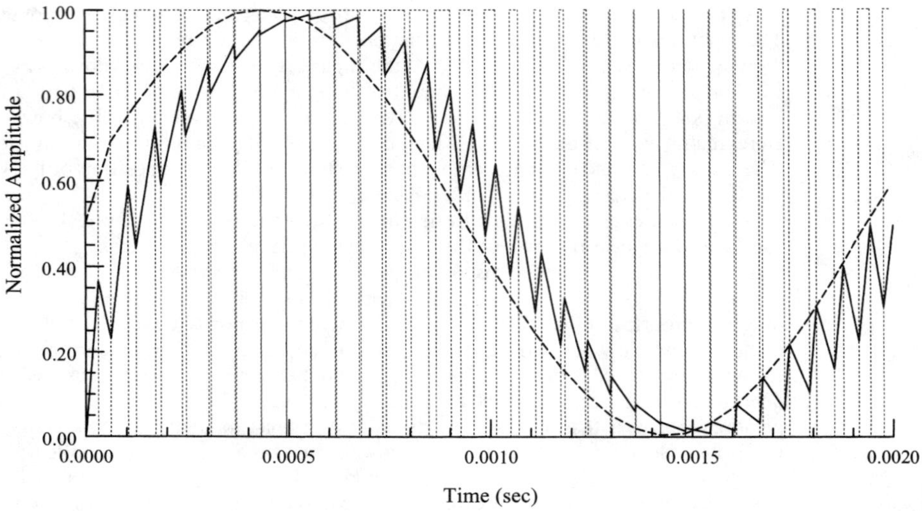

FIGURE 32.24 Plot showing a 500 Hz sine wave, a pulse width modulation (PWM) representation of the sine wave, and the reconstructed sine wave following filtering of the PWM stream by a simple RC filter (jagged waveform). Notice that in the PWM stream the time between pulses (base period) remains constant, however the duration, or the on time (duty cycle) of the pulse changes in proportion to the input sine wave amplitude. The restored sine wave signal has been phase-shifted by the filtering process, and it has a lot of high-frequency components superimposed on the original signal, but an analysis of the power spectra of original sine wave and the reconstructed sine wave reveals the high-frequency components to be low magnitude by comparison with the main signal. The high-frequency components could be further reduced by using a higher order filter, or a higher PWM frequency (1/base period) or both. The benefits of using pulse modulation for motor control are that smaller transistors can be used, possibly without heat sinking, because they are operated either in saturation or off and current pulses are more efficient at overcoming stiction in the motor, which improves overall motor efficiency.

period register is zero, at which time a new value for the duty cycle is loaded, and the process repeats itself (Palmer, 1997).

A variation on PWM is pulse position modulation (PPM), also known as *pulse period modulation* or *pulse frequency modulation* (PFM). In this case, the duty-cycle pulse remains on for a fixed time while the base period is varied. The frequency of the pulses (how close together the pulses are) determines the voltage level. The neuromuscular system is an example of a pulse position modulation system. A muscle is made up of many discrete motor units. A motor unit has an all or nothing response to a nerve impulse in much the same way that a nerve impulse is a nonlinear (thresholded) all-or-nothing event. The level of sustained force output of a motor unit is dictated by the frequency of incidence of the nerve impulses, with the motor units' dynamics [mechanical properties—inertial and damping properties (acts as a mechanical filter)] holding the force output smooth between incoming impulses. The motor unit is pulse frequency modulated by the nervous system.

Finally, it is possible to have a combination of PWM and PPM where one has variable width pulses and variable periods. In all cases, the output PWM stream is a series of pulses in which the dc voltage level is proportional to the ratio of the amount of on time with respect to the amount of off time for the pulses in the stream. Myopulse modulation, as implemented by Childress (1973) in his myoelectric controller, is an example of this form of pulse modulation (Fig. 32.25). This is an elegant proportional controller that can be implemented with a minimum of components. Because percentage myoelectric signal on time (EMG signal level above an arbitrary threshold) is monotonically related to muscle output force level, a proportional signal can be achieved very simply. This

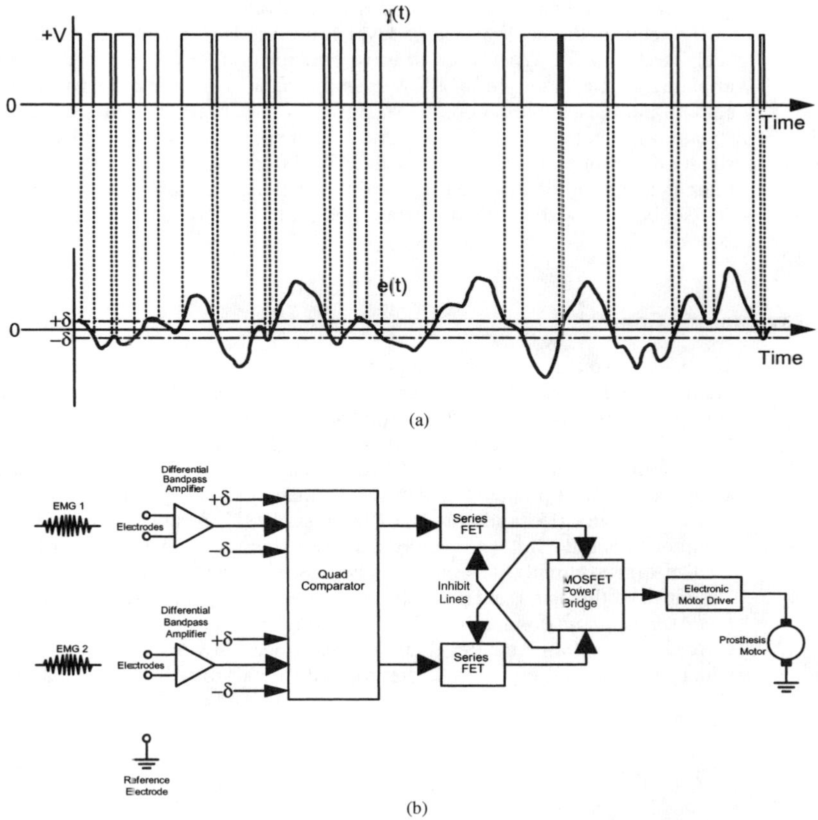

FIGURE 32.25 Myo-pulse modulation (Childress, 1973)—the voltage level seen at the motor is proportional to the ratio of "ON" time of the pulses to "OFF" time. The mean of this ratio is in turn proportional to the intensity of muscle contraction and as such is a very simple means of providing proportional myoelectric control. (*a*) The trace $\gamma(t)$ shows the switching response of a comparator in response to a time-varying EMG signal $e(t)$ and is given by the expression:

$$\gamma(t) = \frac{V}{2}[1 + \text{sgn}(|e(t)| - \delta)]$$

where $|e(t)|$ is the absolute magnitude of the amplified myoelectric signal δ is the threshold, and V is the power supply voltage. The thresholds for switching are $+\delta$ and $-\delta$. One method of choosing these thresholds is to set them at levels corresponding to ± 3 standard deviations of the quiescent (resting) signal amplitude. (*b*) The schematic shows the key components needed to implement this kind of control. The advantages of this EMG processing approach are (1) the simple electronic implementation, (2) No electrical time constant delay (processing delay)–implies faster response, (3) Control signal already in pulse modulation form, and (4) Wide dynamic range of muscle output.

controller simply converts a differentially amplified myoelectric signal directly into a pulse stream through the use of a pair of comparators. This pulse stream is then fed directly to the motor H-bridge without further processing. This controller is able to present a pulse stream to the drive motor in advance of the operator being able to detect motion by the innervated muscle, making it ideally suited for prosthetic mechanisms that require a quick speed of response. This controller is now commercially available through Hosmer-Dorrance Corp.

To recover the analog voltage level from a pulse stream requires the use of a low-pass filter (Palacherla, 1997). When used to drive a motor (via an H-bridge), the inertia and damping of the armature in the motor form a low-pass mechanical filter that provides the filtering necessary to smooth the pulses into a continuous signal. A simple RC circuit can be used to smooth the pulse stream if it is not being fed directly into a dc motor. If using a simple RC circuit, the cutoff frequency should be about two decades below the PWM frequency if a smooth output signal is sought. An RC filter is a first-order filter that has an attenuation of 20 dB/decade. Higher-order filters could, of course, be used with cutoff frequencies closer to the PWM frequency. For motors, a smooth signal is not vital.

32.7.2 Myoelectric Control

Myoelectric control derives it name from the electromyogram (EMG), which it uses as a control input. When a muscle contracts, an electric potential (the EMG) is produced as a by-product of that contraction. If surface electrodes are placed on the skin near a muscle, they can detect this signal (Fig. 32.26). The signal can then be electronically amplified, processed, and used to control a prosthesis. While the intensity of the EMG increases as muscle tension increases, the relationship is a complex nonlinear process that depends on many variables, including the position and configuration of the electrodes (Heckathorne and Childress, 1981). Although the EMG is nonlinear it is broadly monotonic, and the human operator perceives this response as more or less linear.

The first externally powered prosthesis was a pneumatic hand patented in Germany in 1915. Drawings of this hand and possibly the first electric hand were published in 1919 in *Ersatzglieder und Arbeitshilfen* (Borchardt et al., 1919). The first myoelectric prosthesis was developed during the early 1940s by Reinhold Reiter. Reiter published his work in 1948 (Reiter, 1948), but it was not widely known, and myoelectric control had to wait to be rediscovered during the 1950s. Reiter's

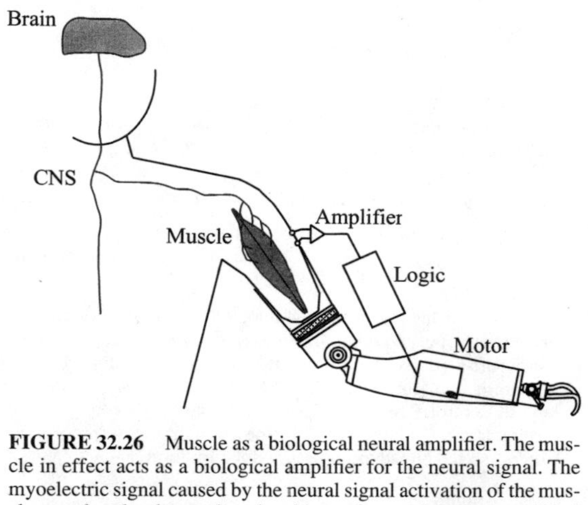

FIGURE 32.26 Muscle as a biological neural amplifier. The muscle in effect acts as a biological amplifier for the neural signal. The myoelectric signal caused by the neural signal activation of the muscle can then be detected at the skin surface and further amplified electronically for use in logic circuitry.

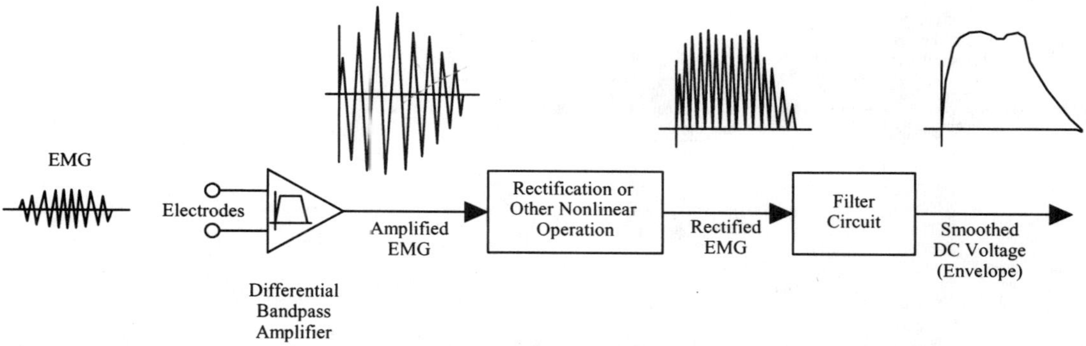

FIGURE 32.27 Schematic showing conventional myoelectric signal processing. A conventional myoelectric processing stream consists of differentially amplifying and bandlimiting the EMG signal. The amplified signal is then changed into a dc signal by rectification, by full wave rectification, squaring, root mean squaring (RMS), or some other appropriate nonlinear processing. The rectified signal is then filtered to obtain the envelope of the EMG signal. This voltage level can then be fed to the motor as a dc voltage level. All these processing steps introduce time delays into the overall control loop. The major delay is introduced by the time constant of the smoothing (envelope) filter circuit. These delays can in turn reduce the bandwidth of the entire motor controller system.

prosthesis consisted of a modified Hüfner hand that contained an electromagnet controlled by a vacuum tube amplifier. The prosthesis was not portable but was instead intended for use at a workstation, although Reiter did hope that one day it might be portable.

The Russian hand was the first semipractical myoelectric hand to be used clinically. This hand also had the distinction of being the first to use transistors (germanium) to process the myoelectric control signal (Childress, 1985). In this country, following World War II, the Committee on Artificial Limbs contracted with IBM to develop several electrically powered limbs. These were impressive engineering feats in their day but never found use outside the laboratory (Klopsteg and Wilson, 1956).

Myoelectric control has received considerable attention since it first appeared during the 1940s, and there is an extensive body of literature describing the EMG's characteristics and properties (see Parker and Scott, 1985; Basmajian and DeLuca, 1985). It was considered to be the cutting edge of technology of the day and was advanced as a natural approach for the control of prostheses since it made it possible for amputees to use the same mental processes to control prosthesis function as had previously been used in controlling their physiological limb (Hogan, 1976; Mann, 1968).

Usually the EMG is amplified and processed (bandlimited, rectified, and thresholded) to provide a dc signal that is related to the force of muscular contraction; this is then used to control the prosthesis (Scott, 1984) [Fig. 32.27]. EMG processing in a typical prosthetic myoelectric control system involves two pairs of differential dry metal electrodes and a reference electrode (Fig. 32.28). Although the electrodes are referred to as dry, the environment inside a prosthetic socket causes the amputee's residual limb to sweat, which creates a conductive interface between the skin and the electrodes (Fig. 32.29). Traditionally, myoelectric control uses electrodes placed on the skin near each of a protagonist-antagonist pair of muscles to control a single degree of freedom. For below-elbow fittings, this usually means electrodes on those muscle groups responsible for flexion and extension of the wrist and fingers (Fig. 32.30). Thinking about flexing or extending the "phantom" fingers controls closing or opening, respectively, of a terminal device.

Surface EMG signal amplitude (RMS) is approximately 100 μv for a moderately contracted forearm muscle. This signal must be amplified to a signal with an amplitude in the range 1 to 10 V before it can used. This implies that a gain of upward of 10,000 is needed. The bandwidth for the surface EMG signals is 10 to 300 Hz, with most of the signals' energy in and around 100 Hz (Childress, 1992). Differential amplifiers are used to amplify the EMG signal because the small EMG signal is often superimposed on large common-mode signals that, at these gain levels, would saturate an amplifier in a single-mode configuration. A differential amplifier can remove the large

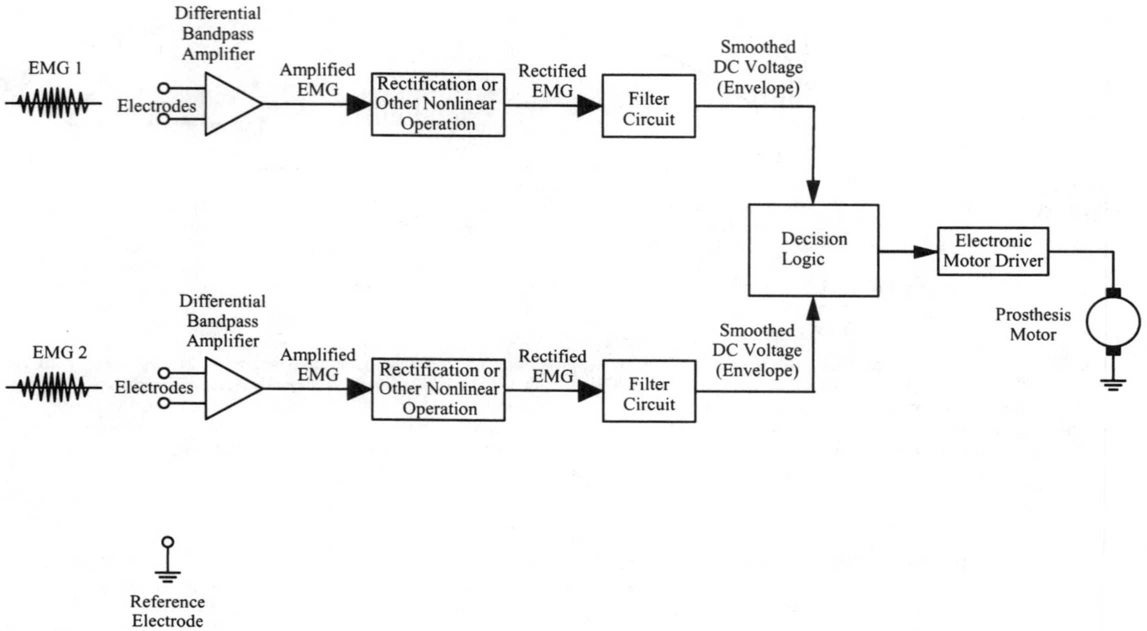

FIGURE 32.28 Schematic of the typical myoelectric processing scheme used in standard commercially available two-site myoelectrically controlled systems. (See also Figs. 32.29 and 32.30). The major difference here from Fig. 32.27 is that there must be some way to decide which of the two incoming signals to use to drive the motor. Usually a drive signal proportional to the magnitude of the difference between the two incoming signals is sent to the motor.

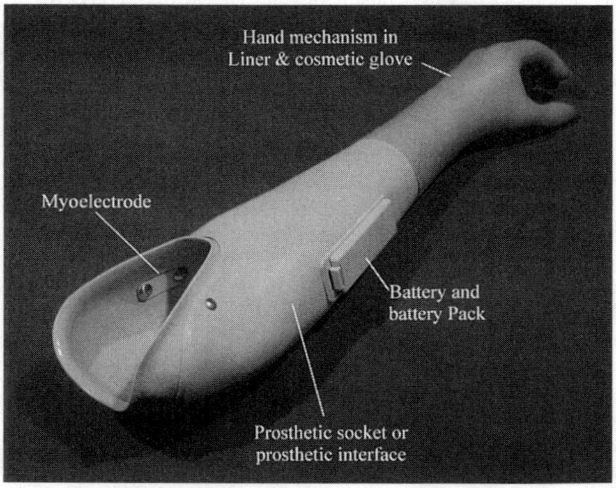

FIGURE 32.29 A standard transradial myoelectric prosthetic interface (socket). The battery pack, wrist unit, and myoelectrodes are all fitted into a custom-made laminated prosthetic socket and forearm. The socket is fabricated after a mold made from the amputee's residual limb.

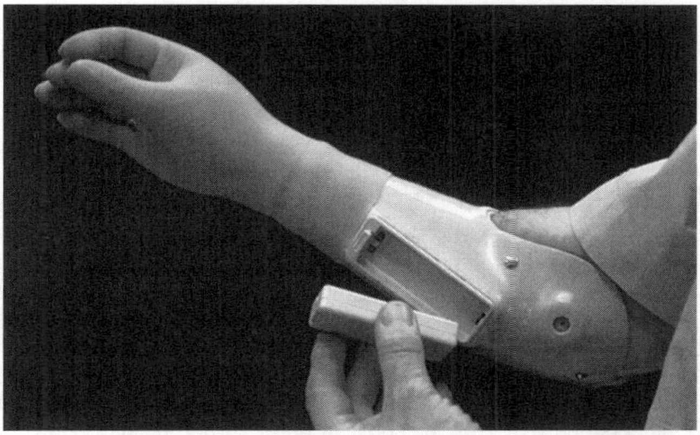

FIGURE 32.30 A person with a transradial amputation wearing a typical myoelectric prosthesis. The user is demonstrating how the battery pack is popped in and out. The battery pack housing is usually located on the medial surface of the prosthesis for appearance reasons and ease of access.

common-mode signals, leaving only the potential difference (EMG signal) between the electrodes to be amplified. This has the effect of most effectively amplifying the EMG signal frequencies around 100 Hz. Because of large gain requirements these differential amplifiers are seldom single op-amps but instead use multiple stages to meet the gain requirements.

Once the EMG signal has been amplified and bandlimited, it is then changed into a dc signal by rectification, by squaring or some other appropriate nonlinear processing. This dc potential is then commonly smoothed with a low-pass filter to remove the pulses and extract the envelope of the signal. For on/off, or switch, control, the smoothed dc voltage is then compared in a logic circuit with a threshold voltage. If the signal is greater than the threshold voltage, then power is supplied to the prosthesis motor, otherwise the power remains off. For proportional control, the smoothed voltage is fed to the motor drive circuit.

When used for proportional control, the EMG signal is usually treated as an amplitude modulated signal, where the mean amplitude of the cutaneous signal is the desired output from the myoelectric processor. However, in order to have accurate estimates of muscle force from EMG signals, a processing system with high signal-to-noise ratio (SNR) as well as short rise time (fast response) is required (Meek et al., 1990). Unfortunately, there is a filtering paradox whereby it is possible to have either fast response or high SNR, but not both (Jacobsen et al., 1984). To overcome this perceived problem, Meek et al. (1990) proposed using an adaptive filter in which the time constant of the low-pass filter used in the final stage of EMG processing (acquire signal envelope) was varied depending on the rate of change of the EMG signal. Their assumption was that an amputee, when moving quickly, will tolerate noise (low SNR) but will not tolerate delays in control. When holding the prosthesis steady or performing slow, dextrous tasks such as threading a needle, the amputee will tolerate slow response as long as there is low noise (high SNR). For the time being, this issue is largely academic. The available components do not have sufficient bandwidth for this issue to matter; however the designer needs to be aware of this filtering paradox should they be involved in the design of high-bandwidth myoelectrically controlled systems.

Processing steps take time! Any delay in the response of the output to a change in the input of greater than about 100 μs is perceptible to the human operator as a sluggish response. Any delay in response reduces the overall system bandwidth. Parker and Scott (1985) suggest that delays between input and output should not exceed 200 μs, and even this can be unacceptable for use with high performance prosthetic components.

Myopulse modulation (Childress, 1973) offers a means of processing myoelectric signals that maximizes the speed of response of an externally powered component. This technique eliminates the delays that are introduced by analog or digital filtering methods, provides proportional control, and keeps the component count to a minimum, thus keeping size and power consumption down. Another alternative method of myoelectric signal smoothing, called *autogenic backlash* (Bottomley, 1965), produced a more or less consistent direct current (DC) output from a fluctuating myoelectric signal while not sacrificing time response.

32.7.3 Comments on Myoelectrodes

A differential amplifier subtracts out those elements of the measured signal that are common to the two inputs of the amplifier and leaves the difference (Fig. 32.31). For prosthetic myoelectric control applications, each input to the amplifier is some form of button, or bar, electrode (Fig. 32.32) that sits on the skin's surface. This pair of electrodes should be oriented along the long axis of the muscle from which the EMG is being measured. This is so because the EMG travels along the muscle's length, creating a difference in potential between the electrodes that forms the input to the differential amplifier. The greater the distance between the electrodes, the larger the potential difference. It should be understood that muscles come in many forms: pennate, bipennate, smooth, etc., but that in the upper limb, the long axis of the superficial muscles, from which the EMGs are generally taken, runs broadly parallel to the underlying skeleton. The situation is complicated at the shoulder disarticulation level, where the direction of the muscles, e.g., the pectoralis or the infraspinatus, are not so obviously aligned to any long bones. In this case electrodes are aligned to the direction of action of the muscle fibers being measured.

The unwanted large common-mode noise signals, on the other hand, are equally present at both electrodes regardless of the orientation of the electrodes. However, these common-mode noise signals are more likely to be the same at both electrodes if the electrodes are located physically close together. However, the potential difference of the actual EMG between the electrodes will be reduced with small spacing. Thus one trades better common-mode rejection (CMRR) for reduced gain. Commercially available myoelectrodes from Otto Bock consist of a pair of bar electrodes spaced about 20 mm apart with a circular reference electrode located between them. The location of Otto Bock's reference electrode presents problems (Fig. 32.32*c*). A better configuration, from a CMRR standpoint is the Hosmer-Dorrance electrode pair which consists of button electrodes 11 mm (7⁄16 in) in diameter spaced about 28 mm (1⅛ in) apart with a separate reference electrode that is located as far away from any electrodes as possible within the confines of the prosthesis (Fig. 32.32). The output from Otto Bock electrodes is not the raw EMG but rather an amplified, bandlimited, rectified and smoothed version of the EMG, i.e., Otto Bock electrodes output a processed version of the EMG signal. Hosmer-Dorrance electrodes output an amplified and bandlimited signal.

A physically separate and remote reference electrode is desirable because the reference electrode is not electrically isolated from the inputs to the differential amplifier. The reference electrode is in contact with the person and consequently is electrically coupled to both inputs of the amplifier by the electrical impedance of the body. This impedance is a function of distance, i.e., the further away the reference electrode is, the greater is the attenuation. The reference electrode forms a floating ground signal rather than a true ground. Furthermore, the noise signal that can be introduced by the reference electrode is not necessarily a common-mode signal that will be removed by the differential amplification process.

The size of these electrodes could be greatly reduced given the availability today of many low-offset, ultrafast single-supply op-amps for the cellular phone industry. The ultralow offset voltage of these op-amps allows for higher gains to be used. This in turn would facilitate the design of smaller electrode pairs with higher common-mode rejection ratios. Physically smaller electrodes would allow more electrodes to be placed on the same residual limb.

A final comment about safety. The input lines from the electrodes in contact with the skin should be capacitively coupled to the inputs of the differential amplifier. The surface electrodes should not be directly attached to the differential amplifier inputs. The capacitive coupling blocks the flow of dc current from the amplifier to the user in the event of an electronic failure. If capacitive elements are not present to protect the user, then the user can suffer burns of the skin under the electrode site similar to those reported by Selvarajah and Datta (2001) for a myoelectric hand using a RSLSteeper single-site electrode.

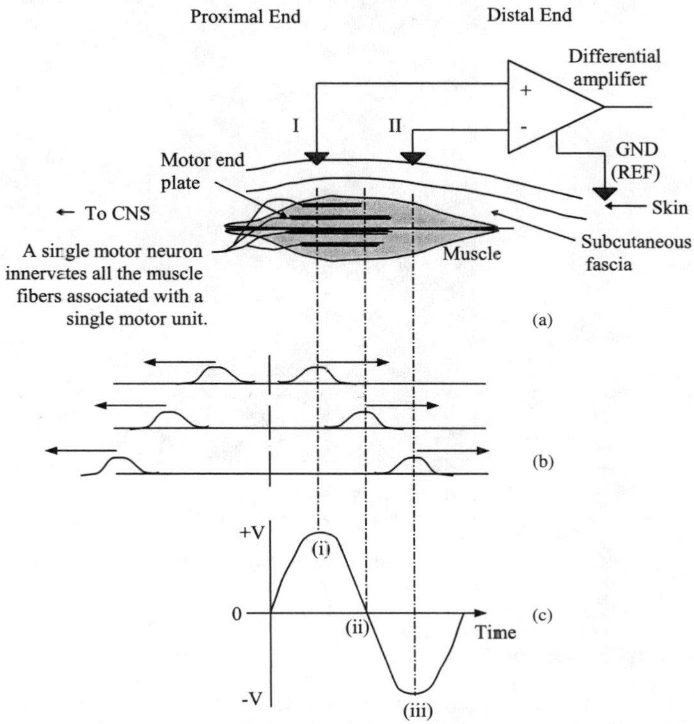

FIGURE 32.31 Schematic of how differential electrodes function. Dry button electrodes I and II are placed over a muscle belly on the skin surface in the line of the long axis of the muscle. These electrodes form the positive (I) and negative (II) inputs of a differential amplifier electrode pair. (*a*) In skeletal muscle, the individual muscle fibers are generally oriented along the long axis of the muscle. A single axon from the central nervous system innervates a single motor unit usually (for skeletal muscle) at the end of the muscle closest to the spinal chord (i.e., the proximal end). The points where the nerve fibers attach to the individual muscle fibers of a motor unit are called "motor end plates." When a nerve sends a signal to an individual muscle fiber to contract, the electric signal travels in both directions away from the motor end plate. For this reason the electrodes of the differential amplifier should be oriented along the long axis of the muscle's fibers. (*b*) Depolarization wave that travels in both directions away from the motor end plate along individual muscle fibers when activated by a motor nerve fiber. (*c*) Typical signal seen at the output of the differential amplifier for a *single* muscle fiber. At (i) the depolarization wave is directly under the positive input (I) with nothing under the negative input (II) to the differential amplifier and consequently registers as a positive going wave at the differential amplifier output; at (ii) the depolarization wave registers equally at the positive (I) and negative (II) inputs and is therefore canceled out, giving a zero at the output of the differential amplifier; at (iii) the depolarization wave is now directly under the negative input (II), with nothing under the positive input (I), so this registers at the differential amplifier output as a negative going wave. In reality, there are thousands of fibers all at different stages in their firing process, which results in the quasi-random signal seen at the skin surface.

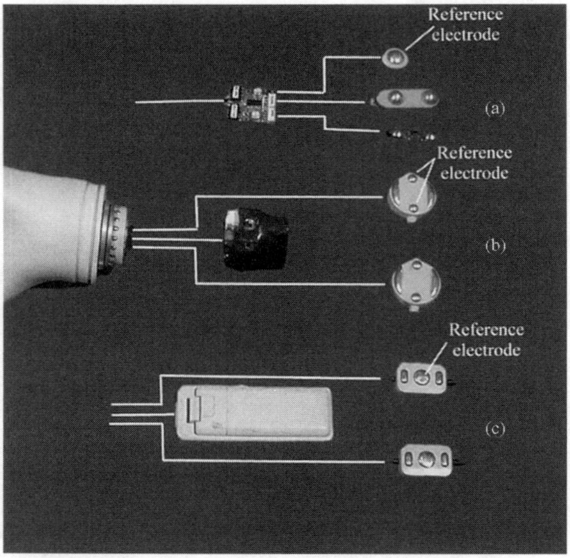

FIGURE 32.32 Photograph of different electrode configurations. (*a*) Hosmer-Dorrance Myopulse modulation electrodes and controller; note physically separate reference electrode to improve the common mode rejection ratio (CMRR). (*b*) and (*c*) Otto Bock electrodes with integrated reference electrodes.

32.7.4 Myoacoustic Control

Myoacoustic signals are auditory sounds created as a by-product of muscle contraction. A myoacoustic control system is very similar in structure to a myoelectric control system, but the elimination of unwanted acoustic noise appears to be a bigger problem than the elimination of unwanted electrical noise in myoelectric systems (Barry and Cole, 1990). It has not gained widespread use.

32.7.5 Prosthesis Control Using Muscle Bulge or Tendon Movement

Tendon or residual muscle movement has been used to actuate pneumatic sensors when interposed between a prosthetic socket and the superficial tendons and/or muscle. These sensors can be used for prosthesis control. The Vaduz hand, which was developed by a German team headed by Dr. Edmund Wilms in Vaduz, Liechtenstein, following World War II, used muscle bulge to increase pneumatic pressure to operate a switch-controlled, voluntary-closing, position-servo hand. This hand can be considered to be a forerunner of the pneumatic Otto Bock Hands of the 1970s and the electrically powered Otto Bock Hands of today.

Simpson (1966) also used muscle bulge to provide a control signal for the proportional control of an Otto Bock gas-operated hand. The width-of-opening of the hand was proportional to the force applied by the muscle's bulge. This device was an example of a force actuated position servomechanism, and because the bulging muscle had to achieve a significant pressure to ensure sufficient force for control-valve operation, a subconscious feedback path existed through the skin's own pressure sensors. This device was a precursor of Simpson's concept of extended physiological proprioception (EPP) (Simpson, 1974).

In 1999 the Rutgers multifunctional hand (Abboudi et al., 1999) received considerable media attention because the developers reported a multifunctional controller that allowed amputees to play the piano in a laboratory setting. This controller used multiple pneumatic sensors that were actuated by the movement of the superficial extrinsic tendons associated with individual finger flexors. For this hand to be clinically viable the developers need to resolve some of the issues that led to the failure of the previous attempts at using pressure/muscle hardness transducers. These issues include the system not being able to differentiate between actual control signals from a tendon and external pressures or impacts from objects in the environment. A prosthesis wearer interacting with the real world will exert forces and moments on the socket, which may actuate the pressure sensors and issue commands to the drive system of the prosthesis.

Otto Bock developed a control system using this idea to control an electric prosthesis in the 1970s (Nader, 1970). Otto Bock's selling points for this system was that it was immune to electrical and magnetic fields, unaffected by changes in skin impedance, and free from a quiescent current drain. This system was used to provide on-off and proportional control. Although the system has been commercially available for a number of years, it's use is not widespread. Members of the Technical University of Delft investigated a similar system to control a pneumatically operated prosthesis (Gerbranda, 1974; Van Dijk, 1976). They found that although the output signal of a muscle hardness transducer is, in theory, proportional to muscle contraction, that external disturbing forces exerted on the prosthesis were too large to practically employ proportional control. Hence this system could be used only for on-off control.

32.7.6 Neuroelectric Control

Neuroelectric control, or the control of a prosthetic device by way of nerve impulses, would appear to be a natural control approach. Although there is, and has been, research concerning prosthesis connections with nerves and neurons (Edell, 1986; Kovacs et al., 1989; Andrews et al., 2001) the practicality of human-machine interconnections of this kind is still problematic. Nerve tissue is sensitive to mechanical stresses, and this form of control also requires the use of implanted systems.

Edell (1986) attempted to use nerve cuffs to generate motor control signals in experimental systems. Kovacs et al. (1989) explored the use of integrated circuit electrode arrays into which the nerve fiber was encouraged to grow. Andrews et al. (2001) reported on their progress in developing a multipoint microelectrode peripheral nerve implant. They used a 100 × 100 grid array of silicon microelectrodes that they inserted into a peripheral nerve bundle using a pulse of air. This group is still working with animal models but claims to be able to identify individual neuron action potentials and also claims good long-term results so far. Concerns about how permanently electrode arrays are fixed have yet to be addressed.

A variation on the use of peripheral neural interfaces is the use of electroencephalogram signals (EEG's). These are electric signals that are detected on the surface of the skull and that result as a by-product of the natural functioning of the brain. Reger et al. (2000) demonstrated a hybrid neurorobotic system based on two-way communication between the brain of a lamprey and a small mobile robot. In this case the lamprey brain was kept alive in vitro and was used to send and receive motor control and sensory signals. The lamprey brain was an in vitro preparation that was used to send and receive motor control and sensory signals. While this is a long way from any practical use in prosthetics it demonstrates a bidirectional interface between nervous tissue and a machine. *IEEE Spectrum* (2001) reported on research at Duke University, where researchers had implanted an electrode array into the cerebellum of a monkey and through the use of appropriate pattern recognition software had it control a remote manipulator at MIT over 1000 km away via the Internet. One of the main practical goals cited for this research is to put paralyzed people in control of their environments. Finally, Kennedy et al. (2000) reported on the use of implanted electrode arrays to provide persons with "shut-in" syndrome control of a computer cursor.

In the area of functional electrical stimulation (FES), Loeb et al. (1998) have designed and built what they consider to be a new class of implantable electronic interfaces with nerve and muscle. These devices, which they call *BIONs*, are hermetically encapsulated, leadless electrical devices that are small enough to be injected percutaneously into muscles (2 mm diameter by 15 mm long). This group's second-generation BIONs will be able to send outgoing telemetry of ongoing movements.

Such devices ought to be able to be used to detect EMG signals and could be used for prosthetic control. Also in the area of FES, FDA approval has been granted to implant FES electrodes for the purpose of providing four-DOF control of an otherwise flail limb (Triolo, 1996).

32.7.7 Multifunctional Myoelectric Control

For prosthetic arms to be more than just position controllers for portable vices, multifunctional mechanisms that have the ability to have multiple degrees of freedom controlled simultaneously (in parallel) in a subconscious manner need to be developed. Current commercially available multifunctional controllers are generally sequential in nature and take the form of two site, three state multifunctional controllers. Motion Control, Inc., in the ProControl hand-wrist controller, uses rapid cocontraction of the forearm extensors and flexors to switch control between hand opening and closing to wrist rotation. Otto Bock uses a similar control strategy in its wrist-hand controller. Motion Control, Inc., in its elbow controller, uses dwell time (parking) to switch from elbow flexion and extension to hand opening and closure and cocontraction of biceps and triceps to switch control from the hand back to elbow.

An alternative to thinking in terms of controlling individual joints is to think in terms of actual grasp patterns to reduce the number of control sites required. To implement grasp pattern control by way of myoelectric signals, a specific grasp pattern of the hand is tied to a specific pattern of EMG signals that the controller identifies. The premise behind this type of control is that an amputee uses phantom limb sensation to visualize a specific hand grasp pattern, such as palmar prehension or lateral prehension, and consequently fires those muscles, or residual muscles, associated with creating that grasp pattern. The controller detects the resulting EMG signals generated by the residual musculature and through the appropriate use of pattern recognition and signal processing techniques extracts key features from these signals. These key features, in turn, enable the controller to recognize the grasp pattern associated with that EMG pattern, enabling the controller to actuate the appropriate motors to generate the desired hand grasp pattern.

In the Philadelphia Arm (Taylor and Finley, 1974) and Sven Hand (Lawerence and Kadefors, 1974) multiple myoelectric signals were used as control inputs, which were processed using adaptive weighted filters. The weights on the filters were adjusted to tailor the filters to a specific individual. In the Philadelphia arm, the choice of location for the myoelectrodes was based on muscle synergies associated with arm movements instead of the phantom limb sensations used in the Sven hand. One of the findings with the Sven hand was that control using myoelectric signals from an intact limb was superior to control using a residual limb and phantom limb sensation. The developers theorized that this was due to the presence of an intact proprioception system.

At present there is ongoing research into multifunctional myoelectric control of artificial hand replacements using time, frequency, and time-frequency identification techniques to identify features in the EMG signals that describe a particular grasp pattern. Hudgins et al. (1993) and Farry et al. (1993) used neural networks to perform the pattern recognition and feature extraction. Farry et al. (1997) used genetic algorithms. Currently, it is the features of Hudgins et al. (1993) that find the most widespread use. But even so a practical multifunctional hand control algorithm has yet to be developed. The processing time required for these complex signal processing algorithms is nontrivial.

32.7.8 Physiologically Appropriate Feedback

Physiologically correct feedback, beyond that provided by vision, is essential if low mental loading or coordinated subconscious control of multifunctional prostheses is to be achieved (Fig. 32.33). When prosthetic arm technology moved to externally powered systems, the control modalities shifted, with the exception of Simpson (1974) and a few others, from the position-based cable control of the body-powered systems to open-loop velocity control techniques (such as myoelectric and switch control). That is, prosthetic technology shifted away from cable inputs, which provide sensory

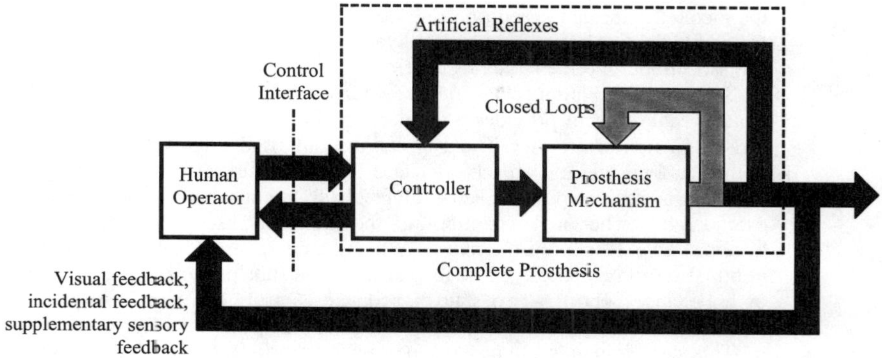

FIGURE 32.33 Feedback in human prosthetic systems. Most powered upper-limb prostheses are currently controlled primarily through *visual feedback* with some assistance from what has been called *incidental feedback*—whine, prosthesis vibration, socket forces, etc., being examples of incidental feedback; i.e., the motor feedback is incidental rather than by design. Attempts have been made to provide *supplementary sensory feedback* (SSF) through the use of vibrations of the skin, electrical stimulation of the skin, auditory and visual signals, and other means. However, because these methods are supplementary to the normal sensory feedback paths of the body they fail to present the feedback information in a physiologically useful manner and consequently have not seen much success. *Control interface feedback* means that the operator receives information concerning the state of the prosthesis through the same channel through which the prosthesis is controlled. Information concerning prosthetic joint position, velocity, and the forces acting on it is available to the operator through the proprioceptors of the controlling joint. Because feedback through the control interface is usually in forms that are easily interpreted by the user, it can be interpreted at a subconscious level, reducing the mental burden placed on the user. *Artificial reflexes* are closed loops within the controller/prosthesis mechanism itself that seek to remove the operator from the control loop, and as a result to also remove the mental burden placed on the user. Such systems use onboard intelligence to automatically respond to some external sensor input. To be effective a user must have confidence in the artificial reflex or else they will not relinquish control to the reflex loop.

and proprioceptive feedback, to techniques that provide little feedback to the operator beyond visual feedback.

Simplicity was probably the primary reason why prosthetics shifted to open-loop velocity control. The actuator of choice, the electric dc motor is an inherently rate-controlled device (i.e., its output speed is directly proportional to the input voltage), and it can be readily controlled with on/off switches. This resulted in the primary use of open-loop, velocity-controlled, externally powered prostheses. In addition, a velocity controlled system does not draw power to maintain a particular position.

Unfortunately, the user must integrate velocity in order to control position when using velocity control. Constant visual monitoring, due to the essentially open-loop nature of myoelectric control, is therefore required for effective operation. For the control of multiple degrees of freedom this places excessive mental load on the user, greatly diminishing any benefits a prosthesis of this complexity might offer. Visual and auditory feedback are slower, less automated, and less programmed than normal proprioceptive feedback and therefore place a greater mental burden on the operator (Soede, 1982).

In contrast, Simpson (1974) advocated augmented cable control. He coined the phrase *extended physiological proprioception* (EPP) to indicate that the body's own natural physiological sensors are used to relate the state of the prosthetic arm to the operator. EPP can be thought of as the extension of one's proprioceptive feedback into an intimately linked inanimate object. Consider a tennis player hitting a ball with a tennis racquet. The player does not need to visually monitor the head of the racquet to know where it will strike the ball. Through experience, the tennis player knows how heavy and how long the tennis racquet is. He or she knows where in space the head of the racquet is located based on proprioceptive cues from his or her wrist and hand; that is, how it feels in his or her hand. The tennis player has effectively extended his or her hand, wrist, and arm's proprioception into the tennis racquet.

The use of locking mechanisms to switch control from one component to the next in a sequential control cable system provides a similar type of control. The prosthesis, when locked, becomes a rigid extension of the user. The user can use this rigid extension to interact with the real world by pulling and pushing on objects. Because the prosthesis is rigid, the user "feels" the forces exerted on the prosthesis and has a sense of feedback about what the prosthesis is doing. The user has extended his or her proprioception into the prosthesis, in a fashion similar to how the tennis racket becomes an extension of the player. When the device is unlocked, it can be easily positioned with minimal effort (Fig. 32.23). The locked/unlocked state of the prosthesis can be thought of as a form of impedance control—two state impedance control. The prosthesis is either in a state of high impedance—rigid or locked, or in a state of low impedance—free or unlocked.

This same principle can be used to provide proprioceptive feedback to a powered prosthetic joint. Consider parking a car that has power steering. The driver feels, through the steering wheel, the interaction between the front wheels and the parking surface, curb, etc. However, the driver does not provide the power to turn the wheels; this comes from the engine. The driver is linked to the front wheels through the steering wheel and has extended his or her proprioception to the front wheels. Essentially, EPP control for externally powered prosthetic components can be thought of as power steering for prosthetic joints.

An ideal EPP system is an *unbeatable position servomechanism.* It is the mechanical linkage between input and output that converts a simple position servomechanism or velocity controller into an *unbeatable* position servomechanism. The mechanical linkage closes the loop and provides the path for the force feedback. The mechanical linkage constrains both the input and output to follow each other. In theory, the input cannot get ahead of (or beat) the output and vice versa. In practice, a small error must exist for the controller to operate. This error can be made very small with an appropriate controller gain. Doubler and Childress, (1982a, b), quantifiably demonstrated the potential effectiveness of applying EPP to the control of upper limb prostheses.

However, after being formalized and implemented by Simpson, control of powered limbs through EPP systems has been slow to gain acceptance. This lack of use is possibly due to the harnessing that is required for these systems. If harnessing is required, why not just use a body-powered system. Until a high bandwidth externally powered multifunctional prosthesis that needs parallel control of a number of degrees of freedom is developed, EPP control is unlikely to be rediscovered. Such a prosthesis might be a Boston Elbow III with a powered humeral rotator. In this case, shoulder elevation and depression would control elbow flexion, whereas shoulder protraction and retraction would control humeral rotation.

32.7.9 Artificial Reflexes

An alternative approach to the problem of the lack of physiologically appropriate feedback associated with today's prosthetic components is to automate more of the control functions of a given system. These *artificial reflexes* strive to reduce both the mental burden placed on the user and the number of control sources required. Artificial reflexes seek to remove the operator from the control loop and use onboard intelligence to automatically respond to some external sensor input and as such, they can be thought of as being similar to an *artificial reflex loop*. By putting more intelligence into the device through the use of embedded microprocessors, etc., more and more of the decision making process can be automated—requiring less attention on the part of the operator. Artificial reflexes are essentially closed loops within the mechanism/prosthesis itself. This trend of putting more onboard intelligence into prosthetic components is likely to increase in importance in the future. A key to the success or failure of an artificial reflex is that a user must have confidence in the artificial reflex or else they will not relinquish control to the reflex loop.

Otto Bock has a hand that has a microprocessor-controlled slip detector (Otto Bock SensorHand). The thumb has a sensor that detects slippage of an object grasped by the prehensor and automatically increases prehension force until it detects that the slippage has stopped. Anecdotal evidence suggests that users of the Otto Bock SensorHand feel more confident that an object they grasp using only visual cues will be grasped properly because the slip detector prevents the object from falling. The Otto Bock transfemoral (above-knee) "C-Leg" prosthesis also serves to reduce the mental burden required of users during walking. This prosthesis uses strain gauges and position transducers coupled with an embedded microcontroller to ensure that the knee is locked at every step before weight is borne on the prosthetic leg. Confidence on the part of the user that knee will not collapse if they stop concentrating on whether the knee is locked at each step will be critical to the success of this device.

Kyberd and Chappell (1994) use a system they call *hierarchical artificial reflexes* to automate the control process. In their multifunctional hand, they take the operator out of the loop and use onboard processing and sensors in the hand to tell the hand what grasp pattern to adopt. The operator only provides a conventional single degree of freedom open or close EMG signal. The idea is that by allowing the processor to take control, it reduces the mental loading on the operator. A major factor in the success or failure of these devices is confidence in the mechanism on the part of the user, so as to relinquish control to the artificial reflex.

32.7.10 Surgical Innovations Required

Surgeons need to perform innovative procedures that create new kinds of human-prosthesis interfaces if major advancements in upper-limb prosthesis control are to occur. They will have to create interfaces between amputee and prosthesis that allow for sensory feedback as well as control signals. *Direct muscle attachment* is one such interface. The concept of direct muscle attachment is not new. It had its origins in Italy at the beginning of the twentieth century and was brought into clinical practice in Germany by Sauerbruch around 1915 (Sauerbruch, 1916). Sauerbruch's technique, called *muscle tunnel cineplasaty*, fashions a skin-lined tunnel through the muscle (released at its insertion) and enables the muscle's power to be brought outside the body. Weir (1995; Weir et al., 2001) has shown the efficacy of tunnel cineplasty control when compared with the control of a conventional above-elbow, body-powered prosthesis. A similar but alternative surgical procedure called *tendon exteriorization cineplasty* (Beasley, 1966) uses tendon transfers combined with skin flaps to bring a tendon loop outside the body.

The subconscious control possible with an EPP controller operating in conjunction with the proprioception of a surgically created muscle tunnel cineplasty presents the intriguing possibility of making independent multifunctional control of a prosthetic hand or arm a reality. Suitable control muscles and EPP controllers in conjunction with powered fingers of an artificial hand would be a step toward achieving the goal of meaningful, independent multifinger control of hand prostheses. In a prototype fitting (Fig. 32.34), a below-elbow amputee with tendon exteriorization cineplasties was fitted with an externally powered hand that utilized the subject's exteriorized tendons as control inputs to an EPP controller (Weir et al., 2001). Movement of the flexor tendon caused the hand to close. Movement of the extensor tendon caused the hand to open. This was the first clinical fitting of a powered-hand prosthesis controlled directly by antagonist muscles (via exteriorized tendons) in a somewhat physiological manner.

The use of muscle/tendon cineplasty procedures would require changes in the way amputation surgery is performed at the time of initial amputation. Preservation of muscle tone and length is of paramount importance if future surgery is to be successful at creating muscle cineplasty interfaces. Muscle tone can be preserved by ensuring that residual muscles retain the ability to develop tension when voluntarily contracted. This requires that either myoplasty and/or myodesis be performed at the time of initial amputation. In myoplasty, agonist-antagonist residual muscle pairs are tied off against each other. In myodesis, the residual muscle is stitched to the bone. In both cases, the residual musculature retains the ability to easily develop tension, thus preventing atrophy during the interval between initial amputation and the revisions necessary to create a muscle/tendon cineplasty control interface. Myoplasty would generally be used on the superficial muscles, whereas myodesis could be used on the deep muscles.

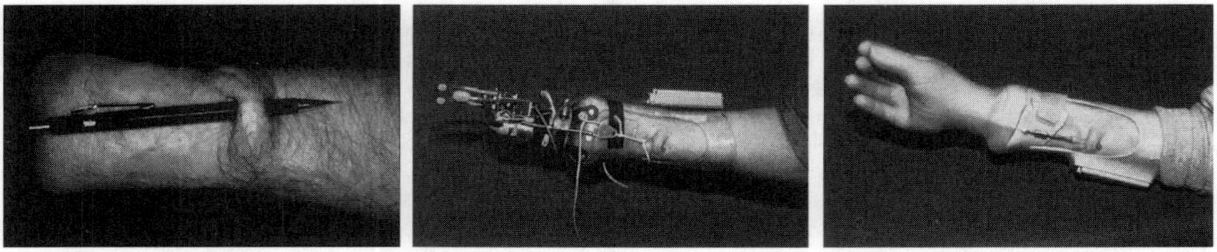

(a) (b) (c)

FIGURE 32.34 Prototype fitting of a person with a transcarpal amputation with tendon exteriorization cineplasties. The subject was fitted with an externally-powered hand that utilized the subject's exteriorized tendons as control inputs to an EPP controller. (*a*) Photograph of a tendon exteriorization cineplasty. A pen is shown passing through the skin-lined tendon loop. (*b*) Prototype version of the proof-of-concept EPP-controlled prosthesis. (*c*) Finished proof-of-concept prosthesis. Contraction of the flexor tendon caused the hand to close. Contraction of the extensor tendon caused the hand to open.

In the area of functional electrical stimulation, skeletal muscle transfers (myoplasty) combined with electrical stimulation have been advocated to provide contractile function that augments or replaces impaired organ function (Grandjean et al., 1996). Clinical investigation of dynamic cardiomyoplasty for the treatment of heart failure and dynamic myoplasty for treatment of fecal or urinary incontinence is already under investigation.

In Sweden, pioneering work in the area of *direct skeletal attachment* has been performed by Brånemark and his team (Brånemark, 1997; Brånemark et al., 2001) The technique is called *osseointegration*, and these surgeons and orthopedic engineers have created interfaces for direct skeletal attachment systems for upper and lower limb amputations. With osseointegration, a prosthesis is attached directly to the skeleton by way of a titanium abutment that protrudes through the skin from the cut end of the bone. Brånemark's techniques appear to have greatly diminished the infection problem that persisted in previous efforts (Hall and Rostoker, 1980). Should direct skeletal attachment prove itself viable, it could revolutionize the prosthetic fitting of some amputation levels.

The work of Kuiken (1995) is another example of the trend toward innovative surgical procedures. He advocates the use of *neuromuscular reorganization* to improve the control of artificial arms. Kuiken observed that although the limb is lost in an amputation, the control signals to the limb remain in the residual peripheral nerves of the amputated limb. The potential exists to tap into these lost control signals using nerve-muscle grafts. As first suggested by Hoffer and Loeb (1980), it may be possible to denervate expendable regions of muscle in or near an amputated limb and graft the residual peripheral nerve stumps to these muscles. The peripheral nerves would then reinnervate the muscles, and these nerve-muscle grafts would provide additional control signals for an externally powered prosthesis.

The nerve-muscle grafts could potentially be directly attached to force sensors using microcineplasty techniques and provide extended physiologic proprioception (EPP). Alternately, the surface EMG from the nerve-muscle grafts could serve as a traditional myoelectric control signal, using existing myoelectric technology. These grafts could provide *simultaneous* control of at least the terminal device and powered elbow/and possibly another degree of freedom such as a wrist rotation or wrist flexion-extension. In a shoulder disarticulation amputee, each of the residual brachial plexus nerves could be grafted to different regions of the pectoralis muscle. The prosthesis of the future might be a multifunctional device mounted into the skeleton using osseointegration and controlled using multiple miniature muscle cineplasties.

32.8 IN CONCLUSION

To put the current state of prosthesis control in the context of the evolution of control in airplanes, automobiles, and remote manipulators, it can be seen that all these fields used similar control methodologies in their early days. However, prosthetics later diverged from the others with the advent of electric power and myoelectric control. In each field, flight surfaces, front wheels, remote manipulators, or prosthetic joints were controlled directly by linking the operator to the device. In early aircraft, control cables connected the joystick to the wing, tail flaps, and rudder. In early automobiles, a rack and pinion mechanism connected the driver directly to the front wheels. In early master-slave manipulators, master and slave were physically connected by cables. In early prostheses, a control cable connected the user to the artificial joint.

In each of these cases, the operator provides both power and control signals and transfers them to the machine via a mechanical linkage that physically interconnects the operator and the machine output. Feedback from the machine is transferred back to the operator via the same mechanical linkage, i.e., feedback is returned to the user via the control interface (Fig. 32.33). The pilot could "feel" what was happening to the aircraft flight surfaces through the joystick. The amputee had a sense of prosthesis state through the control cable.

As aircraft got larger and faster, hydraulic power assist was added to the cable controls. As automobiles advanced, power steering was developed. The control input (the operator at a joystick or steering wheel) remained the same, but instead of the operator providing both the power and control signals, power was now provided by hydraulic systems, which augmented the cable systems of aircraft and the steering of cars. The key is that the same type of feedback through the control linkage was maintained. Position, velocity, and force information was still fed back with these "boosted" systems. A similar pattern of development occurred with remote manipulators (i.e., human-powered cable systems were followed by electric motor boosted cable systems).

However, it was at this point, the advent of externally powered devices, that prosthetics diverged from the others and moved almost exclusively to open-loop velocity control techniques with its associated loss of feedback information. It is the lack of subconscious sensory feedback to the user that has inhibited the use of many externally powered multifunctional prosthetic designs.

Meanwhile, in the area of remote manipulation, electrohydraulic, or electromechanical, bilateral master-slave systems were developed for the nuclear power industry to reduce the mental burden placed on the operator by providing force reflection. Aircraft controls for large planes and some high-performance military aircraft, building on work from the field of teleoperation or remote manipulation, have done away with the direct physical interconnection of the pilot and flight surfaces and now use fly-by-wire systems.

Fly-by-wire systems owe much of their development to Goertz (1951) who built the first bilateral force reflecting master-slave remote manipulators, and Mosher (1967) at General Electric Corp, who built force reflecting manipulators with power augmentation. In these systems, the pilot/operator is connected to the flight surfaces through electrical wire connections, and the systems use bilateral master/slave techniques with force feedback or force reflection to provide the operator with the appropriate feedback or "feel." Automobiles are also beginning to go to steer-by-wire and brake-by-wire systems (DesignFax, 2001).

For the top-of-the-line fighter planes, a stage beyond fly by wire has been reached. These planes are so inherently unstable that a human operator is incapable of controlling them unaided, so now *boosted control* has been added to *boosted power*. Artificial reflex-based systems are the logical conclusion of this trend to move more and more of the control from the operator to automated systems. In the end, the operator is just along for the ride. In fact, operatorless systems are being explored in a new generation of military aircraft and were employed in the U.S. action against Afghanistan in 2001. These aircraft are essentially robot planes that are piloted from the ground. For these aircraft, performance is no longer limited by the bandwidth of the pilot or by what the pilot can withstand in terms of "g" forces.

As for arm prosthetics, the future of what can be achieved depends to a large extent on the interactions of physicians, surgeons, prosthetists, and engineers. If meaningful multifunctional control of prostheses is to be achieved then physicians and surgeons need to perform innovative procedures that can be coupled with the novel components and controllers that engineers and prosthetists create.

ACKNOWLEDGMENT

The author is deeply indebted to Dr. Dudley Childress and Mr. Craig Heckathorne of the Northwestern University Prosthetics Research Laboratory, Chicago, Illinois, for taking the time to review this document.

REFERENCES

Abboudi, R. L., Glass, C. A., Newby, N. A., Flint, J. A., and Craelius, W. (1999). A biomimetic controller for a multifinger prosthesis. *IEEE Trans. on Rehabilitation Engineering*, vol 4, no. 2, pp. 121–129, June.

Andrews, B., Warwick, K., Jamous, A., Gasson, M., Harwin, W., and Kyberd, P. (2001). Development of an implanted neural control interface for artificial limbs. *Proceedings of the 10th World Congress of the International Society for Prosthetics and Orthotics (ISPO), Glasgow, Scotland*, ISPO Publications, Copenhagen, Denmark, July, 1–6 p. TO8.6.

Barry, D. T., and Cole, N. M.(1990). Muscle sounds are emitted at the resonant frequencies of skeletal muscle. *IEEE Transactions on Biomedical Engineering*, vol. BME 37, pp. 525–531.

Basmajian, J., and De Luca, C. (1985). *Muscles Alive*, Williams & Wilkins, Baltimore.

Beasley, R. W. (1966). The Tendon Exteriorization Cineplasty. A Preliminary Report. *Inter-Clinic Information Bulletin (ICIB)*, vol. V, no. 8, pp. 6–8, Committee on Prosthetics Research and Development, New York.

Beasley, R. W. (1983). Surgical treatment of hands for C5-C6 tetraplegia. *Symposium on Rehabilitation After Reconstructive Hand Surgery*, vol. 14, no. 4, pp. 893–904, October.

Beasley, R. W., and de Bese, G. M. (1990). *Prosthesis for the Hand. Surgery of the Musculoskeletal System*, vol. **2**, 2d ed., Evarts, C. M. (ed.), Churchill Livingstone Inc., New York.

Beattie, D., Iberall, T., Sukhatme, G. S., and Bekey, G. A. (1994). EMG control for a robot hand used as a prosthesis. *Proceedings of the Fourth International Conference on Rehabilitation Robotics (ICORR), Wilmington, Delaware*, Bacon, D. C., Rahmin, T., and Harwin, W. S., (eds.), Applied Science & Engineering Laboratories, University of Delaware, A. I. duPont Institute, Wilmington, Delaware, June, 14th–16th, pp. 67–72.

Blair S. J., and Kramer, S. (1981). Partial-hand amputation. *Atlas of Limb Prosthetics: Surgical and Prosthetic Principles*, chap. 10, American Academy of Orthopaedic Surgeons (AAOS.), C. V. Mosby Co., St. Louis, pp. 159–173.

Borchardt, M., Hartmann, K., Leymann, H., Radike, R., Schlesinger, G., and Schwiening, H., (ed.) (1919). *Ersatzglieder und Arbeitshilfen*, Julius Springer, Berlin.

Bottomley, A. (1965). Myo-electric control of powered prostheses. *Journal of Bone and Joint Surgery*, vol. 47-B, no. 3, p. 411.

Bowker, J. H., and Michael, J. W. (eds.) (1992). *Atlas of Limb Prosthetics, Surgical, Prosthetic, and Rehabilitation Principles*, 2d ed., Mosby-Year Book, Inc., St. Louis, Mo.

Brånemark, P-I., (1997): Osseointegration: Biotechnological Perspective and Clinical Modality. In *Osseointegration in Skeletal Reconstruction and Joint Replacement*, P-I. Brånemark, B. L. Rydevik, and R. Skalak, (eds.), Quintessence Publishing Co. Inc., Chicago, pp. 1–24.

Brånemark, R., Brånemark, P-I, Rydevik, B., and Myers, R. R. (2001). Osseointegration in skeletal reconstruction and rehabilitation: a review. *Journal of Rehabilitation Research and Development*, vol. 38, no. 2, pp. 175–182, March/April.

Carlson, L. E., and Primmer, K. R. (1978). Extended physiological proprioception for electric prostheses. In *Advances in External Control of Human Extremities, Proceedings of the 6th International Symposium on External Control of Human Extremities, Dubrovnik, Yugoslavia*, Yugoslav Committee for Electronics and Automation (ETAN), Belgrade, Yugoslavia, Aug. 28th–Sept. 1st.

Childress, D. S. (1973). An Approach to Powered Grasp. In: Advances in External Control of Human Extremities. *Proceedings of the Fourth International Symposium on External Control of Human Extremities, Dubrovnik, Yugoslavia*, August 28th–September 2nd, 1972; M. M. Gavrilovic and A. B. Wilson, Jr., (eds.), Yugoslav Committee for Electronics and Automation (ETAN), Belgrade, Yugoslavia, pp. 159–167, 1973.

Childress, D. S. (1992). Control of limb prostheses, Chapter 6D In *Atlas of Limb Prosthetics, Surgical, Prosthetic, and Rehabilitation Principles*, 2d ed., Bowker, J. H., and Michael, J. W. (eds.), Mosby-Year Book, Inc., St. Louis, Mo., pp. 175–199.

Childress, D. S. (1985). Historical aspects of powered limb prosthetics. *Clinical Prosthetics and Orthotics*, no. 9, pp. 2–13.

Cool, J. C., and Van Hooreweder, G. J. O. (1971). Hand prosthesis with adaptive internally powered fingers. *Medical and Biological Engineering*, vol. 9, pp. 33–36.

D'Souza, S., and Mitra, S. (1997). *Frequency and Resolution Options for PWM Outputs*. MicroChip Application Note AN539, MicroChip Technology Inc., Chandler, Ariz. http://www.microchip.com/Download/appnote//category/17cxx/misc/00539c.pdf

Davies, E. W., Douglas, W. B., and Small, A. D. (1977). A cosmetic functional hand incorporating a silicone rubber cosmetic glove. *Prosthetics and Orthotics International*, no. 1, pp. 89–93.

de Visser, H., and Herder, J. L. (2000). Force-directed design of a voluntary closing hand prosthesis. *Journal of Rehabilitation Research and Development*, vol. 37, no. 3, pp. 261–271, May/June.

DesignFax (2001): By-wire technology creates new car concept, tech upgrade, In: *DesignFax*, May. Adams Business Media, Chicago, pp. 26–27.

Dijk E. J., van, (1976). The Behavior of a Muscle Hardness Transducer in a Prosthetic Forearm Socket, MSc thesis, A-184, Laboratory for Measurement and Control, Technical University of Delft, The Netherlands.

Direct-Link Prehensor (1991a). *NASA Tech Briefs*, vol. 15, no. 12, pp. 78, Dec.

Direct-Link Prehensor (1991b). Technical Support Package, *NASA Tech Briefs*, ARC-11666, NASA, Ames Research Center, Moffett Field, Calif.

Doubler, J. A. (1982). An Analysis of Extended Physiological Proprioception as a Control Technique for Upper-Extremity Prostheses. Ph.D. dissertation, Northwestern University, Evanston, Illinois,

Doubler, J. A., and Childress, D. S. (1984a). An analysis of extended physiological proprioception as a control technique. *Journal of Rehabilitation Research and Development*, vol. 21, no. 1, pp. 5–18.

Doubler, J. A., and Childress, D. S. (1984b). Design and evaluation of a prosthesis control system based on the concept of extended physiological proprioception. *Journal of Rehabilitation Research and Development*, vol. 21, no. 1, pp. 19–31.

Edell, D. J. (1986). A peripheral nerve information transducer for amputees: long-term multichannel recordings from rabbit peripheral nerves. *IEEE Transactions on Biomedical Engineering*, vol. BME-33, no. 2, pp. 203–214, Feb.

Farry, K. A., Fernandez, M. S., Abramczyk, M. E., Novy, B. S., and Atkins, D. (1997). Applying genetic programming to control of an artificial arm. *Proceedings of Myo-Electric Control Symposium '93 (MEC'97), Fredericton, New Brunswick, Canada*, University of New Brunswick, Fredericton, Canada, August 16–20, 1993, pp. 50–55.

Farry, K. A., Walker, I. D., and Sendonaris, A. (1993). Teleoperation of a Multifingered Robotic Hand with Myoelectrics. *Proceedings of Myo-Electric Control Symposium '97 (MEC'97), Fredericton, New Brunswick, Canada*, University of New Brunswick, Fredericton, Canada, August 16–20, pp. 115–122.

Faulhaber Micro Mo Application Notes, http://www.micromo.com/03application_notes.asp MicroMo Electronics, Inc., Clearwater, Fla.

Fletcher, M. J. (1954). New developments in hands and hooks. In *Human Limbs and Their Substitutes*, Klopsteg, P. E., and Wilson, P. D., (eds.), McGraw-Hill, New York, pp. 359–408.

Fraser, C. M. (1998). An evaluation of the use made of cosmetic and functional prostheses by unilateral upper-limb amputees. *Journal of the International Society for Prosthetics and Orthotics(ISPO)*, vol. 22, no. 3, pp. 216–223, Dec.

Fryer, C. M. (1992). Harnessing and Controls for Body-Powered Devices, Chapter 6B. In *Atlas of Limb Prosthetics, Surgical, Prosthetic, and Rehabilitation Principles*, 2d ed. Bowker, J. H., and Michael, J. W., (eds.), Mosby-Year Book, Inc., St. Louis, Mo., pp. 133–151.

Fryer, C. M., and Michael, J. W. (1992). Body-Powered Components, chap. 6A. In *Atlas of Limb Prosthetics, Surgical, Prosthetic, and Rehabilitation Principles*, 2d ed., Bowker, J. H., and Michael, J. W., (eds.), Mosby-Year Book, Inc., St. Louis, Missouri, pp. 107–132.

Gerbranda T. (1974). Feasibility Study of a Muscle Hardness Transducer, M.Sc. thesis, A-144, Laboratory for Measurement and Control, Technical University of Delft, The Netherlands.

Goertz, R. C. (1951). Philosophy, development of manipulators. In *Teleoperated Robotics in Hostile Environments*, (1985), Martin. H. L., and Kuban, D. P., (eds.), Society of Manufacturing Engineers, Dearborn, Michigan, pp. 257–262.

Grandjean, P., Acker, M., Madoff, R., Williams, N. S., Woloszko, J., and Kantor, C., (1996). Dynamic myoplasty: surgical transfer and stimulation of skeletal muscle for functional substitution or enhancement. *Journal of Rehabilitation Research and Development*, vol. 33, no. 2, pp. 133–144, April.

Hall, C. W., and Rostoker, W. (1980). Permanently attached artificial limbs. *Bulletin of Prosthetics Research*, vol. 10, no. 34, pp. 98–100.

Hannaford, B., and Winters, J. (1990). Actuator properties and movement control: biological and technological models. In J. M. Winters and S. L-Y. Woo, (eds.), *Multiple Muscle Systems: Biomechanics and Movement Organization*, Springer-Verlag, Berlin, pp. 101–120.

Hays, M. (2001): Eugene F. Murphy, Ph.D., and early VA research in prosthetics and sensory aids. Guest editorial, *Journal of Rehabilitation Research and Development*, vol. 38, no. 2, pp. vii–viii, March/April.

Heckathorne, C. W., (1992): Components for adult externally powered systems, chap. 6C. In *Atlas of Limb Prosthetics, Surgical, Prosthetic, and Rehabilitation Principles*, 2d ed., Bowker, J. H., and Michael, J. W. (eds.), Mosby-Year Book, Inc., St. Louis, Missouri, pp. 151–175.

Heckathorne, C. W., and Childress, D. S. (1981). Relationships of the surface electromyogram to the force, length, velocity, and contraction rate of the cineplastic human biceps. *American Journal of Physical Medicine*, no. 60, pp. 1–19.

Heckathorne, C. W., Toth, P. J., Childress, D. S., (1995): The role of the non-dominant hand in manipulative tasks. *Proceedings of the 8th World Congress of the International Society for Prosthetics and Orthotics (ISPO), Melbourne, Australia*, ISPO Publications, Copenhagen, Denmark, p. 146.

Herberts, P., Almström, C., and Caine, K. (1978). Clinical application study of multifunctional prosthetic hands. *The Journal of Bone and Joint Surgery*, no. 60, pp. 552.

Hoffer, J. A., and Loeb, G. E. (1980). Implantable electrical and mechanical interfaces with nerve and muscle. *Annals of Biomedical Engineering* 8, pp. 351–360.

Hogan, N. J. (1976). Myoelectric Prosthesis Control: Optimal Estimation Applied to EMG and Cybernetic Considerations for Its Use in a Man-Machine Interface. Ph. D. dissertation, MIT, Boston, Mass.

Hollerbach, J., Hunter, I., and Ballantyne, J. (1991). A comparative analysis of actuator technologies for robotics. In *Robotics Review 2*, pp. 299–342, MIT Press, Cambridge, Mass.

Horowitz, P., and Hill, W. (1995). *The Art of Electronics*, 2d ed., Cambridge University Press, New York.

Hudgins, B. S., Parker, P. A., Scott, R. N. (1993). A new strategy for multifunction myoelectric control. *IEEE Transactions on Biomedical Engineering*, vol. 40, no. 1, pp. 82–94. Jan.

IEEE Spectrum (2001). A mind-internet-machine interaction. News Analysis, *IEEE Spectrum*, p. 33, Jan.

Israelsohn, J. (2001). Listening to Class D. *Electronics Design News (EDN)*, August 30th, pp. 61–72.

Jacobsen, S. C., Knutti, D. F., Johnson, R. T., and Sears, H. H. (1982). Development of the Utah arm. *IEEE Transactions on Biomedical Engineering*, vol. BME-29, no. 4, pp. 249–269.

Jacobsen, S. C., Meek, S. G., and Fullmer, R. R. (1984). An adaptive myoelectric filter. *Proceedings of the 6th Conference of the IEEE Engineering in Medicine and Biology Society (EMBS)*.

Jacobsen, S. C., Smith, F. M., Iversen, E. K., and Backman, D. K. (1990). High Performance, High Dexterity, Force Reflective Teleoperator. *Proceedings of the 38th Conference on Remote Systems Technology, American Nuclear Society (ANS), La Grange Park, Illinois*, vol. 2, pp. 180–185.

Jacobsen, S. C., Wood, J. E., Knutti, D. F., and Biggers, K. B. (1984). The UTAH/M.I.T. dexterous hand: work in progress. *The International Journal of Robotics Research*, vol. 3, no. 4, pp. 21–50, winter.

Kamakura, N., Matsuo, M., Ishii, H., Mitsukoshi, F., and Miusa, Y. (1980). Patterns of static prehension in normal hands. *American Journal of Occupational Therapy*, vol. 34, no. 7, pp. 437–445.

Kato, I., et al. (1970). Multifunctional myoelectric hand prosthesis with pressure sensory feedback system–Wasada Hand-4P. In *Advances in External Control of Human Extremities, Proceedings of the Third International Symposium on External Control of Human Extremities, Dubrovnik, Yugoslavia*, 1969, Yugoslav Committee for Electronics and Automation (ETAN), Belgrade, Yugoslavia, pp. 155–170.

Keller, A. D., Taylor, C. L., and Zahn, V. (1947). *Studies to Determine the Functional Requirements for Hand and Arm Prostheses*, Department of Engineering, University of California at Los Angeles.

Kennedy, P. R., Bakay, R. A. E., Moore, M. M., Adams, K., and Goldwithe, J. (2000). Direct control of a computer from the human Central nervous system. *IEEE Transactions on Rehabilitation Engineering*, vol. 8, no. 2, pp. 198–202.

Kenworthy, G. (1974). An artificial hand incorporating function and cosmesis. *Biomedical Engineering*, vol. 9, no. 12, pp. 559–562, Dec.

Kester, W., and Buxton, J. (1998). Battery chargers. Sec. 5 in *Practical Design Techniques for Power and Thermal Management*, Kester, W. (ed.), Analog Devices, Norwood, Mass., pp. 5.1–5.25.

Klopsteg, P. E., and Wilson, P. D. (eds.) (1954). *Human Limbs and Their Substitutes*, McGraw-Hill, New York.

Klute, G. K., Czerniecki, J. M., and Hannaford, B. (1999). McKibbon artificial muscles: pneumatic actuators with biomechanical intelligence. *IEEE/ASME Conference on Advanced Intelligent Mechatronics, September 19th–22nd, Atlanta.*

Kovacs, G. T. A., Storment, C. W., Hentz, V. R., and Rosen, J. M. (1989). Fabrication techniques for directly implantable microelectronic neural devices. *Proc. RESNA 12th Conf., New Orleans, Louisiana*, RESNA Press, Washington D.C., pp. 292–293.

Krukenberg, H. (1917). *About the plastic reshaping of arm amputation stumps.* Ferdinand Enke - Verlag, Stuttgart.

Kuiken, T. A., Rymer W. Z., and Childress, D. S. (1995). The hyper-reinnervation of rat skeletal muscle. *Brain Research*, vol. 676, pp. 113–123.

Kyberd, P. J., and Chappell, P. H. (1994). The Southampton Hand: an intelligent myoelectric prosthesis. *Journal of Rehabilitation Research and Development*, vol. 31, no. 4, pp. 326–334, Nov.

Lansberger, S., Shaperman, J., Lin, A., Vargas, V., Fite, R., Setoguchi, Y., and McNeal, D. (1998). A small problem: body-powered toddler hand design. *Proceedings of the 9th World Congress of the International Society for Prosthetics and Orthotics (ISPO), Amsterdam, Netherlands*, ISPO Publications, Copenhagen, Denmark, June 28th–July 3rd, p. 222.

Lawrence, P. D., and Kadefors, R. (1974). Classification of myoelectric patterns for the control of a prosthesis. In *The Control Of Upper-Extremity Prostheses and Orthoses, Proceedings of the Conference on the Control of Upper-Extremity Prostheses and Orthoses, Göteborg*, Sweden, October 6–8, 1971, Herberts, P., Kadefors, R., Magnusson, R. I. and Petersén, I., (eds.), Charles C. Thomas, Springfield, Ill., pp. 190–200.

Lee Publications (2001). Gung Ho Grip™ A-Okay, *Lee's Action Figure News & Toy Review (AFN&TR)*, Lee Publications, Monroe, Conn. no. 102, p. 16, April.

Linden. D. (ed.) (1995). *Handbook of Batteries*, 2d ed., McGraw-Hill, New York.

Loeb, G. E., Richmond, F. J. R., Olney, S., Cameron, T., Dupont, A. C., Hood, K., Peck, R. A., Troyk, P. R., and Schulman, J. H. (1998). Bionic neurons for functional and therapeutic electrical stimulation. *Proc. IEEE-EMBS*, vol. 20, pp. 2305–2309.

Lozac'h, Y. (1984). The preferred working plane for an active thumb. *Proceedings of the 2nd International Conference on Rehabilitation Engineering (RESNA 84), Ottawa, Canada, June 17–22*, RESNA Press, Washington D.C., pp. 24–25.

Lozac'h, Y., Madon, S., Hubbard, S., and Bush, G. (1992). On the Evaluation of a Multifunctional Prosthesis. *Proceedings of the 7th World Congress of the International Society for Prosthetics and Orthotics (ISPO), Chicago, June 28–July 3*, ISPO Publications, Copenhagen, Denmark, pp. 185.

Magee, D. J. (1987). *Orthopedic Physical Assessment.* Saunders, Philadelphia.

Mann, R. W. (1968). Efferent and afferent control of an electromyographic proportional rate, force sensing artificial elbow with cutaneous display of joint angle. *Proceedings of the Symposium on the Basic Problems of Prehension, Movement, and Control of Artificial Limbs*, Institution of Mechanical Engineers, London, pp. 86–92.

Mann, R. W., and Reimers, S. D. (1970). Kinesthetic sensing for the EMG controlled "Boston Arm." *IEEE Transactions on Man-Machine Systems*, vol. MMS-11, no. 1, pp. 110–115.

McKenzie, D. S. (1970). Functional replacement of the upper-extremity today. In *Prosthetic and Orthotic Practice*, Murdoch, G., (ed.), Edward Arnold Ltd., London, pp. 363–376.

Meek, S. G., Wood, J. E., and Jacobsen, S. C. (1990). Model-based, multi-muscle EMG control of upper-extremity prostheses. Chap. 22 in *Multiple Muscle Systems: Biomechanics and Movement Organization*, J. M. Winters and S. L-Y. Woo, (eds.), Springer-Verlag, pp. 360–376.

Michael, J. W. (1986). Upper-limb powered components and controls: current concepts. *Clinical Prosthetics and Orthotics*, vol. 10, no. 2, pp. 66–77.

Mosher, R. C. (1967). *Handyman to Hardiman.* S.A.E. Paper 67008, Society of Automotive Engineers (SAE), Warrendale, Pa.

Murphy, E. F. (1955). Engineering—hope of the handless. *Artificial Limbs*, vol. 2, no. 2, p. 10, May 19.

Näder, M., (1970): Experiences and observations with Myobock I and present state of the art in Otto Bock Myostat Systems. *Orthopadie Technik*, vol. 22, no. 12, pp. 337–340.

Napier, J. A. (1956). Prehensile movements of the human hand. *Journal Bone and Joint Surgery*, vol. 38, no. 4, Nov.

Oberg, E., Jones, F. D., Horton, H. L., and Ryffell, H. H. (2000). *Machinery's Handbook*, 26th ed., C. J. McCauley, R. H. Heald, and M. I. Hussain (eds.) Industrial Press, New York.

Palacherla, A. (1997). *Using PWM to Generate Analog Output*. MicroChip Application Note AN538, MicroChip Technology Inc., Chandler, Ariz. http://www.microchip.com/Download/appnote/category/17cxx/misc/00538c.pdf

Palmer M., (1997): *Using the PWM*. MicroChip Application Note AN564. MicroChip Technology Inc., Chandler, Ariz. http://www.microchip.com/Download/appnote/category/17cxx/misc/00564b.pdf

Parker, P. A., and Scott, R. N., (1985): Myoelectric control of prosthesis. *CRC Critical Reviews in Biomedical Engineering*, vol. 13, no. 4, pp. 283–310.

Pillet, J., and Mackin, E. J. (1992). Aesthetic restoration. Chap. 7C in: *Atlas of Limb Prosthetics - Surgical, Prosthetic, and Rehabilitation Principles*, 2d ed., Bowker, J. H., and Michael, J. W., (eds.), American Academy of Orthopaedic Surgeons (AAOS), Mosby Year Book, St Louis, Mo., pp. 227–236.

Plettenburg, D. H. (2002). A Sizzling Hand Prosthesis: On the Design and Development of a Pneumatically Powered Hand Prosthesis for Children. Ph.D. dissertation, Technical University of Delft, The Netherlands.

Razic, D. (1973). Kinematics Design of a Multifunctional Hand Prosthesis. In: Advances in External Control of Human Extremities, *Proceedings of the Fourth International Symposium on External Control of Human Extremities, Dubrovnik, Yugoslavia, Aug. 28–Sept. 2, 1972*, M. M. Gavrilovic and A. B. Wilson, Jr., (eds.), Yugoslav Committee for Electronics and Automation (ETAN), Belgrade, Yugoslavia, pp. 177–183.

Reger, B. D., Fleming, K. M., Sanquineti, V., Alford, S., Mussa-Ivaldi, F. A. (2000). Connecting brains to robots: the development of a hybrid system for the study of learning in neural tissues. *Artificial Life Journal* (MIT Press), vol. 6, no. 4, pp. 307–324, Fall.

Reiter, R. (1948). A new electric hand. *Grenzgebiete der Medizin*, no. 4, pp. 133.

Sarrafian, S. K. (1992). Kinesiology and Functional Characteristics of the Upper-Limb. Chap. 5 in *Atlas of Limb Prosthetics, Surgical, Prosthetic, and Rehabilitation Principles*, 2d ed., Bowker, J. H., and Michael, J. W., (eds.), Mosby-Year Book, Inc., St. Louis, Missouri, pp. 83–105.

Sauerbruch, F. (1916). *The Voluntary Controlled Artificial Hand: A Guide for Surgeons and Technicians*, Julius Springer-Verlag, Berlin.

Schlesinger, G., DuBois-Reymond, R., Radike, R., and Volk, S. (1919). Der mechanische Aufbau der künstlichen Glieder. I.: Der Eratzrarm. In *Ersatzglieder und Arbeitshilfen für Kriegsbeschädigte und Unfallverletzte*, Borchardt M., Hartmann K., Leymann H., Radike R., and Schlesinger, G. (eds.), Julius Springer-Verlag, Berlin, pp. 321–573.

Schulz, S., Pylatiuk, C., & Reischl, M. (2002). A lightweight multifunctional prosthetic hand. Submitted to *Robotica, Special Review: Advance in External Prosthetic Limbs at the Start of the 21st Century*.

Scott, R. N. (1984). An Introduction to Myoelectric Prostheses. In *UNB Monographs on Myoelectric Prostheses*, University of New Brunswick, Institute of Biomedical Engineering, Frederickton.

Sears, H. H., Andrew, J. T., and Jacobsen, S. C. (1989). Experience with the Utah Arm, Hand, and Terminal Device. *Comprehensive Management of the Upper-Limb Amputee*, Atkins, D. J., and Meier, R. H., (eds.), Springer-Verlag, New York, pp. 194–210.

Selvarajah, K., and Datta, D. (2001). An unusual complication of a myoelectric prosthesis. *Clinical Note, Prosthesis and Orthotics International, The Journal of the International Society for Prosthetics and Orthotics (ISPO)*, vol. 25, pp. 243–245.

Simpson, D. C. (1966). Powered hand controlled by "muscle bulge." *J. Scientific Instrumentation*, vol. 43, no. 8, pp. 521–522.

Simpson, D. C., (1969): An externally powered prosthesis for the complete arm. *Biomedical Engineering*, vol. 4, no. 3, pp. 106–110, 119.

Simpson, D. C., (1973): The Control and Supply of a Multi-Movement Externally-powered Upper-limb Prosthesis. *Proceedings of the Fourth International Symposium on External Control of Human Extremities*, Dubrovnik, Yugoslavia, Aug. 28–Sept. 2, 1972, M. M. Gavrilovic and A. B. Wilson, Jr., (eds.), Yugoslav Committee for Electronics and Automation (ETAN), Belgrade, Yugoslavia, pp. 247–254.

Simpson, D. C. (1974). The choice of control system for the multimovement prosthesis: extended physiological proprioception (EPP). In *The Control of Upper-Extremity Prostheses and Orthoses, Proceedings of the Conference on the Control of Upper-Extremity Prostheses and Orthoses, Göteborg, Sweden, October 6–8, 1971*, Herberts, P., Kadefors, R., Magnusson, R. I., and Petersén, I., (eds.), Charles C. Thomas, Springfield, Ills., pp. 146–150.

Soede, M. (1982). Mental control load and acceptance of arm prostheses. *Automedica*, no. 4, pp. 183–191.

Stojiljkovic, Z. V., and Saletic, D. Z. (1975). Tactile pattern recognition by belgrade hand prosthesis. In *Advances in External Control of Human Extremities, Proceedings of the Fifth International Symposium on External*

Control of Human Extremities, Dubrovnik, Yugoslavia, August 1974, Yugoslav Committee for Electronics and Automation (ETAN), Belgrade, Yugoslavia.

Taylor, C. L., (1954). The biomechanics of the normal and of the amputated upper extremity. In *Human Limbs and Their Substitutes*, Klopsteg, P. E., and Wilson, P. D., (eds.), McGraw-Hill, New York.

Taylor, D. R., and Finley, F. R., (1974): Multiple-Axis Prosthesis Control By Muscle Synergies. In *The Control of Upper-Extremity Prostheses and Orthoses, Proceedings of the Conference on the Control of Upper-Extremity Prostheses and Orthoses, Göteborg, Sweden, Oct. 6–8, 1971*, Herberts, P., Kadefors, R., Magnusson, R. I., and Petersén, I. (eds.), Charles C. Thomas, Springfield, Ill., pp. 181–189.

Taylor, D. R., and Wirta, R. W., (1970): Development of a myoelectrically controlled prosthetic arm. In *Advances in External Control on Human Extremities, Proceedings of the Third International Symposium on External Control of Human Extremities, Dubrovnik, Yugoslavia, Aug. 25–30*, 1969, Yugoslav Committee for Electronics and Automation (ETAN), Belgrade, Yugoslavia.

Toth, P. J., (1991). Hand Function Differentiation, Masters thesis, Department of Biomedical Engineering, Northwestern University, Evanston, Ill.

Triolo, R., Nathan, R., Yasunobu, H., Keith, M., Betz, R. R., Carroll, S., and Kantor, C. (1996). Challenges to clinical deployment of upper limb neuroprostheses. *Journal of Rehabilitation Research and Development*, vol. 33, no. 2, pp. 111–122, April.

van Lunteren A., van, Lunteren-Gerritsen, G. H. M., Stassen, H. G., and Zuithoff, M. J. (1983). A field evaluation of arm prostheses for unilateral amputees. *Prosthetics and Orthotics International*, no. 7, pp. 141–151.

Vinet, R., Lozac'h, Y., Beaudry, N., Drouin, G. (1995). Design methodology for a multifunctional hand prosthesis. *Journal of Rehabilitation Research and Development*, vol. 32, no. 4, pp. 316–324, Nov.

Vishay-Siliconix (1992). Low voltage motor drive designs using n-channel dual MOSFETs. In *Surface Mount Packages*. Application Note AN802, (http://www.vishay.com/doc?70592) Vishay Intertechnology Inc., Malvern, Pa.

Weir, R. F. ff, (1995): Direct Muscle Attachment as a Control Input for a Position Servo Prosthesis Controller. Ph.D. dissertation, Dept. of Biomedical Engineering, Northwestern University, Evanston, Ill., June.

Weir, R. F. ff., Heckathorne, C. W., and Childress, D. S. (2001). Cineplasty as a Control Input for Externally Powered Prosthetic Components. *Journal of Rehabilitation Research and Development*, vol. 38, no. 4, pp. 357–363, July/August.

Williams, T. W. (1989). Use of the Boston Elbow for High-Level Amputees. Chap. 19 in *Comprehensive Management of the Upper-Limb Amputee*, Atkins, D. J., and Meier, R. H., (eds.), Springer-Verlag, New York, pp. 211–220.

Yeadon, W. H., and Yeadon, A. W., (2001). *Handbook of Small Electric Motors*. McGraw-Hill, New York.

CHAPTER 33
DESIGN OF ARTIFICIAL LIMBS FOR LOWER EXTREMITY AMPUTEES

M. Barbara Silver-Thorn
Marquette University, Milwaukee, Wisconsin

33.1 OVERVIEW

Artificial limbs for lower extremity amputees are designed and fabricated by prosthetists. These limbs include custom interfaces between the residual limb and prosthesis (prosthetic socket) and commercial components specifically selected for the individual. Biomedical engineers have been involved in the design of some of these commercial components and in the quantitative evaluation of these prostheses and the performance (e.g., walking and running) of the amputee with his or her prosthesis. Biomedical engineers have also been involved in the development of computer-aided design (CAD) and computer-aided manufacture (CAM) of these limbs. Future opportunities for biomedical engineers include continued development and incorporation of strong, lightweight materials in lower extremity prosthetic limbs, high-technology prosthetic components that improve lower extremity amputee performance, and prosthetic sockets and interface materials that minimize risk of dermatological breakdown of residual limb tissues. Additional opportunities involve the enhancement of CAD-CAM systems and technology, assessment of lower extremity prosthetic fit, evaluation of lower extremity amputee function, and development of sensors and technology to more reliably produce comfortable sockets and optimally align prostheses.

33.2 HISTORY OF LIMB PROSTHETICS

The earliest limb amputations generally resulted in death due to blood loss or infection. The amputation was a "guillotine" operation such that all tissues were divided at the same level in one stroke of the blade. Bleeding vessels were immediately cauterized with heated irons or boiling oil. The first recorded successful amputation dates back to 484 B.C., when Hegesistratus reportedly

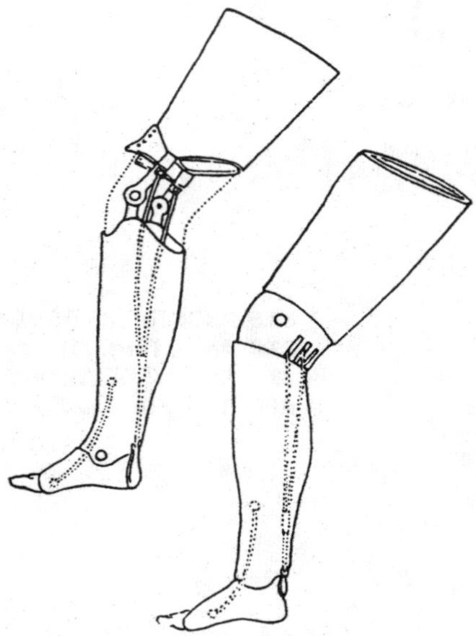

FIGURE 33.1 The Anglesey leg with articulated knee, ankle and toes. (*Adapted from Ref. 5, Fig. 1.4.*)

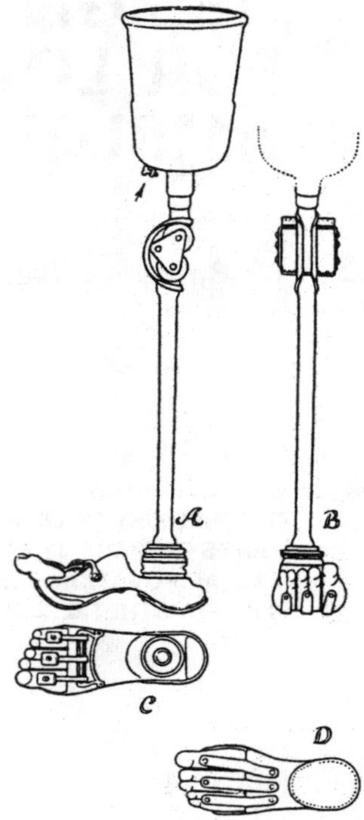

FIGURE 33.2 The Parmalee leg for transfemoral amputees incorporating a suction socket. (*Adapted from Ref. 5, Fig. 1.5.*)

escaped from prison by cutting off one of his feet. (He built himself a wooden foot to compensate for the limb loss.)

The use of anesthesia (ether—Long, 1842; chloroform—Flourens, 1847) and antiseptics (Lister, 1860) significantly improved amputation surgery. Other medical advances contributing to the reduced morbidity of amputation included the use of ligatures or sutures (originally described by Hypocrites, reintroduced by Pare in 1529) to cauterize blood vessels and the tourniquet (von Gersdorff, 1517; Morel, 1674; Faby, 1853).

In addition to resourceful amputees, early prosthetists were blacksmiths, armor makers, and other skilled artisans. Artificial limbs in Europe, and later in America, used metal, wood, and leather. Notable lower extremity prostheses include the first transtibial prosthesis with an unlocked knee, a thigh corset and external hinges introduced in 1696 by Verduyn, a Dutch surgeon and the first transfemoral prosthesis with an articulating ankle and knee, the Anglesey leg (Fig. 33.1), introduced by Potts in 1816. This latter leg was introduced in the United States in 1839. The first artificial limb shop in the United States was opened by Hanger in Richmond, Virginia, during the Civil War. Hanger is also credited with replacing the plantar-/dorsiflexion cords in the Anglesey leg with bumpers. In 1863, Parmalee introduced a transfemoral prosthesis with a suction socket, polycentric knee, and multiarticulated foot (Fig. 33.2). Other historical prosthetic developments include the introduction of the Canadian hip disarticulation prosthesis in the United States (1954, McLaurin), the solid-ankle, cushioned-heel (SACH) foot (1956), and the patellar-tendon-bearing (PTB) prosthesis (1958, University of California at Berkley). The first hydraulic knee, the Hydra-Cadence leg, was introduced by Stewart-Vickers in 1960.

As alluded to earlier, war resulted in many amputees and motivated investment in prosthetic development and training. In 1870 (after the Civil War), Congress passed a law providing artificial limbs to all persons honorably discharged from the U.S. military or navy who had lost a limb in service. This law also entitled these individuals to a new prosthesis every 5 years. In 1945 (after World War II), in response to veterans' demands for more functional prostheses, the National Academy of Sciences (NAS) initiated a study to develop design criteria to improve function for artificial limbs. This Committee on Artificial Limbs (CAL) contracted with universities, industry, and healthcare providers in this initiative. The Veterans Administration established prosthetic laboratories in New York, the University of California at Berkley (lower extremity), and the University of California at Los Angeles (upper extremity) in 1947. In 1949, the American Orthotics and Prosthetics Association developed educational criteria and examinations to certify prosthetists and orthotists. From 1947 to 1976, NAS sponsorship and support from the Veterans Administration, CAL, the Committee on Prosthetics Research and Development (CPRD), and the Committee on Prosthetic-Orthotic Education (CPOE) influenced the development of prosthetics and orthotics. (A detailed history of lower extremity amputation and prosthetics is summarized in Chapter 2 of Sanders' text.[1])

33.3 AMPUTATION SURGERY

Amputation is defined as the removal, usually by surgery, of a limb, part, or organ.[2] Its purpose is to remove dead or diseased tissue, to relieve pain, to obtain healing, and/or to rehabilitate the individual. If the amputation level is properly selected and the surgery is performed well, amputation should be considered not as a salvage procedure but as a rescue procedure that is the first step in rehabilitation.

Causes of amputation include trauma (loss of arterial supply, avulsion, and thermal injury), disease, and congenital deformity.[3] Disease categories leading to amputation include vascular or circulatory disease (e.g., arteriosclerosis, diabetes mellitus, Buerger's disease), cancer, and infection.[4] Since the United States does not have national healthcare, accurate records regarding the incidence and level of amputation are unavailable. However, random polls by the U.S. Public Health Service indicate that there are approximately 1.53 amputees per 1000 population, not including patients in institutions.[4] A 1991 U.S. National Health Survey indicates that the total number of amputees in the United States is 1.5 million, with 75 percent of these amputations (lower and upper extremity) attributed to disease, 23 percent to trauma, and 3 percent to birth defects. Most *lower* extremity amputations are performed secondary to peripheral vascular disease (Fig. 33.3), with primary incidence among the 61- to 70-year age group.[5] Lower extremity amputation is often preceded by attempts at limb salvage through revascularization procedures. Although the data are dated (pre-1975), such data indicate that 10 percent of lower extremity amputees lose their other leg within 1 year, 20 percent within 2 years, and 33 percent within 5 years.[1]

The level of amputation is selected based on the potential for healing and future function. Vascular surgeons, who may or may not have special training in prosthetic rehabilitation, perform most amputations. Orthopedic surgeons, who complete a course in prosthetics as part of their residency, perform amputations necessitated by trauma, malignancy, and other nonvascular causes.[5] Factors influencing the level of vascular amputation include the local blood supply, the nutritional state of the patient, and the probability of successful fitting. Level selection for traumatic amputations or major tumors is based on the nature of the injury and the viability of the remnant tissues.

In the management of acute and chronic ischemia, the vascular status and potential need for amputation are evaluated, followed by assessment of the level of amputation.[6] The primary determinant in assessing the level of vascular amputation is the adequacy of skin blood flow. Assessment measures include segmental blood pressure using Doppler flowmetry, transcutaneous oxygen tension, quantitative skin fluorescence, isotope clearance, and/or laser Doppler and maintenance of normal skin temperature. (Moore,[6] Chaps. 4 through 9, reviews relevant methodological details and

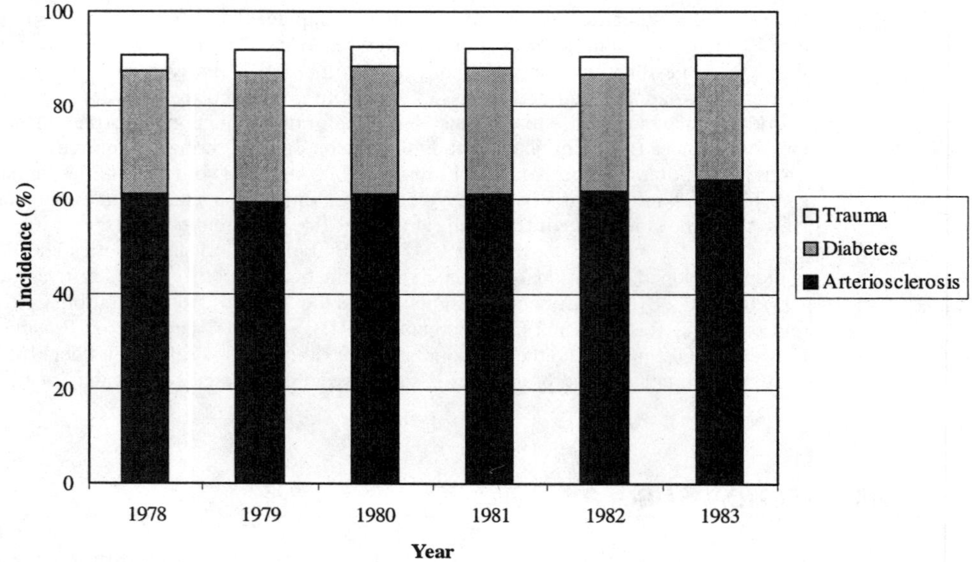

FIGURE 33.3 Trends in lower extremity amputation. These data are from Denmark, but comparable trends are expected in other developed countries. (*Adapted from Ref. 16, Table 1.1 and Figs. 1.3* and *1.4*.)

the diagnostic utility of these techniques and measures.) Amputation levels of lower extremity (Fig. 33.4) include

- Partial foot: transmetatarsal, metatarsal disarticulation (Lisfranc), disarticulation between the talus and the rest of the foot (Chopart), and transcalcaneal amputations (Pirogoff, Boyd)

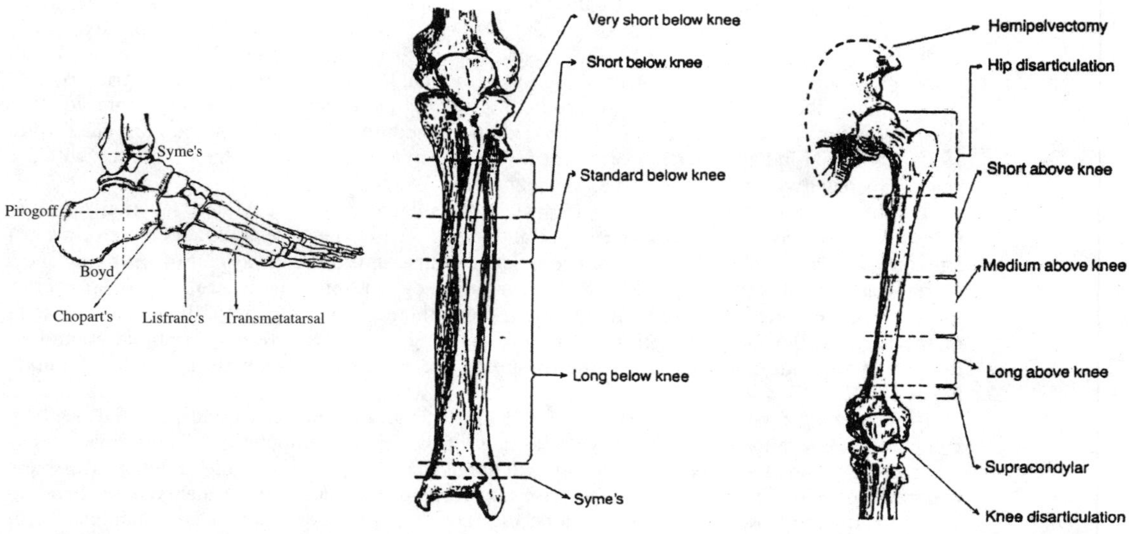

FIGURE 33.4 Lower extremity amputation levels of the foot (*left*), leg (*center*), and pelvis and femur (*right*). (*Adapted from Ref. 12.*)

- Ankle disarticulation or Symes amputation
- Transtibial or below-knee amputation
- Knee disarticulation
- Transfemoral or above-knee amputation
- Hip disarticulation
- Hemipelvectomy

The two primary goals of amputation surgery are the ablation of the diseased or traumatized tissues and reconstruction of the remnant or residual limb.[7] Generally, surgeons want to save as much length as possible while providing a residual limb that can tolerate stresses induced by the prosthesis and ambulation. To minimize the incidence of large, painful neuromas, the major nerves are pulled firmly distally, resected sharply, and allowed to retract into the soft tissue. Amputation is typically a closed technique in which skin flaps are used for primary closure. In cases of infection or when all devitalized tissue has not been removed by debridement, open amputation may be performed. Skin traction is then used to pull the skin and muscle distally over the end of the bone until the wound heals.

When a muscle is severed, it loses its distal attachment. Without attachment at both ends, a muscle is unable to function. If left loose, the muscle will retract, atrophy, and scar against adjacent structures. To improve future muscle function, surgeons have investigated both myodesis and myoplasty to secure the distal muscle. Myodesis involves attaching the distal muscle to the bone. While myodesis attempts to prevent excessive muscle shortening and resulting muscle weakness, it is not commonly practiced because it may "tether" the stump and contribute to bone spur formation. Instead, myoplasty is often performed. The muscles are regrouped about the bone to create distal padding. This procedure secures the distal muscle and improves future muscle function. It also serves to counter the tendency for development of joint contractures.

In addition to the loss of motor function of the amputated joints, the amputee is also deprived of sensory information present in the intact limb. This lack of sensory information is important in the lower extremity but is likely more important in the upper extremity, where lack of sensation is a major factor limiting the effective use of artificial hands and hooks.[3]

Because the artificial limb is attached to the residual limb by a prosthetic socket that encompasses the remnant limb tissues, there is a false joint between the skeletal system and the prosthesis. This insecure attachment often results in perceived instability and uncertainty in control of the prosthesis. The unstable attachment to the skeletal system contributes to the perception that the artificial limb is heavier than the intact limb, even though it may, in fact, be considerably lighter.[3]

The prosthetic socket is fitted over tissues that do not normally bear weight. As such, secondary problems are often observed, including edema from a socket that is too tight proximally, osteoporosis due to reduced skeletal weight bearing, allergic reactions to the socket (or socks, inserts), reduced blood flow, and development of bone spurs, cysts, and/or infections. In addition, many amputees experience phantom sensation (awareness of missing limb, often described as tingling, pressure sensation, or numbness) and/or phantom pain (cramping or squeezing sensation, shooting or burning pain in the missing extremity). Phantom sensation is frequently experienced, whereas phantom pain is reported less commonly.[8,9]

33.4 PROSTHETIC CLINIC TEAM

Following amputation surgery, the patient is often referred to physical therapy for education regarding proper limb positioning and residual limb care and to a prosthetic/amputee clinic. The clinic team typically includes a physician, a physical therapist, and a prosthetist. It may also include a social worker and/or vocational counselor. These multidisciplinary teams were first established after World War II to provide evaluation, prescription, delivery, and follow-up prosthetic services.[10]

33.4.1 Rehabilitation

The earlier the onset of rehabilitation, the greater is the potential for prosthetic success. The longer the delay, the more likely is the development of complications such as joint contractures, general debilitation, and depressed psychological state.[5] The postoperative rehabilitation program includes preprosthetic and prosthetic phases. The preprosthetic or early postoperative residual limb management includes the time between surgery and fitting with a prosthesis. As such, it involves minimizing edema, enhanced healing of the residual limb, prevention of joint contractures and other secondary complications, maintaining or regaining strength in the affected lower extremity, initiating adjustment to the loss of a body part, regaining independence in mobility and self-care, and learning proper care of the intact extremity. The prosthetic phase begins with prosthetic prescription, proper prosthetic fitting and training and follow-up in a prosthetic or amputee clinic.

Preprosthetic Treatment. A postoperative dressing is often used to protect the incision and residual limb and control edema. It may involve an immediate postoperative fitting, a rigid or semirigid dressing, a controlled environment, or a soft dressing. The use of an immediate postoperative dressing greatly limits postoperative edema, thereby reducing postoperative pain and enhancing wound healing. It allows early bipedal ambulation with the attachment of a pylon and foot and allows early fitting of a definitive prosthesis by reducing the length of time to shrink the limb. However, it requires careful application and close supervision during early healing and does not allow daily wound inspection and dressing changes.[5] Soft dressings such as an elastic wrap and/or an elastic shrinker (a socklike garment of knitted rubber-reinforced cotton) are relatively inexpensive, readily available, and washable. Elastic wraps require frequent rewrapping. Shrinkers may be used after the sutures have been removed and drainage has stopped. Semirigid dressings provide better control of edema than the elastic wrap and shrinker but may loosen with time.

The dimensions of the residual limb vary with time (Fig. 33.5). The decrease in stump size results from the reduction of edema, wasting of soft tissues from prosthetic stresses, disuse atrophy of the residual limb musculature, and decrease in fatty tissue with overall weight loss. Initial prosthetic fitting often begins 1 to 2 weeks after surgery, following early compression wrapping. The time

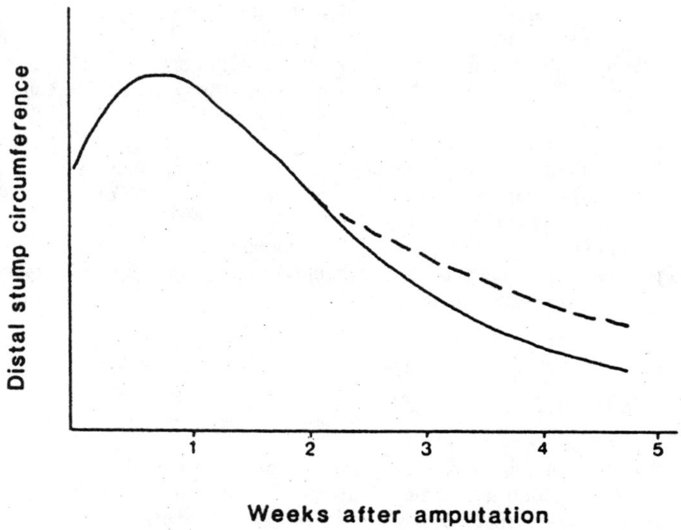

FIGURE 33.5 Residual limb circumference as a function of time. *(Adapted from Ref. 16, Fig. 9.1.)*

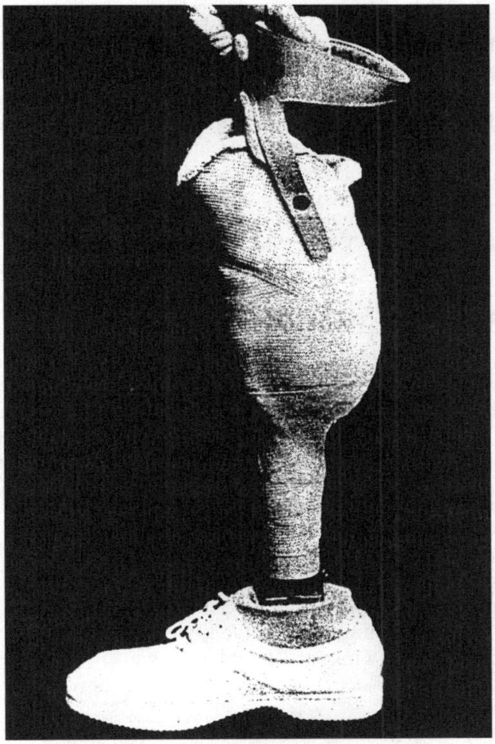

FIGURE 33.6 Preparatory transtibial prosthesis. (*Adapted from Ref. 49, Fig. 3.4.*)

between amputation and definitive prosthetic fitting is at least 6 weeks because definitive fitting requires an approximately stable limb volume for 2 to 3 weeks.[5]

Prosthetic Prescription. The director of the prosthetic clinic (e.g., orthopedist or physiatrist) typically writes the prosthetic prescription, with input from the physical therapist and prosthetist team members. *Prosthetic prescription* refers to specification of the prosthetic socket design, commercial componentry (foot, knee units), suspension, and interface materials (insert, stump socks). Prosthetic selection is influenced by the age and general condition of the patient, his or her skin and vascular status, the presence or absence of disease, and any limitations imposed by such disease.

There are two primary types of prostheses: preparatory and definitive prostheses. The purpose of a *preparatory prosthesis* (Fig. 33.6) is to allow early ambulation at an efficient and safe level and yet allow for rapid changes in limb volume that often occur in the days and weeks postoperatively. Preparatory prostheses may also include temporary prostheses or permanent prostheses that lack the cosmetic cover. Preparatory prostheses are typically endoskeletal, facilitating the interchange of components as may be necessary before finalizing the limb design.

Permanent prostheses may be either endoskeletal or exoskeletal (external support, often used synonymously with crustacean or an outer-shell design). Most permanent endoskeletal prostheses are covered by a soft cosmetic outer covering. Exoskeletal prostheses are less versatile in terms of interchangeability of components and alignment variations but may be indicated for clients with a large build or excessive loading, prior experience with such designs, or use in environments in which the cosmetic cover (foam and stocking) of an endoskeletal prosthesis would be torn or stained.

Due to the limited success of direct skeletal attachment, prosthetic sockets form the interface between the residual limb and prosthesis. The socket fits over and encloses the residual limb tissues. The more intimate the fit, the less need there is for supplemental suspension, and the more control the amputee has over the prosthesis. However, a tightly fitting socket often makes greater demands on the skin, which must be tough enough to tolerate pressure and some slippage and healthy enough to withstand the confining environment. The intimately fitting socket does not merely mirror the residual limb. Rather, its contours are designed so as to provide a comfortable and functional connection between the residual limb and the prosthesis under dynamic loading.

33.5 PROSTHESIS DESIGN

Prosthetic design involves making a replacement of a missing body part of the appropriate shape and size. The prosthesis must be comfortable, functional, and cosmetic (appearance of the prosthetic device, including visual appearance, smell, sound). The prosthesis should take into account the client's general health, weight, activity level, and motivation so as to set realistic goals. The patient's residual limb length, shape, skin condition, circulation, range of motion, and maturation should also be taken into account. Since prostheses are expensive and many insurance companies provide limited reimbursement or a single artificial limb, cost may also be a factor.

The principal function of the residual limb is to serve as the lever to power and control the prosthesis. The prosthetic socket must support the patient's body weight and hold the residual limb firmly and comfortably during all activities. As such, the prosthetic socket should be designed to support the residual limb tissues, facilitate control of the prosthesis during stance and swing, provide suspension during swing, and facilitate alignment of the artificial limb. Near-total contact between the distal limb and socket is required to aid proprioceptive feedback and prevent edema and skin problems.

Since the design of the prosthesis varies with amputation level, prosthetic design will be reviewed for the aforementioned lower extremity amputation levels. Since the primary levels of lower extremity amputation are transtibial (54 percent[11]) and transfemoral (33 percent[11]), the prostheses for these amputation levels will be presented in greater detail.

Partial Foot Amputation Prostheses. The significance of partial foot amputation includes (1) the loss of the anterior lever arm of the foot, thereby affecting the base of support and stability, (2) the functional loss of ankle dorsiflexion, (3) the tendency for the ankle to be fixed in equinus or plantarflexed, (4) a bony anterior residual limb that has poor skin coverage and is therefore difficult to fit and prone to tissue breakdown, and (5) potentially poor cosmetic replacement for prostheses that extend above the shoe. For the more distal levels of partial foot amputations (e.g., transmetatarsal, Lisfranc), prosthetic replacement is minimal, involving the use of a toe filler and arch support. The toe/shoe fillers prevent anterior migration of the foot, provide resistance to creasing of the shoe vamp, and provide shoe-size symmetry. These fillers are typically fabricated from wool, cotton, Silastic foam, synthetic rubber, Plastazote, or Kemblo. More proximal partial foot amputations may have prosthetic replacements that include a slipper or boot, extending to an ankle-foot orthosis (AFO) for Chopart, Boyd, and Pirogoff amputations [12] (Fig. 33.7).

Symes Amputation Prostheses. A Symes prosthesis must compensate for the loss of foot and ankle motion and the loss of leg length (approximately 2 in), as well as provide adequate support during stance and suspension of the prosthesis during swing. The primary difficulty with Symes prostheses is the design of a socket that accommodates the relatively bulbous distal residual limb and yet remains durable and facilitates donning. Symes prostheses include the leather-socket Symes, the posterior-opening Symes, the medial-opening Symes, and the hidden-panel Symes (Fig. 33.8).

FIGURE 33.7 Prostheses for partial foot amputation: transmetatarsal amputation or amputations resulting in an intact functional ankle (*left*), Lisfranc or amputations resulting in an ankle requiring some support (*center*), and Chopart amputation or amputations resulting in an unstable ankle (*right*). (*Adapted from Ref. 4, Fig. 4.3.*)

Transtibial (Below-Knee) Amputation Prostheses. By definition, the residual limb of a functional transtibial amputee includes the tibial tubercle into which the quadriceps tendon inserts so as to retain knee extension capability. The prosthesis for a transtibial amputee, in general, consists of a socket with an optional insert, adapter hardware to attach the socket to the shank, and an artificial foot. In addition, the prosthesis often includes some means of auxiliary suspension.

There are two primary transtibial prosthetic designs: the historic design (dating back to 1696), which incorporates a thigh corset, side joints, and an open-ended wood socket, and the patellar-tendon-bearing (PTB) design (Fig. 33.9). In the historic design, the thigh corset takes load off the residual limb, the side joints prevent knee hyperextension and provide medial-lateral stability, and

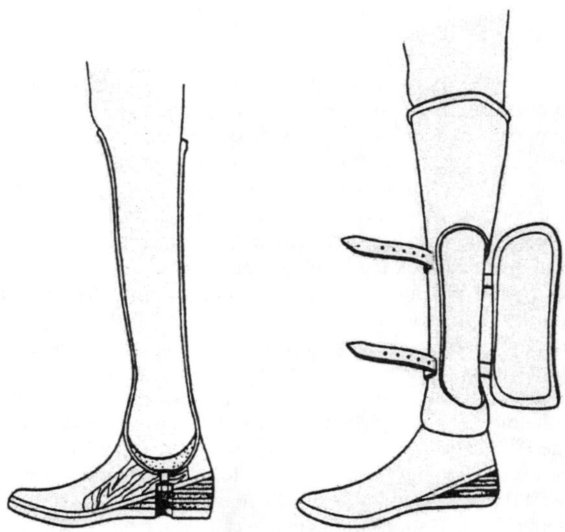

FIGURE 33.8 Hidden-panel (*left*) and medial-opening (*right*) Syme's prostheses. (*Adapted from Ref. 3, Fig. 4.3.*)

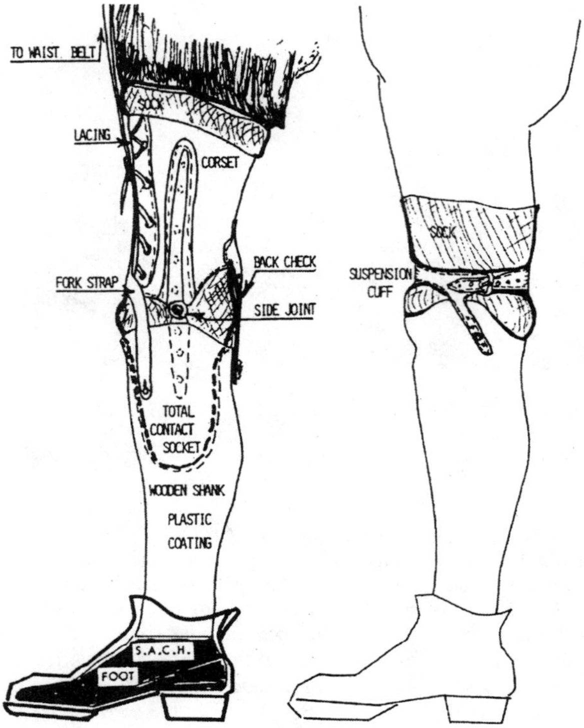

FIGURE 33.9 Transtibial prostheses: historic design (left) and patellar tendon bearing (right). (*Adapted from Ref. 10, Fig. 4.4.*)

the open-ended socket may provide a cooler environment for the residual limb. However, this historic prosthetic design is bulky, heavy, and noncosmetic and may cause distal edema due to the lack of total contact. In addition, this design contributes to thigh atrophy and may result in limb pistoning within the socket. In contrast, the PTB design is nearly total contact. As such, the PTB design may increase proprioception and decrease edema.

The soft tissues of the lower extremity residual limb are not well suited for load bearing. The load-bearing potentials of the residual limb soft tissues are not uniform and vary between individuals. Some tissues, such as the patellar tendon that is relatively avascular, can accept higher loads for a longer period of time without tissue degradation. As such, the patellar tendon is well suited to handling compressive load. The tissues over both the medial and lateral flares of the tibial condyles and shaft are also well suited for load bearing. However, the tissues covering the anterior crest of the tibia cannot assume load without tissue breakdown. Similarly, the tissues over the distal ends of the tibia and fibula cannot tolerate compressive stresses. Finally, due to the presence of the peroneal nerve below the head of the fibula, very low stresses are tolerated in this region.

The PTB socket design, initially developed at the University of California at Berkeley in the late 1950s, accommodates the nonuniform load-bearing tolerances of the residual limb. The basic concept of the PTB socket is to distribute the load over areas of the residual limb in proportion to their ability to tolerate load. Therefore, the majority of the load is to be borne on the patellar tendon (hence the name), medial and lateral flares of the tibia, and the popliteal area. The PTB socket precompresses the residual limb tissues in these load areas so that forces are transmitted comfortably and movement of the socket relative to the skeleton is minimized.[13,14] The socket is thus a replica of the residual limb, with appropriate shape modifications or rectifications such that the pressure-tolerant areas bear the majority of the load and the pressure-sensitive areas are largely relieved of load.

While the PTB socket design remains the default socket design today, several variants exist that enhance medial-lateral stability and/or socket suspension. The nominal suspension for a PTB prosthesis is the supracondylar cuff. (This design can be augmented by a waist belt and modified fork strap for improved suspension.) The PTB prosthesis is therefore indicated for individuals with good ligamentous structure or medial-lateral stability of the knee. Alternative PTB designs include suspension sleeves, the PTB-supracondylar (PTB-SC) socket, the PTB-suprapatellar (PTB/SP) socket, and the PTB-supracondylar-suprapatellar (PTB-SC/SP) socket (Fig. 33.10).

The suspension sleeve may be silicone, latex, neoprene, or elastic. Since this sleeve does not provide medial-lateral stability, it requires inherent knee stability. The sleeve provides excellent suspension and helps mask the prosthetic socket trimlines. However, the sleeve is warm and may induce excessive perspiration and contribute to dermatological problems. As such, suspension sleeves may not be indicated for vascular patients.

The medial and lateral walls of the PTB-SC design are extended proximally so as to enhance medial-lateral stability and provide suspension. This design in indicated for individuals with transtibial amputation who require increased medial-lateral stability and indicate dissatisfaction with the

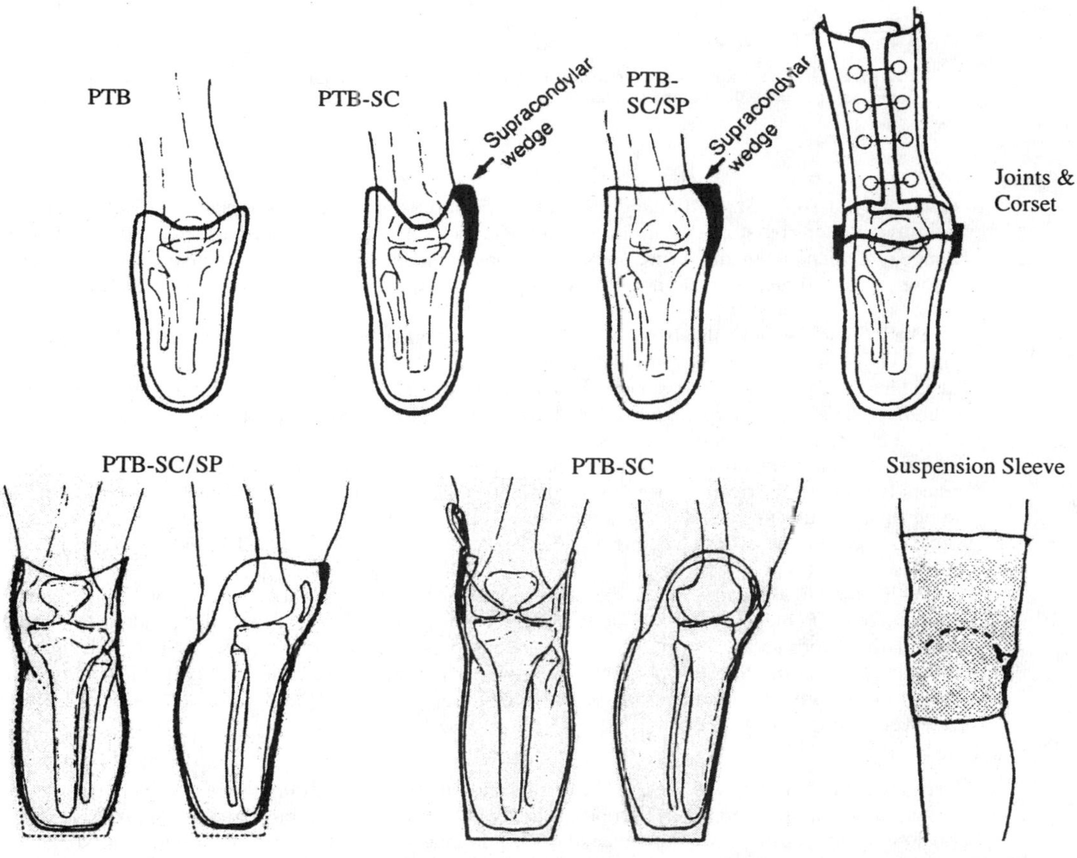

FIGURE 33.10 Alternative means of suspending transtibial prostheses: PTB, PTB-SC, PTB-SC/SP, joints and corset, PTB-SP, removable medial wedge, and suspension sleeve. (*Adapted from Ref. 49, Fig. 5.11, and Ref. 4, Fig. 4.11*).

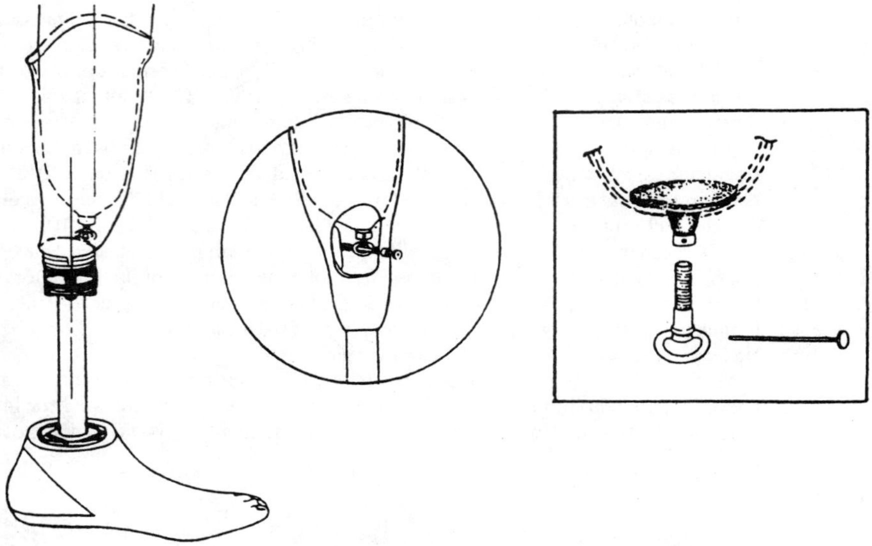

FIGURE 33.11 Suction socket for transtibial amputees. The flexible inner liner is held in place by the outer socket through the use of a locking pin. (*Adapted from Ref. 4, Fig. 4.12.*)

supracondylar strap. Since the anterior socket brimline is lowered, this socket does not provide a rigid hyperextension stop. A variation of the PTB-SC design incorporates a medial supracondylar wedge or detachable medial brim. These variations facilitate donning and suspension for individuals whose femoral dimensions do not otherwise permit donning and suspension via the supracondylar brim.

The PTB socket may also be modified to extend the anterior brimline so as to enclose the patella. This PTB-SP design does not provide inherent medial-lateral stability but provides a hyperextension stop. Finally, the PTB-SC/SP design extends the proximal socket brimlines anteriorly and medial-laterally, providing the hyperextension stop as for the PTB-SP design and the medial-lateral stability offered by the PTB-SC design.

An alternative means of suspension that has become increasingly common for transtibial amputees is suction. Suction has been used routinely for individuals with transfemoral amputation for some time. Its use in transtibial prostheses has increased through the use of a shuttle locking pin, such as incorporated in the Silcone Suction Socket (3S, Durr-Fillauer, Chattanooga, Tenn.) (Fig. 33.11).

While transtibial prostheses may be worn without an insert, such practice is less common. Silicone or foam (e.g., Pelite, Spenco, Plastazote) inserts are often used to cushion the residual limb and potentially absorb and redistribute some of the compressive and shear forces generated during ambulation. In addition, the transtibial amputee typically wears cotton, wool, or synthetic socks, 1 to 15 ply, to absorb or wick perspiration and accommodate residual limb volume fluctuations often observed throughout the day.

Knee Disarticulation Prostheses. The knee disarticulation amputation, while relatively rare in the United States (more common in Europe), is generally fitted as a long transfemoral amputation. Since this long residual limb has a bulbous end due to the retention of the femoral condyles, the socket is similar to that for a Symes amputee. The prosthesis may include a laced leather socket or anterior-opening socket with side joints and a posterior check strap.

Transfemoral Amputation Prostheses. The femur of functional transfemoral amputees must extend distal to the ischial public ramus, providing some lever arm for hip extension and flexion. The prosthesis for a transfemoral amputee includes a prosthetic foot, a shank that may be endo- or exoskeletal in design, a knee unit, a socket, and a means of suspension (Fig. 33.12).

General concepts of transfemoral socket design include (1) proper contouring to facilitate remnant muscle function, (2) stabilized force application so as to apply load to the skeletal structures as much as possible, (3) stretching the hip muscles for improved functionality (i.e., length-tension relationship for muscle[15]), (4) maximized contact area so as to minimize soft tissue pressures, and (5) adequate contact between the distal limb and socket walls so as to prevent edema.

There are three common types of sockets for transfemoral amputees. Plug fit (wood) is the more historic design in which the conical socket interior mirrors that of the residual limb. This design was replaced by the quadrilateral socket,[1] designed by Radcliffe and Foort at University of California at Berkeley in the early 1960s (Fig. 33.13). This socket design provides total contact but not end bearing. This socket design reportedly allows better muscle control and limb comfort by contouring the socket walls and adducting and flexing the femur within the socket. The adducted lateral wall provides support for the femoral shaft and allows the hip abductors to contract effectively during gait. The anterior wall pushes the residual limb onto the posterior ischial seat, whereas the posterior wall provides a major area for weight bearing on the ischial tuberosity and gluteus maximus. The socket also serves to support the femoral shaft so that the hip extensors can contract effectively. The quadrilateral socket design has since evolved to the ischial-containment socket design.

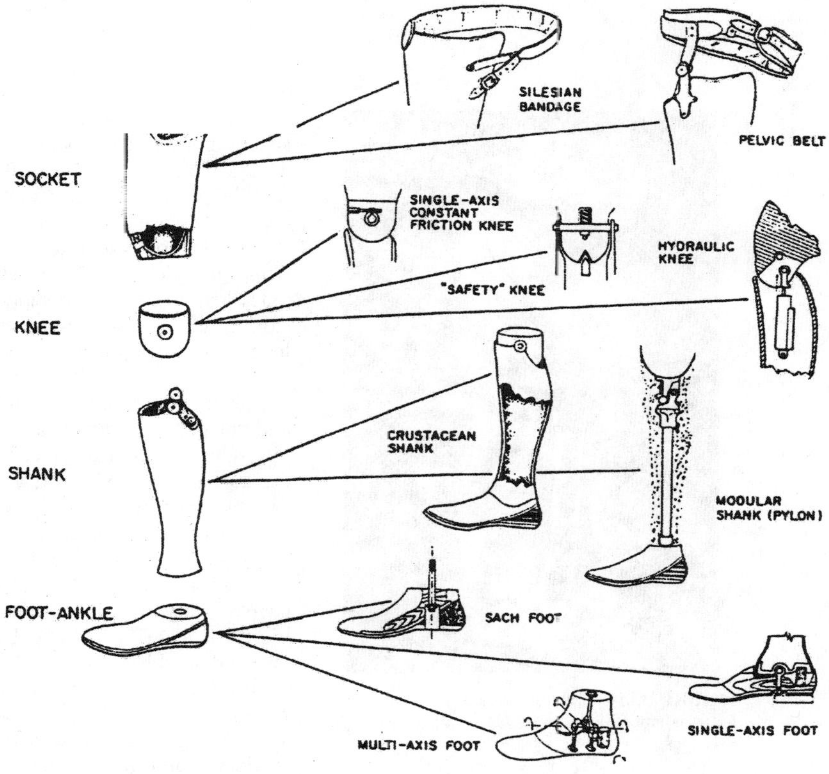

FIGURE 33.12 Prosthetic prescription options for individuals with transfemoral amputation. (*Adapted from Ref. 4, Fig. 4.14.*)

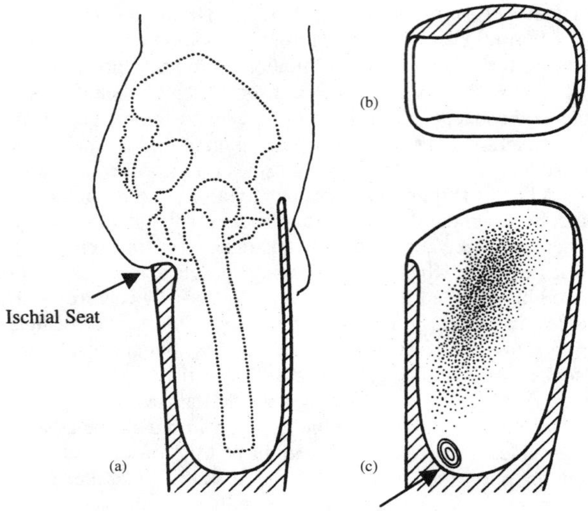

FIGURE 33.13 Quadrilateral ischial seat transfemoral suction socket with total contact: lateral view (*a*), cross section (*b*), and suction valve on medial aspect (*c*). (*Adapted from Ref. 50, Fig. 43.*)

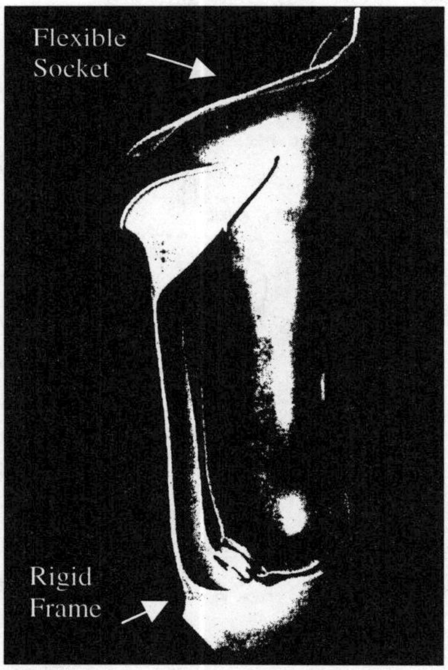

FIGURE 33.14 Flexible socket and rigid frame for transfemoral amputees. (*Adapted from Ref. 16, Fig. 18.1.*)

The ischial-containment socket [e.g., normal shape, normal alignment (NSNA), contoured adducted trochanter controlled-alignment method (CAT-CAM), skeletal CAT-CAM (SCAT-CAM)] contains the ischium within the socket and attempts to provide a bony lock to maintain femoral geometry within the socket. This socket design reportedly provides better mechanical support of the pelvis by restoring the normal pelvic/femoral angle. The ischial lock acts to prevent pelvic shift. In contrast to the quadrilateral socket, the ischial-containment socket is narrower in the medial-lateral dimension, wider in the anterior-posterior dimension, and of course, contains the ischium within the socket. However, this socket design has increased cost, and the fitting and fabrication procedures are not as well documented.

Transfemoral sockets are typically rigid. Flexible sockets fabricated using a malleable thermoplastic [low-density polyethylene and Surlyn (Thermovac)] have recently been incorporated within a rigid or semirigid frame[16] (Fig. 33.14). The potential advantages of these flexible sockets include increased perceived comfort due to the flexible walls, improved proprioception, accommodation of minor volume changes, and enhanced suction suspension.

Suspension of the transfemoral socket and prosthesis is typically achieved through suction (expul-

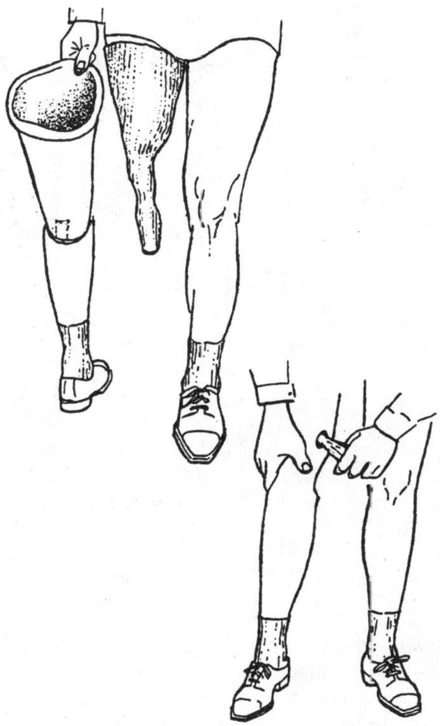

FIGURE 33.15 Common donning procedure for a transfemoral prosthesis incorporating a suction (expulsion valve) socket. (*Adapted from Ref. 4, Fig. 6.1.*)

sion valve or roll-on gel liner with shuttle lock), a total elastic suspension (TES) belt, a Silesian belt, or a hip joint/pelvic band/waist belt. In rare instances, suspenders may also be used. Suction is believed to reduce pistoning of the residual limb within the socket, improve proprioception, provide total contact fit, and result in a lighter prosthesis. However, suction requires that the residual limb maintain constant limb volume and shape. Suction sockets are also hot and, for expulsion systems, may be difficult to don. (As seen in Fig. 33.15, socket donning for an expulsion-valve system requires that a stockinet or nylon stocking be pulled over the residual limb. The residual limb is inserted into the socket. The soft tissues are drawn into the socket by pulling the stockinet out through the valve opening before closing the valve. No sock is worn between the skin and the socket.) The TES belt is a wide neoprene band lined with nylon tricot that can be used with suction or as a primary suspension mechanism. The Silesian belt is typically an auxiliary means of suspension, often augmenting suction. The Silesian belt provides rotational stability but no medial-lateral stability. In contrast, the hip joint/pelvic band/waist belt provides medial-lateral stability and is easy to don, although it is quite cumbersome.

Hip Disarticulation Amputation Prostheses. The Canadian hip disarticulation prosthesis, introduced in the United States in 1954, is still used almost universally today. It consists of a foot, a shank, a prosthetic knee, a "thigh," a hip joint/bumper/control strap, and a socket (Fig. 33.16). The hip disarticulation socket is essentially a bucket, providing a seat for the ischial tuberosity, medial-lateral stability, suspension, and support for weight bearing.

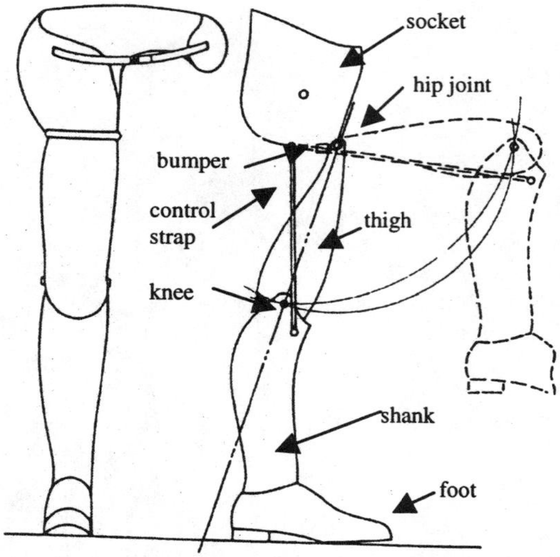

FIGURE 33.16 Alignment principle of the hip disarticulation prosthesis. (*Adapted from Ref. 12.*)

Hemipelvectomy Prostheses. The prosthesis for an individual with a hemipelvectomy is similar to that of the hip disarticulation prosthesis. However, the residuum of an individual with a hemipelvectomy no longer contains the ischial tuberosity. The socket must therefore provide distal support solely through the soft tissues. To minimize soft tissue pressures, the socket may extend more proximally using the iliac crest, distal ribs, and gluteal muscles for support. Suspension may be augmented by shoulder straps. This prosthesis is typically endoskeletal so as to minimize the weight of the prosthesis.

Although the modified Canadian-type hip disarticulation prosthesis can be successfully fitted, many hemipelvectomy amputees prefer crutches for speed and agility in ambulation and use the prosthesis for cosmetic purposes on special occasions.[1]

33.5.1 Prosthetic Socket Fabrication

Lower extremity prostheses are custom devices. While the connective componentry, joints (knees, feet/ankles), and suspension are commercially available, the socket that encases the residual limb is custom-designed. The socket design may or may not involve a computer, as in CAD-CAM.

Many prosthetic facilities now use central fabrication to allow use of complex and expensive CAD-CAM technology without the need for each facility to purchase and maintain such equipment. With central fabrication, patient evaluation, casting, fitting, and delivery continue to be conducted in the prosthetist's office. The fabrication of the socket itself is done off-site and minimizes noise, dust, odors, and potential zoning law difficulties. As such, there is less need to hire technicians and theoretically more efficient use of the practitioner's time. Disadvantages of central fabrication include potential communication problems between the prosthetic and central fabrication facilities, shipping delays, and quality control.[3,7]

PTB Socket for Transtibial Amputees. Prosthetic socket fabrication for transtibial amputees includes some means of residual limb shape sensing (casting or digitizing), socket design (plaster positive or digital representation), and manufacture (vacuum forming or lamination). Specific methodologies for both hand-rectification and CAD-CAM techniques are detailed below.

Sockets for transtibial amputees were originally carved from wooden blocks.[17] However, very few prosthetists use this medium today, and exceedingly few amputees request wooden sockets. Most transtibial sockets are fabricated using thermoplastics (e.g., polypropylene, copolymer) or laminated resins (e.g., acrylic, epoxy, polyester).

Hand-rectified PTB sockets for transtibial amputees involve donning a stockinet over the residual limb, identifying anatomic landmarks (patellar ligament, tibial flares, fibular head) with an ink pen, and casting the residual limb with plaster bandages. This cast is then removed and filled with plaster to create a positive model of the residual limb. The inked bony landmarks are thus transferred to this plaster model. The plaster model is then modified or rectified such that plaster is built up over areas of intended socket relief. A rasp is used to remove plaster over regions of preferential loading. The socket is then vacuum-formed or laminated over this modified plaster positive.

The plaster casting and rectification procedures in current prosthetic practice require skill and experience, as well as considerable time. In an attempt to eliminate factors relating to manual dexterity both in the casting and rectification processes, Murdoch[18,19] developed the Dundee socket in which hydrostatic pressure is used during casting. Unlike the PTB socket design, the entire surface of the residual limb theoretically bears the same load with this design. A similar procedure was recently documented by Lee et al.[20] for use in developing countries.

Successful hand-rectified socket designs cannot be easily duplicated. In addition, such designs do not facilitate documentation of residual limb shape over time. The introduction of CAD-CAM into prosthetics was motivated by the desire to improve the accuracy and quality of socket design, as well as reduce the amount of time for socket manufacture.

CAD-CAM for transtibial prosthetic sockets generally emulates the procedures used in conventional hand-rectified PTB socket fabrication. Starting with a replica of the residual limb, changes are made such that the majority of weight bearing will be borne on the pressure-tolerant areas of the residual limb. As such, the first step of CAD involves obtaining a digital representation of the residual limb, a process known as *shape sensing*. Computer software is then used to modify the digital socket, generally mimicking the rectification procedure of the plaster positive. The final stage, CAM, involves transfer of the proposed socket design to a computerized milling machine so that either a positive mold suitable for vacuum forming is produced or a socket is directly fabricated.

Both contact and noncontact methods of shape sensing of the residual limb have been developed. In general, contact methods are susceptible to errors due to soft tissue deformation. Current commercial shape-sensing technologies include contact techniques such as casting with subsequent cast digitization [Provel d1L ShapeMaker, Seattle Limb Systems (Model+Instrument Development Corporation), Poulsbo, Wash.; CMM and Compression Probe, Active Life Sciences, Troy, Mich.; CANFIT-PLUS, Vorum Research Corporation, Vancouver, British Columbia, Canada], and direct limb shape sensing (TracerCAD, Tracer Corporation, Miami, Fla.; Virtual Casting, BioSculptor Corporation, Coral Gables, Fla.). Noncontact methods typically involve optical or laser scanning techniques [ShapeMaker 3000, Seattle Limb Systems (Model+Instrument Development Corporation), Poulsbo, Wash.; Delta Systems II, Clynch Technologies, Inc., Calgary, Alberta, Canada; CAPOD Systems, CAPOD Systems, South Lyon, Mich.].

The design of the socket is implemented using CAD. In CAD, the sculpting tools of conventional socket rectification are replaced by a computer, graphics terminal, mouse, and on-screen cursor. Modifications typically include scaling, addition or removal of ply, patching or local rectification, and smoothing. Quantification of the rectifications allows direct comparison between various socket designs. In addition, the permanent record of the residual limb geometry provides the capability of monitoring residual limb shape over time (local atrophy or edema, volume changes), as well as the capability of producing duplicate sockets.

As stated previously, CAM of transtibial sockets typically involves use of a computer numerically controlled (CNC) milling machine to carve the equivalent of the positive plaster socket mold. Fabrication of the socket is again performed by vacuum forming or laminating the socket over this mold. A recent development in CAM of prosthetic sockets that obviates the plaster positive is direct fabrication of the prosthetic socket using rapid prototyping techniques (e.g., SQUIRT Shape, Northwestern University Prosthetics Research Laboratory and Rehabilitation Engineering Research Program).[21] Such methods have resulted in successful fitting but are limited to some extent by the materials available for deposition and stereolithography.

Regardless of the rectification and fabrication technique, the prosthetic socket is then trimmed proximally by hand to either allow complete or limited motion of the knee.

Socket Fabrication for Transfemoral Amputees. Fabrication of the transfemoral socket is similar to that for a transtibial socket. Information regarding the residual limb shape is obtained using casts. A Berkeley or Hosmer brim frame is used to shape the proximal socket and provide the ischial seat for quadrilateral sockets. If the patient is unable to stand during casting, CAD-CAM techniques may be used. For transfemoral amputee sockets, CAD involves residual limb shape sensing based on discrete measures (e.g., limb length, limb circumference at specific levels, select anterior-posterior and medial-lateral dimensions). The socket design is then based on scaling of an appropriate socket template. As before, the final socket is vacuum formed or laminated over a plaster positive.

33.5.2 Prosthetic Componentry

The transtibial prosthesis incorporates one of the aforementioned prosthetic socket designs and suspension methods, a spacer (shank), and an artificial foot. The transfemoral prosthesis is similar, with the inclusion of a prosthetic knee unit and an additional spacer for the thigh. The spacer is usually made of wood, plastic, or metal depending on whether the design is exo- or endoskeletal.

Prosthetic Feet. With the exception of partial foot amputees, the prostheses for all lower extremity amputees require a prosthetic foot. The prescription criteria for these feet take into consideration the amputation level, residual limb length, subject activity level, cosmetic needs, and the weight of the individual. Prosthetic feet range from the SACH (solid ankle cushioned heel) foot, which is relatively simple and inexpensive, to dynamic-response or energy-storing feet that are more complicated and considerably more costly. Note that prosthetic feet are often foot and ankle complexes. As such, prosthetic feet may replace plantarflexion/dorsiflexion, pronation/supination, and inversion/eversion. Prosthetic feet are typically categorized in terms of the function(s) they provide or replace and whether or not they are articulated.

Nonarticulated feet are typically the simplest and least expensive. The foot and ankle are combined in a single component, and shock absorption and ankle motion are provided by the materials and structure of the foot. Since these feet are nonarticulated, they are quiet and typically require little maintenance. These feet are also cosmetic, lightweight, and provide good shock absorption and limited inversion/eversion on uneven terrain. Disadvantages of nonarticulated feet are the limited range of plantarflexion/dorsiflexion, difficulty with inclines due to heel compression, lack of adjustability for different heel heights, and little torque-absorption capability.

The SACH foot is the most common nonarticulating foot. As the name implies, this foot contains a rigid wooden keel with a compliant or flexible heel and forefoot (Fig. 33.17). The heel wedge of the SACH foot compresses to emulate plantarflexion; the forefoot flexes to emulate dorsiflexion. This foot is contraindicated for active amputees and amputees who require torque-absorption and/or inversion/eversion capabilities.

Single-axis feet/ankles are articulated to allow plantarflexion and dorsiflexion (see Fig. 33.17). The range of motion is maintained by bumpers or stops and typically ranges from 15 degrees of plantarflexion to 6 degrees of dorsiflexion. The amount of plantarflexion and dorsiflexion can be adjusted by changing the durometer of the bumpers. These feet do not permit inversion/eversion or transverse rotation. As such, ambulation on uneven terrain with these feet results in transmission of torques and shears to the residual limb. This type of articulated foot is typically heavier and less cosmetic than the SACH foot, and the moving parts may become noisy with use and may necessitate bumper replacement.

Multiaxis feet incorporate a "fully" functional ankle (e.g., Greissinger, Otto Bock, Minneapolis, Minn.; College Park, College Park Industries, Inc., Fraser, Mich.; Genesis II, MICA Corp., Kelso,

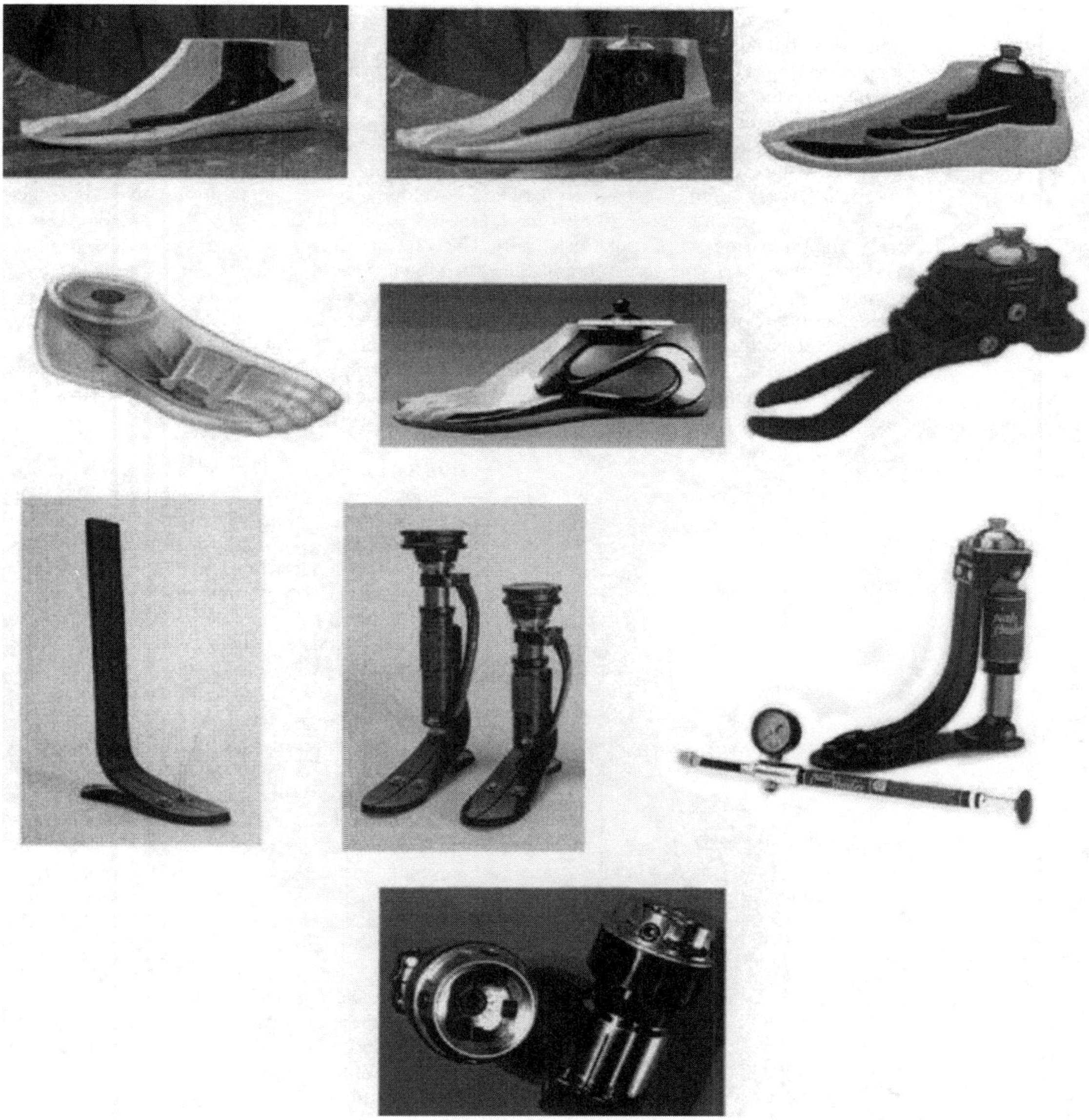

FIGURE 33.17 Prosthetic feet and ankle units: SACH, single-axis, Carbon Copy HP, STEN flexible keel foot, 1D25 dynamic plus foot, TruStep, Modular III, Re-Flex VSP, and Pathfinder. Also shown is a torsion adapter.

Wash.). Degrees of freedom include plantarflexion/dorsiflexion, inversion/eversion, and transverse rotation and may be invoked with mechanical joints and/or composite structures (see Fig. 33.17). While such feet may be bulky, heavy, noisy, and expensive and may require frequent repair, they accommodate uneven terrain and therefore reduce shear and torsional forces that might otherwise be transferred to the residual limb.

While rotators are not explicitly a foot or an ankle, they act to reduce the torque/shear forces on the residual limb due to ambulation over uneven terrain by allowing the socket to rotate independent of foot position. This component may be positioned anywhere in the prosthesis, not just at the ankle, where inertial effects may be problematic (see Fig. 33.17). Since this component adds mass and complexity to the prosthesis, its inclusion may necessitate improved or auxiliary suspension. Another component that is again not a foot or an ankle is shock absorbers. As for the rotators, shock absorbers, which decrease the effective length of the shank during loading, are typically positioned in the shank.

One of the most common areas of prosthetic development is recent years involves dynamic-response or energy-storing feet (e.g., Flex Foot, OSSUR/Flex Foot, Inc., Aliso Viejo, Calif.; STEN, Kingsley Manufacturing Co., Costa Mesa, Calif.; Carbon Copy II, Ohio Willow Wood, Mt. Sterling, Ohio) (see Fig. 33.17). This research and the exceptional performance of disabled athletes using such technology have motivated product development and research assessing the utility (energy-storing capabilities, optimality of energy release) of such technology.[22–36] These feet store energy via heel/forefoot compression during stance, with partial energy release during heel-off and/or toe-off. Some of these feet are multiaxial, accommodating inversion/eversion and rotation. The advantage of these feet include potentially reduced energy expenditure and decreased mass. However, such feet are expensive and may not accommodate children and large adults.

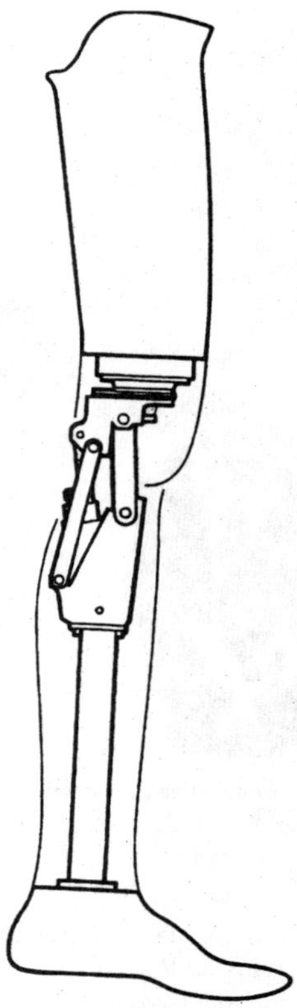

FIGURE 33.18 Four-bar polycentric knee in a transfemoral prosthesis. (*Adapted from Ref. 50, Fig. 36.*)

Prosthetic Knees. During gait, the knee acts as a shock absorber and a support structure and shortens the limb during swing. The purpose of prosthetic knee units is to restore normal function (i.e., quadriceps and hamstring function[37]) and the appearance of gait with minimal energy expenditure. Factors influencing prosthetic prescription include the client's ability to voluntarily control the knee during stance based on their hip musculature and residual limb length and the inherent stability of the knee unit itself. Prosthetic knee units may be classified in terms of their control, such as no mechanical knee control (locked knee), stance phase control only, swing phase control only, and swing and stance phase control.

Stance-phase-control knees include (1) manual locking knees that prevent knee motion until the locking mechanism is manually disengaged, (2) knee units that are alignment controlled such that knee axis is positioned posterior to the weight line from heel strike to midstance such that an inherently stable knee extension moment is provided, (3) friction brakes that "lock" the knee on weight bearing, (4) polycentric linkages, and (5) fluid-resistive devices. A knee unit that is inherently stable requires little voluntary control of the hip musculature for function. Manual-locking knees provide the most inherent stability, followed by many polycentric knees (e.g., Stability Knee, DAW Industries, San Diego, Calif.), weight-activated friction knees, constant-friction knees, and lastly, outside hinges. The manual-locking knee, which can be unlocked for sitting, provides maximum stability during stance but results in an unnatural gait. Polycentric knees (typically four-bar linkages; Fig. 33.18), by definition, have a moving center of rotation that may result in an extremely stable knee. This design is relatively compact, thereby minimizing leg-length discrepancies for knee disarticulation and long transfemoral amputees. Weight-activated stance-control knees (e.g., Stabilized Knee, Ohio Willow Wood, Mt. Sterling, Ohio) function as a single-axis knee during swing and a braked knee during stance. Most weight-activated knees lock as axial load is applied following foot contact and remain locked until the weight has shifted to the sound limb after heel-off, just prior to toe-off. Such knees can lock at 20 to 30 degrees of flexion and may be used in conjunction with a knee-extension assist to initiate swing. Problems such as increased noise, frequent maintenance, and difficulty in jackknifing on stairs have been noted with these knee units. Conventional constant-friction knees are durable, easy to maintain and repair, and relatively inexpensive, but they provide little inherent knee stability. Thus these knees require good voluntary control of the hip musculature. Outside hinges may be used for individuals with knee disarticulation amputation because they do not add length to the femoral section of the prosthesis. They provide little inherent knee stability (but the very long residual limb of a knee disarticulation amputee typically has good hip musculature), with any stability provided via alignment.

Swing-phase-control knees may have constant resistance, variable resistance, or cadence-responsive resistance during swing.[7] The resistance of a constant-resistance swing-phase-control knee, as the name implies, does not vary during swing, regardless of knee angle or cadence. The amount of resistance can be preset by the prosthetist. The resistance of variable-resistance swing-phase-control knees varies as a function of knee flexion angle. In contrast, the resistance of cadence-responsive resistance swing-phase-control knees varies (multiple preset options) as a function of knee angular velocity.

Knee-resistance mechanisms include sliding mechanical friction and fluid mechanisms. The sliding mechanical friction is contact friction, usually applied by a clamp around a knee bolt. It is relatively simple and inexpensive, but since the friction does not vary with cadence, an amputee wearing a prosthesis incorporating this type of knee unit is able to walk at only one cadence with optimal security and ease. The fluid mechanisms consist of a fluid-filled cylinder joined to the knee bolt by a piston, typically located in the posterior aspect of the shank. The resistance to knee flexion during swing is produced as the piston in the cylinder pushes against air or oil. The resistance to swing triggers a knee extension bias that then assists the prosthetic knee into extension. All stance phase control is achieved by alignment and muscle contraction of the hip extensors during stance.

The fluid in these cadence-responsive knee units may be oil (hydraulic) or air (pneumatic). For hydraulic knees, the fluid is incompressible. The resistance to piston motion results from fluid flow through one or more orifices. As such, the resistance is dependent on the fluid viscosity and density, the size and smoothness of the channel, and the speed of movement. In contrast, for pneumatic knees, the fluid is compressible. The resistance is again due to fluid flow through the orifice(s) but is also influenced by fluid compression. Since air is a gas, potential leaks in pneumatic knee units will not result in soiled clothing, unlike what may occur with hydraulic knees. In addition, since air is less dense than oil, pneumatic units tend to be lighter than hydraulic units. However, since air is less dense and less viscous than oil, pneumatic units provide less cadence control than hydraulic units. Note that since viscosity is influenced by temperature, hydraulic (and pneumatic) knee units may perform differently inside and outside in cold weather climates. An example of a hydraulic cadence-responsive knee unit is the Black Max (USMC, Pasadena, Calif.). Additional examples include the Spectrum Ex (pneumatic, Hosmer, Campbell, Calif.), Pendulum (pneumatic, Ohio Willow Wood, Mt. Sterling, Ohio), and Total Knee (hydraulic, Model 2000, Century XXII Innovations, Jackson, Mich.), which combine a cadence-responsive resistance swing-phase-control knee with a four-bar polycentric stance control knee.

33.5.3 Prosthetic Fit

The prosthetist also evaluates prosthetic fit, or the comfort (pressure distribution), stability, suspension, alignment, and function of the prosthesis. The fitting of a prosthesis is an empirical process. The prosthetist has no quantitative information regarding the load distribution of the soft tissues and must rely on experience, feedback from the amputee, and indirect indications of tissue loading to assess socket/prosthesis fit. To assist this process, the prosthetist often fits a test or check socket. This test socket is transparent and is typically fabricated by vacuum forming a polycarbonate sheet over the plaster positive model. After the amputee dons the test socket, the prosthetist provides distal resistance, or the amputee stands in an alignment rig. Skin blanching during loading and subsequent redness (reactive hyperemia) on removal of the prosthesis are used to identify areas of excessive pressure. If a prosthetic sock is worn, areas of excessive pressure are indicated by the compression of the prosthetic sock or sock weave impressions on the residual limb. Other means of estimating tissue loading include the use of chalk, talcum powder, clay, or lipstick to assess distal end bearing and the presence of calluses on the residual limb. These methods provide some basis on which to qualitatively assess comfort and the weight-bearing pressure distribution.

Fitting also includes assessment of prosthetic suspension, the adequacy of the proximal trim lines in terms of comfort and the associated mobility of the proximal joint, and the alignment of the prosthesis itself (see Sec. 33.5.4).

As alluded to earlier, knowledge of the interface stress distribution between the residual limb and the prosthetic socket enables objective evaluation of prosthetic fit. It is this desire for quantitative description of the prosthetic interface stress distribution that has motivated many experimental and numerical investigations of prosthetic interface stress.

Several groups have investigated the stress distribution between the residual limb and prosthetic socket for both transtibial and transfemoral amputees in laboratory and clinical settings. Investigation of the effects of prosthetic alignment, relative weightbearing, muscle contraction, socket liners, and suspension mechanisms on the interface pressure distribution have been conducted (for review, see ref. 38). These experimental stress measures have been limited to specific sites on the limb, since measurements can only be obtained at transducer locations. Comparison of the results of these investigations is difficult because the interface pressure measures are highly dependent on both the means of measurement (transducer type) and the transducer calibration method employed. Ferguson-Pell[39,40] has reviewed several key factors in transducer use and selection relevant to stress measures between human soft tissues and a support structure.

Many of the interface pressure-measurement techniques involve research applications and methodologies that are not appropriate for clinical use. However, the relatively recent development of commercial systems using force-sensitive resistors and capacitive sensors to measure interface loading for seating systems and prosthetic sockets provides a diagnostic tool that can be incorporated into prosthetic fitting. These systems have clinical potential and may facilitate creation of prosthetic databases such that interface pressures, whether measured in research settings or clinical settings or estimated with computer models, may be properly interpreted.

Polliack et al.[41] recently compared the accuracy, hysteresis, drift, and effect of curvature on sensor performance for these commercial prosthetic systems (force sensitive resistors: F-Socket Measurement System, Tekscan, Inc., Boston, Mass.; and Socket Fitting System, Rincoe, Golden, Colo; prototype capacitive sensor, Novel Electronic, Minneapolis, Minn.). These authors concluded that the current systems were more appropriate for static use because the hysteresis, drift, and large standard deviations become more problematic during dynamic and long-term loading.

In contrast to the experimental techniques, computer models of the residual limb and prosthetic socket have the potential to estimate interface pressures for the entire residuum and indeed are not limited to the interface but can also provide information regarding the subcutaneous tissue stresses. Nola and Vistnes[42] and Daniel et al.[43] have found that initial pathological changes in pressure sore formation occur in the muscle directly overlying the bone and then spread outward toward the skin. Therefore, the subcutaneous stresses may be important when evaluating long-term prosthetic success. These subcutaneous stresses are particularly difficult to measure in vivo.

Several groups have used computer models of the residual limb to investigate the residual limb–prosthetic socket interface.[44,45] Many investigators have also used finite-element modeling of the residual limb and the prosthetic socket of lower extremity amputees to investigate residual limb–prosthetic socket biomechanics and to estimate the interface stress distribution (for review, see refs. 38, 46, and 47).

Two primary limitations of these modeling efforts involve the representation of tissue properties across the entire limb and the interface condition between the residual limb and prosthesis. The ability of current finite-element models to estimate prosthetic interface stresses, while performing reasonably well in some cases, has not been highly accurate. Nevertheless, the methodology has potential. Advances in finite-element software enabling nonlinear elastomeric formulations of bulk soft tissue, contact analysis, and dynamic analysis may help address some of the current model limitations. Corresponding advances in pressure-transducer technology will help validate the computer models and facilitate interpretation of the analyses.

Finally, finite-element models have potential applicability in CAD of prosthetic sockets. Current prosthetic CAD systems emulate the hand-rectification process, whereby the socket geometry is manipulated to control the force distribution on the residual limb. Incorporation of the finite-element technique into future CAD would enable prescription of the desired interface stress distribution (i.e., based on tissue tolerance). The CAD would then compute the shape of the new socket that would theoretically yield this optimal load distribution. In this manner, prosthetic design would be directly based on the residual limb–prosthetic socket interface stresses.

33.5.4 Prosthetic Alignment

Prosthetic alignment refers to the orientation of the various components of a lower extremity prosthesis such that the user has optimal physical security, the best possible gait, minimum energy expenditure during walking, and a generally comfortable leg, resulting in good posture without injury to the residual limb even when used for comparatively long periods of time. Good alignment begins with proper fit of the residual limb in the socket (see Sec. 33.5.3). The adapter hardware used to connect the socket to the endoskeletal components, and to the foot and knee, facilitates alignment changes.

Alignment of the prosthesis consists of static and dynamic stages. For transtibial amputees, static alignment places the knee in approximately 5 degrees of flexion to prevent hyperextension and increase pressures on the anterior surface of the limb.[13] The pylon is vertical, and alignment in the sagittal plane is such that a plumb line at the center of the greater trochanter intersects the medial-lateral axis of the knee and aligns with the anterior surface of the pylon so that there is no tendency for the knee to buckle during stance. Additional modifications may be invoked during dynamic alignment based on observation of amputee gait in the frontal and sagittal planes. (Sanders,[1] Chap. 20, reviews static and dynamic alignment of the transtibial prosthesis and identifies malalignment symptoms.)

Zahedi and Spence[52] found that a transtibial amputee can adapt to several alignments, ranging from as much as 148-mm shifts and 17-degree tilts. This tolerance of alignment variability was attributed to the degree of control that an amputee has over the prosthesis. For a transtibial amputee, for example, the retention of the knee joint allows the body to compensate more readily to malalignments. In addition, Zahedi and Spence[52] hypothesized that the nonuniform distribution of alignment data about the mean was indicative of an optimum alignment configuration not readily achieved by current alignment procedures.

For transfemoral amputees, the socket is mounted on an alignment rig. The desired height of the prosthesis is determined, as is the relative position of the knee and foot with respect to the socket/limb. This bench alignment is such that the center of the heel falls just under the ischial seat, with the trochanter-knee-ankle (TKA) line passing through the knee and ankle axes of rotation. As for alignment of transtibial prostheses, this bench alignment is modified based on static and dynamic observations of amputee gait in the frontal and sagittal planes.

In general, prosthetic socket fit and alignment are dependent. Prosthetic alignment depends on the length of the residual limb (lever arm), strength of the remnant musculature, and the amputee's balance and control. For the hip disarticulation prosthesis, alignment *defines* prosthetic stability because there are no knee extensors to prevent the knee from buckling during stance. For stability, the weight line during stance must go through the base of support, be posterior to the hip so as to provide an inherently stable hip extension moment, and be anterior to the knee so as to provide a knee extension moment. As such, the prosthetic hip joint is located distal and anterior to the axis of the anatomic hip (see Fig. 33.16). (The position of the hip joint also determines the length of the thigh when sitting.) The posterior bumper, which controls the amount of hip extension, is mounted well forward of the weight line so as to produce a hip extension moment during stance. The stride-length control strap is positioned posterior to the hip and anterior to the knee. This strap is a critical feature of the hip disarticulation prosthesis, acting as the quadriceps muscle—preventing excess hip extension and excess knee flexion while assisting in knee extension.

33.6 AMPUTEE GAIT

Since one of the goals of lower extremity prosthetics is to replace function, much attention is given to the restoration of gait or ambulation. While some argue that optimal gait for a lower extremity amputee need not be symmetric, symmetry in ambulation is cosmetic. As such, the prosthetist and physical therapists attempt to restore normal, symmetric gait—given constraints such as joint contractures, weak hip/knee musculature, poor balance, and the potential need for an assistive device.

Another primary factor that influences ambulation and prosthetic use is tissue pain. The residual limb soft tissues are routinely loaded during ambulation. As such, these tissues are stressed, unlike prior to amputation. These stresses are the direct result of load transfer from the prosthetic foot through the residual limb soft tissues and subsequently through the skeleton. Thus, while symmetric gait may be desired, tissue sensitivity and pain due to the presence of neuromas and local stress concentrations may mandate altered gait.

The residual limb–prosthetic socket stresses are influenced by the fit of the socket and the alignment of the prosthesis. For transtibial amputees, medial-lateral stability is influenced by foot placement. Foot inset (or outset) may result in varus (or valgus) moments applied to the limb. Similarly, anterior-posterior stability is influenced by the fore-aft position (extension-flexion moment) of the foot, foot plantarflexion/dorsiflexion (extension-flexion moment), and heel durometer (soft heel increases foot plantarflexion). For transtibial amputees with normal knee extensors, knee flexion moments on heel strike are desired, as for individuals without amputation.

Since most prosthetists do not have motion-analysis systems, they cannot quantitatively evaluate lower limb kinematics nor measure lower extremity kinetics. Therefore, prosthetists rely on visual assessment of limb kinematics and indirect measures of tissue loading and joint moments. Joint moments are inferred from the relative position of a weight line with respect to the ankle, knee, and hip axes of rotation. Muscle activity that may be required to oppose such joint moments for stability is similarly inferred. While such moment estimation ignores inertial factors, it provides a visual estimate for evaluation of prosthetic fit and alignment. These visual estimates are only be applied during stance, since stance phase dictates stability and tissue loading. Analysis of swing is restricted to kinematic analysis in the frontal and sagittal planes.

Gait analysis of the transtibial amputee is reviewed in Refs. 7 (Chaps. 14 and 18) and 1. These references highlight common gait deviations, their potential prosthetic and/or physiological causes, and possible corrective measures.

For transfemoral and knee disarticulation amputees, the remnant hip musculature is vital to normal gait and prosthetic stability. As noted previously in the discussion of transfemoral socket design, the thigh is adducted in the socket so as to position the hip abductors at an appropriate functional length. The hip abductors play an important role in stabilizing the pelvis in the frontal plane during single-limb support. Since individuals with a short transfemoral residual limb may have weak hip abductors, their pelvis may dip excessively to the swing side during single-limb support. The hip adductors are important during weight transfer. The hamstrings (hip extensors) act to decelerate the limb during swing in anticipation of heel strike. If the hip extensors are weak, lumbar lordosis may be observed as a common compensatory mechanism. To maximize the efficiency of the hip extensors, the residual femur is routinely flexed within the socket. The amount of hip flexion is dictated by the length of the residual limb, the lever arm for hip extension.

Gait analysis of the transfemoral amputee is reviewed in Ref. 7 (Chap. 14). This reference highlights common gait deviations, their potential prosthetic and/or physiological causes, and possible corrective measures. Note that many individuals with transfemoral amputation choose not to wear a prosthesis and walk with crutches. The use of these assistive devices may facilitate faster and more efficient (per distance traveled) ambulation.

Gait analysis of the hip disarticulation amputee will not be discussed due to the relative rarity of this level of amputation and the fact that the excessive cost of ambulation often dictates that such amputees do not routinely use a prosthesis for ambulation. Many individuals with hip disarticulation prefer to use forearm crutches for ambulation and don their prosthesis for cosmetic purposes only. However, as discussed previously in the design of the prosthesis for individuals with this level of amputation, the stride-length control strap and hip bumper are critical to prosthetic limb stability and function during ambulation.

As alluded to earlier, energy consumption during ambulation is higher for amputees than for nonamputees.[7] While energy cost (energy consumption per distance traveled) is comparable for unilateral traumatic transtibial amputees and normal individuals, energy cost increases substantially as the level of amputation becomes more proximal (Fig. 33.19). These energy costs tend to be higher for vascular amputees than for individuals who have lost their limbs due to trauma. As one might expect, the energy cost of ambulation for individuals with bilateral lower extremity amputation is higher than for unilateral amputees. Finally, the natural velocity of individuals with amputation is typically less than for normal individuals, with the velocity decreasing for the more proximal levels of amputation (see Fig. 33.19). Assessment of amputee performance is further confounded by age, since the majority of lower extremity amputations are due to vascular complications, prevalent in older individuals.

33.7 RECENT DEVELOPMENTS

As indicated throughout this chapter, many advances in lower extremity prosthetics are due to technological advances in materials. This has facilitated fabrication of markedly lighter prostheses, stronger prosthetics sockets, and composite feet with energy-storing capabilities. Prosthetic liners for transtibial amputees have also benefited from such technology. The OrthoGel liners (Otto Bock, Minneapolis, Minn.) incorporate polyurethane gel and provide enhanced comfort for individuals whose residual limb tissues are very sensitive and/or prone to breakdown.

In terms of recent innovations, it is worth acknowledging advances in lower extremity prosthetics with respect to suspension. Historic suspension mechanisms such as hip and waist belts and fork straps and supracondylar cuffs for transfemoral and transtibial amputees, respectively, have been largely replaced by suction. Suction has been induced using neoprene suspension sleeves, TES belts, and shuttle-locking mechanisms.

Sabolich Research and Development (Oklahoma City, Okla.) recently has developed commercial systems for the upper and lower extremities that attempt to supplement sensation lost by amputation. Their Sense of Feel Leg provides sensory feedback to the residual limb. Output from force transducers in the sole of the foot is transmitted to the residual limb via electrodes in the prosthetic socket. The amplitude of the "tingling" sensation is proportional to the force measured at the foot, thereby providing direct feedback to the amputee regarding the relative loading of the forefoot versus the heel. Such information is hypothesized to improve lower extremity balance. Clinical trials are currently underway.

Another new commercial product is the 3C100 C-Leg (Otto Bock, Minneapolis, Minn.) (Fig. 33.20). This transfemoral leg incorporates a microprocessor-controlled hydraulic knee with swing and stance phase control. The controls are adjusted for the individual subject. The knee angles and moments are measured at 50 Hz, and the prosthesis facilitates ambulation at various speeds on inclines, declines, stairs (step over step), and uneven terrain. The rechargeable lithium-ion battery provides sufficient power for a full day (25–30 hours). Similar microprocessor-controlled pneumatic

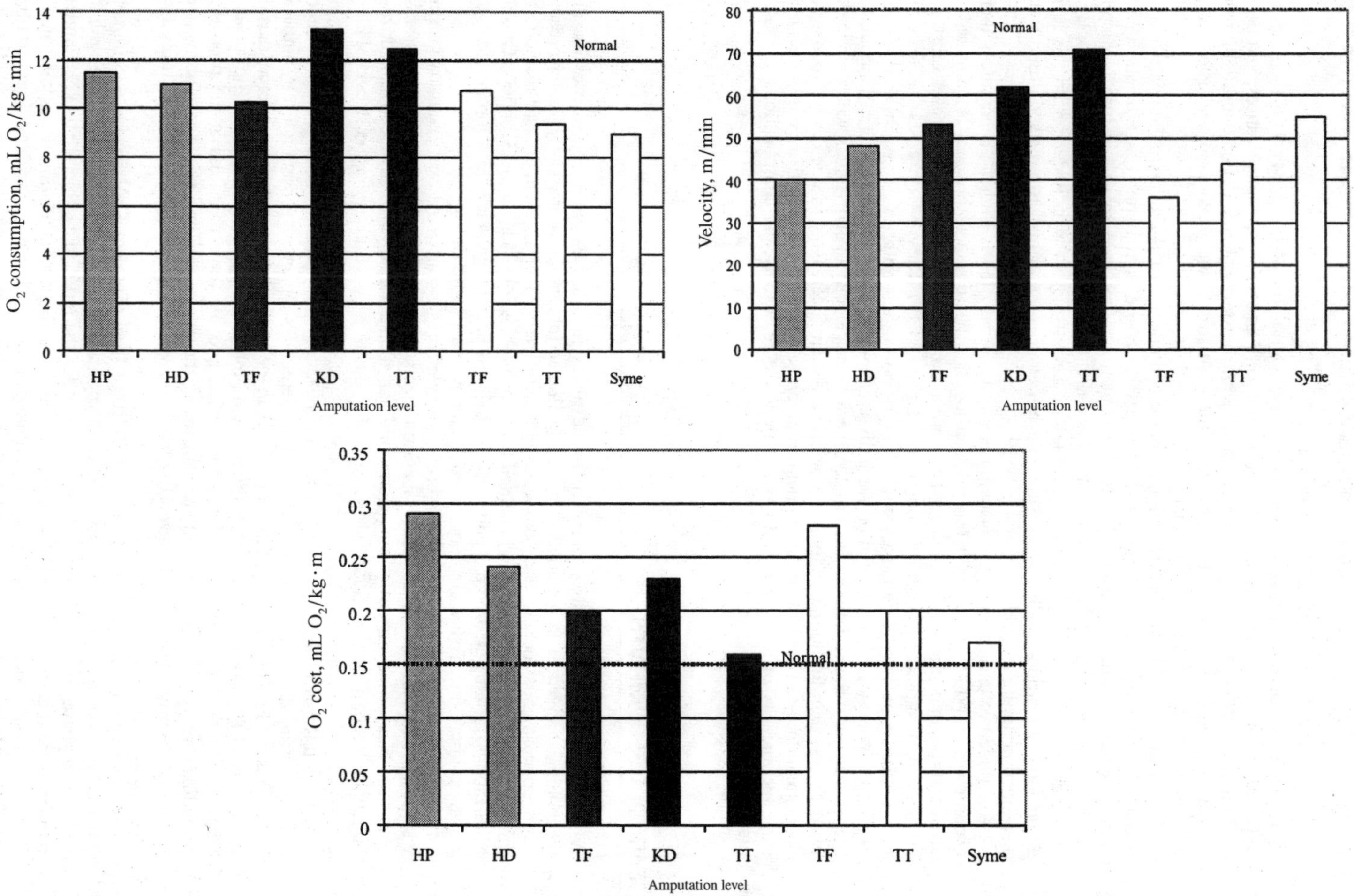

FIGURE 33.19 Rate of oxygen consumption and walking speed for surgical (gray), traumatic (black), and vascular (white) amputees for various levels of amputation (HP = hemipelvectomy, HD = hip disarticulation; TF = transfemoral; KD = knee disarticulation; TT = transtibial). (*From Ref. 6.*)

knees are also available from Endolite (Intelligent Prosthesis Plus, Centerville, Ohio) and Seattle Limb Systems (Power Knee, Poulsbo, Wash.)

One final noteworthy lower extremity development in that of shock-absorbing pylons and shock-absorbing feet (Re-Flex VSP, OSSUR/Flex-Foot, Inc., Aliso Viejo, Calif.) (see Fig. 33.17). These pylons and feet alter the effective length of the shank during loading and introduce new fitting issues and potential problems. The shock absorption or vertical compliance serves to spare the residual limb tissues during high-impact activities.

Some of the latest technology with respect to artificial limbs was recently highlighted in a medical device journal.[48] Many of these developments involve the use of robotics and/or smart materials in lower (and upper) extremity prostheses. A transfemoral prosthesis is under development at Biedermann Motech (Schwennigen, Germany) that incorporates sensors in the prosthetic knee to measure the force and moment exerted on the prosthesis, as well as the angular orientation of the knee. The knee of this prosthesis incorporates a magnetorheological fluid in its damper that reportedly results in improved response times over conventional hydraulic knee units. The BioRobotics Laboratory at the University of Washington in Seattle is investigating use of the McKibben artificial muscle (pneumatically operated actuators) to power a lower extremity prosthesis. After development of the functional prototype of the powered prosthesis, these researchers will compare their active design with conventional limbs in terms of function (gait) and energy expenditure.

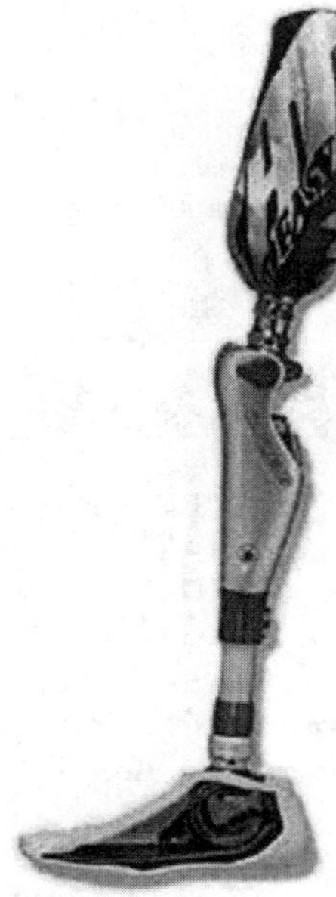

FIGURE 33.20 C-leg system with microprocessor-controlled hydraulic knee with stance and swing phase control.

Prosthetics research is also being conducted in a collaborative effort by researchers at Sandia National Laboratories (Albuquerque, N.M.), the Seattle Orthopedic Group (Poulsbo, Wash.) and the Russian nuclear weapons laboratory at Chelyabinsk-70. A recent project involves the design of a lower extremity prosthetic limb that can adjust to an amputee's gait/environment (incline, decline, and irregular terrain) and compensate for changes in residual limb shape. The first initiative involves the use of microprocessor controls for the hydraulic joints and piezoelectric motors governing the ankle and knee mechanisms. The latter initiative is based on sensors in the socket that monitor the diameter of the residual limb over the course of the day. Difficulties reportedly involve the development of a robust, lightweight power source for the entire system.

As prostheses improve, amputee function improves. Such improved function is often accompanied by increased desire to participate in additional activities and/or further improvements in performance, especially with respect to athletics and recreation. As such, there have been continued developments regarding task-specific prostheses and adaptive equipment. Amputee athletes' prowess with respect to running has received considerable press. These amputees post amazing times, supported by energy-storage prosthetic feet—with dorsiflexed alignment specific for running. Many lower extremity amputees also ski. Individuals with transfemoral amputations often ski without a prosthesis and may use outriggers (adapted ski poles that are a cross between a crutch and a miniski). Transtibial amputees, on the other hand, typically ski with a prosthesis that has been modified (e.g., alignment that accommodates skiing; step-in, rear-entry boots). Individuals with lower extremity amputations may also swim and again may opt to use or not use a prosthesis. Modifications for swimming include a prosthetic socket with a distal fin, peg legs for the beach, and/or a swim leg that does not float but provides some functionality. As for snow skiing, individuals with lower extremity amputation may also waterski, typically using a single ski; they may or may not use a prosthesis. Finally, amputee golfing has gained in popularity. For transfemoral amputees, shear on the residual limb is minimized by the inclusion of a rotator.

Finally, future developments in CAD of prosthetic sockets are also likely to be influenced by alternative shape-sensing methodology and finite-element model development that will enable timely evaluation of residual limb geometry and/or material properties.

REFERENCES

1. G. T. Sanders, B. J. May, R. Hurd, and J. Milani (1986), *Lower Limb Amputations: A Guide to Rehabilitation*, F. A. Davis, Philadelphia.
2. C. L. Thomas (ed.) (1993), *Taber's Cyclopedic Medical Dictionary*, 17 ed, F. A. Davis, Philadelphia.
3. D. G. Shurr and T. M. Cook (1990), *Prosthetics and Orthotics*, Appleton & Lange, East Norwalk, Conn.
4. A. B. Wilson (1989), *Limb Prosthetics*, 6th ed., Demos Publications, New York.
5. B. J. May (1996), *Amputations and Prosthetics: A Case Study Approach*, F. A. Davis, Philadelphia.
6. W. S. Moore and J. M. Malone (1989), *Lower Extremity Amputation*, Saunders, Philadelphia.
7. J. H. Bowker and J. W. Michael (1992), (eds.) *Atlas of Limb Prosthetics: Surgical, Prosthetic, and Rehabilitation Principles*, Mosby–Year Book, St Louis.
8. S. W. Levy, M. F. Allende, and G. H. Barnes, Skin problems of the leg amputee, *Arch. Dermatol.* **85**:65–81.
9. R. A. Sherman (1996), *Phantom Pain*, Plenum Press, New York.
10. S. Banerjee (ed.) (1982), *Rehabilitation Management of Amputees*, Williams & Wilkins, Baltimore.
11. H. W. Kay and J. D. Newman (1975), Relative incidences of new amputations: Statistical comparisons of 6000 new amputees, *Orthot. Prosthet.* **29**(2):3–16.
12. *Lower- and Upper-Limb Prosthetics and Orthotics*, 1992, Northwestern University Medical School, Prosthetic & Orthotic Center, Chicago.
13. W. Barclay (1970), Below-knee amputation: Prosthetics, *Prosthetic and Orthotic Practice* 70–78.
14. J. Hughes (1970), Below-knee amputation: Biomechanics, *Prosthetic and Orthotic Practice* 61–68.
15. T. A. McMahon (1984), *Muscle, Reflexes, and Locomotion*, Princeton University Press, Princeton, N.J.
16. G. Murdoch and R. G. Donovan (eds.) (1988), *Amputation Surgery and Lower Limb Prosthetics*, Blackwell Scientific, Boston.
17. P. E. Klopsteg and P. D. Wilson (1986), *Human Limbs and Their Substitutes*, Hafner Publishing, New York.
18. G. Murdoch (1964), The "Dundee" socket: A total contact socket for the below-knee amputation, *Health Bull.* **22**(4):70–71.
19. G. Murdoch (1968), The "Dundee" socket for below-knee amputation, *Prosthet. Int.* **3**(4–5):15–21.
20. J. G. P. Lee and S. Cheung (2000), Biomechanical evaluation of the pressure cast (PCAST) prosthetic socket for trans-tibial amputee, in *World Congress on Medical Physics and Biomedical Engineering*, Chicago.
21. J. S. Rolock (1998), *Capabilities*, Northwestern University Prosthetics Research Laboratory & Rehabilitation Engineering Research Program, Chicago, pp. **7**(1):1–2, 8–9.
22. D. H. Nielsen, D. G. Shurr, J. C. Golden, and K. Meier (1989), Comparison of energy cost and gait efficiency during ambulation in below-knee amputees using different prosthetic feet - a preliminary report. *J. Prosthet. Orthot.* **1**(1):24–31.
23. A. P. Arya, A. Lees, H. C. Nirula, and L. Klenerman (1995), A biomechanical comparison of the SACH Seattle and Jalpur feet using ground reaction forces, *Prosthet. Orthot. Int.* **19**:37–45.
24. D. G. Barth, L. Schumacher, and S. S. Thomas Gait analysis and energy cost of below-knee amputees wearing six different prosthetic feet, *J. Prosthet. Orthot.* **4**(2):63–75.

25. A. M. Boonstra, V. Fidler, G. M. A. Spits, P. Tuil, and A. L. Hof (1993), Comparison of gait using a Multiflex foot versus a Quantum foot in knee disarticulation amputees, *Prosthet. Orthot. Int.* **17**:90–94.

26. J. M. Casillas, V. Dulieu, M. Cohen, I. Marcer, and J. P. Didier (1995), Bioenergetic comparison of a new energy-storing foot and SACH foot in traumatic below-knee vascular amputations. *Arch. Phys. Med. Rehabil.* **76**:39–44.

27. E. G. Culham, M. Peat, and E. Newell (1986), Below-knee amputation: A comparison of the effect of the SACH foot and single axis foot on electromyographic patterns during locomotion. *Prosthet. Orthot. Int.* **10**: 15–22.

28. N. E. Doane and L. E. Holt (1983), A comparison of the SACH and single axis foot in the gait of unilateral below-knee amputees. *Prosthet. Orthot. Int.* **7**:33–36.

29. Y. Ehara, M. Beppu, S. Nomura, Y. Kunimi, S. Takahashi (1993), Energy storing property of so-called energy-storing prosthetic feet, *Arch. Phys. Med. Rehabil.* **74**:68–72.

30. J. C. H. Goh, S. E. Solomonidis, W. D. Spence, J. P. Paul (1984), Biomechanical evaluation of SACH and uniaxial feet, *Prosthet. Orthot. Int.* **8**:147–154.

31. J. F. Lehmann, R. Price, A. Boswell-Bessette Dralle, and K. Questad (1993), Comprehensive analysis of elastic response feet: Seattle ankle/Lite foot versus SACH foot, *Arch. Phys. Med. Rehabil.* **74**:853–861.

32. J. F. Lehmann, R. Price, A. Boswell-Bessette Dralle, K. Questad, and B. J. deLateur (1993), Comprehensive analysis of energy storing prosthestic feet. Flex foot and Seattle foot versus standard SACH foot. *Arch. Phys. Med. Rehabil.* **74**:1225–1231.

33. M. R. Menard, M. E. McBride, D. J. Sanderson, and D. D. Murray (1992), Comparative biomechanical analysis of energy-storing prosthetic feet, *Arch. Phys. Med. Rehabil.* **73**:451–458.

34. R. D. Snyder, C. M. Powers, and Perry J. Fontaine (1996) The effect of five prosthetic feet on the gait and loading of the sound limb in dysvascular below-knee amputees, *J. Rehabil. Res. Dev.* **32**(4):309–315.

35. L. Torburn, J. Perry, E. Ayyappa, and S. L. Shanfield (1990), Below-knee amputee gait with dynamic elastic response prosthetic feet: A pilot study. *J. Rehabil. Res. Dev.* **27**(4):369–384.

36. J. L. Van Leeuwen, L. A. W. Speth, H. A. Daanen (1990), Shock absorption of below-knee prostheses: A comparison between the SACH and Multiflex foot, *J. Biomechanics* **23**(5):441–446.

37. V. T. Inman, H. J. Ralston, and F. Todd (1981), *Human Walking*, Williams & Wilkins, Baltimore, Md., pp.129–148.

38. M. B. Silver-Thorn, J. W. Steege, and D. S. Childress (1996), A review of prosthetic interface stress investigations, *J. Rehabil. Res. Dev.* **33**(3):253–266.

39. M. W. Ferguson-Pell, F. Bell, and J. H. Evans (1976), Interface pressure sensors: existing devices, their suitability and limitations, (1976), R. M. Kennedy and J. H. Cowden (eds.), *Bedsore Biomechanics*, pp. 189–197, 1976, University Park Press, Baltimore.

40. M. W. Ferguson-Pell (1980), Design criteria for the measurement of pressure at body/support interface, *Eng. Med.* **9**(4):209–214.

41. A. Polliack, S. Landsberger, D. McNeal, R. Sieh, D. Craig, and E. Ayyappa (1999), Socket measurement systems perform under pressure, *Biomechanics* 71–80.

42. G. T. Nola and L. M. Vistnes (1980), Differential response of skin and muscle in the experimental production of pressure sores, *Plast. Reconstr. Surg.* **66**(5):728–733.

43. R. K. Daniel, D. L. Priest, and D. C. Wheatley (1981), Etiological factors in pressure sores, *Arch. Phys. Med. Rehabil.* **62**:492–498.

44. M. Nissan (1977), A simplified model for the short-below-knee stump, *J. Biomechanics* **10**:651–658.

45. D. J. Winarski and J. R. Pearson (1990), First-order model for the analysis of stump stresses for below-knee amputees, *J. Biomech. Eng.* **112**:475–478.

46. S. G. Zachariah and J. E. Sanders (1996), Interface mechanics in lower-limb external prosthetics: A review of finite element methods, *IEEE Trans. Rehabil. Eng.* **4**(4):288–302.

47. M. Zhang, A. F. T. Mak, and V. C. Roberts (1998), Finite element modelling of a residual lower-limb in a prosthetic socket: A survey of development in the first decade, *Med. Eng. Phys.* **20**:360–373.

48. W. Loob (2001), Robotics and electronics research aid building "smart" prostheses, in *Medical Device and Diagnostic Industry*, Jan., p. 64.

49. L. A. Karacoloff (1986), *Lower Extremity Amputation*, Aspen Systems Corporation, Rockville, Md.
50. S. W. Levy and H. Warren (1983), *Skin Problems of the Amputee*, Green, Inc., St. Louis.
51. American Academy of Orthopedic Surgeons, (1981), *Atlas of Limb Prosthetics: Surgical and Prosthetic Principles*, Mosby, St Louis.
52. M. S. Zahedi and W. D. Spence (1986), Alignment of lower-limb prostheses, *J. Rehabil. Res. Dev.* **23**(2): 2–19.

CHAPTER 34
HOME MODIFICATION DESIGN

Blair A. Rowley
Wright State University, Dayton, Ohio

There are many areas of the home both inside and out that may require modification to accommodate individuals with disabilities. This chapter discusses two major indoor areas, the kitchen and the bathroom. The considerations that follow should be used as starting points in meeting the unique needs of each person who has a disability.

34.1 GENERAL CONSIDERATIONS

34.1.1 Electrical

The main electrical consideration is safety. The electrical service in the kitchen and bathroom must incorporate ground-fault monitors as required by local codes. Additional consideration involves the accessibility of switches and outlets. Lighting is also a factor in meeting the needs of the visually impaired.

34.1.2 Lighting

Lighting should be nonglare, and areas of lighting should be nonreflective, using low-sheen laminates. Natural lighting, if available, is preferred. Other lighting sources involve incandescent tungsten, halogen, or fluorescent lighting. The following are considered when using these.

Incandescent Tungsten. This is a harsh light that creates a sharper edge to objects, allowing them to be easily differentiated. The yellow and red characteristics of this light give it a better definition to objects.

Halogen. This is similar to tungsten, except the color characteristics are more constant over the color spectrum. It is also a brighter light and adds dimension if other lighting is used. It is an excellent light to highlight ceilings and walls and to reduce glare without using a diffuser.

Fluorescent. There are choices in selecting fluorescent bulbs with different color lighting. A full-spectrum fluorescent light allows the eyes to see items in a natural state; it is a soft light that may blend objects into the background, making them harder to see. Those that give off blue and green colors may be harder on some eyes yet advantageous for people with color impairments.

34.1.3 Switches

Switches and other type of electrical controls for the disabled are designed for ease of operation and should be placed in an accessible location, where they cannot be accidentally turned on by being bumped into. Depending on the type of disability the user has, these controls can vary greatly and most likely will have to be tailored to each individual's need.

34.1.4 Main Light

The location of all wall switches should not exceed 48 in above the floor. A dimmer switch with a lever handle can control the proper amount of lighting for the main kitchen light fixture and should be placed by the main door on a wall. Additional simple on/off switches can be located by other entrances.

34.1.5 Sink and Range Fluorescent Lights

The best switch locations are a few inches below the overhang of the kitchen counter, roughly at 28 to 30 in of height. Placing these switches at the back wall extends the reach of a seated individual too far, especially over a hot range, so this is *not* a good solution.

34.1.6 Electric Disposal

The disposal switch should be placed where a disabled individual cannot turn it on accidentally by bumping it with his or her wheelchair. The switch should also be located so that no one can operate it and be in contact with the sink at the same time.

34.1.7 Electric Fan

A control knob with a three-speed clicking-type lever handle is a good choice. This variable control can accommodate people with a multitude of disabilities.

34.2 THE KITCHEN

34.2.1 Kitchen Layout

When designing a kitchen for use by persons with mobility impairments, especially those who use wheelchairs, careful layout of the kitchen is crucial to maintaining accessibility. People who are mobility impaired may

- Have walking and standing limitations that require them to sit while working
- Use a mobility aid such as crutches, canes, or walkers
- Use a wheelchair

One of the key issues to consider when designing for persons with mobility impairments is adequate space to maneuver a mobility aid such as a wheelchair or walker.

34.2.2 Maneuvering Space

Space to maneuver close to cabinets, appliances, and work areas must be provided. Each feature must have at least 30 by 48 in of clear floor space arranged for either parallel or perpendicular approach by wheelchair. Clear floor space may extend under the counters and into knee-space areas up to 19 in.

34.2.3 Knee Space

Adequate knee space under counter surfaces is important for people who need to sit while performing kitchen tasks. The space should allow them to pull up under the counter for work areas, sinks, and cook tops. Knee room should be provided beside appliances such as ranges, ovens, and dishwashers. Knee spaces should be at least 30 in wide, 27 in high, and 19 in deep. A width of at least 36 in is preferred because this provides additional turning space, which is especially important in small kitchens.

34.2.4 Turnaround Space

A space large enough for a person to turn around 180 degrees should be provided in the kitchen. If the kitchen is very small, the space can be provided immediately adjacent to the kitchen.

Pivoting Turn. Sufficient space for a complete pivoting turn can be provided with a 5-ft clear diameter floor area. This allows a full turning radius of 360 degrees. The best location for the turning space is away from appliance areas and between walls or cabinets only.

T-Turn. A T-shaped turning space allows a three-point turn to be accomplished. By making one of the necessary kitchen knee spaces 3 ft wide or wider, one leg of the T can be accomplished within the knee space. This arrangement can solve maneuvering problems in very small kitchens.

34.2.5 Laying It All Out

Efficient kitchens are usually designed around a work triangle. This triangle is formed by the location of the refrigerator, sink, and range. The arrangement of the surrounding work center depends on the available space. In general, an L-shaped kitchen provides the best access.

34.2.6 U-Shaped Work Center

Advantages to using a U-shaped work center include

- Plenty of room to maneuver a wheelchair
- Room for two cooks
- Reduced traffic flow problems
- Reduced risk of bumping into appliances

34.2.7 L-Shaped Work Center

Advantages to using an L-shape work center are

- Traffic flow does not interfere with work triangle
- Plenty of room for storage next to each workstation
- Room for two people or a wheelchair

34.2.8 Island and Peninsula Work Centers

A work center with this layout shortens the work triangle, an advantage for people with low vision or those who use walkers or crutches. Open appliance doors may, however, block aisle space needed for a wheelchair.

34.2.9 Corridor and Pullman Work Centers

A corridor work center places appliances across an aisle. A Pullman design has all appliances on one wall. Like island work center designs, these designs shorten the work triangle. The distances between appliances can make working in a kitchen with this type of design tiring for people with mobility impairments. Table 34.1 summarizes the preceding information.

TABLE 34.1 Recommended Work Triangle Dimensions

Appliance/fixture	Standard, ft	Wheelchair, ft	Walker/crutches, ft
Total distance connecting refrigerator, range, and sink	12–22	14–24	10–20
Refrigerator to sink	4–7	6–9	2–5
Sink to range	4–6	6–8	2–4
Range to refrigerator	4–9	6–11	2–7

Source: Whirlpool Home Appliances.

Table 34.2 provides some comfort zones for kitchen dimensions. These are ranges for some kitchen dimensions to maintain usability.

TABLE 34.2 Comfort Zones

Comfort zones	Standing/walking unassisted, ft	Walking with assistance,* ft	Sitting, ft
Minimum aisle space	3	4	4.5
Maximum aisle space between counters	6	6	6.5
Minimum space between workstations			
One cook	4	5	5.5
Two or more cooks	4.5	5.5	6

*Leaning on another person or using a cane, crutches, or walker.
Source: Whirlpool Home Appliances.

34.2.10 Design Notes

- The spaces recommended in Tables 34.1 and 34.2 generally are adequate for most people who use standard-sized manual or electric wheelchairs.
- More space than the minimum is recommended when designing a kitchen for use by more than one person.
- People who use electric scooters for mobility will require more space to maneuver because most scooters are much less maneuverable than a power wheelchair.
- Always consider the specific needs of the person for whom the kitchen is being designed before implementing a standard design.
- Be sure to use nonskid floor coverings.

- Design eating areas with room for round tables for families with deaf members. This provides individuals with a clear view of each other to facilitate communication.
- Keep the pathway for bringing groceries into the house as short and straight as possible.
- Keep the work triangle small for persons with visual impairments.

34.2.11 Refrigerator/Freezer

A refrigerator/freezer in a side-by-side configuration with good door storage spaces, a built-in icemaker, and a cold-water dispenser is recommended.

Dimensions. *Height.* Forty-eight inches is the maximum height a wheelchair-assisted individual should be expected to reach into a refrigerator/freezer.

Width. Between 32 and 48 inches; standard to extrawide dimensions are recommended. Due to a lower reach height, a wider refrigerator/freezer will allow more accessible storage space.

Depth. Standard dimensions are best used along with rollout drawers.

Adaptive Features. *Location.* The best alternative is to position the refrigerator/freezer away from any corners in the kitchen so that doors open 180 degrees. This allows plenty of space to open the doors and for wheelchair accessibility to the interior of the refrigerator/freezer.

Loop Handles. Due to the tight seal and weight of refrigerator/freezer doors, a significant force and space are required to open and close them. Loop handles should be the same dimensions as handrails, 1¼ to 1½ in in diameter for the handgrip, and they should be mounted 1½ in away from the refrigerator door. Sometimes a short leather or nylon loop can be used, e.g., a short dog leash. These are excellent features to ease access for those with degraded motor coordination and for visually impaired individuals.

Powered Doors. A self-opening and closing feature is a good idea for individuals with loss of strength and motor coordination. This can best be accomplished using a custom-designed system with electrical servomotors and a touch-sensitive switch.

Rollout Basket Shelves. Due to standard depth of a refrigerator, the reach must be accommodated for wheelchair-assisted individuals. Simple plastic-covered wire baskets with wheels on a rail allow access to the rear of the refrigerator. A lock-in mechanism should be designed into the shelf at its maximum extension, which should be set at two-thirds its depth.

Side-by-Side Layout of Refrigerator/Freezer. The best configuration is for the two units to sit side by side with their doors opening from the middle. This makes it easier to move items between the two units.

Water Dispenser. Built into refrigerator at 32 in above the ground, this is the average counter top height for wheelchair-assisted individual.

Icemaker. This eliminates the nuisance of filling up and emptying ice trays for mobility-impaired individuals. Location of icemaker dispenser should be next to the water dispenser.

34.2.12 Dishwasher

The assistive kitchen dishwasher is designed for wheelchair-assisted people with varying disabilities but also accommodates people with visual impairments. Since it is impossible to come up with one universal design for all disabilities, adaptations are presented.

Dimensions

- Standard height to fit under the countertop or countertop height for a roll-around unit.
- Standard width of a conventional dishwasher.
- Standard depth of a conventional dishwasher.

Adaptive Features. *Location.* The dishwasher should be located so that it is accessible from either side. It should be raised off the floor 6 to 8 in to ease access.

Controls. Controls for the dishwasher should be designed to require little force to operate and not require gripping, twisting, or fine finger dexterity.

Lever handle or blade control knob. These are excellent controls for setting different types of wash cycles, and their position can be labeled to assist those with visual disabilities.

Electronic touch-pad controls. These are the best controls for those with degraded finger/hand mobility and strength.

Rolling Table. This item is essential for loading and unloading the dishwasher. It should be countertop height with handles for ease of mobility.

Alternative Design. If space is a concern, compact dishwashers are available on the market that fit on top of a countertop with the following dimensions: 19.5 in high by 21.5 in wide by 22.5 in deep. These have the standard drop-down doors with rollout baskets.

Drop-Down Front Door. Most standard dishwashers are equipped with drop-down doors. As an added feature, grip handles are nice for both inside and outside the dishwasher for ease of opening and closing the door.

Roll-Out Basket. Most standard dishwashers are also equipped with this feature, so this should not be an additional need unless absent.

34.2.13 Microwave

Controls. The choice of controls depends on the disability that is being accommodated. It is important to note that many companies offer optional Braille controls for those who are visually impaired.

Dial Controls. *Advantage.*

- More accessible for the visually impaired (versus electric touch controls)

Disadvantages.

- Require grasping and twisting motions
- Difficult for persons with limited motor capabilities to operate

Electrical Touch Controls. *Advantages.*

- Single-touch operation
- Requires little force for operation
- Requires no gripping or twisting
- Does not require fine finger dexterity
- Some manufacturers offer plastic overlay panels with raised tactile openings or labeling to aid the visually impaired

Disadvantage.

- May not be accessible to those with visual impairments

Redundant Cueing/Feedback to the User. Controls should provide redundant cueing or feedback to the user in order to be accessible to persons with visual and hearing impairments.

Examples of Redundant Cueing/Feedback.

- Click stops. These provide distinct audible sound and tactile cues.

- High-contrast labeling.
- Raised markers.

Knee Space. Knee space needs to be available to wheelchair users to allow them access to the microwave. Make sure that the microwave is placed on a countertop that provides adequate knee space.

Minimum Requirements

- *Height*: 27 in
- *Depth*: 19 in
- *Width*: 30 in

Recommended Dimensions

- *Height*: 29 in or greater to allow for wheelchair armrests
- *Width*: 36 in

Reach Ranges. Make sure that the microwave is situated so that its location complies with reach range requirements.

Reach Range for Persons Seated in Wheelchair

- *Down*: 12 to 15 in above the floor
- *Out* (over the counter): 44 in above the floor (maximum)
- *Up*: 48 in above the floor (maximum)

Side Reach for Persons Seated in a Wheelchair

- *Lateral*: 12 in (maximum)
- *Down*: 9 in above the floor (minimum)
- *Up*: 54 in above the floor(maximum)

Reach for Standing Person, Mobility Impaired

- *Up*: 72 in above the floor
- *Down*: 24 in above the floor

Clear Space. A minimum of 10 in of clear space must be available immediately adjacent to the microwave to allow for transfer of foods in and out of the microwave.

34.2.14 Oven

A wall oven is the recommended type of oven to be used when accommodating persons with disabilities. This type of oven can be installed at the most appropriate height for the user, and the controls can be placed within reach of a standing or sitting user. Wall ovens come in several widths (24, 27, and 30 in). There are three types of doors that can be used: drop-front, side-hinged, and swinging doors.

Height. Lowered wall ovens are usually installed 30 to 40 in above the floor. When installing the wall oven, it is important to make sure that its height is appropriate for the user.

Knee Space. Knee space must be available to wheelchair users to allow them to access the oven. For a drop-front door, knee space must be provided on either side of the oven. For a side-hinged or swinging door, knee space must be provided directly under the oven or on the side closest to the door handle.

Minimum Requirements and Recommended Dimensions. The same as for microwaves.

Reach Ranges. The same as for microwaves.

Controls. The type of control chosen should be based on the individual's disability. The following control types can be used for the oven, and the advantages and disadvantages are provided for each type.

Lever Type. Advantages include

- They do not require grasping for operation.
- Their shape provides a natural pointer that indicates the control's position

Blade Knobs. Control with a large straight blade across the center; use the blade to turn the knob. Advantages include

- Blade shape is asymmetrical. It extends on one side, which forms a pointer that helps indicate the control's position.
- The shape acts as a lever. Turning is accomplished with reduced effort.

The disadvantage is that it requires grasping for operation.

Electrical Touch Controls. Advantages include

- Single-touch operation
- Requires little force for operation
- Requires no gripping or twisting
- Does not require fine finger dexterity
- Some manufacturers offer plastic overlay panels with raised tactile openings or labeling to aid the visually impaired

The disadvantage is that they may not be accessible to those with visual impairments.

Redundant Cueing/Feedback to the User. Controls should provide redundant cueing or feedback to the user in order to be accessible to persons with visual and hearing impairments.

Examples of Redundant Cueing/Feedback.

- Click stops. These provide distinct audible sounds and tactile cues.
- High-contrast labeling.
- Raised markers.

Transfer of Foods. The most used oven rack should be placed so that it is at the same height as the adjacent counter space. This facilitates easy sliding of hot pans from the oven to the counter.

Drop-Front Door. Pullout shelf next to the oven, just below the countertop and at the same height as the oven rack.

Side-Hinged Door. Shelf below the oven, 10 in wide, extends the full width of the oven (minimum).

1. Permanent shelf, i.e., the front edge of the counter.
2. Pull-out shelf, located directly under the countertop.

Safety. It is recommended that only electrical ovens be used because (1) there are no products of combustion such as carbon monoxide when using electrical ovens and (2) individuals with an impaired sense of smell will not be able to detect a gas leak.

34.2.15 Range

It is recommended that a cooktop be used because it can be installed at the most appropriate height for the user, its side or front controls are easily reached by most individuals, and counter installation can allow open space below the cooktop for easy access.

Height. It is recommended that the cooktop be installed at a height of 30 to 32 in above the floor. However, to ensure adequate clear space adjacent to the cooktop, make sure that the cooktop is installed at the same height as the adjacent countertop.

Knee Space. Knee space needs to be available for wheelchair users to allow them to access the cooktop. Ideally, space should be available under the cooktop to allow easiest access.

Minimum Requirements and Recommended Dimensions. The same as for microwaves.

Reach Ranges. The same as for microwaves.

Clear Space. Adequate clear space beside burners is required. The recommended minimum amount of clear space is 12 in.

Controls. Controls should never be located at the rear of the unit. Controls should be located on or near the front of the cooking unit. This ensures that there is no need to reach over or around any burners. Also, controls located near or at the front are more accessible for persons with visual impairments. There are several different types of controls that can be used with the cooktop. The choice of control type should be based on the individual's disability. These control types, along with their advantages and disadvantages, are provided below.

Lever Type. Advantages include

- They do not require grasping for operation.
- Their shape provides a natural pointer that indicates the control's position

Blade Knobs. Control with a large straight blade across the center; use the blade to turn the knob. Advantages include

- Blade shape is asymmetrical. It extends on one side and forms a pointer that helps indicate the control's position.
- The shape acts as a lever. Turning is accomplished with reduced effort.

The disadvantage is that they require grasping for operation.

Electrical Touch Controls. Advantages include

- Single-touch operation
- Requires little force for operation
- Requires no gripping or twisting
- Does not require fine finger dexterity
- Some manufacturers offer plastic overlay panels with raised tactile openings or labeling to aid the visually impaired

The disadvantage is that they may not be accessible to those with visual impairments.

Redundant Cueing/Feedback to the User. Controls should provide redundant cueing or feedback to the user in order to be accessible to persons with visual and hearing impairments.

Examples of Redundant Cueing/Feedback.

- Click stops. Provide distinct audible sounds and tactile cues.
- High-contrast labeling.
- Raised markers.

Range Hood. Range hoods come in a variety of widths (30, 36, and 48 in). Controls for the range hood may be located on the lower front panel of the hood to decrease the reach-range requirements for operation. However, it is recommended that these controls be located on the cooktop panel or in nearby base cabinets just below the counter top.

To adapt existing controls located on the range hood itself, a conventional toggle switch can be installed at a lower position as an auxiliary control. The hood controls should be set and left on so that the toggle switch can be used for on/off operation.

Safety. Install the cooking unit near to the sink so that the spray hose can reach some of the burners in case of non-grease-based fires. Cooking units that have flush burners should be specified such that pots and pans can be slid from the cooking surface to the counter without having to be lifted. The counter surface next to the stove should be able to withstand hot items. The burners, cooktop, and counters should be at a smooth common level, with no more than a ⅛-in raised edge, flush is preferred. Burners should be staggered so that the user does not have to reach over a hot burner to get to a rear burner. Placing an angled mirror over the cooktop allows people in wheelchairs to see the contents of pots. It is recommended that only electrical cooktops be used because (1) there are no products of combustion such as carbon monoxide when using electrical cooktops and (2) individuals with an impaired sense of smell will not be able to detect a gas leak.

34.2.16 Kitchen Sink

The adapted kitchen sink is designed for wheelchair-seated people with varying disabilities but also accommodates visually impaired people. Since it is impossible to come up with a universal design for all disabilities, some alternative suggestions are provided.

Design Concept. The adaptive sink design features two bowls with an electric waste disposal in the left sink bowl. Drains are positioned at the center of the left bowl and at the rear inner corner of the right bowl so as to position the plumbing away from the individual's knees in a wheelchair. Since the right basin has its drain positioned at the corner, the bottom of the sink should be sloped toward the drain hole. Lever handles are mounted halfway between the two sink bowls with 300-degree range of motion for maximum flexibility. Another faucet with a flexible rubber hose is provided with a 7-ft reach. A removable sloping protection panel is mounted in front of the plumbing pipe under the sink to protect the knees of a wheelchair-assisted individual.

Dimensions. The recommended countertop height is 27 to 34 in from the finished floor to the underside of the countertop. The upper limit was chosen for the sink height due to the bowl depth being 6.5 in. This allows ample knee space for a wheelchair-assisted individual.

The shallow basin (6.5 in) allows the 27-in minimum knee height from the finished floor to the bottom of the sink necessary for a wheelchair-assisted individual. The total sink dimension (bowl only is 20 in wide by 20 in deep). The maximum depth reach for a wheelchair-assisted person is 44 in, so there is plenty of margin to reach beyond the back of the sink. The width of the right-hand sink plus the counter is 40 in. The wheelchair is generally 26 in wide, so this distance is 14 in wider than the necessary space needed to accommodate a wheelchair-assisted individual's knees under the sink and the counter (40 in).

Adaptive Features

- *Height and reach*: Must be accessible by someone in a wheelchair with limited mobility.
- *Knee space*: Must allow wheelchair to fit under sink to provide maximum access.
- *Bowl depth*: Must be designed for accessible reach
- *Faucet and drain position*: Must be designed for accessible reach and not hinder knee position under sink.

- *Faucet and hose attachment*: Designed to reach countertops.
- *Single-lever faucet*: Adapted to blind individuals or persons with degraded hand coordination.
- *Protective panel*: Plumbing drain pipe is shielded from knees of individual in a wheelchair.

Other adaptive features considered but not used in the preceding adaptive kitchen sink design include

- Hot water dispensers to prevent individuals from having to carry pots of water to and from a range
- Adjustable-height basin that lowers and raises electrically with a touch of a button
- Pedal-operated faucets
- Motion sensors that when tripped activate hot or cold water at a preset rate and temperature for a finite amount of time and shut off automatically

Electric Disposal. A disposer can be installed in any sink that has a full-size drain opening. For an assistive kitchen design, switch location, safety, and disposal location are the main design objectives. The on and off switch should be placed in an accessible area, possibly under the front lip of the countertop. Care should be taken not to position the switch where a wheelchair can accidentally bump the switch while the person is washing the dishes. It should also be located so that a person cannot contact the sink and switch at the same time. The electrical connection should be away from the water line and should be protected with a conduit pipe to eliminate any shock hazard. The disposal should be located away from any wheelchair-accessible area.

34.2.17 Counters

Standard kitchen counters are 36 in high. This is adequate for disabled and nondisabled standing people but too high for people who are seated. Counter heights of 30, 32, and 34 in are more comfortable for a seated person to use for food preparation. This disparity will obviously make the design of a kitchen for use by standing and seated people difficult.

For work such as mixing or beating, a 27-in height is desirable.

NOTE: The usual height for a wheelchair armrest is 29 in. Adequate knee space requires at least 24 in.

Accessible Solutions and Adaptations. *Uniform-Height Counters.* A uniform lowered height is not a recommended solution for a number of reasons:

- This is inconvenient for standing users.
- Appliances such as dishwashers, trash compactors, and ranges are designed for 36-in counter heights.
- Lowered counters may make resale of the house difficult without restoring counters to the standard height.

Dual-Height Counters. A dual-height kitchen includes lowered counter segments to provide work areas for seated people. Each lowered segment must have clear knee space below. Clear vertical space of 30 in at the front of the counter will provide enough clearance for most wheelchairs.

Electrically Adjustable Height. Motor-driven adjustable height counter segments that allow their height to be adjusted at the press of a switch provide a uniquely flexible, highly accessible solution.

Manually Adjustable Segments. A design of this type allows counter height to be adjusted with minimal work. An adaptable design approach such as this is ideal for a rental unit where tenants may change relatively frequently. This can be accomplished in a couple of ways:

Wall-mounted brackets. Counters may be mounted with heavy-duty commercial shelf brackets and standards. Shelving of this type is seen in many retail store shelving units.

Movable wood support. A wooden support strip can be attached to the sides of base cabinets and the countertop to allow for some adjustability.

Breadboards, Chopping Blocks, and Pullout Work Surfaces. Provide a variety of work heights for different jobs. These accessories work best when at a height of 27 in and at least 30 in wide by 24 in deep.

34.2.18 Simple Modifications

The following suggestions are modifications that can be readily and inexpensively made to existing kitchens to make them more accessible.

Carts. A rolling, drop-leaf cart can provide an accessible work surface.

- The leaf can be raised to provide additional workspace.
- The cart can be rolled to the refrigerator or stove to transport food.

A Sturdy Work Table. Providing a heavy-duty kitchen table, which can take the abuse of food preparation work and is located as close as possible to the sink and appliances, is a workable and low-cost solution.

Design Notes

- Use solid surface material for countertops (easy cleaning).
- Round corners on all countertops, especially for visually impaired persons.
- Use contrasting colors on counter edges to increase visibility for those with visual impairments.
- Install drawer organizers.
- Ensure that a fire extinguisher is in easy reach of oven and range and is usable by persons with impaired mobility and dexterity.
- A mirror suspended above the cooking area allows vision into pots for a seated person.
- Include pullout shelving or readily accessible counter space near ovens and microwave to allow for sliding transfer of hot items.
- Be creative when designing a multilevel kitchen. Incorporate desks, eating bars, and tables.

Cabinets and Storage. Storage space is a particularly troublesome issue for people with limited mobility. For many, a significant portion of conventional kitchen storage space is out of reach. In addition, available base cabinet space is reduced when making a kitchen accessible to people who use wheelchairs. By selecting more efficient and accessible storage options, much of this space can be recovered.

Table 34.3 provides some information on shelving height for people with various mobility limitations.

Accessible Storage Solutions. *Full-Extension Drawers.* A deep drawer that extends the full depth of the base cabinet and is mounted on full-extension slides is very useful. These drawers are similar to those found in an office file cabinet (Table 34.4).

TABLE 34.3 Comfort Zones

Comfort zones	Standing/walking unassisted, in	Walking with assistance,* in	Sitting, in
Maximum upper cabinet reach			
Over a counter	68	63	60
Without a counter	77	68	48
Maximum vision for items on a high shelf	61	61	48
Maximum height of storage for daily use	74	65	45

*Leaning on another person or using a cane, crutches, or walker.
Source: Whirlpool Home Appliances.

TABLE 34.4 Recommended Drawer Heights

Purpose	Height, in
Silver, small tools	3–4
Spices	3–4
Linens	3–4
Saucepans	6–7
Canned foods	6–7
Canisters	11–12
Large packaged foods	11–12
Shallow utensils, stored vertically	12–13

Carts. Rolling carts that fit into knee space under countertops can provide additional storage. In addition, they

- Can be easily rolled out to provide knee space.
- Can provide additional workspace.
- Can provide a safe way to transport food and utensils.

Countertop Storage Unit. The space between countertop and upper cabinets can provide easily reachable open storage.

Overhead Cabinet Doors. Traditional swinging wall cabinet doors can be a hazard for blind people. Hardware that allows the cabinet doors to swing up and stay open can reduce this problem.

Concealed Knee Space. Retractable doors can be used to conceal knee space. Special hardware allows the door to be pushed back under the counter after opening.

Pantry. A pantry can provide easily accessible storage space. Height-adjustable shelving can tailor the space to individual needs. Shallow shelves keep items within easy reach. The pantry can be a reach-in unit with storage shelving on the doors or even a walk-in design.

Other Accessible Storage Options. Additional options for accessible storage include

- Storage bins
- Pull-out storage
- Revolving shelves
- Swing-out shelves

Add Storage Bins. The addition of pullout storage bins to shelves and cabinets can help make existing storage space accessible.

Lower Existing Cabinets. Existing overhead cabinets can be lowered as far as the top surface of existing counters if necessary. This provides accessible storage at minimal cost. Cabinets may be lowered to 12 to 15 in above the counters, while keeping the counter surface usable.

Add a Freestanding Storage Cabinet. If floor space is available, a freestanding storage cabinet can provide accessible storage space.

Design Notes. All drawers, doors, and hardware should be selected to provide easy access for people with limited dexterity. Recommended features include the following:

- Sliding or folding doors for cabinets provide for the least interference.
- Magnetic catches should be used on the doors.
- Large loop handles also should be used on doors and drawers.
- Toe space on base cabinets should be at least 9 in high and 6 in deep to allow wheelchair users to maneuver closer to the counters.

34.3 THE BATHROOM

Accessible bathrooms should be designed to accommodate the maximum amount of maneuvering possible for physically challenged persons, their wheelchairs, and the possibility of a second person assisting.

34.3.1 Commode

Location

- Ample space around the commode is necessary for side or front approach by wheelchair or walker user.
- The commode is best located in a corner where a wall is behind and beside the commode to easily install grab bars.
- Clearance of 18 in is needed between center line of commode and side wall for adequate shoulder space.

Seat Height

- An accessible toilet seat height varies from user to user.
- Adjusting the seat height aids people who have difficulty rising.
- A commode seat should be at a height of 15 to 19 in above the floor.
- A good rule of thumb is an 18-in height, which is the same as most wheelchair seats.

If a shower commode wheelchair is used, the chair and commode opening must line up. In general, this requires the commode opening to be 20 in from back wall. Adequate space is needed on both sides of the commode so that there is no obstruction.

Safety Bars

- Mounted onto commode to provide user with grab bars.
- Good for person who has limited range of motion and cannot reach grab bars on walls.

Toilet Paper Dispenser

- Dispenser should be below any grab bars located around the commode area.
- Dispenser should be within easy reach of the person.

34.3.2 Door

Clear Opening

- Doors should have a clear opening of between 32 and 36 in.
- The clear opening is measured from the face of the door in a 90-degree open position to the stop on the opposite jamb; the door itself has to be wider than the clear opening in order to comply.

Level Surface

- The clear opening should lead to a level surface for a minimum of 5 ft in the direction that the door swings.
- A 60- by 60-in clear space is best and complies with all approaches.
- There are exact requirements for given approaches; please see Americans with Disabilities Act Accessibility Guidelines (ADAAG Sec. 4.13 and Fig. 25) for exact requirements.

Threshold

- The threshold maximum height is ½ in.
- There must not be any sharp inclines or abrupt changes.
- A beveled threshold is recommended.

Force to Open Door

- The pressure required to open doors with door closures should be limited to 3 to 6 ft · lb.
- There should be a 3-second minimum closing time to a point 3 in from the latch.
- An automatic door opener may be used in different situations.

34.3.3 Electrical Controls (Wall Switches and Thermostat for Paraplegic Users)

Safety

- Ground fault circuit indicator (GFCI)–protected electrical receptacles, approved by *National Electrical Code*
- Plastic plug inserts for childproofing electrical receptacles whenever toddlers and young children are present in the environment
- Shatterproof light fixtures
- Use of ground wire with wiring and proper installation

Lighting

- Natural
  - Window at natural height

- Glass Blocks that admit light but obscure vision; to install
  1. Reinforced wood framed floor
  2. Bracing at side wall
  3. Provide for trim at top, curbs at floor level, and wall anchors
  4. Set glass blocks with mortar
  5. Caulk around glass wall
- Artificial
  - Above sink with light beams (recommend two minimum) directed toward bowl and away from user's vision field
  - Above and on both sides of grooming mirror/area

In the bathing area, there should be lights above the shower stall or directed into stall above or around the door or curtain.

Electrical Power Outlets

- AC with ground fault interrupters near or within sink reach area for various electrical grooming appliances
- Locate 15 in minimum above floor and 18 in minimum from any corners
- When installed above sink, recommend 15 in above sink top
- Wall switch located at entry 48 in above floor and 3in in from door molding

Infrared or Sunlamp

- Flush mounting in the ceiling
- 115 V ac, 250 to 300 W
- Beam downward just outside shower stall exit
- Function: for warmth and drying comfort

Fan Connection

- 115 V ac, 100 W
- Locate in ceiling, middle of bathroom
- Install concentric with ventilation porting

34.3.4 Flooring

Nonslip Surface. A nonslip surface should be used to prevent slipping and rolling resistance to wheelchairs.

Color and Patterns

- Providing high contrasts allow for visual acuity.
- Patterns and edgings can guide people with low vision.
- Floors should be relatively light in color, with 30 to 50 percent reflectance.
- Examples of color schemes: cool colors (blues and greens), pastel colors, subdued color with intense color accents, or intense colors.

Fire Code. All flooring and materials must meet *National Fire Code*, Class I.

Clear Floor Space for Commode Area

- Ample floor space in front of and beside the toilet fixture is necessary for users to approach the seat and make a safe transfer.
- If the commode can be approached from the front and side and a sink is installed next to it, the floor space must be at least 4 ft by 5 ft, 6 in.
  - The 4-ft dimension extends into room from side wall next to commode fixture.
  - The 5-ft, 6-in measurement is measured from the wall behind toilet to wall in front of toilet.
- If possible, always design the layout of an accessible bathroom to allow both a front and side transfer to the commode.

Clear Floor Space for Shower and Sink Area

- Clear floor space under the lavatory must be a minimum area of 30 in wide by 48 in long that extends a maximum of 19 in under the lavatory.
- Minimum clear floor space for a roll-in shower is 3 ft by 3 ft.

Clear Floor Space for Entire Bathroom

- Adequate floor space is necessary for a person to maneuver a wheelchair or walker around the bathroom without any obstruction.
- At least a 60-in turning radius is recommended.

Clear Floor Space for Door Area

- Clear, unobstructed floor space is necessary at the door area to allow the door to open and close easily.
- Clear floor space includes the area on the hinge side of the door extending the width of the door and the area past the latch side of the door.
- A wider floor space is needed on the pull side of the door to provide space to open the door.
- Less area is needed on the push side of the door.
- Space must be available on the pull side of the door to operate the door latch, pull the door open, and remain out of the way of the swinging door.
- An outward-swinging door is preferable to provide more accessibility.

34.3.5 Grab Bars

Resistance Force. Grab bars need to be capable of resisting at least 250 lb of force. However, very large persons may require more strength in the grab bar.

Diameter of Bars. The bars should have a diameter in the range of 1¼ to 1½ in.

Clear Space. The space between the grab bar and the walls should be at least 1½ in.

Locations of Applicability

- *Roll-in showers*. In 36- by 36-in stalls, bars should be located on the side wall and the wall opposite the seat. In 30- by 60-in stalls, bars should be located on both side walls and the rear wall.
- *Tubs*. Bars should be located on the rear wall, side wall, and side wall opposite controls.
- *Toilet stalls*. Bars should be located on side wall and the wall flush against the commode.

34.3.6 Sink

Type of Sinks

- Wall-mounted and countertop lavatoraties are accessible.
- Countertop sinks are more accessible due to large surface area to place toiletries.
- Countertop space can be adjusted according the consumer's needs.
- A self-supporting sink can have an optional, removable vanity cabinet underneath.

Dimensions

- Depth of the sink (front to back) for persons using a wheelchair to be able to pull underneath sink.
- Accessible sinks should have dimensions of about 20 in wide by 18 to 27 in deep.
- Sink must have at least 29 in of clearance from the floor to the bottom of apron at front of sink.
- Adequate knee space under sink provides clearance for turns as well as space for close approach to sink by wheelchair users.
- Sink bowl should not be any deeper than 6½ in.

Mirror. Mirror must be mounted no higher than 40 in off floor. Examples of accessible mirrors include

- Tilted mirror angled toward floor to accommodate user
- Full-length mirror

Faucet Systems

- Faucets should operate without gripping or twisting of handles.
- For persons who can operate faucet with closed fist, handle should not require more than 5 lb of force.

Accessible faucet handle designs include

- Faucet setup with timer for water to run for period of time and then shut off automatically.
- Faucet that is electronically controlled to eliminate the need to turn handles:
  - Senses the presence of the user's hands.
  - Automatically turns water on and off.
  - Temperature and flow rate of water are preset.
  - Particularly useful for people with severe hand limitations.
- Faucets with single or double loop or lever handle designs.

Plumbing

- Piping underneath sink must not interfere with clear floor space and knee space for wheelchair users and must be insulated to prevent heat injury to legs.
- Wheelchair user needs a clear floor space in front of sink area of approximately 30 by 48 in and knee height to bottom of sink of 22 in.

Reinforcement of Walls. Bathroom walls around sink area may need to be reinforced for proper support of wall-mounted sink.

34.3.7 Storage

Knee Space Underneath Shelves. For a frontal approach to items on shelves, the bottom shelves can be eliminated. This design limits the total shelving space available to user, though.

Depth of Shelves. This should be no greater than 18 in from the front of the closet to the back of the closet. This may vary depending on the size and limitations of person.

Closet Doors. Several options exist:

- Door swings back 180 degrees so that wheelchair users can make a close parallel approach to reach items from the side.
- Eliminate the door, especially if the bathroom is not very large.

Height Range of Shelves

- An adequate reach range for a wheelchair user to make a side reach is 9 to 54 in above the floor
- A good height range for storage is 15 in to 48 in above the floor.

Storage Location and Size

- Placed in an area of bathroom that is easily accessed by disabled user.
- Must not obstruct any of the bathroom appliances.
- Amount of storage necessary varies from individual to individual.
- Users may want room for medical supplies and equipment.
- Others may need only a small amount of storage space.

Pull-Out Drawers or Shelves

- Full-extension drawers can be installed in bathrooms up to 60 in off the floor.
- Higher drawers should be shallow for easy access.
- Lower drawers can be deeper.
- For a built-in storage drawer system, use full-extension drawer slides. These slides allow drawers to be pulled out of the cabinet for easy viewing and reaching to contents. These drawers should not be placed more than 42 in above the floor for wheelchair users.

Handles

- Handles on storage closets and drawers should be accessible.
- Standard round doorknobs should be avoided for users with weak grasp.
- Handles should consist of a single lever or open-loop handle.

ACKNOWLEDGMENTS

I acknowledge the contributions of my graduate students in helping assemble this information. Information was gathered from practical field experience, the literature, and interviews with users and rehabilitation engineers.

BIBLIOGRAPHY

Frechette, Leon A., *Accessible Housing*, McGraw-Hill, New York, 1996.

Germer, Jerry, *Bathrooms: Design-Remodel-Build*, Creative Homeowners Press, Upper Saddle River, N.J., 1995.

Kearne, Deborah S., *The ADA in Practice*, R. S. Means Company, Kingston, Mass., 1995.

Kira, Alexander, *The Bathroom Criteria for Design*, Bantam Books, New York, 1967.

Loversidge, Robert D. Jr., and Laura V. Shinn, *Access for All: An Illustrated Handbook of Barrier Free Design for Ohio*, Schooley Caldwell Associates and the Ohio Governor's Council on People with Disabilities, Cleveland, 1994.

Mace, Ronald L., *The Accessible Housing Design File*, Barrier Free Environments, Inc., Van Nostrand Reinhold, New York, NY, 1991.

Mean ADA Compliance Pricing Guide, R. S. Means Company, and Adaptive Environments Center, Kingston, Mass., 1994.

Wing, Charlie, *The Visual Handbook of Building and Remodeling*, Rodale Press, Emmaus, Pa., 1990.

Wylde, Margaret, et al., *Building for a Lifetime: The Design and Construction of Fully Accessible Homes*, Taunton Press, Newtown, Conn., 1994.

CHAPTER 35
REHABILITATORS

David J. Reinkensmeyer
University of California, Irvine, California

35.1 INTRODUCTION

Millions of people in the United States currently require rehabilitation therapy due to neurologic injury and disease. In the case of stroke alone, there are approximately 600,000 new survivors each year and over 2 million survivors with chronic movement deficits.[1] Recent evidence suggests that intensive therapy improves movement recovery.[2–9] However, such therapy is expensive because it relies on individualized interaction with rehabilitation therapists. The type and quality of therapy also vary greatly between clinics and therapists. Little technology is available to assist patients in practicing therapy on their own. Therapy techniques that require expert feedback or that incorporate manual manipulation of the patient's limbs are often inaccessible once formal therapy is discontinued.

To address these needs, mechatronic and robotic devices are being developed to automate movement therapy after neurologic injury. This chapter discusses the rationale for these rehabilitators, reviews practical design considerations, and discusses their future development.

35.2 RATIONALE FOR REHABILITATORS

35.2.1 Target Populations and Therapies

Stroke survivors are a key target population for rehabilitators. Stroke is a leading cause of severe disability in the United States, with roughly 1 of 100 Americans having experienced a stroke.[1] The incidence of stroke doubles with each decade after age 55. Consequently, as baby boomers age, there will likely be an increase in the number of people experiencing strokes and requiring rehabilitation therapy.

Stroke causes an array of motor impairments.[10] Hemiparesis, or weakness on one side of the body, arises due to destruction of the neural systems and outflow pathways responsible for controlling muscles.[11,12] The degree of hemiparesis varies from complete paralysis to mild weakness depending on the severity of damage and the location of the lesion. Stroke also impairs interjoint coordination, making it difficult to contract muscle groups independently[13–17] and to steer the arm along smooth, coordinated paths.[18,19] In many patients, muscle tone, or the resistance felt when movement is imposed on a passive limb, is increased due to changes in stretch reflex sensitivity (i.e., spasticity) and a loss of soft tissue suppleness (i.e., contracture).[20–22] When combined, these impairments decrease the ability to use the upper and lower extremities for activities of daily living, such as dressing and walking.

Stroke-induced motor impairments are amenable to rehabilitative treatment. Recent evidence from both animal and human research suggests that improved recovery is possible for both the upper and lower extremities with intensive movement practice. In studies of the upper extremity, repetitive-movement practice improved movement ability of the hand and arm after stroke.[5,23,24] This improved movement ability arose at least in part from cortical reorganization. It has been hypothesized that intensive movement practice facilitates "rewiring" of the brain, with undamaged cortical areas assuming control functions previously allocated to damaged ones.[23,25]

A large number of therapeutic techniques besides simple repetitive-movement practice have been developed, many of them requiring manual intervention or detailed feedback from therapists.[10] While none of these techniques has been identified as superior, a number have been shown to produce better recovery when more therapy is given.[2,3,6,7]

Other neurological conditions relevant to rehabilitators are traumatic brain injury (TBI) and spinal cord injury (SCI). About 500,000 people in the United States receive hospital treatment for TBI yearly, and about 20 percent of TBI survivors have long-term impairment.[26] When damage extends to motor regions of the brain, motor impairments similar to those seen in stroke often arise. SCI affects 10,000 new people per year in the United States, and over 200,000 SCI patients are alive today.[26] Displacement of the spinal column crushes the spinal cord, paralyzing muscles that are innervated by nerves below the lesion level. Locomotion is commonly affected since leg muscles are innervated by lower spinal nerves.

Rehabilitative therapy has been shown effective in improving walking ability after both brain and spinal cord injuries. Of special promise here is a technique referred to as *step training with body weight support*.[27,28] In this technique, the patient is suspended over a treadmill, bearing only a fraction of his or her full weight, while therapists provide manual assistance to one or both legs and the torso to assist in locomotion. In one clinical trial, significantly more SCI subjects who received such step training achieved independent overground walking when compared with a group that received conventional physical therapy.[29] Those who could walk before receiving locomotor training were able to walk further and at greater speeds after training. Rhythmical flexor and extensor electromyographic (EMG) activity can be elicited during stepping with manual assistance even in human subjects who have a clinically complete thoracic SCI.[30] Thus even high-level, complete spinal cord injury patients can generate locomotor muscle activity with step training, suggesting that a core of circuitry responsible for locomotion pattern generation resides in the spinal cord. This circuitry apparently "listens" to proprioceptive and cutaneous input from the legs and adapts based on its ongoing sensory experience.

35.2.2 Rehabilitator Designs

A number of devices for providing therapy to the arm and legs after brain and spinal cord injury have been developed. This section introduces three illustrative designs for arm rehabilitators and two for locomotion rehabilitators. The next section reviews clinical results with these devices. For descriptions of other upper extremity rehabilitators, the reader is referred to Refs. 31 to 35, and for other lower extremity, locomotor, and postural rehabilitators, the reader is referred to Refs. 36 to 38. Rehabilitators also have been developed for rodents to facilitate research in animal models of neurologic injury.[39,40]

The *MIT-MANUS* device (*manus*, from the Latin for "hand") was the first rehabilitator to undergo intensive clinical testing.[4,41] This device is a planar, two-revolute-joint, backdrivable robotic device that attaches to the patient's hand and forearm through a brace (Fig. 35.1). The device can assist or

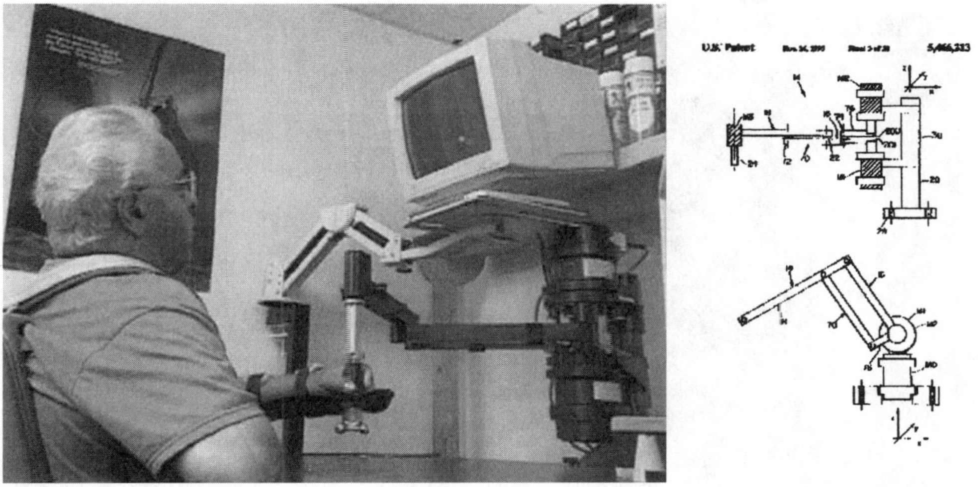

(a) (b)

FIGURE 35.1 MIT-MANUS. (*a*) The patient attaches to the robot through a splint. The patient can move the robot or the robot can move the patient in the horizontal plane. The patient receives feedback of the hand trajectory on the computer screeen. (*From H. I. Krebs, B. T. Volpe, M. L. Aisen, and N. Hogan, Increasing productivity and quality of care: Robot-aided neurorehabilitation, Journal of Rehabilitation Research and Development* ***37****:639–652, 2000. Used with permission.*) (*b*) Details of the mechanism design. MIT-MANUS uses a five bar-linkage design with the elbow and shoulder motors mechanically grounded at the base and their axes colinear. (*From N. Hogan, H. I. Krebs, J. Charnarong, and A. Sharon, Interactive robot therapist, 1995, U.S. Patent No. 5466213.*)

resist in horizontal movement of the arm across a tabletop. It can also accurately measure planar movement of the hand, providing feedback to the patient via a computer screen.

The *MIME* (*Mirror Image Movement Enhancer*) arm rehabilitator incorporates a PUMA 560 industrial robot arm to manipulate the patient's arm[42] (Fig. 35.2) This device can move the patient's arm in three-dimensional space. For hemiparetic patients, the motion of the patient's unimpaired arm can be tracked with a mechanical digitizing stylus, and that motion can be used to control the trajectory of the impaired arm.

The *ARM Guide* (*Assisted Rehabilitation and Measurement Guide*) rehabilitator is a singly actuated, three-degrees-of-freedom device for assisting in reaching movements[43] (Fig. 35.3). The device consists of a linear guide that can be oriented at different yaw and pitch angles. The patient reaches along the linear guide. Like MIT-MANUS and MIME, the device can assist or resist in movement and can measure hand movement.

The *Mechanized Gait Trainer* (*MGT*) is a singly actuated mechanism that drives the feet through a gaitlike trajectory[44] (Fig.35.4). The device consists of two foot plates connected to a doubled crank and rocker system. An induction motor drives the cranks via a planetary gear system. The rear ends of the foot plates follow an ellipsoid-like movement. Different gears can be incorporated to vary stride length and timing. The planetary gear system also moves the patient harness in a locomotion-like trajectory through two cranks attached to suspension ropes. The torque generated by the motor is sensed and displayed online to provide a biofeedback signal to the patient.

The *Lokomat* is a motorized exoskeleton worn by the patients during treadmill walking[45] (Fig. 35.5). This device has four rotary joints that accommodate hip and knee flexion/extension for each leg. The joints are driven by precision ball screws connected to dc motors. Parameters such as the hip width, thigh length, and shank length can be manually adjusted to fit individual patients. The weight of the exoskeleton is supported by a parallelogram mechanism that moves in the vertical direction and is counterbalanced by a gas spring. The hip and knee motors can be programmed to drive the legs along gaitlike trajectories.

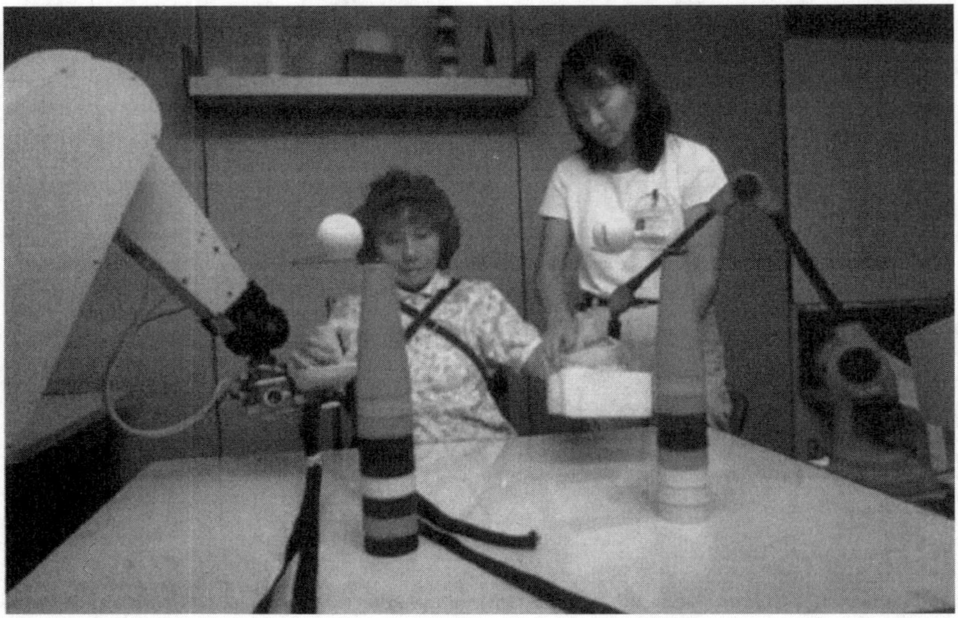

FIGURE 35.2 MIME rehabilitator being used in bimanual mode. When the patient moves her unimpaired arm (left arm), a mechanical digitizing stylus senses the movement. The PUMA 560 robot arm then moves the patient's impaired arm (right arm) along a mirror-symmetrical trajectory. *(From C. G. Burgar, P. S. Lum, P. C. Shor, and H. F. M. Van der Loos, Development of robots for rehabiltitation therapy: The Palo Alto VA/Stanford experience, Journal of Rehabilitation Research and Development, 37:663–673, 2000. Used with permission.)*

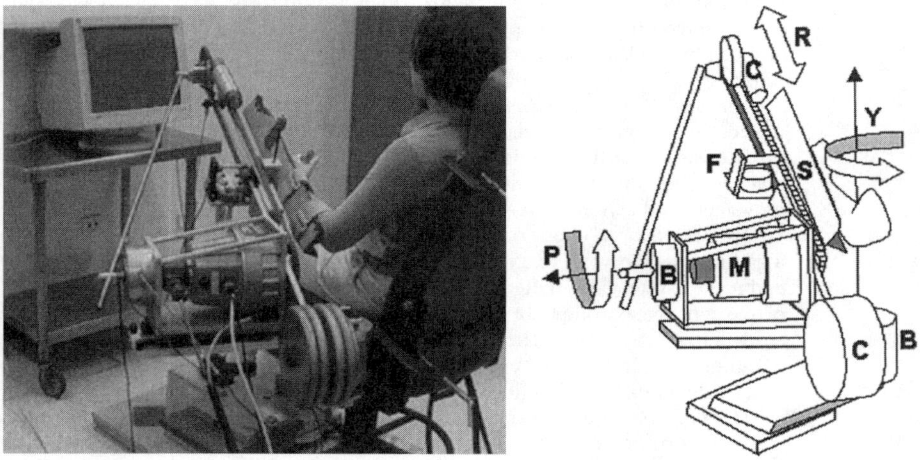

FIGURE 35.3 The ARM Guide. The patient's arm is attached to a hand splint (*S*) that is attached to an orientable linear track. A dc servo motor (*M*) can assist in movement of the subject's arm in the reaching direction (*R*) along the linear track. Optical encoders record position in the reach (*R*), pitch (*P*), and yaw (*Y*) axes. A six-axis force sensor (*F*) records the forces and torques at the interface between the device and the subject. The device is statically counterbalanced with two counterbalance weights (*C*).

(a)

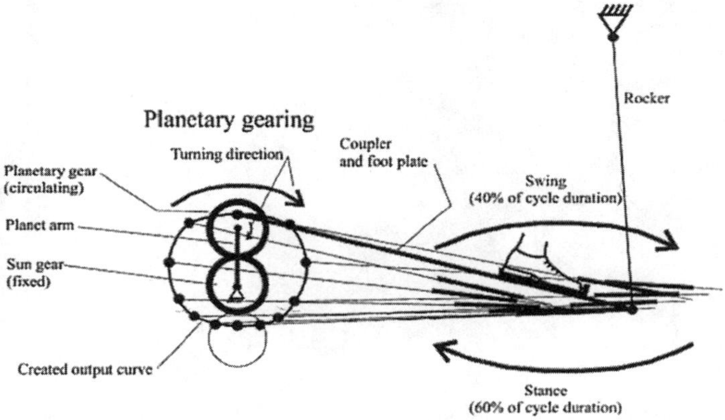

(b)

FIGURE 35.4 The Mechanized Gait Trainer (MGT). (*a*) The patient is supported by an overhead harness as the device drives the feet and torso in a steplike pattern. (*b*) A modified crank and rocker system that includes a planetary gear system simulates the stance and swing phases of gait. (*From: S. Hesse and D. Uhlenbrock, A mechanized gait trainer for restoration of gait, Journal of Rehabilitation Research and Development* ***37****:701–708, 2000. Used with permission.*)

35.2.3 Can Therapy Be Automated?

Recent research using these devices suggests that rehabilitator-based therapy can enhance movement recovery following neurologic injury. In the first clinical trial of robot-aided neurorehabilitation, MIT-MANUS was used to assist acute stroke patients in sliding their arms across a tabletop.[4,41,46] The subjects performed a daily series of movement tasks such as moving to targets and tracing figures. The robot helped complete the movement using an impedance-type controller. It was found that patients who received robot-assisted therapy recovered more than those receiving a sham treatment did, according to a coarse clinical rating scale of arm movement ability. A subsequent study indicated that these relative improvements were maintained in the robot group at a 3-year follow-up.[47]

A therapy study of the MIME device has produced similar results with chronic stroke patients.[42] In this study, the robot was position controlled so as to drive the subject's arm through a desired trajectory, as specified either by the trajectory of the contralateral (unimpaired) arm in real time (bimanual mode) or as measured previously (unimanual mode). Chronic stroke subjects exercised 3 hours per week, performing reaching exercises and tracing shapes. At the end of 2 months, the robot exercise group exhibited enhanced movement ability, as measured by clinical scales similar to those used in the MIT-MANUS study, as well as in terms of measurements of active range of motion and

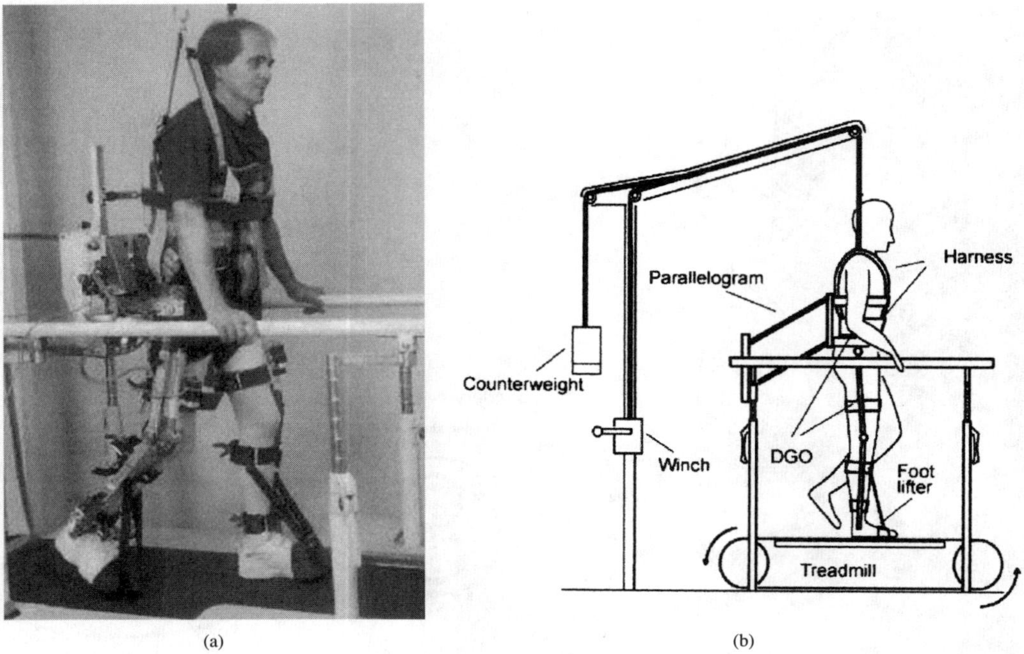

FIGURE 35.5 The Lokomat. (*a*) A motorized orthosis is attached around the upper and lower segment of each leg and drives hip and knee movement in the sagittal plane as the patient walks on a treadmill. (*b*) The patient is suspended from a counterbalanced harness. A parallelogram mechanism with a gas spring supports the weight of the driven gait orthosis (DGO). (*From G. Colombo, M. Joerg, R. Schreier, and V. Dietz, Treadmill training of paraplegic patients with a robotic orthosis, Journal of Rehabilitation Research and Development 37:693–700, 2000. Used with permission.*)

strength. Encouragingly, the improvements in movement ability were comparable with those of a control group that received a matched amount of traditional tabletop occupational therapy exercise.

A clinical trial is also being performed with the ARM Guide.[48] In this trial, one group of chronic stroke patients is receiving mechanically assisted reaching exercise with the ARM Guide. A second group is receiving a matched amount of unassisted, repetitive reaching exercise. All subjects are evaluated using a set of clinical and biomechanical measures of arm movement. The seven subjects who have received therapy with the ARM Guide at the time of writing have shown significant improvement in the measures. However, the amount of improvement in seven unassisted exercise subjects has been comparable. Thus these results also support the concept that rehabilitator-assisted manipulation of the arm following stroke is beneficial but highlight the need to clarify which aspects of that therapy are essential. For example, it may be that the repetitive movement attempts by the patient, rather than the mechanical assistance provided by the rehabilitator, are the primary stimuli to recovery.

Clinical trials with locomotion rehabilitators are just beginning. The MGT has been used to train two patients who were 2 months poststroke.[44] The patients received 4 weeks of gait training with the device, consisting of five 20-minute sessions per week. The patients improved markedly in their overground walking ability. Therapeutic results have not been reported for the Lokomat, although several spinal cord injured patients have tested the device.[45] The device was able to produce gaitlike patterns in the patients, reducing the labor burden on the therapists who were assisting in the step training.

35.3 DESIGN OF REHABILITATORS

This section reviews practical design considerations for rehabilitators using design features of the MIT-MANUS, MIME, and ARM Guide arm rehabilitators and the MGT and Lokomat locomotion rehabilitators for reference.

35.3.1 Connecting the Patient to the Machine

The design of the interface between the patient's limbs and the rehabilitator is a key consideration if the device is to be used comfortably, safely, and with a minimal level of supervision. In the case of therapy for the arm, many stroke patients do not have hand grasp ability and thus cannot grip a handle. During manual therapy with a human therapist, the therapist can compensate for the patient's loss of hand grasp by using his or her own hands to grip the patient's arm. Although replicating the gentle grip of a therapist is difficult, simple approaches can provide safe attachment for many patients.

One of these simple approaches is to make use of existing hand splinting technology. A wide variety of splints are available from rehabilitation therapy supply houses such as Sammons-Preston. These splints are made of thin sheets of thermoplastics that can be heated in a water bath or with a heat gun and then formed into the desired shape. Common configurations for hand splints are ball- , cone- , and U-shaped. Padded hook-and-loop straps can be glued to the splints and secured around the forearm and back of the hand. The MIT-MANUS and MIME devices make use of cone-type splints, whereas the ARM Guide uses a custom-designed grip in which the forearm lies in a padded aluminum trough and the hand wraps around a cylinder that can be slid into the palm and locked.

Many patients also have decreased range of motion of the arm, constraining the set of postures into which they can self-attach their arms to a machine. Commonly affected degrees of freedom are forearm supination and shoulder external rotation. It is thus important to avoid having the patient maneuver his or her hand through a complex attachment geometry. The MIT-MANUS, the MIME, and the ARM Guide allow attachment of the hand in a pronated posture directly in front of the torso.

With respect to therapy for locomotion, therapists typically grasp the patient's leg with both hands. A common technique is to grasp the lower shank with one hand below the knee and with the other hand above the ankle. The therapist may alter contact force during specific phases of the gait cycle in order to stimulate (or avoid stimulation) of tendons and cutaneous reflexes. A gentle touch is essential to avoid skin damage, since decreased skin health and sensation are common after spinal cord injury.

The existing locomotion rehabilitator designs rely on simple attachment schemes. The MGT attaches to the patient's foot via a foot plate, whereas the Lokomat attaches to the thighs and shanks through padded straps. Skin irritation has been reported with the Lokomat if the exoskeleton is not properly adjusted.[45] For both devices, modified parachute or rock-climbing-type harnesses provide an attachment to the torso that allows partial body weight support through an overhead cable.

Cutaneous reflexes are an important consideration in the design of attachment interfaces for locomotion rehabilitators. Cutaneous input significantly modulates muscle activity during stepping, and cutaneous reflexes are often hyperactive following neurologic injury. By virtue of their mechanical contact with the limb, rehabilitators alter cutaneous input to the limb. In SCI rodents, attaching small robotic devices to the paws disrupts stepping, presumably by stimulating cutaneous reflexes.[40] Further research is needed to determine whether attaching rehabilitators to the lower extremities in humans excites cutaneous reflexes and to identify attachment points for which that excitation is minimized or utilized.

35.3.2 Mechanism Design

The design of a kinematic linkage for a rehabilitator is constrained by many factors, including the desired movements to be performed, cost, and safety. Current rehabilitator designs reflect a compromise between these factors.

The MIT-MANUS device relies on a SCARA-like* or five-bar mechanism to achieve two-dimensional movement of the patient's hand in the horizontal plane (see Fig. 35.1). An advantage of this mechanism is that it allows both motors to be affixed at the robot base (or grounded). Thus, instead of placing the elbow motor at the elbow joint (and requiring the shoulder motor to move the mass of the elbow motor), the five-bar linkage allows the elbow motor to be affixed across from the shoulder motor and the linkage to remain lightweight. A variation on this design allows both motors to remain stationary and on the same side of the linkage, which is useful if the device's base is to be placed next to the patient's shoulder.[49] A mechanism that allows the actuators to remain grounded while providing two-degree-of-freedom end-effector movement on the surface of a sphere, or even three-degree-of-freedom spatial motion, also has been conceived.[50]

The MIME device uses an elbow manipulator kinematic configuration with a spherical shoulder, hinge elbow, and spherical wrist, allowing arbitrary positioning and orientation of the end effector with highly geared dc motors (see Fig. 35.2). The device is powerful enough to move a patient's arm throughout the workspace against gravity.

The ARM Guide incorporates a linear bearing that can be pointed in different yaw and pitch directions by rotating it about two revolute axes (see Fig. 35.3). Magnetic particle brakes can lock the bearing in a desired orientation. Movement along the linear bearing can then be assisted or resisted via a motor with a cable chain drive. Thus movement across a large workspace can be achieved using only one actuator. This design can be viewed as analogous to a five-bar mechanism, with the revolute elbow joint replaced with a prismatic joint and the shoulder joint actuated only with a brake.

The MIT-MANUS and MIME devices and the ARM Guide all attach to the patient's hand at their end effectors. Attaching at multiple positions on the upper extremity is also possible and may allow improved control over joint loading.[51] Exoskeleton designs that parallel the degrees of freedom of the upper extremity have been proposed.[52]

With respect to locomotion rehabilitators, the Lokomat device uses an exoskeleton approach to interface with the leg (see Fig. 35.5). This device has rotary degrees of freedom at the hip, knee, and ankle and allows for vertical torso movement via a four-bar linkage. The four-bar linkage also counterbalances the weight of the leg exoskeletons through a gas spring. The MGT device incorporates a single-degree-of-freedom crank and rocker linkage to drive the foot along a fixed steplike trajectory (see Fig. 35.4). Another crank connected to the foot crank also drives the torso periodically in the sagittal plane in a gaitlike pattern.

35.3.3 Backdrivability

An important issue in rehabilitator design is backdrivability, defined as low intrinsic endpoint mechanical impedance[53] or simply as the ability to move a device by pushing on its linkages. Good backdrivability has several advantages for rehabilitator design. It allows the patient to move relatively freely when the actuators are not powered. Thus a backdrivable device can record movements of the patient in order to quantify recovery progress. The MIT-MANUS device has been used in this way to assess the smoothness of reaching movements following stroke, providing insight into possible fundamental subunits of movement.[54] The ARM Guide has also been used in this way to assess the role of abnormal limb tone during reaching movements.[20] Backdrivable machines can also be made to fade to nothing by reducing the amount of assistance they provide as patient recovery improves. Additionally, a backdrivable device can be controlled in such a way that it deviates for the desired path when the patient exerts uncoordinated forces, providing direct and natural kinematic feedback

*Selectively compliant, articulated robot arm.

of movement control errors. In contrast, a nonbackdrivable device must rely on force sensing and visual, tactile, or auditory feedback of the sensed force to provide feedback of movement error. A possible safety advantage is that a properly controlled backdrivable machine can "get out of the way" of the patient if the patient rapidly changes his or her pattern of force development.[54]

Backdrivability with substantial actuator power is difficult to achieve. Modern robotic devices commonly incorporate dc electric motors because of their low cost, ease of operation, and good controllability. However, dc electric motors generate maximum power at high speeds and thus must be geared down to produce high torque at lower speeds. Devices that rely on highly geared motors are in turn difficult to backdrive because of the frictional resistance of the gear train and because the frictional resistance of the motors is amplified by the gear ratio. The PUMA robot used in the MIME device is an example of a geared device that is strong but difficult to backdrive. The Lokomat is another example of a device that is difficult to backdrive because of its use of ball-screw drives.

Backdrivable robots can be made using several approaches. Directly driving the robot linkages using large, nongeared electric motors, such as those offered by PMI, Inc., is a common technique and is used by the MIT-MANUS device. The ARM Guide achieves backdrivability using a cable-chain drive attached to its actuated degree of freedom. A clever design for backdrivability that should be mentioned is the three-degree-of-freedom PHANToM haptic input device manufactured by Sensable Technologies, Inc. This device uses small, coreless dc brushed motors (such as those available from Micromo, Inc., and Maxon, Inc.) with cable and capstan drives in which the motor rolls around the cable to produce excellent backdrivability.

Some backdrivability can be endowed to a nonbackdrivable device by sensing the contact force between the device and the environment and moving the actuators to control that force. However, this approach is intrinsically limited by delays in sensors, processors, and actuators.[55,56] Adding a spring in series with a stiff, position-controlled device and controlling the length of the spring provides instantaneous compliant behavior, enhancing backdrivability and allowing good force control across a wide range of forces.[57]

A related concept to backdrivability is *Z-width*, defined as the range of mechanical impedances that a robotic device can achieve while maintaining stable contact with a human.[58] Counterintuitively, intentionally incorporating a passive, mechanical viscosity into a device can allow higher stiffnesses to be achieved, thus expanding Z-width.[58] When low impedance is desired, the passive viscosity can be actively canceled with the robot actuators.

Minimizing the inertia of the linkages is also important for maintaining good backdrivability and maximizing the Z-width of a machine. Carbon fiber and thin-wall aluminum tubing are common choices for strong, lightweight linkages.

Achieving backdrivability often increases cost and complexity. In direct-drive configurations, larger, more expensive motors are needed to produce high torques. Cabling systems such as those used by the PHANToM robot may add complexity. Designing a machine that can actively drive a limb against gravity over a large workspace while maintaining good backdrivability is a key design challenge. Present designs either sacrifice backdrivability for power, as in the case of the MIME and Lokomat devices, or use passive structural support to support the weight of the arm and maintain backdrivability, as in the case of the MIT-MANUS device, the ARM Guide, and the MGT device.

35.3.4 Controller Design

Early research with simple bimanual rehabilitators indicated that simple proportional position feedback control laws can be used to assist in impaired movements.[33,59] In this scheme, the controller specified an actuator force proportional to the error between a reference trajectory and the actual movement trajectory. The reference trajectory was chosen as the normative or "desired" trajectory for the limb. Thus the more the limb deviated from the desired trajectory, the more force was provided by the rehabilitator to drive the limb toward the desired trajectory. The firmness of the assistance was determined by the gain of the position controller.

In the first clinical trial of a rehabilitator, the MIT-MANUS device used a version of this scheme in which the stiffness and damping of the controller were designed to be isotropic and soft using impedance-control techniques.[46] The ARM Guide has also used a version of position control to deliver therapy.[60] In this approach, a smooth desired trajectory is initialized when the patient initiates a movement from a starting position outside a position threshold window. The gains of the position controller are set to be low at first, producing soft assistance, and then are gradually increased, producing increasingly firmer assistance.

Therapy has also been provided using stiff position control with the MIME device.[42] For this device, a desired movement trajectory is initialized when the patient generates a threshold contact force against the machine. MIME then servos the arm along the desired movement trajectory with a high-gain position controller. In bimanual therapy mode, the sensed movement of the contralateral arm determines the desired trajectory of the rehabilitator (see Fig. 35.2).

Counterbalancing has also been proposed as an assistive technique for therapy.[60] In this approach, the rehabilitator compensates for the gravitational forces resisting movement of the arm. In the case of counterpoise control, the tone of the arm can also be measured and actively counteracted. The patient then uses any residual force-generating ability to move. Such control is an enhanced version of two common clinical devices—the mobile arm support and the overhead sling—both of which act to relieve gravity. A clever arm orthosis that counterbalances its own weight plus the weight of the user's arm using springs instead of inertia-increasing counterweights is described in ref. 61. The benefits of these passive counterbalancing approaches are that they allow patients to reach under their own control and that they are relatively inexpensive. However, some patients may exhibit position-dependent weakness that limits their effectiveness.[60]

With respect to locomotion rehabilitation, the MGT and Lokomat devices have also initially used proportional position-control algorithms to drive the legs along step trajectories. It has been proposed that the nonbackdrivable Lokomat device can be made more responsive to patient movement using the above-mentioned technique of sensing contact forces between the legs and device and moving the device to control those forces.[45] Some work has been done to quantify the magnitude and pattern of forces actually applied by therapists during locomotion training.[62] In addition, dynamic motion optimization has been used as a tool for identifying possible torso movement trajectories for facilitating leg movement during locomotion training.[63]

35.3.5 Safety

Safety is of paramount importance when attaching a robotic device to a human. A sensible approach is to use redundant precautions. At the hardware level, actuator size can be limited. A breakaway connector that separates at a prescribed force/torque level can be used to attach the device to the patient's limb. Sensors can be built into contact locations with the patient such that if contact is lost, power is cut to the device, thus reducing the risk of the device swinging freely and colliding with the patient. The rehabilitator's linkage can be designed so that its workspace matches that of the limb as closely as possible, reducing the chance of moving the limb into harmful postures. Mechanical hard stops can also be used to limit motion, and limit switches at those stops can cut power to the actuators when triggered. Software can also set limits on actuator force, position, and velocity, allowing added flexibility in specifying the form of the limiting relationship. A watchdog timer can be used to continually check that the rehabilitator controller is running on time and has not malfunctioned. Handheld or foot-activated cutoff switches can also be incorporated such that the patient can voluntarily cut power to the device.

Care can also be taken in the design of the control algorithm for the device. As mentioned earlier, a backdrivable device may enhance safety since the device can "get out of the way" of the patient if the patient exerts an uncontrolled force, provided the device is controlled with a suitably soft impedance controller. To date, however, the nonbackdrivable MIME device has been operated safely and has not increased joint pain or decreased joint range of motion in over 10 chronic stroke patients.[64] In addition, Food and Drug Administration (FDA)–approved active dynamometers such as the Biodex machine are nonbackdrivable.

35.3.6 Assessment Procedures

A useful feature of rehabilitators is their ability to measure and assess movement ability and recovery. A variety of assessment procedures have been developed for arm rehabilitators. Range, accuracy, and speed of motion are common measures of movement ability. Smoothness of movement, quantifiable using jerk or the number of movement subunits, has been proposed as a measure of recovery.[46] The amount of assistance needed to steer a limb along a trajectory can also be used to quantify patient progress, and off-axis force generation during steered movement can be used to quantify abnormal coordination.[13,17,33] Imposing movement on the arm with the device, either when the arm is relaxed or during voluntary movement, and measuring the resulting resisting force provides a means to quantify abnormal tone.[20] Similar assessment procedures, along with established gait-analysis techniques, can be applied during locomotion rehabilitation.

35.3.7 Networking

As therapy becomes increasingly automated and accessible from home, development of efficient information-transfer software for communication between patients and medical professionals is needed. The feasibility of rehabilitator-based teletherapy for the arm and hand after stroke was recently demonstrated.[65] In this scheme, a library of status tests, therapy games, and progress charts written in the Java programming language were housed on a Web site. This library was used with a low-cost force-feedback joystick capable of assisting or resisting in movement. The system was used to direct a therapy program, mechanically assist in movement, and track improvements in movement ability.

35.4 THE FUTURE OF REHABILITATORS

Many fundamental questions concerning the nature of rehabilitation and neural recovery remain unanswered. Without answers to these questions, it is impossible to optimize rehabilitator designs. This section briefly summarizes three key unanswered questions and then brashly attempts to predict what future rehabilitators will be like despite these uncertainties. For more discussion along these lines, see ref. 66.

35.4.1 Three Things a Rehabilitator Designer Would Like to Know

One question that a rehabilitator designer would like answered is, "What types of movements should be practiced?" Specifically, in the case of arm rehabilitation, should individual joint movements be practiced, or should functional multijoint movements be practiced? Are large excursions needed, or will smaller movements suffice? Is three-dimensional motion better than planar motion? Is bimanual therapy better than unimanual therapy? Is some combination of these approaches better than any one approach? For locomotion rehabilitation, the general type of movement to be practiced is clearer—i.e., stepping—but what stepping trajectory should be practiced? Should a normative trajectory be practiced? Should variability be incorporated into the practiced trajectory? What torso movements should be practiced? The answers to these questions clearly impact rehabilitator design in terms of size, strength, kinematics, and controller parameters.

A second major unanswered question is, "How should the device mechanically interact with the patient?" Should it assist movement, resist movement, tune itself to the level of the patient, or simply "go along for the ride," measuring the patient's movement. As noted earlier, preliminary clinical results with the ARM Guide suggest that unassisted movement exercise may be as effective as mechanically assisted exercise in promoting movement recovery.[48] Other therapy techniques besides assisting in movement have been developed, many of which could be automated.[10] Rehabilitators with strong actuators, precise sensors, and smart processors offer the possibility of applying novel therapeutic manipulations unachievable by therapists, such as state-dependent force fields that are sculpted to take advantage of implicit learning capability[67] or dynamically optimized torso-shifting strategies for assisting in locomotion.[63] The optimal therapy technique will likely depend on the lesion characteristics, stage of recovery, and impairment profile for both arm and leg rehabilitators, further complicating the problem of designing optimal devices and protocols.

A third major unanswered question is, "What should the rehabilitator quantify?" As mentioned earlier, range, speed, tone, straightness, coordination, and smoothness of movement are all affected after neurologic injury, but it is unclear how each of these relates to recovery progress. Which measure or combination of measures is the most sensitive and reliable assessor of ongoing recovery? Which measures are best at motivating and directing patient effort?

Answering these questions will be difficult because it will require many carefully crafted, controlled clinical trials. However, the answers are essential to improving rehabilitator designs in a rational manner.

35.4.2 What Will Rehabilitators Look Like in the Future?

Despite these uncertainties, it is interesting to try to predict what rehabilitators will look like in the future. A best guess is that future devices will be wearable so that the patient can use them to assist in activities throughout the day. These worn devices will address multiple motor systems, including hand movement, arm movement, posture, and locomotion. If simply assisting in movement is found to be a reasonable therapeutic strategy, future devices will likely seek to infer the motor intent of the patient. Then, if the patient imagines or wills a movement (e.g., the patient simply thinks "reach" or "walk"), the device will complete the movement appropriately, even if the patient is severely impaired. As the patient regains movement ability, the devices will then adapt to the level of the patient. Future rehabilitators will also be embedded with smart sensors and networked so that both the patient and the patient's health providers receive continuous information about the patient's status.

35.4.3 What Technology Is Needed to Improve Rehabilitators?

What design problems must be overcome to move toward this future? First, actuators and energy sources must be developed that are lightweight yet powerful enough to allow wearable devices. Novel actuators and energy sources, such as fuel-cell-powered piezo-hydraulic actuators and monopropellant-powered pneumatic actuators, are being developed for wearable exoskeletons targeted at military applications (see *http://www.darpa.mil/DSO/thrust/md/Exoskeletons/briefings.html*). An attractive possibility from an energetic perspective is to use the patient's muscles themselves as the actuators via electrical stimulation of nerves. Millimeter-sized muscle stimulators are being developed that can be implanted near the motor points of individual muscles and controlled via wireless communication from outside the body.[68] Such functional neuromuscular stimulation could provide the supplemental force needed to complete movements and has already shown promise as a therapeutic technique after stroke.[69] Another possibility is to directly stimulate neural networks in the central nervous system with implantable microelectrode arrays.[70] It is currently possible to drive simple limb movements in frogs,[71] rats,[72] and cats[73] via spinal microstimulation.

Second, better algorithms are needed to infer the intent of the patient for each movement. Inferring motor intent from downstream measures of the patient's residual muscle activation (such as force, movement, and EMG) will likely only go so far. Tapping into upstream signals in the cerebral cortex using implantable microelectrode arrays or electroencephalogram techniques may allow greater control, especially for severely impaired patients. Brain-computer interfaces for completely paralyzed patients have been developed that allow typing at slow rates using thought alone.[74] Better brain-computer interfaces will be possible with implanted microelectrode arrays.[75] It was recently demonstrated that hand trajectories during reaching can be inferred in real time from neural signals recorded from microelectrode arrays implanted in monkey motor cortex.[76] Rehabilitator control algorithms that receive information directly from the cortex will also need to be adaptive so that direct cortical control is reduced as biological motor pathways are restored. It is interesting to note that current brain-computer interfaces require substantial patient practice to achieve control. An analogy can thus be made with rehabilitators in that both technologies allow a patient to learn to make new movements by altering neural pathways. Melding the technologies may be synergistic.

Third, sensors and software need to be developed to interpret and monitor daily activity of patients. Micromachined inertial sensors[77] and vision chips,[78] which are rapidly improving in quality and cost, will likely play a role in generating the raw signals for assessing movement. Processing algorithms are needed to log functional activity throughout the day by categorizing movements (e.g., drinking, opening a door, combing hair, etc.) based on sensor readings. Algorithms are also needed to interpret the quality of each movement (e.g., smoothness, speed, range, etc.) to provide a continuous measure of movement recovery.

35.4.4 Rehabilitators and Neuroregeneration Approaches

This chapter has focused on the design of mechatronic/robotic devices for stimulating nervous system recovery. Clearly, however, the development of new cells and connections to replace dead and diseased ones holds great promise for restoring normal movement. Rehabilitators will likely play key roles in the development of cell- and molecule-based neuroregeneration techniques. For example, rehabilitators may act synergistically with neuroregeneration techniques by maintaining the suppleness and mobility of the arms and legs and by preconditioning residual neural movement control circuitry through repetitive sensory-motor training. Thus, as new connections develop, they will find relatively intact motor systems to control. Also, the repetitive sensorimotor stimulation provided by rehabilitators may provide directional guidance to regenerating fibers so that they make appropriate functional connections. Rehabilitators will also provide a means to more precisely quantify the functional effects of neuroregeneration techniques, thus providing a more rational basis for evaluating and optimizing them.

35.5 CONCLUSION

Rehabilitator design is an evolving art and science. As reviewed earlier, progress has been made in designing patient-machine interfaces, mechanical linkages, control algorithms, safety mechanisms, assessment procedures, and networking software. Even with the current uncertainties in optimal therapeutic parameters, existing rehabilitators have been successful in enhancing movement recovery. As scientific advances are made in understanding neurorecovery mechanisms and technological advances are made in sensors, actuators, and computational algorithms, rehabilitators will continue to improve, making rehabilitation therapy increasingly accessible and effective.

REFERENCES

1. G. E. Gresham, P. W. Duncan, and W. B. Stason (1995), *Post-Stroke Rehabilitation*, AHCPR Publication No. 95-0662, U.S. Department of Health and Human Services, Agency for Health Care Policy and Research, Rockville, Md.
2. D. S. Smith, E. Goldenberg, A. Ashburn, G. Kinsella, K. Sheikh, P. J. Brennan, T. W. Meade, D. W. Zutshi, J. D. Perry, and J. S. Reeback (1981), Remedial therapy after stroke: A randomised controlled trial, *British Medical Journal* **282**:517–520.
3. R. S. Stevens, N. R. Ambler, and M. D. Warren (1984), A randomized controlled trial of a stroke rehabilitation ward, *Age and Aging* **13**:65–75.
4. M. L. Aisen, H. I. Krebs, N. Hogan, F. McDowell, and B. Volpe (1997), The effect of robot-assisted therapy and rehabilitative training on motor recovery following stroke, *Archives of Neurology* **54**:443–446.

5. C. Butefisch, H. Hummelsheim, P. Denzler, and K. Mauritz (1995), Repetitive training of isolated movement improves the outcome of motor rehabilitation of the centrally paretic hand, *Journal of the Neurological Sciences* **130**:59–68.
6. A. Sunderland, D. J. Tinson, E. L. Bradley, D. Fletcher, R. Langton Hewer, and D. T. Wade (1992), Enhanced physical therapy improves recovery of arm function after stroke: A randomized controlled trial, *Journal of Neurology, Neurosurgery, and Psychiatry* **55**:530–535.
7. M. Dam, P. Tonin, S. Casson, M. Ermani, G. Pizzolato, V. Iaia, and L. Battistin (1993), The effects of long-term rehabilitation therapy on poststroke hemiplegic patients, *Stroke* **24**:1186–1191.
8. E. Taub, N. Miller, T. Novack, E. Cook, W. Fleming, C. Nepomuceno, J. Connell, and J. Crago (1993), Technique to improve chronic motor deficit after stroke, *Archives of Physical Medicine and Rehabilitation* **74**:347–354.
9. S. Wolf, D. Lecraw, L. Barton, and B. Jann (1989), Forced use of hemiplegic upper extremities to reverse the effect of learned nonuse among chronic stroke and head-injured patients, *Experimental Neurology* **104**: 125–132.
10. D. J. Reinkensmeyer, J. P. A. Dewald, and W. Z. Rymer (1996), Robotic devices for physical rehabilitation of stroke patients: Fundamental requirements, target therapeutic techniques, and preliminary designs, *Technology and Disability* **5**:205–215.
11. R. W. Bohannon (1997), Measurement and nature of muscle strength in patients with stroke, *Journal of Neurologic Rehabilitation* **11**:115–125.
12. S. C. Gandevia (1993), Strength changes in hemiparesis: Measurements and mechanisms, in A. F. Thilmann, D. J. Burke, and W. Z. Rhymer, (eds.), *Spasticity: Mechanisms and Management*, pp. 111–122, Springer-Verlag, Berlin.
13. D. J. Reinkensmeyer, J. P. A. Dewald, and W. Z. Rymer (1999), Guidance based quantification of arm impairment following brain injury: A pilot study, *IEEE Transactions on Rehabilitation Engineering* **7**:1–11.
14. J. P. Dewald and R. F. Beer (2001), Abnormal joint torque patterns in the paretic upper limb of subjects with hemiparesis, *Muscle and Nerve* **24**:273–283.
15. R. F. Beer, J. D. Given, and J. P. Dewald (1999), Task-dependent weakness at the elbow in patients with hemiparesis, *Archives of Physical Medicine and Rehabilitation* **80**:766–772.
16. J. P. A. Dewald, P. S. Pope, J. D. Given, T. S. Buchanan, and W. Z. Rymer (1995), Abnormal muscle coactivation patterns during isometric torque generation at the elbow and shoulder in hemiparetic subjects, *Brain* **118**:495–510.
17. P.S. Lum, C. G. Burgar, D. Kenney, and H. F. M. Van der Loos (1999), Quantification of force abnormalities during passive and active-assisted upper-limb reaching movements in post-stroke hemiparesis, *IEEE Transactions on Biomedical Engineering* **46**:652–662.
18. M. F. Levin (1996), Interjoint coordination during pointing movements is disrupted in spastic hemiparesis, *Brain* **119**:281–293.
19. R. F. Beer, J. P. Dewald, and W. Z. Rymer (2000), Deficits in the coordination of multijoint arm movements in patients with hemiparesis: Evidence for disturbed control of limb dynamics, *Experimental Brain Research* **131**:305–319.
20. D. J. Reinkensmeyer, B. D. Schmit, and W. Z. Rymer (1999), Assessment of active and passive restraint during guided reaching after chronic brain injury, *Annals of Biomedical Engineering* **27**:805–814.
21. K. L. Harburn and P. J. Potter (1993), Spasticity and contractures, in *Physical Medicine and Rehabilitation: State of the art reviews*, **7**:113–132.
22. N. J. O'Dwyer, L. Ada, and P. D. Neilson (1996), Spasticity and muscle contracture following stroke, *Brain* **119**:1737–1749.
23. R. J. Nudo, B. M. Wise, F. SiFuentes, and G. W. Milliken (1996), Neural substrates for the effects of rehabilitative training on motor recovery after ischemic infarct, *Science* **272**:1791–1794.
24. E. Taub, G. Uswatte, and R. Pidikiti (1999), Constraint-induced movement therapy: A new family of techniques with broad application to physical rehabilitation—A clinical review, *Journal of Rehabilitation Research and Development* **36**:237–251.
25. J. Liepert, H. Bauder, H. R. Wolfgang, W. H. Miltner, E. Taub, and C. Weiller (2000), Treatment-induced cortical reorganization after stroke in humans, *Stroke* **31**:1210–1216.
26. B. H. Dobkin (1996), *Neurologic Rehabilitation*, F. A. Davis, Philadelphia.

27. I. Wickelgren (1998), Teaching the spinal cord to walk, *Science* **279**:319–321.

28. H. Barbeau, K. Norman, J. Fung, M. Visintin, and M. Ladouceur (2000), Does neurorehabilitation play a role in the recovery of walking in neurological populations? *Annals of the New York Academy of Sciences* **860**:377–392.

29. A. Wernig, A. Nanassy, and S. Muller (1999), Laufband (treadmill) therapy in incomplete paraplegia and tetraplegia, *Journal of Neurotrauma* **16**:719–726.

30. S. J. Harkema, S. L. Hurley, U. K. Patel, P. S. Requejo, B. H. Dobkin, and V. R. Edgerton (1997), Human lumbosacral spinal cord interprets loading during stepping, *Journal of Neurophysiology* **77**:797–811.

31. F. Amirabdollahian, R. Loureiro, B. Driessen, and W. Harwin (2001), Error correction movement for machine assisted stroke rehabilitation, in M. Mokhtari (ed.), *Integration of Assistive Technology in the Information Age*, Vol. **9**, pp. 60–65, IOS Press, Amsterdam.

32. G. Arz and A. Toth (2001), REHAROB: A project and a system for motion diagnosis and robotized physiotherapy delivery, in M. Mokhtari (ed.), *Integration of Assistive Technology in the Information Age*, Vol. **9**, pp. 93–100, IOS Press, Amsterdam.

33. P. S. Lum, S. L. Lehman, and D. J. Reinkensmeyer (1995), The bimanual lifting rehabilitator: A device for rehabilitating bimanual control in stroke patients, *IEEE Transactions on Rehabilitation Engineering* **3**:166–174.

34. J. A. Cozens (1999), Robotic assistance of an active upper limb exercise in neurologically impaired patients, *IEEE Transactions on Rehabilitation Engineering* **7**:254–256.

35. R. Rao, S. K. Agrawal, and J. P. Scholz (2000), A robot test-bed for assistance and assessment in physical therapy, *Advanced Robotics* **14**:565–578.

36. S. Okada, T. Sakaki, R. Hirata, Y. Okajima, S. Uchida, and Y. Tomita (2000), TEM: A therapeutic exercise machine for the lower extremities of spastic patients, *Advanced Robotics* **14**:597–606.

37. N. A. Siddiqi, T. Ide, M. Y. Chen, and N. Akamatsu (1994), A computer-aided walking rehabilitation robot, *American Journal of Physical Medicine and Rehabilitation* **73**:212–216.

38. Z. Matjacic, I. Johannesen, and T. Sinkjaer (2000), A multi-purpose rehabilitation frame: A novel apparatus for balance training during standing of neurologically impaired individuals, *Journal of Rehabilitation Research and Development* **37**:681–691.

39. D. J. Reinkensmeyer, W. K. Timoszyk, R. D. de Leon, R. Joynes, E. Kwak, K. Minakata, and V. R. Edgerton (2000), A robotic stepper for retraining locomotion in spinal-injured rodents, in *2000 International Conference on Robotics and Automation*, pp. 2889–2894. San Francisco.

40. W. K. Timoszyk, R. D. de Leon, N. London, R. Joynes, K. Minakata, V. R. Edgerton, and D. J. Reinkensmeyer (2002), Robot-assisted locomotion training after spinal cord injury: Comparison of rodent stepping in virtual and physical treadmill environments, *Robotica* (to appear).

41. B. Volpe, H. Krebs, N. Hogan, L. Edelstein OTR, C. Diels, and M. Aisen (2000), A novel approach to stroke rehabilitation: Robot-aided sensorimotor stimulation, *Neurology* **54**:1938–1944.

42. C. G. Burgar, P. S. Lum, P. C. Shor, and H. F. M. Van der Loos (2000), Development of robots for rehabilitation therapy: The Palo Alto VA/Stanford experience, *Journal of Rehabilitation Research and Development* **37**:663–673.

43. D. J. Reinkensmeyer, L. E. Kahn, M. Averbuch, A. N. McKenna-Cole, B. D. Schmit, and W. Z. Rymer (2000), Understanding and treating arm movement impairment after chronic brain injury: Progress with the ARM Guide, *Journal of Rehabilitation Research and Development* **37**:653-662.

44. S. Hesse and D. Uhlenbrock (2000), A mechanized gait trainer for restoration of gait, *Journal of Rehabilitation Research and Development* **37**:701–708.

45. G. Colombo, M. Joerg, R. Schreier, and V. Dietz (2000), Treadmill training of paraplegic patients with a robotic orthosis, *Journal of Rehabilitation Research and Development* **37**:693–700.

46. H. I. Krebs, N. Hogan, M. L. Aisen, and B. T. Volpe (1998), Robot-aided neurorehabilitation, *IEEE Transactions on Rehabilitation Engineering* **6**:75–87.

47. B. Volpe, H. Krebs, N. Hogan, L. Edelstein, C. Diels, and M. Aisen (1999), Robot training enhanced motor outcome in patients with stroke maintained over 3 years, *Neurology* **53**:1874–1876.

48. L. E. Kahn, M. Averbuch, W. Z. Rymer, and D. J. Reinkensmeyer, Comparison of robot-assisted reaching to free reaching in promoting recovery from chronic stroke, in M. Mokhtari (ed.), *Integration of Assistive Technology in the Information Age*, pp. 39–44, IOS Press, Amsterdam.

49. H. Kazerooni and H. Ming-Guo (1994), The dynamics and control of a haptic interface device, *IEEE Transactions on Robotics and Automation* **10**:453–464.

50. B. D. Adelstein (1998), *Three degree of freedom parallel mechanical linkage*, U.S. Patent No. 5816105, 1998.

51. D. Khalili and M. Zomlefer (1988), An intelligent robotic system for rehabilitation of joints and estimation of body segment parameters, *IEEE Transactions on Biomedical Engineering* **35**:138–146.

52. S. H. Scott (1999), *Kinesiological instrument for limb movements*, U.S. Patent No. 6155993, 1999.

53. H. I. Krebs, B. T. Volpe, M. L. Aisen, and N. Hogan (2000), Increasing productivity and quality of care: Robot-aided neuro-rehabilitation, *Journal of Rehabilitation Research and Development* **37**:639–652.

54. H. I. Krebs, M. L. Aisen, B. T. Volpe, and N. Hogan (1999), Quantization of continuous arm movements in humans with brain injury, *Proceedings of the National Academy of Science USA* **96**:4645–4649.

55. D. A. Lawrence (1989). Actuator limitations on achievable manipulator impedance, in *Proc. 1989 IEEE International Conference on Robotics and Automation*, pp. 560–565.

56. D. A. Lawrence (1988). Impedance control stability properties in common implementations, in *Proceedings of the 1988 IEEE International Conference on Robotics and Automation*, vol. **2**, pp. 1185–1190.

57. D. W. Robinson, J. E. Pratt, D. J. Paluska, and G. A. Pratt (1999). Series elastic actuator development for a biomimetic walking robot, in *1999 IEEE/ASME International Conference on Advanced Intelligent Mechatronics*, pp. 561–568.

58. J. E. Colgate and J. M. Brown (1999). Factors affecting the Z-Width of a haptic display, *Proceedings 1994 IEEE International Conference on Robotics and Automation*, vol. **4**, pp. 3205–3210.

59. P. S. Lum, D. J. Reinkensmeyer, and S. L. Lehman (1993), Robotic assist devices for bimanual physical therapy: Preliminary experiments, *IEEE Transactions on Rehabilitation Engineering* **1**:185–191.

60. D. J. Reinkensmeyer, C. D. Takahashi, W. K. Timoszyk, A. N. Reinkensmeyer, and L. E. Kahn (2000), Design of robot assistance for arm movement therapy following stroke, *Advanced Robotics* **14**:625–638.

61. T. Rahman, W. Sample, R. Seliktar, M. Alexander, and M. Scavina (2000), A body-powered functional upper limb orthosis, *Journal of Rehabilitation Research and Development* **37**:675–680.

62. A. Bejczy (1999), Towards development of robotic aid for rehabilitation of locomotion-impaired subjects, presented at *the First Workshop on Robot Motion and Control* (RoMoCo'99), pp. 9–16, Kiekrz, Poland.

63. C. E. Wang, J. E. Bobrow, and D. J. Reinkensmeyer (2001), Swinging from the hip: Use of dynamic motion optimization in the design of robotic gait rehabilitation, *Proceedings of the 2001 IEEE International Conference on Robotics and Automation*, Seoul, Korea, pp. 1433–1438.

64. P. C. Shor, P. S. Lum, C. G. Burgar, H. F. M. Van der Loos, M. Majmundar, and R. Yap (2001), The effect of robot-aided therapy on upper extremity joint passive range of motion and pain, in M. Mokhtari (ed.), *Integration of Assistive Technology in the Information Age: Proceedings of the 7th International Conference on Rehabilitation Robotics, Institut National des Télécommunication*, pp. 79–83, IOS Press, Amsterdam.

65. D. Reinkensmeyer, C. Pang, J. Nessler, and C. Painter (2001), Java therapy: Web-based robotic rehabilitation, in M. Mokhtari (ed.), *Integration of Assistive Technology in the Information Age*, vol. **9**, pp. 66–71, IOS Press, Amsterdam.

66. D. J. Reinkensmeyer, N. Hogan, H. I. Krebs, S. L. Lehman, and P. S. Lum (2000), Rehabilitators, robots, and guides: New tools for neurological rehabilitation, J. Winters and P. Crago (eds.), in *Biomechanics and Neural Control of Posture and Movement*, pp. 516–533, Springer-Verlag, Berlin.

67. J. L. Patton and F. A. Mussa-Ivaldi (2001), Robot teaching by exploiting the nervous system's adaptive mechanisms, in M. Mokhtari (ed.), *Integration of Assistive Technology in the Information Age: Proceedings of the 7th International Conference on Rehabilitation Robotics, Institut National des Télécommunication*, pp. 3–7, IOS Press, Amsterdam.

68. G. E. Loeb, R. A. Peck, W. H. Moore, and K. Hood (2001), BION system for distributed neural prosthetic interfaces, *Medical Engineering and Physics* **23**:9–18.

69. M. Glanz, S. Klawansky, W. Stason, C. Berkey, and T. C. Chalmers (1996), Functional electrostimulation in poststroke rehabilitation: A meta-analysis of the randomized controlled trials, *Archives of Physical Medicine and Rehabilitation* **77**:549–553.

70. H. Barbeau, D. A. McCrea, M. J. O'Donovan, S. Rossignol, W. M. Grill, and M. A. Lemay (1999), Tapping into spinal circuits to restore motor function, *Brain Research Reviews* **30**:27–51.

71. E. Bizzi, S. F. Giszter, E. Loeb, F. A. Mussa-Ivaldi, and P. Saltiel (1995), Modular organization of motor behavior in the frog's spinal cord, *Trends in Neurosciences* **18**:442–446.

72. M. C. Tresch and E. Bizzi (1999), Responses to spinal microstimulation in the chronically spinalized rat and their relationship to spinal systems activated by low threshold cutaneous stimulation, *Experimental Brain Research* **129**:401–416.

73. V. K. Mushahwar, D. F. Collins, and A. Prochazka (2000), Spinal cord microstimulation generates functional limb movements in chronically implanted cats, *Experimental Neurology* **163**(2):422–429.

74. J. R. Wolpaw, N. Birbaumer, W. J. Heetderks, D. J. McFarland, P. H. Peckham, G. Schalk, E. Donchin, L. A. Quatrano, C. J. Robinson, and T. M. Vaughan (2000), Brain-computer interface technology: A review of the first international meeting, *IEEE Transactions on Rehabilitation Engineering* **8**:164–173.

75. M. A. Nicolelis (2001), Actions from thoughts, *Nature* **409**:403–407.

76. J. Wessberg, C. R. Stambaugh, J. D. Kralik, P. D. Beck, M. Laubach, J. K. Chapin, J. Kim, S. J. Biggs, M. A. Srinivasan, and M. A. Nicolelis (2000), Real-time prediction of hand trajectory by ensembles of cortical neurons in primates, *Nature* **408**:361–365.

77. N. Yzadi, F. Ayazi, and K. Najafi (1998), Micromachined inertial sensors, *Proceedings of the IEEE* **86**: 1640–1659.

78. GVPP, Generic Visual Perception Processor Web Site.

P · A · R · T · 8

CLINICAL ENGINEERING

CHAPTER 36
CLINICAL ENGINEERING OVERVIEW

Alfred M. Dolan
University of Toronto, Toronto, Canada

36.1 BIOMEDICAL ENGINEERING

36.1.1 Historical Background

No discussion of what is now referred to as *clinical engineering* can begin without considering first the development of biomedical engineering and, in turn, engineering. "In the past engineering was defined as the facility to direct the sources of power in nature for man's use and convenience."[1] Today, modern engineering involves the application of scientific techniques, theories, and technology for the solution of societal needs. This definition has implications that all engineers, and clinical engineers in particular must address.

As professionals who wish to consider society's problems, engineers must first address three issues. It is essential that engineers must possess skills that are appropriate to the problem being addressed or else exempt themselves from that work. Second, engineers must ensure that the solutions they propose do not impose new and more serious problems on society. Third, engineers must meet the real needs of society and must be instrumental in defining those needs. These issues are of particular relevance for biomedical engineers and clinical engineers, as we shall see.

Biomedical engineering, rather than limiting the range of engineering brought to bear on a societal problem, limits the range of problems addressed to those in the medical and biological area. Indeed, we increasingly recognize the field more as the integration of engineering with medicine and biology for better understanding and solutions of societal needs.

Although we may feel that this recognition is progressive, history shows that the established boundaries of the various professional and scientific fields, and the medical and engineering sciences in particular, are of modern origin. Indications are that the giants of scientific engineering investigation throughout history either ignored or were insensitive to these boundaries. More likely, they recognized that the same scientific and engineering laws applied to all of natural phenomena whether in the physical or biological world.

The origin of biomedical engineering, while obscure, can be traced to ancient times. We know that Alcmaeon, a student of Pythagoras, in the period around 500 B.C., became interested in applying mathematics to the study of physiology. This propensity for crossing boundaries and recognizing common scientific principles persisted throughout most of recorded history. Both Plato (427–347 B.C.) and Aristotle (384–322 B.C.) investigated, observed, and systematized the world they studied around them. Their study of the human body included detailed explanations of the pumping action of the heart and the functions of other organs. Galen, who was the son of an architect and mathematician in Roman times, developed theories on hemodynamics that persisted to the time of the great physician Maimonides 1200 years later. Certainly, this integration of sciences and mathematics is consistent with modern biomedical engineering.

Leonardo da Vinci, one of the greatest engineers in history, also applied physical principles and experimental analysis to the study of physiology and medicine. Santorio invented an instrument to count the pulse—the "pulsilogium"—that was almost certainly the first recorded medical instrument. In addition, for his study of metabolic physiology, he built a metabolic balance chair to monitor body weight over long periods. Robert Hooke, who was a student of Robert Boyle, studied respiratory phenomena as he studied the relationship between pressures and volumes. Was this work biomedical engineering?

Stephen Hales, an English clergyman and physicist, carried out a classic experiment in 1732 to determine blood pressure. He connected a U tube to the carotid artery of a mare and observed the height that blood rose in the tube. Then, using fluid dynamic principles, he calculated the velocity of blood in the aorta, force of contraction, and stroke volume.[2] This work has been the foundation of modern hemodynamics and was used by Bernoulli in his quite accurate calculation of cardiac output in 1737.

Lavoiser (1747–1794), during his studies of oxygen, extended that work to respiration. Galvani, recognized as a pioneer in electrical engineering, observed in 1780 the contraction of a frog's leg muscle when stimulated electrically. Alessandro Volta, also a physical scientist, worked extensively with the "animal electricity" that Galvani had observed. Was this work biomedical engineering?

In the mid-nineteenth century, several fields of science saw explosive activity. Men like Orstead, Gauss, Weber, Laplace, Lagrange, Carnot, Faraday, Maxwell, and Helmholtz were contemporaries in physics and chemistry. Divergence and segregation of the fields gradually occurred at this time, but some workers, such as Helmholtz, crossed these boundaries.

Helmholtz, with his early interests in physics and mathematics, undertook a career in medicine in 1838. He applied his knowledge of thermodynamics to the study of conservation of energy and metabolic physiology in 1847. He estimated the velocity of nerve transmission, invented the ophthalmoscope, and investigated the physiology and psychology of hearing, a topic of investigation to this day. Surely Helmholtz was a biomedical engineer.

In 1837, Samuel Haughton quite clearly applied mechanical engineering principles and contemporary engineering technology to acquire new knowledge about biomechanics and cardiovascular physiology.[3] In his Preface, he referred to ". . . the mutual advantages obtainable by Anatomists and Geometers from a combination of the Sciences they cultivate." Using tools that we would consider hopelessly inadequate, he quite accurately deduced the velocity of blood in an artery, capillary resistance, blood pressure, cardiac work, and the time required for the circulation of the blood.

These investigators, scientists, and engineers, like Helmholtz and Haughton, recognized that the same laws apply to studies of physics, engineering, or biology—the unity of the sciences. This understanding represents the greatest challenge and potential both on the research and development side and on the applied side of the field of biomedical engineering today. This fact makes clear why developments in fields such as DNA sequencing and communications theory, which we think of as separate fields, can be successfully melded to yield outstanding gains in our understanding and treatment of disease. It also should make clear to us that techniques and technology, and not just instrumentation, that may derive from traditional engineering applications are equally applicable in health care. This application side defines the portion of biomedical engineering that has come to be called *clinical engineering*, where all aspects of engineering and technology come to play directly in the health care field.

36.2 CLINICAL ENGINEERING

36.2.1 Development

Clinical engineering developed in health care facilities around the world over the last four decades of the twentieth century. There was widespread recognition in professional and government circles of the technological explosion that had affected society in general and health care in particular.[4] A series of workshops held in 1972 provided a forum for the discussion of the need for an engineering approach to effect some control on this technology.[5,6] Let us consider three of the factors that can be cited as having had a great influence on the way in which clinical engineering has developed in the hospital:

1. The rapid influx of technology and its resulting instrumentation into the hospital primarily in the 1960s and 1970s
2. The recognition of an electrical safety issue associated with the increase in clinical instrumentation coming in contact with the patient
3. The move to develop a certification process for engineers in hospital clinical settings

These three factors served to encourage the development of the field and at the same time served to define the field itself. The increased prevalence of technology and medical instrumentation in hospitals meant that hospital organizations had to develop ways to take care of these devices. With this rapid proliferation of what were primarily electrical devices in the vicinity of the patient, some assurance of the electrical safety of the patient needed to be provided. Finally, the skills, training, and education of engineers and technologists who were to become involved in these activities needed to be vetted.

Perhaps the start of the new field, that was to become clinical engineering can be attributed to the establishment in January 1942 of the U.S. Army Medical Equipment Maintenance Course.[7] This developed into the Army Medical Equipment School in Denver, Colorado, and the Air Force Training Wing at Sheppard Air Force Base in Texas. Thus maintenance of medical equipment became the first of the defined functions of what was to become the clinical engineering field.

Others recognized other problems in health care that were associated with this rapid influx of technology and that could be amenable to an engineering approach. Robert Rushmer[8] understood the need for engineering involvement not just in maintenance but also in all parts of health care. He emphasized the need for optimization and improvement of the effectiveness, safety, and benefit of existing technologies as a means for their more efficient and appropriate use by health care professionals and the resulting improved patient care. He further recognized the need for medical engineers to help to define the technological applications beyond the critical-care applications of the day. He understood the need for the engineering approach and involvement in all parts of health care from hospitals to home care.

T. D. Kinney[9] in 1974 broadened the role of biomedical engineering in the hospital as he anticipated the potential for savings in chemistry laboratories through the work of biomedical engineering. He estimated that 90 percent of advanced laboratories were automated in some way at that time.

Raymon D. Garret specifically addressed the computing needs of the hospital.[10] While the solutions available in 1973 clearly were inadequate, the analysis of the hospital as a system remains valid today and serves as an excellent description of an opportunity for clinical engineering involvement.

This brief survey illustrates that there was recognition of the problems that existed in health care in that time period and that there was recognition of the potential for biomedical engineering to effect solutions to those problems. It is interesting to note that the initial understanding of the role engineering could play in health care did not come necessarily or solely from engineers. Rather, that broader understanding came from visionaries such as Robert Rushmer, a cardiovascular research scientist, and Cesar Caceres, a clinical cardiologist, whose recognition of what the range of engineering involvement in health care could be defines, in essence, the role of clinical engineering.

Cesar Caceres offered perhaps the most insightful description of clinical engineering: "An engineer who is trained for and works in a health service facility where he or she is responsible for the direct and immediate application of engineering expertise to health and medical care."[11] Except for the constraint this definition places on the practice of clinical engineering to that being within the precincts of the hospital, this definition accurately outlines the role of clinical engineering. It is important to notice that this definition assumes both appropriate engineering expertise and direct benefit to patient care or health care, which are two of the important issues previously cited that engineers must address.

The next step in the beginnings of what has been called clinical engineering can be traced back to the late 1950s and a convergence of the first two factors. Concern arose over electrical safety for patients in hospitals resulting from the proliferation of electronic equipment in the vicinity of the patient devices.[12,13] The results of a number of reports and analyses[14,15] were that the need for biomedical engineering expertise in the hospitals centered for many years on the electrical safety aspect.[16] S. Powers and D. G. Gisser, in discussing the issues related to monitoring of trauma patients in 1974, again brought forward the electrical safety issues.[9] Indeed, some clinical engineering departments were built on electrical safety testing. Even today, the concern for safety sometimes masks other clinical engineering functions at some hospitals.

The other great influence on the development of the field of clinical engineering was provided by the efforts to establish peer recognition of this new group of engineers who practiced in health care institution. In the early 1970s, Cesar Caceres and Tom Hargest coined the name *clinical engineering* to describe the new field of engineering practice that had gradually been developing in hospitals over the previous decade. Since it was directed toward the improvement of patient care in the clinical environment, the somewhat curious name stuck. They also can be considered as pioneers in developing a certification program for vetting the competence of engineers wishing to practice this new field. Modeled after other speciality fellowship training and certification programs in other medical fields such as cardiovascular surgery, a certification commission was set up in 1974, with the Association for the Advancement of Medical Instrumentation agreeing to serve as secretariat.

Initially, two boards of examiners were established that reported to a certification commission. Subsequently, other boards of examiners were established, starting in Canada in 1979. The bylaws and terms of reference of each board, such as the U.S. Board of Examiners for Clinical Engineering and the U.S. Board of Examiners for Biomedical Technicians, were subject to review and approval by the certification commission. The boards in turn had the responsibility of establishing the examination process, setting appropriate examinations, and conducting the examinations for clinical engineering and biomedical technician applicants. Following completion of that process, the boards of examiners made recommendations to the certification commission, and certification was approved or not approved by the commission.

In parallel, the U.S. Board of Examiners for Clinical Engineering set up a similar, albeit much smaller, certification program. The two programs were merged in July of 1983 to form the International Certification Commission (ICC). Currently, the ICC has seven boards of examiners reporting to it.

Certification influenced development of clinical engineering in three ways. First, it established and crystalized the name of the field as *clinical engineering*. Second, it provided some assurance of the competency of clinical engineering practitioners to the health care facilities where they practiced. Third, however, it tended to result in clinical engineers defining themselves in terms of the certification process. The examination process in particular, which in the U.S. boards of examiners was heavily oriented toward electronics, had an effect on both the type of engineering people who went into the field and the educational programs that provided the training. The wide range of management and broad health care issues identified by such people as Rushmer and Caceres tended not to have as heavy an emphasis.

This background provides a basis for the understanding of the role clinical engineering occupies in health care today. Clinical engineering developed with an early large emphasis on the maintenance, electrical safety, and electronics aspects of medical equipment. Some people, such as Scott[12] and

Rushmer,[8] encouraged the consideration of broader safety aspects in health care. Gordon repeatedly emphasized that the assessment, evaluation, and management of health care technology rather than of medical equipment was key. Caceres consistently outlined the broadest possible role for clinical engineering in health care to include all aspects of management of technology throughout the spectrum of health care delivery.[17] Let us then consider that role.

36.3 ROLE OF CLINICAL ENGINEERING

The application of engineering techniques, technology, and theory to the solution of health care problems and a management of technology in health care is, by definition, clinical engineering. The American College of Clinical Engineering emphasises both patient care and management by defining a clinical engineer as "a professional who supports and advances patient care by applying engineering and management skills to health care technology."[18] More than two decades ago, Scott and Caceres separately identified a number of clinical engineering responsibilities that remain valid today.[19,20] Their combined list of clinical engineering responsibilities includes development and management of medical systems, education, maintenance, safety, clinical research and development, and analysis and development for the more effective patient care systems. Betts, in considering the changing role of clinical engineering in the 1980s, emphasized the need for management skills in addition to the requisite technical knowledge.[21] More recently, Bronzino identified technology management, risk management, technology assessment, facilities design and project management, and training as the key functions for a clinical engineering department.[13] The clinical engineer must be directly involved with solutions in any of these problem areas at the delivery level if available solutions are to be effected.

He or she must provide education for nursing, medical, and paramedical staff to facilitate their understanding of present technology and future trends. In consultation with medical and administrative staff, he or she must ensure that equipment purchases and hospital designs and systems are optimal and that technology acquisitions are appropriate; he or she must engage in applied research and development at all levels to improve patient care and make provisions for the safe and effective use of technology. Accordingly, the following functions can be taken as descriptive of the role of clinical engineering.

36.3.1 Education

- Prime responsibility for making provisions for training and education associated with technology and instrumentation used in the hospital
- Education of clinical engineering staff
- Education of health care facility staff

The objective is to provide to all medical equipment users with the understanding and knowledge necessary for the proper use of all patient care equipment, calibration, routines, instruments errors, safety procedures, and possible hazards.

36.3.2 Clinical Research/Development

- Design of new equipment, patient aids, and techniques to aid in patient care
- Assistive devices

While the old adage "Don't build if you can buy" certainly should apply in all clinical engineering departments, it is clear that in most hospitals there is a continuing need for what we could call *customized devices*. A clinical engineering department in a rehabilitation health care facility may be heavily involved in the design or modification of assitive technologies. Other examples arise from the particular type of patient care or particular procedures encountered in a specific hospital. These may be as simple as the need to provided electrical isolation for an amalgam of devices, which would require an understanding of medical electrical system standards. More commonly, it would involve providing assistance to other departments in clinical research projects and the development of solutions to current clinical problems unique to the facility. Whatever the application, it is important that clinical engineering have the facility or be able to serve as an interface so that specialized devices, techniques, or accessories can be made available for more effective patient care.

36.3.3 Computing Applications

- Development and management of hospital and patient information and data-acquisition systems

Certainly the ubiquitous presence of computing in all functions of the hospital is well recognized. These functions range from what we recognize as mainframe computers and servers to word processors or personal computers and handheld computers and finally to processors incorporated within diagnostic and therapeutic devices. It is incumbent on the hospital, therefore, to provide some management of computing in the hospital. Traditionally, this constituted establishment of a department with such a function and starting with the large-scale accounting computing function. Increasingly, it is understood that the overlap of other computing needs such as patient records, patient information systems, medical device data and storage systems, and diagnostic and direct patient care computing systems requires a much broader approach. Such an approach is consistent with the role of clinical engineering.

36.3.4 Facility Planning

- Advising and consulting with administrative and health care staff on matters related to the impact of technological developments in health care facility planning
- Standards and regulations

Botz has emphasised that the application and proper use of technology entail the appropriate management of all resources, including equipment, personnel, supplies, and space.[22] It is obviously important then that the requirements placed on a health care facility by a particular technology become part of the considerations for the technology and related equipment. However, it must also be reflected at a very early stage in the planning process in the planning and design of the facility itself. This may be as simple as planning for an appropriate electrical power supply or adequate heating and ventilation for the installation of a device. Ensuring that relevant standards are met is assumed. It may be the much more intensive planning required to ensure proper integration of the technology and all associated equipment in the architectural design.

36.3.5 Systems Management

- Systems analysis
- Design and evaluation of health care systems
- Quality management

System analysis and synthesis have long been a well-established engineering approach that has been used in many different branches of engineering. This formal way of examining the inputs, outputs, and transfer functions applied to a health care system or portion thereof has provided some important insights and has identified opportunities for improvement.[23] In a similar way, the relatively recent recognition of the applicability of quality-management approaches long used in engineering offers the same potential.[24] Clinical engineering departments are uniquely positioned for involvement in these activities.

36.3.6 Equipment Management

- Consultation with other health care staff in the planning and purchase of equipment
- Prime responsibility for maintenance and modification of equipment

The large number of individual devices in modern hospitals present a unique set of problems in management and represent a major role for clinical engineering. This set can broadly be segregated into:

- *Planning*. Functional program review.
  - New equipment planning
  - Renewal equipment planning
- *Acquisition*. Definition of clinical requirements.
  - Survey of available equipment
  - Specification writing
  - Equipment evaluation
  - Generation of purchase documents
  - Vendor selection
  - Acceptance testing
- *Control*. Inventory management.
  - Maintenance
  - Repair
  - Test and calibration procedures
  - Scheduled inspection
  - Safety program

The technology planning function is one of the most cost-effective roles for clinical engineering.[25] Equipment planning, which includes planning for new and renewal equipment, is an important part of this process. It begins at the functional programming stage of planning. It involves continuing participation in the organizational wide strategic planning process so that equipment-related issues can be appropriately identified. These issues include life cycle, new technology, obsolescence, capital costs, personnel and training requirements, maintenance requirements, operating costs, and facility design considerations. The hospital needs to have all the information to make a decision on the economic and technical viability of either new or existing equipment.

The acquisition of equipment is the next major part of effective equipment management. It follows and is indeed part of the equipment planning phase. Flowing from definition of requirements, specifications are developed. Based on those specifications, a subset of possible equipment is evaluated for conformance to the specifications or standards, and purchase of the most appropriate equipment is initiated. When the equipment arrives at the hospital, the acceptance or incoming inspection of the equipment is a very common function that clinical engineering fulfils.

For all the devices deemed necessary by the hospital to be utilized effectively, an inventory of those devices needs to be in place. Such a system, therefore, however it is maintained, must be able to provide for the hospital the type of device, the capability of each device, the location, the status, and some way to ensure its availability at the point of care. To meet these requirements, in turn, the devices must be adequately maintained and calibrated. All these functions are routine clinical engineering functions. To ensure that this takes place, some sort of inventory, tracking, maintenance and repair log, and calibration activity has to be provided. Typically, safety testing is also part of this process.

36.3.7 Patient Safety/Risk Management

- Prime responsibility for safety of medical equipment and systems in-patient care areas

The management of the safety risk associated with the use of medical devices is an important function for clinical engineering. While in some facilities this may be incorrectly confined to incident investigation, the broader understanding of the function includes identification of hazards, establishing realistic estimates of the risks associated with those hazards, and instituting suitable control measures to minimize those risks. While the concepts of risk management are more than 400 years old, Leeming was the first to specifically apply the concepts to the clinical engineering field.[26] His work in the electrical safety field allowed realistic estimates of the risk associated with the use of multiple electrical devices in the patient care vicinity and provided a basis for a rational deployment of resources and choice of risk-control measures. Much more recently, there is an increasing recognition of the risk-management approach in clinical engineering for technology management as well as for safety applications.[27]

36.3.8 Regulatory Activities

- Codes
- National and international standards
- Regulations and accreditations

Health care in all countries operates subject to a broad variety of standards and regulations. In many cases, these are technological in nature and should come under the purview of clinical engineering. Feinberg emphasized the importance of this point in hospitals in the United States in listing some of the codes and standards pertaining to the clinical engineer.[28] In all jurisdictions, regulations such as municipal, provincial, or national building codes have an impact on health facility design, for example. Furthermore, technical standards affect every device in the hospital. These standards are increasingly international standards but may also be developed by a national or local standards body. Finally, in most countries, such as in the United States, extensive federal regulations govern virtually all aspects of medical devices design, distribution, and application. Clinical engineering, the department responsible for such devices, must be conversant with the appropriate standards and regulations.

Educational programs for clinical engineers and biomedical engineering technicians and technologists reflect the need for training specific to the perceived role of clinical engineering. Figure 36.1 illustrates a set of guidelines and provides a matrix of duties for graduate students during internship periods in one such program. It is provided as an example only.

With this background of the current role of clinical engineering in health care, we can consider the future role for the field and the way it will address the three issues cited earlier:

- Defining and meeting societal needs
- Ensuring that technological solutions are responsible solutions
- Continuing development of required professional skills

36.4 FUTURE DEVELOPMENT OF CLINICAL ENGINEERING

In keeping with the definition of clinical engineering as supporting and advancing patient care through application of engineering management and technology, it is important to recognize that the

Clinical engineering role	Internship activities
Equipment design	Circuit design, fabrication
	Circuit modification
	Mechanical design, modification
	Pneumatic/hydraulic design, modification
	Computer hardware, software
Facility planning	Assessment of applicable standards
Information systems	System design
	Programming
Professional affairs	Associations/societies
	Government
Education	Education CE staff
	Education health care facility staff
Equipment planning	Functional program strategic planning
	Capital planning
	Renewal
Equipment acquisition	Definition of clinical requirements
	Assessment available equipment
	Specification writing
	Equipment evaluation
	Vendor selection
	Acceptance testing
Equipment control	Inventory control
	Test and calibration procedures
	Scheduled inspection
	Repair
Safety program	Hazard report management
	Risk evaluation
	Incident investigation
System analysis	
Regulatory affairs	Codes
	Standards
	Regulations
Research	Literature review
	Experimentation
	Clinical trials
Continuing education	Courses, conferences, site visits
	Literature, journals
	Practice and lab
Departmental management	Policy and services
	Supervision personnel
	Health care facility meetings
	Coordination activities
	Budgeting
Documentation	Reporting
Technology planning	Strategic
	Short term
	Personnel planning
Technology assessment	Components
	Materials
	Methodology
	Systems

FIGURE 36.1 Example of experience matrix guidelines for clinical engineering students in one clinical engineering master's training program. (*From University of Toronto, Institute of Biomedical Engineering, Internship Guidelines, 2001.*)

Context for the hospital of the future	
• Advancing science and technology	• Closer role for research
• Diagnosis and therapy developments	• External health care challenges
• Information technology	• Epidemiological shifts
• New modalities of treatment	• Economics

FIGURE 36.2 Issues affecting design for hospitals in the future. (*Adapted from The Hospital of the Future, Smith Group, 2001.*)

field of clinical engineering will need to continue to develop as health care develops. This will include developments both in sophistication, such as physiological functional imaging or highly integrated information systems, and in scope, such as wellness care or distributed clinical care. Moreover, it will be important that every clinical engineering program or department obey the laws of entropy. These laws describe the natural tendency for systems to run "down hill" rather than evolve to something bigger and better. Very active development by individual departments and by the field will be required for the field to keep abreast of changes and issues in health care. Let us first consider some of these changing health care issues and how they may influence the role of clinical engineering in the future.

In a later chapter in this book J. Currie carefully discusses the impact that technology and other factors have on health care facility design (see Chap. 38). Figure 36.2 lists some of these factors. Advances in science and technology will lead to continuing developments in diagnosis and therapy. The ubiquitous and expanding presence of information-technology applications in health care is widely accepted. New modalities for treatment stemming from current research in areas such as tissue engineering and biomaterials will come to fruition. The incursion of new or newly important disease entities coupled with shifts in population composition will present challenges to health care systems everywhere. Finally, the economic impact of all these factors on health care systems needs to be taken into account.

More than a decade ago, Robert Rushmer described some of these same health technology issues for the twenty-first century.[29] In reviewing research and development at the Imaging Research Center at the University of Washington, he provided a summary of sophisticated new imaging techniques ranging from three-dimensional confocal microscopy for displaying cellular structure to the more common Doppler imaging of flow (Fig. 36.3). These examples in just this one area of technology serve to illustrate three important related health care issues that were also brought out in Currie's work, mentioned previously. These key issues are the rapid development of new technology, ubiquitous incorporation of computing technology, and the communication and information technology imperative.

Imaging technique
Scanning, tunneling microscopy
Biosensors
X-ray microanalysis
High-resolution optical/electron microscopy
3D confocal microscopy
CT, MRI, PET, SPECT
Metabolic and flow imaging
Burn-depth imaging
Ultrasound monitoring trauma
Monolayer analysis
Microtonometry
Cancer karyotyping
Quantitative cytometry
3D cardiac reconstruction
Ultrasound endoscopy
Impedance imaging

FIGURE 36.3 Applications of imaging analysis. (*Adapted from Rushmer, IEEE Engineering in Medicine and Biology, 1990.*)

First each of these new imaging techniques is based on either new or newly applied scientific principles in health care. Information on the structure, chemical composition, physiological function, and metabolic function of organs or cells can be generated. Imaging of the current density distribution in living muscle or neurological tissue provides a representation of the electrophysiologic function of tissues.[30] Health care practitioners at all levels, including clinical engineering practitioners, will need to become conversant with these underlying scientific principles if the new technique is to be introduced and utilized effectively. Current examples can be drawn from the differences in three of the commonly available scanning techniques, magnetic resonance imaging (MRI), positron-emission tomography (PET), and single-positron-emission computed tomography (SPECT). Since each is based on the recording of different emissions from tissues generated in different ways, each therefore has different clinical capabilities. This requires an ongoing educational and training effort by health care facilities, as well as consistent assessment of the suitability of the new technology in each particular health care environment.

Second, the previously cited examples of MRI, PET, or SPECT imaging are obvious examples of technologies that require extensive computing power for transduction, manipulation, and display of information. Less obvious is the ubiquitous presence of digital processors in virtually all medical devices that may be serving as either input or output stages or be running algorithms to generate diagnostic data. The signal from a finger or ear transducer requires the application of an algorithm to convert it to a representation of oxygen saturation. An appreciation by clinical engineering of the computational requirements for any of these devices is essential.

Finally, the data and information generated in all these imaging examples must be converted to usable information and communicated effectively to care providers around the health care facility if patient care is to be affected positively.[31] Information must be accurate, timely, and appropriate to the user. This demands not only that the technological infrastructure be provided for moving and storing information but also that new, innovative user-friendly ways of presenting information must be available.

Mack, in discussing the development of minimally invasive and robotic surgery, brought out similar issues.[32] The development of high-resolution CCD chips, together with high-intensity lighting and high-precision miniaturized handheld surgical instruments, has allowed the application of minimally invasive techniques for a wider variety of procedures. The further development into robotic and computer-assisted surgery will place a great reliance on communications and telemetry in particular. Griffith and Grodzinsky also recognized engineering advances in these same areas of microinstrumentation and virtual surgery techniques.[33] New challenges will be presented for clinical engineering not only in keeping abreast of these new surgical technologies but also in integrating the related communications and computer technology into the hospital environment.

While these changes in health care are occurring, there are other external factors that will also affect the clinical engineering field:

- *Internationalization of science and technology.* The science and technology used in health care are universal. Coupled with global manufacturing and international companies, it is clear that science and technology are worldwide in application.
- *Integration of technology.* In the examples provided by some of the preceding authors, it is evident that health care technology is sophisticated and cross-disciplinary. The application of communications theory to improving the accuracy and speed of DNA sequencing is such an example.[34] In addition, devices are increasingly interdependent and intercommunicate freely as they are combined into systems.
- *Communications technology.* Information is the new currency of modern health care delivery. The data generated at the instrument site must be transformed into information usable for diagnosis and therapy. The health care facility or system must have the ability to move that information to the patient care site. Considerations of information storage and transfer and the protocols for making such transfers are all now part of health care's technology management decisions.
- *Regulation of health care technology.* The health care medical device field is heavily regulated in all countries of the world. The impact this has on the application of technology at an individual hospital level varies with the sophistication of the hospital. However, three examples of international standards that influence the safety and effectiveness of every medical device in the hospital illustrate the pervasive nature of these standards and regulations. These three standards are the medical electrical standard, IEC 60601, the ISO quality-management standard, ISO 13485, and the ISO risk-management standard, ISO 14971. Clinical engineering personnel need to be fully informed about each of these standards and the impact they have on their hospitals.[35–37]

With the changing health care environment both within and without the hospital, what can we say about the future role that clinical engineering should play? Enderle et al.[38] have offered some helpful insight into the possible future role for clincal engineering, including computer support, telecommunications, facilities operation, and strategic planning. It is perhaps most appropriate, however, first to return to the definition of clinical engineering as being the application of engineering techniques, technology, and theory to the solution of health care problems and a management of technology in health care. While the environment is changing, this definition is not. We understand that there is a moving technological horizon in clinical engineering. In the 1960s, the concern was electrical patient safety; in the 1970s and 1980s, it was equipment acquisition; and in the 1990s, planning came more to the fore. Gordon has consistently strongly emphasized that clinical engineering must concentrate on management.[22] Botz voiced a similar theme, stating that management of technological resources necessitated the management of equipment, personnel, supplies, and space.[7] Technology management, then, must be the overriding theme for clinical engineering departments in the future. Within this broad theme we can identify four key roles that clinical engineering must play that are consistent with those of Enderle et al. and that will ensure that the field deals effectively with the factors that have been previously cited in the changing health care field.

- *Strategic planning.* As new scientific advances are considered for implementation, a clear understanding of the resource, economic, and technical implications must be developed by the hospital planning team. Clinical engineering needs to be involved in all these aspects of planning for the hospital system from the generation of the functional program to the final implementation of the technology within the hospital.
- *Technology management.* Once a technology is in place in a health care system, it must be managed in the best interests of the patient, the users, and the hospital. This management function includes all the traditional maintenance and safety considerations but will increasingly include more advanced management functions such as benefit analysis, reliability, risk-management techniques, and sustainability.
- *Information technology.* The overwhelming presence of information and communications technology in health care will only increase. Incorporated in this broad role is computing support, network support, integration of systems, data transmission and storage considerations, and telemedicine. It is imperative that clinical engineering play a part in this key aspect of modern health care.
- *Standards and regulatory activity.* Health will continue to operate in a highly regulated environment. Standards and the regulations and accreditations that implement those standards affect the hospital explicitly or implicitly. Quality management and risk management are important aspects of technology management, which are now very well defined in national and international standards as well as in many accreditation processes. Technical standards now cover virtually every device used in the hospital. Effective use of these standards and regulations as part of the hospital management will be essential.

Health care and the technology supporting health care will continue to develop. This accelerating trend in scientific developments in health care will continue, and this will, in turn, drive the development in clinical engineering. Active development in clinical engineering will be required if the field is to continue to thrive.

REFERENCES

1. Dolan, A. M., Wolf, H. K., and Rautaharju, P. M. (1975), Medical engineering: Past accomplishments and future challenges, *Professional Engineer* **15**(4):11.
2. Stephen Hales, (1733), *Statistical Essays Containing Haemastatics*, Inn and Manley Publishers, London.
3. Samuel Haughton, (1873), *Principles of Animal Mechanics*, Longmans, Green and Company, London.

4. Report of the Ontario Council of Health (1974), *Biomedical Engineering and Biophysics*, Ontario, Canada.
5. Hopps, J. A. (ed.) (1972), First International Biomedical Engineering Workshop Series: I. *Biomedical Equipment Maintenance Service Programs*, American Institute of Biological Sciences, Bioinstrumentation Advisory Council, Washington, D.C.
6. Craig, J. L. (ed.) (1972), First International Biomedical Engineering Workshop Series: I. *Biomedical Equipment Maintenance Service Programs*, American Institute of Biological Sciences, Bioinstrumentation Advisory Council, Washington, D.C.
7. Gordon, G. F. (1990), Hospital technology management: The tao of clinical engineering, *Journal of Clinical Engineering* **15**:111.
8. Robert F. Rushmer, (1972), *Medical Engineering, Projections for Health Care Delivery*, Academic Press, New York.
9. J. H. U. Brown and J. F. Dickson III (eds.) (1974), *Advances in Biomedical Engineering* **4**:113–150.
10. Raymon D. Garret (1973), *Hospitals: A System Approach*, Auerbach Publishers, Philadelphia.
11. C. A. Caceres (ed.). (1980), *Management and Clinical Engineering*, Artech House Books, Norwood, Mass.
12. Scott, R. N., and Paasche, P. E. (1986), Safety considerations in clinical engineering, *CRC Reviews in Biomedical Engineering* **13**(3):201–226.
13. Bronzino, Joseph D. (ed.) (1995), *The Biomedical Engineering Handbook*, CRC Press, Boca Raton, Fla.
14. Dalziel, C. F. (1956), Effects of electric shock on man, *Transactions of the IEEE Professional Group on Medical Electronics* **5**.
15. Weinberg, D. I., Artley, J. L., Whalen, R. I., and McIntosh, H. D. (1962), Electric shock hazards in cardiac catheterization, *Circulation Research* **11**:1004.
16. Todd, James S. (ed.) (1971). Symposium on intensive care units. *Medical Clinics of North America* **55**(5): 1095–1105.
17. Caceres, C. A. (ed.) (1980), *The Management of Technology in Health and Medical Care*, Artech House, Norwood, Mass.
18. Enhancing patient safety; The role of clinical engineering, white paper, American College of Clinical Engineering, Plymouth Meeting, Pa. 2001.
19. Scott, R. N. (1976), *Portrait of Clinical Engineering: The Report of a Study of the Role of the Professional Engineer in Health Care*, CMBES Monograph 1976–1, pp. vi–85, CMBES, Ottawa, Canada.
20. Cacers, C. A. (1981), Clinical engineering: A model development in medical personnel utilization, *Medical Instrumentation* **15**(1):8.
21. Betts, W. (1983), The changing role of the clinical engineer: The need to develop management skills, presented at the AAMI 18th Annual Meeting., Dallas, Texas.
22. Botz, C. K. (1983), Technology resource management in hospitals: A new opportunity, presented at the AAMI 18th Annual Meeting., Dallas, Texas.
23. Gordon, D. B. (1998), A strategic information system for hospital management, Ph.D. thesis, University of Toronto, Canada.
24. White paper (2000), International Workshop Agreement (IWA 1): Quality management systems—Guidelines for process improvements in health service organizations, based on ISO 9004:2000, 2d ed. 2000-12-15, reference number IWA 1:2001(E).
25. Bauer, S., Dolan A. M., and Ramirez, M. R. (1992), Technology planning programs for clinical engineering departments, presented at the 18th CMBEC Conference, Toronto, Canada.
26. Leeming, M. N. (1976), A probablistic approach to safety design: The two-element environment, *Medical Instrumentation* **10**(1–2):2–8
27. Brewin, D. (2001), Design of an optimization system for a medical engineering device maintenance program, M.H.Sc. thesis, University of Toronto, Canada.
28. Feinberg, B. N. (1986), *Applied Clinical Engineering*, Prentice-Hall, Englewood Cliffs, N. J.
29. Rushmer, R. E. (1990), Technologies and health in the 21st century, *IEEE Engineering in Medicine and Biology* **9**(2):50.
30. Joy, M. L. G., Lebedev, V. P., and Gati, J. S. (1999), Imaging of current density and current pathways in rabbit brain during transcranial electrostimulation, *IEEE Transactions on Biomedical Engineering* **46**(9): 1139.

31. Gordon, D. B. (1998), A strategic information system for hospital management, Ph.D. thesis, University of Toronto, Canada.
32. Mack, J. M., (2001), Minimally invasive and robotic surgery, *JAMA* **285**(5):568.
33. Griffith, L. G., and Grodzinsky, A. J. (2001), Advances in biomedical engineering, *JAMA* **285**(5):556–561.
34. Davies, S. W., Eizenman, M., and Pasupathy, S. (1999), Optimal structure for automatic processing of DNA sequences, *IEEE Transactions on Biomedical Engineering* **46**(9):1044–1056.
35. ISO 14971 (2000), Medical devices: Application of risk management to medical devices (reference number ISO 14971:2000E).
36. ISO 13485 (1996), Quality systems: Medical devices—Particular requirements for the application of ISO 9001 (reference number ISO 13485:1996E).
37. IEC 60601-1 (1988), Medical electrical equipment: Part 1. General requirements for safety (ISBN 2-8318-1561-4).
38. Enderle, J. D., Blanchard, S. M., and Bronzino, J. D. (1999), *Introduction to Biomedical Engineering*, Academic Press, New York.

CHAPTER 37
TECHNOLOGY PLANNING FOR HEALTH CARE INSTITUTIONS

John Smith
Hospital for Sick Children, Toronto, Canada

37.1 INTRODUCTION

In health care, in common with many industries, the application of technology in order to advance business objectives is essential to remain competitive. It is clear that technology can enable, and even facilitate, the creation of business objectives; however, unless the business objectives are themselves sound, successful outcomes cannot be guaranteed. Thus the concept that technology alone is sufficient to sustain a business model has serious flaws. This serves to emphasize the importance of technology planning as an integrated component of developing and implementing business strategies.[1]

The same principles apply to health care, which by its nature is very much a social industry catering to the needs of people in times of ill health and the stresses this condition brings. Health care has many facets, many of which deal with need for emotional support, allaying the fears and anxieties that are associated with modern treatments and therapies. Technology often exacerbates these fears, presenting a harsh and unfriendly environment while at the same time providing diagnostic information or enabling the administration of therapies that can dramatically change the course of disease processes. The affordability of technology is a significant factor in its application, and this will vary from country to country[2] and even among jurisdictions within countries. In fact, technology drivers may themselves cause dramatic changes in the way health care facilities are structured and organized.[3]

Health care technology in its broadest sense includes drugs, therapies, surgical procedures, interventions, and devices and systems. As such, health care technology is the domain of a large number of practitioners, including physicians, surgeons, radiologists, pharmacists, laboratory scientists, nurses, and therapists. Clinical engineers have primary expertise in the subset of technology that centers around instrumentation, devices, and systems, and in this context they are usually separated from the patient by an intervening practitioner or caregiver who uses the applicable technology to deliver a treatment or to perform a diagnosis. It is with this model in mind that the concepts and ideas in this chapter are developed.

Successful technology planning in health care on the part of clinical engineers requires definition of the technology domain and a clear understanding of health care processes and technology stakeholders and how such technologies can enable and enhance these processes and the activities of stakeholders. It also requires establishing working relationships with key technology stakeholders such as those with specific expertise and training in technology management, medical physics, information technology, and technology utilization. It is facility with the processes associated with technology planning that is equally if not more important than knowledge and expertise relating to the inner workings of technology. The position that clinical engineers are able to take will determine whether they are drivers of the planning process or participants along with the many other stakeholders.

37.2 STRATEGIC PLANNING

In recent times, the proliferation of medical technology and its associated costs has placed health care institutions in a position where choices are necessary. This is in contrast to earlier times when the availability of technology was such that choices were limited to essential technologies, and decisions were much easier to make. Today, it is necessary to have a framework in which technology decisions can be made, and this framework is most commonly found as a strategic plan. This is developed in a wider organizational context, defining the broad directions of the organization, the communities that will be served, the types of diseases that will be treated, therapies that will be administered, areas for research focus, and the interactions with external organizations and funding agencies, be they governments, insurance providers, or benevolent organizations. Once these fundamental strategies have been defined, it becomes a lot easier to determine the human resources that will be necessary, the facilities that will be required, and the technology that will be necessary to support the strategic goals.

Technology planning cannot be performed in isolation, and a logical first step is to study, understand, and participate in the development and maintenance of the strategic plan. If such a plan exists, actively determine the technology needs that arise from it, and integrate these into a technology plan. If no such plan exists, facilitate the development of such a plan by identifying key technologies, and challenge the organization to define how it can facilitate the achievement of its goals.

Examples of strategic plan components that drive technology are many. For instance, a strategic initiative to be a regional cardiac cardiac center[4] has immediate technology implications. It defines a basic need for a cardiac catheterization facility, recognizing the contemporary shift from diagnostic to interventional procedures. For adults, this means at least single-plane angiography, cardiac echo ultrasound, and an electrocardiogram (ECG) and stress system. In a broader context, it might imply the implementation of emerging cardiac modalities such as cardiac magnetic resonance imaging (MRI), the implementation of information and image archives, communication with related organizations using communication technologies for the real-time transmission of cardiac angiography and ultrasound, and the sharing of common patient records across referring institutions.

As a basic principle, strategy drives technology; however, at the same time, it must be recognized that technology can enable strategy. It is therefore essential that technology planning is an integral part of strategic planning. A careful balance can avoid the trap of technology push, wherein technologies are implemented simply because they exist, primarily driven by market forces.

To be successful, clinical engineers must take time to understand the strategic planning processes of the organization and bring to bear technology planning skills that enhance this process.

37.3 PLANNING PRINCIPLES

Technology planning is a process, of which the assessment of the particular attributes of a technology is only a part. It is clearly distinguishable from technology management, which relates to the acquisition, application, utilization, education, distribution, safety, maintenance, and repair of technologies once they are in place. Technology management is in fact a consequence of the technology planning process. An analogy can be found in the facilities planning area, where planners develop master program plans and the functional plans arising from them. Architects and consulting engineers take these inputs and realize them in the form of physical buildings and infrastructure.

The outputs of the technology management domain serve as valuable feedback into the technology planning process because they represent the experience associated with the implementation of previous technology plans. Important feedback paths involve the replacement and sustainability of existing technologies and a clearer understanding of risks and influences relating to the effectiveness of managing the technologies that result from the implementation of technology plans. As technology managers, it can be argued that clinical engineers are one of the key stakeholders, as well as key participants, in the technology planning process.[5] The relationship between technology planning and technology management is illustrated in Fig. 37.1.

Acceptance of technology planning as a process allows identification of process components. As with any process, these include inputs, outputs, constraints, stakeholders, and transformation components that are internal to the process, which essentially modify inputs to create outputs. Table 37.1 lists the key components of the technology planning process.

The key outputs of the process are specific requirements for each of the four categories, strategic technologies, information technologies, technology infrastructure, and renewal of existing technologies, together with implementation plans that typically involve a project management component. Further details are provided in the following sections.

Strategic technologies are those arising from the initiatives described in the strategic plan. They typically characterize the need for an organization to be leading edge in particular clinical disciplines, thus providing state-of-the-art care. They typically represent a large financial commitment, may involve the renovation or development of facilities, and may address either new opportunities or the renewal of existing strategic equipment and facilities.

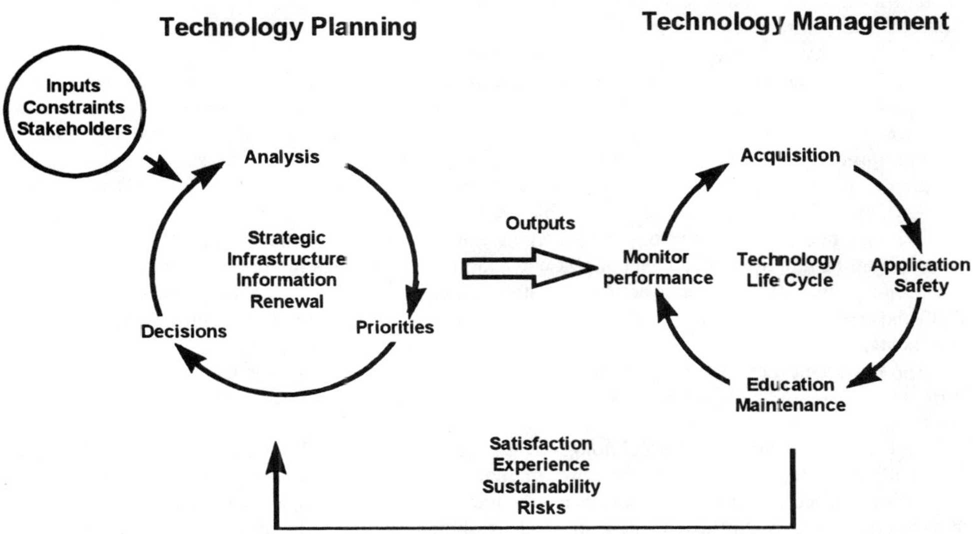

FIGURE 37.1 A schematic representation of the technology planning process and its relationship to technology management.

TABLE 37.1 Key Components of the Technology Planning Process

Inputs	Constraints	Stakeholders	Internal	Outputs
Strategic plan	Government regulation[7]	Corporate	Analysis	Technology plan priorities
Experience: Internal and external	Financial	Providers	Impact	Strategic technologies
Available technologies	Efficacy	Physicians	Setting priorities	Information technologies
Technology assessments	Technology risks	Nurses	Phasing	Technology infrastructure
Priority assessments[6]	Sustainability of current technologies	Clinical engineering	Assessment of benefits and risks	Renewal of existing technologies
Maturity of technologies	Facilities	Allied health professionals	Satisfaction of input criteria	Implementation
Corporate stability		Support services	Satisfaction of stakeholders	Project Management
Standards: Internal and external		Clinical engineering	Impact on workflow	

Information technologies support strategy and are essential for the operation of any contemporary organization. Health care is no exception. In fact, the particular complexities of health care systems, even though they differ from country to country, incorporate a blend of specialized professionals, corporate administration, funding agencies, and insurance providers, each of which presents particular challenges for effective information systems.

Technology infrastructure is a term that applies to the the technologies that provide a foundation for other technologies or that are technologies fundamental to contemporary organizations. As technologies move from stand-alone devices to devices that are integrated into systems, the need for infrastructure becomes more important. In the digital world, networked systems are no longer luxuries, and at a basic level, a good sustainable network infrastructure is essential. At a higher level, the concept of information and communication infrastructures, wherein data can be exchanged among systems, is becoming more and more important. As many institutions strategically strive toward an electronic patient record, the ability to receive and present information from disparate sources in a common context is an essential component. Further, communication systems such as e-mail and voice mail are components of infrastructure.

Funding for the *maintenance of existing technologies* is often neglected in favor of newer technologies until such time as an emergency arises, in which case the problem will be dealt with in the absence of any coordinated planning. Devices on which the daily operation of the organization depend typically include basic surgical tools, patient monitoring devices, life support systems, and laboratory instrumentation through to personal computers that are found on every desktop.

According to this model, the deliverable from the technology planning process is a plan that can be executed in the technology management domain. It is here that the details associated with acquisition, installation, acceptance, and ownership responsibilities relating to maintenance, education, safety, and utilization can be appropriately managed. The strategies and processes associated with these activities are beyond the scope of technology planning and are well known to clinical engineers.

The important message is that technology planning and technology management cannot exist in isolation. There are critical feedforward and feedback links that must be exercised for assurance of success.

Successful execution of the technology planning process requires active executive sponsorship. Typically, this should exist at the vice presidential level in an organization. Even a position of chief technology officer, common in many organizations, without executive-level support may not necessarily be effective. At the executive level, there should be a single champion for all technologies, including both clinical and information technologies. Separation of these two components in contemporary environments will lead to inappropriate emphasis on one or the other. In fact, the demarcation between these two domains is becoming less distinct as information technologies become an integral part of many clinical systems.

In the technology planning context, it is clear that the environment will not adapt to clinical engineers; therefore, clinical engineers must adapt to the environment, an adaptation that is a matter of challenge and choice.

37.4 COMPONENTS OF TECHNOLOGY PLANNING

As described earlier, technology requirements can be subdivided into four primary categories. Each of these will be dealt with in turn. The extent to which these apply will vary according to the environment. For example, the strategic requirements of a large academic organization will be quite different from those of a community-based health care organization.

A key planning concept is to balance strategic needs, information systems requirements, infrastructure needs, and technology renewal needs within a sustainable funding framework. Consideration should be given to the allocation of separate funding categories for each of the components. For example, individual budget line items for each of strategic projects, information technologies, technology infrastructure, and maintenance of existing technologies will avoid the common problem of higher-cost technologies financially overwhelming the maintenance of existing technologies to the extent that the latter are deferred or indefinitely postponed.

37.4.1 Strategic Technologies

Strategic technologies in the domain of devices and systems typically have the following characteristics:

- They are a consequence of the organization strategic plan.
- They are necessary to achieve the strategic goals of the organization.
- They are frequently in the early stage of maturity and dissemination.
- They require a significant investment in either facilities, equipment, or both.
- They require long-term planning cycles.
- They are subject to some risk.

Examples of strategic technologies are provided in Table 37.2. Rather than representing the interests of particular constituencies, they should represent the strategies of the organization.

It is important to note that each of these areas has its own set of stakeholders and specialists whose domain of expertise can individually address each in significant detail. It is unreasonable to expect clinical engineers to have both a broad and in-depth knowledge with respect to all strategic initiatives, and to many, this may at first glance appear to be a significant weakness. However, the strength that clinical engineers can bring to the planning process is that they can operate at a level above the domain experts, bring together the relative strengths and weaknesses of each domain into a broader planning framework, look for common elements, and provide assessments based on the outcomes that particular technologies promise. The role is one of facilitation and requires a special blend of technical and administrative skills. If domain experts are not available within a particular institution, the clinical engineer may be called on to fill this gap to the extent possible or to facilitate the provision of external consulting expertise. In an alternate model, the clinical engineer may fulfill a role as an external consultant, where particular expertise exists.

The planning cycle for strategic technologies incorporates the following components:

- Identification and assessment[15–17] of the applicability of the technology and its efficacy, maturity, reliability, and impact on patient outcomes[18,19] based on, where possible, clinical evidence.

TABLE 37.2 Examples of Contemporary Strategic Technologies

Imaging	Surgery	Clinical laboratories	Monitoring and diagnosis
Positron emission tomography (PET)	Endoscopy	Molecular diagnostic systems	Epilepsy monitoring systems
Magnetoencephalography	Laparoscopy	Automated laboratory systems	Signal processing applied to monitoring
Integrated image archiving systems	minimally invasive surgery and surgical robotics	Electron microscopy	Cardiac mapping systems
Cardiac applications of CT	Image-guided surgery	Mass spectrometry	Telemedicine[14]
Interventional radiology	Tissue engineering[10]	Therapeutic drug monitoring	
Functional MRI and spectroscopy	OR of the future[11]		
Interventional MRI[8,9]	Virtual endoscopy[12]		
Cardiac MRI	Virtual reality[13]		
Radiation oncology			

- Development of a functional plan incorporating the use of the technology. Such a plan includes the patient population that will be served, personnel resources required, impact on workflow,[20] impact on other services, and definition of resources that will be displaced, facility requirements, and generic technology requirements. This plan also will incorporate preliminary budget estimates.
- Development of detailed capital requirements, including equipment specifications, specification of special-purpose facilities, and specification of the costs of implementation.
- Development of a business case that details net cash flows over the life of the equipment based on capital and operating costs.
- Incorporation of the preceding information into a detailed proposal suitable for presentation to key decision makers at the executive level and solicitation of support for the proposal among key stakeholders and decision makers.
- Presentation and justification of the proposal at the appropriate executive level, seeking approval for moving forward with the project. The likelihood of success will to a large extent depend on the strategic importance, the strength of the proposal, and the relevance of the arguments that are presented.

Strategic Technology Example: Interventional Radiology or Image-Guided Therapy. Interventional radiology or image-guided therapy embraces the use of imaging modalities such as x-ray angiography, ultrasound, computed tomography (CT), endoscopy, laparoscopy, and magnetic resonance imaging (MRI). In an appropriate setting, and with the appropriate clinical skills, these technologies can be used for a variety of procedures that are either difficult or impossible with conventional surgical techniques. These include line and shunt placements, biopsies, lineograms, venograms, cecostomies, abscess drainage, tumor removal, correction of vascular malformations, and image-assisted minimally invasive surgery.

The obvious benefits of performing procedures using image-guided techniques include improved patient outcomes, reduced patient trauma when compared with open procedures, reduction in operating room hours allowing reallocation to other procedures, reduction in patient stay, improved administration of intravenous therapies, and a potential reduction in delivery costs. In its broadest sense, image-guided techniques affect the way in which surgical services are delivered and require a new relationship between interventional radiologists and surgeons.

The technology that is required is significant, and it is probable that existing facilities may require significant renovation in order to meet the requirements for the integration of imaging modalities and patient support, especially where sedation, general anesthesia, and recovery are required. This translates into a significant financial commitment.

From a planning perspective, a proposal for the implementation of this technology must include a master and functional plan that addresses technology, facilities, staffing, patient benefits, patient flow, displacement of existing procedures, and quantitative financial projections, all of which must be combined into a coherent and plausible framework for presentation and approval. It is only on approval that detailed technology planning can take place, resulting in a detailed set of requirements for the imaging modalities that can be passed onto the acquisition process.

37.4.2 Information Technologies

Health care information technologies are characterized by their diversity, since many vendors provide niche products directed to meeting the needs of individual professions, whereas others provide comprehensive systems geared to meeting the needs of complete organizations.[21] Quite often the niche products conflict with corporate needs, whereas the comprehensive systems fail to adequately meet niche requirements. Thus a sound management strategy is required. Assuming that in-house software development is not a viable option, there are basically two approaches: one vendor fits all and best of breed in conjunction with integration. Interestingly, this is a moving target, as vendors, through acquisitions, assemble what they consider to be best of breed and present them to the market place as a one-vendor solution. This simply shifts the task of integration from the health care organization to the vendor community and at the same time requires adaptive strategies from a planning perspective.

Information technology serves to automate existing processes and to add value to existing processes by providing data input to quality improvement and management processes, data and facilitation for decision support at the point of physician orders, and data for research and analysis. The diversity of needs, the available systems on the market represented as meeting those needs, and the varying status of legacy systems make it virtually impossible to specify one universal solution. However, planning strategies can be generalized, and these will be summarized in the remainder of this section.

Information technologies are themselves sufficiently complex and diverse that it is wise to develop a specific strategic plan for information technologies that best serve the needs of the organization. Typically, such a plan should address particular organizational unit requirements, including both administrative and clinical, and the data links and relationships that exist between them.

On the clinical side, the plan should specify key organizational units that can benefit from information systems, including laboratories, radiology, cardiology, pharmacy, critical care units, operating rooms, surgery, anesthesia, health records, emergency departments, professional service departments such as physiotherapy, and outpatient or ambulatory departments. Within organizational units, key considerations relate to key stakeholder needs, transactional and analytical requirements, workflow, and desirable data sets. Across organizational units, key considerations are data integration, workflow integration, coordinated scheduling, and data access and presentation. The desirability and scope of an electronic patient record[22,23] and the manner and time frame in which this can be realized should be analyzed, and the results of the analysis should be incorporated into the plan.

A typical information technology plan should be derived from the organization strategic plan and define the information technology that is necessary to support that plan. This will vary from organization to organization; however, there are some fundamental principles that apply. Building on an information technology infrastructure, discussed in Sec. 37.4.3, the plan should define key gatekeeper systems and the hierarchy that links them. Systems should be based on open standards, which allows the integration of new systems with existing legacy systems that have not reached the end of their useful life.

It is not unusual that within organizational units there will be pressure to acquire systems that are geared to particular needs and that these should be implemented before a master plan can be realized. It this case, it is important to include such systems in an integration strategy that allows connection of best-of-breed systems. The most appropriate method for dealing with this problem is to implement an integration engine that can translate data elements within a non–"plug and play" standards environment, such as HL-7, thus ensuring compatibility between applications from different vendors. Internal support for such integration is desirable.

From the time a patient is admitted to the health care organization until the time the patient is discharged, there are many transactions that take place. These may include assessments, medications, laboratory tests, and diagnostic and therapeutic procedures. Each transaction generates a result, which may be in the form of numbers, images, reports, or the administration of a therapy. Transactions may be interdependent, in that subsequent transactions are based on the results of previous transactions. Such processes generate a longitudinal record that profiles the activities and results associated with a particular patient. In a paper world, this is the patient chart, which, if implemented in an electronic context, becomes the electronic patient record.

Transactions arise from an order from qualified practitioners, and each transaction triggers a series of support activities. An order for a medication results in a request to the pharmacy, where it may be screened for correctness, formulated, and delivered to the patient caregiver for administration. An order for an image results in a request to radiology for scheduling the patient to an imaging modality, wherein a series of images is generated, forwarded to a radiologist for reporting, and the resulting report directed back to the responsible physician for assimilation with data from other sources. In these examples, the management of activities within the pharmacy and radiology departments can be met by information systems that address the specific needs of each specialty. In the latter case, this typically comprises a radiology information system (RIS) and a picture archiving and communication system (PACS).[24] In each case, there is a requirement for the bidirectional communication of information between involved systems, and an important part of the planning process is to determine what will be communicated and the means by which it will be communicated.

Systems integration at the point of care is an essential goal, especially where best-of-breed system with integration is the chosen strategy. The concept of a universal clinical workstation, where orders can be placed, results reviewed, and images and reports viewed, all accessed in the same patient context, is the answer to the problem of requiring many workstations running different applications. The evolution of Web technology is perhaps the most significant advance that makes this objective achievable. The planning, specification, and acquisition of any contemporary system should include this functionality.

Transactional systems serve to automate the many transactions that are part of health care delivery. These alone are not sufficient to realize the full potential of information technology, and any comprehensive plan should include an analytical system that allows detailed analysis and queries across patient longitudinal records for the purposes of quality improvement, evaluation of patient outcomes, and other research. Such systems are usually classified as data warehouses and are founded on architectures that facilitate analysis and activities such as data mining in response to ad hoc queries.

Decision-support systems,[25] wherein transactions are screened according to a set of rules, add an additional layer of value to transactional systems. To be effective, such systems require that rules be invoked in real time and draw on data relevant to the transaction. For example, a drug order may be contraindicated based on allergies, or a laboratory order may be contraindicated by the results of previous tests or applicability in the context of the patient's condition as recorded in the electronic chart. The implementation of decision-support technology requires significant effort in the development of acceptable rules and the processing power to process them in real time.

Key operational components of an information technology plan include provision for maintenance of current systems. Software maintenance costs alone can typically run from 18 to 22 percent of software list prices and the qualified human resources necessary to support installed systems. Typically, information technologies in health care systems are either underfunded or place an unreasonable emphasis on administrative systems. An allocation of up to 6 percent of the annual operating budget for information systems is not unrealistic. A summary of contemporary information systems is included in the Table 37.3.

Systems Integration and Standards. An integral part of technology planning is the incorporation and recognition of standards. Standards-based systems improve the probability of compatibility for upgrades, additions, and replacements independent of vendor. Standards for medical devices, such as those relating to safety and performance, are relatively well known in the clinical engineering community; however, information systems standards tend to be better known in the information

TABLE 37.3 Examples of Contemporary Health Care Information Systems Required for Clinical and Administrative Purposes

Clinical systems	Administrative systems	Infrastructure systems
Image archiving systems[26]	Financial systems	Wired network
Radiology information systems	e-Mail	Wireless network
Cardiology information systems	Communications systems	Nurse call systems
Neurology information systems	Voice mail	Central server facility
Laboratory systems	Human resources and payroll	Applications servers
Electronic patient record	Inventory systems	Integration engines
Monitoring networks	Scheduling systems	Desktop computers
Charting systems	Hospital information systems	
Pharmacy systems	Logistics systems	
Telehealth		
Decision-support systems		
Data warehouse		

technology community. Two important standards in this arena at the applications level are Health Level Seven (HL7) and Digital Imaging and Communications in Medicine (DICOM 3.0), the former applying to the communication of patient information and the latter applying to images generated from diagnostic imaging modalities.

Neither standard is plug and play in the sense that a direct seamless connection can be made and interoperability immediately achieved; however, they do define a framework in which prescribed data elements can be exchanged between devices from different manufacturers. In each standard, it is necessary to know message content, as opposed to format, before a reliable interface can be achieved. Once content can be rationalized between two systems, communication can take place. The task of rationalizing content is usually efficiently achieved through an integration engine connecting the two systems. In the case of DICOM 3.0, the task of establishing connectivity involves the matching of DICOM 3.0 conformance statements, statements wherein each device discloses the content of the messages available for communication. Conformance is categorized by class, with separate classes being established for query/retrieve, store, and print, for example.

In the particular case of integration within the radiology environment, one finds radiology information systems that comply with HL7 and PACS that operate with DICOM 3.0. Typically these systems are integrated through a device called a *broker* that serves to translate appropriate messages from the RIS in HL7 to the PACS in DICOM 3.0. An example is the translation of HL7 orders generated in the RIS to DICOM 3.0 work lists that are forwarded to the appropriate modality.

Not all diagnostic imaging modalities comply with all DICOM 3.0 classes, and it is not uncommon to find that contemporary modalities do not provide DICOM 3.0 print class, for example. In this case, third-party systems can be found that effectively provide a DICOM 3.0 print class output from a non-DICOM 3.0 device.

It is essential that any technology planning framework embrace standards and insist that systems comply with appropriate standards to the extent possible. Should noncompliant systems later become compliant through upgrades, integration into the existing infrastructure will be straightforward.

Information Systems Example: Radiology Systems (PACS/RIS). PACS have captured the imagination of technology-oriented radiologists and computer professionals since the early days of computing. It is only recently that the technology of PACS has become sufficiently mature and stable that risks associated with committing an organization to this technology have become acceptable. In addition, the diffusion of digital modalities such as CT, MRI, ultrasound, and x-ray angiography has made the transition to digital imaging easier. Plain films have been a challenge because of the larger dynamic range and higher resolution; however, computed radiography and, more recently, digital radiography address this problem.

Images are the domain of radiologists, and the transition from film-based images to soft-copy images, available throughout the organization on computer screens, requires changes in workflow and work habits. In fact, this transition represents one of the more significant challenges.

From a technology perspective, the establishment of a central server and the reliable and timely distribution of images on demand to networked workstations throughout the organization require a robust network infrastructure. Redundant switching and protection against single points of failure need to be inherent in the network design. The implementation of a physically separate network for PACS is an option; however, this becomes expensive when image and report distribution is required throughout the enterprise.

For an integrated system, PACS must be accompanied by an RIS that essentially manages transactions associated with images, including orders, billing, room scheduling, and reports. Since typical PACS use DICOM 3.0 standards and RIS use HL7 standards, integration of these systems requires a broker that essentially acts as a standards translator. RIS also require a network infrastructure. Traditionally, PACS and RIS have required two separate workstations; however, current trends are to integrate the two systems to the extent that they can be accessed through a common workstation and even a common user interface.

37.4.3 Technology Infrastructure

Technology infrastructure warrants independent planning cycles. It refers to infrastructure that is designed to support the introduction of new technologies in the present and the future. Infrastructure includes computer facilities, telecommunications, information technology networks, and centralized server and database support. Examples of technology infrastructure and listed in Table 37.3.

A large percentage of technology systems acquired today require network communications, allowing the exchange of data among both internal and external stakeholders. Introducing a separate network for each application not only is expensive but also is unsustainable over the long term because of the administrative and support challenges it presents. In the past it was not uncommon for departmental systems to operate on special-purpose proprietary or nonproprietary networks; however, this was generally determined by individual needs rather than those of the organization.

Today, for any sustainable technology plan, investing in, building, and maintaining a state-of-the-art high-speed network is essential. This should include a minimum of category 5 100-Mbps terminal connections, a gigabit backbone incorporating logical segmentation, redundant data pathways, and 7-day by 24-hour support. Every monitored beside, operating room, clinic location, physician's office, laboratory workstation, and general office should have network connectivity even if there is not an immediate need. In fact, a strategy of dual network connections for each location is not excessive.

For example, at a critical-care bedside, network connections may be required on a permanent basis for patient monitoring and a clinical information system and on a temporary basis for connection of an ultrasound machine. In an operating room, similar connections could be expected at each anesthesia location. Unless integrated with the bedside systems, additional network connections will be utilized for connection to institutionwide systems such as hospital information systems and PACS. With an appropriate network infrastructure in place, it becomes much easier to introduce networked systems into the environment and avoids the addition of significant cabling costs for each application. This also means that networked systems not requiring dedicated physical networks and using standard network protocols should be given preference.

In addition to a wired network infrastructure, there is increasing utility to be found in a wireless network infrastructure, providing the advantage of mobility. Applications include wireless patient networks, voice-over Internet Protocol (IP) for mobile communications, and communication with handheld information terminals carried by each physician. A wireless infrastructure requires careful planning, with attention being paid to access points, data rates, and electromagnetic compatibility with medical devices. Current standards such as IEEE 802.11b allowing data rates of 11 Mbps facilitate the seamless coexistence of wireless and wired networks. The use of lower-spread-spectrum techniques in wireless applications reduces problems associated wit medical device interference.

From a planning perspective, if it is not feasible to develop the appropriate network infrastructures at the same time an application is introduced, network design should be such that extension to a larger network system can be done seamlessly. In an integrated information technology environment, the development of a central computer facility is an important part of an infrastructure plan. An adequately sized room with appropriate environmental controls, fire protection, and security, combined with 24-hour by 7-day-per-week operations support, should be considered a minimum requirement. With an appropriate network infrastructure in place, this facility can house applications servers for mission-critical systems, including those for the electronic patient record and more specific clinical systems. This strategy facilitates data backup and data security.

Finally, an important component of the technology infrastructure is the desktop environment, which generally requires supply and maintenance of personal computers at the majority of desks within an organization, as well as those required to run clinical applications at bedsides.

37.4.4 Renewal of Existing Technologies: Capital Equipment

In any organization, technology planning must include the appropriate maintenance of existing technologies. This activity ranges from direct replacement to upgrading of the technologies associated with continuing functional requirements. This category may also include the introduction of new technologies that contribute to operational requirements but do not qualify as strategic initiatives.

It is a fundamental axiom that if stakeholders are polled with respect to their technology needs, the value of requests will exceed any reasonable budget allocations. This is more evident if planning is performed on a renewable annual basis because stakeholders see each year as a new competition.

While funding formulas differ across jurisdictions, two key planning principles can avoid a lot of problems. First, develop a rolling plan that extends over a period between 3 and 5 years, and second, secure a fixed annual budget allocation for maintenance of existing technologies based on the multiyear plan. Once these have been implemented, rational planning that meets both stakeholder expectations and financial constraints becomes possible.

Depending on the type of care provided by the organization, there will always be a need for fundamental technologies. These include patient monitoring, patient thermal support, anesthesia technologies, ventilatory support, infusion therapy, laboratory instrumentation, surgical tools, and diagnostic imaging tools such as ultrasound. Typically these technologies will already exist and will be at some point in their life cycle. An analysis of this situation will allow the development of a replacement plan extending over the selected time frame. The key driver for replacement is obsolescence, which can either be based on the technology or serviceability, both of which can be addressed in the planning and acquisition process. An important resource for the development of replacement strategies is an inventory of existing equipment coupled with a knowledge of the functions expected from the equipment, the risks associated with continuing use, and the status of current technologies providing similar functions. Each of these falls within the knowledge and resource domain of clinical engineering.

Typically, organizations will structure the equivalent of a capital equipment committee to oversee the distribution of funds for the maintenance of existing technologies. Membership of such a committee will comprise key equipment stakeholders, such as physicians from clinical and surgical services, clinical laboratories, and diagnostic imaging. In addition, there will usually be representatives from the administration and financial and purchasing constituencies. The mandate of such a committee can range from the simple allocation of available annual funds based on submitted equipment requests to management of a technology maintenance plan. The latter is obviously more effective and should be an explicit mandate of the group. Clinical engineers can play a variety of roles, from that of a technology consultant to that of a technology planner. They are uniquely qualified to bring a balanced impartial viewpoint and can take a leadership role in facilitating development of the plan through the committee.

Key inputs to the planning process are

- A current inventory of equipment in categories that will be included in the plan.
- An assessment of risks associated with the continuing use of current equipment. Risks include those associated with safe operation, precision, and efficacy when compared with contemporary technologies.

- An assessment of the current point in the life cycle (technological, serviceability). This will be a function of the particular technology and the particular product.
- An assessment of technologies falling within the context of the plan that can enhance operations. New technologies may impact productivity or allow new tests that directly impact patient care to be performed.

Key processes in the development of a plan are

- *Categorization of key stakeholder groups.* It is important that stakeholder groups, usually best defined along organizational lines, have some ownership in the specification and prioritization of their respective internal requirements.
- *Development of equipment and systems groups that can be handled collectively.* For example, an organizationwide patient monitoring plan is more effective than addressing individual requirements and partial replacement strategies.
- *Balancing the plan across a multiyear time frame.* Several iterations of the plan may be necessary to provide a balanced annual cash flow within the set budget allowance.
- *Establishment of an acceptable annual budget allocation to sustain operations.* The plan, once established, balanced, and determined as meeting organizational needs, can be used to determine a proposal for an annual budget.
- *Establishment of a fund for small-cost items and assignment of management to stakeholders.* Responsibility for minor items costing below some predetermined cost threshold can be delegated to organizational units, together with an annual budget allowance. This relieves the committee of the task of prioritizing and approving such items
- *Establishment of a contingency allowance.* A contingency allowance provides for unexpected needs. However, consideration can be given to inserting contingency items into plan priorities, in which case such items would displace planned items.
- *Establishment of a fund for new and innovative technologies.* An allowance for technologies that are innovative and subject to some risk may be included in the plan. Typically, this allowance would be subject to rules of competition.

A key output of the planning process is a plan that

- *Provides flexibility across multiple years.* A multiyear view minimizes the possibility of current year funding cuts because the impact in future years is readily visible.
- *Allows ready consolidation of a plan for the current planning year and coordination with annual operating plans.* Typically, operating plans should be supported by technology, and presenting both the operating plan and the capital plan in the same context reinforces this position.
- *Facilitates discussion and approval at the committee and executive levels.*
- *Can be effectively managed during its implementation.* Execution of the plan involves specification, tender, selection, delivery, acceptance, and payment.
- *Allows metering of cash flow in a particular implementation year.* Cash flow at the time of payment represents the end point of implementation, and the timing of acquisitions can be driven by an annual cash flow plan.
- *Is fair and acceptable to stakeholders.* Stakeholder requirements are often independent; therefore, equity across stakeholders is a desirable outcome.

The multiyear planning process for the maintenance of existing technologies is a framework that has been found to be effective in practice. It avoids traditional problems associated with single-year planning cycles and allows the capital equipment committee to operate at an appropriate executive level, avoiding the need to deal with item-by-item detail. The process is defined independently of particular professional expertise; however, it is apparent that the expertise and experience of clinical engineers are well suited to playing a leadership role. Assertion of this role in a manner consistent with the goals of the process remains a situational challenge in each environment.

It is wise for a planning process such as this to establish an evaluation component so that experience gained during implementation can be fed back into the process in order to introduce improvements. For example, once equipment has been purchased. installed, and used for a period of time, a discussion with users to determine whether the original goals have been achieved can provide valuable information.

Capital Equipment Example: Patient Monitors. Patient monitors are a technology staple for critical-care units. More recently, there has been emphasis on subacute monitoring, which is used for surveillance of patients being cared for outside of critical-care units.

Typically, supply of monitors to these areas has been ad hoc, with no coherent strategy. The introduction of planning principles for these technologies facilitates identification of the rationale for monitoring, selection of the most appropriate parameters for monitoring, selection of the protocols to be used for these parameters, and provision of the education necessary to ensure that maximum benefits are achieved.

For example, if the primary reason for monitoring patients is to determine changes in physiological condition due to cardiac or respiratory difficulties, oxygen saturation monitoring alone is probably appropriate. This parameter, by its nature, is subject to artifact due to movement and low perfusion states, which can be the cause of excessive false-positive alarms. An assessment of the current state of oxygen saturation monitoring reveals that current-generation algorithms that take advantage of advanced signal-processing techniques can reduce false-positive alarm rates, making them suitable for single-parameter monitoring of subacute patients. A decision to use saturation monitoring alone for this purpose precludes the need for ECG monitoring and the application of electrodes, even though this parameter may be available in the device.

A subacute patient monitoring plan should start with requirements based on patient needs and choice of an appropriate technology offering that meets these needs.

37.5 EQUIPMENT PLANNING FOR NEW FACILITIES

The requirement to equip a new facility with medical devices and systems provides an opportunity to begin with a new plan rather than sustain and adapt an existing plan. Despite this, the principles already discussed still apply. It is quite likely that the new facility will be occupied by an existing organization with existing strategic plans or by a restructured organization arising from an amalgamation, for example, in which case one would expect strategies underlying the new structure to already be in place. Thus, from a planning perspective, the primary difference is the fact that the plan is zero-based.

Equipment planners, who may not necessarily be clinical engineers, provide planning services, which typically include pro-forma databases that list typical equipment requirements by category. For example, a critical-care unit will require life-support equipment, thermal-support systems, patient monitors and monitor networks, infusion therapy devices, patient beds and stretchers, sterilizers, and laboratory instruments down to commodity items such as IV poles, suction systems, flowmeters, and resuscitation equipment. Such databases are a key asset of equipment planners and are a key starting point of any new facility planning process.

A typical strategy is for an appropriately structured equipment committee to solicit input from stakeholders in the new facility and to modify and update the database, resulting in equipment lists that represent detailed requirements. At this time, a budget allowance for each line item is also developed. The list is then refined on the basis of available funding and priorities.

As with any technology planning initiative, success depends on the participation of a team of professional with particular areas of expertise. An effective core committee for new facility equipment planning should include a materials management professional, a clinical engineer, an equipment planner, and an architectural planner. Each plays a role as follows:

Materials management. Familiar with all aspects of procurement, including suppliers, pricing, lead times, and logistics.

Clinical engineer. Familiar with performance requirements, detailed specifications, safety, efficacy, and aspects of systems integration.

Equipment planner. Familiar with generic specialty requirements, equipment databases, and list management.

Architectural planner. Familiar with the interface between equipment and facilities. This team member may draw on more detailed information from electrical and mechanical consultants.

Each profession brings a particular expertise and, working together as a team, can produce a more effective result.

37.6 PROJECT MANAGEMENT

The complexity of many technology projects is such that an implementation process is required. Elements are the relationship with facilities, system configuration, integration with existing systems, commissioning, and acceptance procedures. Most vendors will assign a project manager and a project team to coordinate processes from the vendor side and to protect the vendor's interest in the context of the contract. Most responsible vendors expect a similar structure to exist on the customer side, and the establishment of a project team including a project manager is essential. Such a team provides a liaison with the stakeholders in the organization and protects the organization's interest in the context of the governing contract. Project management is a specific skill,[27,28] and if such duties are assumed by a clinical engineer, it is appropriate that the individual has formal project management training or prior relevant experience.

37.7 FACILITY COMPONENTS

Many technology-based systems will have a facility component. This can range from the supply of special-purpose power and network connections to the renovation of a space for a particular need, providing new electrical and mechanical infrastructures. It is important that the facilities component be incorporated into any technology plan.

Facilities can be designed for general or particular use. For example, operating rooms and critical-care units have generic requirements that are usually defined by standards and codes relating to facilities systems. These include electrical and mechanical systems such as normal and emergency power systems, medical gas systems, air-handling systems, and fire-protection systems. This is in contrast to special-procedure rooms, which are generally designed to accommodate special-purpose systems such as x-ray and other imaging systems, display systems, and surgical microscopes.

The implementation of a facilities plan requires the establishment of an internal project team coordinated by a project manager representing the interests of the organization. The project manager can either be an employee of the organization or acquired through an outsourcing contract. The project manager engages a health facilities architect who acts as a coordinator for other design consultants who will become involved in the development of detailed plans.

The project manager assumes responsibility for coordinating and facilitating discussions between the internal stakeholders and the design consultants in order to ensure that needs can be translated into specific requirements that can be incorporated into a preliminary design. Typically, this will be at the architectural level, defining structural constraints, clearances, room layouts, the physical location of key functional elements, traffic flows, and points of access. Completion of the architectural design may take several iterations and, once complete, leads to stakeholder signoff. The project manager keeps detailed minutes of all discussions in case of later disputes.

Once the architectural plans are finalized, the architectural consultants coordinate the detailed design in conjunction with the electrical and mechanical consultants who have been retained for the project. In the case of a renovated facility, this will involve the integration and connection with existing building systems. The end point of this process is a detailed set of design drawings that are suitable for submission to a tender process. Where special-purpose equipment is required, specification of the equipment requirements that impact on the facility and that are not intended to be part of the contract needs to be available for incorporation into design drawings and documents. Equipment that is part of the construction contract is supplied by the contractor and is included in the response to the tender call.

Interested contractors respond to the request for tenders, and from these submissions, a successful general contractor is chosen. A contract is developed between the this bidder and the organization. Typically, this will be done using contract documents that are standardized within the industry.

Once construction starts, the appointed project manager assumes responsibility for the project on behalf of the organization and through the general contractor utilizes the architects and the electrical and mechanical consultants as appropriate for the duration of the project. On completion, a commissioning phase is entered, wherein the installed systems are checked and calibrated according to the design specifications, at which point, subject to minor deficiencies, the facility can be turned over to the owner.

Installation of special-purpose systems can then take place, which, when completed, are handed over to the owner for acceptance testing prior to being placed into clinical use. In practice, some of these processes may be run in parallel, subject to cooperation among the parties involved. Such coordination is the responsibility of the project manager.

It is important for clinical engineers to understand the construction processes and their relationship to the final performance of the installed equipment. There is ample opportunity for involvement during each stage of the project, particularly in the initial planning stages and in the final commissioning stages, and full advantage should be taken of these opportunities to add value to the process. Clinical engineers with special expertise relating to equipment and systems are in a unique position to ensure that the interface with the facility is functional, safe, and effective. The key liaison is with the project manager representing the organization.

37.8 BUSINESS CASES

Assuming that a given project fits with the strategic plan, before approval, the development of a business case[29,30] is an important planning tool that provides a measure of affordability. The validity and structure of business cases will vary from jurisdiction to jurisdiction and the financial basis on which the health care organization is structured. Business cases in systems where revenue is available on a case-by-case basis from insurance providers will differ from those where health care systems are globally funded.

In general, business cases provide an analysis of positive and negative cash flows that can be expected during the lifetime of the technology. This lifetime will depend on the technology under consideration and in today's environment is closely related to the expected lifetime of the platform on which the technology is based. Upgrades to this platform during the lifetime of the platform comprise one component of the negative cash flows. Each year or, for more detail, each month during the life of the technology there will be cash inputs arising from its use and cash outputs necessary to acquire and maintain it. One approach is to calculate a net present value based on the net cash flows over the life of the technology discounted to the present time at an interest rate typical for a secure investment and to compare this value with what the technology is worth in the context of the patients served by the organization. This approach also allows comparisons between technologies where choices must be made. In a strict business environment, the net present value should be positive to represent a financial reward for effort expended. In a globally funded environment, the magnitude of the contributions required from the global budget required to balance the plan becomes an important consideration.

Typical negative cash flows are

- The capital cost, usually occurring in year 1, or in the case of a lease, payments over the lease period
- Maintenance costs, on a quarterly, semiannual, or annual basis
- Cost of technology upgrades as these become available
- Salary costs for operating personnel
- Cost of supplies required to operate the equipment

Typical positive cash flows are

- Revenues earned through use of the technology
- Cost avoidance for personnel and supplies that can be realized through use of the technology
- Cost avoidance relating to reduction in patient stay, for example
- Disposal value of equipment that the technology replaces

Net cash flows, the difference between positive and negative cash flows, can be developed in a spreadsheet, and net-present-value functions available in most spreadsheet applications can be easily applied to calculate the net present dollar value for a particular technology. This technique can also be used to compare different technologies where return on financial investment is a key decision criterion.

While business cases have some value, there will always be technology where the business model does not apply. Typically, these are infrastructure technologies such as networks and computer communications (e-mail, voice mail), and in these cases, investment decisions must be made on the basis of broader organizational criteria.

Business cases are built on assumptions, and one is not complete without an assessment of the risks associated with each assumption. Typically, where risks are significant, mitigating strategies need to be included.

37.9 PRIORITIES AND FUNDING

Technology costs money, and it is clear that among the available technologies, choices will have to be made. Technologies represent an investment, and the criteria that will determine whether such investments should be made and their associated priorities will depend on the type of health care system in which the organization operates. In a for-profit system, decisions are more likely to be influenced by financial business plans driven by return on investment. In a social system, priorities may be set according to community or regional needs or in some cases by a central authority. Whatever the system, the ultimate goal is the delivery of quality patient care, and it is certain that technology will play a role in achieving this goal. Organizations that do not take a proactive approach to technology investment may have an uncertain future.[3]

At the very least, allocating funding for the replacement of the current technology infrastructure is essential. An annual allowance consistent with industry benchmarks for investment in information technology is also a basic necessity. This at least maintains the status quo. For organizations keen to develop new strategic initiatives, funding on a project-by-project basis is necessary.

Allocation of appropriate funding for technology plans will almost always require a technology advocate at the executive level because it is there that the allocation of resources and the setting of priorities are usually decided. To play a meaningful role in technology planning, the clinical engineer needs to have a direct line of communication to the decision makers, and if this does not exist, it should be actively cultivated.

37.10 OUTSOURCING

The complexity of systems, particularly those involving software requiring integration with other systems, places specialized demands on the skills of the individuals charged with the responsibility of managing these systems. Developing an in-house resource requires a critical mass that may only be cost-effective within larger institutions. Under these conditions, outsourcing to a vendor or a third-party supplier is an option. This may be extended to include technology planning and technology management functions and may even involve off-site transaction management and data storage.

The advantages of outsourcing are

- Qualified personnel with appropriate skills should be available.
- Backup resources are available as required.
- Day-to-day management of resources becomes the responsibility of the provider.
- Costs of services are capped for appropriately structured contracts.

The disadvantages of outsourcing are

- Possible loss of control of internal assets, such as data and information.
- Critical systems are in the hands of a third party; their failure is the organization's failure.
- Completion for provider resources may result in the assignment of inexperienced personnel.

Once the risks are understood, problems with provider screening and selection can be minimized, and performance expectations can be contractually developed. Good legal advice in developing a contract is essential. Once a relationship is established, success will depend on the quality of the contract management, and the assignment of a qualified contract manager is highly recommended. Here again, the clinical engineer may have or acquire the skills necessary for this function.

37.11 CORPORATE RELATIONSHIPS

Technologies will generally be acquired from the industrial sector. While make or buy decisions may have been appropriate in the past, in contemporary times with the availability of a broad range of technology from industry, these are no longer particularly relevant, except in the case of special-purpose research applications. For example, an organization that undertakes the development of complex software-based systems assumes the risks associated with continuing support, dependence on particular individuals, adaptation to changing technologies, organizational dependence on the system, and sustainability. Also assumed are the responsibilities in dealing with the regulatory environment.

The contemporary medical industry is currently characterized by consolidation of diverse technologies under a small number of corporate umbrellas. It is now possible to acquire from one vendor technologies ranging from diagnostic imaging systems, monitoring systems, cardiovascular systems, and information systems. Similar consolidation is taking place in technologies such as intravenous therapy systems.

From a technology planning perspective, this environment can provide opportunities in the context of developing strategic corporate partnerships to meet needs across a variety of technology sectors. Such relationships must, however, be approached with appropriate caution. From the point of view of the health care organization, the advantages include definition of a consistent set of generic contractual relationships and performance expectations across products. Within this framework, specific equipment or device specifications and requirements can be developed.

Contractual agreements defining relationships can range from those which require preferential purchasing of specific technologies during the term of the agreement, subject to those technologies meeting clinical requirements, to supply agreements that define the supply of specific goods from a particular vendor over a specified period of time. The degree in which an organization is bound to a particular vendor will define the benefits that accrue. The most frequent examples are favorable pricing structures, access to new and developing technologies, support for research, favorable delivery terms, and favorable access to service options and spare parts. Intangible benefits include the opportunity to standardize on technologies from one vendor and consolidated service agreements and a consistent performance level based on the extent of the relationship.

Criteria on which such agreements should be evaluated include the match between the vendor's offerings and organization requirements, trends for future developments, flexibility with respect to alternative choices, cultural synergies between the organizations, and the extent of favorable contractual conditions. Once established, such agreements require sustained commitment on the part of both parties to make them work. Such agreements should anticipate that issues of disagreement will arise and to the extent possible contain language that addresses these possibilities. In the event of disagreement and dispute, the ability to work these out will be defined by the strength of the agreement and the depth of the respective commitments.

Qualified legal advice in developing the original contract is strongly recommended. Time and effort invested in the development of such agreements prior to signing offsets risks and will yield dividends when difficulties arise.

Such corporate agreements should be driven and executed at the executive level, and the solicited participation of internal stakeholders will depend on management styles that exist at that level. The probability of success is enhanced when internal stakeholders are part of the process. Despite this obvious observation, it will be necessary for clinical engineers to assert their presence in the context of the value that they can contribute to the process along with other stakeholders. Certainly, they will be affected by the outcome and as such should be in a position to influence contributing factors.

37.11.1 Forms of Corporate Agreement

Corporate agreements, should they be appropriate, can take a number of forms. One model that has been found to be effective includes two separate agreements, the first specifying the relationship between parties, such as, for example, purchase obligations and benefits that accrue from the discharge of these obligations. The second agreement, in the form of a master purchase agreement, defines conditions of purchase, including obligations of the vendor with respect to supply and obligations of the purchaser with respect to acceptance and payment. Conditions and remedies for nonperformance are also included in this agreement.

For each individual purchase, a schedule to the master purchase agreement is generated. This schedule details specifications and requirements for the individual item or system and provides the opportunity to override or extend the conditions contained in the master agreement.

If necessary and desirable, a third agreement relating to service of items purchased on individual schedules can be negotiated. Such an agreement contains conditions that would normally be found in any service contract.

37.12 SUMMARY

Technology planning in health care is a process that provides a framework for the identification of technology needs for health care organizations. There are many stakeholders and professionals involved in this process, one of which is clinical engineering. The role that clinical engineers play will be situational and can vary from one of leadership in the process, i.e., as a process facilitator, to one of active participation ensuring that the technology is appropriate, efficacious, and optimized with respect to its environment.

Technology has been subdivided into four distinct categories: strategic technologies, information technologies, technology infrastructure, and renewal of existing technologies. The rationale for each and a discussion of approaches is provided. It is proposed that each be allocated a separate technology budget to avoid the particular problems of large-cost items financially overwhelming lower-cost items. The total technology budget will be a measure of the extent to which an organization is prepared to invest in technology in order to achieve its strategic objectives.

Even though the technology plan is driven by the strategic plan, it is subject to the same scrutiny and challenge as any other plans requiring allocation of resources. It therefore requires careful preparation and advocacy throughout the review process. Technology requires scrutiny in the context of the key end point parameters, of which the most important is the extent to which it improves patient outcomes. Measurement of patient outcomes is a developing science, which in itself requires information technology to provide the necessary data.

A clear distinction is drawn between technology planning and technology management. The domains are interdependent, with clearly defined links between them. The more familiar domain for clinical engineers is that of technology management, and a transition to technology planning requires the adoption of new skills or cooperation with other professionals who already possess these skills.

In times of fiscal restraint, the first casualty is often technology-based capital spending. A well-structured technology plan clearly defines the consequences of such actions and permits rational choices to be made.

REFERENCES

1. Mathews, P. (2000), Leveraging technology for success, *J. Healthcare Manag.* **14**(2):5–12.
2. Weil, T. P. (1995), Comparisons of medical technology in Canadian, German and U.S. hospitals, *Hosp. Health Serv. Adm* **40**(4):524–533.
3. Prince, T. R., and Sullivan, J. A. (2000), Financial viability, medical technology, and hospital closures, *J. Health Care Finance* **26**(4):1–18.
4. Sheldon, W. C. (2001), Trends in cardiac catheterization laboratories in the United States, *Cathet. Cardiovasc. Intervent.* **53**(1):40–45.
5. Patail, B. M., and Arahana, A. N. (1995), Role of the Biomedical Engineering Department in William Beaumont Hospital's technology assessment process, *J. Clin. Eng.* **20**(4):290–296.
6. Singer, P. A., Martin, D. K., Giacomini, M., and Purdy, L. (2000), Priority setting for new technologies in medicine: qualitative case study, *Br. Med. J.* **321**:1316–1318.
7. Perleth, M., Busse, R., and Schwartz, F. W. (1999), Regulation of health related technologies in Germany, *Health Policy* **46**(2):105–126.
8. Manninen, P. H., and Kucharczyk, W. (2000), A new frontier: Magnetic resonance imaging-operating room, *J. Neurosurg. Anesthesiol.* **12**(2):141–148.
9. Jolez, F. A., Nabavi, A., and Kikinis, R. (2001), Integration of interventional MRI with computer assisted surgery, *J. Magn. Reson. Imaging.* **13**(1):69–77.
10. Nerem, R. M. (2000), Tissue engineering: Confronting the transplantation crisis, *Proc. Inst. Mech. Eng. [H]* **214**(1):95–99.
11. Broeders, I. A., Niessen, W., van der Werken, C., and van Vroonhoven, T. I. (2000), The operating room of the future (in Dutch), *Ned. Tijdschr Geneeskd.* **144**(16):773–774.
12. Allan, J. D., and Tolley, D. A. (2001), Virtual endoscopy in urology, *Curr. Opin. Urol.* **11**(2):189–192.
13. Tronnier, V. M., Staubert, A., Bonsanto, M. M., Wirtz, C. R., and Kunze, S. (2000), Virtual reality in neurosurgery (in German), *Radiologe* **40**(3):211–217.

14. Tanriverdi, H., and Iacono, C. S. (1999), Diffusion of telemedicine: A knowledge barrier perspective, *Telemed. J.* **5**(3):223–244.
15. Valk, P. E. (2000), Clinical trial of cost-effectiveness in technology evaluation, *Q. J. Nucl. Med.* **44**(2):197–203.
16. Papatheofanis, F. J. (2000), Health technology assessment, *Q. J. Nucl. Med.* **44**(2):105–111.
17. Crepea, A. T. (1995), A systems engineering approach to technology assessment, *J. Clin. Eng.* **20**(4):297–303.
18. Roberts, G. (2001), Supporting children with serious health care needs: Analysing the costs and benefits, *Eval. Health Prof.* **24**(1):72–83.
19. Thompson, D. I., Sirio, C., and Holt, P. (2000), The strategic use of outcome information, *Joint Comm. J. Qual. Improv.* **26**(10):576–586.
20. Mira, A., and Lehmann, C. (2001), Pre-analytical workflow analysis reveals simple changes and can result in improved hospital efficiency, *Clin. Leadersh. Manag. Rev.* **15**(1):23–29.
21. Austin, C. J., Hornberger, K. D., and Shmerling, J. E. (2000), Managing information resources: A study of ten healthcare organizations, *J. Healthcare Manag.* **45**(4):229–238.
22. Souther, E. (2001), Implementation of the electronic medical record: The team approach, *Comput. Nurs.* **19**(2):47–55.
23. Dansky, K. H., Gamm, L. D., Vasey, J. J., and Barsukiewicz, C. K. (1999), Electronic medical record: Are physicians ready? *J. Healthcare Manag.* **44**(6):440–454.
24. Unkel, P. J., Shelton, P. D., and Inamdar, R. (1999), Picture archiving and communications system planning: A methodology, *J. Digit. Imaging* **12**(3):144–149.
25. Rivers, J. A., and Rivers, P. A. (2000), The ABCs for deciding on a decision support system in the health care industry, *J. Health Hum. Serv. Adm.* **22**(3):346–353.
26. Bryan, S., Buxton, M., and Brenna, E. (2000), Estimating the impact of a diffuse technology on the running costs of a hospital: A case study of a picture archiving and communication system, *Int. J. Technol. Assess. Health Care* **16**(3):787–798.
27. Carr, J. J. (2000), Requirements management: Keeping your technology acquisition project under control, *J. Nurs. Admin.* **30**(3):133–139.
28. Zimmer, B. T. (1999), Project management: A methodology for success, *Hosp. Mater. Manage. Q.* **21**(2):83–89.
29. Abendshien, J. (2001), Managing the future, *Health Forum J.* **44**(1):26–28, 46.
30. Neumann, C. L., Blouin, A. S., and Byrne, E. M. (1999), Achieving success: Assessing the role of and building a business case for technology in healthcare, *Front. Health Serv. Manage.* **15**(3):3–28.

CHAPTER 38
AN OVERVIEW OF HEALTH CARE FACILITIES PLANNING

John Michael Currie
SmithGroup, Inc., Washington, D.C.

38.1 INTRODUCTION

38.1.1 Health Care Facilities Planning

The basic process of planning and design is made up of several sequential steps. Each step builds on the information created and the decisions made in the preceding one. The process is often not linear. Many alternative ideas are tested and the results combined. This chapter concentrates on the work leading up to commencement of the construction activity on site. *This chapter also assumes that the process of planning for and including medical technology and equipment is an intrinsic part of designing a health facility in a responsible manner*. Separate consultants can help in this process, provided they are brought into the project from the beginning. As the work progresses, the opportunity to positively influence its direction with valuable input on medical technology consideration diminishes.

The business of health care facility design is all about the future. We cannot plan well simply to correct today's facility shortcomings. In dealing with the future of health care, the future of technology is critical.

38.1.2 Historical Overview

> "The building or buildings should be simple in style and designed to make a pleasing impression upon the patients. . . . Hospital planning demands the same careful thought that is the foundation of any modern successful business enterprise. . . . The hospital planner must seek to eliminate here all lost motion or unnecessary work."
>
> EDWARD F. STEVENS, *The American Hospital of the Twentieth Century*

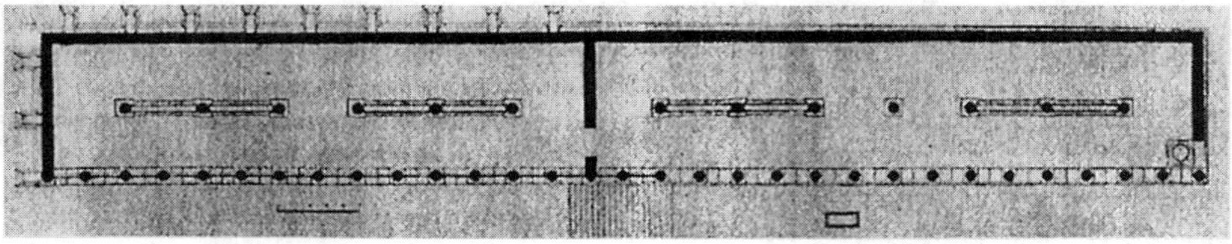

FIGURE 38.1 Stoa plan (Greek Asklepieion).

In a few simple statements, made in 1918, architect Edward F. Stevens has captured ideas that are still central today in hospital planning. While Stevens could not anticipate the enormous growth in medical knowledge, technique, and technology that would come to bear in the fullness of the twentieth century, his words have covered four important goals of hospital facility planning today:

- Simplicity of design
- Focus on the patient
- Understanding of function
- Design to support medical activities

Early hospital architectural designs, the Greek Asklepieion and the Roman Valetudinarium, were forms adapted from buildings already in use. The Asklepieion (Fig. 38.1) was simply a stoa or business arcade put to medical use. The Valetudinarium (Fig. 38.2) was a military barracks building modified for the sick. These plans are derived from other uses, and the planner had few opportunities to introduce design for the specific needs of the medical staff.

Early Christian monasticism and the constant flow of travelers and pilgrims brought the first hospices into being. The Hospice of Turmanin (A.D. 475) in Syria received all travelers, sick or well,

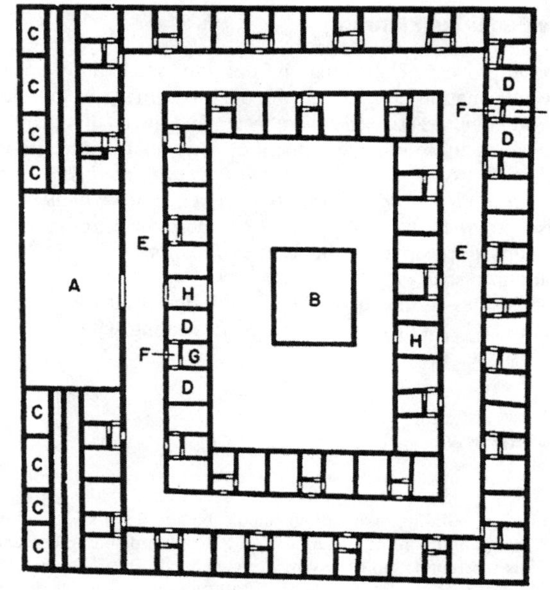

FIGURE 38.2 Valetudinarium plan (Roman).

in the convent building, with nursing care provided for those in need. This building contains an early example of the open ward with deep porticos on all four sides.

The famous Monastery of St. Gall in Switzerland (A.D. 820) had more specific portions of the overall plan devoted to medical care and medical staff. Its plan clearly shows separate facilities for bloodletting, physician housing, infirmary, kitchens, bathing facilities for the sick, and medicinal herb garden.

Many other important European monastic centers cared for patients throughout the Middle Ages. Notable among these are the Great St. Bernard Hospital (A.D. 960) in Switzerland and Cluny (A.D. 1050) in France (Fig. 38.3).

The era of the hospital began in the twelfth century when the sick were to be housed in separate buildings designed for medical care. Examples include St. Bartholomew's Hospital (London, A.D. 1123) and St. Thomas' Hospital (London, A.D. 1200). Both hospitals are in service today, having been refounded in the mid-1500s due to the dissolution of the monasteries in England in A.D. 1536.

Early general hospitals were established throughout Europe during the sixteenth and seventeenth centuries. The pavilion plan was developed in the mid-nineteenth century. This concept was supported by the writings of Florence Nightingale (*Notes on Hospitals*, 1859). These hospitals, arranged into an interconnected series of small patient care buildings, were designed to help control infection, provide separation of various types of patients, and promote natural light and ventilation.

Three main causes influenced the development of specialist hospitals:

1. A growing patient population excluded from general hospitals, such as pregnant women, children, people suffering from contagious disease, and patients requiring a long length of stay

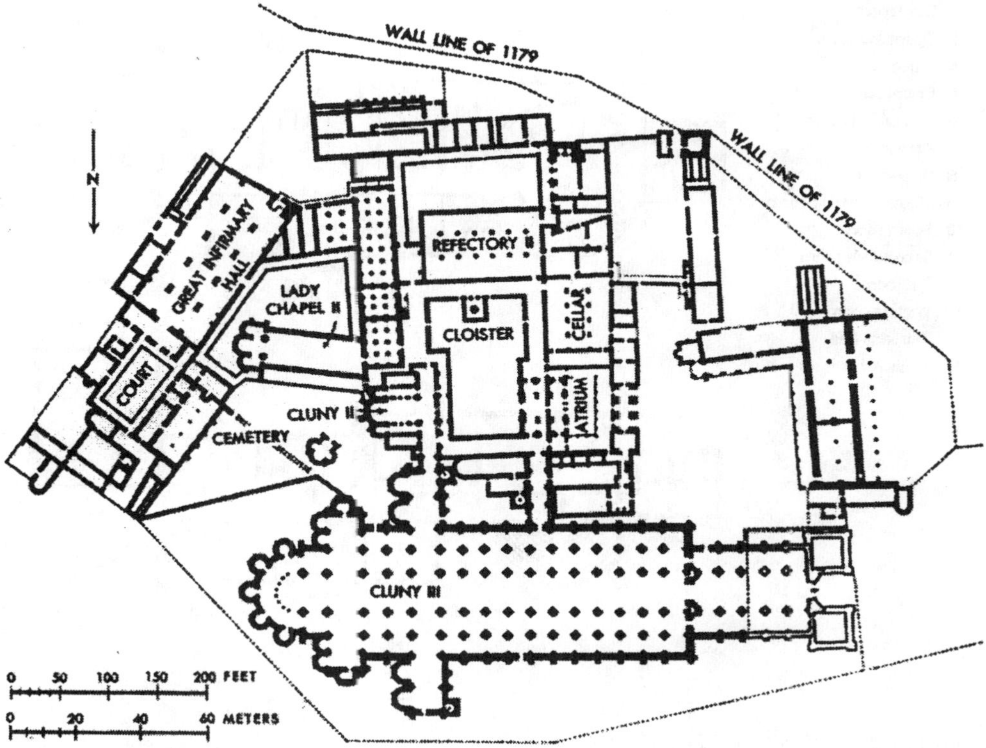

FIGURE 38.3 Cluny Monastery plan.

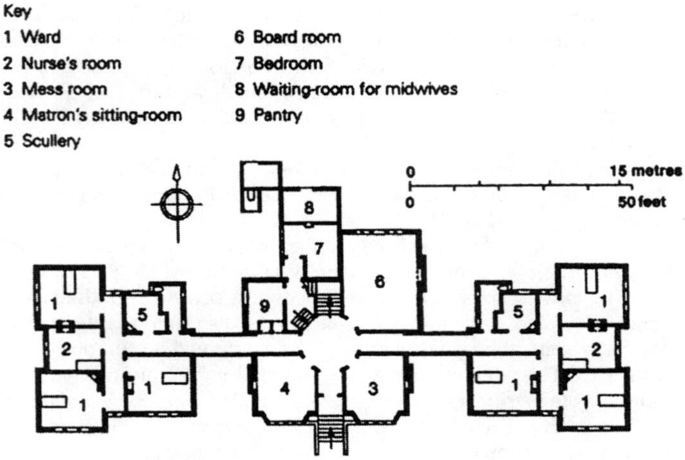

FIGURE 38.4 Plan of Liverpool Lying-In Hospital (1880s).

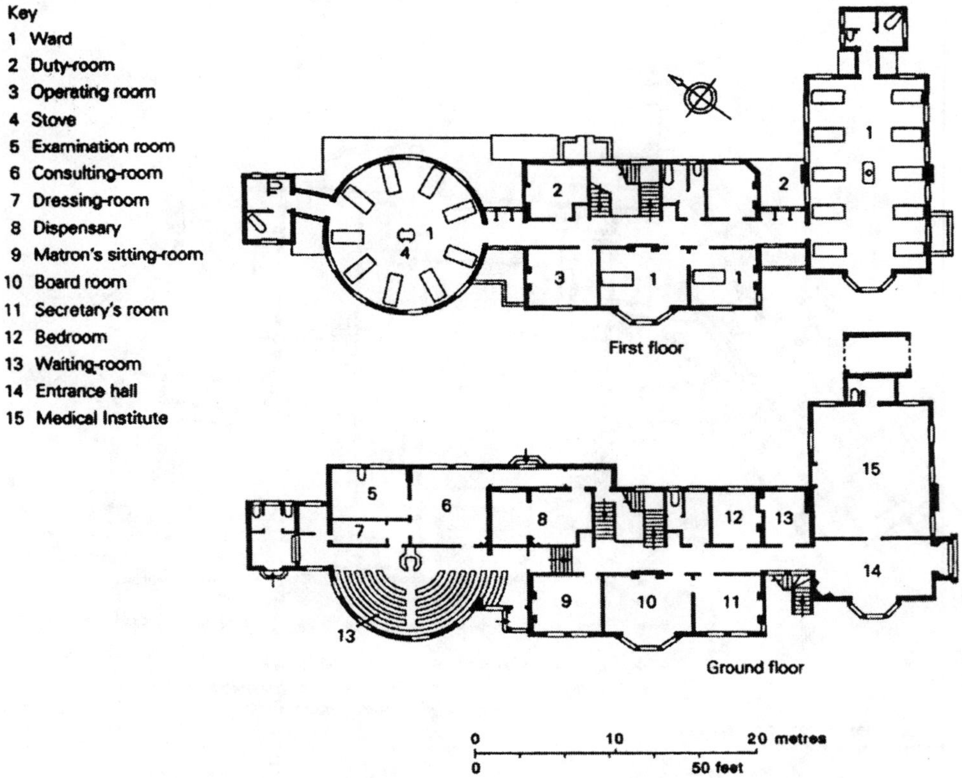

FIGURE 38.5 Plan of Anderson Hospital (London, 1889).

2. Emerging medical and nursing specialties
3. Individual entrepreneurs founding their own hospitals and clinics to promote their private practices

Maternity or lying-in hospitals were among the earliest of specialty hospitals. The Liverpool Lying-In Hospital (Fig. 38.4) was separated from the Hospital for Women to help control infection, again supported by Nightingale's recommendations. One can see from the plan of this building that the architect has created separation between nursing and support functions.

Looking at the plan of the Anderson Hospital in London (1889) (Fig. 38.5), the ward area is seen as a circular room supporting ease of nursing and good visualization of patients but not much privacy for those in bed.

The work of John Shaw Billings, M.D., at Johns Hopkins Hospital in the late 1870s and 1880s is shown in Fig. 38.6. Collaborating with the architect, Billings proposed plans that took into account the control of infection, growth and flexibility, separation, natural light, and ventilation. Billings clearly saw that whatever design was built, continuing scientific advantages would necessitate change:

> . . . no matter what plan is adopted . . . it will appear that it can be improved and a certain amount of funds should be reserved for that purpose. The general principles which I have tried to state . . . are in accordance with the present condition of our knowledge of the

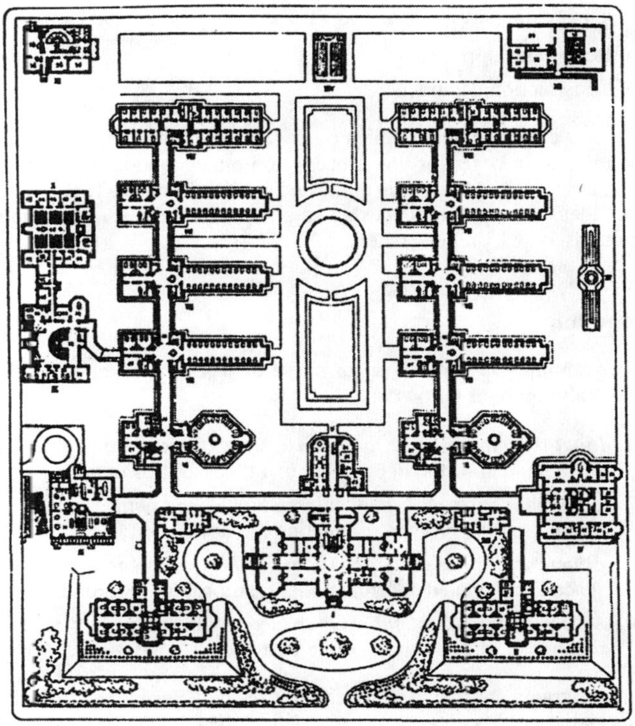

FIGURE 38.6 John Shaw Billings plan for Johns Hopkins Hospital (Baltimore, 1870s).

subject, but that knowledge is imperfect, and too much of the teaching of books on the subject of hospital construction is theoretical only [J. S. Billings, M.D., 1875].

Dr. Billings was right, and today we continue to follow his advice when developing facility plans for health care facilities.

38.2 HEALTH CARE FACILITIES PLANNING AND DESIGN

Health care facilities planning is a unique area of endeavor. The unique character of this work comes from the fact that the patient and family are silent users of the space represented by all the members of the planning and design team. This responsibility to uphold the future well-being of these patients gives the process an added dimension.

It is useful to discuss this process by describing it in phases. These phases or steps are given different names by various groups, but the activities are common to all projects. The American Institute of Architects, as an industry standard, is the source for terms describing the various phases of planning and design through construction and occupancy. In this chapter I will confine my discussion to the work leading up to construction activities.

The health facility planning and design phases are

- Programming
- Schematic design
- Design development
- Construction documents

Each succeeding phase builds on the preceding one as decisions are made and the design solution is refined. In order for this process to produce a good result, all participants should be completely clear in their understanding of the decisions to be made. This is a process of interaction and challenge. All ideas to improve the design solution and all relevant facts must be openly and honestly sought out, discussed, and evaluated.

38.2.1 Programming

Programming is the first major phase of work that the project team undertakes. Programming has several important purposes:

1. *Input from health care facility users* is gathered in this phase. By preparing a thorough program, each element of the department can be described in detail.
2. *Communication with and guidance from the entire team* are recorded to be used and refined throughout the process of creating the new facility.
3. *Adherence to budget, criteria, and other project parameters* can be checked and controlled as space is calculated and functional relationships recorded.
4. *An orientation to the future is ensured* by including new technology considerations and avoiding reliance on the solutions of the past.

Programs are prepared for the use of the team by health facility architects, by consultants who specialize in this area, or by experienced health care users. Regardless of the authorship source, it is imperative that the programmer be a full member of the planning team who stays with the work of creating the project. The program will be refined as the project goes forward, and continuity is important when programming decisions are reconsidered in the light of developing design solutions. Programs consider each space, each department, and each system to be included. Programs describe (1) the activity to be carried out in each space, (2) the people to be accommodated, (3) the technical and support equipment to be included, (4) the furniture and furnishings to be supplied, (5) the physical environment (and environmental controls), (6) critical relationships within and among spaces, (7) the size and makeup of each department, (8) relationships among departments, and (9) the size and makeup of the entire project.

A thoroughly prepared program contains the following components:

- Space listings and area tabulations
- Diagrams
- Room data sheets
- Equipment and furniture
- Technical data sheets on critical equipment systems
- Written functional statements

Space listings and area tabulations are the heart of the program. A number of forms are possible. Commonly, these listings are organized into functional groupings, and then quantities are added. These space listings (Fig. 38.7) must be established using some mutually agreed on forecast of future workload. This can take the form of procedures, visits, or operating room minutes. It is essential that the need for space be directly linked to and driven by a disciplined forecast of future activity.

The areas shown are *net*, i.e., exclusive of walls, doors, structure, and sometimes cabinetry. The areas need to be based on serious discussions among the users plus attention paid to codes and standards that apply to the project. The areas shown need to include full consideration to equipment and other technology that will operate within these rooms. Allowances for departmental *gross* area must be included to provide appropriate circulation, walls, doors, cabinetry, mechanical and electrical

SPACE PROGRAM DRAFT

room #	room name	number of spaces	net area each space	total net area	remarks
	CHEMOTHERAPY				
	Patient/Family Waiting	1	300	300	
	Reception	1	120	120	
	Toilets	2	60	120	
	Patient Treatment - Private	8	100	800	with toilet
	Patient Treatment - 4 Patient	4	280	1120	with toilet
	Patient Treatment - 8 Patient	2	480	960	with toilet
	Nurse Work Area	1	200	200	
	MD Work Area	1	200	200	
	Admin Office	1	120	120	
	Clean Utility	1	100	100	
	Soiled Utility	1	100	100	
	Housekeeping Closet	1	50	50	
	Pharmacy	1	500	500	
	Research Pharmacy	1	500	500	
	Research Office	1	100	100	
	Research Nurse/QA	1	100	100	
	Interview/consult Room	1	100	100	
	subtotal				5490
	Pheresis Unit	1	2000	2000	program to be determined
	subtotal				2000

FIGURE 38.7 Sample space listing.

systems, etc. The remarks column within the space listing spread sheet can contain cues for designers and planners to read further or to refer to functional diagrams included in other sections of the program document.

The inclusion of *room and department diagrams* increases the usefulness of the program document. The facility planning and design process is highly dependent on graphics, drawings, and other visual items and is conducted largely by professionals who use graphics to record and communicate ideas.

Not every space is so demanding that a diagram such as that in Fig. 38.8 is required, but it is clear that words and numbers alone would not completely explain the complexities of the operating room.

Department organization and critical relationships between spaces can be illustrated simply by an adjacency matrix diagram, as shown in Fig. 38.9. This information provides critical guidance for the design team.

Room data sheets are extremely valuable vehicles for carefully recording information about each space. An example, shown in Fig. 38.10, demonstrates the range of facility information that can be shown for the designers and engineers working on the project.

Space programs are concerned with describing rooms/spaces that contain technical equipment. *Equipment and furniture lists* are important to guide designers and to create a complete picture of the project budget. A preliminary list such as shown in Fig. 38.11 can be modified as design and

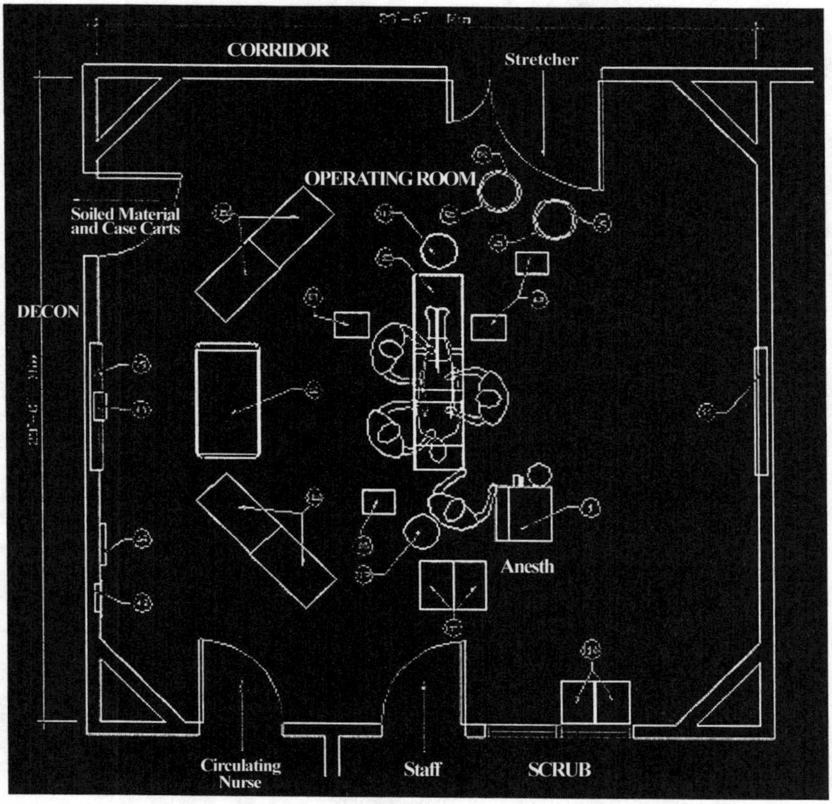

FIGURE 38.8 Sample operating room diagram.

	RECEIVING	BULK STORES	PROCESSED STORES	DECONTAM	LINEN	PACK AND STERILIZATION	PHARMACY
RECEIVING		1	2	3	1	3	1
BULK STORES			1	2	2	1	1
PROCESSED STORES				1	1	1	2
DECONTAM					1	2	2
LINEN						1	3
PACK AND STERILIZATION							3
PHARMACY							

LEGEND	
Directly Adjacent	1
Corridor Connection	2
Indirect	3

FIGURE 38.9 Sample adjacency matrix.

knowledge progress on the project. Including this type of information is an important notice to designers that equipment/systems must not be added in later (or forgotten entirely).

If understanding a particular equipment item is critical to the successful layout of the room or must be considered in developing engineering systems then a *technical data sheet* can be added to this part of the program document. It is unlikely that a specific choice of equipment will be made in final form at the programming stage of a project, but including the technical data sheet as an example will ensure that appropriate space, floor loading, HVAC, and other provisions are made. Manufacturer-supplied material can be used and marked as "*Preliminary—subject to change*."

Each programmed department should be described in writing by including a *written functional statement*. Department health care staff are well equipped to supply this information, which should include

- An overall summary of the department
- Staffing
- Hours of operation
- Workload history and forecasts
- Description of activities, procedures, etc.
- Any unique planning considerations

HEALTH CARE FACILITY ROOM DATA SHEET

PROJECT NAME

Project Number
Date Prepared By

DEPARTMENT
ROOM NAME
ROOM NUMBER

GENERAL

Net Area Occupant Count
Ceiling Height Door Size and Type
Room Function

ROOM FINISHES

Floor Walls
Base Ceiling

Finish Notes

ROOM ACOUSTICS

Level of Speech Privacy Absorption
Wall/Ceiling Performance Other

Acoustics Notes

MECHANICAL REQUIREMENTS

Air Changes/Hour Pressure Relationships Filtration Humidity
O/A Required Exhaust Requirements Temperature Control Other
Equipment Connections

Mechanical Notes

ELECTRICAL and TELECOMMUNICATIONS REQUIREMENTS

Duplex Outlets Special Outlets Special Cabling
Room Illumination Special Lighting Telecom and Data Outlets
Equipment Connections Other

Electrical Notes

PLUMBING AND PIPED SYSTEMS

HW CW Direct Drain Special Waste
Floor Drain
Gas Outlets
Equipment Connections

Plumbing and Piped Systems Notes

EQUIPMENT NOTES

FIGURE 38.10 Sample room data sheet.

SPACE: 24.2004000 INFANT RESUSCITATN QTY. OF SPACES: 2

CAT.NUMBER	ITEM NAME	UNIT COST	QTY	EXT.COST	GRP	CAT	INST.COST
MAS CLK0500	CLOCK ELAPSED TIME	$ 0	1	$ 0	1	E	$ 0
MAS HDW0900	HEADWALL UNIT INTENSIVE	$ 0	1	$ 0	1	B	$ 0
MAS ICC0000	INFANT CARE CENTER	$ 0	1	$ 0	2	B	$ 0
MAS ILL0200	ILLUMINATOR, X-RAY FILM 2 PANEL	$ 0	1	$ 0	2	B	$ 0
MAS LTS0500	LIGHT EXAMINATION	$ 0	1	$ 0	2	B	$ 0
MAS OTS0000	OTOSCOPE/OPHTHALMOSCOPE SET	$ 0	1	$ 0	2	B	$ 0
MAS RSS1600	RESUSCITATOR PEDIATRIC	$ 0	1	$ 0	2	E	$ 0
MAS SCL0900	SCALE INFANT	$ 0	1	$ 0	3	B	$ 0
MAS SNK1900	SINK SCRUB	$ 0	1	$ 0	1	A	$ 0
MAS STD1300	STAND MAYO	$ 0	1	$ 0	3	B	$ 0
MAS STL0100	STOOL ANESTHETIST'S	$ 0	1	$ 0	3	B	$ 0
MAS TBL0900	TABLE INSTRUMENT	$ 0	1	$ 0	3	B	$ 0

FIGURE 38.11 Sample equipment list.

38.2.2 Design Phases

Schematic Design Phase. The American Institute of Architects' *Handbook of Professional Practice* describes the schematic design phase of a project as having the following purpose:

> Schematic Design establishes the general scope, conceptual design, scale, and relationships among the components of the project. The primary objective is to arrive at a clearly defined, feasible concept and to present it in a form that achieves understanding and acceptance. The secondary objectives are to clarify the project program, explore the most promising alternative design solutions, and provide a reasonable basis for analyzing the cost of the project.

Schematic design often begins with the creation of *block diagrams* that address the overall relationships between and among the various departments of the project. Block diagrams (Fig. 38.12) are drawn to a relative scale and show each floor of the project. Major entrances to the building are established for each type of traffic. Circulation routes are studied on each floor to carefully separate traffic. Vertical circulation (stairs, elevations, service lifts, etc.) is planned.

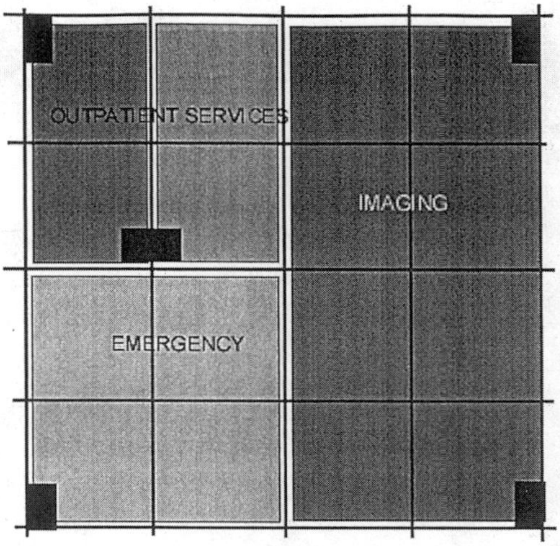

FIGURE 38.12 Sample block plan.

At this very early stage of design, provisions for the communications and data technology must be considered and included within the block diagrams. Allowances for information technology (IT) equipment and cabling pathways within the building must be shown with sufficient accuracy that technology professionals can judge their adequacy. This is not a planning element that can be added in at a later stage of design.

Block diagrams should be drawn using a planning grid that will reflect a structural grid system. This will facilitate planning in later stages and avoid conflicts between desired floor plan arrangements and the structural system.

Block diagrams will also be considered in section showing vertical relationships (Fig. 38.13) from one floor to the next. The building height will be defined in this phase, and issues of floor-to-floor height can be decided. Departments that have high levels of engineered systems can be located to give maximum volume in the overhead spaces, allowing easy routing of ducted and piped systems without conflict.

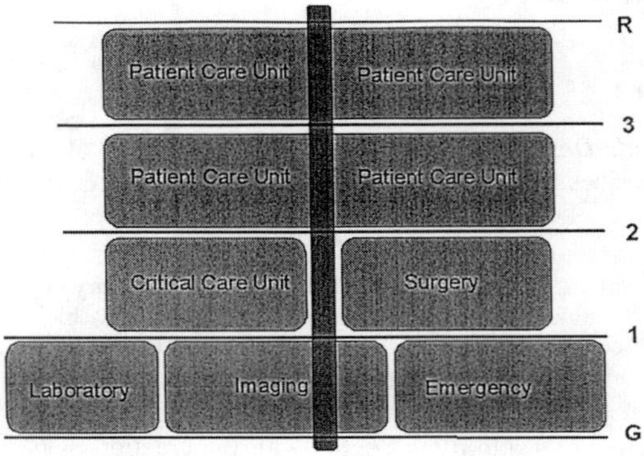

FIGURE 38.13 Sample blocking and stacking plan.

Once all alternative block diagrams have been thoroughly considered and reviewed with hospital personnel, a single direction can be established for more detailed schematic design. Alternative layouts for each department are studied using drawings called *bubble diagrams*.

Figure 38.14 shows a typical bubble diagram. Normally, several alternative arrangements are studied using diagrams such as these. Bubble diagrams can help the team study by showing

- Room-to-room relationships
- Circulation of staff and patients
- The basic size and shape of key spaces
- Provisions for critical support spaces
- Engineering and technology requirements

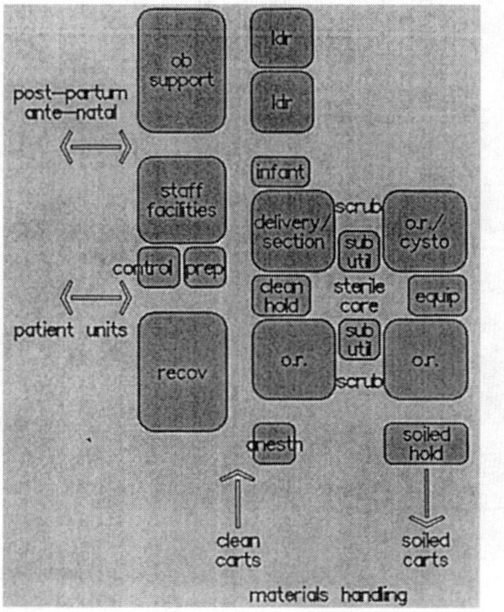

FIGURE 38.14 Sample department bubble diagram.

The success of the planning at this level depends in part on the completeness of the program documents prepared in the programming phase. Program material will guide the team as it studies and evaluates alternative layouts. Bubble diagrams are drawn to scale and also follow the planning grid. The selected diagram layout will be developed in more detail as *single-line schematic drawings*.

Figure 38.15 shows a typical single-line plan. Note that many details are left out, such as door swings and furniture and equipment placement. These important considerations will be relatively easy to add as plans are refined if they have been well described in the program and if those program data are kept in mind as the plan work progresses.

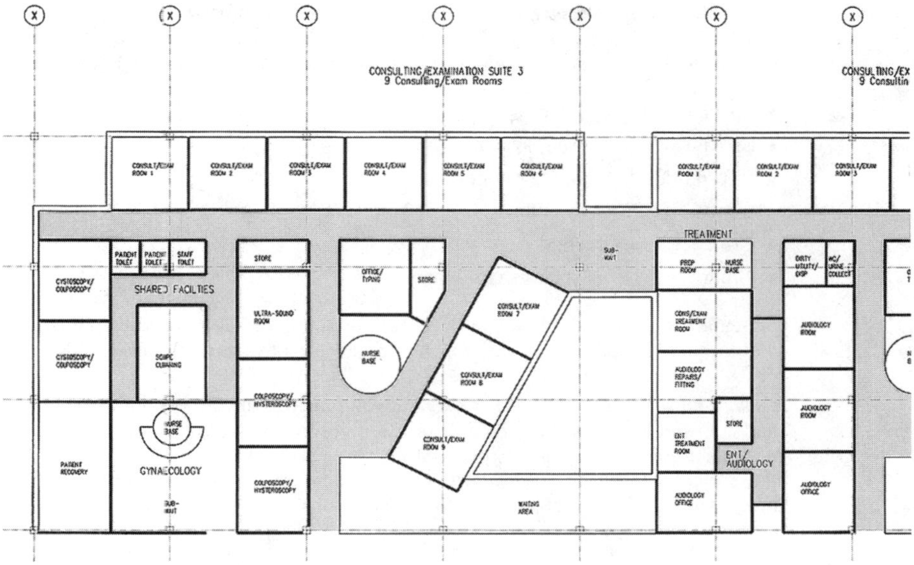

FIGURE 38.15 Sample single-line plan.

Schematic plans are often difficult to fully interpret, and so three-dimensional (3D) sketches are quite useful in judging the success of the plan in meeting the objectives set out in the program. Figure 38.16 shows a 3D sketch drawn with a computer graphics program that very clearly demonstrates the various elements of the plan. Physical models also can be employed to illustrate elements of the schematic plan, but their usefulness is somewhat limited by the comparative lack of detail available.

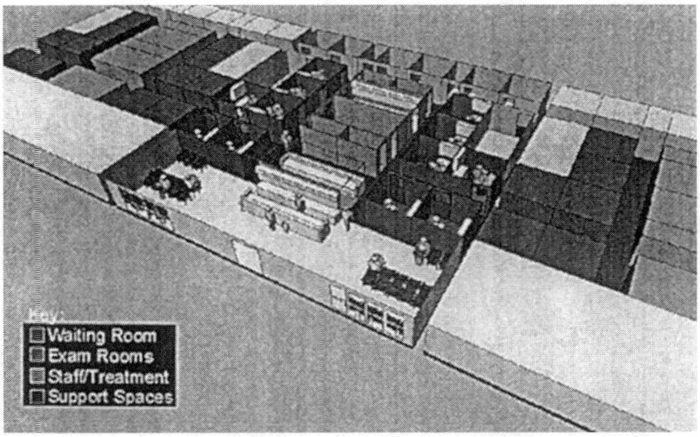

FIGURE 38.16 Sample 3D sketch.

During schematic design, it is also useful to prepare documents called *outline specifications*. Outline specifications describe the various construction contract elements in words citing industry standards, methods, and levels of quality. These documents provide an opportunity to clearly spell out each part of the construction, and they form an important part of the basis for estimating the cost of the project. The program-based equipment listings can be carried forward to be included within the outline specifications so that adequate consideration of the cost of equipment is made a part of the job. Further, these specifications should include clear technical provisions for

- Special piping systems
- Special wiring/cabling systems
- Information technology support
- Communications systems
- Critical environmental controls
- Other fixed items such as casework

Schematic design documents must be submitted together for a formal approval by all the user groups and to be accepted as the basis for moving ahead into the next phase of design.

Design Development Phase. Design development is the project phase in which the design is refined and coordinated among all the disciplines involved on the team. The schematic design work carried out in the preceding phase is brought forward, and detailed information is added. Each element of the project is worked out at a larger scale, and changes are incorporated as the team members see more detail and can arrive at additional decisions.

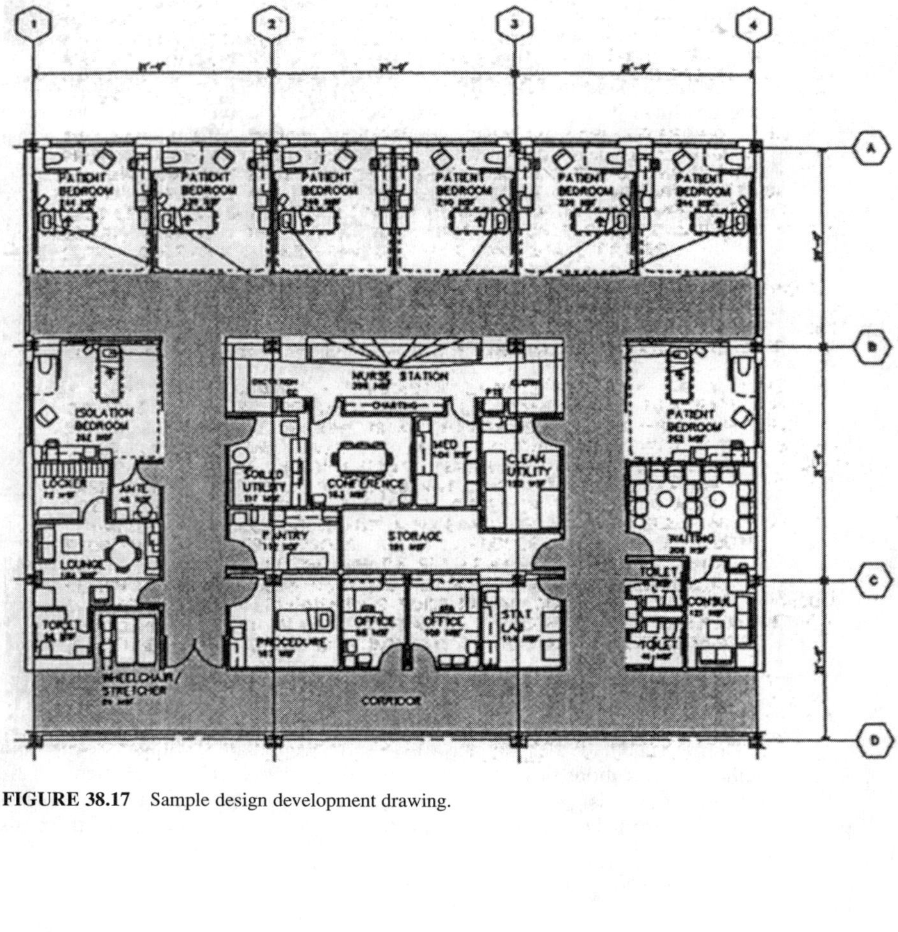

FIGURE 38.17 Sample design development drawing.

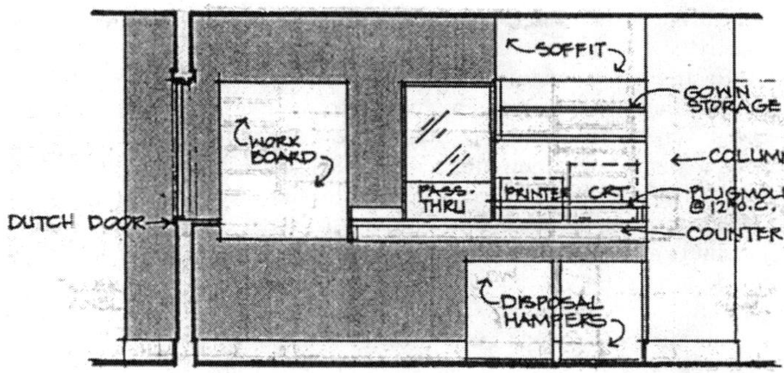

FIGURE 38.18 Sample interior elevation.

Design development begins with plans and sections drawn at increased scale so that users can see the functional and technical details of each space. As information is added, each room in the project should be assigned a unique identification number. This allows the team to track the refinement of the design of each room. Room data sheets prepared during programming can now be brought forward and keyed to these unique numbers. Similarly, equipment lists and technical data sheets are also keyed to the room numbering system. Each room or space will then have a data file that architects and engineers can use for design. Users will use the data file for monitoring the progress of the design and measuring the success of the design of each space in meeting functional and technical needs.

Design development floor plans, such as that shown in Fig. 38.17, contain the details that were not shown in schematic design. These include

- Wall thickness and special wall construction, including shielding or structural support
- Code-required construction for control of smoke spread, fire stopping, etc.
- Doors
- Fixed elements such as plumbing fixtures, cabinetry, etc.
- Equipment placement
- Furniture placement
- Building structure and engineering spaces

Design development also should include drawings that illustrate all wall and ceiling surfaces. *Elevation* drawings of wall surfaces such as shown in Fig. 38.18 will illustrate the placement of electrical, piping, communication, and data outlets to scale, giving mounting heights above the floor and clearances for convenient use by hospital staff. These elevation drawings also show equipment (fixed and movable) to be attached or connected to ensure that the equipment will function and that adequate clearance is provided for service access.

Reflected ceiling plans are useful in controlling the design of the ceiling plane. Figure 38.19 shows that each element (lighting, HVAC, fire-protection sprinkler heads, special systems, and ceiling-mounted equipment) can be installed and operate properly.

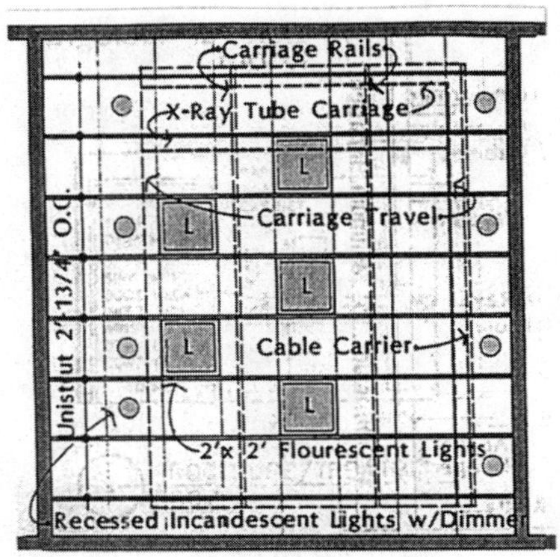

FIGURE 38.19 Sample ceiling plan.

During design development, additional information and detail are created. These are interior finish materials selection and casework and workstation design.

Each space or room will have materials assigned by the design team to be reviewed by users. One method of managing this new information is with a *finish schedule*. Much of these data can come directly from the room data sheets created during programming. This information will now need to be updated and refined. The finish schedule will eventually become an important component of the construction process. The design and placement of casework, workstations, shelving, and other cabinetry are an important part of design development. Each user must be satisfied that the patient care and other work processes to be supported by these items are thoroughly understood by the designers. Each element must be carefully considered. Computers, displays, benchtop equipment, and other critical elements must be drawn to scale and included in the design. It is quite common to build full-sized mockups or models of these workstations so that users can try out the new design before it is built. Workstations should be tested for user comfort, success in accommodating equipment, visual access to patients, etc.

There is no satisfactory substitute for a full-size mockup in discovering and correcting flaws in the design. The same can be said for critical full-room mockups as well. These have been used to excellent results in developing designs of rooms that contain critical new technology features.

Design development also addresses the full coordination of all the design disciplines involved in producing a complete architectural and engineering design for the project. Each department, each space, and all physical elements of the project are brought up to a similar level of design refinement. All systems (HVAC, structural, site/civil) and all specialty areas must be carefully designed and coordinated to avoid conflict. Design development leans heavily on the work done during programming and schematic design.

Design development documents must be published for user and hospital approval. Prior to this approval being obtained as notice to proceed into final working drawings, a cost estimate based on the design development package must be prepared. This is a critical juncture of the project effort, and the completeness of the work done to this point will help to avoid serious problems later in the project.

38.2.3 Construction Documents Phase (Working Drawings)

Most design issues will have been answered during the preceding phases of work. The construction documents phase is principally concerned with the creation of drawings and written instructions to be used by the various building trades in constructing the project. These documents become part of the contract between the owner and the builder. They have important legal consequences. They must be clear, accurate, and free from ambiguity.

The construction documents consist of three basic elements:

1. Specifications (written about in earlier sections)
2. Drawings
3. Written contract provisions

Each element makes reference to the other and must be consistent, using similar language to mean the same in each instance. Each element is briefly discussed here.

Specifications generally fall into two categories. *Descriptive specifications* use words and make reference to industry standards to describe a method of construction or to describe a building product/material. *Performance specifications* use words and make reference to industry standards to make clear how a part of the construction is to perform. Both types are used successfully in health care facility construction. The project's final specifications are developed from the earlier outline specifications prepared during schematic design and design development. Specification writing is a highly specialized endeavor that is the responsibility of the design team and typically is carried out by a design professional with special qualifications in this area.

In medical facility construction, it is very common to have unresolved issues even this late in the design. An example of this is the procurement of certain types of equipment such as medical imaging systems that will be attached to the building but will be supplied by a third-party vendor. Such issues will need to be dealt with in the specifications so that the builder is fully aware that these late decisions will be coming and that funds must be included to accommodate the work to be done.

Certain items to be specified may be covered by *proprietary specifications*. These are items where no competitive alternative exists, and the owner is willing to state an exact make, model, or product name to be used exclusively.

Drawings will be prepared during this phase that will become a legal part of the contract for construction. These drawings are based on the work of earlier phases. The intended audience is the builder and individual trade workers, so the drawings focus on providing the data needed to successfully construct the building. Each element to be constructed is drawn in detail, with all dimensions and explanatory notes shown. In the case cited earlier of a delayed decision regarding equipment (say, for imaging), the drawings must show how the work is to be undertaken to allow for a later decision.

Written documents called *General Conditions, Special Conditions*, and *Supplemental Conditions* are included as important parts of the contract. These set down the various procedures to be followed by the contract parties in constructing the project. The American Institute of Architects publishes a model of these documents (AIA document A201) that is the standard of the construction industry.

38.2.4 Codes, Standards, and Industry Data

There are literally thousands of documents published by more thousands of authorities that provide information, best practices, professional criteria, and otherwise control every part of a project from inception to completion. We will deal here in outline form with those that influence the medical or technical aspects of planning the project.

Building codes have the force of law over the project. The team must comply with the provisions of the building code applied by the authority having jurisdiction over the project. This is usually a building department at the municipal or state level. Building codes often include sections of other documents, most notably the *Life Safety Code* published by the National Fire Protection Association. Three major organizations predominate:

- Building Officials and Code Administrators International (BOCA), publishers of the *BOCA National Building Code*
- International Conference of Building Officials (ICBO), publishers of the *Uniform Building Code*
- Southern Building Code Congress International (SBCCI), publishers of the *Standard Building Code*

These *model codes* form the basis for most building codes in force around the United States.

The International Code Council (ICC), established in 1994 as a nonprofit organization, is dedicated to developing a single set of comprehensive and coordinated national model construction codes. The founders of the ICC are Building Officials and Code Administrators International, International Conference of Building Officials, and Southern Building Code Congress International. The nation's three model code groups created the ICC by developing codes without regional limitations providing one code document series that can be applied from coast to coast in all jurisdictions. There are advantages in combining the efforts of the existing code organizations to produce a single set of codes. Code enforcement officials, architects, engineers, designers, and contractors have access to consistent requirements throughout the United States.

In addition to the state-controlled building codes, a number of other important codes having legal standing are in force. These cover areas of architectural and engineering practice, such as

- Mechanical systems
- Plumbing and piped systems
- Fire protection and life safety
- Energy conservation
- Electrical systems
- Signage

Most code documents refer to a large body of construction industry standards. The American National Standards Institute and the American Society for Testing and Materials publish enormous volumes of important standards to be followed in designing a building, specifying products, and carrying out construction activities.

The National Fire Protection Association (NFPA), headquartered in Quincy, Massachusetts, is an international nonprofit (tax-exempt) membership organization founded in 1896. The mission of NFPA is to reduce the worldwide burden of fire and other hazards on the quality of life by developing and advocating scientifically based consensus codes and standards, research, training, and education. NFPA activities generally fall into two broad, interrelated areas: technical and educational. NFPA's technical activity involves development, publication, and dissemination of more than 300 codes and standards. NFPA codes and standards are developed by nearly 250 technical committees. The Health Care Section of NFPA is comprised of individuals with a focus on the protection of patients, visitors, staff, property, and the environment from fire, disaster, and related safety issues.

Other regulations influence facility design as well. In health care, the requirements for licensure by the states invoke an additional body of criteria to be followed plus other standards by reference. The Joint Commission on Accreditation of Healthcare Organizations conducts surveys of hospitals using criteria that refer to the American Institute of Architects' *Guidelines for Design and Construction of Hospital and Health Care Facilities*. These guidelines, in turn, refer to other sources that govern planning and design. The 2001 edition of the *Guidelines* covers most health care departments and functions and contains a brief chapter on medical equipment.

Federal regulations strongly influence the design of buildings. The provisions of the Occupational Safety and Health Act (OSHA) of 1970 control many of the conditions found on construction sites. Failure to comply with OSHA can involve stiff penalties. In 1990, the Americans with Disabilities Act (ADA) was passed, governing the provision of reasonable access to people with disabilities. This act covers all buildings, both existing and new construction. The Department of Justice (DOJ) administers this act. DOJ has developed accessibility guidelines for use by design professions. Further, the federal government, in regulating the design of its own health facilities, has developed a full panoply of codes, regulations, and other criteria to follow.

Important standards are published in addition to all the preceding by the following organizations that directly or indirectly influence the design of health care facilities:

- Association for the Advancement of Medical Instrumentation (AAMI)
- American Society of Heating, Refrigerating, and Air Conditioning Engineers (ASHRAE)
- Centers for Disease Control and Prevention (CDC)
- College of American Pathologists (CAP)
- Compressed Gas Association (CGA)
- General Services Administration (GSA)
- Illuminating Engineering Society of North America
- National Association of Plumbing, Heating, and Cooling Contractors
- National Bureau of Standards (NBS)
- National Council on Radiation Protection (NCRP)
- National Fire Protection Association (NFPA)
- Underwriters Laboratories, Inc. (UL)

Health care industry data are an important part of this library of criteria that influence design. Each manufacturer or supplier is a useful (and usually willing) source of up-to-date information on products or systems to be included in the project. Some of these have profound effects on the design of spaces and on building systems. During the life of a project, this material must be updated and verified as new items are introduced or as variations are made by the manufacturer.

38.2.5 The Importance of Project Management

The key to successfully mobilizing a health facility planning effort lies in the strength of the work of a team of professionals. Designing a health care facility is a complex task. The team needed to carry out this task is large and multidisciplined.

Project management has emerged as an important area of special experience and training within the architectural profession. Project managers not only pull together the efforts of the architectural and engineering staff but also must coordinate the efforts of the hospital staff and ensure that all the various points of view are included in the effort.

> The Joint Commission [on Accreditation of Healthcare Organizations] expects . . . a collaborative design process. This process should team department direction, medical staff, and individuals having special knowledge. . . .

The project manager needs information in a timely manner and will come to rely on the hospital's technology managers to provide it. No other group is more completely aware of the hospital's capital equipment program. No other group is more completely engaged with the medical equipment industry. This unique knowledge, properly applied, as suggested here, will help ensure a successful, responsive, and flexible design result.

38.3 KEY POINTS OF INFLUENCE: CHECKLIST FOR MEDICAL TECHNOLOGY INPUT INTO THE HEALTH CARE FACILITIES PLANNING PROCESS

38.3.1 During Project Inception

Prior to beginning any formal work, review and evaluate capital purchasing plans for 3 to 5 years into the future to check for possible deferrals. Evaluate leases and maintenance contracts. Evaluate technology staff. Survey technology-intensive departments for upcoming changes or anticipated new technology. Coordinate with IT group.

It would be useful if a technology master plan were in place.

Prepare (or gather together) all *existing* medical equipment listings, showing details useful for programming and later planning. Identify key individuals to serve on programming and design user groups to represent medical technology management.

Organize materials by department or cost center or other hospital organizational entity for ease of use by planners.

38.3.2 During Programming

Direct participation in or authorship of

- Key room sizing and diagramming
- Room data sheets

- Equipment listing
- Manufacturer technical data sheets

Representation on user groups

Review, comment on, or refine programming documents

38.3.3 During Schematic Design and Design Development

Continued representation on user groups and continued responsibility for review and input to design products

Liaison with manufacturers and clinical staff for equipment and systems decision making

Testing critical room layouts with real templates for equipment (plans, elevations, ceiling plans)

Mockup room design and evaluation

38.3.4 During Construction Document Preparation

Continued liaison for design team and clinical staff on equipment and systems questions

Liaison with manufacturers for updated accurate installation templates and diagrams

Prepare a logistics work plan for specification, procurement, receipt, storage, and placement of equipment to be purchased outside of construction contract

Provide coordination for vendor workforces if access to construction site is necessary for rough-in or for equipment placement

REFERENCES

Allison, David J. (ed.) (1997), *Planning, Designing, and Construction of Health Care Environments*, Joint Commission on Accreditation of Healthcare Organizations, Oakbrook, Ill.

Bronzino, Joseph D. (ed.) (1992), *Management of Medical Technology: A Primer for Clinical Engineers*, Butterworth-Heinemann, Boston.

Facilities Guidelines Institute (2001), *Guidelines for Design and Construction of Hospital and Health Care Facilities*, The American Institute of Architects, Washington, D.C.

Galison, Peter, and Thompson, Emily (eds.) (1999), *The Architecture of Science*, MIT Press, Cambridge, Mass.

Haviland, David (ed.) (1994), *The Architect's Handbook of Professional Practice*, Vol. 2, American Institute of Architects Press, Washington, D.C.

Haynes, Marion E. (1996), *Project Management from Idea to Implementation*, Crisp Publications, Menlo Park, Calif.

Howell, Joel D. (1995), *Technology in the Hospital: Transforming Patient Care in the Early Twentieth Century*, Johns Hopkins University Press, Baltimore.

Kostof, Spiro (1985), *History of Architecture: Settings and Rituals*, Oxford University Press, New York.

Lewis, James P. (1997), *Fundamentals of Project Management*, AMACOM, New York.

Marberry, Sara O. (ed.) (1997), *Healthcare Design*, Wiley, New York.

Masson, Madeleine (1985), *A Pictorial History of Nursing*, Hamlyn Publishing, London.

Reiser, Stanley Joel (1978), *Medicine and the Reign of Technology*, Cambridge University Press, Cambridge, England.

Risse, Guenter B. (1999), *Mending Bodies, Saving Souls: A History of Hospitals*, Oxford University Press, New York.

Russell, Louise B. (1979), *Technology in Hospitals: Medical Advances and Their Diffusion*, The Brookings Institution, Washington, D.C.

Stevens, Edward F. (1918), *The American Hospital of the Twentieth Century*, Architectural Record Publishing Company, New York.

Tomasik, Kristine M. (ed.) (1993), *Plant Technology and Safety Management Handbook: Designing the Environment of Care*, Joint Commission on Accreditation of Healthcare Organizations, Oakbrook, Ill.

CHAPTER 39
DEPARTMENT/PROGRAM MANAGEMENT

Ira Tackel
Thomas Jefferson University Hospital, Department of Biomedical Instrumentation, Philadelphia, Pennsylvania

The three keys to success in real estate investing are "location, location, and location." When I think of clinical engineering department management, the one overriding thought that drives to the core of that management is "It's a business." To some, this statement may be obvious, but actually, in what is for the most part a not-for-profit health care delivery world, this obvious statement all too often gets lost in the management of the program. If we keep our sights on this one simple mantra, success can be reached much more easily.

My background starts with a Bachelor of Science Degree in Biomedical Engineering, followed by a Masters of Engineering in Biomedical Engineering (both from Rensselaer Polytechnic Institute, Troy, New York), with an emphasis on and practicum in clinical engineering. It is interesting to note, though, that I never considered myself much of an engineer at all. I viewed my engineering degrees as a stepping-stone to engineering management, more specifically, *technology management*. I always knew that I would not be a very good bench engineer but rather could use my academic background and engineering expertise to solve engineering problems. Through all the physics, thermodynamics, calculus, differential equations, anatomy, and physiology courses, while important in their own right, I was really learning a process by which to solve problems. I truly believe that this aspect of an engineering degree sets it apart from virtually all other academic degrees. Problem solving and a unique perspective to bring to the table in the analysis of solving a problem are what distinguishes an engineer from other professionals.

With this said, the management of a clinical engineering program depends on someone setting the vision, the mission, and the values for the program. In the absence of these three parameters, a program will in all likelihood not be very successful or, perhaps at best, will be in a maintenance mode, not one of growth. In today's health care climate, if we are not progressing, then indeed we are retreating. It is unacceptable to stand still. We must continue to push the envelope and provide additional value for our institutions. The question becomes, How best can we do this?

While some of my colleagues would disagree, I fervently believe that our core competency was, is, and should be managing technology. There are those who would have our profession venture into the information technology (IT) world or expand into other areas to enhance our departmental value. This, I believe, would be a tragic mistake. How many of my peers have any training in IT? The answer is very few. And yet, this is an area in which we should espouse to be experts! Know what you do well, and then exploit it to its fullest. This is exactly what we have done within our own department, first at the Hospital of the University of Pennsylvania for 8 years and then at Thomas Jefferson University Hospital for the past 16 years. Never pretend to be something you are not, and always work to your highest potential.

There is no magic here. I clearly know my limitations. From the beginning, I always wanted to surround myself with the best and the brightest. Let's go back to square one, vision, mission, and values. My job as director of a program is clearly not to do everything. Indeed, that is a path to failure. What is the old saying? If I provide fish for someone for a day, he or she will eat for a day. If I teach him or her to fish, he or she will have fish for a lifetime. If I am to consider myself successful, it is only because I have the very best of professionals working with me, and I have given them the tools they need to make themselves and the program successful.

Another fallacy is that it is unacceptable to make mistakes. I would submit that not only is it acceptable to make mistakes, but also if you are not making mistakes, you are playing the game much too conservatively. Now, making the same mistake over and over again is certainly not acceptable. However, if you learn from your mistakes, then the process gets better and better. It is unfortunate that the health care delivery environment we work in seems more intent on assigning blame when things go wrong than on learning from each failure. One need look no further than patient-related incidents to see some of the worst of how health care assigns blame.

The program at my institution is based on sound principles of technology management. By this I mean that it uses a very traditional organizational structure and reporting relationship. It seems that every time a new management theory comes into vogue, we all scramble to implement it. Whether "management by wandering around (MBWA)" or "self-directed work groups," or "flattening the organizational structure," all have been tried with varying degrees of success. I am not suggesting that any alternative other than a traditional one flat out will not work, but for me a very traditional structured approach has been the road to success. Remember, if you do not care where you are going, any road will take you there. Keep your eye on the target.

Let's not kid ourselves, even a seasoned engineer can miss the mark by a fair amount. Whenever I make a presentation, I start by showing the graph in Fig. 39.1. This material was taken from the *Report of the Presidential Commission on the Space Shuttle Challenger Accident*, June 6, 1986, Washington, D.C.

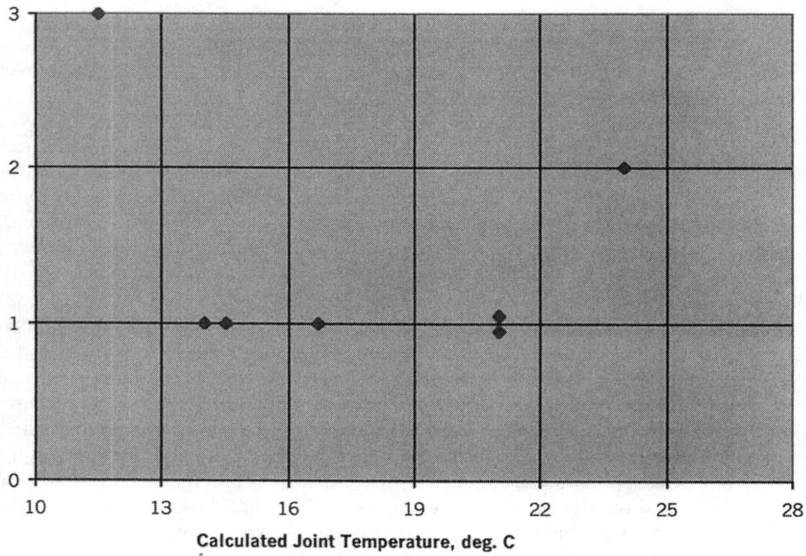

FIGURE 39.1 NASA Shuttle *Challenger* graph 1.

It serves to underscore how analyzing the wrong information, no matter how well intentioned, will result in a wrong conclusion. In this particular case, the analysis resulted in a disastrous outcome. While the circumstances leading up to the *Challenger* disaster are quite complex, the analysis of O-ring thermal distress as a function of launch temperature was certainly a key element to the disaster.

Excerpted from the *Report to the President*, consider the following:

> Temperature Effects
>
> The record of the fateful series of NASA and Thiokol meetings, telephone conferences, notes, and facsimile transmissions on January 27th, the night before the launch of flight 51-L, shows

that only limited consideration was given to the past history of O-ring damage in terms of temperature. The managers compared as a function of temperature the flights for which thermal distress of O-rings had been observed—not the frequency of occurrence based on all flights (Graph 1). In such a comparison, there is nothing irregular in the distribution of O-ring "distress" over the spectrum of joint temperatures at launch between 53 degrees Fahrenheit (≈12 degrees Celsius) and 75 degrees Fahrenheit (≈24 degrees Celsius). When the entire history of flight experience is considered, including "normal" flights with no erosion or blow-by, the comparison is substantially different (Graph 2).

This comparison of flight history indicates that only three incidents of O-ring thermal distress occurred out of twenty flights with O-ring temperatures at 66 degrees Fahrenheit (≈19 degrees Celsius) or above, whereas, all four flights with O-ring temperatures at 63 degrees Fahrenheit (≈17 degrees Celsius) or below experienced O-ring thermal distress.

Consideration of the entire launch temperature history indicates that the probability of O-ring distress is increased to almost a certainty if the temperature of the joint is less than 65 degrees Fahrenheit (≈18 degrees Celsius).

The night before the *Challenger* launch, engineers were analyzing data from previous launches to determine the risk of an O-ring failure due to temperature. If we look at the data in the first graph that the engineers looked at as well, we would probably come to the same conclusion they did, that there was not much correlation between launch temperature and O-ring failure—hence their decision to go ahead with the launch. Unfortunately, they were looking at the wrong data or, at best, incomplete data. If we look at the data in the second graph (Fig. 39.2), we see a very different picture emerging. Now we see all the launches where no O-ring failures occurred, and a statistical model begins to appear, one in which we clearly see that temperature at launch can have a significant effect on O-ring failure. We were not looking at the correct data. Here is one example where failure of any kind was not acceptable, and yet failure occurred. I would like to think that we did learn from our mistake.

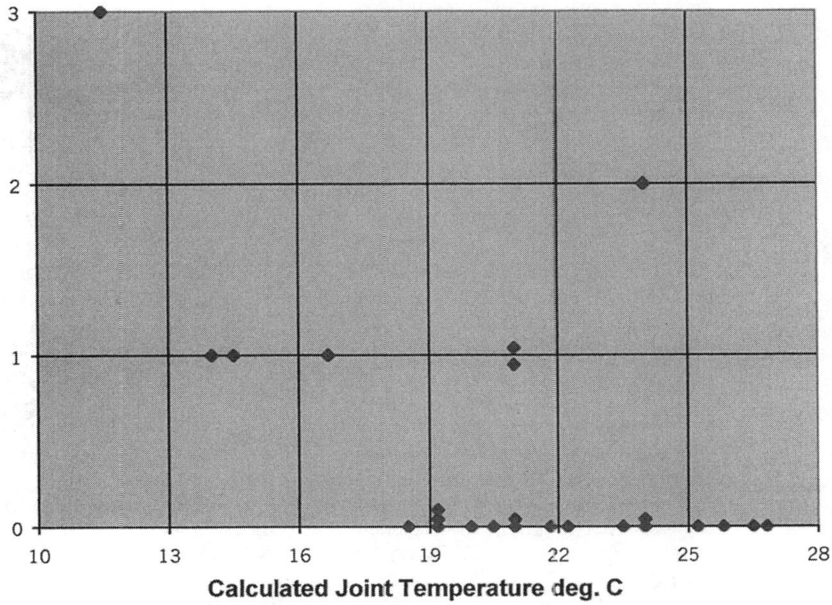

FIGURE 39.2 NASA Shuttle *Challenger* graph 2.

For me, the target was always bringing greater value to the management of technology. In doing so, I had to be the expert in solving problems. If I were to be perceived as creating problems for hospital administration instead of solving problems, my value to the mission of the hospital would not exist. Even while we concentrate on managing the technology of the institution, we must not lose sight of the mission of the hospital. This mission may certainly not be the same for different health care providers. A university-based academic medical center may have a very different mission from that of a community hospital. It is imperative to recognize this mission and identify with it. It will not vary appreciably from year to year. In fact, it may not change much during your entire tenure at the institution. It is the institutional mission that drives all other activities within the hospital. We work under that mission. We work to enhance that mission.

In managing technology to bring greater value to the institution, remember that for the most part biomedical/clinical engineering support programs are cost centers. They do not tend to generate revenue for the hospital. There is a distinct message here. As a cost center, there is only one way to increase value of the program, and that is to decrease costs to the institution. I will talk about becoming a revenue source later in this chapter, but for now, let's concentrate on the obvious—cost reduction. This is our value to hospital administration. It goes without saying that this must be achieved without sacrificing quality. These are the two driving forces that have governed our program: Reduce service costs to the hospital and maintain or increase quality at the same time.

This process must not be accomplished emotionally. When emotion takes over, we lose the battle. I have engaged in numerous conversations where we have been talking about preventive maintenance frequencies, and colleagues begin to speak about technicians/specialists potentially losing their positions if administration sees that we have created "available time." This argument bothers me to the foundation of my professionalism. While we all must be concerned about our employees and strive to preserve their employment, we cannot lose sight of the mission: patient care, technology support. It is a harsh lesson to learn, but if appropriate patient care at the right price means reducing your staff by some number of full-time equivalents (FTEs), then that becomes the correct direction to take. It will not be a popular one, but none the less, it would be the correct one.

I have been very fortunate in the 16 years I have been at the hospital. In a time when other departments have had to cut staff on numerous occasions, I have only had to eliminate one or two individuals over that time frame—that, by the way, in a program that employees 70 FTEs. Again, there is no magic here. I was always able to identify areas I could expand into and thereby take on additional responsibilities. This entailed more work for the staff, with the same amount of time, but the staff in its entirety kept their jobs.

In the very infancy of the program, it became clear to my predecessor that operating a department as a cost center alone had significant limitations. Hence the start of an outreach program where we could provide services that we had perfected in-house to other area hospitals on a for-profit (actually, "for surplus") basis. Again, let's not lose sight of the main objective, providing top-quality technology management services for our own institution. Selling our services should not detract from the quality or performance of the in-house program. It goes without saying that the outreach program also had to show a surplus to the bottom line and not cost the hospital precious cash resources. In a not-for-profit environment, we technically cannot "generate a profit." We can, however, show a surplus of revenue in excess of expenses. When we consider a for-profit health care provider, showing a profit to the bottom line is key to success!

By expanding the program into two complementary segments that worked side by side but keeping the accounting separate, we were able to provide another level of value to the institution. Now economies of scale that could not be achieved by one or the other program became a reality. Both programs were managed by the same administrative staff. We could leverage that staff over a greater program with both an in-house cost center component and a "shared services" outreach program on a (more or less) for-profit basis. We were able to venture into technology-support areas that most other programs simply could not afford. We were able to negotiate from a position of greater strength with manufacturers and vendors by bringing to the table market share that could further drive down costs.

We were fortunate in that we had tremendous administrative support to expand the program. Without that support, expanding the program to include a shared services program (really an independent service organization) would have been very difficult, if not impossible. Once that program was underway, however, had we not performed up to the expectations of our clients and with revenue projections consistent with hospital expectations, our shared services program would have been abandoned very quickly.

The features of a successful health care delivery system and, by extension, a technology management program are

- An eye toward cost containment
- Cost reductions
- Revenue enhancement (where possible)

The dilemma, as you can imagine, is that the current health care environment has

- Enormous duplication of services
- Competition among providers
- Technology changing faster than clinical practice
- Pressure from payers for less costly services
- Pressure on providers to deliver care in a practice environment
- Integration and consolidation
- Pressure for information on the value of new approaches

There appears to be a need for a new paradigm. It is fascinating to listen to Joel Barker, a self-professed futurist, in his video on paradigm shifts (*The Business of Paradigms with Joel Barker*, video). Examples of this phenomenon are the Swiss watch industry before and after 1968, the Japanese economy and manufacturing industry before and after the late 1960s, and his description of a motorist's encounter with another driver and our preconceptions based on limited information. While I will not describe the details of the video, I would encourage you to find it and watch it. It has changed the way I view my business and the way I approach problem solving. The need for a new paradigm in health care is driven by the following:

- New technologies and procedures can improve outcomes and reduce costs.
- Providers do not have unlimited resources.
- Identifying appropriate technologies is essential to ensure that strategic opportunities are not overlooked.
- The cost-based environment discourages use of technology unless it reduces costs.

With this said, therefore, where do you start? You must be prepared to recognize opportunities when presented. Again, there is no magic here. Think a bit out of the box, and view what you are responsible for in a new light. Then seize the opportunity. Be prepared that every opportunity may not be a home run. In fact, some may be strikeouts. Recognize, too, that your environment must we willing for you to strike out every so often. Our industry is basically risk-averse. Typically, we do not take chances. Not all health care administrations are willing to take a calculated risk for significant gain. Some may be willing to take the more traditional approach, with little opportunity for reward but little risk of failure (translated: financial loss). As we consider new opportunities, we must ask ourselves:

- Does the opportunity match the goals of the health system?
- Does it reduce cost?
- Does it improve quality?
- Does it deliver appropriate service at the appropriate site?
- Can it improve the coordination of support?

Now you need to start asking yourself (and your administration) the tough questions:

- Do you possess the ability with existing staff to manage the program?
- Can you cost justify increased staff for the program?
- Do you have the administrative support needed?
- Is this endeavor part of our *core competency*, and do we care?
- Is there an intermediate step between where you are and where you want the program to go or an alternate program that could be structured?

If we approach problems (opportunities) from an emotional perspective, we clearly lose the argument. The only perspective with which to approach the problem is from a dollars-and-cents perspective. Add to this consideration of the quality of the services offered, whether from within your organization or as a service from the outside. Paramount in this soul searching, you must consider whether you have established your program as having substantial credibility. If not, then you are beginning the process from a diminished quality perspective. If your administrator views what you manage and do on a daily basis as his or her headache, he or she will naturally tend to look skeptically on your proposal. If, on the other hand, your administrator views your program as a resource on which he or she can rely and that you are a problem solver, the tendency is to look more favorably on your proposal.

If you possess the requisite skills to direct a new program, you are likely to be successful. If your overall program has demonstrated success and cost containment, you are likely to be successful. If your program has a certain critical mass and cost-effectiveness, you are likely to be successful. However, let's not kid ourselves! Even if you have done all of the preceding and possess all the ingredients in your own mind to ensure the success of a business endeavor, do not be led to conclude that hospital administration will necessarily agree with it or support it. This simply means that you must provide more factual information regarding pricing, quality, response times, user satisfaction, and many of the *value addeds* that make up a complete biomedical support or technology management program. While we would like to think otherwise, politics also plays into the decision-making process. For reasons that may not be clear to you, the cards may be stacked against you from the beginning. You, however, must be objective.

Improper decision making has the following implications:

- Inappropriate technology
- Excess technology
- Unnecessary technology
- Distorted consumer expectations
- *Increased costs*

As we consider new opportunities, we need to keep an eye on the target. Why should we consider involvement or risk in new endeavors at all? The answer to this question should be

- We were able to demonstrate expertise during the assessment phase.
- We were able to contribute to the technology base.
- We were able to contribute to the servicing options assessment.
- We were able to bring dollar savings to the table.
- We were able to generate revenue for the hospital.

In a number of the examples to follow, you may question whether the opportunity we embarked on is part of a traditional biomedical support program. If not, then why should we get involved at all? Consider the following:

- We could enter a new technology arena to expand our departmental horizons.
- As a department, we clearly possess the expertise.

- We could eliminate or at least reduce dependence on the vendor postwarranty.
- We could provide support during the warranty period in concert with the vendor and become familiar with the technology at their expense.
- For a vendor with a good product but poor support, we could add a level of credibility and improved support to a product.

Just because an area of consideration is a nontraditional area of support, it does not mean that we should not be a key player. More than any other hospital department, we possess the technical expertise and probably the negotiating skills to be successful. But remember, it is a business. Any entrepreneurial endeavor must be approached from a business analysis perspective.

I think we can all agree (although I could make the argument to the contrary) that the safest approach to support is by the original vendor or original equipment manufacturer (OEM). I think we can also agree that this approach, for the most part, is the costliest. In an era of dwindling revenue dollars and increased expenses from all areas of patient support, we are all confronted with reducing budgets. The easiest reduction always tends to be reductions in staff. For me, this is the least effective alternative. Our value is in reducing dependence on the OEM. We have already agreed that support by the OEM is the most costly. Thus we now have several alternatives with which to embark on for this cost-containment valuation.

We have experimented with medical instrumentation maintenance insurance (MIMI), ISO contract support, ISO fee-for-service support, and OEM screening (first look) contract support and have arrived at the same conclusion each and every time. While all of the aforementioned approaches has the potential to reduce costs, they only go so far. There is more opportunity to be had. This opportunity can only be achieved with the willingness to assume some risk. Now I would submit that the assumption of risk is very low, and I base that on years of data over numerous health care institutions. Indeed, however, the risk is there. Consider too that the health care business is by and large risk-averse. So this becomes an uphill battle.

We have been successful in transitioning a variety of service areas from contracts to full insourcing or, at the very least, a combined approach that retains a relationship with the OEM but places increased responsibility on the in-house program to achieve cost reductions. Examples of this transitioning includes:

Sterilizer support:

- Vendor support at $178,000/year + parts
- In-house trained specialists, 2 at $130,000 + parts
- Most parts second sourced at 45 percent savings of manufacturers' costs
- Savings of $48,000 (labor) + $54,000 (parts)
- Total savings year 1 = $102,000

General rad/flouro support (for six hypothetical R/F rooms):

- Vendor contract: $90,000
- Time and materials: indeterminate
- Third-party insurance: $76,500
- Equipment support management: $72,000
- In-house support salary and benefits, 50 percent Utilization: $34,375
- Replacement parts budget: $37,625 to $55,625

CT support:

- Vendor X service contract: $140,000
- Medical instrumentation maintenance insurance contract: $110,000
- Actual costs expended: $87,000

MRI support:

- Vendor support at $156,000 per machine per year including cryogen replacement (PMs to be done during normal business hours)
- Third-party support at $96,000 per machine per year including cryogen replacement (PMs to be done after hours; on average, that will free up 11 hours additional for scans per week)

- Total savings of $60,000 per machine per year + added net revenue stream due to available hours for scans

Personal computer support history (1985):

- University opened on-campus "PC Store," computer prescriptions
- Sold hardware and software
- Installed hardware and software
- At the very least, a support mechanism for postwarranty service was needed
- TJUH, TJU combined accounted for over 4000 PCs
- Plus displays and printers
- Additional hardware (routers, modems, etc.)
- Variety of hardware vendors
- Variety of operating systems (DOS, Windows, Mac)

Personal computer support: In-house approach:

- Potential to minimize support costs and equipment down time
- Potential to maximize response time and documentation
- High initial expense to establish program
- Cooperation and acceptance of administrative and clinical staff

Personal computer support: Where are we today?

- Services provided to TJU, TJUH, and Einstein Health System
- Total of over 8000 PCs and peripherals covered
- Proposals to cover four additional health system member hospitals
- Considerable cost savings brought to the bottom line of the hospitals
- Improved response time
- Improved service overall

Flexible endoscope support.

- OEM major repair at $2000
- ISO major repair at $1250
- In-house major repair at $850

Linear accelerator/simulator support (radiation oncology):

- There existed a core group in radiation oncology providing support
- Those three FTEs came under biomed
- The core group expanded to five and began servicing accelerators and simulators outside of TJUH

Many of the vendors and manufacturers we deal with view service as a business unit unto its own. This business unit has its own profit and loss statements and is accountable for sustaining itself. A significant portion of the net profits of any major manufacturer of high-end equipment comes from service, not just sales. If they capture a greater share of the service market, they will reap the financial rewards. It is no wonder, then, that many manufacturers will go to great lengths to protect those precious revenue dollars that are derived from postwarranty support and maintenance. My contention is that we too can profit from this phenomenon. Where the economies of scale exist, it is clear to me that the most prudent course of action, the one where there is the greatest "bang for the buck," is insourcing.

The objective here is not personal survival. What is in the best interests of the organization is in the end in the best interests of you and your department. To look at this any other way is foolish. Determine what is in the best interests of the institution or health system. If your perspective is from any other vantage point, you will likely not succeed.

Success will come to those who develop and prove cost-effectiveness, demonstrate improved outcomes, decrease length of stay, decrease costs, and eliminate complementary or competing technology. While our scope may be limited, we do have a role in effecting these elements of success. *Carpe diem*—seize the day!

INDEX

ABOUT THE EDITOR

MYER KUTZ is president of Myer Kutz Associates, Inc., a publishing and information services consulting firm. Formerly, he was vice president for professional scientific and technical publishing at John Wiley and Sons. Mr. Kutz has been a member of the board of trustees of the Online Computer Library Center and was chair of the ASME publications committee. He has a B.S. degree in mechanical engineering from MIT and an M.S. from RPI. Mr. Kutz resides with his wife in Delmar, N.Y.